RANGELAND WILDLIFE

Paul R. Krausman, Editor

The Society for Range Management
1839 York Street
Denver, Colorado 80206
1996

Rangeland Wildlife

Paul R. Krausman, Editor
Lisa M. Fox, Editorial Assistant
Pat Smith, Copy Editor

First Edition

ISBN 1-884930-05-0

Library of Congress Catalog Card No. 95-072501

Suggested citation formats:

Entire Book:

Krausman, P.R., editor. 1996. Rangeland wildlife. The Society for Range Management, Denver, Colo. 440 pp.

Chapter in Book:

Krausman, P.R., R. Valdez, and J.A. Bissonette. 1995. Bighorn sheep and livestock, Pages 237–243 in P.R. Krausman, ed. Rangeland wildlife. The Society for Range Management, Denver, Colo.

Printed in the United States of America

TABLE OF CONTENTS

DEDICATION

This book is dedicated to all stewards of the land that manage with nature's resources as their top priority.

FOREWORD

The Society for Range Management was formed in 1948—less than 50 years ago. Not even a blip on the geological scale. The Society was founded on the concept of managing rangelands based on ecological principles. In the late forties and early fifties range managers were talking about sustainable systems and they fully believed that they were responsible for managing ecosystems. Those were radical concepts for their time and were largely scorned and believed to be inferior by agricultural and commodity production interests.

What a difference a day makes. Current concepts of ecosystem management differ from earlier concepts but the difference is at the margin, not a radical shift. Rangeland wildlife and their habitats have always been an integral part of rangeland ecosystems and rangeland management. The natural resource disciplines must maintain their individuality in order to develop new knowledge and technology, while continuing to explore new ways of cooperating, working together and working with private industry to advance the concepts of ecosystem management. More will be required than a team comprised of persons each promoting their own interests, areas of specialized knowledge and skills. Future resource management will require an amalgamation of persons with specialized knowledge and skills, planning and acting as a unified whole. The publication of *Rangeland Wildlife* by the Society for Range Management is consistent with these trends. *Rangeland Wildlife* contributes to the amalgamation that is envisioned as being required and there is no denying that the calculus of the integration is headed in the right direction.

Charles B. Rumburg
Executive Vice President
Society for Range Management

PREFACE

Rangelands in the western United States provide habitat for >3,000 species of mammals, birds, reptiles, fish, and amphibians. They are also shared with most of the cattle produced in the 11 western states. Managing livestock and wildlife on public lands in the western United States is one of the most challenging issues for managers of rangelands and answers do not come easily. To complicate issues further, the influences of livestock on flora and fauna are intuitively considered to be detrimental, but these relationships are not well understood. In part, this text was developed to improve the understanding of wildlife on rangelands and their interaction with livestock. Readers will discover very rapidly that many of the answers to questions about the influence of livestock on wildlife will not be uncovered here: the controlled research simply has not been conducted. However, scientific organizations such as the Society for Range Management, The Wildlife Society, and The Society for Conservation Biology call for rangeland management that is governed by scientific inquiry and the application of those results to management problems. For example, The Wildlife Society (The Wildlife Society, Inc., 5410 Grosvenor Lane, Bethesda, MD 20814-2197) recognizes the importance of western rangelands to wildlife. The Wildlife Society Position Statement on Livestock Grazing on Federal Rangelands in the Western United States states "that properly functioning rangeland ecosystems supporting a wide diversity of native plant species are critically important to sustaining wildlife diversity and productivity in the American west. Scientifically sound management plans and practices are key to restoring lands degraded by many years of livestock grazing that damaged soils, water, and plant diversity." The Wildlife Society supports the scientific management of rangelands as exemplified in the first of 17 specific tenants listed in the Position Statement: "...it shall be the policy of the Wildlife Society to support livestock grazing management on federal rangelands in the west that; (a) is based on scientific study and considers all rangeland resources, trends, and interactions as well as the broad spectrum of human values and needs; (b) provides for adaptive management and continued improvement of programs and practices as new knowledge and understanding of rangeland ecosystems become available; (c) includes provisions, support, and criteria for monitoring; and (d) involves effective coordination and cooperation among agencies and affected publics."

The purpose of this book was to provide information about the major vertebrates on rangelands in the western United States and to provide some insight into their interactions with livestock. Not all vertebrates were considered. Collared peccaries were not addressed because of their limited distribution in the United States. Reptiles and amphibians should be addressed in this book but are not. Unfortunately, the author that agreed to prepare the chapter failed to do so.

This book is the cooperative effort of professionals. I am very grateful to all authors who took time to prepare the chapters and subsequent revisions. The reviewers are identified separately and I am grateful for their time and efforts to assist with this volume. Others are also due many thanks. D. E. Padgett assisted with photographs (Chapter 2) and W. M. Pulich, G. Frasier, M. Bombeld, J. Parnell, and W. W. Golder provided help on various matters with Chapter 2. The Caesar Kleberg Foundation for Wildlife Conservation and Texas A & M University provided support for Chapter 6. Outside reviews were provided by C. W. Hanselka, J. L. Holechek, T. E. Fulbright, and J. A. Crawford for Chapter 6. S. E. Peck, W. P. Kuvlesky, and S. King are acknowledged for efforts "beyond the call of duty" in literature acquisition and manuscript assistance for Chapter 7. R. M. Kaminski provided many helpful editorial suggestions for Chapter 7. B. A. Knopf assisted with the editing of Chapter 9. A portion of Chapter 9 originally appeared in *Wild Furbearer Management and Conservation in North America*, Queen's Printer for Ontario, in the Chapter, "Coyote Predation" by W. F. Andelt and is reproduced here with permission of the Ontario Ministry of Natural Resources. B. K. Johnson, M. M. Rowland, D. P. Sheehy, and M. Vavra critically reviewed Chapter 10 and M. M. Rowland reviewed the methods of stocking allocation. H. Heady, R. Kindschy, J. Richardson, M. White, and J. Skovlin reviewed Chapter 13. Numerous professionals assisted with the preparation of Chapter 16, including W. Elmore, L. Meyers, W. S. Platts, S. Schum (riparian ecology); P. C. Marsh and W. L. Minckly (fisheries); J. Fuller (editing); C. D. Zisner (word processing and review); S. Fouty (stream geomorphology and review); L. Meyers (background in northern habitats and agency management); and ranchers willing to share their knowledge and experience in numerous discussions. K. Wethington, T. A. Wethington, and A. V. Zale assisted with the figures and B. E. Coblentz reviewed Table 1 in Chapter 17. V. Catt was instrumental in manuscript preparation and revision, correspondence, and the large task of requesting and coordinating all submissions and reviews from the authors and referees. Numerous Society for Range Management officers and chairs supported this project and were instrumental in bringing it to completion. Special thanks are extended to C. B. Rumburg, J. Stubbendieck, K. D. Sanders, F. Guthery, T. Bidwell, D. Rollins, W. Leininger, G. Tunberg, K. Fulgham, S. Tixier, M. L. Anderson, P. Jackson, and P. Simse. Finally, I thank W. W. Shaw, C. P. P. Reid, and C. C. Kaltenbach of the School of Renewable Natural Resources and College of Agriculture, University of Arizona, for allowing me the time to complete this project. To these and all the other individuals involved, I offer many, many thanks.

Paul R. Krausman
Editor

CHAPTER AUTHORS

William F. Andelt
Dep. of Fishery and Wildlife Biology
Colorado State University
109 Wagar Building
Fort Collins, CO 80523

John A. Bissonette
U.S. National Biological Service
Utah Cooperative Fish and Wildlife Research Unit
Utah State University
Logan, UT 84321-5210

Eric G. Bolen
Dep. of Biological Sciences
University of North Carolina
601 South College Road
Wilmington, NC 28403

Walter M. Boyce
Dep. of Veterinary Pathology, Microbiology and Immunology
School of Veterinary Medicine
University of California
Davis, CA 95616

Stephen C. Bunting
Dep. of Range Resources
University of Idaho
Moscow, ID 83844-1135

W. Dean Carrier
5177 Barnette Loop Road
Shingle Springs, CA 95682

John A. Crawford
Dep. of Fisheries and Wildlife
Oregon State University
Corvallis, OR 97331-3803

Brian Czech
School of Renewable Natural Reources
University of Arizona
325 Biological Sciences East
Tucson, AZ 85721

Steven E. Daniels
Dep. of Forest Resources
Oregon State University
Corvallis, OR 97331-5703

Charles L. Douglas
U.S. National Biological Service
Cooperative Park Studies Unit
Dep. of Biological Sciences
University of Nevada
Las Vegas, NV 89154

Kathleen A. Fagerstone
U.S. Dep. Agric./APHIS
Denver Wildlife Research Center
Denver Federal Center
P.O. Box 25266
Denver, CO 80225-0266

Timothy E. Fulbright
Caesar Kelberg Wildlife Research Institute
Texas A&M University
Campus Box 218
Kingsville, TX 78363

Fred S. Guthery
Caesar Kelberg Wildlife Research Institute
Texas A&M University
Campus Box 218
Kingsville, TX 78363

Harold F. Heady
1864 Capistrano Avenue
Berkeley, CA 94707

Volney W. Howard, Jr.
Dep. of Fishery and Wildlife Sciences
New Mexico State University
Box 4901
Las Cruces, NM 88003

Walter E. Howard
Dep. of Wildlife, Fish and Conservation Biology
University of California
Davis, CA 95616

David A. Jessup
California Dep. of Fish and Game
1701 Nimbus Road, Suite D
Rancho Cordova, CA 95670

Clyde Jones
Dep. of Biological Sciences
Texas Technological University
Lubbock, TX 79409

Robert R. Kindschy
U.S. Department of Interior
Bureau of Land Management
Vale, OR 97918

Fritz L. Knopf
U.S. National Biological Service
4512 McMurry Avenue
Fort Collins, CO 80525-3400

Ben H. Koerth
School of Forestry
Stephen F. Austin State University
P.O. Box 6109
Nacogdoches, TX 75962-6109

Paul R. Krausman
Program in Wildlife and Fisheries Sciences
325 Biological Sciences East
University of Arizona
Tucson, AZ 85721

David M. Leslie, Jr.
U.S. Fish and Wildlife Service
Oklahoma Cooperative Fish and Wildlife Research Unit
Oklahoma State University
Stillwater, OK 74078

Richard W. Manning
Dep. of Biology
Southwest Texas State University
San Marcos, TX 78666

Bart W. O'Gara
215 Red Fox Road
Lolo, MT 59812

Phil R. Ogden
School of Renewable Natural Resources
University of Arizona
325 Biological Sciences East
Tucson, AZ 85721

Robert D. Ohmart
Center for Environmental Studies
Arizona State University
Box 873211
Tempe, AZ 85287-3211

James M. Peek
Dep. of Fish and Wildlife Resources
University of Idaho
Moscow, ID 83843

Craig A. Ramey
U.S. Dep. Agric./APHIS
Denver Wildlife Research Center
Denver Federal Center
P.O. Box 25266
Denver, CO 80225-0266

Robert A. Riggs
Wildlife and Range Science
Boise Cascade Corporation
1917 Jackson Street
LaGrande, OR 97850 3748

Bruce A. Roundy
Dep. of Botany and Range Science
Brigham Young University
Provo, UT 84602

James H. Shaw
Dep. of Zoology
Oklahoma State University
430 Life Sciences West
Stillwater, OK 74078-0459

Wendell G. Swank
2326 Quail Run Road
Cottonwood, AZ 86326

James G. Teer
Welder Wildlife Foundation
P.O. Drawer 1400
Sinton, TX 78387

Jack Ward Thomas
U.S. Dep. Agric./Forest Service
14th and Independence S.W.
P.O. Box 96090
Washington, DC 20090

Raul Valdez
Dep. of Fishery and Wildlife Sciences
New Mexico State University
Las Cruces, NM 88003

Milton W. Weller
Dep. of Wildlife and Fisheries Sciences
Texas A&M University
210 Nagle Hall
College Station, TX 77843-2258

Michael J. Wisdom
U.S. Dep. Agric./Forest Service
Forestry and Range Sciences Laboratory
1401 Gekeler Lane
La Grande, OR 97850

Jim D. Yoakum
P.O. Box 369
Verdi, NV 89439

CHAPTER REFEREES

Morley W. Barrett
Alberta North American Waterfowl Management Plan
401 East Tower, Coronation Plaza
14310-111 Avevnue
Edmonton, AB T5M 3Z7
Canada

Vernon C. Bleich
California Department of Fish and Game
407 West Line Street
Bishop, CA 93514

Michael A. Brogan
U.S. Fish and Wildlife Service
713 Boudler Circle
Fort Collins, CO 80524

John A. Crawford
Dep. of Fisheries and Wildlife
Oregon State University
Corvallis, OR 97331-3803

Don W. DeYoung
University Animal Care
Arizona Health Sciences Center
University of Arizona
Tucson, AZ 85721

Wayne Elmore
U.S. Department of Interior
Bureau of Land Management
Prineville District Office
P.O. Box 550
Prineville, OR 97754

Richard C. Etchberger
Program in Wildlife and Fisheries Sciences
University of Arizona
Tucson, AZ 85721

Jerry L. Holechek
Dep. of Animal and Range Sciences
New Mexico State University
Las Cruces, NM 88003

Thomas Huels
Ecology and Evolutionary Biology
University of Arizona
117 Biological Sciences East
Tucson, AZ 85721

Kendall L. Johnson
Dep. of Range Resources
University of Idaho
Moscow, ID 83843

Ron J. Johnson
Dep. of Forestry, Wildlife and Fisheries
University of Nebraska
202 Natural Resources Hall
Lincoln, NE 68583-0819

Terry B. Johnson
Nongame and Endangered Wildlife
Arizona Game and Fish Department
2221 West Greenway Road
Phoenix, AZ 85023

Richard M. Kaminski
Dep. of Wildlife and Fisheries
Mississippi State University
P.O. Drawer LW
Mississippi State, MS 39762

John Kie
Pacific Southwest Forest Experiment Station
2081 East Sierra
Fresno, CA 93701

David A. King
School of Renewable Natural Resources
University of Arizona
325 Biological Sciences East
Tucson, AZ 85721

Raymond M. Lee
Grame Branch
Arizona Game and Fish Department
2221 West Greenway Road
Phoenix, AZ 85023

David M. Leslie, Jr.
U.S. Fish and Wildlife Service
Oklahoma Cooperative Fish and Wildlife Research Unit
Oklahoma State University
Stillwater, OK 74078

R. William Mannan
Program in Wildlife and Fisheries Sciences
University of Arizona
325 Biological Sciences East
Tucson, AZ 85721

Mary M. Meagher
P.O. Box 126
Yellowstone Park, WY 83020

William H. Miller
School of Agribusiness and Environmental Resources
Arizona State University
Tempe, AZ 85287-3306

John R. Morgart
Yukon Delta National Wildlife Refuge
P.O. Box 346
Bethel, AK 99559

James M. Peek
Dep. of Fish and Wildlife Resources
University of Idaho
Moscow, ID 83843

Bill Platts
Don Chapman Consultants
3653 Rickenbacker, Suite 200
Boise, ID 83705

Dale Rollins
Texas Agriciultural Extension Service
7887 North Highway 87
San Angelo, TX 76901

Keith Severson
South Dakota School of Mines
501 E. St. Joseph Street
Rapid City, SD 57701

Norman S. Smith
26 Avenida de San Ramon
Tucson, AZ 85710

Philip J. Urness
Dep. of Range Science
Utah State University
Logan, UT 84322-5230

Martin Vavra
Eastern Oregon Agricultural Research Center
Sqaw Butte Station
Star Route 1, 4.51 Highway 205
Burns, OR 97720

Michael L. Wolfe, Jr.
Dep. of Wildlife Resources
Utah State University
Logan, UT 84321

Jim D. Yoakum
P.O. Box 369
Verdi, NV 89439

Malcolm J. Zwolinski
School of Renewable Natural Resources
University of Arizona
325 Biological Sciences EAst
Tucson, AZ 85721

Rangeland and Wildlife Resources

Chapter 1
RANGELAND RESOURCES
Harold F. Heady

INTRODUCTION

Rangeland resources include vegetation, wildlife (e.g., game and nongame birds, fish, mammals, amphibians, and reptiles), other animals, parent material, water, and space. None of these can exist without the basic resources of energy and land. *Rangeland* refers to all the resources of the land. In the range management profession, a space or place designation is most often called a range site, ecological site, vegetation type, or habitat type. In wildlife biology, the term is *habitat*. Full understanding of site and habitat must consider biology, the abiotic environment, and the impact of economic, political, and social stresses.

The goals of this chapter are (1) to provide knowledge of rangeland vegetational spatial communities upon which other authors can build; (2) to provide perspective on vegetation changes through time; (3) to describe the primary producer level processes of energy flow, and stability; (4) to bridge that often controversial situation of ineffective knowledge transfer among wildlife biologists, rangeland managers, and public interests; and (5) to outline conceptual changes in condition and trend analysis of rangeland resources.

RANGELAND DEFINED

Rangeland is a type of land supporting grasses, forbs, shrubs, and trees in various combinations that provide habitat to many kinds of native and exotic animals. The vegetation may include native and exotic plants and be temporary after cultivation, harvest of forests, clearing by wildfire, prescribed burns, and other treatments. Rangeland sites and habitats may occur within forests but are essentially missing from cultivated land. Many animal habitats are not related to rangeland.

The connotation of rangeland is neither "forage for livestock" nor "habitat for wildlife" alone, but much more. Accordingly, rangeland resource professionals are equipped by various educational studies and experience to analyze the ecological base and its condition; recommend conservation and preservation measures; predict impact of climate, fire, and use; and manage rangeland on a sustained basis for a stated purpose.

SITE OR HABITAT?

The use of the terms *site* and *habitat* has subtle meaning differences in wildlife biology and range management. *Range site* and *ecological site*, most commonly used by the range management profession, are based on soil and vegetation. *Vegetation type* is a general term. The *habitat type* concept has a special meaning as proposed by Daubenmire (1952).

Habitat for wild animals has been defined in several ways. Ohmann (1992) called it the space occupied by a community of organisms (plants and animals) and the place within it that provides the life needs of an individual. Giles (1978) gave a broad view of habitat that included edges, animal migrations, time, food quality, and managed hunting. Peek (1986) emphasized that climate and substrate were the foremost habitat features because they dictate the availability of nutrients, therefore, animal populations. Morrison et al. (1992) captured the essence of these variations when they separated infinite space characteristics from animal requirements. Habitat for wild animals is a place specific to an individual or species population that allows successful occupancy. It is an area that provides food, cover, water, environment, and absence of predators and competition.

Site and *habitat* have much in common, but are not synonyms. My intent is to use these terms as the professions use them. Definitions that might satisfy several disciplines will not be attempted. Emphasis will be on the primary producer level to minimize misunderstandings among the multiple users of rangeland resources.

SCALE AND MAPPING OF RANGELAND RESOURCES

At the world level, vegetation maps and descriptions are presented for each continent by Dasmann (1976) and many others. The World Resources Institute (1988) listed pasture and rangeland together in North America as 47% of the land area and 50% of the world. These estimates were based on data from the United Nations Production Yearbook for 1986, which is the best estimate available.

Holechek et al. (1989) described the vegetation types in the United States with extensions of some types into Canada and Mexico (Table 1). Their 18 types, based on species composition and general appearance, emphasize division of grasslands and shrublands and the combinination of forest types. The most commonly used vegetation map of the United States is by Kuchler (1964). Typical for states and regions are maps by Franklin and Dyrness (1973). Users of rangeland vegetation types at this scale are teachers who spend a university term leading students through major vegetation types in North America, and scientists interested in the habitat of species with wide distribution.

Shiflet (1994) described 175 rangeland cover types in the United States. The types are arranged in 9 regions that were edited by persons knowledgeable for the areas. Descriptions include distribution, botanical composition, ecological relationships, and variations. Shiflet (1994) gives vegetational descriptions at 2 scales: the regions and within the regions.

A rangeland resource inventory for planning and monitoring purposes consists of classification and measurements. Commonly the maps produced are large scale; i.e., 1 to 50,000. Typical users of large scale maps include vegetation managers who recommend livestock grazing systems according to rangeland sites and those who delineate the specific habitat of threatened and endangered species.

The large- and small-scale approaches to differentiating vegetation types illustrate an important principle. Designation of ecological type, site, and habitat is a tool that supports a program objective, whether finely tuned research or broadscale understanding and management. Efforts to define an ecological type classification system that can be used by numerous agen-

TABLE 1. Area and percent in federal ownership of rangeland types in the United States and adjacent parts of Canada and Mexico (Holechek et al. 1989).

	Area (ha) millions	Federal ownership (%)
Grasslands		
Tallgrass prairie	15	1
Southern mixed prairie	20	5
Northern mixed prairie	30	25
Shortgrass prairie	20	5
Palouse prairie	3	15
California annual grassland	3	6
Alpine grassland	4	99
Desert shrublands		
Sagebrush/grassland	39	65
Salt-desert shrub	35	85
Mojave	14	55
Sonoran	27	55
Chihuahuan	45	55
Woodlands		
Pinyon-juniper	17	65
Mountain shrub	13	70
Western coniferous forest	59	62
Southern pine forest	81	6
Eastern deciduous forest	100	7
Oak woodland	16	4

cies and for many objectives have usually failed. The current approach is establishment of universal data banks that can be retrieved to meet numerous objectives. Stored data in addition to location in space may be productivity of food for animals, the present and potential vegetation composition, landscape characteristics, cover for soil protection and animals, expected variation in response to weather, and many more parameters.

This is a large order for gathering, storing, and organizing information, especially when different programs require retrieval at various scales. Only in the last 2 decades have computers been available that will handle the data mass. For example, the Soil Conservation Service has computer-stored data on an estimated 3,262 range sites in the 20 western and southern states (Joyce et al. 1986). A range site is a manageable landscape subdivision based on similar natural biomass production throughout. Data available on these sites include vegetation composition, erosion hazard, biomass production, livestock carrying capacity, climate, and soil descriptions.

Life zones of rangeland ecosystems were first described by Merriam (1890) for San Francisco Peak near Flagstaff, Arizona. These zones are general, seldom used today, but were the beginning of other systems. For example, the habitat type system designates areas of vegetation and soil with similar characteristics and approximate scale of range sites. It is widely used on forests and rangelands in the Pacific Northwest and broadly through the Rocky Mountain region from Canada to Mexico (Wellner 1989). Habitat types and range sites pay particular attention to potential of climax vegetation and the successional, or present vegetation.

The geographical information system (GIS) stores, analyzes, and displays spatial and nonspatial data on landscapes (Tueller 1989). It creates digital maps and overlays from remote-sensed imagery, such as banks of data on botanical composition, biomass, topography, soil moisture, and amount of solar energy received. The GIS analysis techniques are increasingly useful in measuring short- and long-term change in habitat and site characteristics because imagery can be repeated from satellites. The GIS technology is a relatively new and fast growing approach to inventory of rangeland resources. Cooperation has made the data sets compatible on small, national, and worldwide scales.

If rangeland resource data banks were complete, managers of wild and domestic animals would have access to biomass productivity and other information for calculating stocking rates and carrying capacity. Critical habitats and animal numbers would be identified. Watershed and recreational impacts would be noted. Guidelines for reclaiming abused land would be facilitated. Status and location of threatened and endangered species would be known. The ideal for natural resource data gathering, analyzing, and retrieval has not been attained and may never be, but professionals are working together toward that goal.

Easier collection and handling of extensive data sets and more consistent resource analysis lie ahead. However, resource professionals and scientists must keep in mind that the division of vegetation into types is artificial and of their choosing for communication. While the site and habitat boundaries are important, understanding the ecological processes and changes within the site and habitat is more important.

VEGETATION HABITATS FOR ANIMALS

The rangeland resource manager should be able to answer this question: what is the desired vegetation that best fits the management objective? For a wild animal, that would be the ideal habitat. In livestock grazing, it would be the ideal range condition. For an endangered plant species, it would be the ecological site and uses that foster continued reproduction.

To answer the question of desired habitat and site, assume 5 generalizations. (1) Migratory animals require >1 area (sometimes called >1 habitat) as do those that feed, rest, and reproduce in different places. (2) Habitats favoring 1 animal or plant may disfavor another. (3) Goals in rangeland ecosystem management change over time necessitating preservation of the basic land resource. (4) Wild and domestic animals species differ greatly in the breadth of environmental extremes they will tolerate. (5) The objetive for rangeland resources is multiple use.

HABITAT GUIDES TO MANAGE WILDLIFE

The Blue Mountains of northeastern Oregon support 11 vegetation types described by Kuchler (1964). Because of insufficient fit with animal habitats of interest, Thomas (1979) divided the forest and rangeland ecosystems into 14 plant community habitats and 12 unique habitats that occur within the plant communities. A series of General Technical Reports by the Pacific Northwest Forest and Range Experiment Station described wildlife habitats in managed rangelands of southeastern Oregon (Maser and Thomas 1983). Verner and Boss (1980) in California presented a guide to wildlife and their habitats on the western slope of the Sierra Nevada. The California Wildlife-Habitat Relationships System is an extensive compilation of 52 major vegetation types and >100 habitat elements (Mayer and Laudenslayer 1988). These guides provide information on hundreds of species (Thomas 1979, Verner and Boss 1980).

The complexity of overlapping habitats in relation to rangeland resources was illustrated by loons (*Gavia* spp.) reproducing in 1 habitat and the deer mouse (*Peromyscus maniculatus*) reproducing in 24 of the 26 habitats described. Thomas (1979) also describes 13 tree species available for nesting by 10 species of woodpeckers. Nine of the 10 species nested in fir (*Abies* spp.) and ponderosa pine (*Pinus ponderosa*), but hemlock (*Tsuga* spp.) was used only by Williamson's sapsucker (*Sphyrapicus thyroideus*).

A unique habitat feature may occur in several plant communities. This can cause problems with universal classification systems for habitats, sites, and vegetation. Thomas (1979) showed conclusively that restriction of a species to a special or unique area is uncommon. Habitats of most species extend over a variety of vegetations, some as extensive as continents. Commonly used rangeland vegetation categories (e.g., habitat types, range sites, and ecological sites) do not suffice as precise descriptions of animal habitats; nor do descriptions of animal habitats suffice for overall rangeland ecosystem management. At the present state of knowledge and societal demands, management of rangeland wildlife and vegetation probably should be limited to local planning goals.

The major principle in habitat designation may be that most animal species are more narrowly adapted to vegetation structure than to botanical composition. For example, 28 plant communities were recognized in southeastern Oregon, but on a basis of structural similarity these were grouped into 14 categories (Dealy et al. 1981).

Crosswalks among site and habitat classifications are needed and illustrated by the CALPLAN simulation model that forecasts changes in multiple resources including forage for livestock and acreage for wildlife habitat (Ewing et al. 1988). While breeding, seasonal use, and habitat suitability may be related to plant succession and size of the trees, one must be careful not to imply that increasing tree height describes plant succession.

Rangeland resources for wildlife are more complex than rangeland requirements for domestic species. Where animals are found may not be the habitat of greatest production. Rating of habitat condition, whether good, bad, deteriorating, or improving in terms of potential, is seldom done (Peek 1989).

Land resource inventories and monitoring form a basis of rangeland management. They will be used more as management becomes increasingly intensive. Methods of inventory and monitoring rangeland resources are described in books by Cook and Stubbendieck (1986), Cooperrider et al. (1986), and Bonham (1989).

VEGETATION CHANGES OVER TIME

Common field measurements of vegetation are number of plants per unit area (i.e., density), number of plots in which a species occurs (i.e., frequency), weight of material (i.e., biomass), basal area, and foliage cover. These data are used to calculate species composition, diversity, production, and relative stability that characterize range sites and habitats. Repeated measurements over time (i.e., monitoring) give indications of vegetation trends or stability.

Types of Change

Hourly and daily cycles in plants may seem unimportant in work with vegetation, but considerable change occurs during a day and from day to day. Water content is highest in the morning. At that time, young grass growth may be so high in water that ruminants cannot consume the bulk needed for adequate nutrition. One-day plant cycles include water content, concentration of photosynthates, position of leaves, and flower opening. Peak activity of numerous plant and animal rhythms are keyed to day length and perhaps to lunar cycles. Habitats of animals cycle. Without careful measurement and assessment, weekly change in environment may appear as treatment differences and so may regular morning and afternoon sampling.

Seasonal cycles correlate with warm and cold temperature sequences and with alternating wet and dry periods. These physiological, behavioral, and phenological changes raise no particular argument as to their existence, but a sampling program that spans several days may confound phenological changes in habitats with treatment differences.

Yearly variation in precipitation and temperature results in different growing seasons. Average annual precipitation and temperature are not duplicated in any year. Vegetation responds to these and other abiotic environment variations by never having the same relative composition and biomass from 1 year to the next. Some rodents, insects, and other animals exhibit cycles that span several years. Daily and seasonal changes may be confused with longer cycles and annual variations, especially if measurements are keyed to calendar dates alone. These kinds of changes are superimposed on treatments in research and management. Measurement techniques should separate them.

Immigrations of species and minerals that ride wind, water, animals, and human devices are common. A safe assumption is that immigrations have occurred for as long as there have been organisms to emigrate. The number of exotics everywhere has increased with humans' propensity to travel, and new plant species arrive in the United States every year. Many are well established and spreading before discovery. Becoming established to the extent of reproduction and spread indicates a status of "new native" because their elimination is not practical. Return to pristine conditions is unlikely.

Evolution and drift in macroclimate result in another type of vegetation change. Research into pollen profiles and woodrat (*Neotoma* spp.) middens has indicated that the cool-moist period in the Great Basin region about 12,000 years before present (BP) was followed by a warmer time beginning about 7,000 years BP. Sagebrush (*Artemisia* spp.) became abundant at the expense of conifers and grass. By 4,000 years BP the trend reversed. For the last 3,000 years, aridity and temperatures have been increasing, but with moist-cool periods alternating with dry-warm periods on a 200-500 year basis. Charcoal in the middens increased directly with grass pollen. Charcoal:grass pollen ratios indicate that western juniper (*Juniperus occidentalis*) had reached its northward and upward extent by 6,000 years BP (Mehringer and Wigand 1987). Since 1850, changes in vegetation due to nonhuman causes have been confused with human-related factors including domestic animal grazing, abundant alien invaders, cultivation and abandonment, water diversion, unregulated hunting, seeding of exotic perennial species, and increased frequency of fire.

Grasses and forbs evolved under selective pressure by large herbivores, fire, and climate. Characteristics pointing toward that conclusion are anatomical like thorns, basal meristems, and dormancy with buds in or near the soil; chemicals with poisons and bitter taste; and a short physiological time for rapid growth and seed maturity. Plants in the shortgrass and mixedgrass prairies have these characteristics in abundance and they resist heavy grazing and fire.

Primary Succession

Succession in vegetation is usually defined as directional replacement of the dominant species. The dominants beginning on a bare area where no previous plants existed are likely to be annual broadleaved species. These are followed by annual grass; perennials begin to take over and finally dominate. A point is reached where the number of individuals in each species remains about the same in successive generations. The succession process was described by Clements (1916) for the central United States grasslands and by Redente and DePuit (1988) for reclamation of bare areas left as mine spoil.

The first step in primary succession is invasion of the site by plants and microorganisms, usually transported by wind and animals. The seeding of adapted plants upon a reconstituted landscape speeds the invasion. Inoculation of the soil with mycorrhizae, mulching the soil surface, fertilization, and irrigation may help achieve the second step, establishment. The third step is competition. The fourth is called reaction or species response to the competition. Soil building or development of a soil profile continues through the 4 steps. The last step in succession is stability of vegetation composition. Animal populations show successional responses to the plant succession.

Primary succession after the retreat of glaciers or the development of wide flood plains follows the above steps. A recent example describes forest succession on terraces and permafrost along the Tanana River flood plain in interior Alaska (Viereck 1989). The succession begins with willow (*Salix* spp.), horsetail (*Equisetum* spp.), and rush (*Juncus* spp.) on recent eluvium. That vegetation is invaded by alder (*Alnus tenuifolia*), balsam poplar (*Populus balsamifera*), and a rose (*Rosa acicularis*) understory. White spruce (*Picea glauca*), rose, alder, and evergreen shrubs indicate the beginning of permafrost. The stand becomes open with black spruce (*Picea mariana),* the dominant tree. Finally, thaw ponds develop in the permafrost with sphagnum bogs.

Based on potential vegetation the entire sere would be classed as 1 habitat type. However, definition of stability is difficult because each stage takes from 100 to 200 years for the forest composition to change. Measured in terms of the observer's lifespan all stages in the sere appear stable. Classification based on actual vegetation may be better than on potential vegetation in this example. Primary successions that occur within 1 to 3 decades may be found on raw mine spoil and the thick ash left by the explosion of Mount Saint Helens. Regardless of the time, the manager needs to work with the present vegetation and know the potential to take advantage of the changes that will occur.

Secondary Succession

Secondary succession begins after vegetation has been partially altered, often with little damage to the soil. It follows disturbances such as logging, reduced livestock and wildlife grazing, fire, and abandonment of cultivation. These disturbances cause vegetation changes that differ from the autotrophic time series of primary succession. Invasion may not occur because the flora and fauna are present and expand rapidly through reproduction. The number of species is likely to be similar between early and later secondary succession stages. However, the dominant plant species will change. Replacement by different species of trees is succession. Animal habitat designations may not follow this distinction if they are related more to tree height than to species of tree as implied in habitat classifications used by Thomas (1979) and Verner and Boss (1980).

Deterioration of vegetation may be sudden, as being covered with silt, ashes, and molten lava, or it may be gradual because of excessive grazing by herbivores. In the latter case the order of plant species decrease will not be the reverse of succession. A widely accepted hypothesis is that the most preferred forages will be the first and heaviest grazed, thus giving other plants room to increase. Few data substantiate that hypothesis. A plant response is related to intensity, frequency, and timing of defoliation more than to kind of grazer. Reduced intensity usually results in secondary succession toward a stability, but along different pathways than the deterioration. In several large western rangeland types of vegetation, change is related more to weather than to grazing.

Many departures and alternative theories about succession have been added to the traditional views of Clements (1916). The intensity, frequency, extent, and kind of disturbance cause numerous pathways of changing species, many as yet unpredictable. Continued disturbance may cause apparent stability or prevent development of stable vegetation, hence a view that climax does not exist.

Another view emphasizes the role of chance in individual plant location, therefore species composition of an area. This view contends that composition after disturbance follows nonlinear directions, lacks reversibility, and irregularly pauses for a sufficient time to appear stable. The increased understanding of spatial and temporal changes in vegetation has resulted in renewed examination of the basic tenets of condition and trend of rangeland resources.

Separation of Kinds of Vegetation Change

Separating the types of change is important to understanding the vegetation, defining and mapping spatial units, research results, and management. Changes attributed to time of day, seasonal and yearly climate, evolution, and geologic time cannot be controlled. They influence the results of all applied treatments. Managerial practices yield controllable and uncontrollable results that should not be confused. An example is short-term species composition changes of annual grasses and forbs in Mediterranean climates that may result from heavy grazing intensity, weather, or both. Unfortunately measurements do not always separate types of change in species composition.

ECOSYSTEM PROCESSES

Traditionally, the essentials for livestock and wildlife have been food, water, space, and cover for resting, reproduction, and escape. The food of special interest here is plant material and that conjures such items as primary productivity, transfer of energy, food quality, mineral cycling, plant competition, herbivory, stability, and diversity as interrelated ecosystem processes and conditions.

Primary Food Production (Capture of Energy)

The primary process in ecological systems is the capture of energy in photosynthesis and energy flow through organisms; always closely tied with water, minerals, and nutrients. Green plants combine solar energy, carbon dioxide, water, and minerals into chemical energy, such as protein, carbohydrates, and vitamins. The capture of solar energy is an inefficient process (Williams 1966). Less than half the arriving energy is of wave lengths usable in photosynthesis. The unused portion may be captured as heat by living organisms, thereby favoring environment for photosynthesis and other ecosystem processes. However, lack of water, mineral imbalances in the soil, and unfavorable high or low temperatures may restrict energy capture.

Food Chain, Energy, and Mineral Cycling

Solar energy captured by vegetation constitutes primary production. It is the producer trophic level that begins the flow of energy through food chains. Primary consumers or herbivores of all kinds constitute the second trophic level. Carnivores are secondary consumers and the third trophic level. Organic materials in the 3 trophic levels are eventually mineralized by the decomposers. Part of the energy bound by the primary producers is lost at each level of consumption and is completely dissipated upon release of the minerals.

Increased efficiency in the use of energy is only possible in 2 ways: capturing a greater percent of the solar energy and reducing losses as energy passes through food chains. Of necessity, capture involves green plants and that leads research to techniques for increasing photosynthesis. World increases in atmospheric carbon dioxide and global warming are generally accepted. That should cause increased efficiency of energy capture and greater primary production (Mayeux et al. 1991).

Other means of increasing energy capture include genetic manipulation and controlling plant competition. In practice, the rangeland manager increases biomass production with a wide variety of tools; i.e., seeding higher energy capturing plants, controlling population size of animals, adding soil nutrients, better user distribution, and reduced intensity of defoliation. Most management practices on rangeland attempt to protect the resources and increase productivity.

Social, political, and economic structures often hinder or prevent more energy capture and reduced loss. For example, planting exotic forage plants and releasing biocontrol organisms are undesirable practices according to some viewpoints. Deer (*Odocoileus* spp.) management, where restricted harvest is mandated, leads to excessive winter kill. Locusts in the United States are subject to herbicides, but in parts of Africa they are food for people and cannot be poisoned.

Food quality is another aspect of energy capture and transfer in rangeland plants. There are several well-known and seldom questioned relationships. (1) Young green and growing material is relatively high in protein and soluble carbohydrates. (2) During the plant maturing process, protein and soluble carbohydrates decrease in proportion to complex carbohydrates. (3) Most seeds and fruits are high in fat, starch, and/or protein. (4) Browse maintains soluble carbohydrates and protein longer than most grasses and forbs.

To interpret the above points in the context of animal nutrition requires that the types of digestive systems be considered. Most understanding of rangeland foods has come from the study of domestic animals. Microbial fermentation in Bovidae and Cervidae is pregastric whereby major fiber or cellulose digestion is ruminal with minimal digestion in the colon and cecum. Digestion in these animals is a bulk processing operation.

In nonruminants and postgastric fermenters, including the horse, rabbits (*Leporidae*), and pigs, major digestion occurs in the small intestine, colon, and cecum. Postgastric animals require foods high in nonstructural plant parts, those low in complex carbohydrates such as lignin.

Interest in the habits of rangeland animals has greatly increased since the environmental movement of the 1960s. Diets of hunted species, and those classed as threatened and endangered are fairly well known, but many species have received little study of nutritional requirements beyond the listing of foods in diets. Based on a literature cited of approximately 1,600 publications, Ehrlich et al. (1988) gives the principal dietary items for 646 species of birds in the United States. The diets are broad, many change with the season of year, but nearly all are specific to some degree. Bird nutrition is said to be similar to our needs for proteins, fats, carbohydrates, vitamins, and minerals, but all animals need these food elements. With all this information on diets, the question still remains whether or not the diet represents an ideal supply of foods. Vegetation managers need the answer to that question.

Spaced trees and shrubs, say 50% canopy cover, accumulate nutrients from the adjacent openings through their root systems. Fallen leaves, bark, and fruits deposit the nutrients beneath the canopy and that which is not lost by wind and running water becomes part of the soil. Eventually half the area becomes enriched and half depleted. Botanical composition and plant density often differ beneath the canopy and openings. This redistribution phenomena may be enriched by the action of animals that use the woody plant sites for resting, nesting, and dens. Shift of the nutrient capital into aboveground biomass and to the immediate site implies that managing the density of woody plants can benefit herbivores (Tiedemann 1987).

Data on nutritional content of rangeland forages have been accumulated to determine dietary deficiencies. The knowledge is useful in designing supplementary feeding, but feeding of extra nutrients to wildlife is usually based on too little feed or a shortage of energy. For example, winter feeding of deer has principally been for energy where animals are starving. That practice has not been accepted by many professional managers. Briefly, the large body of nutritional information on rangeland forages has been used for qualitative evaluation of available foods with little emphasis on quantitative intake studies for nutritional evaluation. Quantity and quality of intake by large herbivores on rangeland have been difficult to determine and evaluate according to animal requirements. A listing of food plants by proportion in the diet fails to show dietary deficiencies. Dietary intake without evaluation of the forages available gives little help to the resource manager.

Redistribution of minerals and plants by large herbivores is easily observed in dung accumulations such as those of feral horses and rhinoceros (*Diceros* spp.). Horses and cattle had markedly different influences on pasture botanical composition through nonrandom distribution of dung sites on village "commons" outside London. Time required for the pattern development is unknown, but these commons had been grazed the same way since 1085 (Putman et al. 1991).

Herbivory

Herbivory is the selection and consumption of live plant materials. It is universal; no primary production exists without the possibility of being eaten. Herbivory has developed in almost as many ways as there are herbivores.

The impact of herbivory is a combination of intensity, selectivity, seasonality, and frequency of defoliation. It influences mineral cycling, distribution of other organisms, and animal behavior. The herbivore itself exerts physical pressures on the system. The ultimate challenge is to separate these effects.

Every herbivore through the process of consuming live plant material disturbs that living plant. Studies in rangeland management for domestic animals have attempted to characterize those impacts. The concepts of overuse and proper use have been described and quantitative ways for their measurement suggested. Exploitation of the idea that 1 species of herbivore can be used to develop habitats suitable for another species is largely in the idea stage. However, in Africa, browsers improve the forage resources for grazing animals and vice versa (Heady and Child 1994).

Competition

Competition is a process that occurs when organisms are making a common endeavor to gain requisites in excess of the immediate supply. Darwin (1859) separated the "struggle for life" into 3 parts: physical conditions, predation and parasitism, and competition for food and space.

Short supplies of nutrients and space play a major role in determining which plants survive in plant succession. As plants add organic matter to the soil, their environmental conditions change, such as water holding capacity, structure of soil, temperature amelioration, and increased fertility. Thus, 1 plant species improves conditions for another, and the second wins a dominant place in the plant community.

Competition is among organisms in the same trophic level. This is plant to plant, not plant to environment or plant to animal. For example, drought may cause a shortage of water but competition for water is between organisms. An herbivore does not compete with the plant, however much it influences the plant's competitive ability.

Degree of herbage removal is a factor in rangeland deterioration. Defoliation too early, too often, and before dormancy reduces photosynthate stores (Heady 1975). For decades that explained the physiology of a plant's demise. Understanding the process of food storage and reactivation in plants has been altered by recent work on several grasses showing that new growth is from photosynthesis in a nearby plant part, rather than from stored material in roots and lower stems. Previously stored materials may furnish foods to new growth for only a day or 2. Apparently, food storage in plants is not as important as originally thought (Caldwell and Richards 1986). Not all questions about plant vigor are answered. Much remains to be learned concerning the relative importance of defoliation and competition with neighbors.

Competition is a major factor in the establishment of plants. If more seeds germinate than the environment can supply with minerals, water, and space, many will die. When disturbance such as a fire results in a burst of germination, the net effect may be a short supply of the resources for all. Resource abundance or scarcity has important implications on the sorting of species and survival of individuals within a species (Samuel and DePuit 1987). Competition depends upon the proximity of organisms. The biology of individual growth in isolation will not suffice.

Stability

Stable vegetation refers to self-perpetuating communities of plants in equilibrium with their environment and each other. Darwin (1859) stated it this way, ". . . in the long-run the forces (in nature) are so nicely balanced, that the face of nature remains uniform for long periods of time." However, stability is dynamic and exists under the influence of variable climate, influx of minerals and exotic organisms, and drifts occurring in populations due to evolutionary and geologic changes. Stability is often called climax, and whether it is one particular composition or a dynamic range of compositions has been argued since the term was proposed (Clements 1936, Heady 1973).

The varied time scale for the types of vegetation change needs emphasis in terms of generation time. The shorter generations may be minutes or days; the longer at least centuries. The generation time of scientists and rangeland resource managers approximates 30 years. For practical usefulness, secondary succession and stability probably should take no longer than a decade. When generation times in the flora and fauna are longer than for humans, the use of succession and stability to guide management becomes questionable. Change may be imperceptible and disturbance has more chance to interrupt the sequence of species replacement. One of numerous examples comes from a 37-year-old exclosure in sagebrush/bluebunch wheatgrass (*Agropyron spicatum*) region in southcentral Oregon. It indicated that little increase in grass occurred until the sagebrush was reduced (Sneva et al. 1984).

Few data sets in rangeland evaluation and management are for >20 years and those that do exist often show contrasting changes. The Park Grass fertilizer experiment near London that began in 1856 provided data sets, some as long as 87 years. New equilibria were established in the ratio of floral components on the different fertilizer treatments, subject to variation caused by low or high rainfall (Silvertown 1980).

In climates with rainfall above approximately 400 mm, grassland successions move in a few years from low or intermediate states to stable or climax composition. Responses to habitat manipulation and management are rapid if a seed source is present. In lower rainfalls, succession moves slowly and early composition stages may appear stable. When shrub-dominated vegetation in dry areas is destroyed by fire, succession may not exist if the same shrub species return. Sprouting chaparral species of the Mediterranean climates return immediately, but a ≥5-year succession of forbs and annual grasses covers the land, gradually disappearing as the shrub canopy closes.

Diversity

Biological diversity is an umbrella term for nature's variety (World Resources Institute 1992). It has been divided into genetic diversity covering the intraspecies variability; species diversity including different taxonomic entities of plants and animals; ecosystem diversity commonly defined in terms of space differences; and human cultural diversity that helps people adapt to changing conditions. A taxonomist can destroy diversity and rare organisms by lumping taxonomic categories.

Diversity in rangeland vegetation embodies species richness or number of species, and species evenness or distribution. Many indices of diversity have been suggested and compared (Chambers 1983), but none has been fully accepted. Diversity is usually derived from estimates of numbers, biomass, and cover (Whittaker 1972).

Diversity is commonly included as 1 parameter of an ecosystem. It is used to compare species change within the system over time and differences between ecosystems over space. Thomas (1979) uses length of edges between plant communities in a defined area to calculate diversity. High diversity is desirable and is said to indicate vigorous animal populations through adequate nutrition from mixed diets, stability of vegetation and soil, and sustained biomass production. Optimum white-tailed deer (*Odocoileus virginianus*) habitat is described as the highest floral diversity that nature produces (Bryant and Demarais 1991). Pronghorn (*Antilocapra americana*) do best where many forbs are in the short vegetation (Yoakum 1980).

Silvertown (1980) found in the 87 years of Park Grass fertilizer applications that as biomass increased, diversity decreased when measured by species number. Collins et al. (1987), working in southern Great Plains, claimed that high diversity could only be maintained with livestock grazing that reduced biomass. In understory succession in the grand fir (*Abies grandis*)/myrtle (*Pachistima myrsinites*) habitat type following a fire in north-central Idaho, diversity increased to a stable point early in the sere, then slowly declined (Zamora 1982). Managed herbivory, logging, and fire reduce shade and encourage short species, which are characteristic of high diversity.

Although grazing can increase diversity, it does not always do so. Semiarid vegetation with an evolutionary history of light grazing may lose diversity with low-intensity livestock use. In higher rainfall regions, no grazing often permits large accumulations of mulch and complete dominance by a few tall plant species. On balance, high diversity on rangeland occurs most often at intermediate successional stages attained by grazing.

Measurements and standards of biodiversity are required by law in certain states for bonding release in mined land reclamation. In 1992, 24 federal, state, and university agencies in California agreed to work together on the proposition that "Sustaining the diversity and condition of its natural ecosystems is a prerequisite for maintaining the state's prosperity" (Biodiversity Executive Council, Memorandum of Understanding, the Agreement on Biological Diversity, Univ. California, Oakland, 6pp., 1992). Biodiversity is central to claims and actions by rangeland resource managers, conservationists, and environmentalists. However, biodiversity indices are difficult to interpret and have not been accepted by those who favor simpler measures such as a species list and measures of abundance and distribution (West 1993).

RANGELAND RESOURCES FOR MULTIPLE USERS

Multiple-use is defined as the use of the land for >1 product or service. The different uses may be consumptive (e.g., mining) or nonconsumptive (e.g., wildlife viewing). Multiple-use may be concurrent and cannot be separated like forage and water from the same land. Another type is sequential use in which 1 use follows another. A campground may serve families in the summer, hunters in the fall, and skiers in the winter. A third type is contiguous where uses occur side-by-side, as a concessionaire near a boat ramp. These different situations are compatible, but conflicts occur over increasing demand for space or time. Perhaps the basic problem is lack of procedures that adequately trade off market allocation, competitive allocation, and carrying capacity of the individual uses.

In addition to the biological production of multiple resources, multiple-use has elements of planning, politics, law, economics, and sociology. Conflicts for space and time are usually couched in these subjects, especially on use of public land where public opinion eventually wins. In contrast, multiple-use of private lands depends upon the owner's choice and profit produced. The "best" land use may differ for public and private land.

One problem area is publicly owned wildlife on private land where the landowner furnished food, cover, and water with insufficient or no compensation for the public service. This conflict continues although considerably reduced by such programs as damage control, compensation for damage, owners given partial control over hunter harvest in exchange for habitat improvement, and direct rebates via tags on hunters' licenses.

Multiple-use values may be set in the market place for harvested products such as timber, minerals, and forage. These and all other products from private or public land have social value, and gradually succumb to public pressures. Joyce (1989) described 4 categories of social value. Cultural values are illustrated in the "western way of life." Societal values focus on the community, as relations between the small western townspeople and the surrounding ranchers. Psychological values are illustrated by the tourist visit to a waterfowl refuge or just knowing of its existence. The fourth is physiological, illustrated by health benefits from hiking in clean air. Other societal values include maintenance of diversity, preservation of historic sites, and laboratory sites. Considerable controversy continues over livestock grazing and mining in wilderness.

A discussion of multiple-use on rangelands must include a word on wilderness. This part of the public land has been set aside for those who take nothing and leave nothing. It is roadless, to a degree, and permits no access by any type of motorized vehicle except in emergency. The major purpose is for people, often stated in terms of perpetuation of threatened and endangered species and providing unmanipulated ecosystems in which people can study. A subset of wilderness use is for individual personal growth, therapy, and education (Driver et al. 1990).

Models of tradeoffs among the values must be based on assumptions. These may be worthwhile but most decisions about them will be made in the courts or ballot box. In other models where at least a few market values exist, assumptions may give reasonable answers. Bastian et al. (1991) showed by linear programming of substitution rates that a wide combination of pronghorn and cattle yield more economic benefits than either animal alone.

Multiple-use of certain public lands became federal policy over 100 years ago and has been stipulated in most federal land laws since. Policy acceptance of multiple-use has been slow. The National Environmental Policy Act of 1969 and the resultant Environmental Impact Reports and Statements (EIS) require impact analysis of various alternative management plans. The real effect was analysis of combinations of multiple-uses on rangeland. The EISs were used as strong pressure to change the types and degree of uses. The new policy signaled the increasing importance of wildlife, clean air, clean water, open space, threatened and endangered species, genetic material, diversity, scenic quality, public recreation, protection, and overall environmental quality. Twenty years later, Joyce (1989) wrote that the United States' demand for wilderness, water quantity and quality, threatened and endangered species, and most aspects of outdoor recreation had outstripped demands for domestic animal grazing.

People using public land for a great variety of reasons have increased in number, which conflicts with the traditional products of livestock forage, timber, and minerals. Consequently the thrust of arguments for continued livestock grazing has turned toward management of livestock that does not conflict with other uses and that will maintain and improve wildlife habitat (Severson 1990). Proper livestock management has altered the botanical composition, increased biomass production, improved the nutritive quality of wildlife food, and altered the structure of vegetation. While these results are not universal, they are sufficiently frequent to warrant further development.

A symposium in the 1991 meeting of the American Association of Biological Sciences showed that herbivory by large animals is prehistoric in origin, that plants have evolved to withstand and escape defoliation, and that all grazing is not overgrazing. New importance was attached to the hypothesis that a defoliated plant may react more to competition from nearby undefoliated plants than from the loss of photosynthetic tissue. Return of pristine systems is not a viable rangeland resource objective. Managed livestock grazing is one of the few vegetation management tools available to wildlife habitat management (Heady 1994).

At this point a restatement of consumer management on rangeland is appropriate. The principal consumers are herbivores but for management purposes people may be considered consumers of space, air and water quality, facilities, and wildlife. There are 4 overall guiding principles. First and foremost, consumer numbers must be commensurate with resource carrying capacity. Second, distribution of the consumers should be in harmony with the distribution of resources. Third, use in certain seasons must not harm the resources, Fourth, a mixture of uses in most instances will be more productive of goods and services than a single use.

COORDINATED RESOURCE MANAGEMENT (CRM)

Conflicts in resource use at the local level have fostered a decision-making process known as CRM. Representatives of public and private interests work together to resolve land management conflicts. Coordination is achieved through thoughtful analysis, bargaining, and compromise. That may take a year or more because a dissenting opinion requires further compromise before the decision is made. The CRM is local, often involving a watershed and a single serious problem concerning it. A riparian problem that crosses several ownerships is an example and another is an overused winter deer range. A common goal is apparent and there is no assessment of blame. Successful CRMs usually finish 1 problem before they move to another. That first decision is landmark because it is proof that diverse interests can agree to management practices by bargaining.

Several additional items help sell a CRM and keep it moving. Solving a riparian problem has benefits like clear water and uniform stream flow that are easily identified. A recovered length of stream presents useful guidelines, and it is a location for discussion. The key persons should be few in number and have authority to speak for the landowner or agency they represent. Attainable goals need to be set and actions established on them that keep all parties involved and communicating. Results of monitoring and flexibility to meet indicated changes are necessary before moving to the next problem (Hancock 1989).

Outstanding results of coordinated resource problem solving include the Experimental Stewardship Programs at Challis, Idaho; New Pioneer in western Montana; Modoc/Washoe in northeastern California and adjacent Nevada; the Oregon Watershed Improvement Coalition; The Mulelshoe Ranch Preserve in Arizona; and the Timber, Fish, and Wildlife Agreement in Washington. The City of Fort Collins, Colorado, has included riparian area creation and development in its farsighted plans for open space. Benefits include storm water control and utility, an urban fishery, and numerous recreational facilities (Horak 1989). Planning the use of rangeland resources combines many interests. Cities of all sizes present opportunity for rangeland resource planners to gain public support. Guidelines for CRM planning that have developed over a 40-year period are available (Cleary and Phillippi 1993).

RANGELAND CONDITION AND TREND EVALUATION

This part of the description of rangeland resources is placed last because condition and trend evaluation must be oriented toward the ecosystems of soil, vegetation, and abiotic factors, and the multiple-use objectives, management inputs, and monitoring. Rangeland evaluation must be for stated purposes and grounded in the perspective of all kinds of animal, soil, and vegetation changes.

Rangeland condition is an inventory of resources and an evaluation of that inventory against a standard. For about 40 years it has been a measurement of present vegetation for each specifically named and designated range site. The standard has been the potential and most desirable vegetation for the site. In many situations, the potential vegetation has been the climax of Clementsian ecology, hence the range condition and trend system has been criticized for being based on a theoretical standard, rather than the desired ecosystem.

The range condition system was first applied and later perfected in the tallgrass and mixedgrass region, commonly known as the prairies, of the central United States (Dyksterhuis 1949). This is mostly cultivated private land and subject to severe droughts. By using the system, the Soil Conservation Service helped ranchers attain better management of their land. The goals were good to excellent stable grassland, erosion control, and sustained forage for livestock and wildlife. Two additional results are the overstated viewpoints that range management is applied Clementsian climax theory and is for livestock grazing alone. People trained in range management do not agree with this narrow view.

The validity of the range condition and trend principles has been increasingly questioned along 2 lines. One is that measurements of plant community do not measure or even apply to all the products and resources from rangeland. In opposition, is the view that vegetation at the producer level must be healthy, stable, and protect the soil to support all types of consumer demands. Soil cover, moderate grazing use, and advanced succession attain that level. Secondly, succession and climax lack supporting evidence for universal application. The response is that the system has been adequate in many vegetation types for the purposes of evaluating erosion control and production of forage. Fix it only where it does not apply.

Where it does not apply, major questions center on the causes and speed of vegetation change and on the nature of stability. For example, vegetation composition in the arid and semiarid regions can remain relatively stable for long periods, but change rapidly in response to intense disturbance. Westoby et al. (1989) called this the "state-and-transition" model of vegetation change and gave suggestions for data gathering and analysis. Hopefully, procedures for practical field application of rangeland evaluation based on these ideas will emerge. The following definitions recognize new ecological thinking and multiple-use pressures in evaluating rangeland condition and trend.

Ecological Site. An ecological site is defined as a kind of land in its ability to produce distinctive kinds and amounts of vegetation and in its response to management. The term *range site* connotes grazing by livestock or other named uses. Evaluation of a single site may be based on use condition or on vegetation changes and stability.

Desired Plant Community. Of the many plant communities that may occupy an ecological site, desired plant community is one that has been identified to best meet management objectives. This community conserves the site productivity and produces goods and services as stipulated in the management plan.

Site Conservation Rating. This is a rating of the effectiveness of the present vegetation to protect the site against accelerated erosion by wind and water. Site conservation rating is either good or bad depending upon the Soil Conservation Threshold, which separates geologic and accelerated erosion.

These terms and definitions have been accepted by the Society for Range Management, but have not become common usage in the profession. Considerable discussion and field trial lie ahead in evaluating and monitoring rangeland condition and trend analysis.

SUMMARY

Rangeland is a kind of land including deserts, grasslands, shrublands, savannas, and sometimes open forests and woodlands. Rangeland has many uses and is managed with little control over environmental conditions. Wildlife habitats have been defined by analysis of a single animal species place and with a data base that describes the general habitat requirements of all the resident species.

Rangeland resource evaluation has centered on the condition and trend of vegetation and soil on range sites for more than 6 decades. Following the "environmental 70s," evaluations of range sites have separated conditions for use from those of plant succession and stability.

The approaches for wildlife habitat and rangeland site evaluations have much in common. Some duplication in terminology occurs, and more understanding is needed by each profession of the other's requirements.

Rangeland plant community inventories are of 2 types: 1 based on actual vegetation and the other on potential vegetation. Habitat types and range sites concentrate on potential vegetation with considerable attention to actual vegetation. Major wildlife publications concentrate on actual vegetation with specific habitats further described.

Spatial division of vegetation and habitat requires definition of classification categories and determination of the limits of each unit. Mapping scale varies from 1:1 to 1:several million, depending upon purpose. Both extremes of scale and those between are accomplished by aerial imagery. Hundreds of vegetation maps exist. They are basic to land-use planning and must be made according to the objectives of each project.

A common and uniform vegetation classification system of rangeland resources does not exist for 2 major reasons. Each agency is responsible for different types of vegetation to which different inventory procedures apply. Subject matters (e.g., wildlife management, range management, teaching) require different kinds of information.

Data gathering systems with agreed upon inventory measurements that facilitate information storage and retrieval are evolving, but none has been fully implemented for rangelands. Vegetation changes over time may be hourly, seasonally, year-to-year, geological through evolution and climatic drift, and most important to current understanding and management, change on a directional or replacement basis called plant succession.

Although many persons agree that succession occurs, others show that it is not an orderly process and claim stability at various stages in the successional series. Herbivory, disturbance, competition, energy transfer, mineral cycling, physiological and morphological plant characteristics, change in diversity, and random chance where a seed lands are believed to cause or describe vegetation change and stability.

The concepts of autogenic plant succession leading to a stable climax and the basis of range condition and trend are in a period of intense activity. New terms have been suggested. Further changes will be forthcoming. The major challenge is developing a reporting system to define actual range condition and desired range condition for all users of rangeland. Thereby, trend can be evaluated.

Livestock and wildlife have not been emphasized. Rangeland resource ecology *for all users* is the approach taken. Coordinated Resource Management Planning procedures are suggested as a way to remove the "versus" from livestock and wildlife discussions. These 2 subject areas need each other for rangeland resource use and management. Without cooperation both are likely to fail (Williamson 1992).

LITERATURE CITED

Bastian, C. T., J. J. Jacobs, L. J. Held, and M. A. Smith. 1991. Multiple use of public rangeland: antelope and stocker cattle in Wyoming. J. Range Manage. 44:390-394.

Bonham, C. D. 1989. Measurements for terrestrial vegetation. John Wiley & Sons, Inc., New York, N.Y. 338pp.

Bryant, F. C., and S. Demarais. 1991. Habitat management guidelines for white-tailed deer in south and west Texas. Pages 9-13 *in* R. S. Lutz and D. B. Wester, eds. Vol. 22 Research highlights noxious brush and weed control, range and wildlife management. Texas Tech. Univ., Lubbock.

Caldwell, M. M., and J. H. Richards. 1986. Competitive position of species in respect to grazing tolerance: some perspective on ecophysiological processes. Pages 447-449 *in* P. J. Joss, P. W. Lynch, and O. B. Williams, eds. Rangelands: a resource under siege. Proc. Second Int. Rangeland Congr., Aust. Acad. Sci., Canberra, Australia.

Chambers, J. C. 1983. Measuring species diversity on revegetated surface mines: an evaluation of techniques. U.S. For. Serv. Res. Pap. INT-322. 15pp.

Cleary, C. R., and D. Phillippi. 1993. Coordinated resource management guidelines. Soc. Range Manage., Denver, Colo. 11 units paged separately.

Clements, F. E. 1916. Plant succession: an analysis of the development of vegetation. Carnegie Inst., Washington, D.C. Publ. 242. 512pp.

———. 1936. Nature and structure of the climax. J. Ecol. 24:252-284.

Collins, S. L., J. A. Bradford, and P. L. Sims. 1987. Succession and fluctuation in *Artemisia* dominated grassland. Vegetatio 73:89-99.

Cook, C. W., and J. Stubbendieck, eds. 1986. Range research: basic problems and techniques. Soc. Range Manage., Denver, Colo. 317pp.

Cooperrider, A. Y., R. J. Boyd, and H. R. Stuart, eds. 1986. Inventory and monitoring of wildlife habitat. U.S. Bur. Land Manage. Service Center, Denver, Colo. 858pp.

Darwin, C. R. 1859. On the origin of the species by means of natural selection, or the preservation of favoured races in the struggle for life. J. Murray, London, U.K. 502pp.

Dasmann, R. F. 1976. Environmental conservation. Fourth ed. John Wiley & Sons, Inc., New York, N.Y. 436pp.

Daubenmire, R. 1952. Forest vegetation of northern Idaho and adjacent Washington, and its bearing on concepts of vegetation classification. Ecol. Monogr. 22:301-330.

Dealy, J. E., D. A. Leckenby, and D. M. Concannon. 1981. Wildlife habitats in managed rangelands—The Great Basin of southeastern Oregon: plant communities and their importance to wildlife. U.S. For. Serv. Gen. Tech. Rep. PNW-120. 66pp.

Driver, B. L., A. T. Easley, and J. F. Passineau. 1990. Introductory comments on the benefits of wilderness. Pages 1-10 *in* A. T. Easley, J. F. Passineau, and B. L. Driver, compilers. The use of wilderness for personal growth, therapy, and education. U.S. For. Serv. Gen. Tech. Rep. RM-193.

Dyksterhuis, E. J. 1949. Condition and management of rangeland based on quantitative ecology. J. Range Manage. 2:104-115.

Ehrlich, P. R., D. S. Dobkin, and D. Wheye. 1988. The birder's handbook. Simon and Schuster, Inc., New York, N.Y. 785pp.

Ewing, R. A., N. Tosta, R. Tuazon, L. Huntsinger, R. Marose, K. Nielson, R. Motroni, and S. Turan. 1988. California's forests and rangelands: growing conflict over changing uses. Calif. Dep. For. and Fire Prot., Sacramento. 348pp.

Franklin, J. F., and C. T. Dyrness. 1973. Natural vegetation of Oregon and Washington. U.S. For. Serv. Gen. Tech. Rep. PNW-8. 417pp.

Giles, R. H., Jr. 1978. Wildlife management. W. H. Freeman and Co., San Francisco, Calif. 416pp.

Hancock, J. L. 1989. Selling a successful riparian management program: a public land manager's viewpoint. Pages 1-3 *in* R. E. Gresswell, B. A. Barton, and J. L. Kershner, eds. Practical approaches to riparian resource management: an educational workshop. U.S. Bur. Land Manage., Billings, Mont. 193pp.

Heady, H. F. 1973. Structure and function of climax. Pages 73-80 *in* Arid shrublands—Proc. third workshop of the United States/Australia panel. Soc. Range Manage., Denver, Colo.

———. 1975. Rangeland management. McGraw-Hill, Inc., New York, N.Y. 460pp.

———. 1994. Summary: ecological implications of livestock herbivory in the west. Pages 289-297 *in* M. Vavra, W. A. Laycock, R. D. Pieper, eds. Ecological implications of livestock herbivory in the west. Soc. Range Manage., Denver, Colo.

———, and R. D. Child. 1994. Rangeland ecology and management. Westview Press, Boulder, Colo. 519pp.

Holechek, J. L., R. D. Pieper, and C. H. Herbel. 1989. Range management principles and practices. Prentice-Hall, Englewood Cliffs, N.J. 501pp.

Horak, G. C. 1989. Integrated riparian planning in the urban setting. Pages 41-43 *in* R. E. Gresswell, B. A. Barton, and J.L. Kershner, eds. Practical approaches to riparian resource management: an educational workshop. Bur. Land Manage., Billings, Mont.

Joyce, L. A. 1989. An analysis of the range forage situation in the United States: 1989-2040. U.S. For. Serv. Gen. Tech. Rep. RM-180. 137pp.

———, D. E. Chalk, and A. D. Vigil. 1986. Range forage data base for 20 Great Plains, Southern, and Western States. U.S. For. Serv. Gen. Tech. Rep. RM-133. 17pp.

Kuchler, A. W. 1964. Potential natural vegetation of the conterminous United States. Am. Geogr. Soc., New York, N.Y. Spec. Publ. 36. 116pp.

Maser, C., and J. W. Thomas. 1983. Wildlife habitats in managed rangelands—The Great Basin of southeastern Oregon. U.S. For. Serv. Gen. Tech. Rep. PNW-160. 15pp.

Mayer, K. E., and W. F. Laudenslayer, Jr., eds. 1988. A guide to wildlife habitats of California. Calif. Dep. For. and Fire Prot., Sacramento, Calif. 166pp.

Mayeux, H. S., H. B. Johnson, and H. W. Polley. 1991. Global change and vegetation dynamics. Pages 62-74 *in* L. F. James, J. O. Evans, M. H. Ralphs, and R. D. Child, eds. Noxious range weeds. Westview Press, Boulder, Colo.

Mehringer, P. J., Jr., and P. E. Wigand. 1987. Western juniper in the Holocene. Pages 109-119 *in* R. L. Everett, compiler. Proc. pinyon-juniper conference. U.S. For. Serv. Gen. Tech. Rep. INT-215.

Merriam, C. H. 1890. Results of a biological survey of the San Francisco Mountain region and the desert of the Little Colorado, Arizona. U.S. Dep. Agric., North Am. Fauna 3:1-136.

Morrison, M. L., B. G. Marcot, and R. W. Mannan. 1992. Wildlife-habitat relationships: concepts and applications. Univ. Wisconsin Press, Madison. 343pp.

Ohmann, J. L. 1992. Wildlife habitats of the north coast of California: new techniques for extensive forest inventory. U.S. For Serv. Res. Rep. PNW-440.

Peek, J. M. 1986. A review of wildlife management. Prentice-Hall, Englewood Cliffs, N.J. 486pp.

———. 1989. On furthering the use of habitat classifications in wildlife habitat management in the northwestern United States. Pages 234-237 *in* D. E. Ferguson, P. Morgan, and F. D. Johnson, compilers. Proc. land classifications based on vegetation: applications for resource management. U.S. For Serv. Gen. Tech. Rep. INT-257.

Putman, R. J., A. D. Fowler, and S. Tout. 1991. Patterns of use of ancient grassland by cattle and horses and effects on vegetational composition and structure. Biol. Conserv. 56:329-347.

Redente, E. F., and E. J. DePuit. 1988. Reclamation of drastically disturbed rangelands. Pages 559-584 *in* Vegetation science applications for rangeland analysis and management. Kluwer Acad. Publ., Dordrecht, Netherlands.

Samuel, M. J., and E. J. DePuit. 1987. Competition and plant establishment. Pages 138-148 *in* G. W. Frasier, and R. A. Evans, eds. Proc. of symposium "seed and seedbed ecology of rangeland plants." U.S. Agric. Res. Serv., Wash., D.C.

Severson, K. E., tech. coord. 1990. Can livestock be used as tool to enhance wildlife habitat? U.S. For. Serv. Gen. Tech. Rep. RM-194. 123pp.

Shiflet, T. N., editor. 1994. Rangeland cover types of the United States. Soc. Range Manage., Denver, Colo. 152pp.

Silvertown, J. 1980. The dynamics of a grassland ecosystem: botanical equilibrium in the Park Grass Experiment. J. Appl. Ecol. 17:491-504.

Sneva, F. A., L. R. Rittenhouse, P. T. Tueller, and P. Reece. 1984. Changes in protected and grazed sagebrush-grass in eastern Oregon, 1937-1974. Oreg. State Univ. Agric. Exp. Stn. Bull. 663. 11pp.

Thomas, J. W., tech. ed. 1979. Wildlife habitats in managed forests, the Blue Mountains of Oregon and Washington. U.S. For. Serv. Agric. Handb. 553. 511pp.

Tiedemann, A. R. 1987. Nutrient accumulations in pinyon-juniper ecosystems—managing for future site productivity. Pages 352-359 *in* Proc. pinyon-juniper conference. U.S. For. Serv. Gen. Tech. Rep. INT-215.

Tueller, P. T. 1989. Remote sensing technology for rangeland management applications. J. Range Manage. 42:442-453.

Verner, J., and A. S. Boss, tech. coordinators. 1980. California wildlife and their habitats: Western Sierra Nevada. U.S. For. Serv. Gen. Tech. Rep. PSW-37. 439pp.

Viereck, L. A. 1989. Flood-plain succession and vegetation classification in interior Alaska. Pages 197-203 *in* D. E. Ferguson, P. Morgan, and F. D. Johnson, compilers. Proc. land classifications based on vegetation: applications for resource management. U.S. For. Serv. Gen. Tech. Rep. INT-257.

Wellner, C. A. 1989. Classification of habitat types in the western United States. Pages 7-21 *in* D. E. Ferguson, P. Morgan, and F. D. Johnson, compilers. Proc. land classifications based on vegetation: applications for resource management. U.S. For. Serv. Gen. Tech. Rep. INT-257.

West, N. E. 1993. Biodiversity of rangelands. J. Range Manage. 46:2-13.

Westoby, M., B. Walker, and I. Noy-Meir. 1989. Opportunistic management for rangelands not at equilibrium. J. Range Manage. 42:266-274.

Whittaker, R. H. 1972. Evolution and measurement of species diversity. Taxonomy 21:213-251.

Williams, W. A. 1966. Range improvement as related to net productivity, energy flow, and foliage configuration. J. Range Manage. 19:29-34.

Williamson, L. L. 1992. Identifying shinola. Rangelands 14:81-84.

World Resources Institute. 1992. World resources 1992-1993. Oxford Univ. Press., New York, N.Y. 385pp.

———, and The International Institute for Environment and Development. 1988. World resources 1988-1989. Basic Books, Inc., New York, N.Y. 372pp.

Yoakum. J. 1980. Habitat management guides for the American pronghorn antelope. U.S. Bur. Land Manage. Tech. Note 347. 77pp.

Zamora, B. A. 1982. Understory development in forest succession: an example from the inland Northwest. Pages 63-69 *in* J. E. Means, ed. Forest succession and stand development research in the Northwest. For. Res. Lab., Oreg. State Univ., Corvallis.

Chapter 2
THE BIRDS OF RANGELANDS
Eric G. Bolen and John A. Crawford

INTRODUCTION

Rangelands, from tallgrass prairie to desert grasslands, harbor an avifauna as diverse as the vegetation. Nonetheless, only a few species of rangeland birds are true grazers in those grassland ecosystems where grazing is an otherwise dominant feature of the fauna. In South America, some species of sheldgeese (*Chloephaga* spp.) compete with sheep for forage (Summers and Dunnett 1984), but associations of this sort are uncommon on North American ranges. One of the more interesting aspects of zoogeography concerns the absence of large, cursorial birds on the rangelands of the Northern Hemisphere, whereas each of the southern continents has remarkably similar species; i.e., ratites such as the rhea (*Rhea americana*) of South America and ostrich (*Struthio camelus*) of Africa.

Few studies have addressed the zoogeographical relationships of birds on grazing lands. In North America, Johnsgard (1978) determined that >50% of the birds breeding in the continental United States occur in the area defined as the Great Plains (i.e., about 260 species of birds regularly breed on just 17% of the area within the 48 contiguous United States). Hence, this relatively small area is surprisingly rich in birdlife. On the other hand, the number of grassland-adapted species in the Great Plains is small, typically represented by as few as 34 species (Table 1). Thus, the central grasslands of North America have not played a major role in avian speciation (Fig. 1), perhaps because there are relatively few niches (i.e., reduced vertical stratification, given the general lack of overstory vegetation). Drought tolerance may be another factor contributing to the low diversity of birds in some grassland communities (Zimmerman 1992). In addition, the plains seemingly served as an isolating agent, separating other regions such as boreal forest and southwestern shrublands in which speciation has been much more extensive. Mengel (1970) thus likened the function of the Great Plains to a central sea of open land. Johnsgard (1978: 97-112), expanding on this notion, described the Great Plains as an ocean

> physically separating major faunas to varying degrees, but also acting like a semipermeable barrier to the bird faunas associated with Canada's boreal forest, the deciduous forest of the eastern and southeastern United States, the Rocky Mountain coniferous forests, and the aridlands of the American southwest.

The dominance of eastern forest species in the avifauna of the Great Plains becomes apparent when the geographical and ecological affinities are analyzed (Table 2). As noted by Johnsgard (1978), grassland habitats cover >80% of the Great Plains, but grassland birds comprise only 11% of the total avifauna. Conversely, forest species represent >50% of the region's avifauna, although forest habitats occupy only 15% of the area. Similarly, the importance of wetlands in the Great Plains is equally clear; limnic birds make up 22% of the avifauna even though marshes, rivers, and lakes occupy only 1% of the area. Hawkins (1945:110) emphasized the latter point in relation to playa lakes on the southern Great Plains. He noted "Fewer than forty kinds of birds use the open plains . . . but give the plains a few lakes and it becomes a bird haven, increasing its bird potential by at least eighty species." Similar biogeographical analyses are not available for other regions of North America where grazing has been a dominant use of the land, although likely candidates for

TABLE 1. The North American grasslands avifauna. After Mengel (1970).

Non-passerines	Passerines
I. Primary species (endemic)	
1. Ferruginous hawk (*Buteo regalis*)	1. Sprague's pipit (*Anthus spragueii*)
2. Mountain plover (*Charadrius montana*)	2. Lark bunting (*Calamospiza melanocorus*)
3. Long-billed curlew (*Numenius americanus*)	3. Baird's sparrow (*Ammodramus bairdii*)
4. Marbled godwit (*Limosa fedoa*)	4. Cassin's sparrow (*Aimophila cassinii*)
5. Wilson's phalarope (*Phalaropus tricolor*)	5. McCown's longspur (*Calcarius mccownii*)
6. Franklin's gull (*Larus pipixcan*)	6. Chestnut-collared longspur (*Calcarius ornatus*)
II. Secondary species (more widespread)	
1. Mississippi kite (*Ictinia mississippiensis*)	1. Horned lark (*Eremophila alpestris*)
2. Swainson's hawk (*Buteo swainsoni*)	2. Sage thrasher (*Oreoscoptes montanus*)
3. Marsh hawk (*Circus cyaneus*)	3. Eastern meadowlark (*Sturnella magna*)
4. Prairie falcon (*Falco mexicanus*)	4. Western meadowlark (*Sturnella neglecta*)
5. Greater prairie chicken (*Tympanuchus cupido*)	5. Dickcissel (*Spiza americana*)
6. Lesser prairie chicken (*Tympanuchus pallidicinctus*)	6. Green-tailed towhee (*Pipilo chlorurus*)
7. Sharp-tailed grouse (*Tympanuchus phasianellus*)	7. Savannah sparrow (*Passerculus sandwichensis*)
8. Sage grouse (*Centrocercus urophasianus*)	8. Grasshopper sparrow (*Ammodramus savannarum*)
9. Upland sandpiper (*Bartramia longicauda*)	9. Henslow's sparrow (*Ammodramus henslowii*)
10. Burrowing owl (*Athene cunicularia*)	10. Vesper sparrow (*Pooecetes gramineus*)
11. Short-eared owl (*Asio flammeus*)	11. Lark sparrow (*Chondestes grammacus*)
	12. Sage sparrow (*Amphispiza belli*)
	13. Brewer's sparrow (*Spizella breweri*)
	14. Clay-colored sparrow (*Spizella pallida*)

TABLE 2. Ecological and geographical affinities of the Great Plains avifauna. After Johnsgard (1978).

Ecological affinities	Eastern	Northern	Western	Southern	Pandemic	Endemic	Introduced	Totals	%
Woodland and forest birds	63	37	25	8	31	—	—	164	51
Limnic birds	14	12	17	—	25	2	—	70	22
Grassland birds	4	1	7	—	7	15	—	34	11
Xeric scrub birds	—	—	7	8	—	—	—	15	4
Other birds	6	5	5	8	12	—	5	41	12
Totals	87	55	61	24	75	17	5	324	
%	27	17	19	7	23	5	2		100

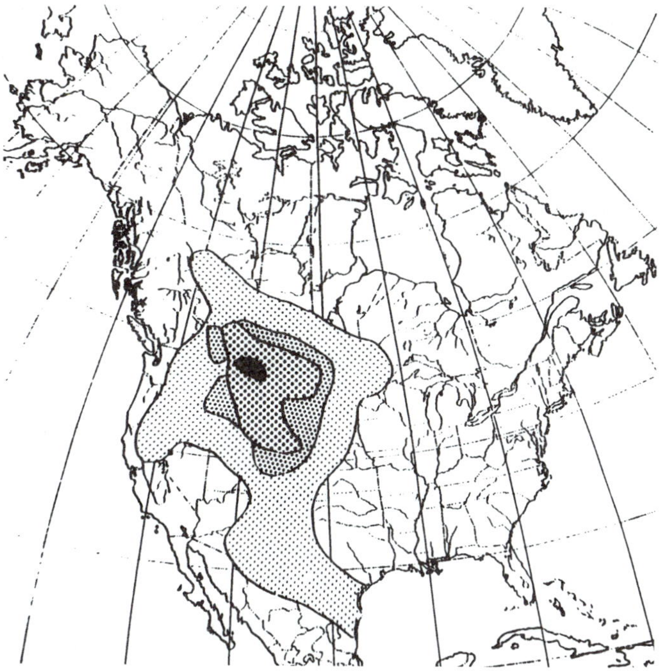

FIGURE 1. Despite the large number of birds associated with a large area of North America, relatively few are endemic to grassland environments. Species richness increases concentrically from 1 to 3 species in the outermost stippled area to 10 to 12 species in the blackened center (see list in Group I, Table 1). After Mengle (1970).

research of this type include such areas as the Palouse Prairie and Great Basin (see Ryser 1985).

Despite the relative paucity of birds specifically adapted to grasslands and other ranges, the avifauna of these environments is represented by a number of orders, some of which are mentioned in the following sections in which we note selected species of interest and highlight some biological features peculiar to each. Our purpose is to provide a broad-brush introduction to the many kinds of birds occurring on the rangelands of North America.

SELECTED GROUPS OF RANGELAND BIRDS

Anseriformes

Prairie wetlands are important to breeding, migrating, and wintering waterfowl, particularly ducks such as mallards (*Anas platyrhynchos*), northern pintails (*A. acuta*), canvasbacks (*Aythya valisneria*), and redheads (*A. americana*). Foremost among these wetlands are the prairie potholes located in a wide belt extending from western Minnesota across the Dakotas into eastern Montana, and including large areas in the southern regions of Manitoba, Saskatchewan, and Alberta. Prairie potholes are popularly known as "duck factories" because of the immense numbers of ducks using these wetlands as breeding habitat. The potholes provide only 10% of the breeding area for waterfowl in North America, but produce 50% of the duck crop in an average year, and even more when conditions are optimal (Smith et al. 1964).

Because periodic drought is a normal phenomenon in grassland ecosystems, it is not surprising that some species of waterfowl adapted to its occurrence. During drought years, for example, northern pintails move farther north, leaving the prairie pothole region for breeding areas where wetlands are more permanent. Smith (1970) thus described a strong inverse relationship between availability of water-filled prairie potholes and the distribution of the breeding population of pintails. He also found a similar relationship between breeding success and the distribution of the pintail population. When the potholes were dry, pintails moved north where they experienced poorer production (Fig. 2).

At least 2 conclusions stem from these relationships. First, breeding pintails seemingly prefer unstable wetlands, but respond to changing conditions and avoid strong attachments to specific sites (i.e., pintails do not remain in drought-stricken areas). Second, production is low during periods of drought, therefore the survival of a viable adult population becomes paramount. The northern areas thus supply the prairie potholes with a reservoir of breeding adults when the drought ends (see also Hansen and McKnight 1964). In contrast, lesser scaup (*Aythya affinis*) faced with drought on the northern prairies often leave their breeding areas early in the season. Of those that remain, only a few attempt to nest; most simply loaf in groups of pairs while the reproductive organs of the females regress (Rogers 1964). In short, drought inhibits reproductive efforts in lesser scaup, a species whose breeding range normally lies north of the drought-prone prairie pothole region.

The playa lakes dotting the Southern High Plains in the Texas Panhandle and adjacent states offer an example of a major wintering area for waterfowl (Bolen et al. 1989*a*). With favorable conditions, some 25,000 of these circular wetlands serve as winter habitat for >1,000,000 ducks each year. Four species form the bulk of the winter population: mallard, northern pintail, American wigeon (*Anas americana*), and green-winged teal (*A. crecca*), although several other species overwinter or visit the playas (Bolen et al. 1989*b*). Wigeon in some years may comprise 30% of the total duck population wintering in the region (Soutiere et al. 1972). Canada geese (*Branta canadensis*) also overwinter in the playa region, which was once a shortgrass prairie but now is one of the more heavily cultivated areas in North America. The racial composition within this population of Canada geese is presently unknown, although it may include most of the subspecies mentioned by Grieb (1970). In recent years, the winter population of lesser snow geese (*Chen caerulescens*) has increased steadily on the Southern High Plains (L. Smith, Texas Tech Univ., pers. commun.).

Bolen and Rylander (1978) compared the lamallar structure of Canada geese (i.e., grazers) with lesser snow geese (i.e., grubbers). The latter species has stouter maxillary and mandibular serrations than the former, which likely reflects adaptations associated with their respective feeding niches. In this regard, the grubbing behavior of snow geese frequently damages the coastal prairies and marshes of Texas and Louisiana, prompting longstanding complaints from local cattlemen (Lynch et al. 1947).

Much has been written about the influences of grazing on waterfowl production. Most research indicated that grazing is

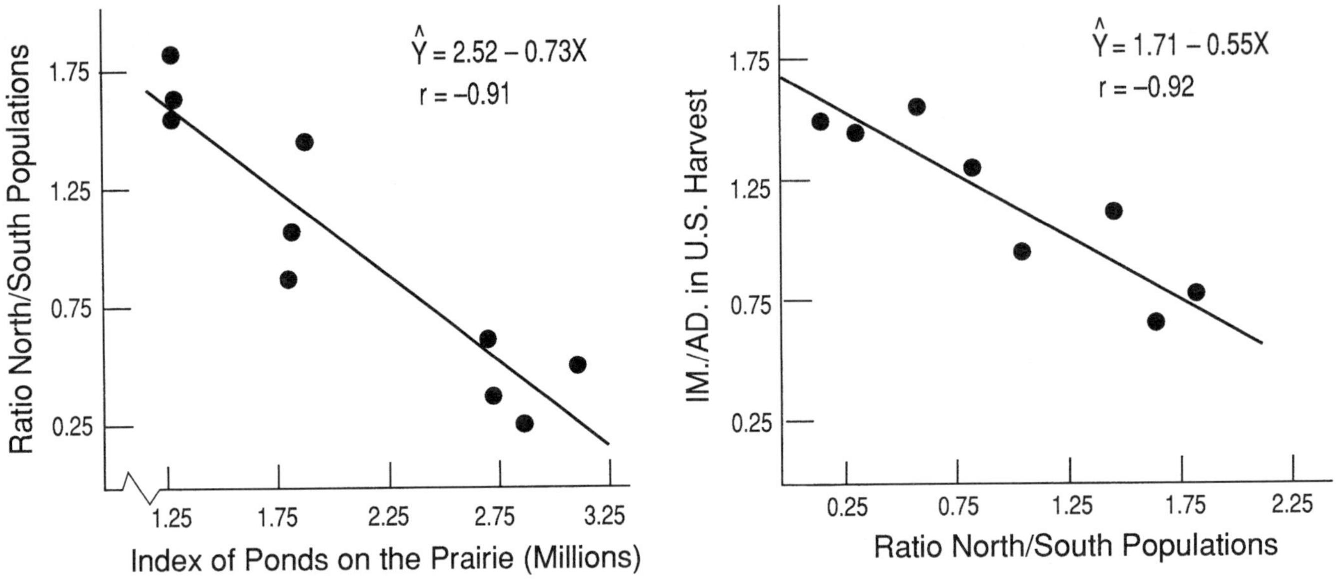

FIGURE 2. Drought displaces pintails to wetlands well north of their usual nesting areas, as shown by the ratio of northern and southern populations in relation to the availability of prime nesting habitat ("prairie potholes") in southern Canada (left). However, production falls significantly under these circumstances (right). After Smith (1970).

detrimental to waterfowl, largely because livestock remove much of the cover that helps ensure the successful nesting of pintails and other upland-nesting ducks. Indeed, after examining the literature on this topic, Braun et al. (1978) noted that ≥55 studies demonstrated that grazing was detrimental to waterfowl production; only 1 study reported otherwise.

Falconiformes

Not many raptors are exclusively associated with rangelands, although a few have relatively strong affinities with rangeland environments in North America. The prairie falcon (*Falco mexicanus*) is aptly named, but like many grassland birds, its habitat is broad and includes more than the prairies for which it is named. Nonetheless, this somewhat uncommon species occurs in grassland systems in western North America from southcentral Canada well into Mexico. Moreover, as pointed out by Mengel (1970), prairie falcons prey heavily on 2 grassland species in much of their breeding range: horned larks (*Eremophila alpestris*) and Richardson's ground squirrels (*Spermophilus richardsonii*). Prairie falcons were among the birds of prey whose reproduction was harmed by exposure to pesticides (Enderson and Berger 1970). Chlorinated hydrocarbons inhibited the function of the enzyme carbonic anhydrase, thereby interrupting the normal transport of calcium to the oviducts of contaminated prairie falcons. Thus, with diminished calcium, the eggshells of falcons contaminated with DDT and its sister compounds and derivatives were abnormally thin and broke during incubation. The reproductive failure in prairie falcons was directly related to this impairment (Table 3).

Golden eagles (*Aquila chrysaetos*) also occupy a broad geographical range, including western rangelands. They feed chiefly on rodents and lagamorphs, but have acquired a reputation among ranchers as predators of sheep and goats. The severity of eagle predation on livestock undoubtedly varies in time, space, and ecological setting, and the subject remains controversial wherever eagles share rangelands with domestic sheep (Bolen 1975). Golden eagles have been persecuted, principally by poisoning and by shooting from aircraft, resulting in heavy fines and other punishment for those who perpetrate such illegal actions.

TABLE 3. Relationships between production of young and eggshell thickness in prairie falcons. After Enderson and Berger (1970).

Eggshell thickness index	Prairie falcon nests		No. young fledged	
	Total	% fledgling ≥ 1 young	Total	Per pair
>1.75	21	75	54	2.6
1.45–1.75	31	50	51	1.6
<1.45	10	10	3	0.3

Two buteos, those heavy-bodied hawks with broad wings and wedge-shaped tails, are especially notable as rangeland raptors. The ferruginous hawk (*Buteo regalis*), with a wingspan of about 1.4 m, is the largest of the North American buteos and shares with the golden eagle the attribute of feathered tarsi. This species breeds over a wide area of the western Great Plains and the Great Basin and adjoining areas. The Swainson's hawk (*B. swainsoni*) also is widely distributed across the plains and open areas of western North America. Both buteos feed extensively on rodents, although Swainson's hawks at times consume large numbers of grasshoppers.

Each of the foregoing species is among the raptors nesting at the Snake River Birds of Prey Area, which is managed by the Bureau of Land Management. The Area contains the largest concentration of noncolonial nesting raptors in the world. More than 1,500 birds of prey breed on this 130-km stretch of canyon in southern Idaho, including >200 pairs of prairie falcons, approximately 5% of the entire world population (Kochert and Pellant 1986).

Final mention goes to the Mississippi kite (*Ictinia mississippiensis*), a sleek-bodied species commonly breeding on the southern plains of Texas, Oklahoma, and Kansas, and as far east and west, respectively, as the Carolinas and Arizona. During the Dust Bowl of the 1930s, the Prairie States Forestry Project established some 32,000 km of shelterbelts, which provided new nesting habitat, and the breeding range for Mississippi kites thereafter shifted westward (Love and Knopf 1978, Bolen and Flores 1989). Mississippi kites are skillful fliers, catching and eating cicadas (Cicadidae) and other insects on the wing. Biologists have sometimes observed Mississippi kites feeding on insects flushed into flight by the movements of ungulates (Jackson 1945), hence we may envision a primal setting in which kites regularly swept over the legions of bison (*Bison bison*) once grazing the southern plains (Bolen and Flores 1993).

Galliformes

Perhaps no group of birds better fulfills an image of rangeland than several species of the subfamily Tetraoninae, for which Johnsgard (1983) offers a fine overview. Specifically, these include the sage grouse (*Centrocercus urophasianus*), sharp-tailed grouse (*Tympanuchus phasianellus*), lesser prairie chicken (*T. pallidicinctus*), and greater prairie chicken (*T. cupido*).

In spring, the males of these 4 species of grouse vigorously court on communal display grounds, called leks, which have long served as convenient sites for conducting annual population surveys (Davison 1940, Lehman 1941, Douglass 1942). Lek behavior offers an interesting contrast between the behavioral adaptations of grassland birds and those related species living in forests where vision is limited (Fig. 3). The grouse of the open range make strong use of visual stimuli by dancing and displaying colorful air sacs, whereas ruffed grouse (*Bonasa umbellus*), a species of the forest interior, rely heavily on sound as the primary means of courtship and territorial defense (Bump et al. 1947, Johnsgard 1973).

FIGURE 3. Sage grouse lek in Wyoming. Note the adaptation of this visual form of courtship behavior to the openness of the habitat, whereas grouse associated with forests, where visibility is limited, rely almost exclusively on long-distance sound for communication. Photo courtesy USDA Forest Service.

Lek sites are used year after year. Wiley (1978) mentioned a sage grouse lek in Wyoming in use for 28 years; this and other leks he studied normally cover about 2 ha, each with a mating center about 10 m in diameter. Some 50–60 male sage grouse display on territories in a typical lek, but extremely large leks, which have several mating centers, may accommodate >400 males (see Johnsgard 1973 for corresponding data in prairie chickens and sharp-tailed grouse). In all, ≥ 90% of the copulations are accomplished by just 10% of the males; the dominance of males increases proportionately for those whose territories are nearer to the mating center. Conversely, essentially all female sage grouse, and other lek-forming galliformes, mate each year. For birds displaying on open rangelands, lek behavior offers the advantage of safety in numbers, which protects courting males from surprise attacks by predators (Wiley 1978). The alternating light and dark bands on the plumage of prairie chickens mimic the colors and shades of grassland vegetation and serve as near-perfect camouflage.

Northern greater prairie chickens (*T. c. pinnatus*) once inhabited mid-continental grasslands from Ohio to Texas and the prairie provinces of Canada. Two other subspecies of the greater prairie chicken occupied coastal grasslands. One of these, the heath hen (*T. c. cupido*), was found from Massachusetts south to Maryland, but intensive human uses of land steadily compressed its range until the last bird died in 1932 (Greenway 1958). The second subspecies is the Attwater prairie chicken (*T. c. attwateri*), once found along much of the Texas Gulf Coast but now highly restricted to several small, isolated populations and officially listed as an endangered species. Lehman (1941) estimated the peak abundance of Attwater prairie chickens at 1,000,000 birds, but unmitigated exploitation and habitat loss soon diminished the population to alarmingly low levels. One account tells of hunting parties of 10–20 men who tallied their kills in piles of 100 birds, which were left to rot at the campsite (Lehman 1941).

The range of the northern greater prairie chicken shifted westward as agricultural practices intensified in the eastern part of their range. Ironically, the development of low-intensity agriculture initially facilitated conditions for these birds on the western prairies, but intensive farming later reduced the availability of suitable habitat and their populations declined. Prairie chickens were common fare in the kitchens of settlers, and later in the 19th century the birds were hunted for market. In an attempt to conserve wildlife resources, Iowa became the first state (in 1878) to institute a bag limit on the harvest of game: 25 prairie chickens per day!

Lesser prairie chickens formerly ranged across the southern parts of the shortgrass prairie, where they were closely associated with bluestem grasses (*Andropogon* and *Schizachyrium*), sand sage (*Artemisia filifolia*), and shinnery oak (*Quercus havardii*). Once again, however, habitat changes imposed by land-use practices, particularly the widespread conversion of rangeland to row crops, diminished the populations of lesser prairie chickens. In Texas, Crawford and Bolen (1976) concluded that areas with <63% rangeland appeared incapable of supporting stable populations of lesser prairie chickens. Additionally, the resulting patchwork of habitat apparently interrupted the migratory behavior of the species (Crawford 1980).

The range of sharp-tailed grouse once extended from northern Michigan and western Quebec to the panhandle regions of Texas and Oklahoma westward to northern California and north to Alaska. In the southern part of this range, sharp-tailed grouse largely associate with grasslands whereas the early seral stages of the boreal forest are preferred by northern birds. The southern populations have diminished, especially in the midwest, intermountain regions, and the Palouse Prairie of the Pacific Northwest.

Sage grouse favor shrub-steppe habitats in much of the intermountain West and Great Basin, where they are closely associated with sagebrush (*Artemisia* spp.) for much of their food and cover. Control of sagebrush, largely with herbicides, thus has adversely affected sage grouse; spraying reduces breeding activities, nesting habitat, and winter habitat. In Idaho, for example, Klebenow (1970) discovered marked reductions in densities of sage grouse nests on sprayed ranges. Similarly, the numbers of male sage grouse declined by 63% during a 2-year period after a large block of sagebrush in Montana was treated with herbicide (Wallestad 1975). Management guidelines prepared by the Western States Sage Grouse Committee proposed the elimination of sagebrush control in several settings: where live sagebrush cover is < 20%; within 3 km of leks; and on critical wintering areas and along riparian zones (Braun et al. 1977).

Northern bobwhites (*Colinus virginianus*) occur in many environments, including those considered rangelands. Grasslands in Texas, for example, provide food and cover for bobwhites (Wilson and Crawford 1987). A western race, the masked bobwhite (*C. v. ridgwayi*), once occurred in the grasslands in southern Arizona and Sonora in northern Mexico, but severe overgrazing so depleted the range that the remaining population faces extinction (Tomlinson 1972). Scaled quail (*Callipepla squamata*), also known as blue quail or simply "cottontops," are familiar grassland species in much of the arid and semiarid Southwest. Farther west, 2 closely related species, Gambel's quail (*C. gambelii*) and California quail (*C. californica*), also are rangeland birds. Gambel's quail occur in the very arid ranges of the Southwest, whereas California quail are found in grassland, shrub, and agricultural areas of Baja, California, and Arizona. California quail also have been introduced into suitable mesic environments in the Pacific Northwest. In the southern parts of Arizona and New Mexico, and the Trans-Pecos of Texas, the harlequin or Montezuma quail (*Cyrtonyx montezumae*) inhabits oak (*Quercus* spp.)-grassland areas.

Only 1 species of wild turkey (*Meleagris gallopavo*) occurs in North America north of Mexico and, for the most part, this species is associated with forested habitats. However, among the various races, the Rio Grande (*M. g. intermedia*) and Merriam's (*M. g. merriami*) wild turkeys use rangelands. Both races have been used widely in reintroduction programs in much of the American West, including locations where no wild turkeys previously occurred (Bailey 1980). In some cases, ecological mistakes were made when the racial stocks were mismatched with the reintroduction sites. For example, the range of the Rio Grande wild turkey is bordered by isohyets of about 41 cm on the west and 81 cm on the east; thus when stocks of this race were released outside these boundaries (e.g., the eastern seaboard), the birds proved ill-adapted to the more humid environment and the efforts failed (Glazener 1967). Lacking such lapses, however, restocking efforts across the country have returned the wild turkey to its former position of prominence among American game birds. The availability of roost sites, especially on rangeland habitats, is among the factors that frequently limit the distribution of wild turkeys (Boeker and Scott 1969).

Two species of gallinaceous birds have been introduced successfully into North American rangelands. Chukars (*Alectoris chukar*) are associated with rocky hillsides, particularly in the Great Basin where the exotic cheatgrass (*Bromus tectorum*) is available as a dietary stable. Chukar stock originated from various locales in Asia minor (e.g., Turkey and Iran). Gray partridges (*Perdix perdix*), introduced from Europe, today occur in many areas of the northern United States and the prairie provinces of Canada, often in association with irrigated agriculture. Commonly known as Hungarian partridges, or "Huns," gray partridges thrive on ranges dominated by bunchgrasses (e.g., *Festuca* spp. and *Agropyron* spp.).

Another exotic, arguably the best known of the gamebirds, cannot be regarded as a rangeland species. Ring-necked pheasants (*Phasianus colchicus*), however, may inhabit grasslands or shrublands adjacent to crops such as wheat and corn. Where pheasants share a common range with prairie chickens, some evidence suggests that the exotics physically dominate the native species. In the Nebraska Sandhills, Sharp (1957) recorded persistent attacks of pheasants that eventually drove many prairie chickens from their booming grounds. Such strife may have contributed to the declining population, possibly even eliminating isolated pockets of prairie chickens from their remaining geographical range. Vance and Westemeier (1979) encountered a similar situation in Illinois, including an instance in which a single pheasant apparently dominated a booming ground with 80 prairie chickens. The Illinois study also revealed instances of nest parasitism that reduced the hatching success of prairie chicken nests from 51% in unparasitized nests to 24% in nests where pheasants had added their eggs.

Charadriiformes

A large number of shorebirds frequent the shallow wetlands associated with the rangelands of North America. Many of these, including a large number of sandpipers (Scolopacidae), cross the plains and prairies during migration. Other shorebirds, such as the American avocet (*Recurvirostra americana*) and black-necked stilt (*Himantopus mexicanus*), breed on the shores of lakes and marshes in the western United States. Among the more interesting birds nesting on prairie sloughs and other wetlands is Wilson's phalarope (*Steganopus tricolor*), which represents a group that is otherwise pelagic. As part of their feeding behavior, these birds whirl about in shallow water, thereby bringing to the surface food that could not be reached by probing (Johns 1969). Phalaropes are also one of the relatively few vertebrate taxa with polyandrous breeding habits (Hohn 1969).

Nearly everyone is familiar with the killdeer (*Charadrius vociferus*) and its entertaining "broken-wing" antics (a classic example of a distraction display) during the nesting season. The eggs of killdeer are spotted and blotched, and blend remarkably with the stony substrates that serve as nest locations. Mountain plovers (*C. montana*), which are not birds of montane environments, have a breeding distribution somewhat similar to the long-billed curlew's (*Numenius americanus*). Upland sandpipers (*Bartramia longicauda*) breed on grasslands as far north as Alaska. Most of the plover group (subfamily Charadriinae) have black head and breast markings, which seem to function as disruptive coloration during nesting and, for at least some species, as reinforcement for courtship and aggressive displays (Graul 1973).

Long-billed curlews also nest on the open grasslands from the Texas Panhandle to the southern edges of prairie Canada (see Allen 1980). Whereas the eggs of this species also are well camouflaged, many nests are placed near cow dung, which suggests the possibility that an incubating curlew may resemble just another "chip" on a closely clipped grassland (King 1978). Curlews also at times flip over dried cow chips in search of insect foods. The underside of dung is often rich in insect life, which flourishes in the microcosm of moisture when surface soils elsewhere are dry (Hayes 1927, Mohr 1943). McDaniel and Balsbaugh (1968), for example, found 10 families of beetles (Coleoptera) under cow chips in South Dakota. Such relationships between birds and dung on rangelands surely evolved during the long association with bison and continue today with the homolog provided by cattle.

Of the various gulls and terns (Laridae), Franklin's gull (*Larus pipixcan*) exhibits a strong association with the continent's interior grasslands. Whereas the species was described early in the last century, it was not until the western plains were settled and plowed that the immensity of their numbers was realized. Sod-breaking settlers knew the species as a "prairie dove," whose flocks arrived each spring when the land thawed and gathered invertebrates exposed in freshly plowed soil (Job 1910). Today, Franklin's gulls still swarm behind even the noisiest of tractors in search of food. Nonetheless, their seasonal movements are erratic, likened by Bent (1921) to the wanderings of Plains Indians. The California gull (*L. californicus*), despite the implications of its name, also is bird of the western plains during the breeding season. In 1848, the species gained fame in the history of Utah by checking a crop-threatening plague of crickets, thereby saving the Mormon settlers from probable starvation. In commemoration, a bronze California gull now tops a prominent monument in Salt Lake City.

Columbiformes

The distribution of mourning doves (*Zenaida macroura*) stretches from coast to coast across North America, including all western rangelands in the United States and Canada. Mourning doves also are numerous, being among the more abundant birds in the United States. Some estimates suggest an autumn population of 500 million birds, and where they are classified as game birds, the harvest of mourning doves runs into the tens of millions of birds each year.

Hence, the species features an unusually high reproductive potential, which in part results from plasticity in its selection of nesting habitat. In addition to nesting in rural environments of nearly every type, mourning doves also adapt well to urban areas (Swank 1955, Leopold and Dedon 1983). Moreover, although mourning doves commonly nest in trees and shrubs, they also nest on the ground. Consequently, doves may fare well on rangelands where fires, either natural or prescribed, remove woody vegetation otherwise providing elevated supports for nests. Indeed, based on work in Texas, ground and tree nests experience similar rates of predation and, overall, production from the ground nests largely compensated for the loss of trees on burned ranges (Soutiere and Bolen 1972, 1976). The densities of ground nests diminished, however, each year after burning, indicating that ground cover played a role in the way mourning doves responded to burned ranges. On cold desert ranges in Idaho, the shading and overhead cover afforded by big sagebrush (*Artemisia tridentata*) is important to the placement of dove nests on the ground; hence practices that maintain shrub cover should be encouraged for the management of mourning doves in these areas (Howe and Flake 1989).

On ranges with few trees or shrubs, the bulk of annual production may in fact result from ground nests. Despite their low density, ground nests account for nearly 60% of the total production of doves in northwestern Oklahoma (Downing 1959). Howe and Flake (1989) recorded an unusually high nesting success of 63% from a small sample of ground nests in Idaho compared with 29% success for ground nests in Oklahoma.

The reproductive strategy of mourning doves provides an interesting contrast with some of the other common game birds associated with rangelands. Quail, for example, typically have only 1 successful nest during each season but they have large clutches (commonly 10–16 eggs), which results in a high reproductive output. In contrast, mourning doves produce small clutches (normally 2 eggs), but they nest 3 to 7 times each breeding season. Quail and doves thus offset their high rates of annual mortality by producing many young, but each accomplishes that end with a different reproductive strategy.

Cuculiformes

Roadrunners (*Geococcyx californianus*) are engaging birds associated with arid and semiarid ranges throughout much of the West and Southwest. Curiously, however, no studies have addressed the influences grazing may exert on the ecology of these interesting birds despite a broad overlap of their geographical distribution and the lands used extensively by livestock.

Roadrunners are predators of lizards, snakes, young birds, and a variety of arthropods, which both parents also secure for their nestlings during the breeding season. The young hatch asynchronously, thus suggesting an adaptation somewhat similar to the hatching strategy of raptors; namely, that the younger nestlings are killed and eaten by the adults when food is not abundant (Ohmart 1973). Moreover, the clutch size, normally 2–8 eggs but at times as many as 12, apparently varies in keeping with the availability of food, at least on highly arid ranges in Arizona, with more eggs at times when food is readily available and fewer in lean times. Conversely, clutch size did not differ by season in southern Texas (Folse and Arnold 1978), perhaps because the setting was more mesic than in the former study and food supplies thus were more stable.

Because lizards, one of their primary foods, are active only during the cooler part of the day, roadrunners are able to hunt without exposing their nestlings to excessive heat loads (Ohmart 1973). That is, the nestlings can be left unshaded in the morning while the parents hunt, whereas during the heat of the afternoon, lizards are no longer active and the adults return to the nest to shade their young. Also, 4–5 days after hatching the black skin of the nestlings permits solar brooding, thereby enabling the parents to hunt freely during the morning hours.

On the Santa Rita Experimental Range in Arizona, Calder (1968) reported that roadrunners sought the shade of mesquite at midday; they also employed gaping, gular flutter, and extended their wings away from their bodies as other means of cooling. Conversely, roadrunners enter into hypothermy during the winter when they use low intensity solar energy to maintain their body temperature (Ohmart and Lasiewski 1971). When sunning for solar energy, the birds expose the black skin and black plumage on their backs, which otherwise remain covered. Roadrunners thus adapt for energy conservation during the winter when the desert's productivity is at its lowest ebb.

Strigiformes

The distribution of several species of owls includes rangeland environments (see Marti 1974). However, because of their interesting ecology in relation to rangeland, 2 species warrant mention here. Burrowing owls (*Athene cunicularia hypugaea*) are uniquely associated with prairie dogs (*Cygnomys ludovicianus*), whose burrows serve as nest sites for these interesting birds of the western plains (Fig. 4). Butts and Lewis (1982) recorded a density of 1 owl/1.9 ha of prairie dog town in Oklahoma. In other locations, the owls adopt the burrows of ground squirrels (*Spermophilus* spp.). Interestingly, an eastern race (*A. c. floridana*) occurs in a highly disjunct population in southern Florida, often in urbanized areas and without the presence of prairie dogs or other burrowing mammals (Wesemann and Rowe 1987). Debate continues as to whether the owls may at times excavate their own burrows, but preferences for certain types of soil are apparent (Butts and Lewis 1982). In addition to their unusual fossorial nesting habits, burrowing owls are active during the day and, much to the delight of birdwatchers, are easily observed peering from the mounds at the entrance of their burrows. Even so, the birds may be more active at night than previously believed (Zarn 1974). The legs of burrowing owls are uncommonly long, which may play a role in the thermoregulation necessary for birds standing on bare ground in hot environments (Coulombe 1971; see also Coulombe 1970).

Burrowing owls are largely insectivorous, although a variety of other foods are consumed (e.g., crayfish and scorpions). A diet of rodents and other vertebrate foods may be more common in the winter, when insects are scarce (Coulombe 1971). The switch to a winter diet dominated by rodents seems correlated with the birds adopting more nocturnal habits (i.e., most rodents are active nocturnally, hence so are the owls). In most raptors, females are larger than males (reversed sexual dimorphism) but this relationship does not hold for burrowing owls (Earhart and Johnson 1970). Many theories, including those associated with the role each sex plays in securing food, have been variously examined, but none satisfactorily explained either the presence of reversed sexual dimorphism in virtually all species of owls or the specific lack of its occurrence in burrowing owls (Mueller 1986). It would be tempting to associate the latter enigma with the fossorial habits of burrowing owls, apparently an unexplored idea, but we cannot offer any rational hypothesis to explain such a relationship.

Burrowing owls collect dried dung as a lining for the entrances, tunnels, and nests of their underground chambers, possibly for insulation and as a means of camouflaging their scent from predators (Martin 1973*a*). Today, the dung of horses, cows, and dogs is used, but no doubt bison once were the primary source of this material. During the nesting season, active burrows may be censused by the presence of dung, which the owls quickly replace whenever it is removed (Martin 1973*a*, Zarn 1974). Moreover, dung beetles (Scarabaeidae) may form a significant item in the diet of burrowing owls (Scott 1940).

While still confined to their nests, nestling burrowing owls employ a singularly distinctive adaptation: when disturbed, their calls mimic the buzz of rattlesnakes (*Crotalus* spp.), a clearly valuable defensive mechanism for helpless birds in a confined space. Martin (1973*b*), in fact, used spectrograms to compare the vocalizations of young burrowing owls with the buzz of prairie rattlesnakes (*C. viridis*), which disclosed "good quantitative evidence demonstrating basic similarity in the sounds." Moreover, Rowe et al. (1986) experimentally determined that ground squirrels whose distribution is sympatric with rattlesnakes indeed responded to the audio mimicry of burrowing owls, whereas ground squirrels from areas where rattlesnakes are absent did not react similarly.

FIGURE 4. The western race of burrowing owl shares a closely knit ecological association with prairie dogs and, to some degree, rattlesnakes. Photo courtesy of Dan L. Flores.

Because prairie dogs are colonial, burrowing owls also are a somewhat colonial species. And because the abundance of burrowing owl populations seems related to the abundance of prairie dogs and ground squirrels, measures aimed at controlling these burrowing rangeland mammals indirectly reduce the numbers of burrowing owls (Butts and Lewis 1982). Accordingly, various public and private conservation agencies at times have expressed their concerns for the species. Other forms of management also may be harmful. Brush control in California, and presumably elsewhere, influences the presence of ground squirrel colonies; with frequent control, the ground squirrel populations are disrupted, and with too little control, the vegetation grows too tall for good habitat. In either case, the burrows of ground squirrels are no longer available for the owls (Zarn 1974). Conversely, measures restricting the control of prairie dogs in areas with black-footed ferrets (*Mustela nigripes*) coincidentally protect burrowing owls. In some settings, where burrows suitable as nesting sites are not available, the owls have readily accepted artificial structures designed for the purpose (Collins and Landry 1977, Collins 1979).

Short-eared owls (*Asio flammeus*) are associated with grasslands, marshes and other open areas across North America. This species also is active during the daytime, typically skimming a few meters above ground in its distinctive, erratic flight pattern. In keeping with their preferences for treeless areas, short-eared owls nest on the ground in slight depressions and with the addition of only spare amounts of grasses or other materials. Their ground-nesting habit, more than any other characteristic, subjects short-eared owls to high rates of nest loss (Clark 1975).

Passeriformes

As might be expected, the largest order of birds (nearly 5,000 species) also includes the largest number of forms whose associations in part or in total include rangelands. Hence, space will limit mention to only a few species.

Of the several sparrows, Brewer's (*Spizella breweri*), clay-colored (*S. pallida*), Cassin's (*Aimophila cassinii*), and sage (*Amphispiza belli*) sparrows are among those having strong affinities with rangelands (Table 1). Similarly, 2 species of longspurs (*Calcarius* spp.) are associated with plains environments, as are the Sprague's pipit (*Anthus spragueii*), lark bunting (*Calamospiza melanocorys*), and horned lark (*Eremophila alpestris*).

The distributions of 2 species of meadowlarks (*Sturnella* spp.) include large areas on the North American prairies and plains. The eastern (*S. magna*) and western (*S. neglecta*) meadowlarks are remarkably alike in appearance and, where their ranges overlap, they are safely separated in the field only by their distinctive, species-specific vocalizations. A gross comparison of the frequencies for each species' songs, measured in vibrations/second, illustrates these differences. Mean, highest, and lowest notes for the eastern meadowlark are 4,400, 6,025, and 3,150, respectively, whereas the corresponding data for the western meadowlark are 2,500, 3,475, and 1,475 (Brand 1938). Lanyon (1957), in a more detailed analysis, concluded that marked differences in the songs of the males determined species recognition, which thereby maintained the biological integrity of the eastern and western meadowlarks as separate species.

Presumably, the 2 species developed during the Pleistocene, when glaciers and the climactic changes they wrought split the then homogeneous continental population of meadowlarks into 2 isolated breeding populations. With the retreat of the glaciers, the once separated populations, by then evolved into 2 species, met along a broad contact zone in the continental interior. Within the contact zone, each species maintains a preference for habitats governed by differences in moisture, with the eastern meadowlark preferring wetter areas compared with the more xeric sites selected by the western meadowlark (Lanyon 1956). Interbreeding of the 2 species may occur at times, but the distinctive songs of each show little evidence of evolutionary convergence (Sutton and Dickson 1965, Johnsgard 1979). When hybridization does occur, hatching success and fledgling rates are low, and the mature hybrids seem infertile (Szijj 1966). Interestingly, the essentially identical meadowlarks of North America themselves bear an extraordinary resemblance to the unrelated African pipits (*Macronyx* spp.); the latter birds also have distinctive yellow underparts with black pectoral bands, white outer tail feathers, and streaked backs. Moreover, *Sturnella* and *Macronyx* are both associated with grasslands, and the species in both genera build nests of similar construction (Friedmann 1946). Such a degree of correspondence thus represents a remarkable example of evolutionary convergence among the world's avifauna.

South of the interior "sea of grass" lie the brushy fringes of the southern Great Plains, specifically the rangelands of the Edwards Plateau, or Texas "Hill Country." Here each year an alarmingly small population of golden-cheeked warblers (*Dendroica chrysoparia*) returns to breed from their wintering grounds in southern Mexico and Central America. Indeed, the current breeding range of this imperiled species is one of the most restricted within the North American avifauna. Pulich (1976) has amassed a considerable volume of information about this rather unique bird, from which the following is derived.

Ashe juniper (*Juniperus ashei*), one of several species forming "cedar brakes" in the Hill Country, is inexorably linked with the ecology of the golden-cheeked warbler: the birds' nests are constructed solely of Ashe juniper, and only mature trees yield the long strips of bark necessary for this crucial necessity. Three other species of *Juniperus* occur in the same area, but these do not suffice as a source of nesting materials. The female alone builds the nest, using cobwebs to secure the bark into an intricately woven bowl lined with feathers from cardinals (*Cardinalis cardinalis*) or other readily available species.

The breeding range of golden-cheeked warblers coincides almost identically with the distribution of Ashe juniper, which in the Texas Hill Country is among the vegetation subject to removal, both for range improvements and housing developments. Secondary growth on overgrazed ranges provides unsuitable habitat for the birds, and urban sprawl from area cities (e.g., Austin and San Antonio) and second-home construction (i.e., "ranchettes") steadily constricts the remaining habitat. Additional juniper is harvested for fence posts; once cut selectively with hand axes, cedar brakes today are clear-cut with power equipment. Accordingly, habitat destruction is paramount among reasons for the declining population of golden-cheeked warblers.

However, in those cases where cedar brakes are not cleared, sheep and goat ranching is fully compatible with the nesting requirements of golden-cheeked warblers. In these and other instances (e.g., state wildlife areas and parks, scout camps) the uncleared land functions well as refuges for the birds. And in a few places cedar brakes have been set aside specifically as warbler habitat, and lands on military areas in the Hill Country are managed appropriately for the birds. Bryant (1991) recently reviewed the ecology and management of wildlife and other natural resources on the juniper-dominated ranges in the Texas Hill Country.

BIRDS AND BARBED WIRE

We conclude with an overview of interactions between barbed wire and birds on the rangelands of North America irrespective of the taxonomic status of the birds involved. Indeed, except for grass and livestock, perhaps there is no image associated more often with grazed ranges than that of barbed wire. Patented in 1874, and at first the bane of ranchers championing open range, barbed wire later became the *sine qua non* of grazing systems (McCallum and McCallum 1965). Soon, too, the iron barbs became an industrial-age analog for the thorns of mesquite (*Prosopis glandulosa*) and other native vegetation on which shrikes (*Lanius* spp.) impale insects, rodents, small birds, reptiles, and amphibians. Because of this unusual behavior, shrikes are colloquially known as "butcher birds." Two species, the loggerhead (*L. ludovicianus*) and northern (*L. excubitor*) shrikes, occur in rangeland areas across much of North America. Both species, although correctly classified as songbirds (Passeriformes: Laniidae), exhibit hawklike behavior and have hooked bills, replete with a tomial tooth not unlike those of many raptors (Fig. 5). Unlike true hawks, however, shrikes lack grasping talons, which has suggested to some observers that shrikes impale their prey because thorns and, later, barbed wire offer a means of anchoring victims during feeding. Another idea suggests that the impaled items represent a food cache, but many items are left to rot without ever being touched again. Sloane (1991), in fact, found that nearly one-half (48%) of the victims impaled by loggerhead shrikes were abandoned untouched. Moreover, if impaling behavior is a means of storing food, then one might expect shrikes to conceal, not prominently display, their fence-line hoard. On the other hand, the northern shrikes Cade (1967) studied often fed on the impaled items, especially during the breeding season, hence the behavior varies in purpose, both by season and species.

Still another theory, discussed long ago by the "father" of American ornithology, Alexander Wilson, proposes that the impaled grasshoppers and other insects act as bait for small birds, on which the shrikes then feed (Brewer 1840). However, Wilson dismissed the idea as "fanciful" on the clear logic that shrikes impale their avian prey as well. Sloane (1991) concluded that the impaled victims, at least in part, are a means by which shrikes advertise their territories, thereby explaining why some of the items, once secured, are never touched again. Whatever the case may be, barbed-wire fencing offered shrikes a unique and widely available analog on which to impale their prey. Regrettably, numbers of loggerhead shrikes are declining by as much as 5%/year in Oklahoma, and the species already is extirpated or rare in much of its eastern distribution (Tyler 1992).

Second, barbed-wire fencing has become a source of mortality for birds (Allen and Ramirez 1990). Birds of widely divergent taxa are involved; these include grebes (Podicipedidae), ducks, geese, and swans (Anatidae), eagles (Accipitridae), owls (Tytonidae and Strigidae), among others. Curiously, the literature currently available on this subject does not mention gallinaceous birds (e.g., Tetraoninae).

The extent of fence-related mortality is difficult to assess; the deaths usually occur singly along seemingly endless lengths of fenceline, thus relatively few occurrences are apparent when compared with accidents such as collisions with television towers where large numbers of moribund birds are encountered (Stoddard and Norris 1967; see also Avery et al. 1980). However, for endangered species such as whooping cranes (*Grus americana*), even a single death represents a substantial loss for the population and the species' gene pool. Unfortunately, whooping cranes have become entangled with barbed-wire fences on a number of occasions, a clear impediment for management programs (Idaho Coop. Wildl. Res. Unit 1974, Braun et al. 1977).

FIGURE 5. Loggerhead shrikes (upper) bear some convergent features with hawks, including hooked bills equipped with tomial teeth. Unlike hawks, however, shrikes often impale their prey on thorns and have adapted this habit to an analog of barbed wire (lower). Photos courtesy of Reuven Yosef.

Because ducks and other waterbirds are among the birds killed by barbed-wire fences, especially those crossing wetlands, even 1 or 2 deaths may set the stage for much greater losses of birdlife. This happens when maggots feeding on the rotting carcasses of wire-killed birds accumulate the toxin produced by *Clostsridium botulinum* and initiate an epizootic of botulism type C in which hundreds or even thousands of birds are killed (Wobeser 1981). Just such an epizootic occurred in Montana, although in this instance the birds collided with a powerline, not barbed-wire fencing (Malcolm 1982). Hence, unneeded barbed-wire and other fencing crossing wetlands should be removed, both to end this form of avian mortality and to enhance the esthetics of wetland sites (Cornwell and Hochbaum 1971). Barring complete removal, hazardous fences left in place should be marked with flagging to increase their visibility to birds (Fitzner 1975). Carcasses of birds killed by fences or other causes should be removed from wetlands to lessen the potential for epizootics of botulism (Reed and Rocke 1992).

SUMMARY

Birds of many kinds are associated with grasslands and other types of ranges, but surprisingly few species are endemic to the Great Plains. These species illustrate various physiological, morphological, and behavioral adaptations to life on rangelands. Some birds, such as prairie chickens, are narrowly adapted in terms of their diet, sexual displays, and plumage. Other species, including waterfowl and many songbirds, are tied less closely with rangeland habitats and consequently reveal broader adaptations. Ways in which birds have adapted to rangeland environments include such wide-ranging features as those for coping with drought (northern pintails) and heat (roadrunners), courtship in open areas (prairie grouse), protection from predators (burrowing owls), specialized nesting material (golden-cheeked warblers), and employment of man-made analogs in lieu of naturally occurring components (shrikes).

The fate of rangeland birds is a function of the quality and quantity of available habitat, which in turn depends largely on human uses of the landscape. Some birds have benefitted from human activities on rangelands (horned larks, for instance, in the Great Basin), but many others have experienced reductions in both distribution and numbers (e.g., greater prairie chickens in the Great Plains). A few species have decreased to the point where they are protected by federal legislation (e.g., Attwater's prairie chickens and masked bobwhite), including some not without long-standing controversy (e.g., golden eagles). For birds and other wildlife resources, conservation depends on the wise use and thoughtful management of rangeland habitats (e.g., sagebrush and sage grouse).

LITERATURE CITED

Allen, J. N. 1980. The ecology and behavior of the long-billed curlew in southeastern Washington. Wildl. Monogr. 73. 67pp.

Allen, G.T., and P. Ramirez. 1990. A review of bird deaths on barbed-wire fences. Wilson Bull. 102:553-558.

Avery, M. L., P. F. Springer, and N. S. Dailey. 1980. Avian mortality at man-made structures: an annotated bibliography (revised). U.S. Fish and Wildl. Serv. Biol. Serv. Program, Nat. Power Team, FWS/OBS-80/54.

Bailey, R. W. 1980. The wild turkey status and outlook in 1979. Pages 1-9 *in* J. M. Sweeney, ed. Proc. Fourth Nat. Wild Turkey Symposium, Nat. Wild Turkey Fed., Edgefield, S.C.

Bent, A. C. 1921. Life histories of North American gulls and terns. Smithsonian Instit. U.S. Nat. Mus. Bull. 113, Washington, D.C. 337pp.

Boeker, E. L., and V. E. Scott. 1969. Roost tree characteristics for Merriam's turkey. J. Wildl. Manage. 33:121-124.

Bolen, E.G. 1975. Eagles and sheep: a viewpoint. J. Range Manage. 28:11-17.

———, G. A. Baldassarre, and F. S. Guthery. 1989*a*. Playa lakes. Pages 341-365 *in* L. M. Smith, R. L. Pederson, and R. M. Kaminski, eds. Habitat management for migrating and wintering waterfowl in North America. Texas Tech. Univ. Press, Lubbock.

———, and D. L. Flores. 1989. The Mississippi kite in the environmental history of the Southern Great Plains. Prairie Nat. 21:65-74.

———, and ———. 1993. Mississippi kite, portrait of a southern hawk. Univ. Texas Press, Austin. 128pp.

———, and M. K. Rylander. 1978. Feeding adaptations in the lesser snow goose (*Anser caerulescens*). Southwest. Nat. 23:158-161.

———, L. M. Smith, and H. L. Schramm, Jr. 1989*b*. Playa lakes: prairie wetlands of the Southern High Plains. BioScience 39:615-623.

Brand, A. R. 1938. Vibration frequencies of passerine bird song. Auk 55:263-268.

Braun, C. E., T. Britt, and R.O. Wallestad. 1977. Guidelines for maintenance of sage grouse habitats. Wildl. Soc. Bull. 5:99-106.

———, K. W. Harmon, J. A. Jackson, and C. L. Littlefield. 1978. Management of national wildlife refuges in the United States: its impacts on birds. Wilson Bull. 90:309-321.

Bryant, F. C. 1991. Managed habitats for deer in juniper woodlands of west Texas. Pages 59-75 *in* J. E. Rodiek and E. G. Bolen, eds. Wildlife and habitats in managed landscapes. Island Press, Washington, D.C.

Brewer, T. M. 1840. Wilson's American ornithology. Otis, Broaders, and Co., Boston (reprinted by Arno & The New York Times [1970], New York). 746pp.

Bump, G., R. W. Darrow, F. C. Edminster, and W. F. Crissey. 1947. The ruffed grouse: life history, propagation, management. New York State Conserv. Dep., Albany. 915pp.

Butts, K. O., and J. C. Lewis. 1982. The importance of prairie dog town to burrowing owls in Oklahoma. Proc. Okla. Acad. Sci. 62:46-52.

Cade, T. J. 1967. Ecological and behavioral aspects of predation by the northern shrike. Living Bird 6:43-86.

Calder, W. A. 1968. The diurnal activity of the roadrunner, *Geococcyx californianus*. Condor 70:84-85.

Clark, R. J. 1975. A field study of the short-eared owl, *Asio flammeus* (Pontoppidan), in North America. Wildl. Monogr. 47. 67pp.

Collins, C. T. 1979. The ecology and conservation of burrowing owls. Pages 6-17 *in* P. P. Schaeffer and S. M. Ehlers, eds. Owls of the West, their ecology and conservation. Western Educ. Cent., Nat. Audubon Soc., Tiburon, Calif.

———, and R. E. Landry. 1977. Artificial nest burrows for burrowing owls. North Am. Bird Bander 2:151-154.

Cornwell, G. and, H. A. Hochbaum. 1971. Collisions with wires—a source of anatid mortality. Wilson Bull. 83:305-306.

Coulombe, H. N. 1970. Physiological and physical aspects of temperature regulation in the burrowing owl, *Speotyto cunicularia*. Comp. Biochem. Physiol. 35:304-335.

———. 1971. Behavior and population ecology of the burrowing owl, *Speotyto cunicularia*, in the Imperial Valley of California. Condor 73:162-176.

Crawford, J. A. 1980. Status, problems, and research needs of the lesser prairie chicken. Proc. Prairie Grouse Symposium. 1:1-7.

———, and E. G. Bolen. 1976. Effects of land use on lesser prairie chickens in west Texas. J. Wildl. Manage. 40:96-104.

Davison, V. E. 1940. An eight year census of lesser prairie chickens. J. Wildl. Manage. 4:55-62.

Douglass, D. W. 1942. A prairie chicken booming ground survey in central Michigan. Wilson Bull. 54:171-172.

Downing, R. L. 1959. Significance of ground nesting by mourning doves in northwestern Oklahoma. J. Wildl. Manage. 23:117-118.

Earhart, C. M., and N. K. Johnson. 1970. Size dimorphism and food habits of North American owls. Condor 72:251-264.

Enderson, J. H., and D. D. Berger. 1970. Pesticides: eggshell thinning and lowered production of young in prairie falcons. Bioscience 20:355-356.

Fitzner, R. E. 1975. Owl mortality on fences and utility lines. Raptor Res. 9: 55-57.

Folse, L. J., Jr., and K. A. Arnold. 1978. Population ecology of roadrunners (*Geococcyx californianus*) in south Texas. Southwest. Nat. 23:1-28.

Friedmann, H. 1946. Ecological counterparts in birds. Sci. Monthly 63: 395-398.

Glazener, W. C. 1967. Management of the Rio Grande turkey. Pages 453-492 *in* O. H. Hewitt, ed. The wild turkey and its management. The Wildlife Soc., Washington, D.C.

Graul, W. D. 1973. Possible functions of head and breast markings in Charadriinae. Wilson Bull. 85:60-70.

Greenway, J. C., Jr. 1958. Extinct and vanishing birds of the world. Am. Comm. for Intern. Wild Life Protection, New York, N.Y. Spec. Publ. No. 13. 518pp.

Grieb, J. R. 1970. The shortgrass prairie Canada goose population. Wildl. Monogr. 22. 49pp.

Hansen, H. A., and D. E. McKnight. 1964. Emigration of drought-displaced ducks to the arctic. Trans. North Am. Wildl. and Nat. Resour. Conf. 29:119-127.

Hawkins, A. S. 1945. Bird life in the Texas Panhandle. Panhandle-Plains Hist. Rev. 18:110-150.

Hayes, W. P. 1927. Prairie insects. Ecology 8:238-250.

Hohn, E. O. 1969. The phalarope. Sci. Am. 220:104-111.

Howe, F. P., and L. D. Flake. 1989. Nesting ecology of mourning doves in a cold desert ecosystem. Wilson Bull. 101:467-472.

Idaho Cooperative Wildlife Research Unit. 1974. Semi-annual report 29(2):27.

Jackson, A.S. 1945. Mississippi kites. Tex. Game and Fish Dep. 3:6-7.

Job, H. K. 1910. The Franklin's gull. Bird Lore 12:124.

Johns, J. E. 1969. Field studies of Wilson's phalarope. Auk 86:660-670.

Johnsgard, P. A. 1973. Grouse and quail of North America. Univ. Nebraska Press, Lincoln. 553pp.

———. 1978. The ornithogeography of the Great Plains. Prairie Nat. 10:97-112.

———. 1979. Birds of the Great Plains, breeding species and their distribution. Univ. Nebraska Press, Lincoln. 539pp.

———. 1983. The grouse of the world. Univ. Nebraska Press, Lincoln. 413pp.

King, R. 1978. Habitat use and related behaviors of breeding long-billed curlews. M.S. thesis, Colorado State Univ., Fort Collins. 69pp.

Klebenow, D. A. 1970. Sage grouse versus sagebrush control in Idaho. J. Range Manage. 23:396-400.

Kochert, M. N., and M. Pellant. 1986. Multiple use in the Snake River Birds of Prey Area. Rangelands 8:217-220.

Lanyon, W. E. 1956. Ecological aspects of the sympatric distribution of meadowlarks in the north-central states. Ecology 37:98-108.

———. 1957. The comparative biology of the meadowlarks (*Sturnella*) in Wisconsin. Publ. Nuttall Ornithol. Club No. 1, Cambridge, Mass. 67pp.

Lehman, V. W. 1941. Attwater's prairie chicken and its life history and management. U. S. Fish and Wildl. Serv. North Am. Fauna 57. 65pp.

Leopold, A. S., and M. K. Dedon. 1983. Resident mourning doves in Berkeley, California. J. Wildl. Manage. 47:780-789.

Love, D., and F. L. Knopf. 1978. The utilization of tree plantings by Mississippi kites in Oklahoma and Kansas. Pages 70-74 *in* Trees, a valuable Great Plains multiple use resource. Proc. 13th Annu. Meeting. Forestry Comm., Great Plains Agric. Council Publ. No. 87. Lincoln, Nebr.

Lynch, J. J., T. O'Neil, and D. W. Lay. 1947. Management significance of damage by geese and muskrats to Gulf Coast marshes. J. Wildl. Manage. 11:50-76.

Malcolm, J. M. 1982. Bird collisions with a power transmission line and their relation to botulism at a Montana wetland. Wildl. Soc. Bull. 10:297-304.

Marti, C. D. 1974. Feeding ecology of four sympatric owls. Condor 76:45-61.

Martin, D. J. 1973*a*. A spectrographic analysis of burrowing owl vocalizations. Auk 90:564-578.

———. 1973*b*. Selected aspects of burrowing owl ecology and behavior. Condor 75:446-456.

McCallum, H.D., and F. T. McCallum. 1965. The wire that fenced the West. Univ. Oklahoma Press, Norman. 285pp.

McDaniel, B., and E. U. Balsbaugh, Jr. 1968. Bovine manure as an overwintering medium for Coleoptera in South Dakota. Ann. Ento. Soc. Am. 61:765-768.

Mengel, R. M. 1970. The North American Central Plains as an isolating agent in bird speciation. Pages 279-340 *in* W. Dort, Jr., and J. K. Jones, Jr., eds. Pleistocene and Recent environments of the Central Great Plains. Spec. Publ. 3, Dep. Geol., Univ. Kansas. Univ. Kansas Press, Lawrence.

Mohr, C. O. 1943. Cattle droppings as ecological units. Ecol. Monogr. 13:275-298.

Mueller, H. C. 1986. The evolution of reversed sexual dimorphism in owls: an empirical analysis of possible selective factors. Wilson Bull. 98:387-406.

Ohmart, R. D. 1973. Observations on the breeding adaptations of the roadrunner. Condor 75:140-149.

———, and R. C. Lasiewski. 1971. Roadrunners: energy conservation by hypothermia and absorption of sunlight. Science 172:67-69.

Pulich, W. M. 1976. The golden-cheeked warbler, a bioecological study. Texas Parks and Wildl. Dep., Austin. 172pp.

Reed, T. M., and T. E. Rocke. 1992. The role of avian carcasses in botulism epizootics. Wildl. Soc. Bull. 20:175-182.

Rogers, J. P. 1964. Effect of drought on reproduction of the lesser scaup. J. Wildl. Manage. 28:213-222.

Rowe, M. P., R. G. Coss, and D. H. Owings. 1986. Rattlesnake rattles and burrowing owl hisses: a case of acoustic Batesian mimicry. Ethology 72:53-71.

Ryser, F. A., Jr. 1985. Birds of the Great Basin, a natural history. Univ. Nevada Press, Reno. 604pp.

Scott, T. G. 1940. The western burrowing owl in Clay County, Iowa, in 1938. Am. Midl. Nat. 24:585-593.

Sloane, S. A. 1991. The shrike's display advertising. Nat. Hist. 6:32-38.

Sharp, W. M. 1957. Social and range dominance in gallinaceous birds—pheasants and prairie grouse. J. Wildl. Manage. 21:242-244.

Smith, A. G., J. H. Stoudt, and J. B. Gollop. 1964. Prairie potholes and marshes. Pages 39-50 *in* J. P. Linduska, ed. Waterfowl tomorrow. U.S. Fish and Wildl. Serv., Washington, D.C.

Smith, R. I. 1970. Response of pintail breeding populations to drought. J. Wildl. Manage. 34:943-946.

Soutiere, E. C., and E. G. Bolen. 1972. Role of fire in mourning dove nesting ecology. Proc. Tall Timber Fire Ecol. Conf. 12:277-288.

———, and ———. 1976. Mourning dove nesting on tobosa grass-mesquite rangeland sprayed with herbicides and burned. J. Range Manage. 29:226-231.

———, H. S. Myrick, and E. G. Bolen. 1972. Chronology and behavior of American widgeon wintering in Texas. J. Wildl. Manage. 36:752-758.

Stoddard, H. L., Sr., and R. A. Norris. 1967. Bird casualties at a Leon County, Florida TV tower: an eleven-year study. Tall Timbers Res. Station Bull. No. 8. 104pp.

Summers, R. W., and G. M. Dunnet. 1984. Sheldgeese and man in the Falkland Islands. Biol. Conserv. 30:319-340.

Sutton, G. M., and G. W. Dickson. 1965. Interbreeding of the eastern and western meadowlarks in central Oklahoma. Southwest. Nat. 10:307-310.

Swank, W. G. 1955. Nesting and production of the mourning dove in Texas. Ecology 36:495-505.

Szijj, L. J. 1966. Hybridization and the nature of isolating mechanisms in sympatric populations of meadowlarks (*Sturnella*) in Ontario. Zeitschrift fur Tierpsychologie 23:677-690.

Tomlinson, R. E. 1972. Current status of the endangered masked bobwhite quail. Trans. North Am. Wildl. and Nat. Resour. Conf. 37:294-311.

Tyler, J. D. 1992. Nesting ecology of the loggerhead shrike in southwestern Oklahoma. Wilson Bull. 104:95-104.

Vance, D. R., and R. L. Westemeier. 1979. Interactions of pheasants and prairie chickens in Illinois. Wildl. Soc. Bull. 7:221-225.

Wallestad, R. 1975. Male sage grouse responses to sagebrush treatment. J. Wildl. Manage. 39:482-484.

Wesemann, T., and M. Rowe. 1987. Factors influencing the distribution and abundance of burrowing owls in Cape Coral, Florida. Pages 129-37 *in* L. W. Adams and D. L. Leedy, eds. Integrating man and nature in the metropolitan environment. Natl. Inst. for Urban Wildl., Columbia, Md.

Wiley, R. H., Jr. 1978. The lek mating system of the sage grouse. Sci. Am. 238:114-125.

Wilson, M. W., and J. A. Crawford. 1987. Habitat selection by Texas bobwhites and chestnut-bellied scaled quail in South Texas. J. Wildl. Manage. 51:575-582.

Wobeser, G. A. 1981. Diseases of wild waterfowl. Plenum Press, New York, N.Y. 300pp.

Zarn, M. 1974. Burrowing owl, *Speotyto cunicularia hypugaea*. Habitat management series for unique or endangered species. Rep. 11, Tech. Note 250. Bur. Land Manage., Denver, Colo. 25pp.

Zimmerman, J. L. 1992. Density-independent factors affecting the avian diversity of the tallgrass prairie community. Wilson Bull. 104:85-94.

Chapter 3
THE MAMMALS
Clyde Jones and Richard W. Manning

INTRODUCTION

The rangelands of the United States occur over a broad geographic area, with many diverse features (Thomas et al. 1990, Chapter 1). As would be expected, these rangelands support diverse mammalian faunas, including the smallest (insectivores) and largest (ungulates) vertebrates that occur in these communities (Table 1). Seasonal and annual fluctuations in climate are common on rangelands; these fluctuations are reflected in extremes of temperatures, availability of cover, and supplies of food. Phenomena such as fire and grazing also contribute to the fluctuations of conditions on rangelands. Mammals that exist in these variable conditions are somewhat "plastic." Species of mammals that inhabit rangelands respond in various ways to these kinds of environmental changes; for example, populations of some species of mammals "pulse" periodically in response to temporal changes in the conditions of the habitats (Kotler 1989). In addition, the mammalian faunas of rangelands are complex; there are some intricate competitive interactions and some complicated predator-prey relationships among the mammals that inhabit the rangelands of North America. The mammalian faunas of rangelands are complicated further by the presence of "islands" that harbor special kinds of mammals; the Black Hills, riparian systems, and marshes on the Great Plains are such "island" habitats. Habitats in upland, xeric, grazed rangelands may support different components of the mammalian faunas than areas of lowland, mesic, ungrazed habitats.

Mammals have important roles in the ecosystems of rangelands. This chapter presents an overview of the diversity and the importance of mammals that inhabit some of the rangelands of North America. In its compilation, we have relied heavily on information available in the published literature on mammals and have also drawn much from our own experiences in conducting studies and observations of mammals on rangelands in North America and elsewhere. A comprehensive survey of the extensive literature of mammals on rangelands, however, is well beyond the scope of this chapter.

HABITATS

Native habitats for mammals on rangelands in North America (Fig. 1, 2) are extant in many areas, such as parts of the Sandhills of Nebraska (Freeman 1989). Besides the summary of mammals that occur on the Sandhills and information about their affinities to different kinds of habitats provided by Freeman (1989), results of other intensive studies of native mammals in natural habitats on rangelands of the Great Plains have been presented by Andersen and Fleharty (1967), Clark et al. (1987), Kaufman et al. (1990), Clark and Kaufman (1991), and Bowles and Copsey (1992).

In many areas of North America in general and on the Great Plains in particular, native habitats of mammals on rangelands have been modified greatly, removed completely, or replaced with other kinds of vegetation by activities of humans. As a result, the native habitats of the mammals in some areas have been changed from homogenous structures to fragmented and mosaic patterns. Studies of native mammals in such disturbed situations reveal that some animals adapted rather remarkably to these changes. Fleharty and Navo (1983) demonstrated that some small mammals used irrigated cornfields, at least for a portion of the year, in western Kansas. Manning and Geluso

TABLE 1. Some ecological characteristics of selected native mammals of the Northern Great Plains (N.D., S.D., Neb.).

Mammals	Food regime	Period of activity
Insectivora		
Soricidae		
Arctic shrew (*Sorex arcticus*)	Insectivorous	Nocturnal, diurnal
Masked shrew (*Sorex cinereus*)	Insectivorous	Nocturnal, diurnal
Merriam's shrew (*Sorex merriami*)	Insectivorous	Nocturnal
Dwarf shrew (*Sorex nanus*)	Insectivorous	Nocturnal
Water shrew (*Sorex palustris*)	Insectivorous	Nocturnal, diurnal
Pygmy shrew (*Sorex hoyi*)	Insectivorous	Nocturnal, diurnal
Northern short-tailed shrew (*Blarina brevicauda*)	Insectivorous	Nocturnal, diurnal
Southern short-tailed shrew (*Blarina carolinensis*)	Insectivorous	Nocturnal, diurnal
Least shrew (*Cryptotis parva*)	Insectivorous	Nocturnal, diurnal
Talpidae		
Eastern mole (*Scalopus aquaticus*)	Insectivorous	Nocturnal, diurnal
Lagomorpha		
Leporidae		
Desert cottontail (*Sylvilagus audubonii*)	Herbivorous	Nocturnal, diurnal
Eastern cottontail (*Sylvilagus floridanus*)	Herbivorous	Nocturnal, diurnal
Mountain cottontail (*Sylvilagus nuttallii*)	Herbivorous	Nocturnal, diurnal
Snowshoe hare (*Lepus americanus*)	Herbivorous	Nocturnal
Black-tailed jackrabbit (*Lepus californicus*)	Herbivorous	Nocturnal, diurnal
White-tailed jackrabbit (*Lepus townsendii*)	Herbivorous	Nocturnal, diurnal
Rodentia		
Sciuridae		
Least chipmunk (*Tamias minimus*)	Herbivorous	Diurnal
Eastern chipmunk (*Tamias striatus*)	Herbivorous	Diurnal
Yellow-bellied marmot (*Marmota flaviventris*)	Herbivorous	Diurnal
Woodchuck (*Marmota monax*)	Herbivorous	Diurnal
Wyoming ground squirrel (*Spermophilus elegans*)	Herbivorous	Diurnal
Franklin's ground squirrel (*Spermophilus franklinii*)	Herbivorous	Diurnal
Richardson ground squirrel (*Spermophilus richardsonii*)	Herbivorous	Diurnal
Spotted ground squirrel (*Spermophilus spilosoma*)	Herbivorous	Diurnal
Thirteen-lined ground squirrel (*Spermophilus tridecemlineatus*)	Herbivorous	Diurnal
Black-tailed prairie dog (*Cynomys ludovicianus*)	Herbivorous	Diurnal
Eastern gray squirrel (*Sciurus carolinensis*)	Herbivorous	Diurnal
Eastern fox squirrel (*Sciurus niger*)	Herbivorous	Diurnal
Red squirrel (*Tamiasciurus hudsonicus*)	Herbivorous	Diurnal
Southern flying squirrel (*Glaucomys volans*)	Omnivorous	Nocturnal
Geomyidae		
Northern pocket gopher (*Thomomys talpoides*)	Herbivorous	Nocturnal, diurnal
Plains pocket gopher (*Geomys bursarius*)	Herbivorous	Nocturnal, diurnal
Heteromyidae		
Olive-backed pocket mouse (*Perognathus fasciatus*)	Herbivorous	Nocturnal
Plains pocket mouse (*Perognathus flavescens*)	Herbivorous	Nocturnal
Silky pocket mouse (*Perognathus flavus*)	Herbivorous	Nocturnal
Hispid pocket mouse (*Chaetodipus hispidus*)	Herbivorous	Nocturnal
Ord's kangaroo rat (*Dipodomys ordii*)	Herbivorous	Nocturnal
Castoridae		
American beaver (*Castor canadensis*)	Herbivorous	Nocturnal, diurnal
Muridae		
Western harvest mouse (*Reithrodontomys megalotis*)	Herbivorous	Nocturnal
Plains harvest mouse (*Reithrodontomys montanus*)	Herbivorous	Nocturnal
White-footed mouse (*Peromyscus leucopus*)	Omnivorous	Nocturnal
Deer mouse (*Peromyscus maniculatus*)	Omnivorous	Nocturnal

Table 1. (continued)

Mammals	Food regime	Period of activity
Northern grasshopper mouse (*Onychomys leucogaster*)	Omnivorous	Nocturnal
Hispid cotton rat (*Sigmodon hispidus*)	Omnivorous	Nocturnal, diurnal
Bushy-tailed woodrat (*Neotoma cinerea*)	Herbivorous	Nocturnal
Eastern woodrat (*Neotoma floridana*)	Herbivorous	Nocturnal
Southern red-backed vole (*Clethrionomys gapperi*)	Herbivorous	Nocturnal, diurnal
Long-tailed vole (*Microtus longicaudus*)	Herbivorous	Nocturnal, diurnal
Prairie vole (*Microtus ochrogaster*)	Herbivorous	Nocturnal, diurnal
Meadow vole (*Microtus pennsylvanicus*)	Herbivorous	Nocturnal, diurnal
Woodland vole (*Microtus pinetorum*)	Herbivorous	Nocturnal, diurnal
Sagebrush vole (*Lemmiscus curtatus*)	Herbivorous	Nocturnal, diurnal
Common muskrat (*Ondatra zibethicus*)	Herbivorous	Nocturnal, diurnal
Southern bog lemming (*Synaptomys cooperi*)	Herbivorous	Nocturnal, diurnal
Zapodidae		
Meadow jumping mouse (*Zapus hudsonius*)	Omnivorous	Nocturnal
Western jumping mouse (*Zapus princeps*)	Omnivorous	Nocturnal
Erethizontidae		
Common porcupine (*Erethizon dorsatum*)	Herbivorous	Nocturnal, diurnal
Carnivora		
Canidae		
Coyote (*Canis latrans*)	Omnivorous	Nocturnal, diurnal
Gray wolf (*Canis lupus*)	Carnivorous	Diurnal
Kit or swift fox (*Vulpes velox*)	Carnivorous	Nocturnal, diurnal
Red fox (*Vulpes vulpes*)	Carnivorous	Nocturnal, diurnal
Common gray fox (*Urocyon cinereoargenteus*)	Carnivorous	Nocturnal, diurnal
Ursidae		
Black bear (*Ursus americanus*)	Omnivorous	Nocturnal, diurnal
Procyonidae		
Common raccoon (*Procyon lotor*)	Omnivorous	Nocturnal, diurnal
Mustelidae		
American marten (*Martes americana*)	Carnivorous	Nocturnal, diurnal
Fisher (*Martes pennanti*)	Omnivorous	Nocturnal, diurnal
Ermine (*Mustela erminea*)	Carnivorous	Nocturnal, diurnal
Long-tailed weasel (*Mustela frenata*)	Carnivorous	Nocturnal, diurnal
Least weasel (*Mustela nivalis*)	Carnivorous	Nocturnal, diurnal
Mink (*Mustela vison*)	Carnivorous	Nocturnal, diurnal
American badger (*Taxidea taxus*)	Carnivorous	Nocturnal
Eastern spotted skunk (*Spilogale putorius*)	Omnivorous	Nocturnal
Striped skunk (*Mephitis mephitis*)	Omnivorous	Nocturnal
Northern river otter (*Lutra canadensis*)	Carnivorous	Nocturnal, diurnal
Felidae		
Mountain lion (*Felis concolor*)	Carnivorous	Nocturnal, diurnal
Lynx (*Lynx lynx*)	Carnivorous	Nocturnal, diurnal
Bobcat (*Lynx rufus*)	Carnivorous	Nocturnal, diurnal
Artiodactyla		
Cervidae		
Wapiti or elk (*Cervus elaphus*)	Herbivorous	Diurnal
Mule deer (*Odocoileus hemionus*)	Herbivorous	Nocturnal, diurnal
White-tailed deer (*Odocoileus virginianus*)	Herbivorous	Nocturnal, diurnal
Moose (*Alces alces*)	Herbivorous	Diurnal
Antilocapridae		
Pronghorn (*Antilocapra americana*)	Herbivorous	Diurnal
Bovidae		
Mountain or bighorn sheep (*Ovis canadensis*)	Herbivorous	Diurnal

(1989) made an intensive study of habitat use by mammals in a man-made forest in the Sandhill region of Nebraska that revealed the overall effects of habitat manipulation on native mammals of this region. Results of some other studies of native mammals of rangelands in central North America in connection with the presence of woodlots and shelterbelts were presented by Gottfried (1979) and Yahner (1982).

Habitats for mammals on rangelands in North America have been subjected to some kinds of modifications for millions of years (Colbert and Morales 1992). Grazing by large herds of American bison (*Bison bison*) and other large ungulates previously caused periodic changes in rangelands. Koford (1958) postulated that intensive grazing by bison resulted in a mosaic pattern of grazed and ungrazed areas. The development of such a pattern undoubtedly affected other mammals in the areas. For example, grazed areas might not support populations of the prairie vole (*Microtus ochrogaster*), which is a diurnal herbivore, but ungrazed areas would harbor these rodents (Manning and Geluso 1989). A recent discussion of hypothetical relationships between bison and some other mammals on rangelands was presented by Stangl et al. (1992). In modern times, grazing by domestic livestock causes similar impacts to these habitats (Fig. 3). Grazing favors infestation by invader plants, usually woody vegetation (Fig. 4, 5), and depletion of plants either preferred by the mammals or easily killed by trampling (Thomas et al. 1990). Overgrazing is followed frequently by erosion (Fig. 6). However, lack of grazing also favors invasion by some kinds of vegetation, especially woody plants. Removal of grass by mechanical methods also impacts habitats for mammals on rangelands. Lemen and Clausen (1984) summarized the results of a study of the effects of mowing on rodents in their native rangeland.

Fires, either caused by natural phenomena or set by humans, are important factors that cause changes in habitats of mammals on rangelands. Some important and recent studies of the results of fires on habitats and mammals on rangelands in the central United States have been conducted by Kaufman et al.

FIGURE 1. Natural rangelands (tallgrass prairie) in the Sandhills, Cherry County, Nebraska. Note the small cluster of cottonwood trees (slightly left of center) that surround a marsh.

FIGURE 3. Impacts of grazing by livestock on rangelands, Thomas County, Nebraska.

FIGURE 2. Natural ungrazed rangelands (tallgrass prairie) in the Sandhills, Cherry County, Nebraska.

FIGURE 4. Results of grazing (left of fenceline) by livestock on rangelands, Thomas County, Nebraska.

FIGURE 5. Overgrazed rangelands, Llano Estacado, Deaf Smith County, Texas.

FIGURE 6. Erosion by winds on overgrazed rangelands, Cherry County, Nebraska.

(1983, 1988, 1989) and Kaufman and Kaufman (1990). In general, suppression of fires on rangelands results in increases of herbaceous plants and woody vegetation.

DIVERSITY OF MAMMALS

Numbers of species of several major groups of native mammals (e.g., Insectivora, Lagomorpha, Rodentia, Carnivora, Artiodactyla) that occur on rangelands in the central part of North America vary in accordance with latitude (Table 2). Although numbers of species vary within these 5 orders of mammals, the numbers of species of these mammals on rangelands in the central part of North America are not highly variable (Jones et al. 1983, Caire et al. 1989), except for Texas (Jones and Jones 1992), which has more species of these mammals than any other of the states listed. Texas is unique in this context, at least in part, because rangelands there include habitats from 4 major physiographic divisions of North America: the Rocky Mountains, Great Plains, Eastern Forests, and Chihuahuan Desert.

Of all terrestrial forms in these 5 orders of native mammals (approx. 310 species) known to occur in North America north of Mexico (Jones et al. 1992), approximately 25% inhabit the rangelands of the central part of the continent. Rangelands in this part of North America harbor about 15% of the species of insectivores, approximately 30% of the kinds of rabbits and hares (excluding pikas), about 19% of the species of rodents, approximately 59% of the species of the extant carnivores, and about 55% of the species of the native ungulates. Rangelands in the central portion of North America are very important for the continued existence of major components of the native mammalian fauna.

Insectivores

The insectivores (i.e., shrews, moles) are important components of the mammalian faunas on rangelands (Table 1, 2). The desert shrew (*Notiosorex crawfordi*) that inhabits arid regions at the southern end of the Great Plains and in the desert regions of the Southwest is associated with short-grass and desert-scrub vegetation. The least shrew (*Cryptotis parva*) is common in upland habitats that may be subjected to grazing. Members of the genera *Sorex* and *Blarina* generally are associated with mesic habitats not subjected to heavy grazing. Moles, which are fossorial mammals, are absent from extremely arid regions and areas with heavily compacted soils. As implied by the name of the order, these are carnivorous mammals that consume mostly insects and other invertebrates. However, these mammals also consume other small animals and carrion, and cannibalism is common, especially among shrews. Removal of insects and other invertebrates is a major contribution of shrews and moles to habitats on rangelands. In addition, the fossorial activities of

TABLE 2. Numbers of species of some native mammals that occur at various latitudes on rangelands in the central part of North America.

Orders	Manitoba	North Dakota	South Dakota	Nebraska	Kansas	Oklahoma	Texas
Insectivora	7	5	8	6	5	5	5
Lagomorpha	4	5	5	4	5	4	4
Rodentia	32	32	33	34	33	44	63
Carnivora	23	23	21	20	19	20	27
Artiodactyla	7	8	6	6	5	4	6
Total	73	73	73	70	67	77	105

moles result in the mixing of humus and aeration of the soils in which they occur (Jones et al. 1983).

Lagomorphs

Leporids (i.e., rabbits, hares) are rather conspicuous members of the mammalian faunas in various habitats on rangelands (Table 1, 2). As a group, these herbivorous mammals generally occur in areas where short grasses and herbs are abundant and clumps of tall grasses or brush are available for cover. Several taxa of lagomorphs are major components of the mammalian faunas of rangelands in North America. The mountain cottontail (*Sylvilagus nuttallii*) occurs on montane "islands"; the desert cottontail (*Sylvilagus audubonii*) is an inhabitant of upland, grazed areas; the eastern cottontail (*Sylvilagus floridanus*) seems restricted mostly to ungrazed, riparian-edge habitats. The pygmy rabbit (*Brachylagus idahoensis*) occurs on some rangelands of the Great Basin. The black-tailed jackrabbit (*Lepus californicus*) ranges widely in semiarid and arid regions and in the well-developed grasslands of the central Great Plains (Fig. 7). The white-tailed jackrabbit (*Lepus townsendii*) inhabits mixed grasslands of the northern Great Plains. Snowshoe hares (*Lepus americanus*) occur on "islands" in montane areas. The white-sided jackrabbit (*Lepus callotis*) frequently is listed as a component of southwestern desert rangelands; however, the current range of this species in the United States apparently is restricted to about 120 km^2 in southwestern New Mexico (Best and Henry 1993). The habitats on rangelands for some of these mammals have been expanded as a result of alterations of the environment by humans.

Rabbits and hares are prolific breeders and some kinds, especially black-tailed jackrabbits, may become abundant at times in areas where they might contribute to overgrazing of grasslands and even become a nuisance to agricultural crops (Jones et al. 1983). In some places, special control programs have been developed to attempt to keep populations of these mammals at levels acceptable to humans. However, rabbits and hares are major dispersers of seeds of some important plants, especially dropseed (*Sporobolus* spp.) and are important components of the food chain in that they serve as major prey species for some carnivores. Lagomorphs are even of some value to humans as food.

Rodents

The order Rodentia represents the most diverse (Fig. 8, 9, 10, 11,12) and largest component of the mammalian faunas in habitats on the rangelands of North America (Table 1, 2). Rodents range in size from the small harvest mice (*Reithrodontomys* spp.) and pocket mice (*Perognathus* spp.) to the large porcupine (*Erethizon dorsatum*) and beaver (*Castor canadensis*). This group of mammals exhibits a variable food regime, ranging from strictly herbivorous to almost carnivorous forms; some species are omnivorous. In activity, rodents are nocturnal, diurnal, and crepuscular. Although most typical rodents are strictly terrestrial, some forms are scansorial and semiarboreal, including some kinds capable of gliding. Some rodents are mostly fossorial and others semiaquatic. These adaptations of rodents enable them to occur in every available habitat on rangelands. Freeman (1989)

FIGURE 8. Ord's kangaroo rat (*Dipodomys ordii*), Thomas County, Nebraska.

FIGURE 7. Black-tailed jackrabbit (*Lepus californicus*), Lubbock County, Texas.

FIGURE 9. Juvenile thirteen-lined ground squirrel (*Spermophilus tridecemlineatus*), Lubbock County, Texas.

FIGURE 10. Black-tailed prairie dog (*Cynomys ludovicianus*), Lubbock County, Texas.

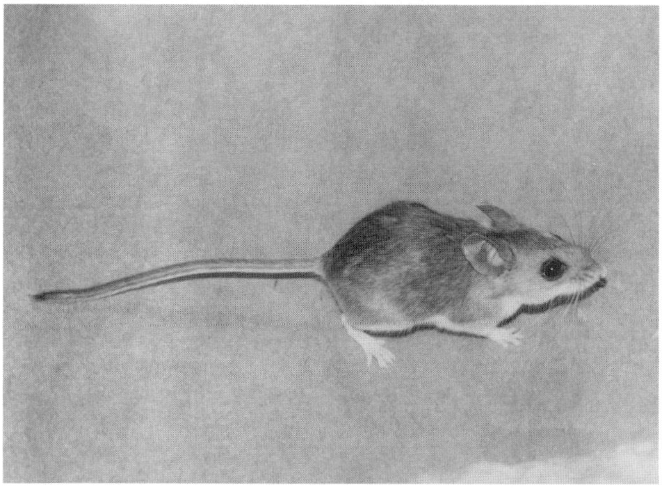

FIGURE 11. Texas mouse (*Peromyscus attwateri*), Randall County, Texas.

describes the affinities of rodents for various habitats on the Sandhills of Nebraska. In Kansas, Kaufman and Fleharty (1974) demonstrated that life form of vegetation (e.g., riparian, tallgrass, shortgrass) affected the distribution of species of rodents more than either the presence or absence of particular species of plants. Some rodents, such as the deer mouse (*Peromyscus maniculatus*), plains harvest mouse (*Reithrodontomys montanus*), and prairie vole (*Microtus ochrogaster*), are somewhat typical of upland, xeric habitats on rangelands. Some other species, such as the white-footed mouse (*Peromyscus leucopus*), western harvest mouse (*Reithrodontomys megalotis*), and meadow vole (*Microtus pennsylvanicus*), occur commonly in association with mesic, riparian habitats. Many of the heteromyid rodents inhabit overgrazed areas and sparcely vegetated areas on sandy soils. However, the hispid pocket mouse (*Chaetodipus hispidus*) occurs frequently in weedy areas of succession. The zapodids (*Zapus* spp.) and bog lemmings (*Synaptomys* spp.) are mostly obligate inhabitants of mesic areas. In the arid Southwest, it seems that the structures of rodent communities are recognizably nonrandom (Findley 1989).

FIGURE 12. Yellow-faced pocket gopher (*Cratogeomys castanops*), Lubbock County, Texas.

Sometimes populations of rodents reach levels that are considered nuisances to the activities of humans, and management practices have been implemented. Long-term control programs have been underway in some areas of North America with regard to certain rodents (Fig. 9, 10, 12), such as prairie dogs (*Cynomys* spp.), ground squirrels (*Spermophilus* spp.), and pocket gophers (*Geomys* spp., *Thomomys* spp., *Cratogeomys* spp.).

Some important contributions of rodents to habitats on rangelands include the dispersal of seeds of plants, consumption and shredding of vegetation that contributes to the deposition of humus, and the mixing and aeration of soils by burrowing activities. In addition, numerous rodents are major sources of food for predators. A few rodents, such as the common muskrat (*Ondatra zibethicus*) and American beaver are of some value to humans for their fur.

Predators

Members of the order Carnivora are diverse and widespread in habitats on rangelands in North America (Table 1, 2). As implied by the ordinal name, these are mostly flesh-eating mammals. However, some members of the group regularly consume carrion and various kinds of plant materials, so they are truly omnivorous.

Some species of carnivores have been extirpated from rangelands of North America as a result of long-term control measures and other activities of humans. For example, wolves (*Canis lupus*, *Canis rufus*), black bears (*Ursus americanus*), grizzly bears (*Ursus arctos*), and mountain lions (*Felis concolor*) no longer are prominent members of the mammalian fau nas on rangelands. On the other hand, a few species of carnivores have expanded their geographic ranges as a result of habitat modifications brought about by humans. For example, coyotes (*Canis latrans*) and badgers (*Taxidea taxus*) now inhabit areas that were not used by them historically. In addition, some members of this order have adjusted to living in urban environments. For example, raccoons (*Procyon lotor*) are considered as nuisances in some areas of human habitations.

Some predator-prey relationships between carnivores and other mammals have fascinated biologists and wildlife managers for years. For example, the closely linked cycles between some members of the family Felidae and rabbits and hares have been studied intensively (Jones et al. 1983). Likewise, the intricate interactions between black-footed ferrets (*Mustela nigripes*) and prairie dogs (*Cynomys* spp.) have been known for some time (Seal et al. 1989). Many of the carnivores are used by humans for their fur.

Ungulates

The ungulates (order Artiodactyla) are conspicuous members (Fig. 13, 14) of the mammalian faunas that occur on rangelands (Table 1, 2). All of these mammals are strictly herbivorous in that they feed on grasses and forbs, such as the pronghorn (*Antilocapra americana*), and the bark, twigs, and leaves of shrubs and trees, such as mule deer (*Odocoileus hemionus*) and white-tailed deer (*Odocoileus virginianus*). It seems that the pronghorn is the ultimate large herbivore on rangelands in the central part of the United States; earlier, pronghorn were nearly as numerous as bison (Jones et al. 1983). As mentioned previously, at least some of the ungulates have had major roles in the structuring of ecosystems on rangelands (Koford 1958, Stangl et al. 1992). The importance of rangelands in the evolution of the ungulates has been emphasized (Colbert and Morales 1992). Coevolution of ungulates and growth forms of certain plants has been a major factor in the evolution of these mammals and in the development of rangelands. Populations of many of these mammals in habitats on rangelands have fluctuated widely for some time because of pressures from hunting, disease, competition with domestic livestock, and manipulations of habitats by humans. Also, as discussed elsewhere, there have been numerous translocations and reintroductions of these mammals by humans in recent times. Certain management activities directed at ungulates have been practiced at various places in North America since at least 1694 (Matthiessen 1987). As a group, the artiodactyls are of some value to humans for reasons of aesthetic appeal; in many areas, these mammals are of great value to humans for hunting for food and recreation.

Bats

Some discussion is warranted of bats (order Chiroptera) that occur on rangelands. In addition to being volant mammals (Fig. 15), most North American bats either migrate or hibernate (or both). Therefore, these unique mammals are either seasonal or transient components of the mammalian faunas of most North

FIGURE 14. White-tailed deer (*Odocoileus virginianus*), Hidalgo County, Texas. Photo courtesy of T. R. Simpson.

FIGURE 13. Pronghorn (*Antilocapra americana*), Brewster County, Texas.

FIGURE 15. Big brown bat (*Eptesicus fuscus*), Crosby County, Texas.

American rangelands. Bats are restricted mostly to areas where appropriate roosting sites, such as crevices, caves, and trees, and man-made structures (e.g., buildings, mines), and water are available. As with many other mammals, numbers of species of bats present correlate somewhat with latitude. For example, only 13 species of bats are reported from Nebraska (Jones et al. 1983), but 32 taxa of bats are known to occur in Texas (Schmidly 1991). The bats in temperate North America forage at night and are mostly insectivorous.

Because many bats consume large numbers of flying insects and a few kinds of nectarivorous bats play a role in the pollination of certain plants, these mammals are important components of the ecosystems in which they occur. In addition, bats are of interest to humans because occasionally they are carriers of paralytic rabies.

Introduced Mammals

Numerous species of mammals not native to North America have been introduced over the years and now occur on rangelands. Introductions of non-native mammals have taken place through either accidental or deliberate actions by humans. Some kinds of introduced mammals have become major components of the mammalian faunas on rangelands in some areas (Kaufman and Kaufman 1990, Traweek 1985). In some cases, the presence of large numbers of feral mammals on rangelands is the source of major management and political concerns (Berger 1986, Brisbin and Mayer 1991). Jones et al. (1992) listed 25 species of exotic mammals that now occur in the wild state in considerable numbers in North America.

In addition to the introductions of species of mammals not native to North America, several species of native mammals have been moved around by humans. These translocations of native species of mammals have been carried out mostly in connection with programs to remove problem mammals, such as bears and mountain lions, or to replenish stocks, such as bighorn sheep (*Ovis canadensis*), reduced by hunting or disease (Schmidt and Gilbert 1978). Also, in recent years, some efforts have been made to produce in captivity some kinds of mammals that have become extinct in the wild, such as the red wolf (*Canis rufus*) and the black-footed ferret, and to reintroduce them into their native habitats (Seal et al. 1989).

SOME INFORMATION NEEDS

This brief survey of some of the major groups of mammals that are important components of the ecosystems of rangelands could not be concluded without some comments about areas of future research and information needs with regard to the perpetuation of these important wildlife resources. Constant vigilance is needed in development of lists and status reports on species of concern. Development of baseline information and determining levels of chemicals present in tissues of mammals is important. Up-to-date biological surveys, inventories, and evaluations of presence and abundance of mammals in habitats on rangelands are vital, including long-term monitoring of populations and related studies of biodiversity. The effects of some modern conservation practices, such as the Conservation Reserve Program, renewed interests in planting of trees, and changes in grazing regimes, should be studied and documented in detail. Development of research activities to determine the impacts of global warming and desertification on mammals in habitats on rangelands should be initiated on a broad scale. A wide array of more intense education programs oriented toward importance and values of rangeland wildlife in general and mammals in particular needs to be developed; management of ecosystems rather than species and appreciation for and wise use of these kinds of resources should be emphasized.

SUMMARY

Mammals that occur in habitats on rangelands in North America are diverse and widespread. These animals are distributed throughout all natural habitats, and in most habitats modified by activities of humans. Mammals have evolved numerous adaptations that enable them to continue to exist in the natural and modified habitats on rangelands. Based on the numbers of mammal species that inhabit rangelands, it is apparent that these kinds of habitats are important to mammalian faunas. Mammals have important roles in the ecosystems of rangelands. However, much information is needed to understand the complex interactions between mammals of all kinds and the continuously changing habitats on rangelands. Rangeland mammals exemplify dynamic biological systems in that some mammals are expanding their geographic ranges, while others are becoming more restricted in their distributions. In addition, the phenomenon of extinction continues, and numbers of kinds of introduced mammals seem to be increasing. Various needs of the constantly increasing human population will continue to impact rangeland ecosystems and the diversity of the mammals that occur there. The mammals that occur on rangelands are of considerable value to humans in general and to those humans that inhabit rangelands in particular.

LITERATURE CITED

Andersen, K. W., and E. D. Fleharty. 1967. Mammalian distribution within biotic communities of northeastern Jewell County, Kansas. Fort Hays Stud. Sci. Ser. 6. 46pp.

Best, T. L., and T. H. Henry. 1993. Lepus callotis. Mammal. Species. 442:1-6.

Berger, J. 1986. Wild horses of the Great Basin: social competition and population size. Univ. Chicago Press, Chicago, Ill. 326pp.

Bowles, J. B., and A. D. Copsey. 1992. Small mammal abundance as a function of herbaceous cover type in south central Iowa. Prairie Nat. 24:109-119.

Brisbin, I. L., and J. Mayer. 1991. Wild pigs in the United States: Their history, morphology, and current status. Univ. Georgia Press, Athens. 336pp.

Caire, W., J. D. Tyler, B. P. Glass, and M. A. Mares. 1989. Mammals of Oklahoma. Univ. Oklahoma Press, Norman. 567pp.

Clark, B. K., and D. W. Kaufman. 1991. Effects of plant litter on foraging and nesting behavior of prairie rodents. J. Mammal. 72:502-512.

———, ———, G. A. Kaufman, and E. J. Finck. 1987. Use of tallgrass prairies by *Peromyscus leucopus*. J. Mammal. 68:158-160.

Colbert, E. H., and M. Morales. 1992. Evolution of the vertebrates: A history of the backboned animals through time. John Wiley & Sons, New York, N.Y. 470pp.

Findley, J. S. 1989. Morphological patterns in rodent communities of southwestern North America. Pages 253-263 *in* D. W. Morris, Z. Abramsky, B. J. Fox, and M. R. Willig, eds. Patterns in the structure of mammalian communities. The Museum, Texas Tech. Univ. Spec. Publ. 266pp.

Fleharty, E. D., and K. W. Navo. 1983. Irrigated cornfields as habitat for small mammals in the Sandsage Prairie Region of western Kansas. J. Mammal. 64:367-379.

Freeman, P. 1989. Mammals. Pages 181-188 *in* A. Bleed and C. Flowerday, eds. An atlas of the sandhills. Inst. Agric. and Nat. Resour., Univ. Nebraska, Lincoln. 238pp.

Gottfried, B. M. 1979. Small mammal populations in woodlot islands. Am. Midland Nat. 102:105-112.

Jones, J. K., Jr., D. M. Armstrong, R. S. Hoffmann, and C. Jones. 1983. Mammals of the Northern Great Plains. Univ. Nebraska Press, Lincoln. 379pp.

———, R. S. Hoffmann, D. W. Rice, C. Jones, R. J. Baker, and M. D. Engstrom. 1992. Revised checklist of North American mammals north of Mexico, 1991. The Museum, Texas Tech. Univ. Occas. Pap. 146. 23pp.

———, and C. Jones. 1992. Revised checklist of Recent land mammals of Texas, with annotations. Tex. J. Sci. 44:53-74.

Kaufman, D. W., E. J. Finck, and G. A. Kaufman. 1990. Small mammals and grassland fires. Pages 46-80 *in* S. L. Collins and L. L. Wallace, eds. Fire in North American tallgrass prairies. Univ. Oklahoma Press, Norman.

———, and E. D. Fleharty. 1974. Habitat selection by nine species of rodents in north-central Kansas. Southwest. Nat. 18:443-452.

———, and G. A. Kaufman. 1990. House mice (*Mus musculus*) in natural and disturbed habitats in Kansas. J. Mammal. 71:428-432.

———, ———, and E. J. Finck. 1983. Effects of fire on rodents in tallgrass prairie of the Flint Hills region of eastern Kansas. Prairie Nat. 15:49-56.

———, ———, and ———. 1988. Influence of fire and topography on habitat selection by *Peromyscus maniculatus* and *Reithrodontomys megalotis* in ungrazed tallgrass prairie. J. Mammal. 69:342-352.

———, ———, and ———. 1989. Rodents and shrews in ungrazed tallgrass prairie manipulated by fire. North Am. Prairie Congr. 11:173-177.

Koford, C. B. 1958. Prairie dogs, white faces, and blue grama. Wildl. Monogr. 3. 78pp.

Kotler, B. P. 1989. Temporal variation in the structure of a desert rodent community. Pages 127-139 *in* D. W. Morris, Z. Abramsky, B. J. Fox, and M. R. Willig, eds. Patterns in the structure of mammalian communities. The Museum, Texas Tech. Univ. Spec. Publ. 28. 266pp.

Lemen, C. A., and M. K. Clausen. 1984. The effects of mowing on the rodent community of a native tall grass prairie in eastern Nebraska. Prairie Nat. 16:5-10.

Manning, R. W., and K. N. Geluso. 1989. Habitat utilization of mammals in a man-made forest in the Sandhill region of Nebraska. The Museum, Texas Tech. Univ. Occas. Pap. 131. 34pp.

Matthiessen, P. 1987. Wildlife in America. Viking Penguin, New York, N. Y. 332pp.

Schmidly, D. J. 1991. The bats of Texas. Texas A&M Univ. Press, College Station. 188pp.

Schmidt, J. L., and D. L. Gilbert, editors. 1978. Big game of North America. Stackpole Books, Harrisburg, Penn. 494pp.

Seal, U.S., E. T. Thorne, M. A. Bogan, and S. H. Anderson, editors. 1989. Conservation biology and the black-footed ferret. Yale Univ. Press, New Haven, Conn. 302pp.

Stangl, F. B., Jr., T. S. Schafer, J. R. Goetze, and W. Pinchak. 1992. Opportunistic use of modified and disturbed habitat by the Texas kangaroo rat (*Dipodomys elator*). Tex. J. Sci. , 44:25-35.

Thomas, G. W., C. H. Herbel, and G. T. Miller, Jr. 1990. Rangeland resources. Pages 340-359 *in* G. T. Miller, Jr., ed. Resource conservation and management. Wadsworth Publ., Belmont, Calif.

Traweek, M. S. 1985. Statewide census of exotic big game animals. Tex. Parks Wildl. Dep., Fed. Aid Wildl. Restor. Proj. W-109, job 21. 40pp.

Yahner, R. H. 1982. Microhabitat use by small mammals in farmstead shelterbelts. J. Mammal. 63:440-445.

Chapter 4
THREATENED AND ENDANGERED WILDLIFE AND LIVESTOCK INTERACTIONS

W. Dean Carrier and Brian Czech

INTRODUCTION

> This region, when grazed by livestock, reverted through a series of more and more worthless grasses, shrubs, and weeds to a condition of unstable equilibrium. Each recession of plant types bred erosion; each increment to erosion bred further recession of plants. The result today is a progressive and mutual deterioration, not only of plants and soils, but of the animal community subsisting thereon. . . . It is quite invisible to the tourist who finds this wrecked landscape colorful and charming (Leopold 1949:206).

A better-trained assessment of the New Mexican desert than Leopold's (1949) would be hard to find, and the deterioration of the animal community is translated today in the lists of threatened and endangered species from rangelands throughout the West.

There are many examples of studies that address the impact of livestock grazing on plant community composition (Klopatek 1992), individual game species (Loft 1989), and groups of selected species (Baker 1989). Fewer studies address the influence of grazing across broad taxonomic scales, such as Barker's (1991) work on nongame birds in North Dakota. Studies that clearly isolate grazing as the factor endangering specific species (Berry 1978) are scarce.

The relative infrequency with which livestock grazing is linked with dangerous declines of species cannot logically be viewed as evidence that grazing is not a problem. Studies designed to detect this linkage are, *inter alia*, logistically difficult, expensive, politically contentious, and statistically indefensible. On the other hand, there is strong circumstantial evidence that grazing is a major problem. That is probably why "Many conservationists claim that livestock has done more damage to the native biodiversity of western North America than all the chainsaws and bulldozers combined" (Noss 1994:613).

While typically the effects of grazing on individual species are neither obvious nor demonstrable, certain related facts are; wildlife occupy ecosystems valued for livestock forage, grazing alters those ecosystems, and many native species associated with those ecosystems have suffered severe population declines. Fleischner (1994) reviewed studies that demonstrated the negative effects of grazing on ecosystem structure, function, and composition.

A discussion of endangered wildlife on western rangelands, then, is logically conducted within the context of "what happens" on those lands with grazing. In this chapter, we: 1) review information on the effects of domestic livestock on North American ecosystems; 2) discuss the Endangered Species Act and its application to rangelands and grazing; and 3) address the impacts, direct and indirect, of grazing on selected listed species.

LIVESTOCK GRAZING: AN HISTORICAL PERSPECTIVE

Even before the flora and fauna of North America felt the effects of the ax and the plow, domestic livestock had modified the continent's ecosystems. Ancestral Equidae had begun their Beringian departure for the Asian steppes millions of years prior, and *Equus* had not been seen in the Americas since the Pleistocene (Nowak 1991). Then, during their expeditions to the Southwest in the 1500s, Coronado and Cortez reintroduced the horse to the continent (Dasmann 1972, Reisner 1986).

As other European cultures populated North America, they brought livestock domesticated from wild European and Asian stock. Horses, cattle, sheep, goats, and swine, used for transportation, plowing, meat, milk, and hides, accompanied the European settlers from the time of their initial explorations. Following their escape, these species altered native ecosystems (Dasmann 1972), but feral stock never became as ecologically problematic as "managed" stock.

In North America, the effects of unrestricted livestock grazing became a public concern by the turn of this century (Cubbage 1994). Many rangelands had been eroded to mineral soil and blowing dust. Exposed rocks covered the ground, rocks that were not to be found in the memories of the earliest ranchers. Ranges were denuded of grass, and a vicious cycle began. Freed from the root systems of native grasses, soils blew or washed away. As soil disappeared, so did the likelihood of recovery for the grasses.

Grazing was not the only activity of the livestock industry that impacted rangeland ecosystems. Livestock growers, hoping to reduce livestock mortality and increase production, have historically sought to eliminate predators. Wolves (*Canis lupus*), coyotes (*Canis latrans*), mountain lions (*Felis concolor*), grizzly bears (*Ursus arctos*), black bears (*Ursus americanus*), and many smaller predators were poisoned, shot, and trapped relentlessly during most of the 1900s. In most areas, the ultimate goal was extirpation, and in most cases (with the perpetual exception of coyotes), the goal was achieved.

The elimination of large predators over vast areas had profound impacts on rangeland ecosystems. In fact, if Savory (1988) was correct in asserting that the pack-hunting predator was an essential element in the maintenance of the world's dry grasslands (by regularly stampeding wild ungulate herds), then the destruction of wolves contributed to the demise of the shortgrass plains. In addition to the ecological impacts, the destruction of American predatory megafauna, viewed in the context of Wilson (1984) and Kellert (1984), has probably damaged the American psyche.

Species regarded as competitors with livestock have also been eradicated to provide more forage for domestic herds. Tales of great shoots and elimination of wild ungulate herds are common in journals written near the turn of the century (Allen 1954, Matthiessen 1987). Small mammals were eliminated with the intent of range enhancement. Expansive prairie dog (*Cynomys* spp.) towns of the Great Plains and high desert states were eliminated via poison.

Wholesale elimination of species must have had an eerie quality to some, due to the swiftness with which an erstwhile keystone could be completely annihilated. For example, on the 750,000 ha San Carlos Apache Reservation in Arizona, blacktail prairie dogs (*Cynomys ludovicianus*) were abundant on the central plateau between the Nantac Rim and the Gila Mountains. Federal control experts came in near mid-century, setting out poison in the prairie dog towns. The following spring, the Indians observed not a single prairie dog emerge from its den, changing the sights and sounds (and the ecological processes) of this high desert grassland forever (J. Stevens, San Carlos Recreation and Wildl. Comm., pers. commun.).

In addition to the extirpation of predators and competitors, wildlife populations suffered considerable losses through indirect effects. Trophic reverberation was widespread, from the bottom up and from the top down. The decline of prairie dog colonies spelled the demise of the black-footed ferret (*Mustela nigripes*) (Clark and Harvey 1988), while elimination of mountain lions contributed to an ecologically devastating irruption of mule deer (*Odocoileus hemionus*) on the Kaibab Plateau (Kormondy 1969). Burk (1973) argues that more than predator removal caused the irruption of deer on the plateau, but in the process implicates grazing itself as the factor keeping deer levels at unnaturally low levels to begin with. In either scenario, then, livestock production was a serious problem.

Diseases and parasites were transmitted by domestic stock. American wild ungulates had never been exposed to these organisms, which were often benign to their familiar domestic hosts. Native species were infected and populations were extirpated. In the case of bighorn sheep (*Ovis canadensis*), this was a particularly devastating process that swept through the species' range (Beuchner 1960). Hoof and mouth disease, liver flukes, and a number of bacterial and viral infections common to domestic species still occasionally run rampant through wild populations (Allen 1954). Brucellosis, tuberculosis, and anthrax threaten the existence of the endangered wood bison (*Bison bison athabascae*) in Canada (Carbyn et al. 1993).

Changes in vegetation composition and structure caused by grazing allowed increases in populations of wildlife compatible to the altered conditions. This is exemplified by the cowbird (*Molothrus ater*), which has steadily increased and is now a primary factor in the decline of many forest and riparian bird species (Line 1993).

Human population growth of the western United States was based as much on production of red meat for eastern and midwestern markets as it was on furs, gold, or agriculture (Wilkinson 1992). The cowboy replaced the mountain man as a symbol of the American wilderness, and the urban population began to view cowboys and livestock as symbolic of and indigenous to the West, just like the elk (*Cervus elaphus*), bison, and Indians.

As the western livestock industry grew into a political power, the Bureau of Land Management (the largest range manager in the nation) became the prototypical captured agency, comprising one third of an "iron triangle" along with western ranchers and senators (Cubbage 1994). The political clout of this triangle has been stronger than the concern for the preservation of species (Barton 1987), thus far. It is not surprising, then, that rangeland reform is high on the agenda of the growing class of scientist-advocate (Noss 1994). Reform has received increasing support from the general public, too. As Zaslowsky (1989:46) stated, "It is hard to be nostalgic about cowboys when the degradation by cattle is right before your eyes."

THE ENDANGERED SPECIES ACT

In 1966, President Johnson signed the Endangered Species Preservation Act, protecting about 70 species of birds and mammals believed to be threatened with extinction (Reffalt 1988, Kohm 1991). The act won broad congressional support and was judged to have little effect on the nation's economic or social activities. It was widely supported by federal lawmakers and the general public (Kohm 1991). The initial legislation required little more than cataloging so-called "rare and endangered species," and provided vague federal direction for their protection. It focused on species that inhabited federal lands, and then only provided protection "insofar as practicable and consistent with . . . [a federal agency's] primary purposes."

The act protected only highly regarded species whose populations were in serious trouble. Aesthetic value probably played an important role in these early determinations. The California condor (*Gymnogyps californianus*), snail kite (*Rostrhamus sociabilis plumbeus*), key deer (*Odocoileus virginianus clavium*), and sea otter (*Enhydra lutra*) were among the target species protected by this initial legislation.

The plight and protection of the bald eagle (*Haliaeetus leucocephalus*), especially, was highly publicized. The bald eagle had suffered population declines throughout its range in the contiguous United States for decades. Despite the Bald Eagle Protection Act of 1940, eagles were victimized by illegal shooting, habitat loss, and environmental contaminants.

One may view the eagle, and American fauna in general, as a political interest group, albeit an unwitting one. In the early years of the endangered species agenda, the eagle would have been classified under Ingram and Schneider's (1994) system as "dependents." They were powerless to protect their own interests but positively constructed; society valued them and would vote to protect them. Who in Congress would deny protection to our nation's symbol?

As the American environmental movement grew, so did public support for protection of listed species. The initial act was re-written in 1969 as the Endangered Species Conservation Act. Then in, 1973, the Endangered Species Act (ESA) was signed by President Nixon. With each revision and re-authorization, the Act became stronger and more specific, yet each was carried with virtually no congressional opposition. The original Act of 1966 merely required the U.S. Fish and Wildlife Service to publish a list of native species threatened with extinction; i.e., the Red Book (Kohm 1991).

Many species found in the Red Book were included based solely on the unfounded testimony of an individual expert. The initial list contained 63 vertebrate species. The second and last publication of the Red Book came early in 1973 and included many more species.

In many states, companion legislation protected species that were declining locally, and the combination of state and federal legislation provided protection to a formidable list of species. Eventually, ESA protected not only vertebrate species of national significance, but also local populations of species common in other states, and invertebrates and plants.

Now, there are thousands of protected species, subspecies, and populations that the public never knew existed, let alone considered protecting. Snail darters (*Percina tenasi*), red-cockaded woodpeckers (*Picoides borealis*), and spotted owls (*Strix occidentalis*), once virtually unknown, are now within the public vocabulary. As the list of protected species has exploded, the "dependent" status of endangered species has been compromised.

Using Ingram and Schneider's (1994) scheme, many endangered species now fit best under under the "deviant" category; they have remained politically powerless, but have assumed a problematic and intrusive construction. The Stephens' kangaroo-rat (*Dipodmys stephensi*), for example, has been implicated in costing taxpayers exorbitant mitigation fees, endangering human lives and properties because of fire management requirements, and devastating property values in several California counties. Although the Department of Interior contests the blame placed on the rat in an unpublished pamphlet entitled "Common Endangered Species Act 'Horror' Stories," the damage of anti-ESA publicity has been done.

Fortunately for the political environment of species preservation, ecosystem management is now in vogue. This is not a new concept to wildlife ecologists; Aldo Leopold described ecosystem management more than 50 years ago, without calling it such (Salwasser 1994, Kaufman et al. 1994). However, assisted by the Forest Service's adoption of ecosystem management under the direction of Dale Robertson in 1992, American awareness that the health of the entire ecosystem is necessary for the preservation of individual species is growing. Individual species should regain their "dependent" political status as the public identifies them as victims to and indicators of ecosystem degradation.

THE ESA AND THE LIVESTOCK INDUSTRY

Endangered species legislation immediately restricted livestock industry activities. The wolf, extirpated from most of its range in the contiguous United States, was protected on federal lands under the initial Act in 1966. However, it was not until the passage of ESA in 1973 that protection of wolves encompassed all lands in the nation. Then it became a federal offense for anyone to kill or even harass wolves, either for sport or for the protection of domestic livestock.

In Minnesota, where the only viable population of wolves in the contiguous United States remained, livestock owners predicted that a lack of intensive wolf control would result in the decimation of their dairy herds. In a historic confrontation between the State of Minnesota and the federal government, the Forest Service prohibited the hunting or trapping of wolves on Superior National Forest. Section 7 of ESA required the Forest Service (along with all federal agencies) to "insure that any action authorized, funded, or carried out by such an agency is not likely to jeopardize the continued existence of any endangered species." Backing the Forest Service was an Eighth Circuit Court decision that prohibited the state from authorizing public taking of a listed species (Kohm 1991), and subsequent events indicate that the Minnesota livestock growers' concerns were unwarranted (Fritts et al. 1992).

Listing of the grizzly bear also drew the ire of the livestock industry in the West. Livestock operators and their associated organizations regarded protection of grizzly bears as a threat to their livelihood. In Idaho, Montana, and Wyoming, the states with most of the grizzlies in the contiguous states, livestock interests had been given discretion to eliminate any bears that threatened to prey on their herds. The listing of the grizzly as "threatened" tightly restricted sport hunting and provided only for the elimination of proven stock killers.

In 1982 a sheepherder shot a grizzly bear he claimed was about to attack his flock. He was charged with a violation of ESA and fined $2,500. In a highly publicized appeal supported by the livestock industry, he argued that he was within his rights because he was protecting his property. However, the Tenth Circuit Court found that ESA did not include the protection of property under its exceptions to compliance (Littell 1992).

While the livestock industry took an adversarial stand against ESA, environmental activists portrayed the livestock industry as profiteers exploiting public resources for private benefit, and their audience was large and interested. Books like *Sacred Cows at the Public Trough* (Ferguson and Ferguson 1983) became best sellers overnight. The rangeland iron triangle was stressed by the heat of Rangeland Reform '94, which promised to make room on the range for market grazing fees and conservation easements. As of 1995, however, the iron triangle has proven formidable, and few changes to federal policy have occurred.

THE EFFECTS OF LIVESTOCK PRODUCTION ON ENDANGERED WILDLIFE SPECIES

Livestock grazing is considered a significant factor in the decline of native wildlife on North American rangelands (Moen 1973, Wilkinson 1992, Fleischner 1994). The overgrazing of native grasslands, vectoring of disease and parasites, pollution of water holes, degradation of riparian areas, and compaction of soils have profound impacts on the native flora and fauna. In addition to grazing, other activities associated with livestock production are detrimental to wildlife. Examples in this section represent a variety of effects of grazing on endangered species.

Harting (1987) described 5 kinds of impacts of livestock production on grizzly bears: 1) direct loss, 2) indirect loss, 3) habitat loss or modification, 4) displacement, and, 5) competition. These categories are equally applicable to the effects of livestock grazing on other native species.

For our purposes, a species is endangered if it is listed by the United States Fish and Wildlife Service or by a state government as a threatened or endangered species or subspecies. We will also consider species that are considered "sensitive" or are proposed for listing, imperiled populations of non-endangered species, and other species or populations that serve to illustrate the principles set forth.

Direct Effects

Livestock production results in the direct mortality of wildlife. Such mortality can be intentional (e.g., predator control) or accidental (e.g., trampling). It is difficult to assess the relative importance of intentional versus accidental mortality. It is logical to speculate that accidental mortality is not as amenable to documentation as is intentional mortality; its occurrence in nature, therefore, is probably much higher than that noted in the literature.

Predator Control. Large predators have found livestock to be an abundant, available, and vulnerable prey source. Livestock owners have responded by eliminating predators from private rangelands (Matthiessen 1987). Reacting to industry pressure, the federal government subsidizes ranchers by paying control agents to kill predators and rodents on public lands. In 1990, the Animal Damage Control branch of the Fish and Wildlife Service (and now of the Animal and Plant Health Inspection Service) killed 91,219 coyotes, 207 mountain lions, and 247 bears in 17 western states. There is no record of the loss of animals that inadvertently succumbed to poison set out for wolves and other large predators (Allen 1974).

No predator has incurred the wrath of the livestock industry more than the wolf. Sometimes basing their view on fictional tales, even educated naturalists such as Seton and Leopold have condoned the extirpation of this species from its native habitat in North America. Lopez (1978:169) quoted a German wolf biologist named Ziman, who commented on the "relentless carnage" of wolf control efforts in North America. "We killed the wolf in Europe . . . and we hated the wolf, but it was not anything like what you have done in America."

Historically, wolves inhabited 46 of the contiguous United States, but by 1950 they had been eliminated from all but Minnesota and Michigan, and possibly Montana and North Dakota (Mech 1970). Wolf bounties were paid by most states, the last being Minnesota in 1965. Bounty records chronicle the systematic decline of the species. From 1833 to 1918, the State of Montana paid bounty on 80,730 wolves. From 1915 to 1942 federal hunters continued to kill wolves, taking 24,132 from Colorado, Wyoming, and Montana (Lopez 1978). Today, only a scattered remnant remains in Montana, including those recently released by the United States Fish and Wildlife Service.

In continuation of its historical war on the wolf, the livestock industry now condemns the proposed reestablishment of wolves in former habitats, even when such reintroduction is the only way to salvage the species as an ecological entity. Reintroduction of gray wolves into the Greater Yellowstone Ecosystem and red wolves (*Canis rufus*) into Texas and Louisiana is severely hampered by the industry's lobby (Alderson 1987).

The fate of grizzly bears and mountain lions parallels that of the wolf. Considered too dangerous to humans and livestock, grizzly bears and lions have been extirpated from most of their native range, in most areas before the wolf had disappeared. In California the last grizzly bear was killed in 1922, although most of the population had been exterminated long before. Grizzly bears were gone from Texas by 1890, New Mexico by 1933, Arizona by 1935, and Colorado by 1952 (Schneider 1977). Even under the protection of ESA, some sheepherders in Montana have admitted killing any bear they see (Knight et al. 1980).

Reduction of Competitors. Concerns regarding competition for forage led early American ranchers to reduce or eliminate wild ungulates from private and public rangelands. For example, in the early 1900s a group of ranchers shot the last bighorn sheep on Cobblestone Mountain in Ventura County, California. Their sole purpose was the extirpation of this herd, to reduce competition for forage with their domestic herds (D. Mills, rancher, pers. commun.).

Rodents are another focus of control efforts, because they eat grass. Millions of kgs of poisoned bait have been applied to rangelands to eliminate prairie dogs in Arizona, Utah, Wyoming, Colorado, Montana, and the Dakotas. Although prairie dogs eat grass, it is simplistic to classify them as competitors; areas cohabited by cattle and prairie dogs have higher plant production than those that lack prairie dogs (Williams 1992).

Prairie dog control has endangered this species in several areas. The blacktailed prairie dog has been extirpated from Arizona, is listed as endangered in Utah, and occupies $< 10\%$ of its former range in the remaining grassland states (Matthiessen 1987, Williams 1992). Seal (1989:xii) wrote, "[Prairie dogs are] systematically poisoned as a matter of public and private land-management policy. Prairie dog towns were once distributed over 40 million hectares; it now appears difficult to allocate even 3,000 to 15,000 hectare patches [for the preservation of prairie dogs and, thus, black-footed ferrets]."

Accidental Mortality. Occasionally, livestock kill endangered species. In the San Joaquin Valley and Mojave Desert of California, cattle have been implicated in causing the deaths of burrowing animals through crushing of individuals or collapsing burrows in which the residents are entombed. Affected species include several species of kangaroo rat (*Dipodomys* spp.), the desert tortoise (*Gopherus agassizii*), and the blunt-nosed leopard lizard (*Gambelia silus*) (T. R. Rado, consultant, pers. commun.). In the Mojave Desert, K. H. Berry (A brief summary of the effects of sheep grazing on desert tortoises in the western Mojave Desert, Bur. Land Manage., Riverside, Calif., 1989, mimeo.) observed that the passage of domestic sheep resulted in the deaths of 11 juvenile tortoises by overturning and crushing them.

It has been suggested that the deaths of 2 whooping crane (*Grus americanus*) chicks at Grays Lake National Wildlife Refuge was due to trampling by domestic cattle, and another chick was killed in a collision with a barbed wire fence (U.S. Fish and Wildl. Serv. 1978). Studies in California indicate that livestock foraging in riparian habitats destroyed southwestern willow flycatcher (*Epidonax trailii extumus*) nests (Tibbetts 1993).

Not only cattle, but predator control efforts can kill other species incidentally. According to R. Jurek (in an unpublished report for the California Department of Fish and Game), an immature California condor, one of only 18 remaining in the wild, was killed in 1983 by a cyanide "coyote getter" placed on private rangelands by public officials.

Indirect Effects

Some species fall victim to livestock production inadvertently and unexpectedly. These species' declines are indirect effects of grazing and other industry activities, effects that are generally more difficult to identify, analyze, and control than are direct mortality factors. These effects illustrate the complexity of challenges to wildlife biology on western rangelands.

Trophic Linkage. A fundamental concept of wildlife ecology is that of trophic interaction. Animals must eat to survive, and if they are dependent on species that are themselves endangered, they become endangered species, *de facto*. Due to the webbed nature of trophic interaction, a problem for a particular species can become problematic for many others.

A clear case is the black-footed ferret, an incidental casualty of the war on prairie dogs. Dependent on prairie dogs as its sole source of prey, the ferret has been virtually extirpated from its range (Clark and Harvey 1988). It is quite possible that the problem is one of quantity and quality. Most of the prairie dog towns are gone, and many of the those remaining have probably been contaminated with poisons.

The California condor has reached a similar fate. It can no longer depend on elk and pronghorn (*Antilocapra americana*) carrion. Ranchers have long supported hunting regimes that reduce wild ungulate (especially elk) populations. The lack of wild carrion alone would not cause the elimination of the condor. In fact, recovery strategies traditionally supported livestock operations on public and private rangelands (Koford 1953, McMillan 1968, Wilbur 1978), because livestock carrion would provide food for the condor. However, when the livestock carrion is laced with poisons intended to control rangeland predators, including strychnine, sodium flouroacetate, and Compound 1080, the condor is no better off (McMillan 1968, Wilbur 1978). Considering only trophic linkages, then, the condor falls victim to the livestock industry in at least 2 ways.

Disease and Internal Parasitism. As with most gregarious species, domestic livestock can transmit disease and parasites. Having inbred immunities to the effects of many of these diseases, they often become vectors. Vaccination, food supplements, and spraying of pesticides have significantly reduced the effects of these diseases on domestic stock, but with no natural immunities, native species can be decimated.

Bighorn sheep are particularly sensitive to livestock-borne disease. Jessup (1985) cataloged 12 populations having exposure to domestic sheep since 1970. Seven of these populations suffered 100% loss, primarily due to disease. The California bighorn sheep (*O. c. californiana*) is listed as a threatened species by the state of California, having been extirpated from many areas by competition with domestic and feral livestock, introduced disease, and hunting (Steinhart 1990).

In an effort to reestablish this species, 4 state and federal agencies transplanted 65 California bighorns to an enclosure on the Lava Beds National Monument in northeastern California. For several years it appeared that the reintroduction was successful. Then, in 1988, domestic sheep from a grazing allotment on the adjacent national forest were allowed into a pasture adjacent to the bighorn enclosure. The bighorn were exposed to a pneumophillic bacteria common in domestic sheep but deadly to bighorn. In a few months the entire bighorn herd of 43 individuals had died (Foreyt and Jessup 1982). This phenomenon is common: when bighorn sheep interact with domestic sheep the former die (Krausman 1995).

External Parasitism. The brown-headed cowbird is a native species that has formed a symbiotic association with domestic livestock. Conditions resulting from livestock grazing have resulted in the increase of this species' population far above those within natural ecosystems. Prior to 1900, cowbirds were absent from most of the habitats they currently occupy and were considered rare, even in their native range. Conversion of natural landscapes to irrigated pastures and the associated increases in livestock triggered an increase in the range and numbers of the cowbird. As early as the 1930s this was reported as a population increase unparalleled by any native bird species (Willet 1933). The cowbird, a brood parasite, is a primary factor in the proposed listing of several riparian species, including the least Bell's vireo (*Vireo belli pusillus*) and southwestern willow flycatcher (Tibbetts 1993).

Chemical Contamination. The persistence of DDT affects the eggshell thickness of many bird species, especially predatory and water-associated species. Rangelands were regularly sprayed to reduce insect populations that may vector livestock disease. From 1950 to 1960, annual spraying of DDT occurred on over 370,000 ha of rangelands within the condor range in the San Joaquin Valley of California (McMillan 1968). Eggshell thinning was indicated as a contributing cause of reproductive failure in the California condor population (Darlington 1987).

Both incidental and secondary poisoning are considered factors in the decline of the endangered San Joaquin Valley kit fox (*Vulpes macrotus mutica*) (U.S. Fish and Wild. Serv. 1983). It feeds on baits deposited for coyote control, and scavenges the carcasses of burrowing rodents that have been poisoned.

Habitat Loss and Degradation

Grazing by domestic livestock alters native plant communities. Smith and Berg (1988), in a survey for the California Native Plant Society, indicated that overgrazing threatened >70 species of native plants in California. The demise of native plant communities has a negative effect on the associated wildlife species.

Perpetual cropping of perennial grasses quickly changes the range composition by reducing or eliminating perennial grass species, allowing invasion of annual broad-leafed plants, forbs, and grasses. Grazing also results in the compaction and baring of fragile topsoils, which are then lost to erosion from wind and water (Ohmart and Anderson 1982). Due primarily to heavy grazing, the perennial grass prairies once found throughout California and other western states have been replaced by native and exotic annual grasslands (Jensen et al. 1990).

Nests of ground-nesting birds, especially those inhabiting wetlands where cattle tend to congregate, are trampled. Furthermore, early spring grazing reduces hiding cover, rendering untrampled nests vulnerable to predators. This has been cited as a significant factor in the decline of the greater sandhill crane (*Grus canadensis tabida*) in Oregon and California (U.S. Fish and Wildl. Serv. 1978, Littlefield 1982).

In the Southwest, the masked bobwhite quail (*Colinus virginianus ridgwayi*) is extirpated by overgrazing (Allen 1974). Formerly found in the Altar and Santa Cruz valleys of southern Arizona (with a wider distribution in northcentral Sonora), the species was first described in 1884, a time when cattle numbers in the Arizona Territory were estimated at 250,000. By 1887 there were over a million cattle grazing on Arizona's desert grasslands. That same year, the last officially documented Arizona masked bobwhite was presented to the *Daily Arizona Citizen* (Allen 1974).

The habitat of the masked bobwhite consisted of wide, grass-covered valleys. By 1970 there was virtually no native grass remaining in the range of this species, and years of overgrazing had resulted in brush invasion (Tomlinson 1972). While the bobwhite might survive in lightly grazed areas, its range has been overgrazed for so long that food and cover have been almost totally eliminated (Mackenzie 1977, Matthiessen 1987).

Efforts to restore the species via protection and habitat rehabilitation have generally been unsuccessful (Gabel and Dobrott 1988). The 45,000 ha Buenos Aires National Wildlife Refuge in Arizona's Altar Valley was established for the bobwhite in 1985. Despite the fact that Buenos Aires is now the largest ungrazed grassland in Arizona, and that approximately 22,000 (>2,000/year) bobwhite have been released over the past 10 years, only about 750 survivors exist. Refuge personnel remain confident that the bobwhite will recover if wild stock from Sonora are released rather than pen-raised birds. They feel that the elimination of cattle grazing was the key to recovery, and that the range is gradually returning to conditions suitable for a self-sustaining bobwhite population (B. Kuvlesky, Jr., refuge biologist, pers. commun.).

The great gray owl (*Strix nebulosa*) is another species adversely affected by grazing. This bird, listed as endangered by the State of California, is found in restricted mountain meadow habitats in the Sierra Nevada range (Steinhart 1990). Heavy cattle and sheep grazing of these small meadows has resulted in the owls' abandonment of nearly all the meadow systems it formerly occupied (Winter 1986). The owl's prey consists of small rodents, especially voles. Grazing reduces the herbaceous cover necessary to provide forage and cover required by the rodents. The only known remaining population is in an area adjacent to and within Yosemite National Park where livestock grazing is prohibited (J. Winter, consultant, pers. commun.).

Overgrazing has contributed to the reduction of many native species in the San Joaquin Valley of California. This area hosts the largest concentration of state and federally listed species of any area within the continental United States. Substantial areas of San Joaquin kit fox habitat have been modified or lost to livestock grazing. Shifts in species composition of small mammal communities from granivores to herbivores such as California ground squirrels (*Citellus beecheyi*) may be a result of excessive grazing. Unfortunately for the fox, the diurnal ground squirrels are not adequate prey (U.S. Fish and Wildl. Serv. 1983).

Overgrazing may also destroy blunt-nosed leopard lizard habitat. It results in soil compaction, damage to rodent burrows (used by the lizard for escape cover), and loss of vegetational cover for the lizard and its prey (U.S. Fish and Wildl. Serv. 1988).

Western riparian ecosystems are of special concern because of the concentration of biodiversity in them (Chaney et al. 1990). About three fourths of the vertebrates in the Southwest, for example, are dependent upon riparian areas for some portion of their life history (Johnson 1989). As Fleischner (1994:635) noted, "Because livestock spend much of their time in riparian communities, and because the ecological stakes are highest here, many of the adverse impacts of grazing are magnified in these habitats."

Destruction of adjacent vegetation has resulted in siltation of stream bed gravels, warming of the water, and elimination of above-ground flow during summer and fall. Aquatic species, especially native fish, are unable to adapt to such changes. Fleischner (1994) called livestock grazing a principal factor contributing to the decline of native trout in the West. Several native fish species have disappeared entirely (Warner and Hendrix 1985). The United States General Accounting Office (1988) reported that degradation of riparian and aquatic areas was the most serious threat to biodiversity in the West.

A number of listed and candidate species reside in riparian ecosystems. An example is the endangered least Bell's vireo found in California and the arid Southwest. The vireo's nesting habitat is restricted to the lower strata of vegetation in riparian ecosystems within its range. Livestock grazing has been the primary factor in the destruction of this strata (Franzreb 1989).

In California's Central Valley, oak (*Quercus* spp.) ecosystems have been drastically altered by human activities, including the grazing of livestock. Continuous livestock occupancy for the past century has resulted in the browsing of nearly all available oak seedlings and acorns (Duncan and Clawson 1979). Reproduction of several species of oak is non-existent. Existing oaks average >100 years in age. Barrett (1979) states that >60 species of mammals use these oaks, and Verner (1979) indicates that 110 species of birds use valley and foothill habitats where oaks make up a significant part of the tree canopy. Many of these species will continue to decline as the oaks disappear.

Livestock grazing in the Southwest has destroyed the cryptogamic crust and its associated microbiota (Fleischner 1994). This alters native plant communities and allows the spread of exotic species. The alteration is a major factor in the listing of the desert tortoise by the State of California and the federal government, and the Mojave ground squirrel (*Spermophilus mohavensis*) by California, according to an unpublished biological opinion on sheep grazing prepared by the Fish and Wildlife Service in 1992.

Native habitats have also been purposefully altered for the "improvement" of rangelands. Vegetation not considered important livestock forage, especially sagebrush (*Artemisia* spp.) and juniper (*Juniperus* spp.), is removed through chaining, burning, or herbicide application. Treated sites are re-seeded to grasses, often exotic species. In the western states, these activities have reduced the original sage grouse (*Centrocercus urophasianus*) habitat by 50% (Call 1974.)

Positive Aspects of Livestock Grazing on Listed Species

Savory (1988) hypothesized that the behavior of herding ungulates, in the presence of large pack-hunting predators, was essential to the prehistoric maintenance of "brittle" (i.e., arid, roughly) grasslands. In this scenario, grazing was necessary to remove dead plant material to stimulate growth, and periodic excited hoof action was required to till the soil, enhancing water retention and seed germination. Edwards (1992) supports this view for the native grasslands of California.

However, other researchers have argued the contrary (Baker 1992, Belsky 1992, Stebbins 1992). Certainly something is awry; wolves (the only large pack-hunting predator in North America) were only common in the Southwest (with its erstwhile productive grasslands) for a brief period in the late 1800s and early 1900s (Brown 1983), and the journals of Lewis and Clark attest to the paucity of ungulates in the pristine Palouse grasslands (DeVoto 1953).

Nevertheless, it seems logical that, where native ungulates have been extirpated or significantly reduced, judicious application of domestic livestock grazing may replicate natural conditions. For example, pronghorn sometimes fare better on rangelands grazed by cattle, rangelands that were once grazed by bison or elk, because the lower foliage of grazed rangelands allows pronghorn does to protect their fawns from coyotes more successfully (J. D. Yoakum, pronghorn biologist, pers. commun.). This benefit only applies as long as grazing does not result in conversion of dry grasslands to shrubby plant communities, which are less secure for pronghorn.

Rarely, livestock have assisted in the management of an endangered species, by acting as a surrogate for native grazing species that have been depleted or eliminated. For the whooping crane, succession of the plant community from grass to brush is detrimental. In a mesic plant community, livestock grazing can reduce brush invasion, maintaining suitable conditions for the cranes. However, this activity must be timed so that cranes are not present when grazing occurs (Olson 1980).

SUMMARY

The livestock industry has come as a shock to many American ecosystems, endangering species in the process. Species have suffered direct and indirect effects from livestock. Direct effects have included predator and rodent control, reduction of competitors, and accidental mortality. Indirect effects have included trophic linkage, disease and internal parasitism, external parasitism, and chemical contamination. Classic examples of species that have suffered the direct effects of livestock production are the wolf, prairie dog, and desert tortoise. Those that have suffered indirect effects include the black-footed ferret, willow flycatcher, and California condor. Some species suffer direct and indirect effects.

The "peaceful coexistence" of species in competition with each other is an oxymoronic ideal, as is the coexistence of domestic livestock and many native species. Ecosystems that appear relatively stable in the managerial time frame are actually dynamic in the evolutionary long term (Botkin 1990). The argument that endangered species will or should evolve to conform with modern conditions is unreasonable, because the molecular clock of evolutionary supply does not run at the speed of industrial demand. As Soule (1983:115) pointed out, "we are dealing with the problem that the rate of environmental change . . . is several orders of magnitude higher than the rate at which genes are substituted in . . . vertebrates."

Hopefully, the livestock industry will soon accept its share of the responsibility to perpetuate native ecosystems, including the full array of native species. After all, industry is a collective form of citizenship, and citizenship "involves an ongoing effort to synthesize questions of 'what is best for the world' with 'what is best for me'" (Landy 1993:20). The American public's desire for the conservation of endangered species is well documented in federal and state legislation. If this desire is not acknowledged by the livestock industry, the public will conclude that livestock interests view public lands as resources devoted to what is best for them; i.e., that the livestock industry is not a good citizen. This may turn the political tide in favor of species conservation, but by then, many more species may be extinct.

Wilkinson's (1992:113) assessment seems appropriate:

> [The] range has long been conceptualized as, and used as, a place to graze cows. We know now that it is much more than that. It is fragile: if its cover is not protected, it can lift off and blow away. It houses other animals: elk, antelope, and [other species] also have grazing needs; trout require clean, cool streams; and wildfowl must have nesting places in green marshlands . . . Ranchers have contributed much to the West, and they deservedly hold a prime place among those with access to public rangeland, but they have no right to a monopoly or to practices that injure the ground. If we look first to the needs of these long-neglected lands, the western range will provide benefits aplenty.

LITERATURE CITED

Alderson, G., editor. 1987. Saving endangered species. Defenders of Wildlife, Washington, D.C. 28pp.

Allen, D. L. 1954. Our wildlife legacy. Funk and Wagnalls Co., New York, N.Y. 422pp.

Allen, T. B. 1974. Vanishing wildlife of North America. Natl. Geogr. Soc., Washington, D.C. 207pp.

Baker, D. L. 1989. Responses of bobwhites, eastern meadowlarks, mourning doves, and their habitat to two intensities of continuous grazing. M.S. Thesis, Texas A&I Univ., Kingsville. 55pp.

Baker, H. G. 1992. Letter to editor. Fremontia 20:32-33.

Barker, W. T. 1991. Influence of grazing treatments on nongame birds and vegetation structure in south central North Dakota. Ph.D. Thesis, North Dakota State Univ., Billings. 164pp.

Barrett, R. 1979. Mammals of California oak habitats- management implications. Pages 275-278 *in* T. R. Plumb, tech. coord. Proc. symposium on the ecology, management and utilization of California oaks. U.S. For. Serv. Gen. Tech. Rep. PSW-44.

Barton, K. 1987. Bureau of Land Management. Pages 3-59 *in* Audubon wildlife report, 1987. Academic Press, Inc., San Diego, Calif. 697pp.

Beuchner, H. H. 1960. The bighorn sheep in the United States, its past, present and future. Wildl. Monogr. 4. 119pp.

Belsky, A. J. 1992. Letter to editor. Fremontia 20:30-31.

Berry, K. H. 1978. Livestock grazing and the desert tortoise. Trans. North Am. Wildl. and Nat. Resour. Conference 43:505-519.

Botkin, D. B. 1990. Discordant harmonies. Oxford Univ. Press, U.K. 241pp.

Brown, D. E. 1983. The wolf in the Southwest. Univ. Arizona Press, Tucson. 195pp.

Burk, C. J. 1973. The Kaibab deer indicent: a long-persisting myth. Bioscience 23:113-114.

Call, M. 1974. Habitat requirements and management recommendations for sage grouse. U. S. Bur. Land Manage. Tech. Note 330. 37pp.

Carbyn, L. N., S. M. Oosenbrug, and D. W. Anions. 1993. Wolves, bison and the dynamics related to the Peace-Athabascan Delta in Canada's Wood Buffalo National Park. Circumpolar Res. Series 4. Canadian Circumpolar Institute. Univ. Alberta, Calgary. 270pp.

Chaney, E., W. Elmore, and W. S. Platts. 1990. Livestock grazing on western riparian areas. U.S. Environ. Prot. Agency. and Northwest Resour. Inf. Center, Inc., Eagle, Colo. 44pp.

Clark, T. W., and A. H. Harvey. 1988. Implementing endangered species recovery policy: learning as we go. Endangered Species Update 5:35-42.

Cubbage, F. W., J. O'Laughlin, and C.S. Bullock III. 1993. Forest resource policy. John Wiley and Sons, Inc. New York, N.Y. 562pp.

Darlington, D. 1987. In condor country. Houghton Mifflin Co., Boston, Mass. 242pp.

Dasmann, R. F. 1972. Environmental conservation. John Wiley and Sons, Inc., New York, N.Y. 473pp.

DeVoto, B., editor. 1953. The journals of Lewis and Clark. Houghton, Mifflin, and Co., Boston, Mass. 504pp.

Duncan, D. A., and W. J. Clawson. 1979. Livestock utilization of California's oak woodlands. Pages 306-313 *in* T. R. Plumb, tech. coord. Proc. symposium on the ecology, management and utilization of California oaks. U.S. For. Serv. Gen. Tech. Rep. PSW-44.

Edwards, S. W. 1992. Observations on prehistory and ecology of grazing in California. Fremontia 20:3-11.

Ferguson, D., and N. Ferguson. 1983. Sacred cows at the public trough. Maverick Publications, Bend, Oreg. 250pp.

Fleischner, T. L. 1994. Ecological costs of livestock grazing in western North America. Conserv. Biol. 8:629-644.

Foreyt, W. J., and D. A. Jessup. 1982. Fatal pneumonia of bighorn sheep following association with domestic sheep. J. Wildl. Dis. 18:163-168.

Franzreb, K. E. 1989. Ecology and conservation of the endangered least Bell's vireo. U.S. Fish and Wildl. Serv. Bio. Rep. 89. 17pp.

Fritts, S. H., W. J. Paul, L. D. Mech, and D. P. Scott. 1992. Trends and management of wolf-livestock conflicts in Minnesota. U.S. Dep. Inter. Res. Publ. 181. 27pp.

Gabel, R. R., and S. J. Dobrott. 1988. Saving the masked bobwhite. Endangered Species Tech. Bull., U. S. Fish and Wildl. Serv., Washington, D.C. 5pp.

Harting, A. L. 1987. Intraspecific relationships: impacts from livestock and other agricultural impacts. Pages 63-78 *in* M. N. LeFranc, Jr., M. B. Moss, K. A. Patnode, and W. C. Sugg, III, eds. Grizzly bear compendium. Natl. Wildl. Fed., Washington, D.C.

Ingram, H., and A. Schneider. 1993. Constructing citizenship: the subtle messages of policy design. Pages 68-98 *in* H. Ingram, and S. R. Smith, eds. Public policy for democracy. The Brookings Inst. Washington, D.C.

Jensen, D. B., M. Thorn, and J. Harte. 1990. In our own hands: a strategy for conserving biological diversity in California. Calif. Policy Semin. Res. Rep., Univ. Calif., Berkeley. 184 pp.

Jessup, D. A. 1985. Diseases of domestic livestock which threaten bighorn sheep populations. Desert Bighorn Counc. Transactions. 29:29-33.

Johnson, A. S. 1989. The thin, green line: riparian corridors and endangered species in Arizona and New Mexico. Pages 35-46 *in* G. Mackintosh, ed. In defense of wildlife: preserving communities and corridors. Defenders of Wildlife, Washington, D.C.

Kaufman, M. R., R. T. Graham, D. A. Boyce, Jr., W. H. Moir, L. Perry, R. T. Reynolds, R. L. Bassett, P. Mehlhop, C. B. Edminster, W. M. Block, and P. S.

Corn. 1994. An ecological basis for ecosystem management. U.S. For. Serv. Gen. Tech. Rep. RM-246. Rocky Mt. For. and Range Exp. Stn. Fort Collins, Colo. 22pp.

Kellert, S. R. 1984. Wildlife values and the private landowner. Am. For. 90:27-28, 60-61.

Klopatek, C. C. 1992. Effects of disturbance on mycorrihizae and microbes in pinyon-juniper ecosystems. Ph.D. Thesis, Ariona State Univ., Tempe. 172pp.

Knight, R. R., B. M. Blanchard, K. C. Kendall, and L. E. Oldenburg. 1980. Yellowstone grizzly bear investigations: annual report of the interagency study team, 1978-1979. U.S. Dep. Inter., Bozeman, Mont. 91pp.

Koford, C. 1953. The California condor. Natl. Audubon Soc. Res. Rep. 4, New York, N.Y. 154pp.

Kohm, K. A., editor. 1991. Balancing on the brink of extinction. Island Press, Washington, D.C. 318pp.

Kormondy, E. J. 1969. Concepts of ecology. Prentice-Hall, Inc., Englewood Cliffs, N.J. 209pp.

Krausman, P. R. 1995. Problems facing bighorn sheep in and near domestic sheep allotments. Pages 60-65 *in* Sustaining Rangeland Ecosystems Symposium. In Press.

Landy, M. 1993. Public policy and citizenship. Pages 19-44 *in* H. Ingram, and S. R. Smith, eds. Public policy for democracy. The Brookings Inst. Washington, D.C.

Leopold, A. 1949. A sand county almanac, and sketches from here and there. Oxford Univ. Press, New York, N.Y. 226pp.

Line, L. 1993. The curse of the cowbird. Natl. Wildl. 32:40-45.

Littell, R. 1992. Endangered and other protected species: federal law and regulation. Bur. Natl. Aff., Washington, D.C. 613pp.

Littlefield, C. D. 1982. The status and distribution of greater sandhill cranes in California, 1981. Calif. Dep. Fish and Game. Adm. Rep. 82-1. 27pp.

Loft, E. R. 1989. Habitat and spatial relationships between mule deer and cattle in a Sierra Nevada forest zone (California). Ph.D. Thesis, Univ. Calif., Berkeley. 151pp.

Lopez, B. H. 1978. Of wolves and men. Charles Scribner's Sons, New York, N.Y. 309pp.

Mackenzie, J. P. S. 1977. Birds in peril. Pagurian Press, Toronto, Ont., Canada. 192pp.

Matthiessen, P. 1987. Wildlife in America. Viking Press, New York, N. Y. 332pp.

McMillan, I. 1968. Man and the California condor. E. P. Dutton and Co., Inc., New York, N.Y. 191pp.

Mech, L. D. 1970. The wolf: the ecology and behavior of an endangered species. Nat. Hist. Press, Garden City, N.Y. 384pp.

Moen, A. N. 1973. Wildlife ecology. W. H. Freeman and Co., San Francisco, Calif. 458pp.

Nowak, R. M., editor. 1991. Walker's mammals of the world, Fifth ed. Johns Hopkins Univ. Press, Baltimore, Md. 1629pp.

Noss, R. F. 1994. Cows and conservation biology. Conserv. Biol. 8:613-616.

Ohmart, R. D., and B. W. Anderson. 1982. North American riparian ecosystems. Pages 33-479 *in* G. L. Bender, ed. Reference handbook on the deserts of North America. Greenwood Press, Westport, Conn.

Olson, D., team leader. 1980. Whooping crane recovery plan. U.S. Fish and Wildl. Serv., Washington, D.C. 206pp.

Reffalt, W. C. 1988. United States listing for endangered species: chronicles of extinction? Endangered Species Update 5:10-13.

Reisner, M. 1986. Cadillac desert. Viking Press, New York, N.Y. 514pp.

Salwasser, H. 1994. Ecosystem management: what's in store for wildlife? Fair Chase, Summer:10-11.

Savory, A. 1988. Holistic resource management. Island Press, Washington, D.C. 564pp.

Schneider, B. 1977. Where the grizzly walks. Mt. Press Publ. Co., Missoula, Mont. 191pp.

Seal, U. S. 1989. Introduction. Pages xi-xvii *in* U. S. Seal, E. T. Thorne, M. A. Bogan, and S. H. Anderson, eds. Conservation biology and the black-footed ferret. Yale Univ. Press, New Haven, Conn. 302pp.

Smith, J. P. Jr., and K. Berg, editors. 1988. Inventory of rare and endangered vascular plants of California. Calif. Native Plant Soc., Spec. Publ. No. 1., Fourth ed., Sacramento, Calif. 168pp.

Soule, M. E. 1983. What do we really know about extinction? Pages 111-124 *in* C. M. Schonewald-Cox, S. M. Chambers, B. MacBryde, and W. L. Thomas, eds. Genetics and conservation: a reference for managing wild animal and plant populations. Benjamin/Cummings Publishing Company, Inc., London, U. K.

Stebbins, G. L. 1992. Letter to editor. Fremontia 20:31-32.

Steinhart, P. 1990. California's wild heritage. Calif. Dep. Fish and Game, Sacramento. 108pp.

Tibbetts, T. J. 1993. Proposed rule to list the southwestern willow flycatcher as endangered with critical habitat. Fed. Register. 58:39495-39522.

Tomlinson, R. E. 1972. Current status of the endangered masked bobwhite quail. Trans. North Am. Wildl. and Nat. Resour. Conf. 37:294-311.

United States General Accounting Office. 1988. Rangeland management: some riparian areas restored but widespread improvement will be slow. June, 1988. GAO/RCED-88-105, Washington, D.C.

United States Fish and Wildlife Service. 1978. Guidelines for management of the Central Valley population of greater sandhill cranes. U.S. Fish and Wildl. Serv., Portland, Oreg. 42pp.

———. 1983. San Joaquin kit fox recovery plan. U.S. Fish and Wildl. Serv., Portland, Oreg. 84pp.

———. 1988. Revised blunt-nosed leopard lizard recovery plan. U.S. Fish and Wildl. Serv., Portland, Oreg. 85p.

Verner, J. 1979. Birds of California oak habitats-management implications. Pages 246-261 *in* T. R. Plumb, tech. coord. Proc. symposium on the ecology, management and utilization of California oaks. U.S. For. Serv. Gen. Tech. Rep. PSW-44.

Warner, R. E., and K. M. Hendrix. 1985. Riparian resources of the central valley and California desert. Calif. Dep. Fish and Game, Sacramento. 235pp.

Wilbur, S. R. 1978. The California condor, 1966-1976: a look at its past and future. U.S. Dep. Inter., North Am. Fauna 72, Washington, D.C. 136pp.

Wilkinson, C. F. 1992. Crossing the next meridian. Island Press, Washington, D.C. 376pp.

Willet, G. 1933. A revised list of the birds of southwestern California. Pacific Coast Avifauna 21:1-204.

Williams, T. 1992. No dogs allowed. Audubon 94:26-35.

Wilson, E. O. 1984. Biophilia, the human bond with other species. Harvard Univ. Press, Cambridge, Mass. 157pp.

Winter, J., 1986. Status, distribution and ecology of the great gray owl (*Strix nebulosa*) in California. M.A. Thesis, San Francisco State Univ., San Francisco, California. 121pp.

Zaslowsky, D. 1989. A public beef. Harrowsmith 4:39-47.

Management of Wildlife on Rangelands

Chapter 5
PERSPECTIVES ON GRAZING NONGAME BIRD HABITATS

Fritz L. Knopf

INTRODUCTION

Nongame birds generally refer to that group of taxonomically diverse orders, families, and species that, collectively, are not usually hunted for food or sport in the United States. This group comprises 772 (93%) of the 831 species recorded in the United States (Chandler 1986). Rangeland wildlife responses to grazing practices for most species and groups of mammals and birds being addressed in this book are relatively straightforward and tightly focused compared to the task of defining responses of so many nongame birds to grazing.

Because a large number of species potentially occur in an avifauna at a given site, each with differential responses to grazing, a single-species approach to managing grazing for nongame birds would be overwhelming and generally cost prohibitive. More importantly, introducing grazing mammals into nongame bird habitats can either directly or indirectly favor populations of some species while depressing others. Grazing alters abiotic and biotic relationships within and among local bird species.

With a growing proportion of the North American public being nonhunters, antihunters, and even antigrazing on public lands surely the focus of range management practices will evolve to focus more and more on the biological diversity (i.e., the sustainable resource) of a locale rather than the harvestable biomass (i.e., commodity) that has historically been emphasized by natural resource agencies. Nongame birds represent a major component of the biological diversity of rangelands, and the objective of this chapter is to provide perspectives on, and an approach to, the conservation of nongame birds on native rangelands managed simultaneously for the production of beef cattle.

THE NONGAME AVIFAUNA

The nongame avifauna of North America, which includes representatives of 180 orders, is taxonomically diverse. However, generalizations about the taxonomic affinities of nongame birds can be risky. Closely related species within a genus, prairie-chickens (*Tympanuchus* spp.) for example, may or may not be hunted even though they are in a taxonomic order (Galliformes) in which most species are hunted. In contrast, game species among the rails (*Rallus* spp.) are in an order (Gruiformes) in which most species are not hunted. In addition, individual species (e.g., mourning dove [*Zenaida macroura*]) may be hunted in one state and protected from hunting in a contiguous state. The 18 orders, however, comprise species that occur in 4 distinct ecological groups: aquatic birds, wetland birds, birds of prey, and terrestrial birds.

Aquatic Birds

Aquatic bird species spend most of their time on open water or use submersed and emergent vegetation in the zone between open water and the shoreline. They include members of the Gaviiformes (i.e., loons), Podicipediformes (i.e., grebes), Pelecaniformes (i.e., pelicans and cormorants), and Coraciiformes (i.e., kingfishers). Aquatic birds generally respond to water presence and quality based on the abundance of their aquatic food base. In addition, they tend to nest in floating mats of vegetation or at sites (as islands) inaccessible to cattle. Grazing, therefore, is rarely considered to have a major direct effect on populations of aquatic birds but may indirectly affect food base availability by altering water quality.

Wetland Birds

Wetland species that spend more time closer to shorelines include members of the Ciconiformes (i.e., herons, ibises, and storks), Phoenicopteriformes (i.e., flamingos), Gruiformes (i.e., rails and cranes), and Charadriiformes (i.e., shorebirds and gulls). Most shoreline birds respond favorably to management actions designed to protect or improve habitats for wetland game birds of the Anseriformes. Many forage and nest in emergent marsh vegetation typically managed to enhance waterfowl nesting (Kirsch and Higgins 1976, Duebbert and Lokemoen 1977, Whyte and Cain 1979). As with waterfowl, the relative effects of grazing generally increase if the nongame species uses drier sites more accessible to cattle (Eddleman et al. 1988). The mountain plover (*Charadrius montanus*) and long-billed curlew (*Numenius americanus*) are examples of species within this group that use drier sites. In the case of plovers, however, habitat quality actually improves as the intensity of grazing increases at some sites (Graul 1975, Bicak et al. 1982). For example, cattle often destroy virtually all vegetation around stock watering tanks on the Pawnee National Grassland of northern Colorado. In June of 1991, I surveyed tanks used by cattle versus those not being used. None of the 28 tank areas unused by cattle were used by plovers. On the other hand, 11 of the 28 used by cattle had plovers present in the vicinity during the survey. The presence of cattle enhances the attractiveness of a site to mountain plovers.

Conservation and management of nongame birds in recent years has emphasized the secondary benefits that wetland enhancement for waterfowl has to species that are not hunted (Knopf 1992*b*). Except for species that frequent drier sites and often prefer heavily grazed areas, specific field practices to favor nongame, wetland bird assemblages follow the general recommendations found in Chapter 7.

Birds of Prey

This group of birds, also known as raptors, includes members of the Falconiformes (i.e., falcons and hawks) and Strigiformes (i.e., owls). Some species, especially those that prey on fish or other bird species (Newton 1979), declined because of organochlorine insecticides in the food chain. Currently, organophosphate insecticides are thought to have major effects upon the health of raptors (Wilson et al. 1991) and possibly other species. Biologists and conservationists have tried to encourage repopulation of historic ranges of some species by promoting legislative regulations to reduce levels of organochlorine compounds in the environment (Wilson et al. 1991), manipulating sites of nesting birds (Phillips and Beske 1984, Smith 1985), or "hacking back" (a historical falconry practice) young birds at sites where the species has been extirpated.

Raptors have received much attention on rangelands since the late 1970s and have been the focus of major management efforts (National Wildlife Federation 1989). Kochert (1989) recently reviewed the responses of raptors to livestock grazing, summarized here. He identified 3 major impacts of livestock grazing on raptor populations. First, cattle can affect the availability of nest sites. Trampling and cropping of standing herbaceous vegetation can have an immediate detrimental effect on cover for ground-nesting species such as the northern harrier (*Circus cyaneus*) and short-eared owl (*Asio flammeus*), similar to impacts cattle have on nesting cover for waterfowl and shoreline birds. Kochert (1989) also identified potential impacts upon overstory nesting substrates, albeit the evidence for long-term negative consequences were highly inferential. The large foraging ranges of most species, coupled with a tendency for many raptors to use isolated trees for nests and perches, argue against the universality of grazing effects upon forest stand regeneration as a specific threat to nesting sites of many raptors. Obviously, effects of cattle and grazing upon nest sites of species that use exceptionally rugged locations (e.g., cliff faces) within the terrain are inconsequential.

Livestock grazing alters raptor habitats primarily by influencing the abundance and availability of prey populations (Kochert 1989). The importance of voles (*Microtus* spp.) as prey for short-eared owls (Clark 1975), Townsend's ground squirrel (*Spermophilus townsendi*) as prey for prairie falcons (*Falco mexicanus*) (Holthuijzen 1990), and fossorial mammals such as prairie dogs (*Cynomys* spp.) and the California ground squirrel (*Spermophilus beecheyi*) as ecosystem associates of burrowing owls (*Athene cunicularia*) (Coulombe 1971, Thomsen 1971, Butts 1973, Martin 1973) are well established. Ironically, voles are relatively intolerant of grazing, ground squirrels tolerate proper grazing (the Soil Conservation Service recommended stocking intensities based upon forage production data), and prairie dogs increase at sites of heavy grazing. Given an understanding of the diets and vertebrate associates of endemic raptor species, grazing and range prescriptions to enhance raptor habitats should be designed specifically to augment prey populations of a target raptor species or assemblage.

Terrestrial Birds

This group includes nongame members of the Galliformes (i.e., grouse) and Columbiformes (i.e., pigeons and doves), plus all species within the Cuculiformes (i.e., cuckoos), Caprimulgiformes (i.e., goatsuckers), Apodiformes (i.e., swifts and hummingbirds), Trogoniformes (i.e., trogons), Piciformes (i.e., woodpeckers), and Passeriformes (i.e., perching birds). Passeriformes is the largest order of bird species in North America; species in some families as the Icteridae (i.e., blackbirds) and Troglodytidae (i.e., wrens) can also be found in wetlands at some locales. Considering the large number of species involved and the evolution of nongame conservation into the broader perspectives of biological diversity conservation (Knopf 1992*a*), these will be the species of most concern to range managers in the future, and the ones that this chapter addresses specifically.

TERRESTRIAL BIRD RESPONSES TO GRAZING

An Introductory Caveat

Individuals within many conservation groups and natural resource management agencies have historically been very critical of grazing (or more specifically the presence of cattle on public lands). However, scientific evidence about grazing effects has often (if not generally) been flawed by (1) poor design (e.g., inadequate controls, replications) of studies, (2) abusively grazed (i.e., "overgrazed") sites carelessly construed to represent proper range management, and (3) investigator advocacy for a fisheries or wildlife resource. None of these flaws is individually more serious than the other. This point has recently been discussed most specifically relative to the waterfowl literature (Kirby et al. 1992); and the grazing and nongame bird literature (like the riparian literature [Knopf 1987]), is dominated by a high proportion of papers not subjected to critical review by peers. Thus, the following discussion is intended to provide perspectives rather than an exhaustive review that synthesizes the literature on grazing/nongame bird relationships. Simply stated, an adequate literature data base is not currently available.

Zoogeographic Perspectives

Species that evolved in heavily grazed ecosystems as the western Great Plains dominated by 30-60 million bison (*Bison bison bison*) (Fig. 1) and prairie dogs or the grasslands of the Central Valley of California that provided habitat for an estimated 600,000 wapiti (*Cervus elaphus*), kangaroo rats (*Dipodomys* spp.), and ground squirrels (*Spermophilus beecheyi*) are often dependent upon grazers and respond favorably to the introduction of livestock into an area. Mengel (1970) identified 26 endemic species of birds that evolved on the North American grasslands. Each of these endemic species can be ordinated along a gradient of increasing herbaceous vegetation at a locale within the grazed landscape (Wiens 1973). Some of these species such as the mountain plover (Graul 1975), Baird's sparrow (*Ammodramus bairdii*), and chestnut-collared longspur (*Calcarius ornatus*) seem dependent upon grazed landscapes, occurring in much higher densities at grazed sites (Kantrud 1981, Renken and Dinsmore 1987).

The behavior of grazing animals favors among-site heterogeneity of vegetation, especially where grasses are also subjected to periodic fire (Glenn et al. 1992). Cattle, like plains bison, exhibit definite patterns of grazing movement and vegetation use although they tend to concentrate more in riparian areas than bison (Leege et al. 1981, Roath and Krueger 1982, Marlow and Pogacnik 1986). Nongame birds of native grasslands select microhabitats from a mosaic of patches with variable densities of standing herbaceous vegetation created by differential grazing pressures within a locale (Graul 1980). In landscapes historically dominated by massive herds of unconfined native grazers, habitats of some nongame birds became degraded when grazing herbivores were removed. Fences create more uniform grazing pressure across landscapes, especially as allotment size decreases and cattle are managed more precisely for vegetation removal. In my opinion, uniformity of grazing management on the Great Plains influences local avian assemblages more than the actual presence of livestock. Alternatively, introducing cattle into areas (e.g., southwestern deserts and riparian associations) that evolved over the last 10,000 years in the absence of massive herds of native grazers can represent a much greater disturbance to a nongame bird assemblage (Szaro 1980, Bock and Bock 1988, but see Janzen 1986), and requires more precision in stocking density and cattle management.

Vegetation Perspectives

Because birds generally respond to cattle impacts on vegetation (Bock and Webb 1984) and not to cattle or grazing, specifically, vegetation at a site can be viewed as habitat for nongame birds and examined for how it is modified by cattle relative to habitat requirements. Ornithologists have long been interested in how birds select habitats, and recent studies (Wiens and Rotenberry 1981, Wiens 1985, Knopf et al. 1990) have helped to define an hierarchical view of the selection process. The view includes selection of landscapes based upon features of vegetative physiognomy, such as a forest canopy, followed by ≥1 specific feature (e.g., certain tree species that occur in a definable patch), and then selecting specific individual microhabitats within that patch based upon site characteristics (e.g., the vigor of the individual plant where the bird will forage or build a nest). To illustrate, a northern flicker (*Colaptes auratus*) might select a riparian forest, then choose a cottonwood tree (*Populus* spp.) over peachleaf willow (*Salix amygdaloides*) because cot-

FIGURE 1. Some species of nongame birds evolved in ecosystems intensively grazed by herbivores as bison. One species, the mountain plover (right) prefers to use sites where cattle severely disturb native vegetative cover.

tonwoods are better suited for nest cavity construction, and finally select an aging cottonwood with a large dead branch of some minimum diameter for the specific cavity location (Sedgwick and Knopf 1990). In a nearby upland, a green-tailed towhee (*Pipilo chlorurus*) might select a sagebrush (*Artemisia* spp.) dominated landscape, then locate a patch where the sagebrush intergrades with chokecherry (*Prunus virginiana*), and finally settle at a site where bushes are more vigorous forming more of a closed canopy than in the larger patch as a whole (Knopf et al. 1990). In the case of this towhee that often nests on the ground, the bird may never actually select an individual plant for any purpose, but chooses appropriate canopy density. The significance of an individual plant also is lost to species such as the yellow warbler (*Dendroica petechia*) that nest in patches of woody vegetation so dense that canopies of the plants intergrade (Knopf and Sedgwick 1992).

FIGURE 2. Riparian areas at lower elevations in western North America are often subjected to abusive grazing practices. Along the South Platte River of eastern Colorado, the shrub vegetation layer has been almost completely removed by a history of improper grazing practices.

Given this avian view of vegetation, how do livestock influence its desirability as potential bird habitats? Obviously, cattle have different impacts upon the varied native-vegetation associations, and even within a biogeographic region the impact can be highly variable (Wiens 1973). In the vertical plane, cattle have relatively little immediate impact upon birds that nest and forage in forest canopies. Alternatively, cattle can have major, relatively immediate impacts upon shrub (Sedgwick and Knopf 1991) and shrub and ground vegetation layers (Mosconi and Hutto 1982). The shrub layer is often completely removed in areas with a history of abusive grazing (Fig. 2)(Klebenow and Oakleaf 1984). Cattle grazing does not universally result in elimination of the shrub layer, however, but modifies the physical structure of the community with time. In high-elevation willow communities, for example, the long-term grazing history of a site is easily viewed in the shapes of the bushes themselves (Fig. 3, 4). Severely damaged woody vegetation recovers slowly (Knopf and Cannon 1982) and grazing-sensitive passerine species such as the willow flycatcher (*Empidonax traillii*) return to nest in damaged areas only after prolonged periods of rest from grazing or implementation of corrective grazing programs (Table 1).

The herbaceous layer of vegetation also is directly impacted by grazing. This stratum has received the most specific attention, primarily because many bird species are dependent upon taller grasses for concealment or support of nests. As grazing pressures increase, these species decline locally while numbers and densities of species that favor shorter herbaceous vegetation increase. The focusing of grazing impact analyses upon this vegetation layer has even given rise to a "density pole" (Robel et al. 1970) for quantifying standing biomass of grasses as potential bird habitats.

Grazing influences more than just vertical structuring of shrub and herbaceous vegetation layers. Grazing practices change the horizontal patterning of the vegetative layers by creating more open shrub stands (Knopf et al. 1988), bare ground

FIGURE 3. A long history of excessive grazing during the growing season may alter the structure of woody vegetation used by birds, especially in the shrub layer. In high elevation riparian landscapes, such a history is immediately obvious in the shapes of willow shrubs. In contrast to the normal, hemispheric shapes of healthy willows on the Arapaho National Wildlife Refuge (left), which have not been abusively grazed, areas as the one along the nearby North Platte River contain shrubs that have been "notched," that is, the lateral branches are removed (right). The left area has a 100-year history of late-season and dormant-season grazing. The North Platte area has historically been, and still is, grazed during the growing season.

FIGURE 4. The "notching" of willow bushes normally would appear to be a consequence of cattle physically breaking branches from a bush (left); however, the notch actually reflects a history of cattle browsing the lateral branches of growing plants, as seen in the smaller shrubs in the right photograph.

interspersed in grasslands (Bock and Webb 1984), and shrub (Ellison 1960) or forb (Kie and Loft 1990) invasions of grasslands. In western North America, grazing appears to have its greatest influence on nongame birds as it immediately or historically alters the horizontal patterning of the lower vegetation layers (Knopf et al. 1988).

Population Perspectives

Populations of terrestrial birds are commonly viewed as increasing or decreasing with the introduction of cattle into their habitats (Ryder 1980). To illustrate his dichotomy, Ryder (1980) pointed out mountain plovers, horned larks (*Eremophila alpestris*), and McCown's longspurs (*Calcarius mccownii*) as species that increase in numbers and lark buntings (*Calamospiza melanocorys*), Brewer's sparrows (*Spizella breweri*), and western meadowlarks (*Sturnella neglecta*) as species that decline in numbers with increasing intensity of grazing on the western Great Plains.

Viewing each species as an increaser or decreaser in response to cattle grazing is an attractive approach locally, but can represent an oversimplification at larger scales of resolution. Whereas some species such as the willow flycatcher predictably decline or disappear in grazed areas in Colorado (Knopf et al. 1988), Montana (Mosconi and Hutto 1982), and Oregon (Taylor 1986), and species as the brown-headed cowbird (*Molothrus ater*) increase everywhere that cows are introduced, other species may increase with grazing pressure at 1 locale and decrease at another. Horned larks and McCown's longspurs that were labeled as increasers in Colorado (Ryder 1980) actually increase and decrease with increasing grazing intensity depending on soil type at a site when viewed more regionally (Kantrud and Kologiski 1982). The latter authors found that the responses of most species of nongame birds to grazing covaried with grazing intensity and soil characteristics as those 2 factors affected herbaceous vegetation. Thus, the habitat requirements of a bird may occur naturally at one locale, require some disturbance or manipulation at a second locale, and may require excessive disturbance at a third locale. Bock and Webb (1984) concluded

TABLE 1. Numbers of individual birds recorded annually for select passerine species on 400 10-minute stationary plot surveys in riparian vegetation along the Illinois River on the Arapaho National Wildlife Refuge (Jackson County, Colorado) 1980, 1983, 1986, 1989. The refuge initiated a deferred, late-season (Aug-Sep) 3-pasture rotation of grazing in 1980 in pastures in attempt to recover nongame bird habitats that had been damaged by a 100-year history of grazing.

	1980	1983	1986	1989
Willow flycatcher	20	36	68	107
Lincoln's sparrow (*Melospiza lincolnii*)	20	47	41	118
White-crowned sparrow (*Zonotrichia leucophrys*)	17	33	33	51
Common yellowthroat (*Geothlypis trichas*)	3	0	1	15
Veery (*Catharus fuscescens*)	0	0	5	14
Fox sparrow (*Passerella iliaca*)	0	0	0	12

that the bare-ground component of grasshopper sparrow (*Ammodramus savannarum*) habitats in Arizona were similar to those reported by Whitmore (1979) for the species in West Virginia. This sparrow occurs on sites that are protected from grazing in Arizona, are lightly grazed in North Dakota (Kantrud and Kologiski 1982), and are highly disturbed by strip-mining activities in West Virginia.

Even at a given site, a species may initially increase with the introduction of cattle into a landscape and subsequently decrease as grazing pressure intensifies. Numbers of northern flickers (*Colaptes auratus*) and American robins (*Turdus migratorius*) increase where cattle graze riparian areas of Montana (Mosconi and Hutto 1982) and Colorado (Schulz and Leininger 1991). Cattle open rank herbaceous vegetation, thereby improving habitats for these ground-foraging species. Both species would obviously be absent, however, in areas where excessive historical grazing has prevented the regeneration of the overstory woody vegetation that these birds use when nesting.

Finally, classifying a bird species as an increaser or decreaser ignores the dynamic nature of ecological systems. Even in the absence of grazing, seasonal precipitation can dramatically influence the avian assemblage present at a site in a given year, primarily as birds respond to changes in the vigor of vegetation and the abundance of local food supplies (Szaro and Balda 1979, Wiens 1981, Hall 1984, Hejl and Beedy 1986). Stocking cattle at a hypothetical density of *D* in an Oklahoma oak (*Quercus* spp.) savannah during a wetter-than-normal growing season would appear totally compatible with shrub- and tree-nesting species. Stocking at density *D* in a drought year, however, could force cattle to browse heavily and have a direct impact on especially the shrub-nesting guild of species. Species that nest in shrubs generally experience the highest rates of nest predation (Martin 1993) and cattle may increase nest losses to predators by exposing concealed nests through reducing foliage densities and opening dense patches of vegetation, enhancing hunting success of predators. The same observations would follow for Brewer's sparrows nesting in sagebrush rangelands or dusky flycatchers (*Empidonax oberholseri*) nesting in deciduous shrubs associated with pine (*Pinus* spp.) forests of the Rocky Mountain and Intermountain states. As with soils, the influence of grazing on nongame birds will obviously covary with seasonal climatic conditions.

Considering the inherent variabilities in vegetation responses to grazing and in population responses of nongame birds to grazing-induced changes in the vegetation, conservation of individual species at a locale will necessarily be dependent upon site-specific information. Such an intensive focus of management time and energy rarely has been justified outside an endangered species scenario. The local variabilities in population and habitat responses have dampened the enthusiasm of natural resource agencies for identifying management "indicator" species, in part because natural resource agencies have generally tried to develop lists of such species at the regional rather than local scale and the ecology of a species is too variable regionally to sustain a simple, cause-and-effect paradigm at the local scale. The best indicator species will change with the soils and vegetation locally (Kantrud and Kologiski 1983).

The Multiple-species Perspective

With the exception of endangered species considerations, wildlife managers are most concerned with the impacts of grazing upon the local assemblage of species on a natural resource management unit. The inherent differences in responses of nongame birds to cattle grazing, among species and among sites and years within species, should not preclude devising a standardized management plan for a local avian community. Wiens (1973) noted that despite clear responses of individual species to grazing at a given locale, no patterns appeared when avifaunas are examined at the regional scale. Range managers cannot realistically or economically devise separate plans for individual species at each locale, yet are required to define specific, defendable objectives in a grazing prescription.

A major component of the grazing and nongame bird literature describes the impact of historical grazing practices upon the number of species (i.e., "species richness") at a site. Species richness, however, has fallen into disfavor as a meaningful characterization of the biological diversity of a site, primarily because ecological generalists substitute equally for those species that are regionally unique to an area (Murphy 1989, Knopf 1992*b*, Martin 1992). To be ecologically meaningful, conservation of nongame birds must be based upon ecological data that retain information on the composition (i.e., quality) rather than number (i.e., quantity) of species at a locale. Overall, the collection of species present within a management unit is often highly dynamic, being best viewed as an oscillating and evolving assemblage within an arbitrarily defined time frame.

The specific impacts of grazing on a local avian assemblage are poorly studied. Cattle grazing may simplify bird assemblages by reducing the number of species present while simultaneously increasing total bird densities in native grasslands (Kantrud 1981, Kantrud and Kologiski 1982). Alternatively, total densities and numbers of species may not change significantly in western riparian areas even though the species composition does undergo dramatic shifts (Mosconi and Hutto 1982). Species extirpated locally by grazing tend to be the rare species within an assemblage (Martin 1984), and they tend to be replaced (not necessarily displaced) by ecological generalist (i.e., eurytopic) species that have a wide geographic distribution.

Relative to minimizing livestock and grazing impacts upon nongame bird habitats, the season of grazing appears to be a more important consideration than the intensity of grazing (Wiens 1973). Grazing late in the growing season or when the vegetation is dormant can drastically reduce physical impacts of cattle on native vegetation (Knopf and Cannon 1982, Kauffman and Krueger 1984) and on the nongame bird community (Kauffman et al. 1982, Sedgwick and Knopf 1987, Knopf et al. 1988). Grazing programs timed to coincide with overstory leaf fall in deciduous forest sites can especially reduce grazing impacts on vegetation because freshly fallen leaves provide an additional, preferred forage (Sedgwick and Knopf 1991). Late-season grazing, in effect, ecologically shifts livestock from grazers to detritivores that actually facilitate organic breakdown of the previous season's primary productivity on site.

Whereas terrestrial bird species vary in their sensitivity to grazing locally, a local avifaunal analysis is critical to designing a grazing program for any management unit. The challenge to range managers becomes how to define the appropriate nongame bird goals that the grazing program will be evaluated against. An example of such an approach is to critique the local avian assemblage as the individual species differentially respond to grazing practices using the concept of differential "response guilds."

Ecological guilds were first defined by Root (1967) as groups of species with similar patterns of resource use at a locale. Bird species have been grouped based upon foraging or nesting habits (Franzreb and Ohmart 1978, Samson 1980, Skagen et al. 1991), and guilds have been proposed for use in natural resource management programs (Johnson 1981, Severinghaus 1981). Szaro

(1986) proposed defining guilds based upon similarities of response to individual perturbations, and such a response-guild approach has been specifically developed for a historical analysis of grazing on the Arapaho National Wildlife Refuge (Knopf et al. 1988). The latter study used multiple lines of inference from data on densities and habitat-use patterns among the common bird species to define which were most sensitive (stenotopic) to grazing on the refuge. That stenotopic response-guild then became the barometer against which grazing programs in riparian areas on the refuge were to be evaluated in the future.

The advantages of a response-guild approach are numerous. Identifying 3-5 grazing-sensitive species narrows the focus of field inventories while increasing the sample sizes in nongame bird databases ultimately used to design and evaluate grazing programs. Most significantly, sensitive species tend to be the relatively rare ones on any area and the cost effectiveness of program evaluation is improved drastically when information is gathered across a number of species with similar responses (Verner 1984).

SUMMARY

Nongame birds are a biologically diverse group of organisms that compose 93% of native birds of the United States. Grazing (1) has relatively minor impacts upon aquatic species, (2) influences birds of prey primarily by affecting densities of prey, (3) affects especially nesting cover of wetlands-associated species, and (4) alters the food and cover of terrestrial species.

Nongame birds generally respond to indirect impacts of cattle on native vegetation. The response is most immediate among species dependent upon the herbaceous and shrub layers of vegetation. Alteration of vegetative communities by livestock grazing especially impacts local avian assemblages by altering the extent and spatial patterning of the grass and shrub layers.

Terrestrial bird species have been viewed as either increasing or decreasing in numbers when cattle are introduced into their respective habitats. The response of a species to cattle, however, can vary drastically between sites, between years within sites, and even within a species population with progression of grazing-induced alterations of the vegetation. Management programs to favor single species are costly, being justified only when a species declines to the point of being in danger of extirpation or extinction.

The number of bird species and the total density often does not change at a locale subjected to cattle grazing. However, the species composition within an assemblage shifts with rare species being extirpated as ecological generalists invade a site. Within proper grazing guidelines, these shifts in species composition appear to be related as much to the seasonal timing of the grazing activity as the intensity of grazing.

Management of cattle grazing for nongame bird habitats requires specific information on the local avian assemblage and the responses of member species to a grazing prescription. Definition of grazing response-guilds for the bird species on a management unit facilitates assessment of grazing programs. The response-guild approach identifies species most sensitive to grazing disturbance at a locale and provides more information, more cost effectively, on rates of change in the avian assemblage relative to the grazing prescription employed.

LITERATURE CITED

Bicak, T. K., R. L. Redmond, and D. A. Jenni. 1982. Effects of grazing on Long-billed curlew (*Numenius americanus*) breeding behavior and ecology in southwestern Idaho. Pages 74-85 *in* J. M. Peek and P. D. Dalke, eds. Wildlife-livestock relationships symposium: proceedings 10. Univ. Idaho, For., Wildl. and Range Exp. Stn., Moscow.

Bock, C. E., and B. Webb. 1984. Birds as grazing indicator species in southeastern Arizona. J. Wildl. Manage. 48:1045-1049.

———, and ———. 1988. Grassland birds in Southeastern Arizona: impacts of fire, grazing, and alien vegetation. Int. Counc. for Bird Preservation Tech. Publ. 7:43-58.

Butts, K. O. 1973. Life history and habitat requirements of Burrowing Owls in western Oklahoma. M.S. Thesis, Oklahoma State Univ., Stillwater. 188pp.

Chandler, W. J. 1986. Migratory bird protection and management. Pages 215-265 *in* R. L. DiSilvestro, ed. Audubon wildlife report 1986. Natl. Audubon Soc. New York, N.Y.

Clark, R. J. 1975. A field study of the Short-eared Owl *Asio flammeus* (Pontoppidan) in North America. Wildl. Monogr. 47. 67pp.

Coulombe, H. N. 1971. Behavior and population ecology of the Burrowing Owl, *Speotyto cunicularia*, in the Imperial Valley of California. Condor 73:162-176.

Deubbert, H. F., and J. T. Lokemoen. 1977. Upland nesting of American Bitterns, Marsh Hawks, and Short-eared Owls. Prairie Nat. 9:33-40.

Eddleman, W. R., F. L. Knopf, B. Meanley, F. A. Reid, and R. Zembal. 1988. Conservation of North American rallids. Wilson Bull. 100:458-475.

Ellison, L. 1960. Influence of grazing on plant succession of rangelands. Bot. Rev. 26:1-78.

Franzreb, K. E., and R. D. Ohmart. 1978. The effects of timber harvesting on breeding birds in a mixed-coniferous forest. Condor 80:431-441.

Glenn, S. M., S. L. Collins, and D. J. Gibson. 1992. Effects of scale and disturbance in tallgrass prairie: local *vs.* regional effects on community heterogeneity. Landscape Ecol. 7:243-251.

Graul, W. D. 1975. Breeding biology of the Mountain Plover. Wilson Bull. 87:6-31.

———. 1980. Grassland management practices and bird communities. Pages 38-47 *in* R. M. DeGraff and N. G. Tilghman, eds. Management of western forests and grasslands for nongame birds. U.S. Dep. Agric. For. Serv. Gen. Tech. Rep. INT-86.

Hall, G. A. 1984. A long-term bird population study in an Appalachian spruce forest. Wilson Bull. 96:228-240.

Hejl, S. J., and E. C. Beedy. 1986. Weather-induced variation in the abundance of birds. Pages 241-244 *in* J. Verner, M. L. Morrison, and C. J. Ralph, eds. Wildlife 2000. Univ. Wisconsin Press, Madison.

Holthuijzen, A. M. A. 1990. Prey delivery, caching, and retrieval rates in nesting Prairie Falcons. Condor 92:475-484.

Janzen, D. H. 1986. Chihuahuan Desert nopaleras: defaunated big mammal vegetation. Annu. Rev. Ecol. and Syst. 17:595-636.

Johnson, R. A. 1981. Application of the guild concept to environmental impact analysis of terrestrial vegetation. J. Environ. Manage. 13:205-222.

Kantrud, H. A. 1981. Grazing intensity effects on the breeding avifauna of North Dakota native grasslands. Can. Field-Nat. 95:404-417.

———, and R. L. Kologiski. 1982. Effects of soils and grazing on breeding birds of uncultivated upland grasslands of the northern Great Plains. U.S. Fish and Wildl. Serv. Wildl. Res. Rep. 15. 33pp.

———, and ———. 1983. Avian associations of the northern Great Plains grasslands. J. Biogeogr. 10:331-350.

Kauffman, J. B., W. C. Krueger, and M. Vavra. 1982. Impacts of a late season grazing scheme on nongame wildlife habitat in a Wallowa Mountain riparian ecosystem. Pages 208-220 *in* J. M. Peek and P. D. Dalke, eds. Wildlife-livestock relationships symposium: proceedings 10. Univ. Idaho, For., Wildl. and Range Exp. Stn., Moscow.

———, and ———. 1984. Livestock impacts on riparian ecosystems and streamside management implications . . . a review. J. Range Manage. 37:430-438.

Kie, J. G., and E. R. Loft. 1990. Using livestock to manage wildlife habitat: some examples from California annual grassland and wet meadow communities. Pages 7-24 *in* K. E. Severson, tech. coord. Can livestock be used as a tool to enhance wildlife habitat? U.S. Dep. Agric. For. Serv. Gen. Tech. Rep. RM-194.

Kirby, R. E., J. K. Ringleman, D. R. Anderson, and R. S. Sojda. 1992. Grazing on national wildlife refuges: Do the needs outweigh the problems? Trans. North Am. Wildl. and Nat. Resour. Conf. 57:611-626.

Kirsch, L. M., and K. F. Higgins. 1976. Upland Sandpiper nesting and management in North Dakota. Wildl. Soc. Bull. 4:16-20.

Klebenow, D. A., and R. J. Oakleaf. 1984. Historical avifaunal changes in the riparian zone of the Truckee River, Nevada. Pages 203-209 *in* R. E. Warner and K. M. Hendrix, eds. California riparian systems. Univ. California Press, Berkeley.

Knopf, F. L. 1987. "On publishing symposia" revisited: When are proceedings no longer timely? Wildl. Soc. Bull. 15:603-605.

———. 1992*a*. Focusing conservation of a diverse wildlife resource. Trans. North. Am. Wildl. and Nat. Resour. Conf. 57:241-242.

———. 1992*b*. Faunal mixing, faunal integrity, and the bio-political template for diversity conservation. Trans. North Am. Wildl. and Nat. Resour. Conf. 57:330-342.

———, and R. W. Cannon. 1982. Structural resilience of a *Salix* community to changes in grazing practices. Pages 198-207 *in* J. M. Peek and P. D. Dalke, eds. Wildlife-livestock relationships symposium: proceedings 10. Univ. Idaho, For., Wildl. and Range Exp. Stn., Moscow.

———, and J. A. Sedgwick. 1992. An experimental study of nest-site selection by Yellow Warblers. Condor 94:734-742.

———, ———, and R. W. Cannon. 1988. Guild structure of a riparian avifauna relative to seasonal cattle grazing. J. Wildl. Manage. 52:280-290.

———, ———, and D. B. Inkley. 1990. Regional correspondence among shrubsteppe bird habitats. Condor 92:45-53.

Kochert, M. N. 1989. Responses of raptors to livestock grazing in the western United States. Pages 194-203 *in* Proc. western raptor management symposium and workshop. Natl. Wildl. Fed. Sci. and Tech. Ser. 12, Washington, D.C.

Leege, T. A., D. J. Herman, and B. Zamora. 1981. Effects of cattle grazing on mountain meadows in Idaho. J. Range Manage. 34:324-328.

Marlow, C. B., and T. M. Pogacnik. 1986. Cattle feeding and resting patterns in a foothills riparian zone. J. Range Manage. 39:212-217.

Martin, D. J. 1973. Selected aspects of Burrowing Owl ecology and behavior. Condor 75:446-456.

Martin, T. E. 1984. Impact of livestock grazing on birds of a Columbian cloud forest. Tropical Ecology 25:158-171.

———. 1992. Landscape diversity and biological diversity. Trans. North Am. Wildl. and Nat. Resour. Conf. 57:283-291.

———. 1993. Nest predation among vegetation layers and habitat types: revising the dogmas. Am. Nat. 141:897-913.

Mengel, R. M. 1970. The North American Central Plains as an isolating agent in bird speciation. Pages 279-340 *in* W. Dort and J. K. Jones, eds. Pleistocene and Recent environments of the central Great Plains. Univ. Kansas Press, Lawrence.

Mosconi, S. L., and R. L. Hutto. 1982. The effect of grazing on the land birds of a western Montana riparian habitat. Pages 221-233 *in* J. M. Peek and P. D. Dalke, eds. Wildlife-livestock relationships symposium: proceedings 10. Univ. Idaho, For., Wildl. and Range Exp. Stn., Moscow.

Murphy, D. D. 1989. Conservation and confusion: wrong species, wrong scale, wrong conclusions. Conserv. Biol. 3:82-84.

National Wildlife Federation. 1989. Proc. western raptor management symposium and workshop. Natl. Wildl. Fed. Sci. and Tech. Ser. 12, Washington, D.C. 320pp.

Newton, I. 1979. Population ecology of raptors. Buteo Books. Vermillion, S.D. 399pp.

Phillips, R. L., and A. E. Beske. 1984. Resolving conflicts between energy development and nesting Golden Eagles. Proc. symposium on issues and technology in the management of impacted western wildlife. Thorne Ecol. Inst. Tech. Publ. 14:214-219.

Renken, R. B., and J. J. Dinsmore. 1987. Nongame bird communities on managed grasslands in North Dakota. Can. Field-Nat. 101:551-557.

Roath, L. R., and W. C. Krueger. 1982. Cattle grazing and behavior on a forested range. J. Range Manage. 35:332-338.

Robel, R. J., J. N. Briggs, A. D. Dayton, and L. C. Hulbert. 1970. Relationships between visual obstruction measurements and weight of grassland vegetation. J. Range Manage. 23:295-297.

Root, R. B. 1967. The niche exploitation pattern of the Blue-gray Gnatcatcher. Ecol. Monogr. 37:317-350.

Ryder, R. A. 1980. Effects of grazing on bird habitats. Pages 51-66 *in* R. M. DeGraff and N. G. Tilghman, compilers. Management of western forests and grasslands for nongame birds. U.S. Dep. Agric. For. Serv. Gen. Tech. Rep. INT-86.

Samson, F. B. 1980. Use of montane meadows by birds. Pages 113-129 *in* R. M. DeGraff and N. G. Tilghman, compilers. Management of western forests and grasslands for nongame birds. U.S. Dep. Agric. For. Serv. Gen. Tech. Rep. INT-86.

Schulz, T. T., and W. C. Leininger. 1991. Nongame wildlife communities in grazed and ungrazed montane riparian sites. Great Basin Nat. 51:286-292.

Sedgwick, J. A., and F. L. Knopf. 1987. Breeding bird response to cattle grazing of a cottonwood bottomland. J. Wildl. Manage. 51:230-237.

———, and ———. 1990. Habitat relationships and nest site characteristics of cavity-nesting birds in cottonwood floodplains. J. Wildl. Manage. 54:112-124.

———, and ———. 1991. Prescribed grazing as a secondary impact in a western riparian floodplain. J. Range Manage. 44:369-373.

Severinghaus, W. D. 1981. Guild theory development as a mechanism for assessing environmental impact. Environ. Manage. 5:187-190.

Skagen, S. K., R. L. Knight, and G. H. Orians. 1991. Human disturbance of an avian scavenging guild. Ecol. Appl. 1:215-225.

Smith, E. D. 1985. Construction of artificial nesting sites for Prairie Falcons. Wildl. Soc. Bull. 13:543-546.

Szaro, R. C. 1980. Factors influencing bird populations in southwestern riparian forests. Pages 403-418 *in* R. M. DeGraff and N. G. Tilghman, compilers. Management of western forests and grasslands for nongame birds. U.S. Dep. Agric. For. Serv. Gen. Tech. Rep. INT-86.

———. 1986. Guild management: an evaluation of avian guilds as a predictive tool. Environ. Manage. 10:681-686.

——— and R. P. Balda. 1979. Bird community dynamics in a ponderosa pine forest. Stud. Avian Biol. 3:1-66.

Taylor, D. M. 1986. Effects of cattle grazing on passerine birds nesting in riparian habitat. J. Range Manage. 39:254-258.

Thomsen, L. 1971. Behavior and ecology of Burrowing Owls on the Oakland Municipal Airport. Condor 73:177-192.

Verner, J. A. 1984. The guild concept applied to management of bird populations. Environ. Manage. 8:1-14.

Whitmore, R. C. 1979. Short-term changes in vegetation structure and its effect on Grasshopper Sparrows in West Virginia. Auk 96:621-625.

Whyte, R. J., and B. W. Cain. 1979. The effect of grazing on nesting marshbird habitat at the Welder Wildlife Refuge, San Patricio County, Texas. Bull. Texas Ornithol. Soc. 12:42-46.

Wiens, J. A. 1973. Pattern and process in grassland bird communities. Ecol. Monogr. 43:237-270.

———. 1981. Single-sample surveys of communities: Are the revealed patterns real? Am. Nat. 117:90-98.

———. 1985. Habitat selection in variable environments: shrub-steppe birds. Pages 227-251 *in* M. L. Cody, ed. Habitat selection in birds. Acad. Press. New York, N.Y.

———, and J. T. Rotenberry. 1981. Habitat associations and community structure of birds in shrubsteppe environments. Ecol. Monogr. 51:21-41.

Wilson, B. W., M. J. Hooper, E. E. Littrell, P. J. Detrich, M. E. Hansen, C. P. Weisskopf, and J. N. Seiber. 1991. Orchard dormant sprays and exposure of Red-tailed Hawks to organophosphates. Bull. Environ. Contam. Toxicol. 47:717-724.

Chapter 6
UPLAND GAMEBIRDS
Fred S. Guthery

INTRODUCTION

Literature on the use of grazing in management of upland gamebird habitat is a potpourri of unsupported conjecture, mixed research results, and conflicting recommendations. Consider these observations on scaled quail (*Callipepla squamata*):

> Grazing, especially by sheep and goats, is particularly serious because it reduces the supply of native foods and often opens the stand of brush to a degree which [scaled] quail cannot tolerate. *Heavy grazing by cattle is not so serious* [emphasis added] because they ordinarily do not eat unpalatable species such as doveweed [*Croton*], wild potato [?], and nightshade [*Solanum*], which furnish seeds for quail. From a long range standpoint, however, heavy grazing by any type of livestock is not recommended (Tex. Parks and Wildl. Dep. 1945:63).

> . . . grazing [by sheep, cattle, and goats] tends to decrease the availability of food and cover plants for scaled quail. . . . grazing has promoted a more diversified weed flora, which, in years of high rainfall, may be considered beneficial to scaled quail. This results in "boom or bust" quail populations with increasingly more busts as time progresses (Wallmo 1956:121).

> *Moderate grazing by cattle* should be encouraged [emphasis added] (Campbell et al. 1973:46).

> . . . scaled quail hunting is best on *moderately grazed ranges*, during and after years of abundant rainfall [emphasis added] (Brown et al. 1978:286).

The observations lack consensus and provide, at best, weak guidelines for the scaled quail manager. The same situation holds for other gamebirds.

Grazing technology in management of upland gamebird habitat is unsophisticated for several reasons. Wildlife managers may react dogmatically through trained incapacities concerning the role of grazing in cover management (Kessler and Bosch 1981:143). A trained incapacity is a reduction in personal flexibility that arises when the training and experience that produce competence in a specialized field also narrow ranges of perception. Kessler and Bosch (1981:143) found that wildlife biologists concerned with habitat management for sharp-tailed grouse (*Tympanuchus phasianellus*) preferred specialized grazing systems over continuous grazing under the assumption that "good range management is good sharptail management." Empirical evidence did not support this assumption. The prevailing recommendation of "moderate" grazing for gamebirds perhaps has become a trained incapacity.

Another reason for a weak technology of grazing management is use of descriptive terms that have vague and imprecise meanings. Classic examples are the terms "light, moderate, and heavy" grazing. Any grazing intensity is a function of space-and time-specific conditions; i.e., a fixed stocking intensity can range from light to heavy, depending on when and where it is applied. Philosophically, "light, moderate, or heavy" grazing has an infinite number of meanings, so in reality the terms have little meaning at all.

Perhaps the main reason for the lack of a more reliable and useful knowledge base on the use of grazing in habitat management for upland gamebirds is lack of management theory (simplified model of reality [Mendenhall et al. 1990:12]) that explains process and cause and effect and predicts outcomes of grazing programs. Theoretical considerations potentially can resolve conflicting results and recommendations that appear in the literature (Spears 1991). Certainly, the search for a unifying theme in the descriptive work on gamebirds and grazing heretofore published could produce useful knowledge, better focus future research, and provide a conceptual framework for grazing management decisions.

Certain constraints and conditions are invoked at the outset so that a few key issues can be addressed in this chapter.

1. The information relates solely to the use of livestock in managing herbaceous cover for gamebirds; other chapters address the effects of other range management practices.
2. Information is presented as if upland gamebirds are the primary management objective.
3. Unless otherwise specified, the word grazing refers to use by cattle.
4. Six of 19 gamebird species addressed are dismissed or casually discussed because grazing is a minor or irrelevant consideration in their management (Table 1). The remaining 13 species may exhibit negative, neutral, or positive responses to grazing, depending upon the habitat context within which grazing is applied and the grazing program itself.
5. The terms gamebird and gallinaceous bird are used interchangeably; only brief mention is made of pigeons and doves.

Given these 5 constraints, I provide basic and applied information and theory on the role of grazing in management of upland gamebird habitat. Because vegetation is the habitat feature of fundamental importance, the upland gamebird manager must hold sound concepts of the response of plant species and communities to grazing. Therefore, the first section discusses vegetation responses germane to gamebird habitat management. The second section provides seminal management theory for application of grazing in habitat management and discusses issues of topical or ecological interest relative to the theory. The last section outlines a process for decision-making.

TABLE 1. Grazing management issues related to major upland gamebirds in the United States and Canada.

Group Species	Grazing management issues
Grouse	
Sage grouse (*Centrocercus urophasianus*)	Contextual[a]
Blue grouse	Peripheral[b]
Spruce grouse (*Dendragapus canadensis*)	Not germane[c]
Ptarmigan (*Lagopus* spp.)	Not germane
Greater prairie-chicken	Contextual
Lesser prairie-chicken (*Tympanuchus pallidicinctus*)	Contextual
Sharp-tailed grouse	Contextual
Ruffed grouse	Peripheral
New World quails	
Mountain quail (*Oreortyx pictus*)	Peripheral
Scaled quail	Contextual
Gambel's quail (*Callipepla gambelli*)	Contextual
California quail (*C. callifornica*)	Contextual
Bobwhite	Contextual
Harlequin quail (*Cyrtonyx montezumae*)	Contextual
Old World partridges and pheasants	
Gray partridge (*Perdix perdix*)	Contextual
Chukar partridge (*Alectoris chukar*)	Contextual
Turkeys	
Wild turkey (*Meleagris gallopavo*)	Contextual
Pigeons and doves	
Band-tailed pigeon (*Columba fasciata*)	Not germane
Mourning dove (*Zenaida macroura*)	Contextual

[a] Grazing by cattle can be negative, neutral, or positive, depending upon the habitat context within which it is applied.

[b] Grazing by cattle is a minor management concern or irrelevant over most occupied range of the species in question, but grazing could have negative, neutral, or positive effects under local conditions.

[c] Because of the distribution of the species in question or the type of habitat it occupies, grazing by cattle is not a relevant habitat management consideration.

VEGETATION CONSIDERATIONS

Dyksterhuis (1949) was among the first scientists to classify herbaceous plants based on their response to grazing. He recognized (1) decreasers (i.e., species preferred by livestock or intolerant to grazing), (2) increasers (i.e., species less preferred or released from competition with decreasers by grazing), and (3) invaders (i.e., species that occur on sites subject to long-term heavy grazing). He classified range condition as poor (<25%), fair (25-50%), good (50-75%), or excellent (>75%) based on the percent composition of decreasers (shown in parentheses).

The Dyksterhuisian system, given the limitations discussed below, is relevant to upland gamebird habitat management because these birds are adapted to attributes associated with seral stages, which in a sense are similar to range condition classes (Pieper and Beck 1990). Greater prairie-chickens (*Tympanuchus cupido*) (Fig. 1) originally occupied tall-grass prairies dominated by big bluestem (*Andropogon gerardii*), little bluestem (*Schizachyrium scoparium*), Indiangrass (*Sorghastrum nutans*), and similar species (Evans 1968). In other words, greater prairie-chickens were associated with decreasers and occupied sites in excellent range condition. Sympatric northern bobwhites (*Colinus virginianus*), conversely, were adapted to lower successional conditions; i.e., sites dominated by increasers such as ragweed (*Ambrosia* spp.) and invaders and perhaps considered in poor range condition (Ellis et al. 1969).

Dyksterhuis's (1949) pioneering work has come under heavy criticism in recent years, because it is an unacceptable model for categorizing the status of rangeland and for describing vegetation response to grazing. The descriptors "poor, fair, good, and excellent" carry connotations that may improperly reflect the productivity of rangeland and the value of the condition classes to wildlife and livestock (Smith 1989). The proper habitat goal of the upland gamebird manager might be poor range condition. For example, diet quality for northern bobwhite chicks is higher on rangeland in poor condition than on rangeland in good condition (Weaver 1984), given the environmental context under which these data were collected. The public (and many wildlife biologists), however, generally use a literal (poor = miserable, pathetic, wretched) as opposed to a technical (poor = <25% climax vegetation) translation of this range condition class. Mistranslation leads to confusion and highlights the need for better descriptors of rangeland status.

The Dyksterhuis model is said to be deterministic, because it presumes a "single persistent state (the climax) in the absence of grazing" (Westoby et al. 1989:266). The model further assumes "succession towards this climax is a steady process. Grazing pressure produces changes which are also progressive and are in the opposite direction to the successional tendency."

Ecologists have collected empirical evidence that vegetation dynamics on rangeland should be considered a stochastic process (Westoby et al. 1989). This means that alternative vegetation responses to grazing are possible (not just a single climax plant community) and that any outcome can be governed by largely random processes, such as precipitation (Fig. 2). For gamebird management, we must assume there exists some successional tendency in plant communities, despite the stochastic nature of the final outcome.

Further, rangeland vegetation can enter stable states that are nonresponsive to grazing pressure or program (Friedel 1991, Laycock 1991). A stable state of considerable importance to the upland gamebird manager is that of shrub-dominated rangeland (Fig. 3). On North American rangeland, shrub-dominated stable states occur in the Great Basin, the shortgrass steppe, and the southwestern desert grasslands (Laycock 1991). There are lower successional stable states dominated by annual grasses in California and southern Idaho.

FIGURE 1. The Attwater's prairie-chicken (*Tympanuchus cupido Attwateri*) and other races of greater prairie-chickens are adapted to high seral stages on prairie. Photo courtesy of Nell Bolen.

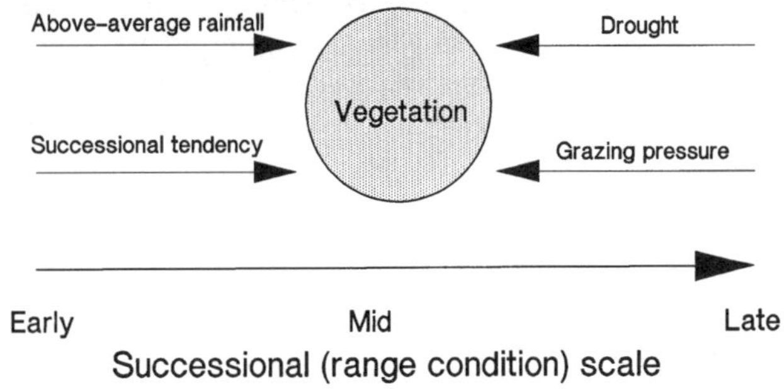

FIGURE 2. Relationships among successional stage of upland gamebird ground cover, precipitation, and grazing pressure (redrawn from Westoby et al. 1989).

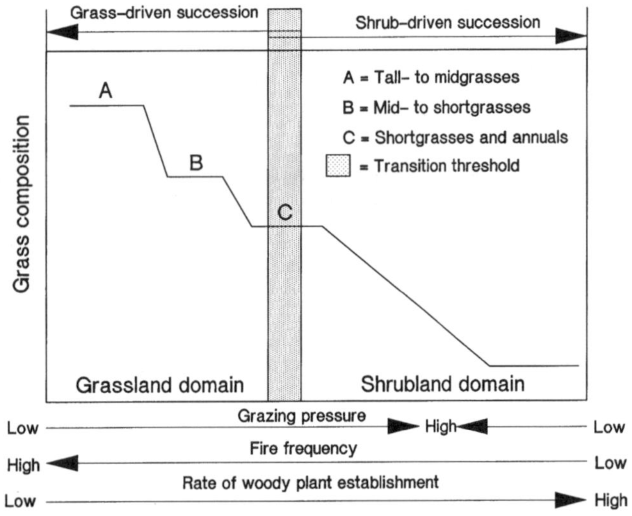

FIGURE 3. When rangeland vegetation crosses a transition threshold from the grassland domain into the shrubland domain, grazing pressure and program become largely irrelevant in gamebird habitat management (redrawn from Archer 1989).

The point for upland gamebird habitat managers, repeated for emphasis, is that the vegetation of stable states does not necessarily respond to alterations in grazing pressure or program. If the shrubland domain (Fig. 3) is reached, substantial cultural or energy inputs may be required to return to the grassland domain and the conversion may be short-lived (Archer 1989). Stable states dominated by shrubs necessitate brush management treatments (see Chapters 19, 20, 21). When the grassland domain is re-established, then grazing program becomes a relevant management consideration.

Given that a site is in the grassland domain (Fig. 3), the domestic grazing animal can be applied to alter the structure and to some extent the composition of herbaceous flora to benefit upland gamebirds birds, if alteration is indicated. Gamebird managers should realize that, given the initial condition of vegetation on a site and the adaptations of any gamebird species, no grazing might be the optimal management decision.

Grazing may increase landscape diversity (e.g., macroscale structural effects) because cattle use space nonuniformly in most rangeland settings, resulting in mosaics of unused, lightly used, moderately used, and heavily used vegetation (Stuth 1991:72). Although cattle avoid steep slopes and other topographic or biotic features that inhibit their presence, which contributes to nonuniform use, it occurs mainly because they are central-place foragers (Stuth 1991:67). The daily behavior of such foragers revolves around some focal habitat feature. The primary focus of activity for cattle is the watering facility, although mineral supplementation points and thermal cover also focus activity.

If cattle used space randomly (a type of uniform grazing), they would occur within 10 m of a water facility for about 3.4 seconds/24 hours, given a radius of movement away from water of 1.6 km. However, individuals spend from 4 to 20 minutes at water (Prasad and Guthery 1986). These times indicate use of watering facilities was 71 to 353 times greater than expected based on random use. Because of higher than expected use, watering facilities on rangeland typically promote concentric zones of plant communities consisting of bare ground, invaders, increasers, and decreasers as distance from the water facility increases (Guthery et al. 1990, Stuth 1991:70). Knowledge of the effects of central-place behavior can be used to disperse or focus grazing pressure in gamebird management.

Grazing also alters the structure of gamebird habitat on a microscale (Fig. 4). Structural changes may occur at levels well above grazable heights through effects on woody vegetation (George and Powell 1979, Schulz and Leininger 1990). Livestock use of a riparian zone was associated with reductions in brush canopy cover (1-2 m tall) from 62 to 23% and in mid-level (2-5 m) woody cover from 31 to 24% (Mosconi and Hutto 1981). These effects apparently occurred because of suppressed survival of woody plant seedlings caused by grazing. Ruffed grouse (*Bonasa umbellus*) density averaged 3.7 birds/10 ha on an ungrazed site and 0.7/10 ha on a grazed site. Reduced grouse density might have resulted from the structural effects of grazing.

Microscale structural effects usually are most pronounced on low level, herbaceous vegetation because of consumption and trampling. As a general guideline, cattle consumption of forage (dry wt) amounts to 2% of body mass/day (Holechek et al. 1989:195). Loss to trampling varies with stocking intensity; it may be nil at higher rates or roughly equal to consumption at lighter rates (Allison et al. 1982).

Consumption and trampling have obvious effects on the structure of herbaceous ground cover (Fig. 4). Generally, bare ground increases while litter, average height of vegetation, foliar cover, and visual obstruction decrease (Whyte and Cain 1981, Watts et al. 1987, Baker and Guthery 1990, Schulz and Leininger 1990).

Although management of structure is the keystone in grazing management decisions, grazing also affects composition. It may increase the diversity of herbaceous flora if it is imposed on a plant community at or near climax or otherwise in some stable state subject to disruption by livestock. Increased diversity is a prediction of the intermediate disturbance hypothesis; i.e., sites subject to intermediate disturbance, as from light to moderate grazing, will have more species than those subject to light or intense disturbance (Petraitis et al. 1989).

FIGURE 4. Overview and closeups of grazing effects on vegetation structure, Umatilla County, Oregon, 1990. Top: Comparison of grazed (left side of road) and ungrazed habitat showing dramatic reduction in woody cover with grazing. Middle: Closeup of grazed area showing increase in bare ground and reduction in stature and foliar cover of herbaceous vegetation. Bottom: Closeup of ungrazed area showing buildup of plant matter, which is inimical to gamebirds.

The process by which intermediate disturbance is presumed to increase diversity can be explained on the basis of Dyksterhuis's (1949) classification of plants. Decreasers outcompete other classes of vegetation and therefore dominate a plant community in the absence of grazing. Decreasers are strong competitors but unable to withstand disturbance. Conversely, invaders are poor competitors but well adapted to disturbance. Light or nil disturbance favors decreasers whereas intense disturbance favors invaders. Moderate disturbance, on the other hand, may maintain limited populations of decreasers, promote populations of increasers, and permit limited populations of invaders. Diversity in the sense of species richness (no. species) therefore maximizes at some intensity of grazing that imposes moderate disturbance. Positive effects of grazing on plant species richness have been observed in Florida (Lewis et al. 1988).

A THEORETICAL APPROACH TO GRAZING MANAGEMENT

Material in the preceding section provided some important concepts concerning the ultrastructure within which grazing management decisions must be made. The theory that follows operates within that ultrastructure.

For theory development, we assume attributes of woody vegetation and geological components of structure are acceptable and that other factors of importance in bird ecology (e.g., food, predation, diseases, parasites, competition, hunting) are held constant. Specifically, we deal with the effects of grazing on herbaceous cover.

Foundation Principles

In developing a theory, we need to start with principles which, though subject to variation, hold on a broad scale. Five such principles pertain.

Principle 1. Any gallinaceous bird is adapted, through natural selection and evolution, to habitat with a certain structure; i.e., any such bird has an optimum structural niche. Structure refers to the height, density, biomass, and dispersion of herbaceous and woody vegetation. Vegetation structure is a key habitat feature for birds in general (Pitelka 1941, MacArthur and MacArthur 1961) and for gamebirds in particular (Leopold 1977:48-49, Kessler and Dodd 1978, Healy 1985, Guthery 1986:115-116).

A corollary of Principle 1 is that optimum structure is more or less constant throughout the occupied range of a gallinaceous bird. This implication is reasonable, because size, behavior, coloration, and physical abilities of the birds are more or less fixed by genetic mechanisms. Climatic influences, such as harsh winters, might make a certain structure optimal that would be suboptimal under milder climate. However, this circumstance does not invalidate the corollary, because suboptimum structure will exact a negative influence independent of climate.

Principle 2. Primary production and hence structure of vegetation vary with climate (i.e., long-term patterns of precipitation and temperature), weather (i.e., short-term fluctuations in precipitation and temperature), and soils. Soil implies type (i.e., sands, loams, clays), depth, fertility, and slope. In general, growth and standing biomass of herbaceous vegetation are greater in rich environments (i.e., longer growing seasons, more precipitation, higher soil fertility) than in poor environments (i.e., shorter growing seasons, less precipitation, lower soil fertility).

Principle 3. Structure and composition of herbaceous vegetation covary; i.e., 1 attribute is confounded with the other. A general principle is that standing crop biomass (a component of structure) increases with seral stage. Late seral stages (e.g., tall perennial grasses) would be expected to support more biomass (kg/ha dry wt) than early seral stages (e.g., annual forbs).

Principle 4. Structure, in contrast with composition, serves as the keystone of grazing management decisions. That gallinaceous birds are associated with (i.e., adapted to) certain seral stages (i.e., composition states) is a powerful theory of habitat management (Grange 1949) and certainly relevant to a theory of grazing management. In certain environments, standing crop biomass is relatively constant as seral stage varies (Tiedeman et al. 1991), in which case the gamebird manager can simultaneously manage for optimum structure and composition of vegetation, if the environment permits optimum structure.

In other rangeland habitats amenable to alteration with grazing, particularly those with a successional tendency towards tall grass communities, the effects of structure and seral stage on gamebird abundance are inseparable (Principle 3). In such situations, the manager can manage for structure or composition, because they are essentially the same variable.

Suppose, however, the target of management is a gamebird that requires structural conditions like those provided by midgrass communities and the successional tendency is towards midgrass communities. The manager has no option but to manage for structure in this situation. Vegetation that provides the optimum structure might not provide the optimum composition, yet management for optimum composition would result in structure that renders habitat unusable (Principle 1). This effect is the site productivity penalty; i.e., marginal habitat may exist despite optimum grazing management, because climate, weather, and soils preclude optimum structure and composition.

If grazing management is based on optimum structure, manifold situations are covered, whereas if it is based on composition, unacceptable situations arise. Hence, structure is the more general variable in grazing management decisions. In essence, grazing is a strong technique for managing the quantity of habitat, but a rather weak technique for managing its quality based on plant composition.

Principle 5. Within a certain domain, gamebird abundance is positively correlated with usable space in time (space-time). Given constancy in other factors, this relationship is considered axiomatic (i.e., held to be true without proof).

Usable space is the area that provides acceptable structure (therefore it is usable) to gamebirds. Time refers to the number of days (or any convenient unit) in a year (or any convenient period) that a given amount of space is usable.

The domain of positive correlation is bounded by minimum space-time required for a viable (i.e., self-sustaining) population and the point at which the relationship between space-time and abundance becomes asymptotic (Fig. 5). An asymptote arises when the rate of change or slope of a relationship approaches zero. The asymptote occurs at maximum possible abundance for the species in question on the area in question.

Space-time can be estimated in discrete or continuous terms. For example, if 1,000 ha were fully usable for a year, then there would be 1,000 ha × 365 days = 365,000 ha-days of space-time. If, because of adverse impacts of grazing, 500 ha were fully usable for a year and 500 ha were fully usable for 100 days, space-time = (500 ha × 365 days) + (500 ha × 100 days) = 232,500 ha-days. In continuous terms,

$$S = \int_a^b s(t)dt$$

where
S = space-time,
a = start of time period,
b = end of time period, and
$s(t)$ is an equation that gives usable space as a function of time.

Theory and Example

The above principles support the following theory of grazing management for gamebirds.

> The optimum approach to grazing in any environment is the approach that maximizes space-time by optimizing structure of herbaceous vegetation for the gamebird in question, given that site integrity is maintained. Because sites differ in primary productivity and vegetation structure, optimal grazing may range from no use to heavy use of forage by livestock.

Research by Spears et al. (1993) on the optimum seral stage for northern bobwhites as influenced by site productivity helps demonstrate the concepts presented above. They recognized that conflicting research results and recommendations on grazing in bobwhite habitat management could be reconciled under the hypothesis that "optimum seral stage for northern bobwhites varies inversely with site productivity." In different words, they hypothesized that bobwhites were mid- to high seral species on less productive sites, while they were low- to mid seral species on more productive sites.

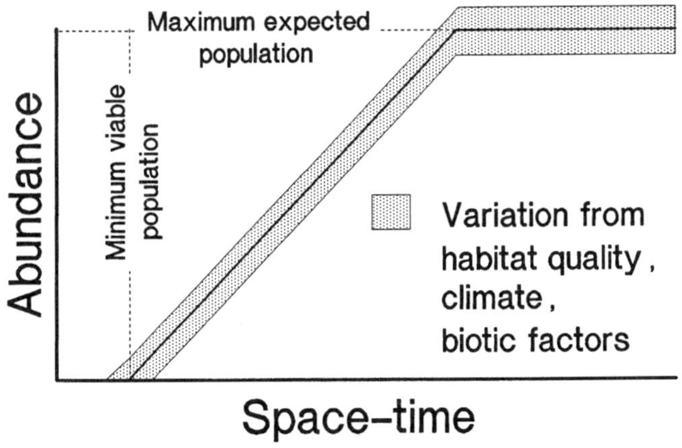

FIGURE 5. Axiomatic relationship between abundance of gamebirds and usable space in time (space-time).

To develop a theoretical construct for the hypothesis, Spears et al. (1993) assumed that bobwhite density varies as a parabolic function of standing crop biomass of grasses at some critical point in the year (Fig. 6A). The parabolic relationship means there can be too little, too much, or proper amounts of grass biomass, consistent with Principle 1. Under the hypothesized model, density would vary positively with biomass until the optimum amount was reached, then it would vary negatively with biomass.

As discussed above, higher seral stages generally have more biomass than lower seral stages on rangeland, especially in the sense of range condition classes. However, absolute amounts of biomass in a given seral stage vary with productivity of a site (Principle 2). Biomass in a high seral stage on a less productive site might equal that in a low seral stage on a more productive site (dotted line, Fig. 6B). If bobwhite abundance varies in a parabolic manner with grass biomass, one might expect their density to be positively associated with seral stage on less productive sites and negatively associated on more productive sites. The hypothesis of Spears et al. (1993) predicts an interaction effect between bobwhite abundance and seral stage on sites of different productivity (Fig. 6C). This effect is tantamount to saying higher seral stages maximize space-time on less productive sites, whereas lower seral stages maximize this variable on more productive sites.

Results from field tests in the Rio Grande Plains of Texas supported the seral stage-site productivity hypothesis (Fig. 6D). The concept needs further testing, but the same type of situation might hold for other gallinaceous birds because of the foundation principles and the proposed theory. The results of Spears et al. (1993) show why the context (Table 1) is important when decisions on managing gamebird habitat with grazing are made. In certain contexts (e.g., less productive sites) a particular gamebird species might be sensitive to grazing pressure. This same species might be robust to grazing in a different context (e.g., more productive sites).

The results of Spears et al. (1993) also highlight the site productivity penalty; observed densities were higher on a low seral stage on a more productive site than on a high seral stage on a less productive site (Fig. 6D). Thus, even though grazing might be optimum on less productive sites, the likelihood remains that gamebird populations will be lower than in a more suitable environment. Differential abundance is a natural outcome of the environment, and not a problem of grazing if proper grazing (defined below) is practiced.

Space-time in Management

Measurement of space-time could be accomplished by sampling herbaceous vegetation structure through time with a visual

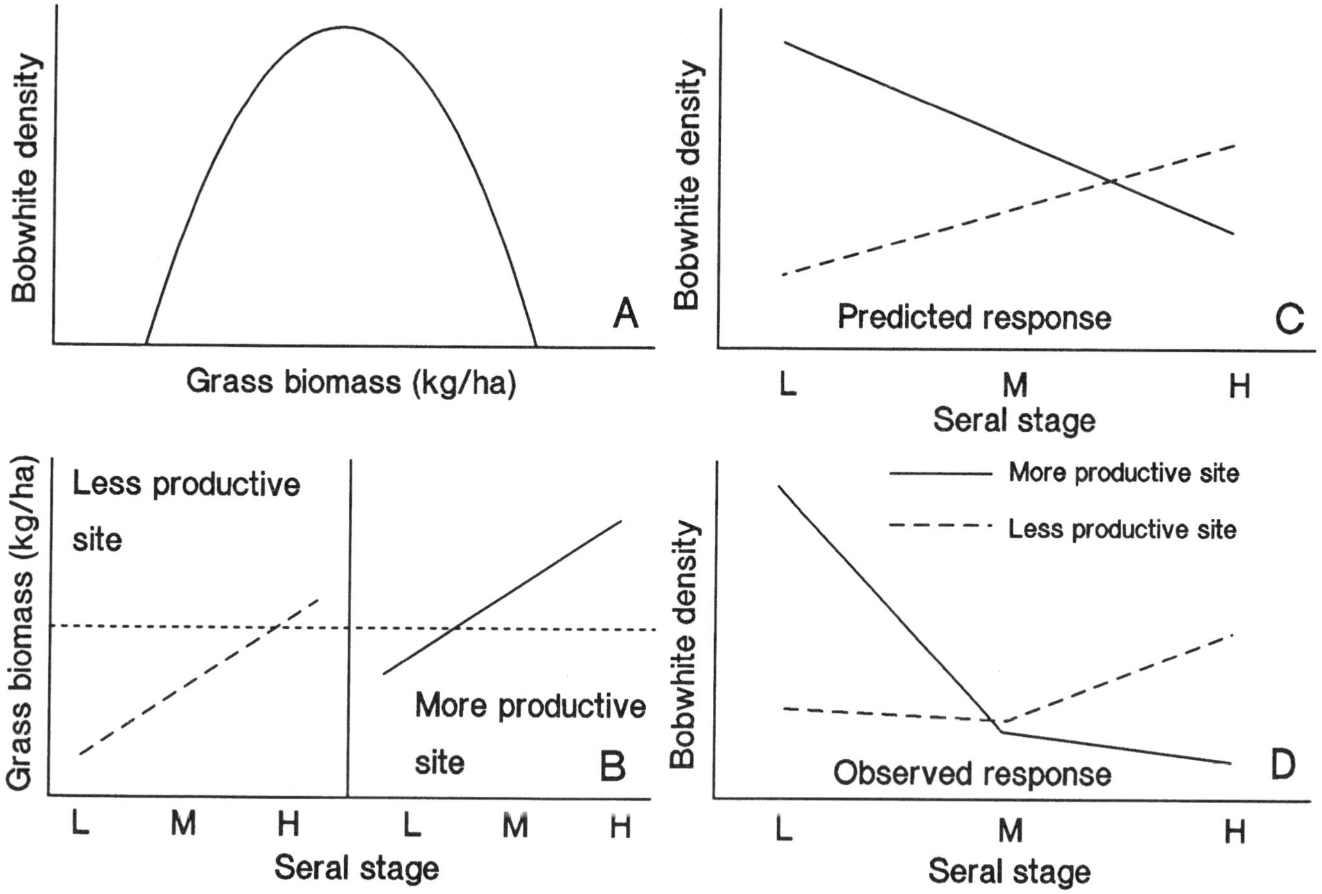

FIGURE 6. Theoretical constructs (A and B) for the relationship between northern bobwhite density and seral stage on rangeland, predicted relationship based on the constructs (C), and observed relationship (D) in the Rio Grande Plains of Texas (Spears et al. 1993).

obstruction pole (Robel et al. 1970, Nudds 1977, Guthery et al. 1981). Knowledge of acceptable structural conditions (e.g., effective height, percent visual obstruction) for the gamebird in question (Klebenow 1981, Mattise et al. 1981) would be necessary for interpretation of the resulting data. As an example, Schulz and Guthery (1988) used a profile pole to quantify vegetation structure at points used by northern bobwhites and compared the results with measurements at random points in 2 different grazing programs.

For management purposes, an easily determined correlate of space-time is useful. The stubble height hypothesis (Holechek et al. 1982:208) provides a convenient management touchstone.

> Whenever use [removal of forage with grazing] is considered, the amount of forage removed is not nearly so important as the amount of residue that remains. If evaluation of grazing intensity were based on stubble height rather than percentage use, managers would have a common reference point for decision making that reflects grazing severity from the standpoint of wildlife, livestock, and vegetation.

In other words, these authors advocated a constant structural condition (stubble height), consistent with Principle 1. Their use of the word "residue" implies structure at the lowest point in the annual cycle of vegetation growth and decline.

Managers should recognize that constant stubble height is an impossible goal if this variable is managed with grazing; there will always be a mosaic of stubble heights unless grazing is so intense that little or no stubble remains. Stubble height is, in fact, a random variable with a mean and a variance, and managers should apply grazing pressure such that some optimum mean height and variability are obtained.

As a unifying concept, assume that optimum mean stubble height for gamebird species i (s_i) is a function of the mean standing height of the species (h_i). In symbols,

$$s_i = f(h_i).$$

If it is assumed that optimum mean stubble height is equal to mean gamebird height, then

$$s_i = h_i.$$

Given the assumption, the research and management problem is to determine the optimum level of variability in stubble height for the gamebird species in question. For species like quail, managers might expect the optimum frequency distribution of stubble heights to be skewed right; i.e., high frequencies of low stubble heights and low frequencies of high stubble heights (high and low defined relative to h_i). Managers might further expect high variability in stubble heights to be optimum. For species like wild turkeys (*Meleagris gallopavo*), one might expect the opposite circumstances to be optimum. Research needs to determine the nature of these relationships.

The stubble height hypothesis also provides a concept for generalizing definitions of grazing intensity. Define a constant stubble height (k_i) as appropriate for gamebird species i at some critical point in time (e.g., the beginning of nesting season) and define observed stubble height at that time as s. These variables lead to the following definitions.

Undergrazing: $s > k_i$ (stubble too high).

Proper grazing: $s = k_i$ (stubble height proper).

Relative overgrazing: $s < k_i$ (stubble too low but character and potential of site maintained).

Relative overgrazing occurs when stubble height falls below recommended levels relative to the gamebird species in question. Absolute overgrazing occurs when grazing intensity is high enough to increase soil erosion beyond acceptable background levels or otherwise damage the character and longterm potential of a site. Because absolute overgrazing destroys the productivity and integrity of environments, it should be considered immoral (Bailey 1984:33). Relative overgrazing, however, is an issue of human values (no principles of logic or philosophy indicate relative overgrazing is right or wrong). Gamebird managers should keep in mind the distinction between absolute and relative overgrazing.

Space-time and Grazing Programs

The concept of space-time is applicable to any approach to grazing, whether that approach is simple (e.g., continuous yearlong or season-long grazing) or complex (e.g., single-herd-multipasture short duration grazing). Space-time is a common denominator of all grazing programs, including no grazing.

Space-time may fall below maximum in the absence of grazing. This could occur in environments incapable of providing acceptable structure, a context within which a gamebird population might be sensitive to grazing. Conversely, in relatively rich environments that provide herbaceous structure unusable because of density, height, or some related factor that inhibits gamebirds, the context results in opportunities for use of grazing in habitat management.

In some cases, grazing has no apparent effects on space-time as indexed by comparative gamebird populations in grazed and ungrazed areas. Zwickel (1972) compared blue grouse (*Dendragapus obscurus*) population attributes between an area grazed at 1.5 AUM/ha and an ungrazed area. He found similar density of adults, percentage of hens with broods, and chicks/brood between areas, despite "striking differences in the vegetative character of the areas" (Zwickel 1972:632). He observed that an apparent increase in shrubs on the grazed area might have compensated for a decline in cover from native bunchgrasses.

Zwickel's (1972) observations may simply reflect site differences, because treatments were unreplicated. However, the observations draw attention to an important consideration in grazing management for gamebirds. The dispersion and structure of woody cover influence the quantity of space-time available, given some stocking rate (cattle/area). Guthery (1986:42-43) observed

> the coverage and diversity of brush affect use of cattle in management of bobwhite cover. Grazing must be lighter in a pasture with 5% brush coverage than in a like pasture with 15% brush coverage. At 5% coverage, herbs (cattle foods) provide most

> cover for travel and escape, so they cannot be grazed. At 15% coverage, shrubs take over some of these functions, so fewer herbs are required.

The upshot is that removal of herbaceous vegetation by grazing can be heavier with more brush but must be lighter with less brush to maximize space-time.

Wildlife biologists have devoted much attention to comparisons between continuous (yearlong or season-long) grazing and one of the specialized grazing systems. Specialized systems have in common the division of a large pasture unit into subunits and rotation of livestock among ≥2 subunits such that nonuse occurs in ≥1 subunit at any point in time during a grazing period.

Over several years, a grazing system might or might not maximize space-time in comparison with continuous grazing if the structure associated with a higher seral stage is desired. In more humid environments, "the most rapid rate of [rangeland plant succession] occurs on badly deteriorated ranges when grazing systems are used which maximize rest periods in relation to grazing periods" (Drawe 1991:28). In more arid environments, conversely, moderate continuous grazing (e.g., 30-45% use) results in higher rates of succession than any known grazing system (J. L. Holechek, New Mexico State Univ., pers. commun.). The more rapidly the proper successional stage and structure are attained and maintained, the greater the space-time available over a period of years. However, use of continuous versus systematic grazing to maximize space-time over the long term depends on the environmental setting.

In the context of near-term (i.e., this year) space-time, there is no particular reason to expect benefits from a grazing system as opposed to continuous grazing in upland gamebird habitat management. The most important consideration in grazing management, the one which has the greatest impact on vegetation, livestock, and wildlife, is the stocking rate (Holechek et al. 1989:173-174). This principle is evident if we consider that a given number of cattle have a fixed requirement for forage (kg dry wt/time period); that requirement does not change appreciably no matter how the herd is distributed in time and space, although efficiency of forage use can be altered with grazing program (C. W. Hanselka, Tex. Agric. Ext. Serv., pers. commun.)

Under certain conditions, grazing systems would be expected to reduce space-time in comparison with continuous grazing. Systems invariably increase the extremes and therefore the range of grazing intensities a unit receives in comparison with continuous grazing. Given a fixed number of cattle, nonuse of ≥1 subunit implies heavy use of ≥1 subunit in comparison with continuous grazing of the unit. If nonuse, heavy use, or both resulted in structurally unusuitable cover, systems could reduce space-time in comparison with continuous grazing (Principle 1). Mattise et al. (1981) advocated season-long as opposed to deferred-rotation grazing of sharp-tailed grouse nesting and brood-rearing areas, because deferred-rotation grazing reduced space-time under the process described above (heavy use in subunits reduced overall space-time).

The advent of short duration grazing (e.g., 1-herd, many pastures, rapid rotation of the herd through pastures) raised a space-time question relative to ground nests. Short duration grazing results in livestock densities 16-90 times higher in subunits than under continuous grazing of the unit at the same overall stocking rate (Savory and Parsons 1980). Concern arose that high stock densities in subunits might accelerate trampling loss of ground nests.

Bryant et al. (1981) argued that the relative danger to trampling loss of ground nests was similar among grazing systems, because while systems resulted in higher subunit livestock densities and increased probability of trampling therein, this was balanced by reduced cattle movement and nonuse of other subunits at any point in time. These results have held in field comparisons between short duration and continuous grazing (Koerth et al. 1983, Bareiss et al. 1986).

At some livestock density, of course, trampling of ground nests must become a management concern. Jensen et al. (1990) showed that cumulative percentage nest trampling (*N*) could be modeled as

$$N = 100\left[1 - \exp\left(-bt\right)\right]$$

where

b = a parameter that governs the rate of increase in trampling loss given stock density (no./ha) and

t = elapsed time of exposure in days.

The parameter increases with stock density, which means the higher the stocking rate, the more rapidly the cumulative percentage trampled increases during the first few days of exposure (Fig. 7).

Implications and Predictions of the Theory

If the theory proposed herein has merit, then one would expect a lack of consensus, if not conflicting recommendations and evaluations of grazing, in the literature on gamebird habitat management. This expectation has been realized, as evidenced

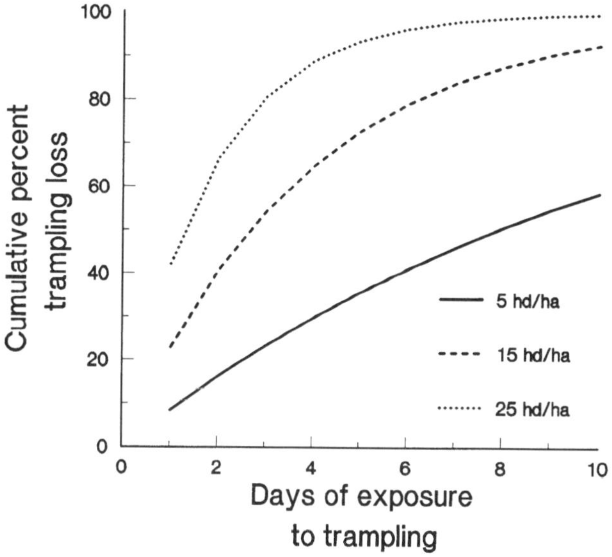

FIGURE 7. Relationship between ground nest trampling and days of exposure at different livestock stocking rates (Jensen et al. 1990).

by the quotations on scaled quail in the introduction to this chapter. The literature also contains conflicting recommendations on grazing in bobwhite management (Spears 1991). These conflicts appear because gamebirds are constant (Principle 1) but habitats vary around them (Principle 2). Thus, in some contexts, gamebirds will be sensitive to if not damaged by grazing, whereas in other contexts grazing or a similar type of disturbance might be essential to population vigor.

The theory and principles developed to support it indicate grazing must be tailored to the environment in which it is practiced. There is no single optimum grazing program, but there are optimum approaches to grazing, given the context, and each optimum approach will maximize space-time. It follows, based on Principles 1 and 2, that optimum grazing will always have 2 properties: great flexibility with respect to stocking rates, and high control over dispersion of livestock use.

The first property is necessary because of variation in weather. The gamebird manager might want to decrease stocking rate in drier years and increase it in wetter years. The manager should not necessarily be concerned about livestock performance, e.g., calving percentages and rate of weight gain. He might want to sacrifice performance to achieve a habitat objective, given humane treatment of the herd. Sacrificing kilograms of beef production might be cheaper than accomplishing the same habitat objective with burning or heavy equipment.

The second property arises because of variation among sites on management areas. This variation indicates dispersion of livestock use must be nonuniform to maximize space-time for the gamebird species in question. This property is directly contrary to a hallowed tenet of range management: uniform use.

The same tools that livestock managers use to improve livestock distribution can be reversed to foster nonuniform use. The primary tools are placement of water, fencing, and supplement (i.e., salt, mineral, and feed) (Holechek et al. 1989:250-261). For example, if relative overgrazing ($s < k_i$) is occurring on portions of a site, this problem can be addressed by fencing a water facility, providing supplement at some distance, or fencing the area from use. Conversely, if undergrazing ($s > k_i$) is occurring, the opposite steps can be taken.

SUMMARY

The information presented in this chapter can be used as a conceptual framework for decision-making in management of gamebird habitat with grazing. The decision process will entail answers to these questions:

1. What is the ecological status of vegetation on the management area? Is grazing of potential value in altering the structure of herbaceous cover for the gamebird species in question, or must we first disrupt a stable state?
2. Given that grazing is useful, how many cattle should be stocked and over what time period to maximize space-time or to preserve acceptable residual cover at some critical point in time? Answering this question will require knowledge of vegetation structure acceptable to the gamebird in question, rate of herbaceous cover removal by cattle, and rate of forage growth.
3. What actions (water, mineral, shade, fencing) need to be taken to focus grazing pressure where it is needed to improve structure and to reduce or eliminate grazing pressure where it is unnecessary or potentially damaging?

Remember that in answering these questions one is under no moral or ethical obligation except for maintaining the integrity of the site and treating the livestock herd humanely.

While I have attempted to unify diverse information and reconcile conflicting recommendations with seminal theory on grazing management, readers should recognize that the theory is perhaps more properly termed an hypothesis. Readers are encouraged to challenge and improve upon the hypothesis, to find nuances of interpretation and application.

LITERATURE CITED

Allison, C. D., M. M. Kothmann, and L. R. Rittenhouse. 1982. Efficiency of forage harvest by grazing cattle. J. Range Manage. 35:351-354.

Archer, S. 1989. Have southern Texas savannas been converted to woodlands in recent history? Am. Nat. 134:545-561.

Bailey, J. A. 1984. Principles of wildlife management. John Wiley & Sons, New York, N.Y. 373pp.

Baker, D. L., and F. S. Guthery. 1990. Effects of continuous grazing on habitat and density of ground-foraging birds in south Texas. J. Range Manage. 43:2-5.

Bareiss, L. J., P. Schulz, and F. S. Guthery. 1986. Effects of short-duration and continuous grazing on bobwhite and wild turkey nesting. J. Range Manage. 39:259-260.

Brown, D. E., C. L. Cochran, and T. E. Waddell. 1978. Using call-counts to predict hunting success for scaled quail. J. Wildl. Manage. 42:281-287.

Bryant, F. C., F. S. Guthery, and W. M. Webb. 1981. Grazing management in Texas and its impact on selected wildlife. Pages 94-112 *in* J. M. Peek and P. D. Dalke, eds. Wildlife-livestock relationships symposium: proceedings 10. Univ. Idaho, For., Wildl. and Range Exp. Stn., Moscow.

Campbell, H., D. K. Martin, P. E. Ferkovich, and B. K. Harris. 1973. Effects of hunting and some other environmental factors on scaled quail in New Mexico. Wildl. Monogr. 34. 49pp.

Drawe, D. L. 1991. Influence of grazing on vegetation and cattle on the Welder Wildlife Foundation Refuge, 1983-1987. Welder Wildl. Found. Contrib. B-11. 31pp.

Dyksterhuis, E. J. 1949. Condition and management of rangeland based on quantitative ecology. J. Range Manage. 2:104-115.

Ellis, J. A., W. R. Edwards, and K. P. Thomas. 1969. Responses of bobwhites to management in Illinois. J. Wildl. Manage. 33:749-762.

Evans, K. E. 1968. Characteristics and habitat requirements of the greater prairie chicken and sharp-tailed grouse: a review of the literature. U.S. Dep. Agric. Conserv. Res. Rep. 12. 31pp.

Friedel, M. H. 1991. Range condition assessment and the concept of thresholds: a viewpoint. J. Range Manage. 44:422-426.

George, J. F., and J. Powell. 1979. Cattle grazing impacts on small cleared areas in dense American elm woodlands. J. Range Manage. 32:78-79.

Grange, W. B. 1949. The way to game abundance. Charles Scribner's Sons, New York, N.Y. 365pp.

Guthery, F. S. 1986. Beef, brush and bobwhites. Caesar Kleberg Wildl. Res. Inst. Press, Kingsville, Tex. 182pp.

———, C. A. DeYoung, F. C. Bryant, and D. L. Drawe. 1990. Using short duration grazing to accomplish wildlife habitat objectives. Pages 41-55 *in* K. E. Severson, tech. coord. Can livestock be used as a tool to enhance wildlife habitat? U.S. Dep. Agric., For. Serv. Gen. Tech. Rep. RM-194.

———, T. B. Doerr, and M. A. Taylor. 1981. Use of a profile board in sand shinnery oak communities. J. Range Manage. 34:157-158.

Healy, W. M. 1985. Turkey poult feeding activity, invertebrate abundance, and vegetation structure. J. Wildl. Manage. 49:466-472.

Holechek, J. L., R. Valdez, S. D. Schemnitz, R. D. Pieper, and C. A. Davis. 1982. Manipulation of grazing to improve or maintain wildlife habitat. Wildl. Soc. Bull. 10:204-210.

———, R. D. Pieper, and C. H. Herbel. 1989. Range management principles and practices. Prentice Hall, Englewood Cliffs, N.J. 501pp.

Jensen, H. P., D. Rollins, and R. L. Gillen. 1990. Effects of cattle stock density on trampling loss of simulated ground nests. Wildl. Soc. Bull. 18:71-74.

Kessler, W. B., and R. P. Bosch. 1981. Sharp-tailed grouse and range management practices in western rangelands. Pages 133-146 *in* J. M. Peek and P. D. Dalke, eds. Wildlife-livestock relationships symposium: proceedings 10. Univ. Idaho, For., Wildl. and Range Exp. Stn., Moscow.

———, and J. D. Dodd. 1978. Responses of coastal prairie vegetation and Attwater prairie chickens to range management practices. Proc. Int. Rangeland Congr. 1:473-476.

Klebenow, D. A. 1981. Livestock grazing interactions with sage grouse. Pages 113-123 *in* J. M. Peek and P. D. Dalke, eds. Wildlife-livestock relationships symposium: proceedings 10. Univ. Idaho, For., Wildl. and Range Exp. Stn., Moscow.

Koerth, B. H., W. M. Webb, F. C. Bryant, and F. S. Guthery. 1983. Cattle trampling of simulated ground nests under short duration and continuous grazing. J. Range Manage. 36:385-386.

Laycock, W. A. 1991. Stable states and thresholds of range condition on North American rangelands: a viewpoint. J. Range Manage. 44:427-433.

Leopold, A. S. 1977. The California quail. Univ. California Press, Berkeley. 281pp.

Lewis, C. E., B. F. Swindel, and G. W. Tanner. 1988. Species diversity and diversity profiles: concept, measurement, and application to timber and range management. J. Range Manage. 41:466-469.

MacArthur, R. H., and J. W. MacArthur. 1961. On bird species diversity. Ecology 42:594-598.

Mattise, S. N., R. L. Linder, and G. D. Kobriger. 1981. Effects of grazing systems on sharp-tailed grouse habitat. Pages 124-132 *in* J. M. Peek and P. D. Dalke, eds. Wildlife-livestock relationships symposium: proceedings 10. Univ. Idaho, For., Wildl. and Range Exp. Stn., Moscow.

Mendenhall, W., D. D. Wackerly, and R. L. Scheaffer. 1990. Mathematical statistics. Fourth ed. PWS-Kent Publ. Co., Boston, Mass. 818pp.

Mosconi, S. L., and R. L. Hutto. 1981. The effects of grazing on the land birds of a western Montana riparian habitat. Pages 221-233 *in* J. M. Peek and P. D. Dalke, eds. Wildlife-livestock relationships symposium: proceedings 10. Univ. Idaho, For., Wildl. and Range Exp. Stn., Moscow.

Nudds, T. D. 1977. Quantifying the vegetative structure of wildlife cover. Wildl. Soc. Bull. 5:113-117.

Petraitis, P. S., R. E. Latham, and R. A. Niesenbaum. 1989. The maintenance of species diversity by disturbance. Q. Rev. Biol. 64:393-418.

Pieper, R. D., and R. F. Beck. 1990. Range condition from an ecological perspective: modifications to recognize multiple use objectives. J. Range Manage. 43:550-552.

Pitelka, F. A. 1941. Distribution of birds in relation to major biotic communities. Am. Midl. Nat. 25:113-137.

Prasad, N. L. N. S., and F. S. Guthery. 1986. Wildlife use of livestock water under short duration and continuous grazing. Wildl. Soc. Bull. 14:450-454.

Robel, R. J., J. N. Briggs, A. D. Dayton, and L. C. Hulbert. 1970. Relationships between visual obstruction measurements and weight of grassland vegetation. J. Range Manage. 23:295-297.

Savory, A., and S. D. Parsons. 1980. The Savory grazing method. Rangelands 2:234-237.

Schulz, P. A., and F. S. Guthery. 1988. Effects of short duration grazing on northern bobwhites: a pilot study. Wildl. Soc. Bull. 16:18-24.

Schulz, T. T., and W. C. Leininger. 1990. Differences in riparian vegetation structure between grazed areas and exclosures. J. Range Manage. 43:295-299.

Smith, E. L. 1989. Range condition and secondary succession: a critique. Pages 103-141 *in* W. K. Lauenroth and W. A. Laycock, eds. Secondary succession and the evaluation of rangeland condition. Westview Press, Boulder, Colo.

Spears, G. S. 1991. Optimum seral stage for northern bobwhites as influenced by site productivity. M.S. Thesis, Texas A&I Univ., Kingsville. 89pp.

———, F. S. Guthery, S. M. Rice, S. J. DeMaso, and B. Zaiglin. 1993. Optimum seral stage for northern bobwhites as influenced by site productivity. J. Wildl. Manage. 57:805-811.

Stuth, J. W. 1991. Foraging behavior. Pages 65-83 *in* R. K Heitschmidt and J. W. Stuth, eds. Grazing management: an ecological perspective. Timber Press, Portland, Oreg. 259pp.

Texas Parks and Wildlife Department. 1945. Principal game birds and mammals of Texas: their distribution and management. Texas Game, Fish and Oyster Comm., Austin. 149pp.

Tiedeman, J. A., R. Beck, and R. V. Ecret. 1991. Dependence of standing crop on range condition rating in New Mexico. J. Range Manage. 44:602-605.

Wallmo, O. C. 1956. Ecology of the scaled quail in west Texas. Texas Game and Fish Comm., Austin. 134pp.

Watts, C. R., L. C. Eichhorn, and R. J. Mackie. 1987. Vegetation trends within rest-rotation and season-long grazing systems in the Missouri River Breaks, Montana. J. Range Manage. 40:393-396.

Weaver, L. J. 1984. Food habits of adopted bobwhite quail chicks in south Texas. M.S. Thesis, Texas A&I Univ., Kingsville. 92pp.

Westoby, M., B. Walker, and I. Noy-Meir. 1989. Opportunistic management for rangelands not at equilibrium. J. Range Manage. 42:266-274.

Whyte, R. J., and B. W. Cain. 1981. Wildlife habitat on grazed or ungrazed small pond shorelines in south Texas. J. Range Manage. 34:64-68.

Zwickel, F. C. 1972. Some effects of grazing on blue grouse during summer. J. Wildl. Manage. 36:631-634.

Chapter 7
BIRDS OF RANGELAND WETLANDS
Milton W. Weller

INTRODUCTION

Although many North Americans associate rangelands with arid western grazing lands, and wetlands with other regions of the country, neither view is completely correct. The extensive western rangelands often have large basin or riparian wetlands, and the wetter areas of North America have wetlands used as rangeland. Water drives the development of wetland vegetation in all areas and attracts birds and other wildlife, but water in arid regions more often comes via distant snowfall and/or rainfall that arrives by flooding rather than by rain and snow on site. Thus, this discussion considers various geographic areas and types of wetlands in North America and cites worldwide examples where grazing of wetlands is a standard practice.

Birds of wetlands often have unique adaptations enabling them to use water and water-based resources, but are mobile enough to search for suitable water areas despite characteristic dynamic water regimes. The most conspicuous, popular, and sought-after birds are waterfowl (i.e., ducks, geese, and swans of the family Anatidae), but other groups of birds are well-adapted and conspicuous, composing a unique community of waterbirds that exploit diverse resources of the wetland (Weller et al. 1958, Weller and Spatcher 1965, Capen and Low 1980, Burger 1985). Rather than focusing only on gamebirds, it is best to consider the community holistically, because any modification of structure or process within the system influences all species. To do so requires an understanding of wetlands, how they function, the resources available within them, and strategies used by birds to exploit resources. While stressing the uniqueness, diversity, and values of wetlands as a biotic community, the interface between wetland and upland must be considered relative to community dynamics, energy flow, productivity, resource use and management, and wildlife use of this edge. Finally, effects of grazing are considered.

WETLANDS AND THEIR CLASSIFICATION

There are 3 major constituents of a wetland: water, water-adapted plants (hydrophytes), and hydric soil or a precursor such as alluvium deposited by water movement (Cowardin et al. 1979). There has been a changing view of whether all 3 components are necessary to meet the definition (Cowardin et al. 1979, Tiner 1984, Federal Interagency Committee for Wetland Delineation 1989, United States Environmental Protection Agency 1991), but in many cases, only 1 or 2 components seem present in aquatic systems that perform wetland functions or have characteristic values of wetlands. In this chapter, I use the original federal interagency definition (Cowardin et al. 1979) that implies that the presence of ≥1 of these components at a given time may designate a wetland.

Additionally, it is important to understand that wetlands are not necessarily in basins, as are ponds or lakes, or in channels, as are streams; they may be a community of hydrophytes in seeps on drainage slopes with no standing water. Moreover, they need not have permanent water; they may be wet during most years only long enough to allow hydrophytes to grow in spring, as is commonly true in riparian zones. Wetlands may be shallow parts of ponds, lakes, or streams rather than the entire entity. Wetlands perform certain community functions, such as productivity, including the support of invertebrates that are exportable resources for other organisms in adjacent areas. Their hydrologic functions such as ground-water recharge, floodwater alteration, sediment stabilization and retention, and nutrient trapping are important ingredients with societal values and influence on wildlife diversity and abundance (Adamus et al. 1991).

Western wetlands are especially known for their variability in water presence and depth due to erratic precipitation patterns. In intermontane basins fed mainly by snow, water usually is seasonally predictable even if volume is not. In dry or even desert areas fed by periodic rainfall, wetlands may be wet only in 1 of many years, and arid basins in Arizona, Nevada, and Texas may be filled only during major subtropical storms. The Prairie Pothole Region also is highly variable in rainfall, with only the larger wetlands considered semipermanent (Stewart and Kantrud 1971) and many smaller ones having water only on a temporary or seasonal basis.

Wetland Delineation

In recent years, delineating the boundary between wetland and upland has become important for regulatory purposes, and considerable effort has been devoted to techniques (Federal Interagency Committee for Wetland Delineation 1989). Delineation is difficult because wetland dynamics make the edge between upland and water mobile over time. But the term delineation also has been applied to determination of whether a plant community contains wetland plant species based on water tolerance. Ephemeral and other shallow wetlands are dynamic and may not be identifiable within or among years. Moreover, the edges of wetlands shift over time with water conditions, and establishing a high-water elevation may not be possible in erratic systems; the high-water mark, however, is most vital because conflicts over usage most often involve tracts between the high- and low-water marks (Odum 1983).

Vegetation Life Form and Structure

Although wetland vegetation may be extremely diverse, relatively few plants tend to become visible dominants in most wetlands. These dominants often are species subject to grazing or to impacts by livestock activities. Emphasis on *vegetation structure* or *life form* is a convenient approach to making generalizations about habitat features of wetlands and identifying potential livestock-vegetation relationships.

From completely aquatic to the more terrestrial, these plant life forms result from adaptation to the presence, duration, seasonality, chemistry, and depth of water. In wetlands confined to basins, tolerance of water depth produces concentric zones of plants of differing species and structure. These form habitats for different bird species. Some plants float on the water's surface (*floating plants*) (e.g., lesser duckweed [*Lemna minor*]) or occasionally in the water column (e.g., star duckweed [*L. trisulca*]), and take their nutrients directly from the water via suspended roots or a suitable cell structure, and thus are less restricted in distribution in a wetland when water is present. Such floating vegetation does not support the weight of even small walking birds, but it often inhibits swimming by larger ones. Other plants have floating leaves but are rooted (e.g., water lilies [*Nuphar* spp., *Nymphaea* spp.]) and are termed *floating-leaf* or *pad plants*, and can flourish in fluctuating or turbid water. Other more aquatic plants are submersed but rooted at some stage and are therefore associated with shallow waters or with the littoral zone of lakes or ponds where light penetration is greatest. These often, like the pondweeds (e.g., sago pondweed [*Potamogeton pectinatus*]), have filiform leaves able to capture light underwater more effectively. Some submersed species may survive floating in the water column when they are torn free from their roots or fragment; still other submergents are unrooted. Much of the great productivity of wetlands comes in the form of rooted herbaceous emergent plants that live in the air/sunlight medium for ideal photosynthesis but also thrive in water and tap the nutrients of the basin soil (Fig. 1). Examples are cattail (*Typha* spp.), sedges (*Carex* spp.), bullrushes (*Scirpus* spp.), rushes (*Juncus* spp.), and spike-rushes (*Eleocharis* spp.), and the more water-tolerant grasses, such as cordgrass (*Spartina* spp.) and whitetop (*Scholochloa festucacea*) that often are favorite livestock foods. Finally, there are woody plants that also may stand in water, including very water tolerant shrubs like buttonbush (*Cephanthus occidentalis*), and early-succession or mud flat species like willows (*Salix* spp.). Other woody shrubs occur along the shoreline slopes of basin wetlands and in stream riparian zones sites where flooding enables germination and survival of willow and alder (*Alnus* spp.). Similar species occur in seeps from snowfields at high altitude, even though there is rarely standing water. Large trees include those of seasonally flooded areas such as cottonwood (*Populus deltoides*) or water oak (*Quercus nigra*) that may be common to riparian zones or bottomland floodplains, respectively, grazed during drier periods.

Wetland Plant Succession

Plant dynamics of wetlands are water-driven, and changes in species composition and dominance are common especially in Midwestern and Western areas of low and variable precipitation. Surprisingly few emergent wetland plants germinate under water, and the drying of basin substrates provides a moist-soil medium ideal for germination. Major plant community changes occur in these settings and numerous workers have observed changes and suggested successional patterns. In herbaceous marshes, seeds are long-lived and seed banks respond to various flooding regimes, allowing both prediction of plant species and the development of management procedures to regulate species composition (Harris and Marshall 1963). Reverse plant zonation may result from germination in formed open water areas during drought periods, and waterbirds and other wildlife respond quickly to such vegetation changes (Weller and Spatcher 1965). These relatively short-term changes are a result of chance events such as rainfall patterns and timing of dewatering, and are best viewed as Gleasonian succession rather than the long-term view of Clementian succession formerly linked to aquatic succession (van der Valk 1981).

Wetland Classification

Considerable effort has been made to group wetlands according to conspicuous features to enhance communication about wetlands and aid in regulatory and conservation programs. Because wetlands are so variable, however, and different specialists emphasize different features, no one system is universal. The most widely accepted standard in the National Wetland Inventory and other surveys is by Cowardin et al. (1979). Their procedure recognizes 5 "systems" based on broad characteristics: (1) marine—saline ocean water with wetlands only in the shallow intertidal shorelines; (2) estuarine—brackish water zone produced by freshwater inflow mixing with ocean-derived saltwater, typically dominated by robust emergents; (3) lacustrine—shallow but sparsely vegetated waters along the more shallow littoral zone (<2 m) of large freshwater lakes or reservoirs where central depths often exceed 2 m (the general limit of emergent plant growth); (4) palustrine—fresh or interior saline water areas of mostly emergent and woody vegetation; and (5) riverine—the more shallow (<2 m) channels of streams where characteristic wetland vegetation may grow. However, adjacent wetlands not in the active channel are considered

FIGURE 1. Emergent vegetation and pools form ideal cover-water interspersion providing suitable feeding and nesting sites, especially for swimming birds. Some species use mainly the emergent vegetation, some the pools, and some the interface between them. Livestock rarely invade such areas when the water is deep and food is available elsewhere. San Bernard National Wildlife Refuge, Brazoria, Texas.

palustrine rather than riverine. Systems are subdivided into hierarchical subsets that may be used at any level of detail necessary: subsystems, classes, subclasses, and dominant vegetation types, with various modifiers relating to water regime and human impacts (Cowardin et al. 1979).

Most wetland habitat examples discussed here will be: palustrine wetlands dominated by rooted, herbaceous emergents at the shallow edges (including interior salt marshes); riparian meadows and shrub or forested zones that technically are considered palustrine wetlands; and coastal estuarine wetlands or "salt" marshes.

BIRDS OF WETLANDS

Wetlands are important habitats for birds because their high productivity provides diverse, abundant, and available resources. Yet, compared with the total number of bird species in a region, only a modest number have adapted to wetlands, and these tend to be represented by large numbers of individuals. As a result, wetlands are known for large concentrations of birds, but often these concentrations come at different times of the year for different species and for different reasons. Colonial nesting wading birds (mostly of the order Ciconiiformes) (Sprunt et al. 1978, Capen and Low 1980) and socially nesting blackbirds (family Icteridae) (Orians 1980), gulls and terns (order Charadriiformes), and grebes (order Podicipediformes)(Burger 1985) are conspicuous breeders, but other families of birds that generally nest in solitary fashion may concentrate as well. Waterfowl like gadwalls (*Anas strepera*) may concentrate on islands to nest (Duebbert 1982), but most waterfowl concentrations are found after breeding. At this time some species like ducks, geese, and American coots (*Fulica americana*) undergo the annual wing molt and become flightless. Subsequently, they flock for migration and move to wintering areas.

The biodiversity among wetland communities commonly is compared using species richness (i.e., no. species) that provides a simple measure of ecological diversity, or diversity parameters that incorporate richness with relative population size of each species (e.g., species composition or species diversity index [James and Rathbun 1981]). In addition, focusing on larger taxonomic categories of birds is useful because orders or families often specialize on unique habitats (e.g., emergents or open water) or foods (e.g., invertebrates or seeds), often with specific adaptations (e.g., a duck's bill or a heron's long legs).

Considering the birds of wetlands as communities or assemblages of species rather than taxonomic categories alone allows comparison between and among wetlands and wetland types. Several ecological approaches elucidate how birds use wetlands and what habitat features are essential to their presence (Fig. 2). We can partition the habitat into subhabitats or zones based on a combination of water presence, depth, and vegetation. Open water areas lack vegetation at the surface, but may have submerged aquatic plant beds just under the surface, and are attractive to swimming and diving birds. Persistent emergent plants (i.e., those lasting to the next season) are characteristic of semipermanent waters, and are habitats for species that use dense vegetation and associated food resources. Vegetated shallow and often temporary waters are characterized by nonpersistent emergent plants (i.e., those deteriorating in several months) such as annuals or fine-leaved perennials found in moist-soil flats or meadowlands. Some seasonally flooded areas form shallow "sheetwater" suitable for temporary use by quite different groups of birds such as feeding shorebirds or waders. At times, these edges and unvegetated basins dry up and form mudflats attractive to diverse foragers in search of invertebrates or seeds. In some areas and flooding regimes, woody vegetation such as shrubs or trees may form the perimeter of basin wetlands or be the dominant vegetation in more permanently flooded flowages. This is especially true in riparian zones, resulting in structural diversity in otherwise low-profile wetland vegetation.

In describing bird use of the various habitats mentioned above, birds may be considered relative to their foods or trophic level, such as herbivores, omnivores, or carnivores, and even a more specific focus, such as insectivores or granivores. Addi-

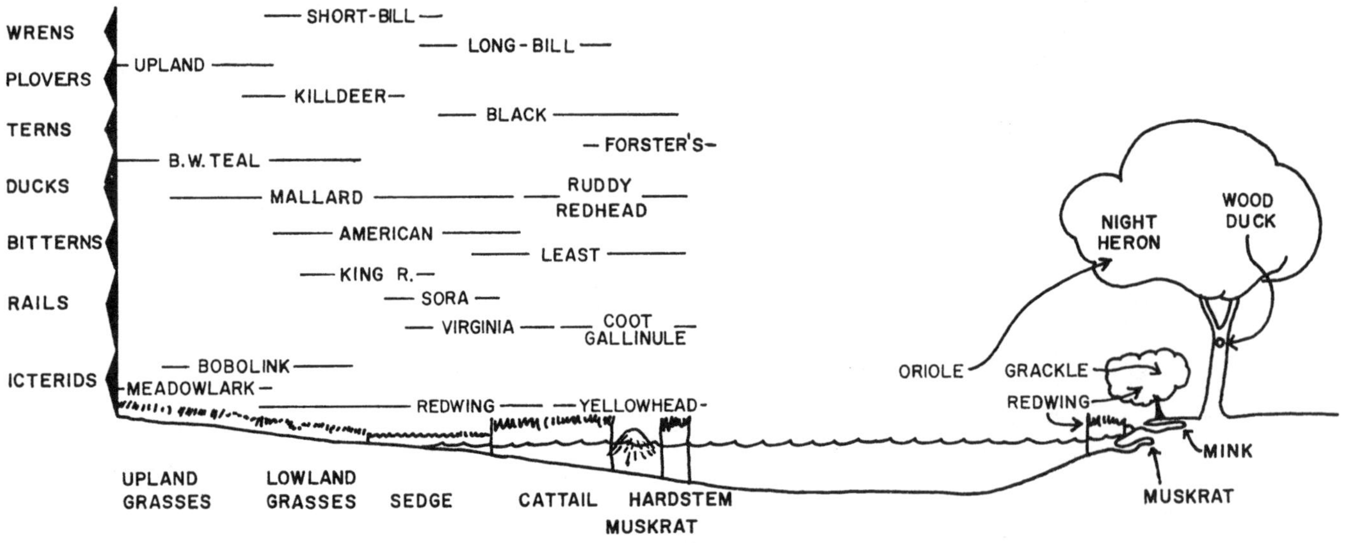

FIGURE 2. General relationship between wetland vegetation and use by major families of birds. After Weller and Spatcher (1965).

tionally, we can separate by seasonal food use, as different life cycle stages require different nutrients (Krapu and Reineke 1992). Finally, such groups of organisms can be grouped by guilds: those species that obtain resources in the same way, regardless of their taxonomy, such as bottom or benthic divers, aerial foragers, or surface gleaners (DeGraaf et al. 1985). Nest sites can also be used to classify guilds, because the vegetation or cover is a necessary resource for successful reproduction, and Short and Burnham (1982) have used matrices between such resources to assess habitat and community complexity.

Open-water Birds

Most birds of deeper, open water are divers that feed on mobile fish, large invertebrates, and occasionally amphibians such as salamanders (*Ambystoma* spp.). Such birds include pelicans (Pelicanidae), cormorants (Phalacrocoracidae), loons (Gaviidae), certain grebes (Podicipedidae), and fish-eating raptors like eagles (*Haliaeetus* spp.) and osprey (*Pandion haliaetus*) (Esler 1988). Some are benthic divers that feed on clams, worms, or insect larvae, such as lesser scaup (*Aythya affinis*), and canvasback (*A. valisineria*) or redhead (*A. americana*) ducks (Bartonek and Hickey 1969, Bergman 1973, Woodin and Swanson 1989) (Fig. 3). A few species, such as northern shoveler (*Anas clypeata*), are planktonic feeders and take tiny crustaceans by straining though special bill structure. They tend to feed in shallow, rich waters in groups or with other birds that stir-up waters and make prey vulnerable.

Birds of Submerged Aquatic Beds

Birds that use submergent plant beds during the nonbreeding period are those like American coots, swans (*Cygnus* spp.), gadwall, American wigeon (*Anas americana*), or canvasbacks and redheads, that feed on the foliage or tubers of plants such as pondweeds. Blue-winged teal (*A. discors*) and mallards (*A. platyrhynchos*) may surface feed on seed heads and associated invertebrates such as snails, especially in more shallow water depths (Swanson and Meyer 1977). Pied-billed grebes (*Podilymbus podiceps*) also feed among the vegetation, seemingly able to move easily through the dense vegetation to capture small fish and large invertebrates that take cover there (Esler 1988). The little snowy egret (*Egretta thula*) may walk on dense mats of such vegetation or use them as stationary feeding perches.

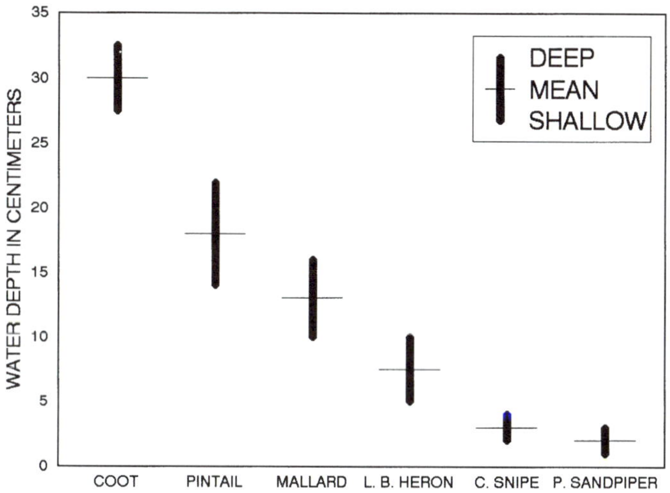

FIGURE 3. General relationship between water depth and use by feeding waterbirds adapted to different foods and water depths. After Fredrickson and Taylor (1982).

Birds of Emergent Vegetation

Emergent vegetation has the greatest concentration of species, presumably because of its diverse food resources and structural opportunities of cover for different species and different functions during individuals' life cycles. Use of emergents is especially prominent during the nesting season, when nests of different species may occur at various sites. Cover-water patterns are important because species often favor edges, and swimming birds move between patches of vegetation via water channels. Moreover, some birds require sizable areas for taking flight. Thus, dense stands of single species are less attractive to most species of birds (Weller and Spatcher 1965, Weller and Fredrickson 1974, Kaminski and Prince 1981, Murkin et al. 1982) and maintenance of balanced cover-water interspersion (i.e., "hemimarsh") is a management target used by various wildlife management agencies. Invertebrates may also be more abundant among these emergents in some seasons (Murkin et al. 1992) and attract birds that use such foods in preparation for egg laying.

Birds attracted to emergent vegetation often differ in use of its biological or physical features resulting from differing resources, water depths, or vegetation types. Some favor the protected but open-water pools within the emergents; some use the dense emergents themselves, but may have preferences in height or density; others still mainly prefer the interface. These habitats within the emergent zone are considered separately below.

Pool Birds. Some birds favor the protection of emergents in preference to extensive open-water areas that may be wind- or wave-influenced, but still use the resources of the small pools in the shallow-water emergent zone. Among these are diving ducks like the ring-necked duck (*Aythya collaris*) and ruddy duck (*Oxyura jamaiensis*) that may nest in emergents but feed in the muddy bottom or dense submergents often found in these pools. Because they are excellent swimmers and divers, they escape predators by diving and move to adjacent open water to take flight. The swimmers among the rail family, including coots using deeper water and common moorhens (*Gallinula chloropus*) of more shallow water, are common in these pools. Black terns (*Chlidonias niger*) feed on invertebrates at the surface and nest on small mats in such openings. Ibis (*Plegadis* spp.) and large rails (*Rallus* spp.) use such pools for feeding, and many ducks rear their broods in such sites where cover is adjacent to food.

Birds of Persistent Emergents. Persistent emergents are favored by conspicuous blackbirds such as the yellow-headed blackbird (*Xanthocephalus xanthocephalus*) (Fig.4) in western marshes, and red-winged blackbirds (*Agelaeus phoenicius*) and several other blackbirds or grackles (Burger 1985, Orians 1980). Their cup-shaped nests of local plant materials are positioned below the tips of vegetation for wind protection. Sedge wrens

FIGURE 4. Male yellow-headed blackbird on territory in deep marsh zone.

FIGURE 5. Least bittern on nest in emergent vegetation.

FIGURE 6. Canada Goose nest in emergent vegetation of a small marsh near Bear River, Utah.

(*Cistothorus platensis*) nest and feed in dense grasses and sedges of shallow waters, whereas marsh wrens (*C. palustris*) use taller vegetation in more permanent water. Both build ball-shaped nests above the flooding zone and find suitable invertebrate food at water level as well as in the upright vegetation. Habitat preference of walking rails like the purple gallinule (*Porphyrula martinica*), Virginia rail (*Rallus limicola*), sora (*Porzana carolina*), and king rail (*Rallus elegans*) grade from shallow water to almost dry ground (Weller and Spatcher 1965) within the emergents. However, all are adaptable, and their precise nest site and feeding location in relation to vegetation structure seems subject to variation in water regime, vegetation density, food availability, season, and local population traditions. Herons, egrets, bitterns, and ibises may nest among or on top of robust emergents like cattail or California bullrush (*Scirpus californicus*), and some like bitterns also feed there during nesting. Such stands of vegetation are excellent waiting sites for stationary feeders like least bitterns (*Ixobrychus exilis*) (Fig. 5), great egrets (*Casmerodius albus*), and great-blue herons (*Ardea herodius*). Canada geese (*Branta canadensis*) (Fig. 6) and other waterfowl may nest on lodges built by muskrats (*Ondatra zibethicus*) of emergent vegetation.

Birds of Emergent Edges. Many birds of persistent emergents and shorter marginal nonpersistent vegetation also are conspicuous at the shallow edges of wetlands or the perimeter of large islands within a wetland. These areas tend to be dominated by walkers rather than swimmers, including king and clapper rails (*Rallus longirostris*), American bitterns *Botaurus lentiginous*), snowy egrets, and little blue herons (*Egretta caerulea*). Redwinged blackbirds prefer the edges and are less abundant in deeper water. Some species, such as aerial foragers like willow flycatchers (*Empidonax traillii*) and water or ground feeders like swamp sparrows (*Melospiza georgiana*) and northern harrier (*Circus cyaneus*) spend as much time in the adjacent uplands as at the wetland edge (Weller and Spatcher 1965, Esler 1988, McCabe 1991). Small trees and shrubs on unflooded upland sites are used for song perches or nest sites. Tree swallows (*Tachycineta bicolor*) and eastern kingbirds (*T. tyrannus*) may nest in trees overwater near shore. Other species find concentrations of important foods like insects, crayfish, and amphibians in this moist interface.

Birds of Meadows and Other Moist-soil Emergents. Seasonally flooded emergent zones typically develop plant communities of low vegetation structure, such as shorter grasses, annual forbs, and spikerush, that are used more by walking birds such as black-necked stilts (*Himanotopus mexicanus*), yellowlegs (*Tringa* spp.), killdeer (*Charadrius vociferus*), and other shorebirds. Grazing waterfowl such as Canada geese or American wigeon also walk in exposed areas foraging on shoots of new vegetation. Ibises, egrets, and herons may flock to areas temporarily flooded if large invertebrate foods, amphibians, or fish are trapped.

Birds of Sheetwater and Mudflats

Shallow sheetwater over nonvegetated or sparsely vegetated soil or sand substrate may result from rainfall, stream, or tidal flooding of mudflats and shorelines. Plowed fields and other agricultural areas may also be flooded under such circumstances, making available different foods. Open-water areas within wetlands form similar new habitats for edge birds when water levels decline below the emergent and submergent zones and expose bare basin bottoms. Resultant mudflats are moist and usually unvegetated areas generally not appreciated by those who do not recognize the distribution of invertebrates and their abundance and role as detritivores in the cycling of nutrients within wetland systems. Organic mud is the favorite habitat of a number of tiny crustaceans, nematodes, and annelids that are eaten by killdeer, American avocets (*Recurvirostra americana*), and other carnivorous foragers (Capen and Low 1980). American green-winged teal (*Anas crecca*), a species often termed "mudder" for its walking and sifting behavior, use such sites in tidal areas, in western alkaline marshes, and on river shorelines and islands. Waders often walk such flats in search of fish, amphibians, reptiles, or large invertebrates in the water.

Open sheetwater areas also may be used during migration or on wintering areas as overnight roosts by cranes and geese, and protection of these habitats is a vital management approach to capitalizing on waste grain and other adjacent food resources attractive to such granivores (Hobaugh 1984).

Birds of Shrub and Forested Wetlands

Wetland shrubs occur around a variety of wetland types in the seasonally flooded zone or in saturated soils in riparian or alpine habitats. Streamsides or drainages where wetland vegetation is apparent or where terrestrial vegetation is more abundant and diverse due to water presence are well known as important riparian areas for birds (Carothers 1977, Hehnke and Stone 1978, Thomas et al. 1979, Wauer 1977, Taylor 1986). Tree or shrub-nesting passerines and flycatchers are prominent there due to the availability of foods from the stream and from the several layers of vegetation.

Water-induced shrubs such as willow and alder occur as extensive stands or patches in cold-temperate climates at high altitude or latitude around the world. Wild and domesticated herbivores browse these shrubs, and they are important habitats for willow flycatchers, yellow warblers (*Dendroica petechia*), and numerous other passerines that exploit insect life of wetland edges or patches.

More water-tolerant shrubs like buttonbush and certain species of willow that survive flooded conditions even during the growing season are among the plant dominants in areas termed shrub-scrub swamps. Herons, egrets, and passerines often use such habitats for nest sites. These plant zones commonly are browsed by livestock in Western riparian zones where extensive vegetation is present because of the increased pools of water associated with stream flooding.

Forested wetlands include bottomland hardwoods seasonally flooded during the winter dormant period and characteristic of backwaters of floodplains along major southeastern river systems, or naturally impounded and trapped waters that are semipermanent and form true swamps. Birds of larger wetland trees commonly include not only the herons, which may nest at the tops of dead trees in or near water, but also woodpeckers that make and use cavities. Cavities in large trees may later be used by wood ducks (*Aix sponsa*) and hooded mergansers (*Lophodytes cucullatus*). Prothonotary warblers (*Protonotaria citrea*) and tree swallows use holes in old snags that must be preserved if the species are to be successful in the area.

Birds of Stockponds, Reservoirs, and Other Constructed Wetlands

Typically, these water bodies are small but are deep enough that emergent vegetation is absent or restricted to a narrow fringe. They sometimes have extensive submerged plant beds that are valuable for waterbirds. They may be important water sources for a variety of birds that feed on insects over the water or swimming birds, and add avian diversity in arid regions (Evans and Kerbs 1977). The larger reservoirs may have extensive littoral zones, especially in the upstream area, and may be excellent waterfowl habitat if they are not subject to rapid and extreme water fluctuations. Because of the abundance of fish, many are excellent areas for cormorants, pelicans, herons, and mergansers. Canada geese use them as rearing areas in summer because they graze in the uplands and seek water for brood protection. Large reservoirs may also be favored by wintering flocks of geese and mallards because they are deep and remain open during most winters, and birds feed in grain fields nearby (Simpson 1988).

WETLAND-LIVESTOCK RELATIONSHIPS

Herbaceous vegetation of wetlands represents an important resource for livestock all over the world (Maltby 1986). Most domestic animals use wetland plants at wetland edges, where productivity is high because of the ideal combination of available nutrients and water. As water levels allow, livestock may wade to feed, and consequently interact with wildlife indirectly and directly. A major influence on wetland use by livestock is water depth. Although moose (*Alces alces*), deer (Cervidae) and other ungulates around the world commonly graze in deep water, cattle seem to do so only to pursue favorite foods. In a freshwater coastal impoundment, I have seen cattle walk through moist-soil stands of seashore paspalum (*Paspalum vaginatum*) to reach chest-deep stands of the same species where new growth may have been more palatable (Fig. 7). Along marine shorelines on the Falkland Islands, I have seen cattle wade into the ocean at low tide to feed on marine algae, apparently deriving nutrients not available in terrestrial plants. The more usual pattern of livestock use of wetlands occurs during dry periods when forage is scarce, or during late summer drydowns when cattle move into the wet edges and graze on diverse annual emergents (Whyte and Silvy 1981), perennial emergents (Fig. 8) (Weller et al. 1958, Chabreck 1968, Neckles et al. 1985), or browse on willows and other shrubs (Taylor 1986, Chaney et al. 1990). In some instances, water may be managed to encourage productivity of

FIGURE 7. Livestock moving from a gradually sloping edge into shallow water of a Coastal Texas impoundment to reach new growth of *Paspalum vaginatum* during late winter, San Bernard National Wildlife Refuge, Brazoria, Texas.

FIGURE 8. Livestock trails into emergent vegetation during low water, Bear River Marsh area, Utah.

hydrophytes and the area then dewatered to allow grazing or mowing (Neckles et al. 1985).

Some Effects of Grazing on Vegetation and Basin Structure

Grazing impacts vegetation structure and can have beneficial effects in some cases and detrimental effects in others. Most commonly, herbaceous vegetation is shorter and less dense (Chabreck 1968), and total biomass is reduced (Reimold et al. 1975). However, in whitetop grass meadows in the northcentral prairies, light grazing opens the vegetation and allows increased production, as does periodic burning or mowing (Neckles et al. 1985). Intense grazing reduces residual litter for the next spring, which can impact nesting cover for waterfowl and other birds (Bue et al. 1952, Gjersing 1975, Mundinger 1976, Braun et al. 1978). Reduced waterfowl nesting cover as a result of grazing near stockponds was noted in South Dakota (Bue et al. 1952)(Fig. 9). Impacts typically are most serious along shorelines (Hoffman and Stanley 1978), where removal of vegetation can increase sediment exposure and potentially increase water turbidity, thus decreasing submergent vegetation important as food and substrate for waterbirds (Robel 1961). Woody species like willows often are impacted in height and volume by grazing (Knopf and Cannon 1981, Taylor 1986). In addition, soil compaction may reduce water availability to streamside plants (Kauffman and Krueger 1984).

Grazing can modify basin depth through compaction and removal of soil on livestock legs and hooves, as noted on trails crossing wetlands (Weller 1967). Trampling by cattle can influence cover-water ratios by breaking up dense cover, a process most wildlife biologists have found beneficial to waterfowl and other waterbirds (Kantrud 1986). In boggy areas where tussock-forming plants may be prominent, intense grazing results in destruction of the tussock heads and conversion to meadow plants (Costello 1936). As a result, there is great danger of losing the physical structure essential for plants or animals with specialized microhabitats.

Grazing Impacts on Plant Species Composition of Wetlands

As with intensive grazing in terrestrial plant communities, there is evidence that grazing influences species composition through modification of processes that influence long-term plant succession (Chabreck 1968, Reimold et al. 1975). In coastal marshes, some plants respond positively to grazing (increasers) and others decrease. Concurrently, invaders may enter the system. A common post-grazing response is the establishment of monocultures, such as common reed (*Phragmites communis*), which then dominate the area for many years (Larsson 1969). In some studies, plant species richness has decreased with grazing, as has litter production and plant distribution patterns (Andresen et al. 1990). In other studies, diversity has increased (Bakker and Ruyter 1981). Whether these effects are positive or negative to wetland birds depends on the species and the time of year.

Woody plants may be eliminated by intensive grazing in some areas, so that grazed lands may be more open. This may impact plant succession and also bird species composition

FIGURE 9. Damage to waterfowl nesting habitat resulting from cattle trampling and feeding in *Scirpus olneyi*, Bear River Marsh area, Utah.

(Ulfvens 1991). Recovery is rapid in some species and areas, with willows becoming reestablished within 10 years after grazing ceases (Rickard and Cushing 1982). In areas where grazing intensity is greater or condition of the willows poorer, recovery of growth rate and form is less apparent (Knopf and Cannon 1981). Rest-rotation and reduced stocking rates help maintain the structure and diversity of riparian areas (Kauffman and Krueger 1984).

Invertebrate Production, Birds, and Livestock

Invertebrates are important foods for prebreeding, nesting, and growing birds (Bartonek and Hickey 1969, Swanson and Meyer 1977, Krapu and Rieneke 1992), and limited data are available on factors that influence invertebrate production, enhancement, use, or reduction. Many major foods of ducks are invertebrate detritivores such as chironomids, amphipods, and snails (gastropoda) more commonly associated with the decomposition of litter (Swanson and Meyer 1977, Kaminski and Prince 1981, Murkin and Kadlec 1986). Birds shift diets and respond to available invertebrates that in turn result from natural wetland decomposition cycles (Swanson and Meyer 1977). Experimental manipulations have shown that invertebrates respond quickly to detritus produced by mowing (Kaminski and Price 1981). Thus, cattle trampling in vegetation might result in increased detrital materials and digestive excrements when water is present, and potentially increased invertebrate populations. This inference is supported by studies of grazing in salt marsh communities in Holland where there was a shift from herbivorous invertebrates to detrital feeders where grazing and trampling modified vegetation (Andresen et al. 1990). Species richness, densities and community diversity of invertebrates also increased when grazing ceased (Andresen et al. 1990). In addition to the functional value of a particular wetland for food resources, these examples point out the importance of retaining complexes of wetlands that may be in different phases of vegetation succession, decomposition, and productivity (Nelson and Kadlec 1984).

WATERBIRD-LIVESTOCK RELATIONSHIPS

Grazing has the potential to modify avian habitats. Birds and other wildlife have evolved with grazing animals like deer, moose, and bison (*Bison bison*) and show little concern except when large numbers invade small areas. Even direct damage to nests seems rarely recorded. Some birds like the recently arrived cattle egret (*Bubulcus ibis*) are commensals that feed with livestock more often than not (Telfair 1983). It is the damage to habitat that is most common and creates the major problems and conflicts in jointly managing wetlands for wildlife and livestock. From the livestock manager's viewpoint, competition for food may be an issue when geese move from wetlands to meadows or even uplands to graze. Considerable work has been done on competition of sheep and sheldgeese (*Chloephaga* spp.) in the Falkland Islands where impacts seem locally important but difficult to control (Summers and Dunnet 1984).

Livestock Impacts on Bird Communities and Populations

Numerous observations of birds on grazed versus ungrazed areas have been reported (Skovlin 1984), and there have been some comparative experimental studies. Most demonstrate sizable differences between the avian community (i.e., diversity, composition, biomass) or population parameter (i.e., abundance, biomass) of grazed versus ungrazed wetlands, but reasons for the differences are not always obvious. In most studies, grazing modifies bird species composition in all wetland types (Larsson 1969, Medin and Clary 1990, Ulfvens 1991). Species richness or a mathematical diversity index may be higher in grazed than ungrazed lands, as noted in riparian habitats in Idaho (Medin and Clary 1990) and on Swedish meadowlands (Larsson 1969), but that pattern was not evident on Finnish islands (Ulfvens 1991) and willow-dominated riparian zones (Taylor 1986). Where habitats differ greatly due to grazing, cause-effect relationships between bird species composition, plant species composition, and the presence and depth of water (Larsson 1969) are not readily obvious. Population size or breeding-bird density also varied in these studies, but it is evident that increases on grazed areas often are a result of a shift of major habitat types (e.g., from woodland to marsh or to meadows with flocking birds). Birds that require taller herbaceous vegetation or dense litter tend to be reduced. Grazing impacts on plant foods used by herbivorous geese and ducks (Owen 1982) and cranes (Hunt 1987) have been observed in several coastal areas and in the playa lakes region of Texas (Cook and Smith 1985). Where rare species are involved, management must favor the bird to the exclusion of grazing; in other cases, dual use seems compatible (Larsson 1969, Holder et al. 1980).

Most work on effects of grazing on breeding ducks has been on Prairie Pothole wetlands of the northern Great Plains and Canada. Early studies suggested that reducing vegetation density by grazing was beneficial to blue-winged teal (Bennett 1937, Glover 1956, Burgess et al. 1965) or at least not harmful. While recognizing this tendency, Kaiser et al. (1979) emphasized that only grazed rangeland in good condition showed increased nest density and success. Other studies have noted lower nest density or nest success on grazed versus ungrazed land (Kirsch 1969, Higgins et al. 1992). Braun et al. (1978) noted that approximately 55 studies had suggested negative effects of grazing on waterfowl populations in some way. Studies on shorelines of wetlands indicated increased nest trampling by livestock (Gjersing 1975) and excessive grazing reduced use of ponds by broods of ducks (Kantrud 1986). However, some positive effects of the removal of shoreline vegetation were noted by Keith (1961). Impacts of grazing differ by duck species, and obviously species that favor dense terrestrial vegetation during nesting are most affected (e.g. dabbling ducks versus inland diving ducks). Insufficient data exist currently to classify species by impact, but overgrazing clearly must be avoided. Moreover, on lands dedicated to wildlife, some workers feel that use of grazing should be viewed from the standpoint of how it contributes to wildlife management goals, rather than accepting the risk of major plant community changes for economic gains (Kirby et al. 1992).

Grazing Management Systems and Wetland Birds

Although relatively little experimental work has been reported on the advantages or disadvantages of various grazing systems for nesting of wetland birds, several studies provide some important insights and suggest directions for future research. Waterfowl production associated with small reservoirs on 2 rest-rotation grazing areas in Montana was greater than on control areas with continuous grazing (Gjersing 1975). Fall grazing was an especially important negative correlate with spring breeding populations, due perhaps to litter removal. Grazing during nesting further impacted nesting efforts and success. In a follow-up study, data on residual vegetation were related to duck pair and brood use. Lack of grazing (rest stage) during the previous year was identified as the most important factor contributing to increased waterfowl use and success (Gjersing 1975).

In an experimental study involving comparisons of 5 grazing systems with ungrazed prairie in North Dakota, more duck nests were found on idle than on grazed land but nesting success usually was higher on 1 of several grazing systems (Sedivec et al. 1990). Waterfowl production also was greater in 1 of 5 specialized grazing systems rather than in season-long grazing, because delaying grazing probably enhanced residual vegetation important to nest success. Twice-over grazing initiated in late May was recommended for optimal production of livestock and ducks and sage grouse (*Centrocercus urophasianus*), and short-duration grazing was second best in duck production in these trials.

Comparisons of different fall and early winter grazing pressures were made in a rest-rotation system at the Monte Vista National Wildlife Refuge in Colorado designed to open up dense cover for nesting waterfowl. Burning was used after grazing to open areas where livestock had not sufficiently reduced cover. Data gathered over a 15-year period were used to calculate nest success in grazed compared to ungrazed areas (D. R. Anderson, U. S. Fish and Wildl. Serv., pers. comm.).

Thus, different conclusions have been reached depending on area, time, and species of waterfowl, and most other environmental variables, like water regimes that cannot be controlled for. The effects of moderate to light grazing on breeding waterfowl will differ among wetlands and with bird species, but some vegetation effects must be expected. Availability of residual vegetation in early spring is especially important for early nesting ducks like northern pintails (*Anas acuta*) and mallards (Keith 1961), and for brood use (Rumble and Flake 1983). Timing and intensity of grazing must be manipulated accordingly (Sedivec et al. 1990). A regular monitoring and evaluation program can alleviate serious impacts from grazing excessively during the dynamic climatic changes common to many grazing areas (Weller 1986).

USE OF LIVESTOCK IN VEGETATION MANAGEMENT

Grazing of wetlands is a common management tool on refuges and other wild lands (Bossenmaier 1964, Weller 1989) in North America where it supplements water regulation as a vegetation control method (Linde 1985). It is an ancient and worldwide custom, and is still very much in use on selected large wetlands in Europe to exploit the resources of wetlands and control vegetation. In a well-known case in the Camargue of France, manipulating water levels and placement and density of grazing horses reduced common reed and greatly increased diversity of waterfowl foods more economically than traditional mowing techniques (Duncan and d'Herbes 1982).

Especially on the Gulf Coast in the United States, and on a smaller scale on the east coast, livestock are used in combination with burning as part of the vegetation management system. Both on private lands and governmental-owned management areas, grazing is used to reduce dense stands of emergents so that snow geese (*Anser caerulescens*) will enter areas and feed on the tubers of three-square bulrush (*Scirpus olneyi*), saltgrass (*Distichlis spicata*), and cordgrass (Glazener 1946, Stutzenbaker and Weller 1989). White-fronted geese (*A. albifrons*) feed also in wet areas on emergents, but use more leafy shoots and fewer tubers.

SUMMARY

Wetlands often are small but functional components of rangelands responsible for great avian diversity and biological productivity. Different bird species and sometimes higher taxa have specialized on various wetland zones and vertical strata, and these habitats dictate avian diversity and abundance. Wetlands also provide food and water resources for livestock. Although many species coexist with livestock, there is good evidence that even moderate grazing can reduce bird species richness, population size, and nest success of waterfowl or other species that depend on wetlands but nest in adjacent upland vegetation directly affected by livestock. Some exceptions suggest that diverse uses such as waterbird production and livestock grazing can be compatible if grazing is moderate and residual cover is left for spring-nesting birds. Recreational uses of rangelands and other grazing areas are increasing, and livestock/waterbird combinations provide hunting and bird viewing/photography opportunities and traditional rangeland uses. Use of grazing as a tool on wetland wildlife areas should be viewed as an alternative considered in relation to goals. In any case, grazing of wetlands must be managed with concern, knowledge, planning, and regular monitoring to protect the integrity of wetland functions and enhance production of waterbirds, endangered species, and other species of concern.

LITERATURE CITED

Adamus, P. R., L. T. Stockwell, E. J. Clairain Jr., M. E. Morrow, L. P. Rozas, and R. D. Smith. 1991. Wetland evaluation technique (WET), Vol. 1: literature review and evaluation rationale. U.S. Army Corps Eng., Environ. Lab., Vicksburg, Miss. 240pp.

Andresen, H., J. P. Bakker, M. Brongers, B. Heydemann, and U. Irmler. 1990. Long-term changes of salt marsh communities by cattle grazing. Vegetatio 89:137-148.

Bakker, J. P., and J. C. Ruyter. 1981. Effects of five years of grazing on a salt-marsh vegetation. Vegetatio 44:81-100.

Bartonek, J. C., and J. J. Hickey. 1969. Food habits of canvasbacks, redheads, and lesser scaup in Manitoba. Condor 71:280-290.

Bennett, L. J. 1937. Grazing in relation to nesting of the blue-winged teal. Trans. North Am. Wildl. Conf. 2:393-397.

Bergman, R. D. 1973. Use of southern boreal lakes by postbreeding canvasbacks and redheads. J. Wildl. Manage. 37:160-170.

Bossenmaier, E. 1964. Cows and cutter bars. Pages 627-634 *in* J. P. Linduska, ed. Waterfowl tomorrow. U. S. Gov. Printing Off., Washington, D.C.

Braun, C. E., K. W. Harmon, J. A. Jackson, and C. D. Littlefield. 1978. Management of national wildlife refuges in the United States: its impacts on birds. Wilson Bull. 90: 309-321.

Bue, I. G., L. Blankenship, and W. H. Marshall. 1952. The relationship of grazing practices to waterfowl breeding populations and production on stock ponds in western South Dakota. Trans. North Am. Wildl. and Nat. Resour. Conf. 17:396-414.

Burger, J. 1985. Habitat selection in temperate marsh-nesting birds. Pages 253-281 *in* M.L. Cody, ed. Habitat selection in birds. Academic Press, Orlando, Fla.

Burgess, H. H., H. Prince, and D. L. Trauger. 1965. Blue-winged teal nesting success in relation to land use. J. Wildl. Manage. 29:89-95.

Capen, D. E., and J. B. Low. 1980. Management considerations for nongame birds in western wetlands. Pages 67-77 *in* Workshop proceedings: management of western forests and grasslands for nongame birds. U.S. Dep. Agric. For. Serv. Gen. Tech. Rep. INT-86.

Carothers, S. W. 1977. Importance, preservation, and management of riparian habitats: an overview. Pages 2-4 *in* Importance, preservation and management of riparian habitat. U.S. Dep. Agric. For. Serv. Gen. Tech. Rep. RM-43.

Chabreck, R. H. 1968. The relation of cattle and cattle grazing to marsh wildlife and plants in Louisiana. Proc. Southeast. Assoc. Game and Fish Comm. 22:55-58.

Chaney, E., W. Elmore, and W. S. Platts. 1990. Livestock grazing on western riparian areas. Northwest Region Inf. Cent. and U.S. Environ. Prot. Agency. 45pp.

Cook, D. G., and L. M. Smith. 1985. Effects of burning and grazing on playa wetland vegetation and associated wildlife use. Dep. Range and Wildl., Texas Tech. Univ., Lubbock. Research Highlights 16:32-33.

Costello, D. F. 1936. Tussock meadows in Wisconsin. Bot. Gaz. 97:610-647.

Cowardin, L. M., V. Carter, F. C. Golet, and E. T. LaRoe. 1979. Classification of wetlands and deepwater habitats of the United States. U.S. Fish and Wildl. Serv. FWS/OBS-79/31. Washington, D.C. 103pp.

DeGraaf, R. M., N. G. Tilghman, and S. H. Anderson. 1985. Foraging guilds of North American birds. Environ. Manage. 9:493-536.

Duebbert, H. F. 1982. Nesting of waterfowl on islands in Lake Audubon, North Dakota. Wildl. Soc. Bull. 10:232-237.

Duncan, P., and J. M. d'Herbes. 1982. The use of domestic herbivores in the management of wetlands for waterbirds in the Camarque, France. Pages 51-66 *in* D. A. Scott, ed. Managing wetlands and their birds. Int. Wildlfowl Res. Bur., Slimbridge, U. K.

Esler, D. 1988. Avian associations with hydrilla on Lake Fairfield, Texas. M.S. Thesis, Texas A&M Univ., College Station. 90pp.

Evans, K. E., and R. R. Kerbs. 1977. Avian use of livestock watering ponds in western South Dakota. U.S. For. Serv. Gen. Tech. Rep. RM-35. 11pp.

Federal Interagency Committee for Wetland Delineation. 1989. Federal manual for identifying jurisdictional wetlands. U.S. Army Corps Eng., U.S. Environ. Prot. Agency, U.S. Fish and Wildl. Serv., and U.S. Dep. Agric. Soil Conserv. Serv., Washington D.C. Coop. Tech. Publ. 76pp.

Fredrickson, L. H., and T. S. Taylor. 1982. Management of seasonally flooded impoundments for wildlife. U. S. Fish and Wildl. Serv. Resourc. Publ. 148. 29pp.

Gjersing, F. M. 1975. Waterfowl production in relation to rest-rotation grazing. J. Range Manage. 28:37-42.

Glazener, W. C. 1946. Food habits of wild geese on the Gulf Coast of Texas. J. Wildl. Manage. 10:322-329.

Glover, F. A. 1956. Nesting and production of the blue-winged teal (*Anas discors* Linnaeus) in Northwest Iowa. J. Wildl. Manage. 20:28-46.

Harris, S. W. and W. H. Marshall. 1963. Ecology of water-level manipulations of a northern marsh. Ecology 44:331-342.

Hehnke, M., and C. P. Stone. 1978. Value of riparian vegetation to avian populations along the Sacramento River system. Pages 228-235 *in* R. R. Johnson and J. F. McCormick, tech. coords. Strategies for protection and management of floodplain wetlands and other riparian ecosystems. Proc. symposium U. S. Dep. Agric. For. Serv. Gen. Tech. Rep WO-12.

Higgins, K. F., L. M. Kirsch, A. T. Klett, and H. W. Miller. 1992. Waterfowl production on the Woodworth Station in south-central North Dakota, 1965-1981. U. S. Fish and Wildl. Serv. Resourc. Publ. 180. 79pp.

Hobaugh, W. C. 1984. Habitat use by snow geese wintering in southeast Texas. J. Wildl. Manage. 48:1085-1096.

Hoffman, G. R., and L. D. Stanley. 1978. Effects of cattle grazing on shore vegetation of fluctuating water level reservoirs. J. Range Manage. 31:412-416.

Holder, G. L., M. K. Johnson, and J. L. Baker. 1980. Cattle grazing and management of dusky seaside sparrow habitat. Wildl. Soc. Bull. 8:105-109.

Hunt, H. E. 1987. The effects of burning and grazing on habitat use by whooping cranes and sandhill cranes on the Aransas National Wildlife Refuge, Texas. Ph.D. Thesis, Texas A&M Univ., College Station. 173pp.

James, F. C., and S. Rathbun. 1981. Rarefaction, relative abundance, and diversity in avian communities. Auk 98:785-800.

Kaiser, P. H., S. S. Berlinger, and L. H. Fredrickson. 1979. Response of blue-winged teal to range management on waterfowl production areas in southeastern South Dakota. J. Range Manage. 32:295-298.

Kaminski, R. M., and H. H. Prince. 1981. Dabbling duck and aquatic macroinvertebrate responses to manipulated wetland habitat. J. Wildl. Manage. 45:1-15.

Kantrud, H. A. 1986. Effects of vegetation manipulation on breeding waterfowl in prairie wetlands: a literature review. U. S. Fish Wildl. Serv., Fish Wildl. Tech. Rep. 3. 15pp.

Kauffman, J. B., and W. C. Krueger. 1984. Livestock impacts on riparian systems and streamside management implications . . . a review. J. Range Manage. 37:430-438.

Keith, L. B. 1961. A study of waterfowl ecology on small impoundments in southeastern Alberta. Wildl. Monogr. 6. 88pp.

Kirby, R. E., J. K. Ringelman, D. R. Anderson, and R. S. Soijda. 1992. Grazing on National Wildlife Refuges: do the needs outweigh the problems? Trans. North Am. Wildl. and Nat. Resourc. Conf. 57:611-626.

Kirsch, L. M. 1969. Waterfowl production in relation to grazing. J. Wildl. Manage. 33:821-828.

Knopf, F. L., and R. W. Cannon. 1981. Structural resilience of a willow riparian community to changes in grazing practices. Pages 198-207 *in* J. M. Peek and P. D. Dalke, eds. Wildlife-livestock relationships symposium: proceedings 10. Univ. Idaho, For., Wildl., and Range Exp. Stn., Moscow.

Krapu, G. L. and K. J. Reineke. 1992. Foraging ecology and nutrition. Pages 1-29 *in* B. D. J. Batt, A. D. Afton, M. G. Anderson, C. D. Ankney, D. H. Johnson, J. A. Kadlec, and G. L. Krapu, eds. Ecology and management of breeding waterfowl and wetlands. Univ. Minnesota Press, Minneapolis.

Larsson, T. 1969. Land use and bird fauna on shores in southern Sweden. Oikos 20:136-155.

Linde, A. F. 1985. Vegetation management in water impoundments: alternatives and supplements to water-level control. Pages 51-60 *in* M. D. Knighton, ed., Water impoundments for wildlife:a habitat management workshop. U.S. For. Serv. Gen. Tech. Rep. NC-100.

Maltby, E. 1986. Waterlogged wealth. Earthscan Paperbacks, Int. Inst. Environ. and Dev., London and Washington D.C. 200pp.

McCabe, R. A. 1991. The little green bird:ecology of the willow flycatcher. Rusty Rock Press, Madison, Wisc. 171pp.

Medin, D. E., and W. P. Clary. 1990. Bird and small mammal populations in a grazed and ungrazed riparian habitat in Idaho. U.S. Dep. Agric. For. Serv. Res. Pap. INT-425. 8pp.

Mundinger, J. G. 1976. Waterfowl response to rest-rotation grazing. J. Wildl. Manage. 40:60-68.

Murkin, E.J. , H. R. Murkin, and R. D. Titman. 1992. Nectonic invertebrate abundance and distribution at the emergent vegetation-open water interface in the Delta Marsh, Manitoba, Canada. Wetlands 12:45-52.

Murkin, H. R., and J. Kadlec. 1986. Responses by benthic macroinvertebrates to prolonged flooding of marsh habitat. Can. J. Zool. 64:65-72.

———, R. M. Kaminski, and R. D. Titman. 1982. Responses by dabbling ducks and aquatic invertebrates to an experimentally manipulated cattail marsh. Can. J. Zool. 60:2324-2332.

Neckles, H. A., J. W. Nelson, and R. L. Pedersen. 1985. Management of white-top (*Scholochloa festucacea*) marshes for livestock forage and wildlife. Delta, Manit., Canada. Delta Waterfowl Res. Stn. Tech. Bull. 1. 12pp.

Nelson, J. W., and J. A. Kadlec. 1984. A conceptual approach to relating habitat structure and macroinvertebrate production in freshwater wetlands. Trans. North Am. Wildl. and Nat. Resour. Conf. 49:262-270.

Orians, G. H. 1980. Some adaptations of marsh-nesting blackbirds. Monogr. in Pop. Biol. 14. Princeton Univ. Press, Princeton, N.J. 295pp.

Odum, E. P. 1983. Commentary: wetlands and their values. Natl. Wetl. Newsl. 5(3):2.

Owen, M. 1982. Management of summer grazing and winter disturbance on goose pasture at Slimbridge, England. Pages 67-72 *in* D. A. Scott, ed. Managing wetlands and their birds, Int. Wildlfowl Res. Bur., Slimbridge, U. K.

Reimold, R. J., R. A. Linthurst, and P. L. Wolf. 1975. Effects of grazing on a salt marsh. Biol. Conserv. 8:105-125.

Rickard, W. H., and C. E. Cushing. 1982. Recovery of streamside woody vegetation after exclusion of livestock grazing. J. Range Manage. 35:360-361.

Robel, R. J. 1961. Water depth and turbidity in relation to growth of sago pondweed. J. Wildl. Manage. 25:436-438.

Rumble, M. A., and L. D. Flake. 1983. Management considerations to enhance use of stock ponds by waterfowl broods. J. Range Manage. 36:691-693.

Sedivec, K. K., T. A. Messmer, W. T. Barker, K. F. Higgins, and D. R. Hertel. 1990. Nesting success of upland nesting waterfowl and sharp-tailed grouse in specialized grazing systems in Southcentral North Dakota. Pages 71-91 *in* K. E. Severson, Coord. Can livestock be used as a tool to enhance wildlife habitat? U. S. Dep. Agric. For. Serv. Gen. Tech. Rep. RM-194.

Short, H. L., and K. P. Burnham. 1982. Technique for structuring wildlife guilds to evaluate impacts on wildlife communities. U.S. Fish and Wildl. Serv. Spec. Sci. Rep. Wildlife 44. Washington, D.C. 34pp.

Simpson, S. G. 1988. Use of the Missouri River in South Dakota by Canada geese in fall and winter, 1953-1984. Pages 529-540 *in* M. W. Weller, ed., Waterfowl in winter. Univ. Minnesota Press, Minneapolis.

Skovlin, J.M. 1984. Impacts of grazing on wetlands and riparian habitat: A review of our knowledge. Pages 1001-1103 *in* Natl. Res. Counc., Natl. Acad. Sci. Developing strategies for rangeland management: a report prepared by the committee on developing strategies for rangeland management. Westview Press, Boulder, Colo.

Sprunt, A. IV, J. C. Ogden, and S. Winckler. 1978. Wading birds. Natl. Audubon Soc. Res. Rep. 7. 381pp.

Stewart, R. E., and H. A. Kantrud. 1971. Classification of natural ponds and lakes in the glaciated prairie region. Bur. Sports Fish. and Wildl. Res. Publ. 92. 57pp.

Stutzenbacker, C. D., and M. W. Weller. 1989. The Texas Coast. Pages 385-405 *in* L. M. Smith, R. L.Pederson, and R. M. Kaminski, eds. Habitat management for migrating and wintering waterfowl in North America. Texas Tech. Univ. Press, Lubbock.

Summers, R. W., and G. M. Dunnet. 1984. Sheld-geese and man in the Falkland Islands. Biol. Conserv. 30:319-340.

Swanson, G. A., and M. L. Meyer. 1977. Impact of fluctuating water levels on feeding ecology of breeding blue-winged teal. J. Wildl. Manage. 41:426-433.

Taylor, D. M. 1986. Effects of cattle grazing on passerine birds nesting in riparian habitat. J. Range Manage. 39:254-258.

Telfair, R. C., II. 1983. The cattle egret: a Texas focus and a world view. Kleberg Studies in Nat. Res. Dep. Wildl. and Fish. Sci., Texas A&M Univ., College Station. 144pp.

Thomas, J. W., C. Maser, and J. E. Rodiek. 1979. Wildlife habitats in managed rangelands: the Great Basin of southeastern Oregon. U.S. Dep. Agric. For. Serv. Gen. Tech. Rep. PNW-80. 18pp.

Tiner, W. W. Jr. 1984. Wetlands of the United States; current status and recent trends. U.S. Fish and Wildl. Serv., Natl. Wetlands Inventory, Washington, D.C. 59pp.

Ulfvens, J. 1991. The shore- and waterbirds on some grazed and ungrazed islands on the Finnish west coast. Ornis Fennica 68:26-32.

United States Environmental Protection Agency. 1991. Federal manual for identifying jurisdictional wetlands, proposed revisions. U.S. Environ. Prot. Agency, U.S. Army Corps Eng., U.S. Fish and Wildl. Serv., and U.S. Soil Conserv. Serv. Washington D.C. 106pp.

van der Valk, A. G. 1981. Succession in wetlands: A Gleasonian approach. Ecology 62:688-696.

Wauer, R. H. 1977. Significance of Rio Grande riparian systems upon the avifauna. Pages 165-174 *in* R. R. Johnson, and D. A. Jones, tech. coord. Importance, preservation, and management of riparian habitat: a symposium. U.S. Dep. Agric. For. Serv. Gen. Tech. Rep. RM-43.

Weller. M. W. 1967. Notes on some marsh birds of Cape San Antonio, Argentina. Ibis 109:391-4.

———. 1986. Marshes. Pages 201-224 *in* A. Y. Cooperrider, R. J. Boyd, and H. R. Stuart, eds. Inventory and monitoring of wildlife habitat. U.S. Bur. Land Manage., Denver, Colo.

———. 1989. Waterfowl management techniques for wetland enhancement, restoration and creation useful in mitigation procedures. Pages 105-116 *in* J. A. Kusler and M. E. Kentula, eds. Wetland creation and restoration: the status of the science. Environ. Prot. Agency Publ. Vol. 2, EPA/600/3-89/038.

———, and L. H. Fredrickson. 1974. Avian ecology of a managed glacial marsh. Living Bird 12:269-291.

———, and C. E. Spatcher. 1965. Role of habitat in the distribution and abundance of marsh birds. Iowa State Univ. of Agric. & Home Econ. Exp. Stn. Spec. Rep. 43. 31pp.

———, B. H. Wingfield, and J. B. Low. 1958. Effects of habitat deterioration on bird populations of a small Utah marsh. Condor 60:220-226.

Whyte, R. J., and N. J. Silvy. 1981. Effects of cattle on duck food plants in southern Texas. J. Wildl. Manage. 45:512-515.

Woodin, M. C., and G. A. Swanson. 1989. Foods and dietary strategies of prairie-nesting ruddy ducks and redheads. Condor 91:280-287.

Chapter 8
RODENTS AND LAGOMORPHS

Kathleen A. Fagerstone and Craig A. Ramey

INTRODUCTION

Historically, large numbers of wild animals ranged freely over western rangelands of North America; however, the westward expansion of European colonists has gradually brought most of this region under varying degrees of agricultural management. Large scale ranching and irrigated farming displaced buffalo (*Bison bison*) and other wildlife in many areas (Conklin 1928, Fuller and Hafen 1957). This has changed wildlife habitat by producing areas of monocultural crops and emphasizing special range management requirements for livestock. Until the 1840's there was no significant livestock grazing in the West. Today grazing by livestock is a common and economically important practice through much of the western United States, with an estimated 70% of the land in 11 western states subjected to grazing at some time during the year (Mackie 1978). Livestock management practices have been in a rapid state of evolution since the 1800s; from simply placing livestock on a range until the forage was gone, to development of increasingly sophisticated systems of pasture deferment, pasture rotation, and habitat modifications that are designed to stimulate plant productivity, improve efficiency in forage utilization, and avoid overgrazing.

For many years, scientists have been investigating these new interactions among man, wildlife, and rangeland, yet there are relatively few definitive studies of the effects of grazing or of selective grazing systems on rodents and lagomorphs. Although these small mammals are significant wildlife components in all rangeland habitats in North America, research on them has lagged behind more obvious faunal components such as big game and game birds. We know relatively little about the distribution and population abundance of small mammals prior to the onset of grazing or about the effects of grazing on small mammal communities. Small mammals have not generally been of concern to range managers, partially because the empirical data on habitat requirements, species interactions, and life history patterns needed for their proper rangeland management are often lacking. Yet, basic information about the relationships of small mammal species to their habitats and vegetational physiognomy is critical to making informed decisions about further habitat modifications, wildlife management, and rangeland production. Such information would aid development of effective grazing programs where small mammals are a management concern. This chapter will provide information on the known impacts of grazing to rodents and lagomorphs.

The chapter will also discuss ways in which small mammals influence rangelands. In most natural systems, direct impact on vegetation by small mammal consumption is relatively slight because, despite high reproduction rates, increases in small mammal population densities are not rapid enough to take full advantage of the seasonal increase in food supplies (Hayward and Phillipson 1979). And not all primary production in an ecosystem is suitable for or available to small mammals. Grodzinski (1968) defined available primary production as "that food which is easy to find and is being chosen and eaten." Herbivores in most ecosystems remove a very small amount (< 10%) of plant production (Chew and Chew 1970, French et al. 1976), but the proportion of the net primary production available as food to rodents is greatest in grassland ecosystems (Hayward and Phillipson 1979), where rodents commonly remove 30 to 50% of aboveground net primary production (Lacey and Van Poollen 1981, McNaughton 1985). However, consumption alone is a poor index of impact, as the consequences of consumption may exceed the actual amount eaten (Petrusewicz and Macfadyen 1970). Hayward and Phillipson (1979) estimated that the impact of small mammals in grasslands due to consumption may be only 5% of their actual total impact. The indirect effects of grazing by small mammals may include stimulation of production, alteration of plant species composition, changes in plant stature and reproduction, and influence on rates of nutrient cycling and decomposition. Which of these alternatives occurs is dependent on the species of small mammal involved and its population density (Batzli 1975). Considerable data are available on a few species of large rodents and lagomorphs in the western United States. Throughout western North America, jackrabbits (*Lepus* spp.), ground squirrels (*Spermophilus* spp.), and burrowing rodents such as pocket gophers (*Thomomys* spp. and *Geomys* spp.) are common rangeland residents. Prairie dogs (*Cynomys* spp.) reach considerable numbers locally. Smaller and less visible rodents such as voles (*Microtus* spp.) are widely distributed on western rangelands, and in the Southwest kangaroo-rat (*Dipodomys* spp.) populations sometimes become large. Less data are available on other smaller species of rodents and lagomorphs.

PRAIRIE DOGS

Prairie dogs are large burrowing rodents classified into the black-tailed and the white-tailed groups. Four species of prairie dogs occur in the United States. Black-tailed prairie dogs (*Cynomys ludovicianus*) have reddish-brown fur and black tips on their tails. The 3 species of white-tailed prairie dogs (i.e., the white-tailed prairie dog [*C. leucurus*], Gunnison's prairie dog [*C. gunnisoni*], and the Utah prairie dog [*C. parvidens*]) have a yellowish coloration to their fur, white tips on their tails, and dark eyebrows. All of the prairie dog species are diurnal, colonial, and social. Female prairie dogs have only 1 estrous cycle/year, in midwinter. Following about a 30-day gestation period 3-5 young are born in March or April and venture above ground at about 4 or 5 weeks of age. Despite their similarities, the ecology, distribution, and associations with rangeland are quite different for black-tailed and white-tailed prairie dogs. Differences can be attributed to differences in population density and habitat occupied by the two groups. Of the 4 species of prairie dogs, only the black-tailed prairie dog has been studied extensively.

Black-tailed Prairie Dogs

Historical Populations. Black-tailed prairie dogs (Fig. 1) occur throughout the Great Plains from the United States-Canada border to northern Mexico and from the Rocky Mountains east to Nebraska, Kansas, and Oklahoma (Fig. 2). They occur at elevations ranging from 915 to 1,830 m. The black-tailed prairie dog is typically associated with short and midgrass prairie; grasses typically found in the vicinity of prairie dog colonies include western wheatgrass (*Agropyron smithii*), buffalograss (*Buchloe dactyloides*), and blue grama (*Bouteloua gracilis*). Prairie dogs dig extensive burrow systems and create large earth mounds 15-20 cm high around the burrow. Burrows form a highly modified environment where temperatures are fairly stable and relative humidity is much higher than on the surface. Black-tailed prairie dogs are highly social and reside in densely populated colonies (King 1955, Koford 1958, Hoogland 1979), where densities average 15 individuals/ha (range = 5-33/ha). Black-tailed prairie dogs do not hibernate in winter, but may remain underground for long periods of time.

Prairie dogs were widespread on the plains through the 1800s. Seton (1929) estimated that in the late 1800s, prairie dog colonies covered 283 million ha and prairie dogs numbered over 5 billion, often occurring in large colonies covering several km^2 (Merriam 1902, Hollister 1916, Bell 1921). Merriam (1902) reported 1 prairie dog colony in Texas that covered 64,750 km^2, and had about 400 million prairie dogs. During this period, prairie dogs probably had a profound effect on rangeland. With settlement of the plains regions, conflicts arose with man as prairie dogs destroyed agricultural crops planted on former rangeland and were thought to be in direct competition with livestock for rangeland forage. Merriam (1902) estimated that the quantity of grass consumed by 32 prairie dogs equaled that required by 1 sheep, and that the amount eaten by 256 prairie dogs was sufficient to support 1 cow. Black-tailed prairie dogs have therefore been considered pests on shortgrass and mixed-grass prairies of central North America since the late 1800s (Merriam 1902).

To control prairie dog numbers, effective rodenticides were developed and in the early 1900s millions of hectares were treated with strychnine and other poisoned grains, significantly reducing prairie dog numbers and eliminating most large colonies (Fig. 3). By 1919, after 20 years of control efforts, the

FIGURE 1. Black-tailed prairie dog (*Cynomy ludovicianus*). Photo by K. Fagerstone.

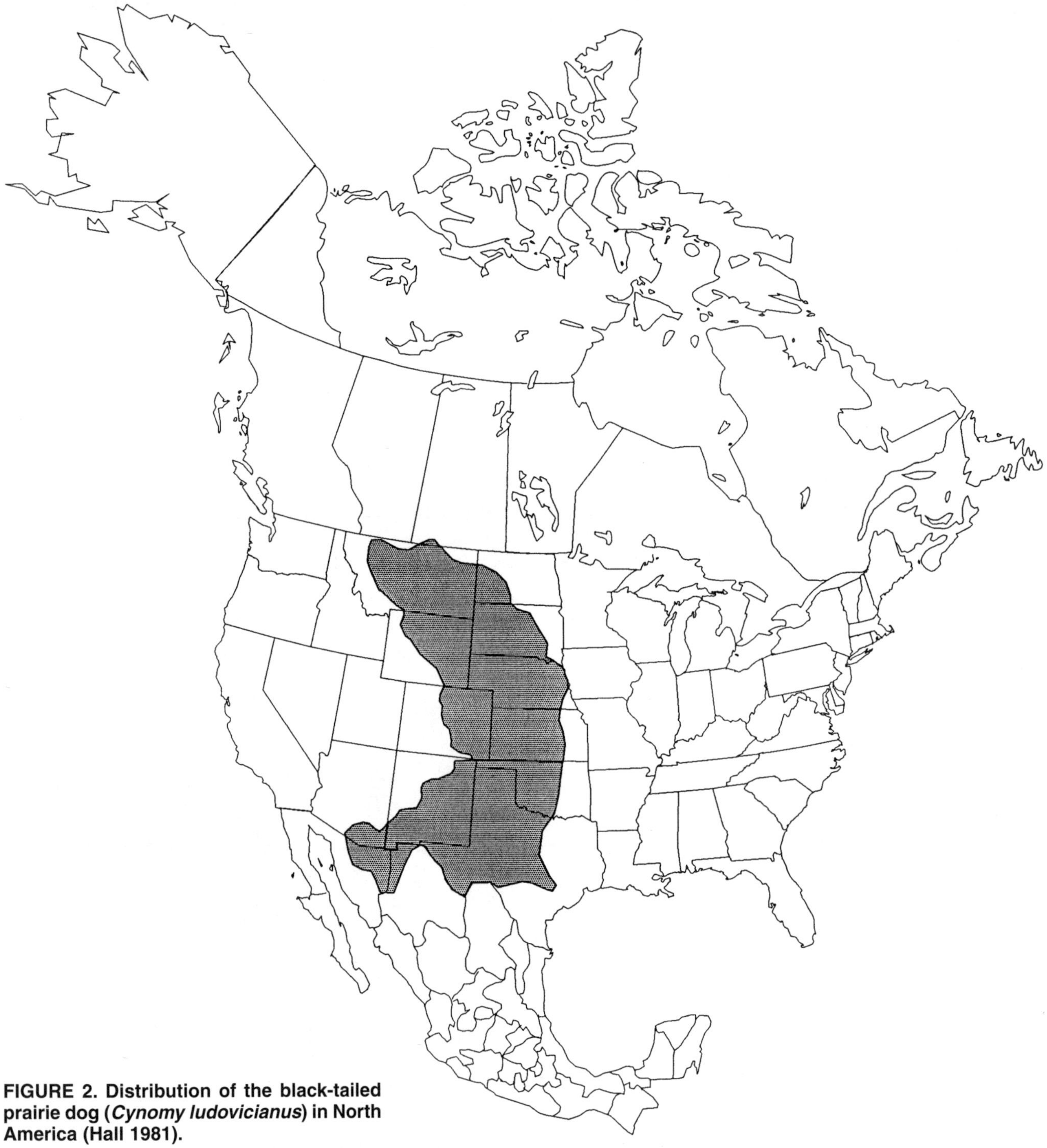

FIGURE 2. Distribution of the black-tailed prairie dog (*Cynomy ludovicianus*) in North America (Hall 1981).

area occupied by prairie dogs was reduced to an estimated 40.5 million ha (Nelson 1919) in the United States and by 1971, they still occupied only 566,000 ha (Cain et al. 1972). The decline was related to federal, state, and local poisoning programs and to changing land use practices that reduced available habitat by converting vast areas of the Great Plains to agriculture and urbanization. Prairie dog populations have increased since 1972, when Presidential Executive Order II 11643 banned the use of secondary poisons on federal lands and thus eliminated most uses of Compound 1080 and strychnine. The largest areas of land in the United States currently occupied by prairie dogs are federally managed lands.

South Dakota serves as an example of the fluctuations that have occurred in prairie dog populations as a result of the social, economic, political, biological, and climatic changes that have occurred on the Great Plains during the last 100 years (Hanson 1988). In South Dakota the first organized control efforts began in 1914, and in 1919 a rodent control law was passed by the South Dakota Legislature that increased prairie dog control efforts. In 1923, prairie dogs occupied almost 809,000 ha. How-

FIGURE 3. Prairie dogs poisoned during the early 1900s, a period when attempts were made to eradicate prairie dogs over most of their range. Photo from Bureau of Biological Survey records.

ever, organized prairie dog control programs continued through the 1920's on private, state, federal, and Indian lands, and by 1930 the prairie dog population had been reduced to widely scattered small towns. In the 1930s, the Dust Bowl forced farmers off agricultural land and depleted existing rangelands, and prairie dog populations increased on these disturbed areas. Utilizing some of the 1930s Federal emergency work programs, control agents pushed prairie dog populations to their lowest recorded level by the 1940s, when only small colonies remained. From 1955 to 1965, prairie dog populations were maintained at about 18,200 ha by use of Compound 1080 baits. The 1972 Executive Order placed most prairie dog control efforts on hold until 1976, when zinc phosphide (that does not have secondary toxicity) was registered for prairie dog control (Tietjen 1976) with the Environmental Protection Agency (EPA). From 1976 to 1979 about 12,000 ha were treated per year (Hanson 1988).

On the Buffalo Grass National Grassland in South Dakota, prairie dog colonies increased about 20%/year after control was discontinued in the early 1970s, from 1,154 ha in 1968 to 14,023 ha in 1978 (K. A. Fagerstone, unpubl. data) (Fig. 4). The increase was so dramatic that a lawsuit was filed in 1980 by the

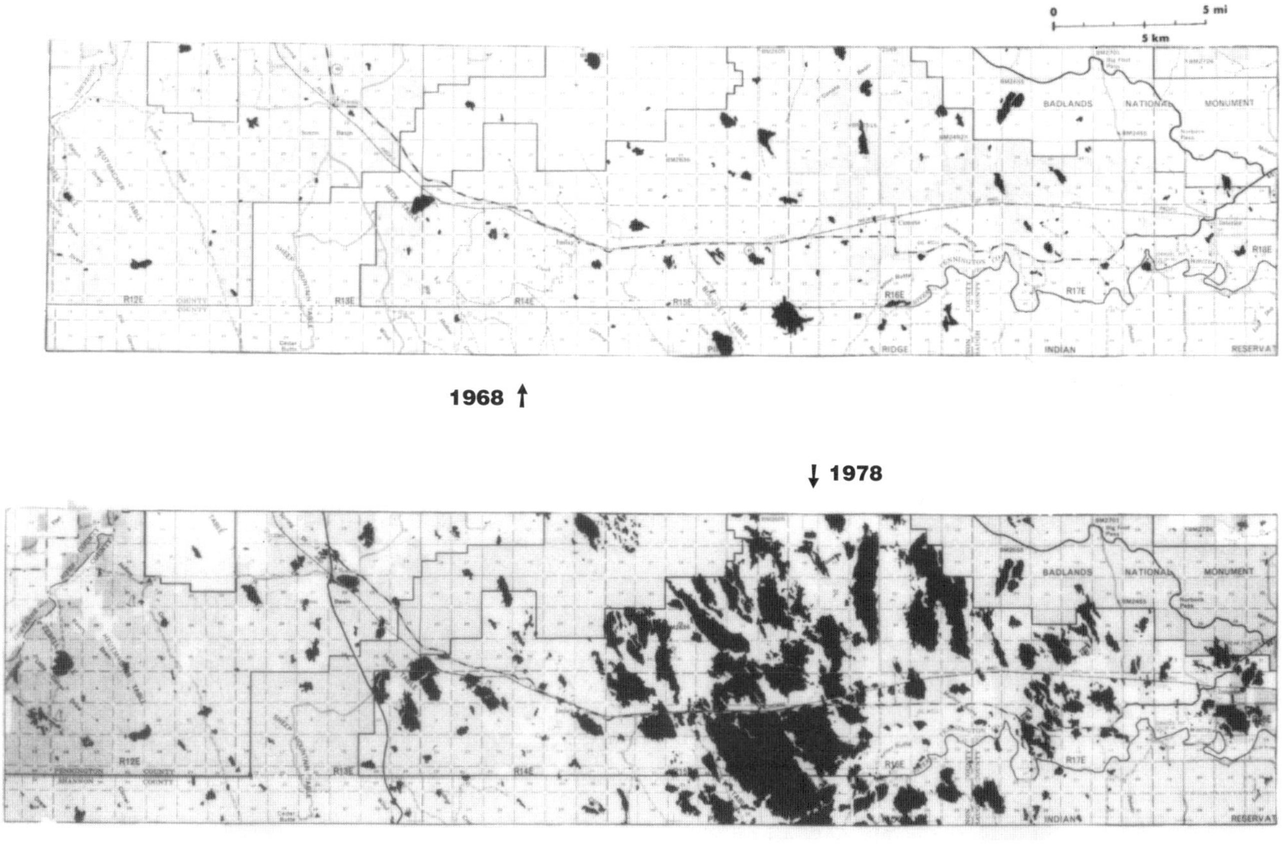

FIGURE 4. Black-tailed prairie dog (*Cynomy ludovicianus*) colony expansion on a 1036 km^2 area of Buffalo Gap National Grassland, South Dakota, between 1968 and 1978. Colonies are shown in black; each square represents 2.6 km^2.

American and South Dakota Farm Bureau Federations and others against the United States Forest Service, National Park Service, and Bureau of Indian Affairs contending that these agencies failed to conduct an effective prairie dog control program on lands administered by them and that adjacent private lands were damaged by prairie dogs originating on federal lands. A large-scale control program was conducted as a result of the lawsuit (Schenbeck 1982), and South Dakota passed a weed and pest law in 1984 that allows county boards the authority to declare lands with prairie dogs a public nuisance and to require prairie dog control in response to complaints from neighboring landowners. Prairie dogs now occupy about 324,000 ha in South Dakota. Because the perception is still prevalent with landowners that prairie dogs compete for forage with livestock, massive prairie dog control programs are still common (Schenbeck 1982).

Effects of Range Condition on Prairie Dogs. Many researchers have suggested that prairie dogs colonize sites that have been overgrazed or otherwise disturbed (Koford 1958, Uresk et al. 1982, Coppock et al. 1983*a*, Uresk and Bjugstad 1983). Black-tailed prairie dogs in particular have been shown to prefer areas with short vegetative cover; low cover apparently allows them to view predators and maintain a complex social system (Slobodchikoff and Coast 1980). Rates of prairie dog colony settlement and expansion have been shown to increase under intense livestock grazing and other human disturbances such as homesteading, fencing, cultivation, and the construction of water impoundments (Osborn and Allan 1949, Uresk et al. 1982, Cincotta 1985, Snell 1985, Cincotta et al. 1987). All of these land management practices reduce height and density of grasses, and provide a desirable environment for prairie dogs to expand and establish new colonies. Prairie dog distribution is correlated to these management practices (Knowles 1982, Cincotta 1985). Uresk et al. (1982) found that prairie dog burrow densities in the Conata Basin of South Dakota increased twice as fast on sites grazed by cattle as on ungrazed sites. Within this area, prairie dog colonies increased at the rate of 20%/year between 1968 to 1978 on the heavily grazed Buffalo Gap National Grassland but only increased 4% on the adjacent ungrazed portions of the Badlands National Monument (K. A. Fagerstone, unpubl. data, Fig. 4).

Effects of Prairie Dogs on Rangeland. Once established, prairie dogs modify the grasslands that they occupy. Because prairie dogs depend on being able to see terrestrial predators from some distance away (Slobodchikoff and Coast 1980), they clip shrubs and other tall vegetation that impede visual detection to a height of only a few centimeters. Vegetation is considerably modified by prairie dog feeding activities and by clipping of unpalatable plants to ground level (King 1955, Koford 1958). In well-established prairie dog colonies (Fig. 5), large areas of bare soil are common (Knowles 1982) and soil temperatures are raised (Archer and Detling 1986). Canopy height in 1 study was decreased by 62% in the first 2 years of colonization (Archer et al. 1987). Prairie dogs are thus able to maintain a low height vegetative community where species composition, biomass, and productivity of vegetation differ from uncolonized areas (Osborn and Allan 1949, King 1955, Bonham and Lerwick 1976, Dalsted et al. 1981, Coppock et al. 1983*b*).

Negative effects of prairie dog colonization on grassland productivity include decreased plant biomass, changes in plant composition from grasses to forbs, a shift toward C4 photosynthetic species, and higher silicon concentrations in forage. Prairie dogs generally reduce total plant biomass, particularly of native perennial grasses (Crocker-Bedford 1976, Hansen and Gold 1977, Crocker-Bedford and Spillett 1977, Knowles 1982, Coppock et al. 1983*b*, Archer et al. 1984, Uresk 1984). They can clip and consume 24 to 90% of the primary production of a site (Osborn and Allan 1949, Hansen and Gold 1977, Crocker-Bedford and Spillett 1977). Koford (1958) found that range condition on prairie dog colonies averaged only 25 to 50% of that of undisturbed range. In South Dakota, Coppock et al. (1980, 1983*b*) found that peak live graminoid biomass in areas colonized for 1-2 years, 3-8 years, and > 25 years was 39, 61 and 97% lower, respectively, than in adjacent uncolonized areas. Coppock et al. (1983*b*) also found that standing dead biomass and litter (mulch) were increasingly reduced in colonies as time since colonization by prairie dogs increased. Archer et al. (1987) found that rapid changes occurred in the first 2 years following colonization, with bare ground increasing from 10 to 35% and litter cover decreasing from 20 to 10%.

High densities of prairie dogs may have a negative influence on native perennial grasses, may cause changes in plant species composition toward shorter grasses and, ultimately, toward annual and short-lived perennial forb species (Severe 1977, Coppock et al. 1983*b*, Lerwick 1974, Uresk 1984, Archer et al. 1987). Often buffalograss is the dominant plant on prairie dog colonies and the taller western wheatgrass and blue grama are most common on uncolonized mixed-grass prairie sites (Fagerstone et al. 1981, Agnew et al. 1986). In areas of greatest prairie dog activity, annual forbs, shrubs and cactus often replace most of the original grass cover (King 1955, Cincotta et al. 1989). Perennial grasses are rapidly displaced by annual forbs within 3 years of colonization by prairie dogs (Coppock et al. 1983*b*, Archer et al. 1987). Coppock et al. (1983*b*) found that

FIGURE 5. Large areas of bare ground are common in well established prairie dog colonies. Photo by K. Fagerstone.

graminoids accounted for 85% of the biomass on uncolonized grassland, 70% on a prairie dog colony colonized for 3 to 8 years, and <3% on a colony >26 years. The formation of forb dominated communities in prairie dog colonies has been attributed to the length of time since colonization and the level of prairie dog activity (Archer et al. 1987, Cincotta et al. 1989); forb domination is usually greatest in the center of the colony.

Forage improvement is very slow on western rangelands that have been heavily grazed for many years and researchers have speculated that total exclusion of herbivores (including cattle) for ≥9 years may be required to increase forage production when the range is in a low condition class (Uresk 1985). Uresk (1985) found that despite 4 years of black-tailed prairie dog removal, production of western wheatgrass did not increase. Buffalograss and needle leaf sedge (*Carex eleocharis*) showed significant decreases in production when cattle were allowed to graze after prairie dog removal, indicating that prairie dog clipping may have stimulated their growth.

Prairie dogs also can cause a shift in the proportion of C3 and C4 plant species (Fagerstone and Williams 1982). There are 2 main photosynthetic pathways used by higher plants: the C3 Calvin cycle and the C4 dicarboxylic acid pathway of photosynthetic CO_2 fixation. Caswell et al. (1973) found that C3 plants are generally nutritionally superior to C4 plants because the concentration of protein in C3 plants is highest in the leaf mesophyll cells, which are more easily broken down than leaf bundle sheath cells, where the concentration of protein is highest in C4 species (Caswell and Reed, 1975, 1976; Akin and Burdick 1977). In black-tailed prairie dog colonies, the percentage of C4 grasses and forbs in the habitat is higher than that predicted for typical short- or mixedgrass prairie (Caswell et al. 1973); the most likely explanation is that C4 species increase with heavy grazing pressure (Fagerstone and Williams 1982).

Prairie dog grazing can cause genetic changes within plant species (Detling and Painter 1983, Detling et al. 1986, Archer et al. 1987). Grasses consumed by prairie dogs contain higher silicon concentrations than those in uncolonized areas (Brizuela et al. 1984), which may make them more difficult to digest. The maximum height of some plant species on prairie dog colonies is also reduced (Detling and Painter 1983, Agnew et al. 1986, Archer et al. 1987), a reduction maintained even in the laboratory. In addition, plants from prairie dog colonies withstand defoliation by grazing better than plants from an ungrazed area, perhaps because those in colonies have a higher leaf blade/leaf sheath ratio and more horizontal leaf angles (Detling and Painter 1983).

Positive effects of prairie dogs on grassland productivity include greater soil aeration, changes in community structure, increased plant species diversity, greater forb production, and better quality food and growing conditions inside prairie dog colonies. Prairie dogs build burrows approximately 12 cm in diameter, 10-30 m long, and 1-5 m deep with 2 or 3 entrances (Sheets et al. 1971). Andelt (1984) and Sieg (1988) postulated that burrowing by prairie dogs decreases soil compaction, increases the intake of water, aerates the soil, and promotes soil formation. Soils in prairie dog colonies are richer in nitrogen, phosphorus, and organic matter than soils in adjacent grasslands (Sieg 1988, Sharps and Uresk 1990).

On the southern Great Plains, prairie dogs can affect the structure and composition of rangeland in a positive manner by destroying mesquite (*Prosopsis* spp.) seedlings before they can grow into nuisance shrubs. Weltzin et al. (1991) showed that honey mesquite (*Prosopsis glandulosa*) seeds disappeared faster and seedling survival was lower on prairie dog colonies than off. They hypothesized that the eradication of prairie dogs may have allowed honey mesquite to increase in abundance over large areas of the Southwest.

Koford (1958) was one of the first to report greater numbers of plants species, primarily forbs, on prairie dog colonies than on native shortgrass prairie sites in Colorado. Since then, numerous other researchers have demonstrated an increase in plant species diversity in black-tailed prairie dog colonies in short- and mixedgrass prairie compared to uncolonized prairie (Lerwick 1974, Bonham and Lerwick 1976, Gold 1976, Hansen and Gold 1977, Severe 1977, Beckstead and Schitoskey 1980, Fagerstone 1982, Archer et al. 1984, Agnew et al. 1986, Archer et al. 1987). Also, forb cover can be significantly greater on prairie dog colonies (Uresk and Bjugstad 1983, Agnew 1983, Agnew et al. 1986).

Prairie dog grazing removes aging leaves and may stimulate growth of new tissue, which usually has a higher nitrogen concentration and greater digestibility than that of an ungrazed plant (McNaughton 1984). Prairie dog colonies, therefore, have been found to contain better quality food and growing conditions than uncolonized areas (Hassien 1976; Beckstead and Schitoskey 1980; Fagerstone 1982; Coppock et al. 1980, 1983*b*, 1983*a*; Detling and Painter 1983, Krueger 1986). Shoot nitrogen concentrations and digestibility are lowest in plants from uncolonized grasslands and increase with the length of time since colonization (Coppock et al. 1983*b*, Krueger 1986). Bison are attracted to prairie dog colonies as grazing sites (Coppock et al. 1983*a*) and use moderately grazed areas near the perimeters of prairie dog colonies because the vegetation contained higher crude protein and nitrogen levels, was more easily digestible, and had less dead matter (Coppock et al. 1983*b*).

Competition Between Prairie Dogs and Cattle. The economic effects of prairie dogs are not entirely clear. Although the black-tailed prairie dog diet is variable (Fagerstone 1982), the degree of dietary competition between prairie dogs and cattle can be high. The majority of plants eaten by prairie dogs are those also desired by livestock (Taylor and Loftfield 1924, Kelso 1939, Koford 1958, Fagerstone et al. 1981). Like cattle, black-tailed prairie dogs eat mainly grasses, followed by forbs and shrubs (Tileston and Lechleitner 1966, Summers and Linder 1978, Fagerstone et al. 1981, Uresk 1984). Similarity in diet between cattle and prairie dogs has been reported to be 64% by Hansen and Gold (1977) and 76% by Kelso (1939). However, prairie dogs feed selectively, so 80% of the biomass they ingest may come from plant parts not highly used by cattle (Crocker-Bedford 1976). Also, potential competition might by minimized by beneficial effects that large herbivores obtain from plants growing in prairie dog colonies because of increases in plant digestibility and nitrogen content (Coppock et al. 1983*b*).

Control of black-tailed prairie dogs has long been considered necessary to increase forage production on rangelands and decrease competition with cattle for rangeland forage. Although experimental studies of competition between prairie dogs and steers failed to show that prairie dogs had a significant negative impact on weight of the steers (Hansen and Gold 1977, O'Meilia et al. 1982), the studies have demonstrated lower weight gains for steers raised on prairie dog colonies amounting to market values of $14 to $24 less/steer.

Eliminating prairie dogs has had little effect on increasing the amount of food available for cattle (Crocker-Bedford 1976, Klatt and Hein 1978, Collins et al. 1984, and Uresk 1985). The increase in livestock forage after prairie dog control in South Dakota was estimated at only 51 kg/ha/year; based on this gain, control of 7.2 ha of prairie dog colonies was required to gain 1 AUM (animal unit months) of grazing (Collins et al. 1984). At a prairie dog repopulation rate of 30%, the control was not economically feasible and annual maintenance costs were greater than the amount of forage gained. Controlling black-tailed prairie dogs on depleted rangelands in western South Dakota did not increase the amount of forage produced after 4 years, whether or not cattle were allowed to graze (Uresk 1985). Total exclusion from prairie dogs and livestock for between 4 and 10 years may therefore be required to increase forage production when the range is in a low condition class. In another study of prairie dog-cattle competition, O'Meilia et al. (1982) found no differences in forb production on pastures with steers only compared to pastures with steers and prairie dogs. However, they did find a significant reduction in availability of blue grama, sand dropseed (*Sporobolus cryptandrus*), and other grasses on pastures with prairie dogs.

Uresk and Paulson (1988) estimated the carrying capacity and forage utilization for cattle in western South Dakota when prairie dogs were present but pastures were maintained in good condition at a near climax stage of mixed perennial cool-season grasses. They found that carrying capacity for cows and for cow-calf units decreased as the number of ha occupied by prairie dog colonies increased; the decrease was approximately 3 AUMs or 2 cow-calf units for every additional 20 ha of prairie dogs. This study showed that when areas with prairie dogs were high, needle leaf sedge (*Carex eleocharis*) and needlegrasses (*Stipa* spp.) could become major limiting factors in determining cow carrying capacity. Western wheatgrass was never a limiting plant species because consumption never exceeded the amount available.

White-tailed Prairie Dogs

Historical populations. There are 3 species of prairie dogs generally grouped under the name "white-tailed prairie dogs" (Fig. 6). These 3 species are found mainly in mountain valleys and plateaus in the Rocky Mountains at higher elevations (1,830-3,660 m) than are black-tailed prairie dogs. Although white-tailed prairie dogs live in colonies, their social organization is loosely structured and more closely resembles that of ground squirrel aggregations than it does the more highly structured organization of black-tailed prairie dogs (Pizzimenti and Hoffmann 1973). White-tailed prairie dogs occur at lower densities, (i.e., 3/ha) (Tileston and Lechleitner 1966, Clark 1977, Hoogland 1979) than black-tailed prairie dogs (approx. 15/ha) (King 1955, Koford 1958, Tileston and Lechleitner 1966, Hoogland 1979). White-tailed prairie dogs construct elaborate burrow systems and mounds; unlike black-tailed prairie dogs, white-tailed prairie dogs hibernate during winter. Habitat of white-tailed prairie dogs is variable and can contain densely vegetated areas, including sagebrush rangeland. As a consequence, visual contact between individuals is often obstructed, and alarm calls are important to the cohesion and survival of a colony (Waring 1970, Slobodchikoff and Coast 1980). Because white-tailed prairie dogs occur at lower densities and will tolerate more densely vegetated areas, they have generally not been considered as destructive to rangeland as black-tailed prairie dogs.

The white-tailed prairie dog has the largest geographic range of the 3 species and is found mainly in western Wyoming, northern Utah and northwestern Colorado (Fig. 7). Gunnison's or Zuni prairie dog occurs in Arizona, New Mexico, Utah, and Colorado (Slobodchikof et al. 1988). The historical populations of these species are not well documented but Taylor and Loftfield (1924) reported that by 1924 heavy overgrazing by livestock caused the disappearance of tall grasses in the Southwest, possibly leading to increases in Gunnison's prairie dog populations over those existing before white men settled the area.

FIGURE 6. White-tailed prairie dog (*Cynomys leucurus*). Photo by K. Fagerstone.

FIGURE 7. Distribution of the white-tailed prairie dog (*C. leucurus*), Gunnison's prairie dog (*C. gunnisoni*), and the Utah prairie dog (*C. parvidens*) in North America (Hall 1981).

The Utah prairie dog has the most restricted range of all the prairie dogs in the United States; it is found only in a few counties in southcentral and southwest Utah and is listed as an endangered species. After the 1920s its range constricted, until by 1975 it inhabited only half of its former range (Collier and Spillett 1975). Climate appears to be the single most important factor influencing the distribution of the Utah prairie dog, as population density has been correlated with the presence of water and populations have disappeared or declined during periods of drought in the 1930s (Collier and Spillett 1975) and in the 1950's (Collier and Spillett 1972). Population numbers have been further reduced in the western portion of its range as a result of drying trends and invasion of shrubby vegetation.

Effects of Range Condition on White-tailed Prairie Dogs. Slobodkichoff et al. (1988) found that Gunnison's prairie dog density was highly correlated with the number of native species in the area and was negatively correlated with the number of introduced-weedy plant species. They concluded that Gunnison's prairie dogs thrived at sites with native species of plants and did not prefer sites with a high proportion of introduced-weedy species. A study of prairie dog diet supported this conclusion, as 60-80% of the diet consisted of native plant species (Shalaway and Slobodchikoff 1988). Introduced weeds did not appear to be favored in Gunnison's prairie dog colonies, even though the soil was disturbed, suggesting that prairie dogs may cause some beneficial restoration of rangeland that has been damaged by grazing.

Gunnison's prairie dog density is positively correlated with the level of grazing (Slobodkichoff et al. 1988), suggesting that Gunnison's prairie dogs may prefer highly grazed sites. However, grazing has not been shown to play as large a role in promoting colony expansion of white-tailed prairie dogs as it does in black-tailed prairie dogs.

Effects of White-tailed Prairie Dogs on Rangeland. Little research has been conducted on the effects of white-tailed prairie dogs on grazing lands and uncertainty remains about their economic impact as competitors of livestock. Taylor and Loftfield (1924) found that destruction of vegetation and burrowing by Gunnison's prairie dogs accelerated sheet erosion, initiated gullies, and damaged primary forage species. A 4-year study on wheatgrass-dominated rangeland showed that Gunnison's prairie dogs utilized 80% of the total potential annual forage production. In blue grama-dominated rangeland, prairie dogs used 83% of the annual production. Gunnison's prairie dog fed on growing grasses and forbs in summer then switched to dead grass and seeds in spring and fall (Shalaway and Slobodchikoff 1988). Kelso (1939) found that plants of value as forage to livestock comprised 84% of the diet of the white-tailed and 75% of the diet of Gunnison's prairie dogs. However, white-tailed prairie dogs relied more heavily than the black-tailed on range plants that are not attractive to livestock, including sage (*Artemisia* spp.), saltbushes (*Atriplex* spp.), and Russian-thistle (*Salsola* spp.), which may offset competition with livestock.

In a Wyoming study, Clark (1977) found few vegetation differences between ungrazed rangeland and white-tailed prairie dog colonies ungrazed by livestock. However, where grazing by both prairie dogs and livestock occurred, grass height was reduced by half and forb abundance was doubled.

Gunnison's prairie dogs decreased plant cover (Slobodkichoff et al. 1988), an effect consistent with that found for black-tailed prairie dogs (Knowles 1982, Archer et al. 1984), but decreased plant species diversity (Slobodchikoff et al 1988, Clements and Clements 1940), which is contrary to findings with black-tailed prairie dogs (Lerwick 1974, Bonham and Lerwick 1976, Gold 1976, Hansen and Gold 1977, Beckstead and Schitoskey 1980, Archer et al. 1984). The difference between plant diversity on black-tailed versus white-tailed prairie dog colonies is probably related to habitat differences associated with the 2 species. Black-tailed prairie dogs clip tall plants, promoting greater numbers of forb species, and increasing diversity. In contrast, white-tailed prairie dogs spend less time clipping forage, and the clipping action that occurs can lower the competitive ability of shrubs, eventually eliminating them from prairie dog colonies and decreasing diversity. The decrease in plant species diversity and plant cover caused by prairie dog grazing apparently comes from a decrease in the introduced weedy plant species, and not from a decrease in native plant species.

Associations With Other Species

Through their modifications of aboveground vegetation, prairie dogs influence the densities, foraging patterns, and nutritional dynamics of other animals. Because these habitat modifications can be extensive, researchers refer to a "prairie dog ecosystem" that consists of the prairie dog community and other associated plants and animals.

Prairie Dog, Ungulate, and Rangeland Interactions. Many observers have reported that bison are attracted to grassland sites altered by prairie dogs (Koford 1958, McHugh 1958, Shult 1972) and suggest that prairie dogs and bison were ecologically fundamental to the functioning of the Great Plains grasslands prior to the coming of the white men (Coppock et al. 1983*b*, Krueger 1985). Koford (1958) suggested a reciprocal ecological relationship between prairie dogs and bison, each of which maintained a shortgrass interspersed with patches of forbs and bare ground, which was ideal habitat for the other. As bison moved into an area and denuded it of taller grasses, they increased visibility for prairie dogs, allowing prairie dogs to invade. Once established, prairie dogs kept their colonies open, which favored use by bison and antelope (*Antilocapra americana*). Coppock et al. (1983*b*) tested these theories and showed that bison actively selected prairie dog colonies over uncolonized grassland. One prairie dog colony was used 90% of the time by bison, even though the colony occupied only 39% of the habitat. Bison partitioned their use of the prairie dog colonies by preferentially resting in the oldest areas of prairie dog colonies (occupied for >26 years) and preferentially grazing in moderately affected areas of prairie dog colonies (occupied by prairie dogs for <8 years). Coppock et al. (1983*a*) found that bison selected for moderately grazed areas near the perimeters of prairie dog colonies; by continuously promoting new growth, prairie dogs modified plants in these areas toward more readily digestible perennial grasses, with lower fiber, higher nitrogen concentrations, and greater accessibility of green tissues, than vegetation from uncolonized areas. Despite the fact that maximum plant biomass occurred on uncolonized areas, bison used these areas least for grazing, perhaps because, as nonselective feeders, they cannot avoid feeding on standing dead vegetation, and prairie dog activities reduced the amount of dead vegetation. Thus, selection by bison for moderately affected sites in prairie dog colonies is a nutritionally advantageous tactic in a mixed-grass prairie. Vanderhye (1985) modeled the nutritional benefits accrued to bison by selectively feeding on prairie dog colonies and suggested that if bison use colonies for summer feeding, mature cows will gain an additional 5 kg of body weight (18% of seasonal weight gain) and yearlings will gain an additional 13 kg (46% of seasonal weight gain) compared to not feeding on colonies.

Elk (*Cervus elaphus*) (Wydeven and Dahlgren 1985) and pronghorn antelope (Krueger 1986) also preferentially use prairie dog colonies for feeding. While bison are attracted to colonies to feed on the regrowth of grasses at colony edges, pronghorn antelope are attracted to the forbs in the core of the colony (Uresk and Bjugstad 1983, Coppock et al. 1983*a*, Knowles 1986, Krueger 1986, Detling and Whicker 1988).

Black-footed Ferret Interactions. Black-footed ferrets (*Mustela nigripes*) have a close association with prairie dogs (Henderson et al. 1969). Their original range corresponded closely to that of prairie dogs and they historically have been observed primarily on prairie dog colonies (Henderson et al. 1969, Hillman and Linder 1973, Campbell et al. 1987). Ferrets depend on prairie dogs for food (Hillman 1968, Fortenbery 1972, Sheets et al. 1972, Campbell et al. 1987), and are dependent on prairie dog burrow systems for denning sites and shelter. Therefore, their decline during this century is probably linked to the reduction in prairie dog numbers (Hillman 1968, Tietjen 1976) and possible secondary poisoning by strychnine and Compound 1080 as the result of eating poisoned animals (Rudd and Genelly 1956, Schitoskey 1975, Tietjen 1976). Although prairie dog populations have increased since 1972 when the use of secondary poisons on federal lands was banned, ferret populations have not increased concomitantly. The last known wild population was found in 1981 near Meeteetse, Wyoming, and was studied until 1985, when canine distemper, a disease fatal to ferrets (Carpenter et al. 1976), was diagnosed in the population following discovery of sylvatic plague (*Yersinia pestis*) in the white-tailed prairie dog colonies. The 18 ferret survivors were taken into captivity and a captive breeding program was begun that has increased their population to about 400 animals. Reintroduction to the wild was attempted during fall 1991 and 1992 in the Shirley Basin of Wyoming with limited survival of a few individuals. Further reintroductions are planned for Montana and South Dakota within 1 or 2 years, with the goal of developing 10 self-sustaining ferret populations. Currently, there are not enough large prairie dog populations to reach this goal so preservation of large prairie dog colonies will be essential for recovery of the black-footed ferret.

Interactions with Other Species. Prairie dog colonies provide a unique habitat that influences the abundance and species composition of birds, small mammals, and large herbivores. Prairie dog colonies provide habitat for many different species of vertebrates other than prairie dogs (Campbell and Clark 1981, Clark et al. 1982, O'Meilia et al. 1982, Agnew 1983). Clark et al. (1982) recorded 107 vertebrate species and subspecies of wildlife on prairie dog colonies in Utah, Colorado, and New Mexico. Sixty-four vertebrate species were recorded by Campbell and Clark (1981) on black- and white-tailed prairie dog colonies in Wyoming. Reading et al. (1989) listed 163 vertebrate species sighted on black-tailed prairie dog colonies and suggested that species richness increases with colony size and density. Sharps and Uresk (1990) determined that 134 species (40%) of western South Dakota wildlife were associated with prairie dog colonies. Agnew et al. (1986) and Deisch et al. (1989) found 5 classes of invertebrates on prairie dog colonies in South Dakota, which may explain why insectivorous rodent species are found in greater numbers on prairie dog colonies than on surrounding rangeland (Agnew et al. 1988).

Prairie dogs serve as an important food source for mammalian predators. The black-footed ferret was historically an important predator, but currently the badger is the main prairie dog predator (Clark 1977). Others include coyotes (*Canis latrans*), bobcats (*Lynx rufus*), foxes (*Vulpes* spp.), and occasionally mink (*Mustela vison*) and long-tailed weasels (*Mustela frenata*) (Hillman and Sharps 1978). Swift fox (*Vulpes velox*) were normally found to have dens within 0.8 km of prairie dog colonies (Hillman and Sharps 1978) because prairie dogs are a major component of swift fox diet in South Dakota (Uresk and Sharps 1986).

Prairie dog burrows serve as homes for rabbits and rattlesnakes, whose numbers are usually higher on prairie dog colonies (Dano 1952, O'Meilia et al. 1982). On a grassland area in Colorado, desert cottontail (*Sylvilagus audubonii*) densities were 0.8 to 1.3/ha on prairie dog colonies and only 0.05 per ha on shortgrass prairie adjacent to the colonies (Hansen and Gold 1977). Colonies also provide habitat for a large number of small mammals. On mixed-grass sites in South Dakota and Oklahoma, small rodent abundance was found to be greater on than off black-tailed prairie dog colonies, but small rodent species richness was significantly lower (O'Meila et al. 1982, Agnew et al. 1986). Prairie dogs act as ecosystem regulators by maintaining habitat with less mulch cover and lower vegetation height than surrounding ungrazed or lightly grazed rangeland. This vegetative manipulation by prairie dogs negatively impacts rodent species associated with dense vegetation of mixed-grass rangeland. However, these vegetation features, combined with high burrow densities, provide quality habitat for other species of small rodents, such as thirteen-lined ground squirrels (*Spermophilus tridecemlineatus*) (Evans and Holdenried 1943, Slade and Balph 1974), deer mice (*Peromyscus maniculatus*), and grasshopper mice (*Onychomys leucogaster*) (Agnew et al. 1986, Uresk 1987), which are associated with sparse, heterogeneous vegetative cover. Unused prairie dog burrows provide nesting and escape cover for these species (Koford 1958, Smith 1967). Maintenance of vegetation in a lower seral stage by prairie dogs is especially favorable to deer mice (Birch 1977), which are adapted to live in open habitat (Baker 1968, Jones et al. 1983, MacCracken et al. 1985*a*, Agnew et al. 1986). Their numbers decrease with increased vegetation height and canopy cover, so after prairie dog control with rodenticides, deer mice populations decrease (Uresk 1985, Deisch et al. 1990) as lack of clipping and grazing by prairie dogs causes increased plant canopy cover and aboveground biomass (Klatt 1971, Potter 1980). Because black-tailed prairie dog colonies contain large numbers of forbs, deer mice may be also be drawn to colonies to feed on seeds, their primary diet (Sieg et al. 1986). In prairie dog colonies an influx of deer mice usually occurs in spring when yearling deer mice establish home ranges (MacCracken et al. 1985*a*), and lower densities occur in August due to dispersal of young-of-the-year (Falls 1968, Metzgar 1980).

On white-tailed prairie dog colonies, where vegetation changes caused by prairie dogs are not as prominent, small mammal populations were not found to differ between prairie dog colonies and uncolonized areas (K. A. Fagerstone, unpubl. data). Clark (1977) also found no differences in population densities of deer mice (about 40/ha), jackrabbits (*Lepus* spp.), and cottontail rabbits among ungrazed rangeland, rangeland grazed by white-tailed prairie dogs only, and rangeland grazed by prairie dogs and livestock. He trapped no ground squirrels on uncolonized areas or areas grazed by prairie dogs only. However, on the area grazed by prairie dogs and livestock, he found high densities (35/ha) of Richardson's ground squirrels (*Spermophilus richardsonii*) and 13-lined ground squirrels (30/ha). On heavily grazed rangeland, prairie dogs, Richardson's ground squirrels, and 13-lined ground squirrels will live together in a complex pattern of overlapping populations.

In a study of fauna associated with black-tailed prairie dog colonies, Agnew et al. (1986) found that bird species diversity and abundance were significantly higher on prairie dog colonies than on mixed-grass sites. They attributed the higher numbers to "patchiness" or structural diversity on prairie dog colonies, to increased forb seed production, and to lower amounts of mulch and lower vegetation height, which may result in greater visibility of macroarthropods and seeds. Species significantly more abundant on prairie dog colonies included horned larks (*Eremophila alpestris*), mourning doves (*Zenaidura macroura*), killdeer (*Charadrius vociferus*), barn swallows (*Hirundo rustica*), and burrowing owls (*Speotyto cunicularia*). The horned lark was the most common species and the shorter grasses on the colonies provided ideal conditions (Apa et al. 1991). Weins (1973) and Skinner (1975) found highest horned lark densities in grazed areas because the larks preferred open areas and low, sparse vegetation (Behle 1942, Skinner 1975) for optimum nesting habitat (DuBois 1935, Giezentanner 1970, Creighton 1974, and Pickwell 1931).

A large number of avian predators feed on prairie dogs, including golden eagles (*Aquila chrysaetos*), bald eagles (*Haliaeetus leucocephalus*), Ferruginous hawks (*Buteo regalis*), red-tailed hawks (*B. jamaicensis*), rough-legged hawks (*B. lagopus*), marsh hawks (*Circus cyaneus*), and other species (Clark 1977, Sharps and Uresk 1990). Burrowing owls use abandoned prairie dog burrows during spring and summer as nesting sites, for escape cover (O'Meilia et al. 1982, MacCracken et al. 1985*b*), and for feeding on the large numbers of insects and arthropods. The continued survival of the burrowing owl will probably depend on the presence of prairie dog colonies.

Prairie Dog Management Programs

It is important for land managers to recognize the positive and negative effects of prairie dogs on rangeland and on other wildlife species. Before developing a comprehensive management program, land managers should answer several questions. For example, what is the primary use of the rangeland: livestock grazing, wildlife value, or both? Are the prairie dog colonies of concern important for other wildlife species? What level of prairie dog occupation will be tolerated by landowners and the public? Is the rangeland in poor or good condition? And what is the purpose of the management plan: to increase, decrease, or maintain current prairie dog colony sizes? Management strategies will differ depending on how these questions are answered.

Most research on prairie dog management has been conducted on black-tailed prairie dog colonies in South Dakota on mixed-grass climax rangelands. This research has focused almost exclusively on development of methods for reducing prairie dog populations and/or preventing increases in their numbers. Traditional techniques include shooting and toxicants. Intensive shooting of prairie dog colonies may successfully control their numbers by disrupting reproductive activities and removing individual animals (Andelt 1984). Toxicants that are legal for prairie dog control include poison grain baits and burrow fumigants (usually used for small areas). Zinc phosphide is a frequently used rodenticide for managing prairie dog colonies to decrease or maintain current colony sizes. Studies have shown that zinc phosphide can be effective in limiting colony expansion. Uresk and Schenbeck (1987) treated 7 colonies on 2 occasions over a 5-year period with zinc phosphide; 5 colonies were not treated. The untreated colonies increased 65% in area, while the treated colonies increased only 1%. They recommended treatment with a rodenticide every 3 years to curtail expansion of prairie dog colonies. Schenbeck (1982, 1986) obtained similar results, with most colonies needing retreatment at least once every 3 years to maintain a desired size.

If rangeland is in poor condition to begin with because of high intensity grazing by livestock and prairie dogs, it may take years after a prairie dog control program for range recovery to occur, even when livestock grazing is reduced. In South Dakota, >4 years of reduced prairie dog densities were needed for an increase in forage production (Uresk 1985). Based strictly on livestock forage response to rodenticide treatments, prairie dog control programs are not economically feasible (Uresk et al. 1982; Collins et al. 1984; Uresk 1985, 1986). Collins et al. (1984) reported that $13.60/ha was spent on prairie dog control while grazing fees generated only $7.30/ha per year. A serious problem with use of rodenticides to curtail colony expansion is the rapid recovery of prairie dog populations and the subsequent costs of frequent retreatments. Prairie dogs will frequently repopulate colonies to initial population numbers within 3 years (Schenbeck 1982, Cincotta et al. 1987).

Prairie dog management needs to be reexamined. Proper range management practices, such as reduction of livestock stocking rates and deferred grazing may be more effective than prairie dog control programs in regulating prairie dog numbers. Prairie dogs prefer habitat managed for warm-season grasses such as blue grama and buffalograss, a habitat favored by heavy livestock grazing (Uresk et al. 1982, Cincotta 1985). Prairie dog colony expansion can be reduced under management plans that use lower stocking rates to maintain a cool-season mid-grass rangeland rather than a short-grass rangeland; light grazing by livestock in mid-grass rangeland can result in greater height, cover, and density of vegetation (Uresk et al. 1982, Cincotta 1985), making habitat less suitable for prairie dog expansion. For example, prairie dog colonies did not significantly expand over 4 years on areas in South Dakota where cattle were excluded (Uresk et al. 1982). Snell and Hlavachick (1980) and

Snell (1985) also reported reduced expansion rates or elimination of prairie dog colonies by using a summer-deferred grazing system. In Nebraska rangeland, deferred grazing for 2 years caused a reduction in natality, population growth, and active areas (Cable and Timm 1988). Deferring a pasture from cattle grazing for 4 consecutive growing seasons reduced a 45-ha colony in southern Kansas by approximately 90% (Snell and Hlavachick 1980); summer grazing on the same colony was deferred again several years later when the colony began to expand, and after just 1 season of deferment, the colony decreased approximately 70% in size (Snell 1985). The average rainfall in the Kansas area was 63.5 cm, so vegetation responses and resulting prairie dog declines would be expected to be more dramatic than in drier areas of the Great Plains.

The most cost-effective control programs for managing prairie dogs to decrease or maintain current colony sizes may be those that combine the use of rodenticides with livestock grazing practices that reduce habitat suitability for prairie dogs by managing for taller grasses. This generally means periodic rest or reduced stocking rates of livestock (Uresk et al. 1982). Expansion and repopulation rates of prairie dog colonies can be controlled by reduced cattle stocking rates to increase height and density of vegetation in combination with rodenticide control programs and mechanical renovation of low quality rangeland (Snell and Hlavachick 1980, Schenbeck 1982, Uresk et al. 1982, Snell 1985).

Management to increase prairie dog colony sizes may become more prevalent in the future as more land managers recognize that prairie dog colonies are important and unique ecosystems. Management of prairie dogs for their positive values will undoubtedly be easier on public lands because private landowner attitudes towards black-tailed prairie dogs are often not favorable. Although many landowners will tolerate some prairie dogs, the 1980 South Dakota Litigation and the 1984 South Dakota weed and pest law are indicative of landowner attitudes in areas where prairie dog populations have been allowed to increase beyond the tolerance limits of local landowners and residents (Schenbeck 1986). In areas where prairie dogs are desired, a moderate level of livestock grazing may be required to enhance colony expansion.

New management techniques need to be developed that can be used in areas where rodenticide or fumigant treatments are not feasible and/or where grazing practices cannot be manipulated. One promising alternative is development of contraceptives that would place the management emphasis on reducing natality rather than on increasing mortality. Because prairie dogs reproduce only once a year, management efforts using immunocontraceptives would be minimal. Previous trials with diethylstilbestrol (DES) (i.e., a hormonal reproductive inhibitor) showed that limiting reproduction successfully limited expansion of a colony (Garrett and Franklin 1982); however, because DES accumulates in tissue and could affect reproduction on predators, it has not been pursued further. Because immunocontraceptives do not accumulate in tissue, they could be especially valuable in urban communities where prairie dog colonies have positive public value.

Less research has been conducted on management of white-tailed prairie dog populations. In areas overgrazed by cattle and prairie dogs, conditions similar to those in black-tailed prairie dog colonies may exist and grazing may need to be curtailed to prevent colony expansion. In other areas it may be beneficial to increase white-tailed prairie dog populations, particularly for the restoration of the endangered Utah prairie dog. Conservation of this species, and of Gunnison's prairie dog, may be enhanced by maintaining habitats with a large component of native vegetation (Slobodchikoff et al. 1988). Slobodchikoff et al. (1988) suggest that Gunnison's prairie dogs may have a beneficial function of promoting native vegetation that could help restore rangeland damaged by grazing.

Important questions still need to be addressed regarding prairie dog influences on rangelands. We know that grazing by prairie dogs decreases plant standing crop but we do not know to what extent that decrease may be offset by the increased nutritive value of the forage around prairie dog colony edges. We also know that a large number of animal species are associated with prairie dog colonies but we do not yet know which of those species are dependent on those colonies. Ideally, management of prairie dog colonies and rangeland should take into consideration a number of factors including range condition and trend, season of livestock use, prairie dog and livestock densities, how much area is available for colony expansion, and potential interactions with other species.

GROUND SQUIRRELS

There are 23 ground squirrel species (*Spermophilus* spp.) and a number of subspecies in North America (Hall 1981). Ground squirrels are distributed widely, with ≥1 species occurring throughout the Western United States. Ground squirrels range in length from about 17 to 52 cm (Fig. 8). Ground squirrels are active in the daytime and some species are social, forming sizable colonies ≥0.5 ha, with complex burrow systems. Ground squirrels feed on a wide variety of green herbage in late winter and spring and, in the more arid West, will often switch to a diet of seeds as green annuals disappear. Most ground squirrel species hibernate in winter; in the dry areas of the West they often have a summer torpor.

Preferred Habitat and Effects on Rangeland

Ground squirrels occupy widely varied habitats. Those species of most concern to rangeland managers primarily inhabit open grassy plains and valleys and generally avoid highly forested or brushy habitats. They are often less numerous where vegetation is tall and dense and their populations are therefore sometimes favored by livestock grazing.

Very little research has been conducted to quantify the effects of ground squirrels on rangeland. Ground squirrels can have positive effects on rangeland. Ground squirrel burrowing activities can contribute to deepening of soils (Grinnell 1923, 1933) by bringing subsoils to the surface where leaching, mixing, and distribution occur. Through their burrowing activities, ground squirrels also loosen the soil, admitting air and water to the roots

FIGURE 8. Several ground squirrel species influence rangeland vegetation, including: upper left, the California ground squirrel(*Spermophilus beecheyi*); upper right, Belding ground squirrel (*S. beldingi*); lower left, Columbian ground squirrel (*S. columbianus*); and lower right, Wyoming ground squirrel (*S. elegans*). Photos by K. Fagerstone, J. Clark, D. Sullivan, and K. Fagerstone, respectively.

of plants. It is estimated that 1 Columbian ground squirrel (*Spermophilus columbianus*) transports 4 to 12.3 kg of soil to the surface annually (Smith and Gardner 1985); newly constructed burrows result in the transfer of 25-50 kg of soil to the surface. Ground squirrel activities may prevent soil erosion by improving the water holding capacity of soil, thus producing greater vegetation cover (Grinnell 1923, Taylor 1935). Alternatively, on bare slopes they may increase erosion by depositing soil brought to the surface downhill from the burrow.

Several ground squirrel species are considered pests on rangelands, including the California ground squirrel (*Spermophilus beecheyi*), Belding's ground squirrel (*S. beldingi*), Richardson's ground squirrel, Wyoming ground squirrel (*S. elegans*), and the Columbian ground squirrel. There is controversy about how much forage ground squirrels consume and the effect of this consumption on rangeland and livestock production. Many range managers contend that ground squirrels consume large quantities of range grasses and, therefore, conduct extensive control programs on rangelands. Although ground squirrels consume grasses that are important as livestock forage, little data have been generated on any species except the California ground squirrel to quantify the degree of conflict with livestock. For most ground squirrel species the primary conflict with man is consumption of cultivated crops rather than rangeland forage.

The California ground squirrel diet and effect on rangeland have been extensively studied. This species inhabits open areas of rangeland from sea level to 2,500 m (Grinnell and Dixon 1918). Populations are small on ranges protected from livestock grazing (Linsdale 1946) and appear to benefit from the vegetation regrowth following disturbances created by grazing. The California ground squirrel has reportedly caused serious economic losses of forage production on livestock ranges in California (Calif. Dep. Agric. 1966). During the 1920s to the 1970s an average of 2.4 million ha/year were treated with poison grain to control ground squirrels, at a cost >$1 million/year. The California ground squirrel competes with livestock primarily during the green forage season on California annual rangeland, where vegetation consists of annual forage plants that sprout from seeds in the fall following the first rains, and reach maturity in March, April, or May (Fitch and Bentley 1949, Howard et al. 1959). Ground squirrels can significantly reduce the amount of green forage available to grazing cattle during winter when plant growth is slow. Grinnell and Dixon (1918) estimated that 200 California ground squirrels would consume as much forage as 1 steer. Studies by Fitch (1948) and Fitch and Bentley (1949) indicated that California ground squirrels can consume a considerable amount of forage in spring; on the basis of an average population of 5.7/ha, they estimated that ground squirrels could consume about 12.4 kg of forage/ha/month. Estimates of the average daily intake of a California ground squirrel range from 20 to 49 g (Grinnell and Dixon 1918, Storer et al. 1944, Fitch 1948, Schitoskey 1978). In contrast to earlier studies, Schitoskey (1978) found forbs, including legumes, the most prevalent items in the diet, and suggested that during the green forage season California ground squirrels and cattle fed largely on different plant species. Schitoskey (1978) also found that intake varied with diet and sex; an adult male ground squirrel consumed 19.6 g of forage/day but only 16.3 g of dry seed/day; an adult female ate 15 g/day when nonreproductive, 19 g/day if pregnant, and 30 g/day if lactating. The ground squirrels could assimilate 83% of the dry matter in forage and 91% of the dry matter in seeds (Schitoskey 1978). This is similar to the 79.9 to 84.5% efficiencies reported for other rodents by Johnson and Groepper (1970) and Harris (1971). During the spring green forage season, ground squirrels feeding on foliage would require 1.5 times their daily energy needs to maintain their weight (Schitoskey 1978). Spring would therefore probably be the peak period of competition with livestock for rangeland forage. During summer and fall seed consumption, squirrels could eat less yet still gain weight, so competition is probably less intense.

Consumption does not wholly express the loss of forage due to ground squirrels because much is cut and left on the ground. California ground squirrels at densities of 4.9/ha reduced the forage by 35%, 10 times more than their forage consumption would indicate (Fitch and Bentley 1949). Howard et al. (1959) found that California ground squirrels can have measurable effects on livestock production. They compared winter weight gains of heifers on California ranges and found that where ground squirrels had been removed cattle weight gain was increased by 0.47 kg/day, equivalent to 2 kg of heifer weight for each ground squirrel removed. During the 2 years that ground squirrels were removed from an experimental pasture, heifers averaged daily gains of 0.5 and 0.3 kg greater than control animals in the pasture where squirrels were present. In contrast, Schitoskey (1978) found that ground squirrels consumed only a small amount of net aboveground plant production.

The Belding's ground squirrel is plentiful on open grasslands of northwestern California and Oregon, southwestern Idaho, and Nevada. It avoids marshy areas and thick brush or rocky slopes but is not adapted to very dry conditions (Hansen 1954). According to Grinnell and Dixon (1918), extensive clearing of sagebrush and seeding to grass and hay benefited this squirrel, which feeds primarily on grasses. Grinnell and Dixon (1918) estimated that the forage eaten on a section of land densely populated with squirrels was sufficient to feed 90 steers throughout the growing season.

The Columbian ground squirrel inhabits wet meadows and grasslands from Montana and Idaho into Canada. Durrant and Hansen (1954) ranked the Columbian second to the Belding's as the ground squirrel least adapted to dry conditions. Density ranges from 12 to 35/ha on rangeland (Elliot and Flinders 1991), with the most densely populated areas situated on well-drained, south facing slopes (Boag and Murie 1981). Columbian ground squirrels eat a variety of rangeland flowers, seeds, bulbs, and fruits (Howell 1938, Manville 1959, Lambeth and Hironaka 1982, Elliot and Flinders 1985, Harestad 1986). The Columbian ground squirrel is considered to be one of the most economically harmful rodent species in western Montana, causing about $800,000 damage in 1 year (Albert and Record 1982) to agricultural crops and rangeland. Much of the damage apparently arises from competition with livestock for forage (Shaw 1921, Fitch 1948, Howard et al. 1959), but good estimates are lacking. Shaw (1920) estimated that Columbian ground squirrels consumed 187% of their weight daily and that consumption by 385 Columbian ground squirrels would be equivalent to 1 cow and 96 to 1 sheep. Because of possible competition with livestock, this species has been the subject of intensive control efforts.

The Wyoming and Richardson's ground squirrels are very similar in habitat requirements, diet, and behavior and until recently were classified as 1 species (Fagerstone 1987). Both squirrels live in small social organizations similar to the clans described for the white-tailed prairie dog. The favored habitat of the Wyoming ground squirrel is well-drained upland slopes covered by dry grassland or sagebrush (Armstrong 1972; Clark 1970*a,b*; Hansen and Johnson 1976; Jones et al. 1983), where heavy clay soils, wet soils, and loose sand are avoided. Richardson's ground squirrels form large colonies in areas of short grasses and rough terrain, especially on heavily grazed slopes (Michener and Michener 1977, Michener 1979, Jones et al. 1983). Both species may be favored by overgrazing of rangeland. Sampson (1952) characterized the range where Wyoming ground squirrels occurred as having moderate to heavy grazing and being in fair to poor condition. He hypothesized that an increased population of squirrels may be a result of depleted range, where the squirrels subsist on secondary succession weedy plants. These squirrels primarily colonize low vegetation, rarely inhabiting sites where tall vegetation restricts their visual field (House 1964, Michener and Koeppl 1985). Hansen and Reed (1969) found that cattle and sheep grazing on their study area in late summer removed about 60% of the aboveground plant biomass, which significantly reduced food for ground squirrels and induced competition for vetch (*Astragalus* spp.), bluegrasses (*Poa* spp.), and sandwort (*Arenaria* spp.). Therefore, although ground squirrels prefer the low vegetation of grazed areas, their populations may at times be adversely impacted in areas of intense livestock grazing.

Wyoming and Richardson's ground squirrels have been considered competitors to cattle on rangelands and reportedly do considerable damage when present at high population densities; however, damage has not been adequately assessed. Wyoming ground squirrels primarily eat green vegetation, with about 60% of their diet consisting of forbs and shrubs and 28% of grasses (Clark 1968, Hansen and Reed 1969, House 1964), so much of their diet consists of plants that are not particularly palatable to cattle. The diet and habitat requirements of Richardson's ground squirrels are similar to those of Wyoming ground squirrels; but Richardson's ground squirrels consume more grasses and fewer forbs and shrubs, so they may come into greater conflict with livestock. Leaves and stems of grasses, and a great variety of forbs, are consumed in spring and early summer, but in late summer and autumn, seeds are their principal foods (Quanstrom 1968). The average Wyoming ground squirrel has been estimated to remove about 1 kg/ha of aboveground vegetation each year, and the density of these squirrels may reach 50/ha (Hansen and Reed 1969). Average annual densities of adult Richardson's ground squirrels are between 1.4 and 5.2/ha (Dorrance 1974, Michener and Michener 1977, and Michener 1983) and on localized areas densities can be as high as 27/ha (Davis 1984); after emergence of young from burrows, Dorrance (1974) noted an average density of 10 juveniles/ha and a maximum of 56/ha.

Several other ground squirrel species are commonly found in association with rangeland but are not significant pests. In most of their range, Uinta ground squirrels (*S. armatus*) typically occur in open grassy meadows several successional stages removed from climax. They reach highest densities in gradually changing habitats (Slade and Balph 1974) and may be helped by moderate grazing. Thirteen-lined ground squirrels have been found to be most abundant in moderately overgrazed pastures, followed by heavily overgrazed and undisturbed areas (Phillips 1936). Jones et al. (1983) reported that thirteen-lined ground squirrels were most commonly found in short grass and Abramsky et al. (1979) typically found them in areas with open vegetation. However, Agnew et al. (1986) captured thirteen-lined ground squirrels most often in mixed-grass sites and only infrequently on shortgrass areas such as prairie dog colonies. Townsend ground squirrel (*S. townsendii*) diets were the same on grazed and ungrazed habitats (Rogers and Gano 1980), not surprising because these ground squirrels are active above ground only during late winter and spring when green plant material is readily available. They remain below ground in a torpid condition the rest of the year when green forage is scarce, thereby precluding intense competition with livestock. Washington's ground squirrels (*S. washingtoni*) inhabit dry, open sagebrush or grassland habitat but are most numerous in areas of high grass cover (Betts 1990), where population densities may range from 120 to 240 individuals/ha (Bailey 1936, Dalquest 1948).

Management

Management for most species of ground squirrels has emphasized population reduction. Various rodenticides are effective for reducing ground squirrel populations, including sodium fluoroacetate and strychnine (no longer available for this use), zinc phosphide, chlorophacinone and diphacinone, bromadiolone and gas cartridges (Record 1978, Albert and Record 1982, Askham 1985, Matschke et al. 1988). Chemical toxicants distributed on grains can result in immediate mortality >70% (Record 1978, Matschke et al. 1983). Generally, however, post-treatment reinvasion by squirrels from surrounding areas rapidly returns populations to pretreatment levels.

Given the paucity of quantitative studies on the effects of ground squirrels on rangeland and livestock, rangeland managers should evaluate carefully before ground squirrel control is undertaken. Ground squirrels have many positive roles in the grassland ecosystem, particularly as an important prey base. Predators of ground squirrels include canids (coyotes and fox), felids (bobcat [*Lynx canadensis*], and mountain lion [*Felis concolor]*), bear (*Ursus* spp.), mustelids (badgers [*Taxidea taxus*], weasels [*Mustela* spp.], marten [*Martes americana*], and skunk [*Mephitis* spp.]), buteos (red-tailed hawks, Ferruginous hawks, and Swainson's hawks [*B. swainsoni]*), marsh hawk, prairie falcon (*Falco mexicanus*), eagles, owls (snowy owl [*Nyctea scandiaca]*, and great horned owl [*Bubo virginianus*]), and black-billed magpies (*Pica pica*) (Michener 1979, Michener and Koeppl 1985, Elliot and Flinders 1991). Domestic cats and dogs also hunt ground squirrels. Long-tailed weasels, badgers, and buteos probably have the greatest impact on ground squirrel populations. Michener (1979) found that predation by long-tailed weasels on unweaned young in the burrow reduced the number of juveniles entering the population > 50%. Ground squirrels are most vulnerable to predation from badgers during fall after squirrels enter hibernation but before the ground is frozen, so badgers can dig them out of burrows (Michener 1979). Ground squirrels can be a very important source of food for predators. During the nestling period for buteos in southern Alberta, Richardson's ground squirrel averaged 90% of the prey biomass for ferruginous hawks, 76% for red-tailed hawks, and 75% for Swainson's hawks (Schmutz et al. 1980). Buteos killed 6% and 15% of the squirrel population in 2 different years.

POCKET GOPHERS

Pocket gophers are fossorial rodents in the family Geomyidae. They are medium-sized, with lengths ranging from about 12.5 to 35.5 cm and color ranging from dark brown to almost white. The great variability in size and color is attributed to a low dispersal rate and limited gene flow that results in adaptation to local conditions. Pocket gophers are found throughout most rangelands in the United States but they are seldom seen, spending most of their lives underground in burrow systems that provide shelter and access to forage. Their many adaptations for life underground include small eyes, inconspicuous ears, powerful forequarters, long claws and incisors for digging, and specialized lips behind the front teeth to prevent soil from entering the mouth. Pocket gophers were named for the large pouches on either side of the face that are used for carrying food. Gophers are active all year round. Usually only 1 gopher occupies a tunnel system except during the breeding season and when females are rearing young. In northern localities pocket gophers have 1 litter/year, usually with 3 to 4 young but in southern areas they may have $\geq$2 litters depending on the locality and climate.

There are 33 species of pocket gophers represented by 5 genera in the western hemisphere (Hall 1981). The 2 primary genera of pocket gophers are *Geomys* and *Thomomys*. *Geomys* are present from the Rocky Mountains east to the Mississippi River, and from southern Canada to southern Texas. The 3 main *Geomys* species in North America are the plains pocket gopher (*Geomys bursarius*) (Fig. 9), the desert pocket gopher (*G. arenarius*), and the Texas pocket gopher (*G. personatus*). The plains pocket gopher is the most widespread (Fig. 10). Adults average about 31 cm in length and weigh 250-450 g and can occur at densities as high as 20/ha.

Thomomys species occur generally in the western states (Hall 1981). The northern pocket gopher (*T. talpoides*) (Fig. 11) is widely distributed from Canada south to northern California and New Mexico, and from the west coast east to the Dakotas (Fig. 12). It is typically 16.5 to 25 cm long with yellowish-brown fur. The southern pocket gopher (*T. umbrinus*) is found from Mexico into southern Oregon and east into Colorado and is slightly larger (13 to 34 cm long) with fur that varies from almost white to black. The other species of *Thomomys* have more limited distribution. Densities are highly variable and may range from 7.5 to 96 animals/ha (Richens 1965, Reid et al. 1966), probably averaging 45/ha.

Pocket gopher burrow systems consist of runways usually 10 to 60 cm underground running generally parallel to the surface. Side tunnels off the main runways are used as exits from the system and for deposition of soil, debris, excess food, and feces. Deeper tunnels, at 1.5 to 3 m, lead to nests and food caches. Tunnels are not open to the surface and gophers will generally plug any openings in the tunnel system within 24 hours. A gopher makes characteristic mounds of soil when digging its tunnel and pushing loosened soil out the tunnel to the surface (Fig. 13). The tunnel system of a single gopher may average 35 to 50 m in length for *Geomys* spp. (Thorne and Andersen 1990, Stubbendieck et al. 1979) and 1 gopher may build 100 mounds and move 1.8 metric tons of soil to the surface in a single year (Stubbendieck et al. 1979). Tunnel systems may range from 25 to 63 m in length for *Thomomys* spp. (Smith 1948, Turner et al. 1973) and occupy about 185 m^2 (Turner et al. 1973). Andersen (1987) and Thorne and Andersen (1990) showed that as much as 69% of the tunnel length excavated was later backfilled, indicating that underground excavation exceeded the amount visible from the surface. During winter, pocket gophers may burrow just under the snow, leaving characteristic castings of soil on top of the ground after the snow melts (Fig. 14).

FIGURE 9. Plains pocket gopher (*Geomys bursarius*) equipped with a radiotelemetry collar. Photo by K. Fagerstone.

Preferred Habitat and Impact of Grazing on Pocket Gophers

Pocket gophers occupy coastal areas to elevations of 3,600 m. They are found in a variety of soil types but reach their highest densities on loose, lightly textured soils with good drainage and high porosity (Davis et al. 1938, Kennerly 1964, Miller 1964, McNab 1966, Turner et al. 1973). Because their tunnels are closed, pocket gophers depend upon diffusion through the soil into the tunnel for gas exchange that is facilitated by light textured, porous soils with good drainage. Wet soils or soils with high clay content diffuse gases poorly and are unsuitable for gophers (Kennerly 1964, McNab 1966). The plains pocket gopher is usually restricted to deep sandy and silty soils (Miller 1964, Turner et al. 1973). Downhower and Hall (1966) found that plains pocket gophers in Kansas occurred only in soils with clay content < 30% and sand content > 40%. In Colorado, they are common in alfalfa (*Medicago sativa*) fields, sandhill rangelands, and river bottoms, but are not abundant in shortgrass prairie or areas with compacted soils (Turner et al. 1973). Beck and Hansen (1966) found plains pocket gophers in eastern Colorado to be more abundant on sandy loam soil compared to dune sand, possibly because of greater abundance of plants used for food on the sandy loam soils. The Mexican pocket gopher (*Pappogeomys castanops*) is a large pocket gopher species (occurring primarily in New Mexico and Texas) that prefers soil conditions similar to the plains pocket gopher but has a broader soil tolerance (Miller 1964, Moulton et al. 1983).

Thomomys species generally have a wider soil tolerance. The valley pocket gopher (*Thomomys bottae*) is a small animal that can live in a wider range of soil types than *Geomys* spp. (Miller 1964). Northern pocket gophers have the widest tolerance, occurring in soils ranging from deep sandy soils of the plains to shallow gravel in mountainous areas (Miller 1964, Hansen and Morris 1968). Local distribution of all pocket gophers can be determined by soil depth (Davis et al. 1938), as they require soils >10 cm deep for their burrows, possibly to prevent cave ins and exposure to extreme temperatures during mid-summer and winter (Howard and Childs 1959, Kennerly 1964, McNab 1966). Pocket gophers live successfully in soils where mean soil moisture ranges from < 10% to > 50% (Hansen and Beck 1968) but are not found in continuously wet soils. Ingles (1949)

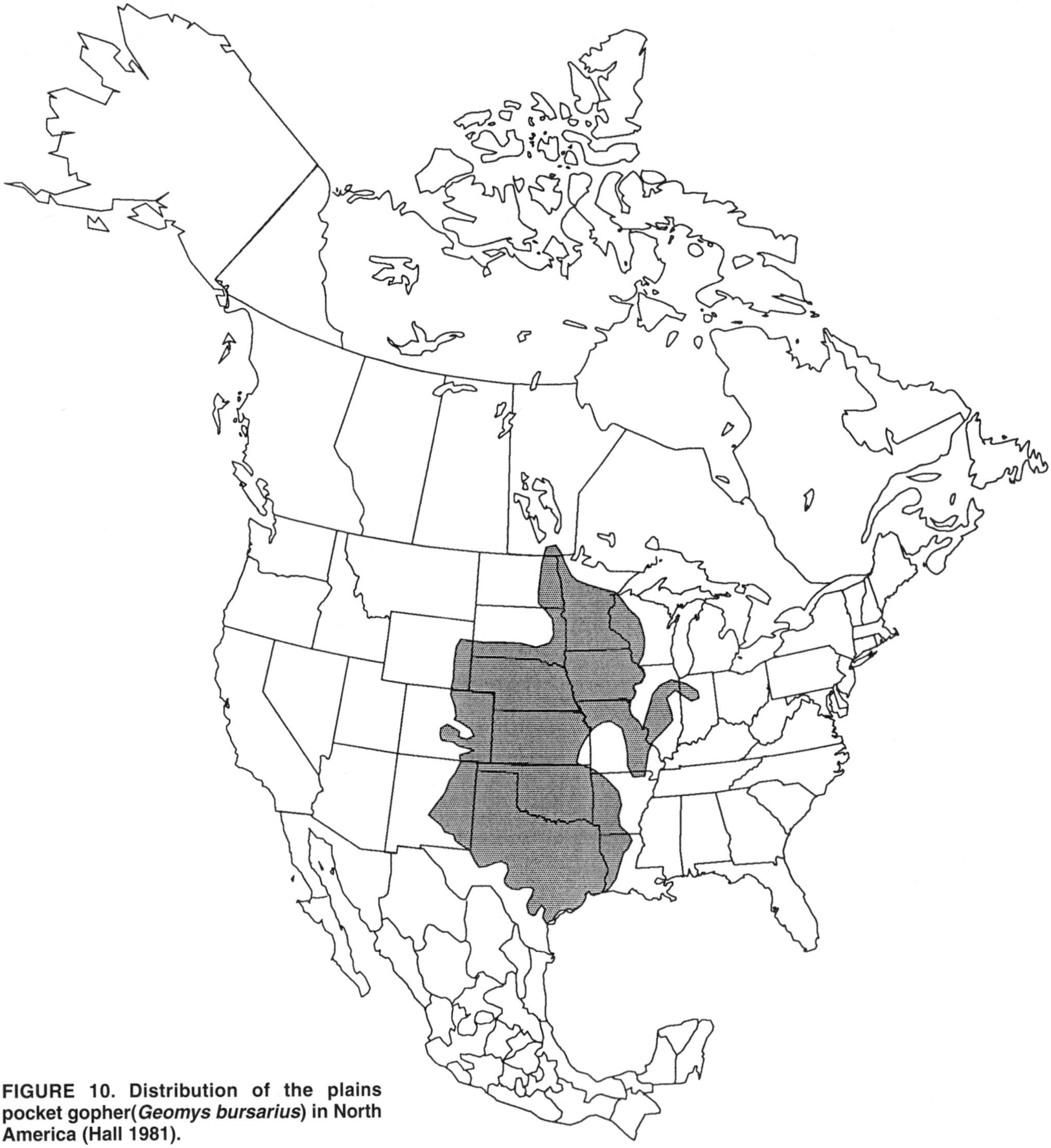

FIGURE 10. Distribution of the plains pocket gopher(*Geomys bursarius*) in North America (Hall 1981).

described how a rising groundwater table forced a local population to vacate suitable habitat.

Pocket gophers are attracted to rangeland in good to excellent range condition, where they use vigorous plants with large root systems. Pocket gopher densities appear to be dependent on plant biomass (Keith et al. 1959, Downhower and Hall 1966, Andersen and MacMahon 1981), particularly biomass of dicotyledonous plants (Williams and Cameron 1990). Tilman (1983) and Reichman and Smith (1985) found that gophers inhabited portions of fields with the highest biomass. Lower plant biomass may require gophers to burrow more extensively to locate food, a theory supported by Reichman et al. (1982). They found that pocket gopher burrows were only half as long in areas with twice the plant biomass. Vleck (1981) observed that once a gopher encountered a resource patch, it exploited that patch rather than constructing additional burrows. Spencer et al. (1985) postulated that on a burned area pocket gopher foraging movement increased in response to the concomitant decrease in forb biomass; similar increases in foraging were not observed on an unburned area.

Grazing has often been alleged to favor large pocket gopher populations (Buechner 1942). However, the impact of grazing on pocket gophers actually appears variable. In 2 studies of northern pocket gopher populations, no significant differences were found among mountain ranges grazed or ungrazed by cattle (Turner 1969, Turner et al. 1973). Other studies have shown higher pocket gopher numbers on ungrazed areas. Moore and Reid (1951) and Ellison and Aldous (1952) found that gophers tended to concentrate inside ungrazed livestock exclosures, and Hansen (1965), Turner et al. (1973), and Hunter (1991) found lower densities of southern pocket gophers and northern pocket gophers in grazed areas. Turner et al. (1973) found a mean of 9 northern pocket gophers/ha on ungrazed ranges compared to only 2/ha on ranges that were grazed lightly, moderately, and heavily by cattle. On Grand Mesa, Colorado, mounds and castings covered 5% more ground surface, and occurred 14% more frequently, on ranges where cattle had been excluded 10 years than on adjacent moderately grazed ranges (Turner 1969). However, other comparisons involving grazing intensity have shown heavily grazed range to have higher gopher densities than lightly grazed range (Laycock 1953). Although these results seem contradictory they may have a biological basis. During the summer grazing season, both sheep and cattle consume large quantities of forbs, which are also the preferred summer foods of northern pocket gophers. Forb availability may be highest for gophers on ungrazed range versus lightly or moderately grazed range and may allow for higher pocket gopher densities on the ungrazed areas. On the other hand, higher pocket gopher densities on heavily grazed range may be an effect of long-term heavy grazing, which can promote greater abundance of forb species than moderate or light grazing.

In contrast to the northern pocket gophers, plains pocket gophers are reported to be most abundant on moderately grazed range, less abundant on ungrazed or lightly grazed range, and absent or least abundant on heavily grazed range (Phillips 1936, Colorado Cooperative Pocket Gopher Project 1960). They seem to prefer the varied vegetation of disturbed grasslands and are more abundant in these areas, provided the soil is suitable for burrowing.

Effect on Rangeland

Pocket gophers affect rangeland in a variety of ways: by burying plants; by transporting nutrients to the soil surface during burrowing and mound formation activities; and by feeding, which decreases biomass of forage plants and alters plant species composition. Pocket gophers compete directly with livestock by consuming range plants, above and below ground. The major food plants of northern pocket gophers are also major food plants of cattle on the same range (Ward and Keith 1962). Consumption of forage by gophers is much higher than for other small mammals because of the higher energetic demands of burrowing (Andersen and MacMahon 1981, Gettinger 1984). Where pocket gophers occur in abundance, they may be the primary non-ungulate consumer of forage, frequently harvesting more vegetation than they actually eat and storing it in underground food caches. Pocket gophers consume primarily belowground vegetation but eat aboveground vegetation during periods when the vegetation is green and succulent (Miller 1964, Andersen and MacMahon 1981, Williams and Cameron 1986*b*).

Although the pocket gopher genera are much alike anatomically and have similar ecological roles, they differ considerably in their diets and in their abilities to use certain foods (Myers and Vaughan 1964). Forbs comprise a much larger portion of the northern pocket gopher diet (67-93%) (Keith et al. 1959, Ward and Keith 1962, Turner et al. 1973, Vaughan 1967) than of the plains pocket gopher diet (30%) (Myers and Vaughan 1964). Keith et al. (1959), Ward (1960), Ward and Keith (1962), and Vaughan (1967) found that forbs are preferred and grasses are a low preference food for northern pocket gophers. Keith et al. (1959) studied the effect of spraying mountain rangeland with 2,4-D. Forb cover was sharply reduced, from 67% of the vegetation and 82% of the pocket gopher diet before spraying to only 30% of the vegetation and 50% of the diet after spraying. Pocket gopher numbers dropped 87% following the spraying because of an apparent inability to survive where forbs were eliminated (Tietjen et al. 1967). Laboratory feeding tests by Tietjen et al. (1967) further demonstrated this inability to survive on a diet high in grasses; only succulent grasses or those bearing corms or rhizomes yielded a subsistence diet, and pocket gophers lost weight even on those grasses. These studies suggest that the optimum habitat for northern pocket gophers is high in forbs.

In contrast to *Thomomys* species, *Geomys* species frequently thrive in grassland areas with few forbs. Myers and Vaughan (1964) studied a population of plains pocket gophers on sandhill rangeland in Nebraska where the summer vegetation was 99% grasses, and the summer diet was 77% grasses, with 6 of 10 preferred plants being grasses. Although grasses constitute the bulk of the diet of plains pocket gophers (Myers and Vaughan 1964, Foster and Stubbendieck 1980, Tilman 1983, Williams and Cameron 1986*b*) and Mexican pocket gophers (Hegdal et al. 1965), succulent forbs are used when available and may be important to pocket gopher survival, as plains pocket gophers feed on some forbs in higher percentages than forbs appear in the habitat. In eastern Colorado, 88% of the vegetation was grasses, but grasses comprised only 64% of the gopher's yearly

FIGURE 11. Northern pocket gopher (*Thomomys talpoides*). Photo from the Denver Wildlife Research Center Archives.

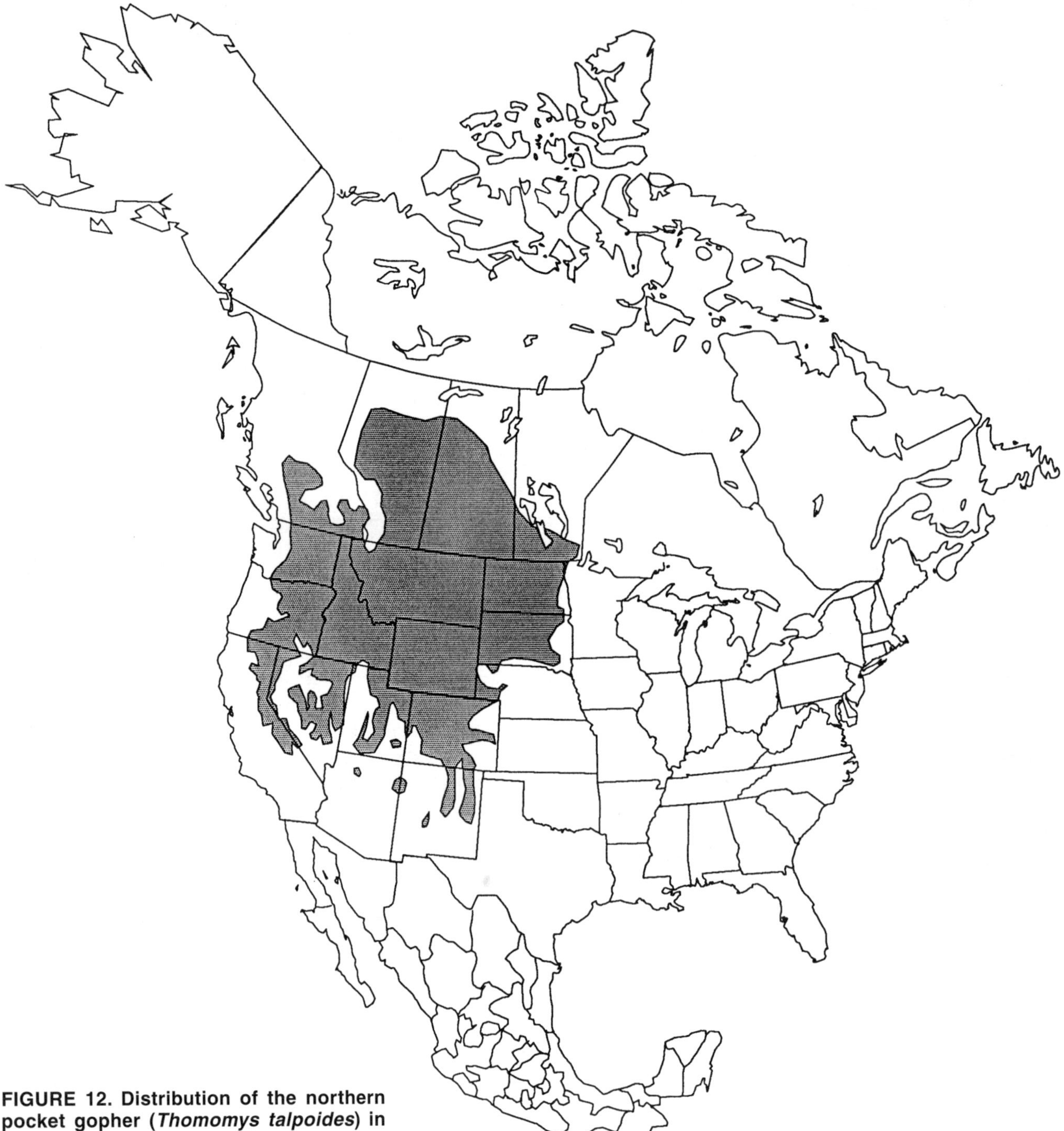

FIGURE 12. Distribution of the northern pocket gopher (*Thomomys talpoides*) in North America (Hall 1981).

diet (Myers and Vaughan 1964). Luce et al. (1980) found that grasses comprised 98% of the vegetation but only 45% of the diet, with forbs comprising 10% and a rush species 15%.

Plains pocket gophers can decrease the forage availability of some desirable livestock grasses on rangeland by their feeding habits. Sand bluestem (*Andropogon hallii*), prairie sandreed (*Calamovilfa longifolia*), needle and thread (*Stipa comata*), western wheatgrass, and blue grama are highly preferred foods for pocket gophers (Myers and Vaughan 1964, Luce et al. 1980). However, not all forage eaten by pocket gophers is of value to livestock; some species of poor forage value to livestock are eaten in small quantities by pocket gophers, including western ragweed (*Ambrosia psilostachya*), thistles (*Cirsium* spp.), and cactus species.

Several studies demonstrated pocket gophers can decrease forage production on rangeland by consumption, clipping, burying litter and vegetation, and reducing plant vigor. However, the amount of the decrease in production varies between short-grass and tall-grass sites and between range condition classes within sites (Stubbendieck et al. 1979, Foster and Stubbendieck 1980,

FIGURE 13. Characteristic mounds of soil made by pocket gophers as they dig tunnels and push soil to the surface. Photo by K. Fagerstone.

FIGURE 14. Soil casts formed by pocket gophers in winter when soil from below ground is pushed into snow tunnels. When the snow melts the soil (tubes) remain on the ground. Photo by K. Fagerstone.

Williams and Cameron 1986*a*). Plains pocket gophers have been shown to have a significant impact on forage production on western Nebraska rangelands, decreasing overall production by 18-49% (Foster 1977, Foster and Stubbendieck 1980) and reducing plant biomass directly over burrows by one-third (Reichman and Smith 1985). In Texas, biomass increased 22% when plains pocket gophers were excluded from rangeland (Williams and Cameron 1986*a*). Fitch and Bentley (1949) observed that southern pocket gophers at a density of 79/ha reduced annual forage production 25% (destroying 284 kg/ha during the growing season) on foothill rangelands in California, where the plant composition was predominantly annual plants. They concluded that pocket gophers had a greater effect on vegetation than ground squirrels and kangaroo rats (*Dipodomys* spp.). Northern pocket gophers reduced herbage production about 20% (224 kg/ha) on mountain rangeland in Colorado (Turner 1969) and 30% in an alpine meadow (Andersen and MacMahon 1981). Alsager (1977) found that forage yields on rangelands in southwestern Alberta, Canada, increased 16% where northern pocket gophers were controlled compared to yields on areas without control. In Utah, following partial control of pocket gophers for 1 to 4 years, herbage production exceeded that on a nearby area on which gophers were not controlled by 560 to 672 kg/ha (Richens 1965).

In addition to changing forage availability, pocket gophers can alter the vegetation species composition of rangeland by feeding upon and burying herbage, and by altering the microenvironment (Branson and Payne 1958, Turner 1969, Turner et al. 1973, Laycock and Richardson 1975, Tilman 1983). Plant species eaten by gophers tend to decrease on rangeland while unpalatable species increase (Myers and Vaughan 1964). Pocket gopher feeding and burrowing activity promotes the presence of annual grasses, annual forbs, and perennial forbs, while decreasing the frequency of perennial grasses (Turner et al. 1973, Foster and Stubbendieck 1980, Kjar and Stubbendieck 1980, Martinsen 1989). These changes in plant composition are related to precipitation, with the most severe changes occurring in areas receiving low precipitation amounts (Kjar and Stubbendieck 1980). Turner et al. (1973) found that pocket gophers caused major changes in vegetation composition on high mountain rangelands, suppressing productivity of some livestock forage species such as common dandelion (*Taraxacum officinale*), lupine (*Lupinus* spp.), agoseris (*Agoseris glauca*), and aspen peavine (*Lathyrus leucanthus*), and increasing production of orange sneezeweed (*Helenium hoopesi*), which is poisonous to sheep and unpalatable to cattle. Two grass species palatable to livestock, slender wheatgrass (*Agropyron pauciflorum*) and mountain brome (*Bromus carinatus*), also benefited from gopher activity. Luce et al. (1980) measured differences in vegetation on rangeland occupied by plains pocket gophers versus unoccupied rangeland and found major changes in basal cover of prairie sandreed and blue grama. Prairie sandreed comprised 25.4% of the basal cover of the vegetation on unoccupied rangeland versus 5.2% on the occupied area; blue grama comprised 16% of basal cover in the unoccupied area and only 7% in the occupied area. A decline in range condition often follows after occupation by pocket gophers as the desirable perennial grasses rapidly decline, accompanied by an increase in annual grass and forbs. This decline may cause gophers to move into previously unoccupied areas and abandon the now weedy areas.

Pocket gopher digging activity can result in decreased biomass and increased bare soil and litter on occupied rangeland. Much of this digging occurs in late summer and fall when young gophers establish their own burrow systems and when adults burrow extensively in search of underground food. However, digging and consequent burial of plants occurs even after snowfall as pocket gophers burrow under the snow. Pocket gopher mounds commonly cover 5 to 15% of the soil surface, with values reported up to 25% (Turner et al. 1973, Stubbendieck et al. 1979, Grant et al. 1980, Spencer et al. 1985) (Fig. 15). Spencer et al. (1985) found that new mounds were formed at the rate of about 58/ha/day on a Texas prairie. Estimates vary regarding

FIGURE 15. Pocket gopher mounds can cover up to 25% of the ground surface in areas of high animal density. Photo by K. Fagerstone.

the amount of soil brought to the surface by pocket gophers. Downhower and Hall (1966) estimated that 1 plains pocket gopher may bring 4,483 kg of soil to the surface annually. The amount of soil brought to the surface each year by *Geomys* spp. in Texas has been estimated at 15,690 kg/ha by Buechner (1942) and between 84,000 and 103,000 kg/ha by Spencer et al. (1985). Gophers in Texas prairie deposited enough soil to cover the entire area in 4 years if no area was covered more than once (Spencer et al. 1985). Estimates for *Thomomys* spp. vary from 11,250 kg (Ellison 1946, Grant et al. 1980) to 85,200 kg (Richens 1966) of soil/ha/year. The deposited soil may bury vegetation and prevent growth of the underlying vegetation (Ellison 1946, Moore and Reid 1951). Plant succession on denuded areas may be slow and may continually provide colonization sites for early successional species (Laycock 1958), many of which are undesirable as livestock forage. Ellison (1946) found that gopher mounds made poor seed beds and Laycock (1953) showed that few plants appeared on northern pocket gopher mounds the season mounds were made; the first plants to appear were usually annuals (Laycock 1958), followed by perennial dicots (Martinsen 1989). Williams et al. (1986) and Martinsen (1989) observed that herbaceous perennial dicots benefit from pocket gopher disturbance by germinating and surviving on mounds in greater numbers than off mounds. Vegetation density on pocket gopher mounds increases rapidly over time as perennial species replace less desirable annual and forb species (Foster and Stubbendieck 1980, Kjar and Stubbendieck 1980). Foster and Stubbendieck (1980) found that vegetation was sparse on gopher mounds <1-year-old and that frequency of perennial grasses increased with mound age while frequency of annual grasses and forbs decreased; the highest increases over 4 years occurred for blue grama (from 2 to 18%), needle and thread (0-63%), prairie junegrass (*Koeleria cristata*) (0-19%), and western wheatgrass (0-24%). After heavy livestock grazing, rangeland improvement through natural plant succession was delayed on areas disturbed by plains pocket gophers (Laycock 1958, Foster and Stubbendieck 1980).

In addition to causing a biomass decrease, pocket gopher mounds may contribute to soil erosion (Foster and Stubbendieck 1980). Northern pocket gophers in Colorado were the principal agent causing erosion (Thorn 1977). At a 11 × 5 m plot, 874 kg of soil had been removed by erosion, equivalent to a plotwide surface lowering of 1.2 cm. Erosion rates caused by pocket gophers were higher than those attributed to other processes in the area.

Unequal displacement of soil by gophers may result in formation of mima mounds (Branson et al. 1965, Murray 1967, Turner et al. 1973, Cox 1990). Such mounds are usually 0.3 to 0.9 m high and 4.5 to 30 m in diameter and are formed over decades as gopher burrowing activities tend to move soil toward the nest (Arkley and Brown 1954, Cox 1984, Cox and Allen 1987). The soil on mima mounds may differ considerably from adjacent soils, having a lower bulk density, higher water permeability, higher organic matter content, and a lack of definite structure in the topsoil (Ross et al. 1968). Stones of the sizes pocket gophers can move are concentrated in mounds (Cox 1984, Cox and Allen 1987) and vegetation on mounds is usually denser and more effective in retarding soil erosion than that off mounds. In a Colorado range seeding project, grasses produced 2 to 5 times more herbage on mounds than between mounds (McGinnies 1960).

Nutrient cycling may be affected by deposition of soil by pocket gophers. Laycock and Richardson (1975) determined that organic matter, nitrogen, and phosphorus increased in areas occupied by pocket gophers. In contrast, other researchers reported reduced nutrients in occupied areas. Grant and McBrayer (1981) measured nutrient content of soil samples from old mounds, new mounds, and away from mounds (controls) and found that old mounds were often significantly lower in nutrient concentrations than new mounds, which were lower than controls. McDonough (1974), Reichman and Smith (1985), and Inouye et al. (1987) reported that gophers reduced the average nitrogen concentration near the soil surface and increased the variability in soil nitrogen. Spencer et al. (1985) found soil deposited by pocket gophers to be lower in nutrient content (i.e., phosphorus, nitrate, and potassium) than randomly collected samples; they theorized that the subsoil deposited on the surface by pocket gophers has had nutrients both leached from it and drawn from it by plant roots. The deficiency of nutrients in mounds may also occur because mounds lack the litter layer that is important in holding moisture and nutrients near the surface. The reduction in average soil nitrogen affects plant species composition and slows succession (Inouye et al. 1987); by creating openings in vegetation that are low in nitrogen and high in light, gopher mounds create conditions typical of the early stages of succession.

Some activities of pocket gophers may be beneficial to rangeland. Gophers loosen compacted soil, allowing better aeration and increased water infiltration, and increase soil fertility by adding excrement and burying vegetation. Soil in newly formed mounds often is more friable than adjacent soil (Turner et al. 1973), so this loose soil may increase the infiltration of water (Hansen and Morris 1968) and retard runoff (Abaturov 1968).

Grant et al. (1980) and Spencer et al. (1985) found that the decrease in biomass caused by mounds is partly compensated for by increased production in areas immediately adjacent to mounds, where production was roughly 5.5% higher than vegetative production 40 or 50 cm from the edge of mounds (Grant et al. 1980, Williams et al. 1986). The researchers hypothesized that increased density of vegetation near mounds is a response to increased nutrient availability caused by leaching of nutrients from mounds into surface soil.

Associations With Other Animals

A variety of animals occupy abandoned pocket gopher burrows, including deer mice (*Peromyscus* spp.), pocket mice (*Perognathus* spp.), kangaroo rats, voles, ground squirrels, and weasels. In addition, numerous reptiles and amphibians have been observed or captured in active and abandoned gopher runways (Howard and Childs 1959, Vaughan 1961, Hansen and Ward 1966). Major predators of pocket gophers include weasels, which pursue the gopher in its tunnel system, and badgers, which are adept at digging them out. Coyotes are common predators but pocket gophers probably make up only a small percentage of the coyote's diet (Sperry 1941). Other predators prey on gophers when they are above ground feeding or constructing their mounds, including bobcats, foxes (*Vulpes* spp. and *Urocyon cinereoargenteus*), skunks (*Mephitis* spp. and *Spilogale* spp.), great horned owls, great gray owls (*Strix robulosa*), barn owls (*Tyto alba*), long-eared owls (*Asio otus*), burrowing owls, red-tailed hawks, Swainson's hawks, goshawks (*Accipiter gentilis*), sparrow hawks (*Falco sparverius*), and snakes (Tryon 1943, Fitch et al. 1946, Evans and Emlen 1947, Fitch 1947*b*, Tryon 1947, Moore and Reid 1951, Craighead and Craighead 1956, Howard and Childs 1959, Hansen and Ward 1966). Dispersing pocket gophers probably are the most vulnerable to predation, as they have no established burrow systems and sometimes travel above ground (Howard and Childs 1959).

Effects of pocket gopher activities on populations of small mammals have not been well studied. However, some general effects can be inferred. Litter can be substantially reduced by pocket gopher activity. Because reduction in litter tends to make the habitat more suitable for deer mice and less suitable for voles, populations of deer mice would tend to increase and voles to decrease. Furthermore, gophers compete directly with voles for food (Ward 1960, Johnson 1964).

Management

Proper range management can favor populations of plains pocket gophers, which are attracted to areas of high and improving range condition, where they use vigorous plants. Once present, pocket gophers interact with grasslands and livestock in ways that can decrease rangeland productivity by 25 to 50%. Range managers therefore need to be aware that the presence of pocket gophers may necessitate some reduction in livestock grazing to maintain good range condition.

Despite lowered rangeland production where pockets gophers occur, many researchers believe that gophers are not a significant problem on well-managed rangelands (Turner et al. 1973), and in some areas, such as high mountain rangeland, the total ecological effects of pocket gopher populations may be beneficial rather than detrimental. Where livestock were excluded, Turner (1969) found the largest grass biomass increase on range with pocket gophers, so it is possible that pocket gophers may actually improve depleted range. Presently, pocket gopher control is rarely practiced on western rangelands and control is more frequently recommended for improving deteriorated rangelands than for maintaining rangelands that are well managed and productive. Where range condition is poor, it may occasionally be advantageous to reduce pocket gopher populations; for example, production on deteriorated mountain rangeland in Colorado increased 290 kg/ha (mainly perennial dicots) after pocket gopher control (Turner 1969). However, even on ranges where gophers are numerous, overgrazing by livestock is frequently the primary cause for deterioration of vegetation cover (Ellison 1946).

Because trapping is extremely slow, time consuming, and practical only in small areas, the most widely used approach to alleviate pocket gopher damage is poisoning. Poisons are distributed either by hand baiting in each burrow system or by a burrow-builder, which creates an artificial burrow and places bait in that burrow (Case 1983). Indirect or ecological control involves habitat modification to make the area less suitable for pocket gopher occupation. Control of forbs, which frequently have large underground storage structures, can be an effective method for minimizing damage to rangeland by northern pocket gophers. Application of 2,4-D herbicide in Colorado reduced pocket gopher populations by 80 to 90%. Success was attributed to decreased forb production (Keith et al. 1959, Hansen and Ward 1966, Tietjen et al. 1967) and resulting starvation of pocket gophers. Where vegetative composition after herbicide treatment remained relatively stable for 5 years (with a grass dominance), repopulation of treated areas was slow. Howard and Childs (1959) also found that herbicide spraying or burning depleted the forb supply and greatly reduced valley pocket gopher numbers. Reduction of forbs is less effective for reducing plains pocket gopher populations, as they survive well on warm-season prairie grasses that have large root to stem ratios.

Other management techniques are needed, particularly for plains pocket gophers. Also, a long-term comprehensive cost:benefit analysis of pocket gopher control needs to be completed that would include reinvasion rates by pocket gophers. Where pocket gophers have been removed from rangeland, it would be advantageous to have a repellent available for use in burrow systems to prevent rapid reinvasion of those burrows by other gophers. In addition, more data are needed on how to manage for good range productivity, in the presence of pocket gophers and after control. For example, what effects do fertilization, reseeding, and manipulation of grazing patterns (such as rest or light grazing) have on rangelands before and after pocket gopher control?

The importance of pocket gophers as a component of grassland ecosystems becomes evident as we consider their role as prey items and their influence on soils, microtopography, habitat heterogeneity, plant species diversity, and primary productivity. Management efforts for rangeland must be based on a balanced ecological approach that takes all of these influences into consideration.

OTHER SMALL RODENTS

General Effects of Grazing on Small Rodents

Variable effects of grazing on small mammals depend on the level of grazing, the type of grassland, and the particular small mammal species involved. Moderate grazing may have little effect or even a positive effect on many species, but overgrazing depresses populations of most small mammals. For example, heavy grazing and repeated fires in sagebrush range caused the establishment of nearly pure stands of annual grasses (cheatgrass [*Bromus tectorum*] and medusa-head [*Taeniatherum asperum*]) over vast areas of southern Idaho (Stewart and Hull 1949, Hironaka 1961). Such sites support only a few deer mice and Great Basin pocket mice (*Perognathus parvus*). In southern Idaho, rodent burrow numbers were significantly higher on ungrazed than on heavily grazed pastures (Anderson 1972), and in a seldom grazed pasture in Arizona, the total rodent population was roughly twice as high as on a heavily grazed pasture (Gallizioli 1979). Carothers et al. (1976), in a study of the impact of burros (*Equus asinus*) on a desert region at the Grand Canyon, found higher total numbers and a greater diversity of small mammals on an area devoid of burros compared to an adjacent heavily grazed area. When range abuse is continued for decades as it has been in certain areas of the West and Southwest, the inevitable severe soil erosion reduces the quality of the habitat for even grazing-tolerant species like kangaroo-rats, prairie dogs, and jackrabbits.

Consistent heavy grazing can reduce the number of species present in an area, but moderate or light grazing may have no effect or can sometimes even promote species diversity. Oldemeyer and Allen-Johnson (1988) studied effects of grazing on small mammal populations on the Sherburne National Refuge, Idaho, where sagebrush and mountain mahogany (*Cercocarpus montanus*)-dominated ecosites were grazed under a deferred-rotation grazing system (grazed mid-Jun to early Aug in 1 year and early Aug through late Oct the next year). They concluded that the grazing regime had no discernible impact on the relative abundance and diversity of small mammals, or on populations of the 2 most abundant species (deer mice and least chipmunks [*Eutamius minimus*]), 4 and 5 years after its implementation. Moulton (1978) and Moulton et al. (1980) found that moderate grazing of cottonwood (*Populus* spp.) riparian woodland provided habitat opportunities for twice the number of species as an ungrazed area. In this case the grazed area had been managed for 17 years at a moderate stocking rate under a spring deferment and a July to December grazing season. This grazing management system promoted microhabitat and edge diversity by allowing patchy habitats to develop and provided niches for a greater variety of small mammals. In contrast, secondary succession on the ungrazed site led toward more uniform plant composition and fewer small mammal species.

Several theories have been tested to explain the presence of small mammals on grasslands, including the size and availability of food and the structural attributes of the habitat. Animal species diversity has often been dependent on habitat structural diversity (Pimlott 1969, Pianka 1973); research by Grant et al. (1982) suggests that this concept is true for grasslands, where the general composition of small mammal communities is determined primarily by habitat structural attributes. Livestock grazing, through selective defoliation of forage species, affects many aspects of grassland ecosystems, including plant cover or biomass, plant species composition and diversity, primary productivity, soil compaction, and soil moisture. These changes can drastically alter the structural attributes of the ecosystem. Plant cover probably has the most influence on small mammal population abundance because it provides food, nests, and protection from predators. Amount of plant cover also influences behavioral interactions such as fighting and dispersal (Warnock 1965, Krebs 1970, Krebs et al. 1973, Myers and Krebs 1974), and moderates ground level humidity, temperature, and soil moisture (Birney et al. 1976). Mulch, a by-product of cover, modifies surface temperatures and retards evaporation (Hopkins 1954). Removal or alteration of cover can cause changes in small mammal communities (LoBue and Darnell 1959, Birney et al. 1976, Geier and Best 1980, Grant et al. 1982). MacCracken et al. (1985*a*) found a significant positive relationship between small mammal abundance and canopy cover in sagebrush-grass rangeland in Montana, where all areas were managed on a rest-rotation grazing system. Geier and Best (1980) found that, of plant life forms, percentage forb cover was most consistently correlated with small-mammal species abundances, grass coverage was of lesser importance, and tree cover was not related to species abundance. They found a significant correlation between the abundance of woody plant debris (e.g., logs, brushpiles, stumps) and small-mammal numbers. These data are consistent with that of Steenbergh and Warren (1977), who reported that rodent abundance and diversity increased with vegetation cover and density and that overgrazing by cattle decreased vegetation complexity. Hanley and Page (1982) observed that grazing altered rodent species diversity through changes in plant species diversity on several habitats in northeastern California.

In assessing grazing impacts on small mammal communities, Hanley and Page (1982) stressed the importance of evaluating effects on a habitat-type basis. Grant et al. (1982) also concluded that the response of small mammals to grazing depends on the site and the original mammal species composition and therefore differs greatly between grassland types; they found differential changes in several small mammal community parameters between grazed and ungrazed sites in 4 Western grassland communities. Where there is sufficient vegetation in ungrazed grasslands to support herbivorous, litter-dwelling species, the small mammal communities are changed significantly by a reduction in vegetation cover caused by grazing. This is true for tallgrass and montane grasslands, which have significantly greater standing vegetation, greater annual net primary production, and greater abundance of such mammals than shortgrass and bunchgrass grasslands (French et al. 1976, Grant and Birney 1979, Grant et al. 1982). The reduction in vegetation cover on tallgrass grazed sites may bring cover below the level required to support dense populations of herbivorous, litter-dwelling small mammals such as voles (Birney et al. 1976, Grant et al. 1982). The result is a decrease in total small mammal biomass, an increase in small mammal species diversity, and a shift from litter-dwelling species with relatively high reproductive rates to surface-dwelling species with relatively low reproductive rates. In studies by Grant et al. (1982), microtines dominated ungrazed tallgrass grassland, with cricetines roughly half as abundant. In contrast, grazed tallgrass grassland was dominated by sciurids and heteromyids. In montane grasslands, grazing-induced reduction in cover resulted in similar decreases in total small mammal biomass and changes in species composition from litter-dwelling species to surface dwelling species, but resulted in a decrease rather than an increase in species diversity. At montane sites microtines dominated the ungrazed area but cricetines dominated the grazed area (Grant et al. 1982). In shortgrass and bunchgrass grasslands (where plant cover was already low) small mammal faunas and abundance were not changed drastically by further reduction in vegetation cover by grazing (Moulton 1978, Grant et al. 1982). Small mammal communities of shortgrass and bunchgrass are frequently composed primarily of surface-dwelling, granivorous and omnivorous species adapted to an open habitat. Grant et al. (1982) found that on bunchgrass grassland sciurids and heteromyids were dominant, and on shortgrass grassland biomass was greatest for cricetines and sciurids, followed by heteromyids. A reduction in cover may actually improve conditions for granivorous rodent species (Baker 1971, Grant et al. 1982), because grazing can increase the abundance of annual grasses and forbs, which produce more seeds than the perennial grasses that are reduced in abundance by grazing. Plant species composition does not appear to be as important a structural attribute as cover in determining rodent species composition. For example, treatment of shortgrass prairie with fertilizer caused a marked increase in vegetation cover and a subsequent immigration of prairie voles (*Microtus ochrogaster*), but did not significantly change plant species composition (Grant et al. 1977). These studies suggest that the effect of grazing on small mammals will be related to the extent to which a habitat can be perturbed. Tallgrass habitats (such as at riparian sites) potentially can be transformed by grazing to midgrass, mixed grass, or shortgrass habitats. Midgrass habitats (i.e., sand sagebrush) can be transformed to shortgrass, but shortgrass remains shortgrass.

Small mammals in riparian habitats can be significantly impacted by grazing. Kauffman et al. (1982) compared small mammal populations before and after late season (late Aug to mid-Sep) grazing. Population estimates decreased in all habitats, from 800 to 83 mammals/ha in hawthorne (*Crataegus* spp.) communities, from 450 to 60 mammals/ha in meadow communities, and from 129 to 42 mammals/ha in cottonwood-mixed conifer communities. Population estimates in ungrazed areas changed during this same time period from 690 to 136/ha in hawthorne communities, from 235 to 463/ha in meadow communities, and from 118 to 254/ha in cottonwood communities. The significant decrease in small mammal populations in grazed areas was apparently caused by a loss of cover due to forage removal, resulting in increased predation and immigration out of grazed habitats into neighboring ungrazed habitats.

In the Great Basin region of the United States, the quality of habitat for small mammals is also strongly dependent upon habitat structural features (Hanley and Page 1982). Overgrazing by domestic cattle and sheep during the late 1800s drastically altered plant communities in the Southwest (Griffiths 1902, Barnes 1936), resulting in severe depletion of vegetation cover by 1900. Much of the Great Basin once dominated by perennial grasses is currently dominated by desert shrubs (Hastings and Turner 1965, Hanley and Page 1982), including creosote (*Larrea tridentata*), tarbush (*Flourensia cernua*), mesquite (*Prosopsis juliflora*), Mormon tea (*Ephedra trifurca*), and acacia (*Acacia* spp.). The effects of livestock grazing on plant life form diversity, and subsequently on rodent species diversity, differ greatly by habitat. Despite long-term protection from grazing, vegetation patterns on desert grasslands have not readily reverted to their original grass cover (Chew 1982, Roundy and Jordan 1988, Heske and Campbell 1991), perhaps because recovery takes a very long time in desert environments or because, once established, woody plants may competitively restrict the reestablishment of herbaceous cover and perennial grasses (Anderson and Holte 1981, West et al. 1984, Daddy et al. 1988). These vegetation changes have caused changes in rodent populations. For example, reduction of herbaceous vegetation and increased abundance of shrubs as a result of livestock grazing have resulted in an increase in plant life form diversity in the most mesic habitats; this is associated with increased rodent diversity (i.e., rodent species richness and evenness), probably because of increased diversity of seed size, seed type, and microhabitat. Generalist granivorous foragers like the least chipmunk, Great Basin pocket mouse, and deer mouse have increased most in mesic habitats. In contrast, reduction of herbaceous vegetation by livestock grazing has resulted in a reduction in plant life form diversity and rodent diversity in xeric communities.

Small Rodent Species Adversely Affected by Grazing

Some rodents may be adversely affected by livestock grazing through depletion of food resources, including wood rats (*Neotoma* spp.) (Steenburgh and Warren 1977), antelope squirrels (*Ammospermophilus* spp.) (Steenbergh and Warren 1977), cotton rats (*Sigmodon hispidus*) (Goertz 1964, Feldhamer 1979), Townsend moles (*Scapanus townsendi*) (Kuhn et al. 1966), voles (Birney et al. 1976), harvest mice (*Reithrodontomys* spp.) (Black and Frischknecht 1971), jumping mice (*Zapus princeps*) (McGee 1982), and masked shrew (*Sorex cinereus*) (McGee 1982). Steenbergh and Warren (1977) found woodrats (*Neotoma albigula*) more common in areas with greater vegetation cover, and away from intensively grazed areas. Vorhies and Taylor (1940) noted that on ranges not overgrazed by domestic stock, the wood rat, by feeding preferentially on woody plants and weedy herbs, can aid in reestablishment of climax perennial grasses. Antelope squirrels eat mostly green grasses, which are also eaten by livestock (Howard et al. 1959, Bradley 1968, Hawbecker 1975). Wood (1969) found that antelope squirrels were most abundant in creosote and climax grassland areas, 2 vegetation types not likely to be overgrazed.

Cotton rats are restricted to the abundant cover of undisturbed grassland (Phillips 1936, Smith 1940, Whitford 1976, Feldhamer 1979). Their populations correlated positively with density and height of perennial grass cover (Goertz 1964) and they do not use heavily grazed areas. Fleharty and Mares (1973) reported that cotton rats preferred habitat with dense undergrowth and protective overstory and Guthery et al. (1979) reported densities 4 times greater on areas with a greater vegetation biomass. Cotton rat populations were low where cattle grazing reduced grass cover; yet where cattle were excluded and there was nearly complete grass cover, populations of cotton rats increased (Whitford 1976).

Cattle grazing impacts from trampling were a significant factor on nestling mortality of Townsend moles in wetland pastures of western Oregon (Kuhn et al. 1966). Normal grazing activities on soft ground early in the season permitted the cattle's hooves to break through the sod and crush the nestlings.

Voles. Most vole species (Fig. 16) select good cover, so increased canopy cover is important in increasing Microtine populations (Zimmerman 1965, Getz 1971, Hodgson 1972, Douglas 1973, Birney et al. 1976, Feldhamer 1979, Grant et al. 1982). Prairie voles generally inhabit areas with dense stands of lush vegetation (Jameson 1947, Carroll and Getz 1976, Abramsky 1978, Abramsky et al. 1979, Fleharty and Navo 1983). They do well in ungrazed or only lightly grazed grasslands but disappear from areas with moderate to heavy grazing (Birney et al. 1976, Moulton 1978). In a study by Agnew et al. (1986), prairie voles were only captured on mixedgrass prairie sites and did not occur on shortgrass sites such as prairie dog colonies. Kauffman et al. (1982) found that the mountain vole (*Microtus montanus*) was also drastically reduced in numbers or disappeared from habitats due to grazing. Studies of vole cycles by Birney et al. (1976) emphasize the importance of cover because at widely distant sites in the western United States, populations of prairie voles and meadow voles (*M. pennsylvanicus*) cycle where the grasslands are dense but not where grasses are kept low by burning or grazing. Birney et al. (1976) showed a significant increase in vole populations after cessation of grazing in a tallgrass prairie, from only 1 individual/ha on a grazed area versus 24/ha on an ungrazed area. Vole populations on a grazed mid-grass prairie in South Dakota were only 2.8/ha, similar to shortgrass prairie populations in Colorado, where prairie voles on an ungrazed area reached a high of only 3.5 individuals/ha, and these individuals may have been transients. Cover, especially litter, was low on both the grazed South Dakota and ungrazed Colorado grasslands and was apparently not sufficient for vole populations to exceed the increase phase of a population cycle. Birney et al. (1976) hypothesized that there may be a cover threshold (between 400 g/m^2 and 600 g/m^2) required before cycles can occur, and that a lower threshold may have to exist before voles can establish resident breeding populations. These researchers postulated that cover provides favorable conditions for population build-ups by providing food, reducing antagonistic contacts between voles, and moderating microhabitat humidity and temperature.

FIGURE 16. Most vole species, such as *Microtus canicaudus*, inhabit areas with dense vegetative cover, and populations decline under heavy grazing. Photo by J. Wolfe.

During normal years, voles have little influence on grasslands, although they may have a direct impact on soil. Golley et al. (1975) pointed out that voles at a density of 200-400/ha probably dislodge 1,000 m^3 of earth/ha/year. Because this activity is restricted to the top 40 cm of soil, such activities influence microtopography and surface water runoff. Voles normally have little effect on vegetation cover because the amount of standing crop vegetation they remove is usually quite small, ranging from 1-35% (Krebs and Myers 1974, Batzli 1975), with the higher figure being unusual. However, Batzli and Pitelka (1970, 1971), after study of the California vole (*Microtus californicus*) in the field, and Krebs and Myers (1974), after review of literature, concluded that during the late increase and peak phases of a population cycle, grazing by voles can have a marked effect on vegetation cover. Foster (1965) found that grazing by microtine rodents removed current-season stem primordia of perennial grasses. And in a series of exclosure experiments, Batzli and Pitelka (1970) showed that grazing by voles kept the habitat open and increased plant species diversity; if voles were excluded, grasses, the preferred food type, increased and became dominant. In some instances, voles can have severe effects on vegetation. Batzli and Pitelka (1970) found that a population of California voles that exceeded 1,500 voles/ha removed 85% of the volume of vegetation for wild oats (*Avena fatua*), Italian ryegrass (*Lolium multiflorum*), and ripgut grass (*Bromus rigidus*). Heavy cropping of plants during reproduction suppressed flowering and caused a 70% seed loss (Batzli and Pitelka 1970). Seed predation by mice, including voles, may be an important regulating factor for some plant species (Janzen 1971). In California annual grasslands, mice at densities of 296/ha consumed 65 to 75% of wild oat seed (Marshall and Jain 1970, Borchert and Jain 1978); these seeds were chosen twice as frequently as those of other plant species, even though they were only 4% of the total plant cover.

During cyclic population peaks, voles (*M. longicaudus*, *M. montanus*, and *M. pennsylvanicus*) can reach densities

≤ 7,400/ha (Spencer 1958). At these high population levels, voles can kill and damage sagebrush and other shrub species (Mueggler 1967, Phillips 1970, Frischknecht and Baker 1972) by stripping bark from plants and girdling stems and branches. Damage is greatest when a dense, ungrazed herbaceous understory exists that favors vole population increases, and when the snowpack persists throughout the winter (Mueggler 1967, Frischknecht and Baker 1972). Usually voles kill only portions of the crowns of individual plants but occasionally entire plants are killed. During 1962-1964, Mueggler (1967) observed an irruption of voles on southwestern Montana rangeland that caused damage to a number of shrubs, including big and silver sagebrush (*Artemisia* spp.), sumac (*Rhus trilobata*), bitterbrush (*Purshia tridentata*), mountain mahogany, and serviceberry (*Amelanchier alnifolia*). Mueggler (1967) recorded crown kills of 35 to 97% of sagebrush on extensive areas, with > 80% of the stand killed in some areas. A similar population explosion of long-tailed voles (*M. longicaudus*) in Utah in 1969 killed 59% of sagebrush plants and damaged another 28% (Frischnecht and Baker 1972). If peaks in population could be predicted, voles could be used to control undesirable shrubs. However, as natives of the sagebrush-grass ecosystem, normal vole populations have little impact on rangeland. Factors responsible for population irruptions and methods of inducing them would have to be identified before voles could be effectively used for biological control of sagebrush.

Sagebrush voles (*Lagurus curtatus*) occur in the most xeric habitats occupied by microtines (O'Farrell 1972) and are 1 of the rarer species of voles. Rickard (1960) and O'Farrell (1972) suggested that the distribution of the sagebrush vole was restricted to areas where big sagebrush (*Artemisia tridentata*) occurs along with associated large bunchgrasses such as bluebunch wheatgrass (*Agropyron spicatum*). Hall (1946) stated that big sagebrush was one of the main foods of sage voles. However, Moore (1943) found little evidence of feeding on sagebrush; and Hansen (1956), Maser (1974), Maser and Strickler (1978), and MacCracken et al. (1985*a*), found that good grass cover, not sagebrush, was the limiting factor for sage vole distribution. Maser (1974) and Maser and Strickler (1978) found voles only in the best rangeland and not on overgrazed, predominantly sagebrush-covered habitat. Oldemeyer et al. (1983) and Oldemeyer and Allen-Johnson (1988) found sage voles more abundant on ungrazed rangeland (with greater plant biomass) than on grazed rangeland in Nevada. Sagebrush voles were found to be beneficial to fescue-dominated (*Festuca* spp.) rangeland in Oregon (Maser and Strickler 1978) because voles deposit vegetation and fecal pellets around the grass that make nutrients available to the fescue plants. Sagebrush voles are not a concern from an economic standpoint; their feeding habits cause little observable permanent damage to rangeland and may help foster grass seed germination (Maser 1974).

Harvest Mice. Western harvest mice (*Reithrodontomys megalotis)* and plains harvest mice (*R. montanus*) usually occur in tall, dense grass habitats (Quast 1950, Maxell and Brown 1968, Armstrong 1972, Larrison and Johnson 1973, Whitford 1976, Ford 1977, Feldhamer 1979). The optimum habitat for western harvest mice is lush vegetation, usually grassy, weedy habitat with 90-99% cover (Whitford 1976, Ford 1977, Abramsky 1978, Abramsky et al. 1979, Fleharty and Navo 1983). Western harvest mice are therefore most abundant on rangeland with no or relatively light grazing (Black and Frischknecht 1971). Moderate grazing tends to diminish their numbers until populations disappear when heavy grazing leads to range depletion and reduced grass cover (Whitford 1976). Western harvest mice were found by Agnew et al. (1986) to occupy mixed-grass prairie sites only; they do not occupy short grass sites such as prairie dog colonies. Plains harvest mice were found by Kaufman et al. (1990) to be the most restricted in distribution of the species studied; they would be severely impacted if their preferred habitat (upland mixed-grass prairie) were altered by grazing.

Species Not Affected or Affected Favorably by Grazing

Populations of a few species of small rodents may be increased where livestock are grazed, including the least chipmunk, deer mice, some pocket mice, and numerous kangaroo rats.

Chipmunk. The least chipmunk is an opportunistic forager and is the most widespread of all North American chipmunks (Hall 1981), suggesting that they adapt rapidly to a variety of habitats (Sullivan 1985). Range depletion favors an increase in least chipmunk density; they were found most common in depleted shadscale (*Atriplex* spp.), where numbers were 10 times greater than in stands in better condition (Larrison and Johnson 1973), and in grazed sagebrush-fescue sites, where they were 4 times more abundant than in ungrazed sites (Hanley and Page 1982). Chipmunks and golden mantled ground squirrels (*Spermophilus lateralis*) have been found to favor the spread and increase of bitterbrush by storing seed in surface caches (Hormay 1943).

Deer Mice. Deer mice (Fig. 17) and white-footed mice (*P. leucopus*) are pioneer species that occur in most vegetation types during most stages of plant succession, but usually not in large numbers (Geier and Best 1980). The deer mouse is a widespread and adaptable species with broad diets, occupying sites ranging

FIGURE 17. Grazing is often beneficial to the white-footed deer mouse (*Peromyscus maniculatus*), a widespread and adaptable species. Photo by E. Kalmbach.

from native prairie to cropfields (Kaufman et al. 1990, Feldhamer 1979). It is sometimes referred to as a weed species, because disturbances that result in early seral stages favor population increases (Williams 1955, Armstrong 1977), and it is usually the most abundant small mammal in severely disturbed areas (Halvorson 1982). Therefore, grazing is generally beneficial to deer mice (Phillips 1936, Harris 1971, Moulton 1978), which select areas with low cover (Black and Frischknecht 1971) and are common in habitats with bare soil surface and open vegetation such as grazed prairie (Phillips 1936, Quast 1950, Grant et al. 1982). Researchers have found that heavy grazing in big sagebrush habitat promotes an increase in deer mice numbers (Black and Frischknecht 1971, Larrison and Johnson 1973). Kaufman et al. (1982) found that the total small mammal population declined in grazed communities, but the density estimates of deer mice increased; they were dominant after the grazing season while they were found in only minor proportions before. Feldhamer (1979) and Oldemeyer and Allen-Johnson (1988) noted that deer mice used microhabitats with greater shrub intercept, which would be the case in grazed areas.

However, range depletion in all habitats does not always favor an increase in deer mouse populations. Hallett (1982) did not find deer mice associated with any measured habitat variable. Geier and Best (1980) found increased deer mice populations with increasing forb cover, and Fitzgerald (1978) found more deer mice in an ungrazed than in a grazed riparian habitat, but fewer deer mice in an ungrazed than in a grazed shortgrass prairie upland. Hanley and Page (1982) concluded that deer mice have differential responses to grazing, decreasing in the most xeric habitats and increasing in mesic habitats.

Deer mice do not normally have an effect on rangeland vegetation. However, seed predation by mice may be an important factor in northwest rangelands. During reseeding efforts, deer mice may consume or cache considerable seed, resulting in poor plant establishment. In contrast, West (1968) concluded that seed caches may result in clumps of seedlings and that 50% of bitterbrush and 15% of ponderosa pines (*Pinus ponderosa*) in Oregon resulted from rodent seed caches.

Effects of Grazing on Desert Rodents

Desert shrub-grasslands in the Southwestern United States are ephemeral environments characterized by seasonal bursts of vegetation productivity after sporadic rains (Reichman 1975). Vegetation is composed of desert shrubs, drought resistant summer-growing perennial grasses, and annual plants. Much of this area is rangeland used primarily for grazing livestock. Seeds persist after green vegetation is gone and are therefore the dominant items in rodent diets. The most specialized granivorous rodents in North American deserts belong to the family Heteromyidae, which includes the pocket mice, kangaroo rats, and kangaroo mice (*Microdipodops* spp.). These rodents feed almost exclusively on seeds and are efficient at locating and harvesting them. The rodents are nocturnal, have external cheek pouches (permitting collection and movement of large quantities of seeds during each foraging bout), and highly efficient kidneys, which enable them to excrete nitrogenous wastes and maintain osmotic balance on a diet containing little free water (Eisenberg 1963, Schmidt-Nielsen 1964). Pocket mice use areas beneath and around shrub canopies (Brown et. al. 1979*b*) where average seed densities are higher and more uniform than in open microhabitats. Bipedal kangaroo rats and kangaroo mice forage primarily in open spaces between perennial vegetation (Brown et al. 1979*b*), where they move rapidly and efficiently between widely spaced seed clumps. Grasshopper mice (*Onychomys* spp.) are commonly found in areas with open vegetation (Abramsky et al. 1979) but are not closely associated with particular habitats in the desert Southwest rangeland, probably because of their insectivorous diet (Hallett 1982).

Pocket Mice. Pocket mice respond differently to heavy grazing depending on the species involved. Most pocket mice species prefer a heavy protective cover of grass and some shrubs (Reynolds and Haskell 1949). In desert Southwest rangeland, pocket mice forage under and around large bushes and clumped vegetation (Rosenzweig 1973, Wondolleck 1978) and are common only if bushes form an important proportion of the vegetation (Rosenzweig and Winakur 1969). Feldhamer (1979) found Great Basin pocket mice resident only in relatively dense cover in sagebrush or greasewood communities. Within this habitat, densities were highest when sand was present, probably because of the limited ability of pocket mice to dig through hard clay soils. These pocket mice are most common in nongrazed dense grass communities or areas of relatively light grazing with heavy cover and seed production, and numbers are reduced in habitats sustaining heavy grazing (Black and Frischknecht 1971, Larrison and Johnson 1973, Hanley and Page 1982). Price (*P. penicillatus*) and Bailey (*P. baileyi*) pocket mice (Reynolds and Haskell 1949) are also most abundant in dense stands of perennial grasses and their numbers are reduced on grazed ranges (Steenbergh and Warren 1977).

The Arizona pocket mouse (*Perognathus amplus*) has been associated with open habitat and increased grazing (Bond 1945, Rosenzweig and Winakur 1969). Guthery et al. (1979) found the greatest densities of silky pocket mice (*Perognathus flavus*) where there was the sparsest grass cover and found them adversely affected by protection of the playa grassland from grazing. Also, hispid pocket mice (*Perognathus hispidus*) are commonly found in areas with open vegetation (Abramsky et al. 1979) or where the prairie is intensely grazed and erosion has removed much of the topsoil (Smith 1940). Hispid pocket mice inhabit a wide variety of habitats, from native prairie to cropfields, and would therefore be expected to be little affected by disturbance (Kaufman et al. 1990, Feldhamer 1979). However, severe disturbance limits their populations, and Agnew et al. (1986) found hispid pocket mice occupying mixed-grass prairie sites but not heavily grazed prairie dog colonies.

Pocket mice are unimportant with regard to their effects on rangeland grasses. Grazing itself, by reducing the height of bunchgrasses, produces a less favorable habitat, and lowers pocket mouse numbers.

Kangaroo Rats. Throughout the southwestern United States, kangaroo rats occur in great numbers in desert shrub-grasslands. Most species of kangaroo rats avoid dense herbaceous vegetation, including the narrow-faced kangaroo rat (*Dipodomus venustus*) (Hawbecker 1940), the Tulare kangaroo rat (*D. heermanni tularensis*) (Fitch 1948), and Heerman's kangaroo rat (*D. h. sayatilis*) (Dale 1939). Ord's kangaroo rat (*D. ordii*), has a wide distribution that includes relatively nonproductive habitats (Rosenzweig and Winakur 1969, Brown 1975, Frye 1983) and is most abundant where disturbed earth provides easy tunneling, such as along roadsides (Larrison and Johnson 1973). The Texas kangaroo rat (*Dipodomys elator*) is restricted to mesquite grassland of northcentral Texas, where it is a threatened species (Stangl et al. 1992). This species shows a preference for heavily grazed short-grass areas with exposed earth resulting from concentrated traffic of vehicles and livestock. Habitat modification by grazing has been proposed to ensure the viability of local populations or to permit the existence of the species elsewhere. Such changes, in fact, may duplicate past habitat modification effects of bison and prairie dogs, and regularly occurring fires on mesquite grasslands.

Populations of Merriam kangaroo rats (*D. merriami*) (Fig. 18) are also favored by grazing. The range of this species, which is widespread in the Southwest, coincides with the distribution of creosotebush, low humidity and rainfall (<24.5 cm annually), high summer temperatures and evaporation rates, and a lightly textured soil favorable to burrowing (sandy or sandy loam) (Reynolds 1958). Most of the lands inhabited by Merriam kangaroo rats are managed primarily as rangelands for cattle grazing (Reynolds 1958). On higher elevation rangelands, maintenance of a perennial grass cover with interspersed shrubs furnishes livestock grazing and reduces erosion; at lower, drier elevations a shrubby cover is all that can be maintained. When rangelands are improperly grazed, there is a gradual downward trend in perennial grass density and a corresponding increase in Merriam kangaroo rats, which avoid sites with dense cover (Stamp and Ohmart 1978, Hallett 1982) and prefer open areas with scattered woody plants and annual grasses. Numbers of Merriam kangaroo rats increase with grazing-induced decreasing grass cover and mesquite invasion (Bond 1945; Reynolds and Glendening 1949; Reynolds 1950, 1958; Rosenzweig and Winakur 1969). When grazing is moderate and percent cover is not decreased, populations of Merriam kangaroo rats have been significantly greater in ungrazed areas (Heske and Campbell 1991); the researchers postulated that cattle had an effect on seed production or plant species composition that was not measured.

FIGURE 18. Populations of Merriam kangaroo rats (*Dipodomys merriami*) are favored by livestock grazing. Photo by E. Kalmbach.

There are interspecific differences in the responses of kangaroo rat populations to range depletion. While the species discussed previously are more abundant on grazed sites (Quast 1950, Reynolds 1958, McCulloch 1962), the chisel-toothed kangaroo rat (*D. microps*) is 2 to 3 times more abundant in healthy shadscale than in stands depleted by grazing (Larrison and Johnson 1973). Shadscale leaves comprise most of its diet (Johnson 1961). The banner-tailed kangaroo rat (*D. spectabilis*) is also restricted to highly productive regions of deserts and arid grasslands.

Effects of Seed Foraging by Desert Rodents on Rangeland

Desert rodents are the primary consumers of seeds of annual and perennial plants and act as predators and dispersers of seeds. In arid rangelands, granivory may have an especially pronounced impact on seed densities and distributions, and thus on the plant community as a whole. Heteromyid rodents, including pocket mice and kangaroo rats, are some of the primary seed eaters (Brown et al. 1979*b*). Up to 95% of the items in kangaroo rat diets are seeds (Reynolds 1958). Heteromyid rodents tend to prefer high density seed patches in their foraging activities (Reichman 1979, 1983; Price and Waser 1985), collecting seeds when abundant and storing them for use when they are scarce. These rodents generally use large seeds >0.25 mg (Price and Reichman 1987). Merriam kangaroo rats forage mainly in open areas (Wondelleck 1978, Rosenzweig 1973) that may have clumped seeds at densities 10-15 times higher than under shrubs (Reichman and Oberstein 1977). Unlike kangaroo rats, pocket mice do not select large seed clumps over small clumps or scattered seeds (Price 1978, Hutto 1978) and they are more efficient at finding smaller, scattered seeds than are kangaroo rats (Reichman and Oberstein 1977). Kangaroo rats escape predators by jumping, and are therefore better suited to open rather than vegetated areas. Pocket mice, however, are better protected from visually hunting predators by remaining in or near vegetation.

Where kangaroo rats live in loose colonies, the plants in the vicinity of the colony are denuded of seeds. Kangaroo rats may harvest > 75% of an entire seed crop (Chew and Chew 1970, Nelson and Chew 1977, Borchert and Jain 1978, Price and Jenkins 1986) and in some years, Merriam kangaroo rats are sufficiently abundant to eat all large perennial grass seed produced (Reynolds 1958). Examination of kangaroo rat burrows by Vorhies and Taylor (1922) and Taylor (1930) showed an average of 1.7 to 1.8 kg of seed material stored in each burrow. With an estimated population of 49 kangaroo rats (≤100)/ha, this would amount to 8.3 kg of seeds/ha.

The abundance of plant species whose seeds are favored foods of these small mammals can be strongly affected by seed predation. Soholt (1973) found that in the Mojave desert, kangaroo rats consumed 95% of the seed of Crane's bill (*Erodium cicu-*

tarium), causing a reduction in the numbers of this plant. Foraging by rodent granivores substantially reduces the standing crop of large-seeded winter annual plants (Brown et al. 1986). After removal of rodents, densities of these plants increased as much as several thousand times. Batzli and Pitelka (1970) and Borchert and Jain (1978) used exclosures to show that rodents can remove the majority of seeds of preferred plant species and that densities of plants can be reduced > 60% by rodent foraging. Brown et al. (1979*a*) removed heteromyid rodents and ants from plots and found seed densities highest on the plot from which ants and rodents were excluded, intermediate on the plots where either ants or rodents had been removed, and lowest on the control plot where all granivores were present. Densities of annuals and biomass of seeds were about 1.5 times greater on plots where rodents had been excluded than on plots where they were present (Brown et al. 1979*a*).

The fate of seeds following harvest by rodents is an important determinant of plant population recruitment. Although rodents consume large amounts of seed, their seed caches are a major source of plant recruitment (West 1968, Evans et al. 1983, McAdoo et al. 1983). Kangaroo rats cache seeds in a centrally located burrow, but also often store seed in small scattered caches just below the soil surface. Unrecovered caches provide for recruitment of new plants throughout rodent home ranges. Seedling establishment of annual plants and small-seeded perennial grasses is influenced by pocket mice (LaTourrette et al. 1971) and seedling establishment of large seeds is influenced by kangaroo rats.

Kangaroo rats can have variable effects on condition of desert or arid rangelands. When a range is in good to excellent condition, Merriam kangaroo rats have little effect on seed dispersal (Reynolds 1958); good quality rangeland is less favorable for kangaroo rats because increased cover is an obstacle to escape from predators and because large-seeded vegetation, the preferred food, is replaced by small-seeded plants as increasing perennial grass density forces out weedy annuals. On rangelands in good to excellent condition, the seed burying habits of heteromyid rodents are probably beneficial to rangeland, as large-seeded perennial grasses and tall shrubby plants have been shown to increase on areas where kangaroo rats were most abundant (Reynolds 1950). During a favorable seed year, and if range condition is at a level where the supply of large-seeded perennial grass seed is in excess of needs of kangaroo rats, much more seed is cached than is ever recovered (Reynolds 1958). Large seeded species have difficulty in germinating and establishing from seed on the soil surface and may require burial in seed caches for seedling establishment to occur (LaTourette et al. 1971, Evans et al. 1983). Because seed left in the ground is in a more favorable environment for germination and early seedling survival than seed lying on top of the ground, the rate of plant restocking may be enhanced by the presence of kangaroo rats (Reynolds 1950, LaTourrette et al. 1971). Rodent caches of antelope bitterbrush, snowbrush (*Ceanothus velutinus*), squawcarpet (*C. prostratus*), green rabbitbrush (*Chrysothamnus viscidiflorus*), cheatgrass, and Indian rice grass (*Oryzopsis hymenoides*) have been reported (West 1968, LaTourrette et al. 1971, McAdoo et al. 1983). These caches provide opportunities for germination of species such as bitterbrush and snowberry (*Symphoricarpos* spp.) on recently burned or denuded pinyon-juniper and sagebrush sites (Ferguson and Basile 1957, West 1968, Everett and Kulla 1976, Evans et al. 1983). Rodents also transport mycorrhizae associated with range plants and therefore could establish plant species and their associated mycorrhizae on denuded range sites (Maser et al. 1988).

McAdoo et al. (1983) observed that heteromyid rodents were very important in maintaining stands of Indian rice grass on rangelands. Kangaroo rats preferentially harvest and cache the largest, most germinable rice grass seeds, rejecting empty seeds and polymorphic forms with reduced germination potential. Germination was greatly enhanced by the rodent's removal of the lemma, palea, and pericarp that induced dormancy, and emergence of seedlings from these caches was the primary means of stand renewal of Indian rice grass.

As rangeland deteriorates to fair or poor condition, woody perennial shrubs increase and perennial grasses decline (Reynolds 1958, Brown et al. 1979*b*), increasing open areas that are believed (Rosenzweig 1973) to allow the kangaroo rat to see and avoid predators (Bartholomew and Caswell 1951). Merriam kangaroo rat populations then increase (Reynolds 1950). Kangaroo rats have been blamed for the increase of mesquite and cholla cactus (*Cylindropuntia* spp.) (Reynolds 1950) by dispersing seed through storage in caches and passage of the seeds through their digestive tracts (Parker and Martin 1952, Reynolds 1954). As grass declines on depleted rangelands, Merriam kangaroo rats increase, more mesquite is favored, followed by a continuing downward cycle of less grass and more rodents. Under these conditions, kangaroo rats can be an important factor in accelerating range deterioration and in very poor rangeland, kangaroo rat activity may be sufficient to prevent range recovery. When the supply of perennial grass seed is sufficiently low, kangaroo rats may consume the entire seed supply and mesquite can continue to increase even without cattle grazing (Reynolds and Glendening 1949). In this case, reducing kangaroo rat populations may be necessary to improve the rangeland. However, if rangelands occupied by mesquite are heavily grazed by cattle, mesquite will increase despite the presence of kangaroo rats, so it will not measurably improve the range to reduce kangaroo rat populations; it will be better to remove the mesquite source of seed. Once the mesquite is gone, perennial grass will increase and kangaroo rats will concomitantly decrease.

Reduction of kangaroo rat populations to increase forage is therefore justified biologically only where the density of perennial grass is low and can be increased by grazing management or range improvement practices. Reduction would also be warranted where artificial reseeding is being attempted with large-seeded plants and is being hampered by kangaroo rats (Reynolds 1950).

Kangaroo rats may prevent rangeland succession by maintaining a sub-climax type of vegetation. Brown and Heske (1990) found that long-term removal of a guild of 3 kangaroo rat species from a Chihuahuan Desert ecosystem led to the conversion of the habitat from shrubland to grassland. Twelve years after removal, density of tall perennial and annual grasses increased approximately 3 times and rodent species typical of arid grassland colonized, including harvest mice. Kangaroo rats were a keystone guild: through seed predation and soil disturbance they had major effects on biological diversity and biogeochemical processes, favoring establishment of annuals and shrubs by selectively foraging on large seeds, and by seed caching and burrowing activities. At this site, the effects of excluding kangaroo rats were much greater than exclusion of cattle. This experimental site was near the natural transition zone from desert to grassland, so the magnitude of changes reported by Brown and Heske (1990) after kangaroo rat removal may not occur in other desert environments.

Management Considerations for Small Rodents

Because of previous overgrazing and concomitant loss of desirable grass, forb and shrub species, millions of hectares of degraded sagebrush-grass, chaparral, and pinyon-juniper rangelands are unable to reach their potential for livestock forage or wildlife habitat (Vallentine 1989). Seeding is commonly used for restoring depleted vegetation (Standley 1988). Efforts to rehabilitate these rangelands by direct seedings have often failed because rodents consume up to 98% of the planted seeds (Nord 1965, Nelson et al. 1970, Sullivan and Sullivan 1982). Reynolds (1958), Brown et al. (1979*b*), Inouye et al. (1980), and Price (1983) found that kangaroo rats prey selectively on large seeds and can be a significant cause of seeding failure when large seeds are planted. If seeds are smaller than 1.1 million/kg, kangaroo rats have no effect on seeding programs. Various methods have been tried for reducing the impact of rodents on seeded areas. The most common method is poisoning, but it often fails because of rapid rodent immigration (Sullivan and Sullivan 1984). Everett and Monsen (1990) recommended adoption of seeding strategies that mimic natural seed predation avoidance mechanisms, such as providing more seed along with sacrifice foods that can be used by the resident rodent population, planting in the spring when rodent populations are low, selecting seeds with low rodent preference, and using chemical repellents on seeds. Many studies show that certain rodents prefer particular species or sizes of seeds (Reynolds and Haskell 1949, Reynolds 1950, Lockard and Lockard 1971, Smigel and Rosenzweig 1974, Everett et al. 1978, Price 1983). Standley (1988) seeded plots with small and large grass seeds and found that large seed was removed but small seeds were still present 36 days after planting. He recommended that when site conditions and management needs allow a choice of plant species to seed, smaller seeds may provide better results. Kangaroo rats operate under an "optimal foraging" strategy where they preferentially harvest seeds from dense patches or clumps rather than dispersed seeds (Pyke et al. 1977, Price and Jenkins 1986). This suggests that seeds should be placed as randomly as possible (Richardson et al. 1986, Everett and Monsen 1990). Evans et al. (1983) reported that only 8% of randomly cached bitterbrush seed was found by rodents compared to complete removal of broadcast seed. Buried seeds are less preferred than seeds on the soil surface (Price and Jenkins 1986) so drilling seed may also discourage seed predation (Everett and Monsen 1990).

Rangeland site conversions can have large impacts on rodent populations. Common range site conversion treatments that reduce cover and food availability for rodents include prescribed burning, spraying with herbicides, chaining, rotobeating, and disking (Vallentine 1989). Following spraying with 2,4-D or rotobeating of black sagebrush (*Artemisia nova*), populations of deer mice and Great Basin pocket mice remained near those of controls, but the least chipmunk disappeared from rotobeaten plots (Zou et al. 1989). Spraying sagebrush-grass ranges with 2,4-D had little effect on density of deer mice, but caused a sharp decline in northern pocket gophers and least chipmunks, and an increase in montane voles (Johnson and Hansen 1969). Density of deer mice and pocket mice dramatically increased the second year following chaining and windrowing of Utah juniper (*Juniper osteosperma*) sites (Baker and Frischknecht 1973). After burning in sagebrush communities (McGee 1982), rodent species that require large amounts of cover such as voles, western jumping mice, and masked shrew decreased in numbers; deer mice populations remained close to preburn conditions. Greater consideration needs to be given to effects of range site conversion on small rodents. A suitable approach to coordinate range and wildlife habitat management would be to determine characteristic small mammal species within grazing allotments, determine cover levels at which each species is reduced in abundance or distribution, and determine the percentage of overall habitats that should be maintained for these sensitive species.

LEPORIDAE: JACKRABBITS AND COTTONTAILS

The order Lagomorpha, with a fossil history dating back to the Oligocene (Hall and Kelson 1959), is composed of 2 modern families: Ochotonidae (i.e., pikas) and Leporidae (i.e., hares and rabbits). The hares and rabbits are found on grazed rangelands and will be considered in this chapter. Two genera of Leporidae occur in North America: *Lepus* (hares) and *Sylvilagus* (rabbits). Hares are larger and have precocial young that are born fully haired, while rabbits are smaller and have altricial young.

Jackrabbits are a prominent rangeland herbivore throughout the West. Two principal species occur on rangeland: the black-tailed jackrabbit (*Lepus californicus*) (Fig. 19) and the white-tailed jackrabbit (*Lepus townsendii*) (Fig. 20). The following descriptions are provided from the work of Jones et al. (1985). The black-tailed jackrabbit is slightly smaller, with a total length of 535-585 mm and weight of 1.8-3.2 kg, compared to the white-tailed jackrabbit, with a total length of 540-640 mm and weight up to 4.5 kg.

The general circadian activity pattern is described for black-tailed jackrabbits by Lechleitner (1958*b*) but is impacted by season (Smith 1990) and ambient temperature (Knowlton et al. 1968). During daylight hours jackrabbits rest in forms, shallow depressions in or under bushes (Vorhies and Taylor 1933). Their activity increases during the crepuscular and early nocturnal period as they forage. They become less active before dawn as they reingest their pellets (Lechleitner 1958*b*, Smith 1990).

Their home ranges (Burt 1943) have been variously reported. Lechleitner (1958b) found that in the Sacramento Valley of Cal-

FIGURE 19. Black-tailed jackrabbit (*Lepus californicus*). Photo by G. Connolly.

FIGURE 20. White-tailed jackrabbit (*Lepus townsendii*). Photo by K. Fagerstone.

ifornia, ranges were small and well defined, usually <20 ha, similar to those later reported by French et al. (1965) in Idaho (16.2 ha) and Tiemeier (1965) in Kansas (16.2 ha). Because home-range size is effected by the ability of the environment to provide food, shelter, and water, and their juxtaposition, it can be as large as several kilometers in diameter in grazing areas of Arizona (Vorhies and Taylor 1933). Orr (1940) reported jackrabbits feeding at distances up to 1.6 km from suitable cover. Donoho (1972) studied the dispersion of white-tailed and black-tailed jackrabbits on the Pawnee National Grasslands in northeastern Colorado and found that the white-tailed jackrabbits had significantly larger home ranges than black-tails. He reported males possessed larger home ranges than females; in contrast, Smith (1990) found no difference between the sexes in the home-range size of black-tailed jackrabbits. Also, seasonal and climatological influences effect the size and shape of jackrabbits' home ranges (Lechleitner 1958*a*, Tiemeier 1965, Gross et al. 1974, Smith 1990). These variables and others such as differences in procedures and analysis make home range generalizations and comparisons among studies difficult.

Jackrabbit densities as estimated by direct observation (Wywialowski and Stoddard 1988), pellet census techniques (Taylor 1930, Arnold and Reynolds 1943), and radiotelemetry (Knowlton et al. 1968) are dependent on vegetation, climate, season, and other factors. Densities in Kansas have been reported from 35/ha in agricultural areas (Bronson and Tiemeier 1959), to 2.3/ha on mixed prairie (Brown 1947*a*), and 0.6/ha in western Kansas (Wooster 1935). Some of the lowest jackrabbit densities (0.01/ha) have been documented in southeastern Colorado (Flinders and Hansen 1973). This information conforms with findings of Fagerstone et al. (1980), who found jackrabbit densities significantly higher near cultivated crops than on isolated rangeland.

Fluctuations in jackrabbit density have been reported in the literature (Keith 1963, Gross et al. 1974) with cycles of approximately 5 to 10 years (Evans et al. 1970, Clark 1975). Populations in local areas can become extremely large during population irruptions (Fig. 21). Some researchers believe that the populations are not actually cyclic, but that drought and food availability (Bronson and Tiemeier 1959) or drought and overgrazing (Taylor et al. 1935) concentrate the hares. Evidence now suggests that the key parameters associated with population fluctuations are much more complex than previously thought. There appear to be geographic trends in jackrabbit frequency fluctuations (Clark 1972) and covariation with many interacting community components (Keith 1979, Johnson 1982, Hansson and Henttonen 1988) and natural phenomena like weather (Stoddard 1985). However, population fluctuations are not believed to be associated with a deviation in a 1:1 sex ratio (Lechleitner 1959, Gross et al. 1974, Griffing and Davis 1976).

Distribution

The black-tailed jackrabbit is the most common jackrabbit in the western and central United States (Fig. 22), ranging from the Pacific coast to western Missouri and Arkansas, and from the prairie and grassland regions of Idaho to South Dakota to the Mexican border (Hall and Kelson 1959, Hall 1981, Dunn et al.

FIGURE 21. Black-tailed jackrabbits (*Lepus californicus*) on Idaho rangeland during a population irruption. Photo from Denver Wildlife Research Archives.

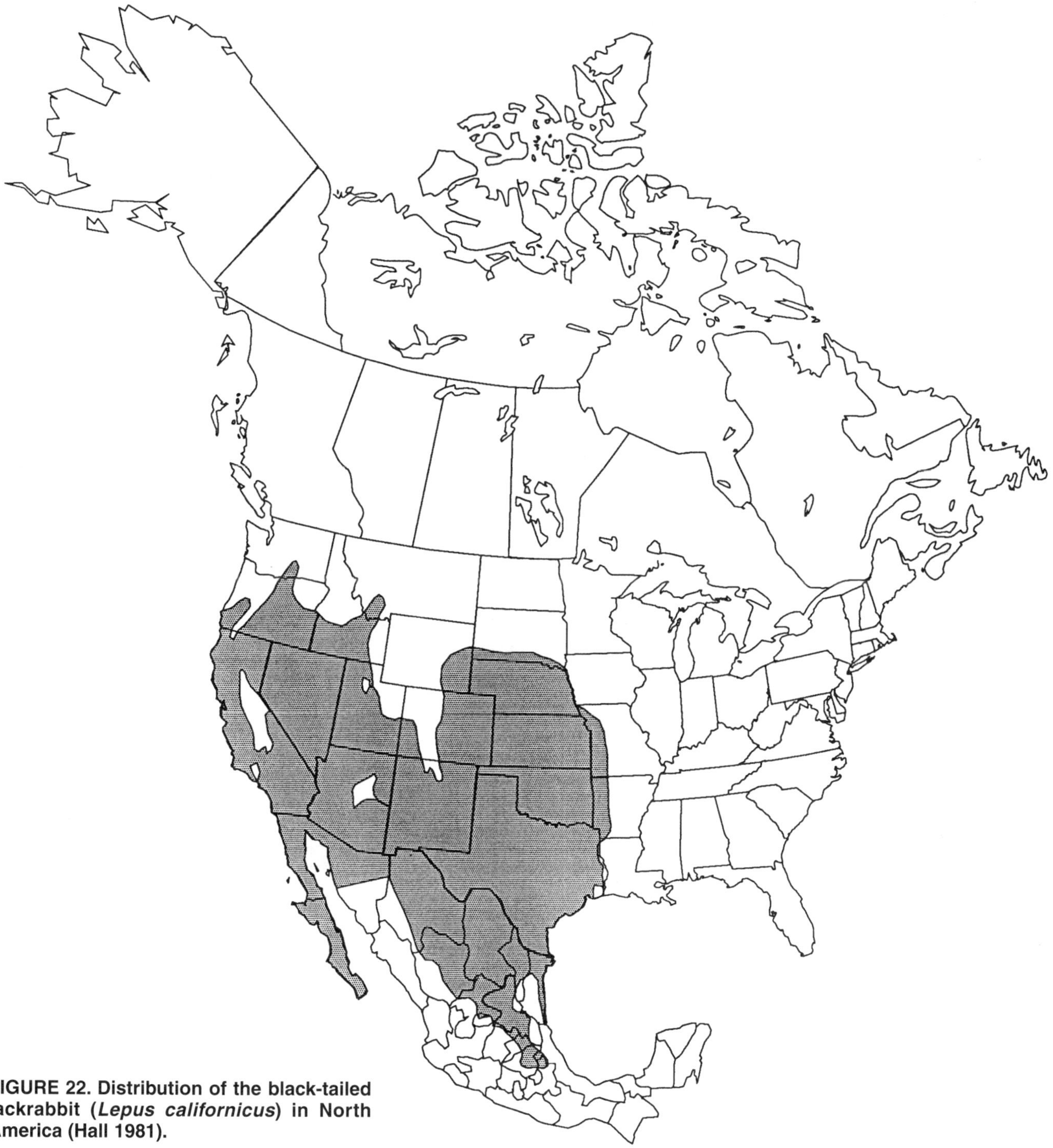

FIGURE 22. Distribution of the black-tailed jackrabbit (*Lepus californicus*) in North America (Hall 1981).

1982, Jones et al. 1985). Although the black-tailed jackrabbit occupies many diverse habitats, it is primarily found in association with shortgrass prairie and open country (Burt and Grossenheider 1959, Dunn et al. 1982). The black-tailed jackrabbit avoids areas of heavy brush or woods, where its principal means of defense, keen eyesight and speed of escape, are reduced (Jones et al. 1985).

The white-tailed jackrabbit occurs in close association with the flora of the northern Great Plains and open areas of the Great Basin (Hall and Kelson 1959, Hall 1981, Dunn et al. 1982, Jones et al. 1985). They range from southern Canada to Colorado and from Michigan to the high mountain slopes of the Rockies, Cascades, and Sierras (Fig. 23). White-tailed jackrabbits once ranged south across the plains states to southern Kansas, but now occur generally north of the Platte River (Jones et al. 1985).

White-tailed jackrabbits prefer large expanses of croplands and pastures with interspersed brush and heavy vegetation in Iowa (Schwartz 1973), both upland and lowland grassland habitat in Colorado (Bear and Hansen 1966), and open flats and rye

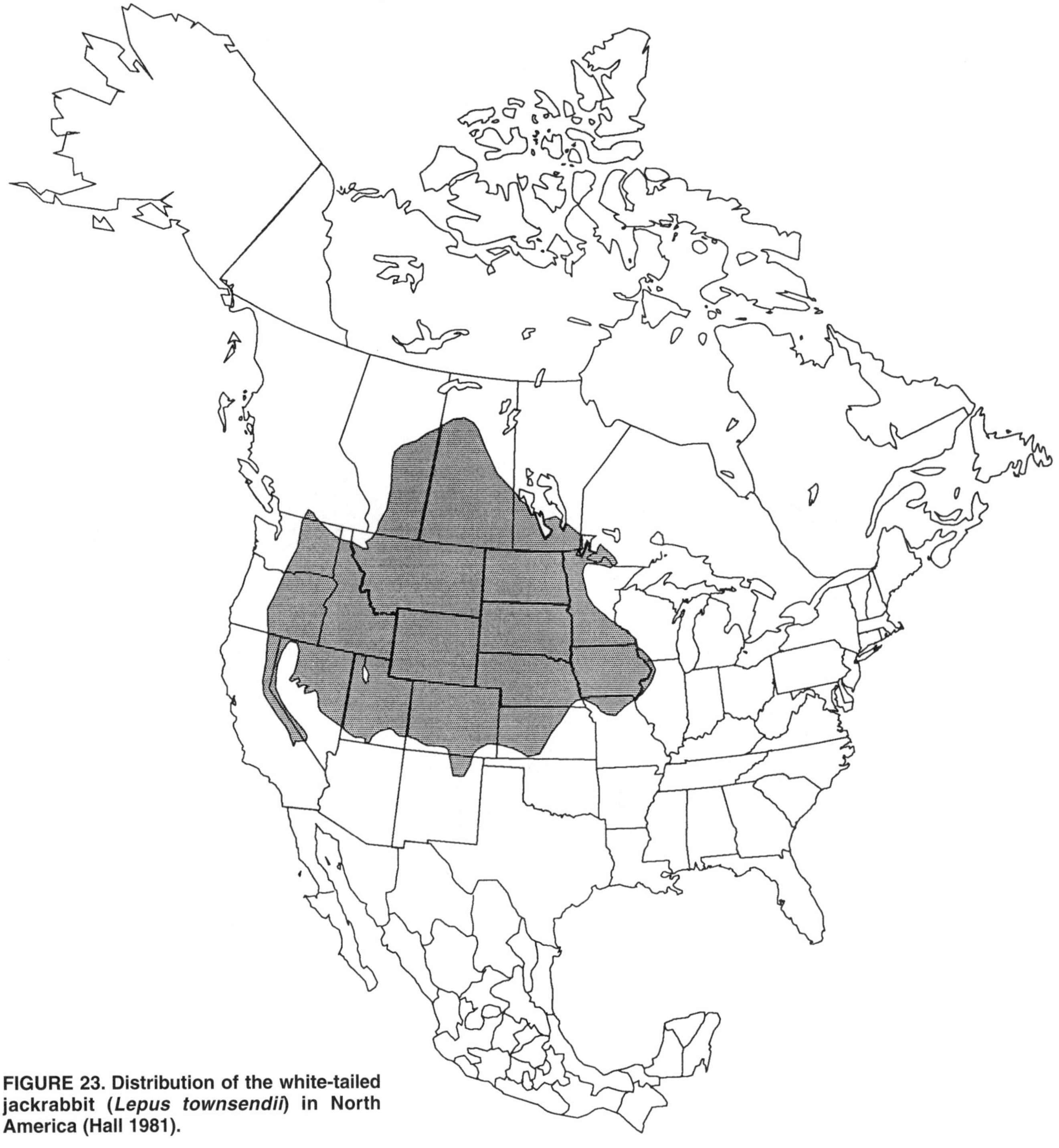

FIGURE 23. Distribution of the white-tailed jackrabbit (*Lepus townsendii*) in North America (Hall 1981).

grass in California (Orr 1940). In general, they avoid forests and woodlands. Leopold (1945) reported that in the early 1900s, the white-tailed jackrabbit was extending its range east into Wisconsin, Iowa, and Missouri, possibly as a result of the disappearance of the original forest cover (Kline 1963, and DeVoes 1964). In recent years, the white-tailed jackrabbit's range has declined, and it is now extirpated from Kansas (Hall 1955) and is rare in Missouri (Watkins and Novak 1973). Jones et al. (1985) suggested that these jackrabbits may be less well adapted to the general climatic warming of the Great Plains and less able to use cultivated areas than the black-tailed jackrabbit. Carter (1939) and Brown (1947*b*) theorized that the expansion of black-tailed jackrabbits into areas formerly occupied by white-tailed jackrabbits was due to the changing habitat brought about by cultivation of the prairie by man. Flinders and Hansen (1972) and Donoho (1972) reported that when both species came into contact on the shortgrass prairie, the white-tailed jackrabbit selected more sparsely vegetated upland habitats and occupied higher elevations. In addition, Hansen and Flinders (1969) suggested that the black-tailed jackrabbit is more efficient than the white-tailed jackrabbit in using feeding sites. Thus, the black-tailed jackrabbit, with its greater adaptability and feeding efficiency, may be able to displace the white-tailed jackrabbit.

Diet

Lechleitner (1958*a*) suggested that jackrabbits prefer to feed in open areas with a supply of succulent young plants, interspersed with patches of dense weeds; the open areas allow them to detect danger and the dense weeds serve as cover. Similarly, Orr (1940) found that jackrabbits prefer the margins of brush tracts adjacent to open areas. Foraging begins during the crepuscular period, increases during early night, and begins to decrease as dawn approaches (Lechleitner 1958*b*, Smith 1990). This general pattern of foraging is influenced by season (Smith 1990), ambient temperature (Knowlton et al. 1968), and phase of the moon (Lechleitner 1958*a*).

Black-tailed Jackrabbits. The diets of black-tailed jackrabbits have been studied by many scientists, especially in the Great Basin and Great Plains (Vorhies and Taylor 1933, Arnold 1942, Riegel 1942, Lechleitner 1958*b*, Currie and Goodwin 1966, Sparks 1968, Hansen and Flinders 1969, Flinders and Hansen 1972, Uresk 1978, Fagerstone et al. 1980, MacCracken and Hansen 1982, Wansi et al. 1992). These studies demonstrated that black-tailed jackrabbits are opportunistic feeders. However, Johnson and Anderson (1984) found that jackrabbits may feed more selectively than previously thought. Although the plant species eaten vary throughout the jackrabbit's range, their diets have some seasonal features in common: they show preferences for shrubs in winter and for grasses and forbs in the spring and summer (Hayden 1966, Westoby 1980). Various cereal crops (Lechleitner 1958*a*) and other cultivated crops (i.e., winter wheat [*Triticum aestivum*], alfalfa, crested wheatgrass [*Agropyron cristatum,*]) (Flinders and Hansen 1972) are preferentially used when available. Fagerstone et al. (1980) documented that plant phenology was a major factor in determining food preferences of rangeland jackrabbits; 85% of their diets was composed of grasses in the spring, grasses and forbs were nearly equal in diets in early summer, and forbs and shrubs increased to 71% of the diets in late summer. In northeastern Colorado, black-tailed jackrabbit seasonal diets consisted mainly of grasses and forbs during summer and changed to shrubs during winter (Hansen and Flinders 1969). Grasses and sedges comprised 49% of their overall diet, with 7 plant species accounting for most of it (64%).

In contrast, Vorhies and Taylor (1933) found, in arid areas of the Southwest, that mesquite made up 54% of the annual diet. Uresk (1978) reported that needle-and-thread grass was preferred in sagebrush habitat, while yarrow (*Achillea lanulosa*) was preferred in bitterbrush habitat. Wansi et al. (1992) found that grasses comprised the largest component of jackrabbit diets in semidesert rangeland, particularly during the summer growing season. Lechleitner (1958*a*) observed that herbaceous weeds were not preferred dietary items. Most investigators agree that jackrabbits select for succulence (Hansen and Flinders 1969, McAdoo and Young 1980), particularly during times of water stress (Westoby 1980). Cooke (1982) and Robertson et al. (1987) reported also that water may be a limiting factor in Australian rabbit populations.

White-tailed Jackrabbits. Bear and Hansen (1966) demonstrated seasonal dietary differences in white-tailed jackrabbits in southern Colorado. They ate 87% shrubs in the spring, including Parry's rabbitbrush (*Chrysothamnus parryi*) (70%) and fringed sage (*Artemisia frigida*) (15%). Summer diets consisted of forbs (70%), grasses (19%), and shrubs (7%). Fall diets were predominantly grasses (43%) and forbs (34%), with higher levels of shrubs (14%). Winter diets were primarily shrubs (76%). Flinders and Hansen (1972) found that white-tailed jackrabbits on the shortgrass prairie of Colorado fed mainly on plants in the prereproductive or early reproductive stages of development that had the greatest nutritive value and contained high proportions of moisture and crude protein. Flinders and Hansen (1972) also noticed seasonal differences in the diet of white-tailed jackrabbits. Early spring diets consisted of western wheatgrass and winter wheat, and western wheatgrass continued to be the most important food item during the summer (21%) and fall (46%). Winter diets were predominantly composed of winter wheat and crested wheatgrass.

Economic Status

Black-tailed jackrabbits have positive and negative relationships with humans. They are used for sport hunting, food, and fur (Jones et al. 1985). They are also an important part of the prairie ecosystem, and constitute a major food base for mammalian predators, particularly the coyote (Clark 1972). They have been reported to influence secondary succession in a positive way on old fields and denuded ranges by dispersing seeds in fecal pellets (Sparks 1968) and by increasing the viability of some seeds in their pellets by their digestive processes (Brown 1947*a*). Negative associations are related to situations where they are reported to cause extensive damage to agriculture (Hegdal 1966, Marsh 1984, Lewis and O'Brien 1990). Fagerstone et al. (1980) reported that cultivated crops adjacent to rangeland are particularly vulnerable to damage by jackrabbits, because the rangeland provides daytime resting areas for them. Jackrabbits have been reported to cause damage to seedling trees (Read 1971), grains and cotton (Vorhies and Taylor 1933), range rehabilitation programs (Wetherbee 1967), and vegetables (Bickler and Shoemaker 1975). Jackrabbit diseases that are transmissible to humans have also been reported, including tularemia (*Pasteurella tularensis*) (McMahon 1965) and (rarely) Rocky Mountain spotted fever (*Rickettsia rickettsi*) (Lechleitner 1959). On western rangelands they may be significant competitors with livestock.

White-tailed jackrabbits have had generally a more positive relationship with humans. Historically, some of the positive benefits of these jackrabbits were associated with sport hunting, food, and fur; their fur has been used commercially to make felt. They are also important to maintenance of ecological balance, biodiversity, and aesthetics on rangeland ecosystems. White-tailed jackrabbits do not cause much damage to agriculture (Jones et al. 1985), but do transmit diseases such as tularemia to humans.

Association with Livestock

The relationship between livestock grazing and jackrabbit population numbers is not well understood (Sanderson 1959, Short 1985) even though the combination can cause severe destruction to rangeland (Allen 1992). Some range scientists believe heavy grazing of rangeland increases suitable habitat for jackrabbits and encourages their presence (Brown 1947*a*, Jones et al. 1985). However, Taylor et al. (1935) reported that jackrabbits prefer moderately grazed pastures, and Norris (1950) stated that they prefer nongrazed rangeland. Short (1987) studied the interactive relationship of herbivores and their pastures and believes that herbivores affect the biomass, growth, and species composition of the pasture. As the density of herbivores increases, a point is reached when plant species of low quality and palatability also increase and make the pasture less suitable for all herbivores, including jackrabbits. Work by Hintz (1969) seems to support this view by concluding that rabbits are only 60% as efficient as ruminants in digesting fiber and therefore must feed on a high quality diet with < 40% fiber (Cooke 1974). Daniel (1991) studied the influence of range condition on the diets of jackrabbits and cattle. He found that: (1) jackrabbit densities were highest on good condition range, (2) range condition influenced the forb and shrub but not the grass diets of jackrabbits, and (3) jackrabbits consumed less grasses (ratio 1:3) and more shrubs than cattle, producing a moderate dietary overlap of 41%. Haskell and Reynolds (1947) reported that food consumption for black-tailed jackrabbits increased until they were 28 to 30 weeks of age then leveled off at an average consumption of 0.12 kg of air-dried native forage/day.

In early studies of competition, scientists estimated the foraging capacity of black-tailed jackrabbits to be: 1 cow equivalent to 148 black-tailed jackrabbits (Vorhies and Taylor 1933) or 62 jackrabbits (Arnold 1942), and 1 sheep equivalent to 6 black-tailed (Currie and Goodwin 1966) or 15 white-tailed jackrabbits (Bear and Hansen 1966). Sparks (1968) found that the greatest direct competition for forage between cattle and black-tailed jackrabbits was in early spring when both species preferred green forage such as western wheatgrass, needle-and-thread, and sunsedge (*Carex heliophila*), and least in late fall and winter.

Vegetation on many rangelands has changed during the last century, partly because of poor grazing practices (Martin 1975). Reports have documented shifts from grassland to shrub-dominated vegetation in New Mexico (Buffington and Herbel 1965), Nevada (Robertson and Kennedy 1954), and central Utah (Christensen and Johnson 1964) as a result of heavy grazing. Rice and Westoby (1978) observed that in Great Basin shrub communities protected for 15 years from domestic sheep, with or without protection from jackrabbits, plant community changes were slow to nonexistent, and that protection from jackrabbits had no effect.

Management

Jackrabbit management in the United States has generally focused on population reduction because of their crop depredation activities (Bronson and Tiemeier 1958, Hegdal 1966). Beginning in the 1840s (Palmer 1896), organized rabbit drives were used to reduce populations; these have given way to issuance of hunting licenses as state game and fish departments began regulating jackrabbits as game species. Concurrently, the federal government employed or investigated a variety of control techniques such as biological control, habitat alteration, disease and parasites, mechanical control, and chemical control (Evans et al. 1970). Natural predators have not been effective in controlling jackrabbits (Palmer 1896, Horn 1941, Pressnall 1950, Lack 1954, Johnson 1964, French et al. 1965) and the introduction of new predators into an ecosystem has not been pursued because of high potential risks to other species (Howard 1967). Although the natural occurrence of enzootics in jackrabbits has been reported (McMahon 1965, Lechleitner 1959), they are not well understood. Disease and parasites have been used for jackrabbit control in other parts of the world. In Australia, the virus myxomatosis (*Myxomatosum*) was introduced in the 1950s for rabbit control (Ratcliffe 1951) with mixed results. Antifertility agents have been under investigation for years (Balser 1964, Howard 1967), but a variety of problems will need to be overcome before these agents can be used in the field, including: (1) current chemicals are not specific to hares, (2) most chemicals are only temporary in effectiveness, (3) carriers and baits are not yet available, and (4) public acceptance will be required (Evans et al. 1970). Several chemicals have been used in the past for jackrabbit control (McCabe 1966, Evans et al. 1970): thallium sulfate, sodium fluoroacetate, strychnine, zinc phosphide, and anticoagulants. All but zinc phosphide have secondary hazards (Evans and Ward 1967, Wetherbee 1967, Kaukeinen 1982, Marsh et al. 1987, McIlroy and Gifford 1992, Ramey et al. 1992, Johnson and Fagerstone 1994). Of these, only zinc phosphide and the anticoagulants are currently registered for jackrabbit control (Johnston 1978, Ramey et al. 1992).

A variety of nonlethal controls like clean farming practices (Allen 1942), fencing (Evans et al. 1970), tree protectors (Johnson 1964), repellents (Johnson 1964, Welch 1967), predator odors (Sullivan and Crump 1986, Sullivan et al. 1988), and buffer crops or vegetative barriers (Lewis 1946) have been used to prevent damage from jackrabbits with some success. Evans et al. (1970) reported that alternate foods or lure crops failed to prevent damage to grain and hay crops during winter, and buffer crops 0.4 km wide or cleared land failed to protect crops when jackrabbit populations were high. However, Borrecco (1976) presented a successful example of habitat manipulation using an herbicide-induced reduction in vegetation cover and availability of summer foods that decreased the clipping damage to Douglas-fir seedlings (*Pseudotsuga menziesii*) by snowshoe hares (*L. americanus*). There is an obvious need for more information about jackrabbit population genetics, ecology, behavior, distributional changes, and livestock grazing interactions (Dunn et al. 1982).

COTTONTAILS

There are 2 primary species of cottontail rabbits: the eastern cottontail (*S. floridanus*) and the desert cottontail (*S. auduboni*) (Fig. 24). The eastern cottontail has a total length of 400-450 mm and a weight of 900-1,500 g. The desert cottontail has a total length of 390-435 mm and a weight of 950-1,375 g (Jones et al. 1985).

Distribution

Cottontails are among the most widely distributed of North American mammals. The eastern cottontail is the most widely distributed cottontail (Fig. 25). It is found from the eastern seaboard west to the Rocky Mountains and from southern Canada south to Costa Rica (Hall and Kelson 1959, Hall 1981, Chapman et al. 1982, Jones et al. 1985). It occurs throughout the plains region primarily in riparian ecosystems (Hall and Kelson 1959) and ubiquitously in the eastern deciduous forests. Lunk (1989) demonstrated a significant preference during their nocturnal foraging activities for farms rather than forested areas. Lord (1963) reported a circadian activity pattern similar to that of hares, but with nonsignificant nocturnal variation; however, Dalke and Sime (1941) have observed 2 pronounced feeding periods: the first, 3 to 4 hours after sunrise; and the second, from sunset to 1 hour later.

Home ranges for eastern cottontails are often sympatric and range from 2 to 6 ha (Chapman et al. 1980, Chapman et al. 1982). Reproductive fecundity is high, with litter sizes up to 12 (Kirkpatrick 1960). Densities as high as 17/ha (Chapman et al. 1982) to 25/ha have been reported (McKay 1978) with 8-10 year cyclic tendencies (Keith 1963, Bailey 1968). Cottontail densities are positively correlated with increased biomass of herbaceous vegetation and with areas ungrazed by livestock (MacCracken and Hansen 1982).

Desert cottontails are distributed widely throughout the arid areas of western North America (Fig. 26), from Montana south to central Mexico, and from the High Plains of Oklahoma, Kansas, and Nebraska to the Pacific Coast (Hall 1981, Jones et al. 1985). Typical habitat in the Great Plains includes weedy margins of fields and pastures, brushy areas, and dry ravines. Home ranges of desert cottontails average 0.4 ha (Trippensee 1934) to 1.2 ha (Allen 1938), with male home ranges much larger than those of females (Ingles 1941). The desert cottontail has a high reproductive potential with an average litter size of 3.6 (Orr 1940), and it is capable of producing 2 to 3 litters annually and more than 1 generation annually (Ingles 1941).

FIGURE 24. Desert cottontail (*Sylvilagus auduboni*). Photo by K. Fagerstone.

Diet

The eastern cottontail has a diet that is cosmopolitan (Fitch 1947*a*), varying with the habitat and seasonal availability of forage (Chapman et al. 1982). Eastern cottontails prefer herbaceous plants when available during the growing season, including bluegrass, orchard grass (*Dactylis glomerata*) (Smith 1950, Dusi 1952, Klimstra and Corder 1957); timothy (*Phleum pratense*); clover (*Trifolium* spp.); and alfalfa (Dalke and Sime 1941). Succulent new growth supplies much of the cottontail's water requirements (Jones et al. 1985). Woody species like apple (*Malus pumila*) and red maple (*Acer rubrum*) (Dalke and Sime 1941, and Smith 1950) are preferred during the dormant season.

The desert cottontail is a crepuscular to nocturnal forager. Desert cottontails can survive droughts by obtaining water from cactus and forbs (Turkowski 1975). Orr (1940) found that they rely seasonally on grasses, sedges, rushes, shrubs (e.g., blackberry [*Rubis allegheniensis*]), and trees (e.g., willow [*Salix* spp.] and oak [*Quercus* spp.]). Hansen (1978) found that their annual diet was very similar to that of prairie dogs and cattle in the kinds of plants eaten, but differed in the relative proportions preferred.

Economic Status

Cottontails have provided a positive relationship with recreationists (Davis 1962) by providing small game hunting. Eastern cottontails are the most widely hunted game mammal in the United States (Jones et al. 1985) and their high reproductive rates allow them to withstand a high hunting pressure. Cottontails are a staple in the diet of various predators including coyotes, foxes, eagles, hawks, and owls (Fitch et al. 1946, Fitch 1947*b*, Capman et al. 1982). Badgers, weasels, and snakes take nestling young. Transmission of cottontail diseases to humans is of some concern (McMahon 1965). Eastern cottontails are responsible for 55% of tularemia cases reported in Americans (McDowell et al. 1964), due to direct contact while skinning and dressing animals. A few cases of plague (*Pasteurella pestis*) have also been reported in cottontail rabbits (Graves et al. 1978). Cottontails are not an important contributor to rangeland overgrazing. Overall, the ecological and economic and recreational benefits from hunting outweigh the minor damage done to crops, nurseries, and orchards (Chapman et al. 1982).

Rabbits have been suggested as seed dispersal vectors (Soriguer 1986) because of their high abundance, intensive use of small annual and perennial herbs, and production of fecal pellets. The literature has scattered reports of germinable seeds present in fecal pellets (Welch 1985, D'Antonio 1990), and Zedler and Black (1992) concluded that seed dispersal by rabbits may influence the distribution and long-term dynamics of some plant species.

Association With Livestock

Crouch (1982) reported that cottontails were significantly greater on the ungrazed bottomlands paralleling the South Platte

FIGURE 25. Distribution of the eastern cottontail (*Sylvilagus floridanus*) in North America (Hall 1981).

River in eastern Colorado and were almost nonexistent on grazed areas. Kundaeli and Reynolds (1972) found that the desert cottontail is negatively impacted when pinyon-juniper habitat is cleared during operations for increased livestock production. Yet, preserving some combination of 175 to 225 downed, dead trees and living shrubs/ha maintained their numbers. Similarly, Green and Flinders (1980*a*) pointed out that the removal of sagebrush for increased cattle production negatively affected the pygmy rabbit (*Brachylagus idahoensis*) because of their critical need for sagebrush (Green and Flinders 1980*b*).

Management

According to Chapman et al. (1982), the optimum habitat for the eastern cottontail in the Midwest is old, weedy, moderately

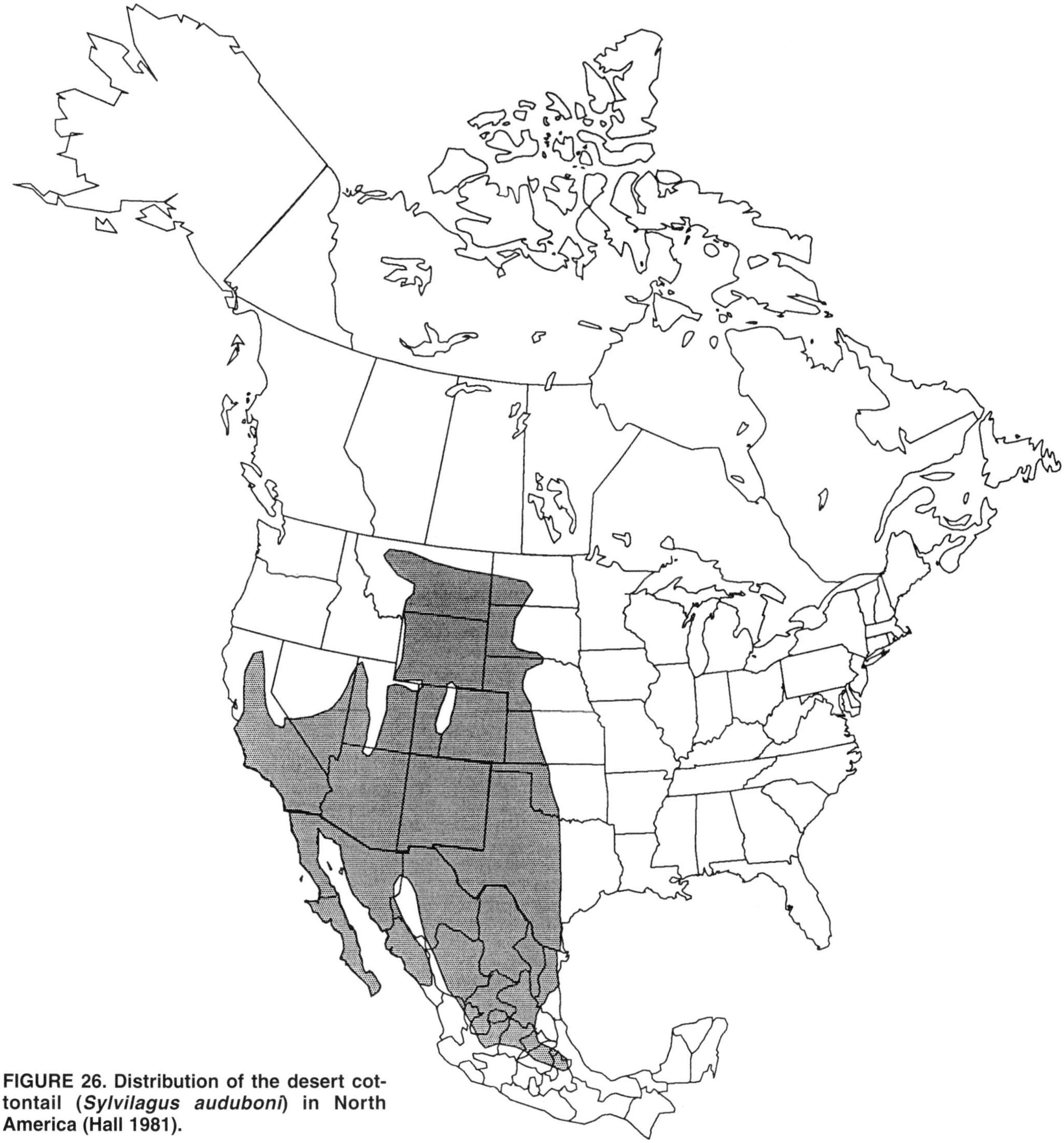

FIGURE 26. Distribution of the desert cottontail (*Sylvilagus auduboni*) in North America (Hall 1981).

grazed, unimproved, native grassland pasture with numerous dense clumps of thorny shrubs and small trees to provide escape cover. Vaughan (1972) documented the cottontail's need for protective cover, and Johnson and Hansen (1979) found them to be most numerous around rock piles. Habitat management to increase populations should emphasize this need for cover, and Madison (1959) suggested building brushpiles. Cottontails utilize primarily successional vegetation; therefore, habitat management techniques like prescribed burning and sharecropping have been suggested by Ellis et al. (1969) to control succession and increase cottontail numbers.

When cottontails arc found in high concentrations, they may damage crops, nurseries, and orchards. One of the principal means used by state game agencies for regulating populations of cottontails is annual hunting seasons (Craven 1983). Other selective controls include exclusionary methods such as tree trunk guards (Marsh et al. 1990), fencing, repellents, habitat modifications, and trapping (Craven 1983).

SUMMARY

Prairie dogs most frequently colonize sites that have been overgrazed or otherwise disturbed. Livestock grazing therefore promotes high prairie dog densities, and colonies tend to expand under heavy grazing. Depending on climatic factors, prairie dog colony expansion rates can be decreased by increasing vegetation cover around colonies through reducing grazing and human disturbance, and prairie dog colony expansion can be increased by increasing grazing or disturbance.

Prairie dogs can greatly alter rangeland. By digging extensive burrow systems they disturb soil and promote growth of disturbance-oriented vegetation. Because prairie dogs have a system of vigilance that depends on being able to see terrestrial predators at a distance, they clip shrubs and other tall vegetation that impedes visual detection. This in turn alters the habitat toward predominantly short grasses and annual forbs, rather than the taller grasses and shrubs characteristic of climax communities. Prairie dogs can cause negative and positive changes in rangeland. Negative changes include decreased forb and grass cover in colonies, higher silicon concentrations in grasses found in areas grazed by prairie dogs, and removal of plant biomass that could be used by cattle. Positive changes include increased plant species diversity, increased production of forbs and grasses caused by clipping, and better quality food and growing conditions inside prairie dog colonies. The economic effects of prairie dogs can be variable, according to the condition of the rangeland. Although considered pests, studies have shown that eradication of prairie dogs has little effect on increasing the amount of food available for cattle and is frequently not economically justified.

Prairie dog colonies provide important habitat for many other species of vertebrates, including the endangered black-footed ferret. Bison are attracted to prairie dog colonies as grazing sites because the vegetation associated with colonies may be more digestible and have a higher nitrogen content than the vegetation at uncolonized sites. Evidence continues to mount that prairie dogs are valuable components of the prairie ecosystem and that one of the management goals for prairie dog colonies should be the maintenance of biological diversity.

Ground squirrels are widely distributed on western rangelands. Their burrowing activities may have positive effects on rangeland by deepening soils, causing soil mixing, and improving the water holding capacity of soils. Many ground squirrel species on rangeland appear to benefit from livestock grazing and are most common on overgrazed ranges, where they can reach high densities. Several species have been considered pests on rangelands, including the California, Belding's, Richardson's, Wyoming, and Columbian ground squirrels, because their consumption and clipping activities can significantly reduce the amount of forage available to livestock. Management of ground squirrels on rangeland has traditionally emphasized use of toxicants; however, ground squirrels are an important part of rangeland ecosystems, and provide a significant part of the predator food base. Given the paucity of information on the effects of ground squirrels on rangeland and on livestock weight gain, further evaluations of rangeland, ground squirrel, and livestock interactions should be undertaken.

Pocket gophers usually prefer rangeland that is in good to excellent condition and may be reduced in numbers on areas heavily grazed by livestock. Like the prairie dogs and ground squirrels, pocket gophers may influence rangelands in positive and negative ways. They may reduce total standing crop biomass by 20% or more, and may alter plant species composition and abundance by differential consumption resulting from dietary preferences. Additionally, by creating soil mounds, feeding tunnels and underground food caches, pocket gophers influence the survival, growth, and biomass of surrounding vegetation. Their tunneling and mound-building activities cause an increase in bare ground, with gopher mounds commonly covering 5 to 15% of the ground surface. Mound building activities may bury vegetation and continually provide germination sites for early successional annual plant species. Mound formation may also be instrumental in maintaining a perennial dicotyledon plant component. Soil deposited by gophers may have a lower nutrient content than surrounding soil, but pocket gopher mound-building activity has a positive effect of increasing the biomass of the vegetation immediately surrounding each mound, partially offsetting lowered rangeland productivity.

The Leporidae (jackrabbits and cottontails) are important herbivores in North American agricultural and rangeland ecosystems. Black-tailed and white-tailed jackrabbits generally prefer open grasslands and agricultural areas and tend to avoid forests. Their relationship with humans have been viewed as positive and negative. Positive values are associated with their role in sport hunting, food and fur production, seed dispersal, ecological balance, biodiversity, and aesthetics. Negative values include their role in crop depredations, declining condition of rangeland, wildlife disease transmission, and competition with livestock for forage. Extensive investigations have been conducted regarding their diets; jackrabbits select succulent prereproductive vegetation when available, and show seasonal preferences for grasses and forbs during the growing season and for shrubs during the remainder of the year. Jackrabbits appear to have a positive effect on the biomass, growth, and species composition of rangeland flora at low densities, but reach a point at which their increasing densities produce lower quality and less palatable vegetation. Their competition with livestock is less than once thought, because recent studies have shown little seasonal and vegetation species overlap. Therefore, management practices now focus less on lethal population control and more on selective control (i.e., lethal and nonlethal) of depredating animals.

Cottontails have a more ubiquitous distribution in North America and a more cosmopolitan diet than jackrabbits. They provide recreation, a staple prey base for predators, seed dispersal, and ecological biodiversity. Overall, these benefits outweigh the minor damage they cause to crops, nurseries, and orchards. They are not considered competitors with livestock, because they prefer ungrazed to moderately grazed pastures with significant amounts of trees and dense shrubs providing escape cover. Management practices have emphasized either increased survival of cottontails through habitat management and the development of successional vegetation, or the selective use of controls such as exclusionary devices, habitat modification schemes, or trapping. The niche for jackrabbits and cottontails on rangelands is presently being debated, with some viewing them as pests needing eradication, while others believe they are essential to global biodiversity in a natural world. Both views are often presented as inflexible alternatives and may be based upon a fundamental lack of knowledge. Additional research on the interactions between rabbits and hares, livestock and rangelands may serve to bring these diverging views closer together towards a common management strategy.

LITERATURE CITED

Abaturov, B. D. 1968. Importance of the digging activity of the mole (*Talpa europaea* L.) on soil cover and vegetation in broadleaf-fir forest. Pedobiologia 8:239-264.

Abramsky, Z. 1978. Small mammal community ecology: changes in diversity in response to manipulated productivity. Oecologia 34:113-123.

———, M. I. Dyer, and P. D. Harrison. 1979. Competition among small mammals in experimentally perturbed areas of the shortgrass prairie. Ecology 60:530-536.

Agnew, W. 1983. Flora and fauna associated with prairie dog ecosystems. M.S. Thesis, Colorado State University, Fort Collins. 47pp.

———, D. W. Uresk, and R. M. Hansen. 1986. Flora and fauna associated with prairie dog colonies and adjacent ungrazed mixed-grass prairie in western South Dakota. J. Range Manage. 39:135-139.

———, ———, and ———. 1988. Arthropod consumption by small mammals on prairie dog colonies and adjacent ungrazed mixed grass prairie in western South Dakota. Proc. Great Plains Wildl. Damage Control Workshop 8:81-87.

Akin, K. E., and D. Burdick. 1977. Rumen microbial degradation of starch-containing bundle sheath cells in warm-season grasses. Crop Sci. 17:529-533.

Albert, W. W., and C. R. Record. 1982. Efficacy and costs of four rodenticides for controlling Columbian ground squirrels in western Montana. Proc. Great Plains Wildl. Damage Control Workshop 5:218-230.

Allen, B. E. 1992. Grazing management of over sown Tussock country: A review of traditional practice experience and research. New Zealand J. Agric. Res. 35:1-6.

Allen, D. L. 1938. Breeding of the cottontail in southern Michigan. Am. Midl. Nat. 20:464-469.

Allen, P. F. 1942. Defensive control of rodents and rabbits. J. Wildl. Manage. 6:122-132.

Alsager, D. E. 1977. Impact of pocket gophers (*Thomomys talpoides*) on the quantitative productivity of rangeland vegetation in southern Alberta: a damage assessment tool. Pages 47-57 *in* W. B. Jackson and R. E. Marsh, eds. Test methods for vertebrate pest control and management materials, ASTM STP 625. Am. Soc. for Testing and Materials, Philadelphia, Pa.

Andelt, W. F. 1984. Managing prairie dogs. Coop. Exten. Serv., Kansas State Univ., Manhattan. 12pp.

Andersen, D. C. 1987. *Geomys bursarius* burrowing patterns: influence of season and food patch structure. Ecology 68:1306-1318.

———, and J. A. MacMahon. 1981. Population dynamics and bioenergetics of a fossorial herbivore, *Thomomys talpoides* (Rodentia: Geomyidae), in a spruce-fir sere. Ecol. Monogr. 51:179-202.

Anderson, J. E., and R. E. Holte. 1981. Vegetation development over 25 years without grazing on sagebrush dominated rangeland in southeastern Idaho. J. Range Manage. 34:25-29.

Anderson, R. D. 1972. Curlew Valley validation site report. USIBP Desert Biome. Utah State Univ., Logan. 123pp.

Apa, A. D., D. W. Uresk, and R. L. Linder. 1991. Impacts of black-tailed prairie dog rodenticides on nontarget passerines. Great Basin Nat. 51:301-309.

Archer, S., and J. K. Detling. 1986. Evaluation of potential herbivore mediation of plant water status in a North American mixed-grass prairie. Oikos 47:287-291.

———, M. G. Garrett, and J. K. Detling. 1987. Rates of vegetation change associated with prairie dog (*Cynomys ludovicianus*) grazing in North American mixed-grass prairie. Vegetation 72:159-166.

———, I. D. Lebreton, M. G. Garrett, and J. K. Detling. 1984. Structural changes in a mixed-grass prairie dog community as a function of prairie dog colonization history. Bull. Ecol. Soc. of Am. 65:162.

Arkley, R. J., and H. C. Brown. 1954. The origin of Mima mound (hogwallow) microrelief in the far western states. Proc. Soil Sci. Soc. Am. 18:195-199.

Armstrong, D. M. 1972. Distribution of mammals in Colorado. Monogr. Mus. Nat. Hist., Univ. Kansas. 415pp.

———. 1977. Ecological distribution of small mammals in the upper Williams Fork Basin, Grad County, Colorado. Southwest. Nat. 22:289-304.

Arnold, J. F. 1942. Forage consumption and preferences of experimentally fed Arizona and antelope jack rabbits. Univ. Arizona Agric. Exp. Stn., Tech. Bull. 98:51-86.

———, and H. G. Reynolds. 1943. Droppings of Arizona and antelope jackrabbits and the "pellet census". J. Wildl. Manage. 7:322-327.

Askham, L. R. 1985. Effectiveness of two anticoagulant rodenticides (chlorophacinone and bromadiolone) for Columbian ground squirrel (*Spermophilus columbianus*) control in eastern Washington. Crop Prot. 4:365-371.

Bailey, J. A. 1968. Region wide fluctuations in the abundance of cottontails. Trans. North Am. Wildl. and Nat. Resource Conf. 33:265-277.

Bailey, V. 1936. The mammals and life zones of Oregon. North Am. Fauna 55:1-416.

Baker, M. F., and N. C. Frischknecht. 1973. Small mammals increase on recently cleared and seeded juniper rangeland. J. Range Manage. 26:101-102.

Baker, R. H. 1968. Habitats and distribution. Pages 98-122 *in* J. A. King, ed., Biology of *Peromyscus* (Rodentia). The Am. Soc. Mammal. Spec. Publ. 2.

———. 1971. Nutritional strategies of myomorph rodents in North American grasslands. J. Mammal. 52:800-805.

Balser, D. S. 1964. Antifertility agents in vertebrate pest control. Proc. Vertebr. Pest Conf. 2:133-137.

Barnes, W. C. 1936. Herds of the San Simon Valley. Am. For. 42:456-457.

Bartholomew, G. A., and H. H. Caswell 1951. Locomotion in kangaroo rats and its adaptive significance. J. Mammal. 32:155-169.

Batzli, G. O. 1975. The role of small mammals in arctic ecosystems. Pages 243-268 *in* F. Golley, K. Petrusewicz, and L. Ryszkowsik, eds. Small mammals: their productivity and population dynamics. Cambridge Univ. Press, London, U.K.

———, and F. A. Pitelka. 1970. Influence of meadow mouse populations on California grassland. Ecology 51:1027-1039.

———, and ———. 1971. Condition and diet of cycling populations of the California vole, *Microtus californicus*. J. Mammal. 52:141-163.

Bear, G. D., and R. M. Hansen. 1966. Food habits, growth, and reproduction of white-tailed jackrabbits in southern Colorado. Colorado State Univ. Agric. Exp. Stn. Tech. Bull. 90. 59pp.

Beck, R. F., and R. M. Hansen. 1966. Estimating plains pocket gopher abundance on adjacent soil types by a revised technique. J. Range Manage. 19:224-225.

Beckstead, M., and F. Schitoskey, Jr. 1980. Assimilation efficiency of the black-tailed prairie dog. Proc. South Dakota Acad. Sci. 59:184-194.

Behle, W. H. 1942. Distribution and variation of horned larks (*Otocoris alpestris*) of western North America. Univ. California Publ. Zool. 46:205-316.

Bell, W. B. 1921. Death to the rodents. U.S. Dep. Agric. Yearbook 1920:421-438.

Betts, B. J. 1990. Geographic distribution and habitat preferences of Washington Ground squirrels (*Spermophilus washingtoni*). Northwest. Nat. 71:27-37.

Bickler, P. E., and V. H. Shoemaker. 1975. Alfalfa damage by jackrabbits in southern California deserts. Calif. Agric. 29:10-12.

Birch, W. L. 1977. Ecological separation of *Peromyscus maniculatus bairdii* and *Peromyscus leucopus novelbaralensis* (Rodentia) in southcentral Ohio. M.S. Thesis, Ohio State Univ., Columbus. 86pp.

Birney, E. C., W. E. Grant, and D. D. Baird. 1976. Importance of vegetative cover to cycles of *Microtus* populations. Ecology 57:1043-1051.

Black, H. L., and N. C. Frischknecht. 1971. Relative abundance of mice on seeded sagebrush-grass range in relation to grazing. U.S. For. Serv., Res. Note INT-147. 7pp.

Boag, D. A., and J. O. Murie. 1981. Population ecology of Columbian ground squirrels in southwestern Alberta. Can. J. Zool. 59:2230-2240.

Bond, R. M. 1945. Range rodents and plant succession. Trans. North Am. Wildl. Conf. 10:229-234.

Bonham, C. D., and A. Lerwick. 1976. Vegetation changes induced by prairie dogs on shortgrass range. J. Range Manage. 29:221-225.

Borchert, M. I., and S. K. Jain. 1978. The effect of rodent seed predation on four species of California annual grasses. Oecologia (Berlin) 33:101-113.

Borrecco, J. E. 1976. Controlling damage by forest rodents and lagomorphs through habitat manipulation. Proc. Vertebr. Pest Conf. 7:203-210.

Bradley, W. G. 1968. Food habits of the antelope ground squirrel in southern Nevada. J. Mammal. 49:14-21.

Branson, F. A., R. F. Miller, and I. S. McQueen. 1965. Plant communities and soil moisture relationships near Denver, Colorado. Ecology 46:311-319.

———, and G. F. Payne. 1958. Effects of sheep and gophers on meadows of the Bridger Mountains of Montana. J. Range Manage. 11:165-169.

Brizuela, M. A., J. K. Detling, and M. S. Cid. 1984. Seasonal silicon content of grasses from sites receiving different grazing pressure by grassland herbivores. Bull. Ecol. Soc. Am. 65:161-162.

Bronson, F. H., and O. W. Tiemeier. 1958. Notes on crop damage by jackrabbits. Trans. Kansas Acad. Sci. 61:226-229.

———, and ———. 1959. The relationship of precipitation and black-tailed jackrabbit populations in Kansas. Ecology 40:194-198.

Brown, H. L. 1947*a*. Coaction of jack rabbit, cottontail and vegetation in a mixed prairie. Trans. Kansas Acad. Sci. 50:28-44.

———. 1947*b*. Why has the white-tailed jackrabbit (*Lepus townsendii campanius*) become scarce in Kansas? Trans. Kansas Acad. Sci. 49:455-456.

Brown, J. H. 1975. Geographical ecology of desert rodents. Pages 314-341 *in* M. L. Cody and J. M. Diamond, eds. Ecology and evolution of communities. Harvard Univ. Press, Cambridge, Mass.

———, D. W. Davidson, J. C. Munger, and R. S. Inouye. 1986. Experimental community ecology: The desert granivore system. Pages 41-62 *in* J. M. Diamond and T. J. Case, eds. Community ecology. Harper and Row, New York, N.Y.

———, ———, and O. J. Reichman. 1979*a*. An experimental study of competition between seed-eating desert rodents and ants. Am. Zool. 19:1129-1143.

———, and E. J. Heske. 1990. Control of a desert-grassland transition by a keystone rodent guild. Science 250:1705-1707.

———, O. J. Reichman, and D. W. Davidson. 1979*b*. Granivory in desert ecosystems. Annu. Rev. Ecol. Syst. 10:201-227.

Buechner, H. K. 1942. Interrelationships between the pocket gopher and land use. J. Mammal. 23:346-348.

Buffington, L. C., and C. H. Herbel. 1965. Vegetational changes on a semidesert grassland range from 1858-1963. Ecol. Monogr. 27:325-349.

Burt, W. H. 1943. Territoriality and home range concepts as applied to mammals. J. Mammal. 24:346-352.

———, and R. P. Grossenheider. 1959. A field guide to the mammals. Houghton Mifflon Co., Boston, Mass. 200pp.

Cable, K. A., and R. M. Timm. 1988. Efficacy of deferred grazing in reducing prairie dog reinfestation rates. Proc. Great Plains Wildl. Damage Control Workshop 8:46-49.

Cain, S. A., J. A. Kadlec, D. L. Allen, R. A. Cooley, M. G. Hornocker, A. S. Leopold, and F. H. Wagner. 1972. Predator control—1971—Report to the Council on Environmental Quality and the Department of the Interior by the Advisory Committee on Predator Control. Inst. for Environ. Quality, Univ. Michigan, Ann Arbor. 207pp.

California Department of Agriculture. 1966. Ground squirrels are nemesis of California farmers. Calif. Dep. Agric. Bull. 55: 195-196.

Campbell, T. M., III, and T. W. Clark. 1981. Colony characteristics and vertebrate associates of white-tailed and black-tailed prairie dogs in Wyoming. Am. Midl. Nat. 105:269-275.

———, ———, L. Richardson, S. C. Forrest, and B. R. Houston. 1987. Food habits of Wyoming black-footed ferrets. Am. Midl. Nat. 117:208-210.

Carothers, S. W, M. E. Stitt and R. R. Johnson. 1976. Feral asses on public lands: An analysis of biotic impact, legal considerations and management alternatives. Trans. North Am. Wildl. and Nat. Resour. Conf. 41:396-406.

Carpenter, J. W., M. J. G. Appel, R. C. Erickson, and M. N. Novilla. 1976. Fatal vaccine-induced canine distemper virus infection in black-footed ferrets. J. Am. Vet. Med. Assoc. 169:961-964.

Carroll, D., and L. L. Getz. 1976. Runway use and population density in *Microtus ochrogaster*. J. Mammal. 57:772-776.

Carter, F. L. 1939. A study in jackrabbit shifts range in western Kansas. Trans. Kansas Acad. Sci. 42:431-435.

Case, R. M. 1983. Pocket gophers. Pages B-13-B-26 *in* R. M. Timm, ed., Prevention and control of wildlife damage. Great Plains Agric. Council Wildl. Res. Comm. and Nebraska Coop. Ext. Serv., Univ. Nebraska, Lincoln.

Caswell, H., and F. C. Reed. 1975. Indigestibility of C4 bundle sheath cells by the grasshopper, *Melanoplus confusus*. Ann. Entomol. Soc. Am. 68:686-688.

———, and ———. 1976. The indigestibility of C4 bundle sheath cells by grasshoppers. Oecologia 26:151-156.

———, ———, S. N. Stephenson, and P. A. Werner. 1973. Photosynthetic pathways and selective herbivory: a hypothesis. Am. Nat. 107:465-480.

Chapman, J. A., J. G. Hockman, and M. M. Ojeda. 1980. *Sylvilagus floridanus*. Mammal. Species 136:1-8.

———, ———, and W. R. Edwards. 1982. Cottontails (*Sylvilagus floridanus* and allies). Pages 124-145 *in* J. A. Chapman and G. A. Feldhamer, eds. Wild mammals of North America. The Johns Hopkins Univ. Press, Baltimore, Md. 1147pp.

Chew, R. M. 1982. Changes in herbaceous and suffrutescent perennials in grazed and ungrazed decertified grassland in southeastern Arizona. 1958-1978. Am. Midl. Nat. 108:159-169.

———, and A. E. Chew. 1970. Energy relationships of the mammals of a desert shrub (*Larrea tridentata*) community. Ecol. Monogr. 40:1-21.

Christensen, E. M., and H. B. Johnson. 1964. Presettlement vegetation and vegetational change in three valleys in Central Utah. Brigham Young Univ. Sci. Bull. Biol. Ser. IV. 16pp.

Cincotta, R. P. 1985. Habitat and dispersal of black-tailed prairie dogs in Badlands National Park. Ph.D. Thesis, Colorado State Univ., Fort Collins. 52pp.

———, D. W. Uresk, and R. M. Hansen. 1987. A statistical model of expansion in a colony of black-tailed prairie dogs. Proc. Great Plains Wildl. Damage Control Workshop 8:30-33.

———, ———, and ———. 1989. Plant compositional change in a colony of black-tailed prairie dogs in South Dakota. Proc. Great Plains Wildl. Damage Control Workshop 9:171-177.

Clark, D. O. 1975. Vertebrate pest control handbook. California Dep. Food and Agric., Sacramento, Calif. 300pp.

Clark, F. W. 1972. Influence of jackrabbit density on coyote population change. J. Wildl. Manage. 36:343-356.

Clark, T. W. 1968. Food uses of the Richardson ground squirrel (*Spermophilus richardsoni elegans*) in the Laramie basin of Wyoming. Southwest. Nat. 13:248-249.

———. 1970*a*. Richardson's ground squirrel (*Spermophilus richardsonii*) in the Laramie Basin, Wyoming. Great Basin Nat. 30:55-70.

———. 1970*b*. Early growth, development, and behavior of the Richardson ground squirrel (*Spermophlus richardsonii elegans*). Am. Midl. Nat. 83:197-205.

———. 1977. Ecology and ethology of the white-tailed prairie dog (*Cynomus leucurus*). Milwaukee Public Mus. Publ. in Biol. and Geol. 3. 97pp.

———, T. M. Campbell, III, D. C. Socha, and D. E. Casey. 1982. Prairie dog colony attributes and associated vertebrate species. Great Basin Nat. 42:572-582.

Clements, F. E., and E. S. Clements. 1940. The biotic significance of disturbance, *in* "Climate, Climax and Conservation." Carnegie Inst. of Washington Year Book 39:174-175.

Collier, G. D., and J. J. Spillett. 1972. Status of the Utah prairie dog, *Cynomys parvidens*. Utah Acad. Sci., Arts and Letters 49:27-39.

———, and ———. 1975. Factors influencing the distribution of the Utah Prairie dog, *Cynomys parvidens* (Sciuridae). Southwest. Nat. 20:151-158.

Collins, A. R., J. P. Workman, and D. W. Uresk. 1984. An economic analysis of black-tailed prairie dog (*Cynomys ludovicianus*) control. J. Range Manage. 37:358-361.

Colorado Cooperative Pocket Gopher Project. 1960. Pocket gophers in Colorado. Colorado State Univ. Agric. Exp. Sta. Bull. 508-S, Fort Collins. 26pp.

Conklin, E. B. 1928. A brief history of Logan County, Colorado. Welch-Haffner Printing Co., Denver, Colo. 354pp.

Cooke, B. D. 1974. Food and other resources of the wild rabbit, (*Oryctolagus cuniculus*). Ph.D. Thesis, Univ. Adelaide, Australia. 142pp.

———. 1982. A shortage of water in natural pastures as a factor limiting a population of rabbits, *Oryctolagus cuniculus* (L.), in arid, north-eastern south Australia. Aust. Wildl. Res. 9:465-476.

Coppock, D. L., J. K. Detling, and M. I. Dyer. 1980. Interactions among bison, prairie dogs, and vegetation in Wind Cave National Park. Final Rep. Natl. Park Serv., Wind Cave Natl. Park, S.D. 195pp.

———, ———, J. E. Ellis, and M. I. Dyer. 1983*a*. Plant-herbivore interactions in a North American mixed-grass prairie. I. Effects of black-tailed prairie dogs on intraseasonal aboveground plant biomass and nutrient dynamics and plant species diversity. Oecologia (Berlin) 56:1-9.

———, J. E. Ellis, J. K. Detling, and M. I. Dyer. 1983*b*. Plant-herbivore interactions in a North American mixed-grass prairie. II. Responses of bison to modification of vegetation by prairie dogs. Oecologia (Berlin) 56:10-15.

Cox, G. W. 1984. The distribution and origin of Mima mound grasslands in San Diego County, California. Ecology 65:1397-1405.

———. 1990. Soil mining by pocket gophers along topographic gradients in a Mima moundfield. Ecology 71:837-843.

———, and D. W. Allen. 1987. Soil translocation by pocket gophers in a Mima moundfield. Oecologia (Berlin) 72:207-210.

Craighead, J. J., and F. C. Craighead, Jr. 1956. Hawks, owls, and wildlife. The Stackpole Co., Harrisburg, Pa., and Wildl. Manage. Inst., Washington, D.C. 443pp.

Craven, S. R. 1983. Cottontail rabbits. Pages D69-D74 *in* R. M. Timm, ed. Prevention and control of wildlife damage. Great Plains Agric. Counc. Lincoln, Neb.

Creighton, P. D. 1974. Habitat exploitation by an avian ground-foraging guild. Ph.D. Thesis, Colorado State Univ., Fort Collins. 154pp.

Crocker-Bedford, D. C. 1976. Food interactions between Utah prairie dogs and cattle. M.S. Thesis, Utah State Univ., Logan. 131pp.

———, and J. J. Spillett. 1977. Home ranges of Utah prairie dogs. J. Mammal. 58:672-673.

Crouch, G. L. 1982. Wildlife on ungrazed and grazed bottomlands on the South Platte river, northeastern Colorado. Pages 186-197 *in* J. M. Peek and P. D. Dalke, eds. Wildlife-livestock relationships symposium: proceedings 10. Univ. Idaho, For., Wildl., and Range Exp. Stn., Moscow.

Currie, P. O., and D. L. Goodwin. 1966. Consumption of forage by black-tailed jackrabbits on salt desert ranges of Utah. J. Wildl. Manage. 30:304-311.

D'Antonio, C. M. 1990. Seed production and dispersal in the non-native invasive succulent *Carpobrotus edulis* (Aizoaceae) in coastal strand communities of central California. J. Appl. Ecol. 27:693-702.

Daddy, F., M. J. Trlica, and C. D. Bonham. 1988. Vegetation and soil water differences among big sagebrush communities with different grazing histories. Southwest. Nat. 33:413-424.

Dale, F. H. 1939. Variability and environmental response of the kangaroo rat, *Dipodomys heermanni saxatilis*. Am. Midl. Nat. 22:703-731.

Dalke, P. D., and P. R. Sime, 1941. Food habits of the eastern and New England cottontails. J. Wildl. Manage. 5:216-228.

Dalquest, W. W. 1948. Mammals of Washington. Univ. Kansas Publ., Mus. of Nat. Hist. 2:1-444.

Dalsted, K. J., S. Sather-Blair, B. K. Worcester, and R. Klukas. 1981. Application of remote sensing to prairie dog management. J. Range Manage. 34:218-223.

Daniel, A. D. 1991. Influence of range condition on density and diet of black-tailed jackrabbits and diet of cattle in southcentral New Mexico (*Lepus californicus*, *Bos taurus*). Ph.D. Thesis, New Mexico State Univ., Las Cruces. 136pp.

Dano, L. 1952. Cottontail rabbit (*Sylvilagus audubonii baileyi)* populations in relation to prairie dog (*Cynomys ludovicianus ludovicianus*) towns. M.S. Thesis, Colorado State Univ., Fort Collins. 132pp.

Davis, L. S. 1984. Behavioral interactions of Richardson's ground squirrels: asymmetries based on kinship. Pages 424-444 *in* J. O. Murie and G. R., Michener, eds. The biology of ground-dwelling squirrels: Annual cycles, behavioral ecology, and sociality. Univ. Nebraska Press, Lincoln.

Davis, W. B., R. R. Ramsey, and J. M. Arendale, Jr. 1938. Distribution of pocket gophers (*Geomys breviceps*) in relation to soils. J. Mammal. 19:412-418.

Davis, W. C. 1962. Values of hunting and fishing in Arizona, 1960. Univ. Arizona, Tuscon. 61pp.

Deisch, M. S., D. W. Uresk, and R. L. Linder. 1989. Effect of two prairie dog rodenticides on ground-dwelling invertebrates in western South Dakota. Proc. Great Plains Wildl. Damage Control Workshop 9:166-170.

———, ———, and ———. 1990. Effects of prairie dog rodenticides on deer mice in western South Dakota. Great Basin Nat. 50:347-353.

Detling, J. K., and E. L. Painter. 1983. Defoliation responses of western wheatgrass populations with diverse histories of prairie dog grazing. Oecologia (Berlin) 57:65-71.

———, ———, and D. L. Coppock. 1986. Ecotypic differentiation resulting from grazing pressure: evidence for a likely phenomenon. Pages 431-433 *in* P. J. Joss, P. W. Lynch, and O. B. Williams, eds. Rangelands: a resource under siege. Australian Acad. of Sci., Canberra.

———, and A. D. Whicker. 1988. Control of ecosystem processes by prairie dogs and other grassland herbivores. Proc. Great Plains Wildl. Damage Control Workshop 8:23-29.

DeVoes, A. 1964. Range changes of mammals in the Great Lakes region. Am. Midl. Nat. 71:210-231.

Donoho, H. S. 1972. Dispersion and dispersal of white-tailed and black-tailed jackrabbits, Pawnee National Grasslands. M.S. Thesis, Colorado State Univ., Fort Collins. 83pp.

Dorrance, M. J. 1974. The annual cycle and population dynamics of Richardson's ground squirrel. Ph.D. Thesis, Univ. Wisconsin, Madison. 137pp.

Douglas, R. J. 1973. Spatial interactions and microhabitat selections of 2 locally sympatric voles, *Microtus montanus* and *Microtus pennsylvanicus*. Ph.D. Thesis, Montana State Univ., Bozeman. 48pp.

Downhower, J. F., and E. R. Hall. 1966. The pocket gopher in Kansas. Mus. Nat. Hist. Misc. Publ. 44:1-32.

DuBois, A. D. 1935. Nests of horned larks and longspurs on a Montana prairie. Condor 37:56-72.

Dunn, J. P., J. A. Chapman, and R. E. Marsh. 1982. Jackrabbits (*Lepus californicus* and allies). Pages 124-145 *in* J. A. Chapman and G. A. Feldhamer, eds. Wild mammals of North America. Johns Hopkins Univ. Press, Baltimore, Md.

Durrant, D. S., and R. M. Hansen. 1954. Distribution patterns and phylogeny of some western ground squirrels. Syst. Zool. 3:82-85.

Dusi, J. L. 1952. The food habits of several populations of cottontail rabbits in Ohio. J. Wildl. Manage. 16:180-186.

Eisenberg, J. F. 1963. The behavior of heteromyid rodents. Univ. California Publ. Zool. 69:1-114.

Elliot, C. L., and J. T. Flinders. 1985. Food habits of the Columbian ground squirrel, *Spermophilus columbianus*, in southcentral Idaho. Can. Field-Nat. 99:327-330.

———, and ———. 1991. *Spermophilus columbianus*. Mammal. Species 372:1-9.

Ellis, J. A., K. P. Thomas, and W. R. Edwards. 1969. Responses of bobwhites to management in Illinois. J. Wildl. Manage. 11:749-762.

Ellison, L. 1946. The pocket gopher in relation to soil erosion on mountain range. Ecology 27:101-114.

———, and C. M. Aldous. 1952. Influence of pocket gophers on vegetation of subalpine grassland in central Utah. Ecology 33:177-186.

Evans, F. C., and J. T. Emlen, Jr. 1947. Ecological notes on the prey selected by a barn owl. Condor 49:3-9.

Evans, F. S., and R. Holdenried. 1943. A population study of the Beechey ground squirrel in central California. J. Mammal. 24:231-260.

Evans, J., and J. L. Ward. 1967. Secondary poisoning associated with anticoagulant-treated nutria. J. Am. Vet. Assoc. 151:856-861.

———, P. L. Hegdal, and R. E. Griffith, Jr. 1970. Methods of controlling jackrabbits. Proc. Vertebr. Pest Conf. 4:109-116.

Evans, R., J. A. Young, G. J. Cluff, and J. K. McAdoo. 1983. Dynamics of antelope bitter brush seed caches. Pages 195-202 *in* Bitterbrush and Cliffrose Symposium. U.S. Dep. Agric. For. Serv. Gen. Tech. Rep. IMT-152.

Everett, R. L., and A. Kulla. 1976. Rodent cache seedlings of shrub species in the southwest. Tree Planters Notes 27:11-12.

———, R. Meeuwig, and R. Stevens. 1978. Deer mouse preference for seed of commonly planted species indigenous weed seed and sacrifice foods. J. Range Manage. 31:70-79.

———, and S. Monsen. 1990. Rodent problems in range rehabilitation. Proc. Vertebr. Pest Conf. 14:186-191.

Fagerstone, K. A. 1982. A review of prairie dog diet and its variability among animals and colonies. Proc. Great Plains Wildl. Damage Control Workshop. 5:178-184.

———. 1987. Comparison of vocalizations between and within *Spermophilus elegans elegans* and *S. richardsonii*. J. Mammal. 68:853-857.

———, G. K. Lavoie, and R. E. Griffith. 1980. Black-tailed jackrabbit diet and density on rangeland near agricultural crops. J. Range Manage. 33:229-233.

———, H. P. Tietjen, and O. Williams. 1981. Seasonal variation in the diet of black-tailed prairie dogs. J. Mammal. 62:820-824.

———, and O. Williams. 1982. Use of C3 and C4 plants by black-tailed prairie dogs. J. Mammal. 63:328-331.

Falls, J. B. 1968. Activity. Pages 543-567 *in* J. A. King, ed. Biology Of *Peromyscus* (Rodentia). Am. Soc. Mammal. Spec. Publ. 2.

Feldhamer, G. A. 1979. Vegetative and edaphic factors affecting abundance and distribution of small mammals in southeast Oregon. Great Basin Nat. 39:207-218.

Ferguson, R. B., and J. B. Basile. 1957. Effects of seedling numbers on bitter brush survival. J. Range Manage. 20:380-382.

Fitch, H. S. 1947*a*. Ecology of a cottontail rabbit (*Sylvilagus audubonii*) population in central California. Calif. Fish and Game. 33:159-184.

———. 1947*b*. Predation by owls in the sierra foothills of California. Condor 49:137-151.

———. 1948. Habits and economic relationships of the Tulare kangaroo rat. J. Mammal. 29:5-35.

———, and J. R. Bentley. 1949. Use of California annual-plant forage by range rodents. Ecology 30:306-321.

———, F. Swenson, and D. F. Tillotson. 1946. Behavior and food habits of the red-tailed hawk. Condor 48:205-237.

Fitzgerald, J. P. 1978. Vertebrate associations in plant communities along the South Platte River in Northeastern Colorado. Pages 73-88 *in* Proc. Loveland river and stream habitat in Colorado, Greeley, Colo.

Fleharty, E. D., and M. A. Mares. 1973. Habitat preference and spatial relations of *Sigmodon hispidus* on a remnant prairie in west-central Kansas. Southwest. Nat. 18:21-29.

———, and K. W. Navo. 1983. Irrigated cornfields as habitat for small mammals in the sandsage prairie region of western Kansas. J. Mammal. 64:367-379.

Flinders, J. T., and R. M. Hansen. 1972. Diets and habitats of jackrabbits in northeastern Colorado. Colorado State Univ. Range Sci. Dep. Sci. Ser. 12. 29pp.

———, and ———. 1973. Abundance and distribution of leporids within a shortgrass ecosystem. J. Mammal. 54:287-291.

Ford, S. D. 1977. Range, distribution and habitat of the western harvest mouse, *Reithrodontomys megalotis*, in Indiana. Am. Midl. Nat. 98:422-432.

Fortenbery, D. K. 1972. Characteristics of the black-footed ferret. U. S. Fish and Wildl. Serv. Resour. Publ. 109:1-8.

Foster, M. A. 1977. Impact of the plains pocket gopher (*Geomys bursarius*) on rangeland. M.S. Thesis, Univ. Wyoming, Laramie. 91pp.

———, and J. Stubbendieck. 1980. Effects of the plains pocket gopher (*Geomys bursarius*) on rangeland. J. Range Manage. 33:74-78.

Foster, R. B. 1965. Effect of heavy winter rodent infestation on perennial forage plants. J. Range Manage. 18:286-287.

French, N. R., R. McBride, and J. Detmer. 1965. Fertility and population density of the black-tailed jackrabbit. J. Wildl. Manage. 29:14-26.

———, W. E. Grant, W. Grodzinski, and D. M. Swift. 1976. Small mammal energetics in grassland ecosystems. Ecol. Monogr. 46:201-220.

Frischknecht, N. C., and M. F. Baker. 1972. Voles can improve sagebrush rangelands. J. Range Manage. 25:466-468.

Frye, R. J. 1983. Experimental field evidence of interspecific aggression between two species of kangaroo rat (*Dipodomys*). Oecologia (Berlin) 59:74-78.

Fuller, H. M., and L. R. Hafen. 1957. The journal of Captain John Bell. Arthur H. Clark Co., Glendale, Calif. 349pp.

Gallizioli, S. 1979. Effects of livestock grazing on wildlife. Calif.-Nev. Wildl. Transactions. 1979:83-87.

Garrett, M. G., and W. L. Franklin. 1982. Prairie dog dispersal in Wind Cave National Park: possibilities for control. Proc. Great Plains Wildl. Damage Control Workshop. 5:185-188.

Geier, A. R., and L. B. Best. 1980. Habitat selection by small mammals of riparian communities: evaluating effects of habitat alterations. J. Wildl. Manage. 44:16-24.

Gettinger, R. D. 1984. Energy and water metabolism of free-ranging pocket gophers, *Thomomys bottae*. Ecology 65:740-751.

Getz, L. L. 1971. Habitat of the meadow vole during a "population low." Am. Midl. Nat. 64:392-405.

Giezentanner, J. B. 1970. Avian distribution and population fluctuations on the shortgrass prairie of north central Colorado. M.S. Thesis, Colorado State Univ., Fort Collins. 113pp.

Goertz, J. W. 1964. The influence of habitat quality upon density of cotton rat populations. Ecol. Monogr. 34:359-381.

Gold, I. K. 1976. Effects of black-tail prairie dog mounds on shortgrass vegetation. M.S. Thesis, Colorado State Univ., Fort Collins. 40pp.

Golley, F. B., L. Ryszkowski, and J. T. Sokur. 1975. The role of small mammals in temperate forest, grasslands and cultivated fields. Pages 223-242 *in* F. B. Golley, K. Petrusewicz, and L. Ryszkowski, eds. Small mammals: their productivity and population dynamics. Cambridge Univ. Press, U.K.

Grant, W. E., and E. C. Birney. 1979. Small mammal community structure in North American grasslands. J. Mammal. 60:23-36.

———, ———, N. R. French, and D. M. Swift. 1982. Structure and productivity of grassland small mammal communities related to grazing-induced changes in vegetative cover. J. Mammal. 63:248-260.

———, N. R. French, and D. M. Swift. 1977. Response of a small mammal community to water and nitrogen treatments in a shortgrass prairie ecosystem. J. Mammal. 58:637-652.

———, ———, and L. J. Folse, Jr. 1980. Effects of pocket gopher mounds on plant production in shortgrass prairie ecosystems. Southwest. Nat. 25:215-224.

———, and J. F. McBrayer. 1981. Effects of mound formation by pocket gophers (*Geomys bursarius*) on old-field ecosystems. Pedobiologia 22:21-28.

Graves, G. N., W. C. Bennett, J. R. Wheeler, B. E. Miller, and D. L. Forcum. 1978. Sylvatic plague studies in southeast New Mexico. Part 2, Relationship of the desert cottontail and its fleas. J. Med. Entomol. 14:511-522.

Green, J. S., and J. T. Flinders. 1980*a*. *Brachylagus idahoensis*. Mammal. Species 125:1-4.

———, and ———. 1980*b*. Habitat and dietary relationships of pygmy rabbit. J. Range Manage. 33:136-142.

Griffing, J. P., and C. A. Davis. 1976. Black-tailed jackrabbits in southeastern New Mexico: population structure, reproduction, feeding and use of forms. New Mexico State Univ. Exp. Stn. Res. Rep. 318pp.

Griffiths, D. 1902. Forage conditions on the northern border of the Great Basin, being a report upon investigations made during July and August, 1901, in the region between Winnemuca, Nevada, and Ontario, Oregon. U.S. Bur. Plant Ind. Bull. 15. 60 pp.

Grinnell, J. 1923. The burrowing rodents of California as agents in soil formation. J. Mammal. 4:137-149.

———. 1933. Native California rodents in relation to water supply. J. Mammal. 14:293-298.

———, and J. Dixon. 1918. Natural history of the ground squirrels of California. Contrib. Univ. California Mus. Vertebr. Zool. 116pp.

Grodzinski, W. 1968. Energy flow through a vertebrate population. Pages 239-252 *in* W. Grodzinski and R. Z. Klekowski, eds. Methods of ecological bioenergetics. Polish Acad. Sci., Warsaw.

Gross, J. E., L. C. Stoddart, and F. H. Wagner. 1974. Demographic analysis of a northern Utah black-tailed jackrabbit population. Wildl. Monogr. 40. 68pp.

Guthery, F. S., T. E. Anderson, and V. W. Lehmann. 1979. Range rehabilitation enhances cotton rats in south Texas. J. Range Manage. 32:354-356.

Hall, E. 1946. Mammals of Nevada. Univ. California Press, Berkeley. 710pp.

Hall, E. R. 1955. Handbook of mammals of Kansas. Univ. Kansas Mus. Nat. Hist. Misc. Publ. 7. 303pp.

———. 1981. The mammals of North America. Second ed. John Wiley and sons, New York, N.Y. 1181pp.

———, and K. R. Kelson. 1959. The mammals of North America, Vol. 1. Ronald Press Co., New York, N.Y. 546pp.

Hallett, J. G. 1982. Habitat selection and the community matrix of a desert small-mammal fauna. Ecology 63:1400-1410.

Halvorson, C. H. 1982. Rodent occurrence, habitat disturbance, and seed fall in a larch-fir forest. Ecology 63:423-433.

Hanley, T. A., and J. L. Page. 1982. Differential effects of livestock use on habitat structure and rodent populations in Great Basin communities. Calif. Fish and Game 68:160-174.

Hansen, C. G. 1956. An ecological survey of the vertebrate animals on Steen's mountain, Harney County, Oregon. Ph.D. Thesis, Oregon State College, Corvallis. 199pp.

Hansen, R. M. 1954. Dispersal of Richardson ground squirrels in Colorado. Am. Midl. Nat. 68:58-66.

———. 1965. Pocket gopher density in an enclosure of native habitat. J. Mammal. 46:508-509.

———. 1978. Prairie dogs are a cottontail's best friend. Colo. Outdoors 27(2):30-31.

———, and R. F. Beck. 1968. Habitat of pocket gophers in Cochetopa Creek drainage, Colorado. Am. Midl. Nat. 79:103-117.

———, and J. T. Flinders. 1969. Food habits of North American hares. Colorado State Univ. Range Sci. Dep. Sci. Ser. 31. 18pp.

———, and I. K. Gold. 1977. Black-tail prairie dogs, desert cottontails, and cattle trophic relations on shortgrass range. J. Range Manage. 30:210-213.

———, and M. K. Johnson. 1976. Stomach content weight and food selection by Richardson ground squirrels. J. Mammal. 57:749-757.

———, and M. J. Morris. 1968. Movement of rocks by northern pocket gophers. J. Mammal. 49:391-399.

———, and L. D. Reed. 1969. Energy assimilation in Richardson ground squirrels. Am. Midl. Nat. 82:290-293.

———, and A. L. Ward. 1966. Some relations of pocket gophers to rangelands on Grand Mesa, Colorado. Colorado State Univ. Agric. Exp. Stn. Tech. Bull. 88. 22pp.

Hanson, R. 1988. A chronology of prairie dog control operations and related developments in South Dakota. Rocky Mountain For. and Range Exp. Stn. U.S. Dep. Agric., For. Serv. Tech. Rep. 154:121-122

Hansson, L., and H. Henttonen. 1988. Rodent dynamics as community processes. Trends Ecol. Evol. 3:195-200.

Harestad, A. S. 1986. Food habits of Columbian ground squirrels: A comparison of stomach and fecal samples. The Murrelet 67:75-78.

Harris, L. D. 1971. A precis of small mammal studies and results in the grassland biome. Pages 213-240 *in* N. R. French, ed. Preliminary analysis of structure and function in grasslands. Range Sci. Dep., Sci. Ser. 10. Colorado State Univ., Fort Collins.

Haskell, H. S., and H. G. Reynolds. 1947. Growth, developmental food requirements, and breeding activity of the California jackrabbit. J. Mammal. 28:129-136.

Hassien, F. 1976. A search for black-footed ferrets in the Oklahoma panhandle and adjacent area and an ecological study of black-tailed prairie dogs in Texas County, Oklahoma. M.S. Thesis, Oklahoma State Univ., Stillwater. 111pp.

Hastings, J. R., and R. M. Turner. 1965. The changing mile. Univ. Arizona Press, Tucson. 317pp.

Hawbecker, A. C. 1940. The burrowing and feeding habits of *Dipodomys venustus*. J. Mammal. 21:388-396.

———. 1975. The biology of some desert-dwelling ground squirrels. Pages 277-303 *in* I. Prakash and P. K. Ghosh. eds. Rodents in desert environments. Junk, The Hague, Netherlands.

Hayden, P. 1966. Food habits of black-tailed jackrabbits in southern Nevada. J. Mammal. 47:42-46.

Hayward, G. F., and J. Phillipson. 1979. Community structure and functional role of small mammals in ecosystems. Pages 135-211 *in* Ecology of small mammals, D. M. Stoddart, ed. Chapman and Hall, London, U.K.

Hegdal, P. L. 1966. Jackrabbit damage in the western United States. Supp. Rep. F-42.2., Denver Wildl. Res. Cent. Cont., Jackrabbit Res. Stn.

———, A. L. Ward, A. M. Johnson, and H. P. Tietjen. 1965. Notes on the life history of the Mexican pocket gopher (*Cratogeomys castanops*). J. Mammal. 46:334-335.

Henderson, F. R., P. F. Springer, and R. Adrian. 1969. The black-footed ferret in South Dakota. South Dakota Dep. Game, Fish and Parks, Tech. Bull. 4:1-37.

Heske, E. J., and M. Campbell. 1991. Effects of an 11-year livestock exclosure on rodent and ant numbers in the Chihuahuan desert, southeastern Arizona. Southwest. Nat. 36:89-93.

Hillman, C. N. 1968. Life history and ecology of the black-footed ferret in the wild. M.S. Thesis, South Dakota State Univ., Brookings. 28pp.

———, and R. L. Linder. 1973. The black-footed ferret. Pages 10-20 *in* R. L. Linder and C. N. Hillman, eds. Proc. black-footed ferret and prairie dog workshop. South Dakota State Univ., Brookings.

———, and J. C. Sharps. 1978. Return of swift fox to northern Great Plains. Proc. South Dakota Acad. Sci. 57:154-162.

Hintz, H. F. 1969. Equine nutrition-comparisons of digestion coefficients obtained with cattle, sheep, rabbits, and horses. The Veterinarian. 6:45-51.

Hironaka, M. 1961. The relative rate of root development of cheatgrass and medusahead. J. Range Manage. 14:264-267.

Hodgson, J. R. 1972. Local distribution of *Microtus montanus* and *Microtus pennsylvanicus* in southwestern Montana. J. Mammal. 53:487-499.

Hollister, N. 1916. A systematic account of the prairie dogs. U.S. Dep. Agric., North Am. Fauna 40:1-37.

Hoogland, J. L. 1979. Aggression, ectoparasitism, and other possible costs of prairie dog (Sciuridae, *Cynomys* spp.) coloniality. Behaviour 69:1-35.

Hopkins, H. H. 1954. Effects of mulch upon certain factors of the grassland environment. J. Range Manage. 7:255-258.

Hormay, A. L. 1943. Bitterbrush in California. Calif. For. and Range Exp. Stn. Res. Note 34. 13pp.

Horn, E. E. 1941. Some coyote-wildlife relationships. Trans. North Am. Wildl. Conf. 6:283-286.

House, W. A. 1964. Food habits of Richardson's ground squirrel in south-central Wyoming. M.S. Thesis, Colorado State Univ., Fort Collins. 57pp.

Howard, W. E. 1967. Biological control of vertebrate pests. Proc. Vertebr. Pest Conf. 3:137- 157.

———, and H. E. Childs, Jr. 1959. Ecology of pocket gophers with emphasis on *Thomomys bottae mewa*. Hilgardia 29:277-358.

———, K. A. Wagnon, and J. R. Bentley. 1959. Competition between ground squirrels and cattle for range forage. J. Range Manage. 12:110-115.

Howell, A. H. 1938. Revision of the North American ground squirrels, with a classification of the North American Sciuridae. North Am. Fauna 56:1-256.

Hunter, J. E. 1991. Grazing and pocket gopher abundance in a California annual grassland. Southwest. Nat. 36:117-118.

Hutto, R. L. 1978. A mechanism for resource allocation among sympatric heteromyid rodent species. Oecologia (Berlin) 33:115-126.

Ingles, L. G. 1941. Natural history observations on the Audubon cottontail. J. Mammal. 22:227-250.

———. 1949. Ground water and snow as factors affecting the seasonal distribution of pocket gophers. *Thomomys monticola*. J. Mammal. 30:343-350.

Inouye, R. S., G. S. Byers, and J. H. Brown. 1980. Effects of predation and competition on survivorship, fecundity and community structure desert annuals. Ecology 61:1344-1351.

———, N. J. Huntley, D. Tilman, and J. R. Tester. 1987. Pocket gophers (*Geomys bursarius*), vegetation, and soil nitrogen along a successional sere in east central Minnesota. Oecologia (Berlin) 72:178-184.

Jameson, E. W., Jr. 1947. Natural history of the prairie vole (mammalian genus *Microtus*). Univ. Kansas Publ., Mus. Nat. Hist. 11:25-151.

Janzen, D. H. 1971. Seed predation by animals. Annu. Rev. Ecol. Syst. 2:465-492.

Johnson, D. R. 1961. The food habits of rodents on rangelands of southern Idaho. Ecology 42:407-410.

———. 1964. Effects of range treatment with 2,4-D on food habits of rodents. Ecology 45:241-249.

———, and K. L. Groepper. 1970. Bioenergetics of North Plains rodents. Am. Midl. Nat. 84:537-548.

———, and R. M. Hansen. 1969. Effects of range treatment with 2,4-D on rodent populations. J. Wildl. Manage. 33:125-132.

Johnson, G. D. and K. A. Fagerstone. 1994. Primary and secondary hazards of zinc phosphide to nontarget wildlife—A review of the literature. USDA/APHIS, DWRC Res. Rep. 11-55-005. 26 pp.

Johnson, M. K., and R. M. Hansen. 1979. Foods of cottontails and woodrats in south-central Idaho. J. Mammal. 60:213-215.

Johnson, R. D. 1982. Relationships of black-tailed jackrabbit diets to population density and vegetal components of habitat. M.S. Thesis, Idaho State Univ., Pocatello. 86pp.

———, and J. E. Anderson. 1984. Diets of black-tailed jackrabbits in relation to population density and vegetation. J. Range Manage. 37:79-83.

Johnson, W. V. 1964. Rabbit control. Proc. Vertebr. Pest Conf. 2:90-96.

Johnston, J. C. 1978. Anticoagulant baiting for jackrabbit control. Proc. Vertebr. Pest Conf. 8:152-153.

Jones, J. K. Jr., D. M. Armstrong, and J. R. Choate. 1985. Guide to mammals of the Plains States. Univ. Nebraska Press, Lincoln. 371pp.

———, ———, R. S. Hoffmann, and C. Jones. 1983. Mammals of the Northern Great Plains. Univ. Nebraska Press, Lincoln. 379pp.

Kaufman, D. W., B. K. Clark, and G. A. Kaufman. 1990. Habitat breadth of nongame rodents in the mixed-grass prairie region of north central Kansas. Prairie Nat. 22:19-26.

Kauffman, J. B., W. C. Krueger, and M. Vavra. 1982. Pages 208-220 *in* J. M. Peek and P. D. Dalke, eds. Wildlife-livestock relationships symposium: proceedings 10. Univ. Idaho, For., Wildl. and Range Exp. Stn. Moscow.

Kaukeinen, D. E. 1982. A review of the secondary poisoning hazard to wildlife from the use of anticoagulant rodenticides. Pest Manage. 1(11):12-14; 1(12):18-19.

Keith, J. O., R. M. Hansen, and A. L. Ward. 1959. Effect of 2,4-D on abundance and foods of pocket gophers. J. Wildl. Manage. 23:137-145.

Keith, L. B. 1963. Wildlife's ten year cycle. Univ. Wisconsin Press, Madison. 201pp.

———. 1979. Population dynamics of hares. Proc. of the world lagomorph conf., Univ. Guelph, Ontario, Can. 440pp.

Kelso, L. H. 1939. Food habits of prairie dogs. U.S. Dep. Agric. Circ. 529. 15pp.

Kennerly, T. E., Jr. 1964. Microenvironmental conditions of the pocket gopher burrow. Texas J. Sci. 16:395-441.

King, J. A. 1955. Social behavior, social organization, and population dynamics in a black-tailed prairie dog town in the Black Hills of South Dakota. Contrib. Lab. Vert. Biol. 67. Univ. Michigan. Ann Arbor. 123 pp.

Kirpatrick, C. M. 1960. Unusual cottontail litter. J. Mammal. 41:119-120.

Kjar, K. J., and J. Stubbendieck. 1980. Natural revegetation of rangeland following the control of *Geomys bursarius*. Proc. Nebraska Acad. of Sci. 19:9.

Klatt, L. E. 1971. A comparison of the ecology of active and abandoned black-tailed prairie dog (*Cynomys ludovicianus*) towns. M.S. Thesis, Colorado State Univ., Fort Collins. 66pp.

———, and D. Hein. 1978. Vegetative differences among active and abandoned towns of black-tailed prairie dogs. J. Range Manage. 31:315-317.

Klimstra, W. D., and Corder, E. L. 1957. Food of the cottontail in southern Illinois. Trans. Illinois State Acad. Sci. 50:247-256.

Kline, P. D. 1963. Notes on the biology of the jackrabbit in Iowa. Proc. Iowa Acad. Sci. 70:196-204.

Knowles, C. J. 1982. Habitat affinity, populations, and control of black-tailed prairie dogs on the Charles M. Russell National Wildlife Refuge. Ph.D. Thesis, Univ. Montana, Missoula. 171pp.

———. 1986. Some relationships of black-tailed prairie dogs to livestock grazing. Great Basin Nat. 46:198-203.

Knowlton, F. F., P. E. Martin, and J. C. Haug. 1968. A telemetric monitor for determining animal activity. J. Wildl. Manage. 32:943-948.

Koford, C. B. 1958. Prairie dogs, white faces, and blue grama. Wildl. Monogr. 3. 78pp.

Krebs, C. J. 1970. *Microtus* population biology: Behavioral changes associated with population cycle in *M. ochrogaster* and *M. pennsylvanicus*. Ecology 51:34-52.

———, M. S. Gaines, B. L. Keller, J. H. Myers, and R. H. Tamarin. 1973. Population cycles in small rodents. Science 179:35-41.

———, and J. H. Myers. 1974. Population cycles in small mammals. Adv. Ecol. Res. 8:267-399.

Krueger, K. 1985. Interactions and activity patterns of bison and prairie dogs at Wind Cave National Park: Implications for managers. Pages 203-208 *in* R. C. Luces, ed. Proc. National Wilderness Resource Conference. Fort Collins, Colo.

———. 1986. Feeding relationships among bisons, pronghorn, and prairie dogs: an experimental analysis. Ecology 67:760-770.

Kuhn, L. W., W. Q. Wick, and R. J. Pedersen. 1966. Breeding nests of Townsend's moles in Oregon. J. Mammal. 47:239-249.

Kundaeli, J. N., and H. G. Reynolds. 1972. Desert cottontail use of natural and modified pinyon-juniper woodland. J. Range Manage. 25:116-118.

Lacey, J. R., and H. W. Van Poollen. 1981. Comparison of herbage production on moderately grazed and ungrazed western ranges. J. Range Manage. 34:210-212.

Lack, D. L. 1954. The natural regulation of animal numbers. Clarendon Press, Oxford, U.K. 343pp.

Lambeth, R. E., and M. Hironaka. 1982. Columbia ground squirrel in subalpine openings in central Idaho. J. Range Manage. 35:493-497.

Larrison, E. J., and D. R. Johnson. 1973. Density changes and habitat affinities of rodents of shadscale and sagebrush associations. Great Basin Nat. 33:255-264.

LaTourrette, J. E., J. A. Young, and R. A. Evans. 1971. Seed dispersal in relation to rodent activities in seral big sagebrush communities. J. Range Manage. 24:118-120.

Laycock, W. A. 1953. An ecological study of the pocket gopher (*Thomomys talpoides tenellus* Goldman) in the Jackson Hole region of Wyoming. M.S. Thesis, Univ. Wyoming, Laramie. 65pp.

———. 1958. The initial pattern of revegetation of pocket gopher mounds. Ecology 39:346-351.

———, and B. Z. Richardson. 1975. Long-term effects of pocket gopher control on vegetation and soils of a subalpine grassland. J. Range Manage. 28:458-462.

Lechleitner, R. R. 1958*a*. Certain aspects of behavior of the black-tailed jackrabbit. Am. Midl. Nat. 60:145-155.

———. 1958*b*. Movements, density, and mortality in a black-tailed jackrabbit population. J. Wildl. Manage. 22:371-384.

———. 1959. Some parasites and infectious diseases in a black-tailed jackrabbit population in the Sacramento Valley, California. Calif. Fish and Game. 45:83-91.

Leopold, A. 1945. The distribution of Wisconsin hares. Trans. Wis. Acad. Sci. 37:1-14.

Lerwick, A. C. 1974. The effect of the black-tailed prairie dog on vegetative composition and their diet in relation to cattle. M.S. Thesis, Colorado State Univ., Fort Collins. 106 pp.

Lewis, J. H. 1946. Planting practice to reduce crop damage by jackrabbits. J. Wildl. Manage. 10:277.

Lewis, S. R., and J. M. O'Brien. 1990. Survey of rodent and rabbit damage to alfalfa hay in Nevada. Proc. Vertebr. Pest Conf. 14:116-119.

Linsdale, J. M. 1946. The California ground squirrel. Univ. California Press, Berkeley. 475pp.

Lobue, J., and R. M. Darnell. 1959. Effect of habitat disturbance on a small mammal population. J. Mammal. 40:425-437.

Lockard, R. B., and J. S. Lockard. 1971. Seed preference and buried seed retrieval of *Dipodomys deserti*. J. Mammal. 52:219-221.

Lord, R. D. 1963. The cottontail rabbit in Illinois. Tech. Bull. Ill. Dep. Conserv. 3:1-94.

Luce, D. G., R. M. Case, and J. Stubbendieck. 1980. Food habits of the plains pocket gopher on western Nebraska rangeland. J. Range Manage. 33:129-131.

Lunk, E. M. 1989. An evaluation of responses of cottontail rabbits to structure and composition of cover: development and testing of habitat-use models for a generalist species (rabbits, North Carolina). Ph.D. Thesis, North Carolina State Univ., Raleigh, 180pp.

MacCracken, J. G., and R. M. Hansen. 1982. Herbaceous vegetation of habitat used by black-tail jackrabbits and Nuttal cottontails in southeastern Idaho. Am. Midland Nat. 107:180-184.

———, D. W. Uresk, and R. M. Hansen. 1985*a*. Rodent-vegetation relationships in southeastern Montana. Northwest Sci. 59:272-278.

———, ———, and ———. 1985*b*. Vegetation and soils of burrowing owl nest sites in Conata Basin, South Dakota. Condor 87:152-154.

Mackie, R. J. 1978. Impacts of livestock grazing on wild ungulates. North Am. Wild. and Nat. Resour. Conf. 43:462-476.

Madison, J. 1959. The cottontail rabbit. Olin Mathieson Chemical Corp., East Alton, Ill. 56pp.

Manville, R. H. 1959. the columbian ground squirrel in northwestern Montana. J. Mammal. 40:26-45.

Marsh, R. E. 1984. An overview of rodent and rabbit pests of agricultural crops in the United States. Pages 57-60 *in* A. C. Dubock, ed. Proc. Conf. Organ. Practices Vertebr. Pest Control. Imperial Chem. Industries, Haslemere, U.K.

———, A. E. Koehler, and T. P. Salmon. 1990. Exclusionary methods and materials to protect plants from pest mammals-a review. Proc. Vertebr. Pest Conf. 14:174-180.

———, R. H. Schmidt, and W. E. Howard. 1987. Secondary hazards to coyotes of ground squirrels poisoned with 1080 or strychnine. Wildl. Soc. Bull. 15:380-385.

Marshall, D. R., and S. K. Jain. 1970. Seed predation and dormancy in the population dynamics of *Avena fatua* and *Avena barbata*. Ecology 51:886-891.

Martin, S. C. 1975. Ecology and management of south-western semidesert grass-shrub ranges: the status of our knowledge. U.S. Dep. Agric., For. Serv. Res. Pap. RM-156. 39pp.

Martinsen, G. D. 1989. Impact of pocket gopher disturbance on a shortgrass prairie: evidence for the non-equilibrium maintenance of species diversity. M.S. Thesis, North. Arizona Univ., Flagstaff. 28pp.

Maser, C. 1974. The sage vole, *Lagurus curtatus* (Cope, 1868), in the Crooked River National Grassland, Jefferson county, Oregon. A contribution to its life history and ecology. Saeugetier Kd. Mitt. 22:193-222.

———, Z. Mazer, and R. Molina. 1988. Small mammal mycophagy in range lands of central and southeastern Oregon. J. Range Manage. 41:309-312.

———, and G. S. Strickler. 1978. The sage vole, *Lagurus curtatus*, as an inhabitant of subalpine sheep fescue, *Festuca ovina*, communities on Steens Mountain—an observation and interpretation. Northwest Sci. 52:276-284.

Matschke, G. H., C. L. Fordham, S. C. Hurlbut, and R. M. Engeman. 1988. Comparative toxicity of strychnine to eight species of ground squirrels. Proc. Great Plains Wildl. Damage Control Workshop 8:75-80.

———, M. P. Marsh, and D. L. Otis. 1983. Efficacy of zinc phosphide broadcast baiting for controlling Richardson's ground squirrels on rangeland. J. Range Manage. 36:504-506.

Maxell, M. H., and L. N. Brown. 1968. Ecological distribution of rodents on the high plains of eastern Wyoming. Southwest. Nat. 13:143-158.

McAdoo, J. K., C. C. Evans, B. A. Roundy, J. A. Young, and R. A. Evans. 1983. Influence of heteromyid rodents on *Oryzopsis hymenoides* germination. J. Range Manage. 36:61-64.

McAdoo, J. D., and J. A. Young. 1980. Jackrabbits. Rangelands 2:135-138.

McCabe, R. A. 1966. Vertebrates as pests: a point of view. Pages 115-134 *in* Scientific Aspects of Pest Control. Natl. Acad. Sci., Nat. Resour. Counc., Washington, D.C.

McCulloch, C. Y., Jr. 1962. Populations and range effects of rodents on the sand sagebrush grasslands of western Oklahoma. Okla. State Univ. Publ. 11. 112pp.

McDonough, W. T. 1974. Revegetation of gopher mounds on aspen range in Utah. Great Basin Nat. 34:267-274.

McDowell, J. W., H. G. Scott, C. J. Stojanovich, and H. B. Weinburgh. 1964. Tularemia. U.S. Dep. H.E.W., Public Health Service, Atlanta, Ga. 81pp.

McGee, J. M. 1982. Small mammal populations in an unburned and early fire successional sagebrush community. J. Range Manage. 35:177-179.

McGinnies, W. J. 1960. Effect of Mima-type microrelief on herbage production of five seeded grasses in western Colorado. J. Range Manage. 13:231-234.

McHugh, T. 1958. Social behavior of the American buffalo (*Bison bison bison*). Zoologica 43:1-40.

McIlroy, J. C., and E. J. Gifford. 1992. Secondary poisoning hazards associated with 1080-treated carrot-baiting campaigns against rabbits *Oryctolagus-cuniculus*. Wildl. Res. 19:629-641.

McKay, D. O. 1978. Estimates of some attributes of a population of Nuttall's cottontails. J. Wildl. Manage. 42:159-168.

McMahon, K. J. 1965. Bacterial and rickettsial diseases. Pages 65-71 *in* The black-tailed jackrabbit in Kansas. Kansas State Univ. Agric. and Appl. Sci. Agric. Exp. Stn. Tech. Bull. 140pp.

McNab, B. K. 1966. The metabolism of fossorial rodents: A study of convergence. Ecology 47:712-733.

McNaughton, S. J. 1984. Grazing lawns: Animals in herds, plant form, and coevolution. Am. Nat. 124:863-886.

———. 1985. Ecology of a grazing ecosystem: The Serengeti. Ecol. Monogr. 55:259-294.

Merriam , C. H. 1902. The prairie dog of the great plains. U.S. Dep. Agric. Yearbook 1901:257-270.

Metzgar, L. H. 1980. Dispersion and numbers in *Peromyscus* populations. Am. Midl. Nat. 103:26-31.

Michener, G. R. 1979. Yearly variations in the population dynamics of Richardson's ground squirrels. Can. Field-Nat. 93:363-370.

———. 1983. Spring emergence schedules and vernal behavior of Richardson's ground squirrels: why do males emerge from hibernation before females? Behav. Ecol. Sociobiol. 14:29-83.

———, and J. W. Koeppl. 1985. *Spermophilus richardsonii*. Mammal. Species 243:1-8.

———, and D. R. Michener. 1977. Population structure and dispersal in Richardson's ground squirrels. Ecology 58:359-368.

Miller, R. S. 1964. Ecology and distribution of pocket gophers (Geomyidae) in Colorado. Ecology 45:256-272.

Moore, A. W. 1943. Notes on the sage mouse in eastern Oregon. J. Mammal. 24:188-191.

———, and E. H. Reid. 1951. The Dalles pocket gopher and its influence on forage production of Oregon mountain meadows. U.S. Dep. Agric. Circ. 884. 36pp.

Moulton, M. P. 1978. Small mammal associations in grazed versus ungrazed cottonwood riparian woodland in eastern Colorado. Pages 133-140 *in* W. D. Graul and S. J. Bissell, eds. Lowland river and stream habitat in Colorado: a symposium. Colo. Chap. Wildl. Soc. and Colo. Audubon Counc.

———, J. R. Choate, S. J. Bissell, and R. A. Nicholson. 1980. Associations of small mammals on the central high plains of eastern Colorado. Southwest. Nat. 26:53-57.

———, ———, and ———. 1983. Biogeographic relationships of pocket gophers in southeastern Colorado. Southwest. Nat. 28:53-60.

Mueggler, W. F. 1967. Voles damage big sagebrush in southwestern Montana. J. Range Manage. 20:88-90.

Murray, D. F. 1967. Gravel mounds at Rocky Flats, Colorado. Mt. Geol. 4:99-107.

Myers, G. T., and T. A. Vaughan. 1964. Food habits of the plains pocket gopher in eastern Colorado. J. Mammal. 45:588-598.

Myers, J. H., and C. J. Krebs. 1974. Population cycles in rodents. Sci. Am. 230:38-46.

Nelson, E. W. 1919. Prairie dogs. Chief Bur. Biol. Survey Rep. 1919:5.

Nelson, J. F., and R. M. Chew. 1977. Factors affecting seed reserves in the soil of a Mojave Desert ecosystem, Rock Valley, Nye County, Nevada. Am. Midl. Nat. 97:300-320.

Nelson, J. R., A. M. Wilson, and C. J. Goebel. 1970. Factors influencing broadcast seeding in bunchgrass range. J. Range Manage. 23:163-170.

Nord, E. C. 1965. Autecology of bitterbrush in California. Ecol. Monogr. 35:307-334.

Norris, J. J. 1950. Effect of rodents, rabbits, and cattle on two vegetation types in semidesert range land. Agric. Exp. Stn., New Mexico College of Agric. and Mech., Art., Bull. 353. 24pp.

O'Farrell, T. P. 1972. Ecological distribution of sagebrush voles, *Lagurus curtatus*, in south-central Washington. J. Mammal. 53:632-636.

O'Meilia, M. E., F. L. Knopf, and J. C. Lewis. 1982. Some consequences of competition between prairie dogs and beef cattle. J. Range Manage. 35:580-585.

Oldemeyer, J. L. and L. R. Allen-Johnson. 1988. Cattle grazing and small mammals on the Sheldon National Wildlife Refuge, Nevada. Pages 391-398 *in* R. C. Szaro, K. E. Severson, and D. R. Patton, eds. Proc. symposium on manage. of amphibians, reptiles and small mammals in North Am. Flagstaff, Ariz.

———, S. J. Martin, and S. G. Woodis. 1983. A preliminary report on the effects of a deferred-rotation grazing system on wildlife at the Sheldon National Wildlife Refuge. Calif.-Nev. Wildl. Transactions. 1983:26-42.

Orr, R. T. 1940. The rabbits of California. California Acad. Sci. 19. San Francisco, Calif. 227pp.

Osborn, B., and P. F. Allan. 1949. Vegetation of an abandoned prairie-dog town in tallgrass prairie. Ecology 30:322-332.

Palmer, T. S. 1896. The jackrabbits of the United States. U.S. Dep. Agric., Bull. 8. 84pp.

Parker, K. W., and S. C. Martin. 1952. The mesquite problem on southern Arizona ranges. U.S. Dep. Agric. Circ. 908. 77pp.

Petrusewicz, K., and A. Macfadyen. 1970. Productivity of terrestrial animals: principles and methods. IBP Handbook 13. Blackwell Scientific Publ., Oxford, U.K.

Phillips, P. 1936. The distribution of rodents in overgrazed and normal grasslands of central Oklahoma. Ecology 17:673-679.

Phillips, T. A. 1970. The status of antelope bitterbrush in the Cassia mountain area of southern Idaho. U.S. Dep. Agric., U.S. For. Serv. Intermountain Region. Range Improvement Notes 15:1-15.

Pianka, E. R. 1973. The structure of lizard communities. Annu. Rev. Ecol. Syst. 4:53-74.

Pickwell, G. A. 1931. The prairie horned lark. Trans. Missouri Acad. Sci. 27:1-153.

Pimlott, D. H. 1969. The value of diversity. Trans. North Am. Wildl. and Nat. Resour. Conf. 34:265-280.

Pizzimenti, J. J., and R. S. Hoffmann. 1973. *Cynomys gunnisoni*. Mammal. Species 25:1-4.

Potter, R. L. 1980. Secondary successional patterns following prairie dog removal on shortgrass range. M.S. Thesis, Colorado State Univ., Fort Collins. 165pp.

Pressnall, C. C. 1950. The predation question-fact versus fancies. Trans. North Am. Wildl. and Nat. Resour. Conf. 15:197-207.

Price, M. V. 1978. Seed dispersion preferences of coexisting desert rodent species. J. Mammal. 59:624-626.

———. 1983. Laboratory studies of seed size and seed species selection by Heteromyid rodents. Oecologia (Berlin) 60:259-263.

———, and S. H. Jenkins. 1986. Rodents as seed consumers and dispersers. Pages 191-235 *in* D. R. Murray, ed. Seed dispersal. Acad. Press, Australia.

———, and O. J. Reichman. 1987. Distribution of seeds in Sonoran Desert soils: implications for heteromyid rodent foraging. Ecology 68:1797-1811.

———, and N. M. Waser. 1985. Microhabitat use by heteromyid rodents: Effects of artificial seed patches. Ecology 66:211-219.

Pyke, G. H., H. R. Pulliam, and E. L. Charnov. 1977. Optimal foraging: a selective review of theory and tests. Q. Rev. Biol. 52:137-153.

Quanstrom, W. R. 1968. Some aspects of the ethoecology of Richardson's ground squirrel in eastern North Dakota. Ph.D. Thesis, Univ. Oklahoma, Norman. 121pp.

Quast, J. C. 1950. Habitat preferences of rodents on grazed and ungrazed foothill pastures in California. M.A. Thesis. Univ. California, Berkeley. 50pp.

Ramey, C. A., E. W. Schafer Jr., K. A. Fagerstone, and S. D. Palmateer. 1992. Back to the future for APHIS's vertebrate pesticides. Proc. Vertebr. Pest Conf. 15:17-21.

Ratcliffe, F. N. 1951. The rabbit problem-a survey of research needs and possibilities. Commonwealth Sci. and Indust. Res. Org. Australia. Leaflet. 1:3-17.

Read, R. A. 1971. Browsing preference by jackrabbits in a ponderosa pine provenance plantation. U.S. Dep. Agric. Rocky Mt. For. and Range Exp. Stn., For. Serv. Res. Note RM-186, 4pp.

Reading, R. P., S. R. Beissinger, J. J. Grensten, and T. W. Clark. 1989. Attributes of black-tailed prairie dog colonies in northcentral Montana, with management recommendations for the conservation of biodiversity. Pages 13-27 *in* T. W. Clark, ed. The prairie dog ecosystem: managing for biological diversity. U.S. Bur. Land Manage., Wildl. Tech. Bull. 2.

Record, C. R. 1978. Ground squirrel and prairie dog control in Montana. Proc. Vertebr. Pest Conf. 8:93-97.

Reichman, O. J. 1975. Relation of desert rodent diets to available resources. J. Mammal. 56:731-751.

———. 1979. Desert granivore foraging and its impact on seed densities and distributions. Ecology 60:1085-1092.

———. 1983. Behavior of desert heteromyids. Great Basin Nat. Memoirs 7:77-90.

———, and D. Oberstein. 1977. Selection of seed distribution types by *Dipodomys merriami* and *Perognathus amplus*. Ecology 58:636-643.

———, and S. C. Smith. 1985. Impact of pocket gopher burrows on overlying vegetation. J. Mammal. 66:720-725.

———, T. G. Whitham, and G. A. Ruffner. 1982. Adaptive geometry of burrow spacing in two pocket gopher populations. Ecology 63:687-695.

Reid, V. H., R. M. Hansen, and A. L. Ward. 1966. Counting mounds and earth plugs to census mountain pocket gophers. J. Wildl. Manage. 30:327-334.

Reynolds, H. G. 1950. Relation of Merriam kangaroo rats to range vegetation in southern Arizona. Ecology 31:456-463.

———. 1954. Some interrelations of the Merriam kangaroo rat to velvet mesquite. J. Range Manage. 1:176-180.

———. 1958. The ecology of Merriam kangaroo rat (*Dipodomys merriami* mearns) on the grazing lands of southern Arizona. Ecol. Monogr. 28:111-127.

———, and G. E. Glendening. 1949. Merriam kangaroo rat: A factor in mesquite propagation on southern Arizona range lands. J. Range Manage. 2:193-197.

———, and H. S. Haskell. 1949. Life history notes on Price and Bailey pocket mice of southern Arizona. J. Mammal. 30:150-156.

Rice, B., and M. Westoby. 1978. Vegetative responses of some Great Basin shrub communities protected against jackrabbits or domestic stock. J. Range Manage. 31:28-34.

Richardson, B., S. Monsen, and D. Bowers. 1986. Interseeding selected shrubs and herbs on mine disturbances in southeastern Idaho. Pages 134-139 *in* Proc. symposium on the biology of *Artemisia* and *Chrysothamnus*. U.S. Dep. Agric. Forest Serv. Gen Tech. Rep. INT-200.

Richens, V. B. 1965. An evaluation of control on the Wasatch pocket gopher. J. Wildl. Manage. 29:413-425.

———. 1966. Notes on the digging activity of a northern pocket gopher. J. Mammal. 47:531-533.

Rickard, W. H. 1960. The distribution of small mammals in relation to the climax vegetation mosaic in eastern Washington and northern Idaho. Ecology 41:99-106.

Riegel, A. 1942. Some observations of food locations of rabbits in western Kansas during periods of stress. Trans. Kansas Acad. Sci. 45:369-375.

Robertson, G., J. Short, and G. Wellard. 1987. The environment of the Australian sheep rangelands. Pages 14-34 *in* G. Caughley, N. Shepherd, and J. Short, eds. Kangaroos: Their ecology and management in the sheep rangelands of Australia. Cambridge Univ. Press, Cambridge, U.K.

Robertson, J. H., and P. B. Kennedy. 1954. Half-century changes on northern Nevada ranges. J. Range Manage. 7:117-121.

Rogers, L. E., and K. A. Gano. 1980. Townsend ground squirrel diets in the shrub-steppe of southcentral Washington. J. Range Manage. 33:463-465.

Rosenzweig, M. L. 1973. Habitat selection experiments with a pair of coexisting heteromyid rodent species. Ecology 54:111-117.

———, and J. Winakur. 1969. Population ecology of desert rodent communities: Habitats and environmental complexity. Ecology 50:558-572.

Ross, B. A., J. R. Tester, and W. J. Breckenridge. 1968. Ecology of Mima-type mounds in northwestern Minnesota. Ecology 49:172-177.

Roundy, B. A., and G. L. Jordan. 1988. Vegetation changes in relation to livestock exclusion and root-plowing in southeastern Arizona. Southwest. Nat. 33:425-436.

Rudd, R. L., and R. E. Genelly. 1956. Pesticides: Their use and toxicity in relation to wildlife. California Fish and Game Bull. 7. 208pp.

Sampson, A. W. 1952. Range management, principles and practices. John Wiley and Sons, Inc., New York, N.Y. 570pp.

Sanderson, H. R. 1959. Relationships between jackrabbit use and availability of forage on native sandhill range. M.S. Thesis, Colorado State Univ., Fort Collins. 85 pp.

Schenbeck, G. L. 1982. Management of black-tailed prairie dogs on the National Grasslands. Proc. Great Plains Wildl. Damage Control Workshop. 5:207-217.

———. 1986. Black-tailed prairie dog management on the northern great plains: New challenges and opportunities. Proc. Great Plains Wildl. Damage Control Workshop. 7:28-33.

Schitoskey, F. 1975. Primary and secondary hazards of three rodenticides to kit fox. J. Wildl. Manage. 39:416-418.

Schitoskey, F., Jr. 1978. Energy requirements and diet of the California ground squirrel. J. Wildl. Manage. 42:373-382.

Schmidt-Nielsen, K. 1964. Desert animals: physiological problems of heat and water. Clarendon Press, Oxford, U.K. 277pp.

Schmutz, J. K., S. M. Schmutz, and D. A. Boag. 1980. Coexistence of three species of hawks (*Buteo* spp.) in the prairie parkland ecotone. Can. J. Zool. 58:1075-1089.

Schwartz, C. 1973. The cottontail and the white-tailed jackrabbit in Iowa, 1963-1972. Iowa Wildl. Res. Bull. 6. 23pp.

Seton, E. T. 1929. Lives of game animals. Doubleday, Doran and Co., New York, N.Y., 4:1-949.

Severe, D. S. 1977. Revegetation of black-tailed prairie dog mounds on short-grass prairie in Colorado. M.S. Thesis, Colorado State Univ., Fort Collins. 92pp.

Shalaway, S., and C. N. Slobodchikoff. 1988. Seasonal changes in the diet of Gunnison's prairie dog. J. Mammal. 69:835-841.

Sharps, J. C., and D. W. Uresk. 1990. Ecological review of black-tailed prairie dogs and associated species in western South Dakota. Great Basin Nat. 50:339-345.

Shaw, W. T. 1920. The cost of a squirrel and squirrel control. Popular Bull. 118, Washington Agric. Exp. Stn. 19pp.

———. 1921. Washington's annual losses from ground squirrels. Popular Bull. 69, Washington Agric. Exp. Stn. 8pp.

Sheets, R. G., R. L. Linder, and R. B. Dahlgren. 1971. Burrow systems of prairie dogs in South Dakota. J. Mammal. 52:451-453.

———, ———, and ———. 1972. Food habits of two litters of black-footed ferrets in South Dakota. Am. Midl. Nat. 87:249-251.

Short, J. 1985. The functional response of kangaroos, sheep and rabbits in an arid grazing system. J. Appl. Ecol. 22:435-447.

———. 1987. Factors affecting food intake of rangelands herbivores. Pages 84-99 *in* G. Caughley, N. Shepherd, and J. Short, eds. Kangaroos, their ecology and management in the sheep rangelands of Australia. Cambridge Univ. Press, U.K.

Shult, M. J. 1972. American bison behavior patterns at Wind Cave National Park. Ph.D. Thesis, Iowa State Univ., Ames. 178 pp.

Sieg, C. H. 1988. Small mammals: Pests or vital components of the ecosystem. Proc. Great Plains Wildl. Damage Control Workshop. 8:88-92.

———, D. W. Uresk, and R. M. Hansen. 1986. Seasonal diets of deer mice on bentonite mine spoils and sagebrush grasslands in southeastern Montana. Northwest Sci. 60:81-89.

Skinner, R. M. 1975. Grassland use patterns and prairie bird populations of Missouri. Pages 171-180 *in* M. K. Wati, ed. Prairie: A multiple view. Univ. North Dakota Press, Grand Forks.

Slade, N. A. and D. F. Balph. 1974. Population ecology of Uinta ground squirrels. Ecology 55:989-1003.

Slobodchikoff, C. N., and R. Coast. 1980. Dialects in the alarm calls of prairie dogs. Behav. Ecol. and Sociobiol. 7:49-53.

———, A. Robinson, and C. Schaack. 1988. Habitat use by Gunnison's prairie dogs. Pages 403-408 *in* R. C. Szaro, K. E. Severson, and D. R. Patton, eds. Proc. symposium on management of amphibians, reptiles, and small mammals in North America. U.S. Dep. Agric. Rocky Mt. For. and Range Exp. Stn. Gen. Tech. Rep. RM-166.

Smigel, B. W., and M. L. Rosenzweig. 1974. Seed selection in *Dipodomys merriami* and *Perognathus penicillatus*. Ecology 55:329-339.

Smith, C. C. 1940. The effect of overgrazing and erosion upon the biota of the mixed grass prairie of Oklahoma. Ecology 21:381-397.

Smith, C. F. 1948. A burrow of the pocket gopher (*Geomys bursarius*) in eastern Kansas. Trans. Kansas Acad. Sci. 51:313-315.

Smith, D. J., and J. S. Gardner. 1985. Geomorphic effects of ground squirrels in the Mount Rae area, Canadian Rocky Mountains. Arctic and Alpine Res. 17:205-210.

Smith, G. W. 1990. Home range and activity patterns of black-tailed jackrabbits. Great Basin Nat. 50:249-254.

———. 1950. Cottontail rabbit investigations. N.Y. Fish and Game, PR Proj. 1-R. Final Rep. 84pp.

Smith, R. E. 1967. Natural history of the prairie dog in Kansas. Kansas Univ. Mus. Nat. Hist. Misc. Publ. 49. 39pp.

Snell, G. P. 1985. Results of control of prairie dogs. Rangelands 7:30.

———, and B. D. Hlavachick. 1980. Control of prairie dogs—the easy way. Rangelands 2:239-240.

Soholt, L. F. 1973. Consumption of primary production by a population of kangaroo rats (*Dipodomys merriami*) in the Mojave desert. Ecol. Monogr. 43:357-376.

Soriguer, R. C. 1986. The rabbit as a plant seed dispenser. Mammal. Rev. 16:197-198.

Sparks, D. R. 1968. Diet of black-tailed jackrabbits on sandhill rangeland in Colorado. J. Range Manage. 21:203-208.

Spencer, D. A. 1958. Biological aspects of the 1957-58 meadow mouse irruption in the Pacific Northwest. U.S. Fish and Wildl. Serv., Denver Wildl. Res. Lab. Spec. Rep. 9pp.

Spencer, S. R., G. N. Cameron, B. D. Eshelman, L. C. Cooper, and L. R. Williams. 1985. Influence of pocket gopher mounds on a Texas coastal prairie. Oecologia (Berlin) 66:111-115.

Sperry, C. C. 1941. Food habits of the coyote. U. S. Dep. Int., Fish and Wildl. Serv. Wildl. Res. Bull. 4. 70pp.

Stamp, N. E., and R. D. Ohmart. 1978. Resource utilization by desert rodents in the lower Sonoran desert. Ecology 59:700-707.

Standley, W. G. 1988. Effect of seed size on removal by rodents. Pages 399-402 *in* R. C. Szaro, K. E. Severson, and D. R. Patton, eds. Proc. symposium on management of amphibians, reptiles, and small mammals in North America. U.S. Dep. Agric. Rocky Mt. For. and Range Exp. Stn. Gen. Tech. Rep. RM-166.

Stangl, F. B., Jr., T. S. Schafer, J. R. Goetze, and W. Pinchak. 1992. Opportunistic use of modified and disturbed habitat by the Texas kangaroo rat (*Dipodomys elator*). Texas J. Sci:44:25-35.

Steenbergh, W. F., and P. L. Warren. 1977. Ecological investigations of natural community status at Organ Pipe Cactus National Monument. Univ. Arizona, Tucson, Coop. Nat. Park Res. Stud. Unit, Tech. Rep. 3:1-152.

Stewart, G., and A. C. Hull. 1949. Cheatgrass (*Bromus tectorum* L.)—An ecological intruder in southern Idaho. Ecology 30:58-74.

Stoddard, L. C. 1985. Severe weather related mortality of black-tailed jackrabbits. J. Wildl. Manage. 49:696-698.

Storer, T. I., F. C. Evans, and F. G. Palmer. 1944. Some rodent populations in the Sierra Nevada of California. Ecol. Monogr. 14:165-192.

Stubbendieck, J., R. Case, K. J. Kjar, and M. A. Foster. 1979. Plains pocket gophers—more than a nuisance. Rangelands 1:3-4.

Sullivan, R. M. 1985. Phyletic, biogeographic, and ecologic relationships among montane populations of least chipmunks (*Eutamius minimus*) in the southwest. Syst. Zool. 34:419-448.

Sullivan, T. P., and D. R. Crump. 1986. Feeding responses of snowshoe hares (*Lepus americanus*) to volatile constituents of red fox (*Vulpes vulpes*) urine. J. Chem. Ecol. 12:729-739.

———, and D. S. Sullivan. 1982. The use of alternative foods to reduce lodgepole pine seed predation by small mammals. J. Appl. Ecol. 19:33-45.

———, and ———. 1984. Influence of range seeding on rodent populations in the interior of British Columbia. J. Range Manage. 37:163-165.

———, ———, D. R. Crump, H. Weiser and E. A. Dixon. 1988. Predator odors and their potential role in managing pest rodents and rabbits. Proc. Vertebr. Pest Conf. 13:145-150.

Summers, C. A., and R. L. Linder. 1978. Food habits of the black-tailed prairie dog in western South Dakota. J. Range Manage. 31:134-136.

Taylor, W. P. 1930. Methods of determining rodent pressure on the range. Ecology 11:523-542.

———. 1935. Some animal relations to soils. Ecology 16:127-136.

———, and J. V. G. Loftfield, 1924. Damage to range grasses by the Zuni prairie dog. U. S. Dep. Agric. Bull. 1227. 16pp.

———, C. T. Vorhies, and P. B. Lister. 1935. The relation of jackrabbits to grazing in southern Arizona. J. For. 33:490-498.

Thorn, C. E. 1977. The magnitude and form of pocket gopher induced erosion in the alpine areas of the Colorado front range. Annu. Meeting Rocky Mt. Sect. Geol. Soc. Am. 30:769.

Thorne, D. H., and D. C. Andersen. 1990. Long-term soil disturbance pattern by a pocket gopher, *Geomys bursarius*. J. Mammal. 71:84-89.

Tiemeier, O. W. 1965. The black-tailed jackrabbit in Kansas. I. Bionomics. Kansas State Univ. Agric. Exp. Stn. Tech. Bull. 140. 37pp.

Tietjen, H. P. 1976. Zinc phosphide—its development as a control agent for black-tailed prairie dogs. U. S. Fish and Wildl. Serv. Spec. Sci. Rep. Wildl. 195. 14pp.

———, C. H. Halvorson, P. L. Hegdal, and A. M. Johnson. 1967. 2,4-D herbicide, vegetation and pocket gopher relationships, Black Mesa, Colorado. Ecology 48:634-643.

Tileston, J. V., and R. R. Lechleitner. 1966. Some comparisons of the black-tailed and white-tailed prairie dogs in north central Colorado. Am. Midl. Nat. 75:292-316.

Tilman, D. 1983. Plant succession and gopher disturbance along an experimental gradient. Oecologia (Berlin) 60:285-292.

Trippensee, R. F. 1934. The biology and management of the cottontail rabbit. M.S. Thesis, Univ. Michigan, Ann Arbor. 217pp.

Tryon, C. A., Jr. 1943. The great gray owl as a predator of pocket gophers. Wilson Bull. 55:130-131.

———. 1947. The biology of the pocket gopher (*Thomomys talpoides*) in Montana. Montana State Coll. Agric. Exp. Stn. Bull 448. 30pp.

Turkowski, F. J. 1975. Dietary adaptability of the desert cottontail. J. Wildl. Manage. 39:748-756.

Turner, G. T. 1969. Responses of mountain grassland vegetation to gopher control, reduced grazing, and herbicide. J. Range Manage. 22:377-383.

———, R. M. Hansen, V. H. Reid, H. P. Tietjen, and A. L. Ward. 1973. Pocket gophers and Colorado mountain rangeland. Colorado State Univ. Agric. Exp. Stn. Bull. 5545. 90pp.

Uresk, D. W. 1978. Diets of the black-tailed hare in steppe vegetation. J. Range Manage. 31:439-442.

———. 1984. Black-tailed prairie dog food habits and forage relationships in western South Dakota. J. Range Manage. 37:325-329.

———. 1985. Effects of controlling black-tailed prairie dogs on plant production. J. Range Manage. 38:466-468.

———. 1986. Food habits of cattle on the South Dakota High Plains. Prairie Nat. 18:211-218.

———. 1987. Relation of black-tailed prairie dogs and control programs to vegetation, livestock and wildlife. Pages 312-323 *in* J. L. Capinera, ed. Westview studies in insect biology: Integrated pest management on rangeland: A shortgrass prairie perspective. Westview Press, Boulder, Colo.

———, and A. J. Bjugstad. 1983. Prairie dogs as ecosystem regulators on the Northern High Plains. Pages 91-94 *in* C. L. Kucera, ed. Proc. seventh North Am. prairie conf. Southwest Missouri State Univ., Springfield.

———, J. G. MacCracken, and A. J. Bjugstad. 1982. Prairie dog density and cattle grazing relationships. Proc. Great Plains Wildl. Damage Control Workshop 5:199-201.

———, and D. D. Paulson. 1988. Estimated carrying capacity for cattle competing with prairie dogs and forage utilization in western South Dakota. Pages 387-390 *in* R. C. Szaro, K. E. Severson, and D. R. Patton, eds. Proc. symposium management of amphibians, reptiles, and small mammals in North America. U.S. Dep. Agric. Rocky Mt. For. and Range Exp. Stn. Gen. Tech. Rep. RM-166.

———, and G. L. Schenbeck. 1987. Effect of zinc phosphide rodenticide on prairie dog colony expansion as determined from aerial photography. Prairie Nat. 19:57-61.

———, and J. C. Sharps. 1986. Denning habitat and diet of the swift fox in western South Dakota. Great Basin Nat. 46:249-253.

Vallentine, J. F. 1989. Range development and improvements. Third ed. Brigham Young Univ. Press, Provo, Ut. 524pp.

Vanderhye, A. V. R. 1985. Interspecific nutritional facilitation: Do bison benefit from feeding on prairie dog towns? M.S. Thesis, Colorado State Univ., Fort Collins. 44pp.

Vaughan, T. A. 1961. Vertebrates inhabiting pocket gopher burrows in Colorado. J. Mammal. 42:171-174.

———. 1967. Food habits of the northern pocket gopher on short-grass prairie. Am. Midl. Nat. 77:176-189.

———. 1972. Mammalogy. W. B. Saunders Co., Philadelphia, Pa. 463pp.

Vleck, D. 1981. Burrow structure and foraging costs in the fossorial rodent, *Thomomys bottae*. Oecologia (Berlin) 49:391-396.

Vorhies, C. T., and W. P. Taylor. 1922. Life history of the kangaroo rat, *Dipodomys spectabilis spectabilis* Merriam. U.S. Dep. Agric. Bull. 1091:1-39.

———, and ———. 1933. The life histories and ecology of jack rabbits (*Lepus alleni* and *Lepus californicus* spp. in relation to grazing in Arizona. Univ. Arizona Agric. Exp. Stn., Tech. Bull. 49:471-587.

———, and ———. 1940. Life history and ecology of the white throated woodrat, *Neotoma albigula albigula* harley, in relation to grazing in Arizona. Bull. Arizona Agric. Exp. Stn. 86:453-529.

Wansi, T., R. D. Pieger, R. F. Beck and L. W. Murray. 1992. Botanical Content of black-tailed jackrabbit diets on semidesert rangeland. Great Basin Nat. 52:300-308.

Ward, A. L. 1960. Mountain pocket gopher food habits in Colorado. J. Wildl. Manage. 24:89-92.

———, and J. O. Keith. 1962. Feeding habits of pocket gophers in mountain grasslands, Black Mesa, Colorado. Ecology 43:744-749.

Waring, G. H. 1970. Sound communications of black-tailed, white-tailed, and Gunnison's prairie dogs. Am. Midl. Nat. 83:167-185.

Warnock, J. E. 1965. The effects of crowding on the survival of meadow voles (*Microtus pennsylvanicus*) deprived of cover and water. Ecology 46:649-664.

Watkins, L. C., and R. M. Novak. 1973. The white-tailed jackrabbit in Missouri. Southwest. Nat. 18:341-357.

Weins, J. A. 1973. Patterns and process in grassland communities. Ecol. Monogr. 43:237-270.

Welch, D. 1985. Studies in the grazing of heather moorland in northeast Scotland U.K. 4. Seed dispersal and plant establishment in dung. J. Appl. Ecol. 22:461-472.

Welch, J. F. 1967. Review of animal repellents. Proc. Vertebr. Pest Conf. 3:36-40.

Weltzin, J. F., R. K. Heitschmidt, and S. Archer. 1991. The historical role of prairie dogs in regulating *Prosopsis glandulosa* (honey mesquite) abundance. Bull. Ecol. Soc. Am. 72:285.

West, N. E. 1968. Rodent influenced establishment of Ponderosa pine and bitterbrush seedlings in central Oregon. Ecology 49:1009-1011.

———, F. D. Provenza, P. S. Johnson, and M. K. Owens. 1984. Vegetation changes after 13 years of livestock grazing exclusion on sagebrush semidesert in west central Utah. J. Range Manage. 37:262-264.

Westoby, M. 1980. Black-tailed jackrabbit diets in Curlew Valley, Northern Utah. J. Wildl. Manage. 44:942-948.

Wetherbee, F. A. 1967. A method of controlling jack rabbits on a range rehabilitation project in California. Proc. Vertebr. Pest Conf. 3:111-117.

Whitford, W. G. 1976. Temporal fluctuations in density and diversity of desert rodent populations. J. Mammal. 57:351-369.

Williams, L. R., and G. N. Cameron. 1986*a*. Effects of removal of pocket gophers on a Texas coastal prairie. Am. Midl. Nat. 115:216-224.

———, and ———. 1986*b*. Food habits and dietary preferences of Attwater's pocket gopher, *Geomys attwateri*. J. Mammal. 67:489-496.

———, and ———. 1990. Intraspecific response to variation in food resources by Attwater's pocket gopher. Ecology 71:797-810.

———, ———, S. R. Spencer, B. D. Eshelman, and M. J. Gregory. 1986. Experimental analysis of the effects of pocket gopher mounds on Texas coastal prairie. J. Mammal. 67:672-679.

Williams, O. 1955. Distribution of mice and shrews in a Colorado montane forest. J. Mammal. 36:3221-231.

Wondolleck, J. T. 1978. Forage-area separation and overlap in heteromyid rodents. J. Mammal. 59:510-518.

Wood, J. E. 1969. Rodent populations and their impact on desert rangelands. New Mexico State Univ. Agric. Exp. Stn. Bull. 555:1-17.

Wooster, L. D. 1935. Notes on the effects of drought on animal populations in western Kansas. Trans. Kansas Acad. Sci., 38:351-352.

Wydeven, A. P., and R. B. Dahlgren. 1985. Ungulate habitat relationships in Wind Cave National Park. J. Wildl. Manage. 49:805-813.

Wywialowski, A. P., and L. C. Stoddard. 1988. Estimation of jackrabbit density: methodology makes a difference. J. Wildl. Manage. 52:57-59.

Zedler, P. H., and C. Black. 1992. Seed dispersal by a generalized herbivore: rabbits as dispersal vectors in a semiarid California vernal pool landscape. Am. Midl. Nat. 128:1-10.

Zimmerman, E. G. 1965. A comparison of habitat and food of two species of *Microtus*. J. Mammal. 46:605-612.

Zou, J., J. T. Flinders, H. L. Black, and S. G. Whisenant. 1989. Influence of experimental habitat manipulations on a desert rodent population in southern Utah. Great Basin Nat. 49:435-448.

Chapter 9
CARNIVORES
William F. Andelt

INTRODUCTION

Carnivores historically had an impact on man by preying upon domestic livestock. This predation caused more problems and controversy for man than any other conflicts with wildlife. Gray wolves (*Canis lupus*) and bears (*Ursus* spp.) were the major carnivores that preyed upon domestic livestock in Europe and Asia. During early settlement of North America, gray wolves, red wolves (*Canis rufus*), grizzly bears (*Ursus arctos*), black bears (*Ursus americanus*), and mountain lions (*Felis concolor*) were the major predators of domestic livestock. Today, with the removal of large predators from extensive areas, coyotes (*Canis latrans*) are the major predator of livestock in North America. Dogs and red foxes (*Vulpes vulpes*), mountain lions, black bears, grizzly bears, gray wolves, and bobcats (*Lynx rufus*) also prey on domestic livestock, but their predation is secondary to that of coyotes. All of the above carnivores have been controlled extensively because of their predation on livestock.

I discuss the extent of carnivore predation on domestic livestock, review past and current control methods, examine the impact of control on predator populations, behavior, and ecology, review public attitudes toward predator control, discuss when and how control should be implemented, and speculate on the future of techniques for prevention and control of predation on livestock. Coyote predation on livestock is emphasized because it causes the majority of conflicts (National Agricultural Statistics Service 1991) and has been researched most intensively. The effect of predation on the population dynamics of ungulates, lagomorphs, and rodents has been reviewed by other authors (Keith 1974, Connolly 1978, Mech 1984, Newsome 1990) and is not discussed here.

EXTENT OF LIVESTOCK LOSSES TO PREDATORS

Numerous studies have been conducted since 1970 to determine the magnitude of livestock losses to predators, particularly coyotes, because the extent of losses was unknown and disputed. The United States Fish and Wildlife Service (1978) and Pearson (1986) summarized several published studies of livestock losses to predators in the 17 western United States, where 88% of the country's sheep (Gee and Magleby 1976) are raised. Pearson (1986) indicated that 2.5% of adult sheep and 9.0% of lambs were lost to all predators, with coyotes taking 74% of the adult sheep and 77.7% of the lambs lost annually to predators. The National Agricultural Statistics Service (1991) reported that coyotes killed 63.7% of the sheep and lambs killed by predators in the United States during 1990. The United States Fish and Wildlife Service (1978) estimated an average annual loss to coyotes of 1-2.5% for ewes and 4-8% for lambs during 1972-78 in the western states. They also reported that livestock losses estimated from biological field studies and questionnaire surveys were similar. Predators were responsible for 25.8% of the adult sheep and 47% of the lambs lost to all causes (Pearson 1986). Pearson (1986) cautioned that many of the published studies in his summary could be considered atypical because many of the biological studies were conducted where predators were a problem and mail surveys might have been biased by greater response rates from producers suffering larger losses. Nass (1977) and O'Gara et al. (1983) reported that coyotes did not feed on 25 and 23% of the domestic sheep that they killed. Most of the above studies summarized by United States Fish and Wildlife Service (1978) and Pearson (1986) were conducted where various types and intensities of predator control were used.

Coyotes have expanded their ranges into the eastern United States where low losses of livestock and fowl recently have been reported (Jones 1987, Slate 1987). Dorrance and Roy (1976) reported that 1.6% of ewes and 2.8% of lambs were lost to predators in Alberta during 1974; coyotes were responsible for 88% of those losses. These mortalities occurred despite assistance from predator control specialists, who used unrestricted controls, including strychnine drop and 1080 baits.

Gee (1978) reported a 5.5% beef-calf loss to all causes in a nationwide United States survey, with predators deemed responsible for 11% of the losses. The United States Fish and Wildlife Service (1978) reported calf losses between birth and weaning to coyotes across the United States at 0.4%, with predation decreasing to nearly zero by weaning time. Dorrance (1982) reported that coyotes, black bears, and wolves were responsible for 35, 31, and 16%, respectively, of the 1,520 confirmed predation losses of cattle in Alberta during 1974-78.

Predators cause substantial losses of domestic goats also. In Texas, where an estimated 1.1 million goats (about 90% of the goats in the United States) are raised (Scrivner et al. 1985), predators were reported to take 18.1% of the adults and 33.9% of the kids in 3 studies (Pearson 1986).

Pearson (1986) stated that predators, particularly coyotes, accounted for losses of hundreds of chickens and turkeys in 14 western states. Andelt and Gipson (1979*a*) reported that a mated pair of coyotes killed 268 domestic turkeys worth $938 (US) between 4 June and 31 August 1976 on 1 farm in Nebraska. Production for the farm was about 130,000 turkeys.

Domestic dogs can be significant predators of livestock and poultry (Denny 1974). Dogs ranked second to coyotes and accounted for 13.6% of the sheep and lambs killed by all predators in the United States during 1990 (National Agricultural Statistics Service 1991). Walton (1990) reported that dogs ranked second behind coyotes in frequency of predation on sheep and goats in his study area in Texas. Schaefer et al. (1981) reported that in Iowa 3% of the sheep owned by questionnaire respondents were allegedly killed by coyotes and 1% were killed by dogs. McAninch and Fargione (1987) reported that 88% of the sheep producers surveyed in New York indicated that dogs were the most harmful predator. Producers in Sonoma and Marin counties, California, reported that dogs were responsible for the majority of predation on sheep (Larson and Salmon 1988).

Livestock losses to mountain lions are not nearly as severe as are losses to coyotes. They accounted for 3.4% of the sheep and lambs killed by all predators in the United States during 1990 (National Agricultural Statistics Service 1991). However, significant impacts on individual operations can occur when a large number of animals are killed by a lion in ≥1 days (Weaver and Sitton 1978). Suminski (1982) estimated that lions kill 0.29% of the sheep annually in Nevada, and reported 59 sheep killed on 1 occasion. Suminski (1982) indicated that lion predation on domestic sheep appears to be more serious in Nevada than in any other western state. Cattle losses to lions appear greatest in Arizona, New Mexico, and Texas with losses generally decreasing northwards within the lion's range (Shaw 1983). Shaw (1977) reported that 6 of 12 mountain lions on a study area in Arizona killed cattle, with calves selected over other prey. Sheep are killed wherever they graze in areas occupied by mountain lions (Lindzey and Wilbert 1989) with most kills occurring during summer.

Bears accounted for 1.6% of the sheep and lambs killed by all predators in the United States during 1990 (National Agricultural Statistics Service 1991). Black bears and grizzly bears killed 1.3 and 0.5% of the sheep grazed on sheep allotments adjacent to Yellowstone National Park during 1976 and 1977 (Johnson and Griffel 1982). Both species killed sheep on the majority of allotments. Davenport (1953) reported that domestic sheep were the primary livestock killed by black bears in Virginia, and that the average annual value of sheep killed from 1941 to 1950 was only 0.09% of the value of sheep sold during 1950. Horstman and Gunson (1982) reported that cattle represented 81% and sheep and swine each represented 9% of the compensation claims submitted for black bear predation on livestock in Alberta; these losses represented 0.02% of the cattle, 0.11% of the sheep, and 0.02% of the swine in the area. Most predation on livestock apparently is by mature and old males (Davenport 1953, Horstman and Gunson 1982, Coolahan 1990).

Gray wolves, in North America, are primarily present in Canada, Alaska, and Minnesota. A few wolves are present in Wisconsin, Michigan, Montana, Idaho, and Washington (Peek et al. 1991). About 1,000 to 1,200 wolves inhabit northern Minnesota (Bailey 1978), where about 12,230 farms were located in 1978 (Fritts 1982). Over 90% of the farms had some livestock, with sheep and cattle present on >80% of the farms (Fritts 1982). Paul (1989) reported that 9 to 38 of 7,200 farms in Minnesota had verified losses to wolves each year from 1976 through 1988. The highest cattle losses to wolves were 0.045% during 1979, and the highest sheep losses were 0.27% during 1981, indicating that wolf predation on livestock in Minnesota is not a large problem except for a few farmers. Tompa (1983) reported that verified losses of all livestock classes to wolves in British Columbia were <0.1% of the respective populations and that the problems are localized. Bjorge and Gunson (1983) and Dorrance (1982) reported that wolves killed calves and yearlings at greater rates than adults and recommended placing only healthy animals on grazing leases in Alberta. The United States Fish and Wildlife Service (1987) reported that only a small fraction of ranchers and permittees in remote wolf country suffered verified livestock losses to wolves, and wolves may live near farms or grazing leases without killing livestock.

Bobcats kill some adult sheep, lambs, and goats (Young 1958, Nass 1977, Coolahan 1990). Bobcats accounted for only 2.8% of the sheep and lambs lost to all predators in the United States during 1990, whereas foxes and eagles accounted for 2.6 and 3.6%, respectively (National Agricultural Statistics Service 1991).

Economic Impacts of Predation on Livestock

Sheep numbers in the United States declined about 80% from 1942 to 1976 (Gee et al. 1977*b*), but stabilized between 1978 and 1989 (Markham 1990). Former sheep producers reported that the principal reasons for leaving the sheep industry included high predation losses, low lamb and wool prices, shortage of good hired labor, and retirement (Gee et al. 1977*b*).

The United States Fish and Wildlife Service (1978) estimated the economic impacts of coyote predation on producers with predator problems, producers without predator problems, and on consumers during 1977. They used an average lamb-loss rate of 4% (267,000 lambs) and a ewe-loss rate of 1.5% (125,000 ewes) to coyotes to estimate an economic loss of $19 million to producers from coyote predation in the 17 western states. The reduced number of sheep and lambs marketed due to coyote predation resulted in a higher market price, which benefited producers by $6 million. The net impact of coyote predation on sheep producers was a loss of $13 million, and the impact on consumers was $4 million in additional costs. The economic impact of coyote predation on calves was estimated at a $20 million loss to producers. However, due to the greater price flexibility of beef compared to sheep, the reduction in beef calves marketed (estimated at 0.4%, 115,000 fewer calves) resulted in a higher price, which benefitted beef producers by $81 million. The net impact of the reduced supply of beef due to coyote predation was a gain of $61 million to beef producers, but it cost consumers an additional $98 million, resulting in an overall loss of $37 million.

Although the average value of livestock losses to predators reflects overall impact on producers, it does not reflect the severity of losses to some individuals. Coyote (Balser 1964, Gee et al. 1977*a*) and bear (Davenport 1953) predation is more serious for some producers than others. Most sheep producers suffer no or minor predator losses, whereas 20-25% of the producers who suffer losses have significantly higher losses than average (United States Fish and Wildlife Service 1978); these losses can drive producers out of business because of low profit margins. Nonfatal injuries and harassment of livestock by predators also can result in reduced weight gain and subsequent reductions in profit.

CHARACTERISTICS OF CARNIVORE PREDATION ON LIVESTOCK

Livestock killed by predators usually can be distinguished from those dying from other causes by the presence of external hemorrhaging; subcutaneous hemorrhaging and tooth punctures; damage to the skin, other soft tissues, and skull; blood on the soil and vegetation; and carnivore tracks, scats, or territorial marks near dead animals. Urgent calling and alert, defensive, and frightened behavior of livestock also suggest that predators may have killed livestock.

Newborn livestock killed by predators and partially consumed can be distinguished from stillborn livestock by characteristics not found in stillborn animals: a blood clot present at the closed end of the navel, pink lungs that float in water, fat around the heart and kidneys, milk in the stomach and intestines, milk fat and lymph in the lymphatic vessels that drain the intestinal tract, a worn soft membrane on the bottom of the hooves, and possibly soil on the bottom of the hooves (Wade and Bowns 1984).

Individual species of predators follow a general pattern of killing and feeding on livestock, but some variation occurs among individuals and some overlap occurs among species. The carnivore species responsible for killing livestock often can be differentiated by the type and location of wounds (e.g., wounds made by teeth, talons, or claws), size and distance between canine punctures, extent of injuries, location of feeding, amount of prey consumed, and if the prey was partly skinned, dragged, or covered. In general, a predator is relatively large compared to its prey if broken bones are common (Wade and Bowns 1984). Bears, coyotes, and foxes scavenge on carcasses, whereas mountain lions and bobcats usually kill their own food and usually do not scavenge on old or spoiled carcasses.

Coyotes generally kill adult sheep and goats by biting the throat just behind the jaw and below the ear (Fig. 1) (Wade and Bowns 1984, Acorn and Dorrance 1990); sheep suffocate an average of 13 minutes after capture (Connolly et al. 1976). The attacks usually leave tooth-puncture marks, subcutaneous hemorrhaging, and external bleeding in the neck region (Davenport et al. 1973, Bowns 1976, Tigner and Larson 1977, Wade and Bowns 1984). The spacing between upper-canine punctures is 2.9-3.5 cm, and the lower canines is 2.5-3.2 cm (Wade and Bowns 1984).

Connolly et al. (1976) found that food deprivation did not have an apparent effect on the prey-killing behavior of coyotes but did influence feeding on kills, suggesting that hunger is not the primary motivation for killing prey. Fox (1969) and Lehner (1976) reported that coyote predatory behaviors, which include identifying, capturing, killing, and consuming prey, are shaped through experience; but Connolly et al. (1976) reported that prey-naive coyotes possessed the inclination and ability to kill sheep. In a pen study, paired males killed more sheep than their mates, whereas 2-year-old males and their mates killed more sheep than yearling males, and unmated females did not attack (Connolly et al. 1976). However, in the wild, all coyotes do not kill sheep (Beasom and Gober 1975, Connolly et al. 1976, United States Fish and Wildlife Service 1978).

Lamb losses to coyotes generally are highest in spring (Wade 1973, Till and Knowlton 1983), which often coincides with lambing and coyote denning season when adults are feeding pups. Another major predation period has been noted during late summer and early fall (Klebenow and McAdoo 1976). Coyotes killed more lambs than ewes (Nesse et al. 1976, Nass 1977, Tigner and Larson 1977, Gluesing et al. 1980), with the most active lambs and those found on the periphery of bedgrounds being selected (Gluesing et al. 1980). Coyotes usually begin feeding on the flank, just behind the ribs, or on the liver, heart, lungs, and mesenteric fat (Wade and Bowns 1984).

Coyotes often attack newborn calves in the flank whereas older calves are attacked in the flank and hindquarters (Acorn and Dorrance 1990). The abdomen of calves usually is opened and the internal organs eaten.

Domestic dogs usually do not attack livestock for food (Wade and Bowns 1984). Their attacks often result in indiscriminate mutilation of prey, with frequent injuries to the hindquarters, shoulders, and nose. Some dogs attack prey in a fashion similar to coyotes, and some coyotes attack prey in an indiscriminate fashion similar to dogs. Dog attacks often result in many dead and wounded animals (Acorn and Dorrance 1990).

Mountain lions usually kill ewes and other large prey by biting them in the back of the neck, although they may also suffocate the animal by biting it in the throat (Shaw 1983, Bruscino 1989). Killed animals often display massive hemorrhaging on the back of the neck and near the base of the skull, and claw marks and rakes along the shoulders and on the back. The upper canine punctures from mature lions range from 4.5 to 5.0 cm apart and lower canine punctures range from 3.0 to 4.0 cm apart (Shaw 1983).

Lions occasionally cover their kills (Shaw 1983, Bruscino 1989). When mass kills occur, such as on sheep bedgrounds, most carcasses are not covered (Shaw 1983). If kills are made

FIGURE 1. Coyotes usually kill domestic sheep by clamping their jaws on the victim's throat and waiting until they die. Photo by G. E. Connolly; courtesy of the U.S. Fish and Wildl. Serv.; published with permission of the *Rangeman's Journal*.

in open habitat, lions usually drag their prey under a low-hanging tree or bush. The presence of drag marks or large tracks (about the size of a large dog) with 3 distinct lobes on the back of the pad and a lack of toenail marks suggests that the kill was made by a lion.

Mountain lions appear to prefer lambs to ewes (Bruscino 1989, Lindzey and Wilbert 1989). They may return for several nights to feed on the carcasses and often kill additional sheep (Shaw 1983). Lions seldom kill calves larger than 136-181 kg (Shaw 1983).

Lions usually enter the carcass at or just behind the rib cage (Shaw 1983, Bruscino 1989). They usually eviscerate the carcass and feed on the lungs, heart, and liver followed by the larger leg muscles and the underside of the legs.

All black bears do not kill cattle or domestic sheep (Murie 1948, Jorgensen 1983); however, grizzly bears appear prone to killing cattle and sheep (Murie 1948, Johnson and Griffel 1982). Black bears and grizzly bears typically kill domestic sheep by biting them on the dorsal side of the neck and less frequently on the frontal or jugal bones of the skull (Griffel and Basile 1981). Grizzly bears typically kill calves and yearling cattle by biting them on the dorsal side of the neck, occasionally on the lumbar region of the spine, and less frequently on the head (Murie 1948). Black bears and grizzly bears apparently do not attack by striking with the paws, but instead seize and hold their victims with their front paws before biting and killing their prey (Murie 1948, Griffel and Basile 1981). Black bears dragged 60% of the sheep carcasses approximately 23-46 m and the remainder <20 m from the kill site. Black and grizzly bear predation can be differentiated by the size of their tracks (Johnson and Griffel 1982).

Black bears and grizzly bears seem to show little preference for killing either lambs or ewes (Griffel and Basile 1981). Black bears and grizzly bears usually return to the kill to finish eating the carcass (Murie 1948, Griffel and Basile 1981). Most sheep and swine predation incidents by black bears were multiple kills (Horstman and Gunson 1982). Johnson and Griffel (1982) reported that grizzly bear predation on sheep occurred only on the bedgrounds during the night or early morning.

Calves were selected over yearlings and adults by black bears (Dorrance 1982, Horstman and Gunson 1982), whereas grizzly bears prefer calves and yearlings to adult cattle (Murie 1948, Acorn and Dorrance 1990).

Most black bears first consume the udder (74%) or the flank (26%) of domestic sheep (Griffel and Basile 1981). Black bears and grizzly bears usually remove the paunch and intestines intact from the body cavity of sheep and cattle (Murie 1948, Griffel and Basile 1981). If additional feeding occurs, black bears usually split the hide over the rib cage and peel it off the more fleshy parts of the carcass; the hide is left intact, which generally distinguishes bear predation. Black bears next eat the costal arch and sternum of sheep, then the front shoulder, and lastly the hindquarters (Griffel and Basile 1981). Sheep that died of causes other than predation, but that were fed upon by bears, lack canine puncture marks on the neck, shoulder, facial area, lack hemorrhaging, and lack lacerations over the back (Griffel and Basile 1981).

Red foxes usually attack the throat of young lambs and kids (Wade and Bowns 1984). They usually begin feeding just behind the ribs and consume the viscera first. Red foxes often carry small carcasses to the den to feed their pups.

Bobcats usually kill small lambs by biting them on the head or back of the neck (Wade and Bowns 1984). The paired upper and lower canine punctures are 1.9-2.5 cm apart. Hemorrhaging from claw punctures often can be found below the skin on the neck, back, sides, and shoulders. Bobcats usually do not attack adult sheep or goats. Bobcats often begin feeding on the viscera after entering behind the ribs. They occasionally cache and cover some kills.

Bjorge and Gunson (1983) reported that wolves primarily attacked cattle on the hindquarters including tail, vulva, lower thigh, and occasionally on the face, behind the front legs, in front of the rear legs, and on the belly. Wolves apparently prey on young, inexperienced, or disabled cattle more frequently than healthy adult cattle (Acorn and Dorrance 1990). Wolves seem to prefer to feed on the viscera and hind legs of large domestic prey (Acorn and Dorrance 1990).

PREVENTION AND CONTROL METHODS

During the past century, predator control methods and philosophies have evolved from an approach of general population reduction to the removal of individual offending animals or the use of nonlethal control techniques. The methods employed for preventing or controlling predation generally depend on the intensity of the problems and the circumstances under which they occur. Most methods fall within nonlethal and lethal controls. Nonlethal methods prevent or control predation without killing predators, whereas lethal controls remove offending animals or suppress predator populations. Advantages of most nonlethal techniques include a minimal level of producer expertise, less reliance on Federal Animal Damage Control field agents, better acceptance by the public, and some techniques (e.g., a good fence) being more permanent solutions than continual population reduction. Disadvantages of nonlethal control techniques include labor and material expenses, maintenance, and occasional lack of success. However, most lethal controls also suffer these disadvantages.

Nonlethal Control Methods

Nonlethal methods used to prevent or control predation on domestic livestock include: various livestock husbandry methods, fencing, guarding dogs, guarding donkeys, and llamas; bonding sheep and goats to cattle; and frightening devices. Other techniques that provide questionable or inconsistent control of livestock depredations include aversive conditioning, repellents, antifertility agents, and electromagnetic and ultrasonic devices. Livestock husbandry practices that can be implemented to reduce sheep losses to predators include confinement, carcass disposal, size of livestock placed on pasture, lambing in confinement, use of herders, and regular surveillance.

Confinement. Confining sheep at night, particularly in predator-proof enclosures, is an effective husbandry practice. Most losses of sheep to coyotes occur at night (Bowns et al. 1973,

Henne 1975) when coyotes are most active (Gipson and Sealander 1972, Andelt and Gipson 1979*b*, Andelt 1985*a*). Higher sheep losses in Kansas were incurred by producers that grazed sheep in pastures day and night whereas intermediate losses occurred when sheep were grazed only during daytime and the lowest losses occurred when sheep were confined day and night (Robel et al. 1981). Although cost-benefit ratios on day and night confinement are needed, confining sheep to corrals during daytime and night likely would be impractical for many producers. Suminski (1982) recommended bedding sheep close to camp or confining them at night to reduce mountain lion predation.

Disposal of Livestock Carcasses. Predators may discover livestock as a source of food by being attracted to and feeding on carcasses (Lehner 1976, Fritts 1982). Todd and Keith (1976) reported higher coyote densities in areas containing carrion than in areas where carrion was removed. Producers in Kansas and Illinois that buried or hauled away dead sheep and swine sustained lower rates of coyote predation than producers that left the carcasses in the pasture or attempted to burn them (Robel et al. 1981, Jones and Woolf 1983). Although attracting predators to the area may be the major reason why the presence of carrion increased predation rates, scavenging on sheep carrion may assist predators to identify sheep as prey (Lehner 1976) and result in increased predation rates.

Size of Livestock Placed on Range. Producers in Kansas that lambed sheep during January to March had higher rates of losses to coyotes than producers that lambed from October to December or throughout the year (Robel et al. 1981). Lambs produced during January to March are more susceptible to predation than those born from October to December because they are smaller when placed on pastures during late spring and early summer, when losses to coyotes are typically highest (Robel et al. 1981). However, it may be important to lamb when forage is abundant to promote better lactation and fast lamb growth.

Shaw (1977) suggested keeping calves out of lion country until they reach 140 kg or converting from a cow-calf operation to a weaner steer operation to reduce lion depredations. However, these practices may not be economically feasible.

Lambing in Confinement. Lambing in sheds or small lots can reduce predator and nonpredator losses (Wade 1973, Boggess et al. 1980). However, this practice requires extra labor, facilities, and feed for the confined sheep. Data are needed on the cost-effectiveness of this technique.

Herders. Herders routinely are employed to attend sheep on open range. Producers using herders generally have lower predation losses than producers without herders (Davenport et al. 1973, Tigner and Larson 1977, Nass et al. 1984). Sheep maintained in tight bands were less susceptible to black bear and grizzly bear predation than were bands allowed to wander freely (Jorgensen 1983). Some producers do not use herders because of difficulties in obtaining adequate numbers of capable herders and increased labor costs.

Record Keeping and Surveillance. By regularly counting sheep and surveying pastures, the onset of livestock losses to predators can be determined. As soon as losses are identified, corrective measures can be implemented. Retaining good records of livestock losses will help to identify loss patterns and problem areas that may require corrective action.

Other Husbandry Practices. Practices that provided inconsistent effectiveness or that may be difficult to manipulate to reduce losses include flock size and various characteristics of pastures. Robel et al. (1981) and Nass et al. (1984) reported that larger sheep operations had lower loss-rates than smaller operations whereas Dorrance and Roy (1976) reported the opposite relationship and Nielson and Curle (1970) reported no relationship. These conflicting results indicate that no conclusions can be drawn between flock size and predation rates.

Robel et al. (1981) reported that producers located >8 km from towns or settlements in Kansas had higher sheep losses to coyotes than producers located <1.6 km from towns, but the opposite trend was observed for sheep losses to dogs. These loss relationships probably reflect higher densities of dogs and lower densities of coyotes near towns (Robel et al. 1981).

Habitat. Robel et al. (1981) reported that coyote predation rates on domestic sheep were higher in larger than smaller pastures, in pastures with taller grass, and in pastures with streams than without streams, but rates were not related to the distance of pastures from residences. These loss relationships may reflect habitat preferences of coyotes. Robel et al. (1981) also reported higher loss-rates in flat versus rough and rolling pastures, but Nass et al. (1984) reported lower predation rates in open pastures. Although the reason for the differences between these 2 studies is unknown, it seems likely that predation rates would be highest in those areas most attractive to coyotes.

Fritts (1982) indicated that calving in forested or brushy pastures was responsible for many instances of wolf predation on livestock in Minnesota. Keeping sheep out of berry patches may reduce predation on sheep by bears (Jorgensen 1983).

Fencing. Properly constructed conventional netwire and electric fences (Fig. 2) can effectively exclude coyotes from pastures containing domestic livestock (deCalesta and Cropsey 1978, Gates et al. 1978, Thompson 1979, Dorrance and Bourne 1980, Linhart et al. 1982, Wade 1982, deCalesta 1983, Shelton 1984, Shelton and Gates 1987). DeCalesta and Cropsey (1978), Dorrance and Bourne (1980), and Linhart et al. (1982) reported that fences were a cost-effective method of preventing predation. Costs of materials for coyote-deterring electric fences were estimated at \$713-\$1,125/km (Gates et al. 1978, Dorrance and Bourne 1980, Linhart et al. 1982) and at \$1,543/km for coyote-deterring conventional fences (deCalesta and Cropsey 1978). One to 3 electrified wires placed outside an existing netwire fence also can reduce entry to pastures by coyotes (deCalesta 1983, Shelton 1984).

Electric fences were reported to complement (Linhart et al. 1982) and reduce demand (Dorrance and Bourne 1980) for lethal control. Other advantages of fencing include elimination of herding, greater control over intensity of grazing, less trampling of vegetation, and reduced parasite infestations due to reduced contact and mixing with adjacent herds (Jones 1938, cited by Wade 1982). Disadvantages of fencing include construction costs; fence maintenance, particularly in washout areas; control of vegetation under electric fences; and obstruction of wildlife (particularly pronghorn antelope [*Antilocapra americana*] and mule deer [*Odocoileus hemionus*]) movements. Howard (1991) reported that a 1.5-m-high electric fence was a barrier to mule deer movements and should not be used in major deer movement/migration corridors. A 1.2-m-high netwire fence restricted fewer mule deer than the 1.5-m-high fence. Predators also may gain access through damaged fences, malfunctioning electric fences, or by digging under or jumping over some fences.

Livestock Guarding Dogs. Livestock guarding dogs (Fig. 3) have been used in the United States to protect sheep from predators since the early 1970s. They have been used to protect small and large flocks of sheep and goats in fenced pastures and large bands of sheep and goats on open range (Green and Woodruff 1988) primarily from coyote predation. Most guarding dogs are members of breeds that have been selectively developed in Europe and Asia to protect livestock from bears and wolves. The most common breeds used in the United States are the Great Pyrenees and Komondor, whereas the Akbash, Anatolian, Kuvasz, Maremma, Shar Planinetz, and mixed breeds are less common (Black and Green 1985, Green and Woodruff 1988). The Akbash was the most common breed used to protect sheep from predators on open range in Colorado (Andelt 1992). Most guarding dogs are large and imposing, weighing 34-45 kg and standing 64 cm or taller at the shoulders. Successful guarding dogs are trustworthy (i.e., will not harm sheep), attentive to sheep, and aggressive toward predators (Coppinger et al. 1983).

Guarding dogs are an effective method of deterring coyote predation of domestic sheep (Linhart et al. 1979; McGrew and Blakesley 1982; Pfeifer and Goos 1982; Coppinger et al. 1983, 1988; Green and Woodruff 1983, 1988; Green et al. 1984; Andelt 1985*b*, 1992; Black and Green 1985). Pfeifer and Goos (1982) surveyed 36 guarding dog owners in North Dakota and reported that dogs reduced predation by 93%. In a survey of 40 owners, Green et al. (1984) reported that guarding dogs saved an average of 68 head of sheep/owner, valued at an average of $3,836 annually. Andelt (1985*b*) reported that 12 producers with their 24 guarding dogs saved $27,000 worth of sheep/year from predators in Kansas. On 1 study site in Montana, guarding dogs, traps, snares, M-44s, and aerial gunning were used to deter coyote predation, but only the dogs successfully stopped predation (O'Gara and Rightmire 1987).

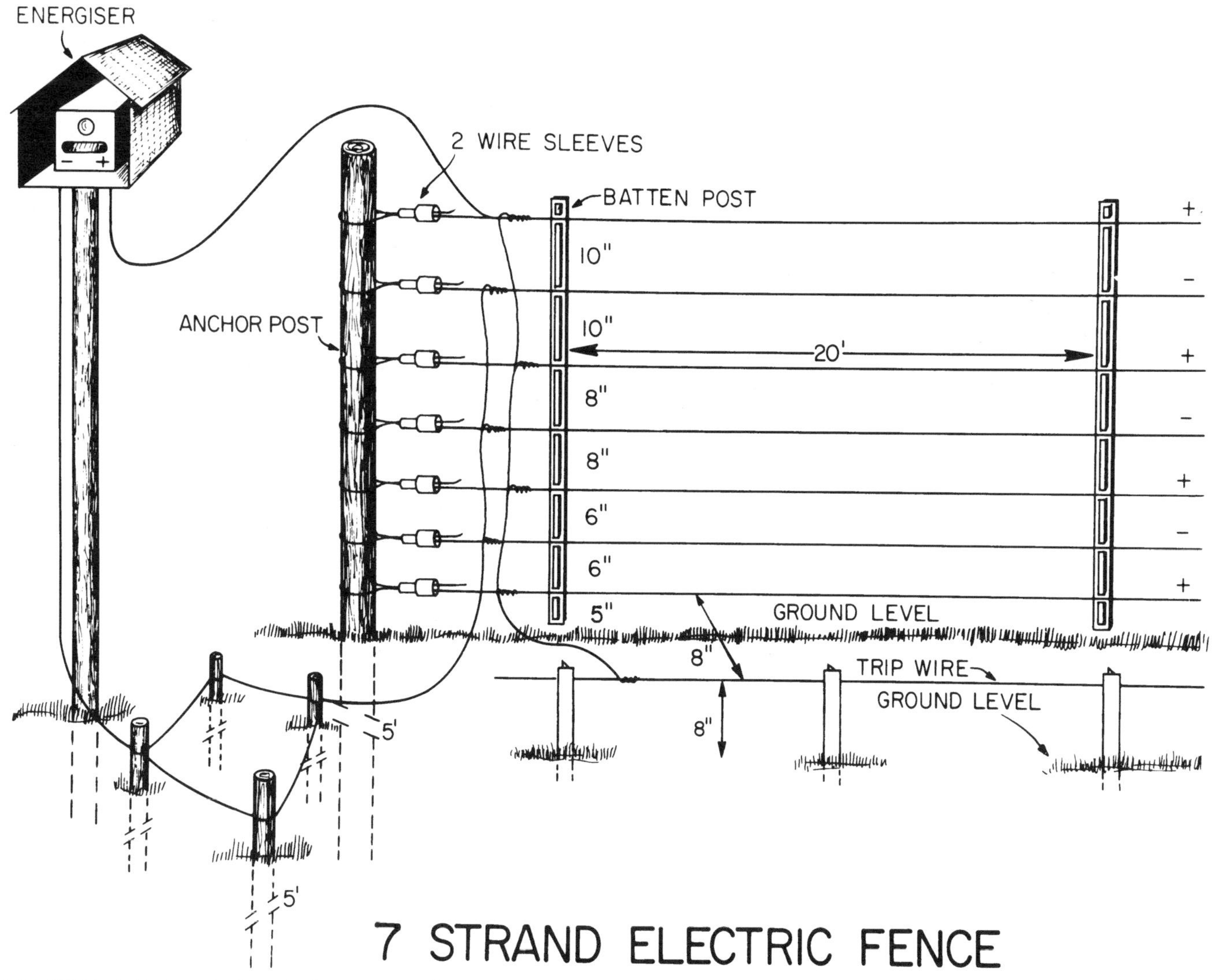

FIGURE 2. A 7-strand electric fence can effectively exclude coyotes from pens and pastures containing livestock. Provided by D. S. deCalesta.

FIGURE 3. Akbash guarding dogs can effectively protect sheep in fenced pastures or on open range. Photo by W. F. Andelt.

Producers in Colorado using guarding dogs lost an average of 0.4% of their ewes and 1.2% of their lambs to coyote predation whereas producers without guarding dogs lost 0.8-1.5% of their ewes and 4.7-9.6% of their lambs (Andelt 1992). Producers in Colorado estimated that each guarding dog saved an average of $3,216 of sheep from predators annually (Andelt 1992).

Green and Woodruff (1988) reported that the rate of success in protecting livestock from predators did not vary among Great Pyrenees, Komondor, Akbash, Anatolians, Maremma, and hybrids, nor was the rate of success different among males and females or intact and neutered dogs. However, dogs that were reared with livestock from ≤2 months old had a significantly higher rate of success than dogs that were >2 months old when placed with livestock.

Green and Woodruff (1988) and Andelt (1992) reported that Komondors were more aggressive toward people than were Akbash, Great Pyrenees, and Anatolians. These differences in aggressiveness should be considered before purchasing guarding dogs. In areas where encounters between guarding dogs and humans are likely, such as on public lands, less aggressive breeds should be considered. Akbash, Anatolian Shepherds, and Komondors are more aggressive than Great Pyrenees and thus might be selected where bears, mountain lions, and wolves are frequent predators (Green and Woodruff 1990).

Purchase prices of guarding dog pups averaged $240 in Kansas (Andelt 1985*b*) and $331 and $458 (depending on breed) in the western United States (Green et al. 1984). Annual maintenance fees (food, veterinary care, miscellaneous costs) averaged $235-$250 (Green et al. 1984, Andelt 1985*b*). The major advantages of using guarding dogs include a decrease or elimination of predation, reduced labor to confine sheep at night, more efficient use of pastures for grazing, and reduced reliance on other predator control techniques. Major disadvantages of guarding dogs include some dogs not staying with or harassing sheep, some dogs being overly aggressive toward people, and the dogs can be subject to injury and premature death. Additional costs and benefits of using guarding dogs are provided in Green et al. (1984) and McGrew and Andelt (1988). Overall, guarding dogs are a cost-effective means of reducing predation (Green et al. 1984).

The effectiveness of guarding dogs for deterring bear, mountain lion, and wolf predation on sheep has not been rigorously evaluated. Green and Woodruff (1989) reported that 15 of 20 encounters between livestock guarding dogs and black bears (17 encounters) and grizzly bears (3 encounters) resulted in bears being chased away without preying on sheep or in bears being shot by shepherds. Guarding dogs apparently were successful in protecting cattle from wolf predation (Coppinger et al. 1988), and were fairly effective in keeping wolves and black bears from carrion feeding sites in Minnesota (Coppinger et al. 1987).

Livestock guarding dogs are not compatible with the use of toxicants to control predators. Some producers have trained guarding dogs to avoid M-44s by allowing them to set off M-44s loaded with pepper; however, any mistakes by the dog likely will be fatal. Guarding dogs that have been tied probably will not be killed if captured in a snare. Most guarding dogs probably will not be injured in traps if removed in reasonable time.

Donkeys. Donkeys (*Equus assinus*) have recently been used with sheep and goats in an attempt to deter predation by coyotes and dogs. Walton and Feild (1989) estimated that 1,000 to 1,800 of 11,000 Texas sheep and goat producers used guarding donkeys in 1989. Green (1989) reported that several ranchers in Virginia and Montana also used guarding donkeys. Donkeys apparently have an inherent dislike for dogs and other canids. They will bray, bear their teeth, run and chase, and attempt to bite and kick an intruder (Green 1989).

Walton and Feild (1989) reported that 40 and 59% of Texas producers rated the effectiveness of donkeys as good or fair for deterring predation by coyotes. The effectiveness of guarding dogs is higher (Green et al. 1984; Green and Woodruff 1988; Andelt 1985*b*, 1992). The effectiveness of donkeys for deterring bear and mountain lion predation of sheep and goats has not been determined. Green (1989) cited 1 donkey raiser who indicated donkeys will flee when bears or mountain lions are in the area.

Donkeys, although apparently less effective than guarding dogs, appear to have some advantages over guarding dogs. Donkeys are relatively cheap (<$250), less prone to accidental death, long-lived, do not require special feeds, stay in the same pasture as sheep, and are not very susceptible to traps, snares, M-44s, and toxic collars.

Green (1989) and Walton and Feild (1989) recommended using only 1 jenny or gelded jack/pasture; intact jacks are too aggressive and ≥2 donkeys might stay together instead of being with the sheep. They also recommended allowing about 4-6 weeks for the donkey to bond with the sheep. Donkeys should be removed during lambing because they might trample lambs or disrupt the ewe-lamb bond. Green (1990) recommended challenging a new donkey with a dog to test its response to canids; donkeys that are not aggressive should not be used. Donkeys are apparently most effective in small open pastures or where sheep are cohesive and graze together. Feeds containing anabolic agents such as monensin (Rumensin) and lasalacid (Bovatec) apparently are poisonous to donkeys.

Donkeys can be obtained from the Bureau of Land Management or United States Forest Service under the Wild Free-Roaming Horse and Burro Act for $75 each. They also can be obtained at stockyard auctions and from breeders for $20 to $250 (Green 1989, 1990; Walton and Feild 1989).

Llamas. During 1990, Franklin and Powell (1993) surveyed 145 producers, primarily in Montana, Wyoming, Colorado, California, and Oregon, to determine the effectiveness of llamas (*Llama glama*) for reducing coyote and dog predation on sheep. These producers reported that they lost an average of 21% of their ewes and lambs annually before acquiring a llama and 7% afterwards. The losses after using a llama were similar to estimated losses for producers without guarding animals in the 17 western states. An average annual savings of $1,253, due to using llamas, was reported by 87 of the producers. Eighty percent of the producers rated their guard llamas as effective or very effective. Markham (1990) also reported that llamas are effective in reducing coyote and dog predation of domestic sheep.

Llamas are naturally aggressive toward coyotes and dogs. Typical responses of llamas to coyotes and dogs are alertness, alarm calling, walking to or running toward the predator, chasing, kicking, or pawing the predator, herding the sheep, or positioning themselves between the sheep and predator. Testimonial accounts indicate that llamas are afraid of mountain lions, and their effectiveness in deterring bear predation is unknown.

Franklin and Powell (1993) reported that the average producer used 1 gelded male llama with 250 to 300 sheep in 250- to 121-ha pastures. One guard llama was more effective than multiple llamas. The effectiveness of gelded males, intact males, and females for deterring predators was similar. However, producers reported more problems with intact (25% of 61 intact males) than gelded males (5% of 135 gelded males) attempting to breed ewes. Producers also reported that aggressiveness by some llamas toward sheep was a problem.

Franklin and Powell (1993) reported that nearly all llamas in their survey were not raised with sheep and were not trained to guard sheep. The initial adjustment period for the llamas and sheep lasted only a few hours for half the llamas, and nearly 80% were adjusted within a week. Sheep that were introduced to llamas in corrals initially sustained lower losses than those introduced in pastures. Otherwise, Franklin and Powell (1993) reported that the success of llamas was not related to age when the llama was introduced, age of llama (after 1 or 2 years old) when guarding, if lambs were present or absent when the llama was introduced, or between open and covered (forested, shrublands, gullies, ravines, etc.) habitat. In contrast, Markham (1990) recommended introducing llamas to sheep just before or at lambing, purportedly because they readily bond to newborn lambs.

Franklin and Powell (1993) reported that gelded male llamas cost $700 to $800, whereas intact males were about $100 less. Most producers reported that daily care for llamas was the same as for sheep and that no special feeds were provided. Average annual expense was $90 for feed (not including pasture) and veterinary costs were about $15. A 73.5-kg gelded llama consumes 3.0 to 4.5 kg of good grass hay/day. Depending on the area, llamas need to be dewormed 2 to 4 times/year. If food is provided for llamas, it should be placed in a feeder high enough to be out of reach of sheep. Llamas offer some of the same advantages as donkeys over livestock guarding dogs for protecting sheep; however, more data are needed on their effectiveness.

For information on llamas or sources of breeders, contact the International Llama Association, P.O. Box 370505, Denver, Colo. 80237, (303) 756-9004 or the Rocky Mountain Llama and Alpaca Association, 593 19-3/4 Road, Grand Junction, Colo. 81503, (303) 241-7921.

Bonding Sheep and Goats to Cattle. Bonding young sheep to cattle (Anderson et al. 1987, Hulet et al. 1987) and goats to sheep and cattle (Hulet et. al 1989) reduced coyote predation. No data are available on the optimum ratio of cattle to sheep or size of bonded herds that are practical for reducing predation. This technique has not been readily adopted by sheep producers, possibly because of the additional labor and expense involved with bonding sheep and goats to cattle.

Frightening Devices. Several frightening devices (Fig. 4) have been used successfully to reduce or prevent coyote predation on domestic livestock. The Denver Wildlife Research Center within the United States Department of Agriculture has developed and tested portable, battery-operated strobe-light and siren devices to reduce coyote predation on pastured sheep.

FIGURE 4. A siren-strobe device, developed by the Denver Wildlife Research Center, can effectively frighten coyotes away from sheep on bed grounds.

The devices consist of an electric timer wired to a strobe light, a warbling-type siren, and a battery. These devices, by emitting a varying and irregular sequence of light and sound stimuli from different locations, should minimize habituation by coyotes and prolong the period of repellency (Linhart 1983, 1984). The original prototype devices provided an average of 53 nights of protection in 10 trials, and a newer device protected pastured sheep for an average of 91 nights in 5 trials (Linhart 1983, Linhart et al. 1984). The siren-strobe devices did not frighten sheep, even when located on the bedgrounds (Linhart 1983). These devices were placed near bedgrounds on open range and reduced sheep losses to coyotes an average of 73% in 10 of 12 trials (Phillips and Fall 1990).

Gas exploders have been used to deter coyotes temporarily from preying on domestic livestock. Gas exploders produce large explosions (similar to rifle or shotgun blasts) that frighten coyotes. Gas exploders are portable, easy to operate, cost around $200, and have relatively low operating costs. Gas exploders deterred coyotes from killing sheep for an average of 31 days on 30 ranches in North Dakota (Pfeifer and Goos 1982) and for 6 weeks on 1 ranch in Saskatchewan (Rock 1978, cited by Linhart 1984). The delay in predation enabled more time to locate offending coyotes that increased the effectiveness of ground-control techniques and aerial hunting (Pfeifer and Goos 1982). The temporary effectiveness of gas exploders should be especially useful around calving operations because calves are most susceptible to predation for a short period after birth. Habituation by coyotes to gas exploders can be delayed by moving the device to various locations in the pasture, by changing the firing rate, or by using rotating guns called Double Johns.

Other methods of frightening coyotes and deterring predation include parking vehicles or playing a radio near areas where predation occurs (Boggess et al. 1980). Robel et al. (1981) reported that producers placing lights over corrals or bells on ≥1 sheep in each corral sustained lower losses than producers that did not use either technique; however, the presence of bells on sheep in pastures did not deter predation.

Aversive Conditioning. Conditioned taste aversion has been proposed and tested as a nonlethal method of preventing coyotes from killing sheep. One method involves placing mutton baits laced with a strong emetic (e.g., lithium chloride) or placing sheep carcasses injected with lithium chloride on the range. Coyotes supposedly ingest the baits, become ill, and develop an avoidance of sheep because they associate sheep with sickness (Gustavson et al. 1974, 1976; Ellins et al. 1977). Gustavson et al. (1974, 1976, 1982), Ellins et al. (1977), and Ellins and Catalano (1980) reported the suppression of attacks upon live prey in pens or reductions in predation in the field after consumption of baits laced with lithium chloride. The experimental designs and conclusions of some of these studies have been questioned (Griffiths et al. 1978, Horn 1983).

Conover et al. (1977), Burns (1980, 1983*a*), Burns and Connolly (1980), and Horn (1983) found that coyotes did not develop an avoidance of live prey after feeding on prey baits or carcasses treated with lithium chloride, and Bourne and Dorrance (1982) reported that distribution of lithium chloride-treated baits on farms did not reduce coyote predation on sheep. Gustavson (1982) attempted to avert wolves from preying on cattle by placing lithium chloride-laced baits on the range, but the effectiveness of the baits could not be determined. Lithium chloride in honey baits did not reduce black bear damage to beeyards (Dorrance and Roy 1978).

Conditioned taste aversion may be difficult to develop in coyotes because coyotes apparently rely more on vision than their other senses during predation (Wells and Lehner 1978) and because only a small proportion of coyotes may ingest the baits (Linhart et al. 1968). Burns et al. (1984) reported that 20 of 21 coyotes that received sublethal doses of toxicants from sheep neck collars subsequently were killed by another toxic collar indicating little potential for the use of repellents or aversive conditioning agents in collars to repel coyotes.

Aversive conditioning currently does not appear to be a viable technique for reducing coyote predation on livestock. However, if coyotes could be trained not to kill sheep, then these coyotes could become valuable assets and should be protected because they would likely prevent other coyotes (that may kill sheep) from entering their territories (Burns 1983*b*).

Repellents. Numerous chemicals that might provide olfactory repellency or that might cause gustatory or taste avoidance in coyotes have been tested (Lehner et al. 1976, Botkin 1977, Lehner 1987). No chemicals consistently repel coyotes while not harming sheep nor do they provide significant efficacy while being practical for producer use (Lehner et al. 1976, Botkin 1977, Linhart 1983). The lack of success with repellents may be related to coyote emphasis on visual cues to locate and attack prey (Wells and Lehner 1978), thus, providing little deterrent effect on the prey-killing behavior of coyotes (Linhart 1983).

Antifertility Agents. The use of antifertility agents for inhibiting coyote (Balser 1964; Linhart et al. 1968; Stellflug et al. 1978, 1984) and red fox (Allen 1982) reproduction has been investigated, but these agents have not been consistent in limiting reproduction in the field because of limited consumption of baits by coyotes and foxes and fairly high consumption by nontarget animals (Linhart et al. 1968). Dogs, crows (*Corvus brachyrhynchos*), and gray foxes (*Urocyon cinereoargenteus*) also consumed baits intended for red foxes (Linhart 1964). Improved attractants and bait delivery systems may increase the success of antifertility agents.

Antifertility agents have been viewed as a method of reducing coyote populations (Balser 1964). However, these agents also may reduce livestock losses because nonreproductive coyotes would not require as much food as reproductive coyotes caring for pups. Use of antifertility agents may provide an acceptable method of reducing coyote densities, providing the delivery system is host specific and that it is used prudently.

Relocation. Capture and relocation of animals has been used to solve some conflicts with wildlife. Weaver and Sitton (1978) reported 5 depredating lions were tagged, removed from the vicinity of livestock losses, and released. None of the relocated lions were retaken on subsequent depredation permits, suggesting that relocation was successful. However, Shaw (1983) reported that 2 relocated lions again killed livestock.

Problems with relocation include high cost, difficulty in finding relocation sites, financial responsibility for moved animals, occasionally lower survival for relocated animals, return of some animals to the capture site, possible introduction of diseases to the relocation site, and the possibility of the animals becoming a problem at the new site. Fritts et al. (1985) found that the annual survival rate of relocated wolves introduced where other wolves were present was 60%, which was similar to nonrelocated wolves. Eight of 104 relocated wolves returned to the capture site from 50 to 65 km, but others that did not return were relocated farther away (Fritts et al. 1984). Fritts et al. (1984) noted that translocation of wolves from areas of livestock depredation was largely unsuccessful in keeping wolves out of livestock production areas.

McArthur (1981) reported that several translocated black bears returned to the capture site and others became a nuisance in another area. A female was likely to return to the capture site if all the cubs were not translocated. Transplants were more successful if they occurred over greater distances, over more ridges, required a greater gain in elevation, and if a physiographic barrier was present.

Compensation. Minnesota and Alberta compensate livestock owners for livestock killed by wolves. Colorado and Wyoming reimburse livestock owners for animals killed by mountain lions and bears. Of the visitors to Yellowstone National Park, 48% felt that the government should pay compensation for livestock lost to wolves, whereas 29% felt conservation groups should pay and only 18% felt that it was the ranchers' burden (McNaught 1987).

Lethal Control Methods

Traditional lethal control techniques include use of steel traps, snares, sodium cyanide ejectors, den hunting, shooting from the air or ground, hunting with dogs, livestock protection collars, and toxic baits. The effectiveness of each method varies with geographic location, and each has advantages and disadvantages. Control methods have been used to stop predation by specific coyotes or to stop recurrences of perennial problems through local population reduction. More than 73,000 coyotes have been killed by various methods used by the Federal Animal Damage Control Program in 15 western states during 1986 (Table 1).

Trapping. Properly set traps can effectively capture offending predators and usually permit release of nontarget animals. Gipson (1975) reported that 33% of the coyotes trapped in response to damage complaints had fed on the items reported damaged. Andelt and Gipson (1979*a*) reported that 6 of 12 coyotes captured near sites of domestic turkey losses showed evidence of having killed turkeys.

Opponents of trapping primarily base their opposition on trapping's perceived or demonstrated lack of selectivity for target species, on foot injuries sustained by captured animals in some types of traps, and on the trauma of restraint (Linhart et al. 1981). Trap selectivity depends upon the type of trap and where and how the trap is set. Trap selectivity can be increased significantly by attaching pan tension devices, which increase the weight required to spring the trap; thus, small animals such as kit foxes (*Vulpes macrotis*), swift foxes (*V. velox*), gray foxes, striped skunks (*Mephitis mephitis*), opossums (*Didelphis marsupialis*), and jackrabbits (*Lepus californicus*) are excluded, but large animals such as coyotes are captured (Linhart et al. 1981,

TABLE 1. Number of coyotes taken by method in 15 animal damage control programs, FY 1986 (Connolly 1988).

State	Shot from aircraft	Trap	M-44	Snare	Den	Called and shot	Other[a]	All methods
Ariz.	752	977	150	22	7	36	77	2,021
Calif.	274	3,776	747	521	499	616	1,054	7,487
Colo.	1,161	126	322	106	461	280	109	2,565
Id.	2,115	737	74	71	136	352	76	3,561
Mont.	2,242	633	542	337	86	0	382	4,222
Nebr.	88	313	491	56	74	96	31	1,149
Nev.	2,378	1,114	58	29	133	82	264	4,058
N.M.	1,172	1,387	1,472	780	95	382	151	5,439
N.D.	977	259	0	41	95	20	29	1,421
Okla.	913	682[c]	720	196	111	557	104	3,283
Oreg.	2,134	2,152	208	551	374	280	257	5,956
S.D.	1,810[b]	187	372	152	28	152	185	2,886
Tex.	2,900	3,478	7,359	3,912	237	854	428	19,168
Ut.	2,120	474	384	71	656	271	140	4,116
Wyo.	3,394	233	58	71	1,103	699	474	6,032
Totals	24,430[b]	16,528[c]	12,957	6,916	4,095	4,677	3,761[a]	73,364
% Total	33.3	22.5	17.7	9.4	5.6	6.4	5.1	100.0

[a]Includes 3,505 shot, 222 taken with dogs, 33 by spotlight, and 1 not specified.
[b]Includes 617 taken by private aircraft under ADC supervision.
[c]Includes 1 taken in live trap.

Linhart 1983, Turkowski et al. 1984). Traps modified with pan tension devices excluded 92-100% of the small nontarget animals whereas unmodified traps excluded 6%. Other methods of reducing capture of nontarget animals include setting traps >8 m from carcasses (Hein 1992), covering baits in dirt-hole sets, and setting traps away from residences.

Padded-jaw traps (Fig. 5) have significantly reduced injuries to the feet of captured coyotes (Linhart 1983; Olsen et al. 1986, 1988; Onderka et al. 1990), gray foxes, red foxes, raccoons (*Procyon lotor*), and bobcats (Olsen et al. 1988), and caused less trauma than unpadded traps for red foxes (Kreeger et al. 1990). Onderka et al. (1990) reported that the limbs of coyotes caught in unpadded traps, compared to padded traps, had a greater tendency to freeze in cold weather.

An earlier version of the padded-jaw trap had lower rates of capturing coyotes (Linhart et al. 1986, 1988; Linscombe and Wright 1988), bobcats, and red foxes (Linscombe and Wright 1988) than standard traps. However, Skinner and Todd (1990) and Linhart and Dasch (1992) reported that newer and improved padded-jaw traps, when properly set, were as efficient as unpadded traps for capturing coyotes. Tullar (1984) reported similar capture rates of red foxes with padded and unpadded traps and Skinner and Todd (1990) reported that the efficiency of padded traps compared to unpadded traps, for capturing coyotes, improved as trappers became more familiar with padded traps. Leghold traps were 3 times as efficient as leg snares for capturing coyotes (Skinner and Todd 1990). Padded leghold traps and Fremont foot snares were superior to standard leghold traps and the Novak snare for holding terrestrial furbearers with minimal injury (Onderka et al. 1990).

FIGURE 5. The padded-jaw trap can reduce injuries to the restrained feet of captured coyotes. Photo by S. B. Linhart; courtesy of U.S. Fish and Wildl. Service.

Pan-tension devices and padded-jaw traps were devised to meet objectionable aspects of leghold traps. Use of smaller traps and daily, early-morning trap checks have reduced injuries to trapped animals (Novak 1987). Drawbacks of these trap modifications include added costs, reduced trap speed, and missed captures (Linhart et al. 1986, Olsen et al 1986). Tranquilizer tabs (Balser 1965) attached to traps reduced injury to the restrained feet of coyotes by 61% (Linhart et al. 1981).

Lindzey (1987) reported that traps set around livestock killed by mountain lions can be effective in removing lions if they return to the kill. Trapping primarily was used to remove depredating wolves in Minnesota (Fritts 1982). Fritts (1982) reported that setting traps near the area of losses and limiting the duration of trapping increased chances that the captured wolf was an offender, however, the extent of trapping necessary to reduce losses was not obvious.

Culvert traps are used extensively in parks and developed areas to capture bears. Culvert traps are expensive and restricted to areas accessible by roads.

Snaring. Snaring (Fig. 6) is an effective method of capturing coyotes in natural runways and in holes under or through woven-wire fences (Young and Jackson 1951). A loop is placed in the snare to encircle the coyote's neck as it passes through, and a locking device holds the loop closed on the neck.

Guthery and Beasom (1978) reported that snares set under fences were easy to deploy and more effective than M-44s but less effective than leghold traps. They also reported that snares were 10 to 12 times more selective than traps; snares could be made even more selective by attaching a stop that would prevent the device from closing below about 5 cm in diameter, thus allowing many small nontarget animals to escape. Phillips et al. (1990) evaluated the tension loads that coyotes, mule deer, domestic calves, and lambs applied to snares so that better breakaway snares could be developed that would allow the release of large mammals. Andelt (1988) recommended that snares should not be set within 50 m of animal carcasses, inside big game wintering yards, on trails traveled by big game, or under fences where deer, antelope, or dogs crawl to avoid capturing nontarget animals.

Snares that capture coyotes by the leg occasionally have been used in the United States and Canada. Coyotes captured in the Fremont foot snare sustained fewer foot injuries than those captured in the Novak foot snare (Onderka et al. 1990). The Fremont foot snare cable is attached to the spring arm and is thicker in diameter than the Novak foot snare cable, which may have reduced foot injuries. Coyotes captured in Novak foot snares sustained less foot injuries when the snares were attached to drags as opposed to stationary objects. Onderka et al. (1990) reported that more ungulates might be captured, held, and injured in foot snares equipped with a spring arm, that throws the snare loop above the hoof, than in traps. Lindzey (1987) reported that foot snares set around livestock killed by mountain lions can be effective in removing lions if they return to the kill.

The Aldrich foot snare and culvert traps have replaced the leghold trap for capturing bears. The Aldrich foot snare is inexpensive, portable, easily hidden, and can be used in a variety of sets (Johnson and Pelton 1980). Prebaiting can be used to locate areas of black bear activity and to eliminate unproductive trap sites (Johnson and Pelton 1980). An automobile hood spring, which acts as a shock absorber, can be attached to snare cables to eliminate virtually all major injuries to bears (Johnson and Pelton 1980). Foot snares do not present a danger to nontarget species (Kolenosky and Strathearn 1987), whereas the leghold trap can cause serious injuries to bears (Stickley 1961, cited by Johnson and Pelton 1980).

Because many black bears and grizzly bears quickly return to the kill to finish eating the carcass (Murie 1948, Griffel and Basile 1981, Johnson and Griffel 1982), depredating bears can be most selectively captured if snares are set only at fresh kills.

Sodium Cyanide Ejectors. Sodium cyanide ejectors (i.e., the Coyote Getter, a 38-cartridge activated device, or the M-44, a spring activated device) have been used by the Federal Animal Damage Control Program in the United States from about 1940 to present, except from 1972 to 1974 (Connolly and Simmons 1984). Coyote Getters have been used in Manitoba and Alberta, but use of the M-44 and Coyote Getter has not been encouraged in Saskatchewan (United States Fish and Wildlife Service 1978).

Sodium cyanide ejectors consist of a hollow metal tube crimped closed at the bottom, a firing mechanism, a sodium cyanide cartridge or capsule holder, and a sodium cyanide capsule or cartridge. The metal tube is driven into the ground and the capsule holder is wrapped with wool or rabbit fur smeared with a lure attractive to canids. A lethal dose of sodium cyanide is ejected into an animal's mouth when the device is pulled. Death occurs within seconds.

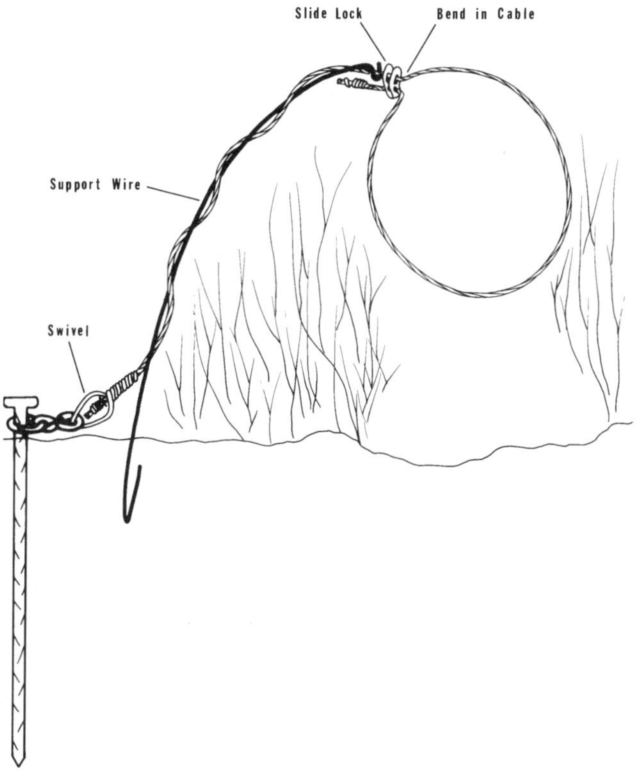

FIGURE 6. Snares can be set in trails and under fences to capture coyotes, foxes, and bobcats (Andelt 1988).

M-44 capsules are registered to control coyotes, red and gray foxes, and wild dogs that depredate livestock and poultry (Connolly 1988). Sodium cyanide ejectors were more selective for capturing coyotes (Robinson 1943, Beasom 1974) and more efficient (Robinson 1943) than steel leghold traps, which were not modified with pan-tension devices. Sodium cyanide ejectors are primarily selective for canids (United States Fish and Wildlife Service 1978), with target species comprising 95% of the animals taken (Dorrance 1980, Connolly 1988).

Denning. Denning, the practice of locating dens of depredating coyotes and destroying the pups and/or adults, has been reported (but not supported with biological data) as a method for coyote population reduction (Young and Dobyns 1945, Gier 1968) and as a method of stopping predation on livestock (Young and Dobyns 1945, Lemm 1973) primarily where the increased food demands of pups cause serious losses. Denning has been criticized as an unselective method of removing offending animals (Defenders Wildlife 1978, Sierra Club 1978, Humane Soc. 1978, cited by Till and Knowlton 1983). However, Till and Knowlton (1983) reported that predation incidents declined 98.2% when pups and adults were removed and 87.7% when only the pups were removed from dens of offending coyotes. Their analysis indicated that denning can be highly selective and cost-effective over a short period of time.

Aerial Hunting. Aerial hunting of coyotes apparently began as early as 1923 and was first officially adopted by the Federal Animal Damage Control Program as an operational control method in 1942 (Wade 1976). Aerial hunting with fixed-wing aircraft and helicopters has been conducted primarily for the protection of livestock and secondarily for the value of furs and bounty payments (Wade 1976). Aerial hunting achieved major importance (concurrent with curtailed use of toxicants in 1972) in the Federal Animal Damage Control Program by 1975, when it accounted for 42% of the coyotes killed (Connolly 1982). Aircraft also can assist in locating den sites (Miner and Quiroz 1974, Wade 1976). Aerial hunting is an expensive although effective method of controlling problem coyotes, especially in open terrain and when control of predation is urgent. Aerial hunting also has been used to control wolf populations (Keith 1983).

Aerial hunting is completely selective for the target species and can be highly selective for offending coyotes (Connolly 1982, Connolly and O'Gara 1987). Connolly and O'Gara (1987) reported that 6 of 11 coyotes taken from a helicopter had recently attacked or fed on collared sheep. Because most coyotes are territorial and usually have home ranges <15 km^2 in size (Camenzind 1978, Andelt 1985*a*), selectivity for offending coyotes can be increased by applying aerial and ground controls near sites of predation.

Ground Shooting. Attracting coyotes within shooting range with predator calls that imitate the sound of an animal in distress is a very selective method of removing individual coyotes (Beasom 1974, Henderson 1986). Producers also have solved many of their own problems by lying in wait and shooting coyotes as they entered sheep corrals and pastures.

Hunting with Decoy and Other Dogs. Some Federal Animal Damage Control Program field agents use decoy dogs to assist in removing coyotes. The agents frequently imitate the howl of a coyote. If coyotes respond, they move to a position that is fairly close to the coyotes and howl again. When coyotes respond, they release 1 or 2 small dogs that run toward the approaching coyote(s). The coyote(s), in an attempt to protect their pups, usually chases the dog(s) back to the field agent who shoots the coyote. This technique is effective from the start of denning until late summer (Rowley and Rowley 1987). Weaver and Sitton (1978) and Suminski (1982) indicated that hunters with dogs were the most frequently used method of removing depredating lions.

Livestock Protection Collars. These devices were developed during the 1970s and 1980s for control of coyote predation on sheep and goats. They were registered by the United States Environmental Protection Agency (EPA) in 1985, and Montana, New Mexico, South Dakota, Texas, and Wyoming have established EPA-approved programs to allow collar use by state-certified applicators as of March 1990 (Connolly and Burns 1990). The collar, consisting of 2 toxicant-filled reservoirs, is positioned on the neck of young and adult sheep and goats to exploit the neck-attacking behavior of coyotes. Coyotes that attack collared livestock usually bite the collars and receive an oral dose of the toxicant, Compound 1080 (sodium fluoroacetate). Generally, 20-50 collared lambs or kids were placed in pastures where predation was occurring and other livestock were removed.

The efficacy of the collars for livestock protection has been studied by the Denver Wildlife Research Center (Connolly 1980, Connolly and Burns 1990), the New Mexico Department of Agriculture (Littauer 1983), the Texas Agricultural Experiment Station (Scrivner 1983, Texas Agricultural Experiment Station 1983), and the Texas Department of Agriculture (Walton 1990). Thirty-eight to 71% of the collars on livestock attacked by predators were punctured in the 4 studies. In 17 of 28 field tests in Texas, Idaho, Montana, and Alberta, predation stopped or declined following use of collars (Connolly 1980). The greatest advantage of the toxic collar is selectivity for livestock-killing coyotes.

Disadvantages include the cost of collars, the labor involved in collaring and managing livestock, the fact that livestock must be sacrificed, and the potential hazards of lost or punctured collars (Connolly 1982). Burns et al. (1988, 1991) concluded that the amount of 1080 residue found on collared sheep killed by coyotes or within coyotes killed by the collars presented minimal primary and secondary hazards to nontarget species.

Scrivner (1983) estimated that the costs of using toxic collars (including collared animals killed or missing, collars punctured or missing, transportation, labor, feed, and miscellaneous) averaged $1,828.78/rancher for a 52-week period in Texas. In New Mexico (Littauer 1983), the cost of collars (excluding labor and transportation) used by ranchers averaged $443/coyote killed when prorating the initial investment in collars over 4 seasons of use and accounting for lost and punctured collars (W. F. Andelt, unpubl. data).

Techniques for using toxic collars are still evolving as experience in their use accumulates. Research on the differential vulnerability of lambs to coyote predation may increase the effectiveness of toxic collars. Lambs with reduced mobility, or lambs from ewes with reduced mobility increased the probability of a lamb being on the periphery of the bedground and thus increased the probability that the lamb would be noticed and attacked by a coyote (Gluesing et al. 1980). Lambs that were most active (Gluesing et al. 1980), orphaned, or recently introduced to a flock (Blakesley and McGrew 1984) were most vulnerable to coyote predation. Placing toxic collars on the most vulnerable lambs should increase the probability of exposing offending coyotes to toxic collars.

Walton (1990) indicated that EPA restrictions resulted in low use of livestock-protection collars. Use of these collars for deterring predation on sheep by bears and mountain lions does not seem feasible because these predators often bite sheep in the back of the neck or through the skull.

Single Lethal-dose Baits. The use of toxic chemicals for predator control in the United States began as early as 1847, when strychnine was introduced (United States Fish and Wildlife Service 1978). Small bite-sized baits containing strychnine were used in the United States until toxic baits were withdrawn from use in 1972 by Presidential Executive Order 11643 and the cancellation of all registrations of predacides by the EPA. Use of strychnine baits by the Federal Animal Damage Control Program increased from 632,187 baits in 1960 to 822,043 baits in 1970, and 1080 baits also were used by an unknown number of governmental hunters (Connolly 1982). Strychnine baits also were used extensively in Alberta in 1974 (Dorrance and Roy 1976). Bjorge and Gunson (1985) reported that the use of bite-sized portions of strychnine-poisoned meat baits were effective in reducing the number of wolves and subsequent mortality of cattle on 1 study area in Alberta.

Data on the hazards and benefits of toxic baits are scarce. Kilgore (1969) and Nunley (1977) reported that increases in small carnivore numbers seemed to coincide with reduced use of strychnine drop baits. Bortolotti (1984) reviewed the causes of deaths of 143 golden eagles (*Aquila chrysaetos*) and 172 bald eagles (*Haliaeetus leucocephalus*) preserved as study skins in museums and reported that 71% (n = 27) of the golden eagles and 2 of 7 bald eagles, where cause of death was known, were killed by poisons (mostly strychnine) or traps. Robinson (1948) indicated that the effectiveness of strychnine baits declined as coyotes learned to detect the toxicant.

Tigner et al. (1981, cited by Connolly 1982) reported that only 9-27% of nontoxic small lard baits placed 10-50 m from animal carcasses ("draw stations") were taken by coyotes; many baits were removed by nontarget animals. Similarly, Linhart et al. (1968) reported that most baits containing chemosterilants were taken by nontarget species. Guthery et al. (1984) indicated that nontoxic baits were taken slightly more frequently by coyotes than by nontarget animals in Texas. Forty-two and 14% of the producers using strychnine baits in Alberta during November-March and April-October 1977 reported that nontarget species (primarily black-billed magpies [*Pica pica*], ravens [*Corvus corax*], and dogs) were killed. Beasom (1974) reported that strychnine meat-baits and strychnine egg-baits were less selective than the M-44 or shooting because they incidentally killed a variety of game animals, raptors, rodents, songbirds, and reptiles.

Selectivity of baits was increased in Texas when employed (1) in December and January, when coyote consumption rates were highest, (2) in dense vegetation to reduce consumption by crows, and (3) on clear nights with a full moon to decrease rodent consumption (Guthery et al. 1984). In Alberta, the uptake of baits by birds was reduced significantly when the baits were covered (M. J. Dorrance, Alberta Agric. Prot. Branch, Edmonton, pers. commun.). Connolly (1982) stated: "It remains to be documented that small, toxic baits can be delivered effectively to coyotes without adverse impact on nontarget species."

Toxic Bait Stations. This method, now banned in the United States, generally consisted of 23-45 kg of livestock meat injected with thallium sulfate or 1 mg of 1080/28.4 g of bait. Thallium sulfate in toxic bait stations was first used for control of coyotes in 1937 and was gradually replaced with compound 1080, starting in 1944 (Robinson 1948), because 1080 was more selective for canid species and safer to apply. The Federal Animal Damage Control Program used 1080 bait stations for coyote control most frequently in the early 1960s when 15,000-16,000 stations were placed each winter in the western United States (Connolly 1982). The use of this technique by the Federal Animal Damage Control Program declined annually after 1964 to 7,289 stations in 1971 (Connolly 1982), and the technique was banned in 1972. Compound 1080 was used extensively in Alberta, British Columbia, Manitoba, and Saskatchewan but its use has significantly decreased recently due to supposed lack of public acceptance (McKay 1975, Dorrance and Roy 1976, United States Fish and Wildlife Service 1978).

Data on the effectiveness and hazards of 1080 bait stations are scarce. Robinson (1948) reported 75-100% reductions in predator losses following early use of toxic bait stations. Lynch and Nass (1981) reported an inverse relationship between the number of toxic bait stations used in the western states and the number of livestock lost to predators on the national forests from 1960 to 1972. Connolly (1982) reported that coyotes fed on 41% of nontoxic (simulated) bait stations and that about 33% of the meat was consumed by coyotes and nontarget species in Wyoming, Idaho, and New Mexico during 1981. Coyotes fed on 94% of the baits in New Mexico but on only 30% in Idaho and 27% in Wyoming; these data suggest that 70% of the bait stations used in Idaho and Wyoming would have had no effect on coyotes (Connolly 1982). Robinson (1948) concluded that although some primary and secondary poisoning of birds and mammals would occur, the selectivity of 1080 was much higher for canids than for other wildlife species.

Toxic bait stations usually were placed at a density not exceeding 1/88.9 km^2 (Robinson 1953*a*,*b*). Considerations of coyote home ranges and territorial behavior may help explain why bait stations did not appear to suppress coyote populations in the southern United States (Wagner 1972). Andelt (1985*a*) found that the home ranges of resident coyotes (87% of the population) in South Texas averaged 4.5 km^2 whereas transient (13% of the population) home ranges averaged >36.7 km^2. Because resident coyotes were territorial and occupied small home ranges (Andelt 1985*a*), 1 bait station/88.9 km^2 probably would be used by only a small proportion of the coyote population. The lower dependence on scavenging in southern areas compared with northern areas likely contributed to the lower effectiveness of treated baits in southern areas (Linhart 1981).

PUBLIC ATTITUDES TOWARD CONTROL OF PREDATORS

Various attitudes toward coyotes and wolves and their control exist. Some ranchers have stated that coyotes are a major threat to the sheep-raising industry (Wagner 1975, Nesse et al. 1976, Gee et al. 1977*a*, Tigner and Larson 1977), whereas animal protectionists have maintained predator losses claimed by sheepmen are exaggerated, control practices pose environmental hazards, and that coyotes are a valuable part of the wildlife resources and need protection. Buys (1975) reported that approximately 92% of sheepmen and 76% of cattlemen believed that predator control is necessary for the survival of their industries. Ranchers and animal protectionists often find it difficult to discuss rationally the issue of coyote control because individuals on both sides are often emotionally charged. Thus, their opinions are often formulated without data on, or knowledge of the effectiveness, hazards, humaneness, and effects of control techniques on individual animals or on the population dynamics of target and nontarget species. Some trapping and sport hunting enthusiasts also do not concur with organized animal damage control activities because they feel these activities compete with their harvest of animals.

Because laws and policy decisions are based upon pressures from special-interest groups, technical and economic information (Arthur et al. 1977), and knowledge of public attitudes, scientific research data are of the utmost importance for educating special-interest groups and the public and for shaping rational policy decisions. Scientific research data should be presented by representatives of recognized nonadvocacy institutions to special-interest groups, with emphasis on fostering mutual understanding and a spirit of compromise between opposing groups (Dorrance 1983).

Public perceptions toward predators, primarily wolves and coyotes, and control measures have been surveyed (Arthur et al. 1977, Stuby et al. 1979, Arthur 1981, Kellert 1985). Approximately one-third of the respondents had not heard of the rancher-environmentalist controversy over the killing of coyotes, and knowledge of coyote habits and population trends was limited. A majority of the public felt that ranchers should have the right to kill those animals that were killing their livestock, but only a minority of the public approved of killing as many coyotes as possible to prevent future problems. The respondents were just about equally concerned over the killing of coyotes and the predatory killing of sheep by coyotes. As sheep and lamb losses to coyotes were hypothetically increased, respondents were more willing to approve of killing coyotes. In general, public attitudes toward wildlife are changing. Kellert

(1976) reported that Americans 18-29 years of age were far more interested in wildlife, more concerned about animal welfare issues, and more likely to oppose hunting than those over 65. Kellert (1976) also found that a humanistic orientation toward animals was shared by about 70% of the public.

A majority of hunters and residents near Glacier National Park, where wolves are recolonizing, visitors to Yellowstone National Park, and residents of Wyoming supported restoration of wolves (McNaught 1987, Bath and Buchanan 1989, Tucker and Pletscher 1989). Kellert (1986) reported that most respondents supported the right of farmers to protect their stock from wolf predation. Most respondents favored control of only the depredating animal and use of nonlethal methods such as relocation, guarding dogs, and improved husbandry.

The public has rated nonlethal control methods (e.g., guarding dogs, repellent chemicals, and birth control) as more acceptable than current lethal control methods, but rancher subsidies and indemnity payments are considered less acceptable than some lethal control methods (Fig. 7) (Arthur et al. 1977, Stuby et al. 1979, Arthur 1981). The public is more concerned about the humaneness (lack of pain and suffering to the animal) and specificity (extent that only offending animals were subject to control) of control methods than about their cost-effectiveness. Trapping and slow-acting poisons were believed to cause more suffering than other lethal control methods and were judged least acceptable, whereas shooting with guns and using fast-acting poisons were regarded as causing the least amount of suffering and were judged most acceptable. Kellert (1985) reported that over 90% of the public objected to the use of poisons.

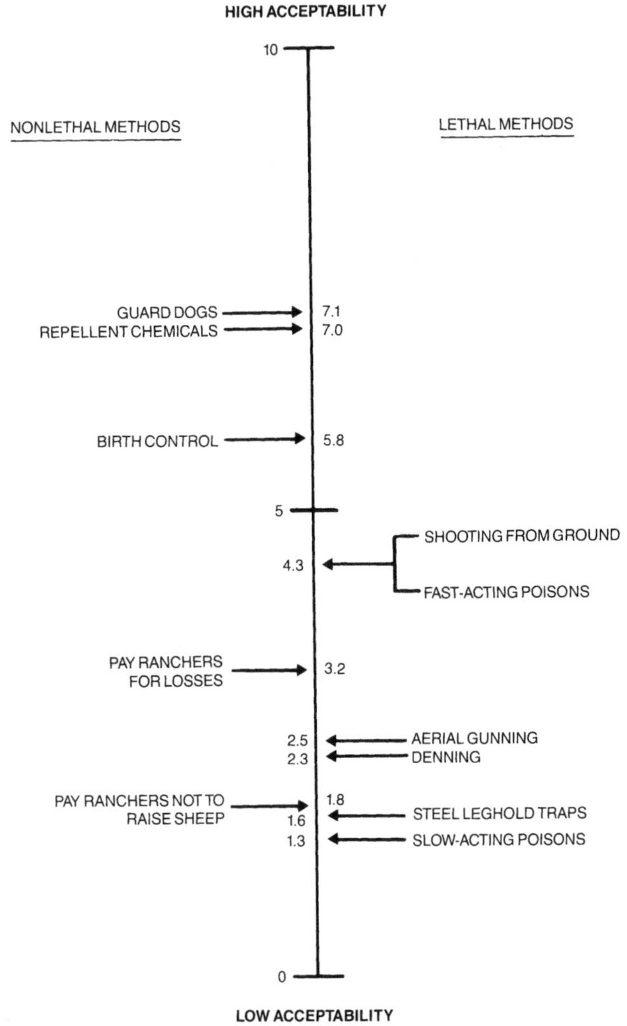

FIGURE 7. Public acceptability of various coyote damage control techniques (modified from U.S. Fish and Wildl. Serv. 1978 and Arthur 1981).

IMPACTS OF PREDATOR CONTROL

The impact of predator control on predator population densities, behavior, and ecology are to a great extent unknown. Few evaluative studies have been conducted, and to assess some of these parameters would require research extending over years, if not decades. Recent control efforts by the Federal Animal Damage Control Program have no significant impacts on target populations at the national level, but target populations may be significantly impacted in localized areas where they are reduced to minimize damage (United States Department of Agriculture 1990). The program removed 76,050 coyotes, 1,226 bobcats, 207 mountain lions, 4,667 red foxes, and 291 black bears during 1988 (United States Department of Agriculture 1990). The control, by approximately 450 Federal Animal Damage Control agents, was conducted on about 11% of the land in 17 western states and the number of coyotes removed annually in 1974 (70,000 to 85,000) represented about 24% of the total harvest and only about 4% of the coyote population (Pearson 1978, United States Fish and Wildlife Service 1978). However, extensive control efforts by the Federal Animal Damage Control Program using thallium sulfate and 1080 bait stations during the 1940s may have temporarily depressed coyote densities around 1950 in some of the western states (Robinson 1953*b*, 1961; Linhart and Robinson 1972; Nunley 1978). Robinson (1948) reported that control of coyotes with toxic bait stations sometimes reached local extirpation in areas where natural foods were scarce during winter. Apparently, coyote densities returned from the 1950 lows to the 1940-41 precontrol levels in Colorado and New Mexico during 1960 and 1970, respectively, but they probably remained depressed in Wyoming over the same period (Linhart and Robinson 1972). However, a large increase in coyote densities after the ban on toxicants in 1972 was not apparent (Roughton 1977). Concurrent with the presumed decrease in coyote densities in Wyoming was an apparent increase in the densities of small nontarget carnivores, particularly the red fox, thus supporting the concept that a decrease in 1 carnivore species leads to an increase in other sympatric carnivores (Linhart and Robinson 1972). Red foxes apparently avoid coyotes with their home ranges abutting or only overlapping the perimeter of coyote home ranges (Voigt and Earle 1983, Major and Sherburne 1987, Sargeant et al. 1987). Similarly, coyotes have been reported to occupy areas between wolf ranges.

Despite considerable man-induced mortality, coyote populations are self-maintaining through behavioral adaptations and biological compensatory mechanisms such as increased rates of reproduction, survival, and immigration (Knowlton 1972, Connolly and Longhurst 1975). Reduced population densities likely result in less competition for limited resources such as food, den sites, and mates and may reduce transmission of diseases and parasites. As coyote populations are reduced, pregnancy rates (Gier 1968), especially for yearling females, and litter sizes (Knowlton 1972) increase, whereas natural mortality rates decrease (Connolly and Longhurst 1975). Connolly and Longhurst's (1975) model suggested that coyotes, through compensatory reproduction, can withstand a 70% annual control level and that 3 coyotes would need to be killed for every animal present at breeding time to hold the density below 50% of the precontrol level. In most areas, coyote numbers likely are controlled by competition for limited resources such as food and by social stress, diseases, and parasites (Connolly and Longhurst 1975).

Connolly and Longhurst's (1975) model indicated that the number of breeding females changed only slightly with increased control intensity but that litter sizes increased. Because livestock losses are relatively serious during spring and summer due to the increased food demand of pups (Wade 1973), control of <75% of the coyotes may actually increase predation due to a greater demand for food by larger litters (Connolly and Longhurst 1975).

The effects of control, sport hunting, and general trapping on coyote behavior and ecology are not well known. However, comparisons of the ecology and behavior of exploited and unexploited coyote populations, in addition to some reported changes in behavior, have suggested that coyotes may be adapting to exploitation either through learning or heredity. Robinson (1948) reported that some coyotes learned to detect strychnine in drop baits or became wary of traps and that others refused to feed on lethal bait stations. Wagner (1975) reported that coyote responses to Coyote Getters dropped off markedly after a few years of use. Andelt et al. (1985) reported that coyotes captured in steel leghold traps and released subsequently avoided scent stations used to determine their relative abundance. Gustavson et al. (1974) and Olsen (1975) reported that lithium chloride-produced aversions in coyotes lasted 2-7 months, and Linhart et al. (1976) reported that electric shock-produced aversions lasted 3-9 months.

Exploitation also may have an effect on the social organization and activity patterns of coyotes. In unexploited areas, most coyotes existed in relatively large groups (Bowen 1978, 1981; Camenzind 1978; Andelt 1985*a*), whereas coyotes in exploited areas generally have been considered to exist in smaller groups (Hibler 1977, Althoff and Gipson 1981). Coyotes have been reported as more active during the daytime in unexploited (Gipson and Sealander 1972, Andelt 1985*a*) areas (Andelt and Gipson 1979*b*). Roy and Dorrance (1985) reported that coyotes avoided open areas near roads during daylight hours in areas where they were hunted. Exploitation likely removed some individuals, causing smaller groups, and also likely selected against diurnally active coyotes that were more visible and thus more susceptible to hunters. Adaptation and selection appear to be occurring.

Red foxes are a major predator of nesting waterfowl. Thus, controlling coyotes to reduce livestock losses may allow an increase in red foxes, which may increase predation on waterfowl.

PREDATOR CONTROL IN PRACTICE

When To Use Prevention And Control Methods

Determining when predator control should be undertaken for livestock protection varies with the type of control methods employed. Nonlethal control methods (i.e., livestock husbandry, fencing, livestock guarding dogs, frightening devices, and herders) are socially acceptable and should be used when they achieve a reduction in predation losses that exceeds the cost of control. The need for lethal control should be based upon aesthetic, social, economic, ecological, political, and administrative considerations (Berryman 1972). Costs of lethal control techniques should be related to the value of livestock saved and equated to social and aesthetic values. Control decisions should be developed cooperatively with other concerned agencies, related to other resource decisions, and based on accurate data. The objective of control programs should be to alleviate the problem, not to destroy offending animals.

The optimal time of year for employing control methods varies with the type of controls used. Nonlethal control techniques, such as frightening devices, should be employed shortly before predation begins (if it is predictable) or immediately after it begins to avoid the establishment of a problem or pattern that may be difficult to disrupt. Frightening devices should be removed as soon as they are not needed because predators are more likely to habituate to them with time. Approved lethal controls for removing specific offending animals should be employed as soon as possible after predation begins to minimize livestock losses. If local populations of predators are removed before predation begins, control efforts should be implemented immediately before predators become a problem because predators quickly move into areas vacated by other predators. Windberg and Knowlton (1988) described the large number of coyotes using small areas and the presence of transient animals available to fill vacant territories. Control applied too long before damage starts likely will be relatively ineffective.

If population suppression over large areas is warranted, control techniques should be employed just prior to whelping, when the population is at a normal low and dispersal has subsided (Knowlton 1972). It does not seem prudent to suppress coyote numbers in the fall, when dispersal is occurring and before the population normally undergoes a large natural reduction within a brief period (Knowlton 1972). Dorrance (1980) suggested that dispersal by coyotes, primarily from mid-February through April, probably negates the effect of preventive control on local coyote populations prior to mid-February in central Alberta.

How To Implement Predator Control

Choosing how to implement predator control involves consideration of who will conduct the control and what methods should be applied. Control measures can be conducted by the general public, sport hunters and trappers, the producer who is suffering damage, private industry, or governmental agencies. The question of who conducts the control requires consideration of the status of the target species, hazards of the control techniques, and the ownership and legal responsibilities for the land where the control measures are to be performed (Berryman 1972). Control should be limited to or supervised by professionals when it might affect sensitive species, requires techniques that threaten human life and nontarget species, or that has permanent adverse environmental effects, and is conducted on public lands (Berryman 1972, Dorrance 1983).

Two governmental approaches to prevention and control of livestock predation involve extension education and governmental animal damage control assistance. The extension approach uses a relatively small number of wildlife specialists who train producers and the general public in nonlethal and lethal methods of controlling predation. After the training period, the producers or volunteers conduct the control themselves. Extension education is successful and is the primary means of animal damage control in the eastern states, Kansas, Missouri, and Alberta, Canada.

The Federal Animal Damage Control Program employs about 450 animal damage control agents who conduct control operations and provide extension education for producers in the western states. The relatively large livestock operations, open-range husbandry practices, and the use of public lands for grazing in the West suggest that using professional animal damage control agents to control predation is more suitable there than it might be in the East (United States Fish and Wildlife Service 1978). The Federal Animal Damage Control Program emphasizes the protection of livestock through nonlethal techniques, the removal of offending animals, and the management of local depredating populations (Fall 1984), replacing earlier attempts at population reduction (Fall 1984).

Future Prevention And Control Techniques

The types of control techniques that will be used in the future likely will be determined by considerations of predator population dynamics, public sentiment, costs, effectiveness, and environmental hazards of techniques. Based upon public sentiment, nonlethal techniques likely will continue to be preferred over lethal techniques directed at offending animals and both approaches will be preferred over lethal methods directed at population reduction. When lethal techniques are used, emphasis will be placed upon using those techniques that are most humane (e.g., using padded instead of conventional leghold traps) and selective (e.g., using pan-tension devices on traps) for the target animals. Livestock producers cannot afford to use illegal techniques (e.g., poisons), especially on public lands, because their use may encourage the public to remove livestock from public lands.

Reducing predator populations over small and large areas is difficult and likely will become less popular. Connolly and Longhurst (1975) indicated that coyote populations probably cannot be substantially reduced over large areas without the use of toxicants. Although toxic drop baits and bait stations are the most economical methods of reducing coyote numbers, they have been relatively unselective for coyotes and may select against carrion-eating coyotes in favor of coyotes that are more prone to kill live prey, including livestock.

The effectiveness of reducing predator populations with some of the current lethal techniques (e.g., trapping, M-44s, shooting from the ground and air) probably decreases in proportion to the decline in the breeding population as control measures increase (Connolly and Longhurst 1975).

Killing coyotes at rates of <75% of the population may stimulate reproduction and aggravate losses by increasing the food demand of larger litters (Connolly and Longhurst 1975). Young and Jackson (1951), Henderson (1972), Lehner (1976), and Windberg and Knowlton (1988) described the difficulty of trying to reduce coyote numbers. Young and Jackson (1951:156) equated reducing coyote populations to "digging a hole in the ocean."

Livestock guarding dogs likely will become the most effective and most popular technique for deterring predation on livestock, particularly sheep and goats. However, lethal techniques will be necessary to remove predators in areas where livestock guarding dogs and other nonlethal techniques may not work. Predators that adapt to guarding dogs and remain a problem will need to be averted from livestock or killed by other techniques.

Antifertility agents likely will be accepted by the public and may become popular if techniques such as the Coyote Lure Operative Device (Marsh et al. 1982, Stolzenburg and Howard 1989) or species-specific contraceptives can be developed so that only the target animals will be affected. Antifertility agents might prevent the compensatory increase in reproduction associated with the killing of coyotes and thus might reduce predation associated with the need to feed pups (Connolly and Longhurst 1975).

Den hunting (Till and Knowlton 1983) may remain an effective control technique even at high rates of control, because the number of females with litters decreases only slightly with increased removal of coyotes (Connolly and Longhurst 1975). However, this technique may not remain viable because of public concern over the removal of pups.

Use of nonlethal and lethal control techniques (particularly those directed at offending animals) that do not affect overall predator numbers may best serve the interests of both ranchers and animal protectionists.

SUMMARY

Predators have been estimated to kill 2.5% of the adult sheep and 9.0% of lambs annually in the western United States. Coyotes kill about 74% of the adult sheep and 77.7% of the lambs lost annually to predators. Dogs, red foxes, mountain lions, black bears, grizzly bears, gray wolves, and bobcats also prey on domestic livestock, but their predation is secondary to that of coyotes.

Nonlethal and lethal control techniques are used to prevent or control predation. The more successful nonlethal methods include various livestock husbandry practices such as confinement, disposal of livestock carcasses, and use of herders; fencing; guarding dogs; and various frightening devices. Lethal methods of deterring predation include trapping, snaring, denning, shooting from the air and ground, and livestock protection collars.

Predators will continue to prey on domestic livestock throughout the foreseeable future. This problem likely will be minimized through effective public education programs, the adoption of nonlethal and more humane lethal control techniques, and the development of new and improved techniques. The trend away from predator population reduction methods is likely to continue, especially when considering current public sentiment, predator population dynamics, costs, and environmental hazards of control techniques. The public controversy over methods of resolving wildlife-human conflicts probably will not diminish until safe, effective, selective, economical, nonlethal methods are developed for preventing predation. Hopefully, innovative research will provide new methods of reducing livestock losses to predators while enabling humans to pursue the recreational and aesthetic opportunities that predators offer.

LITERATURE CITED

Acorn, R. C., and M. J. Dorrance. 1990. Methods of investigating predation of livestock. Alta Agric. Prot. Branch. Edmonton, Can. 36pp.

Allen, S. H. 1982. Bait consumption and diethylstilbesterol influence on North Dakota red fox reproductive performance. Wildl. Soc. Bull. 10:370-373.

Althoff, D. P., and P. S. Gipson. 1981. Coyote family spatial relationships with reference to poultry losses. J. Wildl. Manage. 45:641-649.

Andelt, W. F. 1985*a*. Behavioral ecology of coyotes in South Texas. Wildl. Monogr. 94. 45pp.

———. 1985*b*. Livestock guarding dogs protect domestic sheep from coyote predation in Kansas. Proc. Great Plains Wildl. Damage Control Workshop 7:111-113.

———. 1988. Proper use of snares for capturing furbearers. Coop. Ext. Serv., Colorado State Univ., Fort Collins. 4pp.

———. 1992. Effectiveness of livestock guarding dogs for reducing predation on domestic sheep. Wildl. Soc. Bull. 20:55-62.

———, and P. S. Gipson. 1979*a*. Domestic turkey losses to radio-tagged coyotes. J. Wildl. Manage. 43:673-679.

———, and ———. 1979*b*. Home range, activity, and daily movements of coyotes. J. Wildl. Manage. 44:944-951.

———, C. E. Harris, and F. F. Knowlton. 1985. Prior trap experience might bias coyote responses to scent stations. Southwest. Nat. 30:317-318.

Anderson, D. M., C. V. Hulet, J. N. Smith, W. L. Shupe, and L. W. Murray. 1987. Bonding young sheep to heifers. Appl. Anim. Behav. Sci. 19:31-40.

Arthur, L. M. 1981. Coyote control: the public response. J. Range Manage. 34:14-15.

———, R. L. Gum, E. H. Carpenter, and W. W. Shaw. 1977. Predator control: the public viewpoint. Trans. North Am. Wildl. and Nat. Resour. Conf. 42:137-145.

Bailey, R., ed. 1978. Recovery plan for the eastern timber wolf. U.S. Government Printing Office, Washington, D.C. 79pp. (*Cited from*: Fritts, S. H. 1982. Wolf depredation on livestock in Minnesota. U.S. Dep. Inter., Fish and Wildl. Serv. Res. Publ. 145. 11pp.)

Balser, D. S. 1964. Management of predator populations with antifertility agents. J. Wildl. Manage. 28:352-358.

———. 1965. Tranquilizer tabs for capturing wild carnivores. J. Wildl. Manage. 29:438-442.

Bath, A. J., and T. Buchanan. 1989. Attitudes of interest groups in Wyoming toward wolf restoration in Yellowstone National Park. Wildl. Soc. Bull. 17:519-525.

Beasom, S. L. 1974. Selectivity of predator control techniques in South Texas. J. Wildl. Manage. 38:837-844.

———, and D. R. Gober. 1975. The effectiveness of the M-44 as a tool to curtail sheep losses to predators. Annu. Rep. U.S. Environ. Protection Agency. 49pp.

Berryman, J. H. 1972. The principles of predator control. J. Wildl. Manage. 36:395-400.

Bjorge, R. R., and J. R. Gunson. 1983. Wolf predation of cattle on the Simonette River pastures in northwestern Alberta. Pages 106-111 *in* L. N. Carbyn, ed. Wolves in Canada and Alaska: their status, biology, and management. Can. Wildl. Serv. Rep. Ser. 45.

———, and ———. 1985. Evaluation of wolf control to reduce cattle predation in Alberta. J. Range Manage. 38:483-487.

Black, H. L., and J. S. Green. 1985. Navajo use of mixed-breed dogs for management of predators. J. Range Manage. 38:11-15.

Blakesley, C. S., and J. C. McGrew. 1984. Differential vulnerability of lambs to coyote predation. Appl. Anim. Behav. Sci. 12:349-361.

Boggess, E. K., F. R. Henderson, and C. E. Spaeth. 1980. Managing predator problems: practices and procedures for preventing and reducing livestock losses. Coop. Ext. Serv. Bull. C-620. Kansas State Univ., Manhattan. 19pp.

Bortolotti, G. R. 1984. Trap and poison mortality of golden and bald eagles. J. Wildl. Manage. 48:173-1179.

Botkin, M. P. 1977. The use of aversive agents for predator control. Proc. Great Plains Wildl. Damage Control Workshop 3:13.

Bourne, J., and M. J. Dorrance. 1982. A field test of lithium chloride aversion to reduce coyote predation on domestic sheep. J. Wildl. Manage. 46:235-239.

Bowen, W. D. 1978. Social organization of the coyote in relation to prey size. Ph.D. Thesis, Univ. British Columbia, Vancouver. 230pp.

———. 1981. Variation in coyote social organization: the influence of prey size. Can. J. Zool. 59:639-652.

Bowns, J. E. 1976. Field criteria for predator damage assessment. Utah Sci. 37:26-30.

———, J. W. Davenport, J. P. Workman, D. B. Nielsen, and D. D. Dwyer. 1973. Determination of cause and magnitude of sheep losses in southwestern Utah. Utah Sci. 34:35-37, 52.

Bruscino, M. 1989. Evaluating mountain lion depredation of domestic sheep. Proc. Great Plains Damage Control Workshop 9:32-33.

Burns, R. J. 1980. Evaluation of conditioned predation aversion for controlling coyote predation. J. Wildl. Manage. 44:938-942.

———. 1983*a*. Microencapsulated lithium chloride bait aversion did not stop coyote predation on sheep. J. Wildl. Manage. 47:1010-1017.

———. 1983*b*. Coyote predation aversion with lithium chloride: management implications and comments. Wildl. Soc. Bull. 11:128-133.

———, and G. E. Connolly. 1980. Lithium chloride bait aversion did not influence prey killing by coyotes. Proc. Vertebr. Pest Conf. 9:200-204.

———, ———, and R. E. Griffiths, Jr. 1984. Repellent or aversive chemicals in sheep neck collars did not deter coyote attacks. Proc. Vertebr. Pest Conf. 11:146-153.

———, ———, and P. J. Savarie. 1988. Large livestock protection collars effective against coyotes. Proc. Vertebr. Pest Conf. 13:215-219.

———, H. T. Tietjen, and G. E. Connolly. 1991. Secondary hazard of livestock protection collars to skunks and eagles. J. Wildl. Manage. 55:701-704.

Buys, C. J. 1975. Predator control and ranchers' attitudes. Environ. Behav. 7:81-98.

Camenzind, F. J. 1978. Behavioral ecology of coyotes on the National Elk Refuge, Jackson, Wyoming. Pages 267-294 *in* M. Bekoff, ed. Coyotes: biology, behavior and management. Academic Press, New York, N.Y.

Connolly, G. E. 1978. Predators and predator control in big game management. Pages 369-394 *in* J. L. Schmidt and D. L. Gilbert, eds. Big game of North America. Stackpole Books, Harrisburg, Pa.

———. 1980. Use of compound 1080 in livestock neck collars to kill depredating coyotes. U.S. Dep. Inter., Fish and Wildl. Serv., Denver, Colo. 125pp.

———. 1982. U.S. Fish and Wildlife Service coyote control research. Proc. Great Plains Wildl. Damage Control Workshop 5:132-139.

———. 1988. M-44 sodium cyanide ejectors in the animal damage control program, 1976-1986. Proc. Vertebr. Pest Conf. 13:220-225.

———, and R. J. Burns. 1990. Efficacy of compound 1080 livestock protection collars for killing coyotes that attack sheep. Proc. Vertebr. Pest Conf. 14:269-276.

———, and W. M. Longhurst. 1975. The effects of control on coyote populations: a simulation model. Coop. Extension Serv. Bull. 1872. Univ. California, Berkeley. 37pp.

———, and B. W. O'Gara. 1987. Aerial hunting takes sheep-killing coyotes in western Montana. Proc. Great Plains Wildl. Damage Control Workshop 8:184-188.

———, and G. D. Simmons. 1984. Performance of sodium cyanide ejectors. Proc. Vertebr. Pest Conf. 11:114-121.

———, R. M. Timm, W. E. Howard, and W. M. Longhurst. 1976. Sheep killing behavior of captive coyotes. J. Wildl. Manage. 40:400-407.

Conover, M. R., J. G. Francik, and D. E. Miller. 1977. An experimental evaluation of aversive conditioning for controlling coyote predation. J. Wildl. Manage. 41:775-779.

Coolahan, C. 1990. The north coast animal damage control program. Pages 16-17 *in* G. A. Giusti, R. M. Timm, and R. H. Schmidt, eds. Predator management in north coastal California. Univ. California Hopland Field Stn. Publ. 101.

Coppinger, R., L. Coppinger, G. Langeloh, and L. Gettler. 1988. A decade of use of livestock guarding dogs. Proc. Vertebr. Pest Conf. 13:209-214.

———, R., J. Lorenz, and L. Coppinger. 1983. Introducing livestock guarding dogs to sheep and goat producers. Proc. Eastern Wildl. Damage Control Conf. 1:129-132.

———, ———, and ———. 1987. New uses of livestock guarding dogs to reduce agriculture/wildlife conflicts. Proc. Eastern Wildl. Damage Control Conf. 3:253-259.

Davenport, J. W., J. E. Bowns, and J. P. Workman. 1973. Assessment of sheep losses to coyotes: a problem to Utah sheepmen—a concern to Utah researchers. Utah State Univ., Agric. Exp. Stn. Res. Rep. 7. 17pp.

Davenport, L. B., Jr. 1953. Agricultural depredation by the black bear in Virginia. J. Wildl. Manage. 17:331-340.

deCalesta, D. S. 1983. Building an electric antipredator fence. Pacific Northwest Coop. Ext. Bull. 225. 11pp.

———, and M. G. Cropsey. 1978. Field test of a coyote-proof fence. Wildl. Soc. Bull. 6:256-259.

Denny, R. N. 1974. The impact of uncontrolled dogs on wildlife and livestock. Trans. North Am. Wildl. and Nat. Resour. Conf. 39:257-291.

Dorrance, M. J. 1980. Use of toxicants for coyote control by livestock producers in Alberta. Proc. Vertebr. Pest Conf. 9:209-214.

———. 1982. Predation losses of cattle in Alberta. J. Range Manage. 35:690-692.

———. 1983. A philosophy of problem wildlife management. Wildl. Soc. Bull. 11:319-324.

———, and J. Bourne. 1980. An evaluation of anti-coyote electric fencing. J. Range Manage. 33:385-387.

———, and L. D. Roy. 1976. Predation losses of domestic sheep in Alberta. J. Range Manage. 29:457-460.

———, and ———. 1978. Aversive conditioning tests of black bears in beeyards failed. Proc. Vertebr. Pest Conf. 8:251-253.

Ellins, S. R., and S. M. Catalano. 1980. Field application of the conditioned taste aversion paradigm to the control of coyote predation on sheep and turkeys. Behav. and Neural Biol. 29:532-536.

———, ———, and S. A. Schechinger. 1977. Conditioned taste aversion: a field application to coyote predation on sheep. Behav. Biol. 20:91-95.

Fall, M. W. 1984. Control methods for the future. Pages 27-32 *in* College Agric. Sci., ed. Proc. 1984 agricultural conference days, Oregon State Univ., Corvallis.

Fox, M. W. 1969. Ontogeny of prey-killing behavior in Canidae. Behaviour 35:259-272.

Franklin, W. L., and K. J. Powell. 1993. Guard llamas. Iowa State Univ. Ext. Bull. 12pp.

Fritts, S. H. 1982. Wolf depredation on livestock in Minnesota. U.S. Dep. Inter., Fish and Wildl. Serv. Res. Publ. 145. 11pp.

———, W. J. Paul, and L. D. Mech. 1984. Movements of translocated wolves in Minnesota. J. Wildl. Manage. 48:709-721.

———, ———, and ———. 1985. Can relocated wolves survive? Wildl. Soc. Bull. 13:459-463.

Gates, N. L., J. E. Rich, D. D. Godtel, and C. V. Hulet. 1978. Development and evaluation of anticoyote electric fencing. J. Range Manage. 31:151-153.

Gee, C. K. 1978. Cattle and calf losses in feeder cattle production. U.S. Dep. Agric., Econ. Res. Serv., Agric. Econ. Rep. 345. 15pp.

———, and R. S. Magleby. 1976. Characteristics of sheep production in the western United States. U.S. Dep. Agric., Econ. Res. Serv., Agric. Econ. Rep. 345. 47pp.

———, ———, W. R. Bailey, R. L. Gum, and L. M. Arthur. 1977*a*. Sheep and lamb losses to predators and other causes in the western United States. U.S. Dep. Agric., Econ. Res. Serv., Agric. Econ. Rep. 369. 41pp.

———, ———, D. B. Nielsen, and D. M. Stevens. 1977*b*. Factors in the decline of the western sheep industry. U.S. Dep. Agric., Econ. Res. Serv., Agric. Econ. Rep. 377. 31pp.

Gier, H. T. 1968. Coyotes in Kansas. Kansas State Coll. Agric. Exp. Stn. Bull. 393. 118pp.

Gipson, P. S. 1975. Efficiency of trapping in capturing offending coyotes. J. Wildl. Manage. 39:45-47.

———, and J. A. Sealander. 1972. Home range and activity of the coyote (*Canis latrans frustror*) in Arkansas. Proc. Annu. Conf. Southeast. Assoc. Game and Fish Comm. 26:82-95.

Gluesing, E. A., D. F. Balph, and F. F. Knowlton. 1980. Behavioral patterns of domestic sheep and their relationship to coyote predation. Appl. Anim. Ethol. 6:315-330.

Green, J. S. 1989. Donkeys for predation control. Proc. Eastern Wildl. Damage Control. Conf. 4:83-86.

———. 1990. Reducing predation with guarding animals. Pages 62-68 *in* G. A. Giusti, R. M. Timm, and R. H. Schmidt, eds. Predator management in north coastal California. Univ. California Hopland Field Stn. Publ. 101.

———, and R. A. Woodruff. 1983. The use of three breeds of dog to protect rangeland sheep from predators. Appl. Anim. Ethol. 11:141-161.

———, and ———. 1988. Breed comparisons and characteristics of use of livestock guarding dogs. J. Range Manage. 41:249-251.

———, and ———. 1989. Livestock-guarding dogs reduce depredation by bears. Pages 49-54 *in* M. Bromley, ed. Bear-people conflicts: proceedings of a symposium on manage strategies. Northwest Terr. Dep. Renew. Resour., Canada.

———, and ———. 1990. ADC guarding dog program update: a focus on managing dogs. Proc. Vertebr. Pest Conf. 14:233-236.

———, ———, and T. T. Tueller. 1984. Livestock-guarding dogs for predator control: costs, benefits, and practicality. Wildl. Soc. Bull. 12:44-50.

Griffel, D. E., and J. V. Basile. 1981. Identifying sheep killed by bears. U. S. Dep. Agric., For. Serv. Res. Note INT-313. 3pp.

Griffiths, R. E., Jr., G. E. Connolly, R. J. Burns, and R. T. Sterner. 1978. Coyotes, sheep, and lithium chloride. Proc. Vertebr. Pest Conf. 8:190-196.

Gustavson, C. R. 1982. An evaluation of taste aversion control of wolf (*Canis lupus*) predation in Northern Minnesota. Appl. Anim. Ethol. 9:63-71.

———, J. Garcia, W. G. Hankins, and K. W. Rusiniak. 1974. Coyote predation control by aversive conditioning. Science 184:581-583.

———, J. R. Jowsey, and D. Milligan. 1982. A 3-year evaluation of taste aversion coyote control in Saskatchewan. J. Range Manage. 35:57-59.

———, D. J. Kelly, M. Sweeney, and J. Garcia. 1976. Prey-lithium aversions. I: Coyotes and wolves. Behav. Biol. 17:61-72.

Guthery, F. S., and S. L. Beasom. 1978. Effectiveness and selectivity of neck snares in predator control. J. Wildl. Manage. 42:457-459.

———, W. P. Meinzer, Jr., S. L. Beasom, and M. Caroline. 1984. Evaluation of placed baits for reducing coyote damage in Texas. J. Wildl. Manage. 48:621-626.

Hein, E. W. 1992. Evaluations of coyote attractants and a density estimate on the Rocky Mountain Arsenal. M.S. Thesis, Colorado State Univ., Fort Collins. 57pp.

Henderson, F. R. 1972. Controlling coyote damage. Coop. Ext. Serv. Bull. C-397, Kansas State Univ., Manhattan. 24pp.

———. 1986. How to call coyotes. Coop. Ext. Serv., Kansas State Univ., Manhattan. 8pp.

Henne, D. R. 1975. Domestic sheep mortality on a western Montana ranch. Pages 133-146 *in* R. L. Phillips and C. Jonkel, eds. Proc. 1975 predator symposium. Montana For. Conserv. Exp. Stn., Univ. Montana, Missoula.

Hibler, S. J. 1977. Coyote movement patterns with emphasis on home range characteristics. M.S. Thesis, Utah State Univ., Logan. 125pp.

Horn, S. W. 1983. An evaluation of predatory suppression in coyotes using lithium chloride-induced illness. J. Wildl. Manage. 47:999-1009.
Horstman, L. P., and J. R. Gunson. 1982. Black bear predation on livestock in Alberta. Wildl. Soc. Bull. 10:34-39.
Howard, V. W., Jr. 1991. Effects of electric predator-excluding fences on movements of mule deer in pinyon/juniper woodlands. Wildl. Soc. Bull. 19:331-334.
Hulet, C. V., D. M. Anderson, J. N. Smith, and W. L. Shupe. 1987. Bonding sheep to cattle as an effective technique for predation control. Appl. Anim. Behav. Sci. 19:19-25.
———, ———, ———, ———, C. A. Taylor, Jr., and L. W. Murray. 1989. Bonding of goats to sheep and cattle for protection from predators. Appl. Anim. Behav. Sci. 22:261-267.
Johnson, K. G., and M. R. Pelton. 1980. Prebaiting and snaring techniques for black bears. Wildl. Soc. Bull. 8:46-54.
Johnson, S. J., and D. E. Griffel. 1982. Sheep losses on grizzly bear range. J. Wildl. Manage. 46:786-790.
Jones, E. J. 1987. Coyote damage in the southeastern United States. Proc. Eastern Wildl. Damage Control Conf. 3:320.
Jones, J. M., and A. Woolf. 1983. Relationship between husbandry practices and coyote use of swine in west central Illinois. Wildl. Soc. Bull. 11:133-135.
Jorgensen, C. J. 1983. Bear-sheep interactions, Targhee National Forest. Int. Conf. Bear Res. and Manage. 5:191-200.
Keith, L. B. 1974. Some features of population dynamics in mammals. Trans. Int. Congr. Game Biol. 11:17-58.
———. 1983. Population dynamics of wolves. Pages 66-77 *in* L. N. Carbyn, ed. Wolves in Canada and Alaska: their status, biology, and management. Can. Wildl. Serv. Rep. Ser. 45.
Kellert, S. R. 1976. Perceptions of animals in American society. Trans. North Am. Wildl. and Nat. Resour. Conf. 41:533-546.
———. 1985. Public perceptions of predators, particularly the wolf and coyote. Biol. Conserv. 31:167-189.
———. 1986. The public and the timber wolf in Minnesota. Trans. North Amer. Wildl. Nat. Resour. Conf. 51:193-200.
Kilgore, D. C., Jr. 1969. An ecological study of the swift fox (*Vulpes velox*) in the Oklahoma panhandle. Am. Midl. Nat. 81:512-534.
Klebenow, D. A., and K. McAdoo. 1976. Predation on domestic sheep in northeastern Nevada. J. Range Manage. 29:96-100.
Knowlton, F. F. 1972. Preliminary interpretations of coyote population mechanics with some management implications. J. Wildl. Manage. 36:369-382.
Kolenosky, G. B., and S. M. Strathearn. 1987. Black bear. Pages 443-454 *in* M. Novak, J. A. Baker, M. E. Obbard, and B. Malloch, eds. Wild furbearer management and conservation in North America. Ontario Minist. Nat. Resour. and Ontario Trappers Assoc., Toronto.
Kreeger, T. J., P. J. White, U. S. Seal, and J. R. Tester. 1990. Pathological responses of red foxes to foothold traps. J. Wildl. Manage. 54:147-160.
Larson, S., and T. P. Salmon. 1988. Predators and sheep management practices in Sonoma County, California. Proc. Vertebr. Pest Conf. 13:230-234.
Lehner, P. N. 1976. Coyote behavior: implications for management. Wildl. Soc. Bull. 4:120-126.
———. 1987. Repellants and conditioned avoidance. Pages 56-61 *in* J. S. Green, ed. Protecting livestock from coyotes. U.S. Dep. Agric., Agric. Res. Stn., Dubois, Id.
———, R. Krumm, and A. T. Cringan. 1976. Tests for olfactory repellents for coyotes and dogs. J. Wildl. Manage. 40:145-150.
Lemm, W. C. 1973. Coyote denning as a method of damage control. Proc. Great Plains Wildl. Damage Control Workshop 1:39-44.
Lindzey, F. G. 1987. Mountain lion. Pages 656-668 *in* M. Novak, J. A. Baker, M. E. Obbard, and B. Malloch, eds. Wild furbearer management and conservation in North America. Ontario Minist. Nat. Resour. and Ontario Trappers Assoc., Toronto.
———, and C. Wilbert. 1989. Estimating domestic sheep losses to mountain lions. Proc. Great Plains Wildl. Damage Control Workshop 9:27-31.
Linhart, S. B. 1964. Acceptance by wild foxes of certain baits for administering antifertility agents. N.Y. Fish and Game J. 11:69-77.
———. 1981. Field evaluation of techniques for reducing coyote predation on livestock. Pages 1826-1838 *in* J. A. Chapman and D. Pursley, eds. Proc. worldwide furbearer conference, Frostburg, Md.
———. 1983. Managing coyote damage problems with non-lethal techniques: recent advances in research. Proc. Eastern Wildl. Damage Control Conf. 1:105-118.
———. 1984. Strobe-light and siren devices for protecting fenced-pasture and range sheep from coyote predation. Proc. Vertebr. Pest Conf. 11:154-156.
———, F. S. Blom, G. J. Dasch, R. M. Engeman, and G. H. Olsen. 1988. Field evaluation of padded jaw coyote traps: effectiveness and foot injury. Proc. Vertebr. Pest Conf. 13:226-229.
———, H. H. Brusman, and D. S. Balser. 1968. Field evaluation of an antifertility agent, stilbestrol, for inhibiting coyote reproduction. Trans. North Am. Wildl. and Nat. Resour. Conf. 33:316-327.
———, and G. J. Dasch. 1992. Improved performance of padded jaw traps for capturing coyotes. Wildl. Soc. Bull. 20:63-66.
———, G. J. Dasch, C. B. Male, and R. M. Engeman. 1986. Efficiency of unpadded and padded steel foothold traps for capturing coyotes. Wildl. Soc. Bull. 14:212-218.
———, ———, and F. J. Turkowski. 1981. The steel leg-hold trap: techniques for reducing foot injury and increasing selectivity. Pages 1560-1578 *in* J. A. Chapman and D. Pursley, eds. Proc. third worldwide furbearer conference, Frostburg, Md.
———, J. D. Roberts, and G. J. Dasch. 1982. Electric fencing reduces coyote predation on pastured sheep. J. Range Manage. 35:276-281.
———, ———, S. A. Shumake, and R. Johnson. 1976. Avoidance of prey by captive coyotes punished with electric shock. Proc. Vertebr. Pest Conf. 7:302-306.
———, and W. B. Robinson. 1972. Some relative carnivore densities in areas under sustained coyote control. J. Mammal. 53:880-884.
———, R. T. Sterner, T. C. Carrigan, and D. R. Henne. 1979. Komondor guard dogs reduce sheep losses to coyotes: a preliminary evaluation. J. Range Manage. 32:238-241.
———, ———, G. J. Dasch, and J. W. Theade. 1984. Efficacy of light and sound stimuli for reducing coyote predation upon pastured sheep. Protection Ecol. 6:75-84.
Linscombe, R. G., and V. L. Wright. 1988. Efficiency of padded foothold traps for capturing terrestrial furbearers. Wildl. Soc. Bull. 16:307-309.
Littauer, G. A. 1983. Rancher use of compound 1080 toxic collars in New Mexico. Proc. Great Plains Wildl. Damage Control Workshop 6:92-104.
Lynch, G. W., and R. D. Nass. 1981. Sodium monofluoroacetate (1080): relation of its use to predation on livestock in western national forests, 1960-78. J. Range Manage. 34:421-423.
Major, J. T., and J. A. Sherburne. 1987. Interspecific relationships of coyotes, bobcats, and red foxes in western Maine. J. Wildl. Manage. 51:606-616.
Markham, D. 1990. Llamas the ultimate sheep guard. Llamas: Sept. 101-105.
Marsh, R. E., W. E. Howard, S. M. McKenna, B. Butler, and D. A. Barnum. 1982. A new system for delivery of predacides or other active ingredients for coyote management. Proc. Vertebr. Pest Conf. 10:229-233.
McAninch, J. B., and M. J. Fargione. 1987. Characteristics of predation and losses in the New York sheep industry. Proc. Eastern Wildl. Damage Control Workshop 3:260-268.
McArthur, K. L. 1981. Factors contributing to effectiveness of black bear transplants. J. Wildl. Manage. 45:102-110.
McGrew, J. C., and W. F. Andelt. 1988. Livestock guardian dogs: a new method of reducing livestock losses. Coop. Ext. Serv., Colorado State Univ., Fort Collins. 4pp.
———, and C. S. Blakesley. 1982. How komondor dogs reduce sheep losses to coyotes. J. Range Manage. 35:693-696.
McKay, D. I. 1975. Coyote-livestock interaction in Manitoba. Manitoba Dep. Mines, Resour. and Environ. Manage., Res. Branch, Rep. 75-11. 39pp.
McNaught, D. A. 1987. Wolves in Yellowstone? Park visitors respond. Wildl. Soc. Bull. 15:518-521.
Mech, L. D. 1984. Predators and predation. Pages 189-200 *in* L. K. Hall, ed. White-tailed deer ecology and management. Wildl. Manage. Inst. Stackpole Books, Harrisburg, Pa.
Miner, J. E., and R. T. Quiroz. 1974. Evaluation of the helicopter as a major control method in reducing coyote predation on lambing ranges in Elko County, Nevada. U.S. Dep. Inter., Fish and Wildl. Serv., Reno, Nev. 9pp.
Murie, A. 1948. Cattle on grizzly bear range. J. Wildl. Manage. 12:57-72.

Nass, R. D. 1977. Mortality associated with sheep operations in Idaho. J. Range Manage. 30:253-258.

———, G. Lynch, and J. Theade. 1984. Circumstances associated with predation on sheep and goats. J. Range Manage. 37:423-426.

National Agricultural Statistics Service. 1991. Sheep and goat predator loss. U.S. Dep. Agric., Agric. Stat. Board, Washington, D.C. 12pp.

Nesse, G. E., W. M. Longhurst, and W. E. Howard, Jr. 1976. Predation and the sheep industry in California, 1972-74. Coop. Extension Serv. Bull. 1878. Univ. California, Davis. 63pp.

Newsome, A. 1990. The control of vertebrate pests by vertebrate predators. Trends in Ecol. Evol. 5:187-191.

Nielsen, D. B., and D. Curle. 1970. Predator costs to Utah's range sheep industry. Natl. Wool Grower 60:14-16,22.

Novak, M. 1987. Traps and trap research. Pages 941-969 *in* M. Novak, J. A. Baker, M. E. Obbard, and B. Malloch, eds. Wild Furbearer Management and Conservation in North America. Ontario Minist. Nat. Resour. and Ontario Trappers Assoc., Toronto.

Nunley, G. E. 1977. The effects of coyote control operations on non-target species in New Mexico. Proc. Great Plains Wildl. Damage Control Workshop 3:88-110.

———. 1978. Present and historical bobcat population trends in New Mexico and the West. Proc. Vertebr. Pest Control Conf. 8:177-184.

O'Gara, B. W., K. C. Brawley, J. R. Munoz, and D. R. Henne. 1983. Predation on domestic sheep on a western Montana ranch. Wildl. Soc. Bull. 11:253-264.

———, and W. Rightmire. 1987. Wolf, golden eagle, and coyote problems in Montana. Proc. Eastern Wildl. Damage Control Workshop 3:275-283.

Olsen, A. 1975. Conditioned avoidance of prey in coyotes. M.S. Thesis, Colorado State Univ., Fort Collins. 98pp.

Olsen, G. H., S. B. Linhart, R. A. Holmes, G. J. Dasch, and C. B. Male. 1986. Injuries to coyotes caught in padded and unpadded steel foothold traps. Wildl. Soc. Bull. 14:219-223.

———, R. G. Linscombe, V. L. Wright, and R. A. Holmes. 1988. Reducing injuries to terrestrial furbearers by using padded foothold traps. Wildl. Soc. Bull. 16:303-307.

Onderka, D. K., D. L. Skinner, and A. W. Todd. 1990. Injuries to coyotes and other species caused by four models of footholding devices. Wildl. Soc. Bull. 18:175-182.

Paul, W. J. 1989. Wolf predation on livestock in Minnesota. Annual update of statistics-1988. Proc. Eastern Wildl. Damage Control Workshop 4:101.

Pearson, E. W. 1978. A 1974 coyote harvest estimate for 17 western states. Wildl. Soc. Bull. 6:25-32.

———. 1986. A literature review of livestock losses to predators in western U.S. U.S. Dep. Inter., Fish and Wildl. Serv. Final Rep., Denver, Colo. 20pp.

Peek, J. M., D. E. Brown, S. R. Kellert, L. D. Mech, J. H. Shaw, and V. Van-Ballenberghe. 1991. Restoration of wolves in North America. Wildl. Soc. Tech. Rev. 91-1. 21pp.

Pfeifer, W. K., and M. W. Goos. 1982. Guard dogs and gas exploders as coyote depredation control tools in North Dakota. Proc. Vertebr. Pest Conf. 10:55-61.

Phillips, R. L., F. S. Blom, and R. E. Johnson. 1990. An evaluation of breakaway snares for use in coyote control. Proc. Vertebr. Pest Conf. 14:255-259.

———, and M. W. Fall. 1990. Control methods and their future application in predation management. Pages 81-88 *in* G. A. Giusti, R. M. Timm, and R. H. Schmidt, eds. Predator management in north coastal California. Univ. California, Hopland Field Stn. Publ. 101.

Robel, R. J., A. D. Dayton, F. R. Henderson, R. L. Meduna, and C. W. Spaeth. 1981. Relationships between husbandry methods and sheep losses to canine predators. J. Wildl. Manage. 45:894-911.

Robinson, W. B. 1943. The "humane coyote-getter" vs. the steel trap in control of predatory animals. J. Wildl. Manage. 7:179-189.

———. 1948. Thallium and compound 1080 impregnated stations in coyote control. J. Wildl. Manage. 12:279-295.

———. 1953*a*. Coyote control with compound 1080 stations in national forests. J. For. 51:880-885.

———. 1953*b*. Population trends of predators and fur animals in 1080 station areas. J. Mammal. 34:220-227.

———. 1961. Population changes of carnivores in some coyote-control areas. J. Mammal. 42:510-515.

Roughton, R. D., coord. 1977. Indices of predator abundance in western United States, 1976. U.S. Dep. Inter., Fish and Wildl. Serv. Rep., Denver, Colo. 100pp.

Rowley, G. J., and D. Rowley. 1987. Decoying coyotes with dogs. Proc. Great Plains Wildl. Damage Control Workshop 8:179-181.

Roy, L. D., and M. J. Dorrance. 1985. Coyote movements, habitat use, and vulnerability in central Alberta. J. Wildl. Manage. 49:307-313.

Sargeant, A. B., S. H. Allen, and J. O. Hastings. 1987. Spatial relations between sympatric coyotes and red foxes in North Dakota. J. Wildl. Manage. 51:285-293.

Schaefer, J. M., R. D. Andrews, and J. J. Dinsmore. 1981. An assessment of coyote and dog predation on sheep in southern Iowa. J. Wildl. Manage. 45:883-893.

Scrivner, J. H. 1983. The 1080 toxic collar: economics of field use in Texas. Proc. Eastern Wildl. Damage Control Conf. 1:201-204.

———, D. A. Wade, G. E. Connolly, and L. C. Howard, Jr. 1985. The effects of predation on an Angora goat ranch. Natl. Wool Grower 75:10-13.

Shaw, H. G. 1977. Impact of mountain lion on mule deer and cattle in northwestern Arizona. Pages 17-32 *in* R. Phillips and C. Jonkel, eds. Proc. 1975 predator symposium. Montana For. and Cons. Exp. Stn., Univ. Montana, Missoula.

———. 1983. Mountain lion field guide. Arizona Game and Fish Dep. Spec. Rep. 9. 38pp.

Shelton, M. 1984. The use of conventional and electric fencing to reduce coyote predation on sheep and goats. Texas Agric. Exp. Stn. Bull. MP-1556, 12pp.

———, and N. L. Gates. 1987. Antipredator fencing. Pages 30-37 *in* J. S. Green, ed. Protecting livestock from coyotes. U.S. Dep. Agric., Agric. Res. Serv., Dubois, Id.

Skinner, D. L., and A. W. Todd. 1990. Performance of footholding devices for coyote capture. Wildl. Soc. Bull. 18:166-190.

Slate, D. 1987. Coyotes in the eastern US: status and implications—damage problems and economic losses from coyotes in the northeastern U.S. Proc. Eastern Wildl. Damage Control Conf. 3:325-326.

Stellflug, J. N., N. L. Gates, and R. G. Sasser. 1978. Reproductive inhibitors for coyote population control: developments and current status. Proc. Vertebr. Pest Conf. 8:185-189.

———, C. W. Leathers, and J. S. Green. 1984. Antifertility effect of busulfan and DL-6-(N-2-pipecolinomethyl)-5-hydroxy-indane maleate (PMHI) in coyotes (*Canis latrans*). Theriogenology 22:533-543.

Stickley, A. R., Jr. 1961. A black bear tagging study in Virginia. Proc. Annu. Conf. Southeast Assoc. Game and Fish Comm. 15:43-54. (*Cited from*: Kolenosky, G. B., and S. M. Strathern. 1987. Black bear. Pages 443-454 *in* M. Novak, J. A. Baker, M. E. Obbard, and B. Malloch, eds. Wild furbearer management and conservation in North America. Ontario Minist. Nat. Resour. and Ontario Trappers Assoc., Toronto.)

Stolzenburg, H. W., and V. W. Howard, Jr. 1989. Activation of liquid bait devices by coyotes in Southern New Mexico. Wildl. Soc. Bull. 17:306-312.

Stuby, R. G., E. H. Carpenter, and L. M. Arthur. 1979. Public attitudes toward coyote control. U.S. Dep. Agric., Econ., Statistics, Coop. Serv., Agric. Econ. Rep. ESCS-54. 11pp.

Suminski, H. R. 1982. Mountain lion predation on domestic livestock in Nevada. Proc. Vertebr. Pest Conf. 10:62-66.

Texas Agricultural Experiment Station. 1983. Efficacy of the 1080 toxic collar as a predator damage control method. Texas Agric. Exp. Stn., College Station. 134pp.

Thompson, B. C. 1979. Evaluation of wire fences for coyote control. J. Range Manage. 32:457-461.

Tigner, J. R., and G. E. Larson. 1977. Sheep losses on selected ranches in southern Wyoming. J. Range Manage. 30:244-252.

Till, J. A., and F. F. Knowlton. 1983. Efficacy of denning in alleviating coyote depredations upon domestic sheep. J. Wildl. Manage. 47:1018-1025.

Todd, A. W., and L. B. Keith. 1976. Responses of coyotes to winter reductions in agricultural carrion. Alberta Wildl. Tech. Bull. 5. 32pp.

Tompa, F. S. 1983. Problem wolf management in British Columbia: conflict and program evaluation. Pages 112-120 *in* L. N. Carbyn, ed. Wolves in Canada and Alaska: their status, biology, and manage. Can. Wildl. Serv. Rep. Ser. 45.

Tucker, P., and D. H. Pletscher. 1989. Attitude of hunters and residents toward wolves in northwestern Montana. Wildl. Soc. Bull. 17:509-514.

Tullar, B. F. Jr. 1984. Evaluation of a padded leg-hold trap for capturing foxes and raccoon. N.Y. Fish and Game J. 31:97-103.
Turkowski, F. J., A. R. Armistead, and S. B. Linhart. 1984. Selectivity and effectiveness of pan tension devices for coyote foothold traps. J. Wildl. Manage. 48:700-708.
United States Department of Agriculture. 1990. Animal damage control program: draft environmental impact statement. U.S. Dep. Agric, Animal and Plant Health Inspection Serv., Washington, D.C.
United States Fish and Wildlife Service. 1978. Predator damage in the West: a study of coyote management alternatives. U.S. Dep. Inter., Fish and Wildl. Serv., Washington, D.C. 168pp.
Voigt, D. R., and B. D. Earle. 1983. Avoidance of coyotes by red fox families. J. Wildl. Manage. 47:852-857.
Wade, D. A. 1973. An assessment of the coyote problem in the Great Plains states. Proc. Great Plains Wildl. Damage Control Workshop 1:5-10.
———. 1976. The use of aircraft in predator control. Proc. Vertebr. Pest Conf. 7:154-160.
———. 1982. The use of fences for predator damage control. Proc. Vertebr. Pest Conf. 10:24-33.
———, and J. E. Bowns. 1984. Procedures for evaluating predation on livestock and wildlife. Texas Agric. Exp. Stn. Bull. B-1429. 42pp.
Wagner, F. H. 1972. Coyotes and sheep: some thoughts on ecology, economics, and ethics. Forty-fourth Faculty Honor Lecture. Utah State Univ., Logan. 59pp.
———. 1975. The predator-control scene as of 1974. J. Range Manage. 28:4-10.
Walton, M. T. 1990. Rancher use of livestock protection collars in Texas. Proc. Vertebr. Pest Conf. 14:277-280.
———, and C. A. Feild. 1989. Use of donkeys to guard sheep and goats in Texas. Proc. Eastern Wildl. Damage Control Conf. 4:87-94.
Weaver, R. A., and L. Sitton. 1978. Changing status of mountain lion in California and livestock depredation problems. Proc. Vertebr. Pest Conf. 8:214-219.
Wells, M. C., and P. N. Lehner. 1978. The relative importance of the distance senses in coyote predatory behavior. Anim. Behav. 26:251-258.
Windberg, L. A., and F. F. Knowlton. 1988. Management implications of coyote spacing patterns in southern Texas. J. Wildl. Manage. 52:632-640.
Young, S. P. 1958. The bobcat of North America. Univ. Nebraska Press, Lincoln. 193pp.
———, and H. W. Dobyns. 1945. Coyote control by means of den hunting. U.S. Dep. Inter., Fish and Wildl. Serv. Circ. 7. 8pp.
———, and H. H. T. Jackson. 1951. The clever coyote. Stackpole Books, Harrisburg, Pa., and Wildl. Manage. Inst., Washington, D.C. 411pp.

Chapter 10
ELK

Michael J. Wisdom and Jack Ward Thomas

INTRODUCTION

Elk (*Cervus elaphus*) are an important herbivore on North American rangelands. Their large size, herding behavior, pioneering habits, and high mobility make them an especially conspicuous herbivore in such open areas. Moreover, the potential for elk to compete with livestock makes them an obvious source of controversy between stockgrowers and wildlife advocates.

In this chapter, we describe the history of elk on rangelands. We discuss the competitive interactions between elk and livestock, and identify methods of stocking allocation between both. We also provide concepts, prescriptions, and examples for managing rangelands for elk, particularly through the use of livestock grazing systems.

HISTORY OF ELK ON RANGELANDS

Historical Distribution

Before European settlement, elk were among the most common and widely distributed of the wild ungulates in North America (Seton 1927, Guthrie 1966, Bryant and Maser 1982). Six subspecies inhabited areas from Atlantic to Pacific coasts: eastern elk (*C. e. canadensis*) in eastern North America, Manitoban elk (*C. e. manitobensis*) in the northern lakes and central prairies, Rocky Mountain elk (*C. e. nelsoni*) in the plains and Rocky Mountains, Merriam elk (*C. e. merriami*) in the mountains and chaparrals of the Southwest, tule elk (*C. e. nannodes*) in the valleys and bottomlands of California, and Roosevelt elk (*C. e. roosevelti*) along the Pacific Coast (Bryant and Maser 1982) (Fig. 1).

Considerable controversy exists about whether elk once occupied rangelands of the Great Basin and Desert Southwest. This controversy centers on recent transplants of elk and/or recent colonization by elk into these areas (Bunnell and Hancock 1985, Hess 1985) and the potential for competition with livestock (U.S. Dep. Agric. For. Serv. 1990).

According to Guthrie (1966) and Bryant and Maser (1982), elk originally occupied most grassland biomes throughout western North America, including major areas of the Great Basin and Desert Southwest. However, lack of permanent water likely restricted elk occurrence in the driest portions of these regions (McCabe 1982, O'Neil 1985). These dry areas presumably are outside the historical range of the species (Bryant and Maser 1982) (Fig. 1); this premise agrees with Mack and Thompson (1982), who provided compelling evidence favoring hypotheses that large, wild ungulates historically were absent from or sparsely distributed across large areas of the Intermountain West.

Population Declines and Extirpations

Seton (1927) estimated numbers of North American elk at 10 million before the arrival of European settlers. McCabe (1982) also believed that population size was significant, numbering in the range of at least several million. By contrast, with the exception of the Rocky Mountain and Roosevelt subspecies in the West, elk are now distributed in small, disjunct populations across much of their former range (Bryant and Maser 1982, Witmer 1990) (Fig. 2).

Declines in elk numbers, subsequent reductions in their distribution, and ultimately the extirpation of populations and entire subspecies occurred across large expanses of North America as European settlers moved west in the seventeenth, eighteenth, and nineteenth centuries. By the early 1800s, the eastern subspecies was likely extinct (Boyd 1978, Bryant and Maser 1982). Tule elk, once abundant in valleys and bottomlands of California,

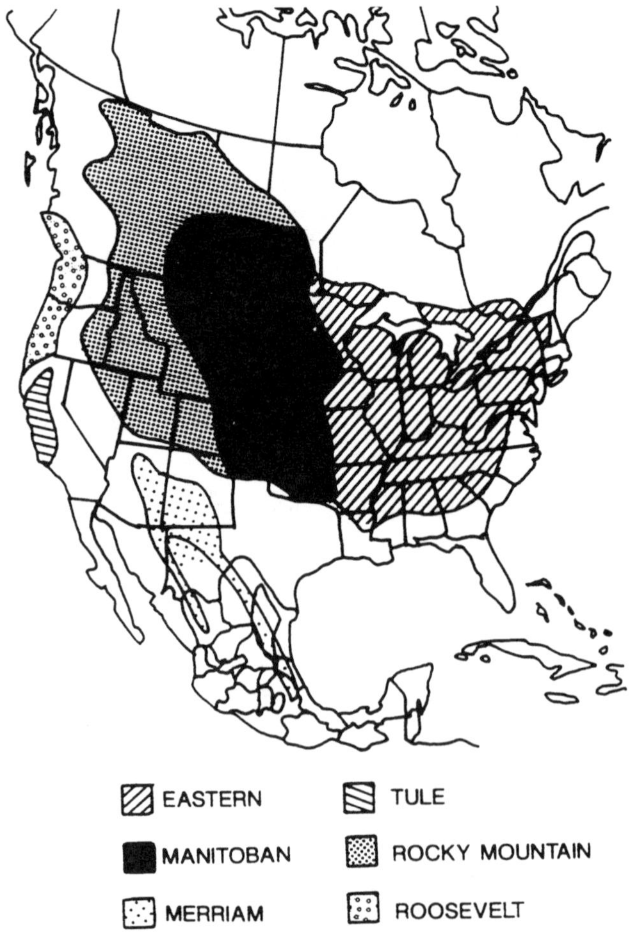

FIGURE 1. Historical distribution of elk in North America (Bryant and Maser 1982:24).

were reduced to ≤100 animals by 1875 (McCullough 1969). The Merriam subspecies was extirpated from the Southwest by 1906 (Byrant and Maser 1982). By the early 1900s, Manitoban elk of the northern prairies were close to extinction (Bryant and Maser 1982). During the same period, populations of Rocky Mountain and Roosevelt elk were reduced to scattered, isolated pockets within their former range (Murie 1951). Unregulated hunting, overgrazing by livestock, and conversion of habitat to agriculture and cities led to these broad-scale declines and extirpations (Boyd 1978, Bryant and Maser 1982, Skovlin 1982).

Current Abundance and Distribution

Jackson (1944) estimated that ≤60,000 elk remained in North America in 1910, with fragmented populations present in 7 western states. By this time, most states had enacted legislation that prohibited the hunting of elk; efforts to recover elk populations were initiated through active programs of translocation. From 1892 to 1939, >5,000 elk were live-trapped and transported from the Yellowstone area for release in 36 states and parts of western Canada (Murie 1951).

By the 1930's, these management changes were showing results. Elk numbers and distribution increased substantially, with the population estimated at 165,764 by 1937 and 236,787 by 1941 (Jackson 1944). By 1942, elk hunting was allowed in 11 states, and translocations continued throughout North America. By the 1960s the elk population in North America had grown to 440,000 (Swanson et al. 1969); it reached 500,000 by the late 1970s (Bryant and Maser 1982) and 600,000 by 1987 (Thomas 1990).

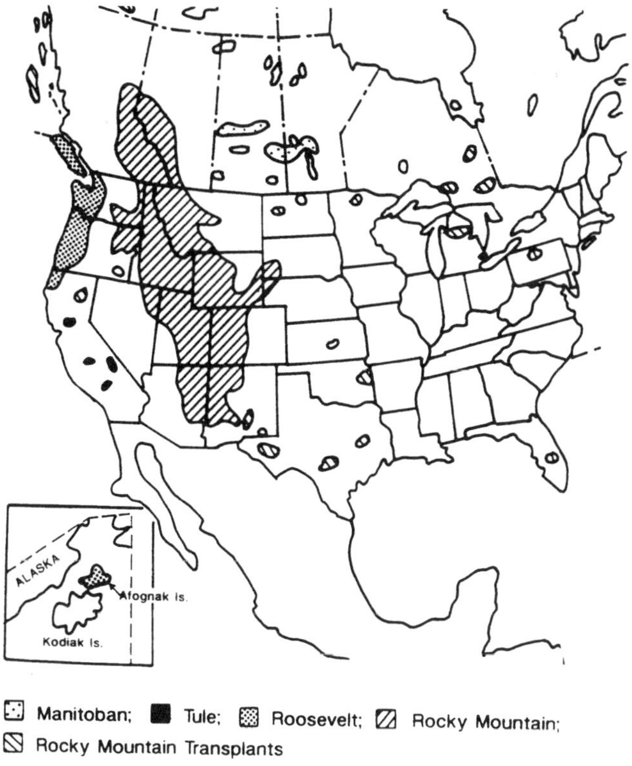

FIGURE 2. Present distribution of elk in North America (Bryant and Maser 1982:25).

Today, we estimate that ≥700,000 elk exist in North America. Our estimate is based on an annual rate of increase of 2% from 1987 to 1995, which is the approximate rate we calculated for increases in elk populations from 1977 (Bryant and Maser 1982) to 1987 (Thomas 1990). Populations are generally increasing and translocations continue. Elk are present in ≥22 states and 6 provinces (Fig. 2); additional states in the Midwest and East are implementing programs of restoration.

ELK-LAND USE ISSUES ON RANGELANDS

Elk on Public Versus Private Lands

Before European settlement, North American elk often made expansive seasonal movements in response to annual changes in weather and the availability of forage (Murie 1951, Adams 1982). Such movements are not possible today. Many seasonal rangelands used historically by elk are now privately owned and managed exclusively for agriculture, livestock, and timber production (Vavra et al. 1989, Vavra 1992). Elk use of these areas is not tolerated or tolerated minimally. In other cases, historic rangelands have been converted to cities, industrial developments, and recreation areas (Henderson and O'Herren 1992). Elk now reside in "ecologically incomplete" or "ecologically compressed" habitats across much of their historic range (Cole 1971).

Patterns of land ownership in the western United States exemplify the problem (Fig. 3). In the West, > 90% of the elk use public land during summer while the majority of winter range is privately owned (Peek et al. 1982, Thomas and Sirmon 1985). Typically, large numbers of elk are produced on publicly owned summer range, where they are desired for hunting and viewing. Animals then migrate to privately owned ranges at lower elevations each winter. On such ranges, elk often encounter severe shortages of habitat features and active programs of population control to minimize competition with livestock and damage to crops (Peek et al. 1982, Thomas 1984).

Efforts to control elk damage on privately owned ranges are intense and varied. Activities include hazing and harassment, live-trapping and transport of elk away from problem areas, fencing to exclude animals from private ranges, feeding to attract and hold elk on public winter ranges, government payments to landowners who experience significant economic damage, special hunts to reduce overall numbers, and leasing or acquisition of additional winter range to maintain winter carrying capacity (Lyon and Ward 1982).

In central and eastern North America, patterns of public and private ownership complicate elk-land use problems even more. Large expanses of public land do not exist. Elk cannot meet their seasonal requirements, let alone their year-round needs, on the fragmented public range. As a result, elk often cause damage to agricultural crops, tree seedlings, and livestock forage on nearby private lands throughout the year. Numbers of elk are thus controlled closely and their distribution is restricted to small areas in and near public land (Witmer 1989, 1990).

The shortage of year-round, ecologically complete habitats for elk and other wild ungulates poses significant management

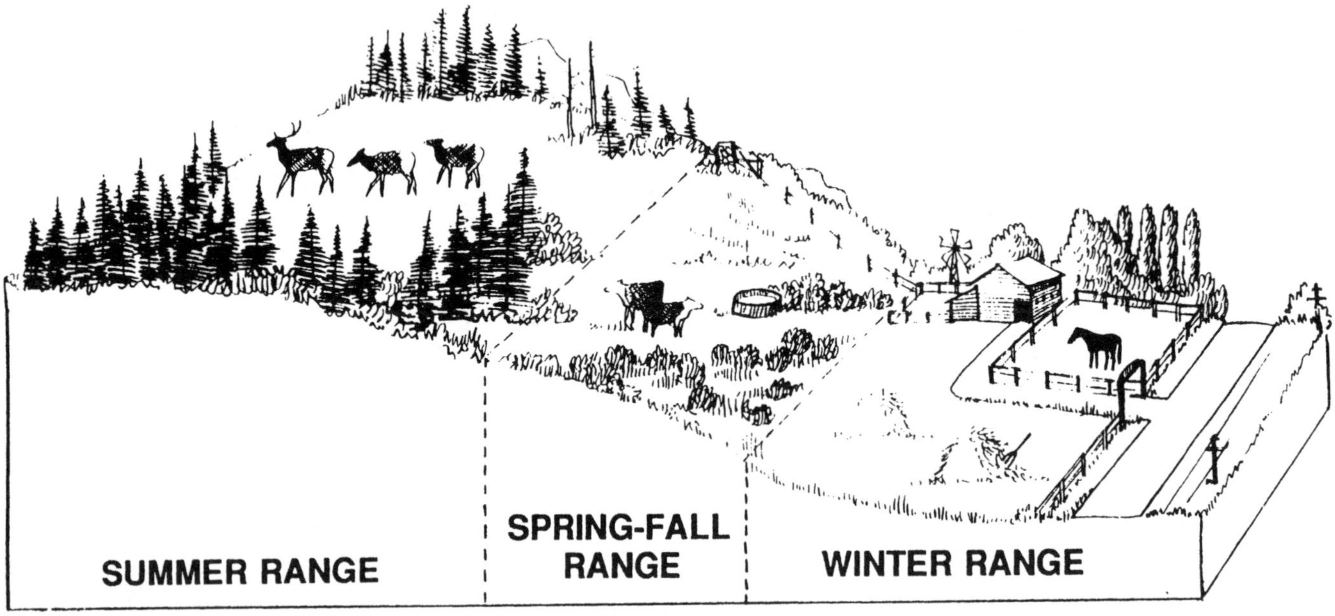

FIGURE 3. Elk produced on publicly owned summer range often face inadequate winter and spring-fall ranges due to conflicts with urban development, livestock management, and crop production on private lands.

problems on North American rangelands (Cole 1971). Many of today's management problems can be attributed to fragmented ownerships and the associated differences in land use goals across the interface of public and private lands (Vavra 1992).

Fee Hunting

Not all private landowners view elk as a problem. Many ranchers earn significant income from entry fees charged to elk hunters (White 1987). Often, such fees supplement income derived from livestock operations, timber management, or farming. In other cases, ranchers use hunting fees as a means of offsetting economic losses caused by elk damage to fences, livestock forage, tree seedlings, and crops (Lacy et al. 1993).

Fee hunting is controversial. Many ranchers manage their rangelands to attract and "hold" elk for purposes of fee hunting. Wildlife agencies and elk users typically equate such management with a commensurate loss of hunting and viewing opportunities on adjacent public lands (Jordan and Workman 1989). Regardless, fee hunting is well established and a growing industry on private rangelands (White 1987). Options exist, however, to make it compatible with management on adjacent public lands. One suggestion has been to implement similar fees on federal lands (Thomas 1984) such as the Sikes Act Program in New Mexico (Morrison 1988).

Elk Ranching

Many rangeland owners have entered the business of game ranching. Elk and red deer (*C. e. elaphus*) are especially lucrative animals for such ventures; sale of meat and antlers from domesticated forms of these animals is a multi-million dollar industry in North America and New Zealand (White 1987). Ecologically, these ungulates may also make better use of North American rangelands compared to livestock that have not coevolved with the vegetation (Peek et al. 1982).

The emergence of elk ranching in North America is fraught with controversy. Domestic elk and red deer have transmitted diseases to wild elk and livestock (Lanka and Guenzel 1991, Lanka et al. 1992). For example, bovine tuberculosis (*Mycobacterium bovis*) was recently discovered in domestic elk on game ranches in Alberta, Colorado, and Montana; infected elk in Alberta subsequently transmitted the disease to cattle (Lanka and Guenzel 1991). Eighty-one people associated with these infected elk also tested positive for tuberculosis (Roffe and Smith 1992).

Domestic elk and red deer also have escaped from game ranches, raising concerns about hybridization with and subsequent genetic effects on wild elk (Geist 1985, Roffe and Smith 1992, Williams 1992). Some states and provinces have enacted strict legislation to control, monitor, and reduce such problems; others have banned game ranching of elk and red deer altogether. Holistic programs of disease control and prevention are needed to make elk ranching compatible with management of wild elk and livestock (Roffe and Smith 1992).

Other Economic and Social Values

Elk hunting and viewing are highly valued experiences, both aesthetically and economically (Duffield and Holliman 1988, Loomis et al. 1988, Brooks et al. 1991, Bolon 1994). Economic benefits of elk are being realized by rural towns and private landowners across western North America (Boyd 1978), as consumptive and nonconsumptive users of elk are drawn from nearby urban areas (Potter 1982).

New sources of revenue from consumptive and nonconsumptive users of elk and other wildlife, combined with growing political influence of such users, are causing shifts in economic and social values that affect management of public lands (O'Toole 1988, Conner 1991, Wondolleck 1991). The growing public advocacy for elk and other wildlife appears to be directed particularly at management of livestock (Ferguson and Ferguson 1983, Gillis 1992, Rennicke 1992). Calls for the elimination of livestock grazing on rangelands (Jacobs 1991, Rifkin 1992) are increasing.

Elk-Livestock Interactions

Most western rangelands occupied by elk are also grazed by livestock. Common grazing by elk and livestock is perhaps the most significant land use issue affecting management of public rangelands in North America today (Williamson 1992). Issues of forage allocation and competition are subjects of unending debate (Vavra et al. 1989, U.S. Dep. Agric. For. Serv. 1990, Cool 1992, Vavra 1992, Wisdom 1992); few solutions seem imminent. Accordingly, relationships of elk and livestock and their management on western rangelands are the subjects of the rest of this chapter.

COMPETITION BETWEEN ELK AND LIVESTOCK

Definitions and Concepts

Elk potentially compete with domestic ungulates for food and space. Dual use of rangelands by elk and livestock, however, does not constitute competition in and of itself.

Birch (1957) postulated that competition can occur in 1 of 2 ways: (1) when individuals of the same or different species use a common resource that is in short supply; or (2) when individuals use a common resource that is readily available but the individuals "nevertheless harm one another in the process."

Miller (1967) and Nelson (1982) referred to definition 1 as "exploitive competition" because it involves direct use, or exploitation, of a limited resource between individuals. Nelson (1982) described definition 2 as "interference competition," behavior that includes any active defense of territory or other spatial resources by 1 animal against another. Definition 2 can also take a more passive form of "disturbance competition," which occurs "when the mere presence of an animal intimidates or annoys another animal into leaving the area" (Nelson 1982).

Wagner (1969) further stated that "for competition to occur between 2 species, each must share a resource that is in short supply and in using the resource, 1 or each species reduces the other's population performance to levels below that which would exist in the absence of the other species." Odum (1971) presented similar concepts.

These definitions have common elements applicable to the issue of elk competition with livestock: (1) common use of a limited resource by 2 different species, (2) disturbance or displacement of 1 species by another, and (3) reduction in population performance of 1 species by another because of competition for a limited resource or because of physical disturbance.

To demonstrate competition, either elements 1 or 2 must occur and result in element 3. A myriad of studies have examined element 1; these studies have focused primarily on food as the limited resource between elk and cattle (Nelson 1982). Likewise, numerous studies have examined element 2 regarding disturbance interactions between elk and livestock (Lyon 1985).

We know of only 1 study (Hobbs et al. In press *a*) that has evaluated element 3 for elk and livestock under field conditions. In that study, commonly used foods were the limited resource (element 1) between elk and cattle; winter grazing by elk was evaluated in terms of its subsequent effect on cattle performance during spring and early summer (element 3). We know of no studies that have documented the effect of disturbance of 1 herbivore by another (element 2) on population performance (element 3). Nelson (1982) concluded that studies of element 3 were probably impossible to conduct for large grazing animals under free-ranging conditions. Schoener (1982) concluded much the same for all studies involving competition among wild species.

Food as a Limited Resource

Many factors influence competition for food among large herbivores (Vavra et al. 1989). The most important include (1) consumption equivalence, (2) dietary overlap, (3) forage quantity and quality, (4) forage use, (5) timing of use, (6) height of foraging reach, (7) density or stocking rate of animals, and (8) spatial and temporal distribution of animals (Nelson 1982). These factors are not independent; they can interact in multiple ways to heighten or minimize the actual degree of competition (Fig. 4).

Space as a Limited Resource

Competition for space and other resources is assumed to be minimal among sympatric species of wild ungulates that have evolved together (Bell 1971; McNaughton 1985, 1991). On the other hand, competition may occur when ≥2 species have not evolved fully as separate species or when an exotic species is introduced to an environment already occupied by an ecological equivalent (Odum 1971, Davies 1979).

The latter is often the case when a domestic herbivore is introduced to a wild herbivore's range (Barnes et al. 1991); many researchers have documented an aversion by wild ungulates to the presence of livestock (Hood and Inglis 1974, Lonner and Mackie 1983, Wallace and Krausman 1987, Loft et al. 1991). We know of no studies in North America, however, that have documented livestock aversion to wild ungulates.

It is important to recognize that nonoverlapping use of rangelands by ≥2 ungulates does not necessarily mean that interference or disturbance competition is occurring. On the contrary, different species use landscapes differently.

RANGELAND RELATIONSHIPS OF ELK WITH OTHER UNGULATES

Elk and Cattle

Elk and cattle often make similar dietary choices. Both can extract high nutritive value from grasses (Janis 1976, Hanley 1982). However, both are dietary opportunists, feeding readily on a variety of grasses, forbs, and shrubs (Kufeld 1973, Heady 1975, Leege et al. 1977, Nelson and Leege 1982, Holechek et al. 1989, Jenkins and Starkey 1992, Ralphs and Pfister 1992).

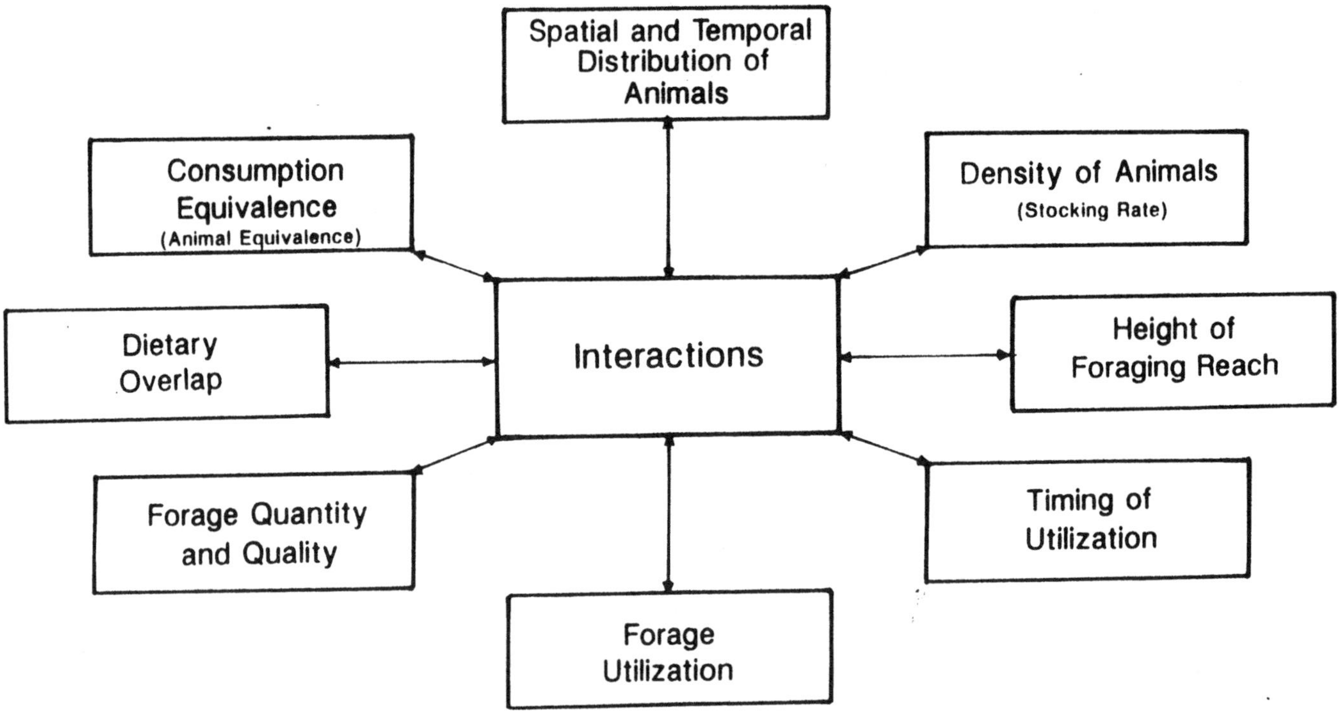

FIGURE 4. Potential interactions among variables that affect ungulate competition for food. Interactions can heighten or reduce the potential for competition in a myriad of ways that are not always predictable.

As a result, dietary overlap between elk and cattle can be high at specific times and places.

Because elk and cattle often share the same ranges and have similar diets, they are among the most likely of ungulate competitors on rangelands of North America (Nelson and Leege 1982, Nelson 1984). We have developed 12 generalizations from the studies of interactions between elk and cattle.

1. The potential for competition between elk and cattle is highest on winter and spring-fall ranges where either forage quantity or quality is limited and where both ungulates commonly share "ecologically compressed habitats" on low-elevation bottomlands or foothills. Elk typically use these ranges during late fall, winter, and early spring. Cattle usually graze these areas during late spring or early summer after elk have migrated upslope, or during fall before elk arrive.

On winter ranges, elk can reduce forage availability for cattle or vice versa. Forage quality is lowest during winter (Fig. 5) (Cook 1972, Hobbs et al. 1979), further compounding limitations in availability. If such areas are grazed heavily by cattle in the fall, insufficient forage may remain for elk during winter (Nagle and Harris 1966). On the other hand, heavy grazing by elk during winter may reduce the forage available to cattle during spring and early summer (Hobbs et al. In Press *b*) and reduce cattle production (Hobbs et al. In Press *a*). Inadequate winter range for elk, a problem common to all western rangelands, further magnifies the potential for competition.

Similar negative interactions can occur between elk and cattle on seasonal rangelands where common grazing occurs during spring and fall; dietary overlap is consistently high during these seasons as elk and cattle feed heavily on grasses (Nelson and Leege 1982). On the other hand, Sheehy (1988) and Hart et al. (1991) found that elk and cattle used habitats on seasonal rangelands that were spatially and temporally separate, thereby reducing the competitive effect of diet similarity.

2. Competition between elk and cattle is usually low on high elevation summer ranges where forage of moderate to high quality is readily available during late spring and summer and where animals have a more expansive land base from which to make optimal grazing choices. This is the period in which forage quality is highest (Fig. 5) (Cook 1972, Hobbs et al. 1979) and grasses are readily available. Furthermore, these ranges are composed largely of public lands that

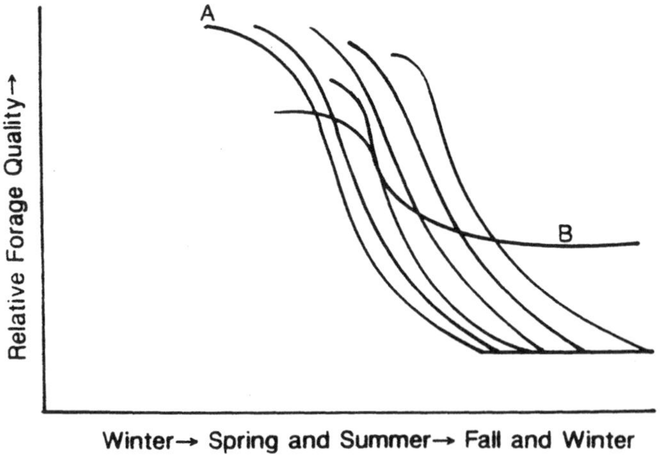

FIGURE 5. Generalized seasonal changes in forage quality of herbaceous and woody forages that are phenologically different (Nelson and Leege 1982:360). A = seasonal changes for various grasses and forbs. B = seasonal changes for a typical deciduous shrub.

encompass expansive areas in which elk can make optimal foraging choices. Elk usually arrive first to such ranges in spring and use many grasses in early stages of growth (Nelson and Leege 1982). Early defoliation of bunchgrasses can stimulate regrowth and extend forage quality into the summer and fall (Pitt 1986), with potential benefits to cattle and elk.

This response is not always predictable, however (Allayne-Chan 1986, Bryant 1993). Westenskow-Wall et al. (1994) found that spring clipping of bluebunch wheatgrass (*Agropyron spicatum*) did not increase forage quality later in the fall. Bryant (1993) found similar results, except during 1 year when sufficient rain occurred in summer and early fall to facilitate regrowth.

Another problem with spring grazing is increased mortality of bunchgrasses (Bryant 1993). If grazing is repeated and prolonged, bunchgrass mortality can be high and production reduced later in the year (Mueggler 1967, 1975; McClean and Wikeem 1985). This could negatively affect the nutritional plane of elk and cattle during late summer and fall, with carryover effects into winter and spring (Bryant 1993).

3. The potential for competition between elk and cattle increases during late summer and fall on high elevation summer ranges following the onset of prolonged seasonal drought and the subsequent decline in forage quality. This is a period in which elk often shift their diet from forbs to grasses (Nelson and Leege 1982), resulting in higher dietary overlap with cattle. It also is a time in which forage quality declines significantly (Fig. 5) (Skovlin 1967, Cook 1972, Hobbs et al. 1979, Baker and Hobbs 1982), sometimes resulting in a shortage of protein for lactating cow elk (Nelson and Leege 1982).

Continued intake of nutritious forage allows elk to put on nutrient reserves essential for proper body functions during winter and early spring, enhancing their ability to minimize depletion of fat and muscle, ensure survival of calves to the yearling age class, and carry fetal development to full-term. Late summer-early fall is a critical period for elk nutrition (Hobbs et al. 1979, Nelson and Leege 1982) and for meat production in cattle (Skovlin 1962, Holechek and Vavra 1983).

4. Elk and cattle often distribute themselves spatially in a manner that minimizes competition. Cattle typically select gentle slopes and areas near water (Holechek et al. 1989, Pinchak et al. 1991, Hart et al. 1993). In contrast to cattle, elk venture farther from water (Nagle and Harris 1966, Nelson and Burnell 1975, Hart et al. 1991), use steeper slopes (Julander and Jefferey 1964, Mackie 1970, Nelson and Burnell 1975, Hart et al. 1991, Yeo et al. 1993) and different aspects (Reynolds 1964, Sheehy 1988), eat a wider variety of foods (Nelson 1982, Nelson and Leege 1982), stay closer to forage-cover edges (Reynolds 1966, Sheehy 1988), and avoid areas near roads (Lyon 1983). Elk also prefer areas where cattle are absent (Skovlin et al. 1968, Komberec 1976, Lonner and Mackie 1983, Wallace and Krausman 1987, Frisina 1992, Yeo et al. 1993).

Not all these factors operate independently. Their combined effect, however, can minimize spatial overlap between elk and cattle; such data can be used to adjust animal unit equivalencies to reflect more realistic, complementary use of rangelands between both ungulates (Nelson 1984).

5. Elk and cattle can be distributed temporally in a manner that minimizes or heightens competition. Elk seek low elevations and foothills during winter while cattle usually graze these areas during spring, early summer, or fall (Nelson 1982, Sheehy 1988, Hart et al. 1991). Separation of ungulates by grazing season may reduce overall grazing pressure (Sheehy 1988). Forage quality may also be extended over multiple seasons from the continued regrowth of defoliated grasses (Allayne-Chan 1986, Pitt 1986). Anderson and Scherzinger (1975) and Alt et al. (1992) presumably observed such a response on elk winter ranges in Oregon and Montana, respectively. Elk use increased during winter following implementation of summer grazing by cattle.

On less productive rangelands, temporal separation of cattle and elk may not reduce competition for forage. On sagebrush grasslands of western Colorado, winter use by elk was manipulated experimentally by varying the density of animals among a replicated set of 32-ha pens; at higher densities of elk, quality of forage available to cattle during spring and early summer increased slightly; the quantity of forage available to cattle, however, was reduced (Hobbs et al. In Press *b*), with a subsequent reduction in cattle production when forage biomass fell below 45 g/m^2 (Hobbs et al. In Press *a*).

6. On ranges where spatial or temporal distributions overlap, differences in the diets of elk and cattle can sometimes minimize competition. When cattle arrive to high elevation ranges in early summer, elk often switch from a spring diet of grasses to a summer diet of forbs, thereby reducing dietary overlap (Nelson and Leege 1982). On many winter ranges, elk feed on browse (Nelson and Leege 1982) or a combination of browse and grasses (Hobbs et al. 1979), when available, in contrast to a cattle diet often dominated by grasses (Holechek et al. 1989). Competition for grasses, however, may be high on commonly grazed winter ranges where browse is not available to elk. The same is true on summer ranges during late summer and fall; elk and cattle feed heavily on grasses during this period.

7. The potential for competition between elk and cattle is high on unproductive rangelands, especially in arid ecosystems. Potential for competition also is high on rangelands grazed to full or maximum use by elk or cattle, and on rangelands experiencing a declining trend. Management for "proper use" (Archer and Smeins 1991) will reduce or minimize competition between elk and cattle. What constitutes "proper use" for dual grazing by elk and cattle, however, is not clear. Holechek et al. (1989:190-191) estimated acceptable levels of grazing use for most major range types in North America. These estimates were based on the premise that moderate livestock use of key forage species would retain sufficient vegetation residues for long-term maintenance of plant vigor, species composition, and site productivity (Smith 1965). Acceptable use of key forage species varied from 40 to 60% of annual biomass on more productive rangelands and from 20 to 40% on more arid rangelands.

We believe these estimates are useful as benchmarks for assessing the potential for competition between elk and cattle. Accordingly, on arid rangelands, we assume that elk or cattle use of perennial grasses at levels >25-30% will heighten forage competition with the other ungulate. On more productive rangelands, use of grasses at levels > 50-60% will likely induce this effect. Use of perennial grasses by elk or cattle at levels > 60-70% will likely bring about forage competition on nearly all sites, unless alternative, woody sources of nutritious forage are available.

Relatively unproductive rangelands, such as those at low elevation in the Great Basin and Desert Southwest, often are in poor condition due to past overgrazing by livestock (Holechek et al. 1989). These ranges cannot support high stocking rates by cattle and maintain low use of key grasses. Many of these ranges were likely ungrazed or grazed lightly by large ungulates before the arrival of livestock, and many key forages are not well-adapted to herbivory (Mack and Thompson 1982). High use of grasses by elk or cattle on these ranges during winter and/or spring and fall likely heightens potential competition between the 2 ungulates.

8. High densities of elk may induce negative effects on forage conditions similar to that caused by high stocking rates of cattle, resulting in lower animal performance. Intraspecific competition among elk, and resultant negative effects on forage, likely are density-dependent and manifested at high population densities. Intraspecific competition can occur within cervid populations that persist at high densities (Klein 1968, McCulloch 1986, Clutton-Brock et al. 1987, Irwin et al. 1994). Intensive herbivory resulting from competition among elk may alter the composition and vigor of key forages (Houston 1982, Edgerton 1987), reduce soil productivity (Tiedemann and Berndt 1972), and lower calf recruitment (Irwin et al. 1994). Similar effects on vegetation and soil have resulted from high stocking rates of livestock (Hart and Norton 1988). High densities of elk, combined with high stocking rates of cattle, can magnify competition between the 2 ungulates and alter plant composition and succession (Krueger and Winward 1974, DeByle 1979, Irwin et al. 1994). These effects are manifested in areas of high dual use such as riparian habitat, where conflicts between ungulate grazing and other resource objectives often are unresolved and intense.

9. Elk show an aversion to the presence of cattle that may or may not restrict their grazing choices. Many studies have shown that elk avoid or decrease their use of areas grazed concurrently by cattle (Nelson and Burnell 1975, Komberec 1976, Sauer 1980, Knowles and Campbell 1982, Lonner and Mackie 1983, Wallace and Krausman 1987, Frisina 1992, Yeo et al. 1993). Avoidance is not total, however. Ward et al. (1973) found no difference in patterns of elk use with or without cattle. They concluded that elk and cattle were socially compatible; both ungulates were observed grazing near each other on numerous occasions. Jeffrey (1963), Burbridge and Neff (1976), Stark (1973), Wallace and Krausman (1987), and Yeo et al. (1993) also observed elk close to cattle.

Such results may seem contradictory to the studies that show elk aversion to livestock (Nelson 1982), but we do not agree. As Sauer (1980:7) explained, "The introduction of cattle coincided with a significant decrease in elk use of the area, but elk that stayed in the area did not seem to change their habits in response to the cattle." Wallace and Krausman (1987) and Frisina (1992) found similar results: after cattle were introduced, most elk moved to adjacent areas without cattle; the elk that remained, however, were observed grazing near cattle.

Frisina (1992) believed that elk in his study area moved away in response to changes in vegetation in the grazed pasture rather than from a social intolerance of cattle; no supporting evidence was presented. However. Yeo et al. (1993) noted that elk avoidance of areas grazed by cattle might have been in response to increased human activities associated with cattle, rather than cattle themselves.

Regardless, the trend seems evident: with the onset of cattle grazing (and a likely increase in associated human activities), elk use shifts to areas unoccupied by cattle. Potential effects of these shifts on the population performance of elk are unknown and likely confounded with effects of elk harvest strategies (Yeo et al. 1993).

10. On productive rangelands that were grazed historically by native herbivores, systems of cattle grazing can be designed to enhance forage or foraging conditions for elk. Likewise, grazing by elk can enhance conditions for cattle.

Grazing systems are being used by state wildlife agencies to manage forage for elk and control elk distribution. Notable examples are those in Oregon (Anderson and Sherzinger 1975; Anderson et al. 1990*a,b*) and Montana (Frisina and Morin 1991, Alt et al. 1992, Frisina 1992). Rest-rotation grazing by cattle is used during late spring and summer to "condition" grasses (i.e., improve forage quality or foraging access) for later use by elk.

The reverse (i.e., elk "conditioning" grasses for later grazing by cattle) also is plausible. Pitt (1986) found that clipping bluebunch wheatgrass during spring extended forage quality into the fall compared to unclipped plots, although this response may not be predictable (Bryant 1993). Moreover, Dragt and Havstad (1987) found no increase in quality of Idaho fescue (*Festuca idahoensis*) and rough fescue (*Festuca scabrella*) during fall after spring grazing by elk and summer grazing by cattle.

Systems of rest-rotation grazing designed specifically for elk appear to work best on relatively productive rangelands such as those of the Pacific Northwest and northern Rocky Mountains. We do not believe these systems would be effective on rangelands in arid ecosystems of the Great Basin and Desert Southwest, due to inherent limitations in site productivity. In these areas, we believe that both the retention of vegetation residue and the regrowth of desirable forage would be *inadequate* to provide the desired benefits to elk and to maintain the desired composition, quantity, and vigor of forages.

11. Competition can be high at a given time and place during 1 year and low or nonexistent in the same place and time in subsequent years. Annual and seasonal variations in weather can dramatically alter forage availability and quality (Holechek et al. 1989). Periods of prolonged drought may induce competition between elk and cattle by reducing the quality or the availability of forage. Moreover, overgrazing during drought years can have lasting negative effects on range health, and thus on animal condition, which cannot be overcome during years of increased precipitation (Heady 1975, Holechek et al. 1989, Frank and McNaughton 1992). Stocking rates of elk and livestock, therefore, must be adjusted to match the carrying capacity of rangelands under drought conditions rather than for periods of average or optimum precipitation.

12. Perception is rarely reality when judging competitive interactions between elk and cattle. All too often, the mere presence of 1 ungulate on the other's range inspires heated debate and calls for reductions in stocking rates of the "culprit" species (Lonner and Mackie 1983). Casual observations of elk-cattle interactions often lead to wild speculation about the effects on vegetation and animal condition, most of which cannot be substantiated by the existing literature or by credible data from the area.

Professional expertise, valid research, and long-term monitoring data are needed to provide objectivity. Mechanisms for conflict resolution among divergent interest groups (Gerrans 1992, Brocci 1993) are also essential.

Elk and Domestic Sheep

Interactions between elk and domestic sheep are often restricted to relatively small areas of high elevation summer range. This is when forage of high quality exists in sufficient quantities over expansive areas. Consequently, the potential for competition between elk and sheep typically is not as pronounced nor as prolonged as that between elk and cattle. Rhodes and Sharrow (1990) found that summer grazing by sheep in western Oregon extended the high quality of grasses well into late summer and fall, with elk as the apparent beneficiaries, much like the effect of summer grazing by cattle claimed by Anderson and Sherzinger (1975), Anderson et al. (1990*a*), Frisina and Morin (1991), Alt et al. (1992), and Frisina (1992).

Situations exist, however, that increase the potential for competition. Sheep often feed heavily on forbs during summer, as do elk (Leege et al. 1977, Nelson 1982). Another is the aversion that elk apparently show to the presence of sheep (Stevens 1966, Ward 1973), much like their aversion to cattle. Finally, where elk and sheep share winter ranges containing browse as the dominant food source, competition can be direct and intense (Nelson 1982). These instances are relatively uncommon, however; options usually exist to reduce winter conflicts by moving sheep to alternate ranges or by supplementing the forage of elk.

Elk and Other Livestock

Little is known about the rangeland relationships of elk with other domestic or feral ungulates. Distributions of elk and ungulates such as goats, horses, burros, and pigs rarely overlap; feral varieties of these animals typically inhabit the driest portions of western rangelands where elk either are not present or are not common. That is changing, however, as elk colonize new areas or reestablish populations in drier areas of the Great Basin and Desert Southwest (Bunnell and Hancock 1985, Hess 1985).

As elk become more common in desert areas, it is likely that questions of competition will arise, especially between elk and feral horses. Horse and cattle diets are highly similar on desert rangelands in southeastern Oregon (McInnis and Vavra 1987); the added interaction of elk may heighten competition. Olsen and Hansen (1977) found high seasonal overlap of diets among elk, horses, and cattle on shared range in Wyoming, as did Hansen and Clark (1977) in Colorado and Salter and Hudson (1980) in Alberta. The dearth of forage and water and the poor condition of many desert rangelands could magnify competitive interactions.

Elk and Other Native Ungulates

Most western rangelands inhabited by elk also are occupied by mule deer (*Odocoileus hemionus*). Distributional overlap can also occur with white-tailed deer (*O. virginianus*), bison (*Bison bison*), moose (*Alces alces*), pronghorn (*Antilocapra americana*), bighorn sheep (*Ovis canadensis*), and caribou (*Rangifer tarandus*). In eastern North America, overlap occurs with white-tailed deer and sometimes with moose and caribou.

It is difficult to generalize about the competitive relationships of elk with these ungulates. Information is either lacking or inconclusive in documenting any measurable degree of competition (Mackie 1981). What is known is summarized below.

Elk, mule deer, bison, bighorn sheep, moose, and pronghorn apparently partition forage and habitat resources to minimize competition in areas of overlapping distribution such as Yellowstone National Park (Houston 1982). Often, patterns of foraging and habitat use are complementary. For example, grazing by elk and bison can enhance forage conditions for pronghorn and mule deer (Houston 1982).

Similar patterns of effective partitioning have been inferred from interaction studies of elk, mule deer, and other native ungulates across many western rangelands (Julander and Jeffrey 1964, Mackie 1970, Edgerton and Smith 1971, Constan 1972, Hudson et al. 1976, Skovlin and Vavra 1979, Hobbs et al. 1983, Sheehy 1988). Results appear to agree with partitioning observed among native ungulates in Africa (Bell 1971, Owen-Smith 1989, McNaughton 1991, Seagle and McNaughton 1992).

Regardless, competition does occur in some situations (Mackie 1981). In contrast to mule deer and bighorn sheep, elk generally are more opportunistic and catholic in their diet selection, more easily digest forage of low quality, are more mobile and wide-ranging, and form larger herds, all of which may result in a competitive edge when resources are limited (du Toit and Owen-Smith 1989). Accordingly, Cliff (1939), Mackie (1981) and Nelson (1982) presumed that elk out-compete mule deer on winter ranges that are limited in size and available forage. Nelson (1982) also believed that mule deer may leave or avoid areas of heavy use by elk, even if forage is available and dietary overlap with elk is low.

Ecologically incomplete habitats may cause similar problems between elk and bighorn sheep. Competition may occur on relatively unproductive alpine and subalpine ranges during summer, and on some bighorn wintering areas (Lyon and Ward 1982). Elk have sometimes overgrazed key forage plants and reduced their availability to bighorn sheep (Nelson 1982). Due to sheer numbers, elk may have a competitive advantage over bighorn sheep when vying for use of limited summer and winter ranges.

Elk are a potential host for meningeal worm (*Parelaphostrongylus tenuis*), a nematode parasite that is fatal to many cervids, including moose, caribou, mule deer, and black-tailed deer (*O. h. columbianus*). Meningeal worm is also fatal to elk, but animals exposed to lower densities of larvae can survive and transmit the disease to other cervids in the same manner as white-tailed deer (Samuel et al. 1992). Although meningeal worm is currently restricted to eastern North America, translocations of infected elk from eastern states and provinces could spread the disease westward. The result could have catastrophic effects on populations of cervids in the West, particularly caribou and moose. The likelihood of infected elk being translocated from eastern to western North America is increasing: hundreds of game-ranched elk are routinely moved across the continent each year (Williams 1992), and no reliable test exists to certify that such animals are not infected (Samuel et al. 1992).

GRAZING MANAGEMENT TRADE-OFFS BETWEEN ELK AND LIVESTOCK

Type of Grazing System

Few data exist about elk response to various systems of livestock grazing. Most research has focused on the effect of livestock grazing per se, with few tests of elk response to different grazing systems under experimental replication. The scant information that is available appears inconclusive regarding elk preference for or aversion to any particular grazing system.

Moreover, no studies have rigorously examined elk performance under various grazing systems. Furthermore, few data from controlled experiments are available regarding the effects of different grazing systems on vegetation parameters considered important to the nutritional needs of elk.

Skovlin et al. (1968) examined elk response to season-long, deferred-rotation, and ungrazed replicates in northeastern Oregon. Elk preferred ungrazed areas and showed no difference in their use of season-long versus deferred-rotation pastures.

Skovlin and Harris (1970), working in the same area, found higher elk use under season-long versus deferred-rotation grazing when cattle stocking was light. They found the opposite preference when stocking was high.

Knowles (1975) monitored elk response to cattle grazing under rest-rotation in Montana in the same area where Mackie (1970) had studied elk and cattle interactions under season-long grazing. Elk preferred pastures in rest and avoided pastures actively grazed by cattle. Knowles (1975) concluded, however, that overall use by elk was not significantly different under rotation grazing than that observed by Mackie (1970) under season-long grazing.

Yeo et al. (1993) found no changes in population trends of elk following a change from season-long to rest-rotation grazing in east-central Idaho. Like Knowles (1975), Yeo et al. (1993) concluded that elk preferred pastures in rest and avoided those actively grazed by cattle.

Effect of Stocking Rate

Stocking rate of livestock, more than any other grazing variable, directly influences the composition, quantity, and nutritive value of forage available to ungulates (Heady 1975, Van Poollen and Lacey 1979, Heitschmidt and Taylor 1991). This premise, combined with the potential for forage competition between elk and cattle, provide a framework for describing the effect of stocking rate on forage conditions for elk.

1. Regardless of the grazing system used, the composition, quantity, and quality of forage available to elk is determined largely by the stocking rate of cattle, interacting with the density of elk, the inherent characteristics of the site, and weather. Effect of stocking rate of domestic sheep, goats, or other livestock on forage conditions for elk is little known.

2. Light stocking rates of cattle can be neutral or positive in their influence on the composition and nutritive value of forage for elk. On more productive rangelands of the Pacific Northwest and Rocky Mountains, light grazing may "condition" grasses, presumably causing regrowth of high nutritive value or improving animal access to nutritious plant parts. In general, cattle use of key forage species at season-end levels of < 40%, using systems described like those of Alt et al. (1992) and Frisina (1992), may induce this effect while maintaining a desirable composition of forage for elk.

3. Moderate stocking rates of cattle, with allowable use of key species between 25 and 60%, can be positive, neutral, or negative in their effect on forage conditions for elk. Much depends on the timing of grazing, the physiological response of key forage plants to grazing, and the inherent productivity of the range and its condition. On relatively unproductive rangelands of the Great Basin and Desert Southwest, it is likely that cattle use > 25% will negatively affect forage conditions for elk, and vice-versa. Intensive monitoring of forage use, composition, quantity, and quality is required to determine specific effects.

4. High stocking rates of cattle invariably heighten the potential for competition with elk for limited forage. Similarly, high densities of elk may induce intraspecific competition for forage (Houston 1982, Irwin et al. 1993), magnifying potential competition with cattle. Regardless, cattle use of key species at levels > 60% may significantly reduce or eliminate key forage plants for elk and cattle on nearly all range types in North America (Holechek 1988, Holechek et al. 1989). The remaining forage will likely be abundant but of low nutritive value, or scarce but of high nutritive value, resulting in reduced animal performance or lower carrying capacity, respectively (Klein 1968, Hobbs and Swift 1985). On desert rangelands, both low quantity and quality of vegetation may result from overgrazing (Huston and Pinchak 1991), lowering both animal performance and carrying capacity.

These effects on elk forages may explain the inverse relationship between stocking rates of cattle and habitat use by elk (Fig. 6), documented by Skovlin et al. (1968) in northeastern Oregon. This relationship generally held true for all types of grazing systems that were studied.

By contrast, Anderson et al. (1990*a*), working in a similar area in northeastern Oregon, documented a long-term increase in stocking rates of elk and cattle with the use of a specially designed system of rest-rotation grazing during summer. However, close inspection of these data (Anderson et al. 1990*a*:106) indicate that elk numbers declined during the last 8 years of this 27-year study, commensurate with the highest stocking rates of cattle.

Methods of Stocking Allocation

Without question, the most confounding and frustrating process facing public range managers today is that of stocking allocation among wild and domestic ungulates. Much of the controversy centers on allocation between elk and cattle.

The allocation procedures we suggest here are generalized; their purpose is to demonstrate how trade-offs in allocation of forage between elk and cattle can be evaluated on a landscape scale such as a range allotment on public land. These procedures are not designed to estimate absolute carrying capacity of elk, nor to estimate the actual number of elk that may be present. Rather, they are designed to estimate realistic stocking rates of cattle within an allotment, given a potential carrying capacity of elk within the area, the expected patterns of distribution overlap between the 2 ungulates, and the potential for forage competition.

Additional fine-tuning is possible by incorporating more variables and by including site-specific data about potential carrying capacities, distribution patterns, and dietary preferences. That would allow range managers to use sophisticated allocation models such as those described by Nelson (1984), Cooperrider and Bailey (1984), Van Dyne et al. (1984), Ranching Systems Group (1992), and Johnson et al. (In Press). Use of our procedures involves the following steps:

1 Calculation of the stocking rate for cattle, based on the biomass of key forage species available, the allowable use of such forage, the forage demand of animals, the length of grazing period, and the spatial distribution of animals within the planning area.

2 Calculation of the stocking rate for elk, based on these same parameters.

3 Judgment of stocking allocation between the 2 species, based on the stocking rates calculated above, the potential overlap in spatial distributions, and the potential for forage competition. This process includes the following assumptions.

Stocking rates of elk and cattle can be estimated based on the biomass of key forage species. Note that biomass of key forage species is *not* equivalent to the total biomass of forage or vegetation that may be present (Smith 1965). Rather, key forage species are those plants that are abundant, productive, and palatable (Smith 1965). Such plants are the most valid indicators on which to base estimates of proper use and stocking (Holechek and Pieper 1992, Holechek et al. 1992). They represent the quantity of plants most likely to be consumed that provide high nutritional value to the consumer; such forages typically are "decreaser" species (Holechek et al. 1989).

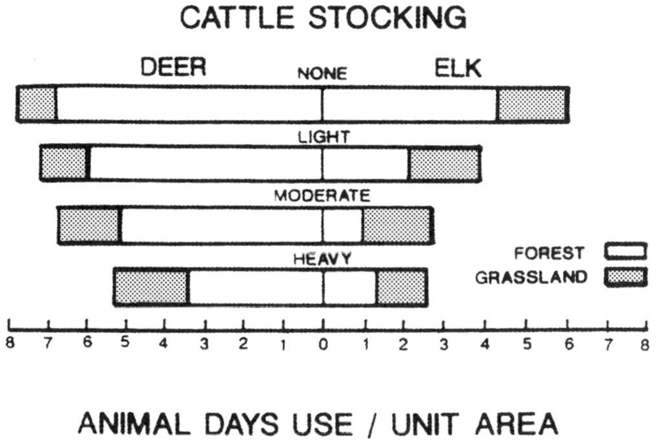

FIGURE 6. Cattle stocking in relation to deer and elk use in northeastern Oregon (Skovlin et al. 1968).

Accuracy of stocking estimates will be improved if the nutritive supply of forage is considered (Hobbs et al. 1982, Hobbs and Swift 1985, Schwartz and Hobbs 1985, Johnson et al. In press). Estimates based on nutrition, however, require intensive, site-specific data about energy and protein levels in forages consumed by ungulates. These data often are not available, and their collection can be time- and cost-prohibitive. Moreover, Klein (1968:363) stated that "With regard to ungulate range, it can perhaps be stated that forage quantity acts primarily to govern population size while quality determines the size of the individual." Schwartz and Hobbs (1985:35) further concluded that "Although the quality of food items has profound influence on the condition of individual herbivores, the quantity of forage ultimately determines how many animals can exist within a habitat." Also, both elk and cattle have similar digestive systems (Janis 1976, Hanley 1982), and in contrast to many smaller ungulates, are better-adapted to foraging in areas where food is high in quantity but low in quality (Wickstrom et al. 1984, Vavra et al. 1989). Biomass of key forage species therefore provides a reasonable basis for calculating a basic stocking rate for elk and cattle.

Stocking rates must be adjusted to reflect the spatial distribution of animals. For example, cattle avoid areas with increasing slope and distance from water (Tables 1, 2). Elk use declines significantly with increasing density of roads open to motorized traffic (Fig. 7). Both ungulates likely avoid forage

TABLE 1. Cattle use of landscapes in relation to topography (Holechek et al. 1989:194).

% slope	% use	% reduction in grazing capacity
0-10	100	None
11-30	70	30
31-60	40	60
> 60	0	100 (consider these slopes ungrazable)

TABLE 2. Cattle use of landscapes in relation to distance from water (Holechek et al. 1989:193).

Distance from water (km)	% use	% reduction in grazing capacity
0-1.6	100	None
1.6-3.2	50	50
> 3.2	0	100 (consider this area ungrazable)

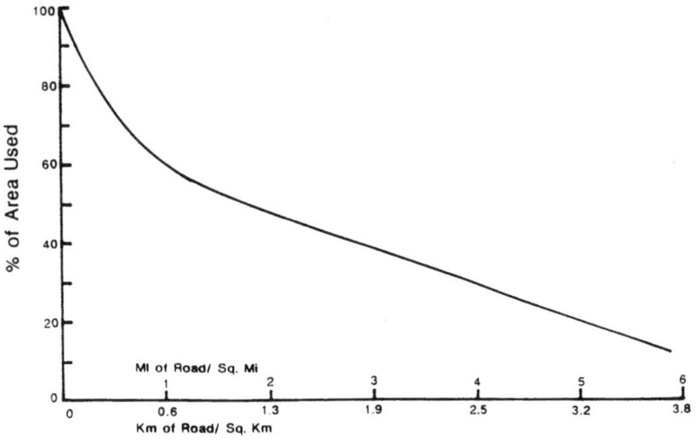

FIGURE 7. Elk use of habitat in relation to the density of roads open to motorized vehicles (Lyon 1983).

patches of low quality, and instead distribute themselves commensurate with the distribution of nutritious forage, as described for ungulates in general (Westoby 1974, Seagle and McNaughton 1992). Deep snow on winter range also can reduce the area available for foraging and alter animal distributions (Lyon and Ward 1982). In addition, elk use declines with increasing distance from cover-forage edges and in certain cover types (Wisdom et al. 1986, Thomas et al. 1988), and with increasing stocking rate of cattle (Skovlin et al. 1968, Painter 1980).

All such variables that induce a nonuniform distribution of ungulates at a specific time and place should be used to adjust stocking rates. One note of caution, however: determination of multi-variable effects on the spatial distribution of ungulates requires knowledge of the interactions among spatial variables. These are largely unknown. Geographic Information Systems (GIS) and Decision Support Systems (DSS) like that used by Ranching Systems Group (1992) and Johnson et al. (In Press) are required to analyze spatial interactions. Our procedures include single variables for purposes of illustration. We assume that a GIS and DSS will be used to analyze spatial variables over extensive areas, especially when analyzing multi-variable effects on animal distribution (Stuth et al. 1990).

Stocking rate, unadjusted for spatial distribution, is calculated as follows:

$$SR_u = \frac{(bf)(as)(au)}{(mw)(dfi)(dg)}$$

where SR_u is the stocking rate (no.) of either cattle or elk, unadjusted for spatial distribution, and where

- *bf* = biomass of key forage species (kg/ha),
- *as* = area size (ha),
- *au* = allowable use of key forages (% of plant wt),
- *mw* = mean weight of animals (kg),
- *dfi* = daily forage intake of animals (kg of forage/kg animal/day, expressed as forage consumed daily as a % of body wt), and
- *dg* = days of grazing (no.).

Stocking rate, adjusted for spatial distribution of animals is then calculated as:

$$SR_a = SR_u \sum_{i=1}^{n} (\text{\% area in category})\ (\text{\% use in category})$$

where SR_a = the stocking rate (no.) of cattle or elk, adjusted for spatial distribution, and where n = number of mapped categories for each spatial variable used, i = each mapped category in which the variable is applied, % area = proportion of area in each category relative to the total area, and % use = % of area or % of time that animals use a category relative to 100% or full use of that category.

Cattle use is that expected in each mapped category due to the effect of slope (Table 1) or distance to water (Table 2). Elk use is that expected in each category based on the density of roads open to motorized vehicles (Fig. 7). Effects of other spatial variables, like those mentioned earlier, can also be included here.

These stocking rates can then be used to judge stocking allocation between elk and cattle. This process includes the following additional assumptions.

If spatial distributions of elk and cattle do not overlap, owing to factors other than the social intolerance of elk for cattle, stocking allocation is additive (complementary) rather than either/or (competitive). If spatial distributions do overlap, stocking allocation is still additive unless forage competition occurs. **Regardless of distributional overlap, stocking allocation between elk and cattle is additive rather than either/or whenever forage is not limiting.**

In areas where elk-cattle distributions overlap, the potential for forage competition must be assessed. If site-specific research or monitoring data are unavailable, we recommend using a decision tree to assess the potential for forage competition, based on range condition and trend (Fig. 8). A better method is to collect site-specific data about elk and cattle use of key forages over time. If results indicate that elk and cattle are using the same key forages, and if total use exceeds that allowed, assume that potential competition for forage is high. If different forages are being used or if total use is at or below that allowed, assume that potential competition is low.

Following is an example of how these procedures can be applied. The example was developed originally by Holechek et

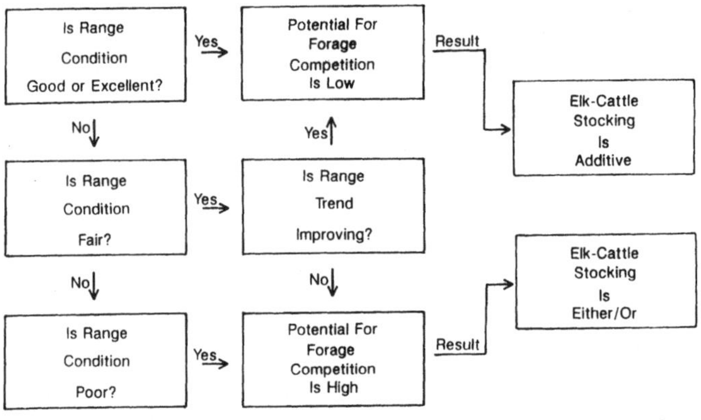

FIGURE 8. Decision tree for assessing the potential for forage competition between elk and cattle in areas where spatial distributions overlap, and where site-specific research or monitoring data about forage availability and dietary overlap are not available.

al. (1989:189-203) to demonstrate the effect of spatial distribution on the calculation of stocking rates for cattle. We extended this process to calculate stocking rates for elk in the same manner. We then estimated a stocking allocation from the 2 stocking rates, based on the overlap in spatial distributions between the 2 ungulates and the assumed potential for forage competition. The assumptions made for calculation of cattle stocking are described by Holechek (1988) and Holechek et al. (1989:189-203).

Example*: *Elk Summer Range in Northeastern Oregon. Cattle graze this hypothetical 1,000-ha allotment for 4 months during summer. Elk use the area for 7 months from late spring through fall. Approximately 200 kg/ha of key forages are available, with allowable use of these forages set at 25% because range condition is poor.

The area is characterized by diverse topography and a high density of roads open to motorized vehicles (Fig. 9A). Thus, the distributions of cattle and elk are clumped (Fig. 9B) due to the effect of slope on cattle use (Table 1) and open road density on elk use (Fig. 7).

No other spatial variables are assumed to affect animal distribution, including social intolerance of elk for cattle. Competition for forage is assumed to be high (i.e., 100% dietary overlap) in areas where cattle-elk distributions overlap because range condition is poor (Fig. 8). Daily forage intake is assumed to be 2% of body weight for both elk and cattle due to the poor condition of range, in contrast to a daily intake of 2.5% or higher that would be expected during summer when forage quality is normally high (Nelson and Leege 1982).

Under these assumptions, the independent stocking rates of each ungulate, unadjusted for effect of spatial variables (SR_u), are 60 elk and 52 cattle within the allotment (Table 3). These stocking rates are then discounted (SR_a, Table 3) to reflect portions of the allotment not used or used less than 100% (Fig. 9B).

As a result of differences in elk-cattle distributions, potential competition is reduced significantly; 40% of the area theoretically is grazed exclusively by cattle, and another 10% is grazed only by elk (categories A and D, Table 3, Figs. 9,10). Hence, stocking allocation is partially additive; that is, cattle and elk can be stocked partially independent of each other to account for 50% of the area where their modeled distributions are mutually exclusive (Fig. 10).

Use of these procedures results in mixed species stocking that is highly complementary and flexible, regardless of which ungulate is emphasized in management (Table 4). By contrast, the traditional method of allocation (Stoddart and Smith 1955) overestimates the number of elk or cattle that can stocked independent of one another (Table 4). This is because the traditional method does not account for clumped distribution of a given

TABLE 3. Independent stocking rates of elk and cattle before (SR_u)[a] and after (SR_a)[b] adjustment for effects of spatial variables on animal distribution within a 1,000-ha allotment in northeastern Oregon.

Mapped category[c]	Cattle		Elk	
	SR_u	SR_a	SR_u	SR_a
A	5	0	6	6
B	16	6	18	18
C	10	7	12	12
D	21	21	24	0
Total	52[d]	34[e]	60[f]	36[g]

[a] $SR_u = \dfrac{bf\,(\text{biomass of forage, kg/ha}) \times as\,(\text{area size, ha}) \times au\,(\text{allowable use, \% of plant weight consumed})}{mw\,(\bar{x}\text{ wt of animals, kg}) \times dfi\,(\text{daily forage intake, kg forage/kg animal/day}) \times dg\,(\text{days of grazing})}$

[b] $SR_a = SR_u \sum_{i=1}^{n} (\text{\% area in category } i)(\text{\% animal use in category } i)$

[c] See Fig. 9

[d] $SR_u = \dfrac{(200)(1000)(0.25)}{(400)(0.02)(120)} =$ 52 cattle (from Holechek et al. 1989:197); no. animals for mapped categories are based on proportion of the biomass of key forages available in each category

[e] SR_a = (52)(0.1)(0.0) + (52)(0.3)(0.4) + (52)(0.2)(0.7) + (52)(0.4)(1.0)
= 0 cattle, category A + 6 cattle, category B + 7 cattle, category C + 21 cattle, category D = 34 cattle
(adjusted for slope, Table 1, Fig. 9)

[f] $SR_u = \dfrac{(200)(1000)(0.25)}{(200)(0.02)(210)} =$ 60 elk; no. animals for mapped categories are based on proportion of the biomass of key forages available in each category

[g] SR_a = (60)(0.1)(1.0) + (60)(0.3)(1.0) + (60)(0.2)(1.0) + (60)(0.4)(0.0)
= 6 elk, category A + 18 elk, category B + 12 elk, category C + 0 elk, category D = 36 elk
(adjusted for roads, Figs. 7, 9)

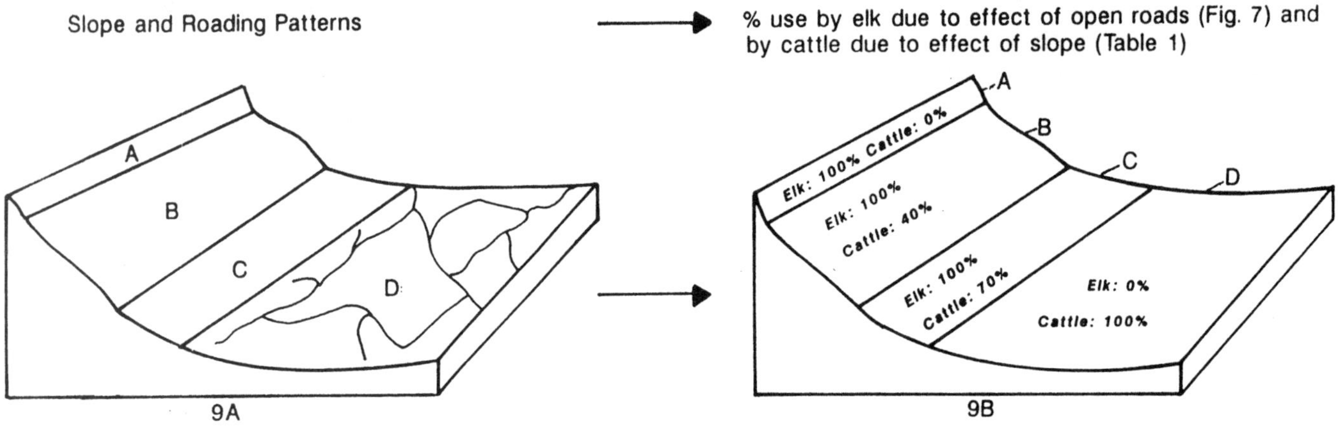

Mapped Categories

A = 10% of area with slopes >60% and road density of 0 Km/sq. Km

B = 30% of area with slopes 31-60% and road density of 0 Km/sq. Km

C = 20% of area with slopes 11-30% and road density of 0 Km/sq. Km.

D = 40% of area with slopes ≤ 10% and road density > 4Km/sq. Km

FIGURE 9. Effect of spatial variables on elk and cattle use of a 1,000-ha allotment in northeastern Oregon.

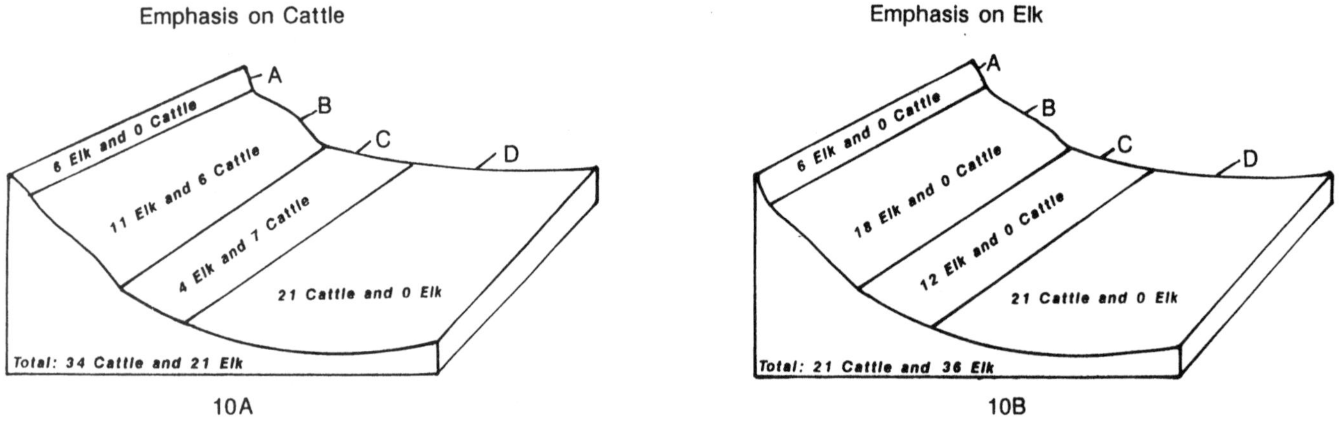

Mapped Categories

A and D = Forage competition low due to little or no overlap of elk and cattle distributions.

B and C = Forage competition high due to high distribution overlap and poor range condition.

FIGURE 10. Stocking allocation with emphasis on cattle (10A) or elk (10B), based on adjusted stocking rates (SRa, Table 3) of each ungulate and the combined effect of distribution overlap and potential competition for key forages in each mapped category of a 1,000-ha allotment in northeastern Oregon. No. elk in 10A is calculated from the biomass of key forage available in mapped categories A, B, and C that is not used by cattle due to effect of slope (Table 3, forage for 18 cattle (SRu – SRa) is available for elk in A, B, and C; no. elk = 1.15 x no. cattle). No. cattle in 10B is the adjusted stocking rate for cattle (SRa, Table 3) in category D, where elk do not occur due to high density of open roads (Table 3).

species, and in essence, overestimates stocking potential whenever a species' distribution is nonuniform (Fig. 9).

The traditional method also calculates stocking of elk as a simple weight ratio, or animal equivalence, of the number of cattle (Table 4), thus not accounting for areas of minimal or no overlap in elk-cattle distributions such as in categories A and D, Fig. 10. Moreover, differences in length of grazing period between elk and cattle in the estimates of animal equivalence are ignored, resulting in highly inaccurate stocking estimates of elk (Table 4). Finally, the traditional method does not account for the biomass of key forages available to 1 species when use of a category by the other ungulate is high but < 100%, such as the forage available to elk with < 100% cattle use of categories B and C, Figs. 9 and 10.

For these same reasons, our procedures allow a higher number of total animals to be stocked in contrast to single species stocking (Table 4). This corroborates conclusions by Nelson (1984:491), who stated that "common use grazing generally results in more efficient use of the forage resources because different herbivores have different habitat and forage preferences."

TABLE 4. Stocking of elk and cattle using different methods of allocation and management objectives for a 1,000-ha allotment in northeastern Oregon.

Allocation method and management objective	Stocking Allocation Cattle (no.)	Elk (no.)	Total no. animals
Traditional stocking			
Maximize cattle	52[a]	0	52
Maximize elk	0	104[b]	104
Single species stocking, adjusted[c]			
Maximize cattle	34	0	34
Maximize elk	0	36	36
Mixed species stocking, adjusted			
Maximize cattle	34	21[d]	55
Maximize elk	21[e]	36	57

[a] SR_u for cattle, Table 3 (unadjusted for effect of spatial variables), per Stoddart and Smith (1955).
[b] Animal equivalence of approximately 1 cow/2 elk (Stoddard and Smith 1955:192).
[c] Numbers for single species stocking are stocking rates adjusted for effects of spatial variables (SR_a, Table 3).
[d] Calculated from biomass of key forage available in mapped categories A, B, and C that is not used by cattle (see Table 3; forage for 18 cattle ($SR_u - SR_a$) is available for elk in A, B, and C; no. elk = 1.15 × no. cattle).
[e] Cattle can graze in category D, where no. elk is assumed to approximate 0 due to high density of roads open to motorized vehicles (see Table 3).

RANGELAND PRESCRIPTIONS FOR ELK

Managing Elk Distribution and Nutrition

By managing for nutritional needs, we can manipulate the spatial distribution of elk and improve animal condition. This is important to rangeland managers who seek to maintain productive elk herds on public land for hunting and viewing, and away from private land where the potential for damage to crops, fences, and livestock forage is high.

Crude protein (CP) and dry matter digestibility (DMD) are typically evaluated as indices of the nutritive value of forages consumed by ruminants (Dietz 1970, Mautz 1978). These values can be used with estimates of other nutritional parameters to derive digestible protein and digestible energy, which indicate the actual amount of protein and energy, respectively, that is available to ruminants for metabolic and other physiological functions (Mautz 1978, Huston and Pinchak 1991).

Nelson and Leege (1982) estimated crude protein requirements of cow elk by season (Fig. 11). These are perhaps the most practical and pertinent estimates of elk nutritional requirements that can be readily used to guide the design of effective land treatments. Note that at a minimum, elk requirements for maintenance are 5-6% CP year-round. This was corroborated by Hobbs et al. (1981), who concluded that "elk diets containing much less than 5% crude protein would fail to meet the animals' metabolic requirements." Hobbs et al. (1981) also cited McCullough (1969), who inferred that a diet of 7% crude protein met elk requirements during winter. Mould and Robbins (1981) found that diets containing < 3.5% CP resulted in negative protein digestibilities in tame elk; they concluded that levels of digestible nitrogen are inadequate for maintenance requirements if CP during winter is < 5% and biomass of forage is limited.

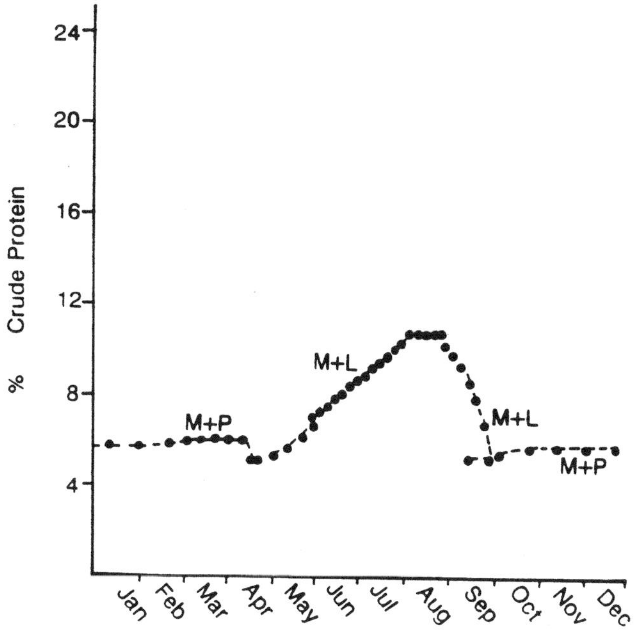

FIGURE 11. Crude protein requirements of cow elk by season, considering needs for maintenance (M), pregnancy (P), and lactation (L), as estimated by Nelson and Leege (1982:365).

Demand by cow elk for protein increases to 10-11% during lactation in spring and summer (Fig. 11). Quality of elk diets appears sufficient in spring and early summer to meet this demand (Baker and Hobbs 1982, Nelson and Leege 1982), but this may not be the case during late summer and fall. Nelson and Leege (1982:365-366) provided an excellent illustration of the importance of early calving in relation to protein availability. For calves born on or before 15 June, their demand for milk coincided with maximum availability of protein in spring and summer forages; these calves usually are weaned before protein content declines rapidly in late summer and early fall on ranges dominated by herbaceous forage. Calves born after June 15 placed protein demands on lactating cows that potentially could not be met in late summer and fall.

Elk requirements for digestible forage have not been studied extensively. Hobbs et al. (1981) found that digestibility of species consumed by elk during winter varied from 23 to 50%, similar to that found by Ward (1971) and Rowland et al. (1983). Hobbs et al. (1981) concluded that elk diets < 50% digestible were substantially submaintenance. They also cited Ammann et al. (1973:200), who concluded that "it is reasonable to assume that a ruminant could not maintain itself on a diet much below 50% digestible energy." This is in agreement with digestibility requirements for maintenance of adult beef cattle, estimated to be 51% (National Research Council 1984). We believe that 55% dry matter digestibility is adequate for maintenance of adult elk if dietary requirements for protein are met (Fig. 11) and daily intake is ≥ 2.5% of body weight.

Assessing the nutritional status of elk, and applying the appropriate mix of land treatments to overcome nutritional deficiencies, is a complex challenge. The following points should be considered:

1. Assessing the nutritional status of elk is most accurate when diet quality is evaluated in tandem with the quality of available forage. Inferring the nutritional status of elk based only on the quality of available forage can be misleading because elk and other ungulates select a diet higher in quality than that generally available (Hobbs et al. 1981, Baker and Hobbs 1982, Rowland et al. 1983, Hobbs and Spowart 1984).

2. Forage quantity and quality often are inversely related (Hobbs and Swift 1985, Hobbs et al. In Press *b*), except on desert rangelands where both quantity and quality can be low (Huston and Pinchak 1991). Management for maximum quality of forage is therefore a strategy to maintain fewer animals in better condition, whereas management for maximum biomass is a strategy to maintain more animals in relatively poor condition (Klein 1968, Hobbs and Swift 1985). Managers should set explicit objectives as to which strategy the range improvements are designed to support.

3. Many range improvements will increase forage quality but reduce quantity in the short-term, and increase forage quantity but reduce quality thereafter. For example, burning may improve the nutritive value of some forages immediately after treatment, but quality usually declines within 2 years (Nelson and Leege 1982, Skovlin et al. 1983, Van Dyke et al. 1991). Continued growth of post-treatment vegetation then results in increased biomass (Leege 1979, Peek et al. 1979, Van Dyke et al. 1991). Maintenance of a patchy mix of habitats, some high in forage biomass and some high in quality, therefore requires diverse temporal applications of range improvements that span multiple seasons and years.

Applying this mix of treatments in a "patchy" manner, temporally and spatially, throughout a seasonal herd range implies a strategy to maintain moderate densities of elk in moderate condition. Moreover, this approach is likely to maintain the desired spatial distribution of animals across public-private ownerships, commensurate with the distribution of range improvements.

4. It is difficult to specify the quantitative, interactive effects of various combinations of range treatments on elk nutrition, due to the lack of research on multi-variable effects. General, single-variable effects of the most common treatments are as follows.

Levels of Stocking

Stocking rates of elk, other wild ungulates, and livestock are rangeland treatments that dramatically affect the nutrition of elk. Proper levels of stocking of all ungulates are essential to maintain desired animal and population performance, and to sustain the range itself. Guidelines for proper stocking and procedures for stocking allocation were given earlier.

Grazing Systems

Livestock grazing can be designed to enhance forage for elk. Moreover, grazing systems can be used to manage the distribution of elk across land ownerships and habitats within a herd range. Rest-rotation grazing by cattle has been used effectively on rangelands dominated by bluebunch wheatgrass and Idaho fescue in Montana and Oregon, as follows.

Cattle graze the Mount Haggin Wildlife Area (8,100 ha) in southeastern Montana from mid-June through mid-October (Fig. 12) (Frisina 1992). Grazing level is 4,000 animal unit months across 3 pastures. During a given year, 1 pasture is grazed early, 1 is grazed late, and 1 is rested, with the order rotated over 3 years (Fig. 12). Grazing in the early pasture begins during the period of maximum growth of grasses (mid-Jun) and generally ends at seedripe (mid to late Aug); grazing in the late pasture is from seedripe to 15 October.

Presumed benefits to elk are continuous availability of preferred grasses in varying stages of phenological development during spring, summer, and fall among the 3 pastures; sustained

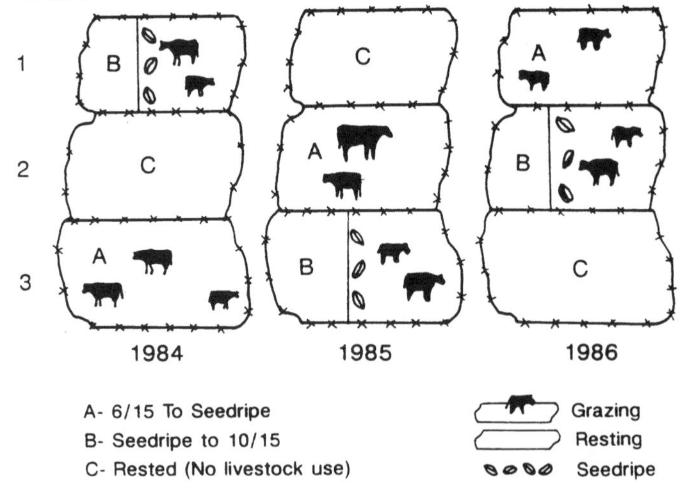

FIGURE 12. A rest-rotation grazing system used to manage forage on elk summer range at the Mount Haggin Wildlife Area in Montana (Frisina 1992:94).

maintenance of grasses of high nutritive value across the Wildlife Area as a whole; improved access to grasses of high quality due to the removal of surrounding litter; adequate periods of grazing rest, allowing preferred grasses to re-seed and/or restore food reserves in roots to maintain their abundance and vigor; and options for elk to graze in expansive areas with little or no interaction with cattle.

Benefits to livestock growers are increased animal unit months in contrast to no grazing or more restricted grazing practices of the past; improved quality of grasses available to cattle; and reduced damage by elk to adjacent, private rangelands (Frisina 1992).

Alt et al. (1992) described a similar system of rest-rotation grazing using cattle at the Wall Creek Wildlife Area in southwestern Montana; the area encompasses a combination of elk summer and winter range. Since its inception in 1984, the grazing system has apparently become increasingly effective in redistributing elk onto the Wildlife Area, and away from adjacent, private lands where damage by elk was significant and unacceptable. Today, the system has become so successful that Alt et al. (1992:15) stated that "The program has eliminated the need for a full-time manager to be housed at Wall Creek since hazing wintering elk off private lands is no longer a major factor." Frisina and Morin (1991), Anderson and Scherzinger (1975), and Anderson et al. (1990*a*) documented similar effective management of elk by using rest-rotation grazing.

These examples employ similar grazing strategies that are presumably responsible for their success: (1) cattle are removed from the early grazed pastures before or just after mid-growing season to allow for sufficient regrowth of grasses; this regrowth is unobstructed by litter and thus provides elk with efficient foraging access to higher quality plant parts during late summer, fall and/or winter; (2) cattle use of grasses is light to moderate, further allowing for regrowth of forage for elk and maintenance of healthy range condition; (3) sufficient grazing rest is provided to facilitate survival and recovery of preferred grasses; (4) ungrazed forage preferred by elk is maintained in the rested pastures; (5) cattle are used primarily to "condition" grasses for elk; beef production is more of a benefit than a goal in itself.

Adopting such grazing strategies demands flexibility from livestock operators in the timing, location, and rate of stocking by season and year. Most conventional systems of livestock grazing do not use such strategies in a similar, holistic manner. Therefore we cannot assume that elk derive similar nutritional benefits on other rangelands where more conventional grazing systems are employed. This is especially true of relatively unproductive rangelands in arid areas of the Great Basin and Desert Southwest. In these areas, it may be difficult to design systems of cattle grazing to achieve the desired forage response for elk, simply because regrowth of vegetation is slow, litter reserves are easily diminished, and precipitation is low and less predictable.

One additional caution: none of the presumed benefits from these examples were documented under a formal experimental design. Moreover, designed experiments have not always produced similar results in like environments (Bryant 1993, Westenskow-Wall et al. 1994). As a result, some of the claimed benefits have been questioned for specific examples (Weurthner 1992) and in general (Bryant 1993).

Prescribed Burning

Fire dramatically changes the composition, quantity, and quality of forage available to ungulates (Wright and Bailey 1982, Bailey 1988). It is 1 of the most common and popular tools used to manipulate forage for elk in the United States today. For example, the Rocky Mountain Elk Foundation, the largest private funder of habitat improvements for elk on public lands, matched >$500,000 with >$1,000,000 from other sources for prescribed burning of approximately 48,000 ha of western rangelands during 1986-1992 (C. Hastings, Rocky Mtn. Elk Found., pers. comm.).

Short-term effects of such broad-scale treatments appear beneficial to forage characteristics (Vallentine 1989) and to foraging conditions (Hobbs and Spowart 1984) for many ungulates, including elk (Lyon and Ward 1982). Burning removes accumulated plant litter, thereby making current, nutritious growth more accessible (Canon et al. 1987, Jourdonnais and Bedunah 1990). Consequently, the regrowth of vegetation up to 2 years after burning is usually desired by elk (Nelson and Leege 1982), presumably due to greater foraging efficiency (Canon et al. 1987). Similarly, burning sometimes improves forage quality immediately after treatment, but nutritive value usually declines within 2 years (Nelson and Leege, Skovlin et al. 1983, Van Dyke et al. 1991). Biomass of herbs and shrubs may increase thereafter (Leege 1979, Peek et al. 1979).

On the negative side, forage quality may not increase after burning, although diet quality may still be enhanced (Hobbs and Spowart 1984). Also, ungulate use usually declines quickly after burning (Wright and Bailey 1982). Moreover, long-term effects on the composition and productivity of forages are highly variable and often unpredictable and/or undesirable (Bailey 1988). For example, burning can significantly reduce nitrogen levels through volatilization, although grazing animals may partially mitigate the effect (Hobbs et al. 1991). Also, improper season of burning can produce unacceptable mortality of key forages. Much depends on the specific objectives and outcomes that are desired, the timing, frequency, and intensity of such fires, the scale of application, the plant communities burned, and the long-term effect on site productivity (Sanders and Durham 1985).

Forage Seeding and Planting

Key forage species are often seeded or planted immediately after a wildfire or prescribed burn to ensure the establishment of nutritious forage for elk. Like burning, such treatments are generally considered beneficial, but few studies have documented specific benefits to elk nutrition.

In general, the effects are likely beneficial if the seedings or plantings provide a more balanced mix of forage classes and species than would be present otherwise. In western Oregon, for example, seeding of grasses and legumes is considered beneficial to elk nutrition because these forage classes complement a naturally high composition of shrubs (Ramsey and Krueger 1986, Wisdom et al. 1986). In eastern Oregon, by contrast, grasses and forbs dominate the forage base (Irwin et al. 1993). Hence, shrubs of high nutritive value like curlleaf mountain-mahogany (*Cercocarpus ledifolius*) and antelope bitterbrush (*Purshia tridentata*) often are planted to provide more diverse phenological stages and composition of forages.

Fertilization with Nitrogen and Macro-minerals

Nitrogen is a critical element that yields microbial protein during fermentation of foods in the rumen. Most rangeland soils are moderately to severely deficient in nitrogen (Holechek et al. 1989, Vallentine 1989). Not surprisingly, nitrogen is the chemical element considered most critical in meeting the nutritional needs of elk (Hobbs et al. 1982, Nelson and Leege 1982). Phosphorus and sulfur also are deficient on many rangelands (Dietz 1970, Vallentine 1989), and both play important roles in physiological functions of ruminants (Mautz 1978, Robbins 1993).

Increasing the content of nitrogen, phosphorus, and/or sulfur in deficient soils can increase the protein content and digestible energy of herbage and its availability to ungulates (Vallentine 1989). Desired changes in elk distribution can be achieved with the application of such fertilizers in deficient areas (Gerrans 1992).

Beneficial effects of nitrogen fertilization, however, can be highly variable between vegetation communities (Mereszcak et al. 1981). Use by elk, and presumably the nutritional benefits to them, are usually limited to ≤2 years unless fertilization is repeated (Skovlin et al. 1983). Repeated applications of nitrogen also can change the species composition of grasses (Vallentine 1989), reduce or eliminate legumes (Duncan et al. 1985), and make some soils more acidic (Berg 1986).

Micro-mineral Supplementation

Trace elements, essential to protein and energy metabolism in ruminants (Huston and Pinchak 1991, Robbins 1993), are sometimes deficient in rangeland soils. For example, selenium is deficient throughout much of the Pacific Northwest and California (Robbins 1993). Masupu (1990) found lower blood levels of selenium in populations of elk, pronghorn, and bighorn sheep than that required by livestock.

Supplemental salting or fertilization with selenium on deficient rangelands is beneficial to livestock (Watkinson 1983, Holechek et al. 1989), and thought to benefit elk (Starkey et al. 1982). Selenium supplementation using metal boluses in female black-tailed deer increased preweaning fawn survival from 0.32 fawns/female to 0.83 fawns/female in a free-ranging population in northern California (Flueck 1994).

Mineral supplementation can improve performance of elk and other ungulates if deficient minerals are identified accurately, deficient minerals are administered at the proper time, rate, and mix, and other minerals not known to be deficient are not included in supplementation efforts.

Forage Supplementation

Winter damage by elk on privately owned farms and rangelands is often predictable and significant unless preventive steps are taken (Lyon and Ward 1982, de Calesta and Witmer 1994).

One approach is supplemental or replacement feeding with alfalfa hay or pellets. Although not as desirable as the acquisition or leasing of additional winter range, feeding is effective in keeping elk away from private land (Emerson 1988, McKeel 1988). This is especially true when feeding is combined with the removal of problem animals on private land, either through live-trapping and transplanting, hazing, or special hunts.

Feeding should meet the nutritional requirements of wintering elk; forage should contain ≥ 5-6% crude protein (Fig. 11) and be ≥ 50-55% digestible. The rate of supplementation depends on the availability and quality of natural forage. If natural forage is limited, alfalfa pellets fed at rates of 4.3 kg/elk/day, or alfalfa hay fed at rates of 5.4 kg/elk/day, appear sufficient (Thorne and Butler 1976). This assumes that alfalfa of high quality is used as the supplement (e.g., 15% crude protein and 66% digestibility of hay used by Thorne and Butler [1976]).

Elk are supplemented at lower rates where natural forage is available but inadequate by itself to meet nutritional requirements. For over 20 years, large numbers of elk have been maintained with a supplementation of 3.2-3.6 kg/animal/day at the National Elk Refuge (Smith and Wilbrecht 1988); here, natural forage is usually available until late January and again by early April.

Feeding is not a panacea for resolving winter range problems (Kimball and Wolfe 1985). Once initiated, it is difficult to reduce elk dependence on such rations. Numbers of elk residing in the area during the non-winter period may increase. Acquisition of additional winter ranges, improvements in the quality of natural forage on such ranges, and reductions in elk numbers to winter carrying capacity all are ecologically superior. Often, however, such approaches are politically unacceptable; supplemental feeding then becomes the likely and often-used alternative.

Water Developments

Water is considered limiting to elk on many arid and semi-arid rangelands (McCabe 1982). Elk may concentrate near water sources in extremely dry areas (McCorquodale et al. 1986). Moreover, some transplants of elk may have been ineffective in part from a shortage of free-standing water (Carpenter and Silvy 1991, Carpenter and Grant 1991). Increasing the distribution and availability of water on many of the driest rangelands will likely enhance elk use of such areas, especially during dry seasons or years.

Further development of water sources on most other elk ranges, however, can be a "double-edge sword" if livestock have access to the water. As discussed earlier, livestock use is usually highest within 1.6 km of water (Table 2); this is the zone of most direct competition between elk and livestock. Improving the distribution of water; i.e., making water more evenly and readily available throughout a pasture, will also result in a more even distribution of livestock (Holechek et al. 1989). This may increase the potential for competition with elk, or reduce elk use in favor of cattle use for 2 reasons.

1) Most water developments have roads associated with them; elk avoid areas near roads open to motorized traffic (Fig. 7). Elk are especially vulnerable to human harassment during hunting

seasons when roads facilitate harvest, and often the over-harvest, of targeted animals (Leptich and Zager 1991, Unsworth et al. 1993). Road densities and traffic will likely increase with an increase in number and distribution of water sources.

2) Elk avoid livestock but return to grazed areas after livestock have left. A more even distribution of livestock, resulting from a more even distribution of water, may hinder grazing choices by elk unless livestock-free areas also are available within a herd's seasonal home range.

Roads

Wherever elk occur, they consistently and dramatically avoid areas near roads that are open to motorized traffic (Marcum 1975, Perry and Overly 1977, Morgantini and Hudson 1979, Witmer 1981, Irwin and Peek 1983). This response was initially quantified by Thomas et al. (1979) and later verified by Lyon (1983) (Fig. 7). Implications of this relationship cannot be underestimated: the positive effects of land treatments like burning, seeding, and fertilization on elk nutrition may be partially or wholly offset by the negative effects of open roads. This is especially true in areas of open rangelands and gentle terrain; here, vegetation and topography provide limited hiding and security from human activities. Aggressive programs of road obliteration and closure are necessary to facilitate elk use of rangelands.

Managing Population Characteristics

Humans are the most effective, dominant, and pervasive predators of elk today. The population dynamics of elk are largely controlled and manipulated through hunting by humans, and less so through predation by other mammals. Moreover, anomalies of weather and disease likely play lesser roles in controlling elk numbers compared to the effects of human predation. Finally, although habitat largely determines carrying capacity, the actual number of elk present in a given area and time, and the sex and age composition of herds, are often more a reflection of hunting regimes than of habitat constraints.

Recognition of these relationships has resulted in a new term that addresses an old paradigm for managing elk demography: that of elk vulnerability to harvest. Lyon and Christensen (1992) define elk vulnerability as a "measure of elk susceptibility to being killed during the hunting season."

Susceptibility to harvest is affected by many variables (Thomas 1991) (Table 5). Those considered to most significantly increase the vulnerability of elk are high density of hunters (Vales et al. 1991), high density of roads (Hurley and Sargeant 1991, Leptich and Zager 1991, Unsworth et al. 1993), loss or absence of large cover blocks (Hillis et al. 1991), and gentle terrain (Edge and Marcum 1991, Unsworth et al. 1993).

Accordingly, elk are highly susceptible to harvest on open rangelands because cover usually is sparse and road densities often are moderate to high. Also, many rangelands are characterized by gentle topography, further increasing vulnerability. To ensure the presence of huntable populations in such areas, tight limits on hunter density, motorized access, and season length are necessary (Table 5). Prudent management of season timing and season type is required. Without such management, the presence and number of elk on open rangelands may be severely limited, during the hunting season and possibly other times of year.

TABLE 5. Management problems that increase elk vulnerability to harvest and the corresponding management remedies (Thomas 1991:319).

Management problems	Management remedies
1. Increasing density of roads.	1. Design roads to minimize impacts. Close roads permanently or temporarily. Enforce road closures.
2. Increasing density of hunters.	2. Restrict hunter numbers.
3. Decreasing amounts of cover.	3. Control stand configuration, juxtaposition, and size through modifications in timber management program.
4. Fragmentation of cover into smaller patches.	4. Retain adequate "escape cover" in the form of stands of several hundred or more acres.
5. No restriction on antler class in bull harvest.	5. Impose regulations on what can be taken—e.g. such as allowing the kill of spike-antlered bulls only.
6. Setting of open seasons that include the rutting period.	6. Insure that open seasons do not include the rutting period.
7. Improving technology.	7. Preclude "modern weapons."
8. Long open seasons.	8. Shorten the open season.
9. Relatively gentle terrain.	9. Decrease road density, maintain more cover, increase size of cover patches, decrease hunter numbers.
10. Increasing number of hunter days.	10. Related to both items 2 and 8 above. Reduce hunter numbers and/or reduce length of hunting season.

CONFLICT RESOLUTION OF ELK-LIVESTOCK PROBLEMS

New Partnerships and Paradigms

Even with the best biological information, conflicts between elk and livestock are rarely resolved without communication, coordination, and consensus among divergent interests. The need for such efforts has been recognized for quite sometime; often, however, the results have been more of frustration than solution (Peek et al. 1982). Nonetheless, more formal partnerships are being forged to resolve long-standing issues.

Two notable partnerships are underway in Colorado (Gerrans 1992) and Oregon (Blue Mountains Elk Initiative 1991, Brocci 1993). Both programs focus on reducing elk damage to livestock and crops on private land; this is accomplished by re-distributing animals away from problem areas through range improvements on nearby public land. Both programs have oversight committees and working groups who help plan, implement, and monitor the land treatments and assess the response of animals. Committees and working groups are composed of state, federal, private, and tribal interests who manage or are affected by the management of elk.

Ecosystem Management of Elk

Elk are mobile, adaptive, and opportunistic; they use a variety of habitats that span many jurisdictions and land ownerships. Their annual movements can encompass hundreds of square kilometers, bringing them into contact with a multitude of landowners and land uses.

At the same time, human development continues unabated on elk ranges. Traditional users of rangelands, such as farmers and livestock growers, are generally intolerant of high elk numbers. Fee hunting operators, hunters, wildlife viewers, and urban settlers on these same rangelands, however, desire more elk for hunting and watching. Given these conditions, elk-land use conflicts will continue to grow in area, frequency, and intensity. The loser, if history is the indicator, is likely to be the elk resource.

Is there hope? We believe so, but only if polarized interests adopt an ecosystem or landscape approach to elk management. Our geographic scale must change from that of an individual pasture, watershed, landowner, or allotment to that of an entire herd range, such as that encompassing ≥10,000 ha. Within each herd range, partnerships like those underway in Colorado (Gerrans 1992) and Oregon (Blue Mountains Elk Initiative 1991, Brocci 1993) must be forged. Such partnerships are politically effective and provide an ecological basis upon which all interests can plan and implement the proper combination of land treatments and hunting regimes necessary to achieve the desired demography and distribution of elk across all land ownerships in time and space.

The restoration of elk to rangelands of North America is likely to continue at a smart pace; with it will come an intense challenge to integrate the management of this special resource in a manner compatible with a myriad of other rangeland uses.

SUMMARY

Before European settlement, elk were among the most common and widely distributed of the wild ungulates in North America. Excessive hunting, overgrazing by livestock, and conversion of habitat to farms and cities in the seventeenth, eighteenth, and nineteenth centuries led to broad-scale population declines and extinctions. Today, population size is increasing and elk are re-colonizing many rangelands they occupied historically.

Existing elk habitat is fragmented along property lines: summer ranges are publicly owned while most winter ranges are in private ownership. Winter ranges continue to be lost to recreation and urban developments. Elk damage on the remaining winter ranges is an increasing problem.

Potential competition between elk and cattle is an issue of intense controversy among rangeland users. On arid rangelands, competition is likely if either ungulate is present at moderate to high densities under common grazing. On more productive rangelands, common grazing by elk and cattle can be complementary. Specially designed systems of rest-rotation grazing are being used by state and federal agencies, in cooperation with private landowners, to improve elk nutrition and control elk distribution across multiple ownerships.

Methods of stocking allocation between elk and cattle must consider differences in the spatial distribution of animals and the potential for forage competition. Stocking of the 2 ungulates is additive (complementary) in areas where spatial distributions do not overlap. Where distributions do overlap, either spatially or temporally, stocking is still additive unless forage competition occurs.

A variety of rangeland prescriptions and practices can be used to effectively manage elk distribution and nutrition. Grazing systems and stocking rates of cattle, prescribed burning, fertilization, forage supplementation, and road management are key practices available to habitat managers.

New partnerships are needed to resolve on-going management problems. Because elk are highly mobile and opportunistic, a landscape-level approach that encompasses multiple ownerships and thousands of hectares is required to effectively manage this resource across rangelands of North America.

LITERATURE CITED

Adams, A. W. 1982. Migration. Pages 301-321 *in* J. W. Thomas and D. E. Toweill, eds. Elk of North America, ecology and management. Stackpole Books, Harrisburg, Pa.

Allayne-Chan, A. C. 1986. Forage conditioning: defoliation and abiotic effects on non-structural carbohydrates, herbage production, and foliar nutrients in bluebunch wheatgrass and rough fescue. B.C. Min. For. Res. Rep. RR85006-HQ, Victoria, B.C., Can. 83pp.

Alt, K. L., M. R. Frisina, and F. J. King. 1992. Coordinated management of elk and cattle, a perspective-Wall Creek Wildlife Management Area. Rangelands 14:12-15.

Ammann, A. P., R. L. Cowan, C. L. Mothershead, and B. R. Baumgardt. 1973. Dry matter and energy intake in relation to digestibility in white-tailed deer. J. Wildl. Manage. 37:195-201.

Anderson, E. W., D. L. Franzen, and J. E. Melland. 1990*a*. Prescribed grazing to benefit watershed-wildlife-livestock. Rangelands 12:105-111.

———, ———, and ———. 1990*b*. Forage quality as influenced by prescribed grazing. Pages 56-70 *in* K. E. Severson, ed. Can livestock be used as a tool to enhance wildlife habitat? U.S. Dep. Agric. For. Serv. Gen. Tech. Rep. RM-194.

———, and R. J. Scherzinger. 1975. Improving quality of winter forage for elk by cattle grazing. J. Range Manage. 28:120-125.

Archer, S., and F. E. Smeins. 1991. Ecosystem-level process. Pages 109-139 *in* R. K. Heitschmidt and J. W. Stuth, eds. Grazing management, an ecological perspective. Timber Press, Portland, Oreg.

Bailey, A. W. 1988. Understanding fire ecology for range management. Pages 527-556 *in* P. T. Tueller, ed. Vegetation science applications for rangeland analysis and management. Kluwer Academic Publ., Boston, Mass.

Baker, D. L., and N. T. Hobbs. 1982. Composition and quality of elk summer diets in Colorado. J. Wildl. Manage. 46:694-703.

Barnes, T. G., R. K. Heitschmidt, and L. W. Varner. 1991. Wildlife. Pages 179-189 *in* R. K. Heitschmidt and J. W. Stuth, eds. Grazing management, an ecological perspective. Timber Press, Portland, Oreg.

Bell, R. H. V. 1971. A grazing ecosystem in the Serengeti. Sci. Amer. 225:86-93.

Berg, W. A. 1986. Effect of 20 years of low N rate pasture fertilization on soil acidity. J. Range Manage. 39:122-124.

Birch, L. C. 1957. The meaning of competition. Am. Nat. 91:5-18.

Blue Mountains Elk Initiative. 1991. Federal, private, state, and tribal partnerships to manage elk in the Blue Mountains of Oregon and Washington, charter for 1991-1995. U.S. Dep. Agric. For. Serv., Pac. Northwest Exp. Stn. 40pp.

Bolon, N. A. 1994. Estimates of the values of elk in the Blue Mountains of Oregon and Washington: evidence from existing literature. U.S. Dep. Agric. For. Serv. Gen. Tech. Rep. PNW-316. 38pp.

Boyd, R. J. 1978. American elk. Pages 11-29 *in* J. L. Schmidt and D. L. Gilbert, eds. Big game of North America. Stackpole Books, Harrisburg, Penn.

Brocci, J. 1993. The vulnerability blues, solving conflicts in the Blue Mountains. Bugle 10:57-65.

Brooks, R., C. S. Swanson, and J. Duffield. 1991. Total economic value of elk in Montana: an emphasis on hunting values. Pages 186-195 *in* A. G. Christensen, L. J. Lyon, T. N. Lonner, comps. Proc. elk vulnerability symposium, Montana State Univ., Bozeman.

Bryant, L. D. 1993. Quality of bluebunch wheatgrass (*Agropyron spicatum*) as a winter forage for Rocky Mountain elk (*Cervus elaphus nelsoni*) in the Blue Mountains of Oregon. Ph.D. Thesis, Oregon State Univ., Corvallis. 147pp.

———, and C. Maser. 1982. Classification and distribution. Pages 1-59 *in* J. W. Thomas and D. E. Toweill, eds. Elk of North America, ecology and management. Stackpole Books, Harrisburg, Pa.

Bunnell, S. D., and N. V. Hancock. 1985. Elk in Utah—past, present, and future. Pages 7-34 *in* G. W. Workman, ed. Western elk management, a symposium. Ut. Agric. Exp. Stn., Utah State Univ., Logan.

Burbridge, W. R., and D. J. Neff. 1976. Conconino National Forest-Arizona Game and Fish Department cooperative roads-wildlife study. Pages 44-57 *in* S. R. Hieb, ed. Proc. elk-logging-roads symposium. Univ. Idaho, Moscow.

Canon, S. K., P. J. Urness, and N. V. DeByle. 1987. Habitat selection, foraging behavior, and dietary nutrition of elk in burned aspen forest. J. Range Manage. 40:433-437.

Carpenter, J., and W. Grant. 1991. The use of computer modeling to examine the population dynamics of a small elk herd in West Texas. Pages 123-125 *in* A. G. Christensen, L. J. Lyon, and T. N. Lonner, comps. Proc. elk vulnerability symposium. Montana State Univ., Bozeman.

———, and N. Silvy. 1991. Movements and habitat relations of a small elk population in West Texas. Pages 64-73 *in* A. G. Christensen, L. J. Lyon, and T. N. Lonner, comps. Proc., elk vulnerability symposium. Montana State Univ., Bozeman.

Cliff, E. P. 1939. Relationship between elk and mule deer in the Blue Mountains of Oregon. Trans. North Am. Wildl. Conf. 4:560-569.

Clutton-Brock, T. H., M. Major, S. D. Albon, and F. E. Guinness. 1987. Early development and population dynamics in red deer, I, density-dependent effects on juvenile survival. J. Anim. Ecol. 56:53-67.

Cole, G. F. 1971. An ecological rationale for the natural or artificial regulation of native ungulates in parks. Trans. North Am. Wildl. and Nat. Resour. Conf. 36:417-425.

Conner, J. R. 1991. Social and economic influences on grazing management. Pages 191-199 *in* R. K. Heitschmidt and J. W. Stuth, eds. Grazing management, an ecological perspective. Timber Press, Portland, Oreg.

Constan, K. J. 1972. Winter foods and range use of three species of ungulates. J. Wildl. Manage. 36:1068-1076.

Cook, C. W. 1972. Comparative nutritive values of forbs, grasses, and shrubs. Pages 303-310 *in* C. M. Mckell, J. P. Blaisdell, and J. R. Goodin, eds. Wildland shrubs-their biology and utilization. U.S. Dep. Agric. For. Serv. Gen. Tech. Rep. INT-1.

Cool, K. L. 1992. Seeking common ground on western rangelands. Rangelands 14:90-92.

Cooperrider, A. Y., and J. A. Bailey. 1984. A simulation approach to forage allocation. Pages 525-560 *in* Developing strategies for rangeland management. Westview Press, Boulder, Colo.

Davies, N. B. 1979. Ecological questions about territorial behaviour. Pages 317-350 *in* J. R. Krebs and N. B. Davies, eds. Behavioural ecology, an evolutionary approach. Blackwell Scientific Publ., London, U.K.

DeByle, N. V. 1979. Potential effects of stable versus fluctuating elk populations in the aspen ecosystem. Pages 13-19 *in* M. S. Boyce and L. D. Hayden-Wing, eds. North American elk: ecology, behavior, and management. Univ. Wyoming Press, Laramie.

deCalesta, D. S., and G. W. Witmer. 1994. Elk. Pages D41-D50 *in* S. Hygnstrom, R. M. Timm, and G. Larson, eds., Prevention and control of wildlife damage. Second ed. Univ. Nebraska Coop. Ext. Serv., Lincoln.

Dietz, D. R. 1970. Animal production and forage quality—definitions and components of forage quality. Pages 1-9 *in* H. A. Paulsen, Jr., and E. H. Reif, eds. Proc. range and wildlife habitat evaluation—a symposium. U.S. Dep. Agric. For. Serv. Misc. Publ. No. 1147.

Dragt, W. J., and K. M. Havstad. 1987. Effects of cattle grazing upon chemical constituents within important forages for elk. Northwest Sci. 61:70-73.

Duffield, J., and J. Holliman. 1988. The net economic value of elk hunting in Montana. Mont. Dep. Fish, Wildlife, and Parks, Helena. 53pp.

Duncan, D. A., N. K. McDougald, and J. R. Larson. 1985. Effects of two nitrogen and a sulfur fertilizer on yield of native clover. California Agri. Technol. Inst., California State Univ., Fresno. 4pp.

du Toit, J., and N. Owen-Smith. 1989. Body size, population metabolism, and habitat specialization among large African herbivores. Am. Nat. 133:736-740.

Edge, W. D., and C. L. Marcum. 1991. Topography ameliorates the effects of roads and human disturbance on elk. Pages 132-137 *in* A. G. Christensen, L. J. Lyon, and T. N. Lonner, comps. Proc. elk vulnerability symposium. Montana State Univ., Bozeman.

Edgerton, P. J. 1987. Influence of ungulates on the development of the shrub understory of an upper slope mixed conifer forest. Pages 162-167 *in* F. D. Provenza, J. T. Flinders, and F. D. McArthur, comps. Proc. symposium on plant-herbivore interactions. U.S. Dep. Agric. For. Serv. Gen. Tech. Rep. INT-222.

———, and J. G. Smith. 1971. Seasonal forage use by deer and elk on the Starkey Experimental Forest and Range, Oregon. U.S. For. Serv. Res. Pap. PNW-112. 12pp.

Emerson, K. 1988. Elkhorn Wildlife Area: artificial feeding of elk. Pages 50-54 *in* M. Zahn, J. Pierce, and R. Johnson, eds. Proc. Western states and provinces 1988 elk workshop. Wash. Dep. Wildl., Olympia.

Ferguson, D., and N. Ferguson. 1983. Sacred cows at the public trough. Maverick Publ., Bend, Oreg. 250pp.

Flueck, W. T. 1994. Effect of trace elements on population dynamics: selenium deficiency in free-ranging black-tailed deer. Ecology 75:897-812.

Frank, D. A., and S. J. McNaughton. 1992. The ecology of plants, large mammalian herbivores, and drought in Yellowstone National Park. Ecology 73:2043-2058.

Frisina, M. R. 1992. Elk habitat use within a rest-rotation grazing system. Rangelands 14:93-96.

———, and F. G. Morin. 1991. Grazing private and public land to improve the Fleecer Elk Winter Range. Rangelands 13: 291-294.

Geist, V. 1985. Game ranching: threat to wildlife conservation in North America. Wildl. Soc. Bull. 13:594-598.

Gerrans, J. 1992. The habitat partnership program in Colorado. Rangelands 14:84-87.

Gillis, A. M. 1992. Should cows chew cheatgrass on commonlands? BioScience 44:668-675.

Guthrie, R. D. 1966. The extinct wapiti of Alaska and Yukon territory. Can. J. Zool. 44:47-57.

Hanley, T. A. 1982. The nutritional basis for food selection by ungulates. J. Range Manage. 35:146-151.

Hansen, R. M., and R. C. Clark. 1977. Foods of elk and other ungulates at low elevations in northwestern Colorado. J. Wildl. Manage. 41:76-80.

Hart, R. H., J. Bisso, M. J. Samuel, and J. W. Waggoner, Jr. 1993. Grazing systems, pasture size, and cattle grazing behavior, distribution and gains. J. Range Manage. 46:81-87.

———, and B. E. Norton. 1988. Grazing management and vegetation response. Pages 493-525 *in* P. T. Tueller, ed. Vegetation science applications for rangeland analysis and management. Kluwer Academic Publ., Boston, Mass.

———, K. W. Hepworth, M. A. Smith, and J. W. Waggoner, Jr. 1991. Cattle grazing behavior on a foothill elk winter range in southeastern Wyoming. J. Range Manage. 44:262-266.

Heady, H. F. 1975. Rangeland management. McGraw-Hill Book Co., New York, N.Y. 460pp.

Heitschmidt, R. K., and C. A. Taylor. 1991. Livestock production. Pages 161-177 *in* R. K. Heitshmidt and J. W. Stuth, eds. Grazing management, an ecological perspective. Timber Press, Portland, Oreg.

Henderson, R. E., and A. O'Herren. 1992. Winter ranges for elk and deer: victims of uncontrolled subdivision? West. Wildlands 18:20-25.

Hess, M. 1985. Elk transplanting programs in Nevada and the status of Nevada elk. Pages 39-43 *in* G. W. Workman, ed. Western elk management, a symposium. Utah Agric. Exp. Stn., Utah State Univ., Logan.

Hillis, J. M., M. J. Thompson, J. E. Canfield, L. J. Lyon, C. L. Marcum, P. M. Dolan, and D. W. McCleerey. 1991. Defining elk security: the Hillis paradigm. Pages 38-43 *in* A. G. Christensen, L. J. Lyon, and T. N. Lonner, comps. Proc. elk vulnerability symposium. Montana State Univ., Bozeman.

Hobbs, N. T., D. L. Baker, J. E. Ellis, and D. M. Swift. 1979. Composition and quality of elk diets during winter and summer: a preliminary review. Pages 47-62 *in* M. S. Boyce and L. D. Hayden-Wing, eds. North American elk: ecology, behavior, and management. Univ. Wyoming, Laramie.

———, ———, ———, and ———. 1981. Composition and quality of elk winter diets. J. Wildl. Manage. 45:156-171.

———, ———, ———, ———, and R. A. Green. 1982. Energy- and nitrogen-based estimates of elk winter range carrying capacity. J. Wildl. Manage. 46:12-21.

———, D. L. Baker, and R. B. Gill. 1983. Comparative nutritional ecology of montane ungulates during winter. J. Wildl. Manage. 47:1-16.

———, and R. A. Spowart. 1984. Effects of prescribed fire on mountain sheep and mule deer during winter and spring. J. Wildl. Manage. 48: 551-560.

———, and D. M. Swift. 1985. Estimates of habitat carrying capacity incorporating explicit nutritional constraints. J. Wildl. Manage. 49:814-822.

———, D. S. Schimel, C. E. Owensby, and D. Ojima. 1991. Fire and grazing in the tallgrass prairie: contingent effects on nitrogen budgets. Ecology 72:1374-1382.

———, D. L. Baker, G. D. Bear, and D. C. Bowden. In press *a*. Ungulate grazing in sagebrush grassland II: effects of resource competition on secondary production. Ecol. Appl.

———, ———, ———, and ———. In press *b*. Ungulate grazing in sagebrush grassland I: mechanisms of resource competition. Ecol. Appl.

Holechek, J. L. 1988. An approach for setting the stocking rate. Rangelands 10:10-14.

———, and M. Vavra. 1983. Drought effects on diet and weight gains of yearling heifers in northeast Oregon. J. Range Manage. 36:227-231.

———, R. D. Pieper, and C. H. Herbel. 1989. Range management principles and practices. Prentice Hall, Englewood Cliffs, N.J. 501pp.

———, and ———. 1992. Estimation of stocking rate on New Mexico rangelands. J. Soil and Water Conserv. 47:116-119.

Hood, R. E., and J. M. Inglis. 1974. Behavioral responses of white-tailed deer to intensive ranching operations. J. Wildl. Manage. 38:488-498.

Houston, D. B. 1982. The northern Yellowstone elk, ecology and management. Macmillan Publ., New York, N.Y. 474pp.

Hudson, R. J., D. M. Hebert, and V. C. Brink. 1976. Occupational patterns of wildlife on a major East Kootenay winter-spring range. J. Range Manage. 29:38-43.

Hurley, M. A., and G. A. Sargeant. 1991. Effects of hunting and land management on elk habitat use, movement patterns, and mortality in western Montana. Pages 94-98 *in* A. G. Christensen, L. J. Lyon, and T. N. Lonner, comps. Proc. elk vulnerability symposium. Montana State Univ., Bozeman.

Huston, J. E., and W. E. Pinchak. 1991. Range animal nutrition. Pages 27-63 *in* R. D. Heitschmidt and J. W. Stuth, eds. Grazing management, an ecological perspective. Timber Press, Portland, Oreg.

Irwin, L. L., J. G. Cook, R. A. Riggs, and J. M. Skovlin. 1994. Effects of long-term grazing by big game and livestock in the Blue Mountains forest ecosystem. U.S. Dep. Agric. For. Serv. Gen. Tech. Rep. PNW-325. 49pp.

———, and J. M. Peek. 1983. Elk habitat use relative to forest succession in Idaho. J. Wildl. Manage. 47:664-672.

Jackson, H. H. T. 1944. Big game resources of the United States. U.S. Dep. Inter. Bur. of Sport Fish. and Wildl. Res. Rep. 8. 56pp.

Jacobs, L. 1991. Waste of the West: public lands ranching. Jacobs Publ., Tucson, Ariz. 602pp.

Janis, C. 1976. The evolutionary strategy of the Equidae and the origins of rumen and cecal digestion. Evolution 30:757-774.

Jeffrey, D. E. 1963. Factors influencing elk and cattle distribution on the Willow Creek summer range, Utah. M.S. Thesis, Utah State Univ., Logan. 46pp.

Jenkins, K. J., and E. E. Starkey. 1992. Food habits of Roosevelt elk. Rangelands 13:261-265.

Johnson, B. K., A. Ager, S. Crim, M. Wisdom, D. Sheehy, and S. Findholt. In press. Allocating forage among wild and domestic ungulates: case example for the Blue Mountains. *In* W. D. Edge and D. Pyke, eds., Proc. sustaining rangelands symposium, Oregon State Univ., Corvallis.

Jordan, L. A., and J. P. Workman. 1989. Overview of fee hunting for deer and elk on private land in Utah. Trans. North Am. Wildl. and Nat. Resour. Conf. 54:190-195.

Jourdannais, C. S., and D. J. Bedunah. 1990. Prescribed fire and cattle grazing on an elk winter range in Montana. Wildl. Soc. Bull. 18:232-240.

Julander, O., and D. E. Jeffrey. 1964. Deer, elk, and cattle range relations on summer range in Utah. Trans. North Am. Wildl. and Nat. Resour. Conf. 29:404-413.

Kimball, J. F., Jr., and M. L. Wolfe. 1985. Elk management opportunities in northern Utah: to feed or not to feed. Pages 191-197 *in* G. R. Workman, ed., Proc., western elk management, a symposium. Utah State Univ., Logan.

Klein, D. R. 1968. The introduction, increase, and crash of reindeer on St. Matthew Island. J. Wildl. Manage. 32:350-367.

Knowles, C. J. 1975. Range relationships of mule deer, elk, and cattle in a rest-rotation grazing system during summer and fall. M.S. Thesis, Montana State Univ., Bozeman. 111pp.

———, and R. B. Campbell. 1982. Distribution of elk and cattle in a rest-rotation grazing system. Pages 46-60 *in* J. M. Peek and P. D. Dalke, eds. Wildlife-livestock relationships symposium: proceedings 10. Univ. Idaho, Wildl. and Range Exp. Stn., Moscow.

Komberec, T. J. 1976. Range relationships of mule deer, elk, and cattle in a rest-rotation grazing system during winter and spring. M.S. Thesis, Montana State Univ., Bozeman. 79pp.

Krueger, W. C., and A. H. Winward. 1974. Influence of cattle and big game grazing on understory structure of a Douglas fir-ponderosa pine-Kentucky bluegrass community. J. Range Manage. 27:450-453.

Kufeld, T. P. 1973. Foods eaten by the Rocky Mountain elk. J. Range Manage. 26:106-112.

Lacey, J. R., K. Jamtgaard, L. Riggle, and T. Hayes. 1993. Impacts of big game on private land in southwestern Montana: landowner perceptions. J. Range Manage. 46:31-37.

Lanka, R. P., and R. J. Guenzel. 1991. Game farms: what are the implications for North American elk? Pages 285-291 *in* A. G. Christensen, L. J. Lyon, and T. N. Lonner, comps. Proc. elk vulnerability symposium. Montana State Univ., Bozeman.

———, E. T. Thorne, and R. J. Guenzel. 1992. Game farms, wild ungulates, and disease in Western North America. West. Wildlands 18:2-7.

Leege, T. A. 1979. Effects of repeated burns on northern Idaho browse. Northwest Sci. 53:107-113.

———, J. R. Nelson, and J. W. Thomas. 1977. Food habits and diet quality of North American elk. Pages 221-241 *in* M. Recheigel, ed. Handbook series in nutrition and food, section G: diets, culture media, and food supplements, volume 1. CRC Press, Cleveland, Oh.

Leptich, D. J., and P. Zager. 1991. Road access management effects on elk mortality and population dynamics. Pages 126-137 *in* A. G. Christensen, L. J.

Lyon, and T. N. Lonner, comps. Proc. elk vulnerability symposium. Montana State Univ., Bozeman.

Loft, E. R., J. W. Menke, and J. G. Kie. 1991. Habitat shifts by mule deer: the influence of cattle grazing. J. Wildl. Manage. 55:16-26.

Lonner, T. N., and R. J. Mackie. 1983. On the nature of competition between big game and livestock. Pages 53-58 *in* B. F. Roche, Jr., and D. M. Baumgartner, eds. Proc. forestland grazing symposium, Washington State Univ., Pullman.

Loomis, J., J. Cooper, and S. Allen. 1988. The Montana elk hunting experience: a contingent valuation assessment of economic benefits to hunters. Montana Dep. Fish, Wildl., and Parks, Helena. 47pp.

Lyon, L. J. 1983. Road density models describing habitat effectiveness for elk. J. For. 81:595-613.

———. 1985. Elk and cattle on the National Forests: a simple question of allocation or a complex management problem? West. Wildlands 11:16-19.

———, and A. L. Ward. 1982. Elk and land management. Pages 443-477 *in* J. W. Thomas and D. E. Toweill, eds. Elk of North America, ecology and management. Stackpole Books, Harrisburg, Pa.

———, and A. G. Christensen. 1992. A partial glossary of elk management terms. U.S. Dep. Agric. For. Serv., INT-288. 6pp.

Mack, R. N., and J. N. Thompson. 1982. Evolution in steppe with few large, hooved mammals. Am. Nat. 119:758-773.

Mackie, R. J. 1970. Range ecology and relations of mule deer, elk, and cattle in the Missouri River breaks, Montana. Wildl. Monogr. 20. 79pp.

———. 1981. Interspecific relationships. Pages 487-507 *in* O. C. Wallmo, ed. Mule and black-tailed deer of North America. Univ. Nebraska Press, Lincoln.

Marcum, C. L. 1975. Summer-fall habitat selection and use by a western Montana elk herd. Ph.D. Thesis, Univ. Montana, Missoula. 188pp.

Masupu, K. V. 1990. Selenium blood levels in free-ranging antelope (*Antilocapra americana*), elk (*Cervus elaphus* spp.), bighorn sheep (*Ovis canadensis* spp.) from California, seven western states, Mexico, and British Columbia, Canada. M.S. Thesis., Univ. California, Davis. 53pp.

Mautz, W. W. 1978. Nutrition and carrying capacity. Pages 321-348 *in* J. L. Schmidt and D. L. Gilbert, eds. Big game of North America, Stackpole Books, Harrisburg, Pa.

McCabe, R. E. 1982. Elk and Indians: historical values and perspectives. Pages 61-123 *in* J. W. Thomas and D. E. Toweill, eds. Elk of North America, ecology and management. Stackpole Books, Harrisburg, Pa.

McClean, A., and S. Wikeem. 1985. Influence of season and intensity of defoliation on bluebunch wheatgrass survival and vigor in southern British Columbia. J. Range Manage. 38:21-26.

McCorquodale, S. M., K. F. Raedeke, and R. D. Taber. 1986. Elk habitat use patterns in the shrub-steppe of Washington. J. Wildl. Manage. 50:664-669.

McCulloch, C. Y. 1986. A history of predator control and deer productivity in northern Arizona. Southwest. Nat. 31:215-220.

McCullough, D. R. 1969. The Tule elk: its history, behavior, and ecology. Univ. California Publ. Zool. Vol. 88. Univ. California Press, Berkeley and Los Angeles. 209pp.

McInnis, M. L., and M. Vavra. 1987. Dietary relationships among feral horses, cattle, and pronghorn in southeastern Oregon. J. Range Manage. 40:60-66.

McKeel, R. 1988. Status of elk winter feeding in Yakima and Kattites Counties, Washington. Pages 26-29 *in* M. Zahn, J. Pierce, and R. Johnson, eds. Proc., Western states and provinces 1988 elk workshop. Washington Dep. Wildl., Olympia.

McNaughton, S. J. 1985. Ecology of a grazing system: the Serengeti. Ecol. Monogr. 55:259-294.

———. 1991. Evolutionary ecology of large tropical herbivores. Pages 509-522 *in* P. W. Price, T. M. Lewinsohn, G. W. Fernandes, and W. W. Bensen, eds. Plant-animal interactions: evolutionary ecology in tropical and temperate regions. John Wiley and Sons, New York, N.Y.

Mereszcak, I. M., W. C. Krueger, and M. Vavra. 1981. Effects of range improvements on Roosevelt elk winter nutrition. J. Range. Manage. 34:184-187.

Miller, R. S. 1967. Patterns and process in competition. Adv. Ecol. Res. 4:1-74.

Morgantini, L. E., and R. J. Hudson. 1979. Human disturbance and habitat selection. Pages 132-139 *in* M. S. Boyce and L. D. Hayden-Wing, eds. North American elk: ecology, behavior, and management. Univ. Wyoming Press, Laramie.

Morrison, B. L. 1988. The Sikes Act in New Mexico: a bright beginning. Proc. West. Assoc. of Fish and Wildl. Agencies 68:106-117.

Mould, E. C., and C. T. Robbins. 1981. Nitrogen metabolism in elk. J. Wildl. Manage. 45:323-334.

Mueggler, W. F. 1967. Response of mountain grassland vegetation to clipping in southwestern Montana. Ecology 48:942-949.

———. 1975. Rate and pattern of vigor recovery in Idaho fescue and bluebunch wheatgrass. J. Range Manage. 28:198-204.

Murie, O. J. 1951. The elk of North America. Stackpole Books, Harrisburg, Pa. 376pp.

Nagle, J. P., and G. A. Harris. 1966. Grazing compatibility of cattle and elk on mountain ranges. Proc. Int. Grassland Congr., Helsinki, Finland. 10:994-997.

National Research Council. 1984. Nutrient requirements of beef cattle. Natl. Acad. Press, Washington, D.C. 90pp.

Nelson, J. R. 1982. Relationships of elk and other large herbivores. Pages 415-441 *in* J. W. Thomas and D. E. Toweill, eds. Elk of North America, ecology and management. Stackpole Books, Harrisburg, Pa.

———. 1984. A modeling approach to large herbivore competition. Pages 491-524 *in* Developing strategies for rangeland management. Westview Press, Boulder, Colo.

———, and D. G. Burnell. 1975. Elk-cattle competition in central Washington. Pages 71-83 *in* Proc. range multiple use shortcourse. Univ. Idaho, Moscow.

———, and T. A. Leege. 1982. Nutritional requirements and food habits. Pages 323-367 *in* J. W. Thomas and D. E. Toweill, eds. Elk of North America, ecology and management. Stackpole Books, Harrisburg, Pa.

Odum, E. P. 1971. Fundamentals of ecology. W. B. Saunders Co., Philadelphia, Pa. 574pp.

Olsen, F. W., and R. M. Hansen. 1977. Food relations of wild free-roaming horses to livestock and big game, Red Desert, Wyoming. J. Range Manage. 30:17-20.

O'Neil, J. 1985. Population status—management strategies and future research on elk in Arizona. Pages 35-43 *in* G. W. Workman, ed. Proc., western elk management, a symposium. Utah Agric. Exp. Stn., Utah St. Univ., Logan.

O'Toole, R. 1988. Reforming the Forest Service. Island Press, Washington, D.C. 250pp.

Owen-Smith, N. 1989. Morphological factors and their consequences for resource partitioning among African savanna ungulates: a simulation modelling approach. Pages 155-165 *in* D. W. Morris, Z. Abramsky, B. J. Fox, and M. R. Willig, eds. Patterns in the structure of mammalian communities. Texas Tech. Univ. Press, Lubbock.

Painter, G. G. 1980. Elk and cattle spatial interactions. Special M.S. Paper, Washington State Univ., Pullman. 93pp.

Peek, J. M., R. J. Pedersen, and J. W. Thomas. 1982. The future of elk and elk hunting. Pages 599-625 *in* J. W. Thomas and D. E. Toweill, eds. Elk of North America, ecology and management. Stackpole Books, Harrisburg, Pa.

———, R. A. Riggs, and J. L. Lauer. 1979. Evaluation of fall burning on bighorn sheep winter range. J. Range Manage. 32:430-432.

Perry, C., and R. Overly. 1977. Impact of roads on big game distributions in portions of the Blue Mountains of Washington, 1972-1973. Appl. Res. Bull. Wash. Dep. Game, Olympia. 39pp.

Pinchak, W. E., M. A. Smith, R. H. Hart, and J. W. Waggoner, Jr. 1991. Beef cattle distribution patterns on foothill range. J. Range Manage. 44:267-275.

Pitt, M. D. 1986. Assessment of spring defoliation to improve fall forage quality of bluebunch wheatgrass (*Agropyron spicatum*). J. Range Manage. 39:175-181.

Potter, D. R. 1982. Recreational use of elk. Pages 509-559 *in* J. W. Thomas and D. E. Toweill, eds. Elk of North America, ecology and management. Stackpole Books, Harrisburg, Pa.

Ralphs, M. H., and J. A. Pfister. 1992. Cattle diets in tall forb communities on Mountain rangelands. J. Range Manage. 45:534-537.

Ramsey, K. J., and W. C. Krueger. 1986. Grass-legume seeding to improve winter forage for Roosevelt elk: a literature review. Oregon State Univ. Agric. Exp. Stn. Rep. 763. 28pp.

Ranching Systems Group. 1992. Grazing land application user's guide. Dep. Rangeland Ecology and Management, Texas A&M Univ., College Station. 378pp.

Rennicke, J. 1992. Sacred cows? Backpacker 20:46-51.

Reynolds, H. G. 1964. Elk and deer habitat use of a pinyon-juniper woodland in southern New Mexico. Trans. North Am. Wildl. and Nat. Resour. Conf. 29:438-443.

———. 1966. Use of ponderosa pine forest in Arizona by deer, elk and cattle. U.S. Dep. Agric. For. Serv. Res. Note RM-63. 7pp.

Rhodes, B. D., and S. H. Sharrow. 1990. Effect of grazing by sheep on the quantity and quality of forage available to big game in Oregon's Coast Range. J. Range Manage. 43:235-237.

Rifkin, J. 1992. Beyond beef: the rise and fall of the cattle culture. Penguin Books, New York, N.Y. 353pp.

Robbins, C. T. 1993. Wildlife feeding and nutrition, second ed. Acad. Press, New York, N.Y. 352pp.

Roffe, T., and B. Smith. 1992. Tuberculosis: will it infect wild elk? Bugle 9:86-92.

Rowland, M. M., A. W. Alldredge, J. E. Ellis, B. J. Weber, and G. C. White. 1983. Comparative winter diets of elk in New Mexico. J. Wildl. Manage. 47:924-932.

Salter, R. E., and R. J. Hudson. 1980. Range relationships of feral horses with wild ungulates and cattle in western Alberta. J. Range Manage. 33:266-271.

Samuel, W. M., M. J. Pybus, D. A. Welch, and C. J. Wilke. 1992. Elk as a potential host for meningeal worm: implications for translocation. J. Wildl. Manage. 56:629-639.

Sanders, K., and J. Durham, eds. 1985. Proc., rangeland fire effects, a symposium. U.S. Bur. Land Manage., Boise, Idaho. 124pp.

Sauer, J. R. 1980. Elk-cattle interactions on elk calving areas. J. Colo.-Wyo. Acad. Sci. 12:7.

Schoener, T. W. 1982. The controversy over interspecific competition. Amer. Sci. 70:586-595.

Schwartz, C. C., and N. T. Hobbs. 1985. Forage and range evaluation. Pages 26-51 *in* R. J. Hudson and R. G. White, eds. Bioenergetics of wild herbivores. CRC Press, Boca Raton, Fla.

Seagle, S. W., and S. J. McNaughton. 1992. Spatial variation in forage nutrient concentrations and the distribution of Serengeti grazing ungulates. Landscape Ecol. 7:229-241.

Seton, E. T. 1927. Lives of game animals, vol. 3 part 1. Doubleday, Page, and Co., Garden City, N.J. 412pp.

Sheehy, D. P. 1988. Grazing relationships of elk, deer, and cattle on seasonal rangeland in northeastern Oregon. Ph.D. Thesis, Oregon State Univ., Corvallis. 269pp.

Skovlin, J. M. 1962. Cow and calf weight trends on mountain summer range in the Blue Mountains. U.S. Dep. Agric. For. Serv. Pac. Northwest Res. Stn. Note 220, Portland, Oreg. 7pp.

———. 1967. Fluctuations in forage quality on summer range in the Blue Mountains. U.S. Dep. Agric. For. Serv. PNW-44. 20pp.

———. 1982. Habitat requirements and evaluations. Pages 369-413 *in* J. W. Thomas and D. E. Toweill, eds. Elk of North America, ecology and management. Stackpole Books, Harrisburg, Pa.

———, P. J. Edgerton, and R. W. Harris. 1968. The influence of cattle management on deer and elk. Trans. North Am. Wildl. and Nat. Resour. Conf. 33:169-181.

———, and R. W. Harris. 1970. Management of woodland grazing resources for cattle, deer, and elk. Int. Grassland Congr. 11:75-78.

———, and M. Vavra. 1979. Winter diets of elk and deer in the Blue Mountains, Oregon. U.S. Dep. Agric. For. Serv., Gen. Tech. Rep. PNW-260. 21pp.

———, P. J. Edgerton, and B. R. McConnell. 1983. Elk use of winter range as affected by cattle grazing, fertilizing, and burning in southeastern Washington. J. Range Manage. 36:184-189.

Smith, A. D. 1965. Determining common use grazing capacities by application of the key species concept. J. Range Manage. 19:196-201.

Smith, B. L., and J. E. Wilbrecht. 1988. Supplemental winter feeding of elk on the National Elk Refuge, Wyoming. Pages 30-41 *in* M. Zahn, J. Pierce, and R. Johnson, eds. Proc. west. states and provinces 1988 elk workshop. Washington Dep. Wildl., Olympia.

Stark, R. H., Jr. 1973. Elk-cattle interactions on the Clockum Creek watershed. M.S. Thesis. Washington State Univ., Pullman. 87pp.

Starkey, E. E., D. S. de Calesta, and G. W. Witmer. 1982. Management of Roosevelt elk habitat and harvest. Trans. North Am. Wildl. and Nat. Resour. Conf. 47:353-362.

Stevens, D. R. 1966. Range relationships of elk and livestock, Cow Creek Drainage, Montana. J. Wildl. Manage. 30:349-363.

Stoddart, L. A., and A. D. Smith. 1955. Range management. McGraw-Hill Book Co., New York, N.Y. 433pp.

Stuth, J. W., J. R. Conner, W. T. Hamilton, D. A. Riegel, B. G. Lyons, B. R. Myrick, and M. J. Couch. 1990. RSPM: a resource systems planning model for integrated resource management. J. Biogeo. 17:531-540.

Swanson, G. A., J. T. Shields, W. H. Olson, E. Decker, R. B. Held, J. A. Strayer, R. Hill, G. Hunter, and G. Perdue. 1969. Fish and wildlife resources on the public lands. U.S. Dep. Comm. Clearinghouse for Fed. Sci. and Tech. Inf. Vol. 1. 233pp.

Thomas, J. W. 1984. Fee hunting on the public lands?—an appraisal. Trans. North Am. Wildl. and Nat. Resour. Conf. 49:455-468.

———. 1990. Wildlife. Pages 175-204 *in* R. N. Sampson and D. Hair, eds. Natural resources for the 21st century. Island Press, Covelo, Calif.

———. 1991. Elk vulnerability—a conference perspective. Pages 318-319 *in* A. G. Christensen, L. J. Lyon, and T. N. Lonner, comps. Proc. elk vulnerability symposium. Montana State Univ., Bozeman.

———, H. Black, Jr., R. J. Scherzinger, and R. J. Pedersen. 1979. Elk. Pages 104-127 *in* J. W. Thomas, ed. Wildlife habitats in managed forests, the Blue Mountains of Oregon and Washington. U.S. Dep. Agric. Handb. 553, Washington, D.C.

———, and J. M. Sirmon. 1985. The National Forests and land-use planning: keys to the future of elk and elk hunting. U.S. Dep. Agric. For. Serv., Pacific Northwest For. and Range Exp. Stn., La Grande, Oreg. 13pp.

———, D. A. Leckenby, M. Henjum, R. J. Pedersen, and L. D. Bryant. 1988. Habitat effectiveness index for elk on Blue Mountain winter ranges. U.S. Dep. Agric. For. Serv. Gen. Tech. Rep. PNW-218. 28pp.

Thorne, T., and G. Butler. 1976. Comparison of pelleted, cubed, and alfalfa hay as winter feed for elk. Wyo. Game and Fish Dep., Laramie. 38pp.

Tiedemann, A. R., and H. W. Berndt. 1972. Vegetation and soils of a 30-year deer and elk exclosure in central Washington. Northwest Sci. 46:59-66.

United States Department of Agriculture Forest Service. 1990. Livestock/big game interaction: activity review, Intermountain and Southwest Region. U.S. Dep. Agric. For. Serv., Washington D.C. 17pp.

Unsworth, J. W., L. Kuck, M. D. Scott, and E. O. Garton 1993. Elk mortality in the Clearwater Drainage of northcentral Idaho. J. Wildl. Manage 57:495-502.

Vales, D. J., V. L. Coggins, P. Matthews, and R. A. Riggs. 1991. Analyzing options for improving bull:cow ratios of Rocky Mountain elk populations in northeast Oregon. Pages 174-181 *in* A. G. Christensen, L. J. Lyon, and T. N. Lonner, comps. Proc. elk vulnerability symposium. Montana State Univ., Bozeman.

Vallentine, J. F. 1989. Range development and improvements. Academic Press, San Diego, Calif. 524pp.

Van Dyke, F. G., J. P. Dibenedetto, and S. C. Thomas. 1991. Vegetation and elk response to prescribed burning in south-central Montana. Pages 163-179 *in* R. B. Keiter and M. S. Boyce, eds. The Greater Yellowstone Ecosystem: redefining America's wilderness heritage. Yale Univ. Press, New Haven, Conn.

Van Dyne, G. M., P. T. Kortopates, and F. M. Smith. 1984. Quantitative frameworks for forage allocation. Pages 289-412 *in* Developing strategies for rangeland management. Westview Press, Boulder, Colo.

Van Poollen, H. W., and J. R. Lacey. 1979. Herbage response to grazing systems and stocking intensities. J. Range Manage. 32:250-253.

Vavra, M. 1992. Livestock and big game forage relationships. Rangelands 14:57-59.

———, M. McInnis, and D. Sheehy. 1989. Implications of dietary overlap to management of free-ranging large herbivores. Proc. West. Sect. Am. Soc. Animal Sci. 40:489-495.

Wagner, F. H. 1969. Ecosystem concepts in fish and game management. Pages 259-307 *in* G. M. Van Dyne, ed. The ecosystem concept in natural resource management. Acad. Press, New York, N.Y.

Wallace, M. C., and P. R. Krausman. 1987. Elk, mule deer, and cattle habitats in central Arizona. J. Range. Manage. 40:80-83.

Ward, A. L. 1971. In vitro digestibility of elk winter forage in southern Wyoming. J. Wildl. Manage. 35:681-688.

———. 1973. Elk behavior in relation to multiple uses on the Medicine Bow National Forest. Proc. West. Assoc. State Game and Fish Comm. 53:125-141.

———, J. J. Cupal, A. L. Lea, C. A. Oakley, and R. W. Weeks. 1973. Elk behavior in relation to cattle grazing, forest recreation, and traffic. Trans. North Am. Wildl. and Nat. Resour. Conf. 28:327-337.

Watkinson, J. H. 1983. Prevention of selenium deficiency in grazing animals by annual topdressing of pasture with sodium selenate. New Zeal. Vet. J. 31:78.

Westenskow-Wall, K. J., W. C. Krueger, L. D. Bryant, and D. R. Thomas. 1994. Nutrient quality of bluebunch wheatgrass regrowth on elk winter range in relation to defoliation. J. Range Manage. 47:240-244.
Westoby, M. 1974. An analysis of diet selection by large generalist herbivores. Am. Nat. 108:290-304.
White, R. J. 1987. Big game ranching in the United States. Wild Sheep and Goat International, Mesilla, N.M. 355pp.
Wickstrom, M. L., C. T. Robbins, T. A. Hanley, D. E. Spalinger, and S. M. Parish. 1984. Food intake and foraging energetics of elk and mule deer. J. Wildl. Manage. 48:1285-1301.
Williams, T. 1992. The elk-ranch boom. Audubon 94:14-20.
Williamson, L. L. 1992. Identifying shinola. Rangelands 14:81-84.
Wisdom, M. 1992. The Starkey Project: new technologies chase old questions about deer and elk management. West. Wildlands 18:32-38.
———, L. R. Bright, C. G. Carey, W. W. Hines, R. J. Pedersen, D. A. Smithey, J. W. Thomas, and G. W. Witmer. 1986. A model to evaluate elk habitat in western Oregon. U.S. Dep. Agric. For. Serv. Rep. R-6-F&WL-216-1986, Portland, Oreg. 36pp.
Witmer, G. W. 1989. Elk and crop damage in Pennsylvania. Proc. East. Wildl. Damage Control Conf. 4:220-224.
———. 1981. Roosevelt elk habitat use in the Oregon Coast Range. Ph.D Thesis. Oregon State Univ., Corvallis. 104pp.
———. 1990. Re-introduction of elk in the United States. J. Pa. Acad. Sci. 64:131-135.
Wondolleck, J. M. 1991. Public lands conflict and resolution: managing National Forest disputes. Plenum Press, New York, N.Y. 261pp.
Wright, H. A., and A. W. Bailey. 1982. Fire ecology of the United States and southern Canada. John Wiley and Sons, New York, N.Y. 499pp.
Wuerthner, G. 1992. Wall Creek Game Range: a dissenting view. Rangelands 14:8-11.
Yeo, J. F., J. M. Peek, W. T. Wittinger, and C. T. Kvale. 1993. Influence of rest-rotation cattle grazing on mule deer and elk habitat use in east-central Idaho. J. Range Manage. 46:245-250.

Chapter 11
GRAZING AND MULE DEER

James M. Peek and Paul R. Krausman

INTRODUCTION

While much concern has been directed at the adverse effects of livestock grazing upon mule deer (*Odocoileus hemionus*) and other wildlife, there is increasing emphasis on adjustments in grazing systems to accommodate deer, innovative management of grazing to enhance mule deer forage and cover, and use of livestock to modify vegetation cover. We reviewed earlier work to illustrate the range of relationships between livestock and mule deer, and emphasize contemporary aspects of active range management that integrate livestock and mule deer management, which points the way to the future.

Mule deer populations have fluctuated extensively over the past 25 years, declining in the 1960s and 1970s and increasing in the 1980s (Flather and Hoekstra 1989) (Fig. 1). Mule deer are the most common big game animal in the Rocky Mountain and Pacific states. Reasons for declines in the 1970s were related to adverse changes in habitat (Wallmo 1978), liberal hunting regulations (Flather and Hoekstra 1988), habitat loss, weather, road development, human harassment, and grazing (Connolly 1981). However, by 1988 populations exceeded 3 million and may still be increasing (Flather and Hoekstra 1988).

Domestic livestock grazing may interact with mule deer by altering plant succession to favor or reduce forage and cover. Changes in plant productivity may also affect mule deer. Disturbance of mule deer by livestock and associated human activity may also occur. Mule deer responses may include changes in distribution patterns and habitat use, modification of activity, or alteration in population density.

Much of the early history of mule deer and livestock relationships may be traced to the famous Kaibab incident of the 1920s (Rasmussen 1941, Leopold et al. 1947, Russo 1964, Caughley 1970). This case history was used extensively as an example of how a deer population may increase to levels that cause damage to range plants when predators are removed and no population controls are implemented. However, Russo (1964) pointed out that changes in plant communities by earlier

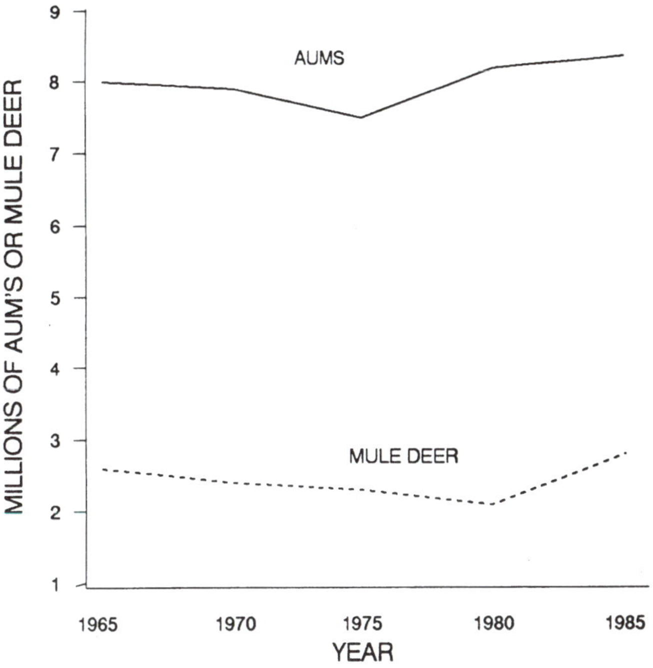

FIGURE 1. Trends in livestock and mule deer in the Rocky Mountain States (Flather and Hoekstra 1988). AUM = animal unit months.

heavy grazing by livestock may have predisposed the mule deer irruption by enhancing important forage plants. After livestock were reduced and hunting pressures on the deer were also decreased, the combination of high quality habitat, favorable climate, and low human exploitation may have allowed the deer to increase (Krausman et al. 1992). Regardless, the history was not well documented, and the correlations were inconclusive; Caughley (1970) concluded that little was to be learned from further examination. Still, the influences of predators and livestock on mule deer and their habitats were indelibly impressed upon the public consciousness (Krausman et al. 1992).

Wagner (1978) noted a general decline in livestock on public rangelands has occurred over the last half-century, even as livestock numbers in the Western states increased, due to more cattle on other land ownerships. In the 1980s, livestock grazing increased on federally administered lands, after lows in 1975 attributable to market conditions (Fig. 1). The mule deer population low of the mid-1970s correlates with the low grazing levels on public lands, and the subsequent increase in mule deer correlates with the increase in livestock grazing. The correlations are spurious because they are too general and the trends are unrelated. However, the common tendency to generalize the relationship between livestock and wildlife is fueled with such generalizations. Obviously, the relationship between mule deer and livestock varies locally, depending on ecological condition, and generalities concerning effects of livestock on mule deer habitat use and populations are not warranted.

FACTORS THAT AFFECT COMPETITION FOR FORAGE

Forages used by the various livestock species and mule deer are an obvious source of potential competition. However, simple comparisons of diet items will not provide adequate information on whether competition is occurring. Cole (1958) listed 4 items that need to be assessed before forage competition can be concluded: (1) both species have to use the same area, although not necessarily at the same time, (2) the diet of the 2 species has to overlap, (3) >1 item in the diet has to be important to both species, and (4) that item has to be shown to be declining in productivity or as part of the plant community as a result of the combined use. Item (1) can be assessed by a variety of means, including direct observation, tracks or pellet and chip counts, and radio-frequency marked or other types of marked animals. The diet can be assessed by direct examination of bite counts, use of conditioned animals, and from rumen or fecal analysis (Wallmo et al. 1973). However, knowledge of what each species is eating on the area of concern is needed, so methodology used must be able to identify foraging patterns by specific plant community. When plant species of concern occupy >1 community type, direct observation and use of experimental animals appears more appropriate than the indirect methods of fecal and rumen analyses, which do not identify the locations where each plant species was fed upon. Finally, an examination of range condition and trend, which includes utilization estimates of the critical forage species by deer and livestock, is necessary before declines in productivity can be demonstrated.

Forage Choice by Mule Deer

Kufeld et al. (1973) synthesized the mule deer diet literature (99 references) into patterns of seasonal composition. Shrubs and trees averaged > 74% of the winter diet (range = 27-100%), forbs averaged 15% (range = 0-43%), and grasses and grasslikes (sedges and rushes) 11% (range = 0-53%). The several species of mountain mahogany (*Cercocarpus* spp.), cliffrose (*Cowania mexicana*), Mormon tea (*Ephreda* spp.), horizontal juniper (*Juniperus horizontalis*), Pachystima (*Pachystima myrsinites*), bitterbrush (*Purshia tridentata*), oak (*Quercus* spp.), and yucca (*Yucca* spp.) received high ratings in ≥2 different studies.

Spring diets included shrubs and trees (49%, range = 6-92%), grasses and grasslikes (26%, range = 4-64%), and forbs (25%, range = 0-43%). Use of grasses was highest in spring, with bromes (*Bromus* spp.), junegrass (*Koeleria cristata*), and bluegrasses (*Poa* spp.) receiving high ratings in ≥2 different studies. These species initiate growth early in spring or in fall if moisture conditions are good, and are a readily available source of digestible nutrients when other species are still dormant. Willms and McLean (1978) reported that Sandberg bluegrass (*Poa secunda*) was an important spring forage for mule deer in south-central British Columbia.

Summer diets included shrubs (49%, range = 12-95%), and forbs (46%, range = 3-77%), with grasses and grasslikes being very low. Important forbs in ≥2 studies included aster (*Aster chilensis*), alfalfa, sweetclover (*Melilotus* spp.), dandelions (*Taraxacum* spp.), salsify (*Tragopogon* spp.), clover (*Trifolium* spp.), and mule's ear (*Wyethia* spp.). The forbs that were palatable are domesticated, latex producing, or large leafed, with deer taking only parts of the leaves.

The fall diet showed an increase in shrubs and trees to 60% (range = 3-97%), while forbs declined to an average of 30% (range = 2-78%), and grasses and grasslikes 9% (range = 0-24%). Important plants included those listed plus quaking aspen (*Populus tremuloides*).

In central Arizona, desert mule deer (*O. h. crooki*) diets vary significantly from season to season, within years and between years due to phenological development and climatic influences on the relative availability of different types of forage (McCulloch 1973). Overall, the vegetation diversity provides ample choices for mule deer to be highly opportunistic feeders. For example, in some seasons and years fruits were common in the diets. In years of poor fruit production other forage is used. Variability in available forage is a result of unpredictable rains, drought, and other climatic factors. One hundred and six species were identified in desert mule deer diets in central Arizona. Farther south in the Lower Sonoran Desert there are fewer plants and Krausman et al. (1989) reported only 28 species were consumed by desert mule deer.

The general diet picture for mule deer emphasizes the high variability across the range depending on what is available. For Rocky Mountain mule deer, the species listed appear to be especially important, and seasonal patterns are apparent. Mushrooms, lichens, acorns, and horsetails may become locally important if present. Burning or grazing that removes standing litter and stimulates early growth of bluebunch wheatgrass (*Agropyron spicatum*) may increase its palatability to mule deer (Willms et al. 1980).

Forage Choice by Cattle

Forage use by livestock changes with intensity of use of a range, between areas, seasons, and years, as much as with mule deer (Vallentine 1990). Grasses generally predominate in the summer diet of range cattle, but forbs may constitute a significant proportion in some areas (Vallentine 1990:227). Forbs appear to be used more in spring and when grazing is light. Highest forb use occurred in late spring on eastern Oregon grasslands (Holechek et al. 1982*a*). Shrubs and trees typically receive light use during all seasons, but may be a significant part of the diet during some years and in some areas. Browse, including ninebark (*Physocarpus malvaceous*), spiraea (*Spiraea betulifolia*), and snowberry (*Symphoricarpos alba*) constituted 23% of the spring-summer-fall diet of range cattle on eastern Oregon forested rangelands (Holechek et al. 1982*b*). Important grass species include buffalograss (*Buchloe dactyloides*), bluestems (*Andropogon* spp.), and switchgrass (*Panicum virgatum*) in southern prairie (Dwyer 1961); Arizona fescue (*Festuca arizonica*), mountain muhley (*Muhlenbergia montana*), and little bluestem on pine-bunchgrass ranges in Colorado (Johnson and Reid 1958); bluebunch wheatgrass and Idaho fescue (*Festuca idahoensis*) on Montana mountain rangeland (Schmautz 1954); and bluebunch wheatgrass, Idaho fescue, junegrass, elk sedge (*Carex geyeri*), and pine grass (*Calamagrostis rubescens*) in eastern Oregon ponderosa pine rangeland (Harris 1954). The contrast between cattle and mule deer forage preferences is best exemplified by the grasses: cattle concentrate on the most abundant grasses that often are dominants while mule deer tend to use subdominant species such as the bluegrasses, fine-leafed species such as fescues, or tips of leaves of species such as crested wheatgrass (Table 1).

Forage Choice by Domestic Sheep

The diet of domestic sheep typically includes more forbs than that of cattle, but again the variation between areas needs to be emphasized (Vallentine 1990:227). Woody plants typically comprise a higher proportion of the diet for sheep than cattle, but grasses are typically significant during all seasons. In western Montana, domestic sheep preferred forbs when grazing an area for the first time, with the proportion of grasses in the diet increasing with subsequent passes (Stevens 1966). Smith and Julander (1953) found that 6 of 10 species prominent in the summer diet of mule deer were also prominent in the diet of domestic sheep. They also reported heavy use of mountain mahogany and bitterbrush in winter by sheep. The similarity of diets between deer and sheep is typically higher than between deer and cattle, but Longhurst et al. (1979) point out that when sufficient forage of all categories is available for deer and sheep, there is not competition, which is consistent with Cole's (1968) criteria for competition. The long-term investigations of Longhurst et al. (1979) at the Hopland Field Station in central California did not indicate any significant competition between deer and sheep except for green forage in fall that is in short supply. In this area, presence of deer and sheep on oak woodland range was concluded to make more efficient use of forage than if only a single species was present.

TABLE 1. Relative values of some common grasses as forage for mule deer and cattle.

Grass species	Rating for mule deer	Rating for cattle
Arizona fescue (*Festuca arizonica*)	Light[1]	Excellent[3]
Bluebunch wheatgrass (*Agropyron spicatum*)	Good[2]	Excellent[2]
Cheatgrass (*Bromus tectorum*)	Moderate	Good[2]
Crested wheatgrass (*Agropyron cristatum*)	Fair[2]	Good[2]
Hairgrass (*Deschampsia caespitosa*)	Fair-good[2]	Excellent[2]
Idaho fescue (*Festuca idahoensis*)	Excellent	Excellent[2]
Kentucky bluegrass (*Poa pratensis*)	Light	Good[2]
Mountain brome (*Bromus carinatus*)	Good[2]	Excellent[2]
Needle and thread (*Stipa comata*)	Moderate	Fair-good[2]
Oatgrass (*Danthonia intermedia*)	Good-excellent[2]	Excellent[2]
Orchardgrass (*Dactylis glomerata*)	Moderate	Excellent[2]
Pinegrass (*Calamagrostis rubescens*)	Light	Fair[2]
Ricegrass (*Oryzopsis hymenoides*)	Good	Good[2]
Sandberg bluegrass (*Poa secunda*)	Moderate	Good[2]
Timothy (*Phleum pratense*)	Good-Excellent	Excellent[2]
Western wheatgrass (*Agropyron smithii*)	Fair	Good[2]

[1] Kufeld (1973).
[2] Stubbendieck et al. (1992).
[3] Humphrey (1970).

Forage Choice by Domestic Goats

Domestic goats are browsers, although high proportions of grasses may occur in the diet in some areas (Vallentine 1990:227). The potential for overlap in use of forage between goats and mule deer for woody plants as forage must be considered high, and management of levels of forage utilization will serve to minimize competition.

Potential Overlaps in Forage Choice: Summary

Reasons for different forage choice between species involve body size, type of digestive system (i.e., caecal or ruminant), rumino-reticular volume to body rate ratio that indicates forage that is most efficiently digested, and mouth size (Hanley and Hanley 1982). Body weights of mule deer (56-74 kg) and domestic sheep (56-74 kg) overlap extensively, but rumino-reticular volume to body weight ratio ranges from 0.160-0.333 in sheep, 0.103-0.105 in mule deer, and 0.150-0.264 in cattle. Weight of fresh rumino-reticular contents as a proportion of body weight was typically less for mule deer than for domestic sheep or cattle. Cattle and domestic sheep, having proportionately larger rumens than mule deer, should be able to digest forage of lower quality than mule deer. Thus, diet selectivity should be highest with the mule deer that requires a higher quality diet than for either of the other 2 species. Hanley and Hanley (1982) reported that the efficient digestion of browse by mule deer is consistent with the relatively small rumen capacity, small mouth, and small body size. Generally, sheep and cattle diets will be more variable than mule deer diets according to the predictions.

Drought and intensity of grazing, however, are significant factors that influence the potential for competition for forage between mule deer and livestock (Julander 1955). For example, Mackie (1970) found that when the highly palatable forb yellow sweetclover (*Melilotus officinalis*) is reduced in abundance because of low precipitation, greater use of shrubs and grass by deer and cattle may result. If grazing becomes heavy, then competition for scarce forage could occur. Diet choice may be expected to vary between years for all species, depending upon what is available and what is palatable. Typically, crude protein levels in cattle and sheep diets when forage is actively growing will occur at 10-12 and 12-16%, respectively (Holechek et al. 1989). Urness (1981) reported that important shrubs in mule deer diets contained crude protein levels ranging from 10 to 14%. Several species may provide the level of nutrients that are needed; flexibility in forage choice is to be emphasized; and species diversity in the plant communities is obviously important. Heavy grazing or drought that reduces plant species diversity or plant productivity, in turn can increase the potential for forage competition among the different species (Skovlin et al. 1968, Willms et al. 1979). Austin and Urness (1985) reported that late summer, when forage was considerably desiccated, was a critical period for forage quality for mule deer. Typically, forage availability in early summer is much higher than later when forages become senescent. While the range manager cannot anticipate the weather, the implications are obviously to keep rangelands productive with a diversity of forage so animals have opportunities to exercise free choice and minimize potential competition.

In winter, forage use and species abundance are influenced by snow depth (Carpenter et al. 1979). Although combinations of forbs, grasses and browse may be sought by mule deer during snow-free periods, snow conditions may preclude use of forbs and grasses that are covered and otherwise would be used during the season. Wallmo et al. (1977) concluded that for a Colorado winter range where grasses and shrubs were typically the usable forage for a 60-day period, duration and severity of winter was critical in determining the length of time deer will survive on that diet. If livestock are grazed on winter ranges prior to snowfall, reductions in available forage should be minimized when mule deer are forced to concentrate on limited areas with naturally sparse forage supplies in winter.

Habitat Use Patterns

Mule deer are capable of including a wide variety of habitats in their natural range. Habitat use patterns are influenced by population density, weather patterns, forage quality and availability, habitat availability, human activities, and by the presence of other ungulate species. Loveless (1967) identified environmental variables in Colorado that influenced mule deer distribution patterns and that may characterize habitat relationships in general. Deer responded to high air temperatures and low atmospheric moisture, low air temperatures associated with high winds or high atmospheric moisture or both, presence or absence of snow, duration of direct sunlight, availability and abundance of preferred forage, and interspersion of sites with these species, and the proximity of feeding areas to locations providing food and cover.

Mackie (1983) characterized mule deer habitats in Montana in terms of complexity, diversity, and stability, indicating that a general relationship to carrying capacity (i.e., deer density) existed. Mountain-foothill habitats containing topographical variation, wide distribution along an elevation gradient, and a variety of plant communities ranging from riparian zones through shrub-grassland to timbered slopes may provide the most stable mule deer habitats, as compared to breaklands and plains. The mountain-foothill regions provide a variety of microsites where plant phenology differs, thus affording mule deer opportunities to select plants of high nutrient content over a longer period than in habitats of less topographical variation where plant phenology is not as variable. In addition, mountain-foothills provide better protection during winter storms, prolonged severe winters, and the attendant late springs than do less complex regions. This conceptual scenario illustrates a pattern of natural intrinsic variation in habitats throughout the range of the mule deer that fundamentally affects population density and distribution, notwithstanding other factors such as predation, hunter harvest, and competition. Habitats with less diversity, stability, and complexity can be expected to sustain deer populations at generally lower levels, but that may also fluctuate more with severe winter weather or summer drought.

Within this concept of habitat quality, distribution and numbers of mule deer will depend upon a variety of interacting proximate factors, that may include livestock grazing intensity, human activity including hunter harvest, predators, and the prevailing weather pattern. Because these factors interact and are variable in time and space, the system is complex and seldom well understood, especially for any period of time. However, mule deer are fundamentally a product of the land, and just as efforts to distinguish between the influence of grazing and the influence of varying precipitation patterns on rangeland vegetation are difficult, so it is difficult to distinguish between the effects of grazing and the effects of varying precipitation patterns on mule deer numbers and distribution. Nevertheless, when drought or prolonged severe winter occurs, livestock grazing may add to the effects of the natural adverse situation. These circumstances can be accentuated in areas of low habitat diversity, complexity, and stability, such as the desert and plains regions.

Differences exist in habitat preferences of cattle and mule deer. Cattle generally prefer to use the most gentle terrain near water, while mule deer will use steeper terrain at distances further from water sources (Julander and Jeffrey 1964, Mackie 1970, Mueggler 1965). However, if forage elsewhere is depleted or terrain is generally steep, cattle will use steep slopes and riparian zones may provide preferred habitats for mule deer (Yeo et al. 1993). In the Southwest, xeroriparian habitat is the most widely used by desert mule deer (Krausman et al. 1985). Julander and Jeffrey (1964) reported that grass-forb communities preferred by cattle were least preferred by mule deer in Utah, although mixed shrub, aspen, and oak communities were used extensively by both species. In eastern Oregon, mule deer made heavy use of forested areas in summer while cattle preferred the grasslands (Skovlin et al. 1968).

Mule deer habitat use patterns are influenced by the intensity of grazing by cattle. Dusek (1975), Griffith and Peek (1989), and Loft et al. (1991) found that deer avoided areas being used by high concentrations of cattle. In the Sierra Nevada, California, overlap in diet and habitat use was low until late summer and early fall on heavily grazed areas (Loft et al. 1991). Wallace and Krausman (1987) documented fewer mule deer on pastures grazed by cattle than on pastures not grazed by cattle on a Ponderosa pine-bunchgrass range in central Arizona. In a southeastern Arizona grass-shrubland, deer used currently ungrazed portions of their home range and dry wash habitats more than expected (Ragotzkie and Bailey 1991). Heavily grazed pastures were used less by mule deer than other areas in eastern Oregon (Skovlin et al. 1968). Kie et al. (1991) found that mule deer altered their behavior by foraging more and resting less on ranges that had increased cattle stocking rates. Hiding cover for fawns was also reduced by these ranges.

Mule deer have shifted habitat use relative to the presence of cattle where high concentrations of livestock or heavy grazing has been observed. Such shifts, probably a gradual and continuing pattern where mule deer and cattle coexist on a range, are not generally of much significance, unless grazing is intensive. Variation in habitat use patterns is expected and attributable to a variety of factors, including the presence of livestock.

Domestic sheep using open rangelands typically occupy upland areas rather than riparian zones. Jensen et al. (1972) reported that spring grazing by domestic sheep on big game winter range was primarily concentrated on forbs, thus minimizing forage competition. Mid-summer grazing on winter ranges increased chances for competition because sheep browsed extensively on bitterbrush and other shrubs. Jensen et al. (1972) recommended that grazing periods be timed so use of shrubs by domestic livestock does not exceed specified levels, forbs would be used sufficiently to minimize competition with shrubs, and time and intensity of use be consistent with other uses and values of the rangelands. Frequently, big game winter ranges are also early-spring ranges because the spring greenup occurs first on the ridgetops and south-facing slopes at low elevations. Since in early spring mule deer prefer grasses and early-season forbs, domestic livestock grazing should be regulated to ensure diversity and vigor of early-season forbs and grasses on these ranges.

PLANT SUCCESSION

Early studies of the relationship between mule deer and livestock questioned whether grazing-induced changes in plant communities favored or diminished their abilities to support mule deer. Concerns today about efforts to increase grass species productivity at the expense of other species through special grazing systems (Mackie 1978) reflect the same issue.

Several approaches to this problem are available, including examination of the effects of livestock grazing on important deer forage plants, and deer use of habitats representing different seral stages induced by grazing. Branson (1985) reviewed vegetation changes on western rangelands that comprise much of the important mule deer habitat. The sagebrush type covers between 35 and 59 million ha, and is one of the more important deer habitats. While disagreement has prevailed over the original composition of much of the area presently covered by sagebrush, Branson (1985) concluded that a general decrease in perennial grasses and associated increase in shrubs had taken place. Widespread establishment of exotics such as cheatgrass (*Bromus tectorum*), medusahead (*Taeniatherum caput-medusae*), halogeton (*Halogeton glomeratus*), Russian thistle (*Salsola kali*), tansy mustard (*Descurainia* spp.), and tumble mustard (*Sisymbrium altissimum)* has occurred also. Young et al. (1976) summarized much prevailing opinion by concluding that increases in native shrubs, reductions in grasses and forbs, and establishment of exotics adapted to intensive grazing have occurred in this type, although Branson (1985) concluded that more recently range conditions have improved. Current efforts to improve range conditions through more intensive livestock management are widespread, so the earlier trends may be reversed as current management becomes more broadly effective.

Sagebrushes may comprise a moderate to heavy part of the winter and spring diet of mule deer (Kufeld et al. 1973), when forbs are unavailable and grasses are cured. Associated native shrubs such as serviceberry, bitterbrush, mountain mahogany, cliffrose, and shiny-leaf ceanothus were all rated as moderately to heavily used by deer. Thus, increases in these shrubs may have benefited mule deer, except that declines in palatable forbs and grasses may offset those benefits. Very often, the winter ranges for mule deer are used during spring after greenup is initiated. Mule deer using less nutritious woody browse forage will switch to green grasses and forbs when they appear in spring without shifting habitats.

Exotics including cheatgrass, tansy mustard, filaree (*Erodium cicutarium*), and tumble mustard are grazed by mule deer. Wyethia, dandelion, salsify (*Tragopogon dubius*), and the eriogonums (*Eriogonum* spp.), which would be expected to increase with cattle grazing pressure, commonly are used by mule deer also. However, balsom-root (*Balsamorhiza* spp.), tall asters (*Aster* spp.), bluebells (*Mertensia* spp.), and geraniums (*Geranium* spp.), which would be expected to decrease with livestock grazing, may be locally important mule deer foods.

A wide variety of conditions resulted from use of the sagebrush type over the past century. Where shrub communities became so dense that understory herbs and grasses important to deer declined, habitats would deteriorate. Where palatable species were replaced by other palatable species that were used in equivalent ecological circumstances, then the situation was not significantly altered. Where palatable species increased, then mule deer habitat would improve.

Variants on this theme are available. Griffith and Peek (1989) reported that bitterbrush-dominated communities that were of high density and in good condition were preferred by mule deer over low density stands with low diversity in the understories in north-central Washington. Bodurtha et al. (1989) suggested that communities in the Palouse grassland type dominated by medusahead were not used by mule deer, and one may assume that sagebrush communities with understories consisting primarily of medusahead would be of low value except as midwinter habitats or for cover.

Shrub communities influenced by earlier heavy grazing have also continued to mature and become senescent (Urness 1990). The net result at this stage is unclear, because declines in shrub overstories resulting from senescence or other factors such as winter-kill may allow understory species to proliferate. If a diversity of plant species favorable to mule deer occurs, then shrub declines, for whatever reason, may benefit. However, the long history of heavy use of winter range shrubs by mule deer that has caused declines in shrub density may not be compensated for by improvements in understory cover and productivity.

The semidesert grasslands of the Southwest have also been subject to extensive modifications due to grazing and reduction of range fire (Branson 1985). A general trend from grass-dominated communities to shrub-dominated communities has occurred in this type, but the trend has occurred in grazed and ungrazed areas. Hastings and Turner (1965) attribute the increases in shrubs to a hotter and drier climatic trend. These grasslands are extremely important mule deer habitats, and probably became more so as shrubs proliferated.

When protection from sheep grazing occurs in California oak woodland, taller grasses increase and native legumes, composites, and filaree decrease (Longhurst et al. 1979). Without sheep grazing, native perennial grasses increase while annuals such as soft chess (*Bromis mollis*) and filaree that are palatable to deer decline. Longhurst et al. (1979) concluded that deer grazing alone does not maintain these grasslands in a productive condition for deer, but may help to minimize establishment of oak seedlings.

Rangeland conditions have generally improved since the drought of the 1930's (Branson 1985). This is typically interpreted to mean that more palatable livestock forage is becoming available, but not necessarily that better mule deer habitat is being created. Increases in livestock forage may mean that ground cover has improved, even if palatable mule deer forage species have declined, and long-term watershed values in relation to site potential are implicated. Mule deer habitat may benefit in some circumstances where plant species diversity is enhanced, and deteriorate in habitats where graminoids dominate and species palatable to deer decline.

Future population levels of mule deer will inevitably be tied to range conditions, and can be expected to vary between regions. Mule deer populations were apparently low prior to and during settlement in much of Utah, California, Nevada, and southern Idaho (Leopold et al. 1947, Leopold et al. 1951), where sagebrush type is the most common habitat. Gruell's (1986) survey of pre-1900 reports of mule deer suggested that populations were locally numerous in Montana, Wyoming, and Idaho. Probably, as range conditions continue to improve, mule deer populations will again exhibit variability depending upon local vegetation conditions, as much as they once did. Most deer herds that were above carrying capacity in the 1940s have been effectively controlled by hunting and habitat manipulation (Krausman et al. 1992). However, the extreme high populations of the 1950s and 1960s are not likely to occur again.

Predictions of future vegetation conditions strictly from current trends related to grazing management are at risk because climatic change is not considered (Svejcar and Brown 1991). Relatively short-term increases in atmospheric CO_2 levels may alter range conditions, favoring shrubs at the expense of grasses (Johnson and Lincoln 1990). While the predictions of climatic change are usually concerned with agriculture, native plant communities and wildlife habitats are obviously involved.

The argument that mule deer population levels will be tied to range conditions presupposes that population management is intensive enough to prevent exploitation that reduces populations below levels that would otherwise be sustained by habitat conditions. It also presupposes that other factors, including human intolerance, human disturbance, habitat changes related to climatic change, and loss of habitat do not adversely influence populations. It certainly indicates, as Mackie (1981) suggested, that population management will become more intensive. It also suggests that purposeful management to retain mule deer habitat, while certainly appropriate, may not take into account climatic variables that may override management and influence plant communities in ways that may or may not alter their capacity to support mule deer.

SPECIAL GRAZING SYSTEMS AND THEIR INFLUENCE ON MULE DEER

Use of deferred grazing; high-intensity, short duration grazing; and rest-rotation grazing systems to manage cattle and rangeland more effectively brings different interactions between mule deer and livestock. Very often these systems entail more fencing and water development. Fences that will permit passage of deer can be constructed to minimize interference with movement between pastures. Generally 4-stranded barbed wire with 30 cm between strands and the lowest 46 cm off the ground are recommended (Rouse 1954). Water developments may be used by deer, and guidelines such as those of Sherrets (1989) accommodate wildlife. However, water development may allow cattle or other ungulate species to graze areas previously not grazed and may displace deer.

Forage use by livestock must be kept at levels that will provide enough for mule deer or will enhance growth of forages palatable to deer following grazing, 2 very different situations. This assumes that the objective of the grazing system is merely to integrate grazing with retention of mule deer habitat, which would be the most common objective. However, if objectives include improvement of habitat by increasing shrub production, then special grazing management is indicated.

Mule deer avoided areas occupied by large numbers of cattle in Arizona (McIntosh and Krausman 1982, Wallace and Krausman 1987), Montana (Dusek 1975), Oregon (Skovlin et al. 1968), Nevada (Rule 1989), Washington (Griffith and Peek 1989), and California (Loft et al. 1991). This means that they will shift to other areas within their home ranges, and this may or may not be detrimental. Cohen et al. (1989) reported shifts by white-tailed deer (*O. virginianus*) as a result of short duration grazing in south Texas, but no direct evidence that this was deleterious to the deer was demonstrated.

In Arizona, mule deer were more abundant in areas ungrazed by cattle. When cattle grazed these same areas the occurrence of mule deer decreased (McIntosh and Krausman 1982, Wallace and Krausman 1987). Deer numbers decreased slowly as cattle were added to the range but they were consistently seen drinking and foraging with cattle. Hungerford (1970) and Ward et al. (1973) also found mule deer and cattle foraging close to each other. Knowles (1975) reported that mule deer either moved from areas inhabited by cattle or used all parts of their home ranges more frequently when cattle were present. Similar results were reported by Rule (1989). Ragotzkie and Bailey (1991) argue for leaving some desert areas ungrazed by livestock to provide a contingency for deer against impacts of cattle during drought.

Dusek (1975) suggested that deer avoided areas occupied by large concentrations of cattle, but argued that "dual use of the range by cattle and mule deer, even when both were utilizing the same forage species, was efficient land use because the combined use was not excessive." As support Dusek (1975) reported an ovulation rate of 2.0. Julander et al. (1961) concluded that good summer range is necessary to approach an ovulation rate of 1.95/female.

In an early study of range use by ungulates Skovlin et al. (1968) concluded that deer use was greater in game-only ranges than in dual use ranges (game and livestock). However, they did not find evidence of competition by cattle and livestock for any particular forage plant. They concluded that "conservative cattle stocking between the light and moderate level would give good overall production from these multiple-use ranges." Loft et al. (1991) reported that mule deer home ranges were larger with moderate and heavy cattle grazing, because of reductions of forage and cover abundance.

Loomis et al. (1991) evaluated the effect of 3 cattle grazing systems on summer range in Sierra Nevada, California, on deer harvest, number of hunters, and economic value of hunting. Continuous moderate grazing provided 82% of the estimated carrying capacity for mule deer when compared to no grazing. A rest-rotation system involving 1 or 2 years of nonuse of each pasture provided 88 or 94%, respectively, of the potential carrying capacity for mule deer. A system using 2 years of heavy use, and 1 year of nonuse reduced deer use to 73% of the no grazing potential. The number of hunters and deer harvest responded similarly to the number of deer in the systems, and suggested that reductions in cattle grazing to accommodate higher numbers of mule deer would result in increased hunter activity and harvest.

Grazing systems may concentrate livestock and affect mule deer distributions and diets. If mule deer redistribute themselves away from concentrations of livestock, adjacent areas that may be within the same or adjacent pasture system may provide satisfactory habitat. Deer may adjust their diet to use the available green regrowth and benefit. When mule deer inhabit the area, a grazing system can be developed with these attributes in mind.

USE OF LIVESTOCK TO ENHANCE MULE DEER HABITAT

An abundance of information illustrates means to coordinate grazing and mule deer habitat needs (Smith 1949; Urness 1982, 1990; Leckenby et al. 1982; Neal 1982; Peek et al. 1978). Also, removal of livestock may not maintain or increase shrub production on mule deer winter ranges. Urness (1990) concluded that efforts to enhance shrub retention and growth would require active manipulation of plant cover in many cases. Fire and grazing management are the most logical tools, because they are relatively economical and a body of experience in their use is available.

Livestock grazing can be used to improve vegetation conditions for mule deer. Early on, observations that browsing stimulated production of additional twigs illustrated the potential for retention or improvement by grazing (Jameson 1963). Gibbens and Schultz (1962) recognized that cattle might be used to alter shrub form and productivity to promote subsequent use by mule deer. Objectives may include removing old growth by grazing to stimulate production of new growth palatable to deer; reducing heights of shrubs to levels that are available to deer; removing old growth that blocks access to palatable forage for deer; and using forage less palatable to deer, reducing competition with the palatable deer forages.

Timing of grazing is critical in accomplishing objectives. On winter ranges of mule deer in Utah, domestic sheep preferred forbs in spring and early summer (Jensen et al. 1972). Shrubs comprised the major portions of the winter diet of mule deer, and sheep grazing tended to improve shrub production. Laycock (1967) reported that heavy late-fall grazing by sheep increased grasses and forbs at the expense of sagebrush while heavy spring grazing caused the opposite trends. On winter ranges where the shrub component is important to deer, spring grazing may be appropriate, while on ranges where the grass-forb component was important for deer, fall grazing would be more appropriate. Longhurst et al. (1979) concluded that sheep grazing maintained oak-woodland vegetation in a productive seral stage for deer by promoting growth and production of important deer forages. Rhodes and Sharrow (1990) found that sheep grazing from May to September may improve forage quality in fall and increase forage quantity in spring for deer in Oregon's Coast Range. Springtime sheep grazing of a Utah winter range which increased bitterbrush current annual growth and reduced the standing dead grass cover allowed mule deer to select a more nutritious diet that included more herbaceous material (Fulgham et al. 1982). Thus, where 2 species may have similar diet preferences, timing of livestock grazing can be adjusted to influence forage production to benefit deer.

Smith and Doell (1968) reported that cattle grazing could be used to drive plant composition in similar ways in Utah. These investigations demonstrated the utility of using livestock to manipulate forage composition and productivity rather than completely excluding livestock.

Horses, which graze primarily grasses, are a logical means to reduce grass competition with shrubs by appropriate controlled grazing. Reiner and Urness (1982) reported that grazing when bitterbrush plants were rapidly growing in early-July produced significant increases in production of this species. All experimental pastures grazed by horses showed increased bitterbrush twig production over the ungrazed control. These results are anticipated by earlier demonstrations that competition between herbaceous plants and shrub establishment can be severe (Holmgren 1956).

Goats may offer unique opportunities to practice integrated grazing management to benefit mule deer (Riggs and Urness 1989, Riggs et al. 1990). In northern Utah oakbrush winter range, goats preferentially grazed bitterbrush and Gamble oak (*Quercus gambelli*), which reduced production of those species but increased production of sagebrush. Mule deer responded by increasing use of sagebrush and rabbitbrush (*Chrysothamnus nauseosus*) when snow depths prevented use of understory plants. This actually enhanced dry matter digestibility of the diet without increasing energy expenditure. When deer could forage on understory plants, browsing-induced changes in forage by goats did not have a significant effect on mule deer diets, although increased herbaceous production occurred.

Long-term changes in vegetation composition that may occur when special grazing treatments are used should be monitored to ensure that trends are detected and modifications of grazing, if necessary, will be timely (Smith and Doell 1968). The conclusion of Longhurst et al. (1979) that sheep grazing promoted deer habitat quality was tempered by the lack of oak seedling establishment, which meant that over time lower mast crops could be expected. McConnell and Smith (1970, 1971) reported that grazing by cattle appeared to slightly increase average crown diameter of bitterbrush without increasing the percentage of dead crown area or altering distribution of plants.

Manipulation of timing and species of livestock to enhance mule deer range has been effectively demonstrated, and will hopefully find broader application in future habitat management programs for mule deer and other wildlife. While such range management programs may require more flexibility and coordination, benefits derived may more than offset the added efforts.

SUMMARY

Enough information is available to describe the patterns that make livestock grazing and mule deer compatible on rangelands. In situations where concentrations of livestock are probable, purposeful planning to provide adjacent cover useful for mule deer is indicated. In situations where livestock are being grazed to alter or maintain plant productivity, timing and intensity of grazing may be adjusted to provide for mule deer. Livestock and mule deer populations will have to be monitored and managed to ensure that range management objectives are being met and that the timing and intensity of grazing is appropriate. Situations occur where livestock may be used to benefit mule deer forage by adjusting timing and intensity of grazing so other forage species are used by the livestock, or so mule deer forage is grazed to improve its quality and/or availability. A knowledge of plant development, livestock and deer food choice, and habitat use patterns can be used to develop reasonable grazing management plans for any given region where mule deer and livestock are grazed. Although distributions and numbers have been affected in local areas, mule deer populations have fluctuated independently of livestock numbers across their range. We can now purposefully manage to benefit livestock, mule deer, and rangeland vegetation by establishing suitable objectives and developing grazing programs that are flexible and tailored to the specific situation.

LITERATURE CITED

Austin, D. D., and P. J. Urness. 1985. Values of four communities for mule deer on ranges with limited summer habitat. J. Range Manage. 38:167-170.

Bodurtha, T. S., J. M. Peek, and J. L. Lauer. 1989. Mule deer habitat use related to succession in a bunchgrass community. J. Wildl. Manage. 53:314-319.

Branson, F. A. 1985. Vegetation changes on western rangelands. Range Monogr. 2. 76pp.

Carpenter, L. H., O. C. Wallmo, and R. B. Gill. 1979. Forage diversity and dietary selection by wintering mule deer. J. Range. Manage. 32:226-229.

Caughley, G. 1970. Eruption of ungulate populations, with emphasis on Himalayan thar in New Zealand. Ecology 51:53-72.

Cohen, W. E., D. L. Drawe, F. C. Bryant, and L. C. Bradley. 1989. Observations on white-tailed deer and habitat response to livestock grazing in south Texas. J. Range. Manage. 42:361-365.

Cole, G. F. 1958. Big game-livestock competition on Montana's mountain rangelands. Montana Wildl. (Apr):24-30.

Connolly, G. E. 1981. Trends in populations and harvests. Pages 225-243 *in* O. C. Wallmo, ed. Mule and black-tailed deer of North America. Univ. Neb. Press, Lincoln.

Dusek, G. L. 1975. Range relations of mule deer and cattle in prairie habitat. J. Wildl. Manage. 39:605-616.

Dwyer, D. D. 1961. Activities and grazing preferences of cows with calves in northern Osage County, Oklahoma. Oklahoma State Univ. Agric. Exp. Stn. Bull. B-588. 61pp.

Flather, C. H., and T. W. Hoekstra. 1988. An analysis of the wildlife and fish situation in the United States: 1989-2040, U.S. Dep. Agric. For. Serv. Gen. Tech. Rep. RM-178.

Fulgham, K. O., M. A. Smith, and J. C. Malachek. 1982. A compatible grazing relationship can exist between domestic sheep and mule deer. Pages 458-478 *in* J. M. Peek and P. D. Dalke, eds. Wildlife-livestock relationships symposium: proceedings 10. Univ. Idaho, For., Wildl. and Range Exp. Stn. Moscow.

Gibbens, R. P., and A. M. Schultz. 1962. Manipulation of shrub form and browse production in game range improvement. Calif. Fish and Game 48:49-63.

Griffith, B., and J. M. Peek. 1989. Mule deer use of seral stage and habitat type in bitterbrush communities. J. Wildl. Manage. 53:636-642.

Gruell, G. E. 1986. Post-1900 mule deer irruptions in the intermountain west: principal cause and influences. U.S. Dep. Agric. For. Serv. Gen. Tech. Rep. INT-206. 37pp.

Hanley, T. A., and K. A. Hanley. 1982. Food resource partitioning by sympatric ungulates on Great Basin rangeland. J. Range Manage. 35:152-158.

Harris, R. W. 1954. Fluctuations in forage utilization on ponderosa pine ranges in eastern Oregon. J. Range Manage. 7:250-255.

Hastings, J. R., and R. M. Turner. 1965. The changing mile: an ecological study of vegetation change with time in the lower mile of an arid and semiarid region. Univ. Arizona Press, Tucson. 317pp.

Holechek, J. L., R. D. Pieper, and C. H. Herbel. 1989. Range management principles and practices. Prentice-Hall, Englewood Cliffs, N.J. 501pp.

———, M. Vavra, and R. D. Pieper. 1982*a*. Botanical composition determination of range herbivore diets-a review. J. Range Manage. 35:309-315.

———, ———, J. Skovlin, and W. C. Krueger. 1982*b*. Cattle diets in the Blue Mountains of Oregon, I. Grasslands. J. Range Manage. 35:109-112.

Holmgren, R. C. 1956. Competition between annuals and young bitterbrush. Ecology 37:370-377.

Humphrey, R. R. 1970. Arizona range grasses. Univ. Arizona Press, Tucson. 159pp.

Hungerford, C. R. 1970. Response of Kaibab mule deer to management of summer range. J. Wildl. Manage. 34:852-862.

Jameson, D. S. 1963. Responses of individual plants to harvesting. Bot. Rev. 29:532-594.

Jensen, C. H., A. D. Smith, and G. W. Scotter. 1972. Guidelines for grazing sheep on rangelands used by big game in winter. J. Range Manage. 25:346-352.

Johnson, R. H., and D. E. Lincoln. 1990. Sagebrush and grasshopper response to atmospheric carbon dioxide concentration. Oecologia 84:103-110.

Johnson, W. M., and E. H. Reid. 1958. Herbage utilization on pine-bunchgrass ranges in Colorado. J. Range Manage. 56:647-651.

Julander, O. 1955. Deer and cattle range relations in Utah. For. Sci. 1:130-139.

———, and D. E. Jeffrey. 1964. Deer, elk and cattle range relations on summer range in Utah. Trans. North Am. Wildl. and Nat. Resour. Conf. 29:404-414.

———, W. L. Robinette, and D. A. Jones. 1961. Relation of summer range condition to mule deer herd productivity. J. Wildl. Manage. 25:54-60.

Kie, J. G., C. J. Evans, E. R. Loft, and J. W. Minke. 1991. Foraging behavior by mule deer: the influence of cattle grazing. J. Wildl. Manage. 55:665-674.

Knowles, C. J. 1975. Range relationships of mule deer, elk, and cattle in a rest-rotation grazing system during summer and fall. M.S. Thesis, Montana State Univ., Bozeman. 111pp.

Krausman, P. R., B. D. Leopold, R. F. Seegmiller, and S. E. Torres. 1989. Relationships between desert bighorn sheep and habitat in western Arizona. Wildl. Monogr. 102. 66pp.

———, K. R. Rautenstrauch, and B. D. Leopold. 1985. Xeroriparian systems used by desert mule deer in Texas and Arizona. Pages 144-149 *in* R. R. Johnson, C. D. Ziebell, D. R. Patton, P. F. Ffolliott, and R. H. Hamre, tech. coord. Riparian ecosystems and their management: reconciling conflicting uses. U.S. Dep. Agric. For. Ser. Gen. Tech. Rep. RM-120.

———, L. K. Sowls, and B. D. Leopold. 1992. Revisiting overpopulated deer ranges in the United States. California Fish and Game 78:1-10.

Kufeld, R. C., O. C. Wallmo, and C. Feddema. 1973. Foods of the Rocky Mountain mule deer. U.S. Dep. Agric. For. Serv. Res. Pap. RM-111. 31pp.

Laycock, W. A. 1967. How heavy grazing and protection affect sagebrush-grass ranges. J. Range Manage. 20:206-213.

Leckenby, D. A., P. Sheehy, C. H. Nellis, R. J. Scherzinger, J. D. Luman, W. Elmore, J. C. Lemos, L. Doughty, and C. E. Trainer. 1982. Mule deer. Wildlife habitats in managed rangelands—the Great Basin of southeastern Oregon. U.S. Dep. Agric. For. Serv. Gen. Tech. Rep. PNW-139. 40pp.

Leopold, A., L. K. Sowls, and D. L. Spencer. 1947. A survey of overpopulated deer ranges in the United States. J. Wildl. Manage. 11:162-177.

Leopold, A. S., T. Riney, R. McCain, and L. Tevis, Jr. 1951. The Jawbone deer herd. Calif. Div. Fish and Game Bull. 4. 139pp.

Loft, E. R., J. W. Menke, and J. G. Kie. 1991. Habitat shifts by mule deer: the influence of cattle grazing. J. Wildl. Manage. 55:16-26.

Longhurst, W. M., G. E. Connolly, B. M. Browning, and E. O. Garton. 1979. Food interrelationships of deer and sheep in parts of Mendocino and Lake Counties, California. Hilgardia 47:191-247.

Loomis, J. R., E. R. Loft, D. R. Updike, and J. G. Kie. 1991. Cattle-deer interactions in the Sierra Nevada: a bioeconomic approach. J. Range Manage. 44:395-398.

Loveless, C. M. 1967. Ecological characteristics of a mule deer winter range. Colo. Game, Fish and Parks Dep. Tech. Publ. 20. 125pp.

Mackie, R. J. 1970. Range ecology and relations of mule deer, elk and cattle in the Missouri River Breaks, Montana. Wildl. Monogr. 20. 79pp.

———. 1978. Impacts of livestock grazing on wild ungulates. Trans. North Am. Wildl. and Nat. Res. Conf. 43:462-476.

———. 1981. Interspecific relationships. Pages 487-507 *in* O. C. Wallmo, ed. Mule and black-tailed deer of North America. Univ. Nebraska Press, Lincoln.

———. 1983. Natural regulation of mule deer populations. Pages 112-125 *in* F. L. Bunnell, D. S. Eastman, and J. M. Peek, eds. Symposium on natural regulation of wildlife populations: proceedings 14. Univ. Idaho, For., Wildl., and Range Exp. Stn. Moscow.

McConnell, B. R., and J. G. Smith. 1970. Some effects of grazing intensity on bitterbrush dispersion. J. Range Manage. 23:92-94.

———, and ———. 1971. Influence of grazing and age on crown characteristics of bitterbrush. U.S. Dep. Agric. For. Serv. Res. Note PNW-146. 4pp.

McCulloch, C. Y. 1973. Seasonal diets of mule and white-tailed deer. Pages 1-37 *in* Deer nutrition in Arizona chaparral and desert habitats. Arizona Game and Fish Dep. Spec. Rep. 3.

McIntosh, B. J., and P. R. Krausman. 1982. Elk and mule deer distributions after a cattle introduction in northern Arizona. Pages 545-552 *in* J. M. Peek and P. D. Dalke, eds. Wildlife livestock relationships symposium: proceedings 10. Univ. Idaho, For., Wildl., and Range Exp. Stn. Moscow.

Mueggler, W. F. 1965. Cattle distribution on steep slopes. J. Range Manage. 18:255-257.

Neal, D. L. 1982. Improvement of Great Basin deer winter range with livestock grazing. Pages 61-73 *in* J. M. Peek and P. D. Dalke, eds. Wildlife-livestock relationships symposium: proceedings 10. Univ. Idaho, For., Wildl. and Range Exp. Stn. Moscow.

Peek, J. M., F. D. Johnson, and N. N. Pence. 1978. Successional trends in a ponderosa pine-bitterbrush community related to grazing by livestock, wildlife, and to fire. J. Range Manage. 31:49-53.

Ragotzkie, K. E., and J. A. Bailey. 1991. Desert mule deer use of grazed and ungrazed habitats. J. Range Manage. 44:487-490.

Rasmussen, D. I. 1941. Biotic communities of Kaibab Plateau, Arizona. Ecol. Monogr. 3:229-275.

Reiner, R. J., and P. J. Urness. 1982. Effect of grazing horses managed as manipulators of big game winter range. J. Range Manage. 35:567-571.

Rhodes, B. D., and S. H. Sharrow. 1990. Effect of grazing by sheep on the quantity and quality of forage available to big game in Oregon's Coast Range. J. Range Manage. 43:235-237.

Riggs, R. A., and P. J. Urness. 1989. Effects of goat browsing on Gambel oak communities. J. Range Manage. 42:354-360.

———, ———, and K. A. Gonzalez. 1990. Effects of domestic goats on deer wintering in Utah oakbrush. J. Range Manage. 43:229-234.

Rouse, C. H. 1954. Antelope and sheep fences. Prelim. Rep. U.S. Fish and Wildl. Serv., Lakeview, Oreg. 22pp.

Rule, M. I. 1989. The effects of cattle grazing on the distribution of mule deer on summer range in northeastern Nevada. M.S. Thesis, Univ. Nevada, Reno. 78pp.

Russo, J. P. 1964. The Kaibab north deer herd. Arizona Game and Fish Dep. Wildl. Bull. 7. 195pp.

Schmautz, J. E. 1954. Grass—cafeteria style in Montana. U.S. Dep. Agric. For. Serv. Intermountain Region Res. Note 12. 4pp.

Sherrets, H. D. 1989. Wildlife watering and escape ramps on livestock water developments: suggestions and recommendations. Idaho Bur. Land Manage. Tech. Bull. 89-4. 30pp.

Skovlin, J. M., P. J. Edgerton, and R. W. Harris. 1968. The influence of cattle management on deer and elk. Trans. North Am. Wildl. and Nat. Res. Conf. 33:169-181.

Smith, A. D. 1949. Effects of mule deer and livestock upon a foothill range in northern Utah. J. Wildl. Manage. 13:421-423.

———, and D. D. Doell. 1968. Guides to allocating forage between cattle and big game on big game winter range. Utah Div. of Fish and Game Publ. 68-11. 32pp.

Smith, J. G., and O. Julander. 1953. Deer and sheep competition in Utah. J. Wildl. Manage. 17:101-112.

Stevens, D. R. 1966. Range relationships of elk and livestock Crow Creek drainage, Montana. J. Wildl. Manage. 30:349-363.

Stubbendieck, J., S. L. Hatch, and C. H. Butterfield. 1992. North American range plants. Fourth ed., Univ. Nebraska Press, Lincoln. 493pp.

Svejcar, T., and J. R. Brown. 1991. Failures in the assumptions of the condition and trend concept for management of natural ecosystems. Rangelands 13:165-167.

Urness, P. J. 1981. Food habits and nutrition. Pages 347-365 *in* O. C. Wallmo, ed. Mule and black-tailed deer of North America. Univ. Nebraska Press, Lincoln.

———. 1982. Livestock as tools for managing big game winter range in the intermountain west. Pages 20-31 *in* J. M. Peek and P. D. Dalke, eds. Wildlife-livestock relationships symposium: proceedings 10. Univ. Idaho, For., Wildl., and Range Exp. Stn. Moscow.

———. 1990. Livestock as manipulators of mule deer winter habitats in northern Utah. Pages 25-40 *in* U.S. Dep. Agric. For. Serv. Gen. Tech. Rep. RM-194.

Vallentine, J. F. 1990. Grazing management. Academic Press, Inc. San Diego, Calif. 533pp.

Wagner, F. H. 1978. Livestock grazing and the livestock industry. Pages 121-145 *in* H. Brokaw, ed. Wildlife and America. Counc. Econ. Qual., Washington, D.C.

Wallace, M. C., and P. R. Krausman. 1987. Elk, mule deer, and cattle habitats in central Arizona. J. Range Manage. 40:80-83.

Wallmo, O. C. 1978. Mule deer and black-tailed deer. Pages 31-41 *in* J. L. Schmidt and D. L. Gilbert, eds. Big game of North America: ecology and management. Stackpole Books, Harrisburg, Pa.

———, L. H. Carpenter, W. L. Regelin, R. B. Gill, and D. L. Baker. 1977. Evaluation of deer habitat on a nutritional basis. J. Range Manage. 30:122-127.

———, R. B. Gill, L. H. Carpenter, and D. W. Reichert. 1973. Accuracy of field estimates of deer food habits. J. Wildl. Manage. 37:556-562.

Ward, A. L., J. J. Cupal, A. L. Lea, C. A. Oakley, and R. W. Weeks. 1973. Elk behavior in relation to cattle grazing, forest recreation, and traffic. Trans. North Am. Wildl. and Nat. Resour. Conf. 38:327-337.

Willms, W., A. W. Bailey, A. McLean, and R. Tucker. 1980. The effects of all grazing or burning bluebunch wheatgrass range on forage selection by deer and cattle in spring. Can. J. Anim. Sci. 80:113-122.

Willms, W., and A. McLean. 1978. Spring forage selected by tame mule deer on big sagebrush range, British Columbia. J. Range Manage. 31:192-199.

———, ———, R. Tucker, and R. Ritcey. 1979. Interactions between mule deer and cattle on big sagebrush range in British Columbia. J. Range Manage. 32:299-304.

Yeo, J. J., J. M. Peek, W. T. Wittinger, and C. T. Kvale. 1993. Influence of rest-rotation cattle grazing on mule deer and elk habitat use in east-central Idaho. J. Range Manage. 46:245-250.

Young, J. A., R. A. Evans, and P. T. Tueller. 1976. Great Basin plant communities-pristine and grazed. Pages 186-216 *in* R. Elston, ed. Holocene environmental change in the Great Basin. Nevada Archaeol. Surv. Res. Pap. 5.

Chapter 12
THE WHITE-TAILED DEER: NATURAL HISTORY AND MANAGEMENT

James G. Teer

INTRODUCTION

Of all the big game animals in North America, members of the deer family, Cervidae, are the most numerous and the most important recreationally and economically. Of all cervids, the white-tailed deer (*Odocoileus virginianus*) is the most widespread and numerous in the New World. It occurs in all states of the United States except California, Nevada, and Utah (Fig. 1) (Southeastern Cooperative Wildlife Disease Study 1982). Thirty-eight races or subspecies range from southeastern Alaska (60° N latitude) through parts of southern Canada into practically all of the contiguous states in the United States through Central America and as far south as northern South America (15° S latitude) (Baker 1984).

Other members of the deer family in North America include the mule deer (*Odocoileus hemionus*), elk (*Cervus elaphus*), moose (*Alces alces*), and caribou (*Rangifer tarandus*). With the exception of the caribou, these species occur sympatrically with white-tailed deer in several regions, but all are much less numerous than whitetails.

In South America, the white-tailed deer is native to Colombia, Venezuela, Guyana, Surinam, French Guiana, Ecuador, Brazil, and Peru (Whitehead 1972). There it shares habitat with 3 species of brocket deer (*Mazama americana*, *M. couazoubira*, and *M. rufina)* in tropical areas, and with 2 species of huemuls (*Hippocamelus antisensis* and *H. bisulcus*) in montane areas of northwestern South America (Walker 1964).

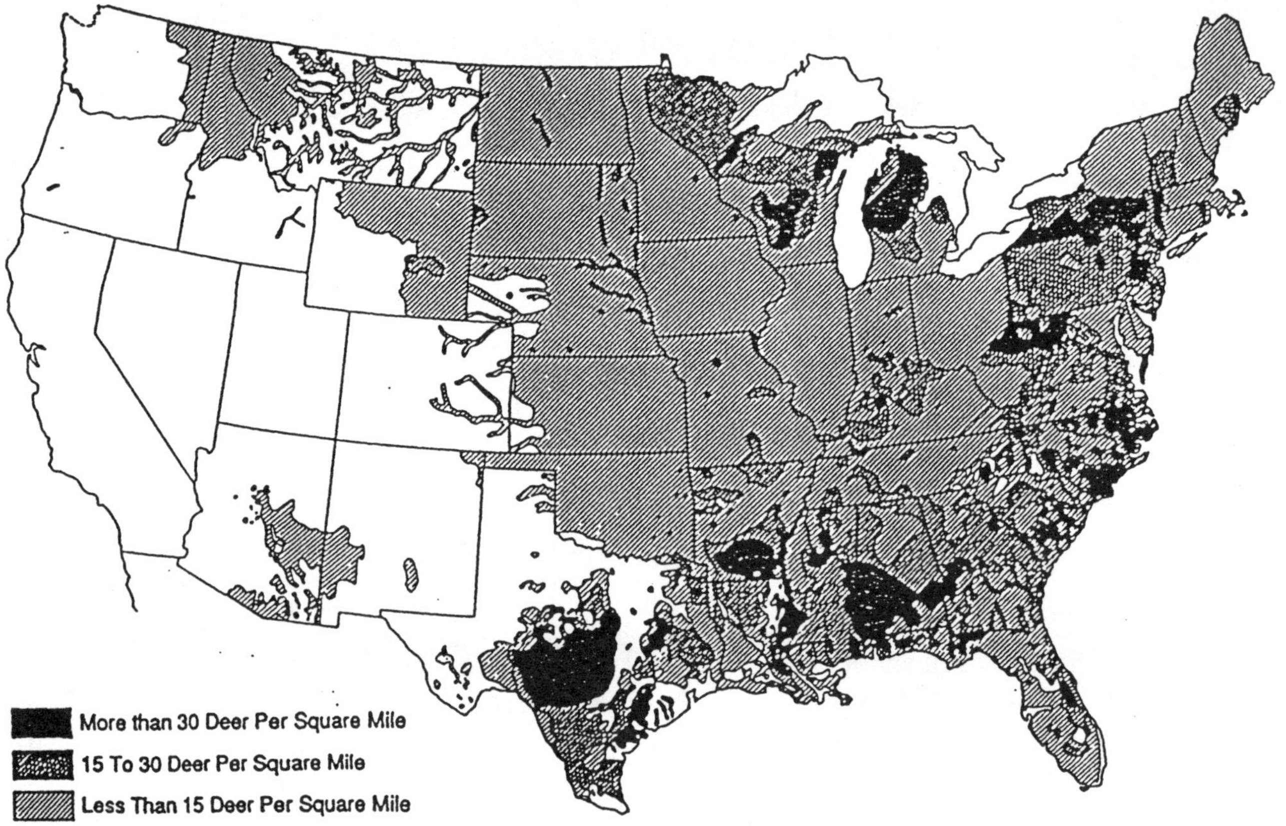

FIGURE 1. Distribution and densities of white-tailed deer in the United States, 1980 (Southeastern Cooperative Wildlife Disease Study 1982).

Other cervids have been translocated from Europe and Asia into North and South America (Teer 1989, Traweek 1989). Most have been released in North America, a few in South America. Among the most important cervid transplants are the red deer (*Cervus elaphus*) in South America; sambar (*C. unicolor*), sika deer (*C. nippon*), axis deer or chital (*Axis axis*), and fallow deer (*Dama dama*) in North America. All have become established in habitats as free-ranging populations and often sympatrically with white-tailed deer.

Only the most experienced taxonomist can separate races of white-tailed deer by morphometric and genetic measurements. Skull characters are the chief criteria and these often are confused by crosses that have occurred through transplants of 1 race into the range of another (Texas Game, Fish and Oyster Commission 1945, Harlow and Jones 1965). Differences in coloration of the pelage and in body sizes and extremities do occur; however, cranial measurements and new technology using mitachrondial DNA are the most reliable for separating subspecies. Deer from more xeric habitats tend to be lighter in color and their extremities appear to be longer. Eastern races are darker and more blocky in conformation.

The ecology and management of the races of white-tailed deer are similar across the various ecosystems they inhabit. Differences in coloration, size, and behavior do occur, and are due primarily to subspecific differences and as adaptations to the species' habitat (Kellogg 1956, Baker 1984). However, a white-tailed deer from any part of its range is recognizable as a white-tailed deer. Management is habitat- and herd-specific and must be adapted to the vegetation and other elements of its habitat.

The purpose of this chapter is to give a brief overview of the natural history and management of white-tailed deer in rangeland habitats. Thus most of the literature used to develop the chapter was selected from rangeland research. Clearly, however, deer are inhabitants of forested areas. I used information on the species from forested regions, particularly the Midwest, Lake States and Northeast. Because much research on habitat management for white-tailed deer has been done in Texas, perhaps the literature cited will not reflect the author's bias more than the amount of work done there.

NUMBERS

Early travelers, settlers, missionaries, military expeditions, and government surveyors described deer as scarce, abundant, or absent in diaries and reports. Venison was described as the main component of many travelers' and settlers' diets. For example, Bailey (1905) reported that with the exception of the Trans-Pecos and part of the Staked Plains, all of Texas was occupied by some form (subspecies) of white-tailed deer.

Even though white-tailed deer were widespread and a staple in the diet of many early settlers and travelers, their numbers were relatively small in comparison to numbers today. Mature ecological stages of the vegetation communities of the time could not support large populations. Vegetation largely occurred as climax communities or habitats made up of seral or successional stages near climax. Mature forests of sawlog and pole timber lacked food because the plants were largely out of reach of deer and understory vegetation was shaded out (Harlow 1984). In grasslands and savannas food sources were seasonally deficient and cover was often lacking.

Numbers of deer increased when the mature forests were logged, when woody species began to invade prairies and plains, and when deer were protected and restocked in areas where they had been extirpated (Swift 1946, Stanton 1963, Mattfield 1984). White-tailed deer densities in the northeastern hardwood and spruce (*Picea*)/fir (*Abies*) forests during pre-colonial times were 3.86 to 5.79/km^2 (Mattfield 1984), a density similar to that reported by Dahlberg and Guettinger (1956) for similar forests in Wisconsin prior to 1800. Huot et al. (1984) reported that white-tailed deer spread northward and eastward in Canada into previously unoccupied range during a period of above-average temperatures concurrent with a general exploitation of forests.

Unlike those in forested habitats, deer numbers in the prairies and plains in the western United States increased following encroachment of woody vegetation. In the Great Plains deer were associated primarily with major streams and their tributaries but began to extend their ranges into upland habitats in the 1940s. In prairie-agricultural habitat in east-central Montana, white-tailed deer selected riparian habitat and its interspersion with cropland and rangeland habitats (Dusek et al. 1988*a*). White-tailed deer on the lower Yellowstone River were associated with grassland sere (Dusek et al. 1989) and used cereal grains, alfalfa and sugar beets in adjacent farmland. Their numbers were positively related to density of riparian cover.

Rangeland habitats in the Edwards Plateau and Rio Grande Plains of Texas were converted to brush-covered ranges after the turn of the twentieth century when cattle grazing and control of wild fires came into widespread use (Johnson 1963, Inglis 1964).

As the habitat became more suited to deer and with increased management and protection of herds, the species reached its zenith in numbers after World War II. Although there are few data for comparison, deer are ≥4 to 5 times as numerous today in many habitats as they were in pre-colonial times.

Grazing lands (i.e., prairies and plains and forests) are the chief habitats of white-tailed deer in the western United States. In most of the West, deer densities are < 5.8/km^2 (Southeastern Cooperative Wildlife Disease Study 1982). However, in some habitats, deer reach densities ≥2 times those reported for landscapes of the eastern and central forests of the United States and Canada. Over 15.4 deer/km^2 were reported on >101,214 km^2 in the Edwards Plateau of Texas in 1980 (Cook 1984). Teer et al. (1965) reported 46.7 deer/km^2 on > 4,048 km^2 of rangeland stocked with cattle, sheep, and goats in the Llano Basin of Texas in 1961. This is perhaps the highest density of deer reported for a large area in North America.

In comparison, Swift (1946) reported 17.4 deer/km^2 in drives of local areas in the northern forest zone of Wisconsin. In the northern parts of white-tailed deer's largely livestock-free range in the Northern Great Lakes and Ontario forests, deer densities ranged between 7.8 and 38.6 deer/km^2 in the mid 1970s. In the southern parts of the species' range in the same region, densities ranged between 38.6 and 116.6 km^2.

Although density or population numbers for white-tailed deer are not available or summarized for the northwest, the annual kill of around 34,000 in the Northern Rocky Mountain Region, that includes Montana, Washington, Idaho, Oregon, and Wyoming, provides some indication of their numbers (Peek 1984).

White-tailed deer are increasing in the West whereas mule deer have declined in many traditional ranges. R. J. Mackie, Univ. Montana (pers. commun.), stated, ". . . white-tails have increased in distribution and number over the past 40-50 years (in Montana) to approach equality with mule deer. When I first came on the scene in the 60's, mule deer predominated in distribution and outnumbered white-tailed deer by perhaps 2:1 or more. Today, we probably run close to 50:50, and certainly no worse than 60:40. . . . White-tails have been strongly favored by development of agriculture and management of riparian environments throughout the region."

HABITAT

The white-tailed deer is extremely adaptable. It thrives in close association with humans and their agricultural and industrial pursuits. The requirements of white-tailed deer are met in practically every ecological type, including grasslands, prairies and plains, mountains, hardwood, coniferous and tropical forests, deserts, and even farmland where it associates with woodlots and riverine habitats (Halls 1984).

Woody vegetation used by deer for cover and often for browse and mast is not an absolute requirement of deer habitats although white-tailed deer are most often associated with brushlands and forested habitats. The species does occur in low densities in open pasturelands and grasslands such as in the Gulf Coast prairies and marshes in Texas (Cook 1984). Usually, however, deer in such open habitats have screening or protective cover nearby for escape. Cover in such habitats may be herbaceous, and woody cover may be available in riparian zones that thread through grasslands and plains. Forests, shrublands, and gallery or riparian forests are the predominant vegetation communities in white-tailed deer habitats.

While not an obligate member, white-tailed deer reach large densities and numbers in disclimax or disturbed vegetation communities. Forbs, browse, and other foods are numerous there. By contrast, climax or seral stages of vegetation often lack seasonal food sources, cover, and shelter.

Deer thrive in areas where grasslands such as those on the Edwards Plateau and Rio Grande Plains of Texas have given way to brushlands (Inglis 1964). However, deer seemed to show little or no selection among the brush communities in south Texas (McMahan and Inglis 1974).

Unlike mule deer that prefer open grassland and parkland types (Wallmo 1981), white-tailed deer prefer to winter in dense, continuous forest (Prescott 1974). Availability of water, great diversity in species composition and structure of plant communities, and long, linear shapes that provide corridors between habitats are important characteristics (Thomas et al.

1979). Clearly, riparian zones and tame pastures are important constituents in white-tailed deer habitat in the West, especially in eastern Montana and the Columbia River Basin in Oregon and Washington (Suring and Vohs 1979, Evans 1984, Peek 1984). Wood et al. (1989) reported white-tailed deer used hardwood draws and mesic shrublands over 62% of the time as determined from aerial surveys in their Cherry Creek Study Area in Montana. Eighty percent of their radiolocationss of white-tailed deer occurred in these habitats during summer. Deer increased their use of harvested grainfields in autumn to feed on waste grain.

Ordinarily, in the mountain states white-tailed deer are found on the lower slopes. The Rocky Mountain mule deer (*O. h. hemionus*) occupies higher elevations during the summer and lower slopes with white-tailed deer in winter. White-tailed deer customarily use riparian zones and valleys where they find food and shelter during cold seasons. However, distribution of the 2 species may be reversed as in Arizona with white-tailed deer on the higher slopes and desert mule deer (*O. h. crooki*) on the lower slopes and desert floor (Anthony and Smith 1977). Krausman and Ables (1981) reported Carmen Mountains white-tailed deer favored habitats > 1,372 m in the Chisos Mountains of Big Bend National Park, Texas, but also shared habitats with desert mule deer at lower elevations.

Migratory mule deer and elk normally move down from higher elevations to valleys and riparian zones to escape snow and cold weather. In some regions, such as the desert Southwest, sympatry occurs in spatial distributions, habitat selection, and diets of white-tailed deer and nonmigratory mule deer (Anthony and Smith 1977).

Lands administered by the public land agencies in the western United States support good numbers of white-tailed deer. These habitats are often a combination of private and public lands. There, livestock grazing is the predominant land-use and thus deer habitats usually consist of grazed forests, open rangelands, and dry and sometimes irrigated haylands. Pastures planted to cereal grains and legumes are often heavily used by white-tailed and mule deer, much to the concern of landowners who plant them (Reed 1981).

NATURAL HISTORY AND ECOLOGY

A swift, graceful animal, the white-tailed deer has great trophy value because of its antlers and large body size. Its shy and elusive habits and venison's fine table qualities are characteristics that make it popular with hunters. Experienced hunters consider the white-tailed deer one of the most elusive big game animals and the most demanding on the hunter of any large ungulate in the New World. Such a claim is defensible when a trophy-class deer is the goal.

Body Size and Antler Development

Large differences in body size occur among races and among populations within races. These differences are predisposed by inherited characteristics and nutritional qualities of the animals' diet and, by the ages at which deer mature physically.

Kellogg (1956) described races of white-tailed deer in northern United States and Canada as "large" or "very large" while ≥3 races in southern United States and all races in Mexico, Central America, and South America were described as "small" or "very small." The largest whitetail on record was taken at Mud Lake, Minnesota in 1926; it field-dressed 196.5 kg (Erickson et al. 1961). A field-dressed deer killed in Sawyer County, Wisconsin, in 1924 weighed 188.7 kg (Bersing 1956). Another killed in 1919 near Trout Creek in the Upper Peninsula of Michigan weighed 173.0 kg field-dressed (Jenkins and Bartlett 1959).

Following Bergman's and Allen's Rules, warm-blooded animals from colder climates tend to be larger in size and have shorter extremities to control heat loss than those in southern environments. Subspecies of white-tailed deer in the north have less body surface in proportion to body mass than do members of the same species farther south. Thus, the largest white-tailed deer in North America are found in the north temperate environments.

Fixed genetically, diminutive races in the southern parts of the species' range in the United States include the Key deer (*O. v. clavium*) in Florida, the Carmen Mountains white-tailed deer (*O. v. carminis*) in the Trans-Pecos Texas, and Coues deer (*O. v. couesi*) in Arizona. Rarely reaching 64.5 kg, they are almost half the size of races of the same species farther north (Table 1). Although some deer attain much larger body sizes than might be expected in populations in southern parts of its range, deer in Mexico, Central, and South America are certainly smaller in stature than in northern areas of its range.

Physical characteristics and population parameters of white-tailed deer respond greatly to nutritional qualities of their diets. Populations of the same race on different ranges often differ significantly in body size and antler development. Simply put, attributes of deer are fixed genetically. However, those populations that are in habitat with all the necessary components including nutritional requirements realize their genetic potential. Conversely, deer in poor range are smaller. Moreover, other biological attributes of their lives such as natality and mortality rates differ.

Males reach physical maturity with the largest body size and antler development at 4.5 to 5.5 years of age (Erickson et al. 1961, Banasiak 1964, Severinghaus 1955, Park and Day 1942, Teer et al. 1965). Maximum antler development may continue through 6 or 7 years of age after which antlers may decrease in size and symmetry. Females reach maximum size at 2.5 to 3.5 years of age in central Texas (Teer et al. 1965).

Physical condition, body weight, and antler size are related to diet quality (French et al. 1956). In temperate zones weights of bucks (especially those in rut) and females with fawns may fluctuate seasonally as much as 25-33% of their normal body size (Sauer 1984). The rut is expensive in terms of energy and fat reserves for males. When high energy foods such as acorns are available, deer will gain weight during late summer and early fall, which sustains them through winter.

Johnson (1937) and Severinghaus (1955) were among the first to report weights of deer related to population density and range conditions. Sanders (1941) compared kill rates and weights of deer on ranges in the Edwards Plateau of Texas and found that

TABLE 1. Comparisons of weights of subspecies of field-dressed white-tailed deer.

Subspecies	Location	Adult males	Adult females	Source
borealis	Minn. (Mud Lake)	166	131	Erickson et al (1961)
borealis	N. Y. (Adirondaks)	130	93	Severinghaus (1955)
borealis	Wisc. (State total)	110	102	Dahlberg and Guettinger (1956)
clavium	Fla. (Everglades)	92		Loveless (1959)
Mixture of Several races	N. C. (Pisgah Natl. Forest)	112		Schilling (1938)
borealis	Me. (> 4.5 years) (Statewide)	185	110	Banasiak (1964)
carminis	Trans-Pecos, Tex.	104 (live)	66 (live)	Krausman (1976)
gymnotis	Venezuela	110 (live)	66 (live)	Brokx (1984)
texanus	Tex. (Llano Basin)	68.1		Teer et al (1965)
texanus	Tex. (Edwards Plateau)	82.3		Teer et al (1965)
texanus	Tex. (Panhandle)	124		DeArmont (1959, 1960)

deer taken from the most dense populations were about 5 kg lighter than those from less dense populations. Hesselton et al. (1965) reported weights of deer and density of the population were inversely related at New York's Seneca Army Depot. The relationship was driven by the quality of the range and competition for forage resources by members of the herd. Teer et al. (1965) reported a similar inverse relationship between density and weights of white-tailed deer in the Llano Basin of Texas.

Home Range, Movements, and Activity Patterns

In the northern parts of their range where winter weather is characterized by cold temperatures with snow frequently covering the landscape, white-tailed deer have different summer and winter ranges. They are not migratory in the true sense of the word. They retire to swamps and other protected areas in the winter where browse plants such as white cedar (*Thuja occidentalis*), willows (*Salix* spp.), and dogwood (*Cornus* spp.) provide forage during an otherwise food-short time and where they can escape deep snow (Dahlberg and Guettinger 1956, Erickson et al. 1961, Banasiak 1964). In the western areas, they usually retire to riparian zones and protected valleys.

In southern climes, white-tailed deer occupy the same range yearlong. Their requirements are met locally and weather conditions are not as severe as they are in the north. While they may move short distances in response to seasonal food availability, they are largely sedentary throughout the year.

Home ranges of whitetails are relatively small. Studies in the south show that most occupy a range of about 5-8 km^2 (Hahn and Taylor 1950, Thomas et al. 1964). Michael (1965) reported home ranges between 24.3 and 137.7 ha for females and 97.2 to 356.3 ha for males on the Welder Wildlife Foundation in south Texas.

White-tailed deer have strong affinities to their home ranges. Dogs, especially free-ranging, untended hounds, chase deer long distances and can temporarily cause deer to leave their ranges (Progulske and Baskett 1958); however, most return after a time. Successful homing of translocated deer to natal areas does not occur frequently. Therefore, choices of release sites need not consider homing tendencies (Glazener 1948, Hamilton 1962).

Food availability has some effect on movement patterns of white-tailed deer. Home ranges of female deer increased from summer to autumn during years with good mast, and males shifted their activities to oak (*Quercus* spp.)-hickory (*Carya* spp.) habitats in response to acorn fall (McShea and Schwede 1993). In times of nutritional stress, more often than not deer choose to remain in their home ranges even to the point of death (Hahn and Taylor 1950, Teer et al. 1991). Adult males are tireless in their movements during the rutting season and will occupy somewhat larger ranges than females (Thomas et al. 1964). Yearlings, especially yearling males, pioneer or leave their natal areas to occupy other ranges. Hawkins et al. (1971) reported 80% of yearling males left natal areas and dispersed to new ranges on Crab Orchard National Wildlife Refuge. Pioneering of young males provides gene flow and also reestablishes numbers when populations are diminished by shooting or when habitat improves to accommodate them.

White-tailed deer are most active in early morning and late evening hours (Michael 1970). They tend to feed during these periods but foraging is not restricted to them. When food is short, as often occurs in the winter or in droughts, deer may feed in all hours of the day. During certain weather conditions and phases of the moon, they may change their activities to accommodate these conditions (Michael 1970).

Diet

White-tailed deer are selective feeders with peaks of feeding activity in early morning (0600) and late afternoon (1800) (Michael 1970). They use many species of browse, forbs, grasses, thallophytes (mushrooms and fungi), and fruiting bodies (mast) of plants. They seem to move constantly while feeding, taking a bite here and there as they walk. As ruminants, they are capable of digesting foods that are low quality and not digestible by monogastric animals.

White-tailed deer use whatever forage is available and have the ability to select the most nutritious and palatable plants or parts of plants (Longhurst et al. 1968). Moreover, they reject plants containing unpalatable terpines and essential oils (Oh et al. 1968). They are heavy users of mast and fruits of plants, preferring them over all other items.

Because of the widespread distribution and differences in habitats throughout the species' range, it is most difficult to characterize diets in specific terms. One can only describe major categories of foodstuffs; i.e., browse, forbs, grasses, mast, and fruits.

They are usually described as browsers. In colder climates, browse is the primary food during winter. Herbaceous foods are primary items in the growing seasons. Green foods, either browse or herbaceous plants, are indicators of the quality of habitats, and are especially important in winter ranges. White cedar (*Thuja occidentalis*) provides such food in yards in the Lake States; Erickson et al. (1961) state that white cedar is the only native browse species that, fed alone, will maintain deer weight in winter for periods ≤60 days.

Diets of whitetails in northwestern and north central habitats range from 60 to 80% browse in the fall and winter seasons (Dalke 1941, Latham 1950, Jenkins and Bartlett 1959, Erickson et al. 1961, Wishart 1984). Browse and mast are supplemented in winter by whatever grasses and forbs are present at the time, and intake of herbaceous foods increases as spring begins. Hollys (*Ilex* spp.), greenbriar (*Smilax* spp.), juniper (*Juniperus* spp.), and other nondeciduous woody plants are usually present and heavily used in winter range in the South and West.

In some rangelands of south Texas, especially in the Coastal Bend with a growing season close to 300 days, forbs are preferred items in deer diets. Browse is second, with grass usually comprising < 15% and then mostly cool-season grasses in mid- to late-winter (Chamrad and Box 1968, Drawe 1968, Drawe and Box 1968).

In more arid habitats in south Texas, mainstays in deer diets are browse and mast. Varner and Blankenship (1985) reported browse and mast in southwest Texas varied from 65% of the diet in the spring to 95% in the winter. Another study in the same region (Arnold and Drawe 1979) reported diet preferences of white-tailed deer were 21.2% cacti, 32.7% browse, 26.6% forbs, 8.3% grass, and 11.3% unknown material. While forbs were preferred when available, browse and mast comprised 50% of foods ingested by whitetails in the same region (Davis and Winkler 1968).

Krausman (1978) reported browse comprised 44% of the diet of white-tailed deer in exclusive white-tailed deer range on Big Bend National Park, Texas. Succulents accounted for 18%; forbs, 14%; and grasses, 3%.

High energy foods such as acorns are important for depositing fat reserves from which white-tailed deer draw in cold weather and during the rut when food intake is reduced. In forested regions of east Texas that may be similar to if not typical throughout southern pine forests, browse provides > 50% and as little as 25% of whitetail diets (Lay 1967). Mast, including fruits and acorns, is more important when available. Herbaceous green forage also is important.

Volume of food intake varies with season, quality of the food, and status of the animals in their life cycles. Nichol (1938) reported that mule deer in Arizona require 1.2 kg of air-dry forage/48.9 kg of body weight/day. A 48.9 kg deer in winter in Michigan requires from 2.0 to 3.4 kg of food/day depending on quality of the food (Jenkins and Bartlett 1959). French et al. (1956:231) reported from penned studies of deer in Pennsylvania that "two pounds (3,600 calories) of high quality food is required daily for a deer weighing 50-60 pounds, 3-4 pounds (6,300 calories) for a 100-pound deer, and 5-6 pounds (9,900 calories) for a 150-pound deer."

Maintenance, growth, and reproductive states require different nutritional intakes. Growth of fawns after weaning requires 14 to 22% (dry-matter basis) protein in their diets (Ullrey et al. 1967). Males have higher requirements than females. Protein requirements for maintenance are lower, 6 to 10% for adults (French et al. 1956). During latter stages of pregnancy, protein requirements are intermediate between that of growth and maintenance (Verme and Ullrey 1984).

Although deer can subsist for long periods on water obtained from vegetation, free water is an important component of deer habitat. Water is usually supplied by natural water areas or can be provided through stock ponds and troughs used to water livestock.

Reproduction and Productivity

Mating habits and the sex cycle of white-tailed deer are well known (Severinghaus and Cheatham 1956, Teer et al. 1965, Robinson et al. 1965, Hirth 1977, Verme and Ullrey 1984). Unlike other species such as elk that are polygamous with a harem arrangement, male and female white-tailed deer do not establish a long-lasting, firm bond. They are promiscuous. The larger males are usually successful.

White-tailed deer do not defend a territory but males and females do establish hierarchical structures in their groups (Hirth 1977, Marchinton and Hirth 1984). During the rut males solicit females and exclude others by threats, pushing matches and, sometimes, fights (Hirth 1977). Pre-rutting "push-fights" occurred throughout the year and were important in establishing a social hierarchy among males in Texas; true fights occurred only during the rut (Michael 1968).

Gonadal activity, shedding of bony antlers, and growth of velvet antlers are related to photoperiod (Nicholls 1978, Goss 1982). Endocrine function, primarily the production of testosterone, regulates the antler cycle (Bubenik 1982). Average testes weights increase more than 5-6 times their quiescent state for the rut (Illige 1951, Robinson et al. 1965). Testosterone production increases accordingly. Cold temperatures seem to stimulate sexual activity after the rut begins and true fighting increases in cold weather (Michael 1968).

Females also become sexually active in response to day length. They experience estrous every 28 days and remain in heat and are receptive to copulation for about 24 hours (Verme and Ullrey 1984). Pre-estrous females may be accompanied by an interested male for 3 or 4 days during the time she is coming to estrous. Does usually conceive during the first estrous even when the sex ratio of the population is quite disparate. If not bred, females can undergo ≤4 estrous cycles (Cheatum and Morton 1946). Normally, even in times of food shortage, most females will conceive but survival of fawns will be low after birth. Females can conceive in their first year of life; males take no part in mating until they are ≥18 months of age (Cheatum and Morton 1946).

The resiliency and recovery of overharvested deer herds is phenomenal and is in part due to high reproductive rates. Precocial breeding of fawns occurs (Cheatum and Morton 1942). Fertility is much lower on poor quality range than on ranges with ample quantity and quality of food. From 75 to 80% of fawns and 100% of older deer breed each fall in deer ranges in intensively farmed midwestern United States (Nixon and Hanson 1992). Deer fawns in Iowa that are primarily deer of riparian and woodlot habitats in association with croplands, are responsible for 30% of the annual increment (Haugen 1975). Deer on poor ranges in western New York had about half the corpora lutea and embryos as did deer on high quality ranges in the Adirondacks (Cheatum and Severinghaus 1950).

Compensatory productivity is displayed by the species. Teer et al. (1965) reported decreases in ovulation and conception rates of white-tailed deer in the Central Mineral Basin of Texas as density increased (Fig. 2, 3). However, Kie and White (1985) could only associate density dependence with reproductive rates of very dense populations (84 deer/km^2) on the Welder Wildlife Foundation, Texas.

Twin pregnancies are usual for adult white-tailed deer. Triplets occur in about 1 out of 100 births; quadruplets occur perhaps 1 in 1,000. Adult females produce an average of 1.5 fawns/doe; yearlings produce an average of about 1.1 fawns/doe (Teer et al. 1965). Twins are unusual in pregnancies of fawns.

Despite the rates of conception and twinning, productivity is related more to survival of fawns after birth. Food conditions must be adequate for the dams to suckle their fawns if increment to the herd is to be substantial. Teer et al. (1965) found that even in periods of food shortage, conception rates of white-tailed deer in the Edwards Plateau and Llano Basin of Texas were high but mortality of fawns after parturition was the greatest source of loss. Reproductive duties; i.e., mating, pregnancy and lactation of females and rutting activities of males, are costly to adults especially on food-short ranges. Dusek et al. (1991) reported white-tailed deer in Swan Valley, Montana committed resources to reproduction at the expense of survival of adults.

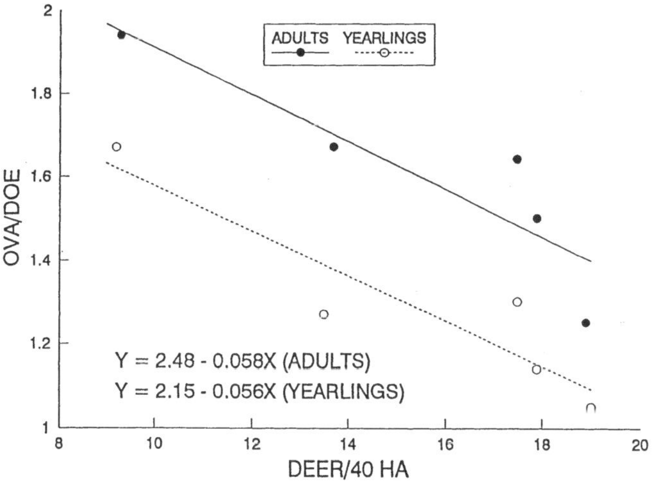

Figure 2. Regression of ovulation rates of yearling and adult white-tailed deer females on density of deer in the Llano Basin of Texas (Teer et al. 1965).

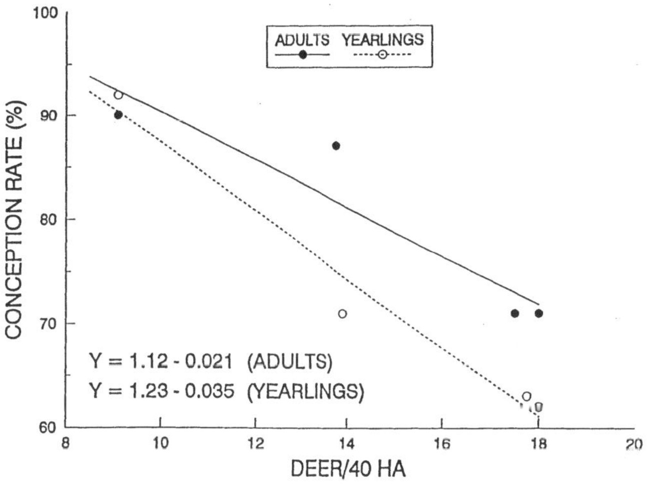

Figure 3. Regression of conception rates of yearling and adult white-tailed deer on density of deer in the Llano Basin of Texas (Teer et al. 1965).

In the Lower Yellowstone herd, deer had sufficient forage resources and higher fawn survival but smaller body size than those in Swan Valley.

DeYoung (1989) demonstrated that managing for older age classes of deer and larger antlers is an inefficient practice because males will die at a rate of 25-29%/year before reaching physical maturity. Post-rut mortality is especially high among bucks. They are debilitated from rutting activities and thus are more susceptible to various sources of mortality including predation by coyotes (*Canis latrans*) and mountain lions (*Felis concolor*) (C.A. DeYoung, Texas A & M, Kingsville, pers. commun.). In most studies of white-tailed deer in the United States, nutrition, predation, and parasites are the primary sources of mortality of neonates and older animals (Halls 1984).

MANAGEMENT STRATEGIES

Incentives for Management of Deer

White-tailed deer are extremely important in recreational hunting. Over 1.7 million were harvested annually in the 1970s in the United States and the harvest continues to increase (U.S. For. Serv. 1981). Texas has the largest population and harvest of all the states with a population of 5,398,874 and a kill of 477,491 (1991-92 hunting season) (Boydston 1992). Only Pennsylvania, New York, Michigan, Wisconsin, and several states in the southeastern United States approach white-tailed deer numbers and kill in Texas.

When hunting opportunities are limited, and public lands are crowded with hunters, many often wish to have exclusive rights to a well-managed deer herd. In North America, hunting rights are sold just as any other product of the land. Animals are not sold outright (i.e., by the head). Rather, access to the land is the product that is sold and bartered.

Economically, the white-tailed deer are an important resource to those who own and manage the land and to local and regional economies. Leasing for hunting privileges is a widespread practice in most livestock ranges in Texas. Commercialization of hunting is increasing throughout the United States. While fees vary widely according to number and quality of the deer in the herd, the average is near $9.88-12.40/ha under a season-lease agreement in Texas (Teer et al. 1983). Many ranches with rangeland habitats lease for white-tailed deer and game birds and the lease price is set accordingly, often reaching $15/ha where trophy deer are available. Leases were purchased by 39% of deer hunters surveyed by Thomas and Adams (1985) in 1982, and as much as 50% of the deer kill comes from such leases. Number of leases and total area under lease have increased in the last decade.

Commercialization of hunting is gathering momentum in the western states as private landowners and government agencies seek funds for personal income and management of wildlife resources (Gartner and Severson 1972, Burger and Teer 1981, Dill et al. 1983, Thomas 1984, Loomis and Fitzhugh 1989). Many western states are opposed and vigorously resist commercialization of hunting recreation. More than 50% of the land area in the western states is made up of public land. Hunters and others in the West feel that charging a fee for hunting compromises their heritage of free access to wildlife and denigrates sport hunting (Teer 1993*a*, *b*). Geist (1993) forcefully argues that commercialization is dangerous to the successful system of wildlife management developed and tuned in North America.

Unless people can benefit from protection and production of white-tailed deer, their management on private lands has little chance of succeeding in the future. Moreover, management is essential if deer are to be produced for recreational and economic purposes. Providence alone cannot provide requirements of deer in habitats where humans use the land for animal and row crop agriculture and for producing wood products. Lands set aside for deer in parks, sanctuaries, and refuges can be an important strategy to produce deer, but protected lands may not be large enough to provide hunting opportunities for even a small number of citizens. Deer management must be done on private lands to take care of hunting demands.

Benefits can be directed to those who use the land and to those who wish to hunt or simply to see deer. While a commercial or fee hunting system may not be applicable to all regions, the deer leasing system has been an extremely important stimulus for landowners to produce deer and other wildlife resources. It is clearly a developing trend in white-tailed deer management throughout the species' range.

Deer Population Management

Management of Herd Numbers. Although habitat loss and deterioration are often cited as the most important factor in deer management, deer are more numerous now than when Europeans reached North America. They have increased to the point that control of their numbers is the most conspicuous and severe problem in deer management in much of North America (Leopold et al. 1947, Marburger and Thomas 1965). Unless natural controls are operative, herbivores outstrip their food resources (Lack 1954). Starvation losses exceed the legal kill in many herds throughout white-tail range.

Krausman et al. (1992) resurveyed the deer ranges identified as overpopulated by Leopold et al. (1947) and found that most herds had been controlled by hunting. Only those herds that had not been hunted at all, or had an inadequate doe harvest, or an inadequate overall harvest were still overpopulated.

Any-sex hunting seasons are not acceptable to many citizens and must be explained and promoted over and over again to the general public. Without proper harvest, the problem of over population is compounded because over-utilization of food results in habitat deterioration. When livestock, especially sheep and goats, are stocked in deer habitats, the problem with habitat deterioration is exacerbated.

The deer manager has 2 options in controlling deer numbers: remove the surplus or do nothing. Surplus deer can be removed by hunting and perhaps by trapping deer to restock other areas. The former is least expensive to those who manage the land and its resources and most useful to hunters. To do nothing allows deer ranges to deteriorate in quality and deer to be wasted through exhaustion of food resources. Starvation is the end product of poor deer herd management or of no management. If the range will not support the herd and if the annual crop of deer is not removed by other means, starvation will remove it. This is an axiom, an inexorable rule, in deer management.

Surplus deer must be removed or the herd will finally exceed its carrying capacity. This has happened repeatedly in every deer range in North America. On parks and reserves where wildlife is protected from use, habitats have been seriously degraded by overpopulations. Natural regulation of herbivore populations is difficult when natural systems have been disrupted by man.

In overpopulated herds, both sexes must be hunted. Natural mortality of males is higher than of females, and when hunting is directed at males only, the sex ratio can be greatly skewed to females. In unhunted herds, the adult sex ratio trends toward a male to 2 females. A male to 10 to 15 females has been reported in herds heavily hunted for bucks in the Edwards Plateau of Texas (Hahn 1945). Similar distorted sex ratios are commonly reported in heavily hunted herds throughout the species' range.

Harvest quotas must be tailored to fit the herd and the objectives of individuals or agencies that are charged with its management. For example, if deer are to be increased, harvest quotas should consider wider sex ratios for productivity. If numbers are to be controlled or reduced, then harvests should encompass both sexes and all ages.

Predator Control. Coyotes, wolves (*Canis lupus*), bobcats (*Lynx rufus*), and mountain lion kill deer. Whether these predators can control numbers of deer has been the subject of long debate. Now, however, considerable evidence has accumulated to show that coyotes can substantially impact, if not control, herd numbers in certain situations (Connolly 1978, Teer et al. 1991).

Coyotes are omnivorous and opportunistic in relation to diet. They take vegetable materials (i.e., fruits, mast) when available and animal prey when abundant and easily caught (Knowlton 1964). Deer fawns comprise a large proportion of coyote diets especially when cover is sparse and other foods are scarce. Approximately 50% of radiocollared fawns on the Welder Wildlife Foundation were taken by coyotes in the first 60 days of their lives; another 20% were suspected to have been taken by coyotes (Teer et al. 1991).

Predators are often cited as the chief causes of deer declines by the general public. Coyotes, mountain lions, bobcats, and other carnivores do kill deer, and their impacts on deer numbers can be important. However, quality of habitat and hunting may be much more important than predation.

Control of predators through some kind of bounty system or other subsidized control program is often the remedy called for by misinformed citizenry. Bounty systems by government agencies have often been abused by those engaged in predator management. Cost and benefit evaluations of the bounty system indicate it is ineffective in preventing predation or encouraging deer numbers.

In the United States at least, predation can be used along with sport hunting to control numbers of deer. Conversely, when deer numbers are to be increased or protected, predator control may be an important activity (Teer et al. 1991). A blanket recommendation should not be made on predator control; each situation must be addressed separately.

Restocking Deer in Suitable Habitat. Restocking deer into habitats where they have been eliminated should be done only after careful evaluation of the habitat. Causes of the loss of deer from areas to be restocked should be examined to see if they continue to exist. Restocking of deer in habitats where their requirements are not met will not succeed.

There are indeed legitimate reasons for restocking. Ranges that have been overgrazed or cleared of brush may improve over time, and logging practices in forested habitats may create habitat for deer. Practically all of the deer in eastern Texas have resulted from translocated stock after the deer were killed out by every means including dogs by the end of the 1930s (Texas Game, Fish and Oyster Commission 1945). In deer restoration programs, the most successful restocking programs are of deer trapped from ranges where they are plentiful and transplanted into ranges where they have been eliminated. Certain conditions must be met before transplants are made in an area. First and foremost, the habitat must be suitable. Careful evaluations of food resources, cover, and water must be made. Of equal importance, the will of the people to protect the transplant must be determined. In some cases, game wardens are assigned to protect deer transplants in their early stages after their release.

A 6,073- to 8,097 ha block of habitat is considered the minimum size to receive a transplant in Texas. An agreement is made between the landowner or a consortium of landowners and the state. In effect, the agreement makes a sanctuary of the deer range. Hunting is prohibited for 5 years, after which an evaluation is made to determine if huntable numbers are present.

Rearing deer in captivity for release into wild habitats has not been as successful as transplanting wild-caught deer. Deer reared in captive herds have low survival; they do not know how to avoid predators or find food and life's other needs. Fawns captured before they are able to outdistance a man can be caught and reared, but they are often imprinted and habituated to humans and are easily killed by predators and hunters.

MANAGEMENT OF LIVESTOCK FOR DEER

Unmanaged, domestic livestock can be inimical to and competitive with white-tailed deer. Managed, livestock and white-tailed deer are compatible and even synergistic. Livestock can change the quantity and quality of the vegetation on which they feed and, by erosion processes and nutrient losses, even the quality of the soil on which the feed is produced (Severson 1990). No other factors in deer ecology and management, outside of poaching and the plow, has influenced the habitat and productivity of deer on rangelands more than livestock (Teer 1987).

Because of tradition and economic worth, cattle, sheep, and goats are ordinarily the animals of choice in rangelands. They are often stocked in combinations on common-use ranges where browse, forbs, and grasses provide preferred foods. They receive priority in management and allocation of resources over other products of rangelands. Until recent years, white-tailed deer and other species of wildlife have not been considered in livestock management; i.e., in stocking rates, grazing systems, and fencing. Changes that have occurred in management of rangeland for wildlife were made largely because of commercialization of hunting recreation.

Further, grazing is the usual land-use priority in rangelands because livestock are tractable and are stocked in habitats where row-crop agriculture cannot be practiced. Moreover, grazing and browsing animals are adapted to harvest forage crops not otherwise harvested or available for human use. Including whitetails in management plans for livestock is practiced much more in the 1990s than certainly in the past 4 or 5 decades throughout the species' range.

Behavioral Relationships with Livestock

Whether whitetails avoid cattle in pastures that are grazed by livestock is not clear. The question is whether they avoid livestock or whether they are present in lower numbers in ranges whose forage quantity and quality are lower from grazing by livestock.

Deer on the Welder Wildlife Foundation in south Texas avoided pastures stocked with cattle and, when present, separated themselves from cattle by 23 to 27 m (Adams 1978). Deer did not alter their home ranges but chose pastures in unstocked portions of their range. Avoidance was most pronounced in heavily stocked, short duration grazing (SDG) systems (Cohen 1985). Kramer (1973) and Ellisor (1969) reported similar avoidance behavior with greater distances between deer and cattle than between deer and cervids.

However, Michael (1967) reported that whitetails ignored cattle and commonly grazed within 27.4 m of them. They also ignored riderless horses but avoided horses with riders, never letting them approach closer than 274.3 m. Merrill et al. (1957*a*) showed avoidance by deer of pastures whose forage supply was decimated by heavy grazing pressures of goats, sheep, and cattle. There did not appear to be an avoidance of livestock per se; lightly or moderately stocked, deferred rotation pastures contained deer in higher numbers than pastures grazed more intensively.

Skovlin et al. (1968) determined that elk use of vegetation decreased as cattle stocking increased; however, stocking rates had little effect on use by deer. That compatibility and avoidance behavior of cervids to other ungulates may be linked to condition of the vegetation is further verified by Yeo et al. (1993). In their study in Idaho, elk preferred rested pastures during June to October but few mule deer used the pastures during the same period. Mule deer selected habitats in the winter that had been grazed previously by cattle.

Competition for Resources

Cattle, sheep, goats, feral horses, burros, swine, and translocated non-native ungulates are often coinhabitants with white-tailed deer of rangeland habitat. Competition for forage occurs to some degree between all herbivores and white-tailed deer.

Depending on rangeland types and forage resources, cattle, sheep, and goats are the most common domestic herbivores. Hardly any rangeland is left ungrazed by domestic animals. Thus forage resources must be managed in the context of grazing by domestic animals but with attention to diets of wild herbivores.

In some western rangelands, exotic or introduced animals are established in white-tailed deer habitat in public and private lands. Three species of deer: Axis deer, Sika deer, and fallow deer, are widespread in central and south Texas (Traweek 1989, Teer et al. 1993). Two bovids, the aoudad sheep (*Ammotragus lervia*) and the nilgai antelope (*Boselaphus tragocamelus*), are also established in large free-ranging, unhusbanded stocks in Texas. The gemsbok (*Oryx gazella*), Siberian ibex (*Capra ibex siberica*), and the aoudad are present as wild and free-roaming populations primarily on public lands in New Mexico (Morrison 1989). Many other large mammals have been translocated but other than those mentioned above, few are established in free-ranging populations. Baccus et al. (1983) and Teer (1993*b*) described relationships between some of these translocated herbivores and domestic livestock.

Competition is important between deer and other ungulates when populations are dense, especially with sheep and goats. Overlaps in diets occur with these species and seasonally there is almost 100% diet overlap. Cattle are less competitive with deer because their diets are primarily grasses and grass-like plants that deer use sparingly.

When ranges are overgrazed, however, competition between all herbivores can occur because all must feed on whatever plant materials are available. It matters little which animals are competitive with each other when undue pressure on range forage is exerted.

Good deer habitat consists of diverse vegetation. Availability of seasonal food sources is imperative in satisfying nutritional requirements. White-tailed deer reach their greatest numbers in rangeland communities where seral stages are below climax. Thus, disturbed land is often the best deer range, because it contains an array of annual and perennial herbaceous plants that serve as food, and woody plants provide food and protection from enemies and weather. Several researchers have attempted to determine forage relationships between livestock and whitetails. Animal unit equivalents proposed by Merrill et al. (1957*b*) on the Edwards Plateau of Texas were 5 sheep, or 6 goats, or 6 white-tailed deer/489 kg cow with calf at side. Davis (1952) reported that 13 deer were equivalent to 1 cow on the King Ranch in South Texas. These ratios are viewed as rough standards for interpreting stocking rates and for developing diet relationships between various herbivores.

Transfer of Diseases and Parasites by Deer and Livestock

Livestock and deer are hosts and vectors of several parasites and diseases that affect both. On ranges stocked with livestock and deer, diseases and parasites may pass easily from 1 to the other. To control or rid livestock of a particular disease is difficult when white-tailed deer (and other wild animals) serve as an intermediate host in the life cycle, or are chronically infected and serve as a reservoir, or serve as a host to the vector of the parasite or pathogen. The reverse may also occur; livestock can be the major source of a particular organism that affects wildlife.

With the exception of the screw worm fly that has now been controlled in the United States, parasites and diseases seldom cause substantial mortality in white-tailed deer. Prior to control of this parasite, a very large percentage of fawns in central and southern portions of the species' range were infested with screw worms in their navels immediately after birth. Females were also often infested in birth wounds. Males also were victims, especially on their ears and backs where ticks had caused an issue of blood or serum.

Salmonella, babesia, anthrax, leptospirosis, and a few other viral and bacterial diseases have been reported in deer. Anthrax occurs rarely but is devastating to a herd. Ticks, deer flies, and keds are common external parasites of deer.

Important parasites common to and passed between whitetails, livestock, and other cervids and bovids are the liver fluke (*Fascioloides magna*), the meningeal worm (*Parelaphostrongylus strongylus*), and several abomasal nematodes (*Haemonchus contortus*, *Trichostrongylus* spp., *Ostertagia* spp.). Several viral and bacterial diseases are also shared by livestock and white-tailed deer. Among the most important are anthrax, blue-tongue virus (BTV), malignant catarrh fever (MTF), leptospirosis, and Texas fever (*Babesia* spp.).

Three examples of parasites will serve to illustrate the action and importance of parasites and diseases in whitetails: the liver fluke, the meningeal worm, and abomasal nematodes.

Liver flukes are important parasites in wet areas of white-tailed deer ranges and cause losses to the cattle industry through liver infestations and perhaps debilitation of the animals (Quereshi et al. 1989). The disease is especially important in cattle ranges of the southeastern United States.

The life cycle of liver flukes in Texas calls for the eggs and larvae to pass through snails, white-tailed deer, and cattle. White-tailed deer are the definitive hosts; a snail (*Lymnae bulimoides*), the intermediate host; and cattle, the dead-end hosts. The cycle may be broken at any stage and thus control of the parasite can be effected by this disruption. Quereshi et al. (1989) controlled liver flukes in a free-ranging population of white-tailed deer and thus in cattle by delivering a pharmaceutical (i.e., Triclabendazole) to deer in medicated corn bait.

White-tailed deer serve as hosts or reservoirs of the meningeal worm but are unaffected by it. However, transferred through the feces to other cervids, the parasite is devastating. Translocated populations of white-tailed deer have carried the parasite and exposed elk, moose (*Alces alces*), mule deer, and exotic deer that died. The parasite is of great concern to game farming of native and exotic cervids in North America. Cattle, on the other hand, are not affected by the parasite.

Abomasal parasites and other organisms can be important when deer are debilitated on food-short ranges. In experiments at the Welder Wildlife Foundation, the number of abomasal parasites doubled after deer numbers had outstripped their food supply (Pederson 1980). Losses of deer were high and were attributed to low food supply, but parasites certainly exacerbated the problem.

Abomasal nematodes are commonly passed between sheep and white-tailed deer. Ruminant hosts can be infected by ingesting larvae on forage or by contact with animals that have infections. Abomasal worms can be debilitating and exacerbate mortality in dense populations of whitetails that are on food-short ranges (Pedersen 1980). Sheep and goats must be treated with phenothiazine and other antibiotics to keep numbers down.

Stocking Rates and Grazing Systems

Grazing by domestic and wild herbivores is important to the quality of a game range (Teer 1987). No use more intimately affects other possible uses of rangeland than does livestock grazing (Stoddart et al. 1975).

The primary use of grazing in management of wildlife is to produce successional stages or seres to which wildlife of 1 kind or another are obligate members or are favored by them. Obviously, although white-tailed deer have wide tolerances and their requirements are met by many ecological types, they are favored by certain seral stages of the vegetation and physical conditions. Vegetation is the key to habitat and is the primary element that can be managed. Livestock grazing can effect changes in vegetation.

In the past, criteria used by range and wildlife managers to judge carrying capacity or quality of the range for livestock had little relationship to quality of white-tailed deer habitats. Obviously, great differences occur in food and cover requirements of various kinds of domestic animals and deer, and these differences were largely ignored in classification systems.

The early classification systems for land capability in North America described wildlife habitat as those lands that were not suited to any other crop. The Soil Conservation Service developed 8 classes of land on the basis of soil characteristics and land form. The best, Type I, was suited to a wide range of use; Type VIII was recommended for wildlife because it had little or no value for conventional row-crop agriculture. It could not be cultivated due to terrain, salinity, alkalinity, wet, or low moisture holding capacity. This system has been largely abandoned because integration and multiple-use are bywords of production systems in rangelands. Land managers now use range site classification systems in which soil nutrients and minerals, topography, elevation, vegetation, and other attributes are judged specifically for livestock and wildlife.

The impact of stocking rates of livestock on whitetails is a 2-pronged relationship. Overgrazing is obviously harmful to deer habitat, and undergrazing can also reduce its quality. Proper (moderate, for the most part) grazing promotes plant vigor and diversity and increases production (Anderson et al. 1990). Undergrazing by livestock permits accumulation of litter and "wolf plants" that inhibit wildlife, including white-tailed deer, from utilizing succulent forage.

Claims of "managing deer and other wildlife habitat with the cow," while not extravagant, have not been well validated by research. Definitive studies for white-tailed deer and livestock are not numerous and few generalities and principles have been produced beyond the concept of "too many or too few."

Connolly and Wallmo (1981:543) lamented the lack of definitive research on livestock grazing and mule deer and black-tailed deer management, stating: "Theory abounds, but supportive documentation of effective results is in short supply. Livestock use probably is beneficial to deer in some circumstances, but intentional manipulation of livestock to enhance deer range is another matter. In this area much remains to be learned." The same may be said of the impacts or benefits of livestock grazing on white-tailed deer, and probably on mule deer and black-tailed deer 12 years after Wallmo's and Connolly's assessment. Urness (1990) called for planned experiments to understand the impacts of grazing on wildlife. Bryant (1982), in a review paper, suggested that research on wildlife-range-livestock interactions is lagging behind development and implementation of grazing systems, and that such research is a fertile area for range and wildlife scientists.

Diets of white-tailed deer vary with kinds of livestock in competition with deer and seasons of the year. They also vary with range conditions (Bryant et al. 1981). At the Ranch Experiment Station near Sonora, Texas, heavy grazing (18.5 animal units/km^2) by cattle, sheep, and goats and by any single kind alone resulted in range deterioration for deer and livestock (Merrill et al. 1957*a*). Angora goats were most competitive with deer for browse, and sheep stocked at heavy intensities also reduced deer numbers in experimental pastures.

However, general management principles and techniques are used by range and wildlife managers in integrating deer and livestock management. Pastures stocked at heavy rates and grazed without rest are often the most or nearly as profitable in livestock production for the short term (Herbel 1971, Drawe 1991, Taylor et al. 1993). Using density as a measure of acceptance by white-tailed deer of an array of grazing systems and regimes, Readon et al. (1978) found deer preferred a 7-pasture rotation system over other systems available to them on the Ranch Experiment Station at Sonora, Texas.

Cows stocked at heavy rates, grazed yearlong in a 4-pasture, deferred rotation system produced more beef/ha and had higher net income/cow than other grazing systems stocked at moderate rates in native rangelands on the Welder Wildlife Foundation in south Texas (Drawe 1991). Aside from movements of deer within and among pastures involved in various grazing systems, results of studies of white-tailed deer responses were largely inconclusive (Drawe 1988).

However, short term gains are offset by damage to the vegetation. The vegetation cannot support high stocking rates of livestock or provide adequate habitat for white-tailed deer over the long term.

Cohen (1985) reported cattle traveled more in the SDG than in the continuous grazed pastures (COT) that were the subject of Drawe's (1988) experiments with grazing systems. Further, key forage species for deer were similar in the systems.

Overgrazing for extended periods by any animal destroys wildlife habitat and creates exploitative competition between herbivores. Perhaps the diversity and site-specific needs in management coupled with vagaries of weather, soil, markets, and land-use goals prevent more than general doctrinal principals and policies. Prescriptions for integrating management of livestock and white-tailed deer are yet to be definitively described.

Grazing systems favored by range and wildlife managers integrate livestock and wildlife requirements with biological and economic efficiency. These usually call for deferment(s) or rest period(s) for vegetation recovery. Drawe (1991:iv) concluded from experiments with cattle and white-tailed deer relationships in several grazing systems at the Welder Wildlife Foundation ". . . that in the short term, heavy continuous or heavy 4-pasture, deferred-rotation grazing will produce the greatest profit, for the long term. However, moderate continuous or moderate 4-pasture, deferred-rotation is recommended with a variable stocking rate which increases in times of high forage production and decreases during times of drought."

Heavy stocking rates in high intensity, low frequency (HILF) or SDG systems, have the real capacity to damage plants that are most valuable to wildlife including browse, forbs, and seed-producing species. Grazed and hedged at the tips of their apical meristems, such plants are more vulnerable to overgrazing than grasses whose meristematic tissues are at the bases of the leaves. Many species of wildlife including white-tailed deer depend on broad-leaved and woody plants rather than grasses.

Many row crops and deer are not compatible. In agricultural regions where they might succeed, deer should not be encouraged because of damage to cereal, vegetable crops, and orchards. Damage claims will surely result if deer are allowed to depredate.

Range Improvements. Range improvements have, until recently, been directed primarily at domestic livestock. Because whitetails and livestock have different requirements, deer were often neglected in management plans. Now, however, commercialization and increased interest in recreational hunting have stimulated consideration for integrating management of all herbivores. When other wildlife species are considered in management plans, management becomes more difficult.

Those species that are to be favored, be they domestic or wild, must be identified and prescriptions made for them. Compromises and trade-offs often must be made to accomplish management for deer and livestock. Some ranches are managed exclusively for livestock, others for deer. Vallentine (1989:40) stated, "In general, ideal big game habitat has been equated with a greater mixture of forage species than needed for livestock, a mosaic of vegetation types, and greater availability of cover than needed for livestock."

Woody plants, some of which are noxious, unwanted, or too dense, dominate many rangelands throughout the white-tailed deer's range (Davis and Spicer 1965, Scifres 1980). Invasions of brush species onto grasslands and prairies created habitat for white-tailed deer in many regions. However, loss of rangeland productivity for livestock and deer is the result of large, unbroken, and unmanaged brush tracts. Competition for nutrients, water, and shade of ground vegetation by woody species reduces forage for both.

The usual prescription for such conditions is removal of brush in some pattern that increases forage production through creation of edge and plant diversity while retaining cover for deer. The result is a mosaic that favors grasses and forbs in cleared areas for cattle and with woody vegetation left in strips or blocks for food and cover for deer. The implication is not that cattle and deer occupy different niches or habitats. Both use cleared and brush strips and both benefit from increased quality and diversity of forage species.

Brush management should consider key or favored food sources. In broken and incised terrain, brush may be cleared on the contour to be aesthetically pleasing, to retain certain plants favored by livestock and deer, and to provide cover in juxtaposition to food, water, and other requirements. Protection of vegetation on ridges and south-facing exposures in north temperate regions may be important considerations in range improvements.

Riparian zones are usually key features in habitats for deer and should be protected in any arrangement or pattern of clearing. Depending on the nature of the zone, generally at least 100 m on the side should be left for deer and other wildlife. Travel lanes are left and positioned to connect brush tracts. These should be ≥30 to 40 m in width or even larger depending on their screening effect.

An ongoing discussion among range and wildlife scientists is the proportion of brush to be cleared in relation to that which is left. The answer is not clear; it probably rests on the kind of animals to be favored. In unbroken brushland habitats, amounts to be cleared and left in any pattern must be site-specific. Soils, vegetation, terrain, and weather patterns are factors that predispose management plans. The Soil Conservation Service's prescription for brush management for white-tailed deer and livestock in Texas ranged between 65 and 90% to be cleared (Rechenthin et al. 1964). With the increased interest in deer management, the amount to be cleared has been reduced in most sites. A rule-of-thumb prescription for favoring deer is a checkerboard or strip pattern in which 40% of the brush is left and 60% is cleared (Fig. 4).

Mechanical and chemical methods, fire, and grazing by domestic livestock are usual methods of controlling woody species. With perhaps the exception of fire, all are expensive and results usually are short-lived. Proper grazing management by livestock and preventive methods are usually less expensive and more effective in the long term. Examples of responses of various treatments follow.

Beasom and Scifres (1977) reported no effects on deer and other species from chemically treating 80% of a mature mesquite (*Prosopis* spp.) brushland with 2,4,5,T + picloram in a pattern of parallel strips alternating with 200 m of cleared and 30 m of brush. However, at 100% coverage, deer numbers were reduced because of loss of forbs. Recovery of deer occurred after 27 months when forb populations recovered.

Root plowing of plant communities in south Texas had very serious long-term impacts on deer habitat (Ruthven et al. 1993). Woody plant species were reduced from 19 to 7 on root-plowed areas. Preferred browse species such as colima (*Zanthoxylum fagara*) and guajillo (*Acacia berlandieri*) were eradicated. Huisache (*Acacia smallii*), an aggressive invader, was increased by 7-fold and forbs were about 2 times that on untreated sites.

The study by Ruthven et al. (1993) suggests that root plowing is a negative influence on white-tailed deer habitat. However, Ruthven et al. (1994) in a following study reported that despite a difference in diets of white-tailed deer on root-plowed and nontreated areas, no long-term effects of root plowing on deer condition, density, reproductive characteristics, and age structure were observed.

Shredding, roller chopping, and discing of woody vegetation is a common practice. While it is often done for livestock, its effects are important to white-tailed deer.

Roller chopping of parallel strips of guajillo- and blackbrush acacia- (*Acacia rigidula*) dominated rangelands and heavy discing of whitebrush- (*Aloysia lycioides*) dominated rangelands in south Texas reduced canopy cover and increased herbaceous and woody forage on both sites (Bozzo et al. 1992*a*). As a result, deer densities were higher on roller chopped and disced areas, probably due to increased forb production (Bozzo et al. 1992*b*).

The content of crude protein and phosphorus was higher in early regrowth of shredded bluewood (*Condalia hookeri*), granjeno (*Celtis pallida*), colima, and lotebush (*Ziziphus obtusifolia*), all important deer browse plants, in south Texas (Everitt 1983).

Whitetails clearly preferred 4.0 to 32.4-ha clearings among openings created by roller-chopping and shredding of various 4.0 to 32.4-ha blocks in south Texas brushlands (Naderman 1979). The smaller were favored over larger clearings primarily due to behavioral patterns of deer in relationship to distance of screening escape cover.

Although fire may kill young plants such as ashe juniper (*Juniperus ashei*), it may only top kill other species such as huisache and mesquite (Rasmussen et al. 1983). Prescribed

Figure 4. Schematic representation of checkerboard pattern of clearings with wooded travelways and protected riparian habitat.

burning can produce very important improvements by eliminating ground vegetation of litter and promoting new growth of herbaceous and woody vegetation relished by deer. Burning can be used to increase forbs (Hansmire et al. 1988). Further, early winter burns favor forbs and late winter burns favor grasses in south Texas.

Cattle readily eat prickly pear (*Opuntia* spp.) after the spines have been removed by burning with backpack butane burners (Vallentine 1989). Although the pads and tunas are important in the diets of whitetails (Knowlton 1964), pastures can be dominated by pear. Removal will increase productivity and diversity of plant communities (Valentine 1989). Low-growing brush and sprout regrowth following mechanical or chemical treatments and fire may be further reduced by the use of goats.

Water, Fences, and Other Physical Features. Water is a requirement of whitetails; however, some populations exist in habitats where free water is scarce. Water is a positive feature and an important factor in deer distribution on rangelands. Deer use water as often as 2-3 times/day in arid environments. Intake is directly related to the physiological state of the animal (i.e., growth, maintenance, lactation) and the amount of free water taken is inversely proportional to the amount in forage (Verme and Ullrey 1984).

Whitetails use free water in natural catchments and also water supplied in troughs and stock tanks. Design of troughs and provision of overflows from windmills are important considerations in providing water to whitetails and other wildlife.

Fences are also important features of deer habitat on rangelands. The normal 1.42-m high stock fence, either of barbed or net wire, is no impediment to adult deer. However, barbed wire fences can cause considerable mortality of whitetails when first built and deer are not accustomed to them. They entangle their hind leg(s) in the topmost 2 wires that become twisted around them. Many die after being released because they usually dislocate their limbs at the hip.

Deer populations confined by high fences, ≥2.44 m, pose special problems in management. While most ranch herds are fenced to produce trophy-sized bucks and to prevent illegal entry by poachers, harvest of females and younger age classes of bucks is often neglected. Such harvest programs lead to overpopulations and distorted herd composition in regions where large numbers of ranches are "deer-proofed." Effects of genetic isolation are also a distant and distinct possibility.

Artificial Feeding and Food Plots. The diets of whitetails can be supplemented in food-short periods by planting grain and forage crops and by putting out grains and other high-energy and protein supplements in feeders. However, feeding and food plots should only be used when natural food sources are in short supply. Plantings of various cereal grains are used to lure deer to the gun.

Cereal grains such as wheat, oats, rye, corn and several clovers are planted in winter to supply green forage to supplement diets of white-tailed deer. In warmer areas of white-tail range, however, where green forage is almost always available, supplemental feeding of green forages may not be used or important.

In forested areas and some rangelands where soils are poor and green forage is seasonally deficient, plantings of cereals provide important food sources in winter. Johnson et al. (1987) found that about 40% of the diet of deer in southern forest habitats was comprised of forages from food plots that totaled <1 % of the habitat. However, fertilized, warm-season plots planted to American jointvetch, red clover, wheat, and subterranean clover failed to produce improvements in deer quality on the Kisatchie National Forest in Louisiana (Johnson and Dancak 1993). Their conclusion was that their ". . . food plot program was not warranted based on biological effects."

Beals et al. (1993) measured standing crops, nutritional quality, and use by deer of several warm-season legumes in food plots in southern Texas. Lablab had higher standing crop and was about equal to cowpeas and soybeans in crude protein. However, cowpeas and soybeans received greater use than lablab. An important attribute of lablab was its tolerance to drought that makes it an important plant for summer food plots in southern Texas.

Corn and sorghum are usually the grains of choice in feeding programs. They are high energy foods but do not contain the 16% protein needed for growth, maintenance, and reproduction.

Grains put out in feeders in Texas, and probably elsewhere, are not usually intended to supplement diets of white-tailed deer. Rather, whole or cracked grains are used to bring deer to the gun where they may be harvested from blinds set nearby, a practice that may not be considered fair chase by some. However, in overstocked ranges, deer must be reduced to prevent overstocking. Food plots and feeders are great aids in accomplishing harvest quotas.

Mineral blocks and protein supplements fabricated into various pelleted foods are on the market for enhancing body and antler development of white-tailed deer. Such foods, if taken by deer, can be an important supplement to native foods. However, wildlife and range managers usually strive to satisfy diet requirements of deer with native vegetation. Salt provided in blocks or in granulated table form is readily used by deer especially on ranges where it is deficient in soils and vegetation.

Control of Illegal Kill. Management of deer herds can be ineffective if poaching of deer gets out of hand. Poaching keeps deer numbers at low levels and, in some instances where dogs are used in hunting and night hunting with spotlights is practiced, deer can be completely eliminated (Jenkins 1952, Marchinton et al. 1970). It is thus absolutely essential in any management program to control illegal kill and to have enforceable regulations on seasons, bag limits, and means and methods of taking deer. Dogs, hunting at night with lights, and traps and snares should be prohibited.

Unless hunting dogs are carefully controlled and penned, they harass deer and take sex and age classes that should not be harvested. Free-ranging dogs owned by rural families and others are particular problems. Females are especially vulnerable during gestation. Fawns <6 weeks of age are easily captured by dogs.

Many ranches employ game guards to assist state and federal game wardens in protecting wildlife resources. The game guards in Texas, or fence riders as they are often termed, are empowered by the state to make arrests, issue citations for violations, and appear in court to testify. They cooperate and patrol private property with government wardens. Their activities are usually confined to the ranches or groups of ranches that employ them. They have been an important force in controlling illegal entry onto private property and therefore in protecting wildlife resources.

SUMMARY

The white-tailed deer is widespread and the most abundant big game mammal in North America. It is important economically and recreationally to hunters and landowners alike. It is primarily a forest-dwelling animal but has adapted to practically all ecosystems including extensive rangelands. Numbers of the species and its distribution have increased as man modified habitats to make them more habitable. Disclimax or disturbed habitats are favored because of the diversity of plants in annual and seasonal abundance.

The species is genetically plastic as evidenced by large differences in body size within and between subspecies. Nutrition is the paramount factor governing the species' ability to achieve its genetic potential. Home ranges of white-tailed deer are small and members of the species have strong affinities to their home ranges. They are not migratory but move seasonally to protected areas and lower slopes in colder climes. Subspecies in the south occupy the same ranges yearlong.

Adaptability is the keynote factor in their diets. When available, they favor forbs, browse, and mast over grasses. Proportions of these foodstuffs taken seasonally by deer differ in relation to their availability. Browse is the primary food in winter in colder areas; forbs and green browse are favored in the south because they are available in winter. High energy foods such as acorns are especially important for laying on fat reserves for the winter season.

The species is promiscuous in its mating system and the sex cycle and antler development are related to daylength as seasons progress. Females are seasonally polyestrous, and may breed first as fawns. Males first breed as long yearlings. Estrous cycles are repeated every 28 days and females can undertake as many as 4 cycles in the breeding season. Most females conceive in their first heat. Twin fawns are usually produced by adult females; singletons by yearlings. Nutrition plays a major part in productivity and annual increment to the herd.

Because of the species' economic value in commercial hunting leases and other fee-hunting systems, integration of management of livestock and wildlife has become a major consideration in range management. Without control of their numbers, deer inexorably exceed the carrying capacity of their ranges. Losses of deer occur frequently where hunting offtake is inadequate and when overuse of forage resources by livestock and other herbivores occurs. Any sex seasons are important in controlling numbers. Natural mortality including predation, parasites, and diseases impact deer numbers but habitat management and hunting are more important to the welfare of white-tailed deer herds. Restocking with wild-caught deer has been an effective method of restoration of deer in ranges where they have been extirpated.

Overgrazing and undergrazing are the two extremes of livestock management that impact deer herds. Goats and sheep are most competitive with deer for range forage; cattle, the least. Large mammals translocated from other regions and continents are also competitors with deer for forage. White-tailed deer avoid close association with livestock, but occupy the same ranges and often use the same forage. Competition between domestic and wild herbivores is most severe on poor ranges.

Grazing systems for livestock can be used to produce successional stages of vegetation (seres) that favor white-tailed deer. However, precise prescriptions for using livestock for creating or maintaining quality deer habitat are difficult to determine and need quantitative research. At present, livestock management for deer habitat is directed at prevention of overgrazing and undergrazing. Deferred rotation grazing systems of 1 form or another with moderate stocking rates are favored over other systems of grazing management for white-tailed deer habitat. Because of variation in vegetation communities, vagaries of weather, economics, and traditions of those who control the land, grazing systems are seldom implemented specifically for white-tailed deer.

Range improvements are usually integrated with increasing carrying capacities for livestock and white-tailed deer. In brush-infested ranges, improvements usually mean partial clearing of woody plants in some pattern that increases diversity of vegetation and amount of forage preferred by livestock and deer. Strip or block clearings are the usual patterns of clearings. The amount of brush to be cleared and amount left varies with the habitat; however, a rule of thumb in ranges heavily infested with brush is to clear 60% and leave 40%.

Protection of key habitats such as riparian zones, wetlands, and particular food plants from overgrazing by livestock and game is an important strategy in habitat management. Chemicals, fire, and mechanical methods are used in managing habitat for white-tailed deer. All have their place. Their long-term impacts must be measured against short-term effects and goals of the manager. Management of native vegetation is the preferred strategy in deer management. Supplemental feeding is often practiced, however. Food plots of especially cereal grains are valuable for providing green forage in winter months. Provision of grain in feeders is not recommended beyond bringing deer to the gun.

Control of illegal kill, uses of dogs in hunting, and the erection of "deer-proof" fences are issues that must be considered and reconciled for herd management to be effective on ranch and timberland habitats.

LITERATURE CITED

Adams, N. E., Jr. 1978. The effects of cattle on white-tailed deer distribution. M.S. Thesis, Texas A&I Univ., Kingsville. 63pp.

Anderson, E. W., D. L. Franzen, and J. E. Melland. 1990. Forage quality as influenced by prescribed grazing. Pages 56-70 *in* K. E. Severson, ed. Can livestock be used as a tool to enhance wildlife habitat? U.S. Dep. Agric. For. Serv. Gen. Tech. Rep. RM-194. 123pp.

Anthony, R. G., and N. S. Smith. 1977. Ecological relationships between mule deer and white-tailed deer in south-eastern Arizona. Ecol. Monogr. 47:255-277.

Arnold, L. A., and D. L. Drawe. 1979. Seasonal food habits of white-tailed deer in the South Texas Plains. J. Range Manage. 32:175-178.

Baccus, J. T., D. E. Harmel, and W. E. Armstrong. 1983. Management of exotic deer in conjunction with white-tailed deer. Pages 213-226 *in* S. H. Beasom and S. F. Robertson, eds. Proc. Game Harvest Manage. Third Int. Symposium. Caesar Kleberg Wildl. Res. Inst., Texas A&I Univ., Kingsville.

Bailey, V. 1905. Biological survey of Texas. North Am. Fauna 25. 222pp.

Baker, R. H. 1984. Origin, classification and distribution. Pages 1-18 *in* L. Halls, ed. White-tailed deer ecology and management. Stackpole Books, Harrisburg, Pa.

Banasiak, C. F. 1964. Deer in Maine. Maine Dep. Inland Fish and Game. Bull. 6. 163pp.

Beals, N. M., T. E. Fulbright, and R. L. Bingham. 1993. Nutritional quality and white-tailed deer use of warm-season forages. Southeast. Assoc. of Fish and Wildl. Agencies 47:In Press.

Beasom, S. L., and C. J. Scifres. 1977. Population reactions of selected game species to aerial herbicide applications in south Texas. J. Range Manage. 30:138-142.

Bersing, O. S. 1956. A century of Wisconsin deer. Wisc. Conserv. Dep. 184pp.

Boydston, G. 1992. White-tailed deer harvest surveys. Tex. Parks and Wildl. Dep. Fed. Aid Wildl. Restor. Proj. W-127-R-1, Job 4. 70pp.

Bozzo, J. A., S. L. Beasom and T. E. Fulbright. 1992*a*. Vegetation responses to 2 brush management practices in south Texas. J. Range Manage. 45:170-175.

———, ———, and ———. 1992*b*. White-tailed deer use of rangeland following browse rejuvenation. J. Range Manage. 45:496-499.

Brokx, P. A. 1984. South America. Pages 525-546 *in* L. K. Halls, ed. White-tailed deer ecology and management. Stackpole Books Press, Harrisburg, Pa.

Bryant, F. C. 1982. Grazing and wildlife: past results, future needs. Pages 59-63 *in* Proc. National conference on grazing management techniques. Texas A&M Univ., College Station. 228pp.

———, C. A. Taylor, and L. B. Merrill. 1981. White-tailed deer diets from pastures in excellent and poor range condition. J. Range Manage. 34:193-200.

Bubenik, G. A. 1982. The endocrine regulation of the antler cycle. Pages 73-107 *in* R. D. Brown, ed. Antler development in Cervidae. Caesar Kleberg Wildl. Res. Inst., Kingsville, Tx.

Burger, G. V., and J. G. Teer. 1981. Economic and socioeconomic issues influencing wildlife management on private land. Pages 252-278 *in* Proc. wildlife management on private lands. Wisc. Chap. The Wildl. Soc., Milwaukee.

Chamrad, A. D., and T. W. Box. 1968. Food habits of white-tailed deer in south Texas. J. Range Manage. 21:158-164.

Cheatum, E. L., and G. H. Morton. 1942. On the occurrence of pregnancy in white-tailed deer fawns. J. Mammal. 23:210-211.

———, and ———. 1946. Breeding season of white-tailed deer in New York. J. Wildl. Manage. 10:249-263.

———, and C. W. Severinghaus. 1950. Variations in fertility of white-tailed deer related to range conditions. Trans. North Am. Wildl. Conf. 15:170-190.

Cohen, W. E. 1985. Deer movement and habitat response to short duration grazing. M.S. Thesis, Texas Tech. Univ., Lubbock. 39pp.

Connolly, G. E. 1978. Predators and predator control. Pages 369-394 *in* J. L. Schmidt and D. L. Gilbert, eds. Big game of North America: ecology and management. Stackpole Books Press, Harrisburg, Pa.

———, and O. C. Wallmo. 1981. Management challenges and opportunities. Pages 537-545 *in* O. C. Wallmo, ed. Mule and black-tailed deer of North America. Univ. Nebraska Press, Lincoln.

Cook, R. L. 1984. Texas. Pages 457-474 *in* White-tailed deer ecology and management. L. K. Halls, ed. Stackpole Books Press, Harrisburg, Pa.

Dahlberg, B. L., and R. C. Guettinger. 1956. The white-tailed deer in Wisconsin. Wisc. Conserv. Dep. Tech. Wildl. Bull. 14. 282pp.

Dalke, P. E. 1941. The use and availability of the more common winter deer browse plants in the Missouri Ozarks. Trans. North Am. Wildl. Conf. 6:155-160.

Davis, R. B. 1952. The use of rumen content data in a study of deer-cattle competition and "animal equivalence." Trans. North Am. Wildl. Conf. 17:448-458.

———, and R. L. Spicer. 1965. Brush control in the Rio Grande Plain. Texas Parks and Wildl. Dep. Bull. 46. 40pp.

———, and C. K. Winkler. 1968. Brush vs. cleared range as deer habitat in southern Texas. J. Wildl. Manage. 32:321-329.

DeArment, R. 1959. To determine game harvest. Tex. Game and Fish Comm. Fed Aid Wildl. Restor. Proj. W-45-R-9. 10pp.

———. 1960. To determine game harvest. Tex. Game and Fish Comm. Fed. Aid Wildl. Restor. Proj. W-45-R-J. 7pp.

DeYoung, C. A. 1989. Mortality of adult male white-tailed deer in south Texas. J. Wildl. Manage.

Dill, T. O., J. Menghini, S. S. Waller, and R. Case. 1983. Fee hunting for Nebraska big game a possibility. Rangelands 5:24-27.

Drawe, D. L. 1968. Mid-summer diet of deer on the Welder Wildlife Refuge. J. Range Manage. 21:164-166.

———. 1988. Effects of three grazing treatments on vegetation, cattle production, and wildlife on the Welder Wildlife Foundation Refuge, 1974-1982. Welder Wildl. Found., Contrib. B-8. Sinton, Tex. 36pp.

———. 1991. Influence of grazing on vegetation and cattle. 1983-1987. Welder Wildl. Found. Contrib. B-11. 31pp.

———, and T. W. Box. 1968. Forage ratings for deer and cattle on the Welder Wildlife Refuge. J. Range Manage. 21:225-228.

Dusek, G. L., A. K. Wood, and R. J. Mackie. 1988. Habitat use by white-tailed deer in prairie-agricultural habitat in Montana. Prairie Nat. 20:135-142.

———, R. J. Mackie, J. D. Herriges, Jr., and B. B. Compton. 1989. Population ecology of white-tailed deer along the lower Yellowstone River. Wildl. Monogr. 104. 68pp.

———, J. G. Mundinger, S. J. Riley, and R. J. Mackie. 1991. Implication of body weight to reproductive strategy in female white-tailed deer. Int. Union Game Biol. 18:555-558.

Ellisor, J. E. 1969. Mobility of white-tailed deer in south Texas. J. Wildl. Manage. 33:220-222.

Erickson, A. B., V. E. Gunvalson, M. H. Stenlund, D. W. Burcalow, and L. H. Blankenship. 1961. The white-tailed deer of Minnesota. Minn. Dep. Conserv. Tech. Bull. 5. 64pp.

Evans, W. 1984. Southern Rocky Mountains. Pages 505-512 *in* L. K. Halls, ed. White-tailed deer ecology and management. Stackpole Books Press, Harrisburg, Pa.

Everitt, J. H. 1983. Effects of plant shredding on nutrient content of four south Texas deer browse species. J. Range Manage. 36:779-781.

French, C. E., L. C. McEwen, J. D. Magruder, R. H. Ingram, and R. W. Swift. 1956. Nutritional requirements for growth and antler development in the white-tailed deer. J. Wildl. Manage. 20:221-232.

Gartner, F. R., and K. E. Severson. 1972. Fee hunting in western South Dakota. J. Range Manage. 25:234-237.

Geist, V. 1993. Great achievements, great expectations: successes of North American wildlife management. Pages 47-72 *In* A. W. L. Hawley, ed. Commercialization and wildlife management: dancing with the devil. Krieger Publ. Co., Malabar, Fla. 124pp.

Glazener, W. C. 1948. Homing instinct of white-tailed deer. Tex. Game and Fish Mag. 6:5.

Goss, R. J. 1982. Control of deer antler cycles by the photoperiod. Pages 1-13 *in* R. D. Brown, ed. Antler development in Cervidae. Caesar Kleberg Wildl. Res. Inst., Kingsville, Tex.

Hahn, H. C. Jr. 1945. The white-tailed deer in the Edwards Plateau Region of Texas. Tex. Game, Fish and Oyster Comm. 52pp.

———, and W. P. Taylor. 1950. Deer movements in the Edwards Plateau. Tex. Game and Fish Mag. 12:4-9.

Halls, L. K., ed. 1984. White-tailed deer ecology and management. Stackpole Books, Harrisburg, Pa. 870pp.

Hamilton, R. 1962. Factors affecting dispersal of deer released in Indiana. J. Wildl. Manage., 26:79-85.

Hansmire, J. A., D. L. Drawe, D. B. Wester, and C. M. Britton. 1988. Effect of winter burns on forbs and grasses of the Texas coastal prairie. Southwest. Nat. 33:333-338.

Harlow, R. F. 1984. Habitat evaluation. Pages 601-646 *in* L. K. Halls, ed. White-tailed deer ecology and management. Stackpole Books, Harrisburg, Pa.

———, and F. K. Jones, Jr. 1965. The white-tailed deer in Florida. Fla. Game and Fresh Water Fish Comm. Tech. Bull. 9. 240 pp.

Haugen, A. O. 1975. Reproductive performance of white-tailed deer in Iowa. J. Mammal. 56:151-159.

Hawkins, R. E., W. D. Klimstra and D. C. Autry. 1971. Dispersal of deer from Crab Orchard National Wildlife Refuge. J. Wildl. Manage. 35:216-220.

Herbel, C. H. 1971. A review of research related to a development of grazing systems on native ranges of the western United States. U.S. Dep. Agric. Jornada Exp. Range Rep. 3. 32pp.

Hesselton, W. T., C. W. Severinghaus, and J. E. Tanck. 1965. Population dynamics of deer at the Seneca Army Depot. N. Y. Fish and Game J. 202:77-107.

Hirth, D. H. 1977. Social behavior of white-tailed deer in relation to habitat. Wildl. Monogr. 53. 55pp.

Huot, J., F. Potvin, and M. Belanger. 1984. Southeastern Canada. Pages 293-304 *In* L. Halls ed. White-tailed deer ecology and management. Stackpole Books, Harrisburg, Pa.

Illige, D. 1951. An analysis of the reproductive pattern of whitetail deer in south Texas. J. Mammal., 32:411-421.

Inglis, J. M. 1964. A history of vegetation on the Rio Grande Plain. Texas Parks and Wildl. Dep. Bull. 45. 122pp.

Jenkins, D. H., and I. H. Bartlett. 1959. Michigan whitetails. Michigan Dep. Conserv. Lansing. 80pp.

Jenkins, J. H. 1952. The extirpation and restoration of north Georgia deer—a sixty-year history. Trans. North Am. Wildl. Conf. 17:472-476.

Johnson, F. W. 1937. Deer weights and antler measurements in relation to population density and hunting effort. Trans. North Am. Wildl. Conf. 2:446-457.

Johnson, M. C. 1963. Past and present grasslands of southern Texas and northeastern Mexico. Ecology 44:456-465.

Johnson, M. K., and K. D. Dancak. 1993. Effects of food plots on white-tailed deer in Kisatchie National Forest. J. Range Manage., 46:110-114.

———, B. W. Delany, S. P. Lynch, J. A. Zeno, S. R. Schultz, T. W. Keegan, and B. D. Nelson. 1987. Effects of cool-season agronomic forages on white-tailed deer. Wildl. Soc. Bull. 15:330-339.

Kellogg, R. 1956. What and where are the whitetails? Pages 31-57 *in* W. P. Taylor, ed. The deer of North America. Stackpole Books, Harrisburg, Pa.

Kie, J. G. and M. White. 1985. Population dynamics of white-tailed deer (*Odocoileus virginianus*) on the Welder Wildlife Refuge, Texas. Southwest. Nat. 30:105-118.

Knowlton, F. F. 1964. Aspects of coyote predation in south Texas with special references to white-tailed deer. Ph.D. Thesis, Purdue Univ., West Lafayette, Ind. 189pp.

Kramer, A. 1973. Interspecific behavior and dispersion of two sympatric deer species. J. Wildl. Manage. 37:288-300.

Krausman, P. R. 1976. Ecology of the Carmen Mountain white-tailed deer. Ph.D. Thesis, Univ. Idaho, Moscow. 160pp. + 26 Appendices.

———. 1978. Forage relationships between two deer species in Big Bend National Park. J. Wildl. Manage. 42:101-107.

———, and E. D. Ables. 1981. Ecology of the Carmen Mountains white-tailed deer. Natl. Park Serv. Sci. Monogr. Ser. 15.

———, L. K. Sowls, and B. D. Leopold. 1992. Revisiting overpopulated deer ranges in the United States. Calif. Fish and Game. 78:1-10.

Lack, D. 1954. The natural regulation of animal numbers. Clarendon Press, Oxford, U.K. 343pp.

Latham, R. M. 1950. Pennsylvania's deer problem. Pennsylvania Game News, Spec. Issue 1. Pa. Game Comm. 48pp.

Lay, D. W. 1967. Deer range appraisal in eastern Texas. J. Wildl. Manage. 31:426-432.

Leopold, A., L. K. Sowls, and D. L. Spencer. 1947. A survey of overpopulated deer ranges in the United States. J. Wildl. Manage. 11:162-177.

Longhurst, W. M., H. K. Oh, M. B. Jones, and R. E. Kepner. 1968. A basis for the palatability of deer forage plants. Trans. North Am. Wildl. and Nat. Resour. Conf. 33:181-192.

Loomis, J. B., and L. Fitzhugh. 1989. Financial returns to California landowners for providing hunting access: analysis and determinants of returns and implications to wildlife management. Trans. North Am. Wildl. and Nat. Resour. Conf. 54:196-201.

Loveless, C. M. 1959. The Everglades deer herd. Life history and management. Fla. Game and Fresh Water Fish Comm. Tech. Bull. 6. 104pp.

Marburger, R. D., and J. W. Thomas. 1965. A die-off in white-tailed deer of the Central Mineral Region of Texas. J. Wildl. Manage. 29:706-716.

Marchinton, R. L., and D. H. Hirth. 1984. Behavior. Pages 129-168 *in* L. K. Halls, ed. White-tailed deer ecology and management. Stackpole Books, Harrisburg, Pa.

———, A. S. Johnson, J. R. Sweeney, and J. M. Sweeney. 1970. Legal hunting of white-tailed deer with dogs: biology, sociology and management. Proc. Annu. Conf. Southeast. Assoc. Game and Fish Comm. 24:74-89.

Mattfield, G. F. 1984. Northeastern hardwood and spruce/fir forests. Pages 305-330 *in* L. K. Halls, ed. White-tailed deer ecology and management. Stackpole Books, Harrisburg, Pa.

McMahan, C. A., and J. M. Inglis. 1974. Use of Rio Grande Plain brush types by white-tailed deer. J. Range Manage. 27:369-374.

McShea, W. J., and G. Schwede. 1993. Variable acorn crops: responses of white-tailed deer and other mast consumers. J. Mammal. 74:999-1006.

Merrill, L. B., J. G. Teer, and O. C. Wallmo. 1957*a*. Reaction of deer populations to grazing practices. Texas Agric. Prog. 3:10-12.

———, G. W. Thomas, W. T. Hardy, E. B. Keng, C. A. Rechenthin, D. C. Langford, T. A. Booker, N. Johnson, J. G. Teer, R. J. Pederson, and J. E. Tatum. 1957*b*. Livestock and deer ratios for Texas ranges. Texas Agric. Exp. Stn. Misc. Publ. 251. 9pp.

Michael, E. D. 1965. Movements of white-tailed deer on the Welder Wildlife Refuge. J. Wildl. Manage. 29:44-52.

———. 1967. Behavioral interactions of deer and some other mammals. Southwest. Nat. 12:156-163.

———. 1968. Aggressive behavior of white-tailed deer. Southwest. Nat. 13:411-420.

———. 1970. Activity patterns of white-tailed deer in south Texas. Texas J. Sci. 21:417-428.

Morrison, B. L. 1989. The introduction of exotics: pro and con. Pages 284-290 *in* First international wildlife ranching symposium. New Mexico State Univ., Las Cruces.

Naderman, J. F. 1979. Deer use of different size mechanically cleared openings in south Texas brush. M.S. Thesis, Texas Tech. University, Lubbock. 77pp.

Nichol, A. A. 1938. Experimental feeding of deer. Ariz. Agric. Exp. Stn. Bull. 75, 39pp.

Nicholls, K. E. 1978. Adenohypophyseal delta cell size in relation to photoperiod and antler growth of mule deer, *Odocoileus hemionus hemionus* (Cervidae). Annu. Proc. North Dakota Acad. Sci. 30:8-19.

Nixon, C. M., and L. P. Hansen. 1992. Habitat relationships and population dynamics of deer in the intensively farmed midwestern United States. Pages 22-29 *In* R. D. Brown, ed. The biology of deer. Springer-Verlag, New York, N.Y.

Oh, H. K., M. B. Jones, and W. M. Longhurst. 1968. Comparison of rumen microbial inhibition resulting from various essential oils isolated from relatively unpalatable plant species. Appl. Microbiol. 16:39-44.

Park, B. C., and B. B. Day. 1942. A simplified method for determining the condition of white-tailed deer herds in relation to available range forage. U. S. Dep. Agric. Tech. Bull. 840. 60pp.

Pederson, W. T. 1980. The relationship of parasite burden to population density of white-tailed deer (*Odocoileus virginianus*) on the Welder Wildlife Refuge. M.S. Thesis, Washington State Univ., Pullman. 105pp.

Peek, J. M. 1984. Northern Rocky Mountains. Pages 497-504 *in* L. K. Halls, ed. White-tailed deer ecology and management. Stackpole Books, Harrisburg, Pa.

Prescott, W. H. 1974. Interrelationships of moose and deer of the genus *Odocoileus*. Can. Nat. 101:493-504.

Progulske, D. R., and T. S. Baskett. 1958. Mobility of Missouri deer and their harassment by dogs. J. Wildl. Manage. 22:184-192.

Qureshi, T., T. M. Craig, D. L. Drawe, and D. S. Davis. 1989. Efficacy of triclabendazole against fascioloidiasis (*Fascioloides magna*) in naturally infected white-tailed deer (*Odocoileus virginianus*). J. Wildl. Dis. 25:378-383.

Rasmussen, G. A., C. J. Scifres, and D. L. Drawe. 1983. Huisache growth, browse quality, and use following burning. J. Range Manage. 36:337-342.

Reardon, P. O., L. B. Merrill, and C. A. Taylor, Jr. 1978. White-tailed deer preferences and hunter success under various grazing systems. J. Range Manage. 31:40-42.

Rechenthin, C. A., H. M. Bell, R. J. Pederson, and D. B. Polk. 1964. Grassland Restoration. Vol. II. Brush Control. U.S. Dep. Agric. Soil Conserv. Serv. Temple, Tex. 39pp.

Reed, D. F. 1981. Conflicts with civilization. Pages 509-535 *in* O. C. Wallmo, ed. Mule and black-tailed deer of North America. Univ. Nebraska Press, Lincoln.

Robinson, R. M., J. W. Thomas, and R. G. Marburger. 1965. The reproductive cycle of male white-tailed deer in central Texas. J. Wildl. Manage. 29:53-58.

Ruthven, D. C. III., T. E. Fulbright, S. L. Beasom, and E. C. Hellgren. 1993. Long-term effects of root plowing on vegetation in the eastern south Texas plains. J. Range Manage. 46:351-354.

———, E. C. Hellgren and S. L. Beasom. 1994. Effects of root plowing on deer condition, population status and diet. J. Wildl. Manage. 58:In Press.

Sanders, E. 1941. A preliminary report on the study of white-tailed deer in the Edwards Plateau of Texas. J. Wildl. Manage. 5:182-190.

Sauer, Peggy R. 1984. Physical characteristics. Pages 73-90 *in* L. K. Halls, ed. White-tailed deer ecology and management. Stackpole Books Press, Harrisburg, Pa.

Schilling, E. A. 1938. Management of white-tailed deer on the Pisgah National Game Reserve (Summary of five-year study). Trans. North Am. Wildl. Conf. 3:248-255.

Scifres, C. J. 1980. Brush management. Principles and Practices for Texas and the Southwest. Texas A&M Univ. Press, College Station. 360pp.

Severinghaus, C. W. 1955. Deer weights as an index to range conditions on two wilderness areas in the Adirondack region. N. Y. Fish and Game J. 2:154-160.

———, and E. L. Cheatum. 1956. Life and times of the white-tailed deer. Pages 57-331 *in* W. P. Taylor, ed. The Deer of North America. Stackpole Books, Harrisburg, Pa.

Severson, K. E. 1990. Can livestock be used as a tool to enhance wildlife habitat? U.S. Dep. Agric. For. Serv., Gen. Tech. Rep. RM-194. 123pp.

Skovlin, J. M., P. J. Edgerton and R. W. Harris. 1968. The influence of cattle management on deer and elk. Trans. North Am. Wildl. and Nat. Resour. Conf. 33:169-181.

Southeastern Cooperative Wildlife Disease Study. 1982. White-tailed deer populations 1980. A map prepared in cooperation with the Emergency Programs, Veterinary Services, Animal and Plant Health Inspection Service, U.S. Department of Agriculture, through cooperative agreement 12-16-5-2230. Univ. Georgia, Athens. 1 page.

Stanton, D. C. 1963. A history of the white-tailed deer in Maine. Maine Dep. of Inland Fish. and Game, Fed. Aid Proj. No. W-37-R. 75pp.

Stoddart, L. A., A. D. Smith and T. W. Box. 1975. Range management. Third ed. McGraw-Hill Book Co. New York, N.Y. 532pp.

Suring, L. H., and P. A. Vohs, Jr. 1979. Habitat use by Columbian white-tailed deer. J. Wildl. Manage. 43:610-619.

Swift, E. 1946. A history of Wisconsin deer. Wisconsin Conserv. Dep. Publ. 323. 96pp.

Taylor, C. A., Jr., N. E. Garza, and T. D. Brooks. 1993. Grazing systems on the Edwards Plateau of Texas: are they worth the trouble? II. livestock response. Rangelands. 15:57-60.

Teer, J. G. 1987. The cow: Its use in managing wildlife. Pages 696-704 *In* L. S. Pope, ed. Beef cattle science handbook. Vol. 21. Int. Stockmen's School. Spilman Press, Sacramento, Calif.

———. 1989. Exotic animals: conservation implications. Int. Union Directors of Zool. Gardens 44:31-46.

———. 1993*a*. Commercial utilization of wildlife: has its time come? Pages 73-83 *in* A. W. L. Hawley, ed. Commercialization and wildlife management: dancing with the devil. Krieger Publ. Co., Malabar, Fla. 124pp.

———. 1993*b*. Dancing with the devil. Boone and Crockett Club News J. 8:9, 10, 34.

Teer, J. G., G. V. Burger, and C. Y. Deknatel. 1983. State-supported habitat management and commercial hunting on private lands in the United States. Trans. North Am. Wildl. and Nat. Resour. Conf. 48:445-456.

———, D. L. Drawe, T. L. Blankenship, W. F. Andelt, R. S. Cook, J. G. Kie, F. F. Knowlton, and M. White. 1991. Deer and coyotes: the Welder experiments. Trans. North Am. Wildl. and Nat. Resour. Conf. 56:550-560.

———, L. A. Renecker and R. J. Hudson. 1993. Overview of wildlife farming and ranching in North America. Trans. North Am. Wildl. and Nat. Resour. Conf. 58:448-459.

———, J. W. Thomas, and E. A. Walker. 1965. Ecology and management of white-tailed deer in the Llano Basin of Texas. Wildl. Monogr. 16. 62pp.

Texas Game, Fish and Oyster Commission. 1945. White-tailed Deer. Pages 91-104 *in* Principal game birds and mammals of Texas: their distribution and management. Texas Game, Fish and Oyster Comm. 149 pp.

Thomas, J. W. 1984. Fee hunting on the public's land: an appraisal. Trans. North Am. Wildl. and Nat. Resour. Conf. 49:455-468.

———, C. Maser, and J. E. Rodiak. 1979. Wildlife habitats in managed rangelands—the Great Basin of Southeastern Oregon. Riparian Zones. Gen. Tech. Rep. PNW-80. 18pp.

———, J. G. Teer, and E. A. Walker. 1964. Mobility and home range of white-tailed deer on the Edwards Plateau in Texas. J. Wildl. Manage. 28:463-472.

Thomas, J. K., and C. E. Adams. 1985. Socioeconomic factors affecting land access to hunt white-tailed deer. Wildl. Soc. Bull. 13:388-394.

Traweek, M. S. 1989. Statewide census of exotic animals. Texas Parks and Wildl. Dep. Fed. Aid Proj. No. W-109-R-12. Job 21. 52pp.

Ullrey, D. E., W. G. Youatt, H. E. Johnson, L. E. Fay, and B. L. Bradley. 1967. Protein requirement of white-tailed deer fawns. J. Wildl. Manage. 31:679-685.

United States Department of Agriculture, Forest Service. 1981. An assessment of the forest and rangeland situation in the United States. For. Res. Rep. 22. 352pp.

Urness, P. J. 1990. Livestock as manipulators of mule deer winter habitats in northern Utah. Pages 25-40 *in* K. E. Severson, ed. Can livestock be used as a tool to enhance wildlife habitat? U.S. Dep. Agric. For. Serv. Gen. Tech. Rep. RM-194. 123pp.

Valentine, J. F. 1989. Range development and improvements. Third ed. Academic Press, Inc., San Diego, Calif. 524pp.

Varner, L. W., and L. H. Blankenship. 1985. Southern Texas shrubs—nutritive value and utilization by herbivores. Pages 108-112 *in* Proc. symposium on plant-herbivore interactions. Snowbird, Utah.

Verme, L. J., and D. E. Ullrey. 1984. Physiology and nutrition. Pages 91-128 *in* L. K. Halls, ed. White-tailed deer ecology and management. Stackpole Books, Harrisburg, Pa.

Walker, E. P. 1964. Mammals of the world. Vol. II. John Hopkins Press, Baltimore, Md. 1500pp.

Wallmo, O. C. 1981. Mule and black-tailed deer distribution and habitats. Pages 1-25 *in* O. C. Wallmo, ed. Mule and black-tailed deer of North America. Univ. Nebraska Press, Lincoln.

Whitehead, G. K. 1972. Deer of the world. Constable and Co., Ltd. London, U.K. 194pp.

Wishart, W. D. 1984. Western Canada. Pages 475-486 *In* L. K. Halls, ed. White-tailed deer ecology and management. Stackpole Books, Harrisburg, Pa.

Wood, A. K., R. J. Mackie, and K. L. Hamlin. 1989. Ecology of sympatric populations of mule deer and white-tailed deer in a prairie environment. Montana Dep. Fish, Wildl. and Parks, Bozeman. 98pp.

Yeo, J. J., J. M. Peek, W. T. Wittinger, and C. T. Kvale. 1993. Influence of rest-rotation cattle grazing on mule deer and elk habitat use in east-central Idaho. J. Range Manage. 46:245-250.

Chapter 13
PRONGHORN ON WESTERN RANGELANDS

J. D. Yoakum, B. W. O'Gara, and V. W. Howard, Jr.

INTRODUCTION

The pronghorn (*Antilocapra americana*) evolved in North America over the last 20 million years (Fig. 1). They are truly North American, occurring no where else in the world (O'Gara 1978). Pronghorn existed with bison (*Bison bison*) in legendary numbers when Lewis and Clark made their historic journey across the continent (Fig. 2). The vernacular name pronghorn will be used throughout this chapter, for some antelope species from Africa now are free-roaming in North America, and in some places, both inhabit the same rangelands (Yoakum and O'Gara In Press).

To better understand this endemic species, first we will discuss its distribution, abundance, and habitat requirements. A review of how pronghorn were almost extirpated in the 1800s, and increased > 3,200% in the twentieth century, is testimony of the species adaptability to living with humans on contemporary rangelands and to changes in wildlife management techniques to benefit pronghorn. Understanding habitat requirements of the pronghorn is the key to managing and perpetuating the species; therefore these components will be discussed in detail.

Today, it is postulated that 98% of pronghorn share their habitat (Fig. 3) with domestic livestock (Yoakum and O'Gara 1990).

FIGURE 1. An adult buck pronghorn with large horns grazes with a small herd of does on rangelands today just as pronghorn have for centuries. Photo by J. D. Yoakum, courtesy of Western Wildlife.

An estimated 60% live on private lands with the remainder on government-administrated lands. Knowledge of the relationships of pronghorn to livestock is consequently of major import, and

FIGURE 2. Historically, pronghorn frequently grazed with bison in the grassland biome, however few herds experience this relationship today. Photo by D. Kitchen.

FIGURE 3. Today, most pronghorn occupy western rangelands with domestic livestock. Photo by George Andrejko, courtesy of the Arizona Game and Fish Department.

we will emphasize these influences. It is necessary to understand basic biological characteristics of pronghorn and various livestock to understand compatibility and competition, foraging characteristics, problems of disease transmittal, and values to modern society. Likewise, managers need to know how rangeland practices such as vegetation manipulation, water development, or fences affect the welfare of pronghorn to plan and implement effective management on private and public rangelands.

Within the past 2 decades, an increased number of technical papers have been published on pronghorn and livestock relationships (Yoakum 1975, 1980; Smith and Beale 1980; Kindschy et al. 1982; Roebuck 1982; Pyrah 1987; Anderson et al. 1990*a*,*b*; Howard et al. 1990; Yoakum and O'Gara 1990; O'Gara and Yoakum 1992; Mosley 1994; U.S. Fish and Wildlife 1994). Others are currently being published (Pyle and Yoakum In Press, Yoakum 1995, Yoakum and O'Gara In Press). Many were generated to provide increased information for species-management plans, land use plans, environmental impact statements, or testimony for litigation in courts. Our objective is to provide a review of currently available literature and major endeavors in progress concerning scientific and management reports documenting the habitat requirements of pronghorn, relationships of pronghorn to livestock, and welfare of pronghorn resulting from management, particularly habitat improvements, and cultural practices common on western rangelands.

PRONGHORN AND RANGELANDS

Distribution

When Euro-Americans began to explore North America, they found pronghorn from the plains of south-central Canada (i.e., Alta., Sask., Manit.), south through most of the western United States, to 160 km of Mexico City, Mexico (Fig. 4) (Nelson 1925). Herds ranged from the Mississippi River to the Pacific Ocean in central California. Further south, herds extended from the Gulf of Mexico in Texas to the Pacific Ocean in Lower California. Pronghorn populations reached greatest densities with bison on grasslands of the Great Plains. Smaller populations occupied intermountain and desert regions. They did not occur east of the Mississippi River in the tallgrass prairies and woodlands (Yoakum 1978).

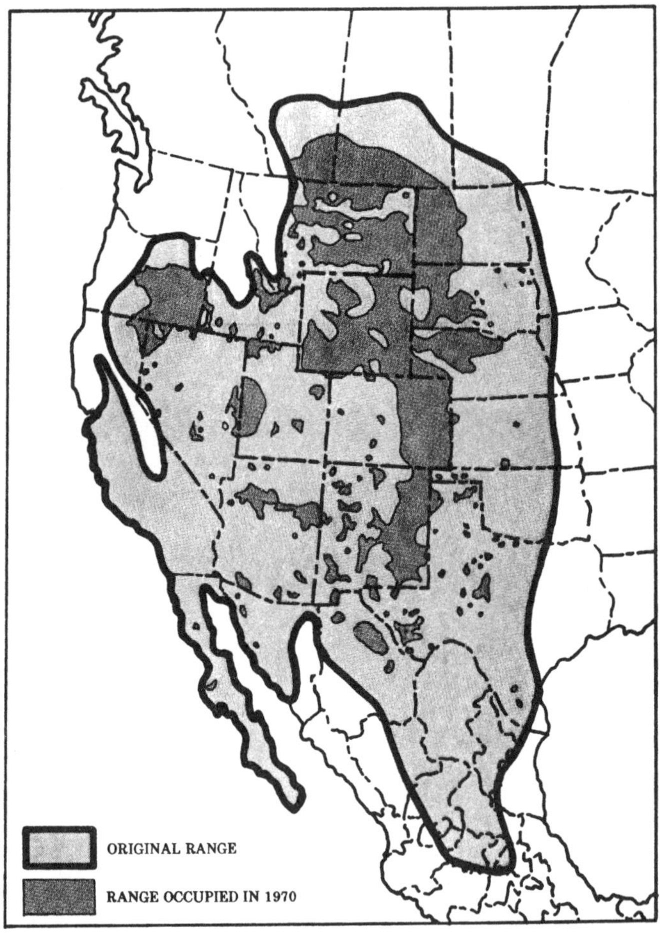

FIGURE 4. Historical and contemporary range of pronghorn. "Original range" adapted from Nelson (1925). Not all areas within these boundaries were occupied. Range occupied in 1970 modified from Yoakum (1978).

Today, pronghorn occupy many of their historic rangelands, but in greatly reduced numbers (e.g., perhaps there is 1 pronghorn today where >60 existed in 1800). More herds now are confined to smaller isolated habitats, unable to make historical seasonal movements because of freeways, railroads, fencing, and other constraints of modern civilization. The endemic pronghorn is the most representative big game species dependent upon western rangelands. Bison, elk (*Cervus elaphus*), and deer (*Odocoileus* spp.) use western rangelands but also occupy forested areas. Only the pronghorn is predominantly limited to American rangelands from Canada to Mexico (O'Gara and Yoakum 1992).

Abundance

Nelson (1925) estimated there were 30 to 60 million pronghorn at the beginning of the nineteenth century. Some reports indicate they were as numerous or possibly more abundant than bison (Seton 1927, Grinnell 1929).

During the late 1800s, herds were hunted relentlessly by sport and commercial hunters who killed animals regardless of sex or age. The biocide led some conservationists to believe the species was doomed to extirpation (Grinnell 1929). Much of the best habitat was lost to the plough, and pronghorn movements became increasingly restricted by sheep-tight fences and other man-made impediments. Pronghorn also were subjected to livestock diseases and parasites to which they had little resistance. Numbers dropped from an estimated 35,000,000 in 1800, to perhaps 13,000 in 1910 (Yoakum 1978). Then a concerned public enacted protective laws and supported conservation and management. Within a decade, the population more than doubled, and their continued recovery has been one of the remarkable success stories of wildlife management in North America.

All regulated sport hunting was curtailed until numbers increased sufficiently to sustain viable populations. Capturing and translocating pronghorn to unoccupied historic rangelands became a major management practice. Water improvements were developed by the thousands as a product of multiple-use in areas where drinking water was scarce. With controlled hunting, rangeland improvements, and wildlife enhancement projects, pronghorn populations prospered. From 1924 to 1984, pronghorn increased > 3,300% (Table 1). They are now second only to deer in abundance and harvest in the United States. A similar story exists for Canada, with accelerated population increases for the last 60 years. However, there has been a continuous decrease in populations in Mexico. Although fully protected by law since the 1920s, herds in Mexico apparently are not effectively protected from illegal hunting and loss of habitat to human occupation. This has resulted in all 3 subspecies in Mexico (*A. a. mexicana*, *A. a. sonoriensis*, and *A.a. peninsularis*) being classified as endangered. Of the total pronghorn population in North America, < 1% is classified as endangered; the populations in Mexico or on rangelands along the Mexico-United States border (Yoakum and O'Gara In Press).

Habitat Requirements

Several recent studies have presented pronghorn distribution by biome (Yoakum 1972, Sundstom et al. 1973, Yoakum and O'Gara In Press). Currently, 68% of the herds inhabit grasslands, 31% shrub-steppes, and 1% deserts.

Pronghorn occupy rangelands from sea level to 3,300 m in elevation. Only small populations inhabit environments at the extreme limits of elevations, whereas the majority occur in habitats from 1,200 to 1,800 m. Highest densities occur on rangelands averaging 25-40 cm of precipitation/year. Some subspecies live in areas of less precipitation, but densities are lower. Most pronghorn rangelands receive some snow; however, when snowfall exceeds >40 cm, pronghorns have difficulty obtaining sufficient forage. Prolonged winters with deep snow are the major cause of mortality for northern herds because of insufficient quality forage, excessive wind chill, and human-made obstacles impeding movement to areas with less snow (O'Gara and Yoakum 1992).

Rangelands maintaining high pronghorn densities have drinking water available every 1.5-6.5 km (Sundstrom 1968, Kindschy et al. 1982, Yoakum In Press *a*). Small herds can exist >8.0 km from water, but Sundstrom (1968) found that 95% of more than 12,000 pronghorn were within 6.5 km of water. Pronghorn usually obtain drinking water from springs, streams, lakes, potholes, water catchments, troughs, or snow. When succulent vegetation is available, about 1 L of water/day appears sufficient. However, during dry, hot summers, an individual may require 4 to 6 L/day (Sundstrom 1968, Yoakum In Press *a*).

The average daily intake of air-dry forage/44 kg pronghorn is 1.0-1.5 kg (Zarn 1981). Pronghorn are opportunistic feeders, selecting palatable, nutritious, succulent forage. They are dainty, selective feeders, taking small bites of preferred leaves, flowers, and terminal parts. Rarely do they feed extensively in one place, but move frequently as they forage. According to Wagner (1978), pronghorn consume < 1% of forage produced on western rangelands in the United States. Similar estimates were made by Kindschy et al. (1982) in Oregon and by Longhurst et al. (1983) in Nevada.

More than 200 diet studies of pronghorn have been reported in the literature. However, only 21 of those presented data regarding diet selection in relation to availability of vegetation yearlong (Yoakum 1990). Nine studies were conducted in grasslands, 11 in shrub-steppes, and 1 for desert. For all biomes, grass was the least preferred forage class (0.2 preference rating). Forbs consistently had the highest preference rating (3.8-5.8). Feeding preferences for shrubs were 1.4-1.6. These analyses indicate that pronghorn prefer forbs, then shrubs, and seldom eat grasses (Fig. 5). Thus, management should promote an abundance of preferred forb species on pronghorn habitats. However, an abundance of less preferred forbs and much reduced grass cover resulting from excessive livestock grazing is detrimental to pronghorn welfare.

The pronghorn has specific habitat requirements. Greater populations occur in grasslands and shrub-steppes with large expanses of flat, or low rolling terrain without major physical barriers to seasonal movements (Table 2). The quality and quantity of vegetation also appear to be major factors affecting pronghorn densities. Habitat quality for pronghorn is directly related to proper percentages, quantities, and distribution of physio-

TABLE 1. Estimated populations of pronghorn in Canada, Mexico, and United States from 1924 to 1984 (Yoakum and O'Gara In Press).

Country	Years				Change 1924-1984		
Province or State	1924[a]	1964[b]	1976[c]	1983-84[d]	No.	%	% population
Canada							
Alberta	1,000	16,000	17,000	21,500	20,500	2,050	2
Saskatchewan	300	4,300	5,300	10,000	9,700	3,250	1
Total	1,300	20,300	22,300	31,500	30,200	2,300	3
Mexico							
Chihuahua	700			307	–393	–56	T
San Luis Potosi	600			15	–585	–97	T
Sonora	600			63	–537	–89	T
Coahuila				12			T
Lower California	500			64	–436	–87	T
Total	2,400	1,200	1,000	461	–1,939	–81	T
United States							
Ariz.	700	10,000	7,300	9,000	8,300	1,200	T
Calif.	1,200	2,700	5,000	6,800	5,700	500	T
Colo.	1,200	15,200	31,000	57,500	56,300	4,900	5
Id.	1,500	4,700	13,300	21,500	20,000	1,300	2
Kans.	10	100	1,100	1,200	1,100	11,000	T
Mont.	3,000	95,000	71,200	161,500	158,500	5,300	15
Nebr.	200	9,000	9,800	9,000	8,800	4,400	T
Nev.	4,300	4,500	6,500	9,800	5,500	130	T
N.M.	1,700	22,500	26,900	30,000	28,200	1,700	3
N.D.	200	14,200	8,100	5,700	5,500	2,700	T
Okla.	20	200	200	400	200	100	T
Oreg.	2,000	8,900	11,300	14,000	12,000	6,000	1
S.D.	700	27,400	35,500	67,000	66,300	9,500	6
Tex.	2,400	9,400	10,500	12,000	9,600	400	1
Ut.	700	1,000	2,600	6,000	5,300	800	T
Wyo.	7,000	140,000	168,000	608,000	601,000	8,600	58
North American Total[e]	30,500	386,300	431,600	1,051,400	1,020,600	3,400	100

[a] All populations rounded to closest 100 except Kansas and Oklahoma (Nelson 1925).
[b] All populations rounded to closest 100 (Yoakum 1968).
[c] Yoakum (1978).
[d] Data for Canada and U.S. are 1983 (Yoakum 1986); Mexico is 1984 (Gonzales and Laffon 1993).
[e] Rounded to closest 100.

graphic and vegetative characteristics; too little or too much of any component may be a major factor limiting pronghorn production and survival. For example, Ellis (1970) compared pronghorn population dynamics for the shrub-steppes of the Great Basin with the grasslands of the Great Plains. Ellis (1970) noted fecundity was 190 fawns/100 producing does for both ecosystems. Fawn survival was twice as high and grass and forb production was higher on the Great Plains than in the Great Basin. Nutritive values (particularly protein) of grasses and forbs were greater than shrubs during late spring and early summer. Ellis (1970) concluded that fawn survival was twice as high on the Great Plains because of abundant, nutritious grasses and forbs during late gestation and early lactation. Thus, the reduced availability of herbaceous forage, partly because of consumption by livestock, apparently resulted in rangelands of lower carrying capacity for pronghorn in the Great Basin.

RELATIONSHIPS WITH OTHER RANGELAND WILDLIFE

Wild animals (including predators, rodents, rabbits, and other herbivores) have coexisted and interacted with pronghorn for centuries. King (1955), Koford (1958), Costello (1970) and Cid et al. (1991) speculated that prairie dogs (*Cynomys* spp.) enhanced grasslands for pronghorn by consuming grasses and disturbing soils, thereby increasing the abundance and variety of forbs. Lovaas and Bromley (1972) found the reverse, report-

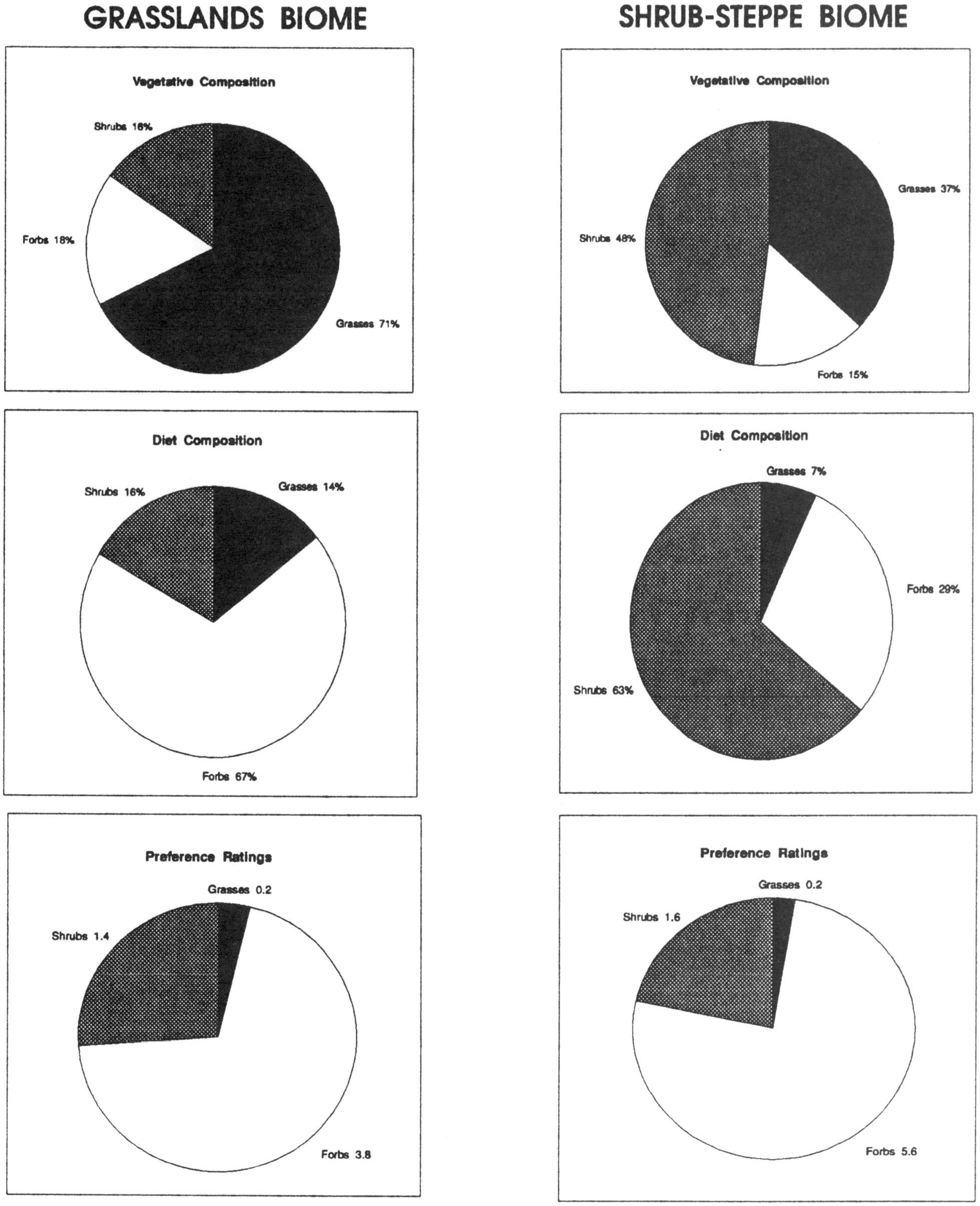

FIGURE 5. Comparison of vegetational composition, diet selection, and preference ratings for pronghorn on grassland and shrub-steppe biomes (modified from Yoakum 1990).

ing that prairie dogs degraded vegetation conditions and competed with pronghorn for preferred forage.

Predatory birds and mammals kill some pronghorn, but predation usually is significant only on marginal rangelands or sites where predator numbers are high in relation to pronghorn numbers. Most predator losses occur among fawns l-3 weeks of age, while separated from their dams (O'Gara and Yoakum 1992). As pointed out by Hornocker (1970), if suitable habitat is not available for a prey species, no amount of predator management will bring about flourishing populations of that prey species.

TABLE 2. Habitat factors and requirements for pronghorn in grassland and shrub-steppe biomes (Yoakum and O'Gara In Press).

Habitat Factor	Pronghorn Requirements: Grassland Biome	Pronghorn Requirements: Shrub-steppe Biome
Abiotic		
1. Physiography	Large, expansive area (40 km minimum), low rolling terrain, no major physical barriers, slopes < 30%	
2. Climate		
Precipitation	25-40 cm	20-30 cm
Snow depth	25-40 cm	25-40 cm
Temperature	Not a major problem, as herds inhabit hot, semi-arid areas to cold alpine steppes	
3. Soils	Not a determining factor except in relation to soil/site vegetation production	
4. Water (drinking)		
Quantity	1.0-5.5 l/day	1.0-5.5 l/day
Distribution	1.5-6.5 km	1.6-6.5 km
Biotic		
1. Vegetation		
Forage consumption	1.0-3.5 kg air-dry forage/day	1.0-3.5 kg air-dry forage/day
Ground cover	60-80% vegetation	30-50% vegetation
Plant spp. composition	50-80% grasses	5-15% grasses
	10-20% forbs	5-10% forbs
	<5% shrubs	10-35% shrubs
Plant spp. diversity	10-20 grasses	5-10 grasses
	20-60 forbs	10-70 forbs
	5-10 shrubs	5-10 shrubs
Height	25-45 cm	25-45 cm
	>65 cm unfavorable for most habitats	
Succulence	The more availability year-round the better for all forage classes	
Communities	Greater variety and diversity important (i.e., meadows, playas, wildfire burns)	
2. Animal		
Wild ungulates	Few competition or compatibility problems	
Predators	Pronghorn may be reduced or limited in areas of low densities or isolated herds	
Mankind	Suitable habitat being usurped. Increased construction of barriers (mainly fences) Predator control, water developments, and alfalfa plantings beneficial. Livestock commensal in grasslands: competitive for forage in shrub-steppe and desert biomes	

Bison and pronghorn lived commensally prior to the arrival of Europeans; both ungulates used the forage and water on the vast grassland prairies (O'Gara and Yoakum 1992). Apparently bison did not coexist with pronghorn historically in the Great Basin or on deserts. Schwartz and Nagy (1976) compared diets of bison and pronghorn in Colorado, and attributed diet differences to the species rather than the rangeland (Table 3). McCullough (1980) studied niche separation for bison and pronghorn in Montana and determined that diet overlap was minimal: bison consumed mostly grasses while pronghorn ate forbs and shrubs. Bison disturbed dominant grass communities by trampling, wallowing, and grazing that resulted in greater production of forbs and shrubs favored by pronghorn.

Range and habitat overlap between elk and pronghorn occurred more in pristine times than it does now. Buechner (1950*a*) observed elk and pronghorn foraging together and noted no acts of aggression and little dietary overlap (Table 3). Bailey and Cooperrider (1982) reported that the greatest seasonal dietary overlap between these 2 species in Colorado was during winter (59%), with less in the spring (33%), summer (38%), and fall (31%). McCullough (1980) found little overlap in the niches occupied by elk and pronghorn in Montana. There was little competition for space, water, or forage. Elk foraged primarily on grasses, whereas pronghorn preferred forbs and shrubs; consequently, dietary overlap was low (11.3%).

Pronghorn occur on rangelands with whitetail and mule deer; however, their distributions usually do not overlap (Buechner

TABLE 3. Dietary overlap for forage classes between pronghorn and bison, white-tailed deer, mule deer, elk, and wild sheep (Yoakum and O'Gara In Press).

Species	Reference	Location	Biome	% dietary overlap			
				Grasses	Forbs	Shrubs	Annual
Bison	Buechner (1950*b*)	Wichita Mountains National Wildlife Refuge, Okla.	Grassland	1.0	1.0	—	2.0
	Schwartz and Nagy (1976)	Pawnee Grasslands, Colo.	Grassland	41.0	3.0	0.0	44.0
	McCullough (1980)	National Bison Range, Mont.	Grassland	2.8	1.4	0.1	4.3
Elk	Buechner (1950*b*)	Wichita Mountains National Wildlife Refuge, Okla.	Grassland	0.1	24.0	0.0	24.1
	McCullough (1980)	National Bison Range, Mont.	Grassland	2.9	4.6	3.8	11.3
	Bailey and Cooperrider (1982)	Trickle Mountain, Colo.	Shrub-steppe	3.5	11.5	45.5	60.5
Whitetail deer	Buechner (1950*b*)	Wichita Mountains National Wildlife Refuge, Okla.	Grassland	0.0	99.0	0.0	99.0
	McCullough (1980)	National Bison Range, Mont.	Grassland	2.9	24.8	18.5	46.2
Mule deer	Barmore (1969)	Yellowstone National Park	Shrub-steppe	16.6	9.2	52.0	77.8
	McCullough (1980)	National Bison Range, Mont.	Grassland	2.3	27.1	33.3	62.7
	Hanley (1980)	Northeast Calif. and Northwest Nev.	Shrub-steppe	2.3	4.2	82.3	88.8
	Bailey and Cooperrider (1982)	Trickle Mountain, Colo.	Shrub-steppe	3.5	10.2	79.5	93.2
	Hansen (1986)	Sheldon National Wildlife Refuge, Nev.	Shrub-steppe	5.0	31.0	39.0	75.0
Wild sheep	McCullough (1980)	National Bison Range, Mont.	Grassland	2.9	12.3	11.4	26.6
	Bailey and Cooperrider (1982)	Trickle Mountain, Colo.	Shrub-steppe	3.5	6.2	56.6	66.3
	Hansen (1986)	Sheldon National Wildlife Refuge, Nev.	Shrub-steppe	5.0	32.0	14.0	51.0

1950*b*, McCullough 1980, and Bailey and Cooperrider 1982). All have similar diets (primarily eating forbs and shrubs) but overlap is minimal because of differences in habitats occupied (Table 3). Behavioral interactions were minimal because pronghorn were generally diurnal while deer were crepuscular and nocturnal.

Three studies reported diets for wild sheep (*Ovis canadensis*) and pronghorn (McCullough 1980, Bailey and Cooperrider 1982, Hansen 1986) (Fig. 6). McCullough (1980) found little overlap in space occupied and little competition for forage between these 2 species in Montana. Bailey and Cooperrider (1982) noted overlap was greatest in Colorado during winter (67%) and least in summer (42%). Hansen (1986) found diets of pronghorn and wild sheep in Nevada to have an annual overlap of 51% with the greatest overlap involving forbs (32.0%).

Decker (1978) reported ≥ 10 species of wild, exotic ungulates were established in the United States. Most were deer, sheep, goats, and pigs occupying forested, mountainous ecosystems. Only the gemsbok (*Oryx gazella*) is a member of the antelope subfamily Antilopineae and inhabits open rangelands similar to pronghorn habitat. Gemsbok were translocated during the 1970s to the White Sands Missile Range in southcentral New Mexico (Pederson 1989) where they are free-roaming and coexist with

FIGURE 6. Pronghorn and bighorn occupy a prescribed burn and non-burned rangelands on the Hart Mountain Antelope Refuge in southcentral Oregon. Photo by Bill Stormont, courtesy of the U. S. Fish and Wildlife Service.

pronghorn. Smith (1994) did not observe gemsbock mingling with pronghorn, although they were frequently within 0.8 km of each other. There was similarity in forage species consumed; however, overlap was only 0.19.

RELATIONSHIPS WITH LIVESTOCK

It has been postulated that 98% of pronghorn populations share rangelands with domestic or feral livestock at some time during a year (Yoakum and O'Gara 1990). Livestock on western rangelands are either domestic animals including cattle, sheep, and horses; or feral stock, primarily horses. Goats, pigs, and burros occur in low numbers and their impact to the pronghorn is minimal.

Interactions between livestock and wild ungulates were documented by Wagner (1978), who compared historic and contemporary populations in the western United States. He calculated that the livestock animal unit months (AUMs) during 50 years in the mid-1900s was the highest in history, while that of wild ungulates (i.e., bison, bighorn, deer, elk, and pronghorn) was < 10% of what it was 125 years ago (Fig. 7).

Livestock grazing may alter pronghorn habitats more than any other human-controlled activity (Leftwich and Simpson 1978, Kindschy et al. 1982) by changing vegetation structure and composition. Pyrah (1987) suggested that livestock grazing in central Montana lowered habitat quality and decreased carrying capacity for pronghorn. When pronghorn and livestock grazed grasslands in southeastern New Mexico, they used many of the same forage species. Pronghorn did not switch to less preferred forage classes and were adversely affected when forbs

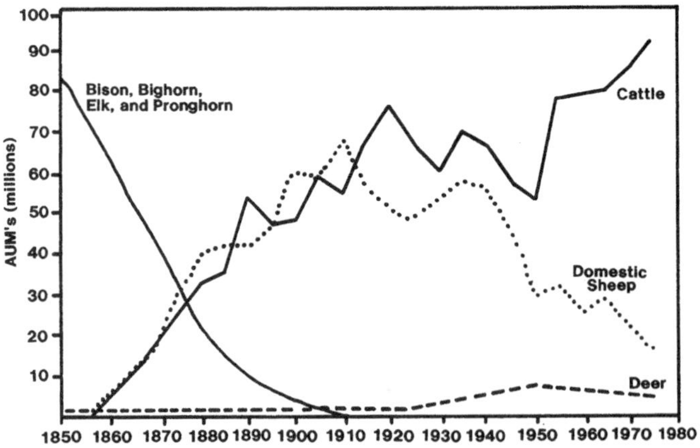

FIGURE 7. Conjectured demand for AUMs of forage by wild and domestic ungulates on western rangelands in the United States from 1850 to 1975 (modified from Wagner 1978).

and shrubs were depleted through drought or heavy livestock grazing (Howard et al. 1990).

Pronghorn use of plants noxious to livestock has been well documented (Einarsen 1948, Buechner 1950*a*, Hoover et al. 1959, Smith and Beale 1980), and pronghorn may be considered beneficial to livestock on rangelands for this reason. Many plants eaten by pronghorn are noxious to livestock because they cause injury or are unpalatable. Spines of bull thistle (*Cirsium* spp.), Russian thistle (*Salsola* spp.), and cacti make these plants less palatable to livestock, but pronghorn frequently eat them. Pronghorn consumed 11% cacti annually in Colorado (Hoover et al. 1959) and 28% in New Mexico (Smith 1994). Plants noxious to livestock but consumed by pronghorn include: woolly locoweed (*Astragalus mollissimus*), larkspur (*Dephinium* spp.), lupine (*Lupinus* spp.), death camas (*Zigadimus* spp.) in Oregon (Einarsen 1948); locoweed (*Astragalus* spp.), woody senecio (*Senecio* spp.), Riddell groundsel (*Senecio* spp.), in Texas (Buechner 1950*a*); cocklebur (*Xanthium* spp.), snakeweed (*Gutierrezia* spp.), bull thistle, Russian thistle, cacti, chokecherry (*Prunus* spp.) in Colorado (Hoover et al. 1959); tarbush (*Flourensia cernua*) in Texas (Hailey 1979); and halogeton (*Halogeton* spp.) in Utah (Smith and Beale 1980). Of 103 plants listed by Buechner (1950*a*) as eaten by pronghorn, 51 were unpalatable and 21 were considered poor forage for livestock, while only 28 were considered poor forage for pronghorn. Most of the plants unsuitable for livestock were relished by pronghorn.

Many references describe pronghorn relations with cattle, sheep, and horses. The effects of these livestock on pronghorn, including compatibility, competition, and disease-parasite factors, are species-specific. Competition, when used in the ecological sense, is an ambiguous term that can denote various meanings and concepts; therefore, we use Wagner's (1978) definition: interspecific competition is an interaction in which 2 species require the same resource, always at the expense of 1 or both.

Cattle

Most authors cited little agonistic behavior between cattle and pronghorn (Einarsen 1948, Buechner 1950*a*, Hoover et al. 1959, Schwartz 1977, Hailey 1979, Salwasser 1980, Kindschy et al. 1982, Roebuck 1982). Most investigations reported a nonaggressive association while feeding, drinking, or resting. However, Pyrah (1987) reported pronghorn generally avoided using pastures concurrently with cattle: 75% of 9,530 observations of pronghorn were in pastures without cattle. Parturient does avoided cattle during the fawning season, resulting in selection of less favorable fawn-production sites in Nevada (McNay and O'Gara 1982).

Competition for forage between cattle and pronghorn generally is minor on grasslands in fair to good ecological condition. Their diets were sufficiently different that little overlap occurred (Table 4) and averaged < 26% for the grassland and shrub-steppes. However, apparently serious competition for forage can occur in spring and summer between cattle and pronghorn in the Great Basin (Ellis 1970). Ellis (1970) reported that habitats grazed by cattle had lesser quantities of forbs and grasses for gestating and lactating pronghorn, resulting in reduced fawn production. Severe competition for forage occurred between pronghorn and cattle in Texas when a drought resulted in intensive use of forage (Hailey 1979). Cattle were given supplemental feed (pronghorn were not), resulting in heavy use of the vegetation. Livestock fences prevented pronghorn from moving to adjacent rangelands, and they turned to nonpreferred, toxic shrubs resulting in large losses and low recruitment. Roebuck (1982) studied pronghorn and cattle forage relations on rangelands in good condition in the Panhandle of Texas. Pronghorn were attracted to pastures with preferred forage regardless of the presence or absence of cattle. Greatest dietary overlap was for forbs during summer. Supplemental feeding of livestock resulted in low spring and winter forage competition. The low overlap for all forage classes and seasons suggested competition was not important on rangelands in good condition.

Exposure to diseases is a concern when pronghorn and cattle share rangelands because their spatial distribution usually overlaps. There is concern that either species may serve as a reservoir for diseases that affect the health of the other (O'Gara In Press). Pronghorn have been repeatedly checked for brucellosis and anaplasmosis but no cases have been confirmed. Blue tongue is probably the most serious disease of pronghorn, and cattle are a primary reservoir for this disease. Cattle do not develop clinical or acute symptoms, but are chronic carriers (Thorne et al. 1983). Leptospirosis causes some mortality in pronghorn, but insufficient evidence exists to implicate either pronghorn or cattle as a primary reservoir of infection (Collins et al. 1981).

Domestic Sheep

Investigators are not in agreement concerning competition between pronghorn and domestic sheep. Einarsen (1948) stated pronghorn usually avoided bands of herded sheep. Gregg (1955) observed pronghorn feeding near isolated, small bands of sheep that were away from herders or dogs. Similarly, Buechner (1950*a*) stated there was no psychological incompatibility between the species. Severson et al. (1968) reported pronghorn and domestic sheep relations in Wyoming and found no apparent stress to either as a result of the other's presence. Both often were seen grazing and drinking together, but pronghorn tended to vacate areas where sheep were herded. This reaction was believed to have resulted from the activities of the herder and dogs because pronghorn returned soon after the herded sheep moved away.

The potential for forage competition between pronghorn and domestic sheep was reported first by Taylor (1936) and subsequently supported by others (Buechner 1950*a*, Hoover et al. 1959, Russell 1964, Taylor 1975, Schwartz and Nagy 1976, Hailey 1979, Smith and Beale 1980, Howard et al. 1990). Buechner (1950*a*) reported intense competition for preferred forbs, supplemental feeding of sheep on rangelands (thus maintaining artificially high populations of sheep), and restricted pronghorn movements because sheep-tight fences limited population increases and distribution of pronghorn. Schwartz (1977) found that pronghorn and sheep ate many of the same species and that crude protein levels of ingested forage for both ungulates were similar (4-12%).

TABLE 4. Forage class dietary overlap among pronghorn and cattle, horses, and domestic sheep (Yoakum and O'Gara In Press).

Class of Livestock	Reference	Location	Biome	% dietary overlap Grasses	Forbs	Shrubs	Annual
Cattle	Buechner (1950*a*)	Trans-Pecos, Tex.	Grassland	4.0	7.0	9.0	20.0
	Campbell (1970)	Southwest, Mt.	Grassland	3.0	20.3	1.6	24.9
	Becker (1972)	Winnett, Mt.	Shrub-steppe	3.0	13.0	0.0	16.0
	Taylor (1975)	Rawlins, Wyo.	Shrub-steppe	7.1	0.2	39.4	46.7
	Schwartz and Nagy (1976)	Pawnee, Colo.	Grassland	46.3	2.0	6.0	54.3
	Hanley (1980)	Northeast Calif. and Northwest Nev.	Shrub-steppe	4.1	3.4	3.7	11.2
	Smith and Beale (1980)	Southwest, Ut.	Shrub-steppe	0.0	0.0	27.0	27.0
	Beasom et al. (1982)	Roswell, N.M.	Grassland	4.0	15.2	1.3	20.5
	Roebuck (1982)	Panhandle, Tex.	Grassland	2.5	19.0	8.5	30.0
	Bailey and Cooperrider (1982)	Trickle Mountain, Colo.	Shrub-steppe	3.5	5.0	14.0	22.5
	McCarty (1982)	Little Lost-Birch Creek, Id.	Shrub-steppe	3.0	2.0	11.0	16.0
	McInnis (1984)	Burns Junction, Oreg.	Shrub-steppe	9.4	4.6	0.2	14.2
	Hansen (1986)	Sheldon Naitonal Wildlife Refuge, Nev.	Shrub-steppe	5.0	9.0	1.0	15.0
Horses	Meeker (1979)	Sheldon National Wildlife Refuge, Nev.	Shrub-steppe	3.0	23.0	2.0	28.0
	Hanley (1980)	Northeast Calif. and Northwest Nev.	Shrub-steppe	2.5	5.2	5.0	12.7
	Bailey and Cooperrider (1982)	Trickle Mountain, Colo.	Shrub-steppe	3.5	2.0	31.0	36.5
	McInnis (1984)	Burns Junction, Oreg.	Shrub-steppe	13.5	2.1	0.3	15.9
	Hansen (1986)	Sheldon National Wildlife Refuge, Nev.	Shrub-steppe	5.0	6.0	0.0	11.0
Sheep	Buechner (1950*a*)	Trans-Pecos, Tex.	Grassland	4.0	19.0	10.0	33.0
	Severson et al. (1968)	Red Desert, Wyo.	Shrub-steppe	3.2	2.6	28.4	34.2
	Campbell (1970)	Southeast, Mt.	Grassland	3.0	27.3	25.3	55.6
	Taylor (1975)	Rawlins, Wyo.	Shrub-steppe	7.1	1.2	39.9	48.2
	Schwartz and Nagy (1976)	Pawnee, Colo.	Grassland	46.3	18.0	3.3	67.6
	Smith and Beale (1980)	Southwest, Ut.	Shrub-steppe	0.0	0.0	46.0	46.0
	McCarty (1982)	Little Lost-Birch Creek, Id.	Shrub-steppe	3.0	14.0	39.0	56.0
	Beasom et al. (1982)	Roswell, N.M.	Grassland	4.0	50.2	6.0	60.2

Percentages of dietary overlap in several studies (Table 4) vary from moderate to heavy (33-67%). A study of pronghorn-sheep forage competition conducted during the early 1980s in New Mexico, determined that livestock used as much as 40% of available forbs (Howard et al. 1990). As forbs declined in number or decreased in moisture content, livestock increased their use of grasses. This suggested that the potential for competition primarily was for forbs. Data on diet similarity supported this premise. Dietary overlap was highest when forbs were most available and lowest when they were least available. Clary and Holmgren (1982) reported moderate use by domestic sheep during the dormant winter period left shrubs unfavorable for pronghorn until spring regrowth occurred on the cold desert of southwestern Utah.

Bever (1957) reported 30-40% losses of pronghorn fawn crops on rangelands heavily grazed by domestic sheep in South Dakota. Bever (1957) also reported that pronghorn had higher parasite loads on rangelands grazed by domestic sheep than rangelands grazed by cattle. In Wyoming, illness and deaths of pronghorn fawns have been attributed to parasitic infections that were prevalent on rangelands grazed heavily by sheep. Blue tongue, a disease fatal to domestic sheep, is also probably the most serious disease of pronghorn; neither species is an important reservoir for this disease, because few animals survive to become carriers (O'Gara In Press).

Horses

Domestic and feral horses occupy rangelands with pronghorn (Yoakum and O'Gara 1990). Two studies in Nevada (Meeker 1979, Berger 1986) reported the degree of aggressive behavior between these ungulates (Table 4). Meeker (1979) noted both animals watered freely together, with pronghorn giving ground only when directly approached by horses. No aggressive action between species was noted during the study. Berger (1986) investigated pronghorn and feral horse relations for several years and recorded various instances when pronghorn were displaced by horses.

Horses predominantly feed on grasses while pronghorn prefer forbs and shrubs. Five studies listed dietary overlap as low to moderate (11-36%), suggesting that competition for food was limited (Table 4). In Idaho, Autenrieth (1982) reported that year-round foraging by feral horses was probably the most important cause of range degradation. Salwasser (1980) noted that high numbers of horses competed with pronghorn for early spring grasses and forbs in Great Basin environments. No reports of disease or parasite relations between pronghorn and horses were found in the literature.

RESPONSES TO RANGELAND PRACTICES AND IMPROVEMENTS

Rangeland management practices and improvements can be beneficial or detrimental to pronghorn populations, depending upon how they are planned and implemented. Techniques that maintain or improve habitat requirements of pronghorn will be beneficial; however, if biological requirements of pronghorn are not met, then any management action can detrimentally affect pronghorn production and survival.

Vegetation Manipulation

After 10 years of restoration, Plummer et al. (1968) reported dominant shrublands and pinyon-juniper (*Pinus* spp.-*Juniperus* spp.) communities can be rehabilitated for pronghorn and mule deer. This requires control of dominant shrub and tree species, followed by seeding a complex mixture (>6 species each) of grasses, forbs, and shrubs. More favorable results for wildlife can be produced with greater plant diversity. The more diverse the plant communities are, the more the biological requirements of pronghorn are met (Yoakum 1980, Kindschy et al. 1982, Yoakum and O'Gara 1990, O'Gara and Yoakum 1992).

The Vale Project on public lands in southeastern Oregon is one of the most extensive rangeland restoration projects carried out to date on pronghorn habitats (Heady and Bartholome 1977, Heady 1988). This large-scale, 11-year, program treated 205,000 ha for shrub control and artificially seeded 108,000 ha. Crested wheatgrass (*Agropyron cristatum*) was the predominant species planted; however forbs, primarily dryland alfalfa, and shrubs were planted in 26 seedings. Fifteen years later, nontreated vegetation areas had a vegetation composition averaging 52% grasses, 3% forbs, and 45% shrubs, and a mean height of 71 cm. Plowed and seeded (with crested wheatgrass and dryland alfalfa) sites had 76% grasses, 11% forbs, and 13% shrubs with an average height of 46 cm. The treated lands met pronghorn habitat requirements better than the untreated areas. This analysis was based upon comparisons of similar lands surrounding the treated areas, and estimates of pronghorn populations obtained from aerial censuses by the state wildlife agency. During the early years of the project (1962-64), the pronghorn herd averaged 1,420/year in the project area. Following treatment practices (1972-74), the herd increased 83% to 2,600, while herds on adjacent nontreated lands increased < 30%. Similar results for treated and untreated lands were reported in other areas of Oregon (Kindschy et al. 1982) and Nevada (Yoakum 1980).

Improvements that change native vegetation composition or structure to large monocultural grasslands of coarse, introduced perennial grasses usually produce poor habitat for pronghorn (Reeher 1969, Yoakum 1980). Pronghorn prefer soft textured grasses, such as Sandberg bluegrass (*Poa sandbergii*). Treatments of large areas require pronghorn to travel long distances to obtain preferred shrubs during the early years of plant succession. Seeded monocultures frequently provide low densities and varieties of forbs and shrubs vital to pronghorn during all seasons.

Vegetation changes resulting from wild or prescribed burns can be beneficial to pronghorn (Deming 1963, Yoakum 1978, Kindschy et al. 1982, Courtney 1989, O'Gara and Yoakum 1992, U.S. Fish and Wildlife Service 1994, Gruell 1995, Pyle and Yoakum In Press). This is especially true when extensive, tall, dense shrub communities are burned, resulting in increased grass and forb production. Courtney (1989) reported frequent use of burned grasslands by pronghorn in Alberta (i.e., Fig. 6).

Water Developments

Pronghorn use water developments with domestic livestock (Sundstrom 1969, Yoakum In Press *a*). Studies in Utah (Beale and Smith 1970) suggested that water developments supported pronghorn where natural water sources were limited, particularly during dry seasons or drought years. Such developments should be placed every 2.5-4 km to meet requirements of pronghorn (Taylor 1972).

Hundreds of small reservoirs have been constructed on public lands through cooperative funding by state wildlife and federal land management agencies, and private sources. Such developments often are natural in appearance and serve a variety of wildlife. Part of each reservoir should be fenced to exclude heavy grazing by livestock. Another water development of high value to pronghorn is the trench reservoir. These are used by pronghorn, especially during late summer and hot weather when vegetation becomes desiccated and physiological requirements for water increase (Yoakum 1980).

Fences

Fences can be major obstacles limiting or restricting pronghorn movement to food and water or to escape from deep snow (Yoakum and O'Gara 1990, Mosley 1994). These impediments to seasonal movements can be disastrous for northern (Spillett 1965) and southern herds (Buechner 1950*a*, Hailey 1979). Most fences on western rangelands are constructed to control livestock. How fences are constructed can have an impact on pronghorn and other wildlife. As early as the 1870s, Caton (1877) reported pronghorn characteristically go under barbed-wire fences rather than through or over them.

Recommendations for wire fences that allow pronghorn easy passage are provided in research reports and agency guidelines (Spillett 1965, Mapston and Zobell 1972, U.S. Bureau Land Management 1985, O'Gara and Yoakum 1992). Specifications for construction of barbed wire fences that allow pronghorn to go under the bottom wire have been established (Fig. 8). Many kms of "sheep-tight" or woven-wire fences, have been built to control domestic sheep. These have become barriers to pronghorn mobility. The "wolf-type" fence constructed with woven and barbed wire to control coyotes on domestic sheep rangelands in the Southwest completely restrict pronghorn movements (Yoakum 1980). The biological and legal implications of the wolf-type fence are well documented in legal hearings held during April 1978 (Gist Ranch, New Mexico 6-78-1, 21 Aug 1978). The conclusion was that wolf-type fences were legal on public lands prior to passage of the Federal Land Use and Management Act of 1976; however, this law mandated multiple-use on rangelands administrated by the U.S. Bureau of Land Management, and such fences now violate that law.

In some areas of the Southwest, stockmen encircle water sources with fences to trap or redistribute livestock (Yoakum 1980). Closed gates restrict pronghorn from the water and sometimes force movements to other rangelands. These enclosures are constructed of woven wire, >6 barbed wires, or snow-

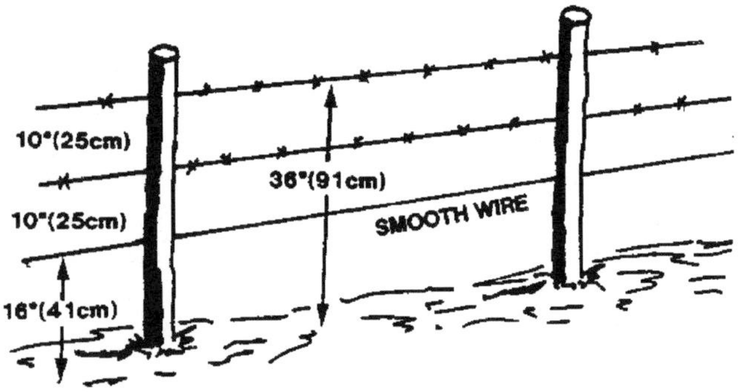

FIGURE 8. Suggested construction specifications for barbed-wire fences on rangelands occupied by pronghorn and cattle (adapted from Kindschy et al. 1982).

control fencing. These fence structures detrimentally affect pronghorn, especially fawns inexperienced in negotiating such barriers. Such fencing of water holes may violate the same basic principle of multiple-use as wolf-type fences.

The feasibility of constructing special facilities that allow pronghorn passage through livestock fences was investigated in Wyoming (Spillett 1965, Mapston and Zobell 1972) and in New Mexico (Howard et al. 1990). A structure called an "antelope pass" was developed that allows some movement by pronghorn. However, the authors were adamant that the structures needed to be improved for safe pronghorn passage because some fawns broke legs in the structures.

Wildlife biologists in Idaho adjusted barbed wire fences to allow seasonal movements of pronghorn on rangelands mutually used with livestock (Anderson and Denton 1980). Height of the lower wire was raised above the ground 46-96 cm. This had special merit in habitats receiving snow depths >30 cm, which can restrict pronghorn movement under fences, and at times entrap individuals.

A second legal test dealing with pronghorn and livestock fences resulted when a wire fence was constructed around approximately 3,885 ha of private and public lands near Rawlins, Wyoming. The fence prevented movement of pronghorn to critical winter habitat; many died of malnutrition when unable to move to adjacent rangelands to obtain forage. The case was tried before the U.S. District Court, and the presiding judge found the woven-and-barbed wire fence in violation of the federal Unlawful Inclosure Act of 1885. The appellant appealed the ruling to the Tenth Circuit Court of Appeals where it was upheld, then appealed again to the United States Supreme Court which recognized the decision of the District Court (O'Gara and Yoakum 1992).

Livestock Grazing Systems

Livestock grazing systems are planned strategies to maintain or improve carrying capacities and systematically manage livestock. Effective management systems should consider the control of livestock, range improvements, determining how many livestock will use a certain rangeland, seasons of use, monitoring studies, physiology of plants, and the effects and needs of other uses on the rangelands. A number of grazing systems have been developed (Stoddart and Smith 1958, Heady and Child 1994, Holechek et al. 1994), and are increasingly used in rangeland management.

Two livestock grazing systems were designed to improve forage for pronghorn and other wildlife in the Great Basin (U.S. Fish and Wildlife Service 1970, 1980; Anderson et al. 1990*a*,*b*). Various designs of deferred livestock grazing were used over a 25-year period. However, subsequent assessments disclosed that apparently these livestock grazing systems did not enhance preferred forage for pronghorn; consequently, they were terminated (United States Fish and Wildlife Service 1994, Pyle and Yoakum In Press, Yoakum 1995).

For rangelands used by pronghorn and livestock, the following guidelines are recommended for design of livestock grazing systems.

1. When allotting forage, the habitat requirements of pronghorn should be considered.
   a. Adequate amounts of preferred plant species should be reserved as forage for pronghorn. These include grasses, forbs, and shrubs determined from diet studies in the same or similar ecosystems. Consideration should be given to proper use of key forbs and shrubs. Make sure that these are not grazed beyond their physiological tolerance levels.
   b. Forage should be reserved for a reasonable number of pronghorn. Reasonable numbers should be based on the average herd population for the past 30 years, the average forage production for the past 15 years, and management objectives for herd size determined by wildlife and land management agencies.
   c. Proper amounts of forage should be apportioned for a reasonable number of animals for specific sites and seasons of use. Special attention should be given to reserving sufficient quality forage for critical sites (i.e., natal areas, winter use areas, movement corridors, and others).
2. Drinking water should be available during all seasons pronghorn are in the area.
3. Fence construction should meet specifications to allow movements year-round for all pronghorn age classes.
4. When livestock grazing systems are designed using the "key plant species" concept, forbs and/or shrubs preferred by pronghorn should be included as key species.
5. Livestock use should be limited on pronghorn natal areas during the fawning season.
6. Livestock grazing systems that restrict, alter, limit or deleteriously affect the habitat requirements of pronghorn should include mitigating measures and alternate procedures for enhancing pronghorn habitat.

Livestock grazing systems on public lands should consider various renewable resources. Each resource is valued differently by various segments of society and decisions should reflect the will of the public. The responsibility of the resource manager is to make sound decisions based on objective analysis of alternatives and consequences.

Animal Equivalents

Managers often need to exchange use of forage for different grazers on the same rangeland. Various methods of calculating exchange ratios (i.e., animal equivalents) have been used, but none have been completely satisfactory (Heady and Child 1994). Heady and Child (1994) were of the opinion that a cow-pronghorn ratio of 1:6 has been used on rangeland more than any other, and appears to be a reasonable expression of the relative impact of these ungulates upon the rangeland.

During the 1988 joint meeting of the Interstate Antelope Conference and the Pronghorn Antelope Workshop, Kniesel (1988) presented a paper reviewing past procedures and practices for using equivalent ratios for pronghorn. He stressed the variation in AUM equivalents currently used by management agencies; e.g., 105 pronghorn = 1 cow in Colorado, 59 pronghorn = 1 cow in Idaho, 38 pronghorn = 1 cow in Texas, 7-14 pronghorn = 1 cow in Oregon, and 5 pronghorn = 1 cow in Montana. He attributed the wide variation in AUM equivalents to different methodologies and information used. Some investigators primarily used weight differentiations, while others included considerations such as dietary overlap, condition of rangelands, and other factors. He concluded that assessing AUM equivalents for pronghorn and livestock is still a problem, because there is little agreement between various state and federal agencies applying exchange ratios in forage allocations for multiple-use programs.

MANAGEMENT RECOMMENDATIONS

Increasing human populations result in ever increasing demands for land resources and products. On western rangelands, this demand requires management practices that produce optimum sustained yields of livestock and wildlife. Management procedures developed over the past 50 years provide techniques to maintain sustained yields of livestock and pronghorn while protecting the rangeland resources. This requires coordinating management practices for both classes of animals. Standards for livestock husbandry and pronghorn enhancement have been developed, but have been poorly coordinated. Despite this, production of both has increased during the last half century. Now, the task is to manage these rangeland resources on a sustaining basis. We have worked with this challenge and suggest the following.

1. Many rangelands can produce concurrent dense populations of pronghorn and livestock. The key is maintaining these rangelands in good ecological condition. Pronghorn and livestock thrive in subclimax habitats, but production decreases for both when excessive livestock grazing produces poor rangeland conditions.
2. Management plans need to incorporate the requirements of livestock and pronghorn simultaneously. Past practices have not always recognized the critical habitat requirements of pronghorn. Managers must be knowledgeable of these requirements and implement practices to meet them, while also providing the much less restrictive requirements of livestock.
3. We cannot emphasize too strongly that, if managers maintain or improve quality habitats, animals have a good chance to maintain healthy condition and numbers. Time is required to discover, understand, and implement techniques to maintain rangeland conditions. Healthy rangelands normally produce healthy animal populations.
4. Based upon our experiences, the following guidelines will help maintain dual use by pronghorn and livestock on western rangelands.
   a. Both animal groups exhibit tolerance and compatibility when they occur together in reasonable numbers on grasslands in fair to good ecological status. However, consumption of grasses and forbs by livestock from March to August is a major competition problem for pronghorn on many shrub-steppes of the Great Basin and desert regions in poor ecological condition.

b. Natural vegetation should be managed to provide an abundance and variety of forage classes. Most livestock graze grasses; pronghorns primarily consume forbs and shrubs. Rangelands producing mixtures of grasses, forbs, and shrubs will best serve livestock and wildlife. The challenge is to maintain existing rangelands in good ecological status with native vegetation. When deteriorated sites require rehabilitation, practices that restore vegetation to natural diversity are more desirable than practices that bring about monocultures and other unnatural conditions.
c. Practices that increase availability of long-term drinking water sites are highly beneficial to both animal groups (see chapter 23). Water should be available every 1.5-6.5 km. Water improvements can be designed in a number of ways; however, those that simulate natural waters are favored and cause fewer problems.
d. Both pronghorn and livestock experience problems with predation, diseases, and parasites. Managers need to recognize how these factors affect the animals and coordinate control techniques beneficial to both groups.

SUMMARY

Pronghorn coexist with wild, domestic, and feral animals on western rangelands. Each group exhibits different degrees of compatibility and competition.

Relative to wildlife, it appears bison and pronghorn occupy similar habitats but have little dietary overlap. Deer, elk, and wild sheep have greater dietary overlap, but differing spatial distribution greatly decreases competition. The recent introduction of exotic antelope to historic pronghorn habitats has received limited study.

Competition with cattle and horses on grasslands in good ecological status does not appear to be a major problem, when these animals are stocked at proper levels. This is primarily the result of food partitioning. However, there appears to be serious competition for grasses and forbs from March to August in the Great Basin and deserts. Cattle may act as a reservoir for bluetongue, a disease highly fatal to pronghorn.

Competition for forage appears to be most likely between pronghorn and domestic sheep, as both consume large quantities of forbs and shrubs. Domestic sheep also carry many diseases and parasites common to pronghorn.

An area of potential conflict between livestock husbandry and pronghorn management is rangeland improvements. Forage manipulation projects that maintain or increase plant diversity and control shrub height can be highly beneficial to pronghorn; however, projects resulting in monocultures of exotic, coarse bunchgrasses have little value. Fences to control livestock can be built to allow pronghorn movements, or they can become barriers. Woven-wire fences commonly used on sheep pastures often are barriers to pronghorn movements. Most water developments have been beneficial to pronghorn.

Livestock grazing systems can be designed to be beneficial or detrimental to pronghorn. Plans that recognize forbs and shrubs as key plants for pronghorn are favored over plans centered on grasses as key species.

Animal equivalents are used for allotting forage on western rangelands. However, the methods and practices for this technique vary and need refinement.

Livestock benefit from pronghorn use of noxious plants. Reciprocal values are received by pronghorn when intensive predator control programs are conducted on rangeland to protect livestock.

The last 50 years have indicated pronghorn, other wild ungulates, and livestock can live together successfully on western rangelands in good ecological status. With enlightened management, few problems of compatibility or competition occur for forage, water, or space. Pronghorn and livestock can be the epitome of sympatric ungulates during contemporary times on managed western rangelands.

LITERATURE CITED

Anderson, E. W., D. L. Franzen, and J. E. Melland. 1990*a*. Prescription grazing to benefit watershed—wildlife—livestock. Rangelands 12:105-111.

———, ———, and ———. 1990*b*. Forage quality as influenced by prescribed grazing. Pages 56-70 *in* K. E. Severson, tech. coord. Can livestock be used as a tool to enhance wildlife habitat? U.S. Dep. Agric. For. Serv., For. and Range Exp. Stn., Fort Collins, Colo. Gen. Tech. Rep. RM-194. 123pp.

Anderson, L. D., and J. W. Denton. 1980. Adjustable wire fences for facilitating big game movement. U.S. Bur. Land Manage., Denver, Colo. Tech. Note 343. 7pp.

Autenrieth, R. E. 1982. Pronghorn fawn habitat use and vulnerability to predation. Pronghorn Antelope Workshop Proceedings 10:112-131.

Bailey, J. A., and A. Y. Cooperrider. 1982. Final report: Trickle Mountain research study. U.S. Bureau of Land Manage., Denver, Colo. 173pp.

Barmore, W. J. 1969. Pronghorn-mule deer-range relationships on the northern Yellowstone winter range (YELL-N-17). Pages 121-137 *in* U.S. Natl. Park Serv., Washington, DC. Office of Nat. Sci. Annu. Rep. 1968.

Beale, D. N., and A. D. Smith. 1970. Forage use, water consumption, and productivity of pronghorn antelope in western Utah. J. Wildl. Manage. 34:570-582.

Beasom, S. L., L. LaPlant and V. W. Howard. 1982. Similarity of pronghorn, cattle, and sheep diets in southeastern New Mexico. Pages 565-572 *in* J. M. Peek and P. D Dalke, eds. Wildlife-livestock relationships symposium: Proceedings 10. Univ. Idaho, For., Wildl. and Range Exp. Stn., Moscow.

Becker, B. W. 1972. Pronghorn-cattle range use and food habits relationships in an enclosed sagebrush control area. M.S. Thesis, Montana State Univ., Bozeman. 57pp.

Berger, J. 1986. Wild horses of the Great Basin. Univ. Chicago Press, Chicago, Ill. 326pp.

Bever, W. 1957. The incidence and degree of the parasitic load among antelope and the development of field techniques to measure parasitism. S.D. Dep. Fish and Game, Pierre. Fed. Aid Wildl. Restor. Proj. 12-R-14. Job A-5.2. 6pp.

Buechner, H. K. 1950*a*. Life history, ecology, and range use of the pronghorn antelope in Trans-Pecos, Texas. Am. Midland Nat. 43:257-354.

———. 1950*b*. Range ecology of pronghorn on the Whichitas Mountains Wildlife Refuge. Trans. North Am. Wildl. Confer. 15:627-644.

Campbell, R. B. 1970. Pronghorn, sheep, and cattle range relationships in Carter County, Montana. M.S. Thesis, Montana State Univ., Bozeman. 87pp.

Caton, J. D. 1877. The antelope and deer of America. Hurd and Houghton, New York, N.Y. 426pp.

Cid, M. S., J. K. Detling, A. D. Whicker, and M. A. Brizela. 1991. Vegetational response of a mixed-grass prairie site following exclusion of prairie dogs and bison. J. Range Manage. 44:100-104.

Clary, W. P., and R. C. Holmgren. 1982. Observations of pronghorn distribution in relation to sheep grazing on the Desert Experimental Range. Pages 581-592 *in* J. M. Peek and P. D. Dalke, eds. Wildlife-livestock relationships symposium: Proceedings 10. Univ. Idaho, For. Wildl. and Range Exp. Stn., Moscow.

Collins, M. T., T. A. Gallego, J. S. Rolf, and W. T. Adrian. 1981. Seroepidemiology of *Leptospira interrogans serovar hardjo* in Colorado antelope and cattle. J. Am. Vet. Med. Assoc. 179:1136-1138.

Costello, D. F. 1970. The world of the prairie dog. J. B. Lippincott Co., New York, N.Y. 160pp.

Courtney, R. F. 1989. Pronghorn use of recently burned mixed prairie in Alberta. J. Wildl. Manage. 53:302-305.

Decker, E. 1978. Exotics. Pages 249-256 *in* J. L. Schmidt and D. L. Gilbert, eds. Big game of North America. Stackpole Books, Harrisburg, Penn.

Deming, O. V. 1963. Antelope and sagebrush. Int. Antelope Confer. Transactions 17:55-60.

Einarsen, A. S. 1948. The pronghorn antelope and its management. Wildl. Manage. Inst., Washington, D.C. 238pp.

Ellis, J. 1970. A computer analysis of fawn survival in the pronghorn antelope. Ph.D. Thesis, Univ. California, Davis. 70pp.

Gonzales, A. y A. Laffon. 1993. Distribution y estado actual del berrendo (*Antilocapra americana*) en Mexico. Pages 409-419 *in* R. A. Medellin y G. Cebellos, eds. Avances en el estudio de los mamiferos de Mexico. Pubicationes Especiales, Vol. 1. Assoc. Mexicana de Mastozoologia, A.C. Mexico, D.F.

Gregg, H. A. 1955. Summer habits of Wyoming antelope. Ph.D. Thesis, Cornell Univ., Ithaca, N.Y. 185pp.

Grinnell, G. B. 1929. Pronghorn antelope. J. Mammal. 10:135-141.

Gruell, G. E. 1995. Historic role of fire on Hart Mountain National Antelope Refuge, Oregon, and Sheldon National Wildlife Refuge, Nevada. U.S. Fish and Wildl. Serv., Lakeview, Oreg. 58 pp.

Hailey, T. L. 1979. A handbook for pronghorn antelope management in Texas. Tex. Parks and Wildl. Dep., Houston. 59pp.

Hanley, T. A. 1980. Nutritional constraints on food and habitat selection by sympatric ungulates. Ph.D. Thesis, Univ. Washington, Seattle. 176pp.

Hansen, M. C. 1986. Dietary overlap between California bighorn sheep, mule deer, pronghorn antelope, feral horses and domestic cattle in northwestern Nevada. Interstate Antelope Conf. Transactions 28.

Heady, H. F. 1975. Range management. McGraw-Hill Book Co., New York, N.Y. 460pp.

———, editor. 1988. The Vale rangeland rehabilitation program: an evaluation. U.S. Dep. Agric. For. Serv., Pacific Northwest For. and Range Exp. Stn., Portland, Oreg. Resour. Bull. PNW-157. 151pp.

———, and J. Bartholome. 1977. The Vale rangeland rehabilitation program: the desert repaired in southeastern Oregon. U.S. Dep. Agric. For., Serv., Pacific Northwest For. and Range Exp. Stn., Portland, Oreg. Resources Bull. RNW-70. 139pp.

———, and R. D. Child. 1994. Rangeland ecology and management. Westview Press, Boulder, Colo. 517 pp.

Holechek, J. L., R. D. Pieper, and C. H. Herbel. 1989. Range management; principles and practices. Prentice-Hall, Engelwood Cliffs, N.J. 501pp.

Hoover, R. L., C. E. Till, and S. Ogilvie. 1959. The antelope of Colorado. Colo. Dep. Fish and Game, Denver. Tech. Bull. 4. 110pp.

Hornocker, M. G. 1970. An analysis of mountain lion predation upon mule deer and elk in the Idaho Primitive Area. Wildl. Monogr. 21. 39pp.

Howard, V. W., J. L. Holechek, R. D. Pieper, K. Green-Hammond, M. Cardenas, and S. L. Beasom. 1990. Habitat requirements for pronghorn on rangeland impacted by livestock and net wire in east central New Mexico. New Mexico State Univ. Agric. Exp. Stn., Las Cruces. Bull. 750. 48pp.

Kindschy, R. R., C. Sundstrom, and J. D. Yoakum. 1982. Wildlife habitats in managed rangelands—The Great Basin of southeastern Oregon: Pronghorn. U.S. Dep. Agric. For. Serv., Northwest For. and Range Exp. Stn., Portland, Oreg. Gen. Tech. Rep. PNW-145. 18pp.

King, J. A. 1955. Social behavior, social organization, and population dynamics in a black-tailed prairie dog town in the Black Hills of South Dakota. Univ. Michigan, Ann Arbor. Contrib. Lab. Vert. Biol. 67. 128pp.

Kniesel, M. 1988. AUM equivalence and its use in rangeland planning. Pronghorn Antelope Workshop Proceedings 13:83-87.

Koford, C. B. 1958. Prairie dogs, white faces, and blue grama. Wildl. Monogr. 3. 78pp.

Leftwich, T. J., and C.D. Simpson. 1978. The impact of domestic livestock and farming on Texas pronghorn. Pronghorn Antelope Workshop Proceedings 8:307-320.

Longhurst, W. M., A. L. Lesperance, M. Morse, R. J. Mackie, D. L. Neal, H. Salwasser, D. Swickard, P. T. Tueller, P. J. Urness, and J. D. Yoakum. 1983. Livestock and wild ungulates. Pages 42-64 *in* J. W. Menke, ed. Proc. Workshop on livestock and wildlife/fisheries relationships in the Great Basin. Agric. Sci. Publ., Univ. California, Berkeley. Spec. Publ. 3301.

Lovaas, A. L., and P. T. Bromley. 1972. Preliminary studies of pronghorn antelope/black-tail prairie dog relations in Wind Cave National Park. Antelope States Workshop Proceedings 5:115-152.

Mapston, R. D., and R. S. Zobell. 1972. Antelope passes: their value and use. U.S. Bur. Land Manage., Denver, Colo. Tech. Note 6508. 11pp.

McCarty, R. S. 1982. Little Lost-Birch Creek antelope management plan. U.S. Bur. Land Manage., Idaho Falls. 35pp.

McCullough, Y. B. 1980. Niche separation of seven North American ungulates on the National Bison Range, Montana. Ph.D. Thesis. Univ. Michigan, Ann Arbor. 226pp.

McInnis, M. L. 1984. Ecological relationships among feral horses, cattle, and pronghorn in southeastern Oregon. Ph.D Thesis, Oregon State Univ., Corvallis. 166pp.

McNay, M. E., and B. W. O'Gara. 1982. Cattle/pronghorn interactions during the fawning season in northwestern Nevada. Pages 593-606 *in* J. M. Peek and G. D. Dalke, eds. Wildlife-livestock relationships symposium: proceedings 10. Univ. Idaho, For., Wildl. and Range Exp. Stn., Moscow.

Meeker, J. O. 1979. Interactions between pronghorn antelope and feral horses in northwestern Nevada. M.S. Thesis, Univ. Nevada, Reno. 101pp.

Mosley, J. D. 1994. Prescribed sheep grazing to enhance wildlife habitat on North American rangelands. Sheep Res. J. Spec. Issue: 79-91.

Nelson, E. W. 1925. Status of the pronghorned antelope, 1922-24. U.S. Dep. Agric., Washington, D.C. Bull. 1346. 64pp.

O'Gara, B. W. 1978. *Antilocapra americana*. Mammal. Species 90. 7pp.

———, and J. D. Yoakum. 1992. Pronghorn management guides. Pronghorn Antelope Workshop, Rock Springs, Wyo. 101pp.

———. In Press. Diseases and parasites. *In*: J. D. Yoakum and B. W. O'Gara. Ecology and management of pronghorn. Wildl. Manage. Inst., Washington, D.C.

Pederson, J. 1989. Quest for the oryx. New Mexico Wildl. 34:6-10.

Plummer, A. P., D. R. Christensen, and S. B. Monsen. 1968. Restoring big game range in Utah. Utah Div. Fish and Game, Salt Lake City. Publ. 68-3. 183pp.

Pyle, W. H., and J. D. Yoakum. In Press. Status of pronghorn management at Hart Mountain National Antelope Refuge. Pronghorn Antelope Workshop Proceedings 16.

Pyrah, D. 1987. American pronghorn antelope in the Yellow Water Triangle, Montana. Mont. Dep. Fish, Wildl. and Parks, Helena. 121pp.

Reeher, J. A. 1969. Antelope use on rehabilitated sagebrush range in southeastern Oregon. West. Assoc. State Game and Fish Comm. Proceedings 49:272-277.

Roebuck, C. M. 1982. Comparative food habits and range use of pronghorn and cattle in the Texas Panhandle. M.S. Thesis, Texas Tech. Univ., Lubbock. 109pp.

Russell, P. 1964. Antelope in New Mexico. New Mexico Game and Fish Dep., Santa Fe. Bull. 12. 103pp.

Salwasser, H. 1980. Pronghorn antelope population and habitat management in northwestern Great Basin environments. Interstate Antelope Conf., Alturas, Calif. 55pp.

Schwartz, C. C. 1977. Pronghorn grazing strategies on the shortgrass prairie, Colorado. Ph.D. Thesis, Colorado State Univ., Fort Collins. 113pp.

———, and J. G. Nagy. 1976. Pronghorn diets relative to forage availability in northeastern Colorado. J. Wildl. Manage. 40:469-478.

Seton, E. T. 1927. Lives of game animals. Vol. 3. Literary Guild of America, Inc. New York, N.Y. 780pp.

Severson, K. E., M. May, and W. Hepworth. 1968. Food preferences, carrying capacities, and forage competition between antelope and domestic sheep in Wyoming's Red Desert. Univ. Wyoming Agric. Exp. Stn., Laramie. Sci. Monogr. 10. 51pp.

Smith, A. D., and D. M. Beale. 1980. Antelope in Utah. Utah Div. Wildl., Salt Lake City. Resour. Pub. 80-13. 88pp.

Smith, C. 1994. Microhistological analysis of diets of exotic and native ungulates in southcentral New Mexico. M.S. Thesis, New Mexico State Univ., Las Cruces. 76pp.

Spillett, J. J. 1965. The effects of livestock fences on pronghorn antelope movements. M.S. Thesis. Utah State Univ., Logan. 138pp.

Stoddart, L. A., and A. D. Smith. 1958. Range management. Third ed. McGraw-Hill Book Co., New York, N.Y. 433pp.

Sundstrom, C. 1968. Water consumption by pronghorn antelope and distribution related to water in Wyoming's Red Desert. Antelope States Workshop Proceedings 3:399-346.

———. 1969. Some factors influencing pronghorn antelope distribution in the Red Desert of Wyoming. West. Assoc. Fish and Game Comm. Proceedings 49:255-264.

———, W. G. Hepworth and K. L. Diem. 1973. Abundance, distribution, and food habits of the pronghorn. Wyoming Game and Fish Comm., Cheyenne. Bull. 12. 61pp.

Taylor, E. 1972. Food habits of the pronghorn antelope in the Red Desert of Wyoming. M.S. Thesis, Univ. Wyoming, Laramie. 90pp.

———. 1975. Pronghorn carrying capacity of Wyoming's Red Desert. Wyoming Game and Fish Dep., Cheyenne. Wildl. Tech. Bull. 3. 65pp.

Taylor, W. P. 1936. The pronghorned antelope in the southwest. North Am. Wildl. Conf. Proceedings 1:652-655.

Thorne, E. T., T. J. Walthall and S. L. Anderson. 1983. The effects of *Leptospira interrogans serovar hardjo* in pronghorn antelope. Pages 38-45 *in* Wyo. Game and Fish Dep. Fed. Aid Wildl. Restor., Proj. CA1, Job BCSWCBG551.

United States Bureau Land Management. 1985. H-1741-1 Fencing manual Release 1-1819. Washington, D.C. 32pp.

United States Fish and Wildlife Service. 1970. Hart Mountain National Antelope Refuge resource management plan. Lakeview, Oreg. 64pp.

———. 1980. Sheldon National Wildlife Refuge renewable resource management plan/final environmental impact statement. Portland, Oreg.

———. 1994. Hart Mountain National Antelope Refuge comprehensive management plan/final environmental impact statement. Lakeview, Oreg.

Wagner, F. H. 1978. Livestock grazing and the livestock industry. Pages 122-145 *in* H. P. Brokaw, ed. Wildlife and America. U.S. Gov. Printing Off., Washington, D.C.

Yoakum, J. D. 1968. A review of the distribution and abundance of American pronghorn antelope. Antelope States Workshop Proceedings 3:4-14.

———. 1972. Antelope-vegetative relationships. Antelope States Workshop Proceedings 5:171-177.

———. 1975. Antelope and livestock on rangelands. J. Animal Sci. 40:985-992.

———. 1978. Pronghorn. Pages 103-121 *in* J. L. Schmidt and D. L. Gilbert, eds. Big game of North America. Stackpole Books, Harrisburg, Pa. 494pp.

———. 1980. Habitat management guides for the American pronghorn antelope. U.S. Bur. Land Manage. Tech. Note 347. 78pp.

———. 1986. Trends in pronghorn populations: 1925-1983. Pronghorn Antelope Workshop Proceedings 12:77-85.

———. 1990. Food habits of the pronghorn. Pronghorn Antelope Workshop Proceedings 14:102-111.

———. In Press *a*. Water requirements for pronghorn. Pronghorn Antelope Workshop Proceedings.

———. 1995. Relationships of pronghorn and livestock in the Great Basin: a review. Pages 170-175 *in* W. D. Edge and S. L. Olson-Edge, eds., Proc. sustaining rangeland ecosystems symposium. Blue Mountain's Nat. Resour. Inst., La Grande, Oreg. 213pp.

———, and B. W. O'Gara. 1990. Pronghorn/livestock relationships. North Am. Wildl. and Nat. Resour. Conf. 55:475-487.

———, and ———. In Press. Ecology and management of pronghorn. Wildl. Manage. Inst., Washington, D.C.

Zarn, M. 1981. Wild ungulate forage requirements—a review. U.S. Bur. Land Manage. Denver, Colo. Spec. Rep. 103pp.

Chapter 14
BISON
James H. Shaw

INTRODUCTION

About 5,000 years ago, the interior of North America became drier. Deciduous forest gave way to grassland and a small species of bison (*Bison bison*) appeared (McDonald 1981). This new species proved able to inhabit most major biomes from the Gulf of Mexico to the Great Slave Lake and eastward as far as the Appalachians, but reached its greatest numbers on prairie grasslands. During the first 70 years of the nineteenth century, there were tens of millions of bison between the Mississippi River and the Rocky Mountains. A single herd near Dodge City, Kansas in 1871 was estimated at 4 million (Roe 1970).

Free-ranging herds numbering in the millions constituted a major ecological force on the prairies. Fires, both anthropogenic and lightning-caused, constituted another force. Herds of bison were attracted to lush grasses on areas recovering from fires. Less attractive unburned regions received little grazing pressure, allowing buildup of greater fuel loads for the next prairie fires. Thus fire and bison combined to create large-scale shifting mosaics of seral stages on the American prairies.

The objectives of this chapter are to review the unique history of bison conservation and how it affects modern management, the ecological effects of bison on both pristine and contemporary grasslands, factors affecting bison populations and productivity, and the future of the American bison as a rangeland resource (Fig. 1).

FIGURE 1. Adult male bison.

BISON DECLINE AND RECOVERY

In 1886, zoologist William T. Hornaday needed specimens of plains bison (*Bison bison bison*) for the National Museum in Washington, D.C.. Because plains bison were rapidly being depleted, he used his extensive network of contacts in the West to search for remnant populations. The reports that Hornaday received were so bleak that he decided to head west earlier than planned; he feared that plains bison would become extinct before the museum's collection could be supplemented. After 8 weeks in the field, Hornaday's expedition collected only 25 bison in a region of Montana that had supported tens of thousands a few years earlier (McHugh 1972).

Hornaday's experience led him to declare what seemed to be the imminent extinction of the largest land mammal native to the western hemisphere. At a time before state, provincial, or federal wildlife programs existed, private individuals undertook the task of saving the plains bison from extinction. Together, these people gathered approximately 75 founders for the 5 foundation herds (Table 1) from which nearly all of today's plains bison are descended. By 1888 Hornaday tallied 184 bison in these 5 foundation herds (Coder 1975).

These early efforts by private individuals accomplished 2 important things. First, they helped to preserve and partially restore plains bison, first on private ranch lands, and later by providing founders for public herds maintained in such places as the Wichita Mountains Wildlife Refuge (WMWR) in Oklahoma, the National Bison Range (NBR) in Montana, Fort Niobrara National Wildlife Refuge (FNNWR) in Nebraska, and Wind Cave National Park (WCNP) in South Dakota. But the nearly complete extirpation of free-living plains bison, and the crucial roles played by private breeders, led to a practice unique to North American wildlife management. In most states and in some Canadian provinces, bison on private land are legally treated as private property, just as if they were livestock.

Indeed, most growth in bison numbers in recent years has occurred in the private herds (Fig. 2). McHugh (1972) estimated that 30,000 bison lived in North America in 1970, roughly 67% of which occurred in the public herds. By the mid-1980s public herds had declined slightly, to just over 17,000 (North American Bison Workshop 1988), and private herds had increased to an estimated 75,000 (Hawley 1989). Public herds in the United States are unlikely to expand because of limited habitat, but prospects for further growth among private herds are substantial. The lower fat content of bison meat compared with beef, the superior hardiness of bison on poorer quality ranges and in more severe climates, and the established tradition of private ownership, should encourage additional increases in private herds.

INFLUENCE OF BISON ON THE LANDSCAPE

Along with fire, bison herbivory has been a major force in maintaining North American grasslands (Larson 1940). Bison have had and may still have localized and large-scale effects on landscapes as a result of their forage selection, movements and herd sizes, wallowing, horning, urine and dung deposition, and seed dispersal. These activities act in concert with other major influences, such as fire and prairie dog (*Cynomys* spp.) herbivory. Each major type of activity is discussed below and compared, where appropriate, with those of domestic cattle.

Forage Selection

American bison are predominantly grazers. On shortgrass prairies, bison diets have been composed of 90-99% grasses and most of the rest of the forage composed of forbs (Table 2). Studies on northern mixed grass prairies revealed similar patterns; summer diets contained 83-98% grasses, 2-17% forbs, and only trace amounts of browse (Plumb 1991). In a shrub-steppe region

TABLE 1. The 5 founding herds for plains bison (Coder 1975).

Herd	Capture Location	Founders		
		Males	Females	Total
McKay-Alloway	Saskatchewan	1	4	5
Goodnight	Texas	—	—	5
Pablo-Allard	Montana	2	2	4
Dupree	Montana	—	—	4-7
Buffalo Jones	Texas	—	—	56
			TOTAL	74-77

FIGURE 2. Private herd of bison.

of Utah, Van Vuren (1984) reported summer diets of up to 99% grasses and 1% forbs. Sedges (*Carex* spp.) occurred in just over 50% the bison diets in Yellowstone National Park (YNP), followed by 35% grasses and 10% rushes (Meagher 1973). In the boreal meadows of the Slave River lowlands (SRL) in the Northwest Territories, bison consumed 51-80% sedges, followed by grasses and willows (*Salix* spp.) (Reynolds and Peden 1987).

Compared with cattle, bison eat more graminoids (grasses and sedges) and less forbs and browse (Table 2). Bison also exhibit a greater preference for warm season grasses than do cattle. On shortgrass prairie between March and October, > 72% of bison diets by dry weight consisted of warm season grasses and 24% of cool-season grasses. Cattle consumed only 38% warm-season grasses and 39% cool-season grasses. Bison diets consisted of < 4% forbs and no shrubs, but cattle consumed 14% forbs and 8% shrubs (Schwartz and Ellis 1981).

An annual comparison of bison and cattle diets on shortgrass prairie also revealed the bison's preference for warm season grasses. Averaged throughout the year, warm-season grasses comprised > 66% of bison diets, compared to 30-38% warm season grasses in the diets of cattle. Warm season and cool season grasses together accounted for an average of 96% of bison diets on lightly grazed sites and 90% on heavily grazed range, compared with 80% and 56% for cattle, respectively (Peden et al. 1974).

Diets of bison and cattle on northern mixed-grass prairies showed similar patterns, except for an increase in the relative importance of cool season grasses, because they were more available than on southern mixed grass prairies. From June through September, bison consumed an average of 6% forbs and < 1% browse, while diets of domestic cattle averaged 11% forbs and 3% browse (Plumb 1991).

TABLE 2. Diets of bison in various biomes and seasons.

Biome	Season(s)	Diets	Reference
Shortgrass	All	90-99% grasses, balance forbs under light grazing, some shrubs under heavy use	Peden et al. (1974)
Shortgrass	Mar-Oct	93-98% grasses, plus 5-8% forbs in summer	Schwartz and Ellis (1981)
Shrub-steppe	Summer	99% grasses, 1% forbs	Van Vuren (1984)
Seeded range	Late summer	96% grasses, 4% forbs	Van Vuren and Bray (1983)
Northern mixed	Summer	83-98% grasses, balance forbs, trace browse	Plumb (1991)
Boreal meadow	All	51-80% sedges, 15-35% grasses, 1-8% willow	Reynolds and Peden (1987)
Mountain meadow	All	51% sedges, 35% grasses, 10% rushes	Meagher (1973)

TABLE 3. Recent population trends in free-ranging bison populations in Canada (Gates 1987, Van Camp 1987, Van Camp and Calef 1987, and C. G. Van Zyll de Jung, Final draft of section on North American bison conservation. North American Section IUCN/SSC Bison Specialist Group, 34pp.).

Location	Recent trends	Probable influences
Wood Buffalo	12-15,000 in 1940s	Mass drowning, severe winter
National Park	4,200 by 1990	Low recuitment, disease
Hook Lake	1,701 in 1971, 183 in 1987	Heavy hunting, predation, disease
Mackenzie Sanctuary	18 in 1963, 2,405 in 1990	"New" disease-free habitat, fires improved habitat
Nahanni River	28 in 1980, 50 in 1990	Chance drowning, poaching, vehicle collision
Slave River Lowlands	2,643 in 1957, 200 in 1990	Disease, human hunting, predation

Bison digest warm season grasses more efficiently than cattle can, particularly during winter. Cattle digest cool-season grasses more efficiently than do bison. In general, bison feed on grasses less selectively than do cattle and exhibit superior digestive abilities on low-protein, low quality forage. These characteristics allow bison to exploit shortgrass prairie more efficiently than cattle (Peden et al. 1974).

Movements and Herd Sizes

The unconfined herds of bison in the early nineteenth century moved over extensive areas of the Great Plains. Historical accounts of movements contain more opinions than specific data. One popular assumption was that bison, like waterfowl, migrated north in spring and south in fall, an interpretation challenged by McHugh (1972). Garretson (1938) inferred that migrations were more east-west, and that bison spent winter on shortgrass prairie and summer on mixed and tallgrass prairies. Hanson (1984) evaluated historical evidence from the northern plains and found that bison were present year round on shortgrass, tallgrass, and mixed-grass regions, which implied that herd movements were localized and flexible. Such movements presumably occurred in response to changing forage quantity and quality, making regular, long-range migrations unnecessary.

Modern studies have often shown predictable seasonal movements between summer and winter ranges (Soper 1941, McHugh 1958, Meagher 1973, Van Vuren 1983, Meagher 1986, Shaw and Carter 1990). Distances over which bison movements have been measured are necessarily shorter than were the original distances, but the rather consistent predictability of seasonal shifts suggests that at least local migrations occurred.

The extent to which bison movements differ from those of cattle can affect the landscape in addition to the effects resulting from differences in forage selection. Bison in the Henry Mountains, Utah, use less accessible sites than do cattle and also forage farther from surface water (Nelson 1965). Nelson (1965) also reported that, compared with cattle, bison were almost constantly on the move, which resulted in less intensive use of range sites.

Bison, especially cow-calf groups, move and forage as closely-knit units. At WCNP, Shult (1972) described feeding activities as progressing in a "wave" fashion; group members spread out laterally behind a leader, typically an older cow. This tendency to forage directionally as a unit was also reported by Nelson (1965).

Groups of mature cows, calves, yearlings, and immature bulls and cows vary in size by season (Shaw and Carter 1988). Shult (1972) reported that the largest cow-calf groups occurred during summer rut at WCNP (Fig. 3). In YNP McHugh (1958) found that cow-calf groups averaged 23.0 during winter and 175.3 in summer. The largest cow-calf groups in the SRL occurred in June and July, and averaged 75.3. During October and November, cow-calf groups averaged 27.4, the smallest of the year (Calef and Van Camp 1987).

Like other ungulates, bison tend to segregate by sex during most of the year. Segregated bulls were seen as singles in 58.6% of Shult's (1972) observations; the remaining 41.4% were observed in groups of from 2 to 17. In general, bull groups are smallest during summer rut and largest during late spring between calving and rut. In the SRL, bull groups averaged 1.3 during rut and 4.2 after calving (Calef and Van Camp 1987). McHugh (1958) reported average sizes for bull groups in Yellowstone of 1.2 during rut and 4.7 after calving.

Wallows

Bull and cow bison wallow, a practice that consists of sniffing, pawing, and then rolling on the ground (McHugh 1958). Repeated wallowings on the same sites constitute small scale intensive and generally continuous disturbance to the microsites (Collins and Barber 1985). Resulting depressions are up to 1 m deep and several meters across. Soil in the bottom of wallows becomes compacted, which causes higher soil moisture levels inside the wallows than out (Polley and Collins 1984). Differences in soil texture and moisture, plus differences in available phosphorus and pH, correlate with changes in vegetation types and remain detectable in unplowed prairies for >100 years after bison leave an area or are removed (Polley and Collins 1984).

Minor topographic and moisture conditions created by wallows also increase overall vegetational diversity following fire. Collins and Uno (1983) found that species richness and diversity inside wallows was less than that outside wallows, but that evenness was greater inside wallows. The combination of patterns from inside and outside wallows resulted in greater overall species richness and diversity following early spring burns (Collins and Uno 1983).

Horning

Horning, particularly by bulls just before and during rut, may affect the landscape by imposing higher mortality rates on tree saplings. In YNP, McHugh (1958) reported that > 50% of lodgepole pine (*Pinus contorta*) in a sample strip along a heavily used meadow had been damaged by bison horning. Of those pines horned, he reported a 14.3% initial mortality plus another 8.6% that were completely girdled and presumably would die soon. Meagher (1973) found that isolated trees and those near forest edges were often killed by horning in YNP.

Pre-settlement bison numbers may have been enough to limit sapling encroachment on to grasslands through horning at least in some regions. Localized areas into which bison are reintroduced might incur long-term ecological effects through horning

FIGURE 3. Adult male bison tending a cow during rut.

that would not occur through cattle grazing, but such effects have not been evaluated.

Seed Dispersal

In some biomes (e.g., tropical moist forests) native mammals play important roles by ingesting fruits and dispersing seeds passed through their digestive tracts. Such roles in grassland are less well documented and in theory less likely (Collins and Uno 1985, Janzen 1984).

Bison also disperse seeds that attach to their hair. McHugh (1958) found thick clots of cockleburs (*Xanthium* spp.) in the leg and head hair of bison at WCNP just as I have seen in the WMWR. Sandbur (*Cenchrus* spp.) is also commonly dispersed by bison. Seeds of Saint Johnswort (*Hypericum perforatum*) cling to bison hair and McHugh (1958) reported that the manager of the NBR thought that bison played an important role in the plant's dispersal there.

FACTORS AFFECTING BISON POPULATIONS

Food Quality and Quantity

Although American grassland ecosystems are highly productive, the quality of forage is generally low compared with that found in forested biomes. That the sole surviving species of *Bison* in North America is the smallest in body size suggests this observation. Belovsky (1986) has, on the basis of optimal foraging models, suggested that herbivores larger than *Bison bison* are not energetically capable of surviving on North American grasslands. Such models seem to suggest that generalist herbivores such as plains bison are, as a general rule, limited by food availability and its apportionment among competing species.

The strong drive to aggregate is an important, if poorly understood, characteristic of bison (M. Meagher, YNP, pers. commun.). Nineteenth century herds were larger and moved farther than contemporary ones, thereby impacting the landscape differently. But even within the spatial limits faced by modern managers, bison productivity and impacts may vary substantially as a function of population size. Five bison confined to 10 ha of mixed grass prairie would not have the same ecological impacts nor the equivalent productivity of 500 bison on 1,000 ha. Given the species' drive to aggregate, total numbers affect ecological relationships.

Winter Severity

For several decades, the relative stability of bison numbers in YNP led Meagher (1973) to conclude that the population appeared to be naturally regulated. Predation and disease seemed insufficient to exert enough pressure to hold numbers in check, so Meagher (1976) concluded that stresses related to winter severity were responsible, particularly in the Pelican Valley herd. Yellowstone's last severe winter occurred during 1975-76. During the mild winters since, aerial counts of bison on the northern winter range increased from 200 in 1975-76 to 594 in 1986-87 (Meagher 1989). Such a rate of increase coinciding with a series of mild winters supports the view that periodic winter severity can exert a major influence on bison populations in colder parts of the species' range. Nonetheless, bison are generally well adapted to all but the most extreme cold conditions. Bison are the sole members of the subfamily Bovinae that have never occurred in the tropics. Compared to domestic cattle, bison exhibit lower metabolic and respiratory rates than those of cattle exposed to the same low temperatures (Christopherson et al. 1979). These comparisons support the contention commonly held by bison breeders that bison are substantially more cold tolerant than are cattle.

Diseases

Bison share diseases and parasites with cattle. The 3 most important infectious diseases are anthrax (*Bacillus anthracis*), bovine tuberculosis (*Mycobacterium bovis*), and brucellosis (*Brucella abortus*). All 3 of these diseases can be zoonotic. More extensive reviews of diseases in bison appear in Broughton (1987) and of parasites and diseases in Reynolds et al. (1982).

Anthrax first entered North America via cattle in Louisiana in the early 18th century (Reynolds et al. 1982). The first known outbreak of anthrax occurring in bison in the SRL was documented in 1962 and has since been confirmed as the causal agent in the deaths of hundreds of bison in the Northwest Territories and Alberta (Broughton 1987). Although vaccination programs have been introduced to combat anthrax, vaccine is difficult to administer to bison in remote areas. The complex and persistent life cycle of anthrax makes it the least likely of the 3 major diseases to be eradicated from a population or region.

Tuberculosis is a serious, chronic, infectious disease. It was first documented in bison at Wainwright Buffalo Park, Alberta in 1923. Canadian authorities transported 6,673 plains bison from the Wainwright herd to Wood Buffalo National Park (WBNP) during the 1920's (Soper 1941). This transfer spread tuberculosis to WBNP, where it remains endemic (Broughton 1987, Redhead 1988), and probably brought anthrax and brucellosis as well. The incidence of tuberculosis in WBNP from 1950 to 1967 averaged 38.8%, based upon post-mortem examinations, although mortality rates were not estimated for the herd (Broughton 1987).

The most controversial disease among North American bison is brucellosis. Agricultural agencies in both the United States and Canada have developed extensive programs to eradicate brucellosis in livestock. They see free-ranging populations of bison as persistent reservoirs of the disease (Fig. 4). Outbreaks of brucellosis can have severe economic consequences because of legal requirements regarding testing for interstate and international shipment of livestock. The complex interplay between biology, regulatory law, economics, and politics has been reviewed by Thorne et al. (1991).

Managers of free-ranging herds point out that it is not technically feasible to round up all members of free-ranging bison herds. Moreover, the serologic testing and vaccinations for brucellosis in cattle are unreliable in bison (M. Meyer, *Brucella abortus* in the Yellowstone National Park bison herd, Rep. Dep. Interior, YNP, 25pp., 1992). Regarding brucellosis in the Yellowstone herd, Meyer concluded that, "A test and slaughter campaign essentially would do nothing but eradicate titers."

FIGURE 4. The author drawing blood from a bison to test for disease.

Natural Predation

There are no significant predators of bison within the continental United States, but gray wolves (*Canis lupus*) are effective predators on some of the publicly owned herds in Canada, generally in areas where there is little, if any, alternative prey. Effects of wolf predation on bison populations are difficult to assess because of difficulties of studying large, mobile mammals in remote regions and the confounding effects of other sources of mortality.

The bison population of the SRL fell from an estimated 2,643 in 1957 (Van Camp and Calef 1987) to approximately 200 in 1990 (C. G. Van Zyll de Jung, Final draft of section on North American bison conservation. North American Section IUCN/SSC Bison Specialist Group, 34pp., 1991). A combination of factors contributed to this decline including hunting pressure, wolf predation, and diseases (Van Camp and Calef 1987). During the winter of 1976-77, field investigations in the SRL focused on the role of wolf predation. By weight, bison made up an approximately 88% of the winter diets of the estimated 64-76 wolves in the SRL (Van Camp 1987). Wolf predation accounted for 31% of adult and subadult mortality and 27% of the calf mortality.

After wolf control removed 72 wolves from the SRL between 1977 and 1979, predation on bison dropped substantially, but the bison population continued to decline (Van Camp 1987). The confounding effects of human hunting and diseases, and the general instability of the bison population in the SRL, rendered the results of the field study inconclusive regarding the effects of wolf predation.

Free-ranging bison populations in northern Canada have fluctuated in recent years. A comparison of recent trends (Table 3) reveals no consistent relationship between wolf predation and rate of increase. Declines in WBNP, Hook Lake, and SRL seem to have resulted from multiple causes, obscuring the role of wolf predation. Rapid increases in the Mackenzie Bison Sanctuary population in the presence of a substantial wolf population, suggest that effects of wolf predation can be overridden by conditions of favorable habitat free of disease.

In the most thorough investigation yet, Carbyn et al. (1993) concluded that wolf predation had a significant impact on bison populations in the Peace-Athalasca Delta of WBNP. Field observations combined with population modeling indicated that bison recruitment would rise quickly without wolf predation.

Wolf predation can affect movements as well as numbers of bison. Bison herds in WBNP, for example, have been known to move up to 86 km following wolf attacks (Carbyn et al. 1993).

Other behavioral changes due to predation have not yet been documented in bison. Berger and Cunningham (1994) compared vigilance (i.e., the amount of time spent scanning their surroundings) in female bison at Badlands National Park where wolves are absent with that at WBNP where wolves were present. They found no significant difference in vigilance.

Hunting

Bison populations were greatly reduced by market hunting in the nineteenth century (Dary 1974, McHugh 1972, Roe 1970). Populations can be hunted on a sustainable basis because in most habitats bison are conspicuous enough to be accurately censused. Quotas for sport hunting can be adjusted for observed population trends and management objectives.

RELATIONSHIPS WITH OTHER HERBIVORES

Diets

Bison are predominantly grazers and their diets closely resemble those of cattle. Bison foraging patterns overlap significantly with those of bighorn sheep (*Ovis canadensis*), although the 2 species rarely share common ranges. Elk (*Cervus elapus*) have such wide feeding niches that they use some of the same forages as bison (Houston 1982). Diets of domestic sheep overlap partially with those of bison (Wagner 1978).

Diets of mule deer (*Odocoileus hemionus*) and white-tailed deer (*O. virginianus*) do not convergence with those of bison under normal conditions. Deer favor browse, forbs, and mast, whereas bison concentrate on graminoids. Deer favor more woody habitats while bison make preferential use of open meadows and grasslands. Pronghorn(*Antilocapra americana*)share common ranges and habitats with bison, but have divergent diets. Pronghorns eat almost exclusively forbs and browse. Horses take a wide range of plant foods and their diets can overlap with those of bison.

Behavioral Relationships

Bison share range with Texas longhorn cattle at the WMWR. The 2 species mingle sparingly, but in general they remain segregated. Hawley and Reynolds (1987) concluded that free-ranging bison populations are not compatible with cattle ranching. They believe that bison would likely drive cattle away from supplemental feed sources, damage fences, and spread infectious diseases.

Habitat relationships among elk, mule deer, pronghorn, and bison were compared at WCNP (Wydeven and Dahlgren 1985). Elk and deer favored sites with predominantly warm-season grasses but bison and pronghorn preferred sites dominated by cool-season grasses. The combination of differential habitat use and different diet preferences suggested low potential for competition among those 4 ungulate species under the observed stocking levels.

A similar pattern emerged between bison and elk at Elk Island National Park (EINP) in Alberta. Bison and elk foraged in different habitats in summer, when both relied principally on herbaceous forage. During winter, both species used similar habitats but their diets diverged, with elk eating mainly browse and bison eating 95% herbaceous plants (Telfer and Cairns 1979).

In YNP, Houston (1982) compared observations of feeding mule deer, bighorn sheep, pronghorn, moose, and bison according to habitat type. Moose (*Alces alces*) were most commonly seen among willows and mule deer on mesic steppe. Bighorn sheep and pronghorn favored xeric steppe. Bison foraged most often in wet meadows and swales.

McHugh (1958) reported that bison charge pronghorns and elk. Even 1-month-old bison calves displaced 6-point bull elk in the Jackson Hole Wildlife Park (JHWP) (McHugh 1958). In YNP, individual bison or herds of bison reportedly displaced groups of elk (McHugh 1958).

Mixed Species Management

Matching bison with other herbivore species is rather simple in terms of diets. Behavioral compatibility could prove to be more of a problem. Under confined conditions of the JHWP, McHugh (1958) observed bison kill 1 8-month old pronghorn, and a 7-month old moose. Two mule deer were attacked and knocked down by bison (McHugh 1958). Presumably bison reared sympatrically with other herbivores would more likely be tolerant of them. Likewise bison confined to large areas would be less likely to injure other species than those kept in smaller pastures. Problems might occur at supplemental feeding areas, with bison displacing other herbivores and under extreme conditions, attacking and injuring them.

DIFFERENT GRAZING SYSTEMS AND PRESCRIBED FIRE

Prescribed Fire

Like other grazers, bison are attracted to rich prairie vegetation during the first growing season following fire (Campbell and Hinkes 1983). Bison use of a 1,860-ha prescribed burn increased from 3.9% in the June the year before the burn to 58.0% in the June following a late February fire in the WMWR (Shaw and Carter 1990).

Frequencies and extent of prairie fires before settlement are difficult to determine. Estimates range from an average fire return interval of once every 3-5 years (Collins and Uno 1983) to once every 5-10 years (Sims 1988).

The effects of fire on natural landscapes are not uniform. Differences in topography and vegetation composition cause fire to create a series of mosaic disturbances. Large, mobile herbivores such as bison forage disproportionately on areas recovering from burns. Thus the combination of fire and herbivory add to overall landscape (gamma) diversity of prairie ecosystems (Collins and Barber 1985, Wright and Bailey 1982).

Grazing Systems

In the early nineteenth century bison herds ranged over long distances in large numbers. The resulting grazing pattern was short-term and intensive. Leopold (1949) used the compass plant (*Silphium* spp.) to illustrate principal differences between the original bison pattern of herbivory and the pattern resulting from herbivory by confined cattle. The compass plant is a highly palatable perennial forb, and cannot endure sustained grazing. The only places it persists are along road rights-of-way and other limited areas from which cattle and other grazers are excluded. Leopold (1949) observed it in an old triangular-shaped cemetery, the corners of which escaped mowers and scythes. The compass plant was widespread and abundant in the prairie states prior to settlement, a pattern that suggests infrequent intensive grazing.

Various systems of rest-rotation grazing have attracted increased attention from land managers in recent years. Perhaps the best known and most controversial is the system developed by Savory (1986, 1988), who attempts to create grazing systems that best duplicate the original animal impacts for a particular region. Savory (1986,1988) maintains that overgrazing is related to time, rather than to animal numbers and that the solution is intensive rest-rotation grazing. In assessing effects of large herbivores on their environment, Savory (1986,1988) distinguishes between those due to grazing (i.e., forage selection) and those due to physical effects, such as trampling, rubbing, horning, urinating, and defecating.

In theory, such a rest-rotation system should work well with bison, because it mimics closely presettlement patterns of bison movements and impacts upon the landscape. However, there have been no published evaluations of bison productivity under rest-rotation conditions. Compared with cattle, bison would require more elaborate and expensive fencing and could be more difficult to move between grazing units.

FACTORS RELATED TO PRODUCTIVITY

Calf production rates vary among the public herds. In the SRL and YNP 50% and 52%, respectively, of the mature cows >3 years gave birth annually (Van Camp and Calef 1987, Meagher 1973). The highest reported rates come from the NBR at 88.2% (Rutberg 1986) and FNNWR and WCNP, both at 78.2% (Haugen 1974). Managers at WCNP, the NBR, and FNNWR keep their bison populations at set levels through the sale of surplus animals. It is thus very likely that the higher productivity is in part due to density-dependent responses within the more intensively managed populations. Bison in YNP and the SRL, in contrast, are not rounded up and their population levels are determined more by natural forces.

Based on research at the WMWR in Oklahoma, Halloran (1968) presented composite samples showing higher calving rates among cows 3, 5, and 7 years of age than among those 4,

6, or 8. He concluded that bison cows typically produce calves in alternate years.

I was unable to confirm any alternate year calving pattern among bison at the WMWR. Calving records of known-aged individual cows over an 8-year period revealed that none of the individuals calved in alternate years (Shaw and Carter 1989). The most likely explanation for this disparity lies in the different stocking rates for the 2 periods. During Halloran's (1968) study, the Refuge bison population averaged 3.46/km^2, compared with 2.30/km^2 during our investigations. If the higher density was placing the bison population under nutritional stress, an alternate year pattern might be expected. Whatever the cause, the generalization that bison produce calves in alternate years should not be uncritically accepted.

Latitude may also affect calf production. If one compares reported calving rates for similarly managed herds in the WMWR (66.9%, Halloran 1968), FNNWR (78.4%, Haugen 1974), WCNP (78.4%, Haugen 1974), and the NBR (88.2%, Rutberg 1986), rates appear to rise with increasing latitude. One explanation for this apparent pattern involves the nutritional value of the grasses in the southern plains in relation to pregnancy and lactation in bison. Crude protein and phosphorus content from grasses in the southern plains are at or below maintenance levels for bison through most of the winter and spring (Speth 1983). Presumably, these conditions impose serious stress upon southern plains bison during the latter phases of pregnancy and most of lactation. This stress would be lower at higher latitudes, accounting for the higher calving rates on northern grasslands.

Attempts to use bison to test the Trivers-Willard hypothesis (Trivers and Willard 1973) have yielded inconsistent results. This hypothesis predicts that in a species with small litter size, strong male-male competition and a correlation between maternal condition and condition of offspring at maturity should produce a greater proportion of male offspring from females in superior condition. At the NBR, Rutberg (1986) found a significantly higher proportion of male fetuses in nonlactating females than in lactating females, which supported the Trivers-Willard hypothesis. In contrast, there was no difference in the sex of calves produced by lactating versus nonlactating bison at the WMWR (Shaw and Carter 1989), which did not support the predictions of the Trivers-Willard hypothesis.

THE FUTURE OF AMERICAN BISON

More than 2 decades ago, Rorabacher (1970) extolled the need for an attitude toward American bison based on recognition of the species as a valuable resource rather than a vanishing symbol of the Old West. Rorabacher (1970) advocated a transition from the impulse to preserve to the challenge for production.

Given the steady rise in the numbers of privately owned bison in North America, it seems apparent that the transition has to some extent been made and that bison will continue to increase as an important commodity on private land (Fig. 5). Hawley (1989) reviewed the outlook for bison ranching in North America and concluded that the best potential for private production will be as a complement to traditional cattle ranching. Because bison are better suited than cattle to areas deemed marginal due to severe winters, predation, or limited availability of water or quality forage, they could be produced on sites of limited value to cattle. The main economic constraints seem to be availability of initial breeding stock and the high initial costs associated with fencing and handling facilities (Hawley 1989).

A more ambitious and controversial plan for American bison, known as the "buffalo commons", has been proposed (F. Popper and D. Popper, Where the buffalo should roam. Oklahoma Academy for State Goals, Oklahoma City, 8 Nov 1990). They developed their ideas based upon their economic and demographic analysis of the Great Plains states. One hundred and nine counties between the 98th meridian and the Rocky Mountains were suffering severe economic distress with no outlook for improvement. The Poppers initially recommended "deprivatizing" these counties and perhaps others as conditions worsened. After entering the public domain, the land could be used to produce bison, with populations kept in check through auctions. The Poppers reasoned that their plan offered a better alternative than laissez-faire collapse, followed by yet another round of resettlement through government subsidies or other public outlays (the 1980s and 1990s constitute the third period of boom and bust for the plains states).

FIGURE 5. A safe way to handle bison when biological samples need to be collected.

Whatever the virtues of the buffalo commons idea, the plan itself triggered outrage among residents of the region. Eastern academics were presuming to tell hard-working farmers and ranchers to abandon their life's work and allow the land to revert to the very conditions that early pioneers sought to tame. Moreover, by recommending deprivatization, the Poppers raised the spectre of government condemnation of private land, an anathema to western landowners.

The Poppers began touring the plains states and listening to views of residents (A. Matthews, The Poppers and the plains. New York Times Mag. 24 Jun. 1990). While retaining the central core of the buffalo commons theme, they refined and clarified some of the means to that end. The Poppers steadfastly maintained that they never advocated forced removal of residents from the plains (J. Klein, plain speaking on the buffalo commons plan. Tulsa World, 8 Nov. 1990). They envisioned a greater role for the private sector, particularly the Nature Conservancy, in obtaining and managing lands.

Although the buffalo commons idea may have been overly ambitious and presented in a less than tactful manner, 2 major elements are essentially irrefutable and together suggest a greater role for bison on the plains. The first element is the deteriorating economic condition of the region, much of which is only marginally suitable for conventional land use, and then only with government subsidies. The second element is the need to maintain the health and vigor of the land by managing a useful product on a sustainable basis.

To the extent that a buffalo commons materializes, it will probably consist mainly of herds owned by a combination of ranchers and nongovernmental organizations who will manage them for production of meat and hides. Other herds will be publicly owned and managed for a variety of objectives. Thus a native species nearly made extinct in the nineteenth century may repopulate parts of the plains in the twenty-first century.

SUMMARY

Extinction of the plains bison was thwarted largely through the efforts of private individuals. These efforts established a unique tradition of private ownership of a native wild species and in the past 20 years, most of the growth in bison numbers has occurred among private herds.

Before settlement, bison influenced the landscape of the Great Plains through grazing, wallowing, horning, and seed dispersal. The spatial distribution of these activities was influenced by fire because bison were drawn to recovering burn sites. Concentrated grazing on burned sites then reduced the amount of standing vegetation or fuel, increasing the chances that the next round of fires would occur elsewhere. Together, fire and bison helped create and maintain a mosaic of vegetation types.

Bison diets consist mainly of grasses and, where available, sedges. Forbs typically make up only a small percentage of the diet and browse even less. Bison thus present little competition for forage with other native ungulates, particularly deer. Among domestic ungulates, cattle have diets overlapping most closely with those of bison, but cattle tend to take more forbs and browse.

Extreme winter conditions may influence bison populations in colder regions, and forage availability can ultimately limit bison numbers locally. Diseases introduced through domestic livestock have exerted pressures on bison populations; they include anthrax, bovine tuberculosis, and possibly brucellosis. Public herds of bison in Canada have declined in recent years, in part due to disease. Concern over bison as reservoirs for brucellosis has led to policy conflicts in the United States and Canada.

The principal natural predator on bison, the gray wolf, presently occurs with bison only in some of the Canadian herds. The effects of wolf predation on bison are not well understood; efforts to assess predation's impact have been confounded by diseases and human predation.

Prescribed fire and rotational grazing systems should both benefit managed herds of bison, but thus far little research has been carried out on these subjects. Bison make heavy use of grassland sites recovering from late winter or early spring burns. Rest-rotational grazing systems should benefit bison and the native prairies through mimicking presettlement conditions.

Several factors influence calf production rates. The highest rates of increase occur among intensively managed herds in which managers use auctions to keep numbers low in relation to carrying capacity and take advantage of density-dependent responses. Latitude may also be an influence, perhaps because of regional differences in forage quality. Recent field studies suggest that bison cows do not produce calves in alternate years as an earlier study had concluded.

The growth in bison ranching during the past 20 years suggests a strong, substantial future for privately owned herds. Bison may also be reintroduced to more public lands and to lands owned by private conservation groups such as The Nature Conservancy. These herds must be kept within the carrying capacities of their ranges, most likely through auctions of surplus animals. The extent to which private and some of the public herds will expand will depend upon market conditions for bison and their products.

LITERATURE CITED

Belovsky, G. E. 1986. Optimal foraging and community structure: implications for a guild of generalist grassland herbivores. Oecologia 70:35-52.

Berger, J., and C. Cunningham. 1994. Bison: mating and conservation in small populations. Columbia Univ. Press. New York, N.Y. 330pp.

Broughton, E. 1987. Diseases affecting bison. Pages 34-38 *in* H. W. Reynolds and A. W. L. Hawley eds. Bison ecology in relation to agricultural development in the Slave River lowlands, Northwest Territories. Can. Wildl. Serv. Occas. Paper 63.

Calef, G. W., and J. Van Camp. 1987. Seasonal distribution, group size and structure and movements of bison herds. Pages 15-20 *in* H. W. Reynolds and A. W. L. Hawley eds. Bison ecology in relation to agricultural development in the Slave River lowlands, Northwest Territories. Can. Wildl. Serv. Occas. Paper 63.

Campbell, B. H., and M. Hinkes. 1983. Winter diets and habitat use of Alaska bison after wildfire. Wildl. Soc. Bull. 11:16-21.

Carbyn, L. N., S. M. Oosenbrug, and D. W. Anions. 1993. Wolves, bisons and the dynamics related to the Peace-Athebasca Delta in Canada's Wood Buffalo National Park. Circumpolar Res. Ser. 4 Can. Circumpolar Inst. Univ. Alberta, Calgary. 270pp.

Christopherson, R. J., R. J. Hudson, and M. K. Christopherson. 1979. Seasonal energy expenditures and thermoregulatory responses of bison and cattle. Can. J. Anim. Sci. 59:611-617.

Coder, G. D. 1975. The national movement to preserve the American buffalo in the United States and Canada between 1880 and 1920. Ph.D. Thesis. Ohio State University, Columbus. 348pp.

Collins, S. L., and S. C. Barber. 1985. Effects of disturbance on diversity in mixed-grass prairie. Vegetatio 64:87-94.

———, and G. E. Uno. 1983. The effect of early spring burning on vegetation in buffalo wallows. Bull. Torrey Bot. Club 110:474-481.

———, and ———. 1985. Seed predation, seed dispersal, and disturbance in grassland: a comment. Am. Nat. 125:866-872.

Dary, D. A. 1974. The buffalo book. Swallow Press, Chicago, Ill. 374 pp.

Garretson, M. S. 1938. The American bison. The N.Y. Zool. Soc., New York, N.Y. 254pp.

Gates, C. C. 1987. Status of bison in northern Canada. Pages 4-5 *in* J. Malcolm ed. North Am. Bison Workshop. U.S. Fish and Wildl. Serv. and Glacier Nat. Hist. Assoc., Missoula, Mont. 74pp.

Halloran, A. F. 1968. Bison (Bovidae) productivity on the Wichita Mountains Wildlife Refuge, Oklahoma. Southwest. Nat. 13:23-26.

Hanson, J. R. 1984. Bison ecology in the northern plains and a reconstruction of bison patterns for the North Dakota regions. Plains Anthrop. 29:93-113.

Haugen, A. O. 1974. Reproduction in the plains bison. Iowa State J. Res. 49:1-8.

Hawley, A. W. L. 1989. Bison farming in North America. Pages 346-361 *in* R. J. Hudson, K. R. Drew, and L. M. Baskin, eds. Wildlife production systems. Cambridge Univ. Press, Cambridge, U.K.. 469pp.

———, and H. W. Reynolds. 1987. Management alternatives for ungulate production in the Slave River lowlands. Pages 63-66 *in* H. W. Reynolds and A. W. L. Hawley, eds. Bison ecology in relation to agricultural development in the Slave River lowlands, Northwest Territories. Can. Wildl. Serv. Occas. Paper 63.

Houston, D. B. 1982. The northern Yellowstone elk: ecology and management. MacMillan Publ. Co., Inc. New York, N.Y. 474pp.

Janzen, D. H. 1984. Dispersal of small seeds by big herbivores: foliage is the fruit. Am. Nat. 123:338-353.

Larson, F. 1940. The role of bison in maintaining the short grass plains. Ecology 21:113-121.

Leopold, A. 1949. A Sand County almanac. Oxford Univ. Press. London, U.K. 226pp.

McDonald, J. N. 1981. North American bison: their classification and evolution. Univ. California Press, Berkeley. 316pp.

McHugh, T. 1958. Social behavior of the American buffalo (*Bison bison bison*). Zoologica 43:1-40.

———. 1972. The time of the buffalo. Alfred A. Knopf, Inc. New York, N.Y. 339pp.

Meagher, M. 1973. The bison of Yellowstone National Park. Natl. Park Serv. Sci. Monogr. Ser. 1. 161pp.

———. 1976. Winter weather as a population regulating influence on free-ranging bison in Yellowstone National Park. Pages 29-38 *in* Research in the parks. Trans. Natl. Park Centennial Symposium. Washington, D.C. 232pp.

———. 1986. *Bison bison*. Mammal. Species. 266:1-8.

———. 1989. Range expansion by bison of Yellowstone National Park. J. Mammal. 70:670-674.

Nelson, K. L. 1965. Status and habits of the American buffalo (*Bison bison*) in the Henry Mountains area of Utah. Utah Fish and Game Publ. 65-2. 142pp.

North American Bison Workshop. 1988. 1987 list and census of North American public bison herds. Pages 2-3 *in* J. Malcolm ed. North American Bison Workshop. U.S. Fish and Wildl. Serv. and Glacier Nat. Hist. Assoc., Missoula, Mont.

Peden, D. G., G. M. Van Dyne, R. W. Rice, and R. M. Hansen. 1974. The trophic ecology of *Bison bison* L. on shortgrass plains. J. Appl. Ecol. 11:489-497.

Plumb, G. E. 1991. Foraging ecology of bison and cattle on a northern mixed prairie. Ph.D. Thesis. Univ. Wyoming, Laramie. 137pp.

Polley, H. W., and S. L. Collins. 1984. Relationships of vegetation and environment in buffalo wallows. Am. Midl. Nat. 112:178-186.

Redhead, B. 1988. The management of brucellosis and tuberculosis in bison in Wood Buffalo National Park. Pages 62-64 *in* J. Malcolm, ed. North American bison workshop. U.S. Fish and Wildlife Serv. and Glacier Nat. Hist. Assoc., Missoula, Mont. 74pp.

Reynolds, H. W., R. D. Glaholt, and A. W. L. Hawley. 1982. Bison. Pages 972-1007 *in* J. A. Chapman and G. A. Feldhamer, eds. Wild mammals of North America. The Johns Hopkins Univ. Press, Baltimore, Md. 1147pp.

———, and D. G. Peden. 1987. Vegetation, bison diets, and snow cover. Pages 39-44 *in* H. W. Reynolds and A. W. L. Hawley, eds. Bison ecology in relation to agricultural development in the Slave River lowlands, Northwest Territories. Can. Wildl. Serv. Occas. Paper 63.

Roe, F. 1970. The North American buffalo. Second ed. Univ. Toronto Press. Buffalo, N.Y. 991pp.

Rorabacher, J. A. 1970. The American buffalo in transition: a historical and economic survey of the bison in America. North Star Press, Saint Cloud, Minn. 142pp.

Rutberg, A. T. 1986. Lactation and fetal sex ratios in American bison. Am. Nat. 127:89-94.

Savory, A. 1986. A solution to desertification and associated threats to wildlife and man. Trans. North Am. Wildl. and Nat. Resour. Conf. 51:116-124.

———. 1988. Holistic resource management. Island Press, Washington. D. C. 564pp.

Schwartz, C. C., and J. E. Ellis. 1981. Feeding ecology and niche separation in some native and domestic ungulates on the shortgrass prairie. J. Appl. Ecol. 18:343-353.

Shaw, J. H. and T. S. Carter. 1988. Long-term associations between bison cows and their offspring: implications for the management of closed gene pools. Pages 50-55 *in* J. Malcolm, ed. North American Bison Workshop. U. S. Fish and Wildlife Serv. and Glacier Nat. Hist, Assoc. Missoula, Mont.

———, and T. S. Carter. 1989. Calving patterns among American bison. J. Wildl. Manage. 53:896-898.

———, and ———. 1990. Bison movement in relation to fire and seasonality. Wildl. Soc. Bull. 18:426-430.

Shult. M. J. 1972. American bison behavior patterns at Wind Cave National Park. Ph.D. Thesis. Iowa State Univ., Ames. 178pp.

Sims, P. L. 1988. Grasslands. Pages 265-286 *in* M. G. Barbour and W. D. Billings, eds. North American terrestrial vegetation. Cambridge Univ. Press. Cambridge, U.K. 434pp.

Soper, J. D. 1941. History, range, and home life of the northern bison. Ecol. Monogr. 11:347-412.

Speth, J. D. 1983. Bison kills and bone counts: decision making by ancient hunters. Univ. Chicago Press, Chicago, Ill. 227pp.

Telfer, E. S., and A. Cairns. 1979. Bison-wapiti interrelationships in Elk Island National Park, Alberta. Pages 114-121 *in* M. S. Boyce and L. D. Hayden-Wing, eds. North American elk: ecology, behavior, and management. Univ. Wyoming, Laramie. 294pp.

Thorne, E. T., M. Meagher, and R. Hillman. 1991. Brucellosis in free-ranging bison: three perspectives. Pages 275-287 *in* R. B. Keiter and M. S. Boyce, eds. The Greater Yellowstone Ecosystem: redefining America's wilderness heritage. Yale Univ. Press, New Haven, Conn. 428pp.

Trivers, R. L., and D. E. Willard. 1973. Natural selection of parental ability to vary the sex ratio of offspring. Science 179:90-92.

Van Camp, J. 1987. Predation on bison. Pages 25-33 *in* H. W. Reynolds and A. W. L. Hawley, eds. Bison ecology in relation to agricultural development in the Slave River lowlands, Northwest Territories. Can. Wildl. Serv. Occas. Paper 63.

———, and G. W. Calef. 1987. Population dynamics of bison. Pages 21-24 *in* H. W. Reynolds and A. W. L. Hawley eds. Bison ecology in relation to agricultural development in the Slave River lowlands, Northwest Territories. Can. Wildl. Serv. Paper 63.

Van Vuren, D. 1983. Group dynamics and summer home range of bison in southern Utah. J. Mammal. 63:329-332.

———. 1984. Summer diets of bison and cattle in southern Utah. J. Range. Manage. 37:260-261.

———, and M. P. Bray. 1983. Diets of bison and cattle on a seeded range in southern Utah. J. Range Manage. 36:499-500.

Wagner, F. H. 1978. Livestock grazing and the livestock industry. Pages 121-145 *in* H. P. Brokaw, ed. Wildlife and America. Counc. on Environ. Quality, Washington, D. C. 532pp.

Wright, H. A., and A. W. Bailey. 1982. Fire ecology: United States and southern Canada. John Wiley & Sons, New York, N.Y. 501pp.

Wydeven, A. P., and R. B. Dahlgren. 1985. Ungulate-habitat relationships in Wind Cave National Park. J. Wildl. Manage. 49:805-813.

Chapter 15
BIGHORN SHEEP AND LIVESTOCK

Paul. R. Krausman, Raul Valdez, and John A. Bissonette

INTRODUCTION

Bighorn sheep in North American belong to one of the most widely distributed genera of ungulates in the world. Bighorn sheep range from Europe (introduced in the Mediterranean Islands of Corsica and Sardinia but now widespread in the continent) through Anatolia, Iraq, Iran, Turkistan, Pakistan and northern India, Central Asia, and Siberia. In North America, bighorn sheep occur from Alaska, south through western Canada and the western United States to northern Mexico and Baja California (Valdez 1982).

Bighorn sheep in North America are divided into 2 basic types: thinhorn and bighorn. Dall's (*O. dalli dalli*) and Stone's (*O. d. stonei*) sheep constitute the thinhorn; Rocky Mountain (*O. c. canadensis*), California (*O. c. californiana*), and desert bighorn sheep (*O. c. mexicana, O. c. nelsoni, O. c. cremnobates, O. c. weemsi*) make up the group known as bighorns (Valdez 1982).

Bighorn sheep have adapted to negotiate rocky, precipitous terrain; they generally avoid areas of tall vegetation that obstruct their vision (Risenhoover and Bailey 1985). Diets of northern forms are primarily grasses and forbs while desert forms subsist mainly on shrubs. Because they are ruminants, sheep regurgitate a cud or bolus of food from the rumen to the mouth and further grind it. If surface water is available during dry periods, sheep may drink regularly; they require approximately 4-5% of their body weight/day of water (Turner 1979*a*,*b*). However, some populations obtain water from moisture in vegetation (Krausman et al. 1985).

People and bighorn sheep have coexisted in North America for ≥ 30,000 years (Hopkins 1967). Bighorn sheep constituted a source of meat protein for aboriginal humans in Asia, and early humans in North America. Seton (1929) estimated there were 1.5 to 2 million bighorn sheep in the contiguous United States prior to the arrival of European humans and another 2 million in Canada and Alaska combined. Seton's (1929) estimate of 4 million sheep in North America is cited often. However, Valdez (1988) doubted that sheep numbers ever exceeded even half a million for all of North America. Bighorn sheep have highly selective habitat preferences and are not distributed uniformly in the mountainous terrain of western North America. In Alaska, where most wild sheep populations still occupy vast regions of habitat undisturbed by man, numbers do not exceed 50,000 (Nichols 1975). Buechner (1960) reviewed historical sheep distributions by state and estimated there were 15,000 to 20,000 wild sheep in the contiguous United States in 1960. In a recent survey by the United States Fish and Wildlife Service (unpubl. data), bighorn sheep biologists representing each state estimated there were 47,750 bighorn sheep in the contiguous United States in 1990. Throughout their range there are over 188,000 bighorn sheep in North America. Dall's and Stone's sheep populations appear to have maintained their historical distributions and numbers. However, bighorn sheep populations of southwestern Canada, the western United States, and northern Mexico have declined because of humans. These sheep have been reduced from relative abundance to one of the rarest ungulates in North America. Unlike their ancestors in the Old World, humans in the New World never domesticated wild bighorn sheep. However, they did bring domesticated livestock to the New World.

The major decline of bighorn sheep populations occurred during the latter half of the 19th century. It was during this period and the early 1900s that cattle and domestic sheep overgrazed much of the northwestern United States and probably southwestern Canada (Honess and Frost 1942, Packard 1946, Jones 1950, Foss 1960, McColm 1963). Heavy grazing in northwestern Mexico and the southwestern United States occurred in the early 1800s (Fradkin 1979). Bighorn sheep declined concurrent with peak cattle numbers in Arizona and Nevada (Gallizioli 1977, McQuivey 1978). Similar trends were reported in Wyoming (McCann 1956). Other factors contributing to bighorn sheep reduction included habitat loss and human disturbances (e.g., dam, canal, fence, and road construction; mining; logging; urban expansion; extensive off-road vehicle use; unregulated hunting during the 1800s and early 1900s; unregulated outdoor recreation; introduction of exotic ungulates; competition with mule deer and elk; mineral, oil, and gas exploration; and the usurpation of water resources for growing agricultural and domestic demands).

Bighorn sheep clearly face a precarious future. They are an ecologically fragile species, adapted to limited habitats that are increasingly fragmented. The remainder of this chapter explores the relationship of bighorn sheep throughout their range with livestock (e.g., cattle and domestic sheep) and reviews recommendations to enhance the survival of bighorn sheep. Three areas of conflict include interactions with cattle, interactions with domestic sheep, and disease transmitted from livestock to bighorn sheep.

INTERACTIONS WITH CATTLE

Cole's (1958) criteria for competition includes the extent that ≥2 species use the same area, extent that ≥2 species use the same forage species, and extent that important plants were in limited supply or deteriorating in production as a result of combined use.

Ranges used by bighorn sheep and cattle usually do not overlap spatially but interactions have been documented (Halloran and Blanchard 1950, Lauer and Peek 1976, King and Workman 1984, Dodd and Brady 1986, Steinkamp 1990). Early reports (Halloran and Blanchard 1950) simply documented the occurrence of both animals but later reports evaluated the relationships between them. Earlier studies of cattle and bighorn sheep interactions (Spencer 1943, Halloran 1949, Couey 1959, Matthews 1960, Arellano 1961) did not demonstrate competition. Habitat preferences for steeper slopes by bighorn sheep and gentler slopes by cattle precluded competition because there was no range overlap. However, Sugden (1961) and Barmore (1962) argued cattle grazing on gentle slopes has precluded the use of those areas by bighorn sheep.

Lauer and Peek (1976) reported bighorn sheep preferred slopes > 30% while cattle preferred slopes ≤ 30% during winter. Both species used grass and grasslikes in > 80% of their diet with bluebunch wheatgrass (*Agropyron spicatun*), the most heavily used species. Both species also used the Wyoming big sagebrush (*Artemisia tridentata wyomingensis*)/bluebunch wheatgrass plant community.

Lauer and Peek (1976) concluded that competition was highest during the initiation of spring growth when cattle and bighorn sheep used bluebunch wheatgrass on low-elevation slopes. Species composition of low-elevation communities was depauperate. Individual plant species were impacted by past use and were adjacent to cover making them especially vulnerable to grazing. Blood (1961) examined competition between cattle and bighorn sheep in Canada where 70% of bighorn sheep winter range was used by cattle. As in Idaho, bluebunch wheatgrass was common in both diets and cattle removed ≥ 35% of vegetation on bighorn sheep range. Blood (1961) concluded that cattle grazing prevented increases in the bighorn sheep population.

King and Workman (1984) reported different associations in southeastern Utah. They found bighorn sheep in higher, steeper, and more rugged talus slopes than cattle that selected lower, gentler slopes and valleys close to roads and developed water sources. In addition, diets of the ungulates were different; cattle diets were dominated by grass but bighorn sheep were browsers. King and Workman (1984) did not demonstrate that cattle and bighorn sheep competed for space or resources; however, they argued that the spatial separation they observed may be from a "social intolerance—avoidance factor." McCann (1956), Barmore (1962), McCullough and Schneegas 1966, Follows 1969, Albrechtsen and Reese 1970, Ferrier and Bradley (1970), Dean (1975), Wilson (1975), Gallizioli (1977), and Van Dyke et al. (1986) argue that bighorn sheep avoid areas used by cattle. Steinkamp (1990) and J. A. Bissonette and M. E. Steinkamp (Utah State Univ., unpubl. data) demonstrated that a translocated population of bighorn sheep clearly avoided cattle. As cattle moved into core areas used by bighorn sheep, sheep moved away. Additionally, the closer cattle grazed, the closer that sheep remained near escape cover. J. A. Bissonette and M. E. Steinkamp (unpubl. data) conducted a field study during August 1989, whereby cattle were moved to within 800 m of sheep. Sheep response was dramatic; they immediately vacated the area and moved >3 km, or 355% further than their mean daily movements during that period.

Social intolerance (Geist 1971) can have serious implications because cattle now graze most rangelands that historically supported bighorn sheep (Mackie 1978); 70% of public lands in the 11 most contiguous western states are grazed at least seasonally (United States Department of the Interior 1986). Livestock grazing, even seasonally, appears to result in habitat fragmentation (Temple 1984; J. A. Bissonette and M. E. Steinkamp, unpubl. data) resulting in the exclusion of sheep from what is otherwise acceptable habitat. Steinkamp (1990) and J. A. Bissonette and M. E. Steinkamp (unpubl. data) demonstrated that social intolerance can be a potent force influencing habitat use by sheep. Their results pertain, however, to newly translocated groups into unoccupied habitat. Whether social intolerance between cattle and bighorn sheep is universal remains in dispute. Resolution of the dispute is clouded by the almost universal disregard for spatial scale. For example, Dodd and Smith (1988) reported an 80% overlap in the distribution of bighorn sheep and cattle in Aravaipa Canyon, Arizona. From their paper, it is clear that they were considering overlap at the level of the home range. Considerations of spatial scale are relevant here. When viewed at a smaller scale (i.e., the level of specific slopes and valley areas), bighorn sheep selected steep slopes with open vegetational associations on southerly aspects while cattle selected level terrain. Dodd and Smith's (1988) finding of no indication of "social intolerance for or avoidance of cattle" by bighorn sheep is defendable only if the scale is explicitly stated. Their conclusion that "It can not be conclusively stated whether bighorn habitat use patterns at Aravaipa reflect solely habitat selection on the part of bighorn or that observed spatial segregation with cattle may represent a subtle form of avoidance," reflects the overt lack of attention to scale that is required to resolve the argument. Lack of consideration of scale effects can have profound implications for management. For example, in 1988-89, the bighorn sheep population in Aravaipa Canyon, Arizona, was reduced by 52%. Mouton et al. (1991) examined the causes of mortality and concluded they were ". . . probably the result of livestock related viral diseases compounded by nutritional stress." Because range overlap has been documented to result in sheep mortality by disease transmission, determination of overlap and the scale at which it occurs is most important. Overlap at the level of home ranges may have very different consequences from overlap on specific slopes or valley floor areas. Additionally, temporal overlap at different scales (e.g., seasonal, annual) would appear to have important ramifications to management.

In other areas of the Southwest grazing by cattle has damaged bighorn sheep habitats (Russo 1956, Gordon 1957, McColm 1963, Riegelhuth 1965, Gallizioli 1977). A recent report by the General Accounting Office (cited by Williamson 1992) "claims that livestock grazing on desert land in the Southwest causes excessive damage to wild plants and animals and is a threat to endangered species." This conclusion supports earlier speculation that grazing of cattle on hot deserts creates long-term damage to highly fragile resources. Low precipitation levels ensure that recovery of ranges will take many years (Jones 1950, Russo 1956) and in some areas damage from livestock grazing may be irreversible. Grazing by cattle has also influenced bighorn sheep habitat in less arid areas (Jones 1950, Buechner 1960, Crump 1971, Geist 1971, Brown 1974) by converting grasslands to shrublands (Demarchi 1970, Morgan 1970).

Bighorn sheep do not do well when they share ranges with cattle. Following the population declines of the late 1800s and early 1900s, bighorn sheep populations did not recover as well as other native ungulates (e.g., mule deer [*Odocoileus hemionus*], elk [*Cervus elaphus*]). Bighorn sheep are not as tolerant as other native North American ungulates to poor range conditions, intraspecific competition, overhunting, and habitat alteration. In addition they are much more susceptible to diseases of livestock than other rangeland wildlife, especially diseases of domestic sheep.

DOMESTIC SHEEP, DISEASES, AND BIGHORN SHEEP

Bighorn sheep are susceptible to many diseases of cattle and domestic sheep. Diseases of cattle that influence bighorn sheep are poorly documented but diseases from domestic sheep have played an important role in bighorn sheep mortality. Throughout the western United States, dieoffs of bighorn sheep and population declines have occurred following the introduction of domestic sheep. Mortality was the result of competition for forage and space and shared diseases (Goodson 1982). According to Goodson (1982), "Co-use of ranges by domestic and bighorn sheep has been consistently linked with declines, dieoffs, and extinctions of bighorn populations from historic to recent times. While much of the evidence for competition between domestic sheep and bighorn is circumstantial, it is sufficiently strong to have prompted management decisions against co-use of ranges by bighorn and domestic sheep by federal land management agencies and state wildlife departments." The Technical Staff of the Desert Bighorn Council (1990) reviewed 24 interactions between bighorn sheep and domestic sheep. Bighorn sheep died as a result of the interactions (Table 1). Recent experimental studies confirmed field observations; when bighorn sheep are exposed to domestic sheep, bighorns die from *Pasteurella haemolytica* (Foreyt 1989, 1990, 1992; Silflow et al. 1993; Foreyt et al. 1994).

The actual mechanisms that kill bighorn sheep after they come in contact with domestic sheep are poorly documented (Jessup 1985) but 2 trends appear clear (Technical Staff Desert Bighorn Council 1990): (1) a large proportion of the bighorn sheep population dies, and (2) domestic sheep do not suffer ill effects because of their contact with bighorn sheep. Bighorn sheep are more susceptible to diseases they share with livestock. Domestic animals have been selectively bred for disease resistance but bighorn sheep have not evolved with the complement of diseases they are now exposed to in the presence of domestic stock. As a result they have not developed effective immunity against these diseases. Silflow et al. (1991) examined lung metabolism between domestic and bighorn sheep and concluded that they had different control mechanisms for lung metabolism, and differences in the metabolites released lead to different regulation of lung defense mechanisms.

Disease

Biologists are not aware of all the factors creating negative interactions between domestic stock and bighorn sheep but scabies, chronic frontal sinusitis, nematode parasites, pneumophilic bacteria, footrot, parainfluenze-III, bluetongue, soremouth, paratuberculosis, and pink-eye are documented decimating factors. Each of these was summarized by Jessup (1985).

Scabies. Scabies is a condition caused by mite (*Psoroptes* spp.) infestation and was not observed in bighorn sheep prior to the introduction of domestic sheep (Buechner 1960). However, with the introduction of domestic livestock onto bighorn sheep ranges in the late 1800s and early 1900s, widespread declines of bighorn sheep were attributed to scabies (Honess and Frost 1942, Couey 1959, Bear and Jones 1973, Lange 1980). This pattern was common throughout the west (Packard 1946, Couey 1959, Buechner 1960, Dean 1977). Scabies has been greatly reduced in domestic livestock, but remains in bighorn sheep populations.

The most recent outbreak occurred in 1978 in San Andres National Wildlife Refuge, New Mexico. When scabies was dis-

TABLE 1. Bighorn sheep declines and die-offs resulting from contacts with domestic sheep.

Location	Cause of die-off	Results	Year(s)	Source
Sun River, Mont.		>70 died	1910-1935	Goodson (1982)
Upper Rock Creek, Mont.		All died	1965-1970s	Goodson (1982)
Thompson Falls, Mont.		All died	1940-1960	Goodson (1982)
Kootenay National Park, B.C., Can.	Pneumonia	Unspecified mortality (UM)	1939	Goodson (1982)
Bull River, B.C., Can.	Pneumonia	96% died	1965	Bandy (1968) in Goodson (1982)[a]
MacQuire Creek, B.C., Can.	Pneumonia	UM	1981-1982	Davidson in Goodson (1982)[a]
Lava Beds National Monument, Calif.[b]	Pneumonia	All died	1980	Blaisdell (1982)
Mormon Mts., Nev.	Pneumonia	50% died	1980	Jessup (1981)
Dinosaur National Monument, Colo.		All died	1950	Barmore (1962) in Goodson (1982)
Rock Creek, Mont.		8 left	1900-1920	Goodson (1982)
Rocky Mtn. National Park, Colo.	Pneumonia	All died	1917-1930	Packard (1939*a*, 1939*b*)[a]
Methow Game Range, Wash.[b]	Pneumonia	13 of 14 died	1979-1981	Foreyt and Jessup (1982)
Warner Mts., Calif.	Pneumonia	All died	1988	Weaver and Clark (1988)
Oregon	Scabies	UM	1936	Lange (1980)
California	Scabies	UM	1870-1879, 1898	Jones (1950)
Grey Bull River, Wyo.			1881	Honess and Frost (1942)
Wyo., Mont.			1885	Hornaday (1901)
Colo.	Scabies	UM	1859-1931	Packard (1946)
Rocky Mtn. National Park, Colo.	Scabies	UM	1878-1903	Lange (1980)
Latir Parks, N.M.	Pneumonia	All died	1978-1982	Sandoval (1988)
Utah State Univ., Logan[c]	Pneumonia	All died	1970s	Spillett in Goodson (1982)[c]
Univ. B.C., Can.[c]	Pneumonia	All died	1970s	Hebert in Goodson (1982)[a]
Colorado St. Univ., Fort Collins[c]	Pneumonia	All died	1970s	Hibler in Goodson (1982)
Utah St. Univ., Logan[c]	Pneumonia	4 of 5 died	1988	T. D. Bunch (Utah State Univ., Pers. commun.)
Washington St. Univ., Pullman[c]	Pneumonia	All died	1987	Foreyt (1989)
Washington St. Univ., Pullman[c]	Pneumonia	All died	1987	Foreyt (1990)

[a] Unpublished report or pers. commun.
[b] Large pen or paddock.
[c] University controlled conditions.

covered in 1978 there were ≥150 bighorn sheep on the refuge. The population declined to <30 by 1982 and has not recovered (Hoban 1990). Scabies was present in all animals during the epizootic but the actual cause of mortality for many individuals was not definitely determined. In addition, attempts to understand mite/host relationships have been inconclusive (Kinzer et al. 1983), which further compounds our understanding of the relationship mites have with bighorn sheep.

Chronic Frontal Sinusitis. Chronic frontal sinusitis is an important cause of mortality in bighorn sheep, especially desert races. The disease is initiated primarily from bacterial decomposition of necrotic nasal bot fly larvae (*Oestrus ovis*) that die from entrapment in the paranasal sinuses (Bunch et al. 1978*a*,*b*, 1985; Paul and Bunch 1978; Allen and Bunch 1982). Progressive debilitation often results in weight loss (≤ 50%), draining lesions on the forehead, and blindness. Little is known about the effects of bot flies in bighorn sheep. Nasal bots are considered minor annoyances of domestic sheep but the sinusitis and subsequent osteonecrosis in desert races of bighorn sheep appear to be a major mortality factor that may have led to the demise of some populations and restricted population increases in others (Bunch et al. 1978*a*,*b*).

Nematode Parasites. Abomasal worms (*Hemonchus*, *Ostertagia*, *Trichostronglyus*), small intestinal worms (*Nematodirus*, *Cooperia*, *Stronggyloides*, *Bunostomum*), and large intestinal worms (*Oesophagostomum*, *Chabertia*, *Trichuris*) are just a few of the internal parasites shared by domestic sheep and bighorn sheep (Jessup 1985). They influence bighorn sheep, usually younger age classes, causing direct blood loss and by contributing to malnutrition. These problems are especially magnified when mountain and domestic sheep graze the same pastures.

Lungworms (*Protostrongylus stelesi*), especially during periods of stress, have been a principal cause of death for Rocky Mountain bighorn sheep (Wishart 1978). This nematode parasite appears to be unique to bighorn sheep (Jessup 1985) while the less common and less pathogenic *P. rushei* is common to mountain and domestic sheep.

Pneumonic Bacteria. When bighorn sheep are held in captivity near domestic sheep, the latter apparently transmit pneumonic bacteria (*Pasteurella multocida*, Foreyt and Jessup 1982, Jessup 1981; and *P. haemolytica*, Foreyt 1989, 1990; Silflow et al. 1993; Foreyt et al. 1994) to bighorn sheep resulting in mortality. In cases where nose-to-nose contact was made with domestic sheep, bighorn sheep died within weeks. This phenomenon has been reported in California (Foreyt and Jessup 1982), Utah, Colorado, Canada, and Wyoming (Jessup 1985) and clinically demonstrated in Washington (Foreyt 1989, 1990; Silflow et al. 1993; Foreyt et al. 1994). Pneumonic bacteria (*P. hemolytica*) causes fatal pneumonia in domestic sheep and cattle and can be a serious source of mortality in wild bighorn sheep.

Viral Diseases. Parainfluenza—III (PI-3), bluetongue and soremouth are only 3 viral diseases that can lead to mortality in bighorn sheep. All 3 are also related to domestic stock; PI-3 is associated with shipping fever and feedlot pneumonias of cattle and is stress related. Bluetongue is a cosmopolitan virus that has been isolated from domestic sheep, goats, cattle, deer, elk, pronghorn (*Antilocapra americana*), and bighorn sheep (Jessup 1985). Bluetongue is transmitted by gnats (*Culicoides* spp.) associated with water in arid areas. Soremouth also is related to stress and can be transmitted from domestic sheep to bighorn sheep. This virus is not as deadly as others but may stunt growth of wild lambs.

Other diseases (e.g., paratuberculosis or Johne's disease, pink-eye, and a *Chylamydia* spp.) have also caused mortalities in bighorn sheep (Jessup 1985). Bighorn sheep did not evolve with these and other diseases of European livestock and they are not "natural" to bighorn sheep. Jessup (1985) stated that bighorn sheep ". . . are possibly the most exquisitely sensitive North American wild ungulate to common livestock diseases and parasites." Because resource managers are not familiar with all of the mechanisms involved in disease transmission and treatment, Jessup (1985) further stated that the "best management strategy is to maintain bighorn herds at optimal nutritional planes, at or below carrying capacity and as widely separated as possible from domestic livestock." Bighorn sheep and livestock do not co-exist well together.

RECOMMENDATIONS TO MINIMIZE MORTALITY OF BIGHORN SHEEP ON RANGELANDS SHARED WITH LIVESTOCK

Because of the fragmentation of bighorn sheep habitat by livestock (Steinkamp 1990; J. A. Bisssonette and M. E. Steinkamp, unpubl. data), social intolerance (Geist 1971), and disease transmission (Jessup 1985), most researchers argue that livestock should not be grazed in bighorn sheep habitat.

Lauer and Peek (1976) proposed a 2-fold approach: modify the livestock grazing system to minimize competition with bighorn sheep and improve the range condition of critical habitats. This supports Jessup's (1985) recommendation to keep livestock and bighorn sheep apart and maintain nutritional quality in the diet of the latter. To minimize avoidance of livestock by bighorn sheep and, hence, avoidance of habitat, livestock and bighorn sheep should not be close to each other (Steinkamp 1990; J. A. Bisssonette and M. E. Steinkamp, unpubl. data). When separation is not possible, efforts should be made to minimize contact (e.g., placement of waters, fencing critical areas), monitor distribution, monitor range conditions, and carefully watch for incidences of disease (McCullough 1982).

Goodson (1982) stated: "On ranges where bighorn sheep are considered an important resource, domestic sheep should not be introduced. Where domestic sheep are currently grazed on bighorn ranges, reduction or elimination of such use is recommended if enhancement of bighorn status is a management goal."

The Technical Staff of the Desert Bighorn Council (1990) recently evaluated problems that arise when there is contact between healthy bighorn sheep and domestic sheep and concluded (from 25 cases [Table 1]) that (1) "There is a preponderance of evidence strongly linking the presence of domestic sheep with subsequent loss of part of all of the affected bighorn population," and (2) "The effects have all been *1 way*—bighorns have died, while domestic sheep never have suffered ill effects because of coming into contact with bighorn." To minimize mortality of bighorn sheep due to contact with domestic sheep the Technical Staff of the Desert Bighorn Council (1990) made 5 recommendations.

1. Domestic sheep in the vicinity of desert races of bighorn sheep habitat should be managed so that bighorn sheep never come in contact with domestic sheep or the disease organisms that the latter carry.
2. Unless topographic features or other barriers prevent interaction between domestic and bighorn sheep, domestic sheep should be excluded from a buffer strip ≥13.5 km wide that surrounds bighorn sheep habitat.
3. Domestic sheep that are trailed or grazed outside the 13.5 km buffer strip but in the vicinity of bighorn sheep range should be closely supervised by competent, capable, and informed herders.
4. Domestic sheep should be trucked to ranges instead of trailing when the latter would bring them closer than 13.5 km to bighorn sheep range. When domestic ewes are in estrus, trailing should never occur.
5. Reintroductions of bighorn sheep should be avoided in areas where grazing has occurred by domestic sheep within

4 years. "The Tech Staff does *not* advocate the co-use of bighorn habitat by both bighorn and domestic sheep."

SUMMARY

Bighorn sheep have coexisted with humans for ≥30,000 years but now face a precarious future. They are an ecologically fragile species, adapted to limited habitats that are increasingly fragmented. Fragmentation of habitats increases when cattle share the same rangelands as bighorn sheep. Domestic sheep pose an even greater threat to bighorn sheep. When these species share the same rangelands, all or most of the bighorn sheep die and domestic sheep do not suffer ill effects because of their contact with bighorn sheep. Common diseases transmitted to bighorn sheep from livestock include scabies, chronic frontal sinusitis, nematode parasites, pneumophilic bacteria, footrot, parainfluenze-III, bluetongue, soremouth, paratuberculosis, and pinkeye. Because of the fragmentation of bighorn sheep habitat by livestock and disease transmission, livestock should be kept off bighorn sheep habitat. The Technical Staff of the Desert Bighorn Council made recommendations to minimize the problem, all of which avoid contact between livestock and bighorn sheep.

LITERATURE CITED

Albrechtsen, B. R., and J. B. Reese. 1970. Problem analysis of habitat management for desert bighorn sheep. Desert Bighorn Counc. Transactions 14:63-65.

Allen, S. D., and T. D. Bunch. 1982. Chronic sinusitis induced cranial lesions in bighorn sheep (*Ovis canadensis*). J. Am. Vet. Med. Assoc. 181:1418-1419.

Arellano, L. M. 1961. Bighorn management in Mexico. Desert Bighorn Counc. Transactions 5:51-52.

Barmore, W. J. 1962. Bighorn sheep and their habitat in Dinosaur National Monument. M.S. Thesis, Utah State Univ., Logan. 134pp.

Bear, G. D., and G. W. Jones. 1973. History and distribution of bighorn sheep in Colorado. Colo. Div. Wildl. Publ. 232pp.

Blaisdell, J. A. 1982. Lava beds wrap-up — what did we learn. Desert Bighorn Counc. Transactions 26:32-33.

Blood, D. A. 1961. An ecological study of California bighorn sheep in southern British Columbia. M.S. Thesis, Univ. British Columbia, Vancouver. 127pp.

Brown, G. W. 1974. Distribution and population characteristics of bighorn sheep near Thompson Falls in northwestern Montana. M.S. Thesis, Univ. Montana, Missoula. 134pp.

Buechner, H. K. 1960. The bighorn sheep in the United States: its past, present and future. Wildl. Monogr. 4. 174pp.

Bunch, T. D., S. R. Paul, and H. E. McCutchen. 1978*a*. Chronic sinusitis and osteonecrosis in desert bighorn sheep (*Ovis canadensis nelsoni*). North Am. Wild Sheep Symposium 3:261-273.

———, ———, and ———. 1978*b*. Chronic sinusitis in desert bighorn sheep (*Ovis canadensis nelsoni*). Desert Bighorn Counc. Transactions 22:16-20.

———, G. Welsh, and R. L. Glaze. 1985. Chronic sinusitis in desert bighorn sheep in northwestern Arizona. Desert Bighorn Counc. Transactions 29:1-3.

Cole, G. 1958. Big game-livestock competition on Montana's mountain rangelands. Montana Wildl. April:24-30.

Couey, F. M. 1959. Rocky Mountain bighorn sheep of Montana. Montana Fish and Game Comm. Bull. 2. 90pp.

Crump, W. 1971. The Wind River bighorn herd—a new approach to sheep habitat management. North Am. Wild Sheep Conf. 1:174-181.

Dean, H. C. 1975. Bighorn investigation in Canyonlands National Park. Desert Bighorn Counc. Transactions 19:7-11.

———. 1977. Desert bighorn in Canyonlands National Park. M.S. Thesis, Utah State Univ., Logan. 86pp.

Demarchi, D. A. 1970. Bighorn sheep and overgrazing in the lower Chilcotin River region, British Columbia. Trans. North. Wild Sheep Counc. 2:39-45.

Dodd, N. L., and W. W. Brady. 1986. Cattle grazing influence on vegetation of a sympatric desert bighorn range in Arizona. Desert Bighorn Counc. Transactions 31:8-13.

———, and R. H. Smith. 1988. Habitat and spatial relationships of sympatric desert bighorn sheep and cattle. Ariz. Game and Fish Dep. Fed. Aid Wildl. Restor. Proj. W-78-R. Final Rep. 25pp.

Ferrier, G. J., and W. G. Bradley. 1970. Bighorn habitat evaluation in the highland range of southern Nevada. Desert Bighorn Counc. Transactions 14:66-93.

Follows, D. S. 1969. Desert bighorn in Canyonlands National Park, Utah. Desert Bighorn Counc. Transactions 13:33-42.

Foreyt, W. J. 1989. Fatal *Pasteurella haemolytica* pneumonia in bighorn sheep after direct contact with clinically normal domestic sheep. Am. J. Vet. Res. 50:341-344.

———. 1990. Pneumonia in bighorn sheep: effects of *Pasteurella haemolytica* from domestic sheep and effects on survival and long-term reproduction. Bienn. Symposium North. Wild Sheep and Goat Counc. 7:92-101.

———. 1992. Failure of an experimental *Pasteurella haemolytica* vaccine to prevent respiratory disease and death in bighorn sheep after exposure to domestic sheep. Bienn. Symposium North. Wild Sheep and Goat Counc. 8:155-163.

———, and D. A. Jessup. 1982. Fatal pneumonia of bighorn sheep following association with domestic sheep. J. Wildl. Dis. 18:163-168.

———, K. P. Snipes, and R. W. Kasten. 1994. Fatal pneumonia following inoculation of healthy bighorn sheep with *Pasteurella haemolytica* from healthy domestic sheep. J. Wildl. Dis. 30:137-145.

Foss, P. O. 1960. Politics and grass. Univ. Washington Press, Seattle. 236pp.

Fradkin, D. L. 1979. Eating of the west. Audubon 81:94, 121.

Gallizioli, S. 1977. Overgrazing of desert bighorn ranges. Desert Bighorn Counc. Transactions 21:21-23.

Geist, V. 1971. Mountain sheep: a study in behavior and evolution. Univ. Chicago Press, Chicago, Ill. 388pp.

Goodson, N. J. 1982. Effects of domestic sheep grazing on bighorn sheep populations: a review. North. Wild Sheep and Goat Counc. 3:287-313.

Gordon, S. P. 1957. The status of the bighorn sheep in New Mexico. Desert Bighorn Counc. Transactions 1:3-4.

Halloran, A. F. 1949. Desert bighorn management. Trans. North Amer. Wildl. Conf. 14:527-536.

———, and W. W. Blanchard. 1950. Bighorn ewe associates with cattle on Kofa Game Range, Arizona. J. Mammal. 31:463-464.

Hoban, P. A. 1990. A review of desert bighorn sheep in the San Andres Mountains, New Mexico. Desert Bighorn Counc. Transactions 34:14-22.

Honess, R. F., and N. M. Frost. 1942. A Wyoming bighorn sheep study. Wyoming Game and Fish Comm. Bull. 1. 127pp.

Hopkins, D. M. 1967. The Cenozoic history of Beringia—a synthesis. Pages 451-483 *in* D. M. Hopkins, ed. The Bering land bridge. Stanford Univ. Press, Palo Alto, Calif.

Hornaday, W. T. 1901. Notes on the mountain sheep of North America with a description of a new species. Annu. Rep. N. Y. Zool. Soc. 5:77-122.

Jessup, D. A. 1981. Pneumonia in bighorn sheep: effects on populations. Cal-Neva Wildl. Soc. Transactions 72-78pp.

———. 1985. Diseases of domestic livestock which threaten bighorn sheep populations. Desert Bighorn Counc. Transactions 29:29-33.

Jones, F. L. 1950. A survey of the Sierra Nevada bighorn. Sierra Club Bull. 35:29-76.

King, M. M., and G. M. Workman. 1984. Cattle grazing in desert bighorn sheep habitat. Desert Bighorn Counc. Transactions 28:18-22.

Kinzer, H. G., W. E. Houghton, and J. M. Reeves. 1983. *Psoroptes ovis* research with bighorn sheep in New Mexico. Desert Bighorn Counc. Transactions 27:6-8.

Krausman, P. R., S. Torres, L. L. Ordway, J. J. Hervert, and M. Brown. 1985. Diel activity of ewes in the Little Harquahala Mountains, Arizona. Desert Bighorn Counc. Transactions 29:24-26.

Lange, R. E. 1980. Psoroptic scabies in the United States and Canada. Desert Bighorn Counc. Transactions 24:18-20.

Lauer, J. L., and J. M. Peek. 1976. Big game—livestock relationships on the bighorn sheep winter range, east fork, Salmon River, Idaho. U.S. Bur. Land Manage. Contract No. 52500-CT5-106. 44pp.
Mackie, R. J. 1978. Impacts of livestock grazing on wild ungulates. Trans. North Am. Wildl. and Nat. Resour. Conf. 43:462-476.
Matthews, W. L. 1960. Management responsibilities and practices of the BLM as they relate to the desert bighorn sheep. Desert Bighorn Counc. Transactions 4:26-27.
McCann, L. J. 1956. Ecology of the bighorn sheep. Am. Midl. Nat. 56:297-325.
McColm, M. A. 1963. A history of the bighorn sheep in central Nevada. Desert Bighorn Counc. Transactions 7:1-11.
McCullough, D. R., and E. R. Schneegas. 1966. Winter observations on the Sierra Nevada bighorn sheep. Calif. Fish and Game 52:68-84.
McCullough, S. A. 1982. Impact of cattle grazing on bighorn sheep, Trickle Mountain, Colorado. M.S. Thesis, Colorado State Univ., Fort Collins. 119pp.
McQuivey, R. P. 1978. The desert bighorn sheep of Nevada. Nevada Dep. Fish and Game Bio. Bull. 6. 81pp.
Morgan, J. K. 1970. Ecology of the Morgan Creek and East Fork of the Salmon River bighorn sheep herds and management of bighorn sheep in Idaho. Idaho Game and Fish Dep. Fed. Aid Wildl. Restor. Proj. W-142-R-1. Final Rep. 155pp.
Mouton, R. J., R. M. Lee, and R. J. Olding. 1991. A desert bighorn sheep decline in Aravaipa Canyon, Arizona. Desert Bighorn Counc. Transactions 35:18-19.
Nichols, L. 1975. Sheep report. Alas. Dep. Fish and Game Fed. Aid Wildl. Restor. Annu. Proj. Rep. Vol. 16.
Packard, F. M. 1946. An ecological study of the bighorn sheep in Rocky Mountain National Park. J. Mammal. 27:3-28.
Paul, S. R., and T. D. Bunch. 1978. Chronic frontal sinusitis and osteolysis in Nelson's desert bighorn sheep (*Ovis canadensis nelsoni*). J. Am. Vet. Med. Assoc. 173:1178-1180.
Riegelhuth, R. 1965. A reconnaissance of Sierra bighorn and bighorn ranges in Sierra Nevada. Desert Bighorn Counc. Transactions 9:35-39.
Risenhoover, K. L., and J. A. Bailey. 1985. Foraging ecology of mountain sheep: implications for habitat management. J. Wildl. Manage. 49:979-804.
Russo, J. P. 1956. The desert bighorn sheep in Arizona. Ariz. Game and Fish Dep. Wildl. Bull. 1. 153pp.
Sandoval, A. V. 1988. Bighorn sheep die-off following association with domestic sheep: case history. Desert Bighorn Counc. Transactions 32:36-38.
Seton, E. T. 1929. Lives of game animals. Vol. III. Doubleday, Page, and Co., New York, N.Y. 780pp.
Silflow, R. M., W. J. Foreyt, and R. W. Leid. 1993. *Pasteurella haemolytica* cytotoxin-dependent killing of neutrophils from bighorn and domestic sheep. J. Wildl. Dis. 29:30-35.
———, S. M. Taylor, W. W. Laegreid, H. D. Liggitt, and R. W. Leid. 1991. Comparison of arachidonate metabolism by alveolar macrophages from bighorn and domestic sheep. Inflammation 15:43-54.
Spencer, C. C. 1943. Notes on the life history of the Rocky Mountain bighorn sheep in the Tarryall Mountains of Colorado. J. Mammal. 24:1-11.
Steinkamp, M. E. 1990. The effect of seasonal cattle grazing on California bighorn sheep habitat use. M.S. Thesis, Utah State Univ., Logan. 57pp.
Sugden, L. G. 1961. The California bighorn in British Columbia with particular reference to the Churn Creek herd. British Columbia Dep. of Rec. and Conserv. 58pp.
Technical Staff of the Desert Bighorn Council. 1990. Guidelines for management of domestic sheep in the vicinity of desert bighorn habitat. Desert Bighorn Counc. Transaction 34:33-35.
Temple, S. A. 1984. Predicting the impacts of habitat fragmentation of forest birds: a comparison of two models. Pages 301-304 *in* J. Verner, M. Morrison, and C. J. Ralph, eds. Wildlife 2000: modeling habitat relationships of terrestrial vertebrates. Univ. Wisconsin Press, Madison.
Turner, J. C. 1979*a*. Adaptive strategies of selective fatty acid deposition in the bone marrow of desert bighorn sheep. Comp. Biochem. Physiol. 62A:599-604.
———. 1979*b*. Osmotic fragility of desert bighorn sheep red blood cells. Comp. Biochem. Physiol. 64A:167-175.
United States Department of the Interior. 1986. Owyhee rangeland program summary. 1981-1986 Prog. Rep. Bur. Land Manage., Boise, Id.
Valdez, R. 1982. The wild sheep of the world. Wild Sheep and Goat Int., Mesilla, N.M. 186pp.
———. 1988. Wild sheep and wild sheep hunters of the New World. Wild Sheep and goat Int., Mesilla, N.M. 270pp.
Van Dyke, W. A., A. Sands, J. Yoakum, A. Poleny, and J. Blaisdell. 1986. Wildlife habitats in managed rangelands—the Great Basin of Southeast Oregon-bighorn sheep. Second ed. Gen. Tech. Rep. PNW-159. 37pp.
Weaver, R. A., and R. K. Clark. 1988. Status of bighorn sheep in California, 1987. Desert Bighorn Counc. Transaction 31:20.
Williamson, L. L., ed. 1992. Desert grazing bad, GAO says. Outdoor News Bull. 46:3-4.
Wilson, L. O. 1975. Report and recommendation of the desert and Mexican bighorn sheep workshop. Pages 110-131 *in* J. B. Trefethan, ed. The wild sheep in modern North America. Winchester Press, New York, N.Y.
Wishart, W. 1978. Bighorn sheep. Pages 161-171 *in* J. L. Schmidt and D.L. Gilbert, eds. Big game of North America. Stackpole Books, Harrisburg, Pa.

Chapter 16
HISTORICAL AND PRESENT IMPACTS OF LIVESTOCK GRAZING ON FISH AND WILDLIFE RESOURCES IN WESTERN RIPARIAN HABITATS

Robert D. Ohmart

INTRODUCTION

What are riparian habitats? How abundant are they? Why should society, and those living in the West, be concerned about their management? How do they function? What is their importance to fish and wildlife? What is their present condition and how can they be better managed? Is domestic livestock grazing contributing to their degradation and, if so, how? Knowledge accumulated on these habitats throughout the West since the turn of the century will provide insight into these questions.

A riparian habitat ". . . is one which occurs in or adjacent to drainageways and/or their floodplains and which is further characterized by species and/or life-forms different from that of the immediately surrounding non-riparian climax" (Lowe 1964:62). In desert areas the transition from verdant floodplain vegetation with an underlying water table to the dry upland supporting creosote bush (*Larrea divaricata*) or big sage (*Artemisia tridentata*) may be less than a meter. Even at elevations where spruce (*Picea* spp.) or grasses dominate the uplands, the transition from the deciduous woodland trees and shrubs to upland species is usually quite distinct.

Soil moisture availability in the floodplain and water table depth are important elements that shape the quantity and quality of the vegetation over the riparian floodplain. The 1-3-year flood event marks a rather distinct boundary between the upland and floodplain vegetation. Upland vegetation is limited from encroaching onto the floodplain by the high water table and prolonged saturated soils from overbank watering during floods. Conversely, the lack of available soil moisture prevents floodplain vegetation from invading the upland.

Studies in hydrology, soils, and geology amply demonstrate that when considered as ecological units, riparian habitats cannot be separated from their watersheds (Hays 1975, Gregory et al. 1991). Therefore, proper management of riparian resources will only be achieved when best management decisions are made for entire watersheds. Management may be applied to individual components of the watershed, but the proper functioning of these ecosystems should always be the ultimate goal. Each species contributes in its own way to the functioning of a healthy ecosystem; the role of some are key or cornerstone to proper functioning of an ecosystem, while others play a more subtle part. Management should always strive to maintain all native species if only for biodiversity and knowledge.

Although there are very important riparian habitats for wildlife along intermittent and ephemeral streams, this chapter will emphasize perennial streams. Intermittent streams have seasonal surface water flow (e.g., from rainfall, snow melt, or water from springs), but there is subterranean flow throughout the year. Ephemeral streams only show surface water as a direct response to heavy rainfall. Though many ephemeral streams are dry most of the time, they have some subsurface flow or seepage that supports a flora different from that of the uplands. There are some drainages with deep alluvium or on bedrock supporting vegetation that may be slightly more robust than the same vegetation (conspecific) in the uplands—these are not riparian habitats.

Because riparian habitats are essentially the upland drainage corridors, they literally are green ribbons anastomosing higher to lower elevations. Though present in all landscapes, their areal or land extent is minuscule when examined as a percent of the total landscape. In California there are 138,800 ha or <0.5% (Sands and Howe 1977). For Arizona, Strong and Bock (1990) estimate 0.5%. On 70,400,000 ha of Bureau of Land Management (BLM) lands there are 40,000 ha of riparian conditions, or <1% (United States Department of the Interior [USDI] 1994). Of the 57,600,000 ha of United States Forest Service (USFS) lands in the West there are 880,000 ha of riparian habitat, or 1%. In the arid Southwest their areal extent is estimated at < 1% (Ohmart and Anderson 1986). At higher elevations with broad, wet meadows they may be ≤ 2% of the landscape, but the mean percent of riparian habitat in the 11 western states may be ≤ 1%.

The importance of riparian areas far exceeds their availability. They are vital to human survival and maintenance of health in the West. They provide drinking water and agricultural waters, which are essential if humans are to persist in the arid West. Functioning riparian systems trap sediments and biodegrade toxic compounds to improve water quality and quantity. The most productive farm lands are on alluvial soils along floodplains. Wilson (1979:82) termed them the "aorta of an ecosystem" because of their importance to the perpetuation of water, fish, wildlife, rangeland, and forest resources. Swanson et al. (1988) termed them one of the most dynamic portions of the landscape.

Gregory et al. (1991) provide an ecosystem perspective modeling spatial and temporal patterns of hydrologic and geomorphic processes, terrestrial plant succession, and aquatic ecosystems in riparian zones. Within the last 3 years numerous articles have appeared on riparian function, structure, energy flow, and landscapes, as well in an excellent and informative book by Malanson (1993).

Streams in riparian habitats may support or provide salmon (*Oncorhynchus* spp.) spawning areas at more northern latitudes, or habitats for the Colorado River squawfish (*Ptychocheilus lucius*) and desert mountain-sucker (*Pantosteus clarki*) in the arid Southwest. Birds reach higher breeding densities in these habitats than any other habitat in the contiguous United States (Carothers et al. 1974), and 60-70% of the total species of western birds are dependent on this habitat to survive and breed. Another 15% use riparian habitat some time in their annual cycle. Knopf (1985) reported 82% of the bird species in northern Colorado were found in riparian areas. Neotropical migrant birds, species that breed in the United States and winter in Mexico, Central America, and South America, make up 60-70% of western riparian breeding birds (Bock et al. 1993). In the Great Basin of southeastern Oregon, of 363 terrestrial species 288 (80%) are directly dependent on them or use them more than other habitats (Thomas et al. 1979). In addition, numerous species of small mammals, amphibians, and reptiles are totally dependent on them. Cross (1985) reported greater species richness and numbers of small mammals in riparian versus upland habitats in Oregon. Large mammals such as deer (*Odocoileus hemonius*, *O. virginianus*), elk (*Cervus elaphus*), desert bighorn sheep (*Ovis canadensis mexicana*), and others use them for water, thermal and hiding cover, as travel corridors, and as a forage source (Thomas et al. 1979, Seegmiller and Ohmart 1981, Krausman et al. 1985).

WHAT IS A PROPERLY FUNCTIONING RIPARIAN SYSTEM?

A recent document (BLM 1993) provides guidelines for assessing proper functioning conditions (PFC) of riparian systems. The PFC are:

> . . . when adequate vegetation, landform, or large woody debris is present to dissipate stream energy associated with high waterflows, thereby reducing erosion and improving water quality; filter sediment, capture bedload, and aid floodplain development; improve flood-water retention and ground-water recharge; develop root masses that stabilize streambanks against cutting action; develop diverse ponding and channel characteristics to provide the habitat and the water depth, duration, and temperature necessary for fish production, waterfowl breeding, and other uses; and support greater biodiversity. **The functioning condition of riparian-wetland areas is a result of interaction among geology, soil, water, and vegetation** (BLM 1993:4).

Another riparian evaluation assessment has a 3-level assessment with the latter being indepth and quantitative (USFS 1992). The final segment covers management implications and interpretations. The review of either of these evaluations provides a good understanding of PFC of riparian habitats.

Photographs of Parley's Fork (Fig. 1) and Red Butte Creek (Fig. 2) in central Utah provide a visual image of streams in PFC. Red Butte Creek has a steeper gradient than the theoretical stream discussed below.

If one assumes an idealized stream running over relatively thick alluvial soils (4.6-6.1 m above bedrock), a moderate or low gradient of 0.4-0.6%, and a floodplain width of 60-90 m on each side of a stream, the channel and vegetation can be generally described. The stream channel itself should be shaped either as a pipe or a U with the vertical legs of the U bent at their tops to provide slightly or well-formed overhanging banks. Overhanging banks exist when a dense root mass provides a physical barrier to the effects of stream velocity and turbulence, creating banks with high surface roughness and high stability (Smith 1976). The width of the stream channel and its depth are a function of the normal stream capacity.

To determine a stream's health, one should evaluate a minimum of 4 km to get a sense of: (1) gradient, (2) geology, (3) types of fluvial material being transported by the stream, (4) condition of the banks and channel, (5) bank material, (6) sinuosity, (7) herbaceous species present, (8) health and condition of the floodplain vegetation, and (9) the distribution of age classes of the major trees and shrubs along the floodplain. Channel geomorphology will probably change over 4 km, so note these changes and why they are occurring.

A very important criterion of PFC is, can the stream access its floodplain in the 1-3-year flood event? The width of the active floodplain, which determines the area available to dissipate stream energy and recharge groundwater, can be judged from the distance the more mesophyllic (i.e., water-loving) vegetation grows from the edge of the stream. Can the stream carry its sediment load? If the stream has more sediment than it can carry, there will be lateral and midchannel bar formation and a decrease in pool depth and numbers as sediments are deposited in pools. Is there an appropriate stream width/depth ratio? In general, a stream in PFC should have a low width/depth ratio, being deep and narrow, except in riffle areas where the stream is changing elevation.

The first indication of a degrading or improving state is usually seen in the condition of the banks. If degrading, they will commonly widen as they erode, changing channel shape from

FIGURE 1. Parley's Fork, Red Butte Canyon, Wasatch Mountains, Utah. Stream depth is approx. 0.8 m, exposed width is 0.2 to 0.3 m. Note the lush herbaceous ground cover, overhanging banks, and the vegetation overhanging the stream. Trees and understory are dense. Photograph by R. D. Ohmart, May 1994.

FIGURE 2. Red Butte Canyon, Wasatch Mountains, Utah. Riffle area with dense *Equisetum* spp. Lush herbaceous ground cover and dense shrub-tree understory are typical along the creek. Photograph by R. D. Ohmart, May 1994.

narrow and deep (low width/depth ratio) to shallow and wide (higher width/depth ratio). Road Creek (Fig. 3) shows a point bar, characteristic of sediment overload, on the inside of the bend, and in flood stage the stream is eroding the outer bank. Bear Creek (Fig. 4), a tributary to Road Creek, shows even worse bank instability and erosion. Road Creek is functioning but at risk, while Bear Creek is not in PFC. The above characteristics apply to floodplains with high rock content. Streams on floodplains with low or no rock content will initially downcut; after which, channel widening establishes a new floodplain at a lower level. In either case, the stream begins to lose its ability to access its floodplain and recharge the water table.

The stream channel may be highly meandering or relatively straight as it makes it way through the floodplain, depending on landform steepness and other variables. In a healthy riparian system, streambanks will be stoutly tied together with roots of trees, shrubs, grasses, and sedges. Many of these species are rhizomatous and/or stoloniferous, providing even greater soil stability. A short distance from the stream, woody vegetation in the form of trees, shrubs, or both begins covering the floodplain. At lower elevations, willow (*Salix* spp.) shrubs and/or trees are common along with Fremont cottonwood (*Populus fremontii*) or narrow-leaf cottonwood (*P. angustifolia*). Other tree species such as ash (*Fraxinus* spp.), box elder (*Acer negundo*), sycamore (*Platanus wrightii*), quaking aspen (*Populus tremuloides*), and big-toothed maple (*Acer grandidentatum*), may be common at moderate to higher elevations. At highest elevations and more northern latitudes, shrubby willows may dominate the floodplain with the stream flowing around individual root masses. All of these vegetation elements are vital to fish and wildlife. The aboveground vegetation forms living strata for wildlife and underground roots stabilize soils and reduce sediment transport.

Numerous researchers have emphasized the importance of the combination of the woody roots of trees, shrubs, sedges, and rushes in providing bank stability during flood events (Platts 1981*a*, Beschta and Platts 1986, Elmore and Beschta 1987, Clifton 1989, Elmore 1992). Elmore (1992:443) wrote:

> Riparian vegetation can withstand high velocities of water and still maintain the positive factors of the bank-building processes. The grasses, forbs, rushes, shrubs, and trees produce a variety of fibrous and woody roots that bind and hold soils in place. The woody roots provide physical protection against the hydraulic forces of high flows and allow the fibrous roots to bind the finer particles. This diversity of plant species is much more effective in promoting bank stability than is any single species alone.

Beschta and Platts (1986) similarly reported the importance of the woody and fibrous mix of roots that created high bank stability during flood stage in small streams. Platts et al. (1985) reported that along Big Creek in Utah where there was good bank structure, that in abnormally large floods well-vegetated banks were trapping sediment and actually building better channel banks. These observations corroborate Smith's (1976) finding that there is an inverse relationship between erosion and the percentage of vegetation roots in streambanks. He observed that as the percent of roots increased, bank erosion decreased. Streambanks containing a 5-cm root mix resisted erosion 20,000 times better than nonrooted streambanks.

Well-developed or mature sedge communities may approach the soil-holding capacity of a woody-fibrous root mix in resisting erosion. Manning et al. (1989) measured root length density (total root length) in a *Carex nebraskensis* community on the Sheldon Antelope Refuge in Nevada and found that in 16 cm^3 there were 15 m of root. This root density extended downward for 10 cm before it began to decline, and root depth was measured to 40 cm. The upper roots contained very little soil and these root length densities exceed any measured for any plant community type. This type of root density and depth combined with the tough fleshy leaves overlaying the roots in flood stage creates a formidable barrier to erosion.

FIGURE 3. Road Creek near Challis, Idaho. Large point bar formed on the inside stream bend. New materials are being eroded away on the outside bend in high flows. Photograph by R. D. Ohmart, May 1994.

FIGURE 4. Bear Creek near Challis, Idaho. The stream is entrenched and undercutting it's bank. In the absence of woody-rooted species the fibrous roots of grasses cannot stabilize the bank. The separation of the stream from its floodplain and a lowered water table has allowed sagebrush to invade. Photograph by R. D. Ohmart, May 1994.

The nonwoody vegetation covering the floodplain is referred to as the herbaceous groundcover and serves 3 very important functions during a flood. One is to be smashed over the floodplain soil and repel erosive forces. As the water subsides, the herbaceous vegetation lifts to provide roughness that slows the water and suspended sediments are trapped on the floodplain. Stems of shrubs and trees act similarly. The herbaceous groundcover is also important in creating a boundary layer over the soil to prevent solar heating and moisture from being swept away by wind. The under, mid, and overstory act similarly at a larger scale.

Stromberg et al. (1993) quantified the responses of a riparian floodplain following a 10-year flood event of 368 m^3/sec along the Hassayampa River, central Arizona, in March 1991. Pole-sized cottonwoods suffered 6% mortality in 1991 on the high floodplain, while those lower and closer to the channel had 40% mortality. The 150-200-m-wide floodplain received good overbank watering and a mean of 8 cm of new sediment. Densely vegetated areas received up to 0.5 m of new soil deposits. An abundance of new seedlings of cottonwoods and willows followed the flood along overflow channels and main channel sediment bars.

Natural floods play a vital role in the functioning and health of riparian systems. Normal 1-3-year floods in functioning systems define the stream channel characteristics and are key in maintaining the health and annual productivity of riparian systems. Usually 1 heavy annual flood occurs in late spring or early summer during snow melt. Some systems may experience late summer floods as well. Annual floods in functioning systems irrigate most, if not all, the floodplain and bring in alluvial soils and organic material for soil enrichment. Floodwaters saturate the overbank soils, hastening detrital decomposition releasing new nutrients. The flooding of the overbanks saturates these soils and this water eventually works its way back to the stream. This slow irrigation leaches surface and subsurface salts to the stream and out of the system.

Stream level at normal flow is at or slightly below the level of the floodplain and establishes the level of the groundwater table. In general, a mound of water parallels the edge of the stream as water is forced from the stream by hydrostatic forces between the water and stream channel interface. Along the outside edge of this mound the water table slopes gently downward following topography. The high water table irrigates the roots of the grasses and sedges keeping them constantly inundated. Some trees and shrubs cannot tolerate constant root inundation and grow farther from the stream where their roots are established in the capillary fringe of the water table.

Studies in the area of biogeochemical interactions in riparian areas demonstrate that these habitats are the physical, biological, and chemical links between upland and aquatic environments (Dahm et al. 1987). They serve as phosphorus sinks where the ions are absorbed to clay particles to become trapped as sediment by the floodplain vegetation. The phosphorus ions are then available for plant or bacterial uptake (Cooper et al. 1987). Floodplains in PFC systems have also been reported as important areas for denitrification for maintenance of high water quality (Jacobs and Gilliam 1985). Green and Kauffman (1989) examined oxidation-reduction potentials in riparian zones and demonstrated the importance of nutrient cycling, especially at the land-water interface. They stress the patterning and diversity of vegetation from the stream's edge along this aquatic cline or gradient and how each plant community contributes to high water quality. In an undisturbed watershed in the Sierra Nevada, Rhodes et al. (1985) reported that over 99% of the incoming nitrate-nitrogen was converted to nitrous oxide or elemental nitrogen. Decoupling of the stream and its banks immediately begins degrading water quality. Streams in PFC produce high quality and quantity water and outside influences that alter the soil-water interface seriously impair the functional integrity of the system.

Occasionally a healthy stream will be subjected to a storm event with heavy loss of trees and shrubs. While an unusually heavy flood event may appear to be destructive, it serves to rejuvenate the system. Older or weak, senescent plant communities may be eroded away and banks lost; new sediment beds will be deposited in their place. As riparian plants evolved with floods, they are highly adapted to pioneering into newly deposited soils left as a flood recedes. Many tree and shrub species are rhizomatous and sucker when the roots are hit or abraded by rocks, and most have wind- and water-carried seeds that ripen and are dispersed prior to and during natural floods. These species usually depend upon the presence of new wet sandbars as nursery sites. Many riparian trees and shrubs can also propagate vegetatively; so if young plants are uprooted in 1 area and buried downstream, they root and begin sending up suckers.

STREAM VELOCITY AND EROSIVE FORCE IN FLOOD STAGE

To appreciate the value of riparian vegetation in spreading and slowing bank overflows and reducing flood damage consider the relationship between stream velocity, resistance to flow, and stream gradient. The equation (Chow 1959) for determining water velocity shows an inverse relationship between stream velocity and resistance to flow provided by riparian vegetation. Thus, if the resistance of flow is doubled (increase vegetation), stream velocity is cut in half. The floodwaters are slowed and spread laterally over the floodplain as the vegetation resists flow.

The erosive force or working power of the stream is proportional to the third power of velocity. Therefore, if water velocity over the floodplain is reduced by a factor of 5 the erosive power of water is reduced by 125. These physical relationships highlight the importance of the vegetated streambanks and the ability of the floodplain trees and shrubs to bend and sway but not break in flood events.

If 2 streams (1 with grassed or smooth rock floodplain, the other with dense willows) experienced a flood of about 142 m^3/sec, which is not unusual for a western small order stream, water velocity in the channel and the floodplain of the rocked or grassed stream would be about 11.3 km/hr, while that over the floodplain of shrubs, trees, and willows would be about 2.3 km/hr. The vegetated floodplain reduced velocity approximately 5 times, which means the erosive power of the flood was reduced by 125.

In the above example, the well-vegetated floodplain receives a slow irrigation, a heavy nutrient and sediment load from the slowed water, and little, if any, scouring damage. The flood height would be less, with longer duration and lower intensity.

Newly formed sandbars with elevated water tables are the seedling germination points and nursery areas for cottonwoods, willows, and other riparian species. A slow decline of the water table in a well-vegetated floodplain would promote new seedling germination and establishment. The converse would occur with a rapid decline of the water table in the highly eroded stream.

In the degraded stream with a broad channel and only fibrous roots of grasses holding the soils, there would be erosion and scouring in the 11.3-km/hr-moving flood. The wide channel might be sufficiently large to contain the flood event entirely within its banks. This could undercut banks and possibly scour out mature trees adjacent to the stream. Should channel capacity be insufficient to contain the flood event, the absence of shrubs and herbaceous groundcover would result in scouring and erosion of the floodplain. The overbank flood event would be rapid, possibly manifesting itself in only a few hours (high intensity and short duration), and once terminated the water table would quickly recede to the normal level of the stream, resulting in very little recharge to the water table.

RIPARIAN FUNCTIONS AND FISH AND WILDLIFE VALUES OF RIPARIAN HABITATS

Riparian functions are the ability of the habitat (i.e., biotic and abiotic elements) to provide the ecological needs of a species to fulfill its annual cycle. This includes the suitability of the habitat for breeding and rearing young, foraging, cover (all types), and overwintering. For terrestrial wildlife this also includes quantity and quality of habitat for migratory stopover. The abiotic component (e.g., water temperature, chemistry) may be more limiting for fisheries than terrestrial wildlife, at least in trout populations (Behnke 1992), which are discussed in the next section.

Fishery and wildlife values are extremely high in riparian habitats, and these values are generally expressed as individual densities, species richness (biodiversity), biomass, and number of uncommon or rare species. A frequent expression for birds is number of breeding pairs per unit area. A commonly used value in fisheries is biomass or standing crop. In general, the higher the number the greater the value for fish and wildlife.

Fishes

Fisheries ecologists in the West more quickly determined which riparian elements were more important for fishes than have ecologists for terrestrial species. For example, Platts (1979) summarized fishery needs, while only 2 years earlier the first riparian symposium was held on wildlife (Johnson and Jones 1977). Perhaps the economic importance, recreational value, a more direct sensitivity of fishes to changes in their environment, and declines in population numbers spurred a greater urgency to gather data sets for management purposes. Few terrestrial wildlife species in riparian habitats have economic value or are hunted. Furthermore, at least in trout populations, each person who fishes is, at some crude level, sampling the adult population and as declines occur management is frequently informed loud and strong. A small mammal, reptile, or amphibian species could decline to extirpation and the passage not be noticed.

A wide variety of native fishes occur in the West, ranging from the tiny desert pupfish (*Cyprinodon macularius*) on the valley floor to trout at higher elevations, and anadromous salmonids in the Pacific Northwest. Obviously, fish are restricted to the aquatic portion of the riparian habitat, but these species show a phenomenonally wide evolutionary diversity through this broad range of temperature and oxygen availability extremes.

Trout, as a group, have relatively narrow tolerances to physical and biological changes in stream condition (Platts 1981*a*). For this reason, they are considered excellent indicators of stream condition as they require uncontaminated cool water high in dissolved oxygen and low in suspended sediments (Behnke 1992).

In many instances trout abundance may be constrained more by physical habitat than by food (Behnke 1992). Some physical factors important to trout are water velocity, water temperature, amount of dissolved oxygen, pool volume (includes number, size, and depth), escape cover, and annual discharge and flow. All these factors are directly related to proper channel shape, bank stability, transport of sediment load, and the ability of the stream to dissipate overflow energy onto its floodplain. Important also is the amount of sediment in the channel as excess can affect pool and riffle quality.

Essentially there are 4 types of habitat that are vital to a healthy trout population: spawning habitat, nursery or rearing habitat, adult habitat, and overwintering habitat. Habitat requirements are variable depending on life cycle stage, species involved, season of year, geographical location, and even time of day.

Spawning habitat is found in the riffles of a stream. They are also sites for re-oxygenation of water, production of insects for food, and they contain the spawning gravels necessary for the incubation of fish eggs. Optimum riffle conditions provide water velocities to clean the gravels of sediment, sweep away metabolic wastes from developing embryos, and supply high levels of dissolved oxygen (Behnke 1992). Spawning gravels may also be selected based on their proximity to high-quality rearing habitat (Platts 1990). Trout fecundity is generally very high and in healthy streams there is a tremendous surplus of young fish that die of natural causes (Behnke 1992).

Quality rearing habitats have adequate protective cover and varying situations such as spring seeps, side channels, and small tributaries where water velocity is low. Without these habitats young fish can be swept away leaving imbalanced age classes (Nehring 1986). An overabundance of rearing habitat can result in excessive recruitment with young and adult fish then placed in competition for a common food supply. In such instances, the population consists mostly of young small fish (Behnke 1992).

Adult habitat is frequently the limiting factor in most streams (Behnke 1992). These habitat components are characterized by narrow deep channels, overhanging banks, logs, rocks, and submerged vegetation. These features allow adults to hide and be blocked from the view of other fish. Trout studies in Arizona consistently showed good bank condition with high standing crop (Clarkson and Wilson 1991). A model predicting trout biomass in Wyoming showed that annual flow regime had the greatest influence on trout biomass (Binns and Eiserman 1979). If low base flows in late summer were adequate to keep adult habitats submerged, trout biomass remained high, but if base flows dropped to levels where shorelines and overhanging banks were exposed, trout biomass declined.

Winter habitat is characterized by deep water with low current velocity and protective cover. The latter consists of deep pools with large boulders and root wads (Bjornn 1971). Deep beaver (*Castor canadensis*) ponds provide excellent winter habitat. Behnke (1992) points out the importance of this habitat to winter survival, but it may be overlooked when a river is evaluated for trout habitat.

Trout biologists have suggested a list of conditions necessary for optimum trout habitats (Armour 1978, Bowers et al. 1979, Oregon-Washington Interagency Wildlife Committee 1979, Reiser and Bjornn 1979). A stream should have: (1) a minimum of 60% shade between 1000 and 1600 hours, (2) inorganic sediment should not exceed a covering of 15% of the gravel/rubble substrate, (3) a minimum of 80% of the streambank should be in stable condition, (4) a minimum of 50% of the streambanks should be overhanging, and (5) a minimum of 50% canopy cover for the entire stream.

Streamside vegetation is very important for optimum trout habitat. The fibrous roots of the herbaceous vegetation and the woody roots of trees and shrubs combine to stabilize banks. The trees and shrubs shade the stream reducing water temperatures. The herbaceous groundcover also insulates the soil in winter months, reducing freeze-thaw cycles, which makes them less vulnerable to erosion from high velocity runoff and ice floes (Bohn 1989).

Riparian vegetation also contributes detritus (i.e., leaves, stems, and other woody materials) to the stream. This energy source is extremely important to a trout fishery because > 99% of the stream energy production comes from this source (Bormann and Likens 1969, Likens and Bormann 1974).

Wildlife

In general, wildlife is more responsive to riparian vegetation components and less to physical factors, but some small mammals may be restricted to certain soil types (Harris 1971, Anderson and Ohmart 1984). Riparian habitats satisfy a wide array of wildlife values (e.g., high densities, high species richness, large number of breeding pairs, species that are rare or generally uncommon, many endangered species, large numbers of overwintering wildlife) (Carothers et al. 1974, Carothers and Johnson 1975, Gains 1977, Johnson et al. 1977, Stamp 1978, Knopf 1985).

Specific vegetation components required to fulfill wildlife needs have not been widely studied, even for birds. During 15 years of ecological studies along the Colorado River in western Arizona, Anderson and Ohmart (1984) systematically examined what were the most important vegetation components required to satisfy the ecological needs of groups such as birds and mammals (Rice et al. 1984). In some instances these components were examined at the species level with the objective of designing and revegetating areas that would contain high wildlife values.

Statistical methods were used to group hundreds of quantified communities throughout the Southwest (ignoring plant species composition) to examine common groups. Six structural groups emerged (Anderson and Ohmart 1986) (Fig. 5). This grouping allows comparison of wildlife values of similar structural types among themselves and between themselves based solely on structure or, when desired, plant species included. It also allowed tracking of changes in wildlife values as young communities changed through time to maturity. Ultimately, these analyses allowed the testing of numerous vegetation variables and which were most strongly correlated with highest wildlife values. For example, foliage volume or dense foliage at any layer always supported more species of birds and greater densities than sparse foliage volumes.

There are indirect data on the importance of foliage volume and the willow community in locations other than the Southwest. Duff (1979) reported a 350% increase in raptors and passerine birds on Big Creek in Utah with the inclusion of willows and increased foliage volume in the midstory. Taylor (1986), on the Blitzen River in southeastern Oregon, reported increases in avian species richness and densities 11-13 times higher in low willow foliage versus high willow foliage understory habitats. Similar avian responses have been reported for Sheep Creek in northwestern Colorado (Schulz and Leininger 1991).

Undoubtedly an important component of willows is their rich and diverse insect fauna. Southwood (1961) reported that the Salicaceae supports one of the richest and most diverse insect faunas found among tree families. This rich food abundance must be very attractive to insectivorous fishes, amphibians, reptiles, birds, and small mammals. Further, arthropod abundance has been demonstrated to be a better predictor of densities of insectivorous birds than either foliage volume or foliage height diversity (Brush and Stiles 1986). Rotenberry (1985) has also suggested that birds may respond more to plant taxa than structure, based on resources provided by the vegetation.

In the absence of definitive wildlife-vegetation data at higher elevations and at northern latitudes, desert riparian habitat data can be used in formulating meaningful wildlife management decisions in riparian habitats in the West. A riparian evaluation guide (USDA 1992), by the Intermountain Region of the USFS, supports this as they stress the importance of vertical and horizontal diversity of riparian forests to support greater animal species richness. Cottonwood and/or willow communities have high wildlife values in desert elevations (Carothers et al. 1974, Hubbard 1977, Johnson 1978, Duff 1979, Taylor 1986), at midelevations (Balda 1975, Knopf and Cannon 1982, Wright et al. 1983, Knopf 1985, McEneaney 1988, Thomas 1989), and in quaking aspen at higher elevations (Winternitz 1980, Winternitz and Cahn 1983).

The exotic saltcedar (*Tamarix chinensis*) that began dominating lowland riparian habitats in the twentieth century is now

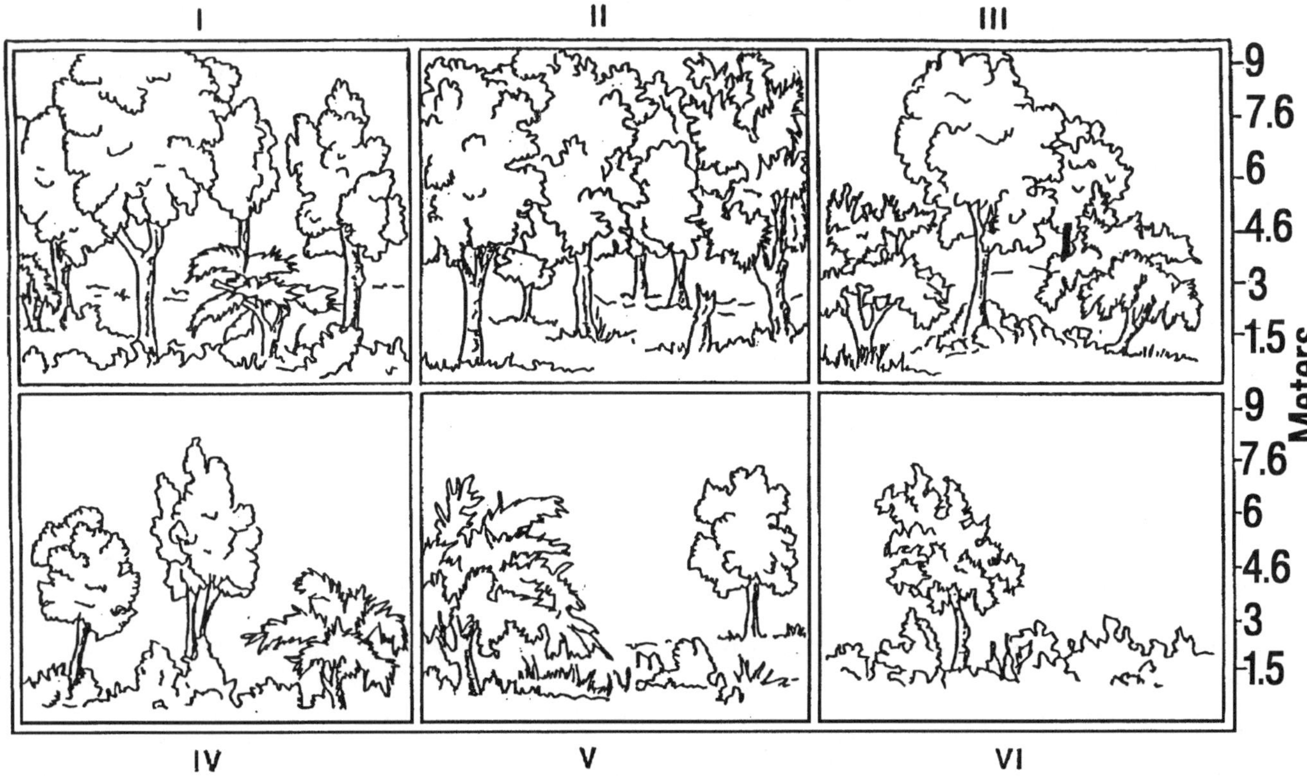

FIGURE 5. Vegetation structural types based on foliage profiles and volumes (Anderson and Ohmart 1984). The paucity of understory vegetation in type II is because of heavy shading by the dense midstory and canopy vegetation. As the forest matures, individuals and groups of trees die, allowing light penetration and an understory to develop.

becoming more common at higher elevations. It generally does not have high wildlife values, but it provides better wildlife habitat than bare soil. Only in exceptional instances have wildlife values in saltcedar begun to approach those of native plant communities (Engel-Wilson and Ohmart 1978, Brown and Trosset 1989). Russian olive (*Elaeagnus angustifolia*), another naturalized exotic, is becoming more abundant in riparian areas in the Intermountain West. Limited information indicates that its wildlife values are not equal that of native riparian trees (Knopf and Olson 1984).

Vegetation components most important to wildlife, in order of importance, are tree species and their densities, foliage height diversity, foliage volume, patchiness, and shrub species and their densities (Ohmart et al. 1988).

Individual Tree Species and Their Densities. The cottonwood-willow component is consistently more important to individual avian species than any of the other vegetation variables (Rice et al. 1984). Avian densities and species richness values in riparian forests are extremely high. Carothers and Johnson (1975) reported 1,059 breeding pairs/40 ha in cottonwood-willow forests on the Verde River in central Arizona; the highest reported values of any habitat in the continental United States.

Foliage Height Diversity. A community with a high foliage height diversity value is tall and structurally complex with high foliage volumes at the herbaceous under, mid, and overstory. This plant community attribute is also important to some arboreal rodents (Anderson and Ohmart 1984) and reptiles (Vitt and Ohmart 1978). Along the lower Colorado River, Ohmart and Anderson (1982) reported 19 breeding bird species associated with the dense canopy layer, 10 with the midstory, and 11 species with the understory. Of all of the possible tree associated communities, the dense mature cottonwood-willow forest has both the important tree species element and the vertical foliage profile, thus providing the 2 most important components in avian habitat selection.

Relative Foliage Volume/m³. The density of the vegetation in the overstory, midstory, understory, and herbaceous layers is extremely important to satisfying the habitat requirements of small mammals, reptiles, amphibians, and breeding birds. Many of the latter are neotropical migrants who tend to be habitat specialists (i.e., foliage gleaners) and as foliage volumes increase in any layer new wildlife species should be added and densities of existing species increased.

Plant Community Patchiness. Patchiness is the unevenness in mixes of different tree species or trees and shrubs horizontally throughout a relatively homogeneous plant community. For example, in the Intermountain West a mix of willow species, when mature, has different heights, providing patchiness. In a cottonwood-willow forest, cottonwoods will generally attain a taller stature at maturity than willows, providing a patchiness component throughout the canopy and midstory layers. In honey mesquite (*Prosopis glandulosa*) communities, quail bush (*Atriplex lentiformis*), wolfberry (*Lycium* spp.), or some other shrub mix provides horizontal patchiness through the community. Patchiness should not be confused with natural edges or ecotones, which are where 2 communities meet. This is considered an intercommunity value, whereas patchiness is an intracommunity value to wildlife.

It has been suggested that patchiness in plant communities provides extra niches or opportunities for bird species to occupy these areas, which would increase avian densities and species richness (Wiens 1989). Comparisons of saltcedar monocultures that have little or no patchiness to very patchy cottonwood-willow habitats, show there are significant differences in avian densities and species richness values (Ohmart and Anderson 1982).

Shrub and Shrub Species Components. Shrubs in desert riparian habitats have received little attention in their importance to wildlife, though willows at more northern latitudes have been reported as important foraging habitats, breeding areas, and as thermal and escape cover for birds (Wright et al. 1983, Krueger and Anderson 1985, McEneaney 1988, Chadde 1989). In desert riparian habitats wolfberry and quail bush are extremely important in the riparian shrub component. Both provide escape cover, but quail bush appears more important as it provides winter thermal cover because it is evergreen, unlike most riparian trees and shrubs. Quail bush's year-round green foliage also supplies a high insect population for foliage-gleaning forms of wildlife; the fruits are consumed as well. A mature quail bush is commonly 2.5 m tall and frequently covers >9.3 m^2. The moist, decomposing leaf litter under these shrubs is replete with detritivorous insects that are heavily fed upon by quail, thrashers, towhees, and small mammals. Along the Colorado River in western Arizona, Ohmart and Anderson (1982) reported that moderate densities of quail bush mixed with exotic saltcedar significantly increased avian species and densities year round.

Patch size or forest community extent is undoubtedly an important wildlife component in broad alluvial floodplains. Most have been modified, fragmented, or so degraded that there has been little opportunity to document the relative importance of this variable to wildlife. Intuitively, a 40-ha riparian forest would fulfill the habitat needs of more species and support greater densities of wildlife per unit area than a 10-ha patch.

For example in a mature stand of willows 1,000 m long and 800 m wide, the outer perimeter (40-60 m) of the stand serves as a buffer area deterring the entry of nest parasites and predators to the core or central portion of the habitat. The core area provides optimum conditions for willow thicket specialists to live and reproduce be they birds, small mammals, amphibians, or reptiles. If this community is fragmented or broken in half by a road or some other interference the core habitat is significantly reduced in size since nest parasites and predators now begin working these new perimeters. This model applies to any expansive plant community type be it a deciduous forest or wetland. Habitat fragmentation can also be highly detrimental to wildlife along streams with narrow bands of vegetation. If the vegetation is destroyed at right angles to the stream so that wildlife populations will not travel across these open areas then gene exchange and dispersal will be stopped until the vegetation regrows to provide cover for movement of individuals.

Relative to migratory wildlife, riparian habitats: (1) provide protective cover and rest areas; (2) supply a rich and abundant insect resource for replenishing fat stores; and (3) serve as wintering habitat for some species. Southwestern riparian habitat importance was summed up by Laymon (1984:595) as "... an essential link for long-distance migrants from the north and are an important wintering ground for many species." Stevens et al. (1977) reported that riparian study plots supported up to 10.6 times as many migrants as paired upland sites. These habitats probably reach their zenith of importance in the Southwest as resting and refueling sites in as they are surrounded by an arid and depauperate upland environment. Terrill and Ohmart (1984) reported that some wood warblers, in an attempt to overwinter as close as possible to the breeding grounds, do so in these habitats as long as winters are mild and insect resources are abundant. Their importance as fueling and stopover sites is explained by studies examining fat reserves, body mass changes, and duration of stay (Cherry 1982, Moore and Kerlinger 1987).

BEAVERS: A KEYSTONE SPECIES IN SMALL-ORDER STREAMS

It is difficult to fathom that before European settlement the beaver population in North America was somewhere between 60,000,000 and 400,000,000 and extended from the Arctic tundra to the deserts of northern Mexico (Seton 1929). Their ability to influence small-order streams is very significant (Naiman et al. 1986, 1988), and yet science is still far from understanding their full role in riparian ecology. Their importance in larger streams as controlling agents may be more significant than currently presumed (see Dobyns 1981).

Naiman et al. (1986, 1988) reported that when beavers remain unexploited they can dramatically alter ecosystem structure and stream dynamics, especially in second- to fifth-order streams. Alteration may be as much as 20-40% by: (1) modifying channel geomorphology and hydrology; (2) retaining sediment and organic matter; (3) creating and maintaining wetlands; (4) modifying nutrient cycling and decomposition dynamics; (5) modifying the plant species composition and physiognomy of plants; (6) influencing the timing, rate, and volume of water and sediment movement downstream; and (7) through the creating of pools and backwaters generating totally new fish and wildlife habitats which results in significant increases in biodiversity. Allred (1980) working in Idaho documented increases in habitat types by beavers and their value to many wetland plants and animals. They may selectively harvest trees to open and modify riparian forest composition and age classes to increase patchiness (Jenkins 1979, 1980). The efficiency of sediment trapping by beaver dams has been reported by Smith (1980), who measured as much as a 90% reduction below dams. Not surprisingly, these habitat alterations are persistent over the riparian landscape for centuries (Rudemann and Schoonmaker 1938, Ives 1942, Neff 1957).

Impounded waters behind dams provide habitat for fish and waterfowl, while emergent and lush vegetation around the pond is favored forage for browsing mammals. Medin and Clary (1991) compared small mammal populations around a willow-dominated beaver pond and an adjacent nonwillow riparian habitat in east-central Idaho. Relative density of small mammals was 3.06 times higher and standing crop biomass was 2.71 times higher in the willow-dominated habitat around the beaver pond. Species richness and diversity were similar between the habitats, but voles (*Microtus* spp.) and shrews (*Sorex* spp.) were more abundant around the beaver pond (Medin and Clary 1991).

To observe a stream in PFC supporting a beaver population is an educational experience, especially with a stream gradient of about 3%. On Rough and Tumbling Creek, Pike National Forest, Colorado, beavers had totally negated the gradient and each dam was a living classroom of hydric to xeric succession. Behind new dams one could see early stages of sediment deposition, older dams showed trapping of sediment by rushes and sedges, others showed stabilization of soils by the woody roots of willows and the forming of backwaters, and ultimately willows and quaking aspen with little surface water. As an interesting exercise, the reader may want to reread the PFC definition (BLM 1993) at the beginning of **What is a Properly Functioning Riparian System** with the beaver in mind.

Some streams may not have the capacity to support beavers for more than a few years (W. Elmore, BLM, pers. commun.). These are streams where shrubby willows dominate the floodplain and the tall deciduous tree element is highly restricted or absent. This may be a food limitation for beavers and after a few years the reduced food supply forces the animals to move to new areas. The importance of beavers in western streams is poorly understood and why they are transient in the above streams needs examination.

HISTORICAL RAMIFICATIONS OF WESTERN LIVESTOCK GRAZING

To give the reader a feel for the evolution and impacts of domestic livestock grazing on western rangeland and riparian habitats, a brief history of this industry's activities seems appropriate. Platts (1979) also gives a thumbnail historical sketch West-wide, and Young and Evans (1989) discuss historical events in Nevada. The West was open and grazing uncontrolled prior to the establishment of the USFS and the various national forests around the turn of the century and the BLM around 1946. A calf could be purchased for $5 and sold a few months later at $65 with the grass and land being free. Arizona (then a territory since statehood was granted in 1912) is used as a model of the consequences of open range and unabated livestock use throughout the 11 western states.

Domestic livestock have grazed portions of the southwestern United States since about 1700. Early Americans did not possess domestic livestock but obtained access to them when the Spaniards brought cattle, horses, sheep, and goats. Simpson (1952) reports that around 1675 there were approximately 200,000 cattle and 2,000,000 sheep on the Central Plateau of Mexico. Within 50 years these numbers would increase to 1,000,000 cattle and 8,000,000 sheep. Ranches were established near the southern fringe of the Sonoran Desert by 1610 (Ewing 1934). By 1694 cattle were grazing the grasslands on the Bavispe River (northern Sonora, Mexico) and headwaters of the San Pedro River (southern Arizona) as reported by Bolton (1948).

Cattle spread rapidly into New Mexico, Arizona, and southern California as each new mission was established. As Father Kino traveled and explored the Pimería Alta, he gave livestock as gifts (Bolton 1948), and in 1701 he made 1,400 animals available to Baja California. The mission in Tucson, San Xavier del Bac, received 700 head in 1702, and by 1703 another 3,500 head were available from Kino's home base in Dolores, Sonora, Mexico. Domestic livestock were extremely important to the new settlers in that they provided a reliable supply of meat, milk, wool, and leather in a harsh and unpredictable environment. By 1750 individual herds of 4,000-5,000 animals were not uncommon (Pfefferkorn 1949).

Many of these cattle became feral as Apaches raided the haciendas and ranches. Bancroft (1883) reported the Apaches preferred horse meat to cattle so raids on ranches were thought to be more for horses and mules, with the cattle being liberated as ranch hands were either killed or abandoned the area. Herds of wild cattle were frequently reported in journals from 1846-1854 (Clarke 1852, Cox 1925, Powell 1931, Durivage 1937, Evans 1945).

The 200-year dominance of the Spanish was essentially terminated at the end of the eighteenth century. Domestic livestock had an important influence on the Indians, and this continued with the Mexicans. Land grants were made along major rivers where water and feed for domestic livestock were most reliable and abundant.

Cooke in 1846 (Bieber 1938), camped near Agua Prieta Creek, wrote that wild cattle were so numerous that the spring had the appearance of a stockyard. Many wild cattle were slain by the officers and an estimated 5,000 watered at the spring (Bieber 1938).

From 1700 to 1850 numbers of domestic livestock grazing in the Southwest were significant and increasing, but stocking rates were much less than those that would be reached between 1850 and 1900. In 1870 there were only 5,000 head of cattle reported in the Arizona Territory (U.S. Bureau of Census 1872:III, 75). Over the next decade this industry grew to provide beef to Army posts, Indian reservations, and growing pioneer settlements. Most cattle brought into Arizona were driven from Texas and Sonora, Mexico. Two drivers brought in over 15,000 head in 4 herds in 1872 (Wagoner 1952). By 1880, 2,500 head were reported east of the San Pedro River in southeastern Arizona; the San Pedro Valley contained 10,000-12,000 head of sheep and about 8,000 cattle. There were 20,000 cattle south of the Gila River and the Arizona Territory contained about 35,000 cattle (Wagoner 1952, U.S. Bureau of Census 1883:III 141-42).

By 1883-1884 in Arizona ". . . every running stream and permanent spring were settled upon, ranch houses built, and adjacent ranges stocked" (Report of the Governor 1896:21). By 1885, there were 435,000 head reported and half were not censused. "This number is being rapidly increased, and within another year it is expected that ranges with living springs and streams will be fully stocked" (Report of the Governor 1885:8).

The 1880s were not a time of tranquility on these open rangelands. Battle lines were drawn and those who controlled water access dictated who grazed the range. Many ranchers recognized the problems of overgrazing but could do little about it. More cattle came while established ranches continued to build their herds. By 1890 it was estimated that >1,000,000 head grazed the territorial ranges of Arizona, and possibly as many as 1,500,000 (U.S. Bureau of Census 1895:I, 29). Peterson (1950) reported >2,500,000 head in Montana, >2,000,000 in New Mexico, and 1,250,000 for Utah and Wyoming in 1890. It is estimated that about 19,000,000 cattle and sheep were grazing the arid West in the late 1880's (General Accounting Office [GAO] 1988). Wilkenson (1992) estimated 26,000,000 cattle and 20,000,000 sheep in the western United States at the end of the century.

Drought struck the cattle industry in Arizona and adjacent states from 1891 to 1893. In 1891 the Governor (Report of the Governor 1896:22) estimated 1,500,000 head of domestic livestock in Arizona. Poor summer rains, coupled with reduced winter moisture, intensified overgrazing to the point that it would be extreme before the drought was over. Cattle died on poorer ranges in the hot dry months of May and June of 1892. Below-normal rains in July and August compounded the problem throughout Arizona. By late spring of 1893 the Governor (Report of the Governor 1896:22) reported the losses as "staggering." Land (1934) stated "Dead cattle lay everywhere. You could actually throw a rock from one carcass to another." J. W. Toumey, Chief Botanist in charge of Grass and Forage Plant Investigations for the Arizona Experiment Station in Tucson in 1891, wrote regarding the southeastern Arizona grasslands, "There are valleys over which one can ride for several miles without finding mature grasses sufficient for herbarium specimens without searching under bushes or in other similar places" (Bahre 1991:113). Livestock mortality estimates were placed at 50-75% (Report of the Governor 1896:22). Wagoner (1952:120-21) supported these mortality estimates for Pima and Cochise counties. Even if mortality rates were only 30% on better rangelands, the ecological destruction of watersheds and riparian habitats in Arizona were easily predictable. Vast areas of rangeland were left barren and unprotected from erosion by wind and rain (Hastings and Turner 1965, Dobyns 1981). In heavy storm events topsoil eroded into the now highly weakened and poorly vegetated riparian habitats. Mature riparian forests were scoured out, leaving more soils vulnerable to erosion from the next storm. With heavily reduced or no groundcover on Arizona watersheds, even small storm events resulted in high surface runoff and heavy soil erosion. Even if flood conditions were conducive to seedling establishment, "continued overuse of riparian bottoms eliminates essentially all reproduction as soon as it becomes established" (Davis 1977:60).

Development of the grazing industry in the other western states does not differ dramatically in timing and consequences to both watersheds and riparian habitats (McArdle et al. 1936, Anderson and Harris 1973, Adams 1975, Behnke 1978, Meehan and Platts 1978, GAO 1988, Chaney et al. 1990). Apparently ranges in northeastern Nevada suffered similarly and during winter of 1889-1890 there was a 95% loss of cattle (Young and Evans 1989). The legend was that one could walk for 161 km along the Mary River, a fork of the Humboldt River, and step from carcass to carcass and never touch the ground (Young and Evans 1989).

Not only did cattle starve, but the resources suffered as well as depicted for Chaco Canyon, New Mexico (Chapman 1933). The first white settlers in 1849 reported the streambed was approximately 3 m wide and 0.6 m deep. In 1924, the stream had entrenched to 10 m and the arroyo was about 100 m wide. The ecological balance had been so ". . . disturbed by overgrazing, erosion has moved a thousand fold more soil in 50 years than in the preceding ten centuries" (Chapman 1933:75). In 1924, Bryan (in Chapman 1933) listed 21 important streams in Arizona, Colorado, New Mexico, and Utah. All streams had floodplains supporting forests of cottonwoods and willows and at that time only supported scattered sage (*Artemisia* spp.), greasewood (*Sarcobatus vermiculatus*), or mesquite.

Deterioration of western riparian systems began with severe overgrazing in the late nineteenth century, and extensive field surveys in the 1980s demonstrate that much of them are in the worst condition in the history of this nation (Chaney et al. 1990). Drought may not have intensified overgrazing as abruptly in the other western states, but the ecological consequences of overgrazing to riparian habitats were similar, throughout western rangelands.

Overgrazing of public lands continued virtually unabated into the twentieth century. Range conditions similar to those in Arizona were reported by Esplin et al. (1928) on lands in Utah, by Keck (1972) in the Great Basin, and by McArdle et al. (1936) when they provided descriptions of unclaimed public lands (now BLM lands). They reported that approximately 84% of these lands had lost more than half of their forage value and forage was depleted on an average of 67% throughout the West.

Overgrazing of National Forest lands "became so critical" (Platts 1981*a*), that the Taylor Grazing Act was passed in 1934 to protect the remaining unclaimed public land and stabilize the livestock industry. Though this action established allotments and adjudicated numbers on these unclaimed public lands, it was, at best, token service to overgrazing on public lands in the 11 western states. On USFS land when permitted livestock numbers were assigned to permittees they were frequently too high. Most USFS lands were not fenced until the 1930s (Bahre 1991). Bahre (1977:27) quotes old timers stating, "The cattle went where the feed was when there was open range, whereas today with fences and supplemental feeding, the cattle stay in pastures for longer than the grass can feed them, ruining the land." The only importance given to riparian habitats during this period was their value in providing extra forage and water for livestock. Up until the late 1960s, riparian habitats were viewed as sacrifice areas. The more valuable grazing allotments contained ≥ 1 perennial streams within their boundaries.

LIVESTOCK IMPACTS TO RIPARIAN HABITATS

This chapter deals with the effects of domestic livestock grazing on riparian habitats, but the reader should be aware that other human activities, both past and present, have destroyed and heavily degraded riparian habitats as well. For example, the virtual elimination of beavers by trapping undoubtedly had a large impact on riparian habitats throughout the West. These animals have major influences on small-order streams and their removal must have radically altered water retention and sediment trapping capabilities of streams. Naiman et al. (1986, 1988), Ehleringer et al. (1992), Elmore and Kauffman (1994) and many others share similar conclusions.

In more recent times, western water management has destroyed and degraded untold thousands of hectares of riparian habitats along major perennial rivers (Stevens et al. 1977, Ohmart et al. 1988). Reservoirs inundate many thousands of hectares of riparian habitat and regulated flows below dams have heavily degraded riparian habitats by stopping or highly altering natural floods. Without natural floods the life cycle of cottonwoods and willows is broken. These moderate-lived tree species persist for many years but eventually disappear because no seedlings are produced as replacements. Along heavily managed streams, channelization and riprapping of banks follow dams and reservoirs. Vegetation is stripped from the banks to place large boulders or riprap, and channels are deepened by dredging. This further decouples the stream from its floodplain and lowers the water table, drying up old oxbows and marsh areas. With the threat of floods eliminated, farming expands on the alluvial floodplain allowing rapid conversion of native habitats to cotton, alfalfa, and other farm crops. Evaporation from reservoirs and leached salts in return irrigation flows to the river increases downstream soil and water salinities, providing optimum conditions for the rapid invasion of saltcedar. Other water management projects such as cutting riparian trees to salvage or save water was undertaken along many perennial streams in the West. Activities such as logging, mining, groundwater pumping, construction of roads, woodcutting, offroad vehicle use, and uncontrolled recreation have also degraded riparian habitats (Busby 1979, Noh 1979, Swan 1979). In general, water management and groundwater pumping has had its greatest impacts to western riparian habitats at lower elevations along most perennial rivers, and domestic livestock grazing has manifested itself ubiquitously at all elevations in the West.

General Considerations

All evidence indicates that virtually all riparian habitats received unmanaged grazing throughout the 11 western states as the livestock industry developed (Elmore and Kauffman 1994). Even Grand Gulch in southeastern Utah, with its vertical sandstone walls of 61 to 122 m, eventually had trails built so that livestock could access the forage (Blackburn 1993). Few western streams with significant forage availability escaped domestic livestock grazing.

Use of the term "unmanaged livestock grazing" refers to the practice of releasing livestock into an area without any planned riparian growing season rest or measures designed to protect the health of the vegetation along the stream or its floodplain. Unmanaged grazing always results in excessive utilization in riparian areas, impairment of plant species vigor, and physical damage to the channel and banks.

Unmanaged grazing of riparian systems has been and continues to be practiced. Today even though most allotments have management plans, all were designed to meet phenological growth requirements of upland vegetation. Watersheds may benefit from these grazing approaches, but riparian habitats are degraded under these plans and will continue to be until management changes are made.

When livestock are put into an allotment or large pasture, they go where they wish or, in many instances, riders drive the animals to wet meadows or other riparian areas where forage and water are abundant. In cow-calf operations the veteran cows know the allotment and where they want to be. Cattle, like most animals, have home ranges, favorite foraging areas that usually include some or all of a riparian habitat, and centers of activity (Martin 1979).

Riparian habitats provide the 4 basic requisites essential to wildlife or domestic livestock: food, water, cover, and space. The attractants of lush vegetation, water, and shade are such that cattle will spend 5-30 times longer in riparian habitats than adjacent uplands, based on areal extent (Skovlin 1984). Cattle congregate in the floodplain in the hotter, drier summer months, imposing heavy use during the heart of the growing season, and in many instances throughout the growing season. Platts and Nelson (1985) reported nearly 100% herbage removal in riparian habitats in the semiarid big sagebrush zone. "Because cattle prefer stream side environments, deterioration of riparian habitats has been significant and much of the deterioration continues" (Platts 1979:48). If grazing use is year-round or even extends into the cooler months, some livestock may disperse into the uplands, but enough will remain in the riparian area to disallow seed development or stored energy reserves for winter. Reduction in livestock numbers is not a management approach to eliminate degradation to riparian habitats.

This was demonstrated in Nevada on Mahogany Creek, where herd size was reduced in efforts to improve trout habitat. Dahlem (1979:34) concluded that

> Based on photographic evidence and data availability, one fact is apparent. The reduction in livestock grazing but continued annual use, had little beneficial effect on riparian habitat along Mahogany Creek. Only after complete removal of livestock use by fencing was significant riparian habitat improvement accomplished along Mahogany Creek.

Gus Hormay related to Olson and Armour (1979:69):

> Vegetation in certain areas, such as meadows and drainage ways, are invariably closely utilized under any stocking rate or system of grazing. Such use may be detrimental to wildlife, esthetic or recreational or other values. Where this is the case, about the only way to preserve values is to fence the area off from grazing. Reducing livestock or adjusting the grazing season usually will not solve such a problem.

The presence of cows (wt ±400 kg each) and/or bulls (wt ±800 kg each) concentrated along streams, foraging along streambanks, and constantly crossing the stream, either season long or year-round, causes extensive physical damage to banks and the channel. That, combined with vegetation removal by each animal (about 350 kg of air dry-forage monthly) for >100 years over most western rangelands has had a devastating effect on riparian systems. Busby (1979) contends things are better in the uplands and he may be correct, but all observations indicate that riparian habitats are highly degraded and generally continue in that state. The Oregon-Washington Interagency Wildlife Council (1979), based on numerous studies in the 11 western states, identified domestic livestock grazing as a major factor in causing serious reductions in wildlife habitat productivity. It was suggested by Bowler (1976) that overgrazing is the largest environmental problem in the United States. Szaro (1989) after extensively surveying riparian conditions in the Southwest reported that livestock may be the major disturbance factor in western riparian habitats. The GAO (1988:11), after talking with agency officials (USFS and BLM) and examining many studies, stated, "Poorly managed livestock grazing is the major cause of degraded riparian habitats on federal range lands." Mosconi and Hutto (1982) working in Montana, suggested that domestic livestock grazing is the major cause of riparian habitat disturbance in the West. Chaney et al. (1990) stated that "extensive field observations in the 1980s suggest riparian areas throughout much of the West are in the worst condition in history." Carothers (1977:3) wrote that ". . . the most insidious threat to the riparian habitat today is domestic livestock grazing."

A GAO (1988) report was very negative on unmanaged livestock grazing and the condition of riparian habitats in the West. It also dealt with restoration of some riparian areas and how these restored areas were highly beneficial to the permittees by providing advantages other than more forage production. These managed areas showed high soil stability and improved range conditions. I have thoroughly reviewed the GAO (1988) document and from my many years of assessing riparian habitats on public lands, interacting with a multitude of USFS and BLM personnel, and working with permittees, I can only say that, it is the most candid and valid assessment of conditions and problems facing riparian restoration. In most instances management knows the problem and generally how to solve it. However, resistance or total opposition by the permittee (and, sometimes agency personnel) and the cost of making changes, severely slows or stops any progress toward better riparian habitat management. Meehan (1991:9) working with salmonid fishes comments on domestic livestock grazing and stream improvement, "Persuasion has been difficult, and change has occurred slowly."

The importance of livestock forage production in riparian habitats is demonstrated in northeast Oregon where 1 ha of moist meadow soils has the potential grazing capacity of 10-15 ha of forested range (Reid and Pickford 1946). These wet meadows represent $\leq 2\%$ of the range and produce approximately 20% of the forage (Roath and Krueger 1982). They further report that because of the way livestock concentrate, the steepness of terrain, and poor water distribution away from the stream, in reality the 2% wet meadow is producing 81% of the practically usable forage in the Blue Mountain grazing allotment.

Most plants in the floodplain are highly palatable to livestock. Sedges in the genus *Carex* maintain a relatively constant level of crude protein throughout the growing season and until the first killing frost (Kauffman and Krueger 1984). Many sedges in riparian habitats in the Pacific Northwest have higher protein and caloric content than key upland forage species (McLean et al. 1963, Skovlin 1967, Paulsen 1969). Not all allotments in the West have the sedge component or broad wet meadows, but the relative value of the riparian forage (plus water availability) to the drier uplands is about the same throughout the West.

Reasons for Management Change

"There is a general acceptance by managers today that most riparian areas are in an unacceptable condition and that approaches to restoration in the past have had limited success" (Elmore and Kauffman 1994:219). The above statement is very true but instead of "most riparian areas" my experiences are that almost all riparian areas are in unacceptable condition. To avoid greater problems that ultimately may exclude grazing in riparian habitats, agencies and permittees should immediately begin to undertake livestock management in riparian habitats.

Important riparian issues loom on the horizon, such as the continued listing of endangered species, more species being considered for listing, water quality, and recreation. If neotropical migrant birds are unquestionably found to be declining in the 11 western states because of domestic livestock grazing, this will elevate the significance of riparian habitat condition to a new level. The affluent and well-educated cadre of birding enthusiasts that pursue this hobby will exert tremendous political pressure on elected officials and federal agencies for immediate legislation to protect riparian habitats. Spring will not be totally silent (Carson 1962), but 60 to 70% of the songbird species breeding in riparian habitats in the western North American are neotropical migrants (Bock et al. 1993).

The willow flycatcher (*Empidonax traillii extimus*), a neotropical migrant, has recently been listed as endangered (U.S. Fish and Wildlife Service 1995). Approximately 718 km or more of streams may be included as critical habitat for this species in the Southwest. Listing packages are in preparation for several other birds (all neotropical migrants) that will only exacerbate user problems in that grazing decisions then must pass Section 7 Consultation under the Endangered Species Act. The gravity of these listings and the rapidity of their occurrence is seen on BLM lands where 10 years ago there were 75 wildlife species federally listed as threatened or endangered. Today, there are 216 and 1,000 more species being readied for listing (Horning 1994). These numbers do not include plants directly affected by livestock grazing. Western livestock growers perceived the Reagan and Bush Administrations as allies, but in reality these elected officials harmed the industry by not enacting slight management alterations over their 12 years that could have avoided drastic management changes today. As it is, permittees may lose use of pastures or possibly entire allotments as new species are federally listed and critical habitats delineated.

What financial burden is being placed on the taxpaying citizens of this country in attempts to recover some of these species that are now endangered from domestic livestock grazing? An indepth cost analysis has not been attempted, but there are some data for mineral extraction activities (Losos et al. 1995). A few examples provide insight into this question. In 1989, BLM, in trying to recover 5 bird species, averaged > \$700,000 per species

(McClure et al. 1991). Two million dollars have been expended over the past 20 years to recover the Gila trout (*Oncorhynchus gilae*), and another million will be expended by the year 2000 (U.S. Fish and Wildlife Service 1993*b*). The species is in greater peril of extinction today than when recovery efforts started because the team wants to avoid controversial issues such as domestic livestock grazing. The U.S. Fish and Wildlife Service (1994) plans to spend $15.5 million over the next 12 years to recover the desert tortoise (*Gopherus agassizii*). Horning (1994) estimates that the BLM total recovery cost for the Lahoutan cutthroat trout (*Oncorhynchus clarki henshawi*) will exceed $14 million, with fencing costs being estimated at $3,000/km. The U.S. Fish and Wildlife Service Office in Phoenix, Arizona, will spend $1.5-2 million to fence Apache trout (*O. apache*) habitat, while cattle continue to overgraze and degrade other streams containing the fish (Horning 1994). These costs have prompted the National Wildlife Federation to press for federal policies that include a thorough cost-benefit analysis to find the most cost-effective approach; graze or not graze riparian habitats (Horning 1994).

Not only are there economic costs but the ecological costs (see Fleischner 1994) from the disruption of ecosystems and the alteration of riparian community structure may well be of greater economic cost than attempting to recover threatened or endangered species. Riparian systems show moderate resistance to grazing and are resilient once livestock are excluded. Systems may return to a semblance of PFC, but can they ever be recovered to their original functioning condition since being so heavily degraded for so long? They are vital to westerners for cheap high-quality and quantity water and to fish and wildlife.

Other driving forces are where sediments are being carried into trout or salmon streams containing listed species (Anderson et al. 1993). Platts (1991) examined the effects of livestock grazing on salmonid fishes and of the 21 studies that he examined all but one had stream and riparian habitats degraded from domestic livestock grazing. All showed habitat improvement when grazing was prohibited. The exception was herded sheep grazing on a well-managed sheep allotment.

There will also be increased restrictions under the Clean Water Act on nonpoint pollution programs with legislation being encouraged by groups such as Mothers for Clean Water. These are but a few of the compelling reasons that managed grazing of riparian habitats is critical if permittees are to continue using them on public lands.

Phases of Pristine Riparian Habitat Degradation with Unmanaged Livestock

Riparian habitat degradation is broken into 3 phases in the hope that it will be easier for the reader to visualize and understand the temporal, physical, and biological changes that occur in each phase. With riparian degradation in 3 phases, along with the knowledge of what biotic and abiotic components are most important to fish and wildlife, it becomes clear when and why certain animal groups began to be stressed by habitat degradation. It is also impressive how long many of these species have managed to persist in spite of this stressor and its duration.

Phase I. Degradation is estimated at 1-10 years. In general, streambanks and channel morphology, herbaceous and understory vegetation, and water quality are changed. The herbaceous groundcover species mix, if not eliminated after a few years, also changes from highly palatable, better soil-holding species to less or even nonpalatable, shallow-rooted annuals and perennials. These changes come about from physical changes of the banks and channel, elimination of herbaceous and understory vegetation, increased erosion from normal and heavy flood events, channel entrenchment, and lowering of the water table.

The concentration of livestock in riparian areas on a yearlong basis or even total growing season use exacerbates the process of bank degradation and stream siltation. As the stream channel deteriorates by widening, more water from each flood event is carried in the channel with greater velocities and erosive force, further widening the channel through in-stream erosion. Channel widening often triggers channel straightening and channel incision, resulting in a dropping water table.

Phase II. Phase II occurs over 100-125 years and as it begins there is a full complement of tree species with high densities, a mature foliage profile, and high foliage volumes at the midstory and canopy layers. In willow-dominated systems without the taller tree element, Phase II may only take 50 or so years with willows being eliminated or becoming highly scattered. Willows managing to persist have a highly modified hourglass physiognomy. Most recruitment of young trees and shrubs ceases and as the youngest trees that escaped the initial grazing mature there are no replacement forests.

A common statement is, "I've lived on this creek all my life and it has always looked the same." In general that statement is true, but after the initial riparian degradation in Phase I, like the aging process, the changes go unnoticed by casual observation. No one living today observed Phase I (but see San Pedro River wildlife consequences) but it did not go unobserved by ranchers (see Bahre 1991). People do not notice themselves aging on a daily basis, but photographs at 5-year intervals show definite changes.

Through the past 100 years deciduous riparian forests, once continuous, have been slowly fragmented leaving small forest islands that have since been subfragmented as individual trees die. Some trees have died of old age, others in blow downs, many have been washed out in more violent floods after watershed and phase I degradation, beavers (where they persist) have girdled and killed many, and others have been left to die with roots perched above declining water tables as a stream downcuts. The slow loss of individual trees through time has progressed to the point today that foliage volumes in the remaining canopy and midstory layers are very low. The decline of the cottonwood-willow gallery forest in Arizona has been so rapid that funds were allocated to quantify the total amount and riparian community types for the state. There are 106,714 ha of floodplain along Arizona's 8,097 km of perennial streams (Valencia et al. 1993). Of the total floodplain, 4.2% or 4,482 ha are remaining cottonwood-willow association. The Arizona Nature Conservancy (1987) reported this community type as the rarest forest type in North America.

In the Zuni Mountains of western New Mexico in the Cibola National Forest streams prior to 1850 were described by hydrologists as, ". . . narrower, deeper and less entrenched. Floodprone areas were broad and densely occupied with hydric and mesic vegetation" (Jackson 1994:4). The author cites extensive clearcutting and extreme overgrazing as being major contributions to the reduction of the original riparian vegetation by 70-90%. Riparian habitat losses are ≥90% along the East Fork of the Gila River in the Gila Wilderness where cattle grazing is the primary stressor (Ohmart In Press).

Phase III. Phase III is the death and collapse of riparian forests in the West and is estimated to take about 50 years. Some streams are in late Phase II, while others are in early Phase III. Upper Black Canyon in the Gila National Forest, New Mexico, is a mid-Phase III. The stream, in the Aldo Leopold Wilderness Area, was once the habitat of the endangered Gila trout. The banks are laid back to predominantly cobbles, the fines having been washed away (Fig. 6), and the stream is entrenched. The stream has become so degraded that it is only marginally suitable for any type of trout.

There are no stands of young cottonwoods represented along the 11 km of Upper Black Canyon that I hiked. There are a few scattered trees (approx. 10-15 years old) but a few scattered trees do not make a forest. Remnant skeletons of mature cottonwood communities are evident along parts of the trail (Fig. 7). There is an occasional line of remaining cottonwoods with an understory of conifers (Fig. 7). However, most cottonwoods are dead and down and the few remaining alive are frequently girdled by beavers (Fig. 8). The gnawed rings are usually 10-15-cm deep and the beavers have begun consuming junipers (*Juniperus* spp.). Willows have been extirpated along Upper Black Canyon except a few decadent hourglass-shaped individuals on the deeded property just outside the wilderness area.

Collapse of decadent quaking aspen communities in phase III may be of shorter duration than 50 years. There is evidence along streams in Idaho that a number of them once supporting willow-aspen mix now only support willows. Once aspens disappear they may or may not pioneer rapidly into the floodplain even with grazing management.

FIGURE 7. Same stream and location as Figure 6. Stream entrenched and banks heavily eroded. The cottonwood forest has begun to collapse. To the back left is a small grove of cottonwoods mixed with conifers. Photograph by R. D. Ohmart, May 1992.

The above is exemplified on a small unnamed stream on the San Felipe Allotment (BLM) near Challis, Idaho, where a 1.5-ha cattle exclosure was constructed about 1988. The only remaining aspens or evidence thereof along this stream are in the exclosure (Fig. 9). The contrast between the grass and sedge-stabilized banks in the exclosure (Fig. 10) is striking against the raw, eroding outside banks. Vegetation in the elk exclosure did not differ from that within the cattle exclosure. Elk pellet groups were inside the cattle exclosure and light utilization of willows was evident but there were no raw or trampled streambanks.

FIGURE 6. Upper Black Canyon, Aldo Leopold Wilderness Area, Gila National Forest, New Mexico. Stream is entrenched at least 1 m, and heavy cobble now forms most banks and the stream channel. Conifers mixed with scattered cottonwoods along the primary floodplain. Photograph by R. D. Ohmart, June 1992.

FIGURE 8. Same location as Figure 6. One of numerous mature cottonwoods showing beaver damage. Note collapsed cottonwood forest around the general area. Photograph by R. D. Ohmart, June 1992.

FIGURE 9. Cattle exclosure on small, unnamed stream on the San Felipe Allotment near Challis, Idaho. The only mature aspen on the stream are within the exclosure as are the only young trees (two in background and black stems in the photograph). Photograph by R. D. Ohmart, May 1994.

FIGURE 10. Fence line contrast with grass and sedge-covered banks within the exclosure contrasted with raw trampled banks outside. Aspen sapling in background of Figure 9 is that on left side. Photograph by R. D. Ohmart, May 1994.

Consequences to Fish and Wildlife

Fish. Most species are sensitive to changes in channel morphology and water quality and quantity, so the immediate physical and biological degradation of streambanks and channels affected this group early in Phase I and continues to do so in most streams today. The above changes in the stream are detrimental to trout populations (Armour 1977, Behnke and Raleigh 1978, Meehan and Platts 1978, Platts 1979). Armour (1979:39) stated ". . . we are concerned about overgrazing adversely impacting thousands of miles of streams associated with federally administered rangeland in the West." Bakke (1977) reported that loss of trout and salmon habitat from overgrazing has been a frustrating problem in Oregon. Behnke and Zarn (1976) identified livestock grazing as the greatest threat to the integrity of trout stream habitat in the West. The physical and biological degradation by domestic livestock grazing of most western streams has prompted fisheries biologists to advocate the abolition of livestock grazing for full stream recovery (Behnke 1979, Dahlem 1979).

Storch (1979:56), working in eastern Oregon, summarizes the problem,

> Uncontrolled livestock grazing has seriously affected the water quality of streams throughout the country. Indiscriminate use of streams by livestock results in breaking down the streambanks, eating and trampling shrubs that shade the streams and/or provide habitat for wildlife, and disturbing stream bottoms. The effects of such use has been erosion of stream banks, higher water temperatures, increased sedimentation, soil compaction, and reduction of the quantity and quality of forage.

The continued deterioration of fisheries habitats on western public rangeland from uncontrolled domestic livestock grazing has prompted the American Fisheries Society to publish a position statement (Armour et al. 1994). The paper has been in preparation a number of years and states, "Overgrazing of riparian and stream ecosystems by domestic livestock has damaged thousands of linear miles in the ecosystems" (Armour et al. 1994:9). Previous position statements and this one point out ". . . overgrazing by domestic livestock was one of the principal factors contributing the damage and loss of riparian and stream ecosystems in the West" (Armour et al. 1994:10).

Hansen (1993:334) observed riparian habitat degradation and stated, "It only takes a few weeks of unauthorized use or overgrazing to set back years of progress in improvements of riparian-wetland systems." Duff (1979) witnessed an area rested for 4 years degrade rapidly after the reintroduction of cattle; overhanging banks were quickly eliminated and after 6 weeks of midsummer grazing the banks fractured and eroded into the stream. Kauffman et al. (1983:683) examined the erosion component in northeastern Oregon and stated, ". . . erosion related to livestock grazing was enough to create significantly greater annual streambank losses when compared to an ungrazed area." Degradation time is rapid when compared to the slowness of the reversal process of 50 years if Wickiup Creek in Oregon is a general indicator of healing time (Clifton 1989). Gregory and Ashkenas (1990), working in Oregon, estimate that with proper management recovery of fish habitat, riparian areas, and water quality may require 25-200 years depending on existing conditions, stream type, and availability of fine sediment for bank rebuilding.

Clarkson and Wilson (1991) examined differences between unmanaged grazing, light, and no grazing during a 4-year study from 243 sampling stations among 75 reaches of 21 high-elevation trout streams in east-central Arizona. The focus of this study was the federally endangered Apache trout. In the data analysis, the amount of ungulate damage to streambanks consistently explained the greatest amount of variation in standing crop of fishes. Clarkson and Wilson (1991) concluded that better livestock management is necessary if the fishery potential of these streams is to be realized.

Banks along some streams may not recover in a lifetime once degraded. In general, these are small-order or headwater streams that carry little, if any sediment load. North Fork Cottonwood Creek may be an example in that there has been no significant change in channel width since livestock exclusion for 24 years (Kondolf 1993). Sediment load is reported to be low, but continual trespass by domestic livestock and exclosure size (0.5 ha or 135 × 35 m) confounds understanding channel response since exclusion (Kondolf 1993).

Numerous studies have examined bank and channel healing after livestock were excluded. Portions of Big Creek in Utah were excluded for 4 years and bank widths of the season-long (May-Oct) grazed area were 173% or almost twice as wide as the rested area (Duff 1979). Sedges and grasses responded rapidly after exclusion, increasing 63% (Duff 1983). Streambanks were initially bare or sparsely covered and within 4 years were described as luxuriant, grassy, and overhanging.

As the protective herbaceous groundcover over the floodplain is heavily grazed and weakened, the inevitable degradation process described earlier begins. The once relatively stable sinuous stream begins to straighten as it erodes its banks. Once incised to a stable point the lowered stream must widen the incised channel to a point that a new a new floodplain can be formed inside the old one. The straightened stream will then begin to reestablish its meander pattern.

Initial vegetation removal generally begins by livestock consuming grasses, sedges, and rushes along the stream and over the floodplain. As this forage resource is depleted, livestock begin browsing young trees and shrubs. If flood events are such that new tree or shrub seedlings germinate they are quickly consumed (Davis 1977), marking the end of tree and shrub recruitment to the riparian community. As stream width increases, large trees near the stream may also be undercut and fall.

Unmanaged grazing extirpates palatable native species and creates opportunities for the establishment and expansion of exotic species that may be undesirable and unpalatable. Cottam and Evans (1945) reported the presence of 10 native grass species in a canyon protected from grazing since the late 1800s (Red Butte), whereas these species were absent in a severely grazed canyon (Emigration) in Utah. Palatable grasses were 5 times greater in Red Butte than Emigration Canyon. Ruderals (unpalatable annuals and perennials, some being exotic such as cheat grass [*Bromus tectorum*]), were 7 times more abundant in Emigration Canyon. Young and Evans (1989) tie deteriorated range condition to the establishment and spread of exotic and noxious weeds in Nevada. Duff (1979) reported that in an exclosure on Big Creek in Utah, the more mesophyllic vegetation along the stream was moving outward from the stream as groundwater reserves increased, while in the grazed portion upland vegetation (i.e., sagebrush) continued invading the floodplain.

Dense shrubs along the stream (e.g., willows) provide shade for the stream, detritus for insect food, and stabilize banks. On Trout Creek in Montana, Marcuson (1977) reported shrub production to be 13 times greater in an ungrazed area as compared to a heavily grazed site. Prior to exclusion of livestock on Big Creek, in Utah, willows were so severely grazed that they were hedged back to basal stems. After exclusion of livestock willows responded slowly, but after 4 years they were 0.5 m tall, and in bend areas mean stem densities were 0.2/1.4 m^2 (Duff 1979). In northern Colorado seasonal grazing practices significantly altered shape, size, volume, and quantities of live and dead willow stems (Knopf and Cannon 1982). Martin (1979) listed livestock tree preference in Arizona as willow, velvet ash (*Fraxinus pennsylvanica* ssp. *velutina*), Arizona alder (*Alnus oblongifolia*), netleaf hackberry (*Celtis reticulata*), and Arizona sycamore, with even the least palatable young trees suffering damage in July. Storch (1979) reported on Camp Creek in eastern Oregon that shrub canopy was < 20% before exclusion of livestock, but 4 years after exclusion it was providing up to 75% shade to the stream. Livestock may remove ≥2 years of willow growth in a summer grazing period (Chaney et al. 1993).

Willows are an extremely important component of riparian areas and probably were one of the first woody elements to decline in the West. A historical literature review covering 1812-1880 reported extensive willow stands throughout western rangelands, but ". . . by the early 1900's, many of these stands were severely damaged or eliminated through cattle overuse" (Kovalchik and Elmore 1992:111). Though willows can withstand heavy browsing and not die, they cease seed production which alters their population dynamics and demography for many generations (Verkaar 1987). Kay and Chadde (1992) studied seed production in 3 willow species in Yellowstone National Park subjected to elk browsing, and willows in exclusions produced a range of 109,000-583,000 seeds/m^2. Browsed willows outside exclosures did not even produce catkins, much less seed. In southeastern Utah there were few, if any, bank or coyote willow (*Salix exigua*) in 5 heavily grazed canyons draining the east side of Cedar Mesa, while on the west side in Grand Gulch (cattle excluded for 20 years), bank willow shoots equaled or exceeded 30/m^2 (Figs. 11 and 12).

As shrubs are overgrazed year after year much of the dense shade component is eliminated. Combined with channel widening, water temperatures increase and oxygen tension levels decline. As willows disappear, the woody roots for stabilizing banks are reduced and may even be lost. Large shifts in water temperature affect fish populations and aquatic insects (Rhodes and Hubert 1991). Platts (1979:41) states: "Streamside vegetation protects streambanks by reducing erosive energy, by helping deposits build the streambanks, and by keeping the streambank from being damaged by ice, log debris or animal trampling." Streams in the Intermountain and Pacific Northwest are frequently icebound in winter. As the ice breaks up in spring it causes shifting dams, which forces the water over the floodplain. If riparian vegetation is not sufficient to protect the banks they can become heavily eroded (Platts 1991).

Comparative water temperatures inside and outside livestock exclosures demonstrate the value of riparian vegetation in depressing water temperatures. After 1 year of livestock exclusion, Van Velson (1979) reported water temperatures were reduced from 24C to 22C in Nebraska. Storch (1979) reported that on Camp Creek in eastern Oregon, mean daily water fluc-

FIGURE 11. Road Canyon in southeastern Utah that drains the east side of Cedar Mesa and has the same physical characteristics as Grand Gulch (Fig. 12). Virtually uncontrolled livestock grazing occurred until December 1993. The stream is entrenched with 2-4 m cut banks with cobble or bedrock bottoms typical along the drainage. Any surface stream flow undercuts banks to further eliminate the remaining riparian trees. No external disturbance occurred except for domestic livestock. Photo by J. Feller, March 1992.

FIGURE 12. Grand Gulch in southeastern Utah that drains the west side of Cedar Mesa and has the same physical characteristics as Road Canyon (Fig. 11). Livestock grazing has been eliminated for 20 years. The dominant woody vegetation along streamside is willow. Some cottonwoods are in the background. Photograph by J. Feller, June 1992.

tuations outside an exclosure were 27C compared to 13C inside the exclosure. Maximum temperatures outside and downstream from the exclosure averaged 11C higher than inside the exclosure. Mean daily water fluctuations were 15C outside the exclosure and 7C inside the exclosure.

Water quality is degraded by sedimentation. Behnke and Raleigh (1978) reported that overgrazing can cause accelerated sedimentation and silt degradation of spawning and insect production areas. Winegar (1977) working on Camp Creek reported sediment loads reduced by 48-79% as it flowed through a 5.6 km exclosure. Accelerated erosion (that caused by grazing) was examined under 3 different grazing levels in Utah to detect sediment transport levels (Croft et al. 1943). They intensively sampled 3 canyons in the Wasatch Mountains and ranked them as to grazing use: lightest (City Creek), moderate (Red Butte), and heavy (Emigration Canyon). Heaviest soil losses were where grazing was heaviest and highly reduced where grazing was lightest. They strongly suggest ". . . grazing management is as much a problem of soil management as of forage management" (Croft et al. 1943:16).

Phillips et al. (1975) reported fine sediments killing fish embryos. Platts (1979) reports that fine sediments cause embryos to receive less oxygen and allow toxic metabolic wastes to accumulate. These sediments also fill spaces in gravel beds, which reduces the protective cover and forces young fish to surface waters where they are more vulnerable to severe winter temperatures and predation. Platts (1978:42) reported that ". . . fish forced to remain in turbid waters may have trouble feeding, using oxygen, and reproducing."

Livestock grazing may also cause chemical and bacterial changes in a stream, but changes may not be manifested immediately. Johnson et al. (1978) did not find any chemical differences between an excluded reach and a grazed reach during the grazing season. However, following the grazing season a significant increase was noted in total dissolved solids, indicating livestock waste entering the stream, possibly from rain showers. In the stream reach where cattle were grazed, there was a significant increase of fecal coliform and fecal streptococci until about 9 days after grazing ceased. Numerous workers have attributed high fecal coli counts in streams to livestock grazing (Kunkle 1970, Darling and Coltharp 1973, Skinner et al. 1974).

Chemical and bacterial changes may, in concert, with physical changes negatively affect fish populations. This was the case in 2 springs in Pahranagat Valley in Nevada (Taylor et al. 1989). Ammonia and nitrate levels became so high that nitrifying bacteria consumed oxygen to levels that fish died. Bacterium populations of *Pseudomonas aeruginosa* and *Aeromonas hydrophila* also increased (Taylor et al. 1989). One fish, the White River springfish (*Crenichthys baileyi baileyi*), was federally endangered and livestock were removed allowing the fish population to recover. Livestock were not removed at Brownie Spring, which supports Pahranagat dace (*Rhinichythys osculus*), and that population has not recovered (Taylor et al. 1989).

Desert fishes were undoubtedly heavily impacted by overgrazing since 1880 (Hastings 1959, Miller 1961, Minckley 1973), but water management activities and introduction of exotic fishes have been more devastating and expedient in eliminating populations (Miller 1961, Minckley 1973). Many species were extirpated before the impacts of domestic livestock were known or fully understood. Hastings and Turner (1965:64, 65, 69, 74) show early photographs (circa 1890) of springs and streams that supported native fishes and these clearly show a highly degraded condition. Possibly many of these springs and small streams supporting native fishes were highly degraded earlier in that Cooke in 1846 described a stream on Agua Prieta Creek with the appearance of a stockyard (Bieber 1938).

Studies of fish populations from streams in PFC demonstrate superior fish habitat conditions. Deeper and narrower streams increase cover, movement areas for trout, and provide a combination of pool types (Raleigh 1982). Fisheries biologists report that lower stream width-depth ratios provide better fish habitat (Behnke and Zarn 1976, Platts 1981*a,b*). Differences in trout standing crop for ungrazed portions of Sheep Creek in Colorado were twice that in the grazed portion (Stuber 1985). In Montana, Gunderson (1968) reported a 30% increase in brown trout (*Salmo trutta*) in an ungrazed stream reach; and Marcuson (1977) reported a brown trout population 3.4 times greater than a grazed reach in Montana. On the Little Deschutes River in Oregon, Lorz (1974) reported trout populations 3.5 times greater in ungrazed versus grazed stream reaches. Similarly, in Rock Creek in Montana, Marcuson (1977) reported brown trout biomass 3.4 times higher in ungrazed stream reaches. Kimball and Savage (1977) reported a 4.25 increase after livestock exclusion for 4 years in Diamond Creek in Utah. Van Velson (1979) reported 88% of a fish population were rough fish while an area was grazed, and after 8 years rest only 1% of the population was rough fish. In Washington, significant reductions in biomass for coho salmon (*Oncorhynchus kisutch*), cutthroat trout, and other salmonids were reported in heavily grazed areas versus ungrazed areas (Chapman and Knudsen 1980).

The validity of some of the above fishery standing crop values have been questioned by Platts (1982). He questions sample size, statistical reliability, lack of controls, and other facets of some of the studies. Some terrestrial studies could be subjected to the same concerns. The inclusion or exclusion of those studies does not change the overall picture of uncontrolled livestock grazing on the degradation of western riparian habitats and their effects on native fishes.

Fifty years of livestock exclusion on Wickiup Creek in the Blue Mountains of central Oregon shows the reversal that occurred when a riparian system was relieved of grazing (Clifton 1989). A 1933 photograph prior to livestock exclusion shows the meadow barren of vegetation, exposed soils, channel banks devoid of vegetation, and banks about 1.3 m high (Clifton 1989). The channel was trapezoidal in shape with outsloping or widened banks. Ten years after exclosure the meadow showed vegetation, the channel had aggraded about 0.6 m, and the channel banks were vegetated. Fifty years after livestock exclusion, the channel had undergone a 94% reduction in cross section and was described as having ". . . thickly vegetated overhanging banks [that] obscure a narrow and deep channel" (Clifton 1989:128). Similar vegetation responses were reported for Sheep Creek on the Roosevelt Forest in north-central Colorado at 2,500 m (Schulz and Leininger 1990). They reported that after 30 years of cattle exclusion there was twice the litter in the protected site, and 4 times more bare ground in the grazed area. Willow canopy was 8.5 times greater in the protected site while Kentucky bluegrass (*Poa pratensis*) was 4 times greater in the grazed site. Fowl bluegrass (*P. palustris*) was 6 times greater in the protected site. Caged plots within the grazed area only produced a peak standing crop of 1,217 kg/ha, while 2,410 kg/ha were produced in the exclosure. I refer the reader to the 1939 repeat photograph of the stream (Schulz and Leininger 1990:296). Twenty years of livestock exclusion in Grand Gulch in southeastern Utah has transformed an entrenched, intermittent stream running on bedrock or heavy cobble (Blackburn 1993) to one aggraded with well-defined banks, there are indications that it may become perennial, and it now supports dense willow-cottonwood communities (Fig. 12).

Wildlife. Structural damage to streambanks along with their denudation and that of the floodplain in Phase I began impacting amphibians, some reptiles, and ground-nesting birds. Data presented by Szaro et al. (1985) from an exclosure in a high-elevation riparian community of alder and willow in New Mexico, demonstrates the importance of the floodplain understory for the wandering gartersnake (*Thamnophis elegans elegans*). In the exclosure (≤10 years protection) both vegetative groundcover and debris accumulated to a level to provide habitat for this snake. Gartersnake density was significantly higher in the ungrazed versus the grazed site (capture rate 5:1). Loss of the herbaceous groundcover and the understory in Phase I probably occurred shortly after bank and channel degradation, and has continued for so long that I suspect many populations of these species were locally extirpated, and if not, significantly reduced in density as was the wandering gartersnake.

The dramatic decline in the herbaceous groundcover and thinning of the understory in Phase I took its toll on all wildlife populations dependent on these layers. Moulton (1978) suggested that species richness in small mammals might increase with grazing because it would create new microhabitats with more diversity. This might be true in some localities, but Medin and Clary (1989) reported the reverse with higher small mammal species richness (11 species vs. 6) in Nevada on a site protected for 11 years compared to a grazed site. They reported a higher standing crop biomass (3.24), species richness (1.83), and species diversity (1.25) on the ungrazed site. Moulton (1978) also reported that grazing may have limited densities of the prairie vole (*Microtus ochrogaster*) which prefers dense groundcover, while improving habitat for mice in the genus *Peromyscus*. Schulz and Leininger (1991) reported trapping 28 small mammals in a grazed site and 41 in a site that had not been grazed for 30 years. The ubiquitous deer mouse (*Peromyscus maniculatus*) dominated the grazed site (15:1) and the western jumping mouse (*Zapus princeps*), preferring dense herbaceous groundcover, dominated the ungrazed site (22:1).

The avifauna inhabiting the understory should be dramatically affected if foliage volume is an important wildlife habitat component. Rucks (1978) stated that understory depletion displaced shrub-nesting species with more generalists that had no preference for nest placement. Taylor (1986) found a significant correlation between increased annual grazing frequency and decreases in bird abundance, shrub volume, and shrub height, as well as between bird abundance and shrub density and height. Numbers of species decreased as intensity of grazing increased and density values were 5-7 times higher on an area ungrazed since 1940 than on 2 areas grazed annually until 1980. His examination of 1930 photos, ". . . showed a tall deciduous upper canopy along the river . . ." and that ". . . cattle grazing

FIGURE 13. San Pedro River, Arizona, approx. 1985. The area was in private ownership with unmanaged livestock grazing. Raw and eroding banks on right, open and wide channel, and absence of herbaceous and understory vegetation are characteristic symptoms of unmanaged grazing. Photo courtesy of BLM Safford District Office, Arizona.

can eliminate or reduce the upper canopy by preventing the establishment of saplings . . ." (Taylor 1986:257).

A 64-km reach of the San Pedro River in southeastern Arizona provides unique insight as to what most perennial desert streams resembled about 1875-1885 (Figs. 13-15) as Phase I was completed. A rare glimpse of this river area before the livestock boom of the 1880s was provided by a pioneer rancher named H. C. Bayless. In 1901, D. A. Griffiths, Chief Botanist over Grass and Forage Plant Investigations for the Arizona Experiment Station in Tucson, sent a circular to a select group of pioneer ranchers in an effort to better understand the role of livestock and the condition of the range prior to and after the 1891-1893 drought (originals not seen *in* Bahre 1991, but also see Hendrickson and Minckley 1984). To a question on entrenchment of the river, Bayless wrote,

> Above twelve years ago the San Pedro Valley consisted of a narrow strip of subirrigated and very fertile lands. Beaver dams checked the flow of water and prevented the cutting of a channel. Trappers exterminated the beavers, and less grass on the hillsides permitted greater erosion, so that within four or five years a channel varying in depth from 3 to 20 feet was cut almost the whole length of the river (Bahre 1991:111).

Bayless' response to the question of whether the current situation was caused by overstocking, drought or both was:

> The present unproductive conditions are due entirely to overstocking. The laws of nature have not changed. Under similar conditions vegetation would flourish on our ranges today as it did fifteen years ago. We are still receiving our average amount of rainfall and sunshine necessary to plant growth. Droughts are not more frequent now than in the past, but mother earth has been stripped of all grass covering. The very roots have been trampled out by the hungry herds constantly wandering to and fro in search of enough food. The bare surface of the ground affords no resistance to the rain that falls upon it and the precious water rushes away in destructive volumes, bearing with it all the lighter and richer particles of the soil. That the sand and rocks left behind are able to support even the scantiest growth of plant life is a remarkable tribute to our marvelous climate. Vegetation does not thrive as it once did, not because of drought, but because the seed is gone, the roots are gone, and the soil is gone (Bahre 1991:112).

The once subirrigated farmland and marshy conditions disappeared on the river as it entrenched and water tables dropped. Somewhere about the turn of the century, cottonwood and willows became established, possibly when livestock numbers were extremely low after the drought and before numbers were reestablished once the range improved. A portion of the San Pedro River described by Bayless came under BLM control as a Riparian National Conservation Area and domestic livestock grazing was eliminated in January 1987. Streambanks, channel, and understory conditions at the time grazing ceased were

FIGURE 14. (Left) San Pedro River, June 1987. The area is now BLM National Riparian Conservation Area (NRCA) and livestock were removed 1 January 1987. This would be typical of southwestern streams about 1885 with a highly modified channel, trampled banks, no herbaceous groundcover, and understory depauperate. Photograph courtesy of BLM Safford District Office, Arizona. (Right) Repeat of (Left) 4 years (June 1991) after livestock were removed. Photopoint moved to left because developed understory disallowed the photo. Note bank vegetation beginning to narrow and deepen the channel. Photograph courtesy of BLM Safford District Office, Arizona.

FIGURE 15. (Left) San Pedro River, June 1987. Different location in the National Riparian Conservation Area but stream conditions are typical of unmanaged grazing along stream courses. Photograph courtesy of BLM Safford District Office, Arizona. (Right) Repeat of (Left), June 1991. Stream being narrowed and deepened with encroachment and sediment trapping of the vegetation. Photograph courtesy of BLM Safford District Office, Arizona.

essentially as I described at the end of Phase I. Within 4 years after livestock exclusion, the understory and bank vegetation had increased significantly (Krueper 1993).

Response of neotropical birds on the San Pedro River during the 4 years after exclusion ranged from virtually unchanged for those species foraging on volant insects to moderate increases of 2-6-fold for those gleaning foliage insects (Krueper 1993). Highly significant density increases were observed in foliage gleaning and understory thicket specialists such as the common yellowthroat (*Geothlypis trichas*) and song sparrow (*Melospiza melodia*) that showed a 25-fold and 61-fold increase, respectively (Table 1). In portions of Sheep Creek in northern Colorado that have been excluded from grazing for 30 years (Schulz and Leininger 1991) showed Wilson's warbler (*Wilsonia pusilla*) and Lincoln's sparrow (*Melospiza lincolnii*), thicket specialists, more common in the restored site. Finch (1986) reported these 2 species dominating healthy subalpine willow communities in southeastern Wyoming.

The ecological contributions that birds make to forest communities are poorly understood but studies over the last decade have focused more attention to the contributions that this group makes to forested ecosystems. Prior to these studies the lay public was highly emotional toward this group (Carson 1962) and birds were perceived more as jeweled forest songsters. Frugivorous forms have been documented as important dispersers of seeds away from the parent tree (Howe and Vande Kerckhove 1979, 1981; Pratt and Styles 1983, Masaki et al. 1994). Insectivorous birds in forests have often been assumed to simply be a small additive factor of mortality to phytophagous or plant-eating insects and Crawford and Jennings (1989) reported a great reduction in densities of spruce budworm by bird predation. The most impressive demonstration of phytophagous insect control

TABLE 1. Increase in bird numbers after removal of cattle from grazing for 5 years. Table adapted from Krueper (1993). NA = data not available.

	Years					
Species (densities are birds/40 ha)	1986	1987	1988	1989	1990	1991
Yellow-billed cuckoo (*Coccyzus americanus*)	6	10	8	6	13	NA
Western wood-pewee (*Contopus sordidulus*)	8	16	22	38	28	29
Brown-crested flycatcher (*Myiarchus tyrannulus*)	21	33	27	36	26	26
Bell's vireo (*Vireo bellii*)	7	11	7	12	15	16
Yellow warbler (*Dendroica petechia*)	29	84	99	227	131	176
Common yellowthroat (*Geothlypis trichas*)	7	24	39	115	110	149
Yellow-breasted chat (*Icteria virens*)	26	44	47	95	100	110
Summer tanager (*Piranga rubra*)	44	84	73	167	94	108
Song sparrow (*Melospiza melodia*)	0	11	14	38	36	61
Northern oriole (*Icterus galbula*)	28	35	28	34	21	32

by birds examined white oak (*Quercus alba*) growth over a 2-year period in a Missouri deciduous forest (Marquis and Whelan 1994). Study trees were compared in a natural setting (controls), caged that allowed insect passage and excluded birds, and trees sprayed with a pesticide. Controls lost 13% leaf area, sprayed trees 6%, and caged plants 25% at the end of the first season (24, 9, and 34%, respectively, the second season). Differences in above ground biomass production (growth) were reduced by one-third in caged trees from sprayed trees with controls having intermediate values. Bird populations significantly controlled insect populations in these studies.

The importance of riparian habitats as nesting and refueling sites for migrating wildlife is frequently mentioned in the literature, but few studies have examined this subject in any depth. Stevens et al. (1977) summarized the literature and from their own data reported that riparian plots contained up to 10.6 times as many migrants per hectare as paired upland sites. More recent and refined studies, many by biologists studying the plight of neotropical migrants, are beginning to provide enough information to indicate the importance of these habitats to migratory wildlife.

Livestock grazing is not the central issue, but, combined with water management, it has contributed heavily to the decline in quality stopover and wintering habitat. Southwestern riparian habitats are an important stopover and wintering area since they are surrounded by arid uplands. As they are degraded and reduced in size their availability and suitability for migrants becomes more limited. Laymon (1984) suspects that riparian forest fragmentation and tiny forest size may now be limiting avian densities nesting to the north. Stanley et al. (1991) contends that they are extremely important areas for migrating birds since they remain green and productive during late summer post-breeding dispersal and in fall migration when there is little upland productivity. The extensive and multiple kilometers of riparian forest along the Sacramento River in California are now only a few trees wide and highly fragmented into patches (Tompson 1980). The same holds true on the lower Colorado River (Ohmart et al. 1988).

Evidence suggests that passage migrants select stopover sites and length of stay based on the intrinsic suitability of the habitat (Moore and Simons 1992). Therefore as riparian habitats continue to be destroyed, fragmented, and degraded in foliage volume and insect productivity, migrant passage or survival in passage could be highly limited as riparian forest size and productivity decline.

A few studies are beginning to indicate the importance of these riparian sites as refueling areas for passage migrants. For example, in the white-crowned sparrow (*Zonotrichia leucophrys*; Cherry 1982) and in wood warblers (Moore and Kerlinger 1987), leaner birds stayed longer and stored larger amounts of fat than those birds with good fat stores. Without quality habitats en route many of the birds in poor fat condition might not finish the migration without rebuilding sufficient fat reserves (Winker et al. 1992).

An interesting data set comes from a 2-ha remnant riparian area in California surrounded by urban and agricultural development along Coyote Creek upstream from where the creek enters the San Francisco Bay. This area was mist netted from 1987-1991 to examine migrant use and body mass changes. Of the 4 species examined in spring and fall migration few (6-18% depending on season and species) stayed on the site >1 day. Most (52-79%) gained or maintained body mass, suggesting the stopover was for refueling. Otahal (ms) calculated flight distance from stored fat for the most extreme specimen, a 10 g yellow warbler (*Dendroica petechia*) which gained 5 g of fat, could then potentially fly 2,848 km on the added fat stores. The continuing loss and degradation of riparian forests may have extensive effects on migratory and overwintering wildlife.

Many western riparian habitats are beginning to approach the threshold where cover and/or foliage availability for insect production for habitat specialists is barely sufficient to sustain populations. Not only vegetation density and distribution, but forest island size may also be a determining factor for some species. Almost all of the most important terrestrial wildlife habitat elements described earlier are essentially gone or highly degraded. The cottonwood-willow community and their tree densities at low and moderate elevations are rapidly disappearing (The Arizona Nature Conservancy 1987). Extensive stands of dense willow have been fragmented and, in some instances, eliminated (Elmore and Kauffman 1994). The foliage profile is now a skeleton of what it was 50-100 years ago and foliage volumes are sparse at all layers. Also, through time the intracommunity plant patchiness element has slowly disappeared. Knopf and Cannon (1982) suggest that in northern Colorado horizontal and vertical structure of the shrub willow community has been eliminated for birds by seasonal grazing over the past 75-100 years. The shrub component may persist along some second terraces, but reduced densities have left a sparse shrub element or it has been converted to agriculture.

Even many wildlife refuges have been subjected to intensive domestic livestock grazing. The 73,200-ha Malheur National Wildlife Refuge in southeastern Oregon was grazed by 40,000 AUMs in the 1930s, by the 1950s this had been increased to over 100,000 AUMs, and in the late 1960s the mean for 9 years was 118,000 AUMs (Taylor 1986). Refuge personnel also used herbicides and grubbing to remove willows to increase livestock forage. Predictably, willow flycatcher and yellow warbler numbers plummeted, but as cattle numbers were reduced in the 1970s the willow element began to recover. Breeding bird surveys showed 7 yellow warblers in 1972 and no willow flycatchers. By 1982 yellow warblers had increased to 56 and willow flycatchers numbered 30. Bird data from transects on the refuge showed similar trends and vegetation data from these transects showed a negative correlation between shrub volume and frequency of cattle use on an annual basis.

Unless grazing management changes are made soon it is predictable that many more species, especially neotropical birds, will be placed on the endangered species list. Horning (1994) reported that of the 76 federally listed plant and animal species on BLM lands where livestock grazing was a significant factor in their decline, 61 species were riparian dependent or associated with riparian habitats.

BEAVERS AND CATTLE

In the western United States it was probably best for the extended longevity of riparian systems that beavers were virtually eliminated prior to the introduction of extensive numbers of domestic livestock to western rangelands. Beavers and unmanaged livestock grazing in stream systems are extremely damaging and together expedite the collapse of riparian forests. Both can be in direct competition for food (i.e., woody and herbaceous) depending on the condition of the riparian habitat (USDA 1992). Livestock crush dams in their efforts to consume the lush forage of sedges, rushes, and willows. They also consume suckering new growth of young trees cut by beavers for dam repair and food, and imbalance the beaver-stream equilibrium that has evolved over the years. As the young trees cut by beavers attempt to put up new shoots these are consumed until the energy reserves are depleted and the tree dies. Repeated growing-season grazing weakens the woody and fibrous-rooted species until they are either washed out in large storm events or die. In the absence of woody and fibrous roots the alluvial soils are then vulnerable to further erosion in each storm event. As the stream widens and downcuts, the water table is lowered, leaving wetland species not eliminated by grazing with soil moisture levels too low to survive. In the damaged system, trampling destroys new dam efforts and with time the elimination of young trees for dams and food begins stressing the beaver population. Beavers are then forced to consume the cambium of the larger deciduous trees expediting the collapse of the mature riparian forest (Figs. 16 and 17). Apparently, once a stream degrades to this level beavers (and livestock) must be removed to expedite the recovery of the stream (L. Meyers, USFS, pers. commun.).

Livestock were not involved in the below example, but it shows how important beaver can be in modifying hydrological and floristic processes. Red Butte Creek in Utah had beavers trapped from the canyon, but they were reintroduced in 1928 (Bates 1963). The Army, then in possession of the canyon, had the beaver removed in 1982, fearing water contamination at Fort Douglas.

Where beavers were active along the stream, the vegetation cover was affected approximately 91 m on either side (Bates 1963). Sediment deposition behind the dams ranged from 0.6-2.4 m deep. Earlier, Scheffer (1938) had reported that 2 dams in the canyon had trapped 4,468 m^3 of silt. In 1983, a year after the beavers were removed, a large storm eroded huge quantities of sediment and incised the stream again, creating a large delta in the reservoir at the mouth of the canyon. In the absence of beavers, 55 plant species have either been extirpated or are highly restricted in their distribution (Ehleringer et al. 1992). Personnel from the USFS claim that flood damage to the canyon would not have been as severe or prolonged had the beaver been active during the floods (Ehleringer et al. 1992).

Beneficial Effects of Livestock Grazing on Riparian Habitat

There is no advantage or benefit to riparian habitats in PFC to be grazed by any large ungulate, be it livestock or elk (Houston 1982, Chadde 1989, Arizona Game and Fish Department 1993). Several years ago when public hearings were held on the transfer of the San Pedro River, Arizona, from private holdings to the BLM, I testified that cattle could be used to economically reduce the fuel load of tall sacaton grasses (*Sporobolus* spp.) growing adjacent to and within cottonwood-willow habitats. Removal of this material would prevent fires that are highly detrimental to these forests. Krueper (1993:323), working on the San Pedro River, stated "Grazing within the riparian zone may be used to reduce dense annual growth . . ." to prevent fires. In this very limited situation, cattle may be useful to help reduce fuel loads and prevent wildfires that are especially detrimental to cottonwood trees.

FIGURE 16. Main Diamond Creek, Gila National Forest, New Mexico. Beaver lodge with sticks of dam visible to back left. The animals are living in the water table and consuming mature cottonwoods around them because there are no young trees for food. Photograph by R. D. Ohmart, May 1994.

FIGURE 17. Same location as Figure 16. Note extent of tree felling and eventual demise of the cottonwoods. Foreground shows cow pies, tracking, and absence of herbaceous groundcover from unmanaged grazing. Photograph by R. D. Ohmart, May 1994.

It has been suggested that cattle might serve to open dense willow thickets and help wildlife in high altitude riparian habitats in southeastern Wyoming (Krueger and Anderson 1985). Willow densities in the study streams were 2,007/ha and 897/ha, the latter stream having a record of overgrazing. Some increased bird densities were recorded in the tunneled willows, but attracted species were habitat generalists and not the specialists that prefer dense willow thickets. Small mammal habitats might be improved by the opening of willows and creating more grass and sedge areas (Krueger and Anderson 1985). Grazing could also be used to create low density willow habitat, but the authors state, "Enough riparian habitat has been overgrazed to create plenty of low density shrub-willow habitat" (Krueger and Anderson 1985:303).

Unquestionably, grazing can be used to enhance habitat for some avian species (Burgess et al. 1965, Kirch and Higgins 1976, Ryder 1980, Crouch 1982, Schulz and Leininger 1991, Clary and Medin 1992). Most of the species added are habitat generalists whose numbers are common in the uplands. Those species thought to be declining and possibly being eliminated are foliage volume or thicket specialists. For example, on a grazed versus ungrazed stream in Colorado, American robin (*Turdus migratorius*) numbers were 30 and 15, Lincoln's sparrows 4 and 13, mountain chickadees (*Parus gambeli*) 8 and 17, and Wilson's warblers 0 and 9, respectively. The ubiquitous American robin increased in heavily grazed riparian habitats, while those species requiring dense habitats to either forage, nest, or for cover declined (Schulz and Leininger 1991). Bock et al. (1993) in reviewing the literature reported that of 43 neotropical migrants, 8 responded favorably to grazing, 17 were negatively affected, and 18 were unresponsive or showed mixed responses. There is no problem with managed grazing or using grazing as a tool to increase biodiversity, but examples of managed grazing are so few and unmanaged grazing so common in riparian habitats that biodiversity is rapidly being lost (Horning 1994).

It's also been advocated that cattle can serve as a tool to modify floodplain terraces to improve groundcover (A. Savory, Holistic Resources Management, pers. commun.). I can only visualize this being true on rivers where the natural flooding process has been stopped by dams. Reduced instream flows below dams with a concomitant decline in the water table converts the higher second terraces to upland habitats. Cattle might be useful in converting decadent shrub communities into grass communities. Otherwise, I fully agree with Chaney et al. (1993:14):

> Some people tout that livestock trampling as a 'tool' to lay back steep or undercut streambanks. The channel of a stream low in sediment could take decades to recover from being 'laid back.'

STATUS OF RIPARIAN AREAS ON PUBLIC LANDS

An early status report on BLM lands (Trout Unlimited 1979) reported that 77% of 30,577 km of streams were in unsatisfactory condition. A 1990 status report on riparian habitats from the USFS in the western national forests estimated that 93,339 km of riparian areas within grazing allotments did not meet and were not moving toward meeting forest plan objectives (GAO 1992). The BLM reported that on 0.5 million ha of riparian-wetland and 78,856 km of riparian streams in 10 BLM state offices, only 7% of the riparian areas were meeting objectives, 8% were not meeting them, and riparian response was unknown in 85% of the areas. The 85% unknown concerns me in that if it is not known what condition they are in then the probability is good that they are not functioning properly. As an example, a BLM report in 1989 by the Gunnison Basin Resource Area Office in Colorado stated,

> . . . that 60 to 100 percent of the riparian areas were being overgrazed. Overgrazing damaged the riparian areas to the extent that forage production was below normal; plant species composition was undesirable; stream channels and stream banks were unstable, causing erosion; soils were compacted, reducing water infiltration; vegetation cover was reduced, resulting in excessive silt from heavy runoffs; groundwater reservoirs were not able to recharge and out of bank heavy runoffs were not slowed down and dispersed (Office of Inspector General 1990).

The most recent (USDI 1994) data estimates that for BLM-managed lands, of the approximately 400,000 ha of riparian habitats 20% are nonfunctioning and 46% are functioning at risk, which means they are threatened by domestic livestock grazing. About 34% are in proper functioning condition. Of the riparian habitats on USFS lands, 63% are meeting objectives while 27% are not. These data very closely agree with data on the Uncompahgre-Gunnison National Forest in Colorado where of 5,885 km of perennial streams, 65% are meeting objectives and 35% are not (R. L. Storch, Forest Supervisor, USFS, pers. commun.). My observations are that of most of the forests in Southwest Region 3 (i.e., Arizona and New Mexico), the number of streams in PFC would be more the opposite. Further, no one in the Regional or Forest Supervisor Offices (Region 3) was aware how the National numbers were obtained, or if they ever were for Region 3. Apparently, Region 3 was not included in the 1994 data set. For example, when the GAO (1988) did its survey they looked indepth at 5 locations and reported that on the Tonto National Forest (Region 3) that 80-90% of the riparian areas were in unsatisfactory condition.

Along important streams for fish, wildlife, or scenic values, which serve as riparian pastures, one would think that the agencies and permittees would have showcased managed-grazing examples. The East Fork of the Gila River, Gila Wilderness Area, Gila National Forest, New Mexico, once a cold-water fishery, is now a warmwater fishery with eroding and caving banks, high sediment loads, and virtually no woody vegetation along the stream. Though a riparian pasture, it is questionable when it will receive better management. The Comb Wash Canyons, Comb Wash Allotment, San Juan Resource Area, BLM in southeastern Utah are entrenched to bedrock or large cobble, support very little riparian vegetation, and provide some of the most scenic riparian habitats in the West. These riparian pastures have only recently received management protection (Rampton 1993).

POSITIVE RESULTS FROM COOPERATING PERMITTEES

Some ranchers have altered their riparian grazing approaches (voluntarily, persuaded, and/or through endangered species restrictions) that has resulted in improved riparian habitats. This has provided increased forage production from these habitats (GAO 1988). The GAO (1988) interviewed many of these permittees and reported savings in reduced feed costs, availability of permanent water supplies where streams had been intermittent, better utilization of upland forage by livestock where they previously had not grazed, and generally better livestock health and calving rates.

A few ranchers have taken the initiative to improve riparian habitats voluntarily. Date Creek in Arizona is a case in point (Fig. 18). After 24 years of dormant-season-only grazing this small stream looks (in 1991) totally incongruous compared to most streams in Arizona. New banks have been formed from trapped sediment and are now matted with grasses, sedges, and rushes. A young and healthy age mix of willows and cottonwoods dominated the floodplain, and in many stream reaches the luxuriant vegetative growth has to be separated to find the 15-20-cm wide banks that encase the 30-40-cm deep riverlet. I visited the stream in July 1991 after heavy March storms had created highly erosive floods throughout Arizona. I expected to see extensive scouring and the possible loss of the 1.6 km-long managed area, but this had not occurred. Figure 18 shows the contrast where there is no management versus the fence line where there is only dormant season use. The obvious difference in standing crop in the fore and background of the photo clearly shows the advantage of increased forage production for both wildlife and livestock. The managed area supports several species of songbirds, nesting zone-tailed hawks (*Buteo albonotatus*), javelina (*Tayassu tajacu*), and mule deer, while only a few songbirds were observed in the few mature trees upstream.

FIGURE 18. Date Creek near Wickenburg, Arizona. A fence separates the unmanaged allotment from the managed downstream area. A major difference in herbaceous groundcover for soil protection and forage availability is obvious in the managed area. Photograph by R. D. Ohmart, June 1992.

Much (approx. 90%) of the floodplain or first terrace of Date Creek was lost in extremely heavy storm events in January and February 1993. A century of January rainfall weather records was broken in 1993, which indicates the magnitude of the event. It should also be kept in mind that the watershed and stream have been degraded for over 100 years and possibly the degradation might have been negligible had they been in good ecological health. The positive side is that management improvements that had accrued over the past 24 years protected the integrity of much of the riparian community. A line of 15-24-year-old willows and cottonwoods had developed along the outer edge of the first terrace and these trees withstood (Fig. 19) and dissipated the erosive force of the flood, keeping most of the scouring in and along the first terrace. In a few places where

FIGURE 19. Date Creek, Wickenburg, Arizona, livestock managed area with trees lining the primary floodplain. Grasses, sedges, and rushes reestablishing streambanks and beginning to stabilize the sandy soils after the 1993 flood event. Photograph by R. D. Ohmart, June 1994.

FIGURE 20. Date Creek managed area showing rapid invasion of knot grass (*Paspalum distichum*) and cattail to begin soil stability, sediment trapping, and stream containment. Photograph by R. D. Ohmart, June 1994.

the tree line did not exist, heavy erosion cut new channels, damaging the second terrace. Though the flood removed much of the first terrace, small clumps of sedges (*Carex* spp., *Cypreus* spp.), rushes (*Juncus* spp.), and cattails (*Typha* spp.) persisted and these are rapidly pioneering into bare areas (Fig. 20).

The allotment upstream that has year-round grazing and no management was devastated by the flood. Channel changes were rampant, mature trees were uprooted near the stream and the second terrace of velvet mesquite (*Prosopis velutina*) (prior to degradation and entrenchment this area was flooded in higher than normal flood events) was heavily eroded (Fig. 21). I repeated the photopoint taken in 1992 (Fig. 18) at the water gap between the 2 properties and though much of the primary floodplain is gone from the managed area, the young trees remain in the background.

I know of a few permittees in Arizona that have, on their own, made financial and personal sacrifices to improve riparian habitats on their deeded lands and leased or public lands. I hope there are more in Arizona and throughout the West who are quietly going about their work and doing similarly. These permittees are to be commended and supported for they will be a standard to those who have not made this commitment. As public concern and litigation rapidly increase over the degraded condition of western riparian habitats, the fate of the livestock industry on public lands may rest on these dedicated permittees. They show how natural resources protection and domestic livestock are compatible with proper grazing management. It can be done but it will take more tax dollars and sacrifices on the permittees' part. But, as riparian systems begin functioning properly, the annual harvestable forage for both wildlife and livestock from these flood-irrigated pasturelands will far surpass a decade of forage production from a degraded system.

GRAZING SYSTEMS AND MANAGEMENT APPROACHES

The best way to manage riparian habitats is not to graze them. If they are to be grazed, the manager must learn how to use the forage resource while maintaining stream PFC. This will be a management challenge, because livestock concentrate in and are highly attracted to riparian areas, these habitats are usually scattered throughout the allotment, and each riparian system has its own set of vulnerable biotic and abiotic components (Elmore and Kauffman 1994). Kinch (1989) reports that the management of domestic livestock grazing in riparian-wetland areas is one of the most difficult and complex issues facing western rangeland management.

Numerous grazing approaches or systems have been developed over the years in an attempt to help deteriorated rangelands and increase forage production. Few, if any of these approaches, consider the condition or grazing impacts on riparian communities (Platts 1981*a*, 1989). Even the most recent treatise on classifying, inventorying, and monitoring rangeland (National Research Council [NRC] 1994) only devotes about 5 sentences to riparian habitats. Present approaches concentrate on forage removal in the uplands, and by the time that grazing level has been achieved most riparian habitats have been heavily overgrazed. For example, Krueger and Bonham (1986) report that cattle are so attracted to riparian areas in the summer that 90-95% of the adjacent uplands receive little or no use. Meyers (1989) examined 34 grazing systems in Montana and 25 (74%) showed no improvement in riparian areas over 10-20 years, while most showed improvement on the watershed. Clary and Webster (1989) in discussions with managers and after reviewing the literature reported there is not a single grazing management approach that has produced consistent improve-

FIGURE 21. Date Creek, before heavy entrenchment this stream once flooded the velvet mesquites on the second terrace to the left of the photograph above the cut bank. Note the absence of rock which would have prevented severe entrenchment. Riparian trees are gone and the sparse understory is dominated by the unpalatable seepwillow (*Baccharis salicifolia*). Photograph by R. D. Ohmart, June 1994.

FIGURE 22. Repeat photograph of Figure 18 after an unusually heavy flood event in January-February 1993. The large 4-trunked tree (visible in 1992) in the center is now covered by the small tree to right of the big tree. The line of trees on either side of the primary floodplain kept the erosive waters contained and dispersed erosive energies. Photograph by R. D. Ohmart, June 1994.

ment of degraded riparian-wetland areas over western rangelands. Elmore and Kauffman (1994) support this conclusion.

The 2 grazing approaches most detrimental to riparian habitats are total growing-season grazing and year-long grazing (Elmore and Kauffman 1994). Cattle are so attracted to riparian habitats that any grazing approach that extends throughout the growing season will insure some cattle in riparian habitats the entire time, unless herded or excluded by fencing. Similar cattle behavior has been displayed in New Mexico (Goodman et al. 1989) and the western Dakotas (Severson and Boldt 1978).

Numerous ecological approaches have been developed for grazing riparian habitats or restoring them (Skovlin 1984, Kinch 1989, Kauffman et al. 1993, Elmore and Kauffman 1994). Elmore (1992) discusses a number of grazing systems and their shortcomings relative to riparian protection and recovery. Selecting a grazing approach in riparian habitats is difficult in that a multitude of variables are involved (Chaney et al. 1993, Elmore and Kauffman 1994). Many workers (Platts 1981*b*, Kinch 1989, Clary and Webster 1989) have examined riparian grazing approaches and no single method has been successful for improving degraded riparian areas. Elmore and Kauffman (1994) summarize grazing approaches based on their experiences and I briefly present what they consider the poorest and the best approaches. Continuous or season-long cattle grazing is the poorest, creating the greatest amount of degradation to the physical and biotic components of the riparian area. This is equal to holding sheep or cattle in riparian areas. Equally poor is short-duration and high-intensity cattle grazing in riparian areas. Winter use with cattle or sheep receives a moderate rating relative to riparian degradation. The best and obvious approach is total closure or rest to riparian areas from all classes of livestock. Two approaches right below exclusion are rest rotation with seasonal preference with sheep, or corridor fencing with either sheep or cattle. Rating high, as well, is fencing the riparian area for prescribed use.

Total closure may be inviting to many but it also has numerous implications. The most obvious is the cessation of domestic livestock grazing on public lands, which I would not support. Secondly, fence all riparian areas, but present fencing on public land is considered more than enough (Jacobs 1991) and costs to fence riparian areas would be overwhelming (current charge $3,000-3,500/1.6 km of fence). Platts (1991) estimated that it would cost $90 million just to fence all of the 24,135 km of fishable streams on BLM lands. These and associated problems such as wildlife entanglement, fences acting as traps and concentrating cattle along streams, operation and maintenance costs, and other problems immediately exclude this option. Herding by the permittee with riparian use constraints appears to be the most viable approach to this point. It is costly to the permittee but less so than the above alternatives.

Interestingly, good riparian healing with high numbers (600-800 head) for a short time (approx. 6 days) in late summer and early fall can occur (R. L. Storch, Forest Supervisor, Grand Mesa, Uncompahgre-Gunnison National Forest, Colorado, pers. commun.). R. L. Storch (pers. commun.) stated that willows and other woody vegetation responded very favorably to this management approach. Fencing of the riparian area includes wide portions of the uplands and cooler temperatures during these times better disperses livestock during the short grazing period.

Some years ago the USFS developed the Integrated Resource Management approach, and I attended one of their training workshops on its use. The team was composed of individuals representing all resource areas. I never observed one of these teams that was not all USFS employees. With range resources being such a dominant part of this agency, I suspect that other resources were not adequately represented or listened to. Further, most team members are low echelon personnel who are knowledgeable of how decisions will be made regardless of their input.

Elmore and Kauffman (1994) have suggested using an interdisciplinary management team (i.e., soils, fishery biologists, botanists, and others) to visit allotments and formulate management plans. They present 6 general recommendations for the team. I observed 1 of these teams, in which all recommendations were carried out except possibly the last and most important action, which was insuring strong compliance recommendations. Further, this team was composed of well-trained, mature, and knowledgeable veteran personnel.

If the time and knowledge of such teams are to be used then the permittee must have an incentive to see that the recommendations are followed. If the team recommends leaving an X-cm tall stubble height over the floodplain for soil protection and sediment trapping in floodstage, it should be the permittee's responsibility to monitor this and report to the agency when limits are being approached. Random measurements by agency personnel would confirm utilization levels and give a date when all livestock must be out of the pasture. If utilization levels were exceeded by the permittee then no utilization would be allowed in the pasture the following year nor would excess numbers be allowed in other parts of the allotment. Attention to detail by the permittee would increase under this approach and the user would share management responsibilities with agency personnel. Most agencies do not have adequate range personnel to properly do their present work load, much less shoulder the total responsibilities of insuring compliance with the recommendations of a management team.

There is strong merit in using a team approach (Elmore and Kauffman (1994). Platts (1979:39) does not exactly suggest a team approach, but alludes to it by saying ". . . no single discipline possesses the skills and knowledge for all problem solving . . ." in riparian management. With an observed team, each discipline had its concerns expressed and the recommendations formulated were more ecologically sound than they would have been if a single person had attempted to encompass all disciplines.

Possibly a set of teams could be established over ecological regions in the West. Each team would be composed of agency and nonagency specialists and would respond to BLM, USFS, or private land managers requesting recommendations on riparian restoration or best grazing management practices. There are excellent people in and out of the agencies whose sole interest is to see range resources improve so that greater conflicts are avoided in the future. Because the permittee would accompany the team, for balance, I recommend that a conservationist who has shown interest in the natural resources on the allotment be invited. Both could enter discussions but *only* the condition of the resource and the team would dictate the management practices to be followed. Time and travel demands on team members may be such that nonagency personnel might have to be employed on a full-time basis. I further suggest these teams report directly to the Secretary or Assistant Secretary of the Interior to remove them from as much agency-permittee politics as possible.

The most striking difference between riparian recovery with total rest and riparian recovery with grazing is the time factor involved. With total rest, most of the systems that I have observed show tremendous change within 8-10 years. These general numbers seem to hold in Colorado (Schulz and Leininger 1990), central Oregon (Clifton 1989), Nevada (Medin and Clary 1989), and numerous stream systems throughout the West (GAO 1988). Other workers have also reported that exclusion provides the most dramatic and rapid rate of recovery in riparian systems (Beschta et al. 1991, Elmore and Kauffman 1994).

With managed grazing riparian healing time is twice (16-20 years) and maybe 4 times (32-40+ years) longer than exclusion. The important question is, can the most degraded riparian areas hang onto their thread of existence for another 30-50 years? They are very weakened and degraded after >100 years of unmanaged grazing, and my experiences are that the agencies and permittees, if willing, could not move rapidly enough with improved management for it to really begin within the next 10-20 years. It might come about faster than this if the managing agencies took a more aggressive role in management as is being done on the Uncompahgre-Gunnison National Forest (R. L. Storch, pers. commun.).

Allotments with improved management approaches ready to be implemented within the next 3-5 years should be enacted. Otherwise, agencies should develop and implement plans to remove cattle from heavily degraded riparian areas for a minimum of 5 years. This suggestion is not new in that many biologists have suggested that there is no grazing plan that allows riparian restoration (Ames 1977, Davis 1977, Behnke 1979, Dahlem 1979, Kindschy 1978, Szaro 1980). If cattle are to be grazed in the floodplain again, riparian habitats should be closely monitored with rigid utilization standards to insure continuing improvement.

KEYS TO BETTER RIPARIAN MANAGEMENT

Most agencies managing public lands in the West have some type of monitoring of upland habitats but have used different methods to evaluate the ecological conditions of western rangelands (NRC 1994). Much of the data are at best trend information, and Box (1990) reported that even these data were lacking for 12% of National Forests lands and 26% of BLM lands.

Monitoring is the key to knowing and documenting riparian improvement. These data are virtually absent for riparian areas in western rangelands. Their collection is essential if management is to recognize changes for improved riparian conditions. A beginning that is cheap, simple, and quick is to establish repeat photographic points at a few representative areas along a stream. A photograph showing the channel, banks, and floodplain at 2-3 year intervals the same month each year along with brief notes on mean channel width and bank and floodplain condition (presence or absence of young trees, percent groundcover of the floodplain) within 2 photographic periods would immediately inform the agency of needed management changes. Funds must be made available to perform these duties.

Too many land managers, both in the USFS and BLM, still believe that presently used grazing systems to improve uplands will also help improve riparian habitats. Related to this is that upper management (typically District Rangers and above and Area Managers and above) have very little knowledge of what management practices are needed to improve riparian areas and of resource conditions on the ground. Too little time is spent in the field by these people to appreciate problems and solutions.

Riparian habitats, unlike uplands, respond very quickly to improved management (usually within 2 years) unless highly degraded. Furthermore, there should be no controversy over what is improvement, stability, or degradation as there is in the uplands. Easily quantifiable and visible objectives can be established allowing the range conservationists and permittee to easily judge if conditions are being met. The consensus of the advisory team, irrespective of the agency person or permittee, would be the sole criterion on meeting riparian objectives.

A second key is to tie specific riparian improvement objectives to land managers' annual performance ratings. These people should be accountable both professionally and monetarily to improved riparian health.

The last key would be to make permittees accountable for riparian health as well. If grazing fees were based on riparian health there would be greater attention and concern to PFC.

Riparian conditions meeting objectives AUM = $3.00
Riparian conditions improving, but not meeting objectives AUM = $5.00
Riparian conditions unsatisfactory and not improving AUM = $15.00

Short courses could be conducted for agency personnel and permittees (attending together) showing the various management targets to be achieved. This approach, I think, would foster a very close, cooperative working relationship to improve riparian conditions between permittees and agency personnel.

SUMMARY

Riparian habitats in western rangelands have exceedingly high values for society, fish, and wildlife. Their resource values far exceed their approximate 0.1% of the land area they cover. They serve to trap and stabilize eroded sediments, detoxify compounds, act as phosphorus sinks for soil enrichment, and serve as denitrification areas to provide high water quality. When functioning they provide bank storage of water and extend the flow regime to perenniality or increase instream flow.

They are vital to fishes when properly functioning by providing uncontaminated cool water, high in dissolved oxygen and low in suspended sediment. These important water quality parameters are all related to proper stream channel shape, bank stability, transport of sediment load, and the relationship of the stream to access its banks. Important physical factors to native salmonids are water velocity, water temperature, amount of dissolved oxygen, pool volume (number, size, and depth), escape cover, and annual discharge and flow. Highest quality fisheries exist under these conditions along with water quality and quantity.

In riparian habitats, vegetative components, in general, fulfill the ecological needs of the greatest array of wildlife species. In their order of importance are tree species and their densities, foliage profile, foliage volumes in the profiles, horizontal patchiness, and shrub species and their densities. The three most important vegetative components are satisfied in the Southwest by mature Fremont or narrowleaf cottonwood intermixed with willow species, in the Pacific Northwest by black cottonwood (*Populus trichocapra*) with willows, and in the Intermountain region by quaking aspen and willow or pure willow habitats. The shrub species and densities element may be most important in the Southwest, but pure willow communities may function similarly. High insect production in willows lends them added importance to wildlife.

Wildlife values in riparian habitats, i.e., density, species richness, biodiversity, number of rare species, number of breeding pairs of birds, and biomass, are extremely high compared to adjacent uplands. Highest fisheries values exist were streams are properly functioning. As knowledge becomes more complete, they may provide some of the most important fish and wildlife habitats in the coterminous United States. Southwestern riparian habitats may be vital to migratory wildlife as migratory corridors and resting, refueling, and wintering habitats in that they provide linear oases when uplands are least productive.

Though knowledge is limited, beavers are unquestionably a keystone species in riparian areas of second- to fifth-order streams. They have the ability to alter habitats by 20-40% by changing channel geomorphology and hydrology; by retaining sediment and organic matter; by creating and maintaining wetlands; by modifying nutrient cycling and decomposition dynamics; by modifying species composition and dynamics of plants; by influencing the movement of water and materials transported downstream; and by creating totally new fish and wildlife habitat that significantly increases biodiversity.

The development of the livestock industry was examined with Arizona being used as an example. Though drought in the 1890's expedited ecological degradation in Arizona, other western states reported similar ecological problems as ranges were overstocked. Riparian habitats suffered the greatest ecological damage by being both highly overgrazed and then experiencing unprecedented flood damage from degraded watersheds.

Livestock are attracted to riparian areas because of lush forage, shade, and water, especially in hotter, arid months. Current management approaches only consider the health and condition of the uplands or watersheds, which does not give grazing relief to riparian habitats. As late as the 1960s riparian habitats were viewed as sacrifice areas. Though fish and wildlife values in riparian habitats are extremely high, there has been little progress in making livestock management changes.

There are many agents of riparian destruction and degradation other than overgrazing by domestic livestock. Most of these have been along major western streams, while unmanaged livestock degradation has been ubiquitous and at all elevations.

Three conceptual phases were used to facilitate and illustrate the consequences of unmanaged livestock degradation of fish and wildlife resources in pristine riparian habitats (minus beavers) from the inception of livestock grazing to present. Completion of the process in Phase I took from 3-10 years depending on livestock density. Though the process was complete the degraded condition of the resource continued and in many instances it has intensified to present. Degradation includes destruction of overhanging banks, overburdened sediment loads, stream channel changes, widening of the channel, virtual elimination of the herbaceous groundcover and the understory, and the cessation of tree and shrub reproduction along the floodplain. Heavier than usual flood events eroded and widened channel widths and depths, and divorced the stream from accessing its floodplain. Sediment loads have increased concomitantly.

Consequences of bank destruction, channel widening, and high sediment loads severely impact fish populations. The elimination of understory trees and shrubs promotes higher water temperatures, lower oxygen levels, and reduces detrital input (the main energy source for small-order streams), further stressing fish populations. Species of small mammals, birds, reptiles, and amphibians requiring dense herbaceous vegetation or dense understory vegetation have experienced density declines and possibly local extirpation as these changes occurred about 1890 or even earlier throughout riparian habitats in the West.

Phase II was initiated about this time and it will take 100-125 years to complete. Beginning Phase II there were extensive and generally continuous riparian forests consisting of cottonwoods and willows, pure willows, or willow-quaking aspen along perennial streams in the West. Tree species abounded in their habitats, the foliage profile was complete (minus the understory), foliage volumes were high in the canopy and midstory, and intracommunity patchiness was high. Phase II is a slow and subtle process of degradation, almost imperceptible without repeat photography or quantified botanical data. Essentially it is the slow process of the youngest trees that escaped being eaten when cattle began heavily grazing these habitats circa 1870 to the point they begin reaching decadence. During the intervening years, older age classes of trees have lived out their life span slowly thinning the forests. This process, in turn lessens foliage volumes at the mid and overstory layers. As these young trees mature and there is no recruitment of young age classes the patchiness element declines as well. There is tree loss through natural mortality, but there are other losses as well, i.e., blow downs, beavers girdling trees, firewood or building material use, trees left perched as water tables declined, and trees washed out with more severe floods from degraded watersheds. Continuous forests over time have been fragmented and thinned, leaving small islands of trees or thin stringers.

The cumulative effects of all human activities in eliminating or degrading riparian habitats is beginning to manifest itself through the reduction in riparian resources in the West. Tree species and their densities, foliage profiles, foliage volumes, patchiness levels, and remaining forest sizes are approaching threshold levels where many more species should show declines in numbers or be locally extirpated.

Phase III is the collapse of the riparian deciduous forest. Most riparian forests are in late Phase II, while some are in early or mid-Phase III. It will take about 50 years to complete this phase. Management options are the (1) immediate use of managed grazing or (2) eventually be forced to use external seed sources with possibly less-adapted ecotypes and/or (3) revegetation efforts which are extremely expensive as compared to grazing management changes.

The management challenge of the twenty-first century will be the continued use of western rangelands while simultaneously healing riparian habitats. Abolition of livestock grazing on public rangelands and fencing are ruled out because of social acceptance and cost. The most viable method at present is herding with stubble height constraints. Strong incentives to both the land manager and permittee to restore proper functioning condition of western streams are key to restoring riparian habitat for optimum social, fish, and wildlife resource values. An approach is suggested.

LITERATURE CITED

Adams, S. N. 1975. Sheep and cattle grazing in forests: a review. J. Appl. Ecol. 12:143-152.

Allred, M. 1980. A re-emphasis on the value of the beaver in natural resource conservation. J. Idaho Acad. Sci. 16:3-10.

Ames, C. R. 1977. Wildlife conflicts in riparian management: grazing. Pages 49-51 *in* R. R. Johnson and D. A. Jones, tech. coords. Importance, preservation and management of riparian habitat: a symposium. U.S. For. Serv. Gen. Tech. Rep. RM-43.

Anderson, B. W., and R. D. Ohmart. 1984. A vegetation management study for the enhancement of wildlife along the lower Colorado River. Final Rep., Bur. Rec., Lower Colo. Reg., Boulder City, Nev.

———, and ———. 1986. Vegetation. Pages 639-660 *in* A. Y. Cooperrider, R. J. Boyd, and H. R. Stuart, eds. Inventory and monitoring of wildlife habitat. U.S. Dep. Inter., Bur. Land Manage. Serv. Cent., Denver, Colo. 858pp.

Anderson, E. W., and R. W. Harris. 1973. References on grazing resources of the Pacific Northwest (Oregon, Washington, British Columbia), 1896-1970. Soc. Range Manage., Pacific Northwest Sect. 103pp.

Anderson, J. W., R. L. Beschta, P. L. Boehne, D. Bryson, R. Gill, S. Howes, B. A. McIntosh, M. D. Purser, J. J. Rhodes, and J. Zakel. 1993. A comprehensive approach to restoring habitat conditions needed to protect threatened salmon species in a severely degraded river. Pages 175-179 *in* B. Tellman, H. J. Cortner, M. G. Wallace, L. F. DeBano, and R. H. Hamre, tech. coords. Riparian management: common threads and shared interests. U.S. For. Serv. Gen. Tech. Rep. RM-226.

Arizona Game and Fish Department. 1993. Elk management operational plan. Public review draft, January. Phoenix, Ariz. 38pp.

Armour, C. L. 1977. Effects of deteriorated range streams on trout. U.S. Bur. Land Manage., Idaho State Off., Boise. 7pp.

———. 1978. Effects of deteriorated range streams on trout. U.S. Dep. Inter., Bur. Land Manage., Boise, Id. 7pp.

———. 1979. Livestock management approaches and the fisheries resource. Page 39 *in* O. B. Cope, ed. Proc. of the forum—grazing and riparian/stream ecosystems. Trout Unlimited, Inc., Denver, Colo.

———, D. Duff, and W. Elmore. 1994. The effects of livestock grazing on western riparian and stream ecosystems. J. Am. Fish. Soc. 19:9-12.

Atlegrim, O. 1989. Exclusion of birds from bilberry stands: impact on insect larval density and damage to the bilberry. Oecologia 79:136-139.

Bahre, C. J. 1977. Land-use history of the Research Ranch, Elgin, Arizona. J. Arizona Acad. Sci. 12(Supple. 2):1-32.

———. 1991. A legacy of change: historic human impact on vegetation in the Arizona borderlands. Univ. Arizona Press, Tucson. 231pp.

Bakke, B. M. 1977. Grazing is destroying our fish. Action Line July:7-9. Trout Unlimited, Portland, Oreg.

Balda, R. P. 1975. Vegetation structure and breeding bird diversity. Pages 59-80 *in* D. R. Smith, tech. coord. Symp. on the management of forest and range habitats for nongame birds. U.S. For. Serv. Gen. Tech. Rep. WO-1.

Bancroft, A. L. 1883. The native races. Vol. I: Wild tribes. A. L. Bancroft and Co., San Francisco, Calif.

Bates, J. W. 1963. The effects of beaver on stream flow. Job Completion Rep., Fed. Aid Proj. W-65-R, Utah Dep. Fish and Game Inf. Bull. 63-13. 87pp.

Behnke, R. J. 1978. Grazing and the riparian zone: impact on aquatic values. Pages 126-132 *in* Lowland river and stream habitat in Colorado: a symposium. Colo. Chap. Wildl. Soc. and Colo. Audubon Counc., Greeley, Colo.

———. 1979. Values and protection of riparian ecosystems. Pages 164-167 *in* G. A. Swanson, tech. coord. The mitigation symposium. U.S. For. Serv. Gen. Tech. Rep. RM-65.

———. 1992. Native trout of western North America. Am. Fish. Soc. Monogr. No. 6. 275pp.

———, and R. F. Raleigh. 1978. Grazing and the riparian zone: impact and management perspectives. Pages 263-267 *in* R. R. Johnson and J. F. McCormick, tech. coords. Strategies for protection and management of floodplain wetlands and other riparian ecosystems. U.S. For. Serv. Gen. Tech. Rep. WO-12.

———, and M. Zarn. 1976. Biology and management of threatened and endangered Western trout. U.S. Dep. Agric., For. Serv. Gen. Tech. Rep. RM-28.

Beschta, R. L., and W. S. Platts. 1986. Morphological features of small streams: significance and function. Water Resour. Bull. 22:369-379.

———, ———, and J. B. Kauffman. 1991. Field review of fish habitat improvement projects in the Grande Rhonde and John Day River basins of eastern Oregon. U.S. Dep. Energy, Bonneville Power Adm., Proj. 91-069, Portland, Oreg. 53pp.

Bieber, R. P., editor. 1938. Cooke's journal of the march of the Mormon Battalion, 1846-1847. Pages 65-240 *in* Exploring Southwest trails, 1846-1854, Vol. VII, Southwest Hist. Ser., Arthur H. Clark Co., Glendale, Calif.

Binns, N. A., and F. M. Eiserman. 1979. Quantification of fluvial trout habitat in Wyoming. Trans. Am. Fish. Soc. 108:215-228.

Bjornn, T. C. 1971. Trout and salmon movements in two Idaho streams as related to temperatures, food, stream flow, cover and population density. Trans. Am. Fish. Soc. 100:423-435.

Blackburn, F. M. 1993. National Wildlife Federation v. BLM. Trans. proceedings UT-06-91-01, Vol. 12, pp. 64-185. U.S. Dep. Inter., Off. Hear. and Appeals, Hear. Div., April 20, 1993.

Bock, C. E., V. A. Saab, T. D. Rich, and D. S. Dobkin. 1993. Effects of livestock grazing on neotropical migratory landbirds in western North America. Pages 296-309 *in* D. M. Finch and P. W. Stengel, eds. Status and management of neotropical migratory birds. U.S. For. Serv. Gen. Tech. Rep. RM-229.

Bohn, C. 1989. Management of winter soil temperatures to control streambank erosion. Pages 69-71 *in* R. E. Gresswell, B. A. Barton, and J. L. Kershner, eds. Practical approaches to riparian resource management: an educational workshop. U.S. Bur. Land Manage., Billings, Mont.

Bolton, H. E., translator and editor. 1948. Kino's historical memoir of Pimería Alta; a contemporary account of the beginnings of California, Sonora, and Arizona. Univ. California Press, Berkeley.

Bormann, F. H., and G. E. Likens. 1969. The watershed-ecosystem concept and studies of nutrient cycles. *In* V. Dyne, ed. The ecosystem concept and natural resource management. Academic Press, New York, N.Y.

Bowers, W., B. Hosford, A. Oakley, and C. Bond. 1979. Wildlife habitats in managed rangelands—the Great Basin of southeastern Oregon: native trout. U.S. For. Serv. Gen. Tech. Rep. PNW-84. 16pp.

Bowler, B. 1976. What future uses can public lands sustain? Pages 49-50 *in* Congress on public lands in the West. Snake River Reg. Studies Cent., College of Idaho, Caldwell.

Box, T. W. 1990. Rangelands. Pages 101-120 *in* R. N. Sampson and D. Hair, eds. Natural resources for the 21st century. Island Press, Washington D.C.

Brown, B. T., and M. W. Trosset. 1989. Nesting-habitat relationships of riparian birds along the Colorado River in Grand Canyon, Arizona. Southwest. Nat. 34:260-270.

Brush, T., and E. W. Stiles. 1986. Using food abundance to predict habitat use by birds. Pages 57-63 *in* J. Verner, M. L. Morrison, and C. J. Ralph, eds. Wildlife 2000: modelling habitat relationships of terrestrial vertebrates. Univ. Wisconsin Press, Madison.

Bureau of Land Management. 1993. Riparian area management: process for assessing proper functioning condition. U.S. Dep. Inter., TR 1739-9. Bur. Land Manage. Serv. Cent., Denver, Colo. 51pp.

Burgess, H. H., H. H. Prince, and D. L. Trauger. 1965. Blue-winged teal nesting success as related to land use. J. Wildl. Manage. 21:89-95.

Busby, F. E. 1979. Riparian and stream ecosystems, livestock grazing, and multiple-use management. Pages 6-12 *in* O. B. Cope, ed. Proc. of the forum: grazing and riparian/stream ecosystems, Trout Unlimited, Inc., Denver, Colo.

Carothers, S. W. 1977. Importance, preservation and management of riparian habitat: an overview. Pages 2-4 *in* R. R. Johnson and D. A. Jones, tech. coords. Importance, preservation and management of riparian habitat: a symposium. U.S. For. Serv. Gen. Tech. Rep. RM-43.

———, and R. R. Johnson. 1975. Water management practices and their effects on nongame birds in range habitats. Pages 210-222 *in* D. R. Smith, tech. coord. Proc. of the symposium on management of forest and range habitats for nongame birds. U.S. For. Serv. Gen. Tech. Rep. WO-1.

———, ———, and S. W. Aitchison. 1974. Population structure and social organization of Southwestern riparian birds. Am. Zool. 14:97-108.

Carson, R. 1962. Silent spring. Fawcett Crest Books, Greenwich, Conn. 304pp.

Chadde, S. 1989. Willows and wildlife of the northern range, Yellowstone National Park. Pages 168-169 *in* R. E. Gresswell, B. A. Barton, and J. L. Kershner, eds. Practical approaches to riparian resource management: an educational workshop. U.S. Bur. Land Manage., Billings, Mont.

Chaney, E., W. Elmore, and W. S. Platts. 1990. Livestock grazing in Western riparian areas. Rep. U.S. Environ. Prot. Agency, Reg. 8, Denver, Colo. 45pp.

———, ———, and ———. 1993. Managing change. Livestock grazing in Western riparian areas. Rep. to U.S. Environ. Prot. Agency, Reg. 8, Denver, CO. 31pp.

Chapman, D. W., and E. Knudsen. 1980. Channelization and livestock impacts on salmonid habitat and biomass in western Washington. Trans. Am. Fish. Soc. 109.

Chapman, H. H. 1933. Influence of overgrazing on erosion and watersheds. J. Civil Eng. 3:74-78.

Cherry, J. D. 1982. Fat deposition and length of stopover of migrant white-crowned sparrows. Auk 99:725-732.

Chow, V. T. 1959. Open-channel hydrolics. Civil Engineering Series, McGraw-Hill, New York, N.Y. 680pp.

Clarke, A. B. 1852. Travels in Mexico and California. Wright and Hasty's Steam Press, Boston, Mass. 143pp.

Clarkson, R. W., and J. R. Wilson. 1991. Evaluation of the U.S. Forest Service fish habitat relationships system in east-central Arizona trout streams. Arizona Game and Fish Dep. Tech. Rep. 8. 39pp.

Clary, W. P., and D. E. Medin. 1992. Vegetation breeding bird, and small mammal biomass in two high-elevation sagebrush riparian habitats. Pages 100-110 *in* W. P. Clary, E. D. McArthur, D. Bedunah, and C. L. Wambolt, compil. Proc. of a symposium on ecology and management of riparian shrub communities. U.S. For. Serv., Intermountain Res. Stn., Gen. Tech. Rep. INT-289.

———, and B. F. Webster. 1989. Managing grazing of riparian areas in the intermountain region. U.S. For. Serv. Gen. Tech. Rep. INT-263.

Clifton, C. 1989. Effects of vegetation and land use on channel morphology. Pages 121-129 *in* R. E. Gresswell, B. A. Barton, and J. L. Kershner, eds. Practical approaches to riparian resource management: an educational workshop. Bur. Land Manage., Billings, Mont.

Cooper, J. R., J. W. Gilliam, R. B. Daniels, and W. P. Robarge. 1987. Riparian areas as filters for agriculture sediment. Soil Sci. Soc. Am. J. 57:416-420.

Cottam, W. P., and F. R. Evans. 1945. A comparative study of the vegetation of grazed and ungrazed canyons of the Wasatch Range, Utah. Ecology 26:171-181.

Cox, C. C. 1925. From Texas to California in 1849. Southwest. Q. XXIX: 128-146.

Crawford, H. S., and D. T. Jennings. 1989. Predation by birds in spruce budworm *Choristoneura fumiferana*: functional, numerical, and total responses. Ecology 70:152-163.

Croft, A. R., L. Woodward, and D. A. Anderson. 1943. Measurement of accelerated erosion on range-watershed land. J. For. 41:112-116.

Cross, S. 1985. Responses of small mammals to forest perturbations. Pages 269-275 *in* R. R. Johnson, C. D. Ziebell, D. R. Patton, P. F. Ffolliott, and R. H. Hamre, tech. coords. Riparian ecosystems and their management: reconciling conflicting uses. Proc. First North Am. Riparian Conf., U.S. For. Serv. Gen. Tech. Rep. RM-120.

Crouch, G. L. 1982. Wildlife on ungrazed and grazed bottomlands on the South Platte River in northeastern Colorado. Pages 186-197 *in* Wildlife-livestock relationships symposium: Proc. 10. Univ. Idaho, For., Wildl. Range Exp. Stn., Moscow, Id.

Dahlem, E. A. 1979. The Mahagony Creek watershed—with and without grazing. Pages 31-34 *in* O. B. Cope, ed. Proc. of the forum: grazing and riparian/stream ecosystems. Trout Unlimited, Inc., Denver, Colo.

Dahm, C. N., E. H. Trotter, and J. R. Sedell. 1987. The role of anaerobic zones and processes in stream ecosystem productivity. Pages 157-178 *in* R. C. Avertt and D. M. McKnight, eds. Chemical quality of water and the hydrolic cycle. Lewis Publ., Chelsea, Mich.

Darling, L., and G. Coltharp. 1973. Effects of livestock grazing on the water quality of mountain streams. Pages 1-8 *in* Water-animal relationships symposium proceedings. U.S. Dep. Agric. Res. Serv., Kimberly, Id.

Davis, G. A. 1977. Management alternatives for the riparian habitat in the Southwest. Pages 59-67 *in* R. R. Johnson and D. A. Jones, tech. coords. Importance, preservation and management of riparian habitat: a symposium. U.S. For. Serv. Gen. Tech. Rep. RM-43.

Dobyns, H. F. 1981. From fire to flood: historic human destruction of Sonoran Desert riverine oases. Ballena Press, Socorro, N. M. 222pp.

Duff, D. A. 1979. Riparian habitat recovery on Big Creek, Rich County, Utah—a summary of eight years of study. Pages 91-92 *in* O. B. Cope, ed. Proc. of the forum: grazing and riparian/stream ecosystems. Trout Unlimited, Inc., Denver, Colo.

———. 1983. Livestock grazing impacts on aquatic habitat in Big Creek, Utah. Pages 129-142 *in* J. W. Menke, ed. Proc. of the workshop on wildlife-fisheries relationships in the Great Basin. Univ. California, Agric. Stn., Sci. Spec. Publ. 3301, Berkeley.

Durivage, J. E. 1937. "Journal," Southern Trails to California in 1849. Pages 159-255 *in* R. P. Bieber, ed., Vol. V. of Southwest Historical Series. Arthur H. Clark Co., Glendale, Calif.

Ehleringer, J. R., L. A. Arnow, T. Arrow, I. B. McNulty, and N. C. Negus. 1992. Red Butte Canyon Research Natural Area: history, flora, geology, climate, and ecology. Great Basin Nat. 52:95-121.

Elmore, W. 1992. Riparian responses to grazing practices. Pages 442-457 *in* R. J. Naiman, ed. Watershed management: balancing sustainability and environmental change. Springer-Verlag, New York, N.Y.

———, and R. L. Beschta. 1987. Riparian areas: perceptions in management. Rangelands 9:260-265.

———, and B. Kauffman. 1994. Riparian and watershed systems: degradation and restoration. Pages 212-231 *in* M. Vavra, W. A. Laycock, and R. D. Pieper, eds. Ecological implications of livestock herbivory in the West. Soc. Range Manage., Denver, Colo.

Engel-Wilson, R., and R. D. Ohmart. 1978. Assessment of vegetation and terrestrial vertebrates along the Rio Grande between Fort Quitman, Texas and Haciendita, Texas. Final Rep., Intl. Boundary and Water Comm., El Paso, Tex.

Esplin, A. C., W. Peterson, P. V. Cardon, G. Stewart, and K. C. Ikeler. 1928. Sheep ranching in Utah. Utah Agric. Exp. Stn. Bull. 204. 60pp.

Evans, G. W. 1945. Mexican gold trail: the journal of a forty-niner. G. S. Dumke. ed., Huntington Library, San Marino, Calif. 340pp.

Ewing, R. C. 1934. The Pima uprising, 1751-52. Ph.D. Thesis, Univ. of California, Berkeley.

Finch, D. M. 1986. Similarities in riparian bird communities among elevational zones in southeastern Wyoming. Pages 105-110 *in* Proc. Wyoming Water 1986 and Streamside Zone Conf. Water Res. Cent., Univ. Wyoming, Laramie.

Fleischner, T. L. 1994. Ecological costs of livestock grazing in western North America. Conserv. Biol. 8:629-644.

Gains, D. A. 1977. The valley riparian forests of California: their importance to bird populations. Pages 57-85 *in* A. Sands, ed. Ecol. Publ. 15, Univ. California, Davis.

General Accounting Office. 1988. Public rangelands: some riparian areas restored but widespread improvement will be slow. GAO/RCED-88-105. U.S. Gen. Accounting Off., Resour., Community, and Econ. Dev. Div., Washington, D.C.

———. 1992. Response to Congressional Requesters. Rangeland management: assessment of Nevada consulting firm's critique of three GAO reports. GAO/RCED-92-178R. U.S. Gen. Accounting Off., Resour., Community, and Econ. Dev. Div., Washington, D.C.

Goodman, T., G. B. Donart, H. E. Kiesling, J. L. Holchek, J. P. Neel, D. Manzanares, and K. Severson. 1989. Cattle behavior with emphasis on time and activity allocations between upland and riparian habitats. Pages 95-102 *in* R. E. Gresswell, B. A. Barton, and J. L. Kershner, eds. Practical approaches to riparian resource management: an educational workshop. U.S. Bur. Land Manage., Billings, Mont.

Green, D. M., and J. B. Kauffman. 1989. Nutrient cycling at the land-water interface: the importance of the riparian zone. Pages 61-68 *in* R. E. Gresswell, B. A. Barton, and J. L. Kershner, eds. Practical approaches to riparian resource management: an educational workshop. U.S. Bur. Land Manage., Billings, Mont.

Gregory, S., and L. Ashkenas. 1990. Management guide Willamette National Forest. Willamette Natl. For., Eugene, Oreg.

Gregory, S. V., F. J. Swanson, W. A. McKee, and K. W. Cummins. 1991. An ecosystem perspective of riparian zones. Bioscience 41:540-551.

Gunderson, D. R. 1968. Floodplain use as related to stream morphology and fish populations. J. Wildl. Manage. 32:507-514.

Hansen, P. 1993. Developing a successful riparian-wetland grazing management plan for the Upper Ruby River Cattle and Horse Allotment in southwestern Montana. Pages 328-335 *in* B. Tellman, H., J. Cortner, M. G. Wallace, L. F. DeBano, and R. H. Hamre, tech. coords. Riparian management: common threads and shared interests. U.S. For. Serv. Gen. Tech. Rep. RM-226.

Harris, L. D. 1971. A précis of small mammal studies and results in the grassland biome. Pages 213-240 *in* N. R. French, ed. Preliminary analysis of structure and function in grasslands. Range Sci. Dep. Sci. Serv. 10, Colorado State Univ., Fort Collins.

Hastings, J. R. 1959. Vegetation change and arroyo cutting in southeastern Arizona. J. Ariz. Acad. Sci. 1:60-67.

———, and R. M. Turner. 1965. The changing mile. Univ. Arizona Press, Tucson.

Hays, H. B. N. 1975. The stream and its valley. Verh. Internatl. Verin. Limno. 19:1-15.

Hendrickson, D. A., and W. L. Minckley. 1984. Ciénegas—vanishing climax communities of the American Southwest. Special Issue. Desert Plants 6:130-175.

Horning, J. 1994. Grazing to extinction: endangered, threatened and candidate species imperiled by livestock grazing on Western public lands. Natl. Wildl. Fed., Washington, D.C.

Houston, D. B. 1982. The northern Yellowstone elk: ecology and management. MacMillan, Publ. Co., New York, N.Y. 474pp.

Howe, H. F.. 1981. Removal of wild nutmeg (*Virola surinamensis*) crops by birds. Ecology 62:1093-1106.

———, and G. A. Vande Kerckhore. 1979. Fecundity and seed dispersal of a tropical tree. Ecology 60:180-189.

Hubbard, J. P. 1977. Importance of riparian ecosystems: biotic considerations. Pages 14-18 *in* R. R. Johnson and D. A. Jones, tech. coords. Importance, preservation and management of riparian habitat: a symposium. For. Serv. Gen. Tech. Rep. RM-43.

Ives, R. L. 1942. The beaver-meadow complex. J. Geomorphol. 5:191-203.

Jacobs, L. 1991. Waste of the West: public lands ranching. Lynn Jacobs, Tucson, Ariz. 602pp.

Jacobs, T. C., and J. W. Gilliam. 1985. Riparian losses of nitrate from agricultural drainage waters. J. Environ. Qual. 14:472-478.

Jackson, F. 1994. Documenting channel condition in New Mexico. U.S. Dep. Agric., For. Serv., Stream Systems Tech. Cent., Stream Notes, Fort Collins, Colo.

Jenkins, S. H. 1979. Seasonal and year-to-year differences in food selection by beavers. Oecologia 44:112-116.

———. 1980. A size-distance relation in food selection by beavers. Ecology 61:740-746.

Johnson, R. R. 1978. The lower Colorado River: a western system. Pages 41-55 *in* R. R. Johnson and J. F. McCormick, tech. coords., Strategies for protection and management of floodplain wetlands and other riparian ecosystems. U.S. For. Serv. Gen. Tech. Rep. WO-12.

———, and D. A. Jones, tech. coords. 1977. Importance, preservation and management of riparian habitat: a symposium. U.S. For. Serv. Gen. Tech. Rep. RM-43.

———, L. T. Haight, and J. M. Simpson. 1977. Endangered species vs. endangered habitats: a concept. Pages 68-99 *in* R. R. Johnson and D. A. Jones, tech. coords. Importance, preservation and management of riparian habitat: a symposium. U.S. For. Serv. Gen. Tech. Rep. RM-43.

Johnson, S. R., H. L. Gary, and S. L. Ponce. 1978. Range cattle impacts on stream water in the Colorado Front Range. U.S. For. Serv. Res. Note RM-359.

Kauffman, J. B., R. L. Beschta, and W. S. Platts. 1993. Fish habitat improvement projects in the Fifteen Mile Creek and Trout Creek Basins of central Oregon. Field Rev. Manage. Reconn., U.S. Dep. Energy, Bonneville Power Admin. Rep., Portland, Oreg. 51pp.

———, and W. C. Krueger. 1984. Livestock impacts in riparian ecosystems and streamside management implications. J. Range Manage. 37:430-438.

———, ———, and M. Vavra. 1983. Impacts of cattle on streambanks in northeastern Oregon. J. Range Manage. 36:683-685.

Kay, C. E., and S. Chadde. 1992. Reduction of willow seed production by ungulate browsing in Yellowstone National Park. Pages 92-99 *in* W. P. Clary, E. D. McArthur, D. Bedunah, and C. L. Wambolt, compil. Proc. symposium on ecology and management of riparian shrub communities. U.S. For. Serv., Intermountain Res. Stn., Gen. Tech. Rep. INT-289.

Keck, W. M. 1972. Great Basin Station—sixty years of progress in range and watershed research. U.S. For. Serv., Intermountain. For. and Range Exp. Stn., Res. Pap. INT-118. 48pp.

Kimball, J., and F. Savage. 1977. Diamond Fork aquatic and range habitat improvements. U.S. For. Serv. Intermountain Reg., Unita Natl. For., Provo, Ut. 19pp.

Kinch, G. 1989. Riparian area management. Grazing management in riparian areas. Tech. Ref. 1737-4, U.S. Bur. Land Manage., Denver, Colo. 44pp.

Kindschy, R. R. 1978. Rangeland management practices and bird habitat values. Pages 66-69 *in* Proc. of the workshop on nongame bird habitat management in the coniferous forests of the western United States. U.S. For. Serv. Gen. Tech. Rep. PNW-64.

Kirch, L. M., and K. F. Higgins. 1976. Upland sandpiper nesting and management in North Dakota. Wildl. Soc. Bull. 4:16-20.

Knopf, F. L. 1985. Significance of riparian vegetation to breeding birds across an altitudinal cline. Pages 105-111 *in* R. R. Johnson, C. D. Ziebell, D. R. Patton, P. F. Ffolliott, and R. H. Hamre, tech. coords. Riparian ecosystems and their management: reconciling conflicting uses. Proc. First North Am. Riparian Conf., U.S. For. Serv. Gen. Tech. Rep. RM-120.

———, and R. W. Cannon. 1982. Structural resilience of a willow riparian community to change in grazing practices. Pages 198-207 *in* Proc. wildlife-livestock relationships symposium. For., Wildl. and Range Exp. Stn., Univ. Idaho, Moscow.

———, and T. E. Olson. 1984. Naturalization of Russian olive: implications to Rocky Mountain wildlife. Wildl. Soc. Bull. 12:289-298.

Kondolf, G. M. 1993. Log in stream channel adjustment to livestock exclosure, White Mountains, California. Restoration Ecol. December: 226-230.

Kovalchik, B. L., and W. Elmore. 1992. Effects of cattle grazing systems on willow-dominated plant associations in central Oregon. Pages 111-119 *in* W. P. Clary, E. D. McArthur, D. Bedunah, and C. L. Wambolt, compil. Proc. symposium on ecology and management of riparian shrub communities. U.S. For. Serv., Intermountain Res. Stn. Gen. Tech. Rep. INT-289.

Krausman, P. R., K. R. Rautenstrauch, and D. B. Leopold. 1985. Xeroriparian systems used by desert mule deer in Texas and Arizona. Pages 144-149 *in* R. R. Johnson, C. D. Ziebell, D. R. Patton, P. F. Ffolliott, and R. H. Hamre, tech. coords. Riparian ecosystems and their management: reconciling conflicting uses. Proc. First North Am. Riparian Conf., U.S. For. Serv. Gen. Tech. Rep. RM-120.

Krueger, H. O., and S. H. Anderson. 1985. The use of cattle as a management tool for wildlife in shrub-willow riparian systems. Pages 300-304 *in* R. R. Johnson, C. D. Ziebell, D. R. Patton, P. F. Ffolliott, and R. H. Hamre, tech.

coords. Riparian ecosystems and their management: reconciling conflicting uses. Proc. First North Am. Riparian Conf., U.S. For. Serv. Gen. Tech. Rep. RM-120.

Krueger, W. C., and C. D. Bonham. 1986. Responses of low cattle densities on large areas of forested pastures. Pages 67-73 *in* Statistical analyses and modeling of grazing systems symposium. Kissimmee, Fla., Soc. Range Manage., Denver, Colo.

Krueper, D. J. 1993. Effects of land use practices on Western riparian ecosystems. Pages 321-330 *in* D. M. Finch and P. W. Stengel, eds. Status and management of neotropical migratory birds. U.S. For. Serv. Gen. Tech. Rep. RM-229.

Kunkle, S. 1970. Sources and transport of bacterial indicators in rural streams. *In* Symp. on Interdisciplinary Aspects of Watershed Management. Montana State Univ., Bozeman. 31pp.

Land, E. 1934. Reminiscences. Manuscript in Arizona Pioneers' Historical Society, Tucson, Ariz.

Laymon, S. A. 1984. Riparian bird community structure dynamics: Dog Island, Red Bluff, California. Pages 587-597 *in* R. E. Warner and K. M. Hendrix, eds. California riparian systems: ecology, conservation, and productive management. Univ. California Press, Berkeley and Los Angeles.

Likens, G. E., and F. H. Bormann. 1974. Linkages between terrestrial and aquatic ecosystems. Bioscience 24:447-456.

Lorz, H. W. 1974. Ecology and management of brown trout in Little Deschutes River. Fishery Res. Rep. 8, Oregon Dep. Fish and Wildl., Portland. 49pp.

Losos, E., J. Hayes, A. Phillips, D. Wilcove, and C. Alkire. 1995. Taxpayer-subsidized resource extraction harms species. Bioscience 45:446-455.

Lowe, C. H. 1964. Part I. Arizona landscapes and habitats. Pages 1-132 *in* C. H. Lowe, ed. The vertebrates of Arizona. Univ. Arizona Press, Tucson.

Malanson, G. P. 1993. Riparian landscapes. Cambridge Univ. Press, Cambridge, U.K. 288pp.

Manning, M. E., S. R. Swanson, T. Svejcar, and J. Trent. 1989. Rooting characteristics of four intermountain meadow community types. J. Range Manage. 42:309-312.

Marcuson, P. E. 1977. The effect of cattle grazing on brown trout in Rock Creek, Montana. Spec. Proj. Rep. F-20-R-21-11-a. Montana Dep. Fish, Wildl., and Parks, Helena. 26pp.

Marquis, R. J., and C. J. Whelan. 1994. Insectivorous birds increase growth of white oak through consumption of leaf-chewing insects. Ecology 75:2007-2014.

Martin, S. C. 1979. Evaluating the impacts of cattle grazing on riparian habitats. Pages 35-38 *in* O. B. Cope, ed. Proc. of the forum: grazing and riparian/stream ecosystems, Trout Unlimited, Inc., Denver, Colo.

Masaki, T., Y. Kominami, and T. Nakashizuka. 1994. Spatial and seasonal patterns of seed dissemination of *Cornus controversa* in a temperate forest. Ecology 75:1903-1910.

McArdle, R. E., D. F. Costello, E. E. Birkmaier, C. Ewing, B. A. Hendricks, C. A. Kutzleb, A. A. Simpson, and A. R. Standing. 1936. The whiteman's toll. Pages 81-116 *in* U.S. Senate, The Western range. 74th Cong., 2nd Sess., Doc. 199, U.S. Gov. Printing Off., Washington, D.C. 620pp.

McClure, B., E. Campbell, T. Cordery, J. Haugh, L. Kaiser, D. Krueper, R. Peck, T. Rick, C. Jacoby, M. Dawson Paladino, and G. Houston. 1991. Nongame migratory bird habitat conservation plan. Fish and Wildlife 2000, Natl. Strategy Plan Ser., U.S. Bur. Land Manage. Public Affairs Off., Washington, D.C.

McEneaney, T. 1988. Birds of Yellowstone. Robert Rhinehart, Boulder, Colo. 171pp.

McLean, A., H. H. Nicholson, and A. L. Van Ryswyk. 1963. Growth productivity and chemical composition of a subalpine meadow in interior British Columbia. J. Range Manage. 16:235-240.

Medin, D. E., and W. P. Clary. 1989. Small mammal populations in a grazed and ungrazed riparian habitat in Nevada. U.S. For. Serv. Intermountain Res. Stn. Res. Pap. INT-413. 6pp.

———, and ———. 1991. Small mammals of a beaver pond ecosystem and adjacent riparian habitat in Idaho. U.S. For. Serv. Intermountain Res. Stn. Res. Pap. INT-445. 4pp.

Meehan, W. R., editor. 1991. Introduction and overview. Am. Fish. Soc. Spec. Publ. 19:1-14.

———, and W. S. Platts. 1978. Livestock grazing and the aquatic environment. J. Soil Water Conserv. 33:274-278.

Meyers, L. H. 1989. Grazing and riparian management in southwestern Montana. Pages 117-120 *in* R. E. Gresswell, B. A. Barton, and J. L. Kershner, eds. Practical approaches to riparian resource management: an educational workshop. U.S. Bur. Land Manage., Billings, Mont.

Miller, R. R. 1961. Man and the changing fish fauna of the American Southwest. Pap. Mich. Acad. Sci., Arts, and Letters 46:365-409.

Minckley, W. L. 1973. Fishes of Arizona. Arizona Game and Fish Dep., Phoenix. 293pp.

Moore, F. R., and P. Kerlinger. 1987. Stopover and fat deposition by North American wood warblers (Parulinae) following spring migration over the Gulf of Mexico. Oecologia 74:47-54.

———, and T. R. Simons. 1992. Habitat suitability and stopover ecology of neotropical landbird migrants. Pages 345-355 *in* J. M. Hagan III and D. W. Johnson, eds. Ecology and conservation of neotropical migrant landbirds. Smithsonian Inst. Press, Washington, D.C.

Mosconi, S. L., and R. L. Hutto. 1982. The effect of grazing on the landbirds of a western Montana riparian habitat. Pages 221-233 *in* L. Nelson, J. M. Peek, and P. D. Dalke, eds.. Proc. of the wildlife-livestock relationships symposium. U.S. For., Wildl., and Range Exp. Stn., Univ. Idaho, Moscow.

Moulton, M. 1978. Small mammal associations in grazed versus ungrazed cottonwood riparian woodland in eastern Colorado. Pages 133-140 *in* Lowland river and stream habitat in Colorado: a symposium. Colo. Chap. Wildl. Soc. and Colo. Audubon Counc., Greeley.

Naiman, R. J., C. A. Johnson, and C. James. 1988. Alteration of North American streams by beaver. Bioscience 38:753-762.

———, M. Melillo, and J. E. Hobbie. 1986. Ecosystem alteration of boreal forest streams by beaver (*Castor canadensis*). Ecology 67:1524-1269.

National Research Council. 1994. Rangeland health. New methods to classify, inventory, and monitor rangelands. Comm. Rangeland Classification, Board on Agric. Natl. Acad. Sci. Press, Washington, D.C.

Neff, D. J. 1957. Ecological effects of beaver habitat abandonment in the Colorado Rockies. J. Wildl. Manage. 21:80-84.

Nehring, R. B. 1986. Fish flow investigations. Colorado Div. Wildl., Stream Fish. Invest., Fed. Aid in Sport Fish Restor., Proj. F-51-R, Job 1, Denver.

Noh, L. 1979. The livestock producers' viewpoint. Pages 3-4 *in* O. B. Cope, ed. Proc. of the forum: grazing and riparian/stream ecosystems. Trout Unlimited, Inc., Denver, Colo.

Office of Inspector General. 1990. Final audit report on survey of selected programs of the Colorado State Office (April 24, 1990). U.S. Bur. Land Manage.

Ohmart, R. D. In Press. Ecological condition of the east fork of the Gila River and selected tributaries: Gila National Forest, New Mexico. Symp. on desired future conditions for southwestern riparian ecosystems. U.S. Dep. Agric., For. Serv., Rocky Mtn. Res. Stn. Gen. Tech. Rep.

———, and B. W. Anderson. 1982. North American desert riparian ecosystems. Pages 433-479 *in* G. L. Bender, ed. Reference handbook on the deserts of North America. Greenwood Press, Westport, Conn. 594pp.

———, and ———. 1986. Riparian habitats. Pages 169-201 *in* A. Y. Cooperrider, R. J. Boyd, and H. R. Stuart, eds. Inventory and monitoring of wildlife habitat. U.S. Dep. Inter. Bur. Land Manage. Serv. Cent., Denver, Colo. 858pp.

———, ———, and W. C. Hunter. 1988. The ecology of the lower Colorado River from Davis Dam to the Mexican-United States International Boundary: a community profile. U.S. Fish Wildl. Serv. Biol. Rep. 85(7.19). 296pp.

Olson, R., and C. Armour. 1979. Economics of improved livestock management approaches. Pages 67-71 *in* O. B. Cope, ed. Proc. of the forum: grazing and riparian/stream ecosystems. Trout Unlimited, Inc., Denver, Colo.

Oregon-Washington Interagency Wildlife Committee. 1979. Managing riparian ecosystems (zones) for fish and wildlife in eastern Oregon and eastern Washington. Riparian Habitat Subcomm. of the Oregon/Washington Interagency Wildl. Comm., March 1979, Portland, Oreg. 44pp.

Paulsen, Jr., H. A. 1969. Forage value on a mountain grassland aspen range in western Colorado. J. Range Manage. 22:102 107.

Peterson, H. V. 1950. The problem of grazing in Western valleys. Pages 404-434 *in* P. D. Trask, ed. Applied sedimentation. John Wiley & Sons, New York, N.Y.

Pfefferkorn, I. 1949. Sonora, a description of the province. T. E. Treutelin, ed., Univ. New Mexico Press, Albuquerque.

Phillips, R. R., L. E. Claire, and J. Moring. 1975. Some effects of gravel mixtures on emergence of coho salmon and steelhead trout fry. Trans. Am. Fish. Soc. 104:461-466.

Platts, W. S. 1978. Livestock interactions with fish and aquatic environments: problems in evaluating. North Am. Wildl. and Nat. Resourc. Conf. 43:499-504

———. 1979. Livestock grazing and riparian/stream ecosystems—an overview. Pages 39-45 *in* O. B. Cope, ed. Proc. of the forum: grazing and riparian/stream ecosystems. Trout Unlimited, Inc., Denver, Colo.

———. 1981*a*. 7. Effects of livestock grazing. W. R. Meehan, tech. ed, Influence of forest and rangeland management on anadromous fish habitat in western North America. U.S. For. Serv. Gen. Tech. Rep. PNW-124.

———. 1981*b*. Sheep and cattle grazing strategies in riparian-stream environments. Pages 82-92 *in* Proc. wildlife-livestock relationships symposium. Univ. Idaho, For. Range Exp. Stn., Moscow.

———. 1982. Sheep and cattle grazing strategies on riparian-stream environments. Pages 251-270 *in* Wildlife-livestock relationships symposium. Proceedings 10. Univ. Idaho, For., Wildl., and Range Exp. Stn., Moscow.

———. 1989. Compatibility of livestock grazing strategies with fisheries. Pages 103-110 *in* R. E. Gresswell, B. A. Barton, and J. L. Kershner, eds. Practical approaches to riparian resource management: an educational workshop. U.S. Bur. Land Manage., Billings, Mont.

———. 1990. Managing fisheries and wildlife on rangelands grazed by livestock: a guidance and reference document for biologists. Nevada Dep. Wildl., Reno.

———. 1991. Livestock grazing. Am. Fish. Soc. Spec. Publ. 19:389-424.

———, K. A. Gebhardt, and W. L. Jackson. 1985. The effects of large stream events on basin-range riparian stream habitats. Pages 30-34 *in* R. R. Johnson, C. D. Ziebell, D. R. Patton, P. F. Ffolliott, and R. H. Hamre, tech. coords. Riparian ecosystems and their management: reconciling conflicting uses. Proc. First North Am. Riparian Conf., U.S. For. Serv. Gen. Tech. Rep. RM-120.

———, and R. L. Nelson. 1985. A stream habitat and fisheries response to livestock grazing and instream improvement structures, Big Creek, Utah. J. Soil Water Conserv. 40:374-379.

Powell, H. M. 1931. The Santa Fe Trail to California, 1849-1852. Book Club of Calif., San Francisco. 272pp.

Pratt, T. K., and E. W. Stiles. 1983. How long fruit-eating birds stay in the plants where they feed: implications for seed dispersal. Am. Nat. 122:797-805.

Raleigh, R. F. 1982. Habitat suitability index models: brook trout. U.S. Fish Wildl. Serv. FWS/OBS-82/10.24. 44pp.

Rampton, Jr. J. R. 1993. Decision dated December 20. National Wildlife Federation vs. Bureau of Land Management, No. UT-06-91-0. U.S. Dep. Inter., Off. Hearings and Appeals, Hearings Div., Salt Lake City, Ut.

Reid, E. H., and G. D. Pickford. 1946. Judging mountain meadow range condition in eastern Oregon and eastern Washington. U.S. Dep. Agric. Circ. No. 748.

Reiser, D. W., and T. C. Bjornn. 1979. Habitat requirements of anadromous salmonoids. U.S. For. Serv. Gen. Tech. Rep. PNW-96. 54pp.

Report of the Governor of Arizona to the Secretary of the Interior. 1885. Gov. Printing Off., Washington, D.C.

Report of the Governor of Arizona to the Secretary of the Interior. 1896. Gov. Printing Off., Washington, D.C.

Rhodes, H. A., and W. A Hubert. 1991. Submerged undercut banks as macroinvertebrate habitat in a subalpine meadow stream. Hydrobiologia 213:149-153.

Rhodes, J., C. M. Skau, D. Greenlee, and D. L. Brown. 1985. Quantification of nitrate uptake by riparian forests and wetlands in an undisturbed headwaters watershed. Pages 175-179 *in* R. R. Johnson, C. D. Ziebell, D. R. Patton, P. F. Ffolliott, and R. H. Hamre, tech. coords. Riparian ecosystems and their management: reconciling conflicting uses. Proc. First North Am. Riparian Conf., U.S. For. Serv. Gen. Tech. Rep. RM-120.

Rice, J., B. W. Anderson, and R. D. Ohmart. 1984. Comparison of the importance of different habitat attributes to avian community organization. J. Wildl. Manage. 48:895-911.

Roath, L. R., and W. C. Krueger. 1982. Cattle grazing influence on a mountain riparian zone. J. Range Manage. 35:100-104.

Rotenberry, J. T. 1985. The role of vegetation in avian habitat selection: physiognomy or floristics? Oecologia 67:213-217.

Rucks, J. A. 1978. Comparisons of riparian communities influenced by grazing. Pages 100-113 *in* Lowland river and stream habitats in Colorado: a symposium. Colorado Wildl. Soc. and Audubon Counc., Greeley.

Rudemann, R., and W. J. Schoonmaker. 1938. Beaver dams as geologic agents. Science 88:523-525.

Ryder, R. A. 1980. Effects of grazing on bird habitats. Pages 51-56 *in* Management of Western forests and grasslands for nongame birds. U.S. For. Serv. Gen. Tech. Rep. INT-86.

Sands, A., and G. Howe. 1977. An overview of riparian forests in California: their ecology and conservation. Pages 91-97 *in* R. R. Johnson and D. A. Jones, tech. coords. Importance, preservation and management of riparian habitat: a symposium. U.S. For. Serv. Gen. Tech. Rep. RM-43.

Scheffer, V. B. 1938. Management studies of transplanted beaver in the Pacific Northwest. Trans. North Am. Wildl. Conf. 6:320-326.

Schulz, T. T., and W. C. Leininger. 1990. Differences in riparian vegetation structure between grazed areas and exclosures. J. Range Manage. 43:294-298.

———, and ———. 1991. Nongame wildlife communities in grazed and ungrazed montane riparian sites. Great Basin Nat. 51:286-292.

Seegmiller, R. F., and R. D. Ohmart. 1981. Ecological relationships of feral burros and desert bighorn sheep. Wildl. Monogr. 78. 58pp.

Seton, E. T. 1929. Lives of game animals. Vol. 4, Part 2, Rodents, etc. Doubleday, Dovan, Garden City, N.Y.

Severson, K. E., and C. E. Boldt. 1978. Cattle, wildlife and riparian habitats in the western Dakotas. Pages 94-103 *in* Management and use of northern Plains rangeland. Regional Rangeland Symposium, Bismarck, N. D.

Simpson, L. B. 1952. Exploitation of land in central Mexico in the sixteenth century. ("Ibero Americana," 36). Univ. California Press, Berkeley. 92pp.

Skinner, Q. J., J. Adams, P. Rechard, and A. Bettle. 1974. Effects of summer use of a mountain watershed on bacterial water quality. J. Environ. Qual. 3:329-335.

Skovlin, J. 1967. Fluctuations in forage quality on summer range in the Blue Mountains. U.S. Dep. Agric. For. Serv., Pac. Northwest Res. Pap. No. 44.

Skovlin, J. M. 1984. Impacts of grazing on wetlands and riparian habitat: a review of our knowledge. *In* Developing strategies for rangeland management. Natl. Res. Coun./Natl. Acad. Sci., Washington, D.C.

Smith, B. H. 1980. Not all beaver are bad; or an ecosystem approach to stream habitat management, with possible software applications. Pages 32-37 *in* Proc. 15th Annu. Meet., Am. Fish. Soc., Colorado-Wyoming Chap., Fort Collins, Colo.

Smith, D. G. 1976. Effect of vegetation on lateral migration of anastomosed channels of a glacier meltwater river. Bull. Geol. Soc. Am. 87:857-860.

Southwood, T. R. E. 1961. The numbers of species of insects associated with various trees. J. Anim. Ecol. 30:1-8.

Stamp, N. E. 1978. Breeding birds of riparian woodlands in southcentral Arizona. Condor 80:64-71.

Stanley, J. T., B. Mori, and D. L. Suddjian. 1991. The importance of riparian vegetation for wildlife in California: a literature synthesis. The Habitat Restoration Group, Scotts Valley, Calif. 19pp.

Stevens, L., B. T. Brown, J. M. Simpson, and R. R. Johnson. 1977. The importance of riparian habitats to migrating birds. Pages 156-164 *in* R. R. Johnson and D. A. Jones, tech. coords. Importance, preservation and management of riparian habitat: a symposium. For. Serv. Gen. Tech. Rep. RM-43.

Storch, R. L. 1979. Livestock/streamside management programs in eastern Oregon. Pages 56-59 *in* O. B. Cope, ed. Proc. of the forum: grazing and riparian/stream ecosystems. Trout Unlimited, Inc., Denver, Colo.

Stromberg, J., B. D. Richter, D. T. Patten, and L. G. Wolden. 1993. Response of a Sonoran riparian forest to a 10-year return flood. Great Basin Nat. 53:118-130.

Strong, T. R., and C. E. Bock. 1990. Bird species distribution patterns in riparian habitats in southeastern Arizona. Condor 92:866-885.

Stuber, R. J. 1985. Trout habitat, abundance, and fishing opportunities in fenced and unfenced riparian habitat along Sheep Creek, Colorado. Pages 310-314 *in* R. R. Johnson, C. D. Ziebell, D. R. Patton, P. F. Ffolliott, and R. H. Hamre, tech. coords. Riparian ecosystems and their management: reconciling conflicting uses. Proc. First North Am. Riparian Conf., U.S. For. Ser. Gen. Tech. Rep. RM-120.

Swan, B. 1979. Riparian habitat—the cattlemen's viewpoint. Pages 4-6 *in* O. B. Cope, ed. Proc. of the forum: grazing and riparian/stream ecosystems. Trout Unlimited, Inc., Denver, Colo.

Swanson, F. J., T. K. Kratz, N. Caine, and R. G. Woodmansee. 1988. Landform effects on ecosystem patterns and processes. Bioscience 38:92-98.

Szaro, R. 1980. Factors influencing bird populations in Southwestern riparian forests. Pages 403-418 *in* R. M. DeGraff, tech. coord. Management of Western forests and grasslands for nongame birds. U.S. For. Serv. Gen. Tech. Rep. INT-86. 535pp.

———. 1989. Riparian forest and scrubland community types of Arizona and New Mexico. Desert Plants 9:69-138.

———, S. C. Belfit, K. Aitkin, and J. N. Rinne. 1985. A preliminary study on the impact of grazing on a riparian garter snake. Pages 359-363 *in* R. R. Johnson, C. D. Ziebell, D. R. Patton, P. F. Ffolliott, and R. H. Hamre, tech. coords. Riparian ecosystems and their management: reconciling conflicting uses. Proc. First North Am. Riparian Conf., U.S. For. Serv. Gen. Tech. Rep. RM-120.

Taylor, D. M. 1986. Effects of cattle grazing on passerine birds nesting in riparian habitat. J. Range Manage. 39:254-258.

Taylor, F. R., L. A. Gillman, and J. W. Pedretti. 1989. Impact of cattle on two isolated fish populations in Pahranagat Valley, Nevada. Great Basin Nat. 49:491-495.

Terrill, S. B., and R. D. Ohmart. 1984. Facultative extension of fall migration by yellow-rumped warblers (*Dendroica coronata*). Auk 101:427-438.

The Arizona Nature Conservancy. 1987. Streams of life. Information pamphlet. Tucson, Ariz.

Thomas, J. W., editor. 1989. Wildlife habitats in managed forests—the Blue Mountains of Oregon and Washington. U.S. Dep. Agric. Handb. 553, Washington, D.C. 512pp.

———, C. Maser, and J. E. Rodiek. 1979. Riparian zones in managed rangelands—their importance to wildlife. Pages 21-31 *in* O. B. Cope, ed. Proc. of the forum: Grazing and riparian/stream ecosystems. Trout Unlimited, Inc., Denver, Colo.

Tompson, K. 1980. Riparian forests in the Sacramento Valley, California. Pages 35-38 *in* A. Sands, ed. Riparian forests in California: their ecology and conservation. Univ. California Press, Davis.

Trout Unlimited. 1979. Proc. of the forum: grazing and riparian/stream ecosystems. Denver, Colo. 94pp.

U.S. Bureau of Census. 1872. Ninth census of the United States: 1870.

U.S. Bureau of Census. 1883. Tenth census of the United States: 1880.

U.S. Bureau of Census. 1895. Eleventh census of the United States: 1890.

U.S. Department of Agriculture. 1992. Integrated riparian evaluation guide. U.S. For. Serv., Intermountain Reg., Ogden, Ut.

U.S. Department of the Interior. 1994. Rangeland reform '94. Draft Environ. Impact Statement. U.S. Bur. Land Manage. and U.S. For. Serv., Washington, D.C.

U.S. Fish and Wildlife Service. 1993*a*. Proposal to list the southwestern willow flycatcher as an endangered species and designate critical habitat. July 23, 1993. Fed. Reg. 58:39495-39522.

———. 1993*b*. Gila trout recovery plan. Albuquerque, N.M. 113pp.

———. 1994. Desert tortoise (Mojave population) recovery plan. U.S. Fish and Wildl. Serv., Portland, Oreg. 77pp + appendices.

———. 1995. Endangered and threatened wildlife and plants: final rule determining endangered status for the southwestern willow flycatcher, February 27, 1995. Fed. Register 60: 10694-10715.

Valencia, R. A., J. A. Wennerlund, R. A. Winstead, S. Woods, L. Riley, E. Swanson, and S. Olson. 1993. Arizona riparian inventory and mapping project. Arizona Game and Fish Dep., Phoenix, Ariz. 138 pp.

Van Velson, R. 1979. Effects of livestock grazing upon rainbow trout in Otter Creek. Pages 53-55 *in* O. B. Cope, ed. Proc. of the forum: grazing and riparian/stream ecosystems. Trout Unlimited, Inc., Denver, Colo.

Verkaar, H. J. 1987. Population dynamics—the influence of herbivory. New Phytologist 106(Suppl.):49-60.

Vitt, L. J., and R. D. Ohmart. 1978. Herpetofauna of the lower Colorado River, Davis Dam to the Mexican border. West. Found. Vertebr. Zool. 2:35-72.

Wagoner, J. J. 1952. History of the cattle industry in southern Arizona, 1540-1940. Univ. Arizona Soc. Sci. Bull. 20.

Wiens, J. A. 1989. The ecology of bird communities. Vol. 2. Processes and variations. Cambridge University Press, New York. 316pp.

Wilkinson, C. F. 1992. Crossing the next meridian. Land, water, and the future of the West. Island Press, Washington, D.C. 376pp.

Wilson, L. O. 1979. Public forum. Pages 77-87 *in* O. B. Cope, ed. Proc. of the forum: grazing and riparian/stream ecosystems. Trout Unlimited, Inc., Denver, Colo.

Winegar, H. H. 1977. Camp Creek channel fencing—plant, wildlife, and soil and water response. Rangeman's J. 4:10-12.

Winker, K., D. W. Warner, and A. R. Weisbrod. 1992. The northern waterthrush and Swainson's thrush as transients at a temperate inland stopover site. Pages 384-402 *in* J. M. Hagan III and D. W. Johnson, eds. Ecology and conservation of neotropical migrant landbirds. Smithsonian Inst. Press, Washington, D.C.

Winternitz, B. L. 1980. Birds in aspen. Pages 247-257 *in* Management of Western forests and grasslands for nongame birds. U.S. For. Serv. Gen. Tech. Rep. INT-86.

———, and H. Cahn. 1983. Nestholes in live and dead aspen. Pages 102-106 *in* Snag habitat management: proceedings of the symposium. U.S. For. Serv. Gen. Tech. Rep. RM-99.

Wright, R., K. Jenkins, B. Butterfield, C. Key, and P. Happe. 1983. Wildlife habitats on riparian zones of the North Fork and mainstem Flathead River, Montana. Final Rep. U.S. Environ. Protect. Agency, Helena, Mont.

Young, J. A., and R. A. Evans. 1989. Silver State rangelands—historical perspective. Rangelands 11:199-203.

Chapter 17
FERAL ANIMALS ON RANGELANDS
Charles L. Douglas and David M. Leslie, Jr.

INTRODUCTION

Release of feral animals from captivity into the wild, whether intentionally or accidentally, has created serious problems for native species on a global scale. Feral animals are domesticated species that have escaped or been released into natural ecosystems. Native or non-native animals could be domesticated, but in this chapter the term "feral" applies only to non-natives. Animals, such as Barbary sheep (*Ammotragus lervia*), that were brought to this country and released for hunting purposes are considered "exotics" and are not discussed. We examine how the presence of primarily large herbivorous feral animals, especially the burro (*Equus asinus*), on western rangelands in the United States changes availability of resources for native species, and how differences in anatomical, physiological, and behavioral adaptations of native wildlife and feral species influence competitive interactions.

Only a few species have become feral and created problems for managers of rangelands in the United States: burros, horses (*Equus caballas*), goats (*Capra hircus*), and pigs (*Sus scrofa*). In some instances, cattle on western rangelands have escaped from, or avoided, roundups and have become feral (at least until they are caught, at which time they revert to being exotic domestics). Likewise, pigs initially escaped captivity in a number of locations and became feral, but now others are imported specifically for stocking some of the same areas for hunting. Those that escaped captivity are properly termed feral, but those brought in for hunting are exotics.

Most wildlife species have clearly limited distributions within a region. These relatively stable distributions reflect a suite of physical and biotic attributes of habitat, including climate, which wildlife have evolved with and adapted to behaviorally and physiologically. We address disruptions of natural systems, including disruptions affecting native wildlife that occur when feral animals are introduced. Wildlife are integral components of natural ecosystems; they should not be managed from a rigid single-species perspective, as if they occurred in a vacuum.

In this chapter, we use the feral burro as a case study of how feral animals interact with native rangeland species of plants and animals. The burro highlights all of the problems affecting native biota that are created by feral animals. Burros are widespread over southwestern rangelands in the United States, have been studied adequately to identify their impacts on some native species, and are a relatively abundant feral herbivore (Table 1).

ADAPTATION OF LARGE HERBIVORES TO RANGELANDS

On an evolutionary scale, grazing animals developed dentition and digestive anatomy and physiology as adaptations to changing availabilities of forage species, especially grasses (Young 1962, Janis 1976). The abrasive nature of silica cells in grasses probably led to evolutionary changes in dentition of Eocene-Pleistocene horses from the low-crowned, generalized

TABLE 1. Feral and exotic large herbivores on western rangelands of the United States.

Type	Common name	Scientific name	General population estimate[a]	Century of first release	States of occurrence
Feral	Horse	*Equus caballus*	50,000	16th	Ariz., Calif., Colo, Id., Mont., Nev., N.M., Oreg., Ut., Wyo.
	Pig	*Sus scrofa*	30,000[b]	16th	Calif.
	Burro	*Equus asinus*	12,000	19th	Ariz., Calif., Nev., N.M., Ut.
	Goat	*Capra hircus*	<500[c]	Unknown	Calif. (islands only), N.M.
	Sheep	*Ovis aires*	<200	Unknown	Calif. (islands only)
Exotic[d]	Axis deer	*Axis axis*	6,000	20th	Calif., Tex.
	Barbary sheep	*Ammotragus lervia*	3,000	20th	N.M., Tex.
	Gemsbok	*Oryx gazella*	>600	20th	N.M.
	Persian wild goat	*Capra aegagrus*	>600	20th	N.M.
	Fallow deer	*Dama dama*	<500	20th	Calif., Tex.
	Siberian ibex	*Capra siberica*	<125	20th	N.M.
	Sambar deer	*Cervus unicolor*	<100	20th	Calif.

[a] Decker (1978), Wagner (1983), Yorks (1989), and Morrison (1985).
[b] Estimate does not include feral pigs in Texas.
[c] Numbers of feral goats on islands off California were ≥ 10,000 prior to recent eradication efforts.
[d] Exotic species that have become free-ranging on western rangelands but are not actively ranched; list does not include other exotic species used in game ranching largely in Texas (see Yorks 1989).

teeth of *Hyracotherium* to high-crowned teeth with intricately folded surface patterns of enamel, cementum, and dentine in Pleistocene and modern horses (Janis 1976) that facilitate grinding this abundant food source. Molarization of the premolars also provided additional grinding surfaces (Colbert 1961).

Ruminants developed a 4-chambered stomach to aid digestion of tough cell walls and to extract additional cell nutrients. Development of ruminant digestion necessitated development of the behavior of regurgitating food items from the rumen to be chewed again into finer particles (Janis 1976). Because of their compartmentalized stomachs, ruminants are limited in their ability to vary the rate of food passage through the digestive tract, and thus must seek a higher quality of forage than nonruminants. Microflora and microfauna of the rumen require micronutrients, obtained from forage, to reproduce and break down cell walls of plants. If ruminal microorganisms do not reproduce, the amount of nutrients that the ruminant host receives is greatly diminished (Van Soest 1982).

Horses and burros, the most common feral species on western rangelands (Table 1), do not have a compartmentalized stomach but can obtain additional nutrients from forage by fermentation in an enlarged caecum (Prosser and Brown 1966, Janis 1976). They have the ability to vary their rate of gut clearance, or passage of food through the digestive tract, by increasing their food intake. This strategy is useful during late summer, or during drought, when nutrient content of forage is reduced. By increasing food intake the individual can increase its intake of nutrients and partly compensate for low forage quality (Janis 1976). The caecum of horses and some rodents is less effective than the rumen in digesting fibers because microorganisms proliferate at a point in the gut beyond (distal to) the site of action of proteolytic enzymes (Prosser and Brown 1966:119).

In general, among related species with similar digestive capabilities, smaller species require better quality food than larger species. Smaller bodied herbivores have higher metabolic rates and higher nutrient requirements per unit of body mass than larger bodied herbivores (Prosser and Brown 1966, Robbins 1983). Young-of-the-year have higher protein and nutrient requirements per unit of body mass than any animals in a population, including lactating females (Robbins 1983). Body size, metabolism, and shape of mouth parts dictate the kinds of strategies animals develop to procure sufficient quantities of good-quality food (Hofmann 1968, Jarman and Sinclair 1979). At one end of the spectrum are small species with delicate mouth parts and high metabolic rates that select the most nutritious or most easily digested parts of plants; at the other end of the spectrum are large species with mouths adapted for rapid ingestion of large quantities of undifferentiated items, possibly of low quality (Jarman and Sinclair 1979). Horses, burros, and cattle tend to feed on a wide variety of plants, depending upon availability.

Diet studies of desert bighorn sheep (*Ovis canadensis* spp.) and feral burros show that they eat a wide variety of plant species, but relatively few species may comprise the majority of the diet (Seegmiller and Ohmart 1981, Ginnett 1982). Large-mouthed animals are unable to be highly selective when grazing, and grasses and various forbs may be consumed together, although grasses are the target species. Relatively large herbivores may coexist on rangelands with ungulate herbivores that are considerably smaller than they are and that select many of the same plant species (Seegmiller and Ohmart 1981, Ginnett 1982). Selective feeders, however, can separate forbs from grasses. A rough rule-of-thumb is that a medium-sized adult herbivore (e.g., sheep or deer) requires approximately its body weight in forage each month (National Academy of Science 1975, Siegmund 1973). Young individuals and lactating females may require 1.5 times their body weight in forage each month. Horses and burros require less than their body weight in forage each month, but because of their relatively large size, total intake greatly exceeds that required by medium-sized ruminants (Janis 1976).

Foraging strategies, anatomy, and physiology determine the interrelationship of herbivores within a system (Hofmann 1968), whether they are native or foreign to that system. Native herbivores evolved with native plants and animals in a given ecosystem and have developed ways to partition forage resources and reduce competition. Non-native species have evolved in different ecosystems and have their own food procuring strategies. Animals using similar food resources in a given habitat must somehow partition those resources or potentially compete with others using the resources. The herbivore guild within an ecosystem is comprised of various genera from grasshoppers, thrips, aphids, and other insect fauna to pocket gophers, mice, rabbits, hares, and ungulates. In a given system, there is a finite amount of yearly forage production that depends on elevation, latitude, soil types, and weather. Primary productivity varies from year to year depending upon a suite of variables such as soil temperature, soil moisture storage, and precipitation prior to and during the critical part of the growing season (Beatley 1969). Additionally, perennial shrubs and trees form bud primordia in the year preceding that in which they open to form leaves and flowers. Therefore, a drought year in which there is too little precipitation to promote production of new shrub growth might be expected to be followed by a second year of modest shrub growth, even though precipitation and weather conditions returned to normal. Deeply rooted shrubs may tap different moisture regimes and respond differently to precipitation events than shrubs with roots near the soil surface. Recharge of soil moisture occurs relatively quickly at shallow depths but takes considerably longer to reach deeper levels. Years of poor forage production will support considerably fewer herbivores than years having good forage production and will affect each species of the herbivore guild in somewhat different ways.

Herbivorous food generalists are better able to use available forage during drought years than food specialists, although they may select alternative species rather than preferred species. Under drought conditions, grazers might select more browse than they would in years with better plant productivity; browsers might select less desirable species or use more energy to search for preferred plants.

There are several ways to look at food availability and shortages on rangelands relative to the potential impact by feral species. In general, native animals have adapted to yearly differences in forage availability and seasonal changes in nutrient content. Nutrient content of rangeland forage species is greatest early in the growing season, as new leaves and stems are developing. As leaves mature and summer progresses, nutrient content decreases markedly, until by autumn forage generally contains only maintenance levels of nutrition. Mammalian herbivores time reproductive events to coincide with seasonal availability of food resources needed to rear their young. At higher elevations and northern latitudes, the growing season is relatively short, thereby constraining reproduction in herbivores to a period of 2-3 months. At lower elevations and more southerly latitudes, the growing season is attenuated, and breeding and birthing periods may extend over 4-5 months (Leslie and Douglas 1979). Drought or late winter storms can disrupt normal timing of plant growth and create serious problems for mammalian herbivores giving birth and nursing young. In drought years, browsing of shrubs may remove several previous years' growth. Overbrowsing may modify the shape of shrubs, or damage them so severely that they are unable to recover. Preferred forage species are the first to be impacted. Feral species can exacerbate seasonally induced reductions in forage availability (Woodward 1976).

FERAL BURROS ON WESTERN RANGELANDS

Domesticated Somalian and Nubian wild asses or burros were introduced into Mexico by the Spaniards in the 1530s and spread northward into what is now the United States as a consequence of Spanish colonization. These hardy, desert adapted animals provided transportation for people and equipment in the United States for >3 centuries and are still used today in developing countries. Domesticated asses were introduced by miners into areas that later became western public lands >100 years ago. Feral burros were released into the wild following the decline of mining, the advent of the railroad, and the availability of motorized vehicles (see Woodward [1976] for an excellent history; Clutton-Brock [1981] for a history of domestication).

Although feral burros represent only a small part of ungulate animal unit months (AUMs) on western rangelands (Fig.1), they have been responsible for extensive habitat damage by overgrazing, selective removal of preferred plants, trampling of plants, soil disturbance leading to erosion (Farrell 1973, Koehler 1974, Carothers et al. 1976, Fletcher and Wauer 1976, Woodward and Ohmart 1976, Norment and Douglas 1977), impact on small vertebrates (Carothers et al. 1976, Yancey and Douglas 1983, Yancey 1984), and competitive interactions with desert bighorn sheep (Seegmiller and Ohmart 1975, 1976; Dunn and Douglas 1982; Ginnett and Douglas 1982). Similar observations have been made for wild horses (Slade and Godfrey 1982, Wagner 1983, Berger 1986). On many, if not most, western rangelands, the biomass represented by feral horses, burros, or cattle exceeds that represented by native ungulates (Fig.1B).

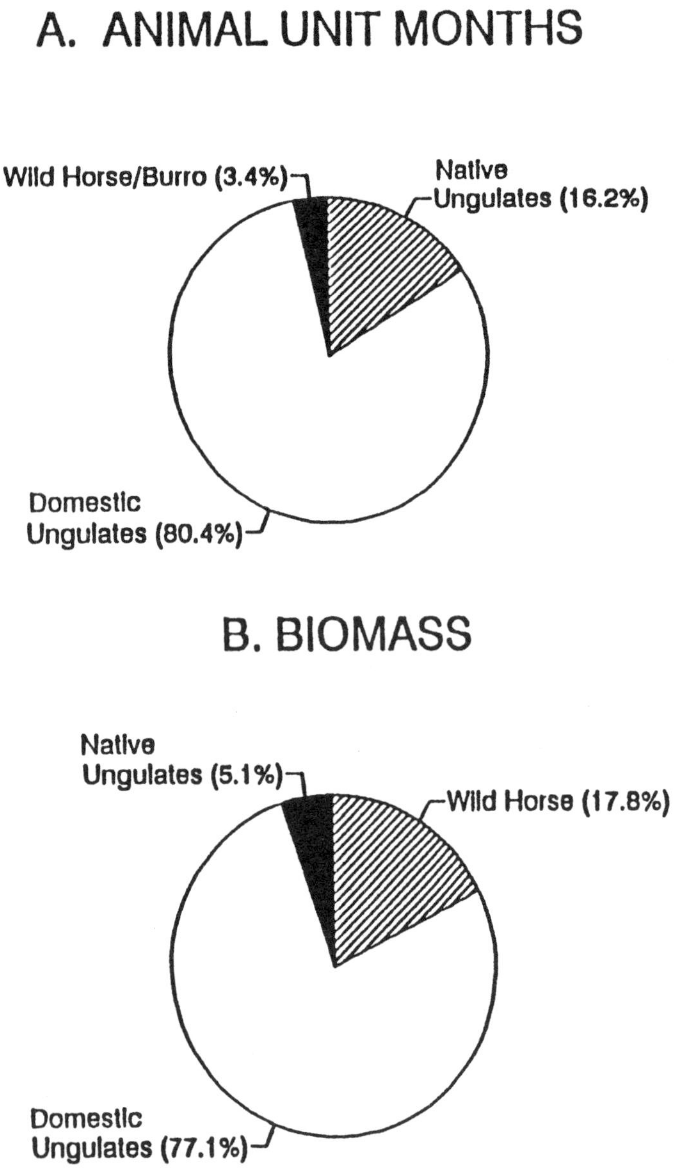

FIGURE 1. Estimates of animal unit months (A) of large herbivorous ungulates on public grazing lands in the western United States (adapted from Wagner 1983) and biomass (B) on public lands in Nevada and Utah (adapted from Berger 1986).

Natural History

Burros live 10-15 years in the wild and longer in captivity. Of 631 burros harvested at China Lake Naval Weapons Center, California, only 12 individuals were >10.5 years old; males lived longer than females. Maximum estimated ages were 15.5 and 20.5 years for females and males, respectively (Johnson et al. 1987). Ruffner and Carothers (1982) reported on age structure, condition, and reproduction of 2 burro populations from Grand Canyon National Park. In the Bedrock Canyon herd, 78% of the animals were between 1 month and 4 years of age, and the remaining 22% were between 5 and 15 years of age. In the Lower Canyon herd, 49% of the individuals were between 1 month and 4 years, and the remaining 51% were from 5 to 15 years of age. Burro populations in the southwestern United States usually have 50% of the individuals >4 years old (Moehlman 1974, Norment and Douglas 1977). However, in the Bedrock Canyon population and in a herd from the Chemehuevi Mountains of southern California, only 22.3 and 36.2% of the individuals, respectively, were >4 years old (Woodward 1976). In a harvested sample of 197 burros from Australia, mean age was 4.0 years ± 3.3 (SD) (McCool et al. 1981), which was ascribed to harvesting by pastoralists, but both males and females >10 years of age were present.

Burros have conservative life strategies. Sexes are segregated for most of the year into male, female, and mixed groups. Only a female and her foal (Fig. 2) form long-term associations (Moehlman 1974). Mixed groups are more common during the breeding season and when animals are dependent upon water. In Death Valley, males began congregating near water sources with females and young in about April, remained in mixed groups through the summer, and segregated again in autumn as cooler temperatures allowed dispersal from water (Norment and Douglas 1977). Breeding occurs in spring and summer when burros are congregated near water. Although females are polyoestrus and can breed throughout the year, most breeding is confined to spring and summer. Some individuals breed and give birth out of synchrony with the majority of the herd; a few young animals are seen throughout the year in Death Valley. Natality in Death Valley appeared to peak in late spring and early summer; the majority of births occur between April and July (Moehlman 1974, Norment and Douglas 1977). A seasonal concentration of natality also is seen in wild asses in African deserts, where breeding occurs following summer rains and greenup of vegetation (Klingel 1977).

Burros have 1 foal/year in years with adequate precipitation; gestation is 12 months (Asdell 1964). In Death Valley, a few animals breed as yearlings, but most breed at 2 years and foal at 3 years of age. Woodward (1976) reported breeding year-round

FIGURE 2. Adult female feral burro and her foal. Such pairs are the only persistent associations among feral burros. Photo by R. D. Ohmart.

along the lower Colorado River; females in that area also reached sexual maturity at 1 year of age. The difference in breeding strategies between areas undoubtedly is related to differences in elevation and climate, and their effect on forage production. Johnson et al. (1987) found that female burros at 1.5 years old were reproductively active and had a 13.8% pregnancy rate on the Naval Weapons Center (NWC), California. Pregnancy rate increased to 78.6% at 3.5 years and varied from 58.3 to 77.8% for females through 10.5 years. Feral burros remain fertile into old age; 2 females at 15.5 years old were pregnant at NWC. The percentage of pregnant or lactating females of reproductive age (76.2%) was similar between the NWC and 2 burro populations in the Grand Canyon (Ruffner and Carothers 1982). Sixty percent of lactating females on the NWC also were pregnant, indicating that burros can give birth in consecutive years (Johnson et al. 1987). Moehlman (1974), however, reported that the inter-foaling interval was typically 2 years in Wildrose Canyon, Death Valley. This observation might have been related to herd density, which was high at the time of Moehlman's (1974) study. Wolfe et al. (1989) reported that mean apparent incidence of pregnancy was 25% in yearling burros and 72.3% in burros >2 years old.

Survival of young burros can be very high; only a few dead foals were seen in 3 different study areas of Death Valley (C. L. Douglas, unpubl. data). White (1980) reported a first year survival rate of 66% in Butte Valley, Death Valley. Moehlman's (1974) data yielded estimates of 72-79% first year survival. Ohmart et al. (Feral burros on the Havasu Resource Area, Colorado River Valley, California-Arizona, unpubl. rep. Bur. Land Manage., Lake Havasu City, Arizona, 33pp., 1975), Morgart (1978), and Seegmiller and Ohmart (1981) found no evidence of foal mortality. Conversely, Ruffner and Carothers (1982) reported 76-89% pregnancy rates in 2 herds in the Grand Canyon but recruitment rates of only 11-17%. Choquenot (1990) reported that > 70% of mature females were pregnant in 2 burro populations in Australia that had experienced earlier herd reductions. Juvenile mortality was 3 times as great at high herd density as at low density and was associated with poorer juvenile body condition and slower growth (Choquenot 1990). In a harvested sample of 197 burros from Australia, 64% of the females were pregnant, and 17% of those were lactating (McCool et al. 1981).

Group integrity is ephemeral in burros except for the group comprised of the female and her foal (Fig. 2), or her foal and yearling. Klingel (1977) reported that there were no permanent bonds between any 2 adult animals. Individuals may be solitary, or occur in a variety of associations. These groups are variable and their composition may change within hours (Moehlman, 1974). Sexual segregation during the nonbreeding season is not rigid; females have been seen with males throughout the year in Death Valley (Norment and Douglas 1977). The female, her foal of the year, and frequently a yearling from the previous year formed the most common group in Death Valley. Young females appeared to remain with their mother for 2-3 years, until they foaled (Moehlman 1974, White 1980). Little is understood about the duration of mother-foal associations, or about how the home range of a young female relates to the home range of her mother. Male foals remain with their mother for most of their first year and then join male groups. Little is understood about the relationship of male home ranges to those of their mothers or to that of the male group they associated with.

Population Dynamics

In rangeland areas having erratic weather patterns and low rainfall, availability of nutritious forage can be a major regulating mechanism of populations. Years having small amounts of precipitation during months critical to plant productivity also have poor survival of young ungulates, especially desert bighorn sheep. Density was secondary to precipitation in regulating bighorn lamb numbers in the Lake Mead area of southern Nevada (Douglas and Leslie 1986). Choquenot (1990) provided evidence that densities of feral burros in northern Australia were imposed by food availability. An inverse relationship between density and percentage of females successfully foaling was suggested for burros in Death Valley (Norment and Douglas 1977). An estimated 24 foals were born into the Wildrose Canyon herd during a 12-month period of high population density; approximately 57% of the adult females foaled. The following year, 90% of the marked females foaled during a 12-month period of lower population density caused by trapping removals (Norment and Douglas 1977).

Because relatively few foals die in most of the herds studied in the western United States, population increases of > 20%/year have been reported (Norment and Douglas 1977). At that rate of increase, the population could double in size every 4 years. Woodward (1976) reported a recruitment rate of about 20% every 18 months in the Chemehuevi Mountains, California. Adult survival apparently is high; Norment and Douglas (1977) found only a 5% loss in adults/year in Death Valley. Reported annual rates of increase for feral burro populations in North America range from 1.2 to 29% (Morgart 1978, White 1980); a finite rate of increase of 23 to 28%/year was estimated for populations in northern Australia following herd reduction of about 40% (Choquenot 1990).

Conley (1980) questioned the high rates of annual increase reported for burro populations in the United States and regarded population increases of 20% as unlikely. Conley (1980) concluded that a 20% rate of increase could only be obtained if: (1) survival rates were high, (2) > 80% of the animals in all age classes bred, (3) age at first parturition was 3 years, and (4) breeding extended beyond 8-10 years of age. All of these conditions have been met by burro herds on southwestern rangelands. Eberhardt's (1982) theoretical modeling of horse population dynamics and observations on 2 Oregon herds supported observations by field researchers that under optimum conditions, some populations of feral equids may have rates of increase >20% per year.

Movements

Mean home ranges of burros range from 2.9 km^2 in Bandelier National Monument, New Mexico (Morgart 1978) to 68 km^2 in Death Valley, California (Norment and Douglas 1977; Table 2).

The unusually small sizes of home ranges at Bandelier were attributed to availability of permanent water and comparatively good forage conditions in contrast to more arid ecosystems. The large home ranges recorded by Norment and Douglas (1977) probably reflected reduced competition for space due to removal of 139 animals by trapping. In Death Valley, burros in Wildrose Canyon ranged over an area of 425 km^2 (Norment and Douglas 1977). Fall-winter ranges had a mean area of 34 km^2 and extremes of 1.0-103.6 km^2; spring-summer ranges averaged 12 km^2 (Norment and Douglas 1977). Moehlman (1974), Woodward (1976), Seegmiller and Ohmart (1976), Norment and Douglas (1977), and Morgart (1978) reported no significant difference in home range size between sexes (Table 2). Burros are faithful to home range areas. Studies in Death Valley showed that herds in 5 different areas differed in distribution of transferrin alleles (Blake 1977, Blake et al. 1981). Herds in adjacent drainages only 8 km apart had significantly different distributions of alleles, indicating insignificant amounts of mixing between herds during the breeding season. Feral burros in Death Valley have wide ranging movements in fall and winter when they are not water dependent (Norment and Douglas 1977, White 1980, Dunn 1984). In summer, 60% of the Wildrose population was restricted to within a 2-km radius of water, and about 98% were restricted to within 4 km of water; 80% were seen >6 km from water in winter (Norment and Douglas 1977).

Diet

Burros require more food and water than native desert ungulates. Much of the success of burros in arid regions stems from their ability to survive on poor quality forage, which is related to their monogastric digestive system and generalized diets. Burros have large faces and mouths, and a simple, nonruminant, digestive system. Their dentition is adapted for grazing, but they also browse and function as food generalists. Burros prefer grasses and forbs when available but readily browse on perennial shrubs. Various reports classify burros as browsers or grazers; however, they are opportunistic, eating grasses and forbs when available and switching to shrubs when necessary (Browning 1960, Hansen and Martin 1973, Woodward 1976, Ginnett 1982, Ginnett and Douglas 1982, Douglas and Hiatt 1987). Results of studies of burro diets vary because of differences in elevation, plant communities from which samples were collected, condition of the plant community, and whether diets were determined by fecal analysis or direct measurement of plant use.

Burros in Wildrose Canyon, Death Valley, functioned almost exclusively as browsers (Moehlman 1974, Norment and Douglas 1977) and have inflicted considerable damage on the shrub community. Most perennial grasses have been removed from the area by overgrazing. Conversely, White (1980) found modest evidence of browsing in Butte Valley, Death Valley, where burro diets consisted largely of red brome (*Bromus rubens*), an abundant exotic grass. Burros used 40 species of plants in Butte Valley, 11 species of which were important throughout the year. The average diet, consisted of 48.3% grasses, 25% shrubs, 19.3% perennial and annual forbs, and 7.3% unknown dicots (Douglas and Hiatt 1987). Browning (1960) analyzed 19 stomach samples of burros from the Cottonwood Mountains, Death Valley, in which he found 10% grasses, 39% forbs, and 51% browse. Ginnett (1982) identified 48 taxa eaten by burros in the Cottonwood Mountains, Death Valley; monthly diets contained 19-31 taxa, depending upon availability of annual forbs. Burros were primarily browsers during winter but switched to perennial grasses in April; the annual diet consisted of 48.4% browse, 41.2% grasses, 2.5% forbs, and 8% unknowns (Ginnett 1982).

In the Bill Williams Mountains of Arizona, Seegmiller and Ohmart (1981) found an annual diet of 40% browse, 22% grasses, and 33% forbs. Woodward and Ohmart (1976) analyzed 89 fecal samples from the Chemehuevi Mountains, California; the annual diet consisted of 61.1% browse, 3.9% grasses, 30.1% forbs, and 4.9% unknown. Hansen and Martin (1973) found that the annual burro diet consisted of 61.0% grasses, 11.5% forbs, and 27.5% browse in the lower Grand Canyon, Arizona. Ruffner et al. (1977) identified 23 species of plants in July diets of burros in the Bedrock Canyon area of the Grand Canyon. Eight species comprised almost 77% of the diet; grasses were the most

TABLE 2. Comparison of home range sizes (km^2) from various feral burro studies.

	Annual		Males		Females			
Location	$\bar{x}$	SE	$\bar{x}$	SE	$\bar{x}$	SE	Density[a]	Source
Tin Mountains, DVNM[b]	7.4	1.2	5.3	1.3	9.6	1.7	4.3	Dunn (1984)
Butte Valley, DVNM	9.9	2.2	8.8	2.4	11.4	4.2	4.4	White (1980)
Wildrose Canyon, DVNM	10.0	1.2	12.3	2.3	8.7	1.0	0.7	Moehlman (1974)
Wildrose Canyon, DVNM	68.1	5.7	72.1	7.6	62.5	8.6	0.5	Douglas and Norment (1977)
Bill Williams Mountains, Ariz.	19.2	3.7	15.2	2.4	24.3	7.9	3.3	Seegmiller and Ohmart (1981)
Lower Colorado River, Ariz.	30.1	6.1	28.7	4.6	30.5	14.3	0.4	Woodward (1976)
Bandelier National Monument, N.M.	2.9	0.4	2.6	0.7	2.7	0.4	2.6	Morgart (1978)

[a] Feral burros/km^2.

[b] Death Valley National Monument, Calif.

important, followed by browse and forbs. Douglas and Hiatt (1987) summarized all information from several diet studies in Death Valley; 119 species were eaten by burros, which reflected their generalized diet. Although burros may prefer grasses and forbs to browse (Woodward and Ohmart 1976), availabilities of forage classes in an area will determine diet composition.

The burro is remarkably adapted to desert environments. Its capacity for dehydration is rivaled by few mammals. It is able to withstand a water loss of 30% of the body weight, which can be replaced in 2-5 minutes by drinking 24-30 L of water (Maloiy 1970). The burro eliminates relatively large amounts of feces because the food is not as well digested as it is in ruminants. Water content of the feces is relatively high, resulting in fecal water loss about 3 times that in the camel (*Camelus dromedarius*) (Schmidt-Nielsen 1979).

IMPACTS OF FERAL BURROS

Vegetation

Because desert plant communities have low productivity, any large herbivore, particularly a feral species, can have a profound impact on individual plants and community composition. Burros are one of the largest grazing and browsing ungulates in the southwestern United States. An average burro requires approximately 4.5 kg of dry forage/day, as determined from weight of stomach contents and theoretical estimates. An adult herbivore requires approximately its own weight in forage each month. The average adult burro in Death Valley weighed about 159 kg (C. L. Douglas, unpubl. data), and as a result, each adult burro must consume about 1,908 kg of forage/year. Burros can modify plant communities in which they forage. The popular and scientific literature is replete with accounts of burro damage to ecosystems throughout the western United States (McKnight 1958; Weaver 1959, 1972*a,b*; C. G. Hansen [Evaluation of burro activity in Death Valley National Monument, Death Valley National Monument, California, 43pp., 1973]; Sanchez 1974; Fletcher and Wauer 1976; Reddick 1981). Seegmiller and Ohmart (1981) reported destructive foraging on ocotillo (*Fouquieria splendens*), smoke trees (*Dalea spinosa*), and little-leaf palo verde (*Cercidium microphyllum*) along the Colorado River. Seegmiller and Ohmart (1981) and Woodward (1976) found that burrobush (*Ambrosia dumosa*) was heavily modified by browsing and was occasionally browsed to the ground.

J. Fisher (Impact of feral asses on community structure in the *Acamptopappus-Grayia* plant community, Panamint Mountains, Death Valley National Monument, Death Valley National Monument, California, 20pp., 1975) measured shrubs inside and immediately outside a burro exclosure established in 1972 in Wildrose Canyon, Death Valley. He resampled in 1974 and reported substantially lower numbers of annual plants outside the exclosure, except for annual buckwheat (*Eriogonum spp.*). Some shrubs outside the exclosure had lower volumes than those inside, but others were more voluminous outside. Ten years later, Yancey (1984) compared total plant volume inside and outside the exclosure and found no significant difference between them. However, she found that the Wildrose exclosure had lower species diversity than other sites in Wildrose Canyon and concluded that the exclosure was not representative of the area (Yancey 1984).

Longshore and Douglas (1988) studied vegetation recovery following burro removal from Wildrose Canyon and Butte Valley, Death Valley; both areas had burro exclosures established in 1972. They found significant differences in species diversity but no differences in mean volumes of perennial shrubs inside and outside the Butte Valley exclosure in 1986 and 1987. Over 3 years of study, there was a general trend toward equality in mean shrub volume inside and outside the exclosure, which suggested rapid recovery from relatively minor browsing by burros. Perennial grasses had been extensively grazed in Butte Valley, where grasses (largely exotics) comprised 48% of the burro diets. Perennial grasses were present inside but not outside the exclosure in Butte Valley in 1985 and 1987. In Wildrose Canyon, significant differences were found in species diversity inside and outside the exclosure in 1986 and 1987, but no differences in mean volumes were detected. Perennial grasses had lower densities outside the Wildrose Canyon exclosure (Longshore and Douglas 1988).

Norment and Douglas (1977) assessed browse impact on plants in Wildrose Canyon, Nemo Canyon, Skidoo, and Emigrant Canyon of Death Valley. They found that 46% of shrubs in Wildrose Canyon showed evidence of browsing. About 30% of plants were heavily browsed (> 75-100% of current year's growth removed and the form of the plant altered), indicating that burros were affecting the structure of the plant community. About 11% of the shrubs were in danger of being removed from the area entirely. A later assessment of browse impact in Wildrose Canyon in 1982-1983, using the method of Norment and Douglas (1977), showed that shrubs reported as being the most heavily browsed in 1977 comprised a smaller percentage of the plant community in 1984, which indicated that burros had changed the vegetational community by selectively removing preferred species (Yancey 1984). During his study of flora and vegetation in the Cottonwood Mountains of Death Valley, Peterson (1984) identified an aberrant shrub community unlike any other in the area; it appeared to have originated from 1 of 2 other shrub communities by plant removal. Peterson (1984) attributed plant removal and modification of the parent community to feral burros.

Bennett et al. (1981) studied effects of burro foraging on 4 vegetation types from the rim of the Grand Canyon to the Colorado River in Grand Canyon National Park. Species diversity and mean importance values were highest at the rim and lowest in communities near the river. Simple communities with <15 species were severely impacted, but more diverse systems showed less impact to vegetation. Carothers et al. (1976) investigated burro impacts to riparian areas in the Grand Canyon, using a control plot and burro impacted plots. They found 28 species of vascular plants on the control plot compared to 19 on the impact plot. Total vegetation cover was 80% on the control plot compared to 20% on the impact plot. Mean area occupied by individual shrubs was reduced on the impact plot. Average absolute density of small mammals was 52 individuals/ha on the control plot and 13/ha on the impact plot (Carothers et al. 1976). Relative densities of small mammals were studied in a burro-free and burro-impacted area of Death Valley (Norment and Douglas 1977); absolute density of shrubs was greater in the control plot, and relative density of small mammals was 260% greater in the control plot than in the impact plot.

Soils

Soil compaction is an aspect of environmental degradation caused by burros and other feral species that has received minimal study. Farrell (1973) examined soil compaction by burros in western Arizona by measuring bulk density and found an average variation of only 0.063 between soils of disturbed and undisturbed areas; minimal compaction was attributed to large percentages of sand and gravel in the soils. Farrell (1973) noted that in concentration areas near waterholes burros had rendered soils fine and powdery like sifted flour. Late summer rainstorms caused soil particles to run together forming crusts. C. L. Douglas and D. Fenn (National Park Service, unpubl. data) studied soil compaction in Death Valley by using bulk density sampling and soil penetrometer readings; treads of burro trails in Butte Valley were as compacted as treads of relatively heavily used jeep roads in the same area. Compaction extended at >15-25 cm below the soil surface. The severity of the soil compaction was surprising; soils in Butte Valley were granitic, relatively coarse, and had poor compressibility.

Heavily used burro trails on steep slopes in Death Valley and the Grand Canyon have contributed to major soil movement and erosion. Shoemaker and Carothers (1976) discussed burro damage in Grand Canyon and illustrated extensive interlacing of burro trails along Mile 124 of the Colorado River. Compacted trails are almost impervious to penetration by water; precipitation tends to run off these heavily compacted soils, leading to erosion. Compacted soils also are resistant to plant colonization. Hundreds of kilometers of burro trails are present in Death Valley and Lake Mead National Recreation Area.

Interaction with Desert Bighorn Sheep

Much has been written about the effects of feral burros on desert bighorn sheep. Many reports have been speculative and unsubstantiated by empirical data. Some concepts in early reports are simply erroneous and have been disproved by later studies.

Because burros eat a wide range of plant species and use habitats similar to those of bighorn sheep (Fig. 3), overlap in food selection between the 2 species would be expected. In the

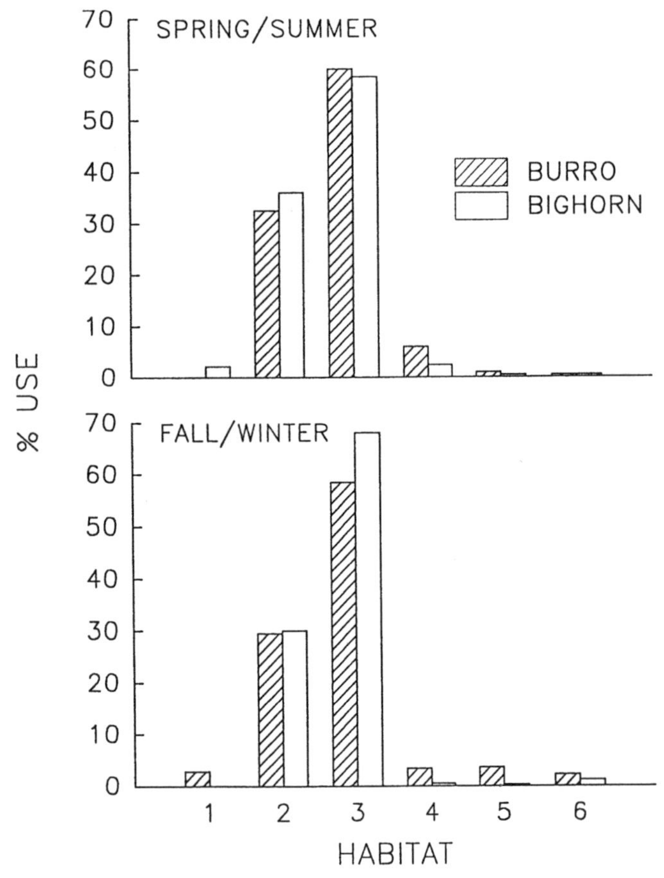

FIGURE 3. Seasonal habitat use of feral burros and desert bighorn sheep in the Cottonwood Mountains of Death Valley National Monument, California. Habitat: 1 = pinyon-juniper, 2 = sagebrush, 3 = shadscale, 4 = desert wash, 5 = blackbrush, and 6 = creosote bush (adapted from Dunn 1984).

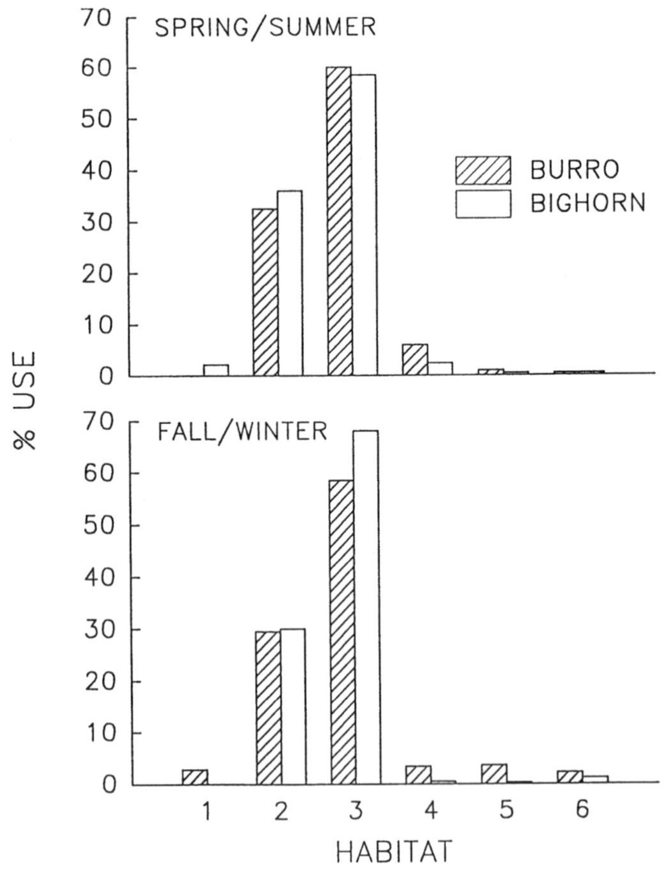

FIGURE 4. Winter and summer diets by forage class of feral burros and desert bighorn sheep in the Cottonwood Mountains of Death Valley National Monument, California (adapted from Ginnett 1982).

Cottonwood Mountains, Death Valley, burros and bighorn sheep exhibited little interspecific resource partitioning in terms of diet composition (Ginnett 1982). Fifty-five forage taxa were identified in annual diets of bighorn and burros, collectively; 67% of these were used in common (Fig. 4). In both species, individual dietary niche breadth was responsible for 79% of the population niche breadth, suggesting that individual variation in diets is of little importance in determining dietary niche breadth in these species (Ginnett 1982). Both species operate as facultative generalists in Death Valley, selecting a narrower range of species in spring, a period of relative abundance, and a broader range of species in the winter. However, as ruminants with slower gut clearance, bighorn must select more nutritious parts of plants than burros.

Walters and Hansen (1978) reported a range from 19.6 to 61.1% similarity between annual bighorn and burro diets from various areas in the Grand Canyon. Seegmiller and Ohmart (1981) studied burro and bighorn diets in western Arizona; similarity values calculated from their data yielded dietary overlaps of 40% during spring to 50% during summer (Ginnett 1982). High similarity of diets between burros and bighorn does not justify a conclusion that the 2 species compete for forage, although competition might exist in some areas. For competition to occur between 2 species, the following must be established: (1) present or past niche overlap, (2) limited resources, and (3) depression in reproductive potential of 1 or both populations. Desert bighorn and burros live in areas having environmental uncertainty. It seems reasonable to assume that desert bighorn are resource limited. Burros also may be resource limited, but because of their digestive strategies, they are less limited by forage resources than bighorn. In deserts of the western United States, habitats of both species have continually fluctuating carrying capacities, being highest in spring and lowest in winter. Primary productivity of desert habitats is low and dependent upon the amount and timing of precipitation. Availability of nutritious forage is thought to be the most limiting resource for herbivores (Norment and Douglas 1977). Foraging strategies of bighorn sheep have evolved in synchrony with their habitats, but feral burros have not coevolved with desert bighorn sheep in southwestern rangelands.

Feral burros have an advantage over bighorn sheep by being monogastric and food generalists. Burros are able to survive on forage having lower nutrient content than ruminants can subsist on. This is accomplished by increasing the rate of gut clearance and processing more volume of forage. Ruminants are restricted in their ability to increase the rate of gut clearance because of the structure of their digestive system (Janis 1976). While they presumably can extract more nutrients from a given amount of forage than burros, bighorn also require forage having a higher nutrient content. Ruminants die in winter with stomachs full of undigested forage because protein content was too low to keep rumen microorganisms alive and reproducing and fiber content too high to permit passage from the rumen (Janis 1976). Ninety adult burros in Death Valley weighed between 147 and 158 kg (C. L. Douglas, unpubl. data); the largest adults weighed about 180 kg. Some of the heaviest individuals were 2-3 year old males; however, most individuals in those age classes had not attained maximum body weight. Adult desert bighorn rams in the Lake Mead area weigh about 40-68 kg and ewes weigh about 34-52 kg; if 45 kg is used as an average bighorn weight, 1 burro equals the same relative mass as 3 to 4 bighorn sheep. A desert ecosystem has a biomass limitation for grazing and browsing organisms, which also includes rabbits, rodents, insect herbivores, and domestic livestock. Because precipitation patterns are highly variable in desert areas, so too is production of plant biomass. The burro can adapt to low quality forage by eating more of it, thereby acquiring at least maintenance levels of nutrition. Bighorn sheep must remain more selective than feral burros and eat the most nutritious parts of plants that are available. Because sheep cannot vary the rate of gut clearance as much as burros and cannot be as general in their food selection, they are at a competitive disadvantage under poor range conditions. Although adults may be able to survive on body stores and maintenance or submaintenance level nutrition, reproduction and postpartum survival of the young will be affected.

Presence of a feral herbivore exacerbates any natural shortage of food resources for native herbivores, when there is a high degree of overlap in their diets. Woodward (1976) makes an interesting analogy between an overpopulation of burros and effects of drought on plant resources. Burro (or any feral or exotic species) overgrazing leads to less plant biomass available for native ungulates and thus a lowered carrying capacity (Woodward 1976).

Competition with Desert Bighorn Sheep for Water

Water is a limiting resource in most rangelands; the potential for burros to usurp water that would otherwise be used by wildlife and to interfere with bighorn obtaining water is a serious management problem. Burros and bighorn may use the same water sources (Welles and Welles 1961, Dunn 1984) at the same time. In Dunn's (1984) study, a bighorn ram and single burro were recorded, by time-lapse camera, drinking from opposite sides of a water puddle about 1 m in diameter. Welles and Welles (1961) recorded sheep and burros drinking together at Lost Spring in Death Valley. At this spring, water forms pools at the top and bottom of a step in the drainage, such that animals drinking at the top source are separated physically from those drinking at the lower source. Dunn and Douglas (1982) found bighorn rams to be less intimidated by burros than ewes. Nevertheless, some rams waited for hours until burros left a spring before going to drink. Ewes generally would not drink if >3 burros were present at the spring. Rams drank at the springs that burros used, but a spring not used by burros received heavier use by ewes than springs used by burros. Burros have the tendency to lounge in groups at water sources, often remaining there for hours, or even entire days. Dunn (1984) regarded this interaction as interference competition; no active aggression was recorded between the species.

Bighorn rams drink as often as ewes, and lactating ewes water more frequently than dry ewes and young animals (Turner and Weaver 1980). Although bighorn can consume large volumes of water, the amount consumed by different individuals is quite variable and can range from a few liters to 20% of an individual's body weight (Turner and Weaver 1980). Drinking rates as high as 2.8 L/minute have been observed. If necessary, bighorn can go for several days without water. Desert bighorn are adapted to desert conditions by having concentrated urine, relatively dry feces, and the ability to rehydrate relatively quickly after drinking (Turner 1973).

Survival of the burro in deserts is enhanced by its ability to withstand high degrees of dehydration and heat (Dill et al. 1979, 1980). Yousef et al. (1970) studied effects of dehydration on body fluids of 2 female adult burros. They reported a decrease of 31.5% in intracellular-fluid-volume, 17% in extracellular-fluid volume, and only 7% decrease in plasma volume. Upon dehydration, burros decrease intracellular fluid volume more than the volume of other water partitions. The small decrease in blood plasma volume suggests that burros maintain a relatively stable blood volume under dehydration. This may help explain their ability to generate saliva and feed on dry forage even when dehydrated.

Burros have a greater need for free water than bighorn because (1) they are larger animals, (2) the amount of feces produced by burros is greater than that of desert bighorn and contains more water, and (3) urine of burros is much less concentrated than that of bighorn. Davis et al. (1978) found an average water turnover rate of 66.1 mL/kg/day ± 20.9 in 4 juvenile and 2 adult burros. Two burros 3 years of age had water turnover rates of 122.5 and 139.7 ml/kg/day. These values represent moisture lost from the body to the environment, and thus daily water requirements. Using these figures, a 158.8 kg burro would require about 20.8 L of water/day (assuming the individual is in a state of water balance).

Body size influences dominance among species of the same trophic level (Berger 1985). Berger (1985) suggested that establishment of exotics in an area must lead to their monopolization of resources used by smaller species, or cause smaller species to shift niches. Burro presence at springs could influence the amount of water bighorn can consume because of apprehension in approaching a spring and remaining there long enough to drink adequately. Dunn (1984) found bighorn ewes in the Cottonwood Mountains of Death Valley used Quartz Spring, which was fenced to exclude burros, significantly more than other unfenced springs in that range. He also found through observation and time-lapse camera records that bighorn would not come to drink at springs when >3 burros were present. The potential decrease in water intake for lactating ewes could result in decreased milk production, which could lead to lower lamb survival. Additionally, limiting use of the range by ewe groups may place limitations on nutrient intake. Nutritional quality of vegetation is strongly correlated with reproductive success (Caughley 1970).

MANAGEMENT OF FERAL BURROS

Because of the pernicious effects of burros on rangelands and native wildlife in the western United States, various control measures have been employed to reduce their numbers or eliminate them altogether. The responsibility for managing feral horses and burros on public rangelands resides with the Bureau of Land Management and United States Forest Service. Management strategies, including trapping, helicopter roundup, fencing, and protection of water sources have been developed by these agencies or their contractors. Unfortunately, political and legal pressures have complicated the ability of wildlife managers to deal directly and effectively with problem populations of burros. For example, the control of burro numbers in California was curtailed in 1953 when the state provided official protection to the feral burro. In 1959, Congress halted the pursuit of wild horses and burros from motorized vehicles by passing the "Wild Horse Annie Act," after the nickname of Mrs. Velma Johnston of Reno, Nevada, who lobbied almost single-handedly for its passage. Public Law 92-195 entitled "Wild and Free Roaming Horses and Burros" was passed in 1971 and provided protection for burros on federal lands administered by the Bureau of Land Management and the United States Forest Service (Wagner 1983). The original legislation proposed by Mrs. Johnston included only horses; she recognized that burros were creating problems on rangelands and required management (V. Johnston, pers. commun.). Nevertheless, burros were added to the bill by Congress, an inclusion that continues to complicate and limit management options. A public-mandated multi-million dollar capture and adoption program was developed and administered by BLM following the above legislation. Studies of birth control and sterilization also are being investigated as a means of population regulation. Groups advocating animal rights of burros and feral horses originated in the wake of these legislative actions.

The National Park Service was specifically exempted from provisions of Public Law 92-195, but burro control was curtailed in most parks until the late 1970s. For a number of years, 2 commercial burro trappers selectively removed burros from Death Valley through an arrangement with the Monument. Burro trapping and direct reduction were terminated in Death Valley during the 1960s, leading to rapid growth in the population and colonization of additional areas until they were removed in the 1980s. Between 1983 and 1986, >6,000 burros were captured and removed from Death Valley, which had the largest known concentration of burros in the United States. Feral burros also have created serious management problems in Bandelier National Monument, Grand Canyon National Park, Lake Mead National Recreation Area, and Virgin Islands National Park.

SUMMARY

Feral animals are properly defined as native or non-native domesticated species that have escaped or been released into natural ecosystems. Feral animals have created serious management problems for native biota on a global scale. Although goats, sheep (islands off California), and pigs have become feral in the western United States, burros and horses have caused more controversy and deleterious impacts to western rangelands than the other species. The feral burro highlights all of the problems affecting native biota that are created by feral rangeland species in general.

Domesticated Somalian and Nubian wild asses or burros were introduced into North America through Mexico by the Spaniards in the 1530's. Burros were released following the decline of mining and advent of the railroad. Although feral burros are a small part of ungulate numbers on western rangelands in the United States, they have been numerous in particular locations, such as national parks and monuments. They have caused (1) extensive habitat damage by overgrazing, which has caused selective removal of preferred plants, (2) soil disturbance and erosion, (3) negative impacts to small mammals, and (4) competitive interactions with desert bighorn sheep.

Feral burros are well adapted to western rangelands in the United States because their monogastric digestive system permits them to vary gut clearance and thereby use forage of poor quality (assuming sufficient quantity can be obtained), and because they can tolerate relatively high levels of dehydration. Their life history strategies are conservative (1 foal/yr), but they are capable of population growth exceeding 20% under favorable conditions.

As a population of feral animals grows, it can exert considerable demand on the resource base and thereby have pernicious effects on native ecosystems. Control measures to limit, reduce, or eliminate populations of feral animals are in place throughout most of the western United States and are clearly warranted.

LITERATURE CITED

Asdell, S. A. 1964. Patterns of mammalian reproduction. Cornell Univ. Press, Ithaca, N. Y. 670pp.

Beatley, J. C. 1969. Dependence of desert rodents on winter annual production and precipitation. Ecology 50:721-724.

Bennett, P. S., A. M. Phillips, G. A. Ruffner, S. W. Carothers, and R. R. Johnson. 1981. Effects of burro foraging on 4 types of Grand Canyon vegetation. Conf. on Sci. Res. in the Natl. Parks 2:20.

Berger, J. 1985. Interspecific interactions and dominance among wild Great Basin ungulates. J. Mammal. 66:571-573.

———. 1986. Wild horses of the Great Basin: social competition and population size. Univ. Chicago Press, Chicago, Ill. 326pp.

Blake, J. G. 1977. Serum protein polymorphisms of the feral ass (*Equus asinus*) in Death Valley National Monument. M.S. Thesis, Univ. Nevada, Las Vegas. 82pp.

———, C. L. Douglas, and L. F. Thompson. 1981. Spatial variation in transferrin allele frequencies among herds of feral donkeys in Death Valley National Monument, California. J. Mammal. 62:58-63.

Browning, B. 1960. Preliminary report of the food habits of the wild burro in the Death Valley National Monument. Desert Bighorn Counc. Transactions 4:88-90.

Carothers, S. W., M. E. Stitt, and R. R. Johnson. 1976. Feral asses on public lands: an analysis of biotic impact, legal considerations and management alternatives. Trans. North Am. Wildl. and Nat. Resour. Conf. 48:396-406.

Caughley, G. 1970. Eruption of ungulate populations, with emphasis on Himalayan thar in New Zealand. Ecology 51:53-71.

Choquenot, D. 1990. Rate of increase for populations of feral donkeys in northern Australia. J. Mammal. 71:151-155.

Clutton-Brock, J. 1981. Domesticated animals from early times. Univ. Texas Press, Austin. 208pp.

Colbert, E. H. 1961. Evolution of the vertebrates. John Wiley and Sons, Inc., New York, N.Y. 479pp.

Conley, W. 1980. The potential for increase in horse and ass populations: a theoretical analysis. Pages 221-234 *in* Nat. Res. Counc., eds. Wild and free-roaming horses and burros: current knowledge and recommended research, phase 1: final report. Natl. Acad. Sci. Press, Washington, D.C. 382pp.

Davis, T. P., M. K. Yousef, and H. D. Johnson. 1978. Partition of body fluids in the burro. J. Wildl. Manage. 42:923-925.

Decker, E. 1978. Exotics. Pages 249-256 *in* J. L. Schmidt and D. L. Gilbert eds. Big game of North America. Stackpole Books, Harrisburg, Pa.

Dill, D. B., L. Soholt, J. Miller, H. Randolph, and T. S. Goudy. 1979. Properties of sweat, burro and man. Comp. Biochem. Physiol. and Comp. Physiol. 62:317- 320.

———, M. K. Yousef, C. R. Cox, and R. G. Barton. 1980. Hunger versus thirst in the burro. Physiol. Behav. 24:975-978.

Douglas, C. L., and H. D. Hiatt. 1987. Food habits of feral burros in Butte Valley, California. Contribution CPSU/UNLV 006/46, Natl. Park Serv. Coop. Park Studies Unit, Univ. Nevada, Las Vegas. 25pp.

———, and D. M. Leslie, Jr. 1986. Influence of weather and density on lamb survival of desert mountain sheep. J. Wildl. Manage. 50:153-156.

Dunn, W. C. 1984. Ecological relationships between desert bighorn and feral burros in Death Valley National Monument, California. M.S. Thesis, Univ. Nevada, Las Vegas. 144pp.

———, and C. L. Douglas. 1982. Interactions between desert bighorn sheep and feral burros at spring areas in Death Valley. Desert Bighorn Counc. Transactions 26:87-96.

Eberhardt, L. L. 1982. Apparent rates of increase for two feral horse herds. J. Wildl. Manage. 46:367-374.

Farrell, J. E. 1973. Behavioral patterns of feral burros as influenced by seasonal changes in western Arizona. M.S. Thesis, Arizona State Univ., Tempe. 35pp.

Fletcher, M. R., and R. H. Wauer. 1976. Feral burro management at Bandelier National Monument, New Mexico. Desert Bighorn Counc. Transactions 20:54-55.

Ginnett, T. F. 1982. Comparative feeding ecology of feral burros and desert bighorn sheep in Death Valley National Monument. M.S. Thesis, Univ. Nevada, Las Vegas. 86pp.

———, and C. L. Douglas. 1982. Food habits of feral burros and desert bighorn sheep in Death Valley National Monument. Desert Bighorn Counc. Transactions 26:81-87.

Hansen, R. M., and P. S. Martin. 1973. Ungulate diets in the lower Grand Canyon. J. Range Manage. 26:380-381.

Hofmann, R. R. 1968. Comparison of the rumen and omasum structure in East African game ruminants in relation to their feeding habits. Symp. Zool. Soc. London 21:179-194.

Janis, C. 1976. The evolutionary strategy of the Equidae and the origins of rumen and cecal digestion. Evolution 30:757-774.

Jarmen, P. J., and A. R. E. Sinclair. 1979. Feeding strategy and the pattern of resource-partitioning in ungulates. Pages 130-163 *in* A. R. E. Sinclair and M. Norton-Griffiths, eds. Serengeti: dynamics of an ecosystem. Univ. Chicago Press, Chicago, Ill.

Johnson, R. A., S. W. Carothers, and T. J. McGill. 1987. Demography of feral burros, (*Equus asinus*), in the Mohave Desert. J. Wildl. Manage. 51:916-920.

Klingel, H. 1977. Observations on social organization and behaviour of African and Asiatic wild asses. Zeitschrift fur Tierpsych. 44:323-331.

Koehler, D. A. 1974. The ecological impact of feral burros on Bandelier National Monument. M.S. Thesis, Univ. New Mexico, Albuquerque. 78pp.

Leslie, D. M., Jr., and C. L. Douglas. 1979. Desert bighorn of the River Mountains, Nevada. Wildlife Monogr. 66. 56pp.

Longshore, K. M., and C. L. Douglas. 1988. Vegetational recovery following burro removal in Death Valley National Monument. Contribution No. CPSU/UNLV 035/04, Natl. Park Serv. Coop. Park Studies Unit, Univ. Nevada, Las Vegas. 36pp.

Maloiy, G. M. O. 1970. Water economy of the Somali donkey. Amer. J. Physiol. 219:1522-1527.

McCool, C. J., C. C. Pollitt, G. R. Fallon, and A. F. Tirmer. 1981. Studies of feral donkeys in the Victoria River-Kimberleys area: observations on behaviour, reproduction, and habitat and some possible control strategies. Aust. Vet. J. 57:444-449.

McKnight, T. L. 1958. The feral burro in the United States: distribution and problems. J. Wildl. Manage. 22:163-179.

Moehlman, P. D. 1974. Behavior and ecology of feral asses (*Equus asinus*). Ph.D. Thesis, Univ. Wisconsin, Madison. 251pp.

Morgart, J. R. 1978. Burro behavior and population dynamics, Bandelier National Monument, New Mexico. M.S. Thesis, Arizona State Univ., Tempe. 94pp.
Morrison, B. L. 1985. Harvest strategies to control exotic ungulate populations in New Mexico. Pages 261-268 *in* S. L. Beasom and S. F. Roberson, eds. Game harvest management. Caesar Kleberg Wildl. Res. Inst., Kingsville, Tex.
National Academy of Sciences. 1975. Nutrient requirements of sheep. Pages 1-72 *in* Subcommittee on sheep nutrition, eds. Nutrient requirements of domestic animals. Natl. Acad. Sci., Washington, D. C.
Norment, C., and C. L. Douglas. 1977. Ecological studies of feral burros in Death Valley. Contribution CPSU/UNLV 006/09. Natl. Park Serv. Coop. Studies Unit, Univ. Nevada, Las Vegas. 132pp.
Peterson, P. M. 1984. Flora and physiognomy of the Cottonwood Mountains, Death Valley National Monument, California. M.S. Thesis, Univ. Nevada, Las Vegas. 241pp.
Prosser, C. L., and F. A. Brown, Jr. 1966. Comparative animal physiology. W. B. Saunders Co., Philadelphia, Pa. 688pp.
Reddick, P. B. 1981. Feral burro management program, Naval Weapons Center, China Lake, California. Pages 1-114, Tech. Appendix 2 *in* U.S. Dep. of Navy. Final environmental impact statement for the Naval Weapons Center, China Lake, California (Kern County).
Robbins, C. T. 1983. Wildlife feeding and nutrition. Academic Press, New York, N.Y. 343pp.
Ruffner, G. A., and S. W. Carothers. 1982. Age structure, condition and reproduction of two *Equus asinus* (Equidae) populations from Grand Canyon National Park, Arizona. Southwest. Nat. 27:403-411.
———, S. W. Carothers, J. W. Jordan, and A. M. Phillips III. 1977. Diets of feral burros (*Equus asinus*) from the Bedrock Canyon area, Grand Canyon National Park, Arizona. Final Rep. Natl. Park Serv., Grand Canyon National Park, Ariz. 14pp.
Sanchez, P. G. 1974. Impact of feral burros on the Death Valley ecosystem. California-Nevada Wildl. Transactions. 13pp.
Schmidt-Nielsen, K. 1979. Desert animals: physiological problems of heat and water. Dover Publ., New York, N.Y. 277pp.
Seegmiller, R. F., and R. D. Ohmart. 1975. Feral burros within desert bighorn habitat. Desert Bighorn Counc. Transactions. 19:45.
———, and R. D. Ohmart. 1976. Feral burro-desert bighorn sheep relations, Bill Williams Mountains, Arizona. Trans. North Am. Wild Sheep Conf. 1:35-37.
———, and ———. 1981. Ecological relationships of feral burros and desert bighorn sheep. Wildlife Monogr. 78. 58pp.
Shoemaker, P. L., and S. W. Carothers. 1976. Burros threaten parts of Grand Canyon. Natl. Park Serv. Newsletter II:1-2, Washington, D. C.
Siegmund, O. H., editor. 1973. The Merck veterinary manual. O. H. Siegmund ed., Merck and Co., Inc., Rahway, N.J., 1600pp.
Slade, L. M., and E. B. Godfrey. 1982. Wild horses. Pages 1089-1089 *in* J. A. Chapman and G. A. Feldhamer, eds. Wild mammals of North America. Johns Hopkins Univ. Press, Baltimore, Md. 1147pp.
Turner, J. C., Jr. 1973. Water, energy and electrolyte balance in the desert bighorn sheep, *Ovis canadensis*. Ph.D. Thesis, Univ. California, Riverside. 138pp.
———, and W. A. Weaver. 1980. Water. pp. 100-113 *in* G. Monson and L. Sumner, eds., The desert bighorn, its life history, ecology, and management. Univ. Arizona Press, Tucson. 370pp.
Van Soest, P. J. 1982. Nutritional ecology of the ruminant. Cornell Univ. Press, Ithaca, N. Y. 373pp.
Wagner, F. H. 1983. Status of wild horse and burro management on public rangelands. Trans. North Am. Wildl. and Nat. Resour. Conf. 48:116-133.
Walters, J. E., and R. M. Hansen. 1978. Evidence of feral burro competition with desert bighorn sheep in Grand Canyon National Park. Desert Bighorn Counc. Transactions. 22:10-16.
Weaver, R. A. 1959. Effects of burro on desert water supplies. Desert Bighorn Counc. Transactions. 3:1-3.
———, 1972*a*. Big game investigations: feral burro survey. California Dep. Fish and Game, Final Rep. Fed. Aid Project W-51-R-17. 14pp.
———, 1972*b*. Conclusion of the bighorn investigation in California. Desert Bighorn Counc. Transactions. 16:56-65.
Welles, R. E., and F. B. Welles. 1961. The bighorn of Death Valley. Natl. Park Serv. Fauna Series 6, Washington, D. C. 242pp.
White, L. D. 1980. A study of feral burros in Butte Valley, Death Valley National Monument. M.S. Thesis, Univ. Nevada, Las Vegas. 124pp.
Wolfe, M. L., L. C. Ellis, and R. MacMullen. 1989. Reproductive rates of feral horses and burros. J. Wildl. Manage. 53:916-924.
Woodward, S. L. 1976. Feral burros of the Chemehuevi Mountains, California: the biogeography of a feral exotic. Ph.D. Thesis, Univ. California, Los Angeles. 178pp.
———, and R. D. Ohmart. 1976. Habitat use and fecal analysis of feral burros (*Equus asinus*), Chemeheuvi Mountains, California. J. Range Manage. 29:482-485.
Yancey, M. J. 1984. A study of burro-small vertebrate interactions in Death Valley National Monument, California. M.S. Thesis, Univ. Nevada, Las Vegas. 90pp.
———, and C. L. Douglas. 1983. Burro-small vertebrate interactions in Death Valley National Monument, California. Desert Bighorn Counc. Transactions. 23:17-24.
Yorks. T. P. 1989. Ranching native and exotic ungulates in the United States. Pages 268-285 *in* R. J. Hudson, K. R. Drew, and L. M. Baskin, eds. Wildlife production systems: economic utilisation of wild ungulates. Cambridge Univ. Press, Cambridge, U.K.
Young, J. Z. 1962. The life of vertebrates. Second ed. Oxford Univ. Press, New York, N.Y. 820pp
Yousef, M. K., D. B. Dill, and M. G. Mayes. 1970. Shifts in body fluids during dehydration in the burro *Equus asinus*. J. Appl. Physiol. 29:345-349.

Other Wildlife Conservation Issues in Range Management

Chapter 18
PRESCRIBED FIRE

Robert A. Riggs, Stephen C. Bunting, and Steven E. Daniels

INTRODUCTION

Research has generated growing support for prescribed fire, and the practice is evolving from a subject of research and demonstration projects into a broadly accepted tool in wildlife habitat management. Public agencies and private institutions, such as the Foundation for North America Wild Sheep and the Rocky Mountain Elk Foundation, have greatly increased resources dedicated to burning projects. As the use of prescription fire grows, practitioners will increasingly be asked to demonstrate that they understand not only fire's potential utility, but also the factors that influence realization of that potential in specific settings. We consider fire's role in natural landscapes and its relevance to modern habitat management, dividing our discussion into 3 parts: (1) a review of fire-vegetation relationships, (2) a review of wildlife responses and the factors that influence them, and (3) a discussion of economic issues that influence the efficiency of burning programs.

FIRE AND VEGETATION

Many early ecologists ignored or underestimated the role that fire can play in shaping biotic communities. Their views may have resulted from a preoccupation with primary successional processes, or simply from a lack of appreciation for secondary succession. Leopold (1920) wrote ". . . light-burning (prescribed fire) reduces the vitality and productiveness of the forage . . ." and ". . . destroys the humus in the soil necessary for rapid tree growth . . . ," adding ". . . now is the time to put the quietus on the agitation for light-burning. . . ." After spending 4 years in southwestern forests, however, Leopold reversed his position and became one of the first ecologists to recognize the importance of fire in biotic communities (Leopold 1924). Clements (1935) wrote that "Under primitive conditions, the great climaxes of the globe must have remained essentially intact, since fires from natural causes must have been both infrequent and localized." More than 2 decades passed before ecologists resumed discussion of fire's role as a determinate of plant communities (Humphrey 1953, Daubenmire 1968).

Ecologists' understanding of the importance of fire in shaping vegetation has increased over the years, yet the early sentiments of Leopold and Clements are still reflected to varying degrees in modern society. Indeed, it is often assumed by lay people that the "pristine" vegetation first viewed by early Euro-Americans in western North America, with its abundant wildlife populations, was also a "climax" vegetation in which fire played no role. On the contrary, numerous studies have established that fire was quite common in North American rangelands (Table 1). Examples include tall grass prairie (Sauer 1950, Wright and Bailey 1982), sagebrush (*Artemisia* spp.) steppe (Gruell 1983, 1986), juniper (*Juniperus* spp.) woodlands (Burkhardt and Tisdale 1976, Young and Evans 1981, Bunting 1987), ponderosa pine (*Pinus ponderosa*) forests (Weaver 1951, Arno 1980, Hall

1980), desert grasslands (Humphrey 1953), mesquite (*Prosopis* spp.) grasslands (Wright and Bailey 1982), aspen (Jones and DeByle 1985), and chapparal (Radtke et al. 1982). Even vegetation of mesic climates was influenced by fire, including "old growth" coastal forests of the Pacific Northwest (Hemstrom and Franklin 1982, Franklin 1988, Agee 1991). Indeed, little vegetation was spared from the influence of fire although the frequency of fire events varied widely and was quite low in some types.

Occurrence of fire during the presettlement period depended on various factors: climate, fine fuel production (primarily herbaceous biomass), vertical and horizontal continuity of fuels, lightning and human ignition sources, and topography. The term "fire regime" was first defined by Heinselman (1978, 1981) as consisting of 3 elements: fire type and intensity, fire size, and fire frequency. To these, Kilgore (1981) added fire severity, season, and pattern. Fire regimes varied among habitat types and geographic locations. Wright and Bailey (1982) estimated that the undulating topography of the tall grass prairie may have had a fire-free interval (FFI) as short as 5-10 years. Southern mixed prairie, in which landscapes were more dissected, may have had longer FFI's of 20-30 years. The FFI's in ponderosa pine forests varied across the dominant species' broad geographic and altitudinal ranges, but several studies indicate that it was <10 years in western Montana (Arno 1976), Arizona and New Mexico (Weaver 1951), central California (Kilgore and Taylor 1979), and eastern Oregon (Hall 1976).

When fire occurs at short intervals its effects on composition of vegetation can be minor because the community becomes dominated by fire tolerant species. Under such regimes, tree seedlings and nonsprouting shrubs can be eliminated or severely reduced, and sprouting shrubs may be top-killed or killed outright when burned at young age (Wright et al. 1976). Fires in grasslands and shrub-steppes generally kill the aerial parts of trees and shrubs except under the lowest of intensities, thus facilitating dominance by fire-tolerant herbaceous species. Sustained short intervals between fires serve to help maintain the dominant herbaceous character of the vegetation and prevent succession to more advanced successional communities.

Short FFI's in forests and woodlands usually result in low intensity surface fires. Under this condition the overstory may be affected at specific locations where fire intensity is high enough to cause some mortality, but most of the stand remains intact. However as the FFI increases, fuel levels often build, thereby increasing the probability that the eventual fire will be of high over-all intensity and initiate replacement of the stand.

Individuals of fire-sensitive species are able to survive wildfire events because of 2 types of variability, which may also occur in the context of prescribed fire. First, fuels may not be continuous and consequently unburned patches remain after many fires. Second, communities also may have considerable variation around the mean FFI. Thus, an unusually long period between 2 successive fires may permit regeneration and development of fire-sensitive plants to a size at which they are able to survive. Long-term survival of such species, however, is further dependent on subsequent variation in fuel continuity and fire frequency.

There are numerous examples of habitat types that are characterized by intermediate FFIs (20-75 years). Many seral ponderosa pine forests in western Montana may have had intermediate-length intervals between fires (Arno 1976). Longer intervals between fires permit some of the more fire-sensitive species, such as Douglas-fir (*Pseudotsuga menziesii*) and grand fir (*Abies grandis*), to establish on these sites. The more mesic big sagebrush steppe may have had natural fire intervals between 20 and 75 years in length (Houston 1973, Wright and Bailey 1982, Bunting et al. 1987). Here the longer FFI's permit fire sensitive mountain big sagebrush (*Artemisia tridentata* subsp. *vaseyana*) to dominate, and conifers such as Douglas-fir, ponderosa pine, or western juniper (*Juniperus occidentalis*) to become established. Under these circumstances, subsequent high-intensity fires result in greater change in plant composition and structure than would occur if the community had been shaped by more frequent, low-intensity fires.

When FFIs are >100 years the effects of fire on communities are often very different from those which typically occur in the context of short- and intermediate-length FFIs. The shorter FFIs

TABLE 1. Examples of pristine fire-free-intervals (FFI) in years for a variety of rangeland vegetation types in western North America.

Vegetation type	FFI	Source
Tallgrass prairie	5-10	Wright and Bailey (1982), Rowe (1969)
Southern mixed prairie	20-30	Wright and Bailey (1982)
Mountain big sagebrush	20-25	Houston (1973)
Wyoming big sagebrush	>100	Wright and Bailey (1982)
Western juniper	<35	Burkhardt and Tisdale (1976)
Southwestern pinyon-juniper	10-30	Leopold (1924)
California chaparral	20-30	Byrne (1978), Radtke et al. (1982)
Desert grassland	10	Leopold (1924)
Coastal plain savannah	2-8	Christensen (1981)

tend to be associated with fire regimes that maintain seral species. Long FFIs, on the other hand, usually are associated with fire regimes characterized by high intensity, overstory replacement fires (Agee 1991). In such cases, initial post-fire vegetation is likely to differ substantially in composition and structure from that which was present immediately prior to the fire. Because of high mortality among species in the pre-fire community, intensely burned sites initially may be dominated by annual, biennial, and short-lived perennial species that can become quickly established on disturbed sites. Examples of rangeland communities with typically long FFIs include Wyoming big sagebrush (*Artemisia tridentata* spp. *wyomingensis*) steppe, and curlleaf mountain-mahogany (*Cercocarpus ledifolius*) woodlands. In rangelands, very long FFIs tend to occur at high elevations, in areas that are moist because of topographic or climatic conditions (Kilgore 1981), or in arid climates where little herbaceous fuel is produced. Conifer forests and woodlands such as subalpine fir (*A. lasiocarpa*) and climax pinyon-juniper are in this category. Infrequency, however, does not necessarily reduce the importance of fire to long-term community function or persistence on the landscape. This is clearly demonstrated by the fire-dependence of several forest vegetations, including lodgepole pine (*P. contorta*), whitebark pine (*P. albicaulis*) (Morgan and Bunting 1990), and coastal Douglas-fir (Agee 1991).

Euro-American land management practices generally have had the effect of increasing FFIs. Important practices in this regard include livestock grazing, fire suppression, and development of roads and agricultural lands. Livestock grazing reduces accumulation of fine fuels, thus limiting the occurrence of fire in grasslands, sagebrush steppe, juniper woodlands, and dry forests. Grazing can greatly limit fire potential in arid and semiarid vegetations, where fine fuel production is usually near or below the minimum capable of sustaining the spread of fire in most years anyway. Grazing also has been important in limiting fire spread in remote rangelands. Here fire spread would normally be rapid and modern fire-suppression capability is typically lower than in less remote areas. Fire suppression activities have most lengthened FFIs in mid- and low-elevation forests, and in areas of high human population. Land development has decreased the continuity of fuels across landscapes.

Increases in FFI associated with Euro-American settlement have resulted in major changes in composition of many communities. Woody species have advanced on to sites formerly dominated by grassland vegetation. Advances of ponderosa pine, oneseed juniper (*J. monosperma*), and mesquite have been documented in the more arid portions of the Great Plains. Quaking aspen (*Populus tremuloides*), eastern redcedar (*J. virginiana*), and numerous oak species (*Quercus* spp.) have advanced in more mesic regions. Ponderosa pine has increased in parts of the mixed prairie and dry mountain meadows. Mesquite has increased in the southern mixed prairie and desert grassland. The sagebrush steppe has been affected by several species including Utah juniper (*Juniperus utahensis*), western juniper, ponderosa pine, and Douglas-fir. Mountain meadows have been invaded by subalpine fir, Douglas-fir, and lodgepole pine (Franklin et al. 1971, Dunwiddie 1977, Butler 1986). Some seral forest types have also been affected by establishment of late-seral tree species that normally would not dominate the pristine communities. Such changes have occurred in areas once dominated by quaking aspen (Loope and Gruell 1973), whitebark pine (Morgan and Bunting 1990), ponderosa pine (Biswell 1972, Arno 1976), Douglas-fir (Habech 1976), and lodgepole pine (Davis et al. 1980).

Compositions of climax forests and woodlands have also been altered by increases in FFIs. Protection of these types from fire commonly changes canopy coverage and vertical structure, as density increases and reproduction by climax species adds the dimension of an uneven-aged stand. Types affected in this regard include pinyon-juniper (West et al. 1975), western juniper (Burkhardt and Tisdale 1976, Johnson and Smathers 1976), ponderosa pine (Johnson and Smathers 1976), Rocky Mountain Douglas-fir (Peet 1988), and southeastern pine forests (Christensen 1981).

Climax shrub communities, such as sagebrush steppe, mountain brush, and chaparral, have responded similarly to the absence of fire. In these vegetations shrubs can increase in the absence of fire until they fully occupy the site. Herbaceous productivity and diversity decline as shrubs achieve dominance (Hanes 1971, Harniss and Murray 1973, Tisdale and Hironaka 1981), and reintroduction of fire enhances the herbaceous strata at the expense of productivity and diversity in the dominant overstory. In Gambel oak (*Quercus gambelii*) shrublands prescribed fire can alter shrub composition more than other management practices such as mechanical or herbicide treatments (Kufeld 1983). In arid and semiarid communities, shrubs continue to dominate without fire and additional community change is minimal once they achieve dominance (West et al. 1984) (Fig. 1).

The density and productivity of understories decline as overstories develop (Gruell et al. 1986, Bunting et al. 1987). This decline has been attributed to changes in light quality, increased competition for water and nutrients, interruption of seedling establishment, and allelopathy. Understory production typically declines curvilinearly as overstories increase (Pase 1958, Arnold et al. 1964, Jameson and Reed 1965, Tisdale et al. 1969, Hanes 1971, Harniss and Murray 1973, Keeley and Zedler 1978, Bojorquez Tapia et al. 1990, Pieper 1990, Ratliff et al. 1991, Scanlan 1992, Moore and Deiter 1992, McDaniel et al. 1993)(Fig. 2).

Alteration of herbaceous structure is often most pronounced in those communities that are dominated by herbaceous species, such as grasslands and wetlands. In grasslands, for example, the previous years' growth is partially or completely consumed. Even in wetlands, fire can markedly alter composition and structure (Schlichtemeier 1967, Vogl 1967, Perkins 1968, Ward 1968, Ewel and Mitsch 1978, Mitsch and Gosselink 1986).

A secondary effect resulting from change in understory is alteration of potential for subsequent fire severity and intensity. In some types of vegetation, such as juniper woodlands, reduction of fine fuel loads over time eventually eliminates the potential for occurrence of all but the most severe fires. When these eventually occur, as either wild or prescribed burns, their greater severity can result in higher mortality of the already depauper-

FIGURE 1. Comparison of early- and late-successional western juniper (*Juniperus occidentalis*) communities, Owyhee Mountains, Idaho. In the absence of fire the late-successional condition on the right persists indefinitely with virtually no change in community composition. Photos by Stephen C. Bunting.

ate understory than would have occurred otherwise (Bunting 1987). Also, the fire resistance of some species declines during prolonged periods of fire exclusion. An example is bitterbrush (*Purshia tridentata*) growing under full ponderosa pine canopy coverage (Bunting et al. 1985); older bitterbrush plants growing under full pine cover are less able to resprout following fire than are younger, more vigorous plants growing in more open conditions. In such cases prescribed fire does not precisely mimic the function of earlier wildfires that occurred at relatively high frequency and low intensity.

Successional changes in the absence of fire do not always reduce fire potential. Establishment of subalpine fir in lodgepole pine or whitebark pine (Morgan and Bunting 1990), or of Douglas-fir or grand fir in ponderosa pine stands (Arno 1976) increases the potential of fire spreading through these types. Similarly, establishment of white fir (*Abies concolor*) and other conifers in the understory of giant sequoia (*Sequoia gigantea*) increases fire potential (Kilgore and Taylor 1979). Because vertically continuous fuels can develop under these circumstances, prescribed fires that follow long FFIs may burn as intense stand replacement fires rather than as cool surface fires. These situations serve as additional examples of circumstances in which reintroduction of fire through prescribed burning may not mimic the function of previous wildfire, at least without intrusive management such as mechanical thinning to reduce vertical fuel continuity prior to ignition.

Human activities have not always increased FFIs. Introductions of exotic annual grasses, particularly cheatgrass and medusahead wildrye, have permanently changed the function of fire in the arid sagebrush steppe and juniper woodlands of the Great Basin, Columbia Basin, and Snake River Plain (Yensen 1981, Young et al. 1987, Pellant 1990). Presence of annuals increases fire potential because they typically cure earlier in summer than perennials, and because they enhance fine fuel continuity. Successive fires reduce sagebrush and increase available habitat for annuals, disrupting the formerly "natural" succession to sagebrush (West 1978, Pellant 1990). Similar changes resulted from introduction of annual grasses into grasslands and oak woodlands of central California. During years with above-average precipitation, cheatgrass may produce adequate biomass to support fires in salt desert shrub communities (Pellant 1990). In these communities low herbaceous biomass production formerly limited fire and many of the species are poorly adapted to survive fire. Once native shrubs are reduced, cheatgrass occupies the site and a cycle similar to that in sagebrush vegetations is initiated.

Effects of fire are closely related to biotic community processes and attributes (Heinselman 1981, Chandler et al. 1983) including: vegetation succession (Hanes 1971, Lyon and Stickney 1976, Gruell 1986, Gruell et al. 1986), nutrient and biomass cycling (Chandler et al. 1983), energy flow (Kucera 1981), species composition and physiognomy (Cattelino et al. 1979), plant regeneration (Cattelino et al. 1979, Lyon and Stickney 1976), forage quality (Hobbs and Spowart 1984), ecosys-

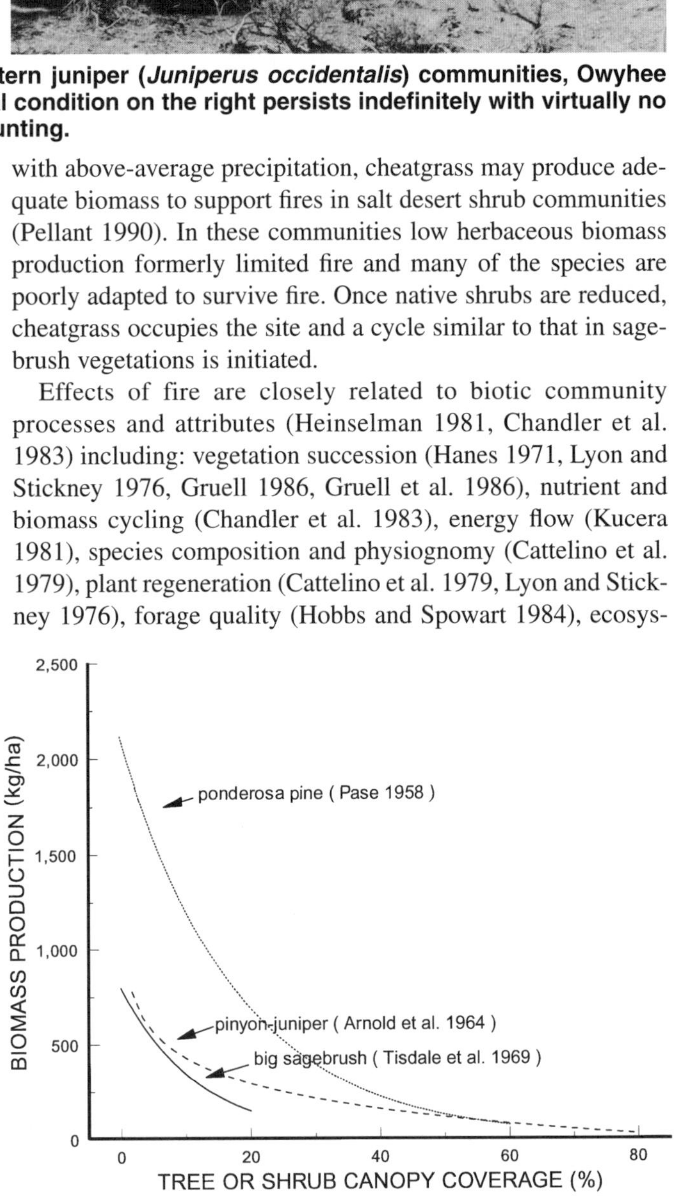

FIGURE 2. Relationships between overstory development and herbage production in understories are generally curvilinear. Light underburns may not enhance the quantity of herbaceous forage substantially. As overstories are reduced more substantially, however, herbaceous production increases disproportionately to the amount of overstory removed. Curves adapted from data in original publications.

tem stability and diversity (Loucks 1970, Taylor 1973, Habeck and Mutch 1973, Habeck 1976, Kilgore 1981), and the development of community type mosaics (Romme and Knight 1982, Lyon et al. 1978). Relationships between wildlife populations and fire are influenced by these processes and by wildlife responses to fire regime.

WILDLIFE RESPONSES

Mechanisms

Several reviews of wildlife-fire relationships are available (Bendel 1974, Kelsall et al. 1977, Lyon et al. 1978, Peek 1986). Studies have been conducted in a wide range of vegetations, climates, and fire histories. Although fire can injure and kill animals (Chew et al. 1959), its most important influences on wildlife are mediated through impacts on plant communities, most notably the food they provide, their architecture, and their relation to communities in the surrounding landscape. In most cases burning changes all 3 interdependently, thereby markedly redefining the choices available to animals. Conversely, animal morphology, physiology, and behavioral plasticity interact with plants and climate in the post-fire environment to define the capacity of animals to adapt to habitat changes. Whether or not fire significantly influences the productivity of wildlife populations often depends on long-term, landscape-scale interactions among plant and animal populations.

Food Resources

Much of the impetus for prescribed burning on behalf of big game at least is rooted in knowledge that wildfires often dramatically enhance forage resources, thereby fueling growth of populations (Young and Robinette 1939, Gaffney 1941, Einarsen 1946, Krefting 1951, Spencer and Hakala 1964, Klein 1982). The main functional aspect of wildfire in this regard is usually its scale. Past wildfires often encompassed thousands of square kilometers, and even today wildfire periodically alters habitats on a grand scale in forests, chaparral, and sagebrush ecosystems. Intensity of wildfires can also be functionally significant, often nearly eliminating decadent, mature, or late-successional overstories. Wildfires burned hundreds of thousands of hectares of the Yellowstone ecosystem in 1988, for example, dramatically altering the character of wildlife habitats in the process. We have yet to demonstrate an ability to manage prescribed fire at such scales, and our ability to mimic the function of such large-scale events, with smaller-scale prescriptions, requires close attention to the relation between animal populations, vegetation, and the various aspects of fire regime (Fig. 3).

Burning can substantially alter the quality of food resources by changing the structure of individual plant canopies, chemical composition of plant tissues, and the botanical composition of communities (Dewitt and Derby 1955, Daubenmire 1968, Pearson et al. 1972, Uresk et al. 1975, Wright and Bailey 1980, Willms et al. 1981, Kalmbacher 1983, Hobbs and Spowart 1984). Within individual plants, concentrations of digestible fiber, minerals, and proteins decline as tissues age with advancing phenology (Rao et al. 1973, Minson 1990). Grass leaves contain less cell wall and fiber than do culms (Goto and Minson 1977). Similarly, shrub leaves are generally less fibrous, and thus more digestible, than twigs or stems. Burning removes old and poorly digestible structures from plant canopies and deposits some of the residues as ash. Light and nutrient availabilities may increase in the surrounding soil along with temperature to facilitate nutrient uptake and increased growth by surviving plants and seedlings (Daubenmire 1968, Pearson et al. 1972, Peet et al. 1975, Woodmansee and Wallach 1981). The secondary walls of plant cells are not formed until after cell enlargement occurs, and thus rapidly growing foliage is characterized by high concentrations of digestible nutrients (Willms et al. 1981). Leaf:stem and live:dead ratios are increased within plant canopies, particularly at the periphery (Willms et al. 1981, Kalmbacher et al. 1985). These responses can occur in shrubs and herbaceous species (Cook and Harris 1950, Willms et al. 1981), and herbivores benefit from them through enhanced efficiencies of food ingestion and digestion.

FIGURE 3. Prescribed fire on Breakneck Ridge, Gros Ventre Range, Wyoming. This 200-ha project failed to achieve its goal of rejuvenating deteriorated aspen clones, and may have hastened their demise, because post-fire herbivory by elk severely suppressed aspen suckers (Bartos et al. 1994). Successful mimicry of wildfire's functions by prescribed fire is dependent upon burning at appropriate frequency, scale, and intensity *and* upon appropriate management of post-fire herbivory. Photograph by N. V. DeByle, U.S. Forest Service.

The effects of range burning on animal nutrition have usually been estimated by analyzing forage quality (Asherin 1973, McAtee et al. 1979, Merrill et al. 1980, Smith et al. 1984, Seip and Bunnell 1985, Wood 1988, Kituku et al. 1992, Soper et al. 1993, Cook et al. 1994), and reports of burn-induced enhancements are common. McAtee et al. (1979), for example, reported that digestible energy content of gulf cordgrass (*Spartina spartinae*) increased 50%, and protein content more than doubled, following burning. Smith et al. (1984) reported that burning improved the protein content of marsh plants as much as 50%, though other nutritional indicators were not greatly altered. However positive effects of burning on forage quality may not be substantially different from those of other management practices (Kituku et al. 1992, Soper et al. 1993), and in shrubs may be reduced or offset somewhat by post-burn increases in secondary plant metabolites that may be toxic, or bind with digestible nutrients thereby rendering them unavailable for metabolism (Bryant and Kuropat 1980, Provenza and Malechek 1984, Pendleton et al. 1992).

While assays of forage quality do provide rough indices to the potential effects of burning on animal nutrition, they do not accurately reflect the benefits that are actually derivable by herbivores. Variation in sample collection, storage, and processing may lead to erroneous conclusions (Van Soest 1965, Jones and Bailey 1972, Goering et al. 1973). Equally important is the fact that plant samples collected for nutritient analysis are usually plucked more or less at random, and thus tend to represent the average of forage items available in a community. Wild herbivores, however, are capable of selecting forage items that are substantially better than the average available to them (Hobbs and Spowart 1984), introducing potential for bias. Diversity of species composition and phenology within communities substantially influences selection of plants and plant parts (Van Dyne and Heady 1965*a*, Carpenter et al. 1979), and selection of chemical constituents to a lesser degree (Van Dyne and Heady 1965*b*). Bias potential exists when sampling either burned or unburned communities, and is not necessarily the same in both circumstances.

Comparisons of actual diet quality between burned and unburned areas are more useful than analysis of random forage samples. Unfortunately, studies of the effects of range burning on herbivore diets are infrequent in the literature. Available data for bighorn sheep (*Ovis canadensis*), mule deer (*Odecoileus hemionus*), and elk (*Cervus elaphus*) (Fig. 4) indicate that herbivores can substantially enhance the quality of their diets by feeding in recently burned plant communities (Hobbs and Spowart 1984, Canon et al. 1987, Cook et al. 1990). Benefits fluctuate seasonally, however, coincident with periods of plant growth and senescence. Svejcar (1989) found that peak benefit for livestock coincided with the period of most rapid plant growth. Similarly, seasonal variation in dietary improvements for wild herbivores (Fig. 4) reflects seasonally enhanced growth and/or tissue composition in burned communities. Benefits of burning persist until rapid growth in burned and unburned communities obviates the burn-induced differences in quality and availability of forage components selected by herbivores in each

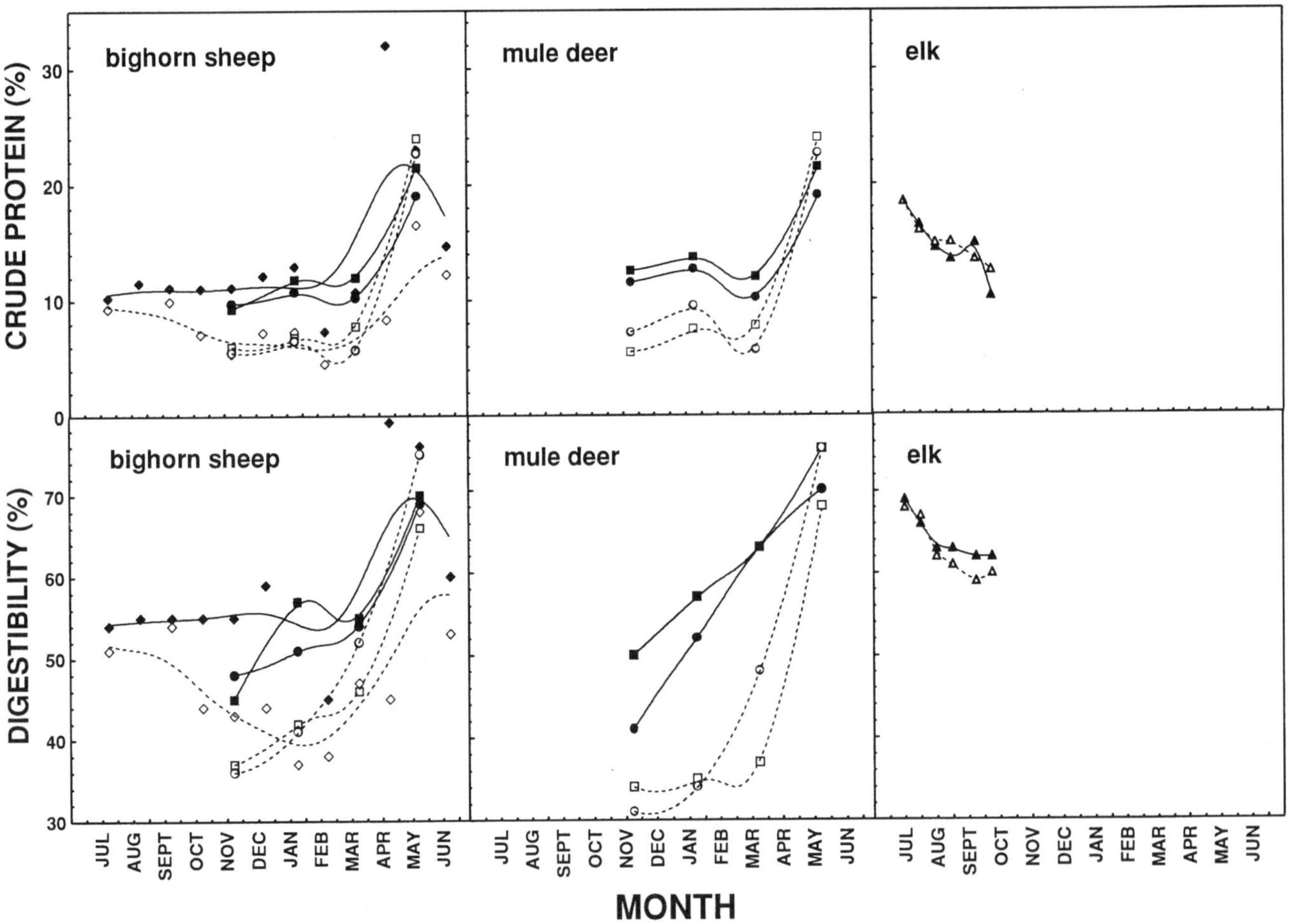

FIGURE 4. Distance-weighted least-squares estimates of dietary crude protein and digestibility for bighorn sheep, mule deer, and Rocky Mountain elk, feeding in burned (b, solid lines) and unburned plant communities (ub, dashed lines). Data are averages reported in the first 2 years following burning of grassland (● = b, ○ = ub), sagebrush-grassland (◆ = b, ◇ = ub), mountain shrubland (■ = b, □ = ub), and aspen/conifer forest (▲ = b, △ = ub). Digestibilities in grassland and mountain shrubland were analyzed on an organic-matter basis (Hobbs and Spowart 1984), whereas those in aspen/conifer forest and sagebrush-grassland were analyzed on a dry-matter basis (Canon et al. 1987, Cook et al. 1990). Adapted from the original publications.

community, or until the potential for accelerated plant growth in burned communities is exhausted (Rao et al. 1973, Hobbs and Spowart 1984). Timing of dietary benefits also is influenced by variation in timing of the burn itself (e.g., spring, summer, fall), topography (e.g., elevation, aspect), weather patterns, and plant type (e.g., warm-season, cool-season).

Nutrient intake by herbivores is not only a function of the nutrient content of food, but also of the amount of food eaten. Benefits of burning on forage and diet quality are widely cited. However, few workers have recognized the potential for rangeland burning to increase the amount of food consumed (Seip and Bunnell 1985). Voluntary consumption of forage can be positively correlated with a number of characteristics that are changed by burning, including solubility and digestibility, organic matter content, protein content, particle size, and trace element concentrations (Van Soest 1982, Arnold 1985, Minson 1990).

Effects of burning on protein content and digestibility of bighorn sheep diets (Fig. 4) provide an example of how burning might enhance forage intake. Bighorn sheep feeding in burned communities consume forage containing about 10% crude protein, on average from October to March, compared to only about 6% in unburned communities. The magnitude of this difference is important in itself, but the scale at which it occurs is also significant because rumen function is impaired when dietary crude protein (N $\times$ 6.25) falls below 7-10% of dry matter (Blaxter et al. 1961, Campling et al. 1962, Elliott and Topps 1963, Milford and Minson 1966). At lower levels, rumen microbial populations are limited by a shortage of available nitrogen. This slows the rate of food digestion and the rate at which digesta pass through the gut, thereby potentially limiting an animal's voluntary intake of dry matter and digestible nutrients. Under these circumstances, increasing dietary nitrogen via commercial supplements, or via range improvements such as burning, can enhance intake of dry matter (Kartchner 1981, Minson 1990). The influence of protein content on forage intake varies as a function of several factors (Elliott and Topps 1963, Milford and Minson 1966, Minson 1990), and is poorly understood for free-ranging wild ruminants. Studies reviewed by Minson (1990) indicate that supplementation of protein-deficient forage can increase voluntary intake of dry matter by domestic ruminants 14-77%. Similar benefits are likely for wild ruminants, but more research is needed to define their significance in specific settings.

Burning also may alter diet quality by influencing community composition because phenological profiles of plant communities are influenced by their botanical composition (Pitt and Wikeem 1990). As burning alters the composition of a plant community it also alters the community's phenological development, and thus the community's time-specific nutritional value. Studies reviewed by Minson (1990) have shown that addition of legumes to nitrogen deficient diets can improve the intake of domestic sheep, sometimes synergistically. Similarly, diversification of plant species and phenologies within communities following fire may improve opportunities for selective foraging. Composition shifts can have negative consequences also, as in the fire-induced conversion of sagebrush-dominated communities to cheatgrass. In enhancing cheatgrass at the expense of sagebrush and perennial herbs, fire alters the temporal pattern of nutrient availability in the community. Springtime greening of forage may occur earlier, but availability of sagebrush is reduced, which may be of critical importance to the nutrition of deer later in the year. Similar seasonal trade-offs are involved in most applications of fire although their functional significance is likely to vary greatly from setting to setting. Generally, the more intense or frequent the burning regime, the more likely communities are to be converted to herbs having relatively short growth periods, which could limit forage options at various times of the year. The importance of fire-induced composition shifts to nutrient limitations depends on their severity, scale of burning, availability of unburned habitat, and seasonal pattern of nutritional constraints on the herbivores involved.

The extent to which fire-induced improvements in forage (quantity and quality), diet quality, and intake actually affect growth and productivity has been difficult for wildlife biologists to demonstrate (Peek 1989, Hobbs and Spowart 1984). However, results of experiments with livestock show that effects can be striking. Svejcar (1989) found that burning tall-grass prairie increased seasonal weight gains of cattle 17%. Angell et al. (1986) observed that overwinter liveweight gains of yearling cattle held in burned pastures of gulf cordgrass were 90% greater than those of cattle held in unburned pastures. Benefits in this range can result in substantially improved energy balance and survival of adults, larger birth weights, better growth and survival of neonates, and ultimately increased recruitment and harvestable yield of wild game animals.

Caveats. While burning can clearly increase carrying capacity of rangeland via enhancement of food resources, there are some practical concerns upon which our expectations should be qualified. First, burn-induced improvements in diet quality can be realized only so long as free-ranging animals are able to select food items of relatively high quality in burned communities. Similarly, increased biomass production is of finite duration. Therefore, the expected duration of burn-induced enhancements should always be a significant variable of concern to managers, for predicting biological effects and for analyzing the economic efficiency of their projects (Daniels and Riggs 1988).

Duration of forage enhancements is a function of the time required for plant growth and succession to restore pre-fire composition and structure of communities. The time required to restructure communities varies from several decades following stand-replacing forest fires to a year or less in the simplest herbaceous communities. In many rangeland settings, however, significant enhancements of forage and diet quality are short-lived (Pearson et al. 1972; Rowland et al. 1983; Thill et al. 1987; Hobbs and Spowart 1984; Cook et al. 1990,1994). Rowland et al. (1983) were unable to detect enhancement of elk diets on burned winter range 3 years post-burn. Declines in nutritional benefits have been reported after 1 year for white-tailed deer (*Odecoileus virginianus*)(Thill et al. 1987), mule deer (Hobbs and Spowart 1984), and bighorn sheep (Hobbs and Spowart 1984; Cook et al. 1990,1994). Repeated cropping and recycling of nutrients may extend benefits on heavily used range sites somewhat (Austin et al. 1983, Miquelle 1983, Willms et al. 1988,

Dick and Urness 1991, Motazedian and Sharrow 1990, Stein et al. 1992, Westenskow-Wall et al. 1994) but with consequent alteration of the community's productivity and successional trajectory possible (Willms et al. 1988). Thus enhancement of forage quality, alone, can be a poor justification for the expense of prescribed burning unless sequential burns are conducted to maintain benefits. Occasional burns, however, can be effective in altering community structure and composition for extended periods, depending of course on the plant community being treated. The benefits of burning are a cumulative function of effects on quality, quantity, structure, and composition, but fire's effects on these may be differentially skewed in time.

A second caveat relates to the concepts of burning scale and population regulation. The area that must be burned to produce or sustain a population effect increases with the size of the herbivore population and with herbivore density. Past wildfires often burned extensive landscapes, and at least earlier in this century this occurred against a background of depleted ungulate populations. Extensive burning and protection from exploitation set the stage for strong population responses. Circumstances are often quite different today. The prevalence of wildfire has declined (Fig. 5), while the scale of prescribed-fire programs remains small (Fig 6). Many big game populations are more strongly influenced by density-dependent nutritional constraints and/or predation than earlier in this century. Where these conditions occur, small-scale prescribed burns may be incapable of increasing population productivity or growth, even when carried out repeatedly. Peek (1989) observed that 20 years of prescribed burning to rejuvenate northern Idaho shrub fields on behalf of elk had yielded no demonstrable increase in herd production. Klinger et al. (1989) did not detect an increase in fawn survival after burning of chaparral even though deer clearly selected recently burned communities. Such observations raise a couple of questions. Has prescribed fire been cost effective in such circumstances? If fire is potentially beneficial, how much burning is required to meet management goals? These questions are not trivial, and they underscore the importance of understanding how various factors influence population dynamics.

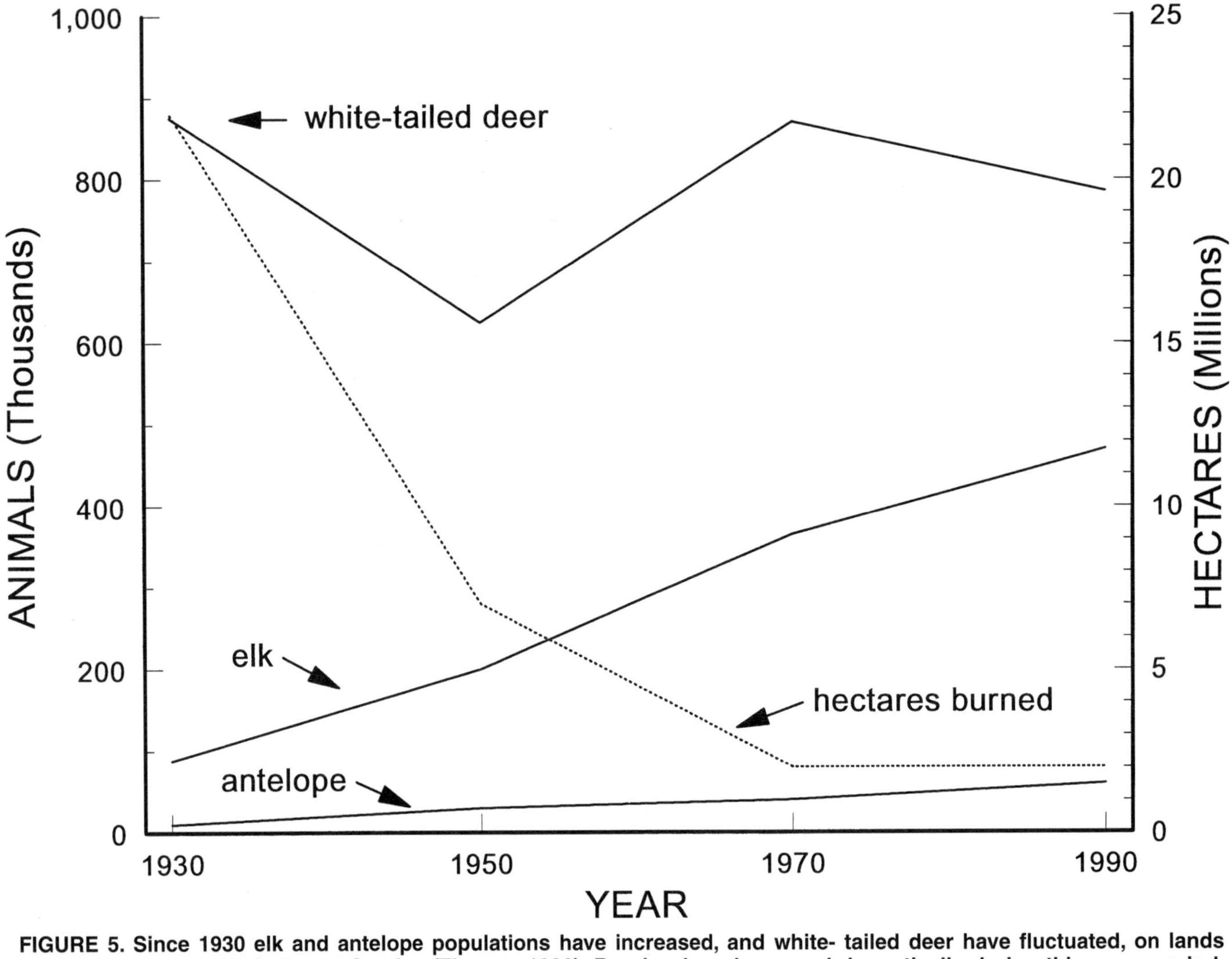

FIGURE 5. Since 1930 elk and antelope populations have increased, and white- tailed deer have fluctuated, on lands administered by the U.S. Forest Service (Thomas 1990). Burning has decreased dramatically during this same period. Game populations are now more likely to be constrained by density-dependent factors than earlier in the century. Responses of game populations and plant communities to prescribed burning are influenced by the intensity of density-dependent constraints.

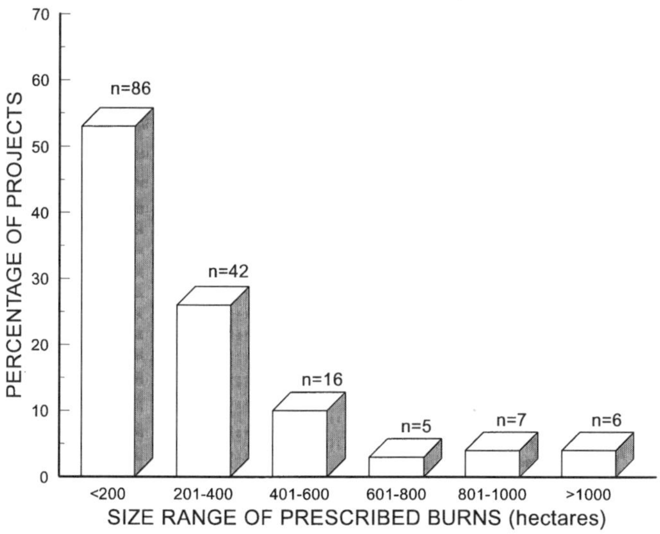

FIGURE 6. Apparently, large-scale prescribed fire is relatively rare in the habitat management programs of public agencies. These data, courtesy of the Rocky Mountain Elk Foundation, illustrate the size distribution for 162 prescribed burns funded cooperatively by the Foundation and various public agencies, November 1986 through April 1993.

A third point is that both the rate and direction of secondary plant succession are influenced by the intensity, timing, and type of ungulate use (Zimmerman and Neuenschwander 1984, Edgerton 1985, Austin et al. 1986, Pastor et al. 1988, Kay 1990, Risser 1990, Urness 1990). Fire's influences on community composition and succession are not independent of herbivory's influences (Hobbs et al. 1991). Yet most research concerning post-fire succession has not emphasized the potential of fire and herbivory to influence succession interactively. Consequently, expectations of the results of fire management may not be realistic if the extant herbivory regime differs markedly from that under which past fire literature was generated (Peek 1989). Furthermore, aggressive management of herbivore populations may be requisite to meeting goals for vegetation management in many circumstances (Bartos et al. 1994).

Because plant-animal interactions alter the outcome of burning programs, they are important to our expectations of the biological effectiveness and economic efficiency of proposed projects. Prescribed burning must be planned, and expectations of its effects tempered, with knowledge that the interactions among animal and plant populations are fundamental to the outcome.

Habitat Use

Big Game. Studies of habitat selection have been the primary means by which the responses of big game to burning programs have been documented. Many examples are available for elk (Skovlin et al. 1983, Canon et al. 1987, Peck and Peek 1991), mule deer (Keay and Peek 1980, Thomas and Tiller 1985), white-tailed deer (Keay and Peek 1980, Ivey and Causey 1984), moose (*Alces alces shirasi*) (Pierce and Peek 1984), bison (*Bison bison*)(Campbell and Hinkes 1983, Coppock and Detling 1986, Shaw and Carter 1990), bighorn sheep (Peek et al. 1979, Riggs and Peek 1980, Bentz and Woodward 1988, Cook et al. 1990), pronghorn antelope (*Antilocapra americana*)(Courtney 1989), and horses (Peck and Peek 1991).

Selection of habitats and foods and their efficiencies of use by herbivores depend on several factors including food biomass (Wickstrom et al. 1984), nutritional quality (Van Soest 1982), secondary chemistry (Bryant and Kuropat 1980, Provenza and Malechek 1984), canopy architecture (Provenza and Balph 1990, Jiang and Hudson 1994), the forager's physiology and morphology (Jarmen and Sinclair 1979, Clutton-Brock and Harvey 1983, Demment and Van Soest 1985), and prior foraging experience (Provenza and Balph 1988). With an appropriate sampling design, selection for or against burned communities can be detected and explained. Preference analyses aid in identifying successional stages of greatest value, seasons of greatest use, and provide clues to the duration of benefits. However, periodic preferences may be exhibited by individuals or groups of animals even if habitat treatments are not sufficient to affect their energy balance or productivity of the greater population. Thus preference for burned areas does not necessarily indicate that burning has resulted in a functionally significant benefit to the population. Conversely, avoidance of burned areas is not indicative of detriment to a population so long as sufficient alternate habitats are available.

Animals may either select or avoid recently burned communities. Ivey and Causey (1984) documented a temporary shift in preference away from dry upland communities, in which most of the understory had been consumed by recent fire, toward wetter bottomland habitats that were not severely burned and contained more residual forage. Gasaway and DuBois (1985) reported a similar response by moose. Short-term responses can be positive as well. Courtney (1989) reported that pronghorn antelope preferred burned sagebrush communities from fall through early spring. Here selection was related to changes in quality and structure of residual vegetation; post-burn greening of herbs attracted pronghorn, but so did prickly pear cactus (*Opuntia polyacantha*) from which spines had been burned, thereby enhancing access to succulent tissues. In both of these studies, selectivity dissipated during the next growing season.

Bighorn sheep may preferentially use burned sagebrush-bunchgrass communities for at least 4 years (Peek et al. 1979, Cook et al. 1990) and seral conifer habitats for several decades (Riggs and Peek 1980). Keay and Peek (1980) observed that mule deer preferred burned habitats in a douglas fir-ponderosa pine winter range complex, while sympatric white-tailed deer avoided these in favor of residual unburned stands. Forage relationships were not implicated in the differential selection, and security characteristics of habitat may be more important to habitat partitioning by these 2 species (Wiggers and Beasom 1986).

Canopy structure, available biomass, and chemistry have implications for acceptance of alternative forages by herbivores and for foraging efficiency (Black and Kenney 1984, Kenney et al. 1984, Wickstrom et al. 1984, Provenza and Malechek 1984, Bunnell and Gillingham 1985). Foraging efficiency of white-tailed deer in thornscrub can vary 6 times depending on the species of forage being eaten (Koerth and Stuth 1991). Rates of forage intake by reindeer (*Rangifer tarandus tarandus*) are affected by biomass and growth form of plants (Trudell and White 1981). Variation in foraging efficiency associated with canopy architecture can influence acceptance of the forages available (Cooper and Owen-Smith 1986). Foraging theory predicts that selection of food and habitat is influenced by such relations (Stephens and Krebs 1986). In theory then, burning can influence habitat selection by altering forage structure alone, independent of effects on quality or quantity.

Canon et al. (1987) compared summer use of burned and unburned aspen forest by Rocky Mountain elk for 2 years following broadcast burning. Burned forest was preferred in both years even though there was no apparent difference in diet quality between habitats. Forage availability approached 2,000 kg/ha in both habitats, the potential intake-rate maxima reported for elk (Hudson and Nietfeld 1985, Wickstrom et al. 1984). Therefore, a strong case could not be made that the habitat preference was motivated by differences in forage quality or quantity. The biting rates of elk did not differ between habitats. Average bite weight, however, was 27% greater when elk fed in burned forest than when they fed in unburned forest (0.71 g/bite vs 0.56 g/bite, respectively). The bite-weight differential between burned and unburned foraging areas was attributed to larger shrub leaves in the burned forest, which could facilitate greater intake rate and perhaps motivate habitat selection (Collins and Urness 1983). In contrast, Skovlin et al. (1983) failed to detect preferential use of burned bunchgrass communities by elk in any of 3 winters following a fall burn.

Selection of burned areas can be a function of vegetation and landscape in the proximity of burns, independent of animal-plant relationships in burned areas. This has been pointed out repeatedly for bighorn sheep (Risenhoover and Bailey 1985, Risenhoover et al. 1988, Bentz and Woodard 1988, Cook et al. 1990). This species' use of burned areas diminishes substantially as distance from escape terrain increases (Bentz and Woodard 1988). Such landscape-level relations can be important to the effectiveness of burning programs conducted on behalf of other species also, particularly in marginal habitats.

Preference for burned areas can help managers redistribute herbivore populations. Shaw and Carter (1990) studied the response of bison to spring burns. Burned areas "held" bison on winter range longer than unburned areas, and use of burned summer range increased 37%. Such responses may be used to advantage in attempts to alter distribution and intensity of ungulate herbivory, whether the goal be vegetation management or range expansion by sedentary populations (Risenhoover et al. 1988, Campbell and Hinkes 1983, Jourdonnais and Bedunah 1990).

Prairie dogs (*Cynomys* spp.) have substantial impacts on composition, biotic diversity, and nutrient dynamics of their colony sites (Coppock et al. 1983*a*, Uresk 1984, Agnew et al. 1986, Cid et al. 1991). Prairie dog colonies attract larger herbivores because forage at their peripheries can be more palatable than that found in the surrounding prairie (Koford 1958, McHugh 1958, Shult 1972, Coppock et al. 1983*a*,*b*). Repeated grazing by bison exacerbates the impacts of prairie dogs on vegetation (Coppock et al. 1983*b*, Cid et al. 1991). Coppock and Detling (1986) hypothesized that prescribed burning might be used to enhance forage in outlying areas and thus attract bison away from colonies. To test their hypothesis they burned 9 of 84 ha surrounding a 36-ha colony. Cow-calf groups reduced their use of vegetation adjoining the colony by 30-63%, depending on the season, and increased their use of the outlying burned prairie by 12 times. Habitat selection by bulls was not affected, however, and the response of cow-calf groups did not extend beyond 1 year. Burning to alter use patterns is frequently touted as a means of alleviating agricultural depredations. Coppock and Detling's (1986) observations suggest that success of such schemes can depend on sequential burning of adjacent rangelands. Judicious hazing of depredating animals may also be required.

Responses to burning can vary substantially depending on the rate of vegetation response and the habitat requirements of the particular wildlife species. The short-term response of bison to burning in tall-grass prairie differs markedly from the long-term response of caribou (*Rangifer* spp.) to burning and regrowth of lichen communities in taiga ranges (Klein 1982, Schaefer and Pruitt 1991). Elk have been shown to prefer early seral stages of forest succession (<20 yr old) substantially more than older seres during winter and early spring (Peck and Peek 1991). Preferential use of seral forest communities on winter range by bighorn sheep over several decades is facilitated by prolonged, fire-induced changes in snow deposition and melt (Riggs and Peek 1980). The benefits of burning to a particular species can be strongly time-dependent, and vary among settings.

Habitat use by sympatric species can influence the response of a target species. Spowart and Hobbs (1985) found that range burning could intensify interspecific competition between mule deer and bighorn sheep through its influences on community composition. Competition for resources at the community level may be reflected in differential use patterns at the landscape level. Roberts and Tiller (1985) noted that post-burn regrowth of chamise (*Adenostoma fasciculatum*) was similar inside and outside of livestock exclosures, but deer apparently made 40% greater use of the exclosures, apparently reflecting an interaction with livestock on the outside.

Weather and climate interact with vegetation change to alter habitat use. In parts of the interior Northwest, moose winter in cedar-hemlock forests in which the chief forage is Pacific yew (*Taxus brevifolia*). These winter ranges typically receive heavy snow packs, and the forest overstories intercept snow, thereby facilitating access by moose to yew in the understory. Clear-cut timber harvesting, followed by broadcast burning, has for some time constituted the preferred silviculture in these forests. These practices have also been encouraged on behalf of big game because voluminous amounts of browse are a predictable result (Leege 1968,1969, Asherin 1973). However these practices do not maintain mature forest canopies or encourage Pacific yew, which is shade-tolerant and fire-intolerant. Therefore, Pierce and Peek (1984) suggested that fire exclusion and selective timber harvesting would be more beneficial to moose than traditional practices.

Similar effects of burning can be important in nonforest habitats. Burning of sagebrush communities may periodically reduce availability of sagebrush, a good source of protein and other nutrients (Kufeld et al. 1981, Welch et al. 1981) to wintering deer. Similarly, shrub reduction may be detrimental in arid landscapes where shrubs are important sources of dietary nutrients during drought (Rominger et al. 1988, Miller and Gaud 1989). Whether shrub removal is beneficial or detrimental depends on the value of shrubs relative to other forages, the timing and length of seasonal nutrient limitations, and the proportion of habitat treated. Where shrubs are critical components of seasonal range and easily removed by fire, judicious livestock grazing may be a useful alternative to burning programs (Urness 1990).

Birds and Small Mammals. Literature regarding fire ecology of large herbivores has focused on forage-mediated relations, but literature for birds and small mammals has focused on structure-mediated relationships. In sagebrush-grasslands, fire's influence on the composition of nongame avifauna appears mediated primarily through its effects on the shrub strata. Rottenberry and Weins (1978) observed that where fire had reduced shrubs by 88%, density of breeding birds was only moderately affected and species composition and richness were unchanged. However the relative abundance of species was altered, with sage sparrows (*Amphispiza belli*) replacing horned larks (*Eremophila alpestris*) as dominant in an apparent response to reduction of sagebrush (Fig. 7).

Castrale (1982) compared bird densities in a 4-year-old burn, a 4-year-old chaining, and a 17-year-old plowed area. All treatments were 16 ha. Density of individual species varied, but neither richness nor total density differed among treatments. Horned lark density was greater in burned and chained areas than in the plowed area, probably because shrubs had more fully recovered in the older plowed area. Brewer's sparrow (*Spizella breweri*) were 50 and 86% less abundant in the burn than in chained or plowed areas, but density in the burn was still 21 territories/km^2. Vesper sparrow (*Pooecetes gramineus*), western meadowlark (*Sturnella neglecta*), and sage thrasher (*Oreoscoptes montanus*) densities were similar among treatments.

Petersen and Best (1987) found that light mosaic burning of sagebrush-dominated stands enhanced species richness and densities of nongame birds. They monitored 4 6.25-ha plots for 2 years, then fall-burned 2 of the plots, excluded livestock grazing, and monitored all plots for an additional 4 years. Burning resulted in vegetation mosaics in which about 45% of the area was burned and dominated by herbaceous species. Richness of the avifauna was consistently higher in burned mosaics than in controls, due largely to colonization of burned areas by horned larks and vesper sparrows. Combined density of all species declined initially, but eventually rebounded to exceed pre-fire levels. Brewer's sparrow declined temporarily, but rebounded strongly after 2 years. Sage sparrows and sage thrashers were not affected, and western meadowlarks increased slightly.

The work of Rottenberry and Weins (1978), Castrale (1982), and Petersen and Best (1987) illustrate a tolerance of rangeland avifaunas to burning, a quality that might be predicted given the relatively high frequency with which their habitats burned naturally. However, responses to fire are not always positive. Nesting success of sage grouse (*Centrocercus urophasianus*), for example, may be locally reduced as a consequence of cover reduction (Connelly et al. 1991, Gregg et al. 1994). Partial kill of sagebrush has little effect on Brewer's sparrow, but total kill of sagebrush can nearly eliminate this species (Best 1972). Thus maintaining alternate, unburned, habitat can be important for some species, and if one goal of burning is to maintain high species richness in small landscapes, burns should be conducted in strips, blocks, or mosaics (Castrale 1982, Petersen and Best 1987). Repeated burning at high frequency, even at light intensity, may severely reduce species that are present because of long-term fire exclusion or where the time required for vegetation to return to pre-fire status is long relative to the FFI. High-frequency burning is required for restoration of some avifaunas following extended FFIs (Fitzgerald and Tanner 1992).

Knowledge of fire-avifauna relations in forests is limited. In Arizona, Blake (1982) censused nonbreeding birds in logged and/or burned Ponderosa pine stands following a large, stand-replacing wildfire. Forty-nine species were detected (Table 2). More of these were found exclusively in unburned forest ($n = 20$) than were found exclusively in burned forest ($n = 13$), and total species observed exclusively in one or the other ($n = 33$) was roughly twice the number found in both ($n = 16$). These data suggest that several species were specialists, either seeking or avoiding burned habitats. Bock and Bock (1983) censused breeding birds in South Dakota following light underburning. They found many more species breeding in both burned and unburned habitats ($n = 21$) than were found exclusively in either habitat ($n = 2$), and none of the observed differences between treatments lasted into the second year following the burn (Table 3).

Several points should be considered when comparing the differing results of these 2 studies. First, differences exist in avifaunal composition between Arizona and South Dakota. Only 8 species were common to both study areas, including northern flicker (*Colaptes auratus*), hairy woodpecker (*Picoides villosus*), white-breasted nuthatch (*Sitta carolinensis*), American robin (*Turdus migratorius*), yellow-rumped warbler (*Dendroica coronata*), dark-eyed junco (*Junco hyemalis*), chipping sparrow (*Spizella passerina*), and the western wood-pewee (*Contopus sordidulus*). Among these only the western wood-pewee was not found in burned and unburned habitats in both studies. Timing of the studies differed; the Arizona study censused nonbreeding populations, and the South Dakota study sampled breeding populations. Both climate and habitat affect habitat use by forest birds (Szaro and Balda 1986), and differences might be expected on this basis. Sampling effort differed greatly between the 2 studies; Blake (1982) sampled 4 times between October and early May, whereas Bock and Bock (1983) sampled 7 times during June alone. More intensive sampling enhances detection of less abundant species, thus reducing the apparent specialization of some species. Fire regimes also differed. Blake (1982) focused on responses to catastrophic wildfire encompassing 12,000 ha in which 4,800 ha of pine forest were burned, with half killed. Bock and Bock (1983), on the other hand, focused on responses to small-scale prescribed fire, which burned cool and left the forest canopy largely intact. In

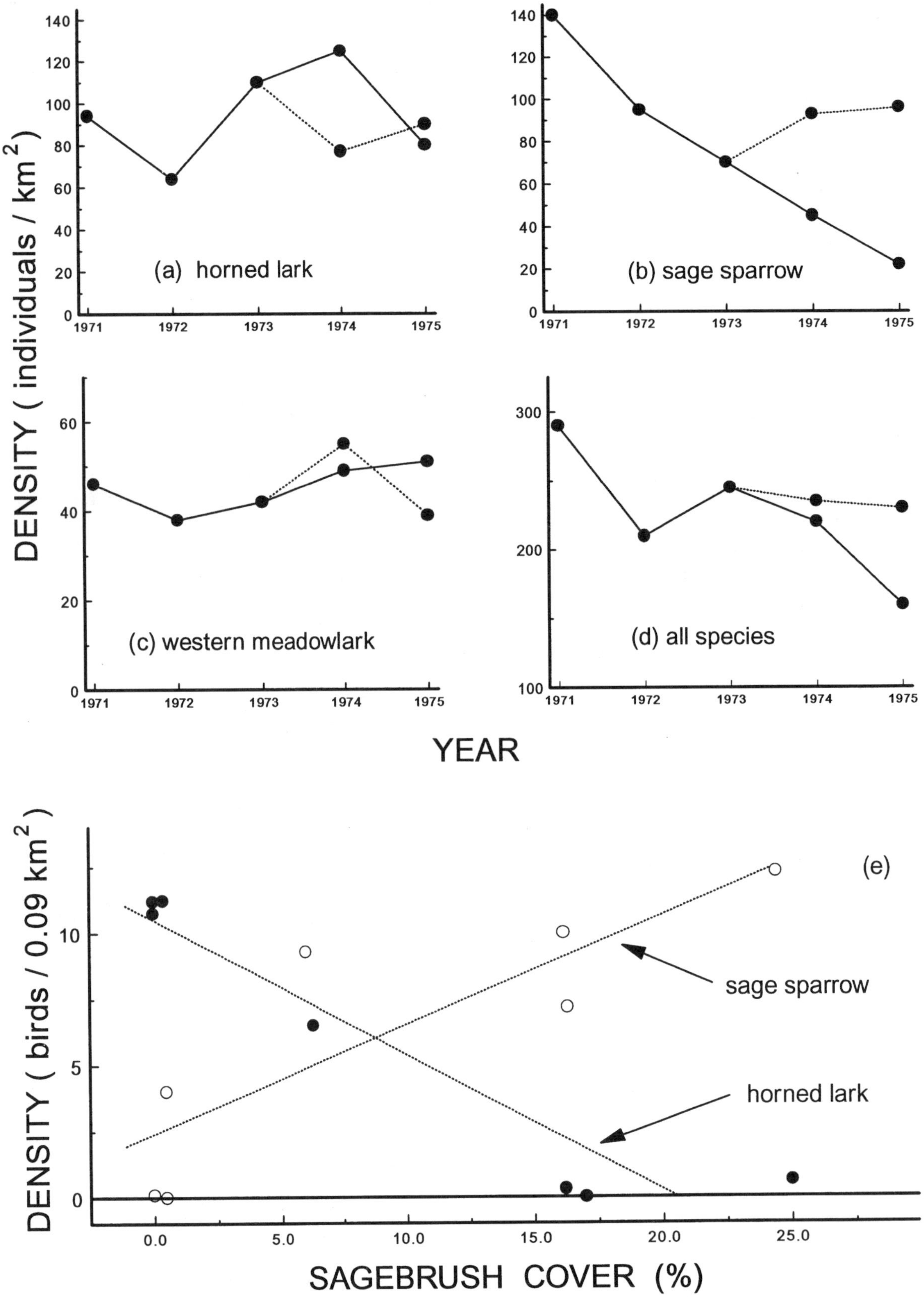

FIGURE 7. Breeding densities of nongame birds in shrub-steppe before and after burning. Some species respond differently to burning in sagebrush grasslands (a,b,c), but density across species increases, at least in the short term (d). Sage sparrows respond negatively to reduction of sagebrush cover, whereas horned larks respond positively (e). Adapted from Rottenberry and Wiens (1978).

TABLE 2. Seasonal composition of bird communities on burned and unburned study sites in Prescott National Forest, Arizona, 1973-74. Species abundances are prominence values (PV).[a] FG = foraging guild. BB1 and TB1 were not logged, BB2 and TB2 were partially logged, and BB3 and TB3 were clear-cut. Reprinted from Blake (1982).

			Burned			Unburned		
Season	Species	FG	BB1	BB2	BB3	TB1	TB2	TB3
Fall								
	Poor-will (*Phalaenoptilus nuttallii*)	H	0.2					
	Western wood pewee (*Contopus sordidulus*)	H		1.0				
	Scrub jay (*Aphelocoma coerulescens*)	O		0.2				
	House wren (*Troglodytes aedon*)	BFI		0.2				
	Hermit thrush (*Catharus guttatus*)	GI		0.2				
	Lesser goldfinch (*Carduelis psaltria*)	G	0.2	0.2	1.3			
	Common flicker (*Colaptes auratus*)	GBI		0.6				0.5
	Hairy woodpecker (*Picoides villosus*)	BI	7.3	6.0	1.0		0.8	0.3
	Mountain chickadee (*Parus gambeli*)	BFI	0.9	0.4		0.2		
	White-breasted nuthatch (*Sitta carolinensis*)	BI	0.8	0.2		4.9	0.6	0.8
	Western bluebird (*Sialia mexicana*)	GI/H	0.6	5.3	1.3		0.8	6.5
	Yellow-rumped warbler (*Dendroica coronata*)	FI/H	1.6	0.2	0.6	1.6	1.0	2.8
	Cassin's finch (*Carpodacus cassinii*)	G		2.1	0.4		0.2	0.8
	Dark-eyed junco (*Junco hyemalis*)	G	0.9	9.1	0.2	2.8	12.5	44.6
	White-crowned sparrow (*Zonotrichia leucophrys*)	G	34.3	21.6	59.4		0.7	19.1
	Band-tailed pigeon (*Columba fasciata*)	G						10.2
	Acorn woodpecker (*Melanerpes formicivorus*)	G						1.2
	Lewis' woodpecker (*Melanerpes lewis*)	G						2.0
	Steller's jay (*Cyanocitta stelleri*)	O				0.5	0.2	0.2
	Pygmy nuthatch (*Sitta pygmaea*)	BFI					0.6	
	Canyon wren (*Catherpes mexicanus*)	GI						0.2
	American robin (*Turdus migratorius*)	GI				2.7		
	Ruby-crowned kinglet (*Regulus calendula*)	FI				5.4	2.9	8.0
	Black-throated gray warbler (*Dendroica nigrescens*)	FI						0.2
	Rufous-sided towhee (*Pipilo erythrophthalmus*)	G						0.5
	Chipping sparrow (*Spizella passerina*)	G						0.2
	Fox sparrow (*Passerella iliaca*)	G						0.2
	No. species		9	14	7	7	10	18
Winter								
	Hairy woodpecker	BI	1.1	1.2	0.9			0.2
	Mountain chickadee	BFI	0.2	0.2		0.4		
	White-breasted nuthatch	BI		0.3		0.4	0.4	
	Western bluebird	GI	3.2	13.5	1.4			17.4
	Cassin's finch	G		8.2			0.4	
	Dark-eyed junco	G	4.6	44.9	0.4		8.5	65.4
	Acorn woodpecker	G						0.2
	Yellow-bellied sapsucker (*Sphyrapicus varius*)	BI			0.2			
	Lewis' woodpecker	G						1.4
	Staller's jay	O						2.0
	Plain titmouse (*Parus inornatus*)	BFI					0.2	
	Pygmy nuthatch	BFI					0.4	
	Ruby-crowned kinglet	FI				7.1	0.6	
	Cedar waxwing (*Bombycilla cedrorum*)	G				2.0		
	No. species		4	6	3	5	6	6

TABLE 2. *(continued)*

Season	Species	FG	Burned			Unburned		
			BB1	BB2	BB3	TB1	TB2	TB3
Spring								
	Cassin's kingbird (*Tyrannus vociferans*)	H			0.2			
	Rock wren (*Salpinctes obsoletus*)	GI		1.0				
	American robin	GI			0.2			
	Solitary vireo (*Vireo solitarius*)	FI	0.2					
	Grace's warbler (*Dendroica graciae*)	FI		0.4				
	Hepatic tanager (*Piranga flava*)	FI	0.2					
	Black-headed grosbeak (*Pheucticus melanocephalus*)	FI	0.4	0.2				
	Cassin's finch	G	0.4					
	Broad-tailed hummingbird (*Selasphorus platycercus*)	N/FI	0.2	0.2	0.5		0.2	
	Common flicker	GBI		0.8	0.4			
	Hairy woodpecker	BI	2.4	0.2			0.2	
	Olive-sided flycatcher (*Nuttalornis borealis*)	H		0.2		0.2		
	Violet-green swallow (*Tachycineta thalassina*)	A		10.3	8.2			
	White-breasted nuthatch	BI		0.2	0.2			
	House wren	BFI	0.8		1.5			
	Western bluebird	GI/H	2.2	5.7	1.0		0.2	
	Brown-headed cowbird (*Molothrus ater*)	G	0.2				0.5	
	Pine siskin (*Carduelis pinus*)	G	2.8	8.0	3.8	0.4	2.9	
	Chipping sparrow	G		7.7	0.8	0.2	8.8	
	Band-tailed pigeon	G						
	Acorn woodpecker	G/H						
	Ash-throated flycatcher (*Myiarchus cinerascens*)	H						
	Empidomax spp.	H				0.4		
	Staller's jay	O						
	Pygmy nuthatch	BFI				2.6		
	Brown creeper (*Certhia familiaris*)	BI				0.5		
	Bewick's wren (*Thryomanes bewickii*)	BFI					0.2	
	Ruby-crowned kinglet	FI				0.5		
	Loggershead shrike (*Lanius ludovicianus*)	GI						
	Virginia's warbler (*Vermicora virginiae*)	FI					0.5	
	Yellow-rumped warbler	FI/H				1.6	0.2	
	Black-throated gray warbler	FI					0.2	
	Townsend's warbler (*Dendroica townsendi*)	FI				1.8		
	Rufous-sided towhee	G					1.3	
	No. species		10	12	10	5	12	

[a] $PV = D \times F^{0.5}$, where D = number of individuals of each species from all census periods on a given site, and F = frequency of occurrence of the species at sample points within the site.

many cases occasional light underburning probably produces few, if any, functionally significant changes in the architecture of forest canopies (Hilden 1965, Emlen 1970, Bock and Lynch 1970, Hanson 1978).

Certain habitat features, however, can be critical and at risk of loss to prescribed fire. Bohn et al. (1980) found that structures of abandoned homesteads were critical to several species not found in surrounding rangelands. Riparian habitats provide structure important to avifaunas in arid habitats (Butorff 1974, Carothers and Johnson 1975, Warner and Hendrix 1984), as do snags and decadent trees in forests (Thomas et al. 1976, Scott et al. 1980, Winternitz 1980, Dickson et al. 1983, McComb et al. 1986, Bull 1987, Horton and Mannan 1988, Land et al. 1989, Milne and Hejl 1989).

Snag dynamics are an interesting case in point. Fire can consume snags and other large woody debris, and also create them by killing live trees. The balance of these 2 processes, in concert with other sources of mortality and natality (e.g., growth rates, insects, disease, silvicultural practices), drives snag dynamics and influences snag-dependent wildlife over time (Fig. 8). After

TABLE 3. Mean numbers of breeding birds on combined burned versus unburned transect ($N = 7$ censuses/treatment/year). Reprinted from Bock and Bock (1983).

	$\bar{x}$ number of birds/census			
	1980		1981	
Species	Burned	Unburned	Burned	Unburned
Mourning dove (*Zenaida macroura*)	0.3	0.4	1.9	1.3
Northern flicker (*Colaptes auratus*)	0.3		0.3	0.1
Hairy woodpecker (*Picoides villosus*)	0.1		0.1	0.1
Western wood-pewee (*Contopus sordidulus*)			0.1	0.1
Gray jay (*Perisoreus canadensis*)		0.7	0.1	0.7
Black-capped chickadee (*Parus atricapillus*)	1.1	1.3	1.1	0.9
White-breasted nuthatch (*Sitta carolinensis*)	0.4		0.1	0.1
Red-breasted nuthatch (*S. canadensis*)	3.4	3.1	0.4	2.0*
Brown creeper (*Certhia americana*)	0.7	0.3		
American robin (*Turdus migratorius*)	10.3	7.6	4.6	10.0*
Townsend's solitaire (*Myadestes townsendi*)			0.6	0.6
Mountain bluebird (*Sialia currucoides*)	1.9*		0.6	0.4
Solitary vireo (*Vireo solitarius*)	2.9*	1.0	2.1	1.1
Ovenbird (*Seiurus aurocapillus*)	0.6	0.3		0.6
Yellow-rumped warbler (*Dendroica coronata*)	6.7*	2.3	4.3	2.3
Western meadowlark (*Sturnella neglecta*)		0.3		0.6
Brown-headed cowbird (*Molothrus ater*)	2.3	0.9	1.3	1.4
Western tanager (*Piranga ludoviciana*)	4.1*	2.3	2.0	2.0
Red crossbill (*Loxia curvirostra*)			1.3	1.6
Grasshopper sparrow (*Ammodramus savannarum*)		0.4		
Vesper sparrow (*Pooecetes gramineus*)	0.1	0.3	1.3*	0.4
Dark-eyed junco (*Junco hyemalis*)	4.9*	1.7	3.6	1.4
Chipping sparrow (*Spizella passerina*)	18.4*	10.9	8.7	9.0
Totals	58.7*	33.7	34.6	36.9

*Greater ($P < 0.05$) than other treatment within year, Mann-Whitney U test.

accounting for fire-induced mortality and natality in ponderosa pine, Horton and Mannan (1988) found that density of large snags (>30 cm dbh) was reduced 50% and that density of mid-size snags (15-30 cm dbh) was reduced 34%, while small snags (<15 cm dbh) increased 20-fold. They observed no short-term changes in avifauna after single burn events that had followed 65 years of fire exclusion, and speculated that regular prescribed fire would not markedly affect bird communities. Other scenarios are plausible, however, depending on the subsequent fire regime's influence on snag dynamics. Snag longevity decreases as some function of size (Bull 1983, Lyon 1977, Rasmussen and Ffolliott 1983), and natality of small snags could be reduced if the FFI between subsequent burns were shortened. Low natality and survival of small snags would result in decreased snag abundance. At the same time, densities of larger snags could be progressively reduced, assuming that surviving trees were to become relatively fire resistant over time and that subsequent ground fires would "kill" existing snags at a constant rate. Thus, depending on the frequency and intensity of prescribed fires, the total density of snags might be reduced and the skewness and kurtosis of their size distribution altered over time. Long-term responses of bird communities to such changes remain unquantified. Some logging regimes apparently can mimic the effects of some fire regimes on avifauna (Blake 1982), and coupled with slash burning and fire control can be an effective alternative to broadcast burning where snag-dependent wildlife are a concern in managed forests.

Prescribed fire can be used to maintain understory vegetation within limits preferred by red-cockaded woodpeckers (*Picoides borealis*)(Hopkins and Lynn 1971). However, red-cockaded woodpeckers almost always excavate their nest cavities in live trees, which develop sap flows in response to the boring activity of the birds. Sap flows increase susceptibility of the tree to fire, and Conner (1979) found it was necessary to rake fine fuels away from nest trees and install fire lines around colony sites. These activities reduce risk, but increase project costs.

The vulnerability of small mammals to predation increases temporarily as a result of reduced protective cover in burned areas (Crowner and Barrett 1979). The rapidity with which small mammals recolonize disturbed areas depends on several factors including the presence of refugia, and the size and severity of the disturbance (Crawley 1983, Kaufman et al. 1990, Archer and Pyke 1991). Kaufman et al. (1990) categorized small mammal species as fire-neutral, fire-negative, or fire-positive, and placed most species in the fire-negative category. Neg-

FIGURE 8. Fire frequency and intensity influence snag natality and mortality. The balance of these processes determine the temporal dynamics of snag density and size distribution. Photograph courtesy of U.S. National Park Service, Yellowstone National Park.

ative responses are typical of species that nest and/or forage in relatively dense vegetation or in litter (microtine rodents, harvest mice, cotton rats and woodrats), while positive responses are more probable for species that forage in relatively open habitat (heteromyid and zapodid rodents, peromyscine rodents, grasshopper mice, and ground squirrels). Species richness is unlikely to be altered much by fire, but the relative abundance of species can change. Duration of fire impacts can be influenced to some extent by the timing and severity of burning, but is typically short because of the rapid revegetation of burned areas and rapid reproduction by small mammal populations.

THE ECONOMICS OF PRESCRIBED FIRE

Economics are increasingly important to wildlife managers. Consider the situation of a wildlife biologist budgeting $25,000 for habitat improvement projects. How should the money be spent? On habitat for game or nongame species? On fire, herbicides, mechanical treatments, or grazing manipulation? Moreover, suppose this biologist chooses prescribed fire; on what basis should specific fire prescriptions be selected? Treat 1 large area or several small ones? Conduct cool underburns to improve forage quality for a few years or hotter burns to alter botanical composition for perhaps several decades? Suppose a year has passed since the $25,000 was received: how can the biologist demonstrate that money was well spent? Consider also the plight of agency administrators, and private foundations, that are inundated each year with proposals for cooperative funding of burning projects. How do they compare the merits of competing proposals and evaluate the effectiveness of their contributions? Applying conventional economic techniques offers some insight into these questions, particularly when economists and biologists collaborate.

This section addresses some of the economic issues that should be addressed when using prescribed fire in wildlife management settings. It is not designed to provide all of the requisite background, but presumes the reader can either perform basic economic calculations or works in an inter-disciplinary environment with someone who can. Texts such as Workman (1986) can help prepare the reader in this regard.

A Theoretical Framework

The expenses associated with prescribed fire should be viewed as investments in wildlife production. The most useful theoretical approach in this area is bioeconomics, which combines biological processes with economic production modeling (Clark 1976, Loomis et al. 1991). Habitat manipulation, when viewed from an ecological perspective, affects what economists refer to as a vertically integrated production process (Fig. 9). Such processes take outputs from one level, and use them in the next. Fire enters this model as an investment into the production process at the soil and plant community levels, in an attempt to ultimately enhance the human value derived from the ecosystem (Fig. 10). Even though the goal is to achieve greater human value, applying fire to the ecosystem does not directly produce that value, except perhaps through improved visual quality. Rather its effects on human value are indirect: processed first through the web of plant responses, then to changes in nutrient balance of herbivores, and thus changes in the size, composition, or growth rate of animal populations.

Thus, economic analysis of prescribed fire as a wildlife management tool requires knowledge in 3 areas: the costs of fire prescriptions (inputs), the effectiveness of fire in wildlife management (production processes), and the valuation of resulting wildlife changes (outputs).

Costs of Prescribed Fire. Cost is important because of obvious budget considerations, but also because any investment should be made in the most cost-efficient manner. Compared to either effectiveness or valuation of benefits, estimating the costs of management though prescribed fire is relatively easy. The major input in prescribed fire is generally labor for planning and on-site fire management, with minor needs for equipment and travel. Aerial ignition is an exception; equipment costs typically predominate in this case.

An important cost consideration is the relationship between cost/ha and project size, which is influenced by fixed-costs.

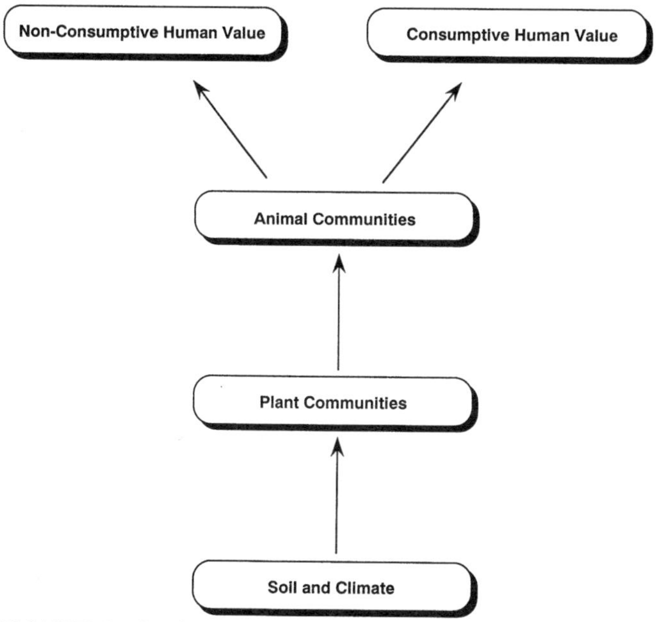

FIGURE 9. A simple ecological view of human use of wildlife.

Expenses that do not vary as a function of project size are fixed costs. When these are relatively high, the total of all costs per hectare is higher for small projects than for large ones. To illustrate this point we plotted average nominal cost/ha against project size for a random sample (n = 33) of 162 burning projects conducted across the West by public agencies, with financial help from the Rocky Mountain Elk Foundation (RMEF), for the period November 1985 through April 1993 (Fig. 11). Average unit cost for the sample projects is \$65/ha, but drops substantially as project size increases. Our sample of the smallest projects (≤ 125 ha, $n = 11$) had average costs of \$124/ha; somewhat larger projects (125-300 ha, $n = 12$) had average costs of \$43/ha; and the largest projects (> 300 ha, $n = 10$) had average costs of only \$25/ha. These results show that a given amount of money treats more area when spent on a single large project than on several small ones, most likely because fixed costs are spread across more area.

Comparing these results across the size distribution of all 162 projects (Fig. 6) raises an interesting point. Apparently, the RMEF's cooperators have focused their burning efforts on more small projects than large ones, but they could have treated substantially more area for the same total cost by focusing on fewer, but larger, projects (Fig. 11). There are a number of reasons why the RMEF and its cooperators (typically agencies such as the Forest Service, Bureau of Land Management, and state or provincial wildlife agencies) have conducted more small burns than large ones: they may be less controversial, they may be the only choices the cooperators brought to the RMEF in some areas, or they may be a more appropriate size for local RMEF chapters to participate in. These are all valid reasons. Certainly average unit cost is not the only important factor in project design (economic theory would focus on marginal cost also). Moreover, whether a single large project is preferable to several small ones could also depend on the relative effectiveness of the

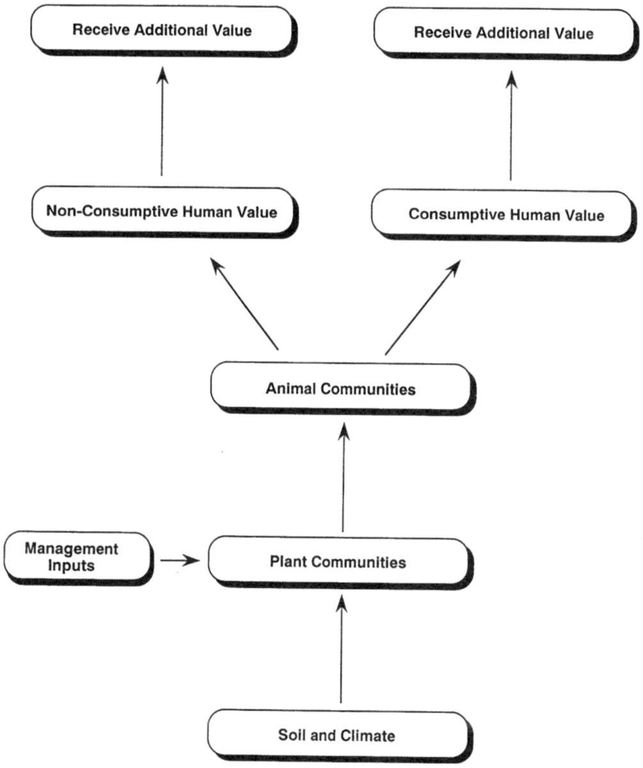

FIGURE 10. A simple bioeconomic model of habitat management. In this model fire is an investment in the vertically integrated production process.

treatments. All other factors being equal, however, large projects are economically more efficient.

Effectiveness of Fire in Habitat Management. Effectiveness refers to the results of any prescribed burning project, in terms of change in habitat and animal populations. Understanding effectiveness requires a more thorough understanding of biological processes than does cost or output valuation. Three of the most difficult considerations regarding the evaluation of the effectiveness of prescribed burning have to do with limiting factors, timing, and risk.

Habitat improvement is called for when habitat conditions are limiting population size, productivity, or distribution. A population's response to a prescribed burn is not simply a direct function of the area burned or of forage response within the burn area, but is also influenced by animal density relative to forage benefits, and further complicated by factors unrelated to the treatment (e.g., weather, predation, reproductive potential). Evaluating the effectiveness of prescribed burning therefore requires a thorough understanding of all the factors limiting the target wildlife population(s).

Timing is a crucial concept in investment analysis and involves 2 separate issues: time's effect on the value of the flow of benefits from the fire treatment, and its effect on the quantity of benefits themselves. The effect of time on the value of benefits embodies the economic concept of time preference, which is handled through the well-accepted process of discounting (Sugden and Williams 1978, Mishan 1976). The second timing issue is that any fire-induced change in vegetation will have nei-

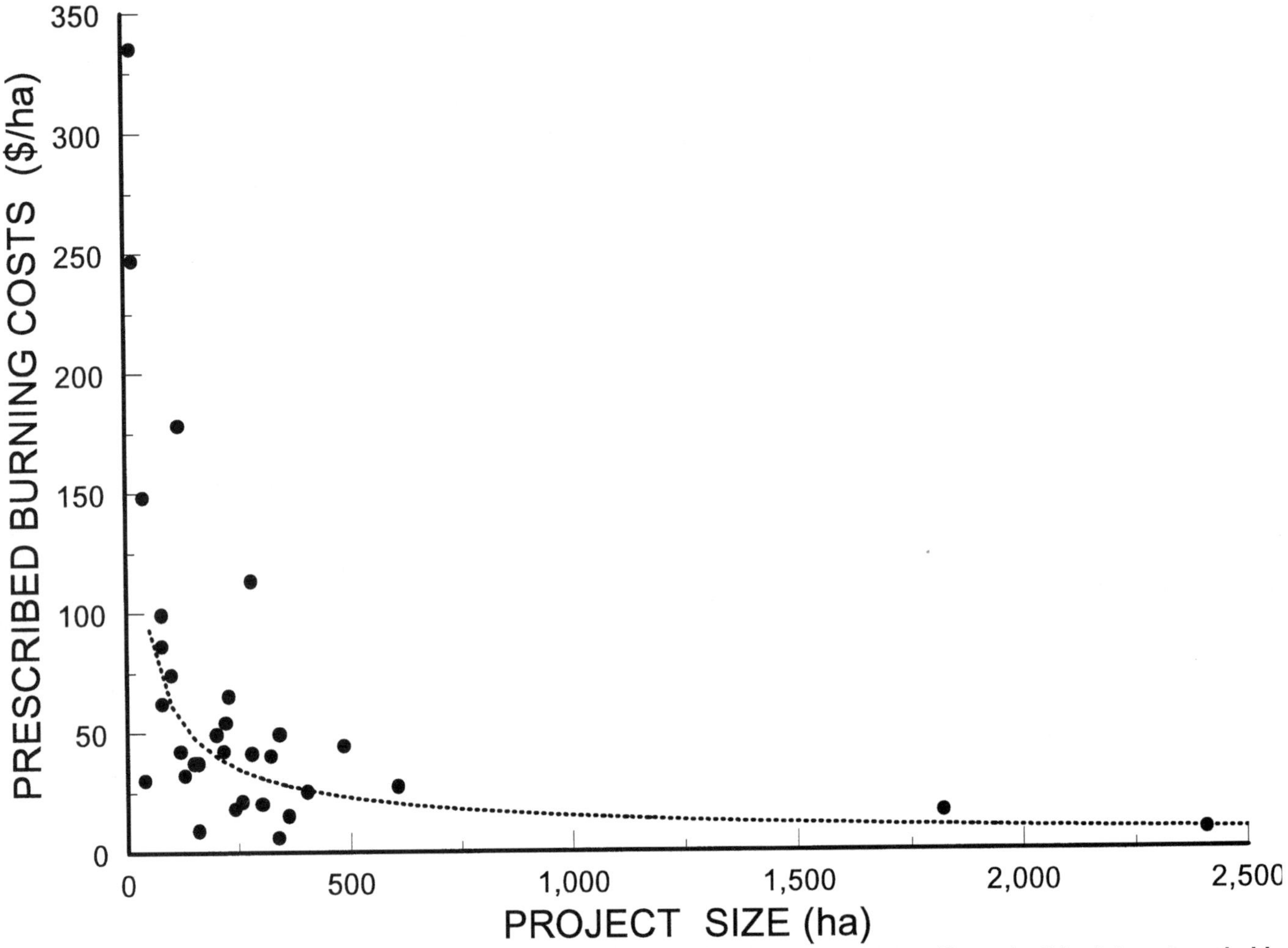

FIGURE 11. The relationship between project size and unit cost ($/ha) is strongly curvilinear in this data set, probably because of fixed-cost effects. Focusing management efforts on large-scale projects can substantially enhance economic efficiency of prescribed burning programs. Data courtesy of the Rocky Mountain Elk Foundation.

ther a permanent nor unvarying effect on wildlife populations. More commonly, there will be a "lifecycle" for the project's effectiveness, that is reasonably predictable based on the particular species' adaptations to fire and the resultant changes in animal and plant populations. Responses such as reproductive performance will usually be most pronounced during the first few years following a burn, but wane as species composition and forage quality return to pre-burn conditions and as herbivore populations grow and fill the regenerating habitat to become subject to density-dependent constraints. In other instances, vegetation changes will be most effective some years after the burn. For example, Lyon (1971) projected peak post-fire development at 20 years for shrubs in a Douglas-fir habitat, while Wittinger et al. (1977) recorded peak post-fire development of tall shrubs in a hemlock (*Tsuga heterophylla*) habitat at 3-5 years. As we have seen earlier, the value of burning to herbivores is a function of the longevity of the production and the composition response, but also the forage quality response. Whereas production and composition responses may be quite long-lived, forage quality responses may be short-lived. This may intensify the skewness of the temporal response to the early years, even though the apparent project life may be substantially longer. All other factors being equal, comparative economic analysis of competing projects will generally favor those having benefits concentrated in the early years.

Unfortunately, when there is not adequate information about the biological lifecycle of habitat improvements, it is very difficult to evaluate economic effectiveness of habitat improvement projects. When the likely path is uncertain, testing the sensitivity of results to changes in the assumed path can be easily undertaken and is worth the additional effort.

Risk is an important consideration when one recognizes that applying prescribed fire in wildlife management is not a precise science. Anyone who has walked through a burned area has seen the seemingly random ways that fire can move across a landscape, consuming vegetation some places, yet leaving it elsewhere. Likewise, anyone who has tried to control or manage fire knows that the best of intentions and planning sometimes go awry, particularly when weather changes unexpectedly. Prior assumptions about the responses of wildlife populations to individual burns can be subject to a great deal of risk, not just because of uncertainty involved with the act of burning but because of ambiguity in biologists' understanding of limiting factors and animal-plant interactions. Consequently, projected economic evaluations of prescribed fires should not be confined to merely deterministic analyses, in which the outcome is assumed to be guaranteed before the project is undertaken. Rather it is more realistic to expect a distribution of likely outcomes from any burning project, and to then use that distribution to construct the likely range of wildlife responses and ultimately that of human values.

Risk is typically handled in investment analyses using the *expected value* method, in which outcomes are weighted by their probability of occurrence. Expected value calculations are not difficult, if the frequency distribution of outcomes is available. When this information is lacking, or does not get conveyed to the economist, the analysis suffers. This is yet another situation where the biological and economic sciences must work together, and where decision makers can benefit from sensitivity analyses. One of the most useful introductory discussions of risk in prescribed fire management is Cleaves and Brodie (1990), which focuses on forest settings.

Valuation of Wildlife Populations. Estimating the value of a change in the size or composition of a wildlife population is the final vexing problem that must be analyzed. The question of what wildlife is worth dates back to the beginning of hunting and fishing. To prehistoric peoples, that worth would have been denominated in caloric balance or perhaps in terms of trade. In an advanced modern society, on the other hand, both the consumptive and nonconsumptive uses of wildlife go far beyond subsistence, and wildlife valuation must therefore extend beyond food or exchange values. Several different measures of value have been put forth for wildlife resources, primarily license receipts, expenditures, net welfare, and intrinsic value. Each of these different measures of value is theoretically correct and may be useful in analyzing the benefits of burning programs, depending on the objective for the investment and the agent undertaking it.

License receipts would be the appropriate value for wildlife if the managing agency had the mandate to maximize the return to the treasury through wildlife use. The agency would therefore take part of the previous year's receipts to fund habitat improvements with the expectation that the increase in future license revenues would more than compensate for the investment. We know of no agency with this charge, and thus do not advocate the use of license receipts as the value of a wildlife population on public lands. The private analog to license receipts is access fees or leases, and these are obviously important to private landowners who are considering prescribed fire as an investment in their operations, which may include fee hunting or wildlife viewing programs.

Expenditure data on hunting and fishing are often viewed as a value of wildlife populations. Clearly, hunting and fishing constitute economic activities created by wildlife resources, but the expenditures associated with them are only a portion of the welfare that is created, and thus may often be an inadequate measure with which to analyze habitat improvement projects. Using expenditures as a measure of wildlife value has some methodological problems also, and has been widely criticized (Knetsch 1977, Sorg and Loomis 1984, McDivitt 1987, Steinhoff et al. 1987). Nevertheless, an expenditures criteria would be correct in the special case of an economic development agency or a chamber of commerce that might invest some money in habitat improvements, like prescribed burning, for the purpose of increasing local economic activity. In that case, their management goal is not the ecological welfare of the wildlife, or benefit to the public; they are interested solely in the portion of the public's benefit that can be extracted by local businesses.

Most public agencies are required to produce value to the public (often with a fairly vague notion of value) while protecting the habitat and the associated animal populations. In this case, the appropriate measure of value is net welfare, which can include all recreational uses as well as nonuse values such as existence and option value. Net welfare is by far the most commonly used measure of value in wildlife valuation research, and in many ways is theoretically most attractive (Steinhoff et al. 1987).

A pervasive philosophical position among lay people and even some professionals is that wildlands, and the associated plants and animals, have some value beyond what can be captured by monetary measures. Moreover, this value is completely independent of any human use, rather it is intrinsic to the ecosystem and the organisms that comprise it. In this context, fire is allowed to burn or is prescribed because it is part of the natural ecosystem and creates outcomes that have worth entirely independent of human enjoyment. How does economics deal with the concept of intrinsic value? The most honest response to this question is "Not ever well. Yet." The emerging discipline of ecological economics offers promise (Costanza 1991, Daly et al. 1994), but no widely accepted methods or results have developed from it. In the meantime, economics is perhaps not the best discipline with which to evaluate intrinsic value. Given that there may not be a strictly rational economic solution to almost any land management issue, debates often move into the political arena. There, however, arguments of intrinsic value run into difficulty because competing intrinsic values are often pitted against each other. Thus intrinsic values are, at best, a risky basis upon which to justify management programs and expenditures.

The question of which economic measure to use is significant; one's choice can determine the apparent value of habitat improvements. For additional discussion of these issues, consider Lenarz (1987), which used licenses and expenditures to evaluate the practice of creating forest openings to improve white-tailed deer habitat, and Daniels and Riggs' (1988) subsequent analysis of the same problem using a welfare criteria. Lenarz (1987) found that created openings were never cost effective using a licenses value, but always were cost effective given an expenditure value; Daniels and Riggs (1988) found that results were more variable, depending on the discount rate used and assumptions about the deer population's temporal response to the habitat improvement.

Two comments summarize this section. First, economic analyses range from the back-of-the-envelope level to sophisticated series of simultaneous nonlinear equations. There is no "right" economic analysis for all situations; the various levels of sophistication have merit to the extent they match the characteristics of the situation at hand and are compatible with analytical goals. Our review of prescribed burning costs is very simple, but it raises some interesting questions regarding the size and cost effectiveness of projects proposed to RMEF and other foundations by competing agencies each year. It also illustrates that economic analysis can enhance the benefits of wildlife management in a fashion analogous to how wildlife management itself enhances the benefits from wildlife populations. Second, economic analyses of prescribed burning projects collectively comprise an area of overlap between economics and wildlife biology. There is arguably no better way to conduct such analyses than through the joint efforts of economists and biologists. It is the rare person who is well versed in the details of wildlife biology and the mechanics of economic analysis. Integrating both perspectives improves on each, and compromises neither.

SUMMARY

Most North American vegetation has evolved and developed to some extent under the influence of fire. Many types of vegetation, especially those of rangelands, are fire-adapted and often dependent upon recurrent burning for their maintenance. Consequently, prescribed fire is a management tool that is potentially beneficial, if not requisite, to enlightened management of rangelands and the wildlife they support.

Prescribed fire is not, however, a panacea, which if practiced indiscriminately ensures the maintenance of "natural" processes or "healthy" wildlife populations or ecosystems. Effects of fire are a function of relationships between fire regime, vegetation, and animal populations, and variance in any one of these factors can influence the outcome of a burning program. The extent to which burning changes plant communities is largely a function of the time elapsed between burns and the severity of the fire. The extent to which prescribed fire influences animal populations is largely a function of the scale, pattern, and duration of vegetation change, the last of these being a function of the interdependent processes of secondary plant succession and animal response. Animal responses involve habitat use, herbivory, and population growth. In some settings the application of prescribed fire, and projection of its effects, may be straightforward. In others, however, interactions between the dominant factors can make prescribed burning an inexact and sometimes risky business, particularly if the practitioner has vague goals, shallow understanding of fire-vegetation-wildlife relations, or incomplete knowledge of the factors that limit wildlife populations.

Because there are risks involved, the practice of prescribed burning to fulfill specific management goals can be a far more challenging endeavor than burning to fulfill some vague notion of "ecosystem management." Where specific goals are stated, these become criteria by which management may be evaluated, held accountable for resources expended, and refined to improve effectiveness and efficiency in the future. Prescribed fire is an economic investment in wildlife management and production just as are other management practices such as grazing, herbicides, and mechanical treatments of rangelands and forests. In this context, individual burning projects and long-term programs should be subjected to economic analysis, in an interdisciplinary context, where the results can benefit from the combined expertise of land managers, wildlife biologists, and economists.

LITERATURE CITED

Agee, J. K. 1991. Fire history along an elevational gradient in the Siskiyou Mountains, Oregon. Northwest Sci. 65:188-199.

Agnew, W., D. W. Uresk, and R. M. Hansen. 1986. Flora and fauna associated with prairie dog colonies and adjacent ungrazed mixed-grass prairie in western South Dakota. J. Range Manage. 39:135-139.

Angell, R. F., J. W. Stuth, and D. L. Drawe. 1986. Diets and liveweight changes of cattle grazing fall burned gulf cordgrass. J. Range Manage. 39:233-236.

Archer, S., and D. A. Pyke. 1991. Plant-animal interactions affecting plant establishment and persistence on revegetated rangeland. J. Range Manage. 44:558-565.

Arno, S. F. 1976. The historical role of fire on the Bitterroot National Forest. U.S. Dep. Agric. For. Serv. Gen. Tech. Rep. INT-187.

———. 1980. Forest fire history in the northern Rockies. J. For. 78:460-465.

Arnold, G. W. 1985. Regulation of forage intake. Pages 81-101 *in* R. J. Hudson and R. G. White, eds. Bioenergetics of wild herbivores. CRC Press, Boca Raton, Fla.

Arnold, J. F., D. A. Jameson, and E. H. Reid. 1964. The pinyon-juniper type of Arizona: Effects of grazing, fire and tree control. U.S. Dep. Agric. Prod. Res. Pap. 84.

Asherin, D. A. 1973. Prescribed burning effects on nutrition, production, and big game use of key northern Idaho browse species. Ph.D. Thesis, Univ. Idaho, Moscow. 96pp.

Austin, D. D., P. J. Urness, and L. C. Fierro. 1983. Spring livestock grazing affects crested wheatgrass regrowth and winter use by mule deer. J. Range Manage. 36:589-593.

———, ———, and R. A. Riggs. 1986. Vegetal change in the absence of livestock grazing, Mountain Brush Zone, Utah. J. Range Manage. 39:514-517.

Bartos, D. A., J. K. Brown, and G. D. Booth. 1994. Twelve years biomass response in aspen communities following fire. J. Range Manage. 47:79-93.

Bendell, J. F. 1974. Effects of fire on birds and mammals. Pages 73-138 *in* T. T. Kozlowski, and C. F. Ahlgren, eds. Fire and ecosystems. Academic Press, New York, N.Y.

Bentz, J. A., and P. M. Woodard. 1988. Vegetation characteristics and bighorn sheep use on burned and unburned areas in Alberta. Wildl. Soc. Bull. 16:186-193.

Best, L. B. 1972. First-year effects of sagebrush control on two sparrows. J. Wildl. Manage. 36:534-544.

Biswell, H. H. 1972. Fire ecology in ponderosa pine-grassland. Proc. Tall Timbers Fire Ecol. Conf. 12:69-96.

Black, J. L., and P. A. Kenney. 1984. Factors affecting diet selection by sheep: II. Height and density of pasture. Aust. J. Agric. Res. 35:565-578.

Blake, J. G. 1982. Influence of fire and logging on nonbreeding bird communities of ponderosa pine forests. J. Wildl. Manage 46:404-415.

Blaxter, K. L., F. W. Wainman, and R. R. Wilson. 1961. The regulation of food intake by sheep. Anim. Prod. 3:51-61.

Bock, C. E., and J. H. Bock. 1983. Responses of birds and deer mice to prescribed burning in Ponderosa Pine. J. Wildl. Manage. 47:836-840.

Bock, C. E., and J. F. Lynch. 1970. Breeding bird populations of burned and unburned conifer forests in the Sierra Nevada. Condor 72:182-189.

Bohn, C., C. Galen, C. Maser, and J. W. Thomas. 1980. Homesteads—manmade avian habitats in the rangelands of southeastern Oregon. Wildl. Soc. Bull. 8:332-341.

Bojorquez Tapia, L. A., P. F. Ffolliott, and D. P. Guertin. 1990. Herbage production-forest overstory relationships in two Arizona ponderosa pine forests. J. Range Manage. 43:25-28.

Bryant, J. P., and P. J. Kuropat. 1980. Selection of winter forage by subarctic browsing vertebrates: the role of plant chemistry. Annu. Rev. Ecol. Syst. 11:261-285.

Bull, E. L. 1983. Longevity of snags and their use by woodpeckers. Pages 64-67 *in* J. W. Davis, G. A. Goodwin, and R. A. Ockenfels, tech. coords. Snag habitat management. U.S. Dep. Agric. For. Serv. Gen. Tech. Rep. RM-99.

———. 1987. Ecology of pileated woodpecker in northeastern Oregon. J. Wildl. Manage. 51:472-481.

Bunnell, F. L., and M. P. Gillingham. 1985. Foraging behavior: Dynamics of dining out. Pages 53-80 *in* R. J. Hudson and R. G. White, eds. Bioenergetics of wild herbivores. CRC Press Inc., Boca Raton, Fla.

Bunting, S. C. 1987. Use of prescribed burning in juniper and pinyon juniper woodlands. Pages 141-144 *in* R. L. Everett, Compiler. Proc. Pinyon-juniper Conf. U.S. Dep. Agric. For. Serv. Gen. Tech. Rep. INT-215.

———, B. M. Kilgore, and C. L. Bushey. 1987. Guidelines for prescribed burning sagebrush-grass rangelands in the northern Great Basin. U.S. Dep. Agric. For. Serv. Gen. Tech. Rep. INT-231.

———, L. F. Neuenschwander, and G. E. Gruell. 1985. Fire ecology of antelope bitterbrush in the northern Rocky Mountains. Pages 48-57 *in* J. E. Lotan and J. K. Brown, compilers. Fire's effect on wildlife habitat, Proceedings of a symposium. U.S. Dep. Agric. For. Serv. Gen. Tech. Rep. INT-186.

Burkhardt, J. W., and E. T. Tisdale. 1976. Causes of juniper invasion in southwestern Idaho. Ecology 57:472-484.

Butler, D. R. 1986. Conifer invasion of subalpine meadows, central Lemhi Mountains, Idaho. Northwest Sci. 60:166-173.

Butorff, R. L. 1974. Cottonwood habitat for birds in Colorado. Am. Birds 28:975-979.

Byrne, R. 1978. Fossil record discloses wildfire history. Calif. Agric. 10:13-14.

Campbell, B. H., and M. Hinkes. 1983. Winter diets and habitat use of Alaska bison after wildfire. Wildl. Soc. Bull. 11:16-21.

Campling, R. C., M. Freer, and C. C. Balch. 1962. Factors affecting the voluntary intake of food by cows. 3. the effect of urea on the voluntary intake of oat straw. Brit. J. Nutr. 16:115-124.

Canon, S. K., P. J. Urness, and N. V. DeByle. 1987. Habitat selection, foraging behavior, and dietary nutrition of elk in burned aspen forest. J. Range Manage. 49:433-437.

Carothers, S. W., and R. R. Johnson. 1975. Water management practices and their effects on non-game birds in range habitats. Pages 190-209 *in* D. R. Smith, tech. coord. Proc symposium on management of forest and range habitats for nongame birds. U.S. Dep. Agric. For. Serv. Gen. Tech. Rep. GTR-WO-1.

Carpenter, L. H., O. C. Wallmo, and R. B. Gill. 1979. Forage diversity and dietary selection by wintering mule deer. J. Range Manage. 32:226-229.

Castrale, J. S. 1982. Effects of two sagebrush control methods on nongame birds. J. Wildl. Manage. 46:945-952.

Cattelino, P. J., I. R. Noble, R. O. Slatyer, and S. R. Kessell. 1979. Predicting multiple pathways of plant succession. Environ. Manage. 3:41-50.

Chandler, C., P. Cheney, P. Thomas, L. Trabuad, and D. Williams. 1983. Fire in, forestry volume I: Forest fire behavior and effects. John Wiley & Sons, New York, N.Y. 450pp.

Chew, R. M., B. B. Butterworth, and R. Grechman. 1959. The effects of fire on the small mammal populations of chaparral. J. Mammal. 40:253.

Christensen, N. L. 1981. Fire regimes in southeastern ecosystems. Pages 112-136 *in* H. A. Mooney, T. M. Bonnicksen, N. L. Christensen, J. E. Lotan, and W. A. Reiners, tech. coords. Fire regimes and ecosystem properties. U.S. Dep. Agric. For. Serv. Gen. Tech. Rep. WO-26.

Cid, M. S., J. K. Detling, A. P. Whicker, and M. A. Brizuela. 1991. Vegetational responses of a mixed-grass prairie site following exclusion of prairie dogs and bison. J. Range Manage. 44:100-104.

Clark, C.W. 1976. Mathematical bioeconomics. John Wiley & Sons, New York, N.Y. 352pp.

Cleaves, D. A., and J. D. Brodie. 1990. Economic analysis of prescribed burning. Pages 271-281 *in* J. D. Walstad, S. R. Radosevich, and D. V. Sandberg, eds. Natural and prescribed fire in Pacific Northwest forests. Oregon State Univ. Press, Corvallis.

Clements, F. 1935. Experimental ecology in the public service. Ecology 16:342-363.

Clutton-Brock, T. H., and P. H. Harvey. 1983. The functional significance of variation in body size among mammals. Pages 632-663 *in* J. F. Eisenberg and D. G. Kleiman, eds. Advances in the study of animal behavior. Am. Soc. Mammal. Spec. Pub. 7.

Collins, W. B., and P. J. Urness. 1983. Feeding behavior and habitat selection of mule deer and elk on northern Utah summer range. J. Wildl. Manage. 47:646-663.

Connelly, J. W., W. L. Wakkinen, A. D. Apa, and K. P. Reese. 1991. Sage grouse use of nest sites in southeastern Idaho. J. Wildl. Manage. 55:521-524.

Conner, R. N. 1979. Effects of a prescribed burn on cavity trees of red-cockaded woodpeckers. Wildl. Soc. Bull. 7:291-293.

Cook, C. W., and L. E. Harris. 1950. The nutritive value of range forage as affected by vegetation type, site, and state of maturity. Ut. Agric. Exp. Stn. Tech. Bul. 344. 45pp.

Cook, J. G., E. B. Arnett, L. L. Irwin, and F. G. Lindzey. 1990. Mountain sheep response to burning in southcentral Wyoming. Dep. Zool. and Physiol., Univ. Wyoming, Laramie. 83pp.

———, T. J. Hershey, and L. L. Irwin. 1994. Vegetation response to burning on Wyoming mountain-shrub big game ranges. J. Range Manage. 47:296-302.

Cooper, S. M., and N. Owen-Smith. 1986. Effects of plant spinescence on large mammalian herbivores. Oecologia. 68:446-455.

Coppock, D. L., and J. K. Detling. 1986. Alteration of bison and black-tailed prairie dog grazing interaction by prescribed burning. J. Wildl. Manage. 50:452-455.

———, ———, J. E. Ellis, and M. I. Dyer. 1983*a*. Plant-herbivore interactions in a North American mixed-grass prairie I. Effects of black-tailed prairie dogs on interseasonal aboveground plant biomass and nutrient dynamics and plant species diversity. Oecologia 56:1-9.

———, J. E. Ellis, J. K. Detling, and M. I. Dyer. 1983*b*. Plant-herbivore interactions in a North American mixed-grass prairie II. Responses of bison to modification of vegetation by prairie dogs. Oecologia 56:10-15.

Costanza, R. 1991. Ecological economics: the science and management of sustainability. Columbia Univ. Press, New York, N.Y. 555pp.

Courtney, R. F. 1989. Pronghorn use of recently burned mixed prairie in Alberta. J. Wildl. Manage. 53:302-305.

Crawley, M. J. 1983. Herbivory. Univ. of California Press, Berkeley and Los Angeles. 437pp.

Crowner, A. W., and G. W. Barrett. 1979. Effects of fire on the small mammal component of an experimental grassland community. J. Mammal. 60:803-813.

Daly, H. E., J. B. Cobb, and C. W. Cobb. 1994. For the common good: redirecting the economy toward community, the environment, and a sustainable future. Second ed. Beacon Press, Boston, Mass. 544pp.

Daniels, S. E., and R. A. Riggs. 1988. Improving economic analysis of habitat management. Wildl. Soc. Bull. 16:452-457.

Daubenmire, R. 1968. Ecology of fire in grasslands. Adv. Ecol. Res. 5:209-266.

Davis, K. M., B. D. Clayton, and W. C. Fischer. 1980. Fire ecology of Lolo National Forest habitat types. U.S. Dep. Agric. For. Serv. Gen. Tech. Rep. INT-79. 77p.

Demment, M., and P. J. Van Soest. 1985. A nutritional explanation for body-size patterns of ruminant and nonruminant herbivores. Am. Nat. 125:641-672.

Dewitt, J. B., and J. V. Derby. 1955. Changes in nutritive value of browse plants following forest fires. J. Wildl. Manage. 19:65-70.

Dick, B. J., and P. J. Urness. 1991. Nutritional value of fresh Gambel oak browse for Spanish goats. J. Range Manage. 44:361-364.

Dickson, J. G., R. N. Conner, and J. H. Williamson. 1983. Snag retention increases bird use of a clear-cut. J. Wildl. Manage. 47:799-804.

Dunwiddie, P. W. 1977. Recent tree invasion of subalpine meadows in the Wind River Mountains, Wyoming. Arctic and Alpine Res. 9:393-399.

Edgerton, P. J. 1985. Influence of ungulates on the development of the shrub understory of an upper slope mixed conifer forest. Pages 162-167 *in* F. D. Provenza, J. T. Flinders, and E. D. McArthur, compilers. Proc. of a symposium on plant-herbivore interactions. U.S. Dep. Agric. For. Serv. Gen. Tech. Rep. INT-222.

Einarsen, A. S. 1946. Management of black-tailed deer. J. Wildl. Manage. 10:54-59.

Elliott, R. C., and J. H. Topps. 1963. Voluntary intake of low protein diets by sheep. Anim. Prod. 5:269-276

Emlen, J. T. 1970. Habitat selection by birds following a forest fire. Ecology 51:343-345.

Ewel, K. C., and W. J. Mitsch. 1978. The effects of fire on species composition in cyprus dome ecosystems. Fla. Sci. 41:25-31.

Fitzgerald, S. M., and G. W. Tanner. 1992. Avian community response to fire and mechanical shrub control in south Florida. J. Range Manage. 45:396-400.

Franklin, J. F. 1988. Pacific Northwest forests. Pages 103-130 *in* M. G. Barbour and W. D. Billings, eds. North American terrestrial vegetation, Cambridge Univ. Press, Cambridge, U.K.

———, W. H. Moir, G. W. Douglas, and C. Wiberg. 1971. Invasion of subalpine meadows by trees in the Cascade Range, Washington and Oregon. Arctic and Alpine Res. 3:215-224.

Gaffney, W. S. 1941. The effects of winter elk browsing, South fork of the Flathead River, Montana. J. Wildl. Manage. 5:427-453.

Gasaway, W. C., and S. D. DuBois. 1985. Initial response of moose, *Alces alces*, to a wildfire in interior Alaska. Can. Field-Nat. 99:135-140.

Goering, H. K., P. J. Van Soest, and R. W. Hemken. 1973. Relative susceptibility of forages to heat damage as affected by moisture, temperature, and pH. J. Dairy Sci. 56:137-143.

Goto, I., and D. J. Minson. 1977. The potential digestibility of leaf and stem fractions in grasses. J. Agric. Sci. 89:143-149.

Gregg, M. A., J. A. Crawford, M. S. Drut, and A. K. DeLong. 1994. Vegetational cover and predation of sage grouse nests in Oregon. J. Wildl. Manage. 58:162-166.

Gruell, G. E. 1983. Fire and vegetative trends in the northern Rockies: interpretations from 1871-1982 photographs. U.S. Dep. Agric. For. Serv. Gen. Tech. Rep. INT-158. 117p.

———, 1986. Post-1900 mule deer irruptions in the Intermountain West: Principal cause and influences. U.S. Dep. Agric. For. Serv. Gen. Tech. Rep. INT-206.

———, J. K. Brown, and C. L. Bushey. 1986. Prescribed fire opportunities in grasslands invaded by Douglas-fir: state-of-the-art guidelines. U.S. Dep. Agric. For. Serv. Gen. Tech. Rep. INT-958.

Habeck, J. R. 1976. Forests, fuels and fire in the Selway-Bitterroot Wilderness, Idaho. Proc. Tall Timbers Fire Ecol. Conf. 14:305-353.

———, and R. W. Mutch. 1973. Fire-dependent forests in the northern Rocky Mountains. Quat. Res. 3:408-424.

Hall, F. C. 1976. Fire and vegetation in Blue Mountains—implications for land managers. Proc. Tall Timbers Fire Ecol. Conf. 15:155-170.

———. 1980. Fire history—Blue Mountains, Oregon. Pages 75-81 *in* M. A. Stokes and J. H. Dieterich, tech. coords. Proc. Fire History Workshop. U.S. Dep. Agric. For. Serv. Gen. Tech. Rep. RM-81.

Hanes, T. L. 1971. Succession after fire in the chaparral of southern California. Ecol. Monogr. 41:27-52.

Hanson, E. E. 1978. The impact of a prescribed burn in a temperate subalpine forest upon the breeding bird and small mammal populations. M.S. Thesis, Central Washington Univ., Cle Elum. 56pp.

Harniss, R. O., R. B. Murray. 1973. Thirty years of vegetation change following burning of sagebrush-grass range. J. Range Manage. 26:322-325.

Heinselman, M. L. 1978. Fire in wilderness ecosystems. Pages 248-278 *in* J. C. Hendee, G. H. Stankey, and R. C. Lucas, eds. Wilderness management. U.S. Dep. Agric. For. Serv. Misc. Publ. 1365.

———. 1981. Fire intensity and frequency as factors in the distribution and structure of northern ecosystems. Pages 7-57 *in* H. A. Mooney, T. M. Bonnicksen, N. L. Christensen, J. E. Lotan, and W. A. Reiners, tech. coords. Fire regimes and ecosystem properties. U.S. Dep. Agric. For. Serv. Gen. Tech. Rep. WO-26.

Hemstrom, M. A., and J. F. Franklin. 1982. Fire and other disturbances of the forest of Mount Rainier National Park. Quat. Res. 18:32-51.

Hilden, O. 1965. Habitat selection in birds: a review. Ann. Zool. Fenn. 2:53-75.

Hobbs, N. T., D. S. Schimel, C. E. Owensby, and D. S. Ojima. 1991. Fire and grazing in the tallgrass prairie: contingent effects on nitrogen budgets. Ecology 72:1374-1382.

———, and R. A. Spowart. 1984. Effects of prescribed fire on nutrition of mountain sheep and mule deer during winter and spring. J. Wildl. Manage. 48:551-560.

Hopkins, M. L., and T. E. Lynn, Jr. 1971. Some characteristics of red-cockaded woodpecker cavity trees and management implications in South Carolina. Pages 140-169 *in* R. L. Thompson, ed. The ecology and management of the red-cockaded woodpecker. Bur. Sport Fish and Wildl. and Tall Timbers Res. Stn., Tallahassee Fla.

Horton, S. P., and R. W. Mannan. 1988. Effects of prescribed fire on snags and cavity-nesting birds in southeastern Arizona pine forests. Wildl. Soc. Bull. 16:37-44.

Houston, D. B. 1973. Wildfires in northern Yellowstone National Park. Ecology 54:1111-1117.

Hudson, R. J., and M. T. Nietfeld. 1985. Effect of forage depletion on the feeding rate of wapiti. J. Range Manage. 38:80-82.

Humphrey, R. R. 1953. The desert grassland, past and present. J. Range Manage. 6:159-164.

Ivey, T. L., and M. K. Causey. 1984. Response of white-tailed deer to prescribed fire. Wildl. Soc. Bull. 12:138-141.

Jameson, D. A., E. H. Reid. 1965. The pinyon-juniper type of Arizona. J. Range Manage. 18:152-153.

Jarmen, P. J., and A. R. E. Sinclair. 1979. Feeding strategy and the pattern of resource partitioning in ungulates. Pages 130-163 *in* A. R. E. Sinclair and M. Norton-Griffiths, eds. Serengeti, dynamics of an ecosystem. Univ. Chicago Press, Chicago, Ill.

Jiang, Z., and R. J. Hudson. 1994. Bite characteristics of wapiti (*Cervus elaphus*) in seasonal *Bromus-Poa* swards. J. Range Manage. 47:127-132.

Johnson, A. H., and G. A. Smathers. 1976. Fire history and ecology, Lava Beds National Monument. Proc. Tall Timbers Fire Ecol. Conf. 15:103-115.

Jones, D. I. H., and R. W. Bailey. 1972. The hydrolysis of cell wall polysaccharides from freeze-dried and oven-dried herbage by rumen and mould carbohydrases. J. Sci. Food Agric. 23:609-614.

Jones, J. R., and N. V. DeByle. 1985. Fire. Pages 77-81 *in* N. V. DeByle and R. P. Winokur, eds. Aspen: ecology and management in the western United States. U.S. Dep. Agric. For. Serv. Gen. Tech. Rep. RM-119.

Jourdonnais, C. S., and D. J. Bedunah. 1990. Prescribed fire and cattle grazing on an elk winter range in Montana. Wildl. Soc. Bull. 18:232-240.

Kalmbacher, R. S. 1983. Total non-structural carbohydrates in burned and unburned creeping bluestem roots, rhizomes and lower stems. Proc. Soil and Crop Sci. Soc. Fla. 42:142-145.

———, F. G. Martin, W. S. Terry, D. H. Hunter, and L. D. White. 1985. Effects of clipping on burned and unburned creeping bluestem. J. Range Manage. 38:531-534.

Kartchner, R. J. 1981. Effects of protein and energy supplementation of cows grazing native winter range forage on intake and digestibility. J. Anim. Sci. 51:432-438.

Kaufman, D. W., E. J. Finck, and G. A. Kaufman. 1990. Small mammals and grassland fires. Pages 46-80 *in* S. L. Collins and L. L. Wallace, eds. Fire in North American tall grass prairie. Univ. Oklahoma Press, Norman.

Kay, C. E. 1990. Yellowstone's northern elk herd: a critical evaluation of the "natural regulation" paradigm. Ph.D. Thesis, Utah State Univ., Logan.

Keay, J. A., and J. M. Peek. 1980. Relationships between fires and winter habitat of deer in Idaho. J. Wildl. Manage. 44:372-380.

Keeley, J. E., P. H. Zedler. 1978. Reproduction of chaparral shrubs after fire: A comparison of sprouting and seedling strategies. Am. Midl. Nat. 99:142-161.

Kelsall, J. P., E. S. Telfer, and T. D. Wright. 1977. The effects of fire on the ecology of the boreal forest, with particular reference to the Canadian north: a review and selected bibliography. Can. Wildl. Serv. Occas. Pap. 32. 58pp.

Kenney, P. A., J. L. Black, and W. F. Colebrook. 1984. Factors affecting diet selection by sheep: III. Dry matter content and particle length of forage. Aust. J. Agric. Res. 35:831-838.

Kilgore, B. M. 1981. Fire in ecosystem distribution and structure: Western forests and scrublands. Pages 58-59 *in* H. A. Mooney, T. M. Bonnicksen, N. L. Christensen, J. E. Lotan, and W. A. Reiners, tech. coords. Fire regimes and ecosystem properties. U.S. Dep. Agric. For. Serv. Gen. Tech. Rep. WO-26.

———, and D. Taylor. 1979. Fire history of a sequoia-mixed conifer forest. Ecology 60:129-142.

Kituku, V. M., J. Powell, M. A. Smith, and R. A. Olson. 1992. Increasing bitterbrush nutrient quality with 2,4-D, mowing, and burning in southcentral Wyoming. J. Range Manage. 45:488-493.

Klein, D. R. 1982. Fire, lichens, and caribou. J. Range Management. 35:390-395.

Klinger, R. C., M. J. Kutilek, and H. S. Shellhammer. 1989. Population responses of black-tailed deer to prescribed burning. J. Wildl. Manage. 53:863-871.

Knetsch, J. L. 1977. Comparisons of methods for recreation evaluation. Pages 450-468 *in* R. Dorfman and N. S. Dorfman, eds. Economics of the environment. Second ed. W. W. Norton and Co., New York, N.Y.

Koerth, B. H., and J. W. Stuth. 1991. Instantaneous intake rates of 9 browse species by white-tailed deer. J. Range Manage. 44:614-618.

Koford, C. B. 1958. Prairie dogs, whitefaces, and blue grama. Wildl. Monogr. 3. 78pp.

Krefting, L. W. 1951. What is the future of the Isle Royale moose herd? Trans. North Am. Wildl. and Nat. Resour. Conf. 16:461-472.

Kucera, C. L. 1981. Grasslands and fire. Pages 90-111 *in* H. A. Mooney, T. M. Bonnicksen, N. L. Christensen, J. E. Lotan, and W. A. Reiners, tech. coords. Fire regimes and ecosystem properties. U.S. Dep. Agric. For. Serv. Gen. Tech. Rep. WO-26.

Kufeld, R. C. 1983. Responses of elk, mule deer, cattle, and vegetation to burning, spraying, and chaining of Gambel oak rangeland. Colo. Div. Wildl. Tech. Publ. 34. 47pp.

———, M. Stevens, and D. C. Bowden. 1981. Winter variation in nutrient and fiber content and *in vitro* digestibility of Gambel oak (*Quercus gambelii*) and big sagebrush (*Artemisia tridentata*) from diversified sites in Colorado. J. Range Manage. 34:149-151.

Land, D., W. R. Marion, and T. E. O'Meara. 1989. Snag availability and cavity nesting birds in slash pine plantations. J. Wildl. Manage. 53:1165-1171.

Leege, T. A. 1968. Prescribed burning for elk in northern Idaho. Proc. Tall Timbers Fire Ecol. Conf. 7:235-253.

———. 1969. Burning seral brush ranges for big game in northern Idaho. Trans. North Am. Wildl. and Nat. Resour. Conf. 34:429-438.

Lenarz, M. S. 1987. Economics of forest openings for white-tailed deer. Wild. Soc. Bull. 15:568-573.

Leopold, A. 1920. "Paiute forestry" vs. fire prevention. Southwest. Mag. 2:12-13.

———. 1924. Grass, brush, timber, and fire in southern Arizona. J. For. 22:1-10.

Loomis, J. B., E. R. Loft, D. R. Updike, and J. G. Kie. 1991. Cattle deer interactions in the Sierra Nevada: A bioeconomic approach. J. Range Manage. 44:395-399.

Loope, L. L., and G. E. Gruell. 1973. Ecological role of fire in the Jackson Hole area, northwestern Wyoming. Quat. Res. 3:425-443.

Loucks, O. L. 1970. Evolution of diversity, efficiency and community stability. Am. Zool. 10:17-25.

Lyon, L. J. 1971. Vegetal development following prescribed burning in Douglas-fir in south-central Idaho. U.S. Dep. Agric. For. Serv. Res. Pap. INT-105. 30pp.

———. 1977. Attrition of lodgepole pine snags on the sleeping child burn, Montana. U.S. Dep. Agric. For. Serv. Res. Note INT-219.

———, H. S. Crawford, E. Czuhai, K. L. Fredericksen, R. R. Harlow, L. J. Metz, and H. A. Pearson. 1978. Effects of fire on fauna: a state-of-the-knowledge review. U.S. For. Serv. Gen. Tech. Rep. WO-6.

———, and P. F. Stickney. 1976. Early vegetal succession following large northern Rocky Mountain wildfires. Proc. Tall Timbers Fire Ecol. Conf. 14:355-373.

McAtee, J. W., C. J. Scifres, and D. L. Drawe. 1979. Digestible energy and protein content of gulf cordgrass following burning and shredding. J. Range Manage. 32:376-378.

McComb, W. C., S. A. Bonney, R. M. Sheffield, and N. D. Cost. 1986. Snag resources in Florida—are they sufficient for average populations of primary cavity-nesters? Wildl. Soc. Bull. 14:40-48.

McDaniel, K. C., L. A. Torell, and J. W. Bain. 1993. Overstory-understory relationships for broom snakeweed-blue grama grasslands. J. Range Manage. 46:506-511.

McDivitt, J. H. 1987. Price and value alternatives for wildlife. Pages 101-108 *in* D. J. Decker and G. R. Goff, eds. Valuing wildlife. Westview Press, Boulder, Colo.

McHugh, T. 1958. Social behavior of the American buffalo (*Bison bison bison*). Zoologica 43:1-40.

Merrill, E. H., H. F. Mayland, and J. M. Peek. 1980. Effects of a fall wildfire on herbaceous vegetation on xeric sites in the Selway-Bitterroot wilderness, Idaho. J. Range Manage. 33:363-367.

Milford, R., and D. J. Minson. 1966. Intake of tropical pasture species. Int. Grassland Congr. 9:815-822.

Miller, G. D., and W. S. Gaud. 1989. Composition and variability of desert bighorn sheep diets. J. Wildl. Manage. 53:597-606.

Milne, K. A., and S. J. Hejl. 1989. Nest-site characteristics of white-headed woodpeckers. J. Wildl. Manage. 53:50-55.

Minson, D. J. 1990. Forage in ruminant nutrition. Academic Press, San Diego, Calif. 481pp.

Miquelle, D. G. 1983. Browse regrowth and consumption following summer defoliation by moose. J. Wildl. Manage. 47:17-24.

Mishan, E. J. 1976. Cost-benefit analysis. Praeger Publ., New York, N.Y. 454pp.

Mitsch, W. J., and J. G. Gosselink. 1986. Wetlands. Van Nostrand Reinhold, New York, N.Y. 539pp.

Moore, M. M., and D. A. Deiter. 1992. Stand density index as a predictor of forage production in northern Arizona pine forests. J. Range Manage. 45:267-271.

Morgan, P., S. C. Bunting. 1990. Fire effects in whitebark pine forests. Pages 166-170 *in* W. C. Schmidt and K. J. McDonald, compilers. Proc. of a symposium on whitebark pine ecosystems: ecology and management of a high mountain resource. U.S. Dep. Agric. For. Serv. Gen. Tech. Rep. INT-270.

Motazedian, I., and S. H. Sharrow. 1990. Defoliation frequency and intensity effects on pasture forage quality. J. Range Manage. 43:198-201.

Pase, C. P. 1958. Herbage production and composition under immature ponderosa pine stand in the Black Hills. J. Range Manage. 11:238-243.

Pastor, J., R. J. Naiman, B. Dewey, and P. McInnes. 1988. Moose, microbes, and the boreal forest. Bioscience 38:770-777.

Pearson, H. A., J. R. Davis, and G. H. Schubert. 1972. Effects of wildfire on timber and forage production in Arizona. J. Range Manage. 25:250-253.

Peck, V. R., and J. M. Peek. 1991. Elk, *Cervus elaphus*, habitat use related to prescribed fire, Tuchodi river, British Columbia. Can. Field-Nat. 105:354-362.

Peek, J. M. 1986. A review of wildlife management. Prentice-Hall, Englewood Cliffs, N.J. 486pp.

———. 1989. Another look at burning shrubs in northern Idaho. Pages 157-159 *in* D. M. Baumgartner, D. W. Breuer, B. A. Zamora, L. F. Neuenschwander, and R. H. Wakimoto, eds. Proc. symposium on prescribed fire in the Intermountain Region, forest site preparation and range improvement. Washington State Univ., Pullman.

———, R. A. Riggs, and J. L. Lauer. 1979. Evaluation of fall burning on bighorn sheep winter range. J. Range Manage. 32:430-432.

Peet, M., R. Anderson, and M. S. Adams. 1975. Effect of fire on big bluestem production. Am. Midl. Nat. 94:15-26.

Peet, R. K. 1988. Forests of the Rocky Mountains. Pages 63-101 *in* M. G. Barbour and W. D. Billings, eds. North American terrestrial vegetation. Cambridge Univ. Press, Cambridge, U.K.

Pellant, M. 1990. The cheatgrass-wildfire cycle—Are there any solutions? Pages 11-18 *in* E. D. McArthur, E. M. Romney, S. D. Smith, and P. T. Tueller, compilers. Proc. of a symposium on cheatgrass invasion-shrub die-off, and other aspects of shrub biology and management. U.S. Dep. Agric. For. Serv. Gen. Tech. Rep. INT-276.

Pendleton, R. L., F. J. Wagstaff, and B. L. Welch. 1992. Winter nutrient content and deer use of Gambel oak twigs in north central Utah. Great Basin Nat. 52:293-299.

Perkins, C. J. 1968. Controlled burning in the management of muskrats and waterfowl in Louisiana coastal marshes. Proc. Tall Timbers Fire Ecol. Conf. 8:269-280.

Petersen, K. L., and L. B. Best. 1987. Effects of prescribed burning on nongame birds in a sagebrush community. Wildl. Soc. Bull. 15:317-329.

Pieper, R. D. 1990. Overstory-understory relations in pinyon-juniper woodlands in New Mexico. J. Range Manage. 43:413-415.

Pierce, J. D. and J. M. Peek. 1984. Moose habitat use and selection patterns in north-central Idaho. J. Wildl. Manage. 48:1335-1343.

Pitt, M. D., and B. M. Wikeem. 1990. Phenological patterns and adaptations in an Artemisia/Agropyron plant community. J. Range Manage. 43:350-357.

Provenza, F. D., and D. F. Balph. 1988. Development of dietary choice in livestock on rangelands and its implications for management. J. Anim. Sci. 66:2356-2368.

———, and ———. 1990. Applicability of five diet-selection models to various foraging challenges ruminants encounter. Pages 423-460 *in* R. N. Hughes, ed. Behavioral mechanisms of food selection. NATOASI Series, Vol. G20. Springer Verlag, Berlin, Germany.

———, and J. C. Malechek. 1984. Diet selection by domestic goats in relation to blackbrush twig chemistry. J. Appl. Ecol. 21:831-841.

Radtke, K. W-H., A. M. Arndt, R. H. Wakimoto. 1982. Fire history of the Santa Monica Mountains. Pages 438-443 *in* C. E. Conrad and W. C. Oechel, tech coords. Dynamics and management of Mediterranean-type ecosystems. U.S. Dep. Agric. For. Serv. Gen. Tech. Rep. PSW-58.

Rao, R. M., L. W. Harbers, and E. F. Smith. 1973. Seasonal changes in nutritive value of bluestem pasture. J. Range Manage. 26:419-422.

Rasmussen, W. O., and P. F. Ffolliott. 1983. A model to predict snag development. Wildl. Soc. Bull. 11:291-292.

Ratliff, R. D., D. A. Duncan, and S. E. Westfall. 1991. California oak-woodland overstory species affect herbage understory: management implications. J. Range Manage. 44:306-310.

Riggs, R. A., and J. M. Peek. 1980. Mountain sheep habitat-use patterns related to post-fire succession. J. Wildl. Manage. 44:933-938.

Risenhoover K. L., and J. A. Bailey. 1985. Foraging ecology of mountain sheep: implications for habitat management. J. Wildl. Manage. 49:797-804.

———, ———, and L. A. Wakelyn. 1988. Assessing the Rocky Mountain bighorn sheep management problem. Wildl. Soc. Bull. 16:346-352.

Risser, P. G. 1990. Landscape processes and the vegetation of the North American grassland. Pages 133-146 *in* S. L. Collins and L. L. Wallace, eds. Fire in North American tallgrass prairies. Univ. Oklahoma Press, Norman.

Roberts, T. A., and R. L. Tiller. 1985. Mule deer and cattle responses to a prescribed burn. Wildl. Soc. Bull. 13:248-252.

Rominger, E. M., A. R. Dale, and J. A. Bailey. 1988. Shrubs in the summer diet of Rocky Mountain bighorn sheep. J. Wildl. Manage. 52:47-50.

Romme, W. H., and D. H. Knight. 1982. Landscape diversity: The concept applied to Yellowstone National Park. Bioscience 32:664-670.

Rottenberry, J. T., and J. A. Wiens. 1978. Nongame bird communities in northwestern rangelands. Pages 32-46 *in* R. M. DeGraaf, tech. coord. Proc. workshop on nongame bird habitat management in coniferous forests of the western United States. U.S. For. Serv. Gen. Tech. Rep. PNW-64.

Rowe, J. S. 1969. Lightning fires in Saskatchewan grassland. Can. Field-Nat. 83:317-342.

Rowland, M. M., A. W. Alldredge, J. E. Ellis, B. J. Weber, and G. C. White. 1983. Comparative winter diets of elk in New Mexico. J. Wildl. Manage. 47:924-932.

Sauer, C. O. 1950. Grassland, climax, fire, and man. J. Range Manage. 3:16-22.

Scanlan, J. S. 1992. A model of woody-herbaceous biomass relationships in ecualypt and mesquite communities. J. Range Manage. 45:75-80.

Schaefer, J. A., and W. O. Pruitt. 1991. Fire and woodland caribou in southeastern Manitoba. Wildl. Monogr. 116. 39pp.

Schlichtemeier, G. 1967. Marsh burning for waterfowl. Proc. Tall Timbers Fire Ecol. Conf. 6:41-46.

Scott, V. E., J. A. Whelan, and P. L. Svoboda. 1980. Cavity nesting birds and forest management. Pages 311-324 *in* R. M. DeGraff and N. G. Tilghman, compilers. Management of western forests and grasslands for nongame birds. U.S. Dep. Agric. For. Serv. Gen. Tech. Rep. INT-86.

Seip, D. R., and F. L. Bunnell. 1985. Nutrition of Stone's sheep on burned and unburned areas. J. Wildl. Manage. 49:397-405.

Shaw, J. H., and T. S. Carter. 1990. Bison movements in relation to fire and seasonality. Wildl. Soc. Bull. 18:426-430.

Shult, M. J. 1972. American bison behavior patterns at Wind Cave National Park. Ph.D. Thesis, Iowa State Univ., Ames. 178pp.

Skovlin, J. M., P. J. Edgerton, and B. R. McConnell. 1983. Elk use of winter range as affected by cattle grazing, fertilizing, and burning in southeastern Washington. J. Range Manage. 36:184-189.

Smith, L. M., J. A. Kadlec, and P. V. Fonnesbeck. 1984. Effects of prescribed burning on nutritive quality of marsh plants in Utah. J. Wildl. Manage. 48:285-288.

Soper, R. B., R. L. Lochmiller, D. M. Leslie, Jr., and D. M. Engle. 1993. Nutritional quality of browse after brush management on cross timbers rangeland. J. Range Manage. 46:399-410.

Sorg, C. F., and J. B. Loomis. 1984. Empirical estimate of amenity forest values: a comparative review. U.S. Dep. Agric. For. Serv. Gen. Tech. Rep. RM-107. 23 pp.

Spencer, D. L., and J. B. Hakala. 1964. Moose and fire on the Kenai. Proc. Tall Timbers Fire Ecol. Conf. 3:10-33.

Spowart, R. A., and N. T. Hobbs. 1985. Effects of fire on diet overlap between mule deer and mountain sheep. J. Wildl. Manage. 49:942-946.

Stein, S. J., P. W. Price, W. G. Abrahamson, and C. F. Sacchi. 1992. The effect of fire on stimulating willow regrowth and subsequent attack by grasshoppers and elk. Oikos 65:190-196.

Steinhoff, H. W., R. G. Walsh, T. J. Peterle, and J. M. Petulla. 1987. Evolution of the valuation of wildlife. Pages 34-48 *in* D. J. Decker and G. R. Goff, eds. Valuing wildlife. Westview Press, Boulder, Colo.

Stephens, D. W., and J. R. Krebs. 1986. Foraging theory. Princeton Univ. Press, Princeton, N.J. 247pp.

Sugden, R., and A. Williams. 1978. The principles of practical cost-benefit analysis. Oxford Univ. Press, U.K. 275pp.

Svejcar, T. J. 1989. Animal performance and diet quality as influenced by burning tallgrass prairie. J. Range Manage. 42:11-15.

Szaro, R. C., and R. P. Balda. 1986. Relationships among weather, habitat structure, and Ponderosa pine forest birds. J. Wildl. Manage. 50:253-260.

Taylor, D. L. 1973. Some ecological implications of forest fire control in Yellowstone National Park, Wyoming. Ecology 54:1394-13

Thill, R. E., A. Martin, Jr., H. F. Morris, Jr., E. D. McCune. 1987. Grazing and burning impacts on deer diets on Louisiana pine-bluestem range. J. Wildl. Manage. 51:873-880.

Thomas, J. W. 1990. Wildlife. Pages 175-204 *in* R. N. Sampson and D. Harn, eds. Natural resources for the 21st century. Island Press, Washington, D.C.

———, R. G. Anderson, C. Maser, and E. L. Bull. 1976. Snags. Pages 60-77 *in* J. W. Thomas, ed. Wildlife habitats in managed forests, the Blue Mountains of Oregon and Washington. U.S. Dep. Agric. Handb. 553.

Thomas, T. A., and R. L. Tiller. 1985. Mule deer and cattle responses to a prescribed burn. Wildl. Soc. Bull. 13:248-252.

Tisdale, E. W., M. Hironaka, and M. A. Fosberg. 1969. The sagebrush region in Idaho: A problem in range resource management. Univ. Idaho Agric. Exp. Stn. Bull. 512. 15pp.

———, ———. 1981. The sagebrush region: A review of the ecological literature. Univ. Idaho For. Wildl. Range Exp. Stn. Bull. 33. 31pp.

Trudell, J., and R. G. White. 1981. The effect of forage structure and availability on food intake, biting rate, bite size, and daily eating time of reindeer. J. Appl. Ecol. 18:63-81.

Uresk, D. W. 1984. Black-tailed prairie dog food habits and forage relationships in western South Dakota. J. Range Manage. 37:325-329.

———, J. F. Cline, and W. H. Rickard. 1975. Impact of wildfire on three perennial grasses in south-central Washington. J. Range Manage. 29:309-310.

Urness, P. J. 1990. Livestock as manipulators of mule deer winter habitats in northern Utah. Pages 25-40 *in* K. E. Severson, tech. coord. Can livestock be used as a tool to enhance wildlife habitat? U.S. Dep. Agric. For. Serv. Gen. Tech. Rep. RM-194.

Van Dyne, G. M., and H. F. Heady. 1965*a*. Botanical composition of sheep and cattle diets on a mature annual range. Hilgardia 36:465-492.

———, ———. 1965*b*. Dietary chemical composition of cattle and sheep grazing in common on a dry annual range. J. Range Manage. 18:78-86.

Van Soest, P. J. 1965. Use of detergents in analysis of fibrous feeds. III. Study of effects of heating and drying on yield of fiber and lignin in forages. J. Assoc. Off. Anal. Chem. 48:785-790.

———. 1982. Nutritional ecology of the ruminant. O&B Books, Inc. Corvallis, Oreg. 374pp.

Vogl, R. J. 1967. Controlled burning for wildlife in Wisconsin. Proc. Tall Timbers Fire Ecol. Conf. 6:47-96.

Ward, P. 1968. Fire in relation to waterfowl habitat of the Delta marshes. Proc. Tall Timbers Fire Ecol. Conf. 8:255-267.

Warner, R. E., and K. M. Hendrix, ed. 1984. California riparian systems. Univ. California Press, Berkeley and Los Angeles, 1035pp.

Weaver, H. 1951. Fire as an ecological factor in the southwestern ponderosa pine forests. J. For. 49:93-98.

Welch, B. L., E. D. McArthur, and J. N. Davis. 1981. Differential preference of wintering mule deer for accessions of big sagebrush and for black sagebrush. J. Range Manage. 34:409-411.

West, N. E. 1978. Basic synecological relationships of sagebrush-dominated lands in the Great Basin and Colorado Plateau. Pages 33-41 *in* The sagebrush ecosystem: a symposium. Utah State Univ., Logan.

———, F. D. Provenza, P. S. Johnson, and M. K. Owens. 1984. Vegetation change after 13 years of livestock grazing exclusion on sagebrush semidesert in west central Utah. J. Range Manage. 37:262-264.

———, K. H. Rea, and R. J. Tausch. 1975. Basic synecological relationships in Pinyon-juniper woodland. Pages 41-52 *in* The pinyon-juniper ecosystem: a symposium. Utah State Univ., Logan.

Westenskow-Wall, K. J., W. C. Krueger, L. D. Bryant, and D. R. Thomas. 1994. Nutrient quality of bluebunch wheatgrass regrowth on elk winter range in relation to defoliation. J. Range Manage. 47:240-244.

Wickstrom, M. L., C. T. Robbins, T. A. Hanley, D. E. Spalinger, and S. M. Parish. 1984. Food intake and foraging energetics of elk and mule deer. J. Wildl. Manage. 48:1285-1301.

Wiggers, E. P., and S. L. Beasom. 1986. Characterization of sympatric or adjacent habitats of 2 deer species in west Texas. J. Wildl. Manage. 50:129-134.

Willms, W. D., A. W. Bailey, A. McLean, and C. Kalnin. 1981. Effects of fall clipping or burning on the distribution of chemical constituents in bluebunch wheatgrass in spring. J. Range Manage. 34:267-269.

———, J. F. Dormaar, and G. B. Schaalje. 1988. Stability of grazed patches on rough fescue grasslands. J. Range Manage. 41:503-508.

Winternitz, B. L. 1980. Birds in aspen. Pages 247-257 *in* R. M. DeGraff and N. G. Tilghman, compilers. Management of western forests and grasslands for nongame birds. U.S. Dep. Agric. For. Serv. Gen. Tech. Rep. INT-86.

Wittinger, W. T., W. L. Pengelly, L. L. Irwin, and J. M. Peek. 1977. A 20-year record of shrub succession in logged areas in the cedar-hemlock zone of northern Idaho. Northwest Sci. 31:161-171.

Wood, G. W. 1988. Effects of prescribed fire on deer forage and nutrients. Wildl. Soc. Bull. 16:180-186.

Woodmansee, R. G., and L. S. Wallach. 1981. Effects of fire regimes on biogeochemical cycles. Pages 649-669 *in* F. E. Clark and T. Rosswell, eds. Terrestrial nitrogen cycles. Ecol. Bull. (Stockholm) 33.

Workman, J. P. 1986. Range economics. Macmillan Publ. New York, N.Y. 217 pp.

Wright, H. A., and A. W. Bailey. 1980. Fire ecology and prescribed burning in the Great Plains—a research review. U.S. For. Serv. Gen. Tech. Rep. INT-77. 60pp.

———, ———. 1982. Fire ecology—United Sates and southern Canada. John Wiley & Sons, Inc., New York, N.Y. 501pp.

———, S. C. Bunting, L. F. Neuenschwander. 1976. Effect of fire on honey mesquite. J. Range Manage. 29:467-471.

Yensen, D. L. 1981. The 1900 invasion of alien plants into southern Idaho. Great Basin Nat. 41:176-183.

Young, J. A., and R. A. Evans. 1981. Demography and fire history of a western juniper stand. J. Range Manage. 34:501-506.

———, ———, and B. L. Kay. 1987. Cheatgrass. Rangelands 9:266-270.

Young, V. A., and W. L. Robinette. 1939. A study of the range habits of elk on the Selway Game Preserve. Univ. Idaho, Moscow, Bull. 34. 48pp.

Zimmerman, G. T., and L. F. Neuenschwander. 1984. Livestock grazing influences on community structure, fire intensity, and fire frequency within the Douglas-fir/Ninebark habitat type. J. Range Manage. 37:104-110.

Chapter 19
CHEMICAL MANIPULATION OF PLANTS

Ben H. Koerth

INTRODUCTION

Blakey (1947:179-180) stated: "Encroachment of brush jungle upon formerly open forest and prairie range is insidious in that it has both good and bad effects upon certain wildlife species, and in some areas has the constant potential for near total exclusion of all valuable forms." Archer (1989) presented evidence that the increase of woody plants on former arid and semiarid grasslands is relatively recent and once established, woody plants likely will persist and become the prominent vegetation without drastic disturbances.

Until recently, brush control methods were used primarily to increase forage for domestic livestock and to facilitate livestock handling. Extensive brush control efforts were often practiced to increase grass production while wildlife considerations were largely ignored (Lehmann 1960). The increasing public demands for use of rangeland for recreational purposes and the potential value of leasing trespass rights for hunting have provoked a critical assessment of the impacts of brush control efforts to capitalize on multiple-use objectives. After decades of controversy, it is now generally accepted that herbicides can be used for wildlife habitat management if positive and negative impacts are considered and proper application techniques and patterns used.

This chapter addresses the use of chemicals in managing rangeland plants for wildlife habitat. Chemicals discussed are compounds exhibiting phytotoxic properties (herbicides). Discussion generally is limited to herbicides used for suppression of woody and herbaceous vegetation on rangeland. This chapter does not give a detailed description of all chemicals or their particular use in control programs, but it provides a basic overview of the mode of herbicide action and methods of application. Effects on wildlife habitat and potential toxic effects of these compounds on animals are addressed. Several case studies are presented examining situations where conditions were improved for wildlife by herbicide treatments and cases where animal populations may have been damaged. Finally, some economic considerations of using herbicides to manage wildlife habitat are discussed.

HISTORY AND CURRENT PRACTICES

Salts, ashes, and waste products have been used for centuries to control undesirable plant growth. Sulfuric acid, iron sulfate, copper nitrate, and potassium salts were used as plant control agents in the late 1800s. In the early 1900s, sodium chlorate and dinitrophenols were used for selective weed control in crops. Sodium arsenate was used as a soil sterilant for industrial or nonselective weed control (Scifres 1980:140). The World War II era initiated a new generation of compounds researched and developed specifically for their properties of regulating plant growth (Fig. 1). The herbicides 2,4-D (Table 1 presents chemical name of herbicides) and 2,4,5-T, first reported in the 1940s, led to the rapid development of the agricultural chemical industry. Since then, thousands of chemicals have been explored for potential herbicidal properties. Of these, about 1% survived the screening process and only about 0.025% survived to the development stage (Scifres 1980:144).

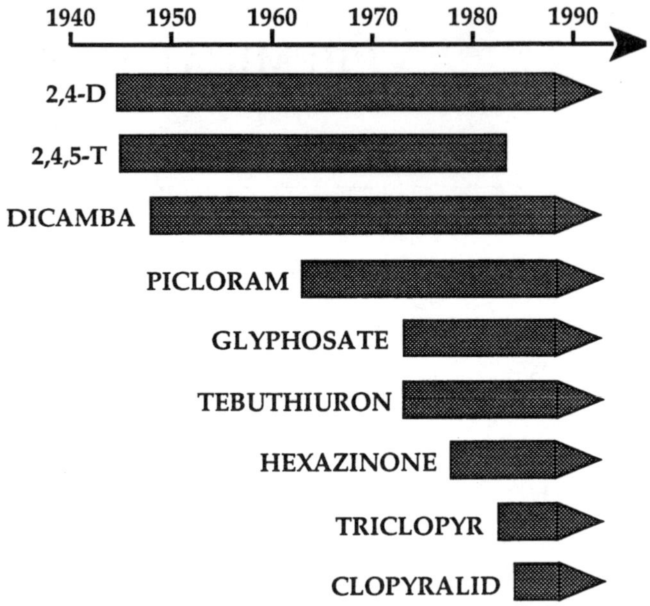

FIGURE 1. Chronology of development of common rangeland herbicides.

Beginning in the 1950s, 2,4,5-T and the closely related silvex were the primary herbicides used for woody plant control. With the introduction of picloram in 1963, mixtures (1:1) of 2,4,5-T and picloram were used commonly. Picloram controlled a broader spectrum of broad-leaved woody and herbaceous plants (Bovey and Scifres 1971) and thereby increased the effectiveness of the initial treatments. Mixtures of dicamba and picloram (1:1) also gave similar levels of control as 2,4,5-T + picloram in complex mixed brush communities (Scifres and Hoffman 1972). Although primarily used for herbaceous weed control, 2,4-D also suppresses some woody species (Scifres 1972*a*, Scifres and Haas 1974).

TABLE 1. Chemical name of herbicides mentioned in the chapter entitled "Chemical manipulation of plants" (Herbicide Handbook Committee 1983).

Common name	Chemical name
Dicamba	3,6-dichloro-o-anisic acid
Clopyralid	3,6-dichloro-2-pyridinecarboxylic acid
Glyphosate	N-(phosphonomethyl)glycine
Hexazinone	3-cyclohexyl-6-(dimethylamino)-1-methyl-1,3,5-triazine-2,4(1H,3H)-dione
Paraquat	1,1'-dimethyl-4,4'-bipyridinium ion
Picloram	4-amino-3,5,6-trichloropicolinic acid
Silvex	2-(2,4,5-trichlorophenoxy)propionic acid
Tebuthiuron	N-[5-(1,1-dimethylethyl)-1,3,4-thiadiazol-2-yl]-N,N'-dimethylurea
Triclopyr	[(3,5,6-trichloro-2-pyridinyl)oxy]acetic acid
Trifluralin	∂,∂,∂-trifluoro-2,6-dinitro-N-N-dipropyl-p-toluidine
2,4-D	(2,4-dichlorophenoxy)acetic acid
2,4,5-T	(2,4,5-trichlorophenoxy)acetic acid

Because of the limited time period during certain phenological stages of plant development when aerial sprays are effective and the concern about potential drift, dry herbicide formulations of extruded pellets (approx. 0.3-cm diam) such as tebuthiuron were developed. Liquid formulations designed for soil application (e.g., hexazinone) have many of the advantages of dry formulations, although they basically are limited to individual plant treatment. Cessation of domestic production and marketing of 2,4,5-T and silvex in 1984 provided the impetus for research and development of alternative herbicides. Subsequently, triclopyr was registered in 1985, and clopyralid was registered in 1987.

BASIS FOR HERBICIDE USE

Woody and Herbaceous Plant Relationships

Structure and composition of the vegetation complex determine the quality of the habitat for various wildlife species. An understanding of the interactions of woody plants and herbaceous forages is necessary in planning a brush management strategy. In simple form, production of warm-season herbaceous plants varies inversely with shrub encroachment (Fig. 2) (Wilson and Tupper 1982). A botanical composition shift also may occur from sun-loving species to shade-tolerant species of lower vigor as shrub cover increases. This woody-herbaceous plant relationship is directly related to animal production with different wildlife species adapted to different levels on the woody-herbaceous continuum.

The woody-herbaceous production relationship may not be as straightforward for cool-season herbaceous species. There are cases where a light-to-moderate canopy cover of woody plants may increase production of some cool-season species, making yearlong herbaceous production greater than on brush-free areas. Scifres et al. (1982) reported production of Texas wintergrass (*Stipa leucotricha*) increased as canopy cover of huisache (*Acacia farnesiana*) increased to about 20% on rangeland in the Texas Coastal Prairie. Production of Texas wintergrass was not reduced until huisache canopy cover exceeded 60%. Brock et al. (1978) reported a similar relationship between honey mesquite (*Prosopis glandulosa*) and cool-season grasses in semiarid areas.

If sufficient numbers of woody plants become established on a former grassland, a new shrub-driven successional process becomes established making it unlikely that the site will revert to a grassland even following drastic disturbances (Hobbs and Mooney 1986, Archer 1989). In general, herbaceous production will increase dramatically following herbicide applications (Fig. 3) with peak production typically occurring from about year 2 to about year 4. A gradual decline in herbaceous production follows until the life of the treatment is exhausted and woody plants again dominate the plant community (Hamilton et al. 1986, Scifres and Hamilton 1989). However, timing and amount of herbaceous response depend upon the amount of woody plant suppression, soil fertility, precipitation, and climate (Scifres 1980:12).

There is an upper limit to woody plant cover required by wildlife (Blakey 1947, Robinette 1972, Inglis 1985). If woody cover exceeds this threshold, it is desirable to suppress woody cover to maintain habitat conditions conducive for the wildlife species in question. For example, wildlife habitat can be described in part as a function of the woody and herbaceous components of the plant community. The "domain of presence" of an animal population would include all combinations of woody and herbaceous cover that will support a population (Fig. 4). Peak animal abundance hypothetically occurs at the optimum combination of woody and herbaceous components.

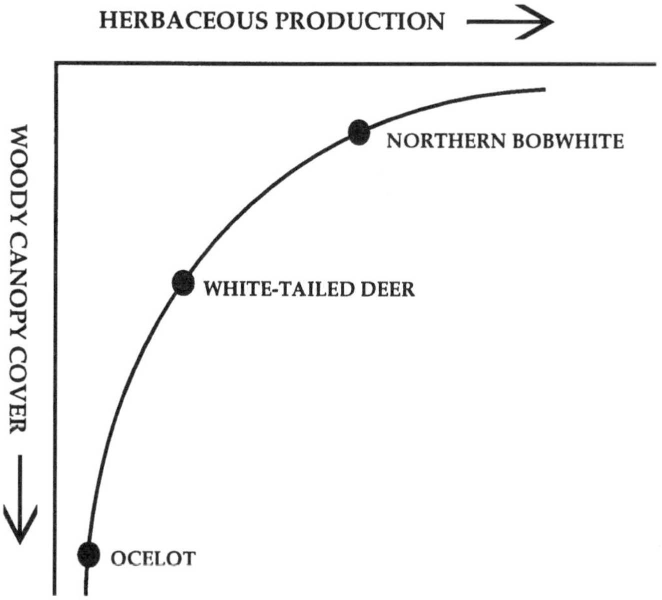

FIGURE 2. Herbaceous-woody plant relationship and examples of wildlife species adapted to different levels of woody cover. Adapted from Wilson and Tupper (1982).

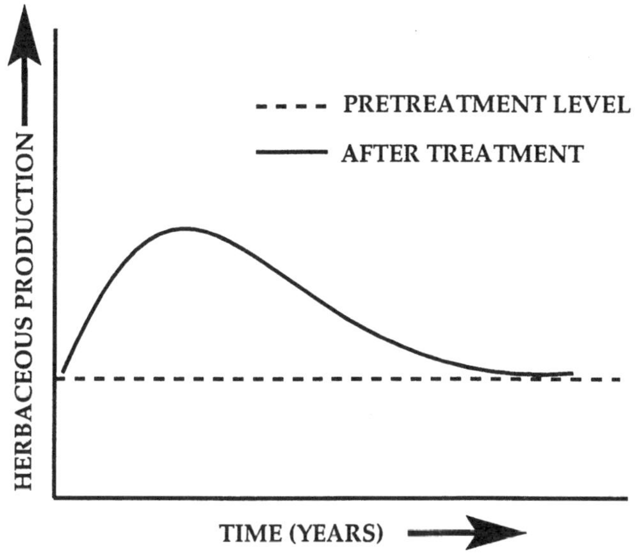

FIGURE 3. Hypothetical response curve for herbaceous production following herbicide treatments.

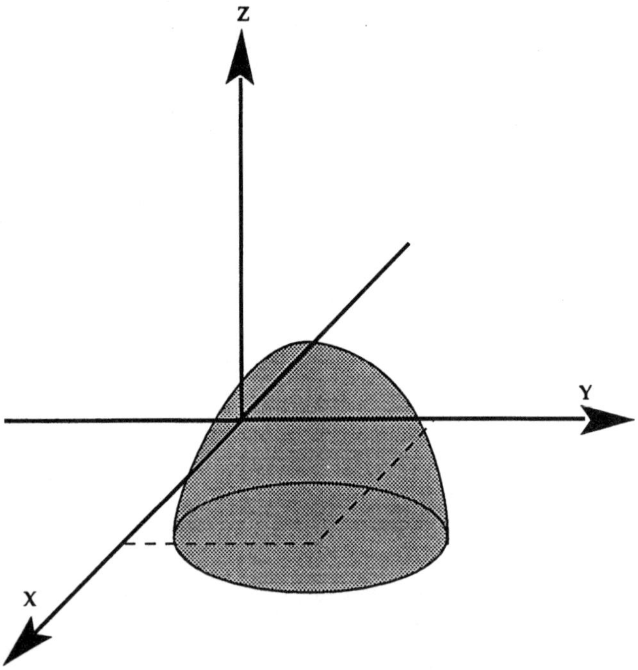

FIGURE 4. Theoretical domain of presence and abundance of an animal population relative to woody and herbaceous components of the habitat. X axis = woody plant cover, Y axis = herbaceous plant cover, and Z axis = animal abundance. As woody cover goes above or below the optimum, animal abundance declines until habitat conditions will no longer support the population. For peak animal populations to occur, the plant community must be manipulated to maintain the requisite habitat conditions.

Because most rangeland supports a variety of wildlife, the requirements of each species need to be understood before treatment. Treatments that benefit some species will likely cause negative responses from other species. Also, animals do not distribute themselves randomly or regularly across an area. They tend to cluster in and around areas that provide specific needs. For a management treatment to do no harm to a population, critical habitat areas must be preserved. For a treatment to benefit populations, useful areas must be developed from suboptimal areas.

Herbicide versus Mechanical Treatments

Alterations of plant communities generally serve to accelerate or retard plant succession. Leopold (1933) listed 4 tools for managing succession: ax, cow, fire, and plow. In a less lyrical sense, and in keeping with present-day technology, there are 4 basic methods of woody plant management: mechanical, biological, pyric, and chemical. Each technique has advantages and limitations. However, broad differences in effects on woody-plant communities can be made by contrasting herbicide treatments with broadcast mechanical treatments.

Vegetation changes with herbicides are more subtle than mechanical treatments. Specific herbicide and plant species susceptibility are of paramount importance as this determines the final extent of defoliation and vegetation change. The most rapid change occurs after application of foliar-active herbicides. Plants generally defoliate within a month. However, branches and trunks of defoliated species remain and provide screening cover and shade (Beasom and Scifres 1977). Defoliation may not be complete for up to 2 years after application of some soil-applied herbicides such as tebuthiuron.

Mechanical treatments are applied more slowly than aerial herbicide applications but result in immediate physical alteration of the habitat (Box and Powell 1965). Mechanical methods reduce screening cover and shade, but those that disturb the soil may increase production of lower successional grasses and forbs (Scifres 1980:264). Mechanical treatments usually provide a more satisfactory seedbed than herbicide treatments if reseeding is being considered (Simanton 1991). Soil disturbance enhances establishment of seeded grasses by burying seeds at desirable depths for emergence, particularly for species with large seeds (Stuth and Dahl 1974, Winkel et al. 1991).

Conversely, foliar-applied herbicides generally reduce forbs for at least a growing season after application (Scifres and Mutz 1978), and establishment of desirable vegetation is dependent upon an adequate seed source in the treated area. Timely precipitation, however, can reduce the length of time that forbs are affected (Tanner et al. 1978).

Scifres (1986) listed the following advantages and disadvantages of using herbicides on rangelands.

Advantages

1. A variety of application methods are available ranging from individual plant treatment to broadcast aerial application.
2. Aerial application is fast and independent of terrain and woody plant growth forms.
3. There is little or no soil disturbance associated with application.
4. Stems and trunks of defoliated plants and foliage of herbicide resistant species remain to provide screening cover and shade for wildlife.
5. Herbicides are effective for suppression of selected woody species.
6. Soil-applied herbicides have minimal drift potential and can be applied during relatively broad periods of the year.
7. Broad-leaved herbaceous plants may be suppressed along with woody species, further reducing competition for establishment of desirable species.

Disadvantages

1. Many broad-leaved herbaceous plants suppressed by herbicides are favored forages of some wildlife species.
2. Foliar sprays of herbicides are usually restricted to certain phenological stages of woody plant development.
3. Application of foliar sprays is restricted to certain environmental conditions (especially wind speed and temperature).
4. The spectrum of species controlled in complex mixed brush associations may not be sufficient to allow maximum herbaceous forage production.
5. Suppression of susceptible species may allow resistant species to increase and form more-difficult-to-manage stands than the original cover.

Applied Use of Herbicides

Selection of brush management technologies may be considered after management objectives are determined. Recommendations for treatment should be based on the characteristics of each problem situation (Fig. 5). Often >1 technique is suitable for each situation, and decision criteria can be overridden by personal preference, economics, or other management considerations. However, different methods should not be considered interchangeable management tools as the subsequent vegetation will differ among treatment methods. After herbicides are identified as a viable option, specific recommendations can be made based on plant species susceptibility, and rate, method, and timing of application.

Species-specific herbicides may be used to thin stands of mixed species of woody cover, thereby reducing competition for more desirable species (Crawford 1960, Krefting and Hansen 1969, Kufeld 1977, Doerr and Guthery 1983, Hurst and Warren 1986, Scifres and Koerth 1989, Newton et al. 1989), and to create and maintain openings in wooded areas (Trumbo and Chappell 1961, McCaffery et al. 1974, McComb and Rumsey 1981).

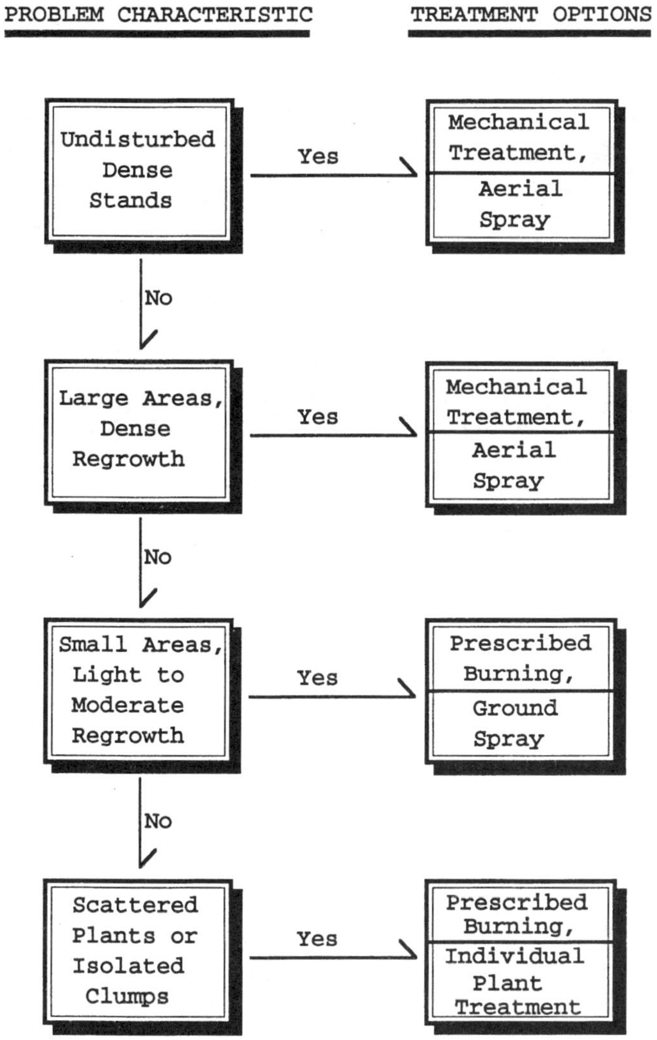

FIGURE 5. Example of a decision making flow chart that might be used in selecting methods for woody plant management.

Specific herbicides also can be used to manage composition of herbaceous vegetation to enhance wildlife food plants (Hamilton and Buckholtz 1953, Bellrose 1956, Coulter 1957, Pletscher and Robel 1979, Guthery et al. 1987).

With careful planning, herbicide treatments and mechanical methods can be used to manage wildlife habitat without degradation. Uncontrolled use of any method can result in requisite habitat needs being decimated and cause the concomitant reduction of some wildlife species (Davis and Winkler 1968, Carr and Glover 1970, Dwernychuk and Boag 1973, Wallestad 1975, Beasom and Scifres 1977, Spencer and Barrett 1980, Rodgers and Sexson 1990, Schulz et al. 1992*a*). A compromise must be reached between the amount of treatment required to meet the objectives and the amount of treatment that would degrade the habitat. The ultimate goal in designing any treatment should be to promote structural and botanical diversity, and to ensure that habitat features for wildlife are provided.

KINDS OF HERBICIDES

Foliar-active

Herbicides can be categorized as foliar- or soil-active (Fig. 6). Foliar-active herbicides are applied directly to leaves and stems in the form of liquids or water-soluble powders. These herbicides can be further divided into contact and translocated herbicides. Contact herbicides kill only the plant tissue directly contacted by the herbicide and are relatively nonselective. Translocated herbicides, on the other hand, must be absorbed by leaf or stem tissue and transported through the phloem to the roots or other organs within the plant (Herbicide Handb. Comm. 1983:xxiv). Factors that alter the photosynthetic activity of the plant affect the rate of translocation (Lym and Messersmith 1991). Extended periods of low light, low soil moisture, or excessive air temperatures tend to reduce physiological activities within the plant and thus diminish translocation. Scifres (1980:144-152) gives a more detailed description of absorption and translocation processes.

Soil-active

Soil-active herbicides are generally applied to the soil surface in liquid or pelleted form. These herbicides depend upon precipitation to move the herbicide into the rooting zone of the

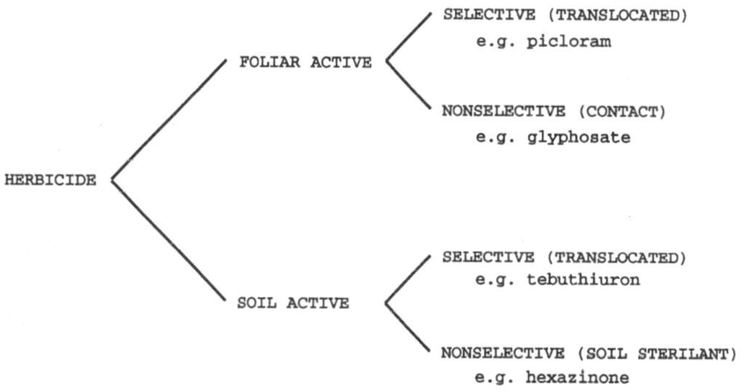

FIGURE 6. General classification of herbicides based on activity, mode of entry, and selectivity. From Scifres (1980).

plants. After incorporation in the soil profile, the herbicide is absorbed by the root system and transported to aerial parts of the plant through the xylem. Soil-applied herbicides may undergo some degradation on the soil surface before being activated by rainfall (Baur et al. 1974), and their activity can be mediated by organic matter and clay content of the soil (O'Connor and Anderson 1974, Chang and Stritzke 1977, Duncan and Scifres 1983). In sandy soil, water soluble herbicides may be leached below the root zone of some plant species (Scifres 1972*b*). Advantages of soil-applied herbicides are reduced drift from aerial applications, negligible volatilization, and long application season (Scifres 1980).

METHODS OF APPLICATION

Aerial

Use of fixed-wing (Fig. 7) and rotary-wing (Fig. 8) aircraft are the most common methods of herbicide application on rangeland. An advantage of using aerial equipment is rapid application to large areas. The method also is relatively unaffected by rough terrain, rocky soil, and densely wooded areas where other herbicide application methods may not be feasible. Herbicides can be applied either as a foliar spray or in a dry formulation such as pellets. The major limitation for fixed-wing aircraft is the necessity of a suitable airstrip for takeoff and landing. Long distance shuttles between the airstrip and the treatment area are costly in time and money. Powerlines, towers, or other tall vertical structures present hazards to low-flying aircraft. Helicopters are best suited for areas remote from landing strips but carry a smaller payload and have slower airspeeds. However, the slower airspeed and greater maneuverability of helicopters may allow more precise application.

Ground

Several ground application systems have been developed and are frequently used on rangeland. The 2 basic types are tractor-mounted (Fig. 9)(usually rubber-wheeled farm tractors) and hand-applied, although spray systems can be mounted on trucks, crawler-type tractors, and other vehicles (Fig. 10).

FIGURE 7. Fixed-wing aircraft for applying herbicides. Photo courtesy of R. Bovey, USDA, ARS, Dep. of Rangeland Ecology and Management, Texas A&M University, College Station.

FIGURE 9. Tractor-mounted boom sprayer. Photo courtesy of T. Welch, Texas Agricultural Extension Service, Texas A&M University, College Station.

FIGURE 8. Rotary-wing aircraft for applying herbicides.

FIGURE 10. Boom sprayer mounted on a 4-wheel drive vehicle. Photo courtesy of T. Welch, Texas Agricultural Extension Service, Texas A&M University, College Station.

Tractor-mounted. Ground application systems are less adapted than aerial application systems for treating large areas of dense brush. However, tractor-mounted systems can be used for broadcast applications in areas with regrowth brush or for herbaceous plant control. Booms or cluster nozzles can be attached to spray tanks for broadcast application of liquids. Pelleted formulations can be broadcast with spreaders typically used for dry fertilizer or seed. Hand-operated nozzles can be used for individual plant treatment when broadcast application is not feasible or is unwarranted (e.g., treatment of fence lines, field borders, or light brush infestations in grassland habitats).

In light infestations of brush or weeds, broadcast applications are relatively inefficient because only 30-50% of the herbicide may contact the target plants (Mayeux and Crane 1983). The hazard of potential drift from sprays is a problem when the proposed area for treatment occurs near herbicide-sensitive crops. The carpeted roller (Fig. 11) is an example of an experimental applicator for individual plant treatment of brush regrowth and for weed control (Mayeux and Crane 1983, 1984, 1985; Messersmith and Lym 1985; Mayeux 1987; Scifres et al. 1988*a*; Crane et al. 1989). The carpeted roller applies herbicide directly to the foliage of target plants, reducing drift and the volume of herbicide solution used compared to broadcast spraying.

Hand-applied. Herbicides can be applied to rangeland by hand in several ways. Most hand application methods are best suited for individual plant treatment, although pelleted formulations can be broadcast with hand-operated spreaders typically used for dry fertilizers or seed. An advantage is the operator can be selective in suppressing only the targeted species or individual plants. Disadvantages are that hand methods are labor intensive and are usually restricted to use on small areas. Where topography and brush density are suitable, some operators have increased speed of application and reduced labor by using all-terrain vehicles or other appropriate transportation to cover ground between plants faster than on foot.

Typical hand application methods include the following.

1. Foliar sprays. Hand operated sprayers are used to apply liquid formulations directly to the foliage of plants.

FIGURE 11. Experimental carpeted-roller applicator for herbicides. Photo courtesy of La Copita Research Area, Texas Agricultural Experiment Station, Alice.

2. Basal bark sprays. Hand operated sprayers are used to apply liquid formulations to the lower 46-61 cm of the trunk.
3. Frill girdle application. A ring is cut through the bark around the trunk near the base. Liquid formulations are applied directly into the cut.
4. Notch application. Notches are cut through the bark near the base of the tree, and liquid formulations are applied directly into the notches.
5. Injector method. Similar to the notch application only an injection system built into the notching tool delivers liquid herbicide into the cuts as they are made.
6. Stump treatment. Herbicides are applied directly to the stump after top removal.
7. Soil applied. Wet or dry formulations are applied directly to the soil at the base of the plant to be controlled.

SHORT- AND LONG-TERM EFFECTS ON HABITAT

Persistence in the Environment

Herbicide residue, as defined by the terminology committee of the Weed Science Society of America, is ". . . that quantity of a herbicide remaining in or on the soil, plant parts, animal tissues, whole organisms, and surfaces" (Herbicide Handb. Comm. 1983:xxii). Often the term residue has negative connotations. The persistence of herbicides is an important factor in the potential for residues to occur in food and water or to contact nontarget organisms (Johnson 1971, Norris 1981). However, residues are not wholly detrimental. To be effective, a herbicide must persist for a time in the environment of the target species to allow for plant uptake and ultimately control. Ideally, the herbicide would then dissipate quickly and completely.

The diversity of vegetation and soils on rangeland complicates determining the fate of herbicide residues (Scifres 1977). Basically, longevity of a herbicide depends upon the chemical attributes of the herbicide and the concentration applied; physical, chemical, and biological attributes of the soil; and climate (Lutz et al. 1973, Radosevich and Winterlin 1977). Herbicides formulated to be soil applied generally persist longer than foliar-applied herbicides. For example, measurable concentrations of tebuthiuron, a soil-applied herbicide, have been found as long as 11 years post-application (Johnsen and Morton 1991). However, concentrations were consistently low and only exceeded limits set by the Environmental Protection Agency for forage at application rates ≥4.5 kg/ha.

Rate of dissipation of most herbicides only can be generalized because loss of herbicidal residue is variable and results from various studies are not directly comparable because of differing experimental procedures and conditions. For example, reported phytotoxic concentrations of 2,4-D in the soil vary from 7 (Klingman 1964) to 98 days (Akamine 1951), dicamba from 30 (Burnside and Lavy 1966) to >365 days (Dowler et al. 1968), and picloram from 71 (Scifres et al. 1971) to >423 days (Youngson et al. 1967). In general, degradation of herbicides accelerates with increasing temperature, soil microflora, and organic matter content of the soil (Johnson 1971). Rainfall also is an important factor influencing the loss of herbicides (Morton et al. 1967) by maintaining soil moisture conditions conducive for microbial decomposition and through leaching, especially for herbicides with high water solubility.

Potential Toxic Effects on Animals

Rangeland herbicides exhibit a lower order of toxicity to mammals and birds than to vegetation (Table 2). However, the low order of toxicity should not be construed to mean that rangeland herbicides can be handled or applied with any less caution than other pesticides. All applicators are legally bound to apply herbicides only according to instructions on the herbicide label.

Some herbicides, especially some formulations of 2,4-D, are relatively toxic to fish (Herbicide Handb. Comm. 1983:133). Contamination of water sources through application, cleaning of equipment, or disposal of wastes should be avoided.

Toxicological data have been gathered on a relatively few species of laboratory animals. For safety sake, it would be prudent to assume that all animals are at least if not more sensitive than the test animals (Scifres 1980:213). A Swedish veterinarian (Erne 1966 *in* Scifres 1980:215) found no clinical effects in rats fed 100-200 mg/kg (mg of herbicidal compound/kg of test animal body wt) of different 2,4-D formulations; however, calves fed the same levels exhibited transient dysphagia, bloating, thirst, and muscular weakness. These levels of intake were far above the exposure of rangeland animals, even at exaggerated levels of application, and are mentioned only as an example of species specific reactions.

TABLE 2. Acute oral toxicity of several herbicides mentioned in the chapter entitled "Chemical manipulation of plants" (adapted from Herbicide Handbook Committee 1983).

Herbicide Species tested	Lethal dose (LD_{50}, mg/kg)
Dicamba	
Rat	1,028
Pheasant	800
Clopyralid	
Rat	>5,000
Mouse	>5,000
Glyphosate	
Rat	5,600
Northern bobwhite	3,850
Hexazinone	
Rat	1,690
Guinea pig	860
Picloram	
Rat	8,200
Rabbit	2,000
Tebuthiuron	
Rat	644
Rabbit	286
Triclopyr	
Rat	713
Rabbit	550
2,4-D	
Rat, Guinea pig, Rabbit	300-1,000[a]

[a] Varies according to formulation or sex.

There is no evidence that under proper application conditions and procedures, any of the herbicides currently labeled for rangeland use will bioconcentrate or pose direct harmful effects to rangeland wildlife or man (Johnson 1971, Leng 1972, Norris 1981). However, other herbicides not as commonly used on rangeland may cause mortality in some wildlife populations. Trifluralin, paraquat, and fosamine ammonium can cause mortality and impaired embryonic growth in exposed mallard (*Anus platyrhynchos*) and northern bobwhite (*Colinus virginianus*) eggs (Hoffman and Eastin 1982, Hoffman 1983, U.S. Fish and Wildlife Service 1982 *in* Robinson and Bolen 1984:165) though these herbicides are listed as posing little hazard to adult mammals and birds at recommended application rates (Herbicide Handbook Committee 1983). There is still much to be learned about the embryotoxic effects of synthetically produced chemicals.

Immediate Plant Defoliation and Mortality

A primary objective in habitat management is to manipulate vegetation so that desired plant species are favored and undesired species are not. Herbicides are more species specific than chemical insecticides. Because vegetation on specific sites varies in composition and different plant species have varying tolerances to different herbicides, the impact of herbicide treatments must be considered on a site by site basis.

Herbicide-site specificity is exemplified by tebuthiuron applications to drainage sites in south Texas that often support a dense understory of whitebrush (*Aloysia gratissima*) with a honey mesquite overstory (Scifres 1980:32). In these situations, whitebrush forms a dense, essentially monotypic community in the understory that precludes production of desirable herbaceous plants for wildlife and livestock. Tebuthiuron effectively suppresses whitebrush and a few associated species while mesquite is tolerant. Application of tebuthiuron followed by prescribed burning or mechanical treatments to remove the dead stems results in a savannah-like appearance.

Cover

Probably the principal acute impact of herbicide treatments in woody plant communities is defoliation and loss of woody canopy cover. With foliar-applied herbicides, leaves of susceptible species typically turn brown within 2 weeks and may completely defoliate within a month. Canopy reduction does not, however, imply plant mortality as mortality in most instances is less than complete. Partial crown survival or "flagging" is common. Resistant plants may be initially defoliated but can refoliate following favorable conditions.

Understory vines and shrubs not controlled by a herbicide treatment may increase following removal of a dominant overstory. Scifres et al. (1987) found stem density of American beautyberry (*Callicarpa americana*) increased to roughly 30 times the density found on untreated areas by the third growing season after application of tebuthiuron at 2.2 kg/ha to an oak (*Quercus* spp.) dominated rangeland. Other understory species increasing in abundance following treatment included saw greenbriar (*Smilax bona-nox*), southern dewberry (*Rubus trivialis*), and peppervine (*Ampelopsis arborea*). Although greenbriar and American beautyberry are favored browse plants for white-tailed deer (*Odocoileus virginianus*) (Kroll 1991:193-194), these secondary woody stands increased to the point of becoming a management problem for livestock grazing and negated increased herbaceous production resulting from the herbicide treatment by the third year.

Even on plants that suffer complete plant mortality, trunks, stems, and all but the smallest twigs persist for some time after treatment. These woody skeletons along with foliage from herbicide resistant species provide the bulk of screening cover at lower canopy levels. Tanner et al. (1978) found no difference in woody plant screening cover at deer height (0.3-0.9 m) between untreated areas and areas that were treated with 0.56 or 1.12 kg/ha of 2,4,5-T + picloram (1:1) although 75-80% of the woody plant canopy had been defoliated by the herbicide treatment. Woody plant screening cover was significantly decreased at higher levels in the canopy primarily from crown defoliation.

Similarly, DeFazio et al. (1988) found no difference in foliage density at the 0.5- to 1.0-m level in the Piedmont Plateau physiogeographic region of Georgia either during summer or winter after application of tebuthiuron at 3.0 kg/ha as a site preparation treatment following clearcutting. In the Coastal Plain Region, foliage density was higher in herbicide-treated areas during summer, primarily from increased grass and sedge production. During the dormant season, no differences in foliage density were found between treated and untreated areas.

Defoliation and mortality of woody overstory may increase the diversity of habitats in some areas. Schulz et al. (1992*b*) found increased snag density, snag basal area, and slash density along with decreased foliar cover on cross timbers rangeland in Oklahoma following treatment 5-6 years previously with 2.2 kg/ha of tebuthiuron or triclopyr. Total nongame bird density, diversity, and species richness did not differ between treated and untreated areas. However, species composition differed. Several species not typically associated with closed woodlands were more prevalent on treated areas presumably because of increased nesting habitat (i.e., snags).

Food

The second major acute impact of herbicide treatments is the potential reduction of food through defoliation of browse plants and suppression of forbs. Like woody plants, forbs also differ in susceptibility to different herbicides. Picloram, for example, controls a broad spectrum of plants and can damage forb populations (Scifres 1980:165-170). Extensive broadcast applications of picloram are seldom recommended for habitat management. However, patterned applications with herbicide mixtures containing picloram have proven effective in wildlife habitat management if sufficient untreated areas are available to supply forbs and browse for wildlife consumption (Beasom and Scifres 1977, Tanner et al. 1978, Scifres and Koerth 1989).

In the northern Lake States, Krefting and Hansen (1969) demonstrated that herbicide treatment of mature forested areas could be used to reduce undesirable forage for deer and increase production of desirable species. Plant species with limited susceptibility to a specific herbicide may only be top-killed. Interruption of apical dominance allows resprouting species to develop vegetative growth along peridermal buds at or below the lowermost level of control. In this way, herbicides can be used to manipulate mature, often decadent woody growth, by bringing nutritious regrowth forage to a level accessible to herbivores (Krefting et al. 1956, Mueggler 1966, Tanner et al. 1978).

In contrast to picloram, many forb species are tolerant of tebuthiuron (Scifres and Mutz 1978). Applications of ≤1.1 kg/ha often maintain forb populations at pretreatment levels or may increase production on some sites after removal of competition from woody species (Scifres and Koerth 1986, Koerth et al. 1989, Fulbright and Garza 1991). Doerr and Guthery (1983) found that application of tebuthiuron at 0.2 or 0.4 kg/ha for control of sand shinnery oak (*Quercus havardii*) increased grass production, forb diversity, and production of seeds useful as lesser prairie-chicken (*Tympanuchus pallidicinctus*) food. Similarly, Olawsky and Smith (1991) found grass and forb composition was greater on sand shinnery oak ranges 3-6 years after treatment with tebuthiuron at 0.56 kg/ha compared to untreated areas.

Loss of forb populations does, nevertheless, occur following application of tebuthiuron and other herbicides. However, all the reduction cannot be attributed to direct effect of herbicides. On some sites, dramatic increases in grass cover following treatment may physically limit some forb production (Beasom et al. 1982; Scifres and Koerth 1986, 1989; DeFazio et al. 1988). Positive impacts of increased production of herbaceous food plants may offset partly the negative impacts of loss of browse.

At very low rates (0.05-0.10 kg/ha), 2,4-D, an auxin-type herbicide, can be used to induce rooting and blossom set and may be useful in establishing some forbs important as food producing plants for wildlife (Hamilton and Buckholtz 1953, Pletscher and Robel 1979). Soil-active herbicides, coupled with autumn soil disturbance, favors establishment of warm-season food plants in the following growing season. Similarly, preemergent herbicides in conjunction with summer soil disturbance could be used to favor cool-season plant growth the following autumn (Guthery et al. 1987). Because the emergent ground cover composition is a predictable function of herbicide by site interaction, the possibility exists to develop very specific plant-response objectives.

Creative use of herbicides may include reducing nuisance wildlife species by selectively controlling their food. Pocket gopher (*Thomomys talpoides*) populations have been effectively reduced by spraying with 2,4-D to control annual and perennial forbs, their major food source (Keith et al. 1959, Tietjen et al. 1967, Johnson and Hansen 1969, Hull 1971). Herbicides also have been used to kill trees to provide snags for cavity-nesting birds in forested situations (Conner et al. 1981, 1983; McComb and Rumsey 1983; Bull and Partridge 1986; Schulz et al. 1992*a*,*b*). Taylor (1979) reported lesser prairie chicken use of man-made leks created with herbicides in sand shinnery oak and sand sagebrush (*Artemisia filifolia*) habitat. Herbicide treatments combined with prescribed burning may modify habitats sufficiently to alter host-parasite relationships resulting in lower infection rates of some intestinal helminths (Boggs et al. 1990*a*,*b*; Boggs et al. 1991*a*,*b*).

Long-term Effects on Habitat

Published information on the long-term effects of herbicide treatments is meager. The majority of studies do not extend beyond the immediate few years after treatment. Even with current technology, herbicide treatments have limited life spans and the first few growing seasons following treatment have the most impact on the habitat. Follow-up treatments are often necessary to maintain effects of the initial treatment as resprouting species and new plants developing from seed begin to recolonize treated areas. Ecologically, the major long-term impact of herbicides on wildlife habitat is alteration of the environment by suppressing certain plant species that are susceptible to the herbicide, thereby reducing the complexity of the vegetation. Fulbright and Beasom (1987) reported similar conditions following root-plowing in south Texas.

The long-term effect of changing the availability of food and cover may be beneficial or detrimental, depending on the species involved and the management objectives for the area. Herbicide treatments do not, however, always result in reduced biodiversity. Selective treatments can create relatively stable shrub communities and somewhat less stable herbaceous communities in mature, mixed woodlands (Niering and Goodwin 1974, Scifres and Koerth 1986, Lochmiller et al. 1991).

Responses by Specific Animals

Sage Grouse. Certain wildlife species have obligate relationships with particular plant species or communities and any program that harms or removes that plant species is decidedly damaging to the welfare of that wildlife population. One example is the specialized relationship of sage grouse (*Centrocercus urophasianus*) and sagebrush (*Artemisia* spp.). Reduction of sagebrush with herbicides can restrict use of treated areas by sage grouse (Johnson 1969, Klebenow 1970, Martin 1970, Wallestad 1975). However, wholesale avoidance of the use of herbicides on ranges containing sagebrush is not necessary if management is required to meet additional objectives. Klebenow and Gray (1968) and Peterson (1970) found that some plant species favored as food by juvenile sage grouse occurred frequently on herbicide treated areas, thus compensating somewhat the loss of other foods. Braun et al. (1977) proposed a series of treatment recommendations that, with careful planning, would allow treatment of sagebrush ranges while minimizing damage to sage grouse populations.

Small Mammals and Birds. Eurytypic wildlife species also may be reduced following herbicide treatments that reduce habitat complexity. In particular, resident avifauna and small mammal populations that have small daily ranges of movement may suffer from changes in successional stage of the habitat or from changes in food and cover quality (Schroeder and Sturges 1975; Savidge 1978; Drew 1988; Santillo et al. 1989*a*,*b*). Changes in cover can also affect predation rates and thermal moderating characteristics of the foliage. However, population reductions often can be ameliorated by appropriate patterning in application whereby untreated areas are interspersed with treated areas creating a habitat mosaic favorable to a number of species (Santillo et al. 1989*a*,*b*).

Some wildlife species that prefer lower successional vegetation stages may successfully adapt to changes following treatment of mature communities (Slagsvold 1977; Schulz et al. 1992*a*,*b*). Soutiere and Bolen (1976) found mourning doves (*Zenaida macroura*) compensated for the loss of trees as nesting sites by nesting on the ground and that ground nests were more successful than tree nests. Other studies have suggested that population changes of small mammals and birds following herbicide treatments were either short-lived or not detectable (Bramble et al. 1984, Morrison and Meslow 1984, Gruver and Guthery 1986, D'Anieri et al. 1987, Sullivan 1990). Therefore, speculative comments on the response of wildlife species to herbicide treatment cannot be generalized. Each case must be examined separately relative to plant composition, the specific herbicide used, timing and method of application, and the particular wildlife species in question.

Large Mammals. Maximum white-tailed deer densities have been estimated to occur at the point where woody canopy becomes dense enough to begin competing with herbaceous vegetation for light, water, and soil nutrients (Inglis 1985). Thinning of excess woody cover leads to increased production of the herbaceous component providing a more diverse food supply (Thompson et al. 1991).

At 2 years after spraying Gambel oak (*Quercus gambellii*) with 2,4,5-TP, Kufeld (1977) found elk (*Cervus canadensis*) and mule deer (*Odocoileus hemionus*) use of the sprayed sites increased 73 and 16%, respectively, over pretreatment use. At 5 years after spraying, elk use was 11% over pretreatment levels, but mule deer use was 21% below. The trend of elk and deer use basically followed the pattern of grass production following partial removal of the oak. The increased grass production essentially changed the forage composition, creating a more diverse food supply than found in the untreated state.

Halls and Crawford (1965) found aerial spraying reduced production of preferred deer browse for the first 2 years in the Ozark Mountains of Arkansas. However, production of preferred species increased in succeeding years. By 8 years after treatment, production of preferred browse species was 6 times greater on treated than on untreated areas.

CASE STUDIES

Strip and Block Applications

Because aerial spraying of 2,4,5-T did not effectively control many common woody species in mixed-brush communities, repeat applications were sometimes practiced in an attempt to eliminate woody cover (Scifres et al. 1978). However, mechanical methods were the most widely used for brush suppression. With the advent of herbicides that controlled a broader spectrum of woody plants (e.g., picloram), aerial spraying for range improvement became widely accepted. Beasom and Scifres (1977) and Beasom et al. (1982) investigated complete and partial spraying of mixed-brush communities in south Texas for the potential of improving rangeland for livestock grazing without detriment to wildlife populations.

Beasom and Scifres (1977), aerially applied 2,4,5-T + picloram (1:1) at 1.1 kg/ha in alternating strips to achieve 80% coverage of a mature honey mesquite-dominated brushland. Another similar area was sprayed to achieve 100% coverage as a broadcast treatment. Game population responses were investigated by helicopter counts at 3, 9, 15, 21, and 27 months post spray and were compared to an adjacent untreated area.

The 80% strip treatment showed no net population changes of white-tailed deer, nilgai antelope (*Boselaphus tragocamelus*), wild turkeys (*Meleagris gallopavo*), or feral hogs (*Sus scrofa*) (Fig. 12). Complete (100%) treatment apparently exceeded the threshold of suitability for all animals surveyed with the exception of nilgai. The lack of response by nilgai was attributed to their apparent ability to shift dietary intake relative to plant availability more readily than the native animal species. Nilgai readily adapt to a grass-dominated diet if browse and forbs become limited (Sheffield et al. 1983).

Approximately 44% of the variation in white-tailed deer density was accounted for by the variation in forb production following treatment. With the restoration of production and species diversity of forbs at 27 months after treatment, deer numbers were not different among treatments (Fig. 12).

Estimates of turkey populations were highly variable but tended to increase on the untreated area and decrease on the treated areas (Fig. 12). However, turkeys continued to use the treated areas for some of their daily activities throughout the sampling period.

Feral hogs were unaffected by the strip treatment but tended to decrease their use of the block-sprayed area (Fig. 12). Although feral hogs were essentially unaffected, collared peccaries (*Dicotyles tajacu*) showed a dramatic negative response in the strip- and block-sprayed treatments. The response by collared peccaries was attributed to the virtual elimination of prickly pear (*Opuntia* spp.), their most important food (Everitt et al. 1981), after application of picloram.

Broadcast Application

Because vegetation response to herbicide treatments is site specific, managers tend to treat only those sites with the greatest potential for herbaceous production to increase carrying capacity more than indiscriminate strip-treatment of an entire pasture. In much of south and west Texas, honey mesquite-dominated drainages offer the greatest potential for herbaceous forage production but are also considered one of the most important habitats for white-tailed deer (McMahan and Inglis 1974, Darr and Klebenow 1975, Tanner et al. 1978). Consequently, managers often refrained from treating drainages to avoid possible damage to deer habitat.

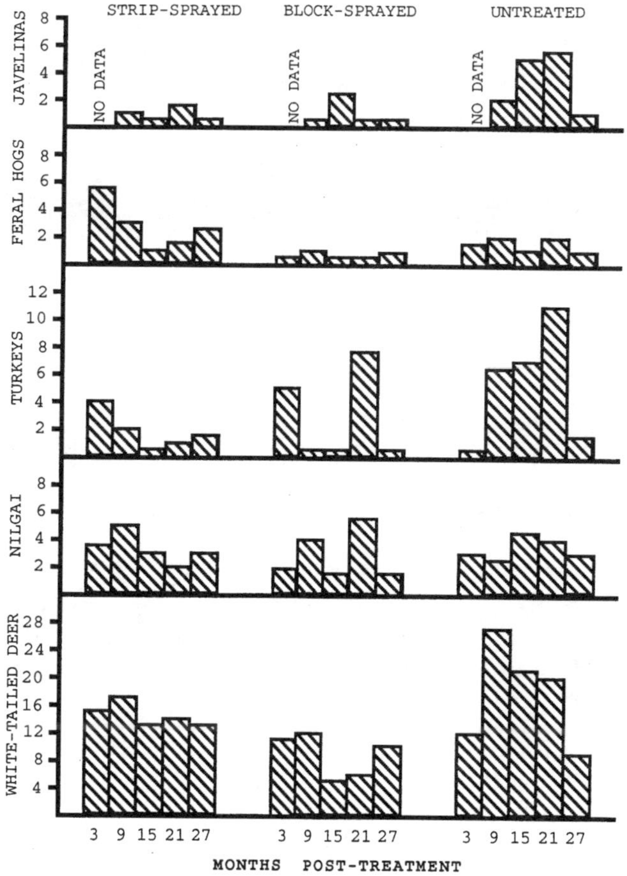

FIGURE 12. Average density (animals/100 ha) of white-tailed deer, nilgai antelope, feral hog, wild turkey, and javelina during various census periods following aerial application from May 11 to 15, 1973, of 1.12 kg/ha 2,4,5-T + picloram to 80% of the brushland in alternating strips (strip sprayed), complete treatment (block sprayed), or no treatment of south Texas mixed brush near Raymondville. From Beasom and Scifres (1977).

Beasom et al. (1982) investigated the effects of site discriminate treatment on white-tailed deer by broadcast-spraying 70% of a drainage (20% of the pasture) that was bordered by upland sites of lesser potential productivity. Based on fecal pellet group accumulations, they concluded that discriminate treatment of the drainage site did not cause consistent differences in deer use of the habitat and did not change deer use of the pasture containing the treatment. A 3- to 10-fold increase in grass production in the treated area also may have caused a shift in livestock diets away from preferred deer foods. Therefore, at least a portion of preferred habitat of white-tailed deer can be broadcast-sprayed if adjacent habitat is left undisturbed.

Variable Rate Patterning

Because large scale block-spraying is usually considered detrimental to wildlife habitat, strip applications alternating with untreated strips of vegetation have been considered the best design for herbicide applications (Fig. 13). A single herbicide dosage is usually applied in the strip to suppress the maximum

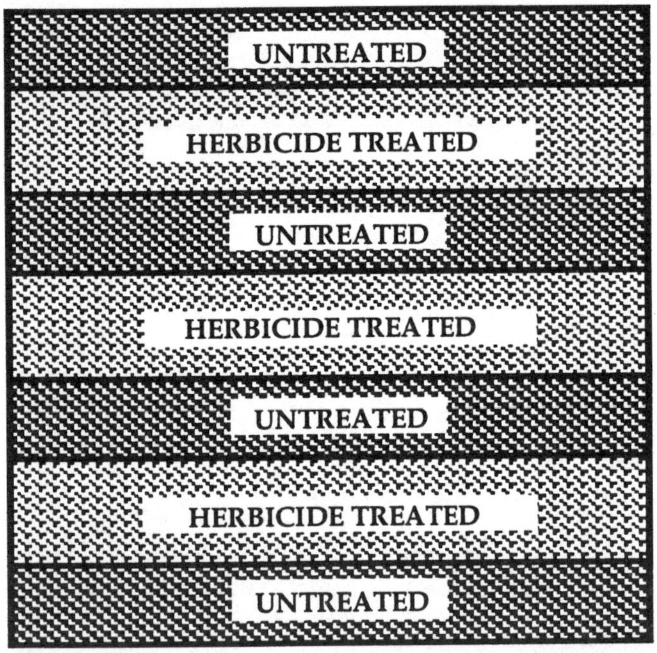

FIGURE 13. Example of herbicides applied in a strip pattern.

proportion of trees and shrubs. Scifres and Koerth (1986) proposed an alternative to strip patterning generically referred to as variable rate patterning (VRP). The simplest VRP can be installed by applying herbicide at half the recommended rate in treated and untreated strips in 2 directions, with the second set of strips applied over the same area but at right angle to the first (Fig. 14). This produces a 3-dosage VRP that is theoretically a mosaic of (1) undisturbed blocks where untreated strips intersect, (2) herb-shrub blocks where the single dose suppresses shrubs and promotes herbaceous growth, and (3) herbaceous-dominated blocks where the full dose at herbicide strip intersections is lethal to most woody species.

The checkerboard effect from VRP is somewhat misleading given the asymmetry that results in complex plant associations (Scifres and Koerth 1986). Ordination analysis of vegetation cover data from a 5-dosage VRP receiving tebuthiuron at 0, 1.1, 2.2, 3.3, and 4.4 kg/ha showed sites that were dissimilar in the untreated state may become functionally similar after treatment (Fig. 15). Structurally and botanically diverse patches emerged because of the various combinations of brush and herbaceous cover created by herbicide and site interactions. This asymmetry also functions to amplify edge compared to strip and block treatments. By 27 months after treatment, distance to closure of the cover screen was significantly increased as herbicide dosage exceeded 2.2 kg/ha. However, with the application pattern used, average distance to closure at deer height was ≤38 m.

Botanical composition and canopy cover of herbaceous vegetation varied with herbicide dosage, range site, and time after treatment (Scifres and Koerth 1986). The greatest change was an increase in grass cover on treated areas. Forb cover and diversity varied with herbicide dosage, range site, and rainfall. Forb populations remained unchanged on sites with shallow, gravelly soil regardless of evaluation date or treatment. On upland sites with deeper soils and in drainages, forb populations were initially decreased as herbicide dosage exceeded 2.2 kg/ha. However, forb availability for wildlife was assured by adjacent untreated patches and was generally greater on treated than untreated sites by the second and third growing season following application of ≤2.2 kg/ha.

Herbicide treatments may cause deer to temporarily leave treated areas, presumably because of the negative effects on forbs. Deer numbers generally return to pretreatment levels or greater after recovery of forbs and appearance of woody plant regrowth (Beasom and Scifres 1977, Tanner et al. 1978). Based on spotlight counts, Taylor (1989) concluded that at 4 years after application of a 3-dosage VRP using 0, 1.1, and 2.2 kg/ha of 2,4,5-T + picloram, there were no differences in deer abundance or distribution between treated and adjacent untreated pastures. However, carrying capacity for livestock in the treated pastures increased 20% in response to decreased woody cover and increased grass and forb production (Taylor 1989).

Application costs can be lower with VRP compared to other approaches. Based on 1987 herbicide and contract application costs of tebuthiuron, a 3-dose VRP using 0, 1.1, and 2.2 kg/ha cost 23% less than application of 2.2 kg/ha in alternating strips to a comparable land area, and 48% less than a broadcast treatment at 2.2 kg/ha (Scifres et al. 1988*b*).

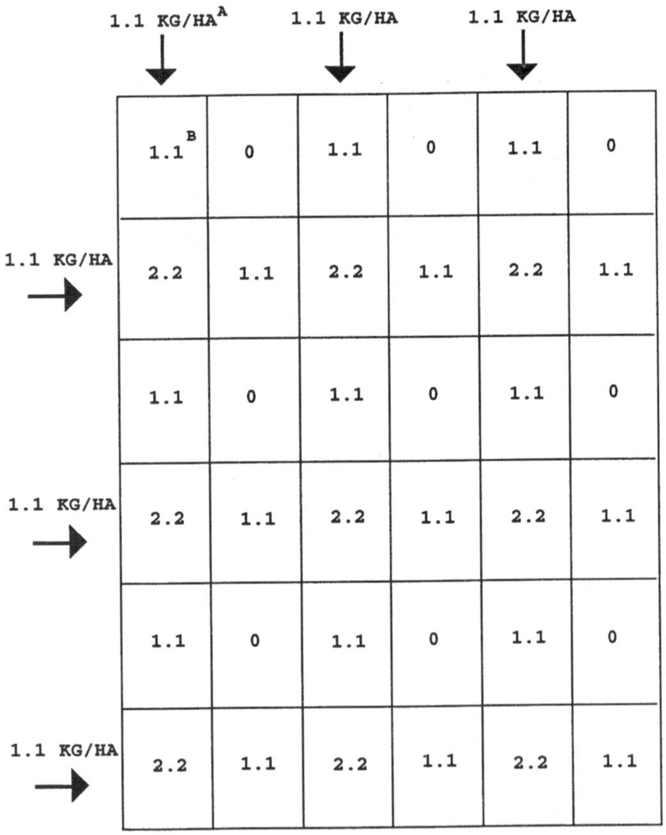

FIGURE 14. Example of 3-dosage variable rate pattern for herbicide application (from Scifres and Koerth 1989).
[A]Herbicide dosage applied in that strip.
[B]Herbicide dosage applied in that block.

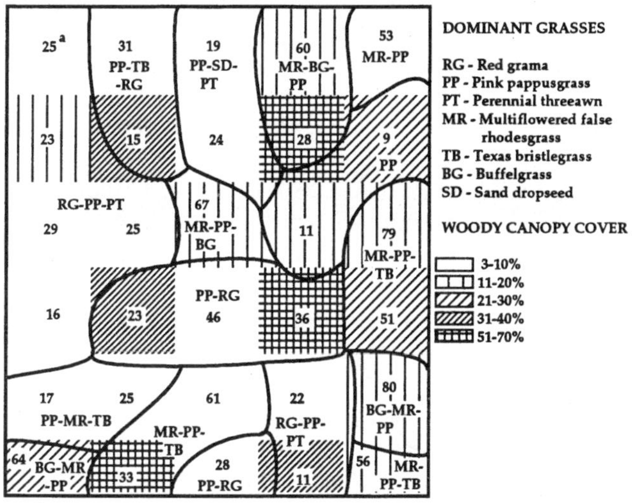

[a] Values indicate average grass cover (%) in that treatment block.

FIGURE 15. Distribution of herbaceous associations, average percent grass cover, and alterations in shrub canopy covers in autumn 1985 after application of tebuthiuron in a 5-dosage variable rate pattern in the spring 1983 near Laredo, Texas. From Scifres and Koerth (1986).

Forest Site Preparation

Mechanical treatments followed by planting have been a typical procedure following clearcutting in pine (*Pinus* spp.) plantations, but mechanical treatments are expensive and intensive treatments can disturb soil and degrade the habitat (Hebb 1971). Herbicides have become an increasingly acceptable alternative to suppress competing vegetation and hasten reforestation of commercial timber species (Morrison and Meslow 1984). Because broadscale use of herbicides in forests is relatively new, long-term impacts of herbicides on forest wildlife habitat and biodiversity is only beginning to receive research attention (Schulz et al. 1992*b*).

In concurrent studies, Santillo et al. (1989*a,b*) investigated the influence of glyphosate-treated clearcuts on breeding songbird and small mammal populations in north-central Maine. Most grasses, forbs, and deciduous trees and shrubs are susceptible to glyphosate, thereby increasing the probability of reforestation by planted conifer seedlings. Herbicide-induced alterations of the habitat and food resources were evident for up to 3 years post-treatment. The principal impact on the habitat was a reduction in vegetative cover and diversity. As intended, treatment with glyphosate changed the vegetative community from deciduous brush dominated to conifer dominated.

Vegetation changes also significantly affected the invertebrate community (Santillo et al. 1989*b*). Herbivorous insects were reduced 89% at 1 year after treatment and were still reduced 25% at 3 years after treatment compared to untreated areas. No trend was found for predatory insects. The reduction in insect numbers and changes in the plant community resulted in changes in abundance of both small mammals and songbirds. Insectivorous birds and mammals were reduced for up to 3 years after treatment. Herbivorous mammals were reduced up to 2 years but by the third year there were no differences between treated and untreated areas. Omnivorous birds and mammals were essentially unaffected by the treatment.

Basically, the untreated areas had more vegetation diversity and contained the highest vertebrate species richness and diversity. It is likely that the differences in species composition would persist, if not increase, as vegetation succession progressed. In contrast, application of herbicides intermixed within larger areas of mature oak forests in Oklahoma increased habitat diversity (Lochmiller et al. 1991) and overall species richness of breeding nongame birds (Schulz et al. 1992*a,b*). In this case, herbicide treatments provided a mosaic of habitats and possibly increased food and nesting habitat availability.

Santillo et al. (1989*a,b*) suggested options to maintain vertebrate diversity in treated clearcuts including intentionally leaving untreated areas where conifer seedlings usually experience severe damage and mortality (e.g., small drainages, wet swales, road edges, and unstable soil on steep slopes). Areas in the treated clearcuts under study that were unintentionally missed during treatment maintained biodiversity equal to the untreated control areas. Similar observations were found by Morrison and Meslow (1984) in Oregon. Another option was to stagger treatment of portions of large clearcuts 3 years apart. This 3-year interval would allow treated areas to attain some degree of structural complexity from partial recovery of woody and herbaceous vegetation before the next portion of the clearcut was treated.

ECONOMIC IMPLICATIONS

Acceptance or rejection of any management activity is commonly based upon economics. Projected monetary gain or loss frequently indicates if a project should be undertaken. Effective evaluation of a treatment requires that the goals and objectives for the management area be clearly defined. Reliable data on costs, risks, and projected benefits resulting from the treatment must be obtained. Contract application costs for most brush management techniques are usually readily available. However, changes in plant communities and in production of salable products from livestock, hunting opportunities, and recreation are more variable and estimates may have to be derived from experienced personnel.

Because the relationship between production and income is based upon a structured marketing system (Conner 1985), domestic livestock are a readily observable economic entity. Tangible values for wildlife and recreation are less straightforward but not impossible to obtain. Commensurate values for hunting opportunities and recreation have been quantified using consumer willingness to pay (Loomis et al. 1991). Values for individual game and nongame animals have been estimated by the cost associated with replacing animals lost or displaced by a management activity (Norman et al. 1976).

Whitson et al. (1977) provided an example economic assessment of the combined benefits of livestock production and lease hunting opportunities following partial spraying (80% in a strip pattern), complete treatment, and no treatment of mixed brush in south Texas (Beasom and Scifres 1977). Assuming a situation where investment capital is limited, partial spraying in alternating strips yielded a higher rate of return than complete treatment or no treatment. Hunting opportunities more than compensated for reduced livestock production where quality wildlife habitat was retained. Similar viewpoints on the economic benefits of multiple use of rangeland have been presented by Robinson and Bolen (1984:180), Bastian et al. (1991), and Loomis et al. (1991).

SUMMARY

While the use of chemicals to suppress plant growth cannot be considered new, the development of commercial production and widespread use of herbicides has basically occurred since the 1950s. Production of warm-season herbaceous plants decreases as woody canopy increases. Some cool-season herbaceous plants may increase production with light-to-moderate woody canopy cover. Infestations of woody plants can create a shrub-driven successional process making it unlikely that the area can be reverted to a grassland. Plant communities must be manipulated to maintain the proper proportions of woody and herbaceous components to provide requisite habitat for wildlife.

Compared to mechanical treatments, herbicide treatments create more subtle changes in the plant community. A negative aspect of herbicide applications is that forb populations are usually suppressed for at least a growing season. However, herbicides can be used to manage wildlife habitat as long as appropriate patterned applications are used.

General categories of herbicides are foliar- and soil-active. Foliar herbicides are applied directly to the foliage of the plants and depend upon absorption by leaf or stem tissue. Soil-active herbicides are applied to the soil surface and depend upon rainfall to move the herbicide into the rooting zone of the plants. In the soil profile, the herbicide is absorbed by the roots.

Aerial application methods are relatively insensitive to soil and terrain factors, and woody plant densities that may make other methods unfeasible. A disadvantage is the necessity of a suitable airstrip for fixed-wing aircraft. Rotary-wing aircraft do not need an airstrip but carry smaller payloads than fixed-wing aircraft.

Application systems mounted on tractors or other vehicles are effective for treatment of regrowth brush and weeds. Size and density of woody plants and rough terrain are factors that limit use of vehicles. Hand-applied systems can be extremely selective in suppressing target species or plants but are limited by time and labor constraints.

Persistence of a herbicide in the environment depends upon the particular chemical used, soil, and climatic factors. Soil-applied herbicides generally persist longer than foliar-applied herbicides. Herbicides generally degrade faster with increasing temperature, soil moisture, soil microflora, and soil organic matter content.

Rangeland herbicides are relatively nontoxic to mature mammals and birds when applied at recommended rates. Less is known about the effects on embryonic stages. Applicators should strictly follow the herbicide label for application particulars. Care should be used in cleaning equipment and disposing wastes to avoid contaminating water sources. There is no indication that rangeland herbicides will bioconcentrate.

Principal acute impacts on wildlife habitat from herbicide treatments are loss of woody cover and potential reduction of food through defoliation of browse plants and suppression of forbs. Increased production of herbaceous plants after removal of woody plant competition for nutrients and light may partially offset losses of browse. Herbicides, coupled with mechanical soil disturbance, can be used to manipulate herbaceous plant composition in favor of warm- or cool-season species.

The major long-term impact on wildlife habitat is a reduction in habitat complexity. However, on a large scale, selective treatments may enhance biodiversity by creating shrub and herbaceous dominated openings in mature woodlands. Wildlife species that have obligate relationships with certain plant species can be harmed by treatments that remove that plant from the habitat. Likewise, wildlife species with limited mobility can be harmed by changes in the habitat. Highly mobile species are less likely to be harmed and may respond favorably to proper habitat manipulation. Thinning of excessive woody cover changes the forage composition and may create a more diverse food supply than found in the untreated state.

Treatment of 80% of a mature honey mesquite-dominated rangeland in alternating strips in south Texas produced no appreciable changes in numbers of native or exotic animals with the exception of collared peccaries. Reduced use by collared peccaries was attributed to the virtual elimination of their most important food. Complete block-spraying of a pasture exceeded the threshold of suitability for all native species. However, at least a portion of preferred habitat for deer can be broadcast-sprayed as long as adjacent, though less preferred, habitat is left undisturbed.

An alternative to strip spraying referred to as variable rate patterning (VRP) has been proposed. A VRP is installed by applying herbicide strips in 2 directions, with the second set over the same area but at right angle to the first. The resultant vegetation pattern is a unique, asymmetrical mosaic that is a function of the particular site by herbicide interactions on the treated area.

Herbicides effectively suppress competing hardwood and herbaceous vegetation and facilitate reforestation by planted conifers. Herbivorous insects declined on treated areas causing the concomitant decline in insectivorous birds and mammals. Herbivorous mammals were initially reduced following treatment but regained pretreatment numbers as herbaceous vegetation began to recover. No changes were detected in numbers of omnivorous birds and mammals. Biodiversity may be maintained on treated clearcuts by intentionally leaving untreated areas or by staggering treatment of portions of large clearcuts 3 years apart.

Before selecting a management option, define goals and objectives for the management area and obtain reliable data on costs, risks, and benefits. Economic benefits can be derived from managing rangeland for multiple use when wildlife and recreation are given real value.

LITERATURE CITED

Akamine, E. K. 1951. Persistence of 2,4-D in Hawaiian soils. Bot. Gaz. 112:312-319.

Archer, S. 1989. Have southern Texas savannas been converted to woodlands in recent history? Am. Nat. 134:545-561.

Bastian, C. T., J. J. Jacobs, L. J. Held, and M. A. Smith. 1991. Multiple use of public rangeland: Antelope and stocker cattle in Wyoming. J. Range Manage. 44:390-394.

Baur, J. R., R. W. Bovey, and H. G. McCall. 1974. Thermal and ultraviolet loss of herbicides. Pestic. Abstr. 8:217.

Beasom, S. L., J. M. Inglis, and C. J. Scifres. 1982. Vegetation and white-tailed deer responses to herbicide treatment of a mesquite drainage habitat. J. Range Manage. 35:790-794.

———, and C. J. Scifres. 1977. Population reactions of selected game species to aerial herbicide applications. J. Range Manage. 30:138-142.

Bellrose, F. C. 1956. The control of aquatic weeds in waterfowl habitats. Proc. North Central Weed Control Conf. 13:26.

Blakey, H. L. 1947. The role of brush control in habitat improvement on the Aransas National Wildlife Refuge. Trans. North Am. Wildl. and Nat. Resour. Conf. 12:179-185.

Boggs, J. F., R. L. Lochmiller, S. T. McMurry, D. M. Leslie, Jr., and D. M. Engle. 1991*a*. *Cuterebra* infestations in small-mammal communities as influenced by herbicides and fire. J. Mammal. 72:322-327.

———, S. T. McMurry, D. M. Leslie, Jr., D. M. Engle, and R. L. Lochmiller. 1990*a*. Influence of habitat modification on the intestinal helminth community ecology of cottontail rabbit populations. J. Wildl. Dis. 26:157-169.

———, ———, ———, ———, and ———. 1990*b*. Parasitism of cottontail rabbits (*Sylvilagus floridanus*) by *Obeliscoides cuniculi* in response to habitat modification in the cross timbers of Oklahoma. J. Helminthol. Soc. 57:146-152.

———, ———, ———, ———, and ———. 1991*b*. Influence of habitat modification on the community of gastrointestinal helminths of cotton rats. J. Wildl. Dis. 27:584-593.

Bovey, R. W., and C. J. Scifres. 1971. Residual characteristics of picloram in grassland ecosystems. Texas Agric. Exp. Stn. Bull. 1111. 24pp.

Box, T., and J. Powell. 1965. Brush management techniques for improved forage values in south Texas. Trans. North Am. Wildl. and Nat. Resour. Conf. 30:285-296.

Bramble, W. C., W. R. Byrnes, and M. D. Schuler. 1984. The bird population of a transmission right-of-way maintained by herbicides. J. Aboric. 10:13-20.

Braun, C. E., T. Britt, and R. O. Wallestad. 1977. Guidelines for maintenance of sage grouse habitats. Wildl. Soc. Bull. 5:99-106.

Brock, J. L., R. H. Haas, and J. C. Shaver. 1978. Zonation of herbaceous vegetation associated with honey mesquite in north-central Texas. Proc. Int. Rangeland Congr. 1:187-189.

Bull, E. L., and A. D. Partridge. 1986. Methods of killing trees for use by cavity nesters. Wildl. Soc. Bull. 14:142-146.

Burnside, O. C., and T. L. Lavy. 1966. Dissipation of dicamba. Weeds. 14:211-214.

Carr, H. D., and F. A. Glover. 1970. Effects of sagebrush control on sage grouse. Trans. North Am. Wildl. and Nat. Resour. Conf. 35:205-215.

Chang, S. S., and J. F. Stritzke. 1977. Sorption, movement, and dissipation of tebuthiuron in soils. Weed Sci. 25:184-188.

Conner, J. R. 1985. Range improvements: The decision process. Pages 174-185 *in* Proc. International Ranchers Roundup. Tex. Agric. Extension Serv., Laredo. 489pp.

Conner, R. N., J. G. Dickson, and B. A. Locke. 1981. Herbicide-killed trees infected by fungi: potential cavity sites for woodpeckers. Wildl. Soc. Bull. 9:308-311.

———, J. C. Kroll, and D. L. Kulhavy. 1983. The potential of girdled and 2,4-D injected southern red oaks as woodpecker nesting and foraging sites. South. J. Appl. For. 7:125-128.

Coulter, L. L. 1957. The role of herbicides in wildlife production through creation and stabilization of habitats. Down-To-Earth 13:4-6.

Crane, R. A., C. J. Scifres, B. H. Koerth, and R. C. Flinn. 1989. Control of honey mesquite, huisache, and blackbrush acacia with herbicides applied by carpeted roller. Pages 9-22 *in* W. T. Hamilton and T. G. Welch, eds. Management of south Texas mixed brush with herbicides. Tex. Agric. Exp. Stn. Bull. 1623.

Crawford, H. S. 1960. Effect of aerial 2,4,5-T sprays on forage production in west-central Arkansas. J. Range Manage. 13:44.

D'Anieri, P., D. M. Leslie, Jr., and M. L. McCormack, Jr. 1987. Small mammals in glyphosate-treated clearcuts in northern Maine. Can. Field-Nat. 101:547-550.

Darr, G. W., and D. A. Klebenow. 1975. Deer, brush control, and livestock on the Texas Rolling Plains. J. Range Manage. 28:115-119.

Davis, R. B., and C. K. Winkler. 1968. Brush vs. cleared range as deer habitat in south Texas. J. Wildl. Manage. 32:321-329.

DeFazio, J. T., Jr., A. E. Stone, and R. J. Warren. 1988. Effects of tebuthiuron site preparation on white-tailed deer habitat. Wildl. Soc. Bull. 16:12-18.

Doerr, T. B., and F. S. Guthery. 1983. Effects of tebuthiuron on lesser prairie-chicken habitat and foods. J. Wildl. Manage. 47:1138-1142.

Dowler, C. C., W. Forestier, and F. H. Tschirley. 1968. Effect and persistence of herbicides applied to soil in Puerto Rican forests. Weeds. 16:45-50.

Drew, G. S. 1988. Coyote-prey interactions on an intensively managed south Texas ranch. M. S. Thesis, Texas A&M Univ., College Station. 88pp.

Duncan, K. E., and C. J. Scifres. 1983. Influence of clay and organic matter of rangeland soils on tebuthiuron effectiveness. J. Range Manage. 36:295-297.

Dwernychuk, L. W., and D. A. Boag. 1973. Effect of herbicide induced changes in vegetation on nesting ducks. Can. Field-Nat. 87:155-165.

Erne, K. 1966. Studies on the analytical chemistry and toxicology of phenoxy herbicides. Natl. Vet. Inst., Stockholm, Sweden.

Everitt, J. H., C. L. Gonzalez, M. A. Alaniz, and G. V. Latigo. 1981. Food habits of the collared peccary on south Texas rangelands. J. Range Manage. 34:141-144.

Fulbright, T. E., and S. L. Beasom. 1987. Long-term effects of mechanical treatments on white-tailed deer browse. Wildl. Soc. Bull. 15:560-564.

———, and A. Garza, Jr. 1991. Forage yield and white-tailed deer diets following live oak control. J. Range Manage. 44:451-455.

Gruver, B. J., and F. S. Guthery. 1986. Effects of brush control and game-bird management on non-game birds. J. Range Manage. 39:251-253.

Guthery, F. S., T. E. Shupe, L. J. Bareiss, and C. E. Russell. 1987. Responses of selected plants to herbicide treatment of disturbed soil. Wildl. Soc. Bull. 15:247-251.

Halls, L. K., and H. S. Crawford. 1965. Vegetation response to an Ozark woodland spraying. J. Range Manage. 18:338-340.

Hamilton, K. C., and K. P. Buckholtz. 1953. Use of herbicides for establishing food patches. J. Wildl. Manage. 17:509-516.

Hamilton, W. T., J. R. Conner, J. W. Stuth, G. L. McBryde, and A. J. Vega. 1986. Incorporating precipitation-induced variation in annual forage production into economic analyses of range improvement practices. Tex. Agric. Exp. Stn. Bull. 1519. 16pp.

Hebb, E. A. 1971. Site preparation decreases game food plants in Florida Sandhills. J. Wildl. Manage. 35:155-162.

Herbicide Handbook Committee. 1983. Herbicide handbook of the Weed Science Society of America. Fifth ed. Weed Sci. Soc. Am., Champaign, Ill. 515pp.

Hobbs, R. J., and H. A. Mooney. 1986. Community changes following shrub invasion of grassland. Oecologia. 70:508-513.

Hoffman, D. J. 1983. Effects of Krenite brush control agent (fosamine ammonium) on embryonic development in mallards and bobwhite. Environ. Toxic. and Chem. 7:69-75.

———, and W. C. Eastin. 1982. Effects of lindane, paraquat, toxaphene, and 2,4,5-trichlorophenoxyacetic acid on mallard embryo development. Arch. Environ. Contam. Toxicol. 11:79-86.

Hull, A. C., Jr., 1971. Effect of spraying with 2,4-D upon abundance of pocket gophers in Franklin Basin, Idaho. J. Range Manage. 24:230-232.

Hurst, G. A., and R. C. Warren. 1986. Deer forage on pine plantations after a herbicide application for pine release. Proc. South. Weed Sci. Soc. 39:238.

Inglis, J. M. 1985. Wildlife management and IBMS. Pages 35-40 *in* C. J. Scifres, ed. Integrated brush management systems for south Texas: Development and implementation. Tex. Agric. Exp. Stn. Bull. 1493.

Johnsen, T. N., Jr., and H. L. Morton. 1991. Long-term tebuthiuron content of grasses and shrubs on semiarid rangelands. J. Range Manage. 44:249-253.

Johnson, D. R., and R. M. Hansen. 1969. Effects of range treatment with 2,4-D on rodent populations. J. Wildl. Manage. 33:125-132.

Johnson, J. E. 1971. The public health implications of widespread use of the phenoxy herbicides and picloram. Bioscience 21:899-905.

Johnson, W. M. 1969. Life expectancy of a sagebrush control in central Wyoming. J. Range Manage. 22:177-182.

Keith, J. O., R. M. Hansen, and A. L. Ward. 1959. Effect of 2,4-D on abundance and foods of pocket gophers. J. Wildl. Manage. 23:137-145.

Klebenow, D. A. 1970. Sage grouse versus sagebrush control in Idaho. J. Range Manage. 23:396-400.

———, and G. M. Gray. 1968. The food habits of juvenile sage grouse. J. Range Manage. 21:80-83.

Klingman, G. C. 1964. Weed control: as a science. John Wiley & Sons, New York, N.Y. 421pp.

Koerth, B. H., C. J. Scifres, R. C. Flinn, and R. A. Crane. 1989. Influence of pre- and post-application management practices on efficacy of pelleted herbicides. Pages 37-44 *in* W. T. Hamilton and T. G. Welch, eds. Management of south Texas mixed brush with herbicides. Tex. Agric. Exp. Stn. Bull. 1623.

Krefting, L. W., and H. L. Hansen. 1969. Increasing browse for deer by aerial applications of 2,4-D. J. Wildl. Manage. 33:784-790.

———, ———, and M. H. Stenlund. 1956. Stimulating regrowth of mountain maple for deer browse by herbicides, cutting, and fire. J. Wildl. Manage. 20:434-441.

Kroll, J. C. 1991. A practical guide to producing and harvesting white-tailed deer. Center for Applied Studies in Forestry, School of Forestry, Stephen F. Austin State Univ., Nacogdoches, Tex. 590pp.

Kufeld, R. C. 1977. Improving gambel oak ranges for elk and mule deer by spraying with 2,4,5-TP. J. Range Manage. 30:53-57.

Lehmann, V. W. 1960. Problems of maintaining game on ranges subjected to brush control. Proc. World For. Congr. 5:1807-1809.

Leng, M. L. 1972. Residues in milk and meat and safety to livestock use of phenoxy herbicides in pastures and rangeland. Down-To-Earth 28:12-20.

Leopold, A. 1933. The conservation ethic. J. For. 31:634-643.

Lochmiller, R. L., J. F. Boggs, S. T. McMurry, D. M. Leslie, Jr., and D. M. Engle. 1991. Response of cottontail rabbit populations to herbicide and fire applications on cross timbers rangeland. J. Range Manage. 44:150-155.

Loomis, J. B., E. R. Loft, D. R. Updike, and J. G. Kie. 1991. Cattle-deer interactions in the Sierra Nevada: A bioeconomic approach. J. Range Manage. 44:395-399.

Lutz, J. F., G. E. Byers, and T. J. Sheets. 1973. The persistence and movement of picloram and 2,4,5-T in soils. J. Environ. Quality 2:485-488.

Lym, R. G., and C. G. Messersmith. 1991. Correlation of environment and root carbohydrate content to picloram translocation in leafy spurge. J. Range Manage. 44:254-258.

Martin, N. S. 1970. Sagebrush control related to habitat and sage grouse occurrence. J. Wildl. Manage. 34:313-320.

Mayeux, H. S., Jr. 1987. Application of herbicides on rangelands with a carpeted roller: shrub density influences volume, dosage and rate. Weed Sci. 35:444-448.

———, and R. A. Crane. 1983. The brush roller—an experimental herbicide applicator with potential for range weed and brush control. Rangelands 5:53-56.

———, and ———. 1984. Application of herbicides on rangelands with a carpeted roller: control of goldenweeds (*Isocoma* spp.) and false broomweed (*Ericameria austrotexana*). Weed Sci. 32:845-849.

———, and ———. 1985. Application of herbicides on rangelands with a carpeted roller: evaluation of four herbicides for control of honey mesquite. J. Range Manage. 38:233-237.

McCaffery, K. R., L. D. Martoglio, and F. L. Johnson. 1974. Maintaining wildlife openings with picloram pellets. Wildl. Soc. Bull. 2:40-45.

McComb, W. C., and R. L. Rumsey. 1981. Habitat characteristics of forest clearings created by picloram herbicides and clearcutting. Proc. Annu. Conf. S. E. Assoc. Fish and Wildl. Agencies 35:174-184.

———, and ———. 1983. Characteristics and cavity-nesting bird use of picloram-created snags in the central Appalachians. South. J. Appl. For. 7:34-37.

McMahan, C. A., and J. M. Inglis. 1974. Use of Rio Grande Plain brush types by white-tailed deer. J. Range Manage. 27:369-374.

Messersmith, C. G., and R. G. Lym. 1985. Roller application of picloram for leafy spurge control in pastures. Weed Sci. 33:258-262.

Morrison, M. L., and E. C. Meslow. 1984. Response of avian communities to herbicide-induced vegetation changes. J. Wildl. Manage. 48:14-22.

Morton, H. L., E. D. Robison, and R. E. Meyer. 1967. Persistence of 2,4-D, 2,4,5-T and dicamba in range forage grasses. Weeds. 15:268-271.

Mueggler, W. F. 1966. Herbicide treatment of browse on a big-game winter range in northern Idaho. J. Wildl. Manage. 30:141-151.

Newton, M., E. C. Cole, R. A. Lautenschlager, D. E. White, and M. L. McCormack, Jr. 1989. Browse availability after conifer release in Maine's spruce-fir forests. J. Wildl. Manage. 53:643-649.

Niering, W. A., and R. H. Goodwin. 1974. Creation of relatively stable shrublands with herbicides: arresting "succession" on rights-of-way and pastureland. Ecology 55:784-795.

Norman, R. L., L. A. Roper, P. D. Olson, and R. L. Evans. 1976. Using wildlife values in benefit/cost analysis and mitigation of wildlife losses. Colo. Div. Wildl., Denver. 18pp.

Norris, L. A. 1981. The behavior of herbicides in the forest environment and risk assessment. Pages 192-215 *in* Proc. J. S. Wright Forestry Conf., weed control in forest management. Purdue Univ., West Lafayette, Ind.

O'Connor, G. A., and J. U. Anderson. 1974. Soil factors affecting the adsorption of 2,4,5-T. Proc. Soil Sci. Soc. Am. 38:433-436.

Olawsky, C. D., and L. M. Smith. 1991. Lesser prairie-chicken densities on tebuthiuron-treated and untreated sand shinnery oak rangelands. J. Range Manage. 44:364-368.

Peterson, J. G. 1970. The food habits and summer distribution of juvenile sage grouse in central Montana. J. Wildl. Manage. 34:147-155.

Pletscher, D. H., and R. J. Robel. 1979. Possible use of 2,4-D in management of sunflower and ragweed. Wildl. Soc. Bull. 7:48-53.

Radosevich, S. R., and W. L. Winterlin. 1977. Persistence of 2,4-D and 2,4,5-T in chaparral vegetation and soil. Weed Sci. 25:423-425.

Robinette, W. L. 1972. Browse and cover for wildlife. Pages 69-76 *in* C. M. McKell, J. P. Blaisdell, J. R. Goodin, eds. Wildland shrubs - their biology and utilization. U.S. Dep. Agric. For. Serv. Gen. Tech. Rep INT-1.

Robinson, W. L., and E. G. Bolen. 1984. Wildlife ecology and management. Macmillan Publ. Co., New York, N.Y. 478pp.

Rodgers, R. D., and M. L. Sexson. 1990. Impacts of extensive chemical control of sand sagebrush on breeding birds. J. Soil and Water Conserv. 45:494-497.

Santillo, D. J., P. W. Brown, and D. M. Leslie, Jr. 1989*a*. Response of songbirds to glyphosate-induced habitat changes in clearcuts. J. Wildl. Manage. 53:64-71.

———, D. M. Leslie, Jr., and P. W. Brown. 1989*b*. Response of small mammals and habitat to glyphosate application on clearcuts. J. Wildl. Manage. 53:164-172.

Savidge, J. A. 1978. Wildlife in a herbicide-treated Jeffrey pine plantation in eastern California. J. For. 76:476-478.

Schroeder, M. H., and D. L. Sturges. 1975. The effect on the Brewer's sparrow of spraying big sagebrush. J. Range Manage. 28:294-297.

Schulz, C. A., D. M. Leslie, Jr., R. L. Lochmiller, and D. M. Engle. 1992*a*. Autumn and winter bird populations in herbicide-treated cross timbers in Oklahoma. Am. Midl. Nat. 127:215-223.

———, ———, ———, and ———. 1992*b*. Herbicide effects on cross timbers breeding birds. J. Range Manage. 45:407-411.

Scifres, C. J. 1972*a*. Herbicide interactions in control of sand shinnery oak. J. Range Manage. 25:386-389.

———. 1972*b*. Sand shinnery oak response to dicamba granules and picloram pellets. J. Range Manage. 25:155-156.

———. 1977. Herbicides and the range ecosystem: Residues, research, and the role of rangemen. J. Range Manage. 30:86-91.

———. 1980. Brush management. Principles and practices for Texas and the southwest. Texas A&M Univ. Press, College Station. 360pp.

———. 1986. Integrated management systems for improvement of rangeland. Pages 227-259 *in* M. A. Sprague and G. B. Triplett, eds. No-tillage and surface-tillage agriculture: The tillage revolution. John Wiley & Sons, Inc., New York, N.Y.

———, R. A. Crane, B. H. Koerth, and R. C. Flinn. 1988*a*. Roller application of clopyralid for huisache (*Acacia farnesiana*) control. Weed Technol. 2:317-322.

———, D. L. Drawe, J. L. Mutz, O. E. Bontrager, and J. W. McAtee. 1978. Range improvement practices for the coastal prairie: needs, present applications, and research advances. Proc. Welder Wildl. Found. Symposium 1:15-24.

———, and R. H. Haas. 1974. Vegetation changes in a Post Oak savannah following woody plant control. Tex. Agric. Exp. Stn. Misc. Publ. 1136. 12pp.

———, R. R. Hahn, J. Diaz-Colon, and M. G. Merkle. 1971. Picloram persistence in semiarid rangeland soils and water. Weed Sci. 19:381-384.

———, W. T. Hamilton, B. H. Koerth, R. C. Flinn, and R. A. Crane. 1988*b*. Bionomics of patterned herbicide application for wildlife habitat enhancement. J. Range Manage. 41:317-321.

———, and ———. 1989. Factors affecting economic performance of herbicides. Pages 49-59 *in* W. T. Hamilton and T. G. Welch, eds. Management of south Texas mixed brush with herbicides. Tex. Agric. Exp. Stn. Bull. 1623.

———, and G. O. Hoffman. 1972. Comparative susceptibility of honey mesquite to dicamba and 2,4,5-T. J. Range Manage. 25:143-145.

———, and B. H. Koerth. 1986. Habitat alterations in mixed brush from variable rate herbicide patterns. Wildl. Soc. Bull. 14:345-356.

———, and ———. 1989. Patterned herbicide applications. Pages 45-48 *in* W. T. Hamilton and T. G. Welch, eds. Management of south Texas mixed brush with herbicides. Tex. Agric. Exp. Stn. Bull. 1623.

———, and J. L. Mutz. 1978. Herbaceous vegetation changes following applications of tebuthiuron for brush control. J. Range Manage. 31:375-378.

———, ———, R. E. Whitson, and D. L. Drawe. 1982. Interrelationships of huisache canopy cover with range forage on the coastal prairie. J. Range Manage. 35:558-562.

———, J. W. Stuth, and B. H. Koerth. 1987. Improvement of oak-dominated rangeland with tebuthiuron and prescribed burning. Tex. Agric. Exp. Stn. Bull. 1567. 24pp.

Sheffield, W. J., B. A. Fall, and B. A. Brown. 1983. The nilgai antelope in Texas. The Caesar Kleberg Res. Progr. in Wildl. Ecol., Texas A&M Univ., College Station. 100pp.

Simanton, J. R. 1991. Revegetation of semiarid rangelands: Problems, procedures, and probabilities. Rangelands 13:129-132.

Slagsvold, T. 1977. Bird population changes after clearance of deciduous scrub. Biol. Conserv. 12:229-244.

Soutiere, E. C., and E. G. Bolen. 1976. Mourning dove nesting on tobosa grass-mesquite rangeland sprayed with herbicides and burned. J. Range Manage. 29:226-231.

Spencer, S. R., and G. W. Barrett. 1980. Meadow vole population response to vegetational changes resulting from 2,4-D application. Am. Midl. Nat. 103:32-46.

Stuth, J. W., and B. E. Dahl. 1974. Evaluation of rangeland seedings following mechanical brush control in Texas. J. Range Manage. 27:146-149.

Sullivan, T. P. 1990. Influence of forest herbicide on deer mouse and Oregon vole population dynamics. J. Wildl. Manage. 54:566-576.

Tanner, G. W., J. M. Inglis, and L. H. Blankenship. 1978. Acute impact of herbicide strip treatment on mixed-brush white-tailed deer habitat on the northern Rio Grande plain. J. Range Manage. 31:386-391.

Taylor, M. A. 1979. Lesser prairie chicken use of man-made leks. Southwest. Nat. 24:683-714.

Taylor, T. T. 1989. White-tailed deer response to habitat attributes in the Rio Grande Plains of south Texas. Ph. D. Thesis, Texas A&M Univ., College Station. 154pp.

Thompson, M. W., M. G. Shaw, R. W. Umber, J. E. Skeen, and R. E. Thackston. 1991. Effects of herbicides and burning on overstory defoliation and deer forage production. Wildl. Soc. Bull. 19:163-170.

Tietjen, H. P., C. H. Halvorson, P. L. Hegdal, and A. M. Johnson. 1967. 2,4-D herbicide, vegetation, and pocket gopher relationships, Black Mesa, Colorado. Ecology 48:634-643.

Trumbo, H. A., and W. E. Chappell. 1961. Techniques involved in the use of chemicals for establishing wildlife clearings. Proc. Annu. Conf. S. E. Assoc. Fish and Wildl. Agencies 13:34-38.

United States Fish and Wildlife Service. 1982. Spraying of treflan may pose a hazard to reproduction in birds. Research Information Bull. 82-20, Patuxent Wildl. Res. Center, Laurel, Md. 1pp.

Wallestad, R. 1975. Male sage grouse responses to sagebrush treatment. J. Wildl. Manage. 39:482-484.

Whitson, R. E., S. L. Beasom, and C. J. Scifres. 1977. Economic evaluation of cattle and white-tailed deer response to aerial spraying of mixed brush. J. Range Manage. 30:214-217.

Wilson, A. D., and G. J. Tupper. 1982. Concepts and factors applicable to the measurement of range condition. J. Range Manage. 35:684-689.

Winkel, V. K., B. A. Roundy, and D. K. Blough. 1991. Effects of seedbed preparation and cattle trampling on burial of grass seeds. J. Range Manage. 44:171-175.

Youngson, C. R., C. A. I. Goring, R. W. Merkle, H. H. Scott, and J. D. Griffith. 1967. Factors influencing the decomposition of Tordon herbicide in soils. Down-To-Earth 23:3-11.

Chapter 20
MECHANICAL MANIPULATION OF PLANTS

Timothy E. Fulbright and Fred S. Guthery

INTRODUCTION

Mechanical manipulations of rangeland usually are implemented to increase forage production for domestic livestock. These manipulations include brush management and other practices used to shift existing plant species composition to a plant community desired by the range manager. Woody plants are common targets of manipulation because shrubs and trees compete with grasses and other herbaceous vegetation, resulting in decreased forage production for livestock.

Woody plant density greatly increased on rangelands in the western United States following settlement by European man. A goal of many resource managers during the first half of the twentieth century was to eradicate woody plants and return rangeland to the grassland or savannah present before the perturbations caused by humans and livestock. Early brush control efforts included hand grubbing of individual plants. Development of power equipment to manage brush began in the 1930s (Fisher et al. 1973). Mechanical brush management became prevalent following World War II with development of heavy equipment such as the root plow (Fig. 1) in 1949. From the 1950s to the early 1970s, range managers and researchers emphasized mechanical manipulations (Holechek et al. 1989:433).

The increase in brush, perceived as a nemesis for livestock, had mixed effects on wildlife populations. Although Berlandier (1980) reported seeing an abundance of white-tailed deer (*Odocoileus virginianus*) on open prairie in southern Texas during his travels in the 1830s, the increase in white-tailed deer populations in Texas during the twentieth century has been attributed to increased brush densities (Hahn 1945) among other factors. Conversely, brush invasion contributed to reduction of the range and abundance of prairie or savannah species (e.g., Attwater's prairie-chicken [*Tympanuchus cupido attwateri*]) (Lehmann 1941).

Historically, wildlife was ignored in the planning and implementation of range improvement programs for livestock (Scifres 1980:262). Yoakum (1980) postulated that < 1% of all range improvements were designed to improve wildlife habitat. However, Adams et al. (1992) surveyed hunting lease operators in Texas and reported that 21% of the respondents controlled brush as a wildlife habitat management technique. Browse rejuvenation (i.e., top-growth removal of woody plants to stimulate sprouting), which temporarily increases nutritional quality and availability of browse (Scifres 1980:272, Vallentine 1980:57, Yoakum et al. 1980:342), is an example of how brush management may be applied to improve wildlife habitat. Likewise, suppression of brush may be used to increase the habitat for prairie species and to increase edge (Leopold 1933:132) for species that use both brushland and grassland habitats.

FIGURE 1. Root plow attached to a crawler tractor.

We emphasize brush management in this chapter. Many range managers and researchers recognized by the 1960s that returning rangeland to the grassland or savannah landscape that formerly existed via large-scale mechanical, chemical, and pyric treatments was an unrealistic goal. Terminology changed from brush eradication to brush control. Brush management became the modern term with recognition of the need to increase productivity of rangelands while minimizing damage to the environment and reducing negative public perception (Scifres 1980:xi).

Planning brush management to benefit wildlife or to improve grazing capacity for livestock without reducing wildlife densities is a complex problem. An infinite array of possible brush management patterns exists and numerous variables affect habitat and wildlife species' responses to mechanical treatments. Consequently, recommendations in the literature regarding the optimum amount of brush to treat to increase or maintain various wildlife species and the proper practices to use are mostly anecdotal. A considerable body of information on wildlife species responses to various mechanical manipulations exists (Kozicky and Fulbright 1991). The inability of managers to provide more resolute recommendations regarding species' responses reflects the lack of a solid theoretical base for decision-making.

Our objectives are to review the common types of mechanical brush manipulations used on rangelands in North America, review empirical evidence on habitat and wildlife population responses to the treatments, and present conceptual models for decision-making in planning range manipulations. Our hope is to provide a conceptual basis for development of habitat manipulation theory on a landscape level.

METHODS OF MECHANICAL MANIPULATION AND IMPACTS ON HABITAT

Categories of Mechanical Manipulations

Mechanical methods of brush manipulation can be divided into major categories: top-growth and plant removal treatments (Table 1). Vallentine (1980) and Scifres (1980) described brush management methods, the regions in which they are applied, modifications of the techniques, and expected benefits for livestock.

Top-growth Removal

The major forms of top-growth removal are roller chopping and shredding. A roller chopper is a large, cylindrical metal drum with blades parallel to its axis that is pulled behind a crawler

TABLE 1. Categories and methods of mechanical brush management.

Category	Method
Top-growth removal	Roller chopping
	Shredding
	Drag-type
	Hydro-ax
	Scalping
Plant removal	Railing
	Cabling
	Chaining
	Tree crusher
	Discing
	Bulldozing
	Grubing
	Root plowing

FIGURE 2. Roller chopper pulled by a crawler tractor in thornscrub in southern Texas. A seeder box attachment is mounted above the roller chopper.

FIGURE 3. A front-mounted brush shredder.

FIGURE 4. Resprouting guajillo 2 months after roller chopping in southern Texas.

tractor (Fig. 2). The chopper severs and crushes the stems of woody plants. Two or 3 roller choppers may be linked to increase width of the strips. The drum may be filled with cement or it may be hollow and the weight increased by filling with water or sand.

The 2 major types of shredders are the hydro-ax (Fig. 3), in which the mower blades are attached to the front of the power source, and the drag-type shredder, which is pulled behind the power source. Drag-type shredders are used on brush with basal stem diameters <6.4 cm. Heavy-duty, hydraulically operated shredders are used on shrubs with trunk diameters >10 cm (Welch et al. 1985). Roller choppers can be used on larger woody plants and in rougher topography than shredders (Scifres 1980:128).

Brush species that resprout from roots, crowns, and stem bases (e.g., mesquite [*Prosopis* spp.], *Acacia* spp., gambel oak [*Quercus gambelii*]) are seldom killed by top-growth removal (Fig. 4). Single-stemmed shrubs usually produce multiple resprouts from crowns and stem bases, and top removal may result in stands that are more dense than were present before treatment. Individual plant density of honey mesquite (*P. glandulosa*) and other woody legumes may increase dramatically following roller chopping (Fulbright and Beasom 1987). Forb and grass canopy cover may increase following top-growth removal (Bozzo et al. 1992*a*); but the increase is temporary because of rapid brush regrowth, and treatment every 2-3 years is recommended when the objective of management is to increase yield of herbaceous vegetation (Welch et al. 1985). Woody species diversity of thornscrub communities that reestablished following roller chopping or repeated shredding was lower than before treatment in southern Texas (Fulbright 1987, Fulbright and Beasom 1987). Conversely, plant species richness and diversity increased following roller chopping in a longleaf (*Pinus palustris*) -slash pine (*P. elliottii*) forest in north Florida (Lewis et al. 1988).

Roller chopping, shredding, railing, crushing, clipping, and pruning stimulate browse regrowth (Scotter 1980). Browse rejuvenation has been applied in chaparral vegetation in California (Yoakum et al. 1980:342); bitterbrush (*Purshia tridentata*) (Yoakum et al. 1980:343) and gambel oak stands in the western United States; curlleaf mountain mahogany (*Cercocarpus ledifolius*) stands in Utah (Vallentine 1980:61); and mixed-brush in southern Texas (Powell and Box 1966, Everitt 1983, Fulbright et al. 1991, Reynolds et al. 1992). Brush regrowth may be higher in crude protein, phosphorus, and digestibility than growth of mature plants (Everitt 1983, Fulbright et al. 1991, Reynolds et al. 1992) and is more palatable and accessible (Box and Powell 1965, Powell and Box 1966). The increase in nutritional quality lasts from 2 to 13 months (Everitt 1983, Asah et al. 1987).

Scalping with a blade mounted on the front of a crawler tractor is used also to remove top-growth of woody plants. The blade shears off woody plants nearer the ground than roller choppers or shredders and causes more soil disturbance (Mutz et al. 1978).

Plant Removal

Root Plowing. A root plow is a large, V-shaped blade pulled behind a crawler tractor. The plow severs crowns and roots of woody species. Root plowing is often followed by raking with a large brush rake that removes debris from the soil (Figs. 5, 6). The treatment disturbs the soil so severely that reseeding may be required to reestablish herbaceous vegetation. Root plowing causes greater woody-plant mortality than any other mechanical method. Usually 10-20 years elapse before the shrub community has reestablished (Welch et al. 1985).

Root plowing often causes a long-term reduction in woody plant species richness and diversity. Huisache (*Acacia smallii*) and mesquite composed 95% of the woody plants on rangeland in the Texas Coastal Bend that was root-plowed and raked 14 years earlier, compared to 25% on nontreated rangeland (Mutz et al. 1978). In the Texas Rio Grande Plain, mesquite density was 2.7-3.6 times greater on rangeland root plowed 25-30 years earlier than on nontreated range (Fulbright and Beasom 1987). Density of important browse plants for white-tailed deer, such as bluewood (*Condalia obovata*) and spiny hackberry (*Celtis pallida*), was much lower on root-plowed rangeland than on nontreated rangeland. Untreated areas in the eastern Rio Grande Plain of Texas contained an average of 19 woody species compared to 7 woody species on areas root-plowed and raked 16-17 years earlier (Ruthven et al. 1993).

Species richness and Shannon's diversity index values of woody plants in ephemeral drainages in the western Rio Grande Plain chained about 40 years earlier or chained (about 40 years earlier) and then root plowed (about 30 years earlier) were similar to those on nontreated rangeland (Nolte et al. 1994). However, beta diversity, a measure of dissimilarity among plots within a treatment (Biondini et al. 1989), was lower on chained and root-plowed areas than on chained or untreated areas. Thus, root plowing resulted in a more homogeneous pattern of species composition.

Variable forage-yield responses to root plowing have been reported. In some cases root plowing and reseeding have resulted in increased forage yield; in others reduced forage yield has resulted. Information on long-term effects of root plowing on herbaceous vegetation is limited. Fourteen years after root plowing and raking in the Texas Coastal Bend, relative percent botanical composition of grasses of good to excellent forage value for livestock was 17% compared to 44% on nontreated rangeland (Mutz et al. 1978). In the eastern Rio Grande Plain, herbaceous species richness and diversity were similar on untreated areas and areas root-plowed and raked 16-17 years earlier (Ruthven et al. 1993). Canopy cover of herbaceous vegetation averaged across sampling dates was 49% on untreated areas, 81% on root-plowed areas.

Chaining and Cabling. These methods employ heavy anchor chains or cables that are dragged in a *U*- or *J*-shape

FIGURE 5. A root rake used to remove roots and debris following root plowing.

FIGURE 6. Recently root plowed and raked rangeland in southern Texas.

between 2 crawler tractors to uproot shrubs (Fig. 7). Shrubs with flexible stems are more difficult to uproot and are less susceptible to damage than shrubs with stiff stems. Chains or cables are sometimes dragged both directions (double-chaining or cabling) to uproot additional woody plants. Chaining and cabling are commonly used in controlling pinyon pines (*Pinus* spp.) and junipers (*Juniperus* spp.) in the Intermountain Region. Chaining is also used in oak- and juniper-dominated rangeland and in mesquite-dominated rangeland in Texas.

FIGURE 7. Anchor chain dragged between crawler tractors to uproot shrubs. Photo courtesy of Dale Rollins.

FIGURE 8. Disc plow pulled by crawler tractor being used to manage thornscrub in southern Texas.

Chaining may result in a long-term reduction in pinyon and juniper density, whereas shrubs may increase in abundance (Skousen et al. 1989). Shrub density on pinyon-juniper rangeland cabled 24 years earlier or double-chained 14 or 20 years earlier was greater than on nontreated rangeland in Utah.

Several researchers have reported increased herbaceous vegetation following chaining. Relative canopy cover of herbaceous plants was greater on cabled or double-chained than on nontreated rangeland (Skousen et al. 1989). Grass and forb standing crop was 55% greater 22 months after chaining Ashe juniper (*J. asheii*)-oak-dominated rangelands in central Texas (Rollins and Bryant 1986).

Discing. Heavy disc plows pulled by crawler tractors (Fig. 8) are used to manage sagebrush (*Artemisia* spp.), rabbitbrush (*Chrysothamnus* spp.), greasewood (*Sarcobatus vermiculatus*), low mountain brush, creosotebush (*Larrea divaricata*), tarbush (*Flourensia cernua*), and southern Texas thornscrub (Vallentine 1980:95, Montemayor et al. 1991, Bozzo et al. 1992*a*). Heavy disc plows sever the roots and crowns of shallow-rooted shrubs, but do not sever the stems of most sprouting species deeply enough to prevent resprouting from lateral and adventitious buds in the crown and stem bases (Fig. 9). Brush density increases following initial treatment and woody plants dominate in about 10 years (Welch et al. 1985).

FIGURE 9. A shrub resprouting 3 months after treatment with a heavy disc.

Heavy discing can be used to selectively reduce abundance of shallow-rooted shrub species and increase abundance of more deeply rooted species. Discing dense whitebrush (*Aloysia lycioides*) colonies reduced cover of whitebrush, a shallow-rooted shrub that forms dense colonies and is poor browse for white-tailed deer, and increased relative cover of more palatable and deeply rooted browse species such as spiny hackberry and guayacan (*Guaiacum angustifolium*) (Bozzo et al. 1992*a*).

Discing once is not as detrimental to shrub species diversity or abundance of preferred white-tailed deer browse species as root plowing. Shrub species diversity and richness were similar 10 or 16 years after treatment in thornscrub disced once and in nontreated thornscrub in southern Texas (Montemayor et al. 1991). Forb canopy cover and species richness were similar or greater, depending on sampling date, on disced areas than on nontreated areas in southern Texas (Bozzo et al. 1992*a*).

Discing is used also to disturb soil to stimulate early-successional forbs to provide herbage and seeds for game such as deer and quail (Webb and Guthery 1983, Nurdin and Fulbright 1988). Light disc plows pulled by farm tractors are commonly used. In Florida, discing promoted a variety of valuable food plants for northern bobwhites (*Colinus virginianus*) (Landers and Mueller 1986:21-23). Discing northwest Texas rangeland promoted several plant species important for wildlife, but forb production and density on the disced strips were relatively low (Webb and Guthery 1983).

Bulldozing and Grubbing. Bulldozing and power grubbing of individual plants are less economical than other mechanical methods and are seldom used for treating extensive areas. Bulldozing (i. e., uprooting woody plants with a bulldozer blade) has been used to thin stands of pinyon-juniper. Density of pinyons and junipers was less on New Mexico rangeland bulldozed 18 years earlier than on nontreated rangeland (Severson 1986*b*). Power grubbing with devices mounted on the front or rear of crawler or row-crop tractors is generally used as a maintenance practice rather than as a primary treatment (Welch et al. 1985).

Bulldozing and grubbing provide the opportunity for highly selective shrub control, which may be desirable in wildlife management. Grubbing is particularly selective in that individual plants of selected woody species can be removed. Individual species removal generally can not be accomplished with other mechanical methods, although stands or individuals of certain species can be left untreated.

IMPACTS OF MECHANICAL MANIPULATIONS ON WILDLIFE

White-tailed Deer

Mechanically clearing extensive tracts of rangeland without leaving portions of the range with adequate woody cover is devastating to populations of white-tailed deer. Inglis et al. (1986:67) suggested that mechanical treatments can produce a chronic reduction in carrying capacity if blocks of rangeland ≥400 ha are cleared. Root plowing large areas of rangeland is probably the most destructive method relative to white-tailed deer habitat, on a short- and long-term basis. The initial effect of root plowing is loss of screening cover, thermal cover, and browse. Deer use of root-plowed areas greatly declines during drought because browse is not available, although the areas may be used when forbs are present (Davis and Winkler 1968). Clearing large areas by chaining (Darr and Klebenow 1975) and other mechanical methods also reduces deer densities until brush reestablishes.

The long-term decline in browse diversity and abundance following root plowing may have little or no impact on white-tailed deer populations in habitats in which forbs are consistently abundant and available. Rangeland in the eastern Rio Grande Plains in southern Texas 17-18 years following root plowing was dominated by huisache and had low brush species diversity, but white-tailed deer densities on root-plowed range were similar to densities on more diverse nontreated rangeland (Ruthven et al. 1994). Indices of deer nutritional status and reproduction were also similar between root-plowed and more diverse areas. In contrast to findings of Ruthven et al. (1994), use by mature male white-tailed deer of rangeland that had been root-plowed and reinvaded by brush was lower than use of nontreated range in the more arid western Rio Grande Plain (Pollock 1989:53).

Mechanically treating 2- to 16-ha patches (Holechek 1981) or strips (Bozzo et al. 1992*a*, *b*) interspersed in woody plant-dominated rangeland may result in greater deer populations. Bozzo et al. (1992*b*) roller-chopped 20% of a 400-ha area each year for 4 years in separate but adjacent roller-chopped strips alternating with nontreated strips. Averaged over a 6-year period beginning the first year of treatment, white-tailed deer densities were 2.4 times greater in the treated area than in a nontreated area. White-tailed deer densities increased following 50% and 70% clearing by double chaining of Ashe juniper and live-oak (*Q. virginiana*)-dominated rangeland on the Edwards Plateau of Texas, but declined on areas that were 80% cleared (Rollins et al. 1988). Chaining was done in 180- × 450-m clearings with 160-, 70-, or 40-m residual brush strips between clearings at the 50, 70, and 80% intensities, respectively.

Reasons for increased deer densities following mechanical manipulation of patches or strips of brush are multiple. Rollins et al. (1988) and Bozzo et al. (1992*b*) speculated that increased deer use of mechanically treated areas resulted from greater abundance of forbs. In the Texas Coastal Bend, Naderman (1979) attributed increased deer use of roller-chopped and shredded openings to a behavioral factor, such as increased visibility of predators, rather than a response to forage availability.

Drawe (1981) postulated that mechanical top-growth removal on a portion of the range each year would provide a continuing supply of nutritious, readily available browse for white-tailed deer. Fulbright et al. (1993*a*, 1993*b*) tested the hypothesis that roller chopping 20% of an area dominated by guajillo (*Acacia berlandieri*) and blackbrush acacia (*A. rigidula*) in separate portions each year to provide browse regrowth would improve the nutritional plane of white-tailed deer. Blood urea nitrogen, kidney fat, and femur marrow fat of deer in the treated area and in a nontreated area were similar. Treatment effects were possibly confounded by greater deer density in the roller-chopped area than in the nontreated area. Also, shrub regrowth may be high in secondary compounds, which may circumvent increased crude protein and other chemical components by reducing digestibility (Haggerman et al. 1992). In support of the hypothesis that greater secondary compounds in regrowth may reduce digestibility, Reynolds et al. (1992) found that in vitro organic matter digestibility of blackbrush acacia browse was lower than that of browse from nontreated plants 2 months after roller chopping.

Mule Deer

Considerable research on mule deer (*O. hemionus*) responses to mechanical manipulation has been conducted in the pinyon-juniper ecosystem. Mechanical treatments, including cabling, chaining, dozing individual trees, and hand chopping or sawing, are widely used in pinyon-juniper habitats (Barnitz et al. 1990). Mechanically clearing extensive areas of pinyon-juniper habitat is detrimental to mule deer populations (Short et al. 1977). Conversely, mechanically treating small patches of rangeland interspersed with pinyon-juniper woodland may maintain mule deer populations and improve habitat (Short and McCulloch 1977:8). Barnitz et al. (1990) compared mule deer use on small (2.6-11.1 ha) patches of pinyon-juniper rangeland in New Mexico 10, 20, and 31 years following 2-way cabling and concluded that cabling did not have a long-term impact on use of the areas by mule deer because of rapid woody plant regrowth and small size of the treated areas. Mule deer use during summer, fall, and winter 1975-82 was similar on nontreated pinyon-juniper woodland in New Mexico and woodland cabled in 1954 or 1975 (Howard et al. 1987). During spring, mule deer use was greater on areas cabled in 1975. Big game use of sites cabled or double-chained 2 to 24 years earlier and ranging in size from 30 to 310 ha was similar to use of nontreated pinyon-juniper sites in Utah (Skousen et al. 1989). Terrel and Spillett (1975) recommended that mechanical clearing of pinyon-juniper habitat be limited to a maximum of 25% of the total winter range in any 25-year period to maintain mule deer populations and openings should be <0.32 km wide.

In the Arizona chaparral, mule deer spent 25 to 50% as much time on root-plowed pastures 32 ha in size seeded to weeping lovegrass (*Eragrostis curvula*) as in adjacent brushland (Urness 1974). However, preferred forbs were more abundant in root-plowed pastures than in brushland. Urness (1974) suggested small root-plowed areas interspersed in brushland might improve the habitat. Mule deer preferred gambel oak-dominated areas in Colorado sprayed with 2,4,5- TP ([2,4,5-tricholorophenoxy] acetic acid + 4-amino-3,5,6-trichloropicolinic acid) more than chained areas (Kufeld 1983).

Elk

Pinyon-juniper and gambel oak-dominated rangeland are the primary habitats occupied by elk (*Cervus elaphus*) that are commonly impacted by brush management. Clearing pinyon-juniper woodland in New Mexico reduced elk use (Short et al. 1977). Short and McCulloch (1977) suggested that elk habitat would be improved by providing many small cleared areas interspersed with natural pinyon-juniper woodland. In gambel oak-dominated habitat, elk preferred burned areas more than chained areas (Kufeld 1983). Kufeld (1983) suggested that to maintain elk populations burning should be used rather than chaining to manipulate gambel oak.

Pronghorn

Woody plants, particularly sagebrush (*Artemisia* spp.), comprise a major portion of pronghorn (*Antilocapra americana*) diets in Wyoming and Alberta (Sundstrom et al. 1973, Mitchell 1980:39), whereas forbs are predominant in pronghorn diets in the southern and western portions of its range (Russell 1964:64, Schwartz and Nagy 1976, Hailey 1979, Edwards and Ohmart 1981). Conversion of large blocks of rangeland to grass monocultures by brush control and seeding is detrimental to pronghorn populations because of reduced availability of browse (Yoakum 1980:52). Extensive areas dominated by big sagebrush (*A. tridentata*), especially if the shrubs are >76 cm tall, support low pronghorn densities (Yoakum 1979). Mechanical treatments such as brushland plowing and chaining can be used to improve vegetation structure (Yoakum 1983). Treated areas should be <405 ha in size and 5-10% shrub cover should remain (Yoakum 1979).

Bighorn Sheep

Bighorn sheep (*Ovis canadensis*) primarily occupy rangeland that is not amenable to mechanical manipulation because of rough topography. However, increased abundance of gambel oak has reduced availability of preferred habitat and has degraded migration corridors between seasonal bighorn sheep ranges in Waterton Canyon, Colorado. Hand cutting followed by herbicide application was effective in creating openings that facilitated bighorn movement (Rominger 1983).

Upland Game Birds

Wild Turkeys. A 63% reduction in Merriam turkey (*Meleagris gallopavo merriamii*) populations occurred following chaining and burning of pinyon-juniper rangeland in Arizona that isolated roost sites ≥300 m from cover (Scott and Boeker 1977). Scott and Boeker (1977) recommended that cleared areas be ≤90 m wide and that strips of cover should be retained as travel lanes to established roost areas to maintain turkey habitat.

Rio Grande turkeys (*M. g. intermedia*) avoided a 228-ha upland area in north-central Texas that was grubbed to remove brush and avoided a 1,298-ha area that was treated with a herbicide, chained, then retreated with a herbicide (Quinton et al. 1980). The authors recommended limiting brush clearing to small, patterned areas to maintain turkey populations.

Northern Bobwhites. Because of their diet, behavior, and weak flight abilities, northern bobwhites are not adapted to dense, mature brush communities. Mechanical brush management is considered an effective method of habitat improvement for these birds if some brush is preserved for loafing and escape cover.

Optimum coverage by brush canopies in unaltered communities for bobwhites ranges from 5 to 15% (Guthery 1986:115, Rice 1991). This range implies mechanical brush management can be beneficial if brush coverage exceeds 15%, and that 85-

95% of an area can be treated if untreated brush is properly dispersed. Because bobwhites seldom venture >200 m from woody cover (Stoddard 1932:181, Lehmann 1946, Murphy and Baskett 1952), strip patterns with 400-m-wide cleared strips and 7- to 10-m-wide preserved strips have been recommended in bobwhite management (Jackson 1969). This pattern results in 97.7% treatment. Guthery (1986:22) recommended halving the width of treated strips to 200 m, which yields 95% treatment and reduces the maximum distance to woody cover by 50% (100 m). The more conservative pattern (200-m cleared strips) would be more robust to variation in weather, grazing pressure, and type of mechanical treatment. This robustness would accrue because the distance that bobwhites forage from woody cover decreases as the quality of screening cover decreases (Guthery 1980). Thus, broader cleared strips would be more likely to inhibit use of clearings than would narrower cleared strips.

In hotter, drier portions of bobwhite range, preservation of taller brush plants and mottes is recommended. Bobwhites seek taller brush for thermal cover during summer but use lower brush during cooler periods (Johnson and Guthery 1988). In southern Texas, summer coverts averaged 2.8-4.5 m tall, whereas winter coverts averaged 1.8-2.4 m tall.

Gambel's Quail. Gambel's quail (*Callipepla gambelii*) were twice as abundant on velvet mesquite- (*P. velutina*) dominated rangeland in Arizona that was chained in patches ranging from 2.8 to 30.4 ha as on nontreated rangeland (Germano et al. 1983). Spot clearing also benefited scaled quail (*C. squamata*).

Nongame Birds

Mechanical brush management may have short- or long-term and positive or negative effects on composition and abundance of nongame birds. The general expectation is a decline in species adapted to woody plant communities and an increase in species adapted to herbaceous plant communities (Best 1972, Renwald 1978, Kruse et al. 1979, O'Meara et al. 1981, Rollins 1983, Vega and Rappole 1994).

Mechanical treatment effects on bird community variables (e.g., total density [all birds], species richness [no. species], and diversity) generally cannot be predicted. Brush manipulation effects on these variables depend on the richness and abundance of bird species in the pretreatment community; the richness and abundance of bird species potentially attracted to the treated community; the nature of the pretreatment brush community (e.g., impacts in structurally and compositionally simple communities may be more dramatic than impacts in complex brush communities); the type of mechanical treatment applied because of effects on structure and diversity of brush communities; and the pattern of treatment.

Research in relatively simple pinyon-juniper and juniper habitats illustrates the lack of predictability of mechanical brush treatments on bird community variables. In Colorado, chaining pinyon-juniper plots as small as 16 ha has reduced bird species richness and diversity (Sedgewick and Ryder 1987); chaining extensive areas has suppressed total bird density for up to 15 years (O'Meara et al. 1981). In contrast, chaining small Ashe juniper plots in central Texas lowered total bird density but increased richness and diversity (Rollins 1983). Bulldozed pinyon-juniper habitat in Arizona had higher total bird density but lower richness and diversity than unchained sites (Kruse et al. 1979).

The only consensus that has emerged from previous research on nongame bird communities and brush management is that size of treated areas should be limited and that untreated brush should be preserved to benefit nongame bird communities or minimize damage thereto. O'Meara et al. (1981) recommended limiting clearing width to 200 m to maximize use by birds that nest in woodland but include chained vegetation in their territory. Castrale (1982), dealing with sagebrush communities, advocated 100-m-wide treated strips alternated with 100-to 200-m-wide untreated strips. We have found that discing in a complex, mesquite-mixed brush community in southern Texas had little effect on diversity, richness, or total density when 137-m-wide strips were preserved every 200 m and 2 27-m-wide strips were preserved in the 200-m strips.

Small Mammals

Rodent populations may increase following mechanical brush clearing. Uprooting trees with a bulldozer and chaining pinyon-juniper woodland resulted in greater small mammal populations (Turkowski and Reynolds 1970, Baker and Frischknecht 1973, O'Meara et al. 1981, Severson 1986*a*). Baker and Frischknecht (1973) reported that small mammals were more numerous for the first 3 years after chaining pinyon-juniper woodland, but 12-13 years after treatment numbers were similar to those on untreated rangeland. Root-plowed rangeland in southern Texas supported greater cotton rat (*Sigmodon hispidus*) densities than nontreated rangeland (Guthery et al. 1979). Greater cotton rat densities were attributed to greater biomass of bristlegrasses (*Setaria* spp.), sumpweed (*Iva* spp.), and western ragweed (*Ambrosia psilostachya*) on root-plowed rangeland. Powell (1968) also reported greater rodent densities on root-plowed than on nontreated rangeland in southern Texas, but root-plowed and raked areas supported lower rodent densities than nontreated areas. A landscape mosaic of small mechanically manipulated and seeded patches interspersed in pinyon-juniper and sagebrush-dominated habitat may support greater small mammal abundance and biomass than continuous woodland because of differences in habitat preferences among small mammal species (Smith and Urness 1984).

Rodent species composition may change following mechanical manipulation. Bulldozed pinyon-juniper rangeland supported a greater proportion of rodents characteristic of grassland habitats and rodents that preferred woodland habitats were reduced in abundance (Kruse et al. 1979, Severson 1986*a*). Complete clearing may be detrimental to white-throated woodrat (*Neotoma albigula*) (Turkowski and Watkins 1976) and brush mice (*Peromyscus boylii*) populations (Turkowski and Reynolds 1970, Short and McCulloch 1977:8), although both species increased in abundance as slash accumulation increased in a New Mexico study (Severson 1986*a*). Pinyon mice (*P. truei*) and rock mice (*P. difficilus*) were more abundant where more slash remained after control, but only on areas where the pinyon-juniper overstory was relatively intact. Deer mice (*P. maniculatus*) comprised 85% of the total rodents captured on a 16-ha chained pinyon-juniper site in Colorado, whereas pinyon mice were captured only on the nontreated area (Sedgewick and Ryder 1987).

Rodent species diversity may be lower on mechanically manipulated areas (O'Meara et al. 1981). However, Sedgewick and Ryder (1987) reported greater small mammal species richness on a 16-ha chained pinyon-juniper site in Colorado during the first 4 years after treatment than on a nontreated area.

Rollins (1983) determined eastern cottontail (*Sylvilagus floridanus*) use of live-oak dominated rangeland on which none, 30, 50, 70, and 80% of the brush was removed in strips by chaining 180- x 450-m openings. On the 50% cleared site, cottontails preferred untreated brush over cleared areas but showed no distinct preferences on other sites. Black-tailed jackrabbits (*Lepus californicus*) preferred cleared areas relative to brush areas. Uprooting, piling, and burning downed trees depressed desert cottontail (*S. auduboni*) use, but use was not affected if 173-222 downed trees or living shrubs/ha remained (Kundaeli and Reynolds 1972). Lagomorphs preferred pinyon-juniper woodland cabled in 1954 or 1975 more than nontreated areas during spring, summer, and fall in New Mexico during 1975 to 1982 (Howard et al. 1987). Black-tailed and antelope (*L. alleni*) jackrabbits were more abundant on velvet mesquite-dominated rangeland in Arizona that was undisturbed or chained in patches ranging from 2.8 to 30.4 ha than on mesquite-free range (Germano et al. 1983).

PLANNING BRUSH MANAGEMENT PATTERNS

Two tenets emerge from the preceding descriptions of wildlife responses to woody plant management: clearing woody plants from larger areas reduces density of wildlife species that require woody cover; and a landscape mosaic of smaller, mechanically manipulated patches interspersed within a woodland matrix may support greater wildlife density, diversity, and biomass than homogeneous shrubland or woodland, in part because of differences in adaptations among wildlife species (Smith and Urness 1984). Bailey (1984:226) criticized widespread application of the second tenet by wildlife managers because of the assumption that all species respond similarly to habitat disturbance under all situations, which is false because certain wildlife species are not favored by disturbance. This is a valid argument against mechanical manipulation on certain public and private lands, but the amount of rangeland left in a natural (i.e., untreated) state is limited by political, social, and economic constraints. Further, brush communities on rangeland may reach stable states (Friedel 1991, Laycock 1991) that are irresponsive to successional trends and can only be altered with cultural inputs such as mechanical brush management (Archer 1989). In short, situations will arise where some type of brush management is required, and the second tenet is germane to such management.

Mechanical methods provide infinite opportunities for managing woody plants such that tenet 2 is fulfilled. Possible patterns include openings bounded by untreated brush, untreated islands surrounded by treated areas, preserved and treated strips of fixed or variable width and linear or curvilinear orientation, and any or all combinations of the above.

The infinity of options creates problems in decision-making. How do managers evaluate a pattern before it is applied? Why select 1 pattern over another? What biological and sociological variables are germane to the planning process? What negative effects should be considered? The following sections provide a simple, conceptual framework for addressing these questions and provide ecological perspectives on the decision model.

The Principle of Edge

Leopold's (1933:132) Principle of Edge, as generalized by Guthery and Bingham (1992), serves as a conceptual basis for decision-making in brush-pattern planning. The general principle as it applies to edge-benefitted species is

$$D = a + kS,$$

where

D = potential density of the species in question,
a = density in the absence of edge (untreated brush), $a \geq 0$,
k = a constant of proportionality that relates density to edge length, and
S = length of biologically meaningful edge.

The principle specifies a simple linear relationship between animal density and length of edge for edge-obligate ($a = 0$) or edge-benefitted species ($a > 0$).

The constant of proportionality can be considered negative for monotype species (Guthery and Bingham 1992). For interior species (those adapted to mature woody plant communities), the Principle of Edge can be expressed

$$D = a - kS,$$

with all variables as defined above and maximum density occurring at $S = 0$ (no edge). The principle for prairie species is similar; i.e.,

$$D = b - kS,$$

where b = density in a prairie situation.

Leopold (1933) recognized that for edge-benefitted species the Principle of Edge applied only "within ordinary limits," which he did not define. Guthery and Bingham (1992) showed that these limits could occur at small and large scales of resolution. The small scale relates to openings, whereas the large scale relates to patterns that cause habitat fragmentation; decisions on brush management patterns are effective only at intermediate scales of resolution.

Openings

At the level of a single opening in a mature woody plant community, traditional thought has held that the greater the amount of edge associated with the opening, the greater its value to edge-benefitted wildlife. However, it is possible to create redundant edge, i.e., edge that has no effect on density of edge-benefitted species (Guthery and Bingham 1992).

Consider a mechanical brush management program that creates treated and untreated brush. Hereafter, we call these resource and refuge cover, respectively. Edge-benefitted wildlife might use treated brush for resources (e.g., food, nesting cover) and untreated brush for refuge (e.g., predator avoidance, thermal stress). The arbitrary definitions do not preclude the possibility that a species obtains refuge from resource cover or resources from refuge cover, nor do they preclude the possibility that a species is monotypic, i.e., restricted to treated or untreated brush.

Edge-benefitted species have a maximum radius of movement away from refuge cover (Sumner 1935, Urness 1974, Inglis 1985, Longland 1991). Define the radius of full use (r) as the radius away from refuge cover such that use of resource cover remains random as distance from refuge cover increases.

If an opening is configured such that the maximum distance to refuge cover is $\leq r$ at any point in the opening, then the opening is defined as being *fully usable* and its shape is not necessarily a management concern. Given a fully usable opening of size A, nonredundant edge (i.e., the length of edge [S] that applies in the generalized Principle of Edge) is $3.545A^{1/2}$. This is the circumference of a circle, the geometric figure that minimizes length of edge given area (Giles 1978), with radius $\leq r$. Redundant edge is simply observed edge $- 3.545A^{1/2}$. The point for managers to keep in mind is that, contrary to general thought, edge-benefitted wildlife is not necessarily responsive to the total amount of edge associated with an opening if the opening is fully usable (as defined).

The Operational Scale

Concepts. At intermediate scales of resolution, it is convenient to think in terms of habitat area as opposed to edge length. Because edge cannot exist with space devoted to cover types, Leopold's (1933:132) principle implies areas (Giles 1978). The problem is contriving a decision model that relates areas of refuge and resource cover to animal density.

Assume there exists some optimum ratio of resource cover *area* to refuge cover *area*; i.e., a certain amount of refuge cover is required to support a given amount of resource cover and vice versa. Define this constant ratio as

$$c = W/R,$$

where

W = the area of refuge habitat required for R units of resource habitat, and
R = the area of resource habitat required for W units of refuge habitat.

Further, assume that w/r is a reasonable approximation of c where w = the minimum linear breadth of refuge cover required by a given species (Inglis 1985) and r = the radius of full use.

With respect to edge-benefitted species, mechanical brush management potentially creates areas where (1) contiguous refuge and resource cover areas are in the proper ratio and density reaches the maximum value of D (i.e., D_m) for the general principle of edge, (2) refuge cover exceeds the amount required given the availability of resource cover and density is approximately the intercept in the general Principle of Edge, or (3) resource cover is unusable because the distance to refuge cover exceeds the maximum acceptable distance for the species in question. If the function relating probability of use to distance from refuge cover were known, the fully usable area could be estimated with integral calculus. Lacking knowledge of this function for any species, we will arbitrarily set habitat value at 0 if distance to refuge cover exceeds the radius of full use, r.

The above definitions lead to the Principle of Edge expressed in areas instead of lengths for edge-benefitted species:

$$D' = (D_m A_1 + aA_2)/A,$$

where

D' = average potential density,
D_m = the maximum value of $a + kS$ for the species in question,
A_1 = the area where contiguous refuge and resource cover are in the proper ratio and all resource cover is fully usable,
a = animal density in untreated brush,
A_2 = the area in redundant refuge cover (untreated brush in excess of needs), and
A = $A_1 + A_2$ = total area.

Note that A_3, that portion of an area beyond the radius of full use from refuge cover, does not enter into the equation because density on such areas was arbitrarily set at 0.

Monotypic species require different expressions for the Principle of Edge expressed as a function of areas:

$$D' = aU/A$$

for interior species and

$$D' = b(A - U)/A$$

for prairie species where U = the total area in untreated brush.

Pattern Evaluation. For monotypic species, pattern evaluation is straightforward. Given any pattern, potential density is simply the proportion of the area in the single required type times the potential density in the type. All areas of the required type must be large enough to receive use and within dispersal distance of the species in question.

For edge-dependent species, 2 situations may arise with respect to refuge and resource cover, where these habitat types are contiguous: availability of resource cover exceeds requirements or availability of refuge cover exceeds requirements. If refuge habitat is limiting, as might occur if islands of brush are preserved, then the amount of supportable resource habitat is approximated as

$$R = Wr/w$$

(W is fixed) whereas if resource habitat is limiting, as might occur with small openings, then the amount of supportable refuge habitat is approximated as

$$W = Rw/r.$$

(R is fixed).

Analytical determination of an optimum brush management pattern usually is possible for simple patterns with regular, repeated, treated-untreated configurations. For example, given w and r for the species in question, 1 optimum pattern is indefinite repetition of preserved strips of width w and cleared strips of width $2r$, resulting in the pattern $w{:}2r{:}w{:}2r{:}w{:}2r{:}w{:}2r\ldots$ (The radius of full use is doubled because an animal can seek refuge in 2 directions). This pattern maximizes density according to the generalized Principle of Edge and its homologue based on areas (given a large management area).

The optimum pattern cannot be analytically determined with unconstrained variation in sizes and shapes of treated and untreated areas. However, the best pattern for edge-benefitted species among a set of options can be identified by applying the concepts described above. The process is to measure the areas (A_1, A_2) based on known or estimated biological properties (D_m, a, r, w) of the target species, calculate expected density based on the spatial homologue of Leopold's Principle of Edge, and select the pattern that results in the highest weighted estimate of density.

The above approaches (analytical or iterative) will provide the approximate optimum pattern for a single, edge-benefitted species. If the optimum pattern for a suite of s species is desired, including edge-benefitted and edge-damaged species, then ≥2 approaches to pattern selection are reasonable. The first approach is to select the pattern that maximizes the sum of D' values for the i species in question, i.e.,

$$\text{maximize } \sum D_i',\ i = 1, 2, \ldots s.$$

The first approach assumes each of the s species is of equal value ecologically, socially, or economically. If species are not of equal value, then the second approach is to maximize the sum of values for species i (v_i) times D_i'; i.e.,

$$\text{maximize } \sum v_i D_i',\ i = 1, 2, \ldots s.$$

Undoubtedly, numerous approaches to pattern optimization are possible; the above 2 approaches are merely suggestions.

As a last comment on pattern optimization within the operational scale of brush management, we observe that livestock have requirements for refuge and resource cover (Fig. 10). Further, their density can be modeled using the principles of edge described above, even though density is a management decision. Therefore, the formulas provided and the methods of identifying optimum patterns can be used with livestock in the suite of s species.

Fragmentation

At some point, creation of additional edge with mechanical brush management may result in habitat fragmentation and with neutral or negative effects on an edge-benefitted species. Guthery and Bingham (1992) suggested that optimum length (L) of edge on a given area could be approximated as

$$L = 2A/(w + 2r),$$

given a repeated linear pattern ($r{:}w{:}2r{:}w \ldots w{:}r$). The estimate of L can be used as a guideline for determining the point at which edge potentially becomes damaging to an edge-benefitted species.

MECHANICAL MANIPULATIONS: AN ECOLOGICAL PERSPECTIVE

A Theoretical Basis for Mechanical Habitat Manipulation

The simple decision model presented in the previous section does not take into account a host of ecological considerations germane to brush management decisions. Many variables besides pattern may affect the optimum amount of woody plant cover required by a wildlife species, including woody plant density, canopy cover, community structure, and species composition. Different wildlife species require more woody cover than others; e.g., white-tailed deer require more than bobwhites.

Better definition regarding the optimum amount of woody plant cover to clear for a wildlife species or wildlife community cannot be achieved by traditional research approaches because there is an infinite array of variables and interactions that may affect wildlife responses. We propose that when managers desire to optimize dual livestock-wildlife responses (i.e., want to clear as much land as possible to increase forage for livestock while maintaining or increasing wildlife populations), brush management decisions must be based on unifying theory that will provide a conceptual framework for decision-making. In this section, we provide concepts relevant in applied brush management decisions and theory development.

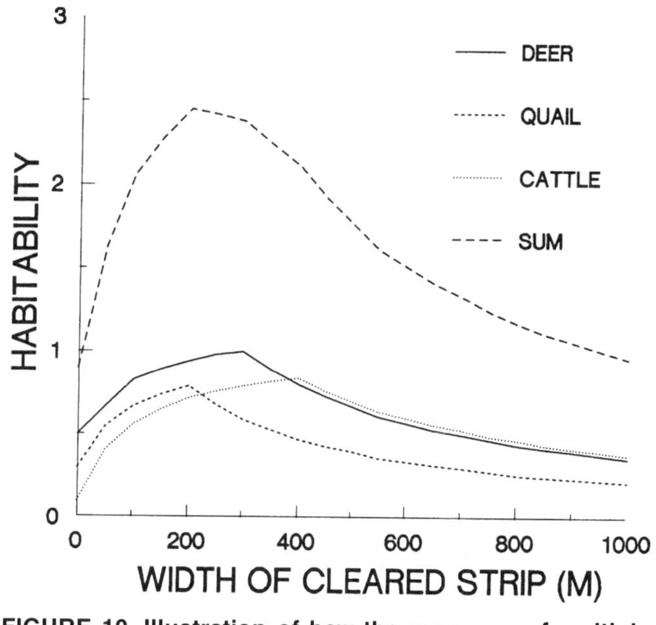

FIGURE 10. Illustration of how the response of multiple species to mechanical brush management can be modeled for decision-making. The pattern evaluated is linear with a constant width of uncleared strip (90 m for deer, the most sensitive species) and variable width of clearing between strips. The habitability index (*y*- axis) is based on the ratio of treated and untreated habitat and the proportion of an area that is fully (randomly) usable. The optimum width of cleared strip, given the fixed requirement for deer, varies with density in untreated brush (*a*) and with the radius of full use (*r*). Habitabilities can be summed as one of the many possible methods of determining optimum for the suite of species examined.

Disturbance and Patch Dynamics

Ecological theories of disturbance and patch dynamics (Pickett and White 1985) may provide a basis for developing management theory with which mechanical manipulations can be better conceptualized and planned, and with which wildlife responses to manipulations can be predicted. According to these theories, individual growth and death are not the sole factors that contribute to ecosystem dynamics (White and Pickett 1985). Natural disturbances, such as fire, drought, or insect damage, play a major role in ecosystem dynamics. Most natural disturbances produce heterogeneous and patchy effects across the landscape. Disturbance and patch dynamics concepts have not been widely applied to management problems (Pickett and White 1985).

The relationship between patch size, patch dispersion, frequency of disturbance, and intensity of disturbance and wildlife population responses must be known before further application of patch theory relative to wildlife populations. Size, shape, and diversity of patches influence patterns of plant and animal species abundance (Turner 1989). Relationships between patch size and use by white-tailed deer (Naderman 1979) and small mammals (Foster and Gaines 1991) have been researched. Once these relationships are understood, managers should be able to use landscape-level models for brush management (Fig. 11) to predict wildlife responses to mechanically created combinations of patch sizes, shapes, dispersion, and frequency and intensity of disturbances.

Functional Ecology

Mechanical manipulations affect multiple ecosystem processes including energy flow, hydrological cycling, and nutrient cycles that may directly or indirectly affect individual wildlife species and communities. Impacts of mechanical manipulations usually have been studied from the standpoint of structural ecology with emphasis on relatively simple parameters such as wildlife population responses or vegetation responses.

The functional role in the ecosystem of the plant species to be manipulated is often ignored or not understood by resource

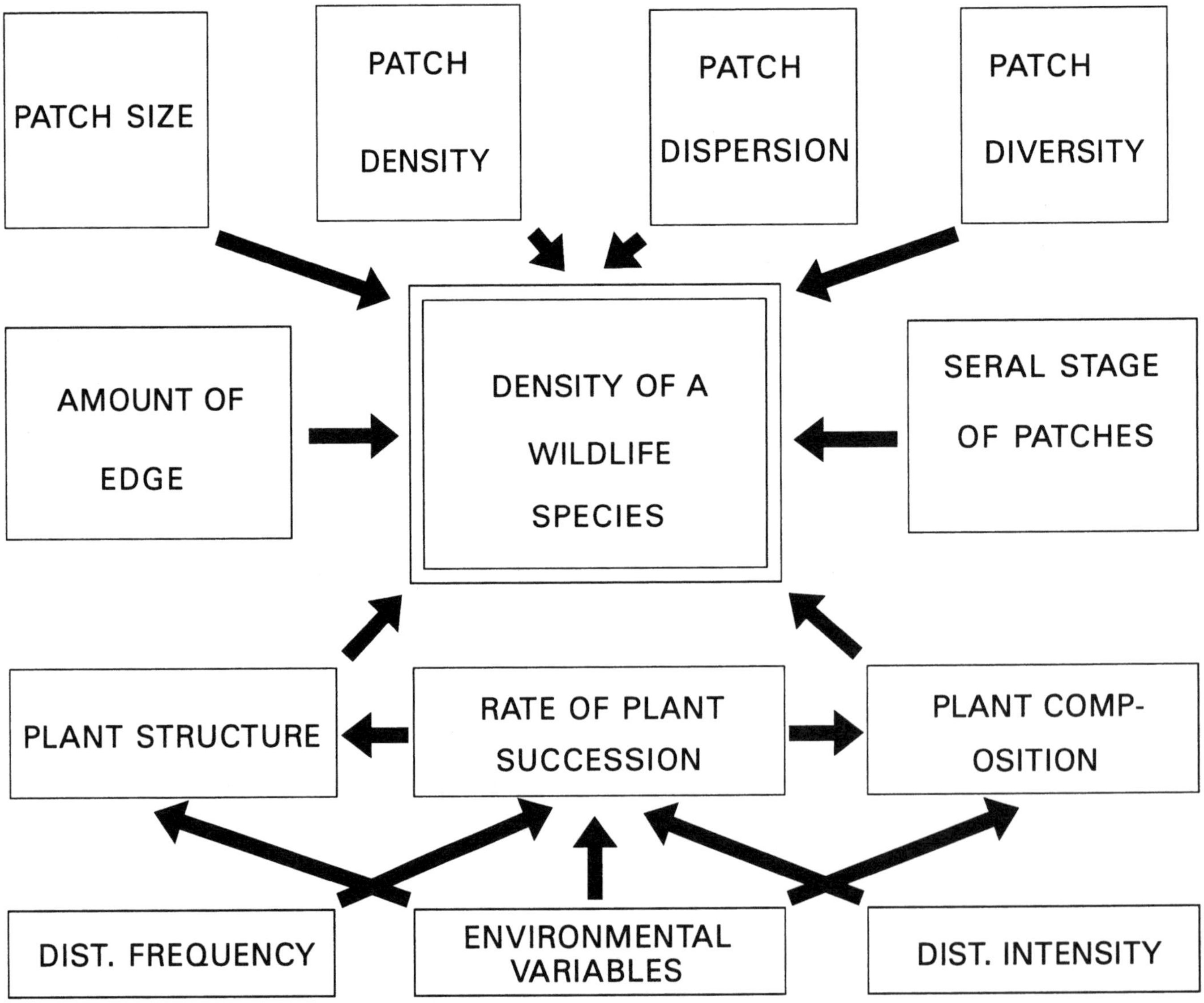

FIGURE 11. A landscape-level model for brush management decisions based on patch dynamic theory.

managers. For example, honey mesquite is commonly targeted as an undesirable species, but it (1) is a nurse plant for many shrubs that are important food and cover species for wildlife (Archer et al. 1988); (2) forms a symbiotic association with nitrogen-fixing microorganisms resulting in significant nitrogen inputs into the ecosystem (Johnson and Mayeux 1990); (3) provides critical habitat for warm-season grasses and Texas wintergrass (Heitschmidt et al. 1986); (4) serves as a nurse plant for ground cherry (*Physalis* spp.), a forb highly preferred by white-tailed deer; (5) provides mesquite pods, extremely important in wildlife and cattle diets and especially important during drought; and (6) provides shade for wild and domestic animals. Heitschmidt et al. (1986) hypothesized that mesquite canopies ameliorate the xeric environment characteristic of most rangelands and that lack of forage response to mesquite control during dry years may occur because the less xeric environment under mesquite canopies may enhance grass production.

Mesquite-dominated ecosystems often consist of a landscape mosaic of grassland and shrub clusters with mesquite as the nucleus and subordinate shrub species and herbaceous plants under its canopy (Archer et al. 1988). Soil nutrient concentrations are greater under mesquite canopies than in grassland areas between mesquites (Tiedemann and Klemmedson 1973). Wild and domestic animals that browse and graze in the shrub clumps and defecate in the grass-dominated interstitial areas between shrubs may redistribute N and other nutrients to the grassland. Thus, the shrub clumps may serve as an important point source of nutrient input into the interstitial grassland areas.

Resource managers should consider the functions of woody species in the ecosystem before mechanical manipulations are planned. The argument that leaving woody plants on rangeland following manipulation will provide a seed source for reinfestation is not valid because once an area has supported a woody plant community the soil already contains an abundant seed source for recolonization.

Ecological Succession

Succession and Diversity. Mechanical habitat manipulation entails manipulating biotic succession. The response of an animal species to mechanical manipulations depends on how the manipulations affect successional dynamics of the mosaic of plant communities in the landscape with which the animal interacts and the successional stages to which the animal is adapted.

The importance of plant species diversity to wildlife is usually associated with the idea that a greater variety of plant species provides a more stable and nutritious food and cover supply. Pollock et al. (1994) suggested that greater use by mature male white-tailed deer of areas with diverse woody vegetation might be related to greater structural diversity as well as greater food diversity. Impacts of mechanical manipulations on vegetation structure relative to the requirements of wildlife species may be as important as impacts of manipulations on plant species composition and stage of succession in many cases.

The short- and long-term effects of mechanical manipulations on plant species diversity and the relationship of plant species diversity to wildlife populations need to be more clearly understood. A prediction of the intermediate disturbance hypothesis is that species richness will be greater in communities experiencing an intermediate level of disturbance and species richness will be lower if the community is undisturbed or intensely disturbed (Pickett and White 1985, Petraitis et al. 1989). The intermediate disturbance hypothesis provides a theoretical basis for testing hypotheses and making predictions about the effects of mechanical treatments on plant species diversity. For example, a hypothesis might be that roller chopping is an intermediate disturbance and results in development of a richer community, whereas root plowing is an intense disturbance that reduces species richness (Fig. 12). At the landscape level, intermediate disturbance rates create a broad range of successional habitats, with some recently disturbed, some of intermediate time since disturbance, and some late successional (Tilman and Pacala 1993). This allows numerous species to coexist. Almost all areas within the landscape are newly disturbed with high rates of disturbance, whereas low rates of disturbance result in habitats that are almost entirely late successional. Planning mechanical manipulations to produce intermediate disturbances may be more desirable if maintaining or increasing plant species richness is a management goal.

Mechanisms of Succession. The ability to predict the patterns of succession following mechanical manipulation would enable managers to determine if mechanical manipulation in a given habitat would produce a plant community that meets the compositional or structural requirements of a particular wildlife species or group of species. Such prediction requires the development of mechanistic theory (Tilman 1990). There are several theoretical approaches to the mechanisms of succession, including viewing succession as a result of facilitation and inhibition mechanisms (Connell and Slatyer 1977), underlying plant population dynamics (Peet and Christensen 1980), or environmental constraints and resource allocation-based tradeoffs (Tilman 1990).

These theories might provide also the basis for post-mechanical treatment manipulations. For example, according to the

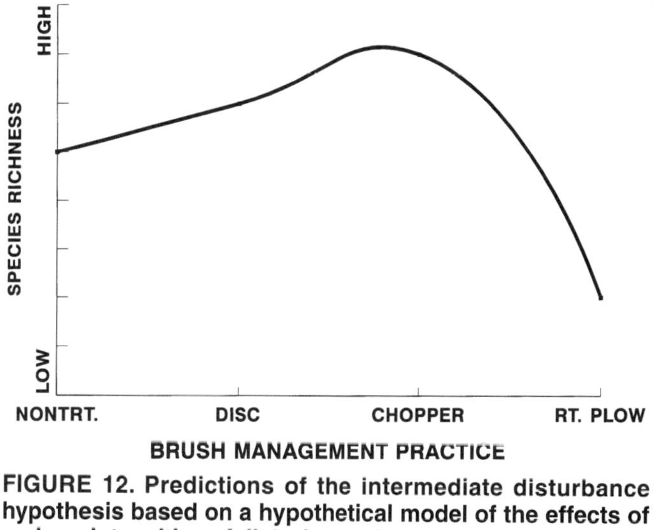

FIGURE 12. Predictions of the intermediate disturbance hypothesis based on a hypothetical model of the effects of various intensities of disturbance by mechanical manipulations (discing [least intense], roller chopping, and root plowing [most intense]) on wildlife and plant species richness.

nutrient-light ratio hypothesis, secondary succession on nutrient-impoverished soils results from a temporal gradient in the relative availabilities of a limiting soil resource and light (Tilman 1985, 1990). A prediction based on this hypothesis is that fertilization with the proper nutrients following mechanical manipulation may speed the rate of succession and inhibit recolonization by early successional shrubs such as mesquite that are adapted to low-nutrient, high-light environments.

SUMMARY

Mechanical treatments temporarily suppress woody plants and the rapidity of woody plant reestablishment depends on the method of treatment and environmental variables such as rainfall, soil fertility, and post-treatment livestock grazing management. Wildlife species that are reduced in abundance following large-scale treatments may increase as woody plants reestablish. However, mechanical manipulations may result in undesirable long-term changes in wildlife habitat. Plant species composition and structure of biotic communities in the seral stages that follow treatment may differ from the communities originally present. Woody plant species richness and diversity of communities that reestablish following mechanical manipulation, particularly root plowing, are generally lower than in unaltered habitat. In many cases, mechanical manipulations have exacerbated rather than slowed the increase of woody plants and have not resulted in increased forage yield. Decisions on whether mechanical manipulations are warranted should be based on the cost of treatment maintenance and long-term impacts on plant species composition in addition to expected increases in forage.

Large-scale clearing is detrimental to habitat for wildlife species obligate on woody plant communities, including mixed-successional species such as white-tailed deer. Conversely, mechanically clearing small areas interspersed within landscape mosaics dominated by woody plants is not detrimental to most wildlife species and may increase wildlife species density or diversity. Wild and domestic herbivores tend to concentrate foraging on cleared areas. Too little clearing may result in decreased forage production because of spot overgrazing. The number, dispersion, and size of cleared areas should be carefully planned to minimize spot overgrazing.

Management-level decisions regarding mechanical manipulation of wildlife habitat have been hampered by the lack of a theoretical basis for decision making. Possible clearing patterns are infinite and decision models are essential for predicting optimum patterns. A host of ecological variables may affect wildlife responses to brush management patterns on a community and landscape level. The ecological concepts of disturbance and patch dynamics may provide the theoretical basis for refining simple decision models and incorporating empirical information in the models.

LITERATURE CITED

Adams, C. E., J. K. Thomas, and C. W. Ramsey. 1992. A synopsis of Texas hunting leases. Wildl. Soc. Bull. 20:188-197.

Archer, S. 1989. Have southern Texas savannas been converted to woodlands in recent history? Am. Nat. 134:545-561.

———, C. Scifres, C. R. Bassham, and R. Maggio. 1988. Autogenic succession in a subtropical savanna: conversion of grassland to thorn woodland. Ecol. Monogr. 58:11-127.

Asah, H. A., T. E. Fulbright, and M. L. Land. 1987. Effect of simulated browsing on spiny hackberry after top removal. Tex. J. Agric. and Nat. Resour. 1:16-18.

Bailey, J. A. 1984. Principles of wildlife management. John Wiley and Sons, New York, N. Y. 373pp.

Baker, M. F., and N. C. Frischknecht. 1973. Small mammals increase on recently cleared and seeded juniper rangeland. J. Range Manage. 26:101-103.

Barnitz, J. A., Jr., V. W. Howard, Jr., and G. M. Southward. 1990. Mule deer and rabbit use on areas of pinyon-juniper woodland treated by two-way cabling. New Mexico Agric. Exp. Stn. Bull. 752. 40pp.

Berlandier, J. L. 1980. Journey to Mexico during the years 1826 to 1834. Vol. 1, 2. Translated from French by S. M. Ohlendorf, J. M. Bigelow, and M. M. Standifer. The Tex. State Hist. Assoc. and Cent. for Studies in Tex. Hist., Univ. Tex., Austin. 672pp.

Best, L. B. 1972. First-year effects of sagebrush control on two sparrows. J. Wildl. Manage. 36:534-544.

Biondini, M. E., A. A. Steuter, and C. E. Grygiel. 1989. Seasonal fire effects on the diversity patterns, spatial distribution and community structure of forbs in the Northern Mixed Prairie, USA. Vegetatio 85:21-31.

Box, T. W., and J. Powell. 1965. Brush management techniques for improved forage values in south Texas. Trans. North Am. Wildl. and Nat. Resour. Conf. 30:297-296.

Bozzo, J. A., S. L. Beasom, and T. E. Fulbright. 1992*a*. Vegetation responses to two brush management practices in south Texas. J. Range Manage. 45:170-175.

———, ———, and ———. 1992*b*. White-tailed deer use of rangeland following summer browse rejuvenation. J. Range Manage. 45:496-499.

Castrale, J. S. 1982. Effects of two sagebrush control methods on nongame birds. J. Wildl. Manage. 46:945-952.

Connell, J. H., and R. O. Slatyer. 1977. Mechanisms of succession in natural communities and their role in community stability and organization. Am. Nat. 111:1119-1144.

Darr, G. W., and D. A. Klebenow. 1975. Deer, brush control, and livestock on the Texas Rolling Plains. J. Range Manage. 28:115-119.

Davis, R. B., and C. K. Winkler. 1968. Brush vs. cleared range as deer habitat in southern Texas. J. Wildl. Manage. 32:321-329.

Drawe, D. L. 1981. Wildlife responses to range management practices: the Welder Wildlife Refuge experience—implications for the future. Beef Cattle Sci. Handb. 18:86-92.

Edwards, C. L., and R. D. Ohmart. 1981. Food habits of the Sonoran pronghorn. Pages 34-44 *in* the Sonoran pronghorn. Special Rep. 10, Arizona Game and Fish Dep. Phoenix.

Everitt, J. H. 1983. Effects of plant shredding on nutrient content of four south Texas deer browse species. J. Range Manage. 36:779-781.

Fisher, C. E., H. T. Wiedemann, C. H. Meadors, and J. H. Brock. 1973. Mechanical control of mesquite. Pages 46-52 *in* Mesquite. Res. Monogr. 1, Texas A&M Univ., Tex. Agric. Exp. Stn., College Station.

Foster, J., and M. S. Gaines. 1991. The effects of a successional habitat mosaic on a small mammal community. Ecology 72:1358-1373.

Friedel, M. H. 1991. Range condition assessment and the concept of thresholds: a viewpoint. J. Range Manage. 44:422-426.

Fulbright, T. E. 1987. Effect of repeated shredding on a guajillo (*Acacia berlandieri*) community. Tex. J. Agric. and Nat. Resour. 1:32-33.

———, and S. L. Beasom. 1987. Long-term effects of mechanical treatments on white-tailed deer browse. Wildl. Soc. Bull. 15:560-564.

———, J. P. Reynolds, and S. L. Beasom. 1993*a*. Effects of browse rejuvenation on selected blood serum characteristics of white-tailed deer. Tex. J. Agr. and Nat. Resour. 6:49-53.

———, ———, and ———. 1993*b*. Effects of browse rejuvenation on white-tailed deer diets and nutrition. Tex. J. Agr. and Nat. Resour. 6:41-48.

———, ———, ———, and S. Demarais. 1991. Mineral content of guajillo regrowth following roller chopping. J. Range Manage. 44:520-522.

Germano, D. J., R. Hungerford, and S. C. Martin. 1983. Responses of selected wildlife species to the removal of mesquite from desert grassland. J. Range Manage. 36:309-311.

Giles, R. H., Jr. 1978. Wildlife management. W. H. Freeman and Co., San Francisco, Calif. 416pp.

Guthery, F. S. 1980. Bobwhites and brush control. Rangelands 2:202-204.

———. 1986. Beef, brush and bobwhites. Caesar Kleberg Wildl. Res. Inst. Press, Texas A&I Univ., Kingsville. 182pp.

———, T. E. Anderson, and V. W. Lehmann. 1979. Range rehabilitation enhances cotton rats in south Texas. J. Range Manage. 32:354-356.

———, and R. L. Bingham. 1992. Comments on Leopold's principle of edge. Wildl. Soc. Bull. 20:340-344.

Haggerman, A. E., C. T. Robbins, Y. Weerasuriya, T. C. Wilson, and C. McArthur. 1992. Tannin chemistry in relation to digestion. J. Range Manage. 45:57-62.

Hahn, H. C., Jr. 1945. The white-tailed deer in the Edwards Plateau region of Texas. Mimeo. Report. Texas Game, Fish, and Oyster Comm., Austin. 52pp.

Hailey, T. L. 1979. A handbook for pronghorn antelope management in Texas. Texas Parks and Wildl. Dep. Booklet 7000-46, Austin. 59pp.

Heitschmidt, R. K., R. D. Schultz, and C. J. Scifres. 1986. Herbaceous biomass dynamics and net primary production following chemical control of honey mesquite. J. Range Manage. 39:67-71.

Holechek, J. L. 1981. Brush control impacts on rangeland wildlife. J. Soil and Water Conserv. 26:265-270.

———, R. D. Pieper, and C. H. Herbel. 1989. Range management, principles and practices. Prentice Hall, Englewood Cliffs, N. J. 501pp.

Howard, V. W., Jr., K. M. Cheap, R. H. Heir, T. G. Thompson, J. A. Dimas. 1987. Effects of chaining pinyon-juniper on nongame wildlife. Pages 541-551 *in* R. L. Everett, compiler. Proc. pinyon-juniper conference. U. S. Dep. Agric., For. Serv., Intermountain Res. Stn., Gen. Tech. Rep. INT-215.

Inglis, J. M. 1985. Wildlife management and IBMS. Pages 35-40 *in* Integrated brush management systems for south Texas: development and implementation. Tex. Agric. Exp. Stn. Bull. 1493.

———, B. A. Brown, C. A. McMahan, and R. E. Hood. 1986. Deer-brush relationships on the Rio Grande Plain, Texas. Tex. Agric. Exp. Stn., College Station. 80pp.

Jackson, A. S. 1969. Quail management handbook for west Texas Rolling Plains. Tex. Parks and Wildl. Dep. Bull. 48. 77pp.

Johnson, D. B., and F. S. Guthery. 1988. Loafing coverts used by northern bobwhites in subtropical environments. J. Wildl. Manage. 52:464-469.

Johnson, H. B., and H. S. Mayeux. 1990. *Prosopis glandulosa* and the nitrogen balance of rangelands: extent and occurrence of nodulation. Oecologia 84:176-185.

Kozicky, E. L., and T. E. Fulbright. 1991. An annotated bibliography on the interaction of range management (livestock grazing, brush management and prescribed burning) or nonmanagement with wildlife habitat and wildlife. Tex. Parks and Wildl. Dep., Fed. Aid and Wildl. Restor. Proj. W-124-M, Austin. 293pp.

Kruse, W. H., R. P. Balda, M. J. Simono, A. M. Macrander, and C. D. Johnson. 1979. Community development in two adjacent pinyon-juniper eradication areas twenty-five years after treatment. J. Environ. Manage. 8:237-247.

Kufeld, R. O. 1983. Responses of elk, mule deer, cattle, and vegetation to burning, spraying, and chaining of gambel oak rangeland. Colorado Div. of Wildl., Fort Collins. Tech. Pub. 34. 47pp.

Kundaeli, J. N., and H. G. Reynolds. 1972. Desert cottontail use of natural and modified pinyon-juniper woodland. J. Range Manage. 25:116-118.

Landers, J. L., and B. S. Mueller. 1986. Bobwhite quail management: a habitat approach. Tall Timbers Res. Stn., Tallahassee, Fla. Misc. Publ. 6. 39pp.

Laycock, W. A. 1991. Stable states and thresholds of range condition on North American rangelands: a viewpoint. J. Range Manage. 44:427-433.

Lehmann, V. W. 1941. Attwater's prairie chicken: its life history and management. U. S. Fish and Wildl. Serv., North Am. Fauna 57. 65pp.

———. 1946. Mobility of bobwhite quail in southwestern Texas. J. Wildl. Manage. 10:124-136.

Leopold, A. S. 1933. Game management. Charles Scribner's Sons, New York, N.Y. 481pp.

Lewis, C. E., B. F. Swindel, and G. W. Tanner. 1988. Species diversity and diversity profiles: concept, measurement, and application to timber and range management. J. Range Manage. 41:466-469.

Longland, W. S. 1991. Risk of predation and food consumption by black-tailed jackrabbits. J. Range Manage. 44:447-450.

Mitchell, G. J. 1980. The pronghorn in Alberta. G. J. Mitchell, Univ. Regina, Sask. 165pp.

Montemayor, E., T. E. Fulbright, L. Brothers, B. Schat, and D. Cassels. 1991. Long-term effects of rangeland disking on white-tailed deer browse. J. Range Manage. 44:246-248.

Murphy, D. A., and T. S. Baskett. 1952. Bobwhite mobility in central Missouri. J. Wildl. Manage. 16:498-510.

Mutz, J. L., C. J. Scifres, D. L. Drawe, T. W. Box, and R. E. Whitson. 1978. Range vegetation after mechanical brush treatment on the Coastal Prairie. Texas Agric. Exp. Stn., Bull. 1191. Texas A&M Univ., College Station. 16pp.

Naderman, J. F. 1979. Deer use of different size mechanically cleared openings in south Texas brush. M. S. Thesis, Tex. Tech. Univ., Lubbock. 77pp.

Nolte, K. R., T. M. Gabor, M. W. Hehman, M. A. Asleson, T. E. Fulbright, and J. C. Rutledge. 1994. Long-term effects of brush management on vegetation diversity in ephemeral drainages. J. Range Manage. 47: In Press.

Nurdin, and T. E. Fulbright. 1988. Disking of improved rangeland to increase wildlife food plants. Tex. J. Agric. and Nat. Resour. 2:32-34.

O'Meara, T. E., J. B Haufler, L. H. Stelter, and J. G. Nagy. 1981. Nongame wildlife responses to chaining of pinyon-juniper woodland. J. Wildl. Manage. 45:381-389.

Peet, R. K., and N. L. Christensen. 1980. Succession: a population process. Vegetatio 43:131-140.

Petraitis, P. S., R. L. Latham, and R. A. Niesenbaum. 1989. The maintenance of species diversity by disturbance. Q. Rev. Biol. 64:393-418.

Pickett, S. T. A., and P. S. White. 1985. Patch dynamics: a synthesis. Pages 371-384 *in* S. T. A. Pickett and P. S. White, eds., The ecology of natural disturbance and patch dynamics. Academic Press, Inc., New York, N. Y.

Pollock, M. T. 1989. Winter and spring habitat selection by mature, male white-tailed deer in southern Texas. M. S. Thesis, Texas Tech. Univ., Lubbock. 69pp.

———, D. G. Whittaker, S. Demarais, and R. E. Zaiglin. 1994. Vegetation characteristics influencing site selection by male white-tailed deer in Texas. J. Range Manage. 47:235-239.

Powell, J. 1968. Rodent numbers on different brush control treatments in south Texas. Tex. J. Sci. 20:69-76.

———, and T. W. Box. 1966. Brush management influences preference values of south Texas woody species for deer and cattle. J. Range Manage. 19:212-214.

Quinton, D. A., A. K. Montei, and J. T. Flinders. 1980. Brush control and Rio Grande turkeys in north-central Texas. J. Range Manage. 33:95-99.

Renwald, J. D. 1978. The effect of fire on woody plant selection by nesting nongame birds. J. Range Manage. 31:467-468.

Reynolds, J. P., T. E. Fulbright, and S. L. Beasom. 1992. Mechanical rejuvenation to dampen seasonal variation in chemical composition of browse. J. Range Manage. 45:589-592.

Rice, S. M. 1991. Models for estimating bobwhite habitat quality on subtropical, subhumid rangeland. M.S. Thesis, Texas A&I Univ., Kingsville. 69pp.

Rollins, D. 1983. Wildlife responses to different intensities of brush removal on the Edwards Plateau of Texas. Ph. D. Thesis, Texas Tech. Univ., Lubbock. 71pp.

———, and F. C. Bryant. 1986. Floral changes following mechanical brush removal in central Texas. J. Range Manage. 39:237-240.

———, ———, D. D. Waid, and L. C. Bradley. 1988. Deer response to brush management in central Texas. Wildl. Soc. Bull. 16:277-284.

Rominger, E. M. 1983. Bighorn sheep food habits and gambel oak manipulations, Waterton Canyon, Colorado. M. S. Thesis, Colorado State Univ., Fort Collins. 124pp.

Russell, T. P. 1964. Antelope of New Mexico. N.M. Dep. Game and Fish, Bull. 12. 103pp.

Ruthven, D. C., III, E. C. Hellgren, and S. L. Beasom. 1994. Effects of root plowing on white-tailed deer condition, population status, and diet. J. Wildl. Manage. 58:59-70.

———, T. E. Fulbright, S. L. Beasom, and E. C. Hellgren. 1993. Long-term effects of root plowing on vegetation in the Eastern South Texas Plains. J. Range Manage. 46:351-354.

Schwartz, C. C., and J. G. Nagy. 1976. Pronghorn diets relative to forage availability in northeastern Colorado. J. Wildl. Manage. 40:464-468.

Scifres, C. J. 1980. Brush management, principles and practices for Texas and the Southwest. Tex. A&M Univ. Press, College Station. 360pp.

Scott, V. E., and E. L. Boeker. 1977. Responses of Merriam's turkey to pinyon-juniper control. J. Range Manage. 30:220-223.

Scotter, G. W. 1980. Management of wild ungulate habitat in the western United States and Canada: a review. J. Range Manage. 33:16-28.

Sedgewick, J. A., and R. A. Ryder. 1987. Effects of chaining pinyon-juniper on nongame wildlife. Pages 541-551 *in* R. L. Everett, compiler, Proc. pinyon-juniper conference. U. S. Dep. Agric., For. Serv., Intermountain Res. Stn., Gen. Tech. Rep. INT-215. 581pp.

Severson, K. E. 1986*a*. Small mammals in modified pinyon-juniper woodlands, New Mexico. J. Range Manage. 39:31-34.

———. 1986*b*. Woody plant reestablishment in modified pinyon-juniper woodlands, New Mexico. J. Range Manage. 39:438-442.

Short, H. L., W. Evans, and E. L. Boeker. 1977. The use of natural and modified pinyon-juniper woodlands by deer and elk. J. Wildl. Manage. 41:543-559.

———, and C. Y. McCulloch. 1977. Managing pinyon-juniper ranges for wildlife. U. S. For. Serv. Gen. Tech. Rep. RM-47. 10pp.

Skousen, J. G., J. N. Davis, and J. D. Brotherson. 1989. Pinyon-juniper chaining and seeding for big game in central Utah. J. Range Manage. 42:98-104.

Smith, C. B., and P. J. Urness. 1984. Small mammal abundance of native and improved foothill ranges, Utah. J. Range Manage. 37:353-357.

Stoddard, H. L. 1932. The bobwhite quail—its habits, preservation and increase. Charles Scribner's Sons, New York, N.Y. 558pp.

Sumner, E. L., Jr. 1935. A life history study of the California quail, with recommendations for its conservation and management. Calif. Fish and Game 21:275-342.

Sundstrom, C., W. G. Hepworth, and K. L. Diem. 1973. Abundance, distribution, and food habits of the pronghorn. Wyo. Game and Fish Comm. Bull. 12. 61pp.

Terrel, T. L., and J. J. Spillet. 1975. Pinyon-juniper conversion: its impact on mule deer and other wildlife. Pages 105-109 *in* The pinyon-juniper ecosystem: a symposium. Utah State Univ., Logan. 194pp.

Tiedemann, A. R., and J. O. Klemmedson. 1973. Nutrient availability in desert grassland soils under mesquite (*Prosopis julifora*) trees and adjacent open areas. Proc. Soil Sci. Soc. Am. 37:107-110.

Tilman, D. 1985. The resource-ratio hypothesis of plant succession. Am. Nat. 125:827-852.

———. 1990. Constraints and tradeoffs: toward a predictive theory of competition and succession. Oikos 58:3-15.

———, and S. Pacala. 1993. The maintenance of species richness in plant communities. Pages 13-25 *in* R. E. Ricklefs and D. Schluter, eds. Species diversity in ecological communities. Univ. Chicago Press, Chicago, Ill.

Turkowski, F. J., and H. G. Reynolds. 1970. Response of some rodent populations to pinyon-juniper reduction on the Kaibab plateau, Arizona. Southwest. Nat. 15:23-27.

———, and R. Watkins. 1976. White-throated woodrat (*Neotoma albigula*) habitat relations in modified pinyon-juniper woodland of southwestern New Mexico. J. Mammal. 57:586-591.

Turner, M. G. 1989. Landscape ecology: the effect of pattern on process. Annu. Rev. Ecol. and Syst. 20:171-197.

Urness, P. J. 1974. Deer use changes after root plowing in the Arizona chaparral. U. S. For. Serv. Res. Note RM-255. 8pp.

Vallentine, J. F. 1980. Range developments and improvements, Second ed. Brigham Young Univ. Press, Provo, Ut. 545pp.

Vega, J. H., and J. H. Rappole. 1994. Effects of scrub mechanical treatment on the nongame bird community in the Rio Grande Plain of Texas. Wildl. Soc. Bull. 22:165-171.

Webb, W. M., and F. S. Guthery. 1983. Response of wildlife food plants to spring discing of mesquite rangeland in northwest Texas. J. Range Manage. 36:351-353.

Welch, T. G., R. P. Smith, and G. A. Rasmussen. 1985. Brush management technologies. Pages 15-24 *in* Integrated brush management systems for south Texas: development and implementation. Tex. Agric. Exp. Stn. Bull. 1493. 71pp.

White, P. S., and S. T. A. Pickett. 1985. Natural disturbance and patch dynamics: an introduction. Pages 3-13 *in* S. T. A. Pickett and P. S. White, eds. The ecology of natural disturbance and patch dynamics. Academic Press, Inc., New York, N. Y.

Yoakum, J. D. 1979. Managing rangelands for pronghorns. Rangelands 1:146-151.

———. 1980. Habitat management guides for the American pronghorn antelope. U. S. Dep. Interior, Bur. of Land Manage., Denver Serv. Cent., Denver. 77pp.

———. 1983. Managing vegetation for pronghorns in the Great Basin. Pages 189-194 *in* Managing intermountain rangelands—improvement of range and wildlife habitats. U. S. For. Serv. Gen. Tech. Rep. INT-157.

———, W. P. Dasmann, H. R. Sanderson, C. M. Nixon, and H. S. Crawford. 1980. Habitat improvement techniques. Pages 329-403 *in* S. D. Schemnitz, ed. Wildlife management techniques. Fourth ed. The Wildl. Soc., Washington, D. C.

Chapter 21
REVEGETATION OF RANGELANDS FOR WILDLIFE

Bruce A. Roundy

INTRODUCTION

Rangeland revegetation technology has been developed in the western United States to improve forage for livestock and to increase watershed stability. Because most revegetation projects were not conducted primarily for wildlife habitat improvement, they have had differential effects on wildlife populations. Rangeland revegetation principles include careful consideration of how revegetation will help meet management objectives, selection of appropriate plant materials and establishment methods, and habitat management after revegetation. Much of the literature on rangeland revegetation to improve wildlife habitat focuses on establishing shrub and forb species on mule deer (*Odocoileus hemionus*) winter ranges in the West. Using this work as a primary model, I will discuss the historical background, special considerations, and effects of rangeland revegetation on wildlife habitat improvement.

HISTORICAL PERSPECTIVE

Rangeland revegetation approaches to improve wildlife habitat in the western United States have developed predominantly in response to a dynamic series of interactions between livestock and mule deer populations and changes in plant community composition. The influx of large domestic livestock populations as the West was colonized in the latter half of the nineteenth century led to a loss of grass dominance. Reduced competition from grasses, the lack of fire associated with suppression efforts, and limited fine fuels, led to an increase in populations of woody species (Smeins 1983, Young et al. 1983). Periodic severe winters and droughts in the Intermountain area (Young et al. 1979) and droughts in the Southwest (Cox et al. 1982) devastated livestock herds and signaled a need for management of grazing. Severe flooding and associated damage to foothill urban areas in Utah (Forsling 1931) generated interest in management and rehabilitation of watersheds (DeByle and Hookano 1973). Range management developed as a profession with an emphasis on grazing management and vegetation improvements as a result of these early impacts (Holechek et al. 1989, Vallentine 1989).

More subtle than the starvation of livestock or the siltation of hydroelectric plants resulting from loss of grass was the shift in woody plant dominance, which many believe supported increases in mule deer populations (Workman and Low 1976, Longhurst et al. 1977). Conservative hunting regulations in the early 1900s allowed mule deer to increase dramatically from 1930 to 1950 (Urness 1990). High deer mortality associated with deep snows during periodic severe winters from the 1930s, through the 1950s, and overbrowsing of winter ranges by excessive deer populations stimulated concern for deer habitat (Julander and Low 1976).

Research on range revegetation in the 1930s and 1940s had mainly been conducted by the United States Forest Service and had emphasized establishment of adapted grasses for livestock forage and watershed rehabilitation. The most successful grasses in research trials and those eventually most extensively seeded were grazing-tolerant introduced species (Roundy and Call 1988).

In the 1940s and 1950s Forest Service and state game and fish departments began to develop research programs to characterize and improve mule deer and other big game habitat. Some of the supporting funds came from the Federal Aid to Wildlife Restoration Act or Pittman-Robertson Act of 1937 (Peek 1986). Early research in many of the western states emphasized establishment of Rosaceae shrubs, especially antelope bitterbrush (*Purshia tridentata*), because of the observed dominance of these shrubs in the diets of mule deer and other wildlife species (Ferguson 1983, Tiedemann and Johnson 1983). Bulletins published from these research efforts in California (Hubbard 1964), Idaho (Holmgren and Basile 1959), Utah (Plummer et al. 1968), and Washington (Brown and Martinsen 1959) recommended species for different sites and discussed the challenges and techniques of brush establishment (Ferguson 1983, Medin and Ferguson 1972). Research in the Southwest concentrated on establishing four-wing saltbush (*Atriplex canescens*) (Springfield 1970).

Actual game winter range improvement projects by 1972 included only 4,000 ha in Nevada, 6,000 ha in Idaho, > 30,000 ha in Oregon, and 48,000 ha in Utah (Medin and Ferguson 1972). Pinyon (*Pinus* spp.) and juniper (*Juniperus* spp.) woodlands were considered priority vegetation types for improvement due to their large geographic dominance (> 30,000,000 ha in western North America), their importance as mule deer winter range (54% of the winter range in Utah), and their potential for increased productivity and diversity after tree control and revegetation (Terrel and Spillett 1975). Actual benefits of these treatments for mule deer and other wildlife populations were not consistent, partly because many projects involved revegetation with grasses predominantly for livestock (Tueller 1976).

Predominance of wildlife revegetation work in Utah is a product of an initial cooperative agreement between the Utah Division of Fish and Game and the United States Forest Service Intermountain Research Station initiated in 1955 (Plummer et al. 1968). Cooperative research by these agencies eventually resulted in establishment of the Shrub Sciences Laboratory in Provo, Utah. This effort probably represents the longest continued cooperative commitment to rangeland revegetation for wildlife in the West. In conjunction with the Shrub Research Consortium, the Shrub Science Laboratory continues to hold and publish proceedings of symposia on use and management of shrubs and other important species for wildlife habitat.

The classic publication of Plummer et al. (1968) summarized much of the early cooperative research and established initial guidelines for range revegetation for wildlife in Utah. A substantial update to that information contributed by many vegetation scientists in the Intermountain area will present the most detailed treatment of ecology, management, and revegetation guidelines for plants for wildlife habitat available in the West (Monsen and Stevens In Press).

In addition to the revegetation literature initiated by concern over western game ranges, information relevant to wildlife habitat improvement has been compiled relative to the Conservation Reserve Program (CRP) for seeding erodible cropland (Mitchell 1987); mineland reclamation (Institute for Land Rehabilitation 1978, 1979; Dickson and Vance 1981, Wasser and Shoemaker 1982) and other disturbances such as military testing (Wood 1980); and roadcut revegetation (Hungerford 1984). Game managers in the southern United States have used food plantings since 1935 to concentrate game animals and compensate for habitat loss associated with increased timber production (Halls and Stransky 1968). Legumes received research attention because of their ability to fix nitrogen and provide high protein forage for domestic livestock. Legumes and other forbs are important species for wildlife (Hermann 1966), therefore literature on wildland plantings of legumes (Rumbaugh 1983) is especially relevant to wildlife habitat revegetation.

CONSIDERATIONS FOR WILDLIFE HABITAT REVEGETATION

Revegetation for wildlife habitat improvement involves consideration of objectives, site potential, plant materials, establishment, and management.

Determining Objectives

Because revegetation is an intensive and expensive practice, it should be justified in terms of specific objectives. Objectives for revegetation can range from creating habitat for maintaining or increasing specific wildlife populations to general goals of improving vegetation diversity and production for encouraging faunal diversity.

Featured species habitat management focuses on habitat requirements for specific species, but also considers the requirements of other wildlife species in the area (Peek 1986). Revegetation for specific species has traditionally been justified mainly for game animals, but could also be justified for threatened and endangered plants and animals. As land treatment objectives shift from production goals to goals of increasing biodiversity and ecological restoration, difficult questions of economics and evaluation of success arise. Capital costs of large scale revegetation are high and raise the questions of how and by whom they will be paid (Godfrey and Nielsen 1980, Young 1990). Re-establishing or restoring natural, self-sustaining plant and animal communities as a goal is problematic due to difficulties in defining "natural", major physical or biological changes on site such as presence of many exotic species, and the prohibitive economics and possibly ecology of large scale projects (Diamond 1987).

Because of their expense, wildlife habitat restoration projects realistically need to be integrated into the overall management of other resources such as forage for livestock and timber (Longhurst et al. 1982, Flather and Hoekstra 1989). A good example of this integration is the use of revegetation species to benefit wildlife on CRP lands in the Great Plains (Mitchell 1987). Multiple use approaches to land management support the maintenance of diversity in the form of vertical, horizontal, and special components of wildlife habitat (Flather and Hoekstra 1989). But just how much economic return for unused resources society is willing to forgo, or what price society is willing to pay for habitat restoration to maintain or promote certain wildlife species is uncertain.

There is a growing awareness that the management of wildlife predominantly involves management of habitat. In a recent survey, state and federal biologists ranked habitat degradation and loss as the 2 most important wildlife and fish management issues (Flather and Hoekstra 1989). Habitat management requires, as Leopold recognized in 1933 (Krausman 1990), an understanding of the "... specific definitions of the environment needed by each species...." Natural resource scientists and managers have produced considerable general, and, for some species, specific information defining habitats for rangeland wildlife. Such information was summarized by the Biological Services Program (Institute for Land Rehabilitation 1978) and can be found in various chapters in this book. Habitats have been defined by observed use of certain landscapes and plant communities and by dietary studies (Longhurst et al. 1982). Game species have received more attention than nongame species. Patton (1992) has reviewed the development and use of wildlife habitat models for management decision making. Payne and Bryant (1994) discuss components of habitat and faunal diversity.

The specificity of objectives in revegetation defines the specificity of information needed about wildlife habitat requirements. If the objective is to maintain or increase a particular population of animals, then the vertical and horizontal scales and specific components of the seasonal habitats and dietary requirements of that population must be defined to guide revegetation for habitat improvement. For example, masked bobwhite quail (*Colinus virginianus ridgewayi*) lack winter food sources on the northern edge of their range at the Buenos Aires Wildlife Refuge in southern Arizona. Restoration of this quail at the refuge may be facilitated by revegetation with white ball acacia (*Acacia angustissima*), which gradually sheds its seeds over winter.

Sometimes the condition and extent of critical habitats are perceived to be limiting to many populations of a particular species. For example, as mule deer populations grew in the West, increased deer productivity was associated with high quality and quantity of forage on summer ranges (Julander et al. 1961). Quality of summer forage influences ovulation and fawning rates (Swank 1958). Deer populations had high mortality during severe winters on some winter ranges but not on others (Plummer et al. 1968). Mortality on winter ranges was related to available forage (Robinette et al. 1952). Revegetation research and projects were directed primarily at winter range plant communities including pinyon-juniper, mountain brush, and big sagebrush (*Artemisia tridentata*), and secondarily at summer range vegetation including aspen (*Populus tremuloides*) and associated conifers and subalpine and aspen openings (Plummer et al. 1968).

Habitat improvement for mule deer was only a part of management, which included increased harvesting to balance deer populations with available habitat (Julander and Low 1976). This exemplifies the fact that because some wildlife populations can increase faster than we can or want to manipulate vegetation or to revegetate to increase habitat for them, revegetation and other manipulations should only be a part of the management plan, which must include other controls to balance available habitat and population size. Revegetation without consideration of an overall management plan may result in alleviating a habitat constraint to a wildlife population in 1 season and lead to increased population size and overuse of habitat in another season. For example, pushing juniper trees and seeding crested wheatgrass (*Agropyron* spp.) in the 1950s on the Blue Ridge Ranger District in Arizona has provided nutritious critical late winter-early spring forage for reintroduced elk (*Cervus elaphus*) and may be contributing to recent population increases resulting in overgrazing of summer ranges.

In some cases, goals of revegetation for disturbed or degraded rangelands are to *reclaim* them to "... be habitable to organisms originally present in approximately the same composition and density ..." as before disturbance or degradation, or *rehabilitate* them by returning the site "... to a form and productivity in conformity with ... prior use ..." (Box 1978). The Surface Mining Control Reclamation Act of 1977 or more stringent state laws set reclamation guidelines for disturbance by coal mining (Chambers and Brown 1983). Where goals are set at the level of reclamation or rehabilitation and where the specific biology of wildlife species are not well known, revegetation is usually planned around establishing the plant communities known to be associated with animal species that inhabit similar undisturbed sites. Wildlife species associations with rangeland vegetation types have been listed for many of the western states as a guide to reclamation (Institute for Land Rehabilitation 1978).

Often, actual revegetation and fire rehabilitation projects include fairly general goals that include improvement of watershed stability, and increasing forage for livestock and wildlife. Actual revegetation projects are usually constrained by seed availability but seed mixes are usually defined to include some plants considered important for wildlife. In general, native species, especially shrubs (Robinette 1972, Urness 1989) and forbs (Hermann 1966), are important for wildlife food and cover.

Also, high diversity in the horizontal and vertical scales of structure, and in kinds of plants available for food, is considered necessary for a diversity of wildlife species (Holechek et al. 1989). High vegetation diversity is often associated with successional, rather than climax vegetation conditions. Lack of understory diversity in climax pinyon-juniper stands, for example, has led to extensive manipulation of this type for wildlife and livestock habitat improvement (Terrel and Spillett 1975).

Traditional revegetation practices to provide forage for domestic livestock often decreased vegetation diversity for wildlife. Seeding of large blocks of big sagebrush (*Artemisia tridentata*) rangelands to a few wheatgrass species reduces diversity of wildlife (McAdoo et al. 1986). Seeding mixtures of species are recommended to avoid monoculture pests and provide a better forage base for all classes of animals (Keller 1979, McKell 1986). Four-wing saltbush and various legumes grown with crested wheatgrass increased total forage and grass yields, and protein yield (Rumbaugh et al. 1982) (Fig. 1). This indicates potential to improve forage quality and quantity for wildlife with mixed seedings. However, attempts to establish mixed communities have often been unsuccessful due to the competitiveness and vigor of some of the seeded species (Roundy and Call 1988). Exotic grasses are often more easily established than native grasses, shrubs, and forbs and therefore may dominate revegetated ranges despite seeding with a mixture of species (Leckenby and Toweill 1983*a,b*).

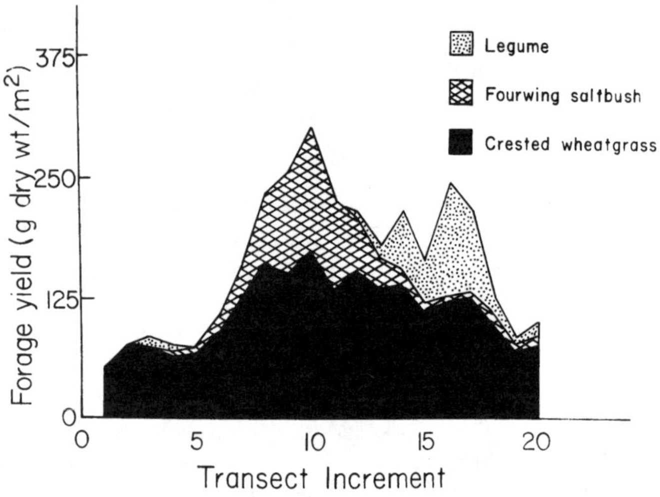

FIGURE 1. Forage yields of crested wheatgrass, fourwing saltbush, and legumes distributed spatially across a rangeland site (Rumbaugh et al. 1982). Each transect increment is equal to 0.3 m.

Guidelines have been developed for revegetation of arid and semiarid rangelands, and specific recommendations have mainly emphasized establishment of grasses (Roundy and Call 1988). The following discussion will emphasize considerations for the selection and establishment of shrubs and forbs based on their importance to wildlife habitat.

Site Potential

Site potential for rangeland revegetation is generally defined in terms of climatic conditions and soil characteristics. Native species are adapted as mature plants to the climate and edaphic conditions of specific sites. Descriptions of the biota of western rangelands (Brown 1982, West 1983) and descriptions of the distribution of specific plants and communities (Tew 1983, West et al. 1975, McArthur and Welch 1984) are a beginning to determining associations between site potential and adapted species (McKell 1990). Soil Conservation Service range site descriptions also list what is thought to be the plant species composition and production potential of undisturbed range sites (Shiflet and MacLauchlan 1986). Payne and Bryant (1994) list appropriate site conditions for numerous plants of known wildlife value. Most revegetation guidebooks categorize land areas by environmental zones based on the normally occurring plant community and climatic conditions (Jordan 1981, Plummer et al. 1968).

Degree of aridity is the most important factor in determining revegetation potential on many rangelands. Aridity is defined by effective soil moisture for plant growth and is a function of annual and seasonal precipitation and evaporative demand (United Nations Educational, Scientific, and Cultural Organization 1979). Although mature plants are adapted to arid areas (< 200-300 mm annual precipitation) natural establishment is generally episodic (MacMahon 1987, Romney et al. 1989*a*), occurring only during unusually wet years. Therefore, revegetation by direct seeding is usually recommended in areas receiving ≥ 300 mm annual precipitation (Roundy and Call 1988). However, successes have been reported in direct-seeding shrubs under lower annual rainfall conditions in the Mojave and Sonoran deserts (Graves et al. 1978, Jackson et al. 1991), and also in the deserts of central Asia (Nechaeva 1985).

A number of authors have discussed site and soil factors that affect revegetation potential and species that indicate potential for direct seeding (Vallentine 1989, Roundy and Call 1988, DePuit and Redente 1988). Changes in physical and biological conditions of a site, such as erosion of surface soil horizons, loss of the original nutrient cycling system and soil fertility, and influxes of weedy species, may reduce the potential of a site so that high successional species native to the site can no longer establish. Conventional recommendations are to revegetate sites with the highest potential first. These sites generally have gentle topography and deep, fertile, moderately textured soils, while those sites with lower potential due to shallow, rocky, coarse or fine-textured soils are often most in need of improvement (Medin and Ferguson 1972).

Stevens et al. (1974) conducted a rare study of site productive potential for a mixture of revegetated shrubs, forbs, and grasses in Utah. For pinyon-juniper and sagebrush-grass communities, they correlated production of seeded big sagebrush, black sagebrush (*Artemisia nova*), and various grasses and forbs with annual and May precipitation, slope, and various soil factors such as percent rock, soil depth, and depth to calcium carbonate accumulation. Regression equations were developed to categorize sites into productivity potential groups based on the site factors for each of the revegetation species. Detailed research of this nature is needed to better quantify the biomass productive potential for different sites and revegetation species.

Plant Materials

Ideally, species are selected for revegetation projects to meet management objectives and because they can establish and persist on the site of interest. In practice, revegetation projects often use species that are considered adapted, have proven establishment ability, and have seeds that are commercially available. The establishment success of many introduced grasses plus the ease with which the seeds of many of them can be produced, harvested, cleaned, and sown continues to make them the least expensive and most available species for revegetation. While still valuing appropriate grasses, scientists oriented toward improvement of game ranges emphasized the use of shrubs and forbs and mixed seedings. Although species for sagebrush, pinyon-juniper, and mountain brush types were most studied because of their importance to deer habitat, species for higher and lower elevation plant communities have also been studied (Plummer et al. 1968). Numerous species trials on many different sites have generated considerable information and species recommendations have been published for different climatic, edaphic, and management situations (Hafenrichter et al. 1968, Plummer et al. 1968, Institute for Land Rehabilitation 1979, Keller 1979, Jordan 1981, Cox et al. 1982, Thornburg 1982, Wasser and Shoemaker 1982, Monsen and Shaw 1983*b*, George et al. 1983, McGinnies et al. 1983, Brown and Wiesner 1984, Tiedemann et al. 1984, Everett 1987, Mitchell 1987, Vallentine

1989, Monsen and Stevens (In Press). In addition, a Plant Information Network (PIN) has been established by Colorado State University and the United States Fish and Wildlife Service to provide an array of computerized information on biology, distribution, revegetation potential, management, and value of native plant species for livestock and wildlife (Vories and Sims 1977). Payne and Bryant (1994) have compiled extensive information on value and use of specific plants for wildlife habitat improvement throughout the United States.

Shrubs. Shrubs provide critical cover and structure for protection of animals from hot and cold temperatures, and protection from predators, especially during breeding and reproduction (Urness 1989). Some shrubs provide important seasonal forage for wildlife. For example, some shrubs provide higher levels of protein, phosphorus, and carotene than grasses and forbs, and can meet the nutritional gestation requirements of ruminants during dormant winter periods (Welch 1989), while others have physical and chemical characteristics that decrease their availability and use as forage (McKell 1989).

Many North American shrubs have genetic flexibility associated with rapid evolution in changing environments (Stutz 1989). Many shrubs have the genetic potential to adapt to specific environments, a characteristic that results in occurrence of specific ecotypes, but also permits opportunities for artificial selection and plant improvement. The ecotypic nature of fourwing saltbush, for example, has led to a number of cultivars selected for specific geographic ranges (Carlson 1984). Use of local seed sources is recommended for many native species (Ferguson 1983), because range of adaptability may not be known (Dunne 1989, McKell 1990).

Many shrubs have characteristics that limit commercial seed availability including low seed production, seeds with coverings and appendages that make them difficult to harvest, clean, and sow, and a general growth form that makes seed harvesting difficult. Young and Young (1986), Dunne (1989), and Lippitt et al. (1994) described methods of collecting and processing wildland plant seeds, while Van Epps (1989) provided guidelines for producing shrub seeds in plantations. Major use of shrubs and other native species for large-scale revegetation for wildlife is constrained by lack of a native seed industry and, therefore, seed supply. Although seeds of frequently recommended species are usually available, use of less common species usually requires contracting at least a year in advance to allow time for special collection (McKell 1990).

Forbs and Legumes. Forbs occur in all rangeland plant communities and many are important seasonal food sources for wildlife (Hermann 1966). Forbs maintain higher protein and phosphorus through the summer than grasses that have flowered (Cook 1983). Forbs add to the ecological, nutritional, and visual diversity of rangeland plant communities. Their forage, seeds, and fruits are important food for livestock and many classes of wildlife. For example, forbs are an important dietary component for sage grouse (*Centrocercus urophasianus*) in the spring and summer (Klebenow and Gray 1968, Wallestad et al. 1975). Shaw and Monsen (1984*a*) and Stevens et al. (1985) reviewed the benefits of nonleguminous forbs, the reasons for their limited use in revegetation, and major species used for revegetation in the Intermountain area. Use of nonleguminous forbs in rangeland revegetation has been limited mainly by low seed availability. Hand collection from native stands is constrained by erratic seed crops; indeterminant flowering and seed set, which results in seeds being ready for harvest at different times on the same plant; and low density populations. Increasing interest in biodiversity, and the decision of the Federal Highway Administration to require revegetation of federal-aid highways with native wildflowers (Scully 1987), may help increase demand and the commercial production and availability of forbs.

The use of leguminous forbs for wildland plantings has been reviewed by Rumbaugh (1983) and Rumbaugh and Townsend (1985). Legumes are considered important forage components on livestock ranges because their nitrogen-fixing ability can increase yield and quality of associated grass species and improve the seasonal forage production and quality of the pasture as a whole (Fig. 1).

Establishment of legumes has the potential to increase site productivity for wildlife by providing diversity and high-protein forage and seeds during critical reproductive periods (Yoakum 1979). Criteria for selecting legumes for revegetation include availability of seeds and *Rhizobium* inoculum, ability to establish and complement associated species, nontoxicity and forage quality, and persistence of perennials or ability of annuals and biennials to reestablish (Rumbaugh 1983). Another important consideration is low preference by rodents, grasshoppers, and other insects (Hewitt et al. 1982). Most frequently seeded legumes in rangeland revegetation include alfalfa (*Medicago sativa* and *M. falcata*) and sweetclover (*Melilotus alba* and *M. officianalis)*, but sainfoin (*Onobrychis viciaefolia*), species of sweetvetch (*Hedysarum*) and milkvetch (*Astragalus*), and other legumes show promise for rangeland plantings.

Plant Improvement. Rangeland revegetation research has always emphasized identification of better plant materials, both among and within species. Traditional selection programs involve comparing establishment, forage yield and quality, and other characteristics in a common location for ecotypes or accessions collected from different natural populations (Shiflet and MacLauchlan 1986). Additional selection criteria for plant selection or more sophisticated plant breeding programs might include the ability to resprout after fire (Pellant 1990); palatability and absence of undesirable oils, digestibility, energy and nutritional content; beneficial establishment characteristics, such as seed germination and seedling vigor; and various adaptive characteristics such as tolerance to salinity, temperature, and drought stress; resistance to defoliation, diseases and insects; and rooting and growth habit (McArthur et al. 1985, McArthur 1988, McKell 1990). Potential for plant materials improvement and promising selections have been discussed for shrubs in the Compositae (McArthur et al. 1985), Chenopodiaceae (Stutz and Carlson 1985), and Rosaceae (Monsen and Davis 1985) families. Shrubs in these families have substantial genetic variability, and high potential for selective improvement. Stutz and Carlson (1985) have proposed that for the erratic environments of wildlands, shrub selections should be genetically rich.

McKell (1990) and McArthur (1988) listed major range plant cultivars developed by mass selection and other methods and discussed more advanced plant improvement methods. Hybridization breeding has been conducted mainly for range grasses, but also for some chenopod shrubs (Stutz and Carlson 1985). McKell (1990) noted that information on shrub breeding systems, such as that from Pendleton et al. (1989), will facilitate shrub improvement programs. McKell (1990) and Carman (1989) discussed the tremendous potential recently developed molecular biology and cellular approaches have for genetic construction of wildland plants, not only for enhanced germplasm, but also as a research tool to better understand interactions of physiology and morphology of plants with the environment.

Although perennial legumes have substantial potential for increasing forage quantity and quality on rangelands (Rumbaugh and Townsend 1985), their use has been limited in revegetation due to their potential to cause bloat in ruminants, and their lack of persistence. Johnson et al. (1981) felt that these problems could be overcome by appropriate grazing management and development of improved legume cultivars. Over 25 improved cultivars of legumes are currently available for humid and semiarid rangelands, but development of cultivars for more arid sites is still needed (Rumbaugh and Townsend 1985).

Stevens et al. (1985) discussed adaptation and use of 29 improved nonleguminous forb varieties. A recognition of the value and appropriate use of these varieties may stimulate their demand in revegetation projects and lead to commercial seed production necessary to create a consistent seed supply and minimize costs.

Establishment

Although many plants desirable for wildlife habitat on semiarid rangelands are adapted as mature plants to specific sites, establishing them on a given year on those same sites by direct seeding continues to be the greatest challenge to the science and application of revegetation. Establishment success is greatly dependent on the physical and biological environmental context of the site, in relation to the germination and seedling growth requirements of the species of interest. Factors that may be controlled to some degree to enhance success include site selection, species-seed selection to obtain certain germination and seedling attributes, seed pretreatments to improve germination, modification of the physical aspects of the seedbed to improve conditions for germination and seedling growth, modification of existing species populations, season of sowing, and location of seeds in the seedbed. Thus, conventional guidelines for direct seeding include recommendations for appropriate site selection, sowing germinable seeds of adapted species, controlling existing plant competition, preparing the seedbed, and sowing seeds in an appropriate way and in the proper season.

General and specific guidelines and considerations for establishing rangeland species are presented in a number of books, bulletins, and articles including Payne and Bryant (1994), Vallentine (1989), Roundy and Call (1988), Plummer et al. (1968), Jordan (1981), Institute for Land Rehabilitation (1978), and Redente and DePuit (1988). The most comprehensive treatment for the Intermountain area is that of Monsen and Stevens (In Press). Traditional recommendations have proven especially effective for establishing large-scale exotic grasslands. Their application to the establishment of biologically diverse plant communities, including native grasses, shrubs, and forbs for improvement of wildlife habitat, has been less successful. Advances in the science of seedbed ecology are needed to better understand establishment requirements and refine guidelines for establishment of plants and communities for wildlife habitat (Call and Roundy 1990, 1991; Pyke and Archer 1991; Archer and Pyke 1991).

An understanding of germination characteristics is important in determining bulk seeding rates for direct seeding and also for propagating plants in the nursery for transplanting. Conventionally, recommendations for direct seeding rates are based on numerous field trials and are given in terms of pure live seed (PLS) or the portion of the seed lot that is germinable seed. Seed germination and ecology of domestic and wildland species has been studied extensively and summarized by a number of authors: Kozlowski 1972; Heydecker 1973; McDonough 1977; Koller and Hadas 1982; Mayer and Poljakoff-Mayber 1982; Bewley and Black 1982, 1983; Fulbright et al. 1982; Redente et al. 1982; Wasser and Shoemaker 1982; Fenner 1985, 1992; Young and Young 1986, 1992; Wester 1991.

Hansen (1989) listed problems associated with the use of shrub seeds in revegetation, including high diversity among seed collections, dormancy, specialized seed coats, polymorphism in size and germinability, greater water availability requirements for germination than for established plants, and unpredictable seed production. Germination of many wildland species can be enhanced by specific treatments, but these treatments often do not consistently enhance germination for different collections of the same species. For example, different collections of true mountain mahogany (*Cercocarpus montanus*) differ in germination after stratification or prechilling (Monsen and Davis 1985) while scarification to break the seedcoat enhances germination of some four-wing saltbush sources and decreases germination of others (Springfield 1970).

Use of many pretreatments to enhance germination, such as prechilling, moist heat, or chemical treatments, are most practical for plant propagation in a nursery, although dry treatments such as mechanical scarification or dry heat may be practical prior to direct seeding in the field. Seeds are often tested for germination in relation to an array of environmental variables in the laboratory such as salinity, light, constant and alternating temperatures, and water potential (Koller and Hadas 1982). Laboratory germination responses of many species to these variables are correlated with field environmental conditions where these species or ecotypes naturally occur and can be used to indicate appropriate soils, sites, and time of seeding (Young 1988, Meyer and Pendleton 1990, Meyer and Monsen 1990). However, laboratory conditions are often not representative of actual environmental dynamics in the seedbed, and, therefore, may not indicate meaningful biological responses to these variables. For example, germination is often tested in relation to abrupt temperature alternations in the laboratory and different static water potentials. However, seedbeds exhibit diurnal temperature curves, not abrupt alternations, and seedbed water potentials can be highly dynamic (Roundy et al. 1992). Models that accurately estimate seedbed soil water and temperature for different climatic conditions coupled with appropriate biological response models could help define establishment requirements and successful establishment characteristics of wildland species (Call and Roundy 1991). This could permit selection for seed and seedling characteristics to maximize establishment in certain environments.

Many wildland plants have seed appendages that may serve useful dispersal and anchoring functions for natural establishment, but may complicate seed harvesting, cleaning, and sowing for revegetation projects (Booth 1987). Conventional equipment such as the hammermill can damage seeds during the cleaning process. New equipment for harvesting and cleaning seeds with appendages without damaging them, and drills that sow trashy seeds are becoming available (United States Department of Agriculture 1991).

Before sowing, legumes should be inoculated with the appropriate strain of *Rhizobium* bacteria to ensure nodulation and nitrogen fixation (Roughley 1985). Propagules of actinomycete for nitrogen fixation of many nonleguminous woody species (Torrey 1978), and of mycorrhizal fungi to improve host plant water and nutrient uptake (Trappe 1981), are not readily adapted to inoculate seeds at the time of sowing, but can be used to inoculate transplants in a nursery (Roundy and Call 1988).

Seedbed preparation and sowing techniques are used to modify the environmental conditions of seedbeds and place seeds in most favorable conditions to enhance germination and establishment. Often methods such as chaining and plowing are used to control existing plants and prepare the seedbed or help cover sown seeds. Equipment for plant control, seedbed preparation, and planting is described in detail by Vallentine (1989) and Larson (1980). Drilling of seeds is preferred when possible but broadcasting has also been successful when used with chaining, rootplowing, or other mechanical plant control methods that help bury seeds at a shallow depth (6-12 mm). Treatments such as furrowing, imprinting, pitting, and mulching may prolong favorable moisture and temperature conditions in the seedbed and increase establishment of some species on some sites and under some weather conditions, but will not ensure success in dry years (Springfield 1972, Roundy and Call 1988, Roundy et al. 1992, Jackson et al. 1991). Spot seeding of shrubs into pits left from chaining pinyon and juniper trees has been successful (Plummer et al. 1968, Ferguson 1983).

Ironically, one of the biggest deterrents to successful revegetation to improve wildlife habitat has been seed and seedling depredation by wildlife species (Medin and Ferguson 1972, Ferguson 1983, Monsen and Shaw 1983a, Standley 1988). Vallentine (1989) advised that high populations of rodents must be controlled or excluded or shrub seedings will fail. He also discussed methods of control and exclusion.

There is potential to distribute and sow seeds by other than mechanical means. The use of animals (Archer and Pyke 1991) and the use of ephemeral arroyo floods (Barrow 1992) are being researched and developed as ways to disseminate and establish desirable wildland plants.

Prior to most rangeland revegetation projects, competing, undesirable vegetation on the site must be controlled to allow seedling establishment of desirable species. Competing species on rangelands include all life forms from annual weeds to mature trees, thereby requiring an array of biological, mechanical, and chemical methods to control them, and the use of prescribed fire. These methods are described in detail by Scifres (1980), McDaniel (1983), Wright and Bailey (1982), and Vallentine (1989), and briefly summarized by Roundy and Call (1988). Monsen and McArthur (1985) discussed factors affecting seedling establishment both naturally and after revegetation on burned rangelands.

Although sowing of mixtures of species is recommended to create diverse plant communities for wildlife and to increase chances of establishment in a variable environment (Monsen 1975, Plummer 1977), interspecific competition may reduce establishment of some species. Forbs and shrubs are generally slow to establish and are usually out-competed by aggressive seeded grasses (Hubbard 1957, Nord et al 1971, Monsen 1975). Possible solutions to this dilemma are to seed forbs, shrubs, and grasses in alternate rows or strips (Keller 1979, Monsen and Shaw 1983), or to interseed or transplant shrubs and forbs into established grass stands (Stevens et al. 1981, Petersen et al. 1986, McKell 1989) (Fig. 2). Seeding or transplanting in straight, evenly spaced rows may not achieve the desired stand heterogeneity found under natural conditions. Patch seeding or transplanting species at different rates and in different patterns would increase heterogeneity (Brown and Hallman 1984), but may leave the stand open to weed invasion or possible undesirable successional trajectories. Mechanical plant control methods are especially versatile for creating vegetation heterogeneity and edge in association with revegetation for habitat improvement (Urness 1979). However partial control or leaving strips of aggressive species with prolific seed production like sagebrush can result in a rapid return to the previous composition (Vallentine 1989). Determination of beneficial plant associations and patterns that maximize diversity, productivity, and stability will greatly increase the potential for wildlife habitat improvement (Call and Roundy 1991).

FIGURE 2. Bitterbrush strips established from transplants on the Boise Front Range in Idaho.

Transplanting of wildland shrubs is an alternative to direct seeding when seed availability is limited, when direct seeding is risky due to germination problems or in harsh or competitive environments, and when establishment of specific genetic material is desired (Plummer et al. 1968, Petersen et al. 1986, Ferguson 1983, Hansen 1989). Transplanting has also been frequently used to establish woody species in highly dynamic riparian areas (Briggs 1992). Methods of culturing and propagating shrubs for transplanting and transplanting techniques have been summarized by Hansen (1989), and Alder and Ostler (1989). Recommendations for specific species important to rangeland wildlife habitat improvement have been given by Ferguson and Monsen (1974), Ferguson (1983), Aldon (1984), McArthur et al. (1984*a,b*), Shaw and Monsen (1984*b*), Shaw (1984), Landis and Simonich (1984) and Romney et al. (1989*b*). Of course, the high cost and labor intensive nature of transplanting constrains its application to relatively small-scale projects.

Management

Revegetation should be considered as part of an overall management plan. Grazing of rangeland seedings for livestock is commonly deferred for 2 growing seasons to allow grass to fully establish (Vallentine 1989). Although this may be too much deferment for grazing-tolerant grasses like crested wheatgrass, it is probably a good recommendation for revegetation with native shrubs and forbs. Moderate browsing by big game may reduce growth and survival of establishing shrubs but is generally not permanently damaging (Ferguson 1968, Medin and Ferguson 1972), while heavy browsing may necessitate replanting (Carson and Edgerton 1989). Livestock grazing is relatively easily deferred by use of fencing. Revegetation projects can be protected from big game animals by the use of special fences (Carson and Edgerton 1989), by reducing populations by special hunts or capture and transplanting, or by seeding large areas (Vallentine 1989). Adherance to principles of grazing management (Vallentine 1990) is necessary to maintain desired rangeland plant communities, whether grazed by wildlife and/or domestic livestock. Management of wildlife populations is more problematic but just as necessary as is management of domestic stock to prevent overgrazing.

Effects of Revegetation on Wildlife

Most of the research on revegetation relevant to wildlife concerns techniques for establishment and evaluation of revegetation species known to be important for wildlife habitat. For example, increases in herbaceous forage and browse production after revegetation (Medin and Ferguson 1972) and persistence of species valuable to wildlife (Monsen and Shaw 1983*a*) demonstrate the potential for using revegetation to improve habitat for wildlife. Other studies have quantified or estimated the increases in forage nutrients for specific wildlife species that have been realized or could be expected from revegetation with certain species (Eckert et al. 1973, Welch 1989). Hubbard (1962) calculated expected increases in deer carrying capacity after revegetation with bitterbrush. Shaw et al. (1984) rated growth and habit of various shrubs in relation to their value for upland game bird habitat. A knowledge of habitat requirements of various game and nongame wildlife species allowed Schramm et al. (1987) to recommend revegetation species for the CRP to help control erosion on croplands and to improve wildlife habitats. Predominately grass seedings on CRP lands have benefited a number of bird species, especially pheasants (*Phasianus colchicus*). Inclusion of shrubs and forbs in future CRP plantings is considered necessary to benefit other wildlife species (Schenck and Williamson 1991).

Only a few studies report effects on wildlife of revegetation practices conducted specifically to benefit wildlife. The approach of manipulating rangeland vegetation for wildlife is straightforward. It involves defining management goals, assessing the current habitat conditions for wildlife, prescribing management actions to meet the goals, and monitoring effects of actions taken to see if desired habitat and wildlife goals are met (Kie and Thomas 1988). However, costs and other resource uses have generally limited direct habitat improvement projects for wildlife to a small scale, while vegetation manipulations for major land uses such as livestock and timber production have been of a comparatively large scale (Kie and Thomas 1988).

Benefits to wildlife of revegetation of rangelands primarily for soil conservation and livestock forage are highly specific to the species, site, and management situation. The effect of any vegetation conversion on wildlife is dependent on where it is done and what is being replaced with respect to seasonal habitat use and value.

Urness (1986) has synthesized numerous studies of the effects of crested wheatgrass seedings on big game. Crested wheatgrass seedings done predominantly on broad valleys of the Intermountain Region provide forage for livestock away from a narrow belt of mountain toe slopes and bajadas that are of importance to mule deer. Crested wheatgrass is less desirable than native grasses to mule deer but can be highly nutritious and an important supplement to browse of lower quality on some rangelands from fall to mid-spring before forbs become available. Crested wheatgrass seedings that replace or exclude native browse and grass species in mid-elevational ranges important to mule deer, such as in the pinyon-juniper type, are not recommended. Emphasis on establishing mixed forb, grass, shrub communities in most revegetation projects should improve habitat for wildlife; however, inclusion of vigorous introduced grasses in the seeding mix may lead to their replacement of natives and dominance over time (Harris and Dobrowolski 1986, Leckenby and Toweill 1983*a,b*). Mid-successional vegetation communities created by vegetation manipulation or revegetation are beneficial to many wildlife species (Terrel and Spillett 1975, Griffith and Peek 1989, Holechek et al. 1989, Vallentine 1989). However, maintenance of these communities may require continued inputs due to their lack of natural stability.

The composition of forbs in a seeding seems to be an important factor determining relative value to pronghorn (*Antilocapa americana*) (Urness 1986). More "open" wheatgrass seedings or those that include forbs in the seeding mixture receive most pronghorn use. Wheatgrass seedings in general do not appear to have a negative effect on pronghorn populations, especially if resident forbs are not controlled by herbicides prior to seeding.

Elk and bison (*Bison bison*) sometimes use seeded exotic wheatgrasses. The use and impacts of these seedings on elk, bison, and bighorn sheep (*Ovis canadensis*) are limited by the limited application of these seedings on the species' ranges.

Rangeland bird habitat is defined mainly in terms of vegetation structure (Wiens and Dyer 1975, Buttery and Shields 1975). McAdoo et al. (1986) have shown that the structure of sagebrush communities before and after conversion to crested wheatgrass determines use by nongame birds. Shrub-nesting species are favored by communities dominated by sagebrush while ground-nesting species are favored by conversion to wheatgrass. Wheatgrass stands reinvaded by sagebrush supported a balance of shrub and ground nesters and were considered to be most beneficial to raptors by providing a diversity of prey species. Loss of critical sagebrush cover associated with conversion of sagebrush to seeded grasses is considered to negatively affect sage grouse (Urness 1989).

Seeding exotic lovegrasses (*Eragrostis* spp.) in semidesert grassland of the Southwest resulted in lower plant and animal diversity than in an adjacent native grass stand (Bock et al. 1986). This comparison is important because Lehmann lovegrass (*Eragrostis lehmanniana*) has the ability to invade some native grasslands (Anable et al. 1992, McClaran and Anable 1992). However, exotic grasses are rarely seeded on native grasslands but are usually seeded on degraded rangelands with plant communities that initially lack vegetation diversity anyway, due to dominance by highly competitive woody or annual herbaceous species. Benefits of revegetation to wildlife should more commonly be determined by comparing wildlife populations and use on degraded sites that have been seeded to similar sites that have not been seeded (Ruyle and Roundy 1990).

Exotic grasses also continue to be the species that are most dependably established on drier semiarid rangelands (Leckenby and Toweill 1983*a*,*b*). As such, they may have value in protecting associated plant communities important to wildlife from invasion by weedy species. Greenstrips of exotic grasses strategically placed can protect mixed sagebrush/grass communities from high fire frequency that can lead to dominance by cheatgrass (*Bromus tectorum*) and medusahead (*Taeniatherum asperum*) (Pellant 1990). Lack of high diversity within these strips themselves for wildlife is of secondary importance compared to their value in protecting much larger areas of habitat from a much greater potential loss of diversity.

A few studies have measured the effect of revegetation or reclamation for wildlife on wildlife populations. Small mammal populations on revegetated minelands compared to unmined areas are highly dependent on the density, structure, and composition of the established vegetation. Diversity of small mammals has been reported to be higher (Hingtgen and Clark 1984), lower (Ireland et al. 1990), or similar (LaRue et al. 1990), to that of unmined lands.

Some of the few "project scale" reports of wildlife response to revegetation are those for pronghorn in Oregon (Kindschy et al. 1982, Yoakum 1983). Alfalfa was aerially seeded on > 27,000 ha of sagebrush rangelands previously plowed and seeded mainly to crested wheatgrass. Alfalfa constituted 10% of the vegetation composition on most of the seedings after 10 years and improved overall forb composition from 2% on untreated areas to 7% on seeded areas. More pronghorn does and fawns were observed on seeded than nonseeded areas in 1976. Similar seedings of crested wheatgrass and dryland alfalfa and other vegetation improvements in Bear Valley, Oregon, over a 40-year-period converted high stature sagebrush rangelands to plant communities with a 57% grass, 47% forb, and 2% shrub composition. Eventually the pronghorn population using the area increased to >600 animals and had one of the highest doe:fawn ratios in Oregon.

SUMMARY

Historically, rangelands have been predominately revegetated with introduced grasses to restore forage for livestock and watershed stability. Increases in mule deer populations in the Intermountain West associated with decreases in grasses and increases in woody species as a result of early overstocking of livestock, led to concern for deer summer and winter ranges. Forest Service and state game and fish departments developed cooperative research programs to study deer habitat use and improvement. The longest and most intensive research program has been in Utah and produced the classic guidebook for restoring big game ranges (Plummer et al. 1968). Since then, numerous symposia and other publications have discussed the details of revegetation with species considered important for wildlife habitat. Revegetation for wildlife habitat improvement requires careful consideration of objectives, selection of plant materials and establishment techniques, and appropriate management after revegetation.

Perhaps the most difficult part of this process is determining realistic and feasible objectives that fit within the framework of an overall management plan. Because of expense, large-scale revegetation projects will probably continue to be constructed for general range improvements to meet multiple-use objectives, rather than to specifically improve habitat for wildlife. These projects in the past, which include mainly replacing sagebrush or pinyon-juniper communities with grasses, have often been constructed without emphasizing wildlife habitat improvement, and have had differential effects on wildlife populations (Severson and Medina 1983). Structuring vegetation control and seeding mixtures of shrubs, grasses, and forbs to increase diversity could improve value of these projects for wildlife.

Direct seeding for habitat improvement has often been constrained by difficulties in producing, harvesting, cleaning, and sowing seeds and establishing seedlings of forbs and shrubs considered valuable for wildlife. Guidelines have been produced for determining site potential, selecting species and ecotypes, and establishing plants on various sites. Research has emphasized shrubs in the Rosaceae, Chenopodiaceae, and Compositae families and leguminous forbs. Selection of adapted species and ecotypes, planting in the appropriate season, minimizing competition among existing plants and those in the seed mixture, and modifying the seedbed to create favorable soil moisture and temperature conditions are critical for successful plant establishment. For example, seeding competitive species in alternate rows or in patches may allow concurrent establishment of a diversity of species. Research on compatibility and competitiveness of different species is needed to test the stability produced by these techniques and to guide the establishment of diverse but stable plant communities to benefit wildlife in the future.

Considerable potential exists for improvement of plant materials for revegetation by conventional selection and plant-breeding methods and by use of advanced techniques in molecular biology. High costs and lack of economic incentives constrain wildland plant improvement, not biological potential.

Propagation and transplanting techniques have been developed for numerous wildland plants. This approach avoids the risk of direct seeding and can be very useful for small scale improvement projects. Small-scale revegetation projects are done much more frequently than large-scale projects for habitat improvement. This trend will probably continue as specific sites are targeted for habitat improvement for threatened and endangered species.

Revegetation projects for wildlife require special management just as do those for livestock. Deferment of use until plants are well established and continued application of principles of grazing management are necessary to prevent overuse and loss of desirable species. Management of wildlife populations is more difficult than that of domestic stock but habitat improvements can be maintained by special fencing or by animal population control by capture and transplanting or special hunts. The overall seasonal habitat requirements of wildlife populations should be considered so that improvement of habitat for some seasons does not result in increased populations that overuse and degrade other seasonal habitat.

Value of revegetation projects to wildlife have usually been assessed by determining establishment success of plant species of known habitat value, and by measuring wildlife use of revegetated areas. Studies of direct effects of revegetation projects for wildlife on specific wildlife populations are lacking. However, seeding alfalfa into crested wheatgrass stands in Oregon has been shown to be associated with increased use of specific sites by pronghorn and increases in the population in the general area (Yoakum 1983). More information at this scale would help guide revegetation projects in the future.

LITERATURE CITED

Alder, G. M., and W. K. Ostler. 1989. Native shrub propagation and nursery stock production. Pages 535-552 *in* C. M. McKell, ed. The biology and utilization of shrubs. Acad. Press, Inc., San Diego, Calif.

Aldon, E. F. 1984. Methods for establishing fourwing saltbush on disturbed sites in the Southwest. Pages 265-268 *in* A. R. Tiedemann, E. D. McArthur, R. Stevens, and K. L. Johnson, compilers. Symp. on the biology of *Atriplex* and related chenopods. U.S. Dep. Agric. For. Serv. Gen. Tech. Rep. INT-172.

Anable, M. E., M. P. McClaran, and G. B., Ruyle. 1992. Spread of introduced Lehmann lovegrass *Eragrostis lehmanniana* Nees. in southern Arizona, USA. Biol. Conserv. 61:181-188.

Archer, S., and D. A. Pyke. 1991. Plant-animal interactions affecting plant establishment and persistence on revegetated rangeland. J. Range Manage. 44:558-565.

Barrow, J. R. 1992. Use of flood water to disperse grass and shrub seeds on native arid lands. Pages 167-169 *in* W. D. Clary, E. D. McArthur, D. Bedunah, and C. Wambolt, compilers. Symp. on ecology and management of riparian shrub communities. U.S. Dep. Agric. For. Serv. Gen. Tech. Rep. INT-289.

Bewley, J. D., and M. Black. 1982. Physiology and biochemistry of seeds in relation to germination. 2. Viability, dormancy, and environmental control. Springer-Verlag, New York, N. Y. 375pp.

———, and ———. 1983. Physiology and biochemistry of seeds in relation to germination. 1. Development, germination and growth. Springer-Verlag, New York, N. Y. 306pp.

Bock, C. E., J. H. Bock, K. L. Jepson, and J. C. Ortega. 1986. Ecological effects of planting African lovegrasses in Arizona. Natl. Geogr. Res. 2:456-463.

Booth, D. T. 1987. Diaspores of rangeland plants: ecology and management. Pages 202-211 *in* G. W. Frasier and R. A. Evans, eds. Seed and seedbed ecology of rangeland plants, U.S. Dep. Agric. Res. Serv., Natl. Tech. Inf. Serv., Springfield, Va.

Box, T. W. 1978. The significance and responsibility of rehabilitating drastically disturbed land. Pages 1-10 *in* F. W. Schaller and P. Sutton, eds. Reclamation of drastically disturbed lands. Am. Soc. Agron., Madison, Wis.

Briggs, M. A. 1992. An evaluation of riparian revegetation efforts in Arizona. M.S. Thesis, Univ. Arizona, Tucson. 229pp.

Brown, D. E., editor. 1982. Biotic communities of the Southwest-United States and Mexico. Desert Plants 4:map.

Brown, D., and R. G. Hallman. 1984. Reclaiming disturbed lands. U.S. Dep. Agric. For. Serv. Equipment Cent., Missoula, Mont.

Brown, E. R. and C. F. Martinsen. 1959. Browse planting for big game. Washington State Game Dep. Biol. Bull. 12. 63pp.

Brown, G. A., and L. E. Wiesner. 1984. Selecting species for revegetation: a guide for disturbed lands in the western coal region. Montana Agric. Exp. Stn. Spec. Rep. 3. 44pp.

Buttery, R. F., and P. W. Shields. 1975. Range management practices and bird habitat values. Pages 183-189 *in* D. R. Smith, tech. coord. Symp. on management of forest and range habitats for nongame birds. U.S. Dep. Agric. For. Serv. Gen. Tech. Rep. WO-1, Washington, D.C.

Call, C. A., and B. A. Roundy. 1990. New perspectives in rangeland revegetation. Pages 1-44 *in* C. A. Call and B. A. Roundy, eds. Perspectives and processes in rangeland revegetation. Soc. Range Manage., Denver, Colo.

———, and ———. 1991. Perspectives and processes in revegetation of arid and semiarid rangelands. J. Range Manage. 44:543-549.

Carlson, J. R. 1984. *Atriplex* cultivar development. Pages 176-182 *in* A. R. Tiedemann, E. D. McArthur, R. Stevens and K. L. Johnson, compilers. Symp. on the biology of *Atriplex* and related chenopods. U.S. Dep. Agric. For. Serv. Gen. Tech. Rep. INT-172.

Carman, J. G. 1989. Cellular and molecular approaches to the ecophysiology and genetic improvement of shrubs. Pages 31-36 *in* A. Wallace, E. D. McArthur and M. R. Haferkamp, compilers. Symp. on shrub ecophysiology and biotechnology. U.S. Dep. Agric. For. Serv. Gen. Tech. Rep. INT-256.

Carson, R. G., and P. J. Edgerton. 1989. Creating riparian wildlife habitat along a Columbia River impoundment in North-central Washington. Pages 64-69 *in* A. Wallace, E. D. McArthur, and M. R. Haferkamp, compilers. Symp. on shrub ecophysiology and biotechnology. U.S. Dep. Agric. For. Serv. Gen. Tech. Rep. INT-256.

Chambers, J. C., and R. W. Brown. 1983. Methods for vegetation sampling and analysis on revegetated mined lands. U.S. Dep. Agric. For. Serv. Gen. Tech. Rep. INT-151. 57pp.

Cook, G. W. 1983. "Forbs" need proper ecological recognition. Rangelands 5:217-220.

Cox, J. R., H. L. Morton, T. N. Johnsen, Jr., G. L. Jordan, S. C. Martin, and L. C. Fierro. 1982. Vegetation restoration in the Chihuahuan and Sonoran Deserts of North America. U.S. Dep. Agric. Res. Serv., Western Region, ARM-W-28, Oakland, Calif.

DeByle, N. V., and E. Hookane, Jr. 1973. Research related to the Davis County Experimental Watershed: an annotated bibliography. U.S. Dep. Agric. For. Serv. Gen. Tech. Rep. INT-4. 16pp.

DePuit, E. J., and E. F. Redente. 1988. Manipulation of ecosystem dynamics on reconstructed semiarid lands. Pages 162-204 *in* E. B. Allen, ed. The reconstruction of disturbed arid lands. Am. Assoc. for the Adv. of Sci., Washington, D.C.

Diamond, J. 1987. Reflections on goals and on the relationship between theory and practice. Pages 329-336 *in* W. R. Jordan, III, M. E. Gilpin and J. D. Aber, eds. Restoration ecology. Cambridge Univ. Press, New York, N. Y.

Dickson, K. L., and D. Vance. 1981. Revegetating surface mined lands for wildlife in Texas and Oklahoma. U.S. Dep. Inter. Fish and Wildl. Serv., U.S. Gov. Printing Off., Washington, D.C.

Dunne, C. G. 1989. Seed collecting, processing, and storage. Pages 511-518 *in* C. M. McKell, ed. The biology and utilization of shrubs. Acad. Press, Inc., San Diego, Calif.

Eckert, R. E., Jr., A. D. Bruner, G. I. Klamp, and F. F. Peterson. 1973. Mountain meadow improvement through seeding. J. Range Manage. 26:200-203.

Everett, R. L., compiler. 1987. Proc. pinyon-juniper conf. U.S. Dep. Agric. For. Serv. Gen. Tech. Rep. INT-215. 581pp.

Fenner, M. 1985. Seed ecology. Chapman and Hall, New York, N. Y. 151pp.

———, editor. 1992. Seeds. The ecology of regeneration in plant communities. C.A.B. International. Wallingford, U.K. 373pp.

Ferguson, R. B. 1968. Survival and growth of young bitterbrush browsed by deer. J. Wildl. Manage. 32:769-772.

———. 1983. Use of Rosaceous shrubs for wildland plantings in the Intermountain West. Pages 136-149 *in* S. B. Monsen and N. Shaw, compilers. Managing Intermountain rangeland-improvement of range and wildlife habitats. U.S. Dep. Agric. For. Serv. Gen. Tech. Rep. INT-157.

———, and S. B. Monsen. 1974. Research with containerized shrubs and forbs in southern Idaho. Pages 349-359 *in* R. W. Tinus, W. I. Stein, and W. E. Balmer, eds. Proc. North American containerized forest tree seedling symposium. Great Plains Agric. Counc. Publ. 68. U.S. Gov. Printing Off., Washington, D.C.

Flather, C. H., and T. W. Hoekstra. 1989. An analysis of the wildlife and fish situation in the United States: 1989-2040. U.S. Dep. Agric. For. Serv. Gen. Tech. Rep. RM-178, Fort Collins, Colo. 146pp.

Forsling, C. L. 1931. A study of the influence of herbaceous plant cover on surface runoff and soil erosion in relation to grazing on the Wasatch Plateau in Utah. U.S. Dep. Agric. Tech. Bull. 220. 71pp.

Fulbright, T. E., E. F. Redente, and N. E. Hargis. 1982. Growing Colorado plants from seed: a state of the art. Vol. II: Grasses and grasslike plants. U. S. Fish and Wildl. Serv., Washington, D.C. 113pp.

George, M. R., T. E. Adams, Jr., and W. J. Clawson. 1983. Seeded range plants for California. Leaflet 21344. Univ. California, Berkeley. 23pp.

Godfrey, E. B., and D. B. Nielsen. 1980. Economic incentives and range improvements on federal rangelands. Pages 9-12 *in* P. E. Packer and J. O. Klemmedson, eds. Constraints to improved management of public rangelands. Soc. Am. For., Washington, D.C.

Graves, W. L., B. L. Kay, and W. A. Williams. 1978. Revegetation of disturbed sites in the Mojave Desert with native shrubs. Calif. Agric. 32:4-5.

Griffith, B., and J. M. Peek. 1989. Mule deer use of seral stage and habitat type in bitterbrush communities. J. Wildl. Manage. 53:636-642.

Hafenrichter, A. L., J. L. Schwendiman, H. L. Harris, R. S. MacLauchlan, and H. W. Miller. 1968. Grasses and legumes for soil conservation in the Pacific Northwest and Great Basin states. U.S. Dep. Agric. Agric. Handb. 339, Washington, D.C. 69pp.

Halls, L. K., and J. J. Stransky. 1968. Game food plantings in southern forests. Trans. North Am. Wildl. and Nat. Resour. Conf. 33:217-222.

Hansen, D. J. 1989. Reclamation and erosion control using shrubs. Pages 459-478 *in* C. M. McKell, ed. The biology and utilization of shrubs. Acad. Press, Inc., San Diego, Calif.

Harris, G. A., and J. P. Dobrowolski. 1986. Population dynamics of seeded species on northeast Washington semiarid sites, 1948-1983. J. Range Manage. 34:46-51.

Hewitt, G. B., A. C. Wilton, and R. J. Lorenz. 1982. The suitability of legumes for rangeland interseeding and as grasshopper food plants. J. Range Manage. 35:653-656.

Hermann, F. J. 1966. Notes on western range forbs: Cruciferae through Compositae. U.S. Dep. Agric. Handb. 293, Washington, D.C. 365pp.

Heydecker, W., editor. 1973. Seed ecology. Pennsylvania Univ. Press, University Park, Pa.. 578pp.

Hingtgen, T. M., and W. R. Clark. 1984. Small mammal recolonization of reclaimed surface-mined land in Wyoming. J. Wildl. Manage. 48:1255-1261.

Holechek, J. L., R. D. Pieper, and C. H. Herbel. 1989. Range management. Prentice Hall, Englewood Cliffs, N.J. 501pp.

Holmgren, R. C., and J. V. Basile. 1959. Improving southern Idaho deer winter ranges by artificial revegetation. Id. Dep. Fish and Game Wildl. Bull. 3. 61pp.

Hubbard, R. L. 1957. The effects of plant competition on the growth and survival of bitterbrush. J. Range Manage. 10:135-137.

———. 1962. The place of browse seeding in game range management. Trans. North Am. Wildl. and Nat. Resour. Conf. 27:394-401.

———. 1964. A guide to bitterbrush seeding in California. U.S. Dep. Agric. For. Serv. Pac. Southwest For. and Range Exp. Stn. and Calif. Dep. Fish and Game. 30pp.

Hungerford, R. D. 1984. Native shrubs: suitability for revegetating roadcuts in northwestern Montana. U.S. Dep. Agric. For. Serv. Res. Pap. INT-331, Ogden, Ut. 13pp.

Institute for Land Rehabilitation. 1978. Rehabilitation of western wildlife habitat: A review. Utah State Univ., Logan, and U. S. Fish and Wildl. Serv., Fort Collins, Colo. 238pp.

———. 1979. Selection, propagation, and field establishment of native species on disturbed arid lands. Utah State Univ., Logan. 49pp.

Ireland, T. T., S. D. Schemnitz, and G. L. Wolters. 1990. Reinvasion of small mammals, reptiles, amphibians, and insects on a reclaimed coal strip mine. Pages 74-85 *in* F. F. Munshower and S. E. Fisher, Jr., tech. coords. Proc. symposium on disturbed land rehabilitation. Montana State Univ., Reclamation Res. Unit Publ. No. 9003, Bozeman.

Jackson, L. L., J. R. McAuliffe, and B. A. Roundy. 1991. Desert restoration: Revegetation trials on abandoned farmland. Restoration and Manage. Notes 9:71-80.

Johnson, D. A., M. D. Rumbaugh, and K. H. Asay. 1981. Plant improvement for semi-arid rangelands: Possibilities for drought resistance and nitrogen fixation. Plant and Soil 58:279-303.

Jordan, G. L. 1981. Range seeding and brush management on Arizona rangelands. Arizona Agric. Exp. Stn. Bull. T81121. Univ. Arizona, Tucson. 88pp.

Julander, O., and J. B. Low. 1976. A historical account and present status of the mule deer in the West. Pages 3-19 *in* G. W. Workman and J. B. Low, eds. Mule deer decline in the west: a symposium. Utah State Univ., Logan.

———, W. L. Robinette, and D. A. Jones. 1961. Relation of summer range condition to mule deer herd productivity. J. Wildl. Manage. 25:54-60.

Keller, W. 1979. Species and methods for seeding the sagebrush ecosystem. Pages 129-163 *in* The sagebrush ecosystem: a symposium. Utah State Univ., Logan.

Kie, J. G., and J. W. Thomas. 1988. Rangeland vegetation as wildlife habitat. Pages 585-605 *in* P. T. Tueller, ed. Vegetation science applications for rangeland analysis and management. Kluwer Acad. Publ., Boston, Mass.

Kindschy, R. R., C. Sundstrom, and J. D. Yoakum. 1982. Wildlife habitats in managed rangelands—the Great Basin of southeastern Oregon: Pronghorns. U.S. Dep. Agric. For. Serv. Gen. Tech. Rep. PNW-145. 18pp.

Klebenow, D. A., and G. M. Gray. 1968. Food habits of juvenile sage grouse. J. Range Manage. 21:80-83.

Koller, D., and A. Hadas. 1982. Water relations in the germination of seed. Pages 402-431 *in* O. L. Lange, P. S. Nobel, C. B. Osmond, and H. Ziegler, eds. Physiological plant ecology II. Water relations and carbon assimilation. Encyclopedia of plant physiology. New Ser., Vol. 12B. Springer-Verlag, New York, N. Y.

Kozlowski, T. T., editor. 1972. Seed biology, Vols. I, II, III. Acad. Press, Inc. New York, N.Y.

Krausman, P. R. 1990. Foreword. Page V *in* P. R. Krausman and N. S. Smith, eds. Managing wildlife in the Southwest. Arizona Chap., The Wildl. Soc., Phoenix.

Landis, T. D., and E. J. Simonich. 1984. Producing native plants as container seedlings. Pages 16-25 *in* P. M. Murphy, compiler. The challenge of producing plants for the Intermountain area: proceedings Intermountain Nurseryman's Association 1983 conference. U.S. Dep. Agric. For. Serv. Gen. Tech. Rep. INT-168.

Larson, J. E. 1980. Revegetation equipment catalog. U.S. Dep. Agric. For. Serv. Equipment Cent., Missoula, Mont. 198pp.

LaRue, C. T., G. C. Bateman, and B. P. Dunfee. 1990. Population characteristics of small mammals on reclaimed surface-mined lands and native sagebrush shrubland. Pages 126-135 *in* P. R. Krausman and N. S. Smith, eds. Managing wildlife in the Southwest. Arizona Chap., The Wildl. Soc., Phoenix.

Leckenby, D. A., and D. E. Toweill. 1983*a*. Response of forage species seeded for mule deer in western juniper types of south-central Oregon. J. Range Manage. 36:98-103.

———, and ———. 1983*b*. Response of selected plant species seeded on mule deer winter range. J. Range Manage. 36:312-316.

Lippitt, L., M. W. Fidelibus, and D. A. Bainbridge. 1994. Native seed collection, processing and storage for revegetation projects in the western United States. Restor. Ecol. 2:120-131.

Longhurst, W. M., A. L. Lesperance, M. Morse, R. J. Mackie, D. L. Neal, H. Salwasser, D. Swickard, P. T. Tueller, P. J. Urness, and J. D. Yoakum. 1977. Livestock and wild ungulates. Pages 42-64 *in* J. W. Menke, ed. Proceedings of the workshop on livestock and wildlife-fisheries relationships in the Great Basin. Sparks, Nev.

———, R. E. Haferfield, and G. E. Connolly. 1982. Deer-livestock interrelationships in the western states. Pages 409-420 *in* J. M. Peek and P. D. Dalke, eds. Wildlife-livestock relationships symposium. Univ. Idaho, Moscow.

MacMahon, J. A. 1987. Disturbed lands and ecological theory: an essay about a mutualistic association. Pages 221-237 *in* W. R. Jordan, III, M. E. Gilpin, and J. D. Aber, eds. Restoration ecology. Cambridge Univ. Press, New York, N. Y.

Mayer, A. M., and A. Poljakoff-Mayber. 1982. The germination of seeds. Third ed. Pergamon Press, New York, N. Y. 211pp.

McAdoo, J. K., R. A. Evans, and W. S. Longland. 1986. Nongame bird responses to type conversion of sagebrush communities. Pages 155-162 *in* K. L. Johnson, ed., Crested wheatgrass: Its values, problems, and myths; symposium proceedings. Utah State Univ., Logan.

McArthur, E. D. 1988. New plant development in range management. Pages 81-112 *in* P. T. Tueller, ed. Vegetation science applications for rangeland analysis and management. Kluwer Acad. Publ., Boston, Mass.

———, A. Clyde Blauer, and G. L. Noller. 1984*a*. Propagation of fourwing saltbush by stem cuttings. Pages 261-264 *in* A. R. Tiedemann, E. D. McArthur, R. Stevens, and K. L. Johnson, compilers. Symp. on the biology of *Atriplex* and related chenopods. U.S. Dep. Agric. For. Serv. Gen. Tech. Rep. INT-172.

———, S. E. Stranathan, and G. L. Noller. 1984*b*. 'Rincon' fourwing saltbush—proven for better forage and reclamation. Rangelands 6:62-64.

———, and B. L. Welch, editors. 1984. Symp. on the biology of *Artemisia* and *Chrysothamnus*. U.S. Dep. Agric. For. Serv. Gen. Tech. Rep. INT-200. 398pp.

———, B. L. Welch, and D. L. Nelson. 1985. Developing improved cultivars of sagebrushes and other composite shrubs. Pages 188-196 *in* Proc. selected papers presented at the 38th Annu. Meet. Soc. Range Manage. Soc. Range Manage., Denver, Colo.

McClaran, M. P., and M. E. Anable. 1992. Spread of introduced Lehmann lovegrass along a grazing intensity gradient. J. Appl. Ecol. 29:92-98.

McDaniel, K. C., editor. 1983. Proc. brush management symposium. Soc. Range Manage., Denver, Colo. 105pp.

McDonough, W. T. 1977. Seed physiology. Pages 155-184 *in* R. E. Sosebee, ed. Rangeland plant physiology. Soc. Range Manage., Denver, Colo.

McGinnies, W. J., W. G. Hassell, and C. H. Wasser. 1983. A summary of range seeding trials in Colorado. Colorado State Univ. Spec. Ser. 21. 283pp.

McKell, C. M. 1986. The role of shrubs in diversifying a crested wheatgrass monoculture. Pages 109-115 *in* K. L. Johnson, ed. Crested wheatgrass: its values, problems and myths: symposium proceedings. Utah State Univ., Logan.

———. 1989. Shrub palatability. Pages 267-282 *in* C. M. McKell, ed. The biology and utilization of shrubs. Acad. Press, Inc., San Diego, Calif.

———. 1990. Improved plant materials for revegetation. Pages 104-132 *in* C. A. Call and B. A. Roundy, compilers. Perspectives and processes in rangeland revegetation. Soc. Range Manage., Denver, Colo.

Medin, D. E., and R. B. Ferguson. 1972. Shrub establishment on game ranges in the northwestern United States. Pages 359-368 *in* C. M. McKell, J. P. Blaisdell, and J. R. Goodin, tech. coords. Wildland shrubs: their biology and utilization. U.S. Dep. Agric. For. Serv. Gen. Tech. Rep. INT-1.

Meyer, S. E., and S. B. Monsen. 1990. Seed-source differences in initial establishment for big sagebrush and rubber rabbitbrush. Pages 200-208 *in* E. D. McArthur, E. M. Romney, S. D. Smith, and P. T. Tueller, compilers. Symp. on cheatgrass invasion, shrub die-off, and other aspects of shrub biology and management. U.S. Dep. Agric. For. Serv. Gen. Tech. Rep. INT-276.

———, and R. L. Pendleton. 1990. Seed germination biology of spineless hopsage: Between population differences in dormancy and response to temperature. Pages 187-192 *in* E. D. McArthur, E. M. Romney, S. D. Smith, and P. T. Tueller, compilers. Symp. on cheatgrass invasion, shrub die-off, and other aspects of shrub biology and management. U.S. Dep. Agric. For. Serv. Gen. Tech. Rep. INT-276.

Mitchell, J. E., editor. 1987. Impacts of the Conservation Reserve Program in the Great Plains: symposium proceedings. U.S. Dep. Agric. For. Serv. Gen. Tech. Rep. RM-158. 134pp.

Monsen, S. B. 1975. Selecting plants to rehabilitate disturbed areas. Pages 76-90 *in* R. S. Campbell and C. H. Herbel, eds. Improved range plants. Range Symp. Ser. 1, Soc. Range Manage., Denver, Colo.

———, and J. N. Davis. 1985. Progress in the improvement of selected western North American rosaceous shrubs. Pages 201-209 *in* Proc. selected papers presented at the 38th Annu. Meet. Soc. Range Manage. Soc. Range Manage., Denver, Colo.

———, and E. D. McArthur. 1985. Factors influencing establishment of seeded broadleaf herbs and shrubs following fire. Pages 112-124 *in* K. Sanders and J. Durham, eds. Rangeland fire effects: a symposium. U.S. Bur. Land Manage., Boise, Id.

———, and N. Shaw. 1983*a*. Seeding antelope bitterbrush with grasses on south-central Idaho rangelands: a 34-year response. Pages 126-136 *in* A. R. Tiedemann and K. L. Johnson, compilers. Research and management of bitterbrush and cliffrose in western North America. U.S. Dep. Agric. For. Serv. Gen. Tech. Rep. INT-152.

———, and ———, compilers. 1983*b*. Managing Intermountain rangelands: improvement of range and wildlife habitats. U.S. Dep. Agric. For. Serv. Gen. Tech. Rep. INT-157. 194pp.

———, and R. Stevens. In Press. Restoration of western range and wildlands. U.S. Dep. Agric. For. Serv. Gen. Tech. Rep.

Nechaeva, N. T., editor. 1985. Improvement of desert ranges in Soviet Central Asia. Harwood Acad. Publ., New York.

Nord, E. C., P. F. Hartless, and W. D. Nettleton. 1971. Effects of several factors on saltbush establishment in California. J. Range Manage. 24:216-223.

Patton, D. R. 1992. Wildlife habitat relationships in forested ecosystems. Timber Press, Inc., Portland, Oreg. 392pp.

Payne, N. F., and F. C. Bryant. 1994. Techniques for wildlife habitat management of uplands. McGraw-Hill, Inc., New York, N.Y. 840pp.

Peek, J. M. 1986. A review of wildlife management. Prentice-Hall, Englewood Cliffs, N.J. 486pp.

Pellant, M. 1990. The cheatgrass-wildfire cycle: are there any solutions? Pages 11-18 *in* E. D. McArthur, E. M. Romney, S. D. Smith, and P. T. Tueller, compilers. Symp. on cheatgrass invasion, shrub die-off, and other aspects of shrub biology and management. U.S. Dep. Agric. For. Serv. Gen. Tech. Rep. INT-276.

Pendleton, R. L., B. K. Pendleton, and K. T. Harper. 1989. Breeding systems of woody plant species in Utah. Pages 5-22 *in* A. Wallace, E. D. McArthur, and M. R. Haferkamp, compilers. Symp. on shrub ecophysiology and biotechnology. U.S. Dep. Agric. For. Serv. Gen. Tech. Rep. INT-256.

Petersen, J. L., D. N. Ueckert, and R. L. Potter. 1986. Cultural practices for establishing fourwing saltbush within perennial grass stands. J. Range Manage. 39:460-463.

Plummer, A. P. 1977. Plants and treatments for revegetation of disturbed sites in the Intermountain area. Pages 155-173 *in* R. A. Wright, ed. The reclamation of disturbed arid lands. Univ. New Mexico Press, Albuquerque.

———, D. R. Christensen, and S. B. Monsen. 1968. Restoring big game range in Utah. Utah Div. Fish and Game Publ. 68-3. 183pp.

Pyke, D. A., and S. Archer. 1991. Plant-plant interactions affecting plant establishment and persistence on revegetated rangeland. J. Range Manage. 44:550-557.

Redente, E. F., and E. J. DePuit. 1988. Reclamation of drastically disturbed rangelands. Pages 559-584 *in* P. T. Tueller, ed. Vegetation science applications for rangeland analysis and management. Kluwer Acad. Publ., Boston, Mass.

———, P. R. Ogle, and N. E. Hargis. 1882. Growing Colorado plants from seed: a state of the art. Vol. III; Forbs. U.S. Fish and Wildl. Serv., Washington, D.C. 141pp.

Robinette, W. L. 1972. Browse and cover for wildlife. Pages 69-76 *in* C. M. McKell, J. P. Blaisdell, and J. R. Goodin, tech. eds. Wildland shrubs: their biology and utilization. U.S. Dep. Agric. For. Serv. Gen. Tech. Rep. INT-1.

———, O. Julander, J. S. Gashwiler, and J. G. Smith. 1952. Winter mortality of mule deer in Utah in relation to range condition. J. Wildl. Manage. 16:289-299.

Romney, E. M., A. Wallace, and R. B. Hunter. 1989*a*. Pulse establishment of woody shrubs on denuded Mojave Desert land. Pages 54-57 *in* A. Wallace, E. D. McArthur, and M. R. Haferkamp, compilers. Symp. on shrub ecophysiology and biotechnology. U.S. Dep. Agric. For. Serv. Gen. Tech. Rep. INT-256.

———, ———, and ———. 1989*b*. Transplanting of native shrubs on disturbed land in the Mojave Desert. Pages 50-53 *in* A. Wallace, E. D. McArthur, and M. R. Hafercamp, compilers. Symp. on shrub ecophysiology and biotechnology. U.S. Dep. Agric. For. Serv. Gen. Tech. Rep. INT-256.

Roughley, R. J. 1985. Nitrogen fixation by the legume-*Rhizobium* symbiosis: The role of the bacteria. Pages 140-145 *in* R. F. Barnes, P. R. Ball, R. W. Brougham, G. C. Martin, and D. J. Minson, eds. Forage legumes for energy-efficient animal production. U.S. Dep. Agric. Res. Serv., Washington, D.C.

Roundy, B. A., and C. A. Call. 1988. Revegetation of arid and semiarid rangelands. Pages 607-635 *in* P. T. Tueller, ed., Vegetation science applications for rangeland analysis and management. Kluwer Acad. Publ., Boston, Mass.

———, V. K. Winkel, H. Khalifa, and A. D. Matthias. 1992. Soil water availability and temperature dynamics after one-time heavy cattle trampling and land imprinting. Arid Soil Res. and Rehabilitation 6:53-59.

Rumbaugh, M. D. 1983. Legumes—their use in wildland plantings. Pages 115-122 *in* S. B. Monsen and N. Shaw, compilers. Managing Intermountain rangelands—improvement of range and wildlife habitats. U.S. Dep. Agric. For. Serv. Gen. Tech. Rep. INT-157.

———, D. A. Johnson, and G. A. Van Epps. 1982. Forage yield and quality in a Great Basin shrub, grass, and legume pasture experiment. J. Range Manage. 35:604-609.

———, and C. E. Townsend. 1985. Range legume selection and breeding in North America. Pages 137-147 *in* Proc. selected papers presented at the 38th Annu. Meet. Soc. Range Manage. Soc. Range Manage., Denver, Colo.

Ruyle, G. B., and B. A. Roundy. 1990. Effects of vegetation manipulation on wildlife on semidesert rangelands. Pages 230-238 *in* P. R. Krausman and N. S. Smith, eds. Managing wildlife in the Southwest. Arizona Chap., The Wildl. Soc., Phoenix.

Schneck, E. W., and L. L. Williamson. 1991. Conservation Reserve Program effects on wildlife and recreation. Pages 37-42 *in* L. A. Joyce, J. E. Mitchell, and M. D. Skold, eds. The Conservation Reserve Program—yesterday, today, and tomorrow: symposium proceedings. U.S. Dep. Agric. For. Serv. Gen. Tech. Rep. RM-203.

Schramm, H. L., Jr., L. M. Smith, F. C. Bryant, R. R. George, B. C. Thompson, S. A. Nelle, and G. L. Vallentine. 1987. Managing wildlife with the Conservation Reserve Program. Texas Tech. Univ., Lubbock. Manage. Note 11. 6pp.

Scifres, C. J. 1980. Brush management principles and practices. John Wiley and Sons, New York, N.Y. 360pp.

Scully, J. 1987. Sowing seeds for federal landscapes. Environment 29:24.

Severson, K. E., and A. L. Medina. 1983. Deer and elk habitat management in the Southwest. J. Range Manage. Monogr. 2. 64pp.

Shaw, N. 1984. Producing bare root seedlings of native shrubs. Pages 6-15 *in* P. M. Murphy, compiler. The challenge of producing plants for the Intermountain area. Proc. Intermountain Nurseryman's Assoc. 1983 Conf. U.S. Dep. Agric. For. Serv. Gen. Tech. Rep. INT-168.

———, and S. B. Monsen. 1984. Nonleguminous forbs for rangeland sites. Pages 123-131 *in* S. B. Monsen and N. Shaw, compilers. Managing Intermountain rangelands: improvement of range and wildlife habitats. U.S. Dep. Agric. For. Serv. Gen. Tech. Rep. INT-157.

———, and ———. 1984. Nursery propagation and outplanting of bareroot chenopod seedlings. Pages 251-260 *in* A. R. Tiedemann, E. D. McArthur, R. Stevens, and K. L. Johnson, compilers. Symp. on the biology of *Atriplex* and related chenopods. U.S. Dep. Agric. For. Serv. Gen. Tech. Rep. INT-172.

———, A. Sands, and D. Turnipseed. 1984. Potential use of fourwing saltbush and other dryland shrubs for upland game bird cover in southern Idaho. Pages 206-214 *in* A. R. Tiedemann, E. D. McArthur, R. Sevens, and K. L. Johnson, compilers. Symp. on the biology of *Atriplex* and related chenopods. U.S. Dep. Agric. For. Serv. Gen. Tech. Rep. INT-172.

Shiflet, T. N., and R. S. MacLauchlan. 1986. Selecting plants for revegetation of rangelands in the United States. Pages 295-296 *in* P. J. Joss, P. W. Lynch, and O. B. Williams, eds. Rangelands: a resource under siege. Proc. Second Int. Rangeland Cong. Australian Acad. Sci., Camberra.

Smeins, F. E. 1983. Origin of the brush problem: a geological and ecological perspective of contemporary distributions. Pages 5-16 *in* K. C. McDaniel, ed. Proc. brush management symposium. Soc. Range Manage., Denver, Colo.

Springfield, H. W. 1970. Germination and establishment of fourwing saltbush in the Southwest. U.S. Dep. Agric. For. Serv. Res. Pap. RM-55. 48pp.

———. 1972. Using mulches to establish woody chenopods. Pages 382-391 *in* C. M. McKell, J. P. Blaisdell, and J. R. Goodin, tech. coords. Wildland shrubs: their biology and utilization. U.S. Dep. Agric. For. Serv. Gen. Tech. Rep. INT-1.

Standley, W. G. 1988. Effect of seed size on removal by rodents. Pages 399-402 *in* R. C. Szaro, K. E. Severson, and D. R. Patton, tech. coords. Management of amphibians, reptiles, and small mammals in North America. U.S. Dep. Agric. For. Serv. Gen. Tech. Rep. RM-166.

Stevens, R., W. L. Moden, and D. W. McKenzie. 1981. Interseeding and transplanting shrubs and forbs into grass communities. Rangelands 3:55-58.

———, A. P. Plummer, C. E. Jensen, and B. C. Giunta. 1974. Site productivity classification for selected species on winter big game ranges of Utah. U.S. Dep. Agric. For. Serv. Res. Pap. INT-158.

———, N. Shaw, and C. G. Howard. 1985. Important nonleguminous forbs for Intermountain ranges. Pages 210-220 *in* Proc. selected papers presented at the 38th Annu. Meet. Soc. Range Manage. Soc. Range Manage., Denver, Colo.

Stutz, H. C. 1989. Evolution of shrubs. Pages 323-340 *in* C. M. McKell, ed. The biology and utilization of shrubs. Acad. Press, Inc., San Diego, Calif.

———, and J. R. Carlson. 1985. Genetic improvement of saltbush (*Atriplex*) and other chenopods. Pages 197-200 *in* Proc. selected papers presented at the 38th Annu. Meet. Soc. Range Manage. Soc. Range Manage., Denver, Colo.

Swank, W. G. 1958. The mule deer in Arizona chaparral. Ariz. Game and Fish Dep. Bull. 3. 109pp.

Terrel, T. L., and J. J. Spillett. 1975. Pinyon-juniper conversion: its impact on mule deer and other wildlife. Pages 105-118 *in* The pinyon-juniper ecosystem: symposium. Utah State Univ., Logan.

Tew, R. K. 1983. Bitterbrush distribution and habitat classification on the Boise National Forest. Pages 32-36 *in* A. R. Tiedemann and K. L. Johnson, compilers. Research and management of bitterbrush and cliffrose in western North America. U.S. Dep. Agric. For. Serv. Gen. Tech. Rep. INT-152.

Thornburg, A. A. 1982. Plant materials for use on surface-mined lands in arid and semiarid regions. U.S. Dep. Agric. Soil Conserv. Serv., SCS-TP-157, Washington, D.C. 88pp.

Tiedemann, A. R., and K. L. Johnson, compilers. 1983. Research and management of bitterbrush and cliffrose in western North America. U.S. Dep. Agric. For. Serv. Gen. Tech. Rep. INT-152. 279pp.

———, E. D. McArthur, R. Stevens, and K. L. Johnson, compilers. 1984. Symp. on the biology of *Atriplex* and related chenopods. U.S. Dep. Agric. For. Serv. Gen. Tech. Rep. INT-172. 309pp.

Torrey, J. G. 1978. Nitrogen fixation by actinomycete-nodulated angiosperms. Bioscience 28:586-592.

Trappe, J. M. 1981. Mycorrhizae and productivity of arid and semiarid rangelands. Pages 581-599 *in* J. T. Manassah and E. J. Briskey, eds. Advances in food-producing systems for arid and semiarid lands. Part A. Kuwait Found. for Adv. of Sci., Acad. Press, New York, N. Y.

Tueller, P. T. 1976. Deer range improvement and mule deer management. Pages 55-66 *in* G. W. Workman and J. B. Low, eds. Mule deer decline in the West: a symposium. Utah State Univ., Logan.

United Nations Educational, Scientific, and Cultural Organization. 1979. Map of the world distribution of arid regions. Explanatory note. Union Typographique, Villeneuve-Saint-Georges, France. 54pp.

United States Department of Agriculture. 1991. Rangeland technology equipment council. 1990. Annual report. U.S. Dep. Agric. For. Serv. Tech. and Dev. Cent., Missoula, Mont. 83pp.

Urness, P. J. 1986. Value of crested wheatgrass for big game. Pages 147-153 *in* K. L. Johnson, ed. Crested wheatgrass: its values, problems, and myths; symposium proceedings. Utah State Univ., Logan.

———. 1979. Wildlife habitat manipulation in sagebrush ecosystems. Pages 169-178 *in* The sagebrush ecosystem: a symposium. Utah State Univ., Logan.

———. 1989. Shrubs as habitat for wildlife. Pages 441-458 *in* C. M. McKell, ed. The biology and utilization of shrubs. Acad. Press, Inc., San Diego, Calif.

———. 1990. Livestock as manipulators of mule deer habitats in Utah. Pages 25-39 *in* K. E. Severson, tech. coord. Can livestock be used as a tool to enhance wildlife habitat? U.S. Dep. Agric. For. Serv. Rocky Mt. Stn. Gen. Tech. Rep. RM-194.

Vallentine, J. F. 1989. Range development and improvements. Third ed. Acad. Press, Inc., San Diego, Calif. 524pp.

———. 1990. Grazing management. Acad. Press, Inc., San Diego, Calif. 533pp.

Van Epps, G. A. 1989. Seed production from plantations. Pages 519-533 *in* C. M. McKell, ed. The biology and utilization of shrubs. Acad. Press, Inc., San Diego, Calif.

Vories, K. C., and P. L. Sims. 1977. The plant information network volume I: A user's guide. Colorado State Univ., Fort Collins. 56pp.

Wallestad, R. O., J. G. Peterson, and R. L. Eng. 1975. Foods of adult sage grouse in central Montana. J. Wildl. Manage. 37:628-630.

Wasser, C. H., and J. Shoemaker. 1982. Ecology and culture of selected species useful in revegetating disturbed lands in the West. U. S. Fish and Wildl. Serv., Fort Collins, Colo. 347pp.

Welch, B. L. 1989. Nutritive value of shrubs. Pages 405-424 *in* C. M. McKell, ed. The biology and utilization of shrubs. Acad. Press, Inc., San Diego, Calif.

West, N. E. 1983. Overview of North American temperate deserts and semi-deserts. Pages 321-472 *in* N. E. West, ed. Temperate deserts and semi-deserts. Ecosystems of the world 5. Elsevier Sci. Publ. Co., New York, N.Y.

———, K. H. Rea, and R. J. Tausch. 1975. Basic synecological relationships in juniper-pinyon woodlands. Pages 41-52 *in* The pinyon-juniper ecosystem: a symposium. Utah State Univ., Logan.

Wester, D. B. 1991. A summary of range plant seed germination research. ICASALS Pub. No. 91-2. Texas Tech. Univ., Lubbock. 112pp.

Wiens, J. A., and M. I. Dyer. 1975. Rangeland avifaunas: Their composition, energetics, and role in the ecosystem. Pages 146-181 *in* D. R. Smith, tech. coord. Symp. on management of forest and range habitat for nongame birds. U.S. Dep. Agric. For. Serv. Gen. Tech. Rep. WO-1, Washington, D.C.

Wood, S. L., editor. 1980. Soil-plant-animal relationships bearing on revegetation and land reclamation in Nevada deserts. Great Basin Naturalist Memoirs No. 4. Brigham Young Univ., Provo, Ut.

Workman, G. W., and J. B. Low, editors. 1976. Mule deer decline in the West: a symposium. Utah State Univ., Logan. 134pp.

Wright, H., and A. Bailey. 1982. Fire ecology in the United States and Canada. Wiley Interscience, New York, N. Y. 501pp.

Yoakum, J. D. 1979. Managing rangelands for pronghorns. Rangelands 1:146-148.

———. 1983. Managing vegetation for pronghorns in the Great Basin. Pages 189-194 *in* S. B. Monsen and N. Shaw, compilers. Managing Intermountain rangelands: improvement of range and wildlife habitats. U.S. Dep. Agric. For. Serv. Gen. Tech. Rep. INT-157.

Young, J. A. 1988. Seedbeds as selective factors in the species composition of rangeland communities. Pages 171-188 *in* P. T. Tueller, ed. Vegetation science applications for rangeland analysis and management. Kluwer Acad. Publ., Boston, Mass.

———. 1990. Revegetation technology. Pages 92-103 *in* C. A. Call and B. A. Roundy, compilers. Perspectives and processes in rangeland revegetation. Utah State Univ., Logan.

———, R. E. Eckert, Jr., and R. A. Evans. 1979. Historical perspectives regarding the sagebrush ecosystem. Pages 1-13 *in* The sagebrush ecosystem: a symposium. Utah State Univ., Logan.

———, R. A. Evans, and D. W. McKenzie. 1983. History of brush control on western U. S. rangelands. Pages 17-25 *in* K. C. McDaniel, ed. Proc. brush management symposium. Soc. Range Manage., Denver, Colo.

———, and C. G. Young. 1986. Collecting, processing, and germinating seeds of wildland plants. Timber Press, Portland, Oreg. 236pp.

———, and C. G. Young. 1992. Seeds of woody plants in North America. Descordes Press, Portland, Oreg. 407pp.

Chapter 22
FENCES, WATERHOLES, AND OTHER RANGE IMPROVEMENTS

Robert R. Kindschy

INTRODUCTION

Rangelands often require development of facilities to improve values for livestock. For example, fencing to enable better distribution of stock and drinking water for added flexibility in season of use are typical of facilities in western North America.

Through the course of development of these improvements for livestock production, certain modifications and developments may be made. The implication to wildlife of typical livestock management facilities is discussed in this chapter. Many such improvements may be modified or managed to benefit wildlife. Planning and foresight can greatly increase wildlife habitat values of fences, waterholes, and other range improvements.

FENCES

Implications for Wildlife

Management of livestock on rangelands normally requires subdivision of the range into pastures through fencing (Heady 1975). Barbed wire fences are commonly used throughout western United States rangelands, as in southeastern Oregon, where 4,300 km of barbed wire fence is actively used to manage 2 million ha of public land (Heady and Bartolome 1977).

Fences have a negative and a positive impact on wildlife. Positive benefits to wildlife occur through the management of livestock to attain a desired status of vegetation (Kimball 1957, Scotter 1980, Holechek et al. 1982, Kindschy 1986, Kie and Loft 1990, Severson 1990, Urness 1990).

An example of livestock grazing to improve habitat for a species of wildlife is moderate to close utilization of crested wheatgrass (*Agropyron cristatum*) during the spring and early summer to enable succulent regrowth and autumn growth availability for mule deer (*Odocoileus hemionus*). Attainment or maintenance of a target successional stage of vegetation to provide optimum habitat for a preferred wildlife species or associated groups of species is also feasible in many rangeland ecosystems (Anderson et al. 1990).

Barriers to Access

Fences, however, often disrupt normal movement patterns for big game which, under extreme situations, may result in death from collisions, entanglement, or entrapment. Howard (1991) documented that although 1.2 m high net wire fence inhibited only young mule deer in pinyon-juniper (*Juniperus* spp.) habitat, 7 strand electric fence 1.5 m high proved to be a barrier to 97% of deer. In northern Arizona, Neff and Woolsey (1979) describe pronghorn (*Antilocapra americana*) losses caused by fence entrapment during severe winter conditions.

Fence Design

Sanderson et al. (1990) list the following types of fences in common use in western North America as barbed wire, woven wire, wooden, and electric.

Barbed-wire fence, the most common type of fencing used on North American rangelands, has several variations. The standard post and wire fence is normally used on arid lands. Steel posts are commonly used in the major portion of the fence; however, wooden posts are used in stretch panels, corners, and gates. Suspension fence is similar but posts are much farther apart and wire tension often greater. Suspension fences are less visible due to the reduced number of posts and, hence, greater danger to pronghorn and horses.

Design of fencing is important. Big game accessibility may be accommodated through modification of standard barbed wire fence design. Such standards have been developed for most big game species (Fig. 1).

Woven wire fencing is associated with domestic sheep ranching. Such fence can be a major barrier to big game if height exceeds 1.4 m (Yoakum et al. 1980). Pronghorn are especially restricted as they normally pass under or through a fence. One or 2 strands of barbed wire are often placed above the woven wire. Deer typically jump fences and may entangle their hind legs in the barbed wire when attempting to clear the obstacle. Prolonged entrapment eventually resulting in death is the usual outcome.

Electric fences have become more popular for temporary fencing in remote areas since solar powered chargers were perfected. One or 2 strands of wire, normally smooth but sometimes barbed, are supported by light weight fiberglass rods. Such fence is difficult to see and collisions by big game and wild horses are frequent. White cloth strips tied on the top wire reduce the frequency of collisions (Kindschy et al. 1982). Wind will blow the cloth down smooth wire causing it to collect at the support rods. Consequently the cloth should be secured in place on smooth wire.

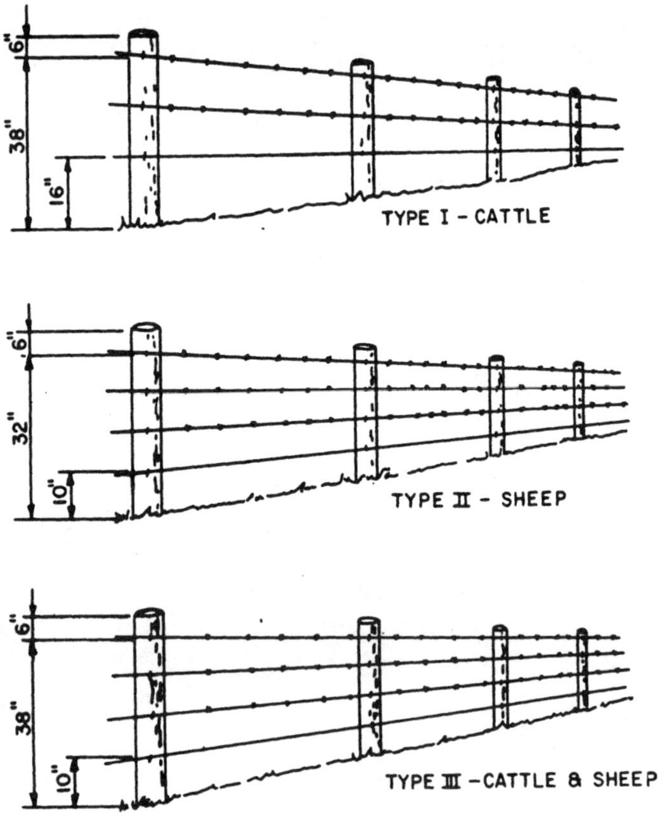

FIGURE 1. Specifications for livestock fencing on pronghorn ranges (Yoakum et al. 1980).

Birds can be electrocuted when they encounter hot and ground wires together. Proper fence design will minimize this loss. Electric fences using steel fence posts with insulators pose a greater risk to passerine birds because each post is an electrical ground. Fiberglass and hard rubber rods and posts are consequently preferred for temporary electric fencing.

Buck-and-pole fences can be major obstacles for big game due to the combination of height and width of these structures (Sanderson et al. 1990, Scott 1992). Fortunately such wooden fences are not popular on western rangelands due to the great expense of materials and construction and the high risk of loss from fire. The same may be said of log-worm and block-and-pole fences. Yet, such wooden fencing is picturesque and, consequently, increasingly used around campgrounds and within wilderness and primitive areas.

Location of the fence is critical to the welfare of many big game animals. Elk (*Cervus canadensis*) trails, for example, should have a lay-down fence provided during the time of maximum elk activity. Fences built on steep slopes become barriers for mule deer during winter snow conditions (Table 1) (Anderson 1980).

A deer is disadvantaged when jumping up-slope (Fig. 2); traction is reduced and the jumping point is farther from the fence, especially when compared to jumping down-slope. As deer approach the limit of jumping ability, the probability of failure increases, especially if deer are in a weakened condition.

Clearance beneath a fence is also affected by slope. Where fence specifications require a 41 cm height of the bottom wire

TABLE 1. Relation of slope of land to height of fence. From Anderson (1980).

Percent slope	Height of jump necessary to clear a standard 1 m high fence
10	1.2
20	1.4
30	1.6
40	1.7
50	1.9
100	2.7

above the ground, a pronghorn whose nose is 50 cm up slope from a fence on a 10% grade confronts an apparent space of 34 cm beneath the fence.

New fence should be flagged with white ribbon or cloth to alert wildlife to the new barrier (Kindschy et al. 1982). A flag every second or third span is normally sufficient. By the time the flagging has deteriorated, wildlife will have become accustomed to the new fence.

Height of the bottom wire for fences is important for pronghorn (Yoakum et al. 1980). Forty-one cm is normally recommended as an above-the-ground clearance. This wire should be non-barbed or smooth. An occasional calf may pass under such a fence but can cross back to rejoin the mother cow and not become trapped.

With special reference to pronghorn, Kindschy et al. (1982) summarized intensive studies of field biologists concerning fencing with the following recommendations.

1. Use of woven or net-wire fences should be minimized.
2. Areas encircled by fences that prohibit pronghorn movement should be large enough to include year-long habitat needs.
3. Pathways and migration routes should have low fences, lay-down panels, pass structures, or adjustable spacing of the wire (Yoakum et al. 1980).
4. Barbed wire fences <3 strands can be negotiated by pronghorn. The bottom wire should be smooth (non-barbed) and >40 cm above the ground. Remaining wires should be spaced at 25-cm intervals. The total height of the fence should not exceed 90 cm, and there should be no stays between posts.
5. White-topped steel fence posts increase the visibility of the fence and thus lessen the likelihood of collision.

Although pronghorn normally pass under or through fences, some jump fences <80 cm high. Of the designs tested by Spillet (1965), the least detrimental to pronghorn passage were net or woven wire 80 cm high without wire above them, or net or woven wire 66 cm high with a barbed wire 10 cm above the net wire.

Modifications to Enhance Wildlife Habitat Values

Certain fence designs provide values to wildlife, especially nongame species, which have not been recognized. Maser et al. (1979) illustrated a number of wildlife habitat enhancements that have been developed for fences. Nest boxes, for example, may be placed in favored habitats along the fence and raptor roosts erected at sites favored by hawks and eagles (Fig. 3). Rock-jack and rock-crib structures afford habitat for smaller rodents and reptiles when of the proper design.

The location of new fence should take into consideration property boundaries, vegetation, wildlife migration routes, available water sources, slope, snowdrift areas, stream crossings, areas of heavy livestock use, gate locations, and road crossings (Sanderson et al. 1990). Unfortunately, many livestock fences are built on survey lines, such as "section" boundaries, with no regard to local topography, location of water, or other biological considerations. The result, in many instances, is poor livestock distribution and forage utilization. Big game are more likely to encounter obstruction to movement when fences contour around steep hill slopes. All wildlife suffers when portions of improperly fenced range are constantly stripped of vegetation.

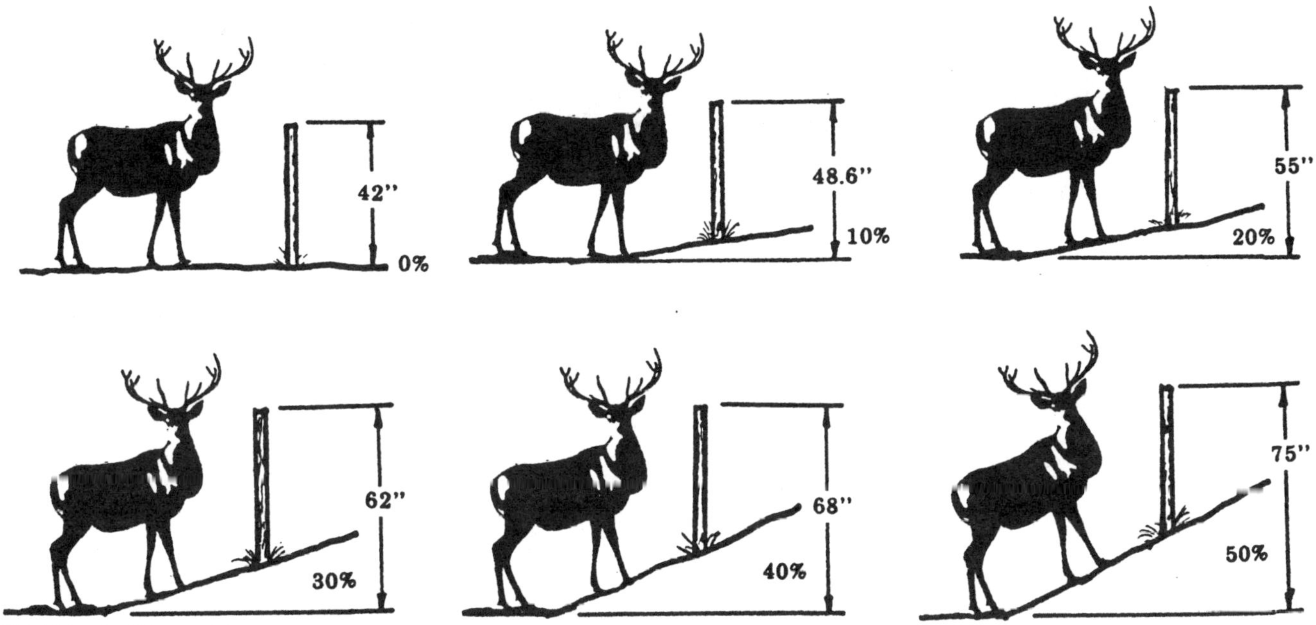

FIGURE 2. Effect of increasing ground slope on barrier height of any fence (Duffy 1988).

FIGURE 3. Placement of perching structures and nesting boxes along fences provides often unique habitat features for wildlife (Maser et al. 1979).

WATERHOLES

Development of Water Sources in Rangelands

Water is typically scarce on rangelands. Yet, an adequate supply of reliable drinking water, fairly uniformly distributed within livestock grazing pastures, is essential for intensive livestock management (Vallentine 1971). Streams, lakes, and springs originally supplied the only sources of open water, except for brief periods of rain or snow accumulation. Management activities, primarily centered around requirements for livestock, have created additional water through the construction of surface water entrapments, such as reservoirs, and the drilling of wells (Bleich et al. 1982).

Water is equally essential for most species of wildlife, both as a drinking source and, in the case of fish and waterfowl, as primary habitat. Water developments, therefore, are an important habitat component in rangeland ecosystems.

Spring Developments. Water discharge from most springs is insufficient to water many livestock at a time. Consequently the flow must be accumulated in a storage site for periodic high volume consumption. Metal, concrete, or fiberglass troughs or stock tanks of several thousand liters capacity are often installed. Surplus water overflows onto the ground at the site of the trough, which creates mud holes. Trampling, in turn, causes the tank height to increase relative to drinking animals. No improvement of vegetation is possible due to the constant trampling and grazing by stock. Unfortunately the original riparian zone associated with the spring is often destroyed following development, which collected all available water. The spring source should generally not be fenced. Such protection from grazing enables willow (*Salix* spp.) and other wetland plants to grow on the site. Eventually roots from these plants plug the perforated collection pipe causing failure of the development. Rather, the overflow of the trough should be piped to an adjacent site that can then be developed as a riparian community. This discharge site, when properly selected, offers an opportunity to mitigate the loss of the original riparian zone and to create enhanced habitat values through the introduction of suited plants. The best sites are those that spill onto a gentle slope >15 to 20 m from the trough. The stream of water may be contoured in a shallow ditch to provide irrigation of the site, the extent of which will be governed by the volume of water and lay of the land. Such sites normally must be fenced to exclude livestock access (Fig. 4) (Taylor and Littlefield 1986). Suited grasses, forbs, shrubs, and trees may be planted within the area of irrigation influence. Introduction of plants such as cottonwood (*Populus* spp.) and fruit trees afford enhanced habitat diversity attractive to many birds as feeding and breeding habitat (Faanes 1984, Szaro and Jakle 1985, Stinnett and Klebenow 1986).

Although seeps provide a limited amount of water for wildlife when undeveloped, the development to accumulate and store water can be positive for wildlife. The collection trough should be designed for easy accessibility by large and small species, and preferably the top should be <40 to 50 cm above ground level (Fig. 5). Escape ramps built of metal grating (Fig. 6) are desirable additions to steep-sided tanks. Small animals can often escape drowning and thus avoid fouling the water (Wilson and Hannans 1977).

Wells and Pipeline Systems. Wells or bores provide water in areas where springs do not occur and impoundments are not possible (Bleich et al. 1982). Often a pipeline system distributes water to float-valve controlled troughs. Such waters are available to big game during times of pipeline function. The persistence or reliability of water is critical for maximum values to wildlife. An on and off system along a pipeline, for example, is less effective in meeting wildlife requirements than a source that persists year-long or, at the least, during the dry season.

Main storage tanks can be left full of water at the time livestock leave a pasture. A sufficient supply of drinking water is thus afforded resident wildlife. Such consideration is especially important in areas where no alternative water sources exist. Other troughs along these pipelines may be deactivated to reduce evaporation and thereby extend the supply availability of water in critical areas.

FIGURE 4. Water trough height <50 cm assures access to more wildlife (Yoakum et al. 1980).

Escape Devices in Troughs. Limited use is also made by wildlife other than big game; however, float valve controlled water levels often are inaccessible to smaller birds and mammals. Large rocks placed in the trough will help alleviate this problem and provide a foothold for young ungulates that occasionally get into troughs.

All steep sided stock tanks and troughs should be equipped with an escape ramp to provide small birds and mammals a means of crawling from the water. Wilson and Hannans (1977), Yoakum et al. (1980), Sherrets (1989), and Sanderson et al. (1990), developed a number of designs for escape ramps (Fig. 6).

Height of the rim of the tank above the ground surface is crucial for water access of young ungulates. Sides should be no higher than 40 to 50 cm (Fig. 5). Concentrations of cattle at troughs often compact and erode soil to the extent that the effective height of the tank rim is increased. Rock of "rip-rap" size may be placed surrounding the trough to prevent soil loss through trampling and pedestaling.

Birdbaths. An adaption to pipeline systems that has proven quite successful is the birdbath (Fig. 7), a small concrete basin that is fed by the pipeline, often through a secondary storage tank. Two buried septic tanks storing 2,650 L, for example, are filled by the main pipeline and supply water to the birdbath when the main line is dry. In eastern Oregon and southern Idaho, such systems been used by upland game birds and small mammals for >25 years.

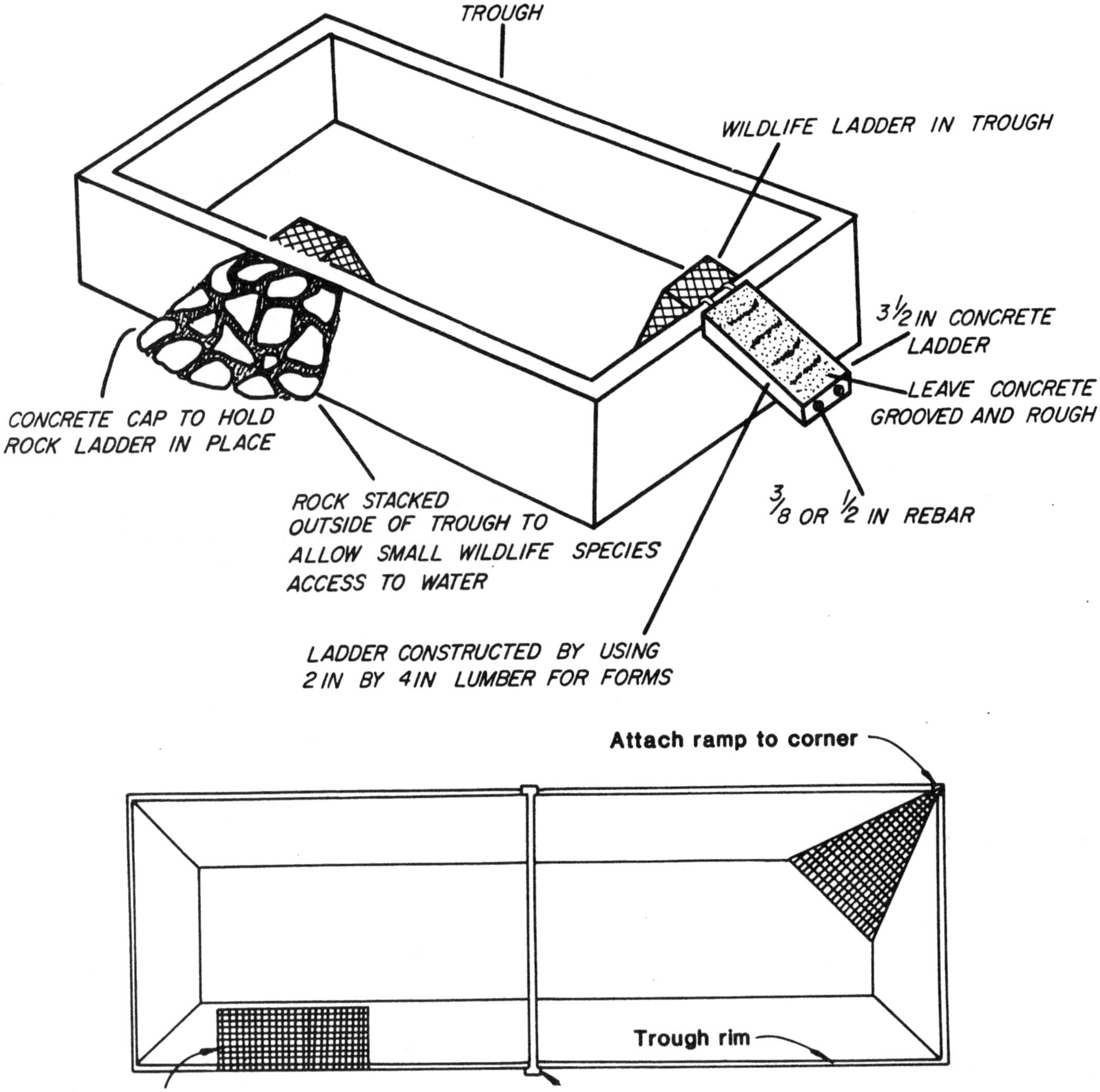

FIGURE 5. Ramps within water troughs enable small wildlife a means of escape and thereby also maintain higher water quality for all rangeland animals (Wilson 1977).

RESERVOIRS

Earthen dams are common throughout the western United States for entrapment and storage of spring snow-melt. Damming of ephemeral streams provides water for livestock and wildlife. Habitat values for wildlife vary by specific site. Impoundments that are perennial may provide habitat for fish and waterfowl. Shoreline vegetation is uniquely important for game and nongame birds and mammals (Thomas et al. 1979, Faanes 1984, Johnson and Dinsmore 1986, Medin and Clary 1990). Protective fencing or a livestock grazing system that enables the development and persistence of such vegetation is necessary for maximizing habitat features (Lokemoen 1973, Kaiser et al. 1979, Candelaria and Wood 1981, Whyte and Cain 1981, Kindschy 1987, Uresk and Severson 1988).

Noodle-bowls

Noodle-bowls (Fig. 8) are a relatively new stock water development that has married spring developments to retention reservoirs. One or often more springs are piped down country to a circular earthen pit that has a water capacity of >1,000,000 L.

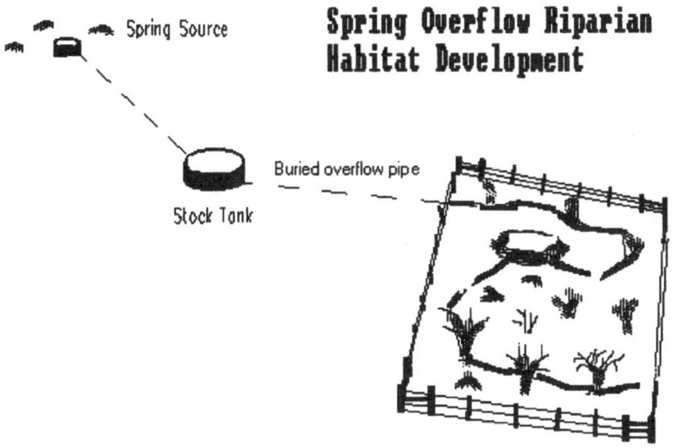

FIGURE 6. Enhancement of spring development through piping of overflow water into an exclosure. Suitable grasses, forbs, shrubs, and trees can be planted within area of irrigation influence.

A gravel packed filter system on the pond bottom discharges water to a series of gravity fed pipelines, each of which is controlled by valves in a manifold located immediately below the dam. Gravity supplies all energy requirements. Spring flows of only a few L/minute accumulate great amounts of water during an entire year. The pond is deep to minimize evaporation. Noodle-bowls are always fenced to exclude cattle. In southern Oregon this type of water development has been highly successful for >25 years, and of high value to wildlife as a desert oasis in a sea of rangeland (Szaro and Jakle 1985). Many sites support put-and-take trout fisheries. All provide habitat for waterfowl and shorebirds plus many passerines that nest in the associated trees (Faanes 1984).

Site Potential for Riparian Features

Stock pond reservoirs exhibit a wide variety of potentials for wildlife habitat depending upon the site (Crouse and Kindschy 1984). As a general rule 4 conditions must be met.

FIGURE 7. A satellite water source for small animals served by a buried pipeline. Such "birdbaths" provide reliable drinking water for wildlife even when the main pipeline is not active if associated storage tanks are installed.

1. Water should be present year-long.
2. Soil must be present.
3. Shore slope should be gradual.
4. Livestock usually must be excluded.

Modifications for Wildlife Enhancement

In a typical fenced reservoir habitat, water is piped to livestock by an underground pipeline fitted with a turn-off valve at the base of the dam (Fig. 9). Trough fill is con-trolled by a standard float valve. Ideally, the pipe should be placed through the dam at the time of construction. Existing dams, however, have been successfully cut for pipe placement and then refilled. Proper compaction is mandatory when refilling the dam materials. Placement of a series of collars along the pipe within the dam further assures that water will not seep along the outside of the pipe. Siphons over the dam are generally not successful due to the interruptions of water flow caused by the float valve control.

Fencing of Stock Ponds

Four strand, barbed wire fences are recommended when fencing a reservoir due to stress exerted by livestock near the water. Fence should be located as far from the water edge as practicable, but no closer than 15 m. The entire impoundment should be enclosed. Water-gaps into the pond are seldom effective due to constant maintenance requirements. Height of the bottom wire above the ground should be 40 to 45 cm. Such clearance will allow passage of pronghorn and young calves (but not the mother cows). The calves will soon return to their mothers. Lower fences often allow calf entry at occasional dips in the land surface, a dry wash for example, but the calf then fails to rediscover the site for exit.

Occasionally it is necessary to fence across the points of inflow and discharge with structures other than the standard "flood gate." One successful structure is formed through the hanging of wooden panels from a cable stretched across the stream channel and anchored to large rock jacks a couple meters back from the stream bank (Duffy 1988) (Fig. 10). High water flow and debris can pass as the panels swing out of the way. Cattle do not seem to press the panels during the grazing period. In fact, these panels are effective in excluding cattle from exclosures and are superior to the traditional flood gate that requires frequent maintenance or replacement.

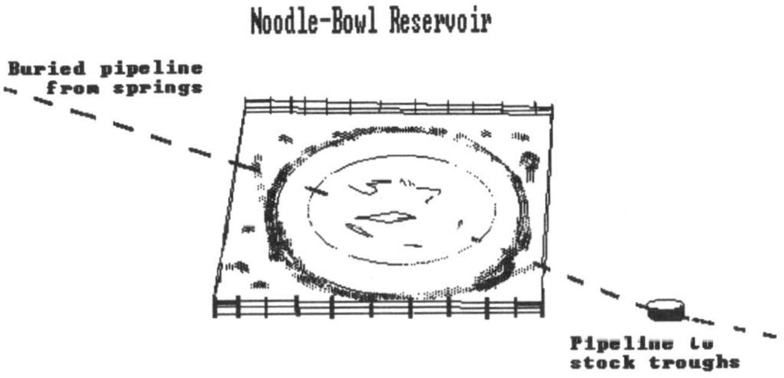

FIGURE 8. "Noodle-bowl" storage pits or reservoirs accumulate piped spring water throughout the year by gravity flow. Water is then piped to stock troughs at lower elevations. Such sites require fencing to protect the pipeline manifolds thus enabling riparian vegetation to develop and persist.

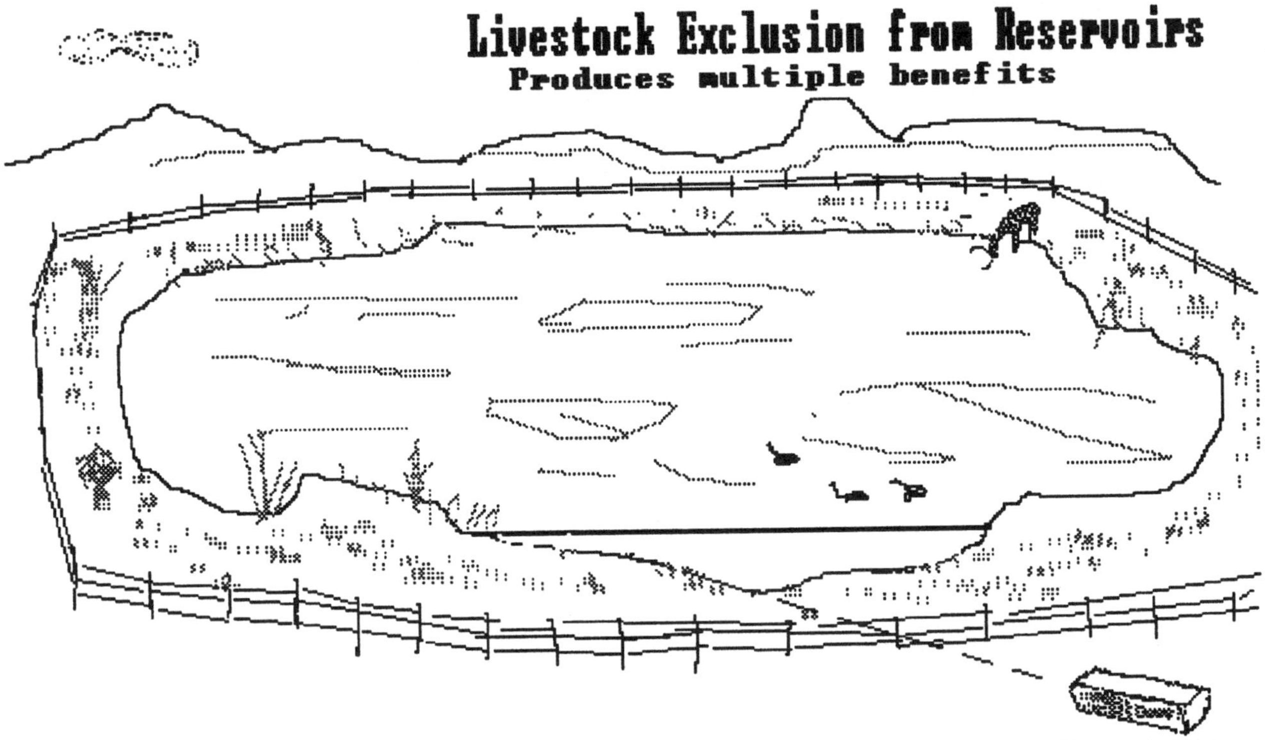

FIGURE 9. Stock pond reservoirs may provide outstanding wildlife habitat when fenced to exclude livestock. High quality water is piped by gravity flow to float controlled stock troughs.

Islands

Waterfowl nesting can be enhanced through the development of small islands within the reservoir (Jones 1975). These may be created at the time of construction through cutting off points that project into the impoundment (Fig. 11). Islands may also be built by piling bottom material to project above the high water line (Fig. 12). This may be either accomplished at the time of initial construction or during maintenance or clean-out operations. A word of caution must be injected for areas where removal of reservoir bottom materials might expose fractures in the substrata which would result in leaks in the pond. Much of the lava country of the northwestern United States has a threat of this problem.

Giroux (1981) found that islands of approximately ⅛ ha were of great value to nesting Canada geese (*Branta canadensis*) and many duck species. These values were maximized when adequate vegetation screening was present. Studies by Duebbert et al. (1983) resulted in recommendations that a mixture of grasses and legumes be planted on islands. My observations in south-

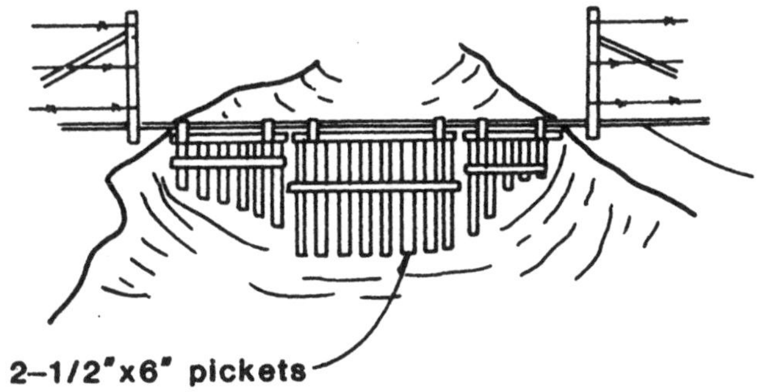

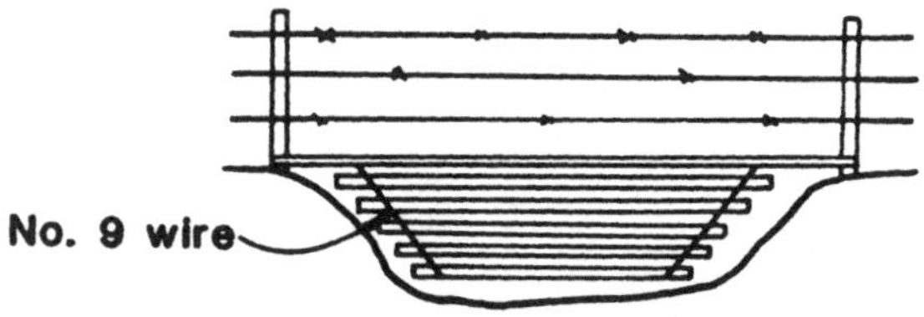

FIGURE 10. Swing gates across streams minimize maintenance following high water yet are effective as cattle barriers (Sanderson et al. 1990).

FIGURE 11. Islands may be created in reservoirs either at the time of construction or during maintenance of existing impoundments. Islands are constructed by mounding earth from the reservoir basin. Such islands are valuable to nesting waterfowl and shorebirds.

eastern Oregon are that the addition of some shrubbery and small trees is also of value. Roots of such larger plants tend to stabilize the shore from wave erosion while the tree structure affords excellent habitat for many nongame birds.

Small impoundments of <0.5 ha may have floating rafts anchored near the center for waterfowl nesting and resting (Yoakum et al. 1980) (Fig. 12). Annual maintenance is usually required to renew nesting materials and prevent winter ice damage.

Goose nesting platforms may be installed above the high water line around larger impoundments (Hacky et al. 1988). In areas of severe winter cold, wind driven floating ice will severely damage structures placed below the high water line. Various designs for nesting platforms are often available through state or provincial wildlife agencies.

Pond ecosystems that have become well advanced present opportunities for the placement of nesting boxes for bluebirds (*Sialia* spp.) and other cavity nesting species (Fig. 13). Often local conservation clubs or scout troops volunteer for this type of project.

Vegetation Introduction

Waterfowl (Mack and Flake 1980) and shorebirds (Ryan et al. 1984) are among the greatest benefactors of pond riparian vegetation. Protection from livestock use is generally advisable if the objective is to optimize waterfowl nesting potential, although grazing systems designed to favor riparian vegetation may function adequately (Mundinger 1976).

Evans and Kerbs (1977) working in South Dakota observed that 82% of broods produced during a 7 year study were on ponds that had been protected the previous 2 years from livestock grazing during the spring.

Vegetation establishment can be either natural or provided at the time of initial pond fill. Often ducks and other birds bring in seed from willow, cottonwood, cattail (*Typha* spp.), and other fine-seeded plants. However, many additional plants may be introduced into the new riparian community to improve wildlife habitats. Again, it is essential to give the preferred plants an opportunity to accomplish their primary life functions of growth and reproduction. This may be accomplished through protective fencing from livestock or through a grazing system that meets these requirements. Trees add important

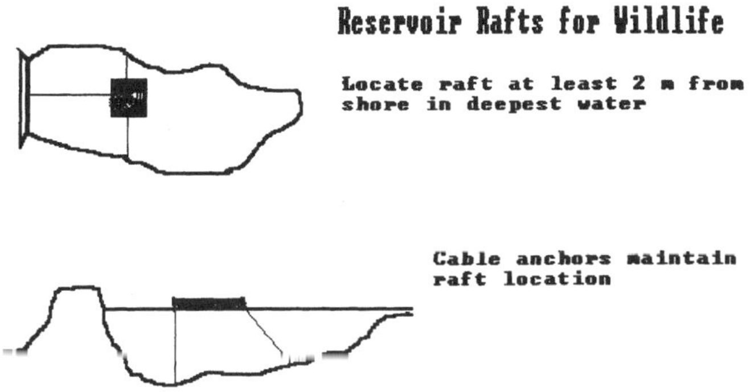

FIGURE 12. Placement of floating raft in smaller ponds for waterfowl nesting and resting. Such structures require annual maintenance and should be removed from water during winter to prevent water-logging and ice damage.

FIGURE 13. Bird houses may be placed in riparian zones around protected reservoirs. This nesting box was built and placed by Scouts from Jordan Valley, Oregon.

structure to rangeland and are important habitat for nesting birds. Lombardy poplar (*Populus nigra italica*) is an excellent choice for introduction as it cannot reproduce and become a pest, a problem with plants such as Russian olive (*Elaeagnus angustifolia*)(Knoph and Olson 1984), Chinese elm (*Ulmus parvifolia*, and salt-cedar (*Tamarix* spp.). Lombardies and box elder (*Acer negundo* are rapid growing and limb structure is exceptionally well suited to bird nesting.

Aquatic plants also enhance habitat features for waterfowl (Hudson 1983) and mammals such as muskrat (*Ondatra zibethicus*) and beaver (*Castor canadensis*). Introduction of species native to the area speeds the process of habitat successional advancement. Plants worth considering include sago pond-weed (*Potamogeton pectinatus*), deep water duck potato (P. *amplifolius*), wapato duck potato (*Sagittaria cuneata*), bur-reed (*Sparganium* spp.), smartweeds (*Polygonium* spp.), and bulrush (*Scirpus* spp.). Duck-weed (*Lemna minor*) and cattail should be considered where habitat objectives and conditions are suitable. Excessive submergent vegetation can adversely effect fish following freeze-up, especially when deep snow covers the ice.

Where dense stands of cattails, bulrush, or other wetland vegetation have developed, it is often desirable to create openings in the stands. Muskrat may be introduced to the site. They will create numerous openings in the emergent aquatic vegetation. Muskrat lodges afford excellent nesting sites for waterfowl and shorebirds. Muskrat also provide a prey base for larger carnivores.

Fish

Most reservoirs that retain water during all but the more severe droughts have a potential for a fisheries. The presence of fish is a recreational value and provides a source of prey for a variety of birds and mammals.

Trout (*Oncorhynchus* spp.) seldom reproduce in ponds. Occasionally an entering stream will allow limited spawning opportunity; however, such reproduction is not necessary and often not desirable due to potential over-population. It is important to stock no more fish than will be able to rapidly grow to desirable size. Over-stocking will cause slow growth and small size that is not attractive to recreationists. Under-harvest further complicates the problem of over-stocking. It is far better to under-stock initially. Ponds that occasionally go dry (i.e., once every 8 or 10 years), have an advantage in totally eliminating fish without the expense of treatment. Restocking is a normal part of a "put and take" trout fisheries program.

Ponds <3 m deep are of low potential for trout due to excessive heating during summer and freezing during winter. Both conditions can result in dissolved oxygen levels below the limit tolerated by trout. Dense stands of underwater vegetation may further deplete oxygen levels during calm summer nights or when winter ice and snow cover prevent sunlight penetration of the water.

Excellent fish populations that reproduce naturally are possible in most perennial reservoirs (Mraz and Cooper 1957, Welch and Ball 1966) through the stocking of black bass (*Micropterus* spp.)(Bennett and Childers 1972), crappie (*Pomosis* spp.), bluegills (*Lepomis macrochirus*), and bullheads (*Ameiurus nebulosus*). Bullhead catfish provide a prey base for some birds (i.e., heron [*Ardea herodias*], terns [*Sterna foresteri*], bittern [*Botaurus lentininosus*], pelican [*Pelcanus erythrorhynchos*], and gulls [*Larus* spp.]. Warm water fish are often prolific and can rapidly over-populate resulting in a stunting of size. While not a problem for fish-eating wildlife, such stunting does greatly reduce the recreational fishing demand. A mixture of largemouth black bass and black crappie or bluegills is recommended over a single species stocking. The species that achieves greatest size first will prey on the other species and thus keep populations in balance with the pond's carrying capacity.

Wetlands are especially used by sensitive wildlife; e.g., nesting use by rare waterfowl or shorebirds. A recreational fisheries at such a site is often incompatible with the welfare of such nesting birds due to constant disturbance by recreationists. In such cases nongame fish, rather than a sports species, may be introduced into the waters to provide forage for the wide array of birds and mammals that prey on fish.

Precipitation Catchments

Precipitation interception and storage structures, often termed "guzzlers," are widely used as drinking water sources for wild birds and mammals.

Prior to initiating a costly program of guzzler construction on a rangeland seemingly devoid of drinking water, it should be determined that water is actually a factor limiting wildlife. Carnivores such as kit fox (*Vulpes macrotis*) obtain fluids from prey such as kangaroo rats (*Dipodomys* spp.) that, in turn, metabolize moisture from carbohydrates in the seeds they consume. Other wildlife may only use areas during a wet season when water is widely available.

Designs are many and varied (Fig. 14). All require an apron to intercept precipitation and some form of water storage container. A greater quantity of high quality water results when stored water is protected from sunlight and other elements.

Plan of guzzlers must first recognize how much water is needed at that site. An apron can then be designed to intercept precipitation based upon the minimum precipitation for the site, rather than the average or maximum, to prevent guzzler failure during drought (Yoakum et al. 1980). A good rule is: 10 $L/m^2/cm$ of rainfall.

Yoakum et al. (1980) and Tsukamoto (1990) present in depth discussions of various guzzler designs, materials, construction techniques, and other considerations. On occasion, natural rock surfaces provide collection surfaces that can be dammed and diverted with a pipe to underground tanks. Water is then supplied to a float-controlled drinking fount. Strong fiberglass tanks are well suited to long-term burial in rangeland soils. Fount should be partially buried in soil to prevent winter freezing damage of the float-valve mechanism (Fig. 15).

Control of water use is often necessary. Cattle, for example, could exhaust a supply in a few days that would water hundreds of birds. Fencing is an obvious solution to such a problem, yet many, perhaps most precipitation catchments are located in areas so remote from water that cattle seldom use the range.

OTHER RANGE DEVELOPMENTS

Roads

Bladed back-country roads are used by humans and by many types of wildlife as less resistant travel routes. Borrow-pits may be seeded to a variety of plants suitable for wildlife cover and forage. The crowned nature of the roadbed affords drainage into the adjacent borrow area thus making it more mesic than the norm. Dryland alfalfas, sweet clovers, and other forbs, grasses, and shrubs do well in such environments. In addition, their presence inhibits the establishment of noxious weeds that readily populate such disturbed areas.

Rehabilitation of disturbed sites, whether road borrow-pits, mining reclamation, fire rehabilitation, or other disturbances, present an opportunity to increase the vegetation diversity and

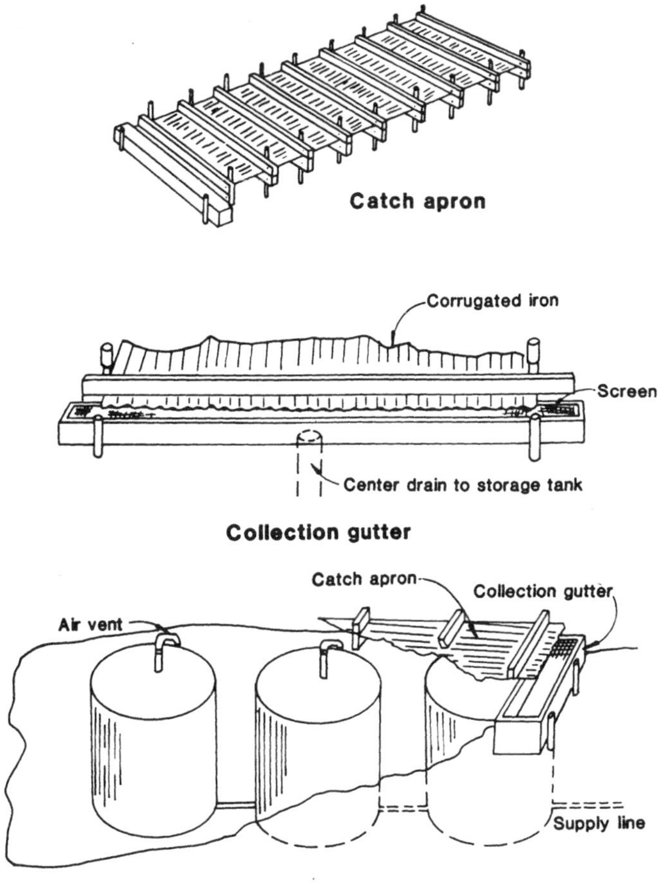

FIGURE 14. Precipitation catchments or "guzzlers" are of many designs but all can provide drinking water to rangeland animals where no other possibilities exist (Sanderson et al. 1990).

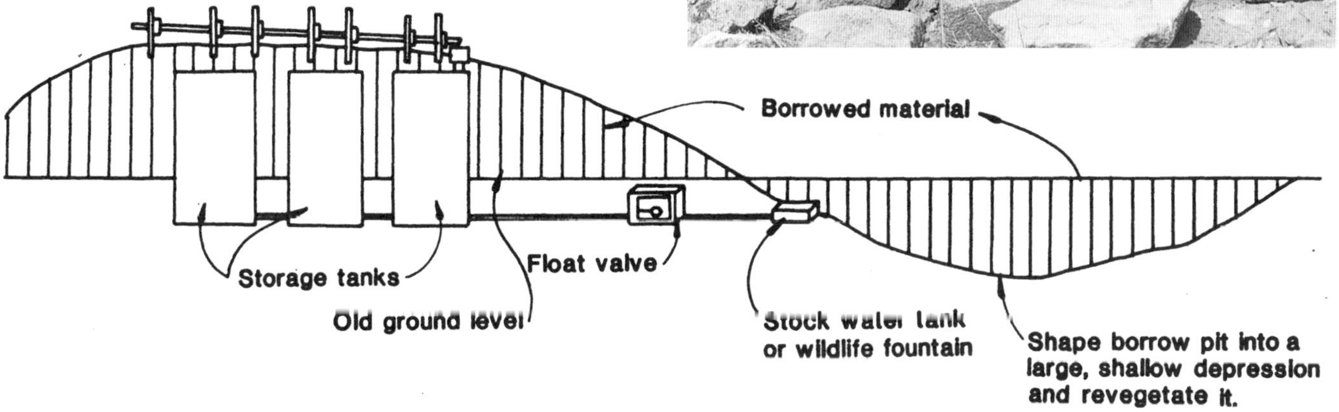

FIGURE 15. Drinking water is provided to animals in a small basin or fount. A float-valve device is buried under sufficient soil to prevent freezing damage (Sanderson et al. 1990).

species composition of an area (Monsen and Shaw 1983). Such change in plant diversity is normally followed by increased animal diversity. Recent horticultural development of native plants plus a variety of beneficial exotic plants enables a rich selection of plant materials for seeding or planting.

Electric Powerlines

Powerlines serving stock wells, solar powered arrays, and electric fences are typical modern uses of electricity on rangelands of western North America. Large birds, especially raptors, are subject to electrocution when perching on powerline poles and cross-arms. The Edison Electric Institute funded research concerning this problem which, in 1975, resulted in an excellent publication entitled *Suggested Practices for Raptor Protection on Powerlines* (Miller et al. 1975). Sixteen variations of powerline configuration were presented. Several of these diagrams have been republished (Nelson and Nelson 1976). Most powerlines of major electric distributors now meet raptor safety standards. However, problems can develop with new construction, especially at drop poles near wells where birds congregate due to water and prey abundance.

Raptor roosts (Fig. 16) on top of select power poles provide a safe perch for large birds (Nelson and Nelson 1976). It is amazing how much use is made of power and telephone poles, especially when paralleling a highway. Such are ideal resting and foraging sites.

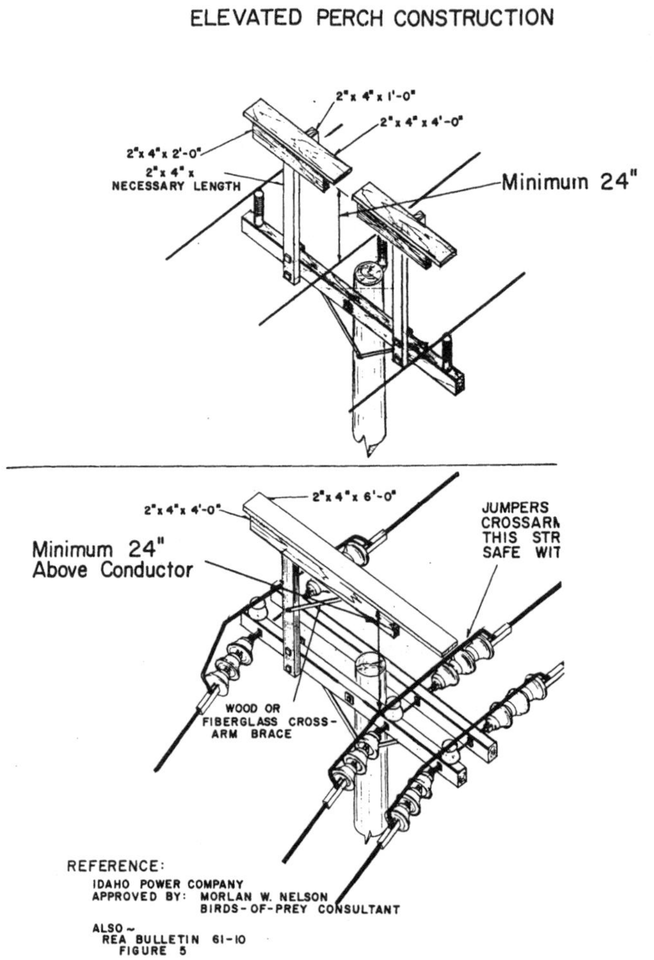

FIGURE 16. Electric powerline configuration has been designed to minimize hazard to large birds (Miller et al. 1975).

SUMMARY

This chapter discussed considerations and modifications of traditional rangeland habitat developments to enhance increased habitat features for wildlife. Food, cover, water and open space, the fundamental components of habitat for all life, are most productive when available in optimal spacing and timing. This principle has been a major factor in rangeland development for livestock production. Such is also the case for wildlife habitat enhancement.

Increasing the variety and number of plant types on a site by including forbs, grasses, shrubs and, where possible, trees, provides habitat for many more species of animals. Management of vegetation must assure that plant growth and reproduction may regularly occur.

Water is normally scarce on rangelands. Such sites are often the most productive habitats for wildlife. Present rangeland management practices afford a wide variety of opportunities for enhancement of wildlife habitat through management and modification of water developments.

Open space needs vary with the requirements of each species of wildlife. Perhaps large ungulates are most affected by livestock fence designs that restrict critical movement among habitats. Modifications of design are possible to reduce such restrictions.

Rangelands offer a multitude of values that include forage for livestock and habitat for wildlife. Optimization of these features will become increasingly more important to humanity.

LITERATURE CITED

Anderson, E. W., D. L. Franzen, and J. E. Melland. 1990. Forage quality as influenced by prescribed grazing. J. Range Manage. 43:56-70.

Anderson, L. D. 1980. Adjustable wire fences for facilitating big game movement. U.S. Bur. Land Manage. Tech. Note 343. 7pp.

Bennett, G. W., and W. F. Childers. 1972. Thirteen-year yield of smallmouth bass from a gravel pit pond. J. Wildl. Manage. 36:1249-1253.

Bleich, V. C., L. J. Coombes, and J. H. Davis. 1982. Horizontal wells as a wildlife habitat improvement technique. Wildl. Soc. Bull. 10:324-328.

Candelaria, L. M., and M. K. Wood. 1981. Wildlife use of stockwatering facilities. Rangelands 3:194-196.

Crouse, M. R., and R. R. Kindschy. 1984. A method for predicting riparian vegetation potential. Pages 18-24 *in* Proc. 1984 Pacific Northwest range management short course: range watersheds, riparian zones, and economics. Pendleton, Oreg.

Duebbert, H. F., J. T. Lokemoen, and D. E. Sharp. 1983. Concentrated nesting of mallards and gadwalls on Miller Lake Island, North Dakota. J. Wildl. Manage. 47:729-740.

Duffy, B. 1988. Fences. U.S. Dep. Agric. For. Serv., U.S. Bur. Land Manage., U.S. Dep. Agric. For. Serv. Tech. and Dev. Center. 210pp.

Evans, K. E., and R. R. Kerbs. 1977. Avian use of livestock watering ponds in western South Dakota. U.S. Dep. Agric. For. Serv. Rocky Mt. For. and Range Exp. Stn. Gen Tech. Rep. RM-35. 11pp.

Faanes, C.A. 1984. Bird use of riparian trees in prairies. Am. Birds 38:3-6.

Giroux, J. 1981. Use of artificial islands by nesting waterfowl in southeastern Alberta. J. Wildl. Manage. 45:669-679.

Hacky, D. L., W. C. Matthews, Jr., and I. J. Ball. 1988. Elevated nest structures for Canada geese. Wildl. Soc. Bull. 16:362-367.
Heady, H. F. 1975. Rangeland management. McGraw-Hill, New York, N.Y. 460pp.
———, and J. Bartolome. 1977. The Vale rangeland rehabilitation program: the desert repaired in southeastern Oregon. U.S. Dep. Agric. For. Serv. Res. Bull. PNW-70. 139pp.
Holechek, J. L., R. Valdez, S. D. Schemnitz, R. D. Pieper, and C. A. Davis. 1982. Manipulation of grazing to improve or maintain wildlife habitat. Wildl. Soc. Bull. 10:204-210.
Howard, V. W., Jr. 1991. Effects of electric predator excluding fences on movements of mule deer in pinyon/juniper woodlands. Wildl. Soc. Bull. 19:331-334.
Hudson, M. S. 1983. Waterfowl production on three age-classes of stock ponds in Montana. J. Wildl. Manage. 47:112-117.
Johnson, R. R., and J. J. Dinsmore. 1986. Habitat use by breeding Virginia rails and soras. J. Wildl. Manage. 50:387-392.
Jones, J. D. 1975. Waterfowl nesting island development. U.S. Dep. Inter. Bur. Land Manage. Tech. Note 260, Denver, Colo. 17 pp.
Kaiser, P. H., S. S. Berlinger, and L. H. Fredrickson. 1979. Response of blue-winged teal to range management on waterfowl production areas in southeastern South Dakota. J. Range Manage. 32:295-298.
Kie, J. C., and E. R. Loft. 1990. Using livestock to manage wildlife habitat: some examples from California annual grassland and wet meadow communities. J. Range Manage. 43:7-24.
Kimball, T. L. 1957. The economic aspects of livestock-big game relationships as viewed by a game administrator. J. Range Manage. 10:67-70.
Kindschy, R. R., C. Sundstrom, and J. D. Yoakum. 1982. Wildlife habitats in managed rangelands: the Great Basin of southeastern Oregon. U.S. Dep. Agric. For. Ser. Gen. Tech. Rep. PNW-145. 17pp.
———. 1986. Rangeland vegetative succession: implications to wildlife. Rangelands 8:157-159.
———. 1987. Riparian reminiscences. Rangelands 9:71-74.
Knoph, F. L., and T. E. Olson. 1984. Naturalization of Russian-olive: implications to Rocky Mountain wildlife. Wildl. Soc. Bull. 12:289-298.
Lokemoen, J. T. 1973. Waterfowl production on stock-watering ponds in the northern plains. J. Range Manage. 26:179-184.
Mack, G. D., and L. D. Flake. 1980. Habitat relationships of waterfowl broods on South Dakota stockponds. J. Wildl. Manage. 44:695-700.
Maser, C., J. W. Thomas, I. D. Luman, and R. Anderson. 1979. Wildlife habitats in managed rangelands—The Great Basin of Southeastern Oregon: Manmade habitats. U.S. Dep. Agric. For. Serv. Gen. Tech. Rep. PNW-86. 40pp.
Medin, D. E., and W. P. Clary. 1990. Bird populations in and adjacent to a beaver pond in Idaho. U.S. Dep. Agric. For. Serv. Res. Pap. INT-432. 6pp.
Miller, D., E. L. Boeker, R. S. Thorsell, and R. R. Olendorff. 1975. Suggested practices for raptor protection on powerlines. Raptor Res. Found. 19pp.
Monsen, S. B., and N. Shaw, editors. 1983. Managing Intermountain rangelands: Improvement of range and wildlife habitats. U.S. Dep. Agric. For. Serv. Gen. Tech. Rep. INT-157. 194pp.
Mraz, D., and E. L. Cooper. 1957. Reproduction of carp, largemouth bass, bluegills and black crappies in small rearing ponds. J. Wildl. Manage. 21:127-133.
Mundinger, J. G. 1976. Waterfowl response to rest-rotation grazing. J. Wildl. Manage. 40:60-68.
Neff, D. J., and N. G. Woolsey. 1979. Effect of predation by coyotes on survival of antelope fawns on Anderson Mesa. Ariz. Game and Fish Dep. Spec. Rep. 8. 36pp.
Nelson, M. W., and P. Nelson. 1976. Power lines and birds of prey. Idaho Wildl. Rev. 28:3-7.
Ryan, M. R., R. B. Renken, and J. J. Dinsmore. 1984. Marbled godwit habitat selection in the northern prairie region. J. Wildl. Manage. 48:1205-1218.
Sanderson, H.R., T. M. Quigley, E. E. Swan, and L. R. Spink. 1990. Specifications for structural range improvements. U.S. Dep. Agric. For. Serv., Pac. Northwest Res. Stn., Gen. Tech. Rep. PNW-250. 120pp.
Scott, M. D. 1992. Buck-and-pole fence crossings by 4 ungulate species. Wildl. Soc. Bull. 20:204-210.
Scotter, G. W. 1980. Management of wild ungulate habitat in the western United States and Canada: A review. J. Range Manage. 33:16-27.
Severson, K. E. 1990. Summary: Livestock grazing as a wildlife habitat management tool. J. Range Manage. 43:3-6.
Sherrets, H. D. 1989. Wildlife watering and escape ramps on livestock water developments: suggestions and recommendations. U.S. Bur. Land Manage., Idaho Tech. Bull. 89-4. 30pp.
Spillett, J. J. 1965. The effects of livestock fences on pronghorn antelope movements. M.S. Thesis, Utah State Univ., Logan. 138pp.
Stinnett, D. P., and D. A. Klebenow. 1986. Habitat use of irrigated lands by California quail in Nevada. J. Wildl. Manage. 50:368-372.
Szaro, R. C., and M. D. Jakle. 1985. Avian use of a desert riparian island and adjacent shrub habitat. Condor 87:511-519.
Taylor, D. M., and C. D. Littlefield. 1986. Willow flycatcher and yellow warbler response to cattle grazing. Am. Birds 40:1169-1173.
Thomas, J. W., C. Maser, and J. E. Rodiek. 1979. Wildlife habitats in managed rangelands—the Great Basin of southeastern Oregon: Riparian zones. U.S. Dep. Agric. For. Serv. Gen Tech. Rep. PNW-80. 18pp.
Tsukamoto, G. K. 1990. Proc. of the wildlife water development symposium, Las Vegas, Nevada; November 1988. U.S. Bur. Land Manage., Nev. Dep. Wildl., and Nev. Chap. Wildl. Soc. 192pp.
Uresk, D. W., and K. Severson. 1988. Waterfowl and shorebird use of surface-mined and unfenced livestock water impoundments on the northern Great Plains. Great Basin Nat. 48:353-357.
Urness. P. J. 1990. Livestock as manipulators of mule deer habitats in northern Utah: Can livestock be used as a tool to achieve wildlife habitat? U.S. Dep. Agric. For. Serv. Gen. Tech. Rep. RM-194. 40pp.
Vallentine, J. F. 1971. Range development and improvements. Brigham Young Univ. Press, Provo. 516pp.
Welch, E. B., and R. C. Ball. 1966. Food consumption and production of pond fish. J. Wildl. Manage. 30:527-536.
Whyte, R. J., and B. W. Cain. 1981. Wildlife habitat on grazed or ungrazed small pond shorelines in south Texas. J. Range Manage. 34:64-68.
Wilson, L. O., and D. Hannans. 1977. Guidelines and recommendations for design and modification of livestock watering developments to facilitate safe use by wildlife. U.S. Bur. Land Manage. Tech. Note 305. 20pp.
Yoakum, J., W. P. Dasmann, H. R. Sanderson, C. M. Nixon, and H. S. Crawford. 1980. Habitat improvement techniques. Pages 329-410 *In* S. D. Schemnitz, ed. Wildlife management techniques manual, Fourth ed. The Wildlife Society, Inc., Washington, D.C.

Chapter 23
DAMAGE TO RANGELAND RESOURCES

Walter E. Howard

INTRODUCTION

Most wild animals that compete with livestock for range forage or actually damage rangelands do so because they benefit from changes in the plant communities brought about by livestock grazing. When habitats are altered to favor livestock grazing, some native vertebrate species almost inevitably become ecological dislocates (Howard 1985). However, the rodent and rabbit species that become pests do so because they can adapt to the modified habitats. These species also have high resilience and promptly recover after control operations.

In disturbed environments, management is necessary to improve the welfare of some species, and to protect wildlife species that have had their habitats altered. Such management is needed to prevent resource depredations, and for other benefits of humans.

The impact of excessive grazing of livestock on the vegetation and soil of rangelands is well known, but less understood is the ecology of wild animals on rangelands that are also grazed, hence modified, by livestock. I attempt to show how some native animals which have attained a natural balance with the vegetation and soils of pristine rangelands can become a liability when grazing by domestic livestock modifies such lands or when exotic plants are intentionally or accidentally introduced. I will also discuss the best procedures for correcting the undesirable results of these habitat alterations.

Wildlife damages rangelands by directly competing with livestock for forage, causing soil erosion, transmitting diseases (Fowler 1985, Greenhall, 1985), and producing excessive livestock predation. Early homestead regulations required tilling a portion of every claim, regardless of the suitability of the soil for such practices. As a result, the worst range problems were from wind erosion rather than from rodents, although rodents caused additional water erosion (Fig. 1). Too often the desperate economic struggle of landowners to survive prevented any consideration of good stewardship of the land. Also, overgrazing, especially during the drought years of the 1920s and 1930s devastated much land.

Humans have modified plant and animal communities, at times unintentionally, by introducing exotic animals and plants that have not had time to evolve an adaptive coexistence, and it

FIGURE 1. Once native trees have been removed, some soils are susceptible to subsurface erosion when rainwater gets channeled down either ground squirrel (*Spermophilus* spp.) or pocket gopher (*Thomomys* spp.) burrows. Once the top caves in gully formation occurs with the next storm.

is the composition of the vegetation that determines the kind and density of vertebrate fauna that can live there. Pests, by definition, do not exist in natural areas, only where such animals cause problems to people. For example, when Europeans introduced domestic livestock into New Zealand, as well as European rabbits (*Oryctolagus cuniculus*), 9 species of deer, the Australian brush-tailed possum (*Trichosurus vulpecula*), and other mammals, they caused extreme environmental instability, requiring costly animal control that continues today (Howard 1958, 1965). Some tree species were largely eliminated from forests and national parks because this endemic vegetation of New Zealand had evolved in absence of any grazing or browsing mammals. Environmental stability depends on the maintenance of a balance between the vegetation and its consumption by herbivorous animals, including livestock. Because many of the plants and animals found today on ranges are not native to rangelands, it is necessary that they be managed responsibly.

ECOLOGY OF WILDLIFE CONFLICTS

Rangelands are often seeded with non-native plants to raise the carrying capacity for livestock and to help prevent erosion where the range is grazed. Similarly, grazing by livestock can alter the makeup of plant communities, which can favor certain species of rodents and rabbits that can then further alter the composition of the forage (Bond 1945) and damage the range.

Any change in species composition of the vegetation, especially if it includes exotic plants, may have profound effects on the densities and species makeup of rodents and other wildlife populations. An ecosystem has many hierarchical levels and is like a quasiorganism with constant changes in the production, storage, modification, utilization, and loss of energy. However, when humans modify the environment, most vertebrate species find the new habitat conditions less suitable and decline, at least temporarily, in numbers, often even becoming locally extinct. But a few may quickly adapt and become pests if they have sufficient genetic plasticity to do so.

If conditions are favorable for species of rodents and lagomorphs capable of significantly competing with livestock, the carrying capacity for livestock will of course be reduced by their activities. Grazing with livestock (Williamson 1992), the introduction of exotic animals (Neele 1992), and the establishment of exotic plants are usually the key causes of environmental disruptions. California grasslands illustrate these phenomena as > 79% of the herbaceous grassland vegetation is now comprised of introduced bromes and other grasses. To maintain the forage species considered more desirable to livestock, the ranges must be grazed by livestock (Bentley and Talbot 1951). If left ungrazed, the forage will be dominated by tall grasses, such as ripgut brome (*Bromus rigidus*) and slender wildoats (*Avena barbata*), whereas under light to close grazing by cattle an earlier stage of succession is maintained with more clovers (*Medicago* spp. and *Trifoluium* spp.) and filaree (*Erodium* spp.) (Bentley and Talbot 1948).

Pristine conditions are not always necessary to obtain the greatest biological diversity in rangeland ecosystems. Species enrichment can be obtained by modifying habitats to accommodate a greater diversity of native and exotic species. There is a big difference between the terms "biological diversity" and "natural biological diversity." The latter refers to original native occupants of a particular area. Due to introductions, all land masses today have a greater biological diversity than a few centuries ago despite the tragedy of many recent extinctions.

Experimental evidence indicates that distantly related taxa (e.g., desert rodents and harvester ants [*Veromessor* spp.]) can compete for limited seed resources, and thus play a major role in the organization of grazed ecological communities (Brown et al. 1979). Rodents and birds, such as quail (*Lophortyx* spp.) and crowned sparrows (*Zonotrichia* spp.), sometimes influence the effects of prescribed burning by consuming all the germinating annual forage for up to about 1 m beyond the outer edge of brush plants that provide cover. Often these bare strips of ground between patches of brush prevent grass from carrying the fire to other brush plants. However, whenever there is a late spring rain, the growth of the exotic annuals becomes too much for the animals to consume.

Most species of large and small native animals do not die from wild or controlled fires (Howard et al. 1959). The main factors responsible for changes in population densities after a fire are in response to alterations of habitat (Lawrence 1966). Because most rodents and harvester ants survive range fires, any reseeding should be timed to keep the interval between seeding and germination as short as possible or rodents and ants may gather the seeds. Animal damage is also much less with drilled seedings than when seeds are broadcast. Many species of small rodents, birds, and harvester ants can prevent successful broadcast reseeding of rangelands following fire, especially if climatic conditions are not favorable for the seeds to germinate shortly after they are broadcast (Howard 1950). The degree of damage by rodents and birds was demonstrated by placing cages made of 6.4-mm mesh hardware cloth at various sites within seedings. In some chaparral areas only seeds under the sire cages germinated (Fig. 2). Wire cages that had openings large enough for rodents to enter also had the seeds taken. The cages, of course, were not effective where harvester ants were present (Howard 1962*a*).

Once a habitat is modified, invaders can more easily move in. Exotic immigrants may flourish and spread rapidly as happened on California rangelands and with cheatgrass (*Bromus tectorum*) in the Great Basin region. In contrast to the spring and summer rains in the Great Plains that produce rhizomatous sod grasses, elsewhere the western bunchgrasses go dormant by mid-July. In the Intermountain West, large hooved herbivores like buffalo (*Bison bison*) obviously were never common, because this vegetation could not tolerate being trampled heavily. Also, it was easily plowed in contrast to the sod of the Great Basin (Mack 1984).

In the high mountain meadows in Utah, Shupe and Brotherson (1985) found secondary succession of increaser and invader plant species shifted differentially when areas were grazed by sheep or cattle. The cattle allotment was predominantly mountain bluebell (*Mertensia ciliate*) with tall forbs, whereas on the sheep allotments vegetation was predominantly smallwing sedge (*Carex microptera*) with low forbs, perennial grasses, and sedges.

FIGURE 2. After a broadcast seeding of grasses and legumes in a stand of chamise (*Adenostoma fasciculatum*) at the University of California Hopland Research and Extension Center, the only seeds which grew into plants were those protected from small rodents and birds by being covered with a wire-mesh cage (Pasture and Range Research Techniques, Comstock Publ. Assoc., New York, N.Y. 1962). Photo by W. E. Howard.

By comparing a 3,160-ha range in southeastern Arizona where livestock have been excluded since 1968 with an adjacent grazed area, there was a significant difference in the number of species of birds and rodents between the areas. Grazing appeared to favor a greater number of birds but there were fewer rodent species. The difference in habitat conditions determined species preference (Bock et al. 1984).

Even though most rangelands by necessity must be managed for multiple use, dual use of rangeland by livestock and wildlife is often compatible when livestock grazing is carefully managed and wildlife needs are considered (Holechek et al. 1982). Where game populations are managed for profit, it then is often financially easier to reduce the number of livestock.

Any modification of livestock numbers may affect rodent numbers. An excellent example of how deferred grazing can profoundly reduce black-tailed prairie dog (*Cynomys ludovicianus*) numbers in Kansas is provided by Snell (1985). When grazing was eliminated during June, July, and August from 1976 to 1980, the increase in plants reduced the size of the dog town from 44.52 ha to 4.86 ha. When a few cattle (5% of stocking capacity) were allowed to graze the area during the plant growing season in 1981 and 1982, the dog town grew to 8.1 ha. In 1983, with deferred grazing again in that pasture, by fall the population of the prairie-dog town declined to 2.31 ha.

SMALL MAMMALS

Rodents (order Rodentia), moles (Insectivora), and rabbits and hares (Lagomorpha) are probably the world's major direct wildlife competitors of livestock for forage (Marsh 1985), although ungulates can locally also do considerable damage on the range, in pastures, and to haystacks. In addition to competing with livestock for forage, rodents create problems by burrowing. The mounds cover forage and the burrows create hazards to animals and equipment and often contribute to soil erosion (Chase et al. 1982).

"Although rodent activities can markedly affect species composition and may induce erosion by channeling water, cultivation of the soil by rodents may also be beneficial. Overgrazing by livestock emphasizes the negative effects of rodents . . ." but good erosion control includes controlling rodents (Ratcliff 1985). What is often overlooked is that if humans do not want to use the forage that is taken by rodents and other animals, then "damage" is of course going to be nil. The condition of a range is defined as the current productivity of a range relative to what that range is naturally capable of producing. All plants are harvested one way or another, and it is the charge of the range manager to use the renewable resources responsibly.

Concerns about small mammal damage to the productivity and well-being of grazed rangelands are still much as people viewed them 60 years ago.

> Rodents are an important part of the biotic communities which make up our grazing ranges. Often, unfortunately, rodents do not fit into the requirements of good range management. Since some

> forage-destroying rodents may consume up to 80 percent of range forage, it will be necessary on many ranges to determine how much forage is being consumed by uncontrolled rodents before the carrying capacity by livestock can be satisfactorily determined. In sample plot work, in artificial reseeding, and in natural range revegetation, the rodent pressure is of widely varying significance depending on the season. In years of deficient moisture it is far more serious than in good years (Taylor 1930).

The damage effects of rodents on rangelands also grazed by livestock are common in all parts of the world. This seems especially true where rainfall is low. According to Gupta (1978), when native rodents in India occur at densities >400 ha, they can consume all the grass on the arid lands.

As Taylor (1930) further states: "We lack methods of determining the true significance of rodent pressure on the soil. The burrowing species stir up the soil, increase its porosity, tend to decrease the size of its particles, augment aeration and water percolation, incorporate large quantities of organic matter, tend to deepen worked portions, deposit feces, urine, and their own dead bodies." The ecological significance of these factors, however, is not known.

Pocket Gopher

The principal burrowing mammals in the United States are pocket gophers (*Thomomys* spp., *Geomys* spp., *Pappogeomys castanops*) of the order Rodentia and 5 genera of moles of the order Insectivora. An interesting question concerning soil formation in northeastern United States is how much has the absence of pocket gophers been a factor? Can uneaten vegetation, prolific plant roots, the myriad of bacteria, protozoans, worms, crustaceans, arachnids, insects, and other small animals together be more effective in soil formation than pocket gophers (Howard and Childs 1959, Chase et al. 1982)? The effect of gophers varies with density, nature of habitat, season, and land use practices (Buechner 1942, Moore and Reid 1951, Ward and Keith 1962, Richens 1965, Hansen and Ward 1966, Vaughan 1967, Turner 1973, Turner et al. 1973, Laycock and Richardson 1975).

Pocket gophers can be responsible for accelerated soil erosion, especially on overgrazed ranges (Ellison 1946). On hillsides or certain soil types, once the woody vegetation has been removed, the slopes are susceptible to land slippage during above-normal wet weather. The hypothesis that water being channeled through rodent burrows, mainly those of pocket gophers and ground squirrels, contributes to these land slips has apparently not been confirmed. But in these types of soils once water does get channeled down a ground squirrel (Longhurst 1957) or pocket gopher burrow, subsurface erosion results (Fig. 1). Some of these tunnels become large enough for a person to enter. However, once the top caves in, overland flow is then responsible for further erosion and gully formation.

Prairie Dog

Koford (1958) and Bonham and Lerwick (1976) found that black-tailed prairie dogs favor a heterogeneous vegetation mixture in the general vicinity of a prairie dog town. Bonham and Lerwick (1976) observed that the number of plant species was greater in the 5 towns and that even though prairie dogs altered the species composition to some extent, ". . . it cannot be said that prairie dogs are always destructive to rangelands . . ." because they encouraged ". . . an increase in plants which are more tolerant to their grazing . . ." although they did decrease the primary productivity. In the same area in Colorado, Hansen and Gold (1977) found that "The amount of above ground herbage eaten and made unavailable because of soil distribution by prairie dogs and cotton tails (*Sylvilagus auduboni*) was about 24% of the total potential production."

In an intensive study of black-tailed prairie dog diets and forage relationships in western South Dakota, Uresk (1984) demonstrated that the highest preference was for uncommon perennial grasses, such as ring muhly (*Muhlenbergia torrevi*), green needlegrass (*Stipa virdula*), and sand dropseed (*Sporobolus cryptandrus*). High densities of prairie dogs could have a high negative impact on grass and grasslike plants, as they make up 87% of the prairie dog's diet from March through September.

In a comparison of forage between pastures with steers and prairie dogs and steers only, O'Meilia et al. (1982) found blue grama (*Bouteloua gracilis*) and sand dropseed (*Sporobolus cryptandrus*) were significantly reduced on pastures with prairie dogs. However, there was no difference in forb production between the pastures.

Uresk (1987) provides a short review of the relation of black-tailed prairie dogs and control programs to effects on vegetation with and without livestock. In an economic analysis of black-tailed prairie dog control in the Contra Basin of South Dakota, there was a negative annual benefit in AUMs gained. The increase in cattle forage from prairie dog control was estimated at 51 kg/ha/year, with the largest increases in blue grama (*Bouteloua gracilis*)-buffalograss (*Buchloe dactyloides*), and western-crested wheatgrass (*Agropyron smithii*) (14 and 18 kg/ha, respectively). This means 7.2 ha of initial prairie dog control was required to gain 1 AUM/year for the life of the treatment (Collins et al. 1984).

In western South Dakota, Uresk (1985) showed that controlling prairie dogs did not result in a positive increase in forage production without or with cattle grazing, even after 4 years. Uresk (1985) concluded that total exclusion from all herbivores may be necessary for >9 years to increase forage production when the range is in a low condition class.

California Ground Squirrel

On open range and in pastures, high populations of ground squirrels (*Spermophilus* spp.) can become serious competitors with livestock for forage in California (Grinnell and Dixon 1918). They calculated that 200 ground squirrels consume the same amount of range forage as a 454-kg steer. Fitch and Bentley (1949), studying the effects of range rodents upon the exotic forage cover at the San Joaquin Experimental Range in California, found that 6 male ground squirrels (*S. beecheyi*) confined to a 0.2 ha enclosure decreased potential forage yield by 240 kg, >10 times the amount the squirrels may have eaten. The results also suggested that natural field populations of ground squirrels, pocket gophers, and kangaroo rats (*Dipodomys heermanii*) in some pastures of the station might well be reducing the annual herbaceous forage crop by >33%.

Fitch (1948) found at the San Joaquin Experimental Range that ground squirrels are highly selective in their diet, feeding part of the year exclusively on plants used by cattle. After seed germination in the fall and through the winter months, the plants grow slowly, and competition for forage between rodents and livestock on the California annual-type forage is most critical. Two pastures, 12.76 and 19.59 ha, after being adjusted in size during 2 years of calibration to be equal in forage production, were selected to measure the degree that the squirrel populations competed with livestock for the exotic, shortgrass forage during the critical winter green forage period (Howard et al. 1959). By removing the squirrels after 2 years of calibration, 10 heifers on the pasture without ground squirrels during the next 2 years averaged a greater daily gain of 0.47 and 0.34 kg, respectively, than the 10 heifers on the pasture with squirrels. This is an average gain of 33.57 kg/heifer during the winter short-grass season after ground squirrels were removed. Without any grazing by livestock, the dominant plant species become less attractive to ground squirrels. Ground squirrels thrive on grazed California rangelands, the vegetation of which is composed almost entirely of Mediterranean annual plants, hence a highly modified environment compared to most grazing lands in the United States (Bentley and Talbot 1948).

Kangaroo Rat

Sometimes the population density of most or all of the rodent species present will fluctuate in synchrony due to some major environmental factor that affects plant production, such as a drought. At other times the rodent species will fluctuate independently, as some factors have different effects on different species. Often there is some overlap in selection of diet but a difference in staple diet, thus allowing a more complete use of the food resources (Wood 1969). For this reason, if there are >1 rodent species present and in significant biomass, their combined effect can be quite serious in the degree of competition to livestock. Wood (1969) found that the Ord's, Merriam's, and banner-tailed kangaroo rats (*Dipodomys ordi*, *D. merriami* and *D. spectabilis*), in climax areas, kept 10% of the area out of vegetation production by denuding the areas around their burrow mounds.

The seed-collecting and caching by kangaroo rats can accentuate the loss of forage. In a 0.1-ha enclosure study at the San Joaquin Experimental Range, California, 79 kangaroo rats (*D. heermanni*)/ha decreased the forage production over a 7-month growing season by 5.26 kg dry weight/rat (Fitch and Bentley 1949). Vorhies and Taylor (1922) found that each banner-tailed kangaroo rat stored about 1.8 g of grass seeds, crowns, and other edible portions during summer and fall. When the density of large-seeded perennial grasses is low, most can be consumed by these rats (Reynolds 1950).

Voles (Meadow Mice)

Voles (*Microtous* spp.), which are primarily diurnal, require dense, lush cover for protection. They are rarely abundant on grazed lands. Extensive destruction of big sagebrush (*Artemisia tridentata*) occasionally results from vole outbreaks. I witnessed it in northeastern California in 1957-58 and Mueggler (1967) reported the same in southwestern Montana during winter 1963-64. During an irruption voles can reach densities of >7,000-12,000/ha, but only in agricultural crops, which they then decimate. The number of voles on rangelands during peak years is much less so range plants, even though damaged, usually survive. Jameson (1958) calculated that a population of 247 meadow mice/ha, which is not a high population, will consume 74 kg equivalent of dry alfalfa/ha/month, or about 519 kg during a 7-month growing season.

Woodchuck (Marmot)

The genus *Marmota* has 6 species in North America. They are the largest of the squirrel family (Sciuridae). The eastern species (*M. monax*) is called woodchuck, while the 5 western species are known as marmots.

Woodchucks and marmots are significant pests of cultivated pastures and other growing crops, consuming 0.45 kg of green vegetation a day (Eadie 1954). As with moles and pocket gophers, their soil mounds cover vegetation and may damage mowing equipment.

Moles (Insectivora)

Moles belong to the order Insectivora, along with shrews, and are not a gnawing animal like the order Rodentia. They are a true fossorial animal and feed mostly on insects, and they cannot burrow in dry, hard soils. Their main problem to livestock grazing occurs in hayfields and pastures. In Oregon Kuhn (1979) found that they can reduce forage production by 50%. In pastures that are mowed, mole mounds break or dull the mowers or cutting equipment. Their tunnels can channel water, causing erosion.

Hares and Rabbits (Lagomorpha)

Like some species of rodents, jackrabbits also seem to be more abundant, but not restricted to, pastures that have been grazed by livestock (Vorhies and Taylor 1933, Taylor et al. 1935, Arnold 1942, Evans et al. 1970). Vorhies and Taylor (1933) calculated that 148 black-tails (*L. californicus*) ate as much forage as 1 cow. Goodwin (1960) estimated that 7 jackrabbits equaled 1 ewe, whereas Currie and Goodwin (1966) found that 6 black-tailed jackrabbits could destroy as much forage as 1 sheep. Bear and Hansen (1966) estimated that 15 white-tailed jackrabbits (*L. townsendi*) consumed as much forage as 1 sheep. Jackrabbits are extremely destructive to rangeland forage when the population reaches a plague density at the peak of its population cycle. Damage to pastures and rangelands by European rabbits (*Oryctolagus cuniculus*) in Australia, New Zealand, and other parts of the world is classical. Like the California ground squirrel, European rabbits do not thrive in dense herbaceous vegetation (Fig. 3). They thrive best with some bare ground or turf conditions (Howard 1957). Thus, they are favored by habitat modification from burning, grazing, rural roads, railroads, shallow soil, or some other factors that create open areas.

FIGURE 3. The bare patch indicates presence of the last European rabbit(s) to survive in this New Zealand paddock. After it was fenced from sheep the rank grass crowded out the rabbits, which apparently can not survive in wet lush forage (Howard 1958).

One of the reasons rabbits can quickly become abundant on grazed land is because they are so prolific in favorable habitat, with females becoming pregnant while still nursing, producing 25 or more young/year (Howard 1958).

DEER

According to Kie et al. (1991), in an intensive study of the interrelationships of cattle and mule deer (*Odocoileus hemionus*) in the Sierra Nevada of California, deer spent only 24% of their time feeding where cattle had not grazed that season, 31% where there was moderate cattle stocking, and 44% where there were heavy stocking rates. As herbaceous forage levels declined in meadow- riparian and aspen habitats from cattle grazing and trampling, deer had to spend more time feeding. As Kie (1991) points out, consumption of forbs by deer declined as did overall diet diversity and also the availability of hiding cover for fawns. The deer also made habitat shifts (Loft et al. 1991). Bowyer and Bleich (1984) found significantly fewer mule deer where cattle grazing occurred. Dusek (1975) found significant dual usage of vegetation types only during late summer when both used the silver sagebrush type in the bottom of major drainages, but specific sites on this type used by deer seldom were used simultaneously by cattle.

Many winter deer ranges in the foothills of the northern Rocky Mountains are close to nonirrigated wheat fields or fields recently seeded to perennial grasses and forbs through the Conservation Reserve Program. These private ranges are critical factors for mule deer and elk (*Cervus elaphus*), which make extensive use of them (Thomas and Irby 1991). In winter big game can compete with livestock for forage and raid haystacks.

Axis (*Axis axis*) and fallow deer (*Dama dama*) were released on Point Reyes Peninsula, California, in the 1940s. By 1967, before it became part of Point Reyes National Seashore, ranchers began shooting the animals because of competition with cattle for forage. After studying the deer-cattle relationship, Wehausen and Elliot (1982) concluded that both species of deer actually benefited from vegetation changes associated with livestock grazing. Local ranchers had noted a shift in distribution of the deer and both species expanded their range just into areas grazed by livestock. Fallow deer left areas once cattle grazing was stopped (Wehausen and Elliot 1982). By closely observing tame deer and examining rumen contents of fistulated cattle, mule deer and cattle were shown to compete very little for forage on a central Colorado ponderosa (*Pinus ponderosa*) and bunchgrass range during the spring-summer-fall grazing season (Currie et al. 1977). McMahan (1964) used a tame white-tail deer (*Odocoileus virginianus*) to compare forage selections with those of livestock. Following a decrease in available browse, deer became competitive with sheep for forbs. Competition for grass between deer and livestock appeared important only when grass was in a succulent stage in the spring and no other forage was available in quantity. Deer, like other animals, can modify species composition of the vegetation (Fig. 4).

FERAL ANIMALS

Many kinds of animals have escaped from captivity and are considered feral if they have well-established breeding populations. Many introduced exotic game animals, or other species not originally living in a particular area, have become established, especially on game ranches in New Mexico and Texas. As for any real or potential damage to rangelands from feral animals, some of the important troublesome exotic species are horses (*Equus calallus*), burros or asses (*Equus asinus*), hogs or pigs, goats, European rabbits (Howard and Marsh 1986), and cervids, especially in New Zealand. Unrestrained and vagrant domestic dogs are often a troublesome predator, especially to sheep.

Horses and Burros

Wild (feral) horses and burros have been present in America since the 1950s. Few question how damaging they have been to

FIGURE 4. When elk were excluded in 1936, aspen (*Populus tremloides*) seedlings soon appeared. Later, half of the enclosure was removed and a new portion protected from elk. The results clearly illustrate that this area would be an aspen grove if there had not been an unnaturally high population of elk present. Yellowstone National Park, 1950.

rangelands. The same is true in parts of southern Argentina where horses and burros run free. In many areas, because horses pull up herbaceous plants they feed upon, high populations have been devastating to the vegetation, causing a deterioration of the entire biota. They cause soil damage, accelerate erosion, destroy vegetation, damage springs and water holes used by other native wildlife (Dunn and Douglas 1982), and compete with native wildlife for food and space.

In a study of range relationships of feral horses with wild ungulates and cattle in western Alberta, competition was greatest between horses and cattle, although there was little contemporaneous spatial overlap of horses and cattle (Salter and Hudson 1980). In southern Argentina I saw where feral horses seriously damaged rangelands, when they pulled plants out by the roots.

Under the protection afforded them by the Wild Free-Roaming Horse and Burro Act (Public Law 92-195) enacted in 1971, populations increased rapidly. Most wild horses are found on public lands in Nevada, but other states with >1,000 are California, Colorado, Idaho, Montana, Oregon, Utah, and Wyoming. About 95% of these animals occur on federal land, principally USDI-Bureau of Land Management (BLM) land, but also USDA-Forest Service land.

Goats

Feral goats, especially on many islands, have been destructive to vegetation (Nass et al. 1984, Severson and Debano 1991). I consider an uncontrolled goat population to be like a slow bulldozer. In southern Argentina, after much of the rangelands had been grazed so heavily with sheep that it became uneconomic to continue, goats were incorporated. Once goats removed the remaining shrubs, huge blow-outs were created by the strong prevailing winds off the Andes, which blew all the soil, often to depths >1 m, into the Atlantic Ocean. In many parts of Asia and the Mediterranean I have also observed where goats have contributed to desertification.

Pigs

Feral hogs or pigs are the most prolific, large, free-living mammal in the United States. They can do considerable environmental damage by their rooting and in some areas they cause considerable concern as a disease reservoir for swine brucellosis (*Brucella suis*) and pseudorabies (Wood and Barrett 1979). In addition to destroying so much livestock vegetation by their rooting, they cause accelerated erosion and in New Zealand I noted they often kill ewes.

PREDATORS

The principal livestock predators found on rangelands in the United States are the coyote (*Canis latrans*), domestic dog, bobcat (*Lynx rufus*), golden eagle (*Aquila chrysaetor*), red fox (*Vulpes fulva*), gray fox (*Urocyon cinereoargentens*), and less commonly, grizzly bear (*Ursus arctor*), black bear (*Ursus americanus*), mountain lion (*Felis concolor*), jaguar (*Felis onca*), gray wolf (*Canis lupus*), Canada lynx (*Lynx canadensis*), ocelot (*Felis pardalis*), and feral pig.

Under natural conditions before civilization arrived on the scene and modified the plant-animal relationships on rangelands, predators played the leading role in preventing wildlife from "damaging" the vegetation and causing soil erosion. The natural stability between native predators and prey had reached a degree of balance. The damage to rangelands commonly attributed to wildlife is the change brought about by humans entering the scene and trying to create a new balance coupled with the habitat modifications necessary to support people. The crux of the damage problems is how best to manage the species composition and population densities of both native and exotic flora and wildlife, including domestic livestock.

The economic benefit of controlling coyotes to protect wildlife (Smith et al. 1986) and domestic livestock can be appreciable (Scrivner and Conner 1984, Terrill 1986). Concerning the frequent obvious need for predator control, those who object to controlling predators have a difficult dilemma to resolve. Is it not callous that we have selectively bred most of the innate escape mechanisms from livestock and then literally "toss them to the wolves" to be fed on, even while alive, by coyotes (Fig. 5) or bald eagles (*Haliaeetus leucocephalus*) as observed by McEneaney and Jenkins (1983)? If people take away the natural predator defenses from domestic animals, they have a responsibility to protect them from being killed by predators (Howard 1962*b*, 1991). Surplus killing of prey (i.e., where a predator makes multiple killings in close succession and some of the kills are abandoned without being utilized) is not uncommon with many predators including coyotes and wolves (Miller et al. 1985). When humans provide suitable livestock as prey for predators, we must recognize this means humans may have to control some predators to provide a new balance (Howard 1983*b*). Coyotes can even be a threat to children (Carbyn 1989).

Carnivores primarily feed on flesh of animals they have killed and generally contribute far more carrion to an ecosystem than they consume. The amount of flesh they leave from uneaten parts of prey they have killed greatly exceeds the carrion they may eat that died from other causes (Howard 1983*b*). To think

FIGURE 5. Coyotes attack fleeing sheep by the neck and crush their trachea. Once a sheep is short of oxygen the coyote is able to throw it to the ground. Typically, the coyote immediately starts eating the sheep's small intestines.

that coyotes, wolves, lions, bobcats, and foxes cannot learn to feed on domestic animals is doing these cunning beasts a discredit. What many people do not seem to understand is that these predators are ecologically important in nature to sustaining their prey populations at higher levels than would prevail if there were no predators (Howard 1980). For example, if all the predators of rodents were removed, rodents would then become over-abundant and then markedly decline from other limiting forces such as starvation, diseases, or habitat destruction.

The most sensible explanation concerning the famous irruption of the Kaibab Plateau deer herd during the 1910-20s is that it was the consequence of extreme control, the almost complete elimination of predators, primarily mountain lions, plus also a possible release of food supply for deer following the drastic reduction of overgrazing by sheep, cattle, and feral horses, and protection from traditional hunting (Mitchell and Freeman 1993). Clearly the deer irruption was not just the consequence of predator control as postulated by Leopold (1943), a theory challenged by Caughley (1970).

MANAGING RANGELAND ANIMAL PESTS

In the limited space available no attempt will be made to provide details on how to manage troublesome wild animals. The reader is referred to the local county, state, or federal animal damage control official. Probably the best general reference on animal control methodology is the revised edition of Prevention and Control of Wildlife Damage (Hygnstrom et al. In Press). It is important to note that regulations concerning the methodology of animal control are highly variable in different parts of the country and are frequently changed.

The pocket gopher burrow builder, developed at the University of California, Davis (Kepner and Howard 1960, Kepner et al. 1965, Marsh and Cummings 1977) and later modified for Rocky Mountain Region (Ward and Hansen 1960), revolutionized pocket gopher control on pasture land and some rangelands. However, more economical control methods are still needed for most rangelands where pocket gophers are a serious competitor.

With effective toxicants and baits, prairie dogs are fairly easy to control. If black-footed ferrets (*Mustela nigripes*) may be present, shooting is the best control option available. Shooting, trapping, and using fumigants are usually too expensive, except as a follow-up control after poisoning. Bait acceptance is usually best when green feed is not available. Bait acceptance tests should be done first and prebaiting is necessary when zinc phosphide is the rodenticide used.

Many people would like to find ways of encouraging natural predators to help control troublesome field rodents. Hall et al. (1981) researched establishing artificial raptor perches. Howard et al. (1985) were unable to demonstrate any measurable reduction of rodents even though hawks and owls quickly accepted the perches.

On rangelands the only economical way of controlling ground squirrels is with acute rodenticides on acceptable grain bait applied by hand, from horseback, vehicle, or airplane (Marsh 1967). Fumigants, trapping, shooting, and use of anticoagulant rodenticides in bait boxes are effective but usually too costly for protection of rangelands. When controlling ground squirrels, timing is critical and in spring it is necessary to be sure all squirrels are out of hibernation. In warmer climates where green forage is plentiful when squirrels come out of hibernation, grain baits may have to be delayed until late spring, summer, or fall when seeds are an important part of their diet. But baiting during summer estivation of mature squirrels must be avoided.

On grazed rangelands there seldom is need to control voles (meadow mice), but poison grain is indicated when needed. Woodchuck (marmot) control can be achieved with fumigants, shooting, and poisoned chopped alfalfa, pieces of apple, carrots, or sweet potato.

Moles, due to their primarily insectivorous diet, are not easy to control with poison bait, but Elshoff and Dudderar (1990) report success controlling the eastern mole (*Scalopus aguaticus*) and starnosed mole (*Condylura cristata*) using a chlorophacinone pellet called Orco Mole Bait (Federal registrations pending in 1993).

To control jackrabbits on rangelands, poison baits, hunting during open seasons, and rabbit drives are the only practical methods. Catch or buffer crops around pastures have not been effective. Fencing is good but expensive. When poisoning there is an additional hazard to nontarget species because jackrabbits do not seek out isolated kernels of grain or small pieces of bait. The bait must be placed in small handfuls. Cottontails and brushrabbits (*Sylvilagus* spp.) are usually not a problem and are easier to poison. Every plausible control method has been tried to control European rabbits in Australia and New Zealand (Howard 1958). With the effectiveness of the disease myxomatosis waning, I noted an increase of problems from rabbits in England and Scotland in 1993.

Establishing more liberal hunting seasons is the most economical way of controlling deer and bovids. Fencing, though more costly, is the best and is often used to exclude large herbivores from hay stacks. New Zealand once tried to eliminate all introduced deer, and fencing and shooting proved to be the most effective control measures (Howard 1965).

The control of wild horses and burros is an emotional issue. These animals would be relatively easy to control if hunting were allowed and carcasses could be sold as pet food. For humane reasons and for the welfare of rangelands and other competing native wildlife, the density of these exotic animals in the West should be reduced. They can be rounded up by cowboys and helicopters for removal. Research on birth control is good public relations, but an economic way of using this approach is probably far off, although their breeding potential is low, having a population growth rate of only 18% annually (Garrott and Taylor 1990).

Feral hogs or pigs and feral goats are difficult and costly to control. Hunting can help keep their numbers down. If these animals have much cover, their control can be very difficult. With goats, a combination of toxicants, shooting, and dogs is used (Parks 1990). Where costs permit, subdividing with fences helps (Van Vuren 1992). Pigs are often more difficult because they are so fecund, and using dogs, shooting, poisoning, and trapping are employed (Hone and Stone 1989, McIlroy and Saillard 1989, Choquenot et al. 1990, Hone 1990, Peine and Farmer 1990).

Predator control is too complex to do more than summarize in this chapter. When predation problems develop it is best to seek professional help from federal, state, county, or private wildlife damage control organizations. Many predators are also game animals, hence involve regulations of state wildlife or game and conservation organizations. The national body responsible for predator control is Animal Damage Control (ADC), Animal and Plant Health Inspection Service (APHIS) of the United States Department of Agriculture.

Several breeds of guarding dogs have been successfully used to protect sheep from coyotes (Coppinger et al. 1988, Green and Woodruff 1988). Their use, unfortunately, is not the solution to all coyote problems (Green et al. 1984), and conflicts can occur concerning their behavior to deer and other wildlife, and the incompatibility of such dogs with other predator damage control methods (Timm and Schmidt 1989).

Varying degrees of success in protecting livestock have been obtained by using net-wire and electric fences to exclude coyotes (Shelton and Gates 1987, Nass and Theade 1988). A livestock protection (LP) collar using a solution of sodium monofluoroacetate (1080) has been registered by the United States Environmental Protection Agency (EPA), and it shows minimal primary and secondary hazard to nontarget species (Burns and Savarie 1988). One of the most effective coyote control devices in addition to traps is the M-44 sodium cyanide ejector (Connolly 1988). Using strobe-light/siren devices to repel coyotes has had some positive benefit (Linhart et al. 1984). Trapping and moving depredating eagles is not a practical operational procedure for protecting lambs (O'Gara 1978).

A device for selectively orally delivering toxicants or other substances to coyotes is called the coyote lure operative device (CLOD), using trimethylammonium deconate (TMAD) or other lures (Marsh et al. 1982, Fagre and Ebbert 1988, Stolzenburgh and Howard 1989). Much research was involved in developing the lure TMAD (Scrivner et al. 1985). Attempts to use taste aversion to condition coyotes not to kill sheep was not successful (Burns and Connolly 1985).

The control of coyotes received a severe setback when 1080 was federally banned by Executive Order 11643 in 1972, and no promised alternative control method ever appeared. A good review of the history of predator damage control from 1964 to February 1986 is provided in the Appendix of Wade (1986). The great coyote-1080 conspiracy was perpetuated during the "Coyotegate Years" of 1971-73, but the aborted attempt to drive livestock off federal lands failed (Howard 1983*a*) and the true selectivity of 1080 for coyotes and some rodents was not revealed (Howard and Schmidt 1984).

SUMMARY

That there will be a change in species diversity when a range is grazed is obvious. The balance of nature is always altered when new species (livestock and exotic plants) are introduced, or the habitat is manipulated mechanically. Whenever humans modify the environment, they have a responsibility to manage the ensuing species composition in a responsible way. If humans create pest problems, because some unwanted species can capitalize on the changed environmental conditions, then humans must alter their influences, for example, grazing intensity. We often overlook the fact that, in natural environments, the species present, or their ancestors, have successively eliminated by survival of the fittest all of their major competing species.

Grazing by livestock, or the seeding of exotic plant species, will have a profound effect on which natural species can adapt to these modifications of the environment. When humans wish to use the natural renewable resources of rangelands, we must accept the fact that it is not a gift without responsibility on our part. The California foothill grasslands provide an excellent example, for practically all of the herbaceous plants are now non-native Mediterranean species of annuals. The original perennial bunch grasses are essentially nonexistent. All rangelands in the United States have been altered a great deal, maybe not to the same degree as in California, but any change tends to affect the different species of birds, mammals, and other animals and plants to some degree. The important point is that humans should not expect a grazed rangeland to have the same fauna and flora as occurs on comparable ungrazed sites. By providing an adequate harvesting scheme, for example with hunting, Texas manages quite well a greater than natural diversity of exotic ungulates (Nelle 1992).

Maintaining favorable rangeland habitats is the key to successful livestock grazing. This can be accomplished by rotation grazing, deferred grazing (Snell 1985), changing the density of animals, seeding more desirable species, resting the range, and controlling competitive species. Such livestock and range management schemes can often be the key that prevents significant wildlife damage to rangelands that would otherwise result from habitat modification due to livestock grazing. On other occasions, what is often now needed to prevent wild animal damage, resulting from grazing rangelands, is more effective animal damage control, whether the species are rodents, lagomorphs, predators, or ungulates. When a habitat is modified, as from livestock grazing or seeding, some wild animal species may benefit, even to where they become pests, but most species are affected deleteriously. This is why humans have an obligation to manage the species composition and fauna densities in environments they have altered, and to do what they can toward establishing land harmony where they have altered nature.

Excessive livestock grazing, with no concern for wildlife and other resource values, as was once common (Williamson 1992), now rarely exists on rangelands, because not only is it ecologically destructive, it is unprofitable to overgraze private land and it is no longer tolerated as it once was on most government land. However, a recent report by the General Accounting Office, RCED-92-12, November 1991 (Anonymous 1992), that many may not agree with, claims that livestock grazing on desert lands in the Southwest causes excessive damage to wild plants and animals and is a threat to endangered species. It noted that grazing on the hot dry lands of the Mohave, Sonoran, and Chihuahuan deserts in southern California, Nevada, Utah, Arizona, and New Mexico is not critical to local economies and offers scant reward to the permittees involved (Anonymous 1992). These areas are administered by the Bureau of Land Management.

When humans manipulate the balance of nature to their advantage, it is axiomatic that they will come into conflict with some of the inevitable changes that will occur in the composition and population density of the native fauna and flora. The presence of weeds, rodents, and predators does not necessarily mean the range has been overgrazed by livestock. Sometimes it is because biologists have not yet learned how the rancher can increase productivity without creating these pest problems. Land use is habitat alteration, and when the habitat is changed it usually is inevitable that there will be undesirable factors along with the good. Fortunately, most rodents, insectivores, wild ungulates, and other native vertebrates are not damaging to properly grazed rangelands, and these species, if managed properly, can probably help nature's balance and even help make a livestock operation better off economically.

LITERATURE CITED

Anonymous. 1992. Rangeland management: BLM's hot desert grazing program merits reconsideration. Outdoor News Bull. 46:3-4.

Arnold, J. F. 1942. Forage consumption and preferences of experimentally fed Arizona and antelope jackrabbits. Univ. Arizona Agric. Exp. Stn. Tech. Bull. 98. 86pp.

Bear, G. D., and R. M. Hansen. 1966. Food habits, growth, and reproduction of white-tailed jackrabbits in southern Colorado. Colorado State Univ. Agric. Exp. Stn. Tech. Bull. 90. 39pp.

Bentley, J. R., and M. W. Talbot. 1948. Annual-plant vegetation of the California foothills as related to range management. Ecology 29:72-79.

———, and ———. 1951. Efficient use of annual plants on cattle ranges in the California foothills. U.S. Dep. Agric. For. Serv. Circular 870. 52pp.

Bock, C. E., J. H. Bock, W. R. Kenney, and V. M. Hawthorne. 1984. Responses of birds, rodents, and vegetation to livestock exclosure in a semidesert grassland site. J. Range Manage. 37:239-242.

Bond, R. M. 1945. Range rodents and plant succession. North Am. Wildl. Conf. 10:229-234.

Bonham, C. D., and A. Lerwick. 1976. Vegetation changes induced by prairie dogs on short grass range. J. Range Manage. 29:221-225.

Bowyer, R. T., and V. C. Bleich. 1984. Effects of cattle grazing on selected habitats of southern mule deer. Calif. Fish and Game 70:240-247.

Brown, J. H., D. W. Davidson, and 0. J. Reichman. 1979. An experimental study of competition between seed-eating desert rodents and ants. Am. Zool. 19:1129-1143.

Buechner, H. K. 1942. Interrelationships between the pocket gopher and land use. J. Mammal. 23:346-348.

Burns, R. J., and G. E. Connolly. 1985. A comment on "coyote control and taste aversion." Appetite 6:276-281.

———, and P. J. Savarie. 1988. Large livestock protection collars effective against coyotes. Vertebr. Pest Conf. 13:215-219.

Carbyn, L. N. 1989. Coyote attacks on children in western North America. Wildl. Soc. Bull. 17:444-446.

Caughley, G. 1970. Eruption of ungulate populations, with special emphasis on Himalaya thar in New Zealand. Ecology 51:53-72.

Chase, J. D., W. E. Howard, and J. T. Roseberry. 1982. Pocket gophers. Pages 239-255 *in* J. C. Chapman and G. A. Feldhamer, eds. Wild mammals of North America. Johns Hopkins Univ. Press, Baltimore, Md.

Choquenot, D., B. Kay, and B. Lukins. 1990. An evaluation of warfarin for the control of feral pigs. J. Wildl. Manage. 54:353-359.

Collins, A. R., J. P. Workman, and D. W. Uresk. 1984. An economic analysis of black-tailed prairie dog (*Cynomys ludovicianus*). J. Range Manage. 37:358-361.

Connolly, G. 1988. M-44 sodium cyanide ejectors in the animal damage control program, 1976-1986. Vertebr. Pest Conf. 13:220-225.

Coppinger, R., L. Coppinger, G. Laugeloh, and L. Gettler. 1988. A decade of use of livestock guarding dogs. Vertebr. Pest Conf. 13:209-214.

Currie, P. O., and D. L. Goodwin. 1966. Consumption of forage by black-tailed jackrabbits on salt desert ranges of Utah. J. Wildl. Manage. 30:304-311.

———, D. W. Reichert, J. C. Malechek, and 0. C. Wallmo. 1977. Forage selection comparisons for mule deer and cattle under managed ponderosa pine. J. Range Manage. 30:352-356.

Dunn, W. C., and C. L. Douglas. 1982. Interactions between desert bighorn sheep and feral burros and spring area in Death Valley. Desert Bighorn Counc. Transactions 26:87-96.

Dusek, G. L. 1975. Range relations of mule deer and cattle in prairie habitat. J. Wildl. Manage. 39:605-616.

Eadie, R. W. 1954. Animal control in field, farm and forest. MacMillan, Publ. Co., New York, N.Y. 257pp.

Ellison, L. 1946. The pocket gopher in relation to soil erosion on mountain range. Ecology 27:101-114.

Elshoff, D., and G. Dudderar. 1990. The effectiveness of Orco Mole Bait in controlling mole damage. Eastern Wildl. Damage Control Conf. 4:205-209.

Evans, J., P. L. Hegdal, and R. E. Griffith, Jr. 1970. Methods of controlling jackrabbits. Vertebr. Pest Conf. 4:109-116.

Fagre, D. B., and S. M. Ebbert. 1988. Development and testing of the coyote lure operative device. Vertebr. Pest Conf. 13:235-240.

Fitch, H. S. 1948. Ecology of the California ground squirrel on grazing lands. Am. Wildl. Nat. 39:513-596.

———, and J. R. Bentley. 1949. Use of California annual-plant forage by range rodents. Ecology 30:306-321.

Fowler, M. E. 1985. Livestock diseases caused by wild mammals. Pages 509-535 *in* S. M. Gaafar, W. E. Howard and R. E. Marsh, eds. Parasites, pests and predators. Elsevier Sci. Publ., Amsterdam, Netherlands.

Garrott, R. A., and L. Taylor. 1990. Dynamic of a feral horse population in Montana. J. Wildl. Manage. 54:603-612.

Goodwin, D. L. 1960. Seven jackrabbits equal one ewe. Farm Home Sci. 21:38,39,51.

Green, J. S., and R. A. Woodruff. 1988. Breed comparisons and characteristics of use of livestock guarding dogs. J. Range Manage. 41:249-251.

———, ———, and T. T. Tueller. 1984. Livestock-guarding dogs for predator control: Costs, benefits, and practicality. Wildl. Soc. Bull. 12:44-50.

Greenhall, A. M. 1985. Effects of vampire bats and paralytic rabies on livestock production. Pages 537-553 *in* S. Gaafar, W. E. Howard, and R. E. Marsh, eds. Parasites, pests and predators. Elsevier Sci. Publ., Amsterdam, Netherlands.

Grinnell, J., and J. Dixon. 1918. Natural history of the squirrels of California. Calif. State Comm. on Horticulture Mon. Bull. 7(11 and 12):597-708.

Gupta, R. K. 1978. Management of range resources in the Indian arid zone with reference to rodent control. Indian Natl. Sci. Acad. 44(B-5):312-319.

Hall, T. R., W. E. Howard, and R. E. Marsh. 1981. Raptor use of artificial perches. Wildl. Soc. Bull. 9:296-298.

Hansen, R. M., and A. L. Ward. 1966. Some relations of pocket gophers to rangelands on Grand Mesa, Colorado. Colorado State Univ. Agric. Exp. Stn. Tech. Bull. 88. 22pp.

———, and I. K. Gold. 1977. Blacktail prairie dogs, desert cottontails and cattle trophic relations on shortgrass range. J. Range Manage. 30:210-214.

Holechek, J. L., R. Valdez, S. D. Schemnitz, R. D. Pieper, and C. A. Davis. 1982. Manipulation of grazing to improve or maintain wildlife habitat. Wildl. Soc. Bull. 10:204-210.

Hone, J. 1990. Predator-prey theory and feral pig control, with emphasis on evaluation of shooting from a helicopter. Australian Wildl. Res. 17:123-130.

———, and C. P. Stone. 1989. A comparison and evaluation of feral pig management in two national parks. Wildl. Soc. Bull. 17:419-425.

Howard, W. E. 1950. Wildlife depredations on broadcast seedlings of burned brushlands. J. Range Manage. 3:291-298.

———. 1957. Ecology of rabbits in New Zealand compared with California rodents. New Zealand Ecol. Soc. 5:13-14.

———. 1958. The rabbit problem in New Zealand. New Zealand DSIR Inf. Serv. 16. 47pp.

———. 1962*a*. Rangeland insect and rodent pests. Pages 191-193 *in* G. O. Mott, ed. Pasture and range research techniques. Joint Commission, Comstock Publ. Assoc. Cornell Univ. Press, Ithaca, N.Y.
———. 1962*b*. Means of improving the status of vertebrate pest control. Trans. North Am. Wildl. and Nat. Resour. Conf. 27:139-150.
———. 1965. Control of introduced mammals in New Zealand. New Zealand DSIR Inf. Serv. 45. 96pp.
———. 1980. Teaching the complex biological problems of wild vertebrate populations. Pages 205-212 *in* T. S. Bakshi and Z. Naveh, eds. Environmental Education. Plenum Publ., New York, N.Y.
———. 1983*a*. Viewpoint: The coyote-1080 conspiracy: an aborted attempt to drive livestock off federal lands. Rangelands 5:134-135.
———. 1983*b*. Livestock predators and the balance of nature. Pages 106-113 *in* Yearbook of agriculture, 1983. U.S. Gov. Printing Off., Washington, D.C.
———. 1985. Vertebrate predators, competitors and disease carriers of livestock: introduction and overview. Pages 389-395 *in* S. M. Gaafar, W. E. Howard and R. E. Marsh, eds. Parasites, Pests and Predators. Elsevier Sci. Publ., Amsterdam, Netherlands.
———. 1991 (revised). Animal rights vs. nature. Howard Press, Davis, Calif. 229pp.
———, and H. E. Childs, Jr. 1959. Ecology of pocket gophers with emphasis on *Thomomys bottae mewa*. Hilgardia 29:277-358.
———, R. L. Fenner, and H. E. Childs, Jr. 1959. Wildlife survival on brush burns. J. Range Manage. 12:230-234.
———, and R. E. Marsh. 1986. Implications and management of feral mammals in California. Vertebr. Pest Conf. 12:226-229.
———, ———, and C. W. Corbett. 1985. Raptor perches: Their influence on crop protection. Acta Zool. Fennica 173:191-192.
———, and R. H. Schmidt. 1984. Biological rational for 1080 as a predacide. Vertebr. Pest Conf. 11:138-145.
———, K. A. Wagnon, and J. R. Bentley. 1959. Competition between ground squirrels and cattle for range forage. J. Range Manage. 12:110-115.
Hygnstrom, S. E., R. M. Timm, and G. E. Larson, editors. In Press. Prevention and control of wildlife damage. Univ. Nebraska Coop. Ext. Serv. Lincoln.
Jameson, E. W., Jr. 1958. Consumption of alfalfa and wild oats by *Microtus californicus*. J. Wildl. Manage. 22:433-435.
Kepner, R. A., M. W. Cummings, and W. E. Howard. 1965. Mechanical application of gopher bait. Am. Soc. Agric. Eng. 8:335-337.
———, and W. E. Howard. 1960. Mechanical gopher-bait applicator for pastures and open fields. Calif. Agric. 14:7,14.
Kie, J. G. 1991. Wildlife and livestock grazing alternatives in the Sierra Nevada. Trans. Western Sect., The Wildl. Soc. 27:17-27.
———, C. J. Evans, E. R. Loft, and J. W. Menke. 1991. Foraging behavior by mule deer: The influence of cattle grazing. J. Wildl. Manage. 55:665-674.
Koford, K. B. 1958. Prairie dogs, whitefaces and blue grama. Wildl. Monogr. 3. 78pp.
Kuhn, L. W. 1979. Mole control. Oregon State University, Coop. Ext. Serv. Circular 987. 4pp.
Lawrence, G. E. 1966. Ecology of vertebrate animals in relation to chaparral fire in the Sierra Nevada foothills. Ecology 47:278-291.
Laycock, W. A., and B. Z. Richardson. 1975. Long-term effect of pocket gopher control on vegetation and soils of a subalpine grassland. J. Range Manage. 28:458-462.
Leopold, A. 1943. Deer irruptions. Wisconsin Conservation Bulletin. Reprinted in Wisc. Conserv. Dep. Publ. 34:3-11.
Linhart, S. B., R. T. Sterner, G. J. Dasch, and J. W. Theade. 1984. Efficacy of light and sound stimuli for reducing coyote predation upon pastured sheep. Prot. Ecol. 6:75-84.
Loft, E. R., J. W. Menke, and J. G. Kie. 1991. Habitat shifts in mule deer: The influence of cattle grazing. J. Wildl. Manage. 55:16-24.
Longhurst, W. M. 1957. A history of squirrel burrow gully formation in relation to grazing. J. Range Manage. 10:182-184.
Mack, R. N. 1984. Invaders at home on the range. Natural History 93(2):40-47.
Marsh, R. E. 1967. Aircraft as a means of baiting ground squirrels. Vertebr. Pest Conf. 3:2-6.
———. 1985. Competition of rodents and other small mammals with livestock in the United States. Pages 485-508 *in* S. M. Gaafar, W. E. Howard and R. E. Marsh, eds. Parasites, pests and predators. Elsevier Sci. Publ., Amsterdam, Netherlands.
———, and M. W. Cummings. 1977. Pocket gopher control with mechanical bait applicator. Univ. California, Div. Agric. Sci., Leaflet 1699. 7pp.
———, W. E. Howard, S. M. McKenna, B. Butler, and D. A. Barnum. 1982. A new system for delivery of predacides or other active ingredients for coyote management. Vertebr. Pest Conf. 10:229-233.
McEneaney, T. P., and M. A. Jenkins. 1983. Bald eagle predation on domestic sheep. Wilson Bull. 95:694-695.
McIlroy, J. C., and R. J. Saillard. 1989. The effect of hunting with dogs on the numbers and movements of feral pigs, *Sus scrofa*, and the subsequent success of poisoning exercises in Namadgi National Park, A.C.T. Australian Wildl. Res. 6:353-363.
McMahan, C. A. 1964. Comparative food habits of deer and three classes of livestock. J. Wildl. Manage. 28:798-808.
Miller, F. L., A. Gusen, and E. Broughton. 1985. Surplus killing as exemplified by wolf predation on newborn caribou. Canadian J. Zool. 63:295-300.
Mitchell, J. E., and D. R. Freeman. 1993. Wildlife-livestock-fire interactions on the North Kaibab: A historical review. U.S. Dep. Agric. For. Serv. Gen. Tech. Rep. RM-222. 12pp.
Moore, H. W., and E. H. Reid. 1951. The Dalles pocket gopher and its influence on forage production of Oregon mountain meadows. U.S. Dep. Agric. Circular 884. 36pp.
Mueggler, W. F. 1967. Voles damage big sagebrush in southwestern Montana. J. Range Manage. 20:88-91.
Nass, R. D., G. Lynch, and J. Theade. 1984. Circumstances associated with predation rates on sheep and goats. J. Range Manage. 37:423-426.
———, and J. Theade. 1988. Electric fences for reducing sheep losses to predators. J. Range Manage. 41:251-252.
Nelle, S. 1992. Exotics: at home on the range in Texas. Rangelands 14:77-80.
O'Gara, B. W. 1978. Trapping and moving depredating eagles is not a practical operational procedure for protecting lambs. Vertebr. Pest Conf. 8:206-213.
O'Meilia, M. E., F. L. Knopf, and J. C. Lewis. 1982. Some consequences of competition between prairie dogs and beef cattle. J. Range Manage. 35:580-585.
Parks, J. P. 1990. Feral goat control in New Zealand. Biol. Conserv. 54:335-348.
Peine, J. D., and J. A. Farmer. 1990. Wild hog management programs at Great Smoky Mountains National Park. Vertebr. Pest Conf. 14:221-227.
Ratcliff, R. D. 1985. Meadows in the Sierra Nevada of California: State of knowledge. U.S. Dep. Agric. Gen. Tech. Rep. PSW-84. 52pp.
Reynolds, H. G. 1950. Relationship of Merriam's kangaroo rats to range vegetation in southern Arizona. Ecology 31:456-463.
Richens, V. B. 1965. An evaluation of control of the Wasatch pocket gopher. J. Wildl. Manage. 29:413-425.
Salter, R. E., and R. J. Hudson. 1980. Range relationships of feral horses with wild ungulates and cattle in western Alberta. J. Range Manage. 33:266-271.
Scrivner, J. H., and J. R. Conner. 1984. Costs and returns of Angora goat enterprises with and without coyote predation. J. Range Manage. 37:166-171.
———, W. E. Howard, R. Teranishi, and D. B. Fagre. 1985. Towards a more effective coyote lure. Rangelands 7:52-54.
Severson, K. E., and L. F. Debano. 1991. Influence of Spanish goats on vegetation and soils in Arizona chaparral. J. Range Manage. 44:111-117.
Shelton, M., and N. L. Gates. 1987. Antipredator fencing. Pages 30-37 *in* J. S. Green, ed. Protecting livestock from coyotes. U.S. Dep. Agric. Res. Serv. Sheep Exp. Stn., Dubois, Idaho.
Shupe, J. B., and J. D. Brotherson. 1985. Differential effects of cattle and sheep grazing on high mountain meadows in the Strawberry Valley of central Utah. Great Basin Nat. 45:141-149.
Smith, R. H., D. J. Neff, and N. G. Woolsey. 1986. Pronghorn response to coyote control—a benefit:cost analysis. Wildl. Soc. Bull. 14:226-231.
Snell, G. P. 1985. Results of control of prairie dogs. Rangelands 7:30.
Stolzenburgh, H. W., and V. W. Howard, Jr. 1989. Activation of liquid bait devices by coyotes in southern New Mexico. Wildl. Soc. Bull. 17:306-312.
Taylor, W. P. 1930. Methods of determining rodent pressure on the range. Ecology 11:523-542.
———, C. T. Vorhies, and P. B. Lister. 1935. The relation of jack rabbits to grazing in southern Arizona. J. For. 33:490-498.

Terrill, C. E. 1986. Trends of predator losses of sheep and lambs from 1940 through 1985. Vertebr. Pest Conf. 12:347-351.

Thomas, T. R., and L. R. Irby. 1991. Winter habitat use by mule deer with access to wheat fields and planted forb-grassland. Wildl. Soc. Bull. 19:155-162.

Timm, R. M., and R. H. Schmidt. 1989. Management problems encountered with livestock guarding dogs on the University of California Hopland Field Station. Great Plains Wildl. Damage Workshop 9:54-58.

Turner, G. T. 1973. Responses of mountain grassland vegetation to gopher control, reduced grazing, and herbicide. J. Range Manage. 22:377-383.

———, R. M. Hansen, V. H. Reid, H. P. Tietjen, and L. A. Ward. 1973. Pocket gophers and Colorado mountain rangeland. Colorado State Univ. Exp. Stn. Bull. 554S. 90pp.

Uresk, D. W. 1984. Black-tailed prairie dog food habits and forage relationships in western South Dakota. J. Range Manage. 37:325-329.

———. 1985. Effects of controlling black-tailed prairie dogs on plant protection. J. Range Manage. 38:466-468.

———. 1987. Relation of black-tailed prairie dogs and control programs to vegetation, livestock, and wildlife. Pages 312-323 *in* J. L. Capinera, ed. Integrated pest management of rangeland: A shortgrass prairie perspective. Westview Press, Boulder, Colo.

Van Vuren, D. 1992. Eradication of feral goats and sheep from island ecosystems. Vertebr. Pest Conf. 15:377-381.

Vaughan, T. A. 1967. Food habits of the northern pocket gopher on shortgrass prairie. Am. Wildl. Nat. 77:176-189.

Vorhies, C. T., and W. P. Taylor. 1922. Life history of the kangaroo rat, *Dipodomys spectablilis spectablilis* Merriam. U.S. Dep. Agric. Bull. 1091. 40pp.

———, and ———. 1933. The life history and ecology of the jackrabbits *Lepus alleni* and *Lepus californicus* in relation to grazing in Arizona. Univ. Arizona Agric. Exp. Stn. Tech. Bull. 49. 117pp.

Wade, D. A. 1986. Predation damage control: 1980 to 1986. Vertebr. Pest Conf. 12:369-386.

Ward, L. A., and R. M. Hansen. 1960. The burrow-builder and its use for control of pocket gophers. U.S. Fish and Wildl. Serv. Spec. Sci. Rep. Wildl. 47. 7pp.

———, and J. 0. Keith. 1962. Feeding habits of pocket gophers on mountain grassland, Black Mesa, Colorado. Ecology 43:744-749.

Wehausen, J. D., and H. W. Elliot, III. 1982. Range relationships and demography of fallow and axis deer on Point Reyes National Seashore. Calif. Fish and Game 68:132-145.

Williamson, L. L. 1992. Identifying shinola. Rangelands 14:81-84.

Wood, J. E. 1969. Rodent populations and their impact on desert rangelands. New Mexico State Univ. Agric. Exp. Stn. Bull. 555. 17pp.

Wood, G. W., and R. H. Barrett. 1979. Status of wild pigs in the United States. Wildl. Soc. Bull. 7:237-246.

Chapter 24
DISEASES OF WILD UNGULATES AND LIVESTOCK

David A. Jessup and Walter M. Boyce

INTRODUCTION

During the latter half of the nineteenth century, as the western United States was explored and developed, cattle, sheep and goats were introduced to western rangelands. The external and internal parasites, bacteria, viruses, and other disease agents they harbored came with them. The modern importation and quarantine laws that may slow or stop spread of disease-causing organisms between countries or continents were not in place and knowledge about treating livestock for diseases and parasites was limited. In the western United States, these animals were grazed on public and private rangeland where they came into contact with native wild ungulates, including white-tailed deer (*Odocoileus virginianus*), black-tailed and mule deer (*O. hemionus*), bison (*Bison bison*), pronghorn (*Antilocapra americana*), elk (*Cervus elaphus*), bighorn sheep (*Ovis canadensis*), and to a lesser extent, moose (*Alces alces*), reindeer or caribou (*Rangifer tarandus*), mountain goat (*Oreamnos americana*), and musk ox (*Ovibos muschates*). These species had their own endemic parasites, bacteria, and viruses. As a consequence, in a variety of locations and under a variety of environmental conditions, disease-causing agents were introduced to new susceptible hosts. In some cases, die-offs of livestock occurred, in other cases wildlife die-offs or local extinctions were recorded. Most of these die-offs occurred before science and veterinary medicine had even discovered and/or named many of the parasites and disease organisms.

In the latter half of the twentieth century, interest developed in ranching exotic wild ungulates in the West. The private land on which this ranching occurs, or is proposed, is often adjacent to public lands that support native wildlife. Once released into suitable habitat, exotic ungulates like axis (*Axis axis*), red (*Cervus elaphus*), and fallow deer (*Dama dama*); mouflon (*Ovis musimon*) and barbary sheep (*Ammotragus lervia*); and oryx (*Oryx gazella*) may be very difficult to control. Although managers may recognize many of the potential disease-causing organisms that may be shared by livestock, native, and exotic wildlife, they have only a few effective medications and methods to control diseases and parasites. The vastness of the land, the limitations of treatment technology, the complexity of the potential host and organism interactions, and legal and financial complications all seem to conspire to limit our disease management options for wildlife.

In the West, land ownership patterns are complex. Much of the land is controlled by various federal agencies such as the United States (U.S.) Forest Service, U.S. Fish and Wildlife Service, Bureau of Land Management, National Park Service, Department of Defense, and the Department of Energy, all of whom have different philosophies and priorities. For practical purposes, the wildlife agencies in the western states share management authority and responsibility for wildlife with the federal agencies that manage wildlife habitat.

Historically, it is fairly clear that some livestock diseases (notably foot and mouth disease, scabies, and several respiratory pathogens shared by wild and domestic ovids) have damaged some wildlife populations. It is also clear that some diseases thought to have evolved with North American wildlife (i.e., anaplasmosis and epizootic hemorrhagic disease) have caused financial losses to livestock owners. Today, some vocal citizen conservation groups blame livestock, occasionally without factual basis, for most of the ills of wildlife. There are strong pressures to limit or eliminate grazing on public lands and also strong financial and social incentives not to do that.

Livestock owners, private game ranchers, and wildlife and land management agencies would generally profit by preventing the introduction and spread of diseases in and between wild, exotic, and domestic animals. Unfortunately, few resources have been devoted to research, preventive medicine, improved diagnosis, and cooperative approaches to management of disease problems. Our objective in this chapter is to improve the understanding of the diseases that may be shared by native and exotic wild ungulates and livestock. We have selected the most important, controversial, and potentially damaging bacterial, viral, and parasitic diseases shared by wild ungulates and livestock. They are organized by type of organism and presented in rough order of currently perceived importance.

Our presentation is based upon personal experiences and the following references (Davis et al. 1981*a*, *b*; Davidson et al. 1981; Foreyt and Jessup 1982; Thorne et al. 1982; Thorne et al. 1992; Onderka and Wishart 1988; Onderka et al. 1988; Foreyt 1989; Silflow et al. 1989; Boyce et al. 1991; Callan et al. 1991; Jessup et al. 1991; Miller et al. 1991; Castro and Heuschele 1992; Drew et al. 1992; Jessup 1993).

BACTERIAL DISEASES

Tuberculosis

Tuberculosis (TB) is a chronic bacterial disease caused by the *Mycobacteria* spp. (mainly *M. bovis*). Tuberculosis can cause infections in cattle and can be a serious health problem in captive members of the deer family. In many species, the disease process is slowly progressive and tubercles are formed as the body attempts to wall off the infection, thus the name. Tuberculosis is a difficult disease to eradicate and small pockets of infection remain in the United States.

Distribution and Hosts. Tuberculosis is found throughout the world and afflicts a very wide variety of animals. In many underdeveloped countries, TB in cattle and subsequent human infection with *M. bovis* is still common. Bovine TB has nearly been eradicated from cattle in the United States by a long standing program of routine test and slaughter. Many TB outbreaks have occurred in captive wildlife. Of cervids, white-tailed deer, fallow, roe (*Capreolus capreolus*), axis (*Axis axis*), sika (*C. nippon*), red deer, and elk have become infected. Zoos certified by the American Association of Zoo Parks and Aquariums (AAZPA) observe strict testing procedures, but unfortunately many small zoos, private game ranches, and auctions do not. As a result there have been a number of recent tuberculosis outbreaks in bison, elk and other cervids, and exotic bovids on game ranches. Outbreaks have occurred in Colorado, Iowa, Wisconsin, Texas, New York, California, Montana, and several Canadian provinces. Where range animals become infected, resident small carnivores (notably badgers [*Taxidea taxus]* in England and possums in Australia) may also become infected and serve as reservoirs.

Transmission, Pathogenesis, and Signs. Tubercle bacteria are shed into exhaled air, in saliva, feces, urine, milk and other bodily fluids, and in the draining exudates from TB abscesses (Davis et al. 1981*a*). These bacteria are hardy and may survive in soil for years. Bacteria usually enter a host by inhalation and ingestion and replicate locally and in regional lymph nodes. Lymph nodes of the neck and mesentery are most commonly infected. In ungulates, pneumonia and tubercles that cause a chronic moist cough and other respiratory signs are seen. The chronic course of the disease may result in emaciation, anorexia, and low grade fever but many infected animals appear outwardly normal (Miller et al. 1991).

Lesions and Diagnosis. Tubercles are often thick walled with a layered, onion skin appearance, but can look like abscesses caused by a variety of common bacteria. Either a Ziehl-Nielsen of Truant's rhodamine-auramine stain is used to identify bacteria as acid-fast bacilli (Thorne et al. 1982). Post mortem diagnosis is based on gross and microscopic lesions and acid-fast staining bacteria that should be confirmed by isolation and characterization of the bacteria.

Diagnosis and screening of populations is done by dermal sensitivity testing. A thickening of skin at the site of an intradermal inoculation with tuberculin at 72 hours post inoculation is considered a positive test. Although the caudal fold test is adequate for cattle, a single cervical test or a comparative cervical test is required for deer, elk, and other wildlife. Even under the best of circumstances the skin tests are not extremely sensitive or specific (Miller et al. 1991). Animals early in infection or very late in infection may give false negatives. False positives are less common, but can result from exposure to *M. avium* or other mycobacteria. A complete post mortem examination should be conducted on all animals that are culled due to dermal sensitivity to tuberculin. A battery of tests, including the ELISA antibody test, are used to screen red deer in New Zealand. The applicability of these screening procedures for North American wildlife is being explored (Thorne et al. 1992).

Management and Control. Elimination of positive herds, and in some cases test and slaughter, remain the only management tools for TB control and eradication. Carcasses should be disposed of on-site, usually by incineration. Herds with tuberculin reactors may be quarantined and retested.

Importance. Tuberculosis is a very persistent and infectious disease. It can cause serious but treatable disease in human beings. Exposure can cause permanent dermal sensitivity conversion. Millions of dollars have been spent to eradicate it from livestock. It is only rarely seen in wildlife and keeping it from spreading to livestock or wildlife is of extreme importance. Many state and federal agencies are tightening regulations on captive-owned wildlife and restricting their movement. If TB becomes more widespread, it will be extremely difficult and expensive to control (Thorne et al. 1992).

Bang's Disease (Brucellosis)

Many species of animals, including man, are susceptible to infection with the various types of *Brucella spp.* bacteria. It causes a high fever that periodically reoccurs (undulant fever of humans), joint disease, abortion and damage to the reproductive system, and in some cases death (Davis et al. 1981*a*). Since 1934, a Cooperative State-Federal Brucellosis Eradication Program has been attempting to eradicate *Brucella abortus* using serologic tests to detect infection, slaughter of infected cattle, and Strain 19 vaccination of calves (Thorne et al. 1992). The incidence of infection in cattle is < 0.5 % nationally. However, brucellosis in herds of elk and bison and a few herds of range cattle make the eventual eradication of this disease problematic.

Other forms of brucellosis occur in reindeer and wild pigs (Drew et al. 1992). If ignored, these remaining pockets of infection could spread and negate the years of effort and millions of dollars spent thus far.

Distribution and Hosts Affected. *Brucella abortus* type 1 is primarily a disease of cattle, but there are foci of infection involving bison and elk, in northwestern Wyoming and Yellowstone National Park. Infection rates of approximately 30% are reported from elk wintering on the National Elk Refuge and Grey's River Feed Ground (Thorne et al. 1982). The bison herds on the National Bison Range in Montana, in Wood Buffalo National Park and Elk Island National Park in Canada are infected also. Infections in moose in Minnesota, Montana, and Alberta, white-tailed deer in Ohio, and coyote (*Canis latrans*) in Texas have been reported (Thorne et al. 1982).

Brucella suis types 1 and 3 have been recovered from wild pigs in California, Georgia, and Florida. In California, several deer from an infected premises were also seropositive, suggesting their potential exposure. *Brucella suis* type 4 infects reindeer in several areas of Alaska and has recently been shown to be transmissible to cattle kept with infected reindeer.

Brucella melitensis is primarily a disease of domestic sheep and goats, and infections of captive or free-ranging wildlife are rather rare. *Brucella ovis* is a venereally transmitted reproductive disease of domestic sheep. The latter 2 types of brucellosis will not be discussed further in this section. Humans can become infected with any type of brucella by consumption of undercooked meat, unpasteurized milk, inoculation into tissue such as when cleaning a carcass, or through contact of the organism with mucous membranes (eyes, nose, and throat).

Transmission, Pathogenesis, and Signs. The oral route of infection (e.g., licking of the fetus, ingestion of fetal membranes and vaginal secretions) appears to be most important for cattle, elk, and presumably for bison. Unfortunately, elk in Wyoming often abort in winter when concentrated on winter feed grounds, exacerbating transmission. Venereal transmission can occur also. In areas where infection rates are high, coyote can become infected by consumption of fetal tissues and may play some role in the maintenance and spread of this disease.

Brucella suis is believed to be transmitted primarily by venereal contact, although the omnivorous nature of wild boar make it possible that oral transmission may also occur. Dead end infections may occur in other scavenging species like raccoon (*Procyon lotor*). A carnivore (primarily Arctic fox [*Alopex eagopus*]) to reindeer cycle occurs in Alaska.

The number of organisms and strains of bacteria to which an animal is exposed, its age, and innate resistance, determine whether exposure results in infection. The bacteria penetrates the local mucous membranes, localizes in regional lymph nodes, and may then spread to other organs and joints causing local, sometimes severe inflammation. In pregnant female animals that produce high levels of erythritol (a carbohydrate) in the uterus, *Brucella* tend to localize and proliferate there (Thorne et al. 1982). This results in damage to the uterine lining, and the placenta, and abortion. In male animals, the bacteria may localize in the testes.

Abortion, premature birth, and neonatal mortality are the most important signs of brucella infection in elk and cattle, but only occasionally are subsequent pregnancies affected. Chronically infected elk and reindeer may show lameness and signs of joint infections, most commonly the fetlock and carpal joints. Bison cows abort or lose calves and male bison may have grossly swollen testicles. Reproductive effects are reported to occur in deer. Moose infected with *B. abortus* become weak, emaciated and die of generalized infection. Wild sows infected with *B. suis* may abort or have nonviable piglets.

Lesions and Diagnosis. In moose with generalized infections, fibrinous pericarditis, pleuritis, peritonitis, and orchitis are the most prominent lesions, although degeneration and necrosis of liver, kidney, spleen, and lymph nodes are seen also. In bison, swollen pus-filled scrotums and degenerated testicles are prominent lesions. The testicular lesions in elk are less severe and may be seen microscopically as mixed to mononuclear inflammatory reaction and focal necrosis.

The lesions in elk, reindeer, and cattle are generally less striking than in bison and moose and have been described extensively. A purulent exudate covering a thickened placenta can be visible following abortion. Microscopically, a necrotizing placentitis with mixed inflammatory cells and sometimes epithelioid and giant cells are visible. This often resolves to a mild chronic endometritis. Lymph nodes of the head and pelvic region may be swollen and inflamed. Swellings in soft tissues over bony prominences (hygromas) with red-brown linings and containing straw colored fluid are also reported. Chronic inflammatory changes are seen in infected joints.

The lesions of brucellosis are characteristic but not definitive. Culture of the organism from animals with typical lesions allows a firm diagnosis. Brucella organisms are fastidious and require enriched media, 10% CO_2, and often bacterial inhibitors. Carcasses and tissues can be frozen to inhibit bacterial overgrowth.

Serologic tests developed for cattle (i.e., the rapid card test or standard plate agglutination) can be used with caution to make a diagnosis. A reaction at 1:50 in elk is interpreted as indicating infection. However, a battery of tests including the Rivanol and complement fixation test has been recommended for screening wildlife. Serosurveys of wild animals captured for a variety of reasons allow wildlife agencies to determine if infected populations exist in their state, and are a relatively inexpensive way for wildlife and livestock interests to cooperate toward the elimination of brucellosis.

Control and Management. Once free-ranging elk, bison, reindeer, or wild pigs become infected, eradication or control of this disease becomes an expensive and difficult, if not impossible, proposition. Infected cattle should not be allowed to use open range where deer, elk, or bison may come into contact with them. Control of brucellosis in wildlife is only possible if eradication of brucellosis in cattle succeeds. The same quarantine procedures that are used for infected cattle should be applied to affected game ranches.

Infected herds of free-ranging elk and bison may need to be culled. Winter feeding may have to be slowly reduced or eliminated. Biobullet vaccination of elk and bison may also be useful management tools. Brucellosis in free-ranging moose and deer appears to be rare and self limiting, so probably does not merit management efforts.

Caribou or reindeer from infected states should be relocated or exported only from closed, negative herds and after 2 negative serologic tests for brucella antibody. Localized shooting programs will be necessary to reduce infection in wild swine populations, but eradication may prove quite difficult.

Importance. Brucellosis can be a serious and life threatening disease of humans. It is an economic disease in the livestock industry and millions of dollars have been spent to eradicate it. To the extent that wildlife may threaten the success of eradication efforts, and that people exposed to infected wildlife may contract the disease, wildlife brucellosis is rather important. Because the infection rate is low, and it is often not fatal in many species of wildlife, it is not a disease that is likely to destroy herds or populations. Brucellosis must not be accidentally spread through relocation of infected wildlife or movement of infected privately owned livestock or wild animals. Routine testing prior to shipment and isolation of new stock prior to releasing them into a herd will help assure this.

Johne's Disease (Paratuberculosis)

Paratuberculosis is an infectious disease of livestock (primarily dairy cattle) that may infect beef cattle, sheep, goats, llamas (*Llama llama*), bighorn sheep, mountain goats (*Oreamnos americanus*), elk, reindeer, and many other wild and domestic ruminants (Thorne et al. 1982). It is caused by the bacteria *Mycobacterium paratuberculosis* and is a chronic disease of the gastrointestinal tract that causes persistent diarrhea and emaciation. Attempts to eliminate it or to limit its spread have met with limited success. The wide variety of hosts it may infect, its persistence in nature (for years in some soils), and long time course from infection, to manifestation, to death, make this a disease of concern. On the other hand, it is not a highly infectious organism and has limited potential to spread widely or rapidly unless moved within infected animals.

Distribution and Hosts Affected. Although it is most often seen in cattle, a wide variety of species can be infected. In free-ranging wildlife, paratuberculosis has been diagnosed in a herd of bighorn sheep, and mountain goats in Colorado. The calves of Tule elk (*C. elaphus nannodes*) released onto a ranch within a National Seashore in California became infected with paratuberculosis (Jessup 1993). Culling of infected elk, and dispersion of the remainder, and removal of cattle appear to have limited the prevalence of disease within the herd, although occasional cases occurred in following years. Fallow and axis deer on this same National Seashore harbored *M. paratuberculosis* and shed it in feces without manifesting signs of disease. A similar situation was reported in white-tailed deer using infected pastures in the eastern United States. Cases in ranch-raised red deer were recently reported under circumstances that could have exposed hundreds of other red deer to infected feces. Captive moose and white-tailed deer have become infected. In zoos and wild animal parks, once established, paratuberculosis has been extremely difficult to diagnose and eliminate, and is capable of infecting a wide variety of species.

Transmission, Pathogenesis, and Signs. Paratuberculosis is transmitted primarily via ingestion of bacteria shed in feces. However, the bacteria have been recovered from milk and congenital infection of domestic calves has been reported. Young animals in contaminated environments are most susceptible. They ingest the bacteria that then multiply in the intestinal mucosa, local lymphoid tissues, and phagocytic cells for several months. As few as 1,000 bacteria may establish an infection in domestic lambs. In individuals and species resistant to infection, no signs of disease are seen and diagnostic tests may be of limited value. Other animals develop mild localized lesions and remain clinically normal, but continually shed organisms into the environment. More susceptible species and individuals , particularly those infected <6 months of age, develop signs of disease from several months to a year or more following initial infection. Signs are related to thickening of the intestinal wall, blockage of lymph channels, and massive enlargement of regional lymph nodes due to chronic inflammation. Fluids and nutrients cannot be absorbed resulting in chronic watery diarrhea without straining, emaciation, and other signs related to malabsorption, and malnutrition. Diarrhea is not as prominent in sheep, goats, and bighorn. Dissemination to the liver and other vital organs may eventually occur. The entire process from infection to death may take as long as 1.5 to 2 years.

Lesions and Diagnosis. The gross changes in organs and tissues are variable. Animals with severe clinical disease may not have severe lesions. Well-developed gross lesions include thickened and folded walls of the distal small intestine, enlarged pale or mineralized lymph nodes, subserosal and mesenteric edema, and thickening and cording of mesenteric lymphatics. Microscopic lesions are primarily those of a chronic diffuse mononuclear inflammatory response, sometimes with giant cells and granuloma formation, but foci of necrosis with mineralization are more common in sheep and goats. Acid-fast stains can be quite dramatic, showing all phagocytic cells stuffed with bacteria. The Ziehl-Nielsen acid-fast stain is relatively common and inexpensive, while the Truant's auramine-rhodamine technique is more dramatic (Thorne et al. 1982). A presumptive diagnosis of paratuberculosis can be reached by finding clumps of acid fast organisms, sometimes within sloughed cells, in fecal smears from live animals that are emaciated and have diarrhea.

Stains of fecal smears or tissue impressions may not detect subclinical carriers or animals with clinical disease that are not shedding large numbers of organisms. A positive culture of feces or tissues (mesenteric lymph nodes, ileocecal junction) is the most sensitive way to reach a definitive diagnosis. Culture must be done on a mycobactin-enriched media (i.e., Harold's egg yolk) and may take 6 to 12 weeks. A commercial polymerase chain reaction technique to detect mycobacteria in feces was developed for use on cattle and can give results in several days. It is considerably less sensitive than culture, however. Enzyme-linked immunosorbent assay (ELISA) and radioimmunoassay (RIA) show some promise as serologic screening tests. Complement fixation and intradermal Johnin skin tests are unreliable in species other than cattle.

Once clinical manifestations are evident, paratuberculosis is invariably fatal. There is no effective therapy or treatment. Mortality rates in infected herds of cattle vary from 1-25%/year, and although such data are not developed for wild animals, similar mortality rates may occur.

Control and Management. Paratuberculosis is a difficult disease to control in captive animals. Animals in early stages of infection and subclinical carriers contaminate the environment with large numbers of relatively robust bacteria. The lack of a rapid and sensitive diagnostic test further complicates management. Culling clinically infected adults, raising young animals away from infected adults, and good sanitation are only partially successful at reducing infection rates. Depopulation, disinfection of the premises, followed by restocking within 1 to 3 years has been effective at some ranches and zoos.

When wildlife become infected, management options are even more limited. Depopulation is seldom realistic as infected and carrier animals can not be easily identified. An infected population of bighorn in Colorado has not been culled, but has declined in numbers and affected individuals are still occasionally reported now, 20 years after the first cases. The gregarious nature of bighorn and their traditional use of bedding, lambing, and foraging areas could help maintain infections. Only clinically affected Tule elk were culled in California and the rate of infection apparently declined to <1%/year but the herd is still infected some 10 years later.

Importance. Prevention of infection is extremely important with Johne's disease. Infected wild animals should not be relocated. Paratuberculosis is difficult to control once established, and owners of wild animals, and stockmen using public rangeland should be encouraged to keep herds free of this disease. Shipment of infected captive wildlife between ranches or zoos should be strongly discouraged. The relatively low infectivity of the organism and low yearly mortality rates make it unlikely that a wild animal population could be severely impacted, but decreased productivity and physical condition, and increased susceptibility to predation could be seen. Feces and carcasses of infected animals should be treated as infectious waste. All applicable state and federal laws and policies for livestock should serve as a basis for wildlife and exotic animal management decisions. This would include basic health management procedures like holding newly acquired animals in pens or corrals and separate from the herd for observation for several weeks.

Pneumophilic Bacteria (*Pasteurella*)

Pasteurella bacteria, primarily *Pasteurella hemolytica* and *Pasteurella multocida*, are major pathogens of wild and domestic hoofstock. There are numerous serotypes and biotypes and *Pasteurella* infections tend either to take a pneumonic or septicemic form. Two basic theories of disease are that *Pasteurella* are present in the tonsils or other tissues and cause pneumonia or septicemia under stressful conditions, and that most species have their own *Pasteurella* and disease generally occurs when they are challenged with more pathogenic *Pasteurella* from another species. These theories are not mutually exclusive.

Distribution and Hosts. *Pasteurella* are common worldwide and most, if not all, species are apparently susceptible to disease, and carrier states appear to be common. Pneumonias and septicemias and more chronic localized infections occur in cattle, domestic sheep, and goats. Pronghorn antelope (*Antilocapra americana*) and bighorn sheep appear to be quite susceptible to *Pasteurella* infections, particularly acute fatal pneumonias. Mule deer and elk appear to be less commonly affected although *Pasteurella* septicemias are reported in elk, deer, bison, and many wild ungulate species.

Die-offs of bighorn sheep from acute bronchopneumonia following association with domestic sheep were first reported in 1982 (Foreyt and Jessup 1982). In two incidents, healthy bighorn sheep herds were in enclosures for long periods of time before deaths occurred. This suggested that the pneumonia was not the result of capture stress or acclimation to new areas. In each case, nose-to-nose contact with domestic sheep was followed by a die-off within weeks. At Lava Beds National Monument, California, all 43 bighorn sheep died. In Washington at the Methow Game Range, 1 ewe of 14 bighorn sheep survived. No sickness or mortality was reported in the domestic sheep. Circumstantial evidence suggested that apparently healthy domestic sheep transmitted pneumophilic bacteria (*Pasteurella multocida* was isolated in the California outbreak) to the bighorn sheep resulting in mortality (Foreyt and Jessup 1982).

Six apparently healthy domestic sheep were placed in with 6 pen-reared bighorn sheep (Foreyt 1989). On the day of introduction, 4 of the 6 domestics had *Pasteurella hemolytica* in their nasal passages, which were not present in the bighorn sheep. All 6 bighorn sheep died over 2 months of *Pasteurella hemolytica* pneumonia, but the domestics remained healthy. This experiment has been repeated under slightly different conditions with the same results. In separate experiments in Canada, researchers also found that *Pasteurella* of domestic sheep were fatal when given to bighorn sheep either by contact or by inoculation (Onderka et al. 1988; Onderka and Wishart 1988). Utah researchers reported that bighorn exposed to domestic and exotic sheep routinely died of *Pasteurella* pneumonias despite attempts at treatment (Callan et al. 1991).

In winter 1987, a major *Pasteurella* pneumonia die-off occurred in the Lostine drainage of eastern Oregon. Contact with domestic sheep was thought to be a factor. Approximately 60% of the Rocky Mountain bighorn sheep (*Ovis canadensis canadensis*) herd died in 3 months. In subsequent years lamb survival was considerably reduced due to pneumonia, only improving gradually over several years. This pattern has been observed in other bighorn herds following pneumonia die-offs.

In winter 1988, an entire herd (approx. 60 animals) of California bighorn sheep (*O. c. californiana*) died of *Pasteurella* pneumonia over 2 months in the Warner Mountains of northeastern California. Again, contact with a stray domestic sheep occurred prior to the incident. Other possible contributors may have been weather stress, and contact with feral goats from which a *Pasteurella* was isolated.

The pulmonary macrophages that guard the upper respiratory tract of bighorn sheep are less capable of killing and processing bacteria than the macrophages of domestic sheep (Silflow et al. 1989). These and other inherited differences in host resistance factors may account for some of the differences in susceptibility between domestic and bighorn sheep.

It is not unusual to recover *Pasteurella* from the respiratory tract of normal, healthy bighorn sheep. These are often an apparently nonpathogenic, nonhemolytic strain. Idaho researchers have presented evidence that bighorn may get pneumonia in the absence of domestic sheep and that the *Pasteurella* cultured from these cases are the same as those they harbor in their own tonsils.

Transmission, Pathogenesis, and Signs. *Pasteurella hemolytica* is capable of causing progressive fatal pneumonia in domestic sheep and cattle once established in the lung. *Pasteurella multocida* may require predisposing viral infections and/or stress mediated immune suppression to cause fatal pneumonia. Once an animal becomes infected large numbers of bacteria are shed in saliva and droplets from the upper respiratory tract. Under moist and mild conditions *Pasteurella* may survive in the environment for weeks to months.

Respiratory infections often affect the anterior lobes of the lung. Some infections are rapidly progressive, the lungs become edematous, the infection becomes systemic, and the animal dies. Other infections progress slower and may become walled off. Abscesses may form and fibrin deposition is prominent. Animals with pneumonic infections often have labored breathing, coughing, and nasal discharge. Some animals may survive long enough to show signs of emaciation. Animals with acute septicemic pasteurellosis show few signs and die suddenly.

Lesions and Diagnosis. Lesions of septicemic infections consist of multiple hemorrhages on serous surfaces; red, wet lungs with froth in the trachea and swollen hemorrhagic lymph nodes of the head, neck, and chest (Davis et al. 1981*a*). The pneumonic forms of infection involve 1 or several lung lobes and may show distinct margins. Abscesses and areas of cavitation and necrosis may be seen in the more anterior and ventral lobes, fibrin exudation may be extensive (classic bronchopneumonia). Bacteria may be deposited in the liver or spleen causing abscess formation.

Although the gross pathologic changes caused by *Pasteurella* are fairly characteristic, isolation of the organism on artificial media is needed for a definitive diagnosis. Typing or "fingerprinting" of the bacteria may allow comparison with bacteria originating from other animals. On smears of affected tissues bipolar shaped rods may be quite evident, but other species of bacteria can also have this appearance.

Management and Control. By the time *Pasteurella* pneumonias are recognized in wildlife, they are very difficult to treat. If given early and at effective doses, penicillins or tetracycline can be an effective treatment in livestock, and in some bighorn sheep. However, some *Pasteurella* are resistant to these and other common antibiotics. New third generation cephalosporins hold promise for treatment of *Pasteurella* pneumonias, and appear to be quite safe.

Many vaccines have been developed, but have not proven to be very valuable, particularly in wildlife. Separation of extremely susceptible species from potential carrier species and maintenance of wild ungulates at or below carrying capacity of their range are the most applicable management tools currently available. The Desert Bighorn Council has documented 25 bighorn sheep die-offs that have occurred subsequent to contact with domestic sheep and have recommended 15 km buffer zone between domestic and wild sheep.

Importance. *Pasteurella* spp. have caused major die-offs of wild animals, in some cases whole herds of bighorn sheep have been destroyed. It is not entirely clear whether these die-offs are due to transmission between species or the result of stressors or both. Probably no disease has caused more conflict between wildlife and livestock interest groups. It does not appear that the *Pasteurella* of cattle cause bighorn sheep die-offs. Contact between domestic sheep and goats and bighorn sheep should be minimized.

Caseous Lymphadenitis (*Corynebacterium pyogenes*)

Corynebacterium pyogenes is a small gram positive rod shaped bacteria. It causes abscesses and suppurative infections in wild and domestic animals.

Distribution and Hosts. This bacteria and the diseases it causes are found throughout the world wherever large numbers of animals are raised. Infections and abscesses due to *C. pyogenes* are common in domestic sheep, goats, and cattle and occasionally seen in mule deer, white-tailed deer, pronghorn antelope, elk, moose, reindeer, and bighorn sheep (Thorne et al. 1982).

Transmission, Pathogenesis, and Signs. *C. pyogenes* is apparently a normal inhabitant of the bacterial flora of many if not all species of grazing ungulates. Inoculation into tissues via puncture wounds, lacerations, ulceration, or bites appear to be the most common routes of entry. Overt disease may result when the host can not localize the initial infection. Local bacterial replication is followed by spread through lymphatic channels. This is often a slow process, with each lymph node along the chain abscessing. Joints may also be infected. If *C. pyogenes* enters the vascular system and replicates in sufficient numbers, metastatic abscesses may occur in a variety of organs and/or may cause valvular endocarditis. Abscesses in the lymph nodes of the peritoneal or thoracic cavities, and liver or kidneys, are fairly common.

Pneumonias from which *C. pyogenes* can be cultured are particularly common in sheep, goats, bighorn sheep, and deer. It is usually thought to be a secondary invader of tissues devitalized by *Pasteurella spp.* or other pneumonic agents.

Animals that die from *C. pyogenes* infections are usually very thin as the disease course is often long and debilitating. Draining abscesses from the site of lymph nodes, swollen joints filled with thick pale green pus, chronic infections of the jaw and mouth are common outward signs of this bacterial infection.

Lesions and Diagnosis. Abscesses due to *C. pyogenes* may be dry (caseated) and well walled off or may contain large amounts of creamy pus. Joints are quite swollen and may be deformed. Pneumonias may be fibrinous or fibrous depending on the time course.

The gross lesions and characteristic pus can lead to a presumptive diagnosis, but these organisms can be confused with

TB as they retain some acid-fast stain. Infection should be confirmed, if possible, by isolation of the organism. Inoculation onto blood agar is sufficient for primary isolation.

Management and Control. This organism is ubiquitous and no form of management is very effective. Isolation of infected animals to a "sick pen", removal of pus, and sanitation of contaminated pens and instruments will reduce spread. Prompt and aggressive treatment of infections by draining abscessed lymph nodes and antibiotic therapy can be helpful in treatment of owned animals.

Importance. This disease is common in wildlife and livestock. It is important to be able to distinguish it from more serious bacterial diseases that may cause similar lesions such as tuberculosis, brucellosis, and pasteurellosis. *Corynebacterium pyogenes* abscesses are a common source of questions from hunters as to edibility of a carcass. If the animal is not emaciated and the infection only localized to a few lymph nodes, the carcass may be trimmed and consumed.

Leptospirosis

Leptospires are small spiral shaped bacteria that can live for variable periods outside the body. They cause a disease that is transmissible between animals of the same and different species, and from animals to man. Manifestations of infection range from subclinical, to a vague flu-like illness, to death. Dairy cattle are fairly commonly infected and range cattle, various species of wildlife, and humans are only rarely infected. Although there are many (over a hundred) serovars, only 1 serotype (*L. interrogans*) is recognized.

Distribution and Hosts Affected. Leptospirosis has a worldwide distribution, but dominant serovars may be a local phenomenon. A wide range of species may become infected and many become inapparent carriers, shedding the organism in their urine. Wild rodents including rats, and raccoon, opossum (*Didelphus virginianus*) and skunk (*Mephitis* spp.) are proven reservoirs (Davis et al. 1981*a*).

Cattle and wildlife may become infected by drinking from contaminated water sources. Illness and limited numbers of deaths of free-ranging pronghorn antelope in Wyoming were attributed to leptospirosis in 1966 (Thorne et al. 1982). Occasional single fatal cases of leptospirosis in black-tailed and mule deer have been seen in California from 1977 through 1992. Infected white-tailed deer have been identified in New York, Pennsylvania, Louisiana, and Ontario (Davidson et al. 1981). Red deer have suffered natural infections and the disease has been experimentally transmitted to moose. Domestic sheep appear to be relatively resistant.

Transmission, Pathogenesis, and Signs. Urine from infected, often inapparent carrier animals is the most common source of infection. Leptospires can survive from days to weeks in stock ponds and slightly alkaline aquatic environments, but do not survive freezing weather well. They may penetrate abraded skin, and mucous membranes of the eye, nose, or mouth. In the western United States contaminated water holes in late summer are suspected to be important in transmission between wildlife and livestock. Venereal and transplacental transmission has occurred.

After penetrating skin or mucous membranes, leptospires replicate in the blood stream for a week. During this time a significant fever will be seen that may cause acute abortion. If infection results in sufficient specific antibodies, clinical illness will be limited. This is the most common sequelae. Some strains produce a toxin that breaks down red blood cells, causing a generalized yellow coloration of all white tissues in the body (jaundice), bloody-colored urine, and anemia. These signs may be confused with anaplasmosis. If the organism persists in the kidney or liver, damage to those organs will produce a variety of signs and usually death. Nervousness, incoordination, aggression, dullness, inappetence, emaciation, diarrhea, and abortion have all been reported.

Lesions and Diagnosis. Deer in terminal stages of infection are extremely anemic with packed cell volumes around 10%. Dark yellow to red wine colored urine and generalized jaundice are classic gross lesions in most species. The kidneys may be soft, pale and swollen and have pale streaks from capsule to cortex. Microscopically this is seen as chronic diffuse interstitial nephritis. The liver may be swollen and soft and areas of coalescing central necrosis may be evident microscopically.

Leptospires are very difficult to isolate in artificial media and diagnosis is often based on typical gross and microscopic lesions. Observation of leptospires by dark field examination of urine or kidney smears, and in silver stains or florescent antibody preparations of liver and kidney are diagnostic.

Serology is often used to diagnose infection in a herd of animals or to screen populations. A 4 fold rise in titer from paired serum samples is used in livestock, but is seldom applicable to wildlife. A single positive microscopic agglutination test at dilutions of 1:100 or greater is diagnostic for previous exposure, and is most often used for screening wildlife.

Control and Management. Supplying water in troughs or drinkers rather than from stock ponds and water holes will greatly reduce the likelihood of transmission between wildlife and livestock. Water sources should be cleaned periodically. Concentration of livestock and wildlife at point sources of water is undesirable for a variety of reasons; multiple sources of water separated by some distance will help. Leptospirosis may be a density dependent disease and maintaining populations of wild animals at or below carrying capacity will reduce chances of transmission or maintenance of this disease. Vaccination of cattle will prevent clinical illness and abortion for 6 to 12 months, and is a good and rather inexpensive method to prevent spread to other cattle or wildlife. Infected cattle can be treated with antibiotics. Neither vaccination nor treatment are practical for free-ranging wildlife, but they may be useful for wildlife on game ranches.

Importance. Leptospirosis in humans, beginning as a severe flu-like illness, although often self limiting, can be fatal. In descending order the most common sources of human infections are water, livestock, domestic pets, and wildlife.

Leptospirosis is not an extremely common disease of free-ranging wild ungulates and is probably not an important population limiting disease. However, individuals from heavily infected herds should not be moved to locations where that particular serovar is not present. For this reason a number of western states routinely test wild ungulates to ascertain patterns of infection.

Good preventive programs for cattle and putting water in troughs instead of ponds can limit wildlife exposure, and thus potential livestock and wildlife loses, and should be encouraged. Vaccination, treatment, and water management for game ranched wildlife are also sound investments.

Anthrax

Anthrax is caused by a gram positive bacteria (*Bacillus anthracis*) and primarily affects ruminants, and only occasionally humans and other animals.

Distribution and Hosts. Anthrax occurs worldwide and has been reported from most areas of the United States where soils are neutral or alkaline and under appropriate weather conditions. Cattle are the species most often affected, but white tailed deer, mule deer, and moose have died of anthrax (Thorne et al. 1982). In Africa, kudu (*Tragelaphus* spp.) and roan antelope (*Hippotragus equinus*) are most susceptible. A number of species of exotic wildlife on Texas game ranches have died of anthrax. Given sufficient contact with the infectious stage of the bacteria, almost any species of animal may succumb.

Transmission, Pathogenesis, and Signs. Decomposing vegetation following flooding creates the appropriate conditions for bacterial replication. Some endemic foci are well known. A lack of good forage that forces cattle or wildlife to graze close to the ground increases the opportunity for grazers to ingest large numbers of organisms. Infection can be via ingestion, inhalation of spores, or inoculation. *Bacillus anthracis* spreads via lymphatics, rapidly becoming septicemic.

Once an outbreak occurs, large numbers of biting insects help spread infection from animal to animal. Toxemia and massive necrosis of vital organs is the cause of death in susceptible individuals. Intense localized infections occur in resistant animals. Aggressiveness, agitation, depression, and death occur very rapidly. The blood often does not clot and bloody fluid may leak from the rectum, nose, or mouth.

Lesions and Diagnosis. It is dangerous and inappropriate to do a post mortem examination on an animal that is suspected to have died of anthrax as this spreads the spores and exposes humans. Aspiration from or removal of an eye for culture is recommended. Nicking an ear to get a blood smear or culture sample is another common diagnostic procedure. Rapid post mortem decomposition and gasification, thick tarry blood, edema, and hemorrhage in almost any or all organs are compatible lesions. It may be difficult to distinguish anthrax from several of the *Clostridial* diseases like gangrene, blackleg, malignant edema, and from viral hemorrhagic diseases (epizootic hemorrhagic disease and bluetongue).

In endemic areas the signs and lesions described are sufficient for a presumptive diagnosis. Blood smears should be Gram stained and examined for characteristic organisms before a postmortem exam is done. Presence of other bacilli may make this difficult. Inoculation of blood or ocular fluid into enriched media like trypticase-soy will allow the typical medusa head vegetative state bacteria to develop.

Management and Control. Cattle may be vaccinated with the Sterne spore vaccine. This is not usually practical for wildlife, although thousands of roan, kudu, and other species have been vaccinated around outbreak areas in South Africa. Carcasses should be completely burned, but should not be moved or scattered.

Importance. The disease is relatively rare but outbreaks of anthrax can kill large numbers of wild and domestic animals. Overgrazing following prolonged wet weather or in endemic areas may start an outbreak. As it is a potentially lethal human disease, hunters and people working around wildlife or cattle carcasses, should be aware of the potential of exposure. Only very early aggressive treatment with antibiotics is successful.

Plague (Yersiniosis)

Plague or bubonic plague is an infectious disease caused by the bacteria *Yersinia pestis.* If untreated, plague is often fatal.

In scattered endemic areas in the Western United States plague is primarily a disease of rodents. It occasionally infects carnivores, and rarely infects humans, and wild or domestic ruminants. Human cases apparently increased in the 1970s, possibly due to more recreational use of rural areas. Only 2 cases have been reported in deer.

Plague is a relatively rare disease that is not of significance for wild ungulates or livestock. However, wildlife workers and ranchers in the western United States should recognize its presence in the environment in which they work.

RICKETTSIAL DISEASES

Anaplasmosis

Anaplasmosis is a tick-transmitted rickettsial disease of cattle, sheep and a number of wild ruminants. It is usually a mild disease in wild animals, and deer may be carriers. In cattle anaplasmosis causes anemia and lost production, running into the hundreds of millions of dollars annually.

Hosts and Distribution. Various species of anaplasmas are found throughout the world. The 2 found in North America are *A. marginale* and *A. ovis*, which infect cattle and sheep, respectively. Black-tailed deer (*O. hemionus columbianus*), mule deer, white-tailed deer, and to a lesser extent elk and antelope may become infected with *A. marginale*, and bighorn sheep may become infected with *A. ovis*.

Transmission, Pathogenesis, and Signs. The *Anaplasma rickettsia* live in red blood cells. Ticks of the genus *Dermacentor* appear to be the most important transmitters of these rickettsia, which they ingest with a blood meal (Davis et al. 1981*b*). Interrupted feedings or multiple host feedings allow transmission, but transovarial transmission may rarely occur. *Dermacentor occidentalis*, found in much of California, is a particularly efficient transmitter as it feeds on deer and cattle at several stages, and both transstadial and transovarial transmission also occur. Recent observations suggest that where bighorn sheep are infested with *D. hunteri*, they may also be infected with *A. ovis*. Horse flies and other biting flies may transmit the organism mechanically. Multiple use of needles or reuse of dehorners or surgical instruments may also allow transmission. Disinfection or sterilization of surgical instruments and needles will prevent this.

There is an approximately 3 week period of silent infection following inoculation. Anemia is seen when the number of parasitized red blood cells evident on a smear reaches 1%. The anemia becomes progressively worse for a week or more until the production of new red blood cells in the bone marrow can begin to make up for those destroyed. Infected animals may get jaundiced (yellow-colored). Unlike leptospirosis, blood does not appear in the urine (hemoglobinuria).

Affected animals may be weak, unwilling to eat and show labored breathing. Although deer and other wild animals often show no signs of infection, in the Sierra foothills of California some severely anemic deer, with packed cell volumes as low as 8% have been found. Many recovered animals become carriers of the organism, some essentially for life.

Lesions and Diagnosis. The blood may be watery and the liver and spleen may be somewhat enlarged and pale. Internal lymph nodes may be enlarged and contain phagocytized red blood cells. Normally, white tissues may appear quite yellow. A diagnosis is made upon demonstration of marginal bodies in the cytoplasm of red blood cells with a Giemsa stain. Several serologic tests developed for use on cattle, notably the card agglutination test, must be modified for use on deer and are inaccurate on other species of wildlife (Thorne et al. 1982). Recently, a very specific indirect immunofluorescent antibody test that shows great promise for use in a wider variety of species has been developed.

Management and Control. Cattle can be vaccinated. Needles and other instruments that come in contact with blood should be disinfected between animals. Treatment of clinically infected animals with tetracyclines may help, however physical restraint for treatment may cause the animals to collapse and die. The disease is usually so mild in wildlife that treatment is not considered necessary. Tick control in heavily infested areas may also reduce the number of infected animals.

Importance. Deer are a common carrier of anaplasmosis that infects cattle. However, if the appropriate tick vectors are present, it is possible that anaplasmosis can be transmitted to and from many species of wild and domestic ruminants.

VIRAL DISEASES

Influenza Viruses

There are several respiratory viruses in the influenza virus family that may affect domestic or wild ungulates. The two most common are parainfluenza-III virus (PI-3) and respiratory syncytial virus (RSV).

Distribution and Hosts. Both PI-3 and RSV viruses have a worldwide distribution and are relatively common causes of upper respiratory disease in cattle and sheep. Virus neutralizing antibodies to PI-3 were found in 95, 92, and 88%, respectively, of normal newborn, weanlings, and adult domestic sheep in Idaho. Although there have been a number of serologic surveys conducted on captive and free-ranging wildlife, these data have not been summarized, the specificity of the serologic testing methods have not been verified, and in many cases it is unclear whether the animals sampled were in contact with livestock. A respiratory disease syndrome associated with common livestock influenza viruses has not been reported from deer, elk, moose, pronghorn antelope or many species of game ranched hoof stock.

In southern California, poor lamb survival and bronchopneumonia in lambs from 2 to 5 months old has been reported. Early in the course of clinical disease when radiographs show consolidation, PI-3 virus can be isolated from some lambs. The virus has been isolated from tissues of a few lambs dying of acute pneumonia. However, other lambs with pneumonia show no evidence of PI-3 infection, and PI-3 virus has been isolated from clinically normal and pneumonic adult bighorn sheep. Isolates of PI-3 from these bighorn are not similar to type strains of PI-3 from cattle, and the source of the PI-3 virus is unknown.

A report from Wyoming, documented PI-3 pneumonia in 10 adult captive Rocky Mountain bighorn sheep. The virus was isolated from 3 animals, and 9 eventually died from pneumonia.

Respiratory Syncytial Virus (RSV) appears to be an important initiator of pneumonia in domestic cattle and sheep. It may act similarly to PI-3 virus, and be a part of a multifactorial respiratory disease complex. Serologic surveys suggest it is now fairly common in bighorn sheep. Seroprevalence does not appear to correlate with clinical disease. Attempts to produce pneumonia in bighorn sheep by challenge with live RSV failed. However, recent reports from Colorado, including isolation of the virus and observation of typical lesions, suggest that RSV may initiate pneumonias of captive bighorn sheep.

Transmission, Pathogenesis, and Signs. Influenza viruses are transmitted in droplets and respiratory secretions and are fairly infectious, although they do not survive outside the body very well. Upon entering the respiratory tract they infect the ciliated epithelial cells and the pneumocytes. During replication they kill the cells that line the respiratory tract and reduce the responsiveness of lymphocytes. Bacteria are then free to colonize the respiratory tract.

Animals with viral respiratory infections may be depressed and have nasal or ocular discharge, and may cough and sneeze. By the time the pneumonia becomes severe and the discharge becomes purulent the virus is often no longer present.

Lesions and Diagnosis. The lesions caused specifically by the influenza viruses are subtle and transient. Microscopically the epithelial cells will be hyperplastic and mononuclear inflammatory cells infiltrate around infected airways. Occasionally characteristic fused epithelial (syncytial) cells can be found. Often, however, secondary bacterial invasion will overshadow any viral changes and the lesions are similar to those caused by *Pasteurella's*. The characteristic microscopic lesions and specific florescence from tissues or swabs allow a presumptive diagnosis. Definitive diagnosis can be made only by isolation of the virus. Serology may be of some assistance in that animals that survive infection will usually show a rise in antibody titer over a several weeks following infection.

Management and Control. There is no treatment for infected animals except to control bacterial infections that may follow the viral infection. The procedures discussed in the section on Pasteurella's are appropriate. However, cattle and captive wildlife can be vaccinated for both of these viruses as there are commercial vaccines for PI-3 and RSV. These appear to be safe in bighorn and domestic sheep, and may be considered if viral pneumonia appears to be a problem. Modified live PI-3 vaccine in biobullet form has been used on penned and free-ranging bighorn sheep (Jessup et al. 1991). These are relatively common viruses and their actual role in causing any disease process in wild animals is not confirmed. It is also not clear if the influenza viruses that have been isolated from wild animals came from livestock or are indigenous to wildlife.

Importance. Influenza viruses are blamed for outbreaks of pneumonia in some herds of bighorn sheep. Some groups have tried to tie this to interactions with livestock. The evidence is not consistent and not very convincing at this time.

Hemorrhagic Disease (bluetongue [BT] and Epizootic Hemorrhagic Disease [EHD])

Bluetongue and EHD cause very similar disease syndromes in a variety of wild and domestic ungulates. They can only be distinguished by isolating the causative virus or by serial blood sampling for specific antibodies, so both diseases are described together under a name descriptive of the signs, gross and microscopic pathology, "hemorrhagic disease". There are 4 serotypes of BT virus (10, 11, 13 and 17) in the United States and 2 types of EHD virus (i.e., 1 and 2). These diseases are transmitted by biting gnats, most notably *Culicoides variipennis*, and are not directly transmitted between hosts.

Hosts and Distribution. The viruses of hemorrhagic disease occur in livestock across most of the United States, with the possible exception of the extreme northeast and northwest coastal areas. The disease is most often seen in wildlife in the southeastern United States, California and portions of the Rocky Mountain states. Occasional epizootics occur as far north as the Dakotas and British Columbia. Some of the other 17 serotypes of BT virus are found in various areas of South America, Asia and Africa and the disease was first described in sheep in Africa around the turn of the century.

Although BT is primarily recognized as a disease of domestic sheep; sheep, white-tailed deer, pronghorn antelope, black-tailed or mule deer, bighorn sheep and cattle, in that approximate order of sensitivity can become infected (Thorne et al. 1982). Epizootic hemorrhagic disease is primarily recognized as a disease of white-tailed and black-tailed deer, but pronghorn antelope, sheep, bighorn sheep, elk and cattle may be infected (Davidson et al. 1981). In zoos and on game farms an even wider variety of wild ungulates, including most deer species and many African antelope and wild bovids, have become infected by these viruses.

Transmission, Pathogenesis, and Signs. An outbreak of hemorrhagic disease requires a number of susceptible hosts, a sufficiently pathogenic strain of virus, and sufficient populations of biting midges or gnats. The *Culicoides variipennis* gnat is most commonly thought to be the primary vector in the United States, but recent entomological studies suggest other species of gnats may be important vectors in the desert Southwest. Gnats are very small and tend to be seasonal. They become infected when they feed on a viremic host and in 10 to 14 days can transmit the virus to another host. Gnats prefer muddy, fecal contaminated pond edges for breeding, and concentrating susceptible host around such ponds creates perfect conditions for an outbreak. In most areas of the United States late summer or early fall are the gnat season, but in the desert Southwest late winter and early spring may favor emergence. In central California spring gnat blooms and subsequent disease transmission occasionally occur.

Cattle appear to be potential reservoirs as they seldom show overt signs of disease and the virus persists much longer in their blood than in other species of wildlife or sheep. Viremias range from 2 days to 2 weeks in most of the latter species but can be >30 days in cattle. Apparent transmission from a cow elk to its calf has been reported.

Once inoculated into the host by the bite of the gnat the virus reproduces in the cells lining blood vessels. Most of the signs and lesions of hemorrhagic disease are caused by damage to and rupture of small blood vessels. Blood vessels of the oral cavity, tongue, pillars of the rumen and coronet of the foot are for some reason preferentially infected. As the cells lining blood vessels die and expose collagen in the vessel walls, a clotting cascade occurs. The clotting factors may be consumed allowing multiple hemorrhages and leakage of fluids. Tissues of the mouth, rumen and hoof overlying the affected vessels may die and slough. In the blood stream the virus is closely associated with red blood cells.

Infection with hemorrhagic disease viruses often cause no overt signs of disease. However, animals with acute, severe disease may be depressed and die within 24 hours. These animals may appear to be blind, have swollen ears, eye lids, lips or tongue, and they may have froth exuding from the nostrils. Large numbers of dead animals are found around water holes, as they apparently become very thirsty and febrile near death.

Animals with less severe signs of disease may have ulcers of the mouth, the tongue may be dark (blue) or ulcerated, or necrosis of the coronary band may lead to sloughing of the hooves and secondary foot infections (sometimes confused with foot rot). Some animals with less severe signs survive but tend to remain emaciated.

Abortion and neonatal deaths occur in deer when infection takes place during the last third of pregnancy. Reduced pronghorn antelope fawn production has been reported in the year following an epidemic.

Lesions and Diagnosis. Hemorrhages occur in various organs, most prominently on serosal and mucosal surfaces. Ruminal or intestinal ulcers may lead to bloody diarrhea. Edema fluid may fill the lungs and pleural or peritoneal cavities. In wildlife, the classic lesions described in sheep, medial necrosis of the great vessels of the heart and myocardial necrosis, are uncommon. In domestic and bighorn sheep and some experimentally infected pronghorn antelope secondary pneumonias are commonly reported.

Microscopically, hemorrhage, vasculitis and thrombosis in small arteries and associated necrosis are prominent. Diagnosis may have to be based only on a compatible case history, gross and microscopic lesions, as isolation of the virus, particularly from carcass material, is very difficult. Rising antibody titers to one of the strains of EHD or BT in surviving animals may implicate that virus. Without virus isolation EHD and BT can not be distinguished. Also, hemorrhagic diseases may resemble a variety of other diseases including foot and mouth and other vesicular diseases, malignant catarrhal fever, and other bovine viral diseases.

Management and Control. There is no good treatment for infected animals. A killed type 10 BT vaccine and, in California, modified live type 11 and 17 vaccines are available for sheep. There is no vaccine for BT type 13 or EHD. Safety and efficacy of these and other vaccines for cattle and various species of wildlife have yet to be proven.

Gnat breeding habitat can be reduced by keeping the ponds sides steeply sloped and by reducing the organic content of water. Troughs or spring boxes are less likely than stock ponds to serve as gnat breeding sites. During gnat seasons water sources should be kept clean, animals as dispersed as possible, and gnat breeding habitat minimized. Unfortunately, these precautions are not terribly easy to implement or likely to be completely effective. Research on a variety of gnat larvicides and pond treatments failed to reveal a treatment that was practical and cost effective. Control of these diseases will only be possible when we better understand the ecology of the insect vectors and the pathogenesis of the disease.

Importance. In some areas these diseases have caused major die-offs of sheep, deer, antelope, and exotic wild animals. However, the full impact of an outbreak may not be estimated by carcass count alone as birth rates and neonatal survival may also be effected. In some locations these diseases may be population regulating, in others they may be quite common and apparently benign. The conditions under which hemorrhagic disease outbreaks occur are complex and difficult to manage. This is another disease that has caused conflicts between wildlife and livestock interests. Most importantly BT and EHD can resemble malignant catarrhal fever, foot and mouth disease, and a number of other very serious diseases.

Malignant Catarrhal Fever (MCF)

Although it is rather rare in the United States and not highly contagious, malignant catarrhal fever is an often fatal viral disease. Infected cattle have a very high fever and the upper respiratory tract, eyes, and intestinal tract are affected. In Africa, a herpes virus causes a wildebeest (*Connochaetes* spp.) associated form of MCF. At least 1 outbreak of MCF in cattle was associated with an adjacent wildlife park where wildebeest were calving. In the United States, the virus causing MCF has often proven illusive. Some of the MCF in cattle and captive wildlife in the United States appears to be associated with contact with domestic sheep (so called "sheep associated MCF").

Distribution and Hosts. MCF has been reported from cattle in all regions of the United States. Infections in a variety of captive exotic wild ungulates including Pere David's deer (*Cervus davidensis*), red deer, axis deer, and several species of antelope, and in captive native wildlife including white-tailed deer, mule deer, bison, and moose have been reported (Castro and Heuschele 1992). Only 2 cases in free-ranging deer, 1 black-tailed deer in California and 1 white-tailed deer in the eastern United States, have been reported.

Transmission, Pathogenesis, and Signs. The virus that causes MCF is associated with white blood cells and it takes a rather large transfusion of infected blood to transmit the disease experimentally. The mode of natural transmission and pathogenesis are unknown, but thought to be via aerosol and then spread from the respiratory tract via the blood stream.

There are 4 recognized forms of MCF based on the organ system most affected (i.e., acute, head and eye, intestinal, and mild); however, the signs may overlap (Davis et al. 1981*a*). High fever, catarrhal to mucopurulent inflammation of the nasal passages, ocular mucous membranes, photophobia, corneal opacity, dark and fetid diarrhea, dehydration, emaciation, and enlargement of lymph nodes are all typical signs (Castro and Heuschele 1992).

Lesions and Diagnosis. The lesions of MCF are difficult to distinguish from some of those caused by foot and mouth disease or hemorrhagic disease. These include erosions of the lips, oral mucous membranes, tongue, and pharynx. Additionally tracheobronchitis, sinusitis, corneal opacity, generalized lymph node enlargement, hemorrhagic ulcers of the abomasum, bloody or watery diarrhea, and lesions of the heart, liver, or kidney may be seen.

Microscopically the lesion that differentiates MCF from other diseases is fibrinoid necrotizing vasculitis. Mononuclear infiltration around affected vessels is seen also. The best organs to find these characteristic lesions in are the brain and abdominal organs (i.e., liver, kidney, and adrenal). A second common, but less characteristic, lesion is depletion or necrosis of small lymphocytes in the lymph nodes and/or lymphoid hyperplasia.

The diagnosis of MCF is difficult and usually based on compatible signs, gross and microscopic lesions, particularly fibrinoid necrosis of blood vessels. Serology is useful for diagnosing wildebeest associated MCF in a herd or group of animals. Virus isolation from post mortem specimens of non-wildebeest associated cases of MCF has not been successful.

Management and Control. With so little understood about this disease there are few practical methods of control. Separation of susceptible species from wildebeest, especially during calving periods, and/or from domestic sheep are suggested. In some states the importation and private ownership of wildebeest is prohibited.

Importance. It does not appear that wildlife are commonly exposed to MCF or that it is an important disease to free-ranging populations. However, its high case fatality rate has caused a number of states to discourage the ownership of wildebeest. The fact that it looks so much like foot and mouth disease and other foreign animal diseases (FADs) is a cause for concern.

Foot and Mouth Disease

Many species of cloven-hoofed animals including cattle, sheep, goats, deer, antelope, and elk are susceptible to infection with foot and mouth disease (FMD) virus. This disease is caused by a very small RNA virus. It is considered to be a foreign animal disease, and an outbreak would likely cause the declaration of a national disease emergency. Infected animals are subject to quarantine and slaughter. The most recent outbreak in England in 1967-68 resulted in the slaughter of 450,000 livestock. The last outbreak in North America was in California in 1928 and >22,000 deer were slaughtered in the Stanislaus National Forest.

Hosts and Distribution. Domestic livestock and most species of North American wildlife are susceptible. Many species of African antelope held on game ranches are also susceptible, but both Asian and South American camelids are somewhat resistant. The only areas of the world free of FMD are North and Central America, Australia, New Zealand and Oceania, Japan, and portions of Europe.

Transmission, Pathogenesis, and Signs. The FMD virus is shed in large amounts in the oral secretions and survives in manure, blood, and excreta for long periods. Infected animals and recovered animals often become carriers. The disease is highly infectious and transmission by aerosol over long distances is possible.

The virus replicates in the superficial layers of the mucous membranes and coronary band causing the formation of blisters (vesicles). These vesicles rupture releasing large numbers of virus particles. Ulcers and secondary bacterial infections follow. The virus also causes a systemic infection with a high fever that often damages the heart muscle.

Animals infected with FMD are usually depressed, anorectic, sometimes emaciated, they are often lame and salivate profusely due to painful blisters. Many infected animals will survive but seldom regain full health and can spread the disease.

Lesions and Diagnosis. Clinical signs and gross lesions of FMD are not adequate for diagnosis but should trigger immediate investigation. The prominent vesicular lesions and focal myocardial necrosis of FMD can be histologically distinguished from the vascular based lesions of hemorrhagic disease and MCF. They can not be distinguished from other vesicular diseases.

A diagnosis of FMD is based on the presence of virus specific proteins, or antibodies in serum, detected by 1 of several tests, with the ELISA being the most sensitive, or by isolation of the virus.

Control and Management. There is no treatment for FMD. Trade in live animals and uncooked animal products with countries that have FMD is forbidden. In North America any infection that occurs will be handled by quarantine, test, and slaughter.

Importance. It is important that the lesions of FMD are well known and that any disease resembling it is reported immediately. Other vesicular diseases, particularly vesicular stomatitis, do occasionally occur in North America and must be immediately distinguished from FMD. Other (e.g., MCF, BT, and EHD) lesions can also be confused with FMD. These similarities make it all the more important that wildlife and livestock diseases are monitored and diagnosed correctly.

Sore Mouth (Contagious Ecthyma)

A large double-stranded DNA (parapox) virus with a propensity to infect cells in the basal layer of the skin causes contagious ecthyma (CE). Common names for this disease include scabby mouth, sore mouth, and orf (when it infects humans). It was first described in Europe over 200 years ago.

Distribution and Hosts. This disease has a worldwide distribution and sheep and goats are most commonly infected. Since 1954 periodic outbreaks in free-ranging bighorn sheep and mountain goats have been documented in Alaska, British Columbia, Alberta, Colorado, California, and New Mexico. Natural infections in musk-ox, deer, and reindeer have occurred and the disease has been experimentally produced in elk, pronghorn antelope, mule deer, and white-tailed deer.

Transmission, Pathogenesis and Signs. The infection is spread primarily by direct contact between animals, but the virus may survive for >20 years in scabs. Small breaks in the skin such as occur when eating dry coarse forage may allow entry of the virus. The mucous membranes of the nose, mouth, and rectum are most often effected, but also eyes, vagina, teats, and locations subject to trauma or inoculation (e.g. thigh or lower legs). Salt blocks and alfalfa hay on which infected animals have previously fed have been the point sources of outbreaks in bighorn sheep.

The virus begins to replicate in the epithelial cells, and within 2 days a papule is seen. A vesical and pustule follow as bacteria invade the open lesion, and a scab is usually visible by the end of one week. The lesions may get quite large and proliferative but usually diminish within 3 weeks and heal spontaneously in a month or so.

Contagious ecthyma apparently causes intense itching and pain, and infected sheep are nervous, lick their lips and nostrils, and rub affected areas. Grazing and suckling are difficult and lambs are often rejected by their dams. If an outbreak occurs during lambing season, lamb losses may be high.

Lesions and Diagnosis. The early lesions of CE are rarely noticed. When the tissues beneath the scabs become thick and cauliflower-like, they bleed profusely. Microscopically the epidermal cells in the basal layer can be seen undergoing ballooning degeneration with extensive rete-peg formation down into the dermis (Castro and Heuschele 1992). The initial inflammatory response is mononuclear (mostly lymphocytes) but becomes mixed and then predominantly neutrophilic as secondary bacterial invasion proceeds.

The species affected, time course and progression of lesions, their gross appearance, and histology are all characteristic and, in combination, are sufficient to make a diagnosis. Virus isolation and electron microscopy are used to confirm.

Control and Management. A virulent live virus vaccine is used to inoculate lambs and kids on the scarified inner thigh to induce an infection and subsequent immunity. This is not applicable to free-ranging wildlife. Also, the vaccine virus can be spread into the environment, where it is quite persistent, as noted. Phenolic disinfectants and steam can be used to clean contaminated clothes and equipment. There is no effective treatment, but the disease is usually self limiting and by the end of a month, most adult animals will recover spontaneously. Animals in active stages of infection should not be mixed with susceptible populations. Hay from fields that have been grazed by sheep and goats should not be fed to highly susceptible species.

Importance. This disease is transmissible to humans. Four of 65 people handling or working near infected bighorn sheep during a capture developed lesions. It is painful, unsightly and can leave some scaring. Implements, clothes and equipment must be disinfected thoroughly or discarded. This disease is probably not an important population limiting factor, although some mortality may occur. It is not uncommon in free-ranging bighorn sheep herds in the western United States. Infected populations tend to show recurrent signs when other stressing factors like malnutrition, overpopulation or prolonged extreme weather are present.

Ovine Progressive Pneumonia (OPP) and Caprine Arthritis Encephalitis (CAE)

These 2 diseases are caused by very similar slow-growing lentivirus. They share many of the same biological, chemical, structural, and serologic properties, but differ in some nucleic acid sequences.

Distribution and Hosts. These diseases are primarily found in North America. There is cross-susceptibility of sheep and goats to both viruses, but as the names suggest, OPP is generally a disease of sheep and CAE of goats. Exotic sheep, goats, and bighorn sheep may be susceptible. Although extensive inoculation trials have not been conducted on a wide variety of species, cattle and native and exotic wild ungulates do not appear to be susceptible. In goats, CAE is very common in the United States, and in many areas >50% of the flocks are infected. There are no cases of diseases resembling OPP or CAE in cattle, deer, elk, moose, pronghorn antelope, or other species of wildlife. The only well developed cases of a disease similar to OPP/CAE in bighorn sheep occurred in 2 hand-reared male lambs fed goat colostrum and goats' milk. Both animals developed signs of a wasting disease at just over a year of age and eventually died, and brain lesions compatible with CAE were noted at post mortem. Both animals were seropositive for OPP/CAE antibodies in later stages of disease, having been seronegative at about three to four months of age. No herds of free-ranging bighorn sheep in the western United States had been shown to have antibodies to OPP/CAE.

Pathogenesis and Signs. Ovine progressive pneumonia is a highly fatal disease of sheep; however, the onset is very slow and loss of weight and difficulty breathing are the only outward signs in adults. Emaciated sheep will have a normal appetite and are ≥2 years old by the time they get ill. Chronic lameness and thickened udder with decreased milk flow is reported in sheep. This can contribute to high lamb mortality. Many sheep carry and will shed the virus to lambs via the colostrum and rarely in utero. Also, the virus can sometimes be transmitted via the respiratory route. Caprine arthritis and encephalitis is a crippling disease of goats that is eventually fatal. It most commonly causes joint disease rather than pneumonia, and it causes progressive brain damage in young animals born to infected nannies or fed on infected colostrum, but otherwise is similar to OPP.

Lesions and Diagnosis. Although diagnosis can often be made by post mortem examination of an infected animal and demonstration of typical lesions and/or virus, extensive exposure of a flock is likely to have occurred by this time. A serologic test for the antibody allows screening for evidence of prior exposure. This procedure will also allow selection of uninfected goat flocks to use as a source for colostrum and milk to feed valuable orphaned exotic sheep and goats. The OPP/CAE antibodies cross react in the AGID and ELISA tests, so 1 test may pick up exposure to either virus. More sensitive tests have been developed recently. Some sheep do not start to shed antibody until six months or more after infection.

Control and Management. Although not highly infectious, these diseases are difficult to eliminate from infected flocks. Thus, keeping valuable seronegative flocks of angora goats, mouflon, aoudad, bighorn sheep, and other rare breeds of sheep separated from untested stock is a wise management option. Although attempts to eradicate OPP from domestic sheep are ongoing, it is still found in many states. Goats milk or colostrum for feeding orphaned wildlife, lambs, and kids should be taken from CAE/OPP negative flocks.

Importance. Both OPP and CAE are slow, progressive diseases for which there is no vaccination or treatment. Every effort should be made to keep wild sheep and goat herds free of these diseases. Mixing sheep and goats from many sources without testing may result in problems years later.

INTERNAL AND EXTERNAL PARASITES

The parasites covered in this section were selected because of their demonstrated or potential importance in the ecology and management of the interface involving wildlife and domestic animals. However, these selected parasites represent only a fraction of the array of parasites that occur in or on wildlife and domestic animals. For more information on specific topics and problems, interested readers should examine the listed references and consult appropriate experts such as parasitologists, entomologists, and wildlife veterinarians.

Virtually every animal is infected with ≥1 (typically several) different parasites including protozoa, nematodes, tapeworms, flukes, and arthropods. Some parasites occur in both wildlife and domestic animals, while in other cases it is not clear whether or not different hosts share identical parasites. In many situations, infected animals do not exhibit any outward signs of infection or disease. However, if an animal is weakened by some preexisting stress or condition such as malnutrition, then parasites may "gain the upperhand" and cause clinical disease and death. In contrast, some parasites are so pathogenic that they can be primary causes of morbidity and mortality in the absence of any detectable preexisting condition.

Another important fact is that parasites often tend to be over-dispersed in their hosts, that is, most of the individual parasites of a given species occur in only a small fraction of the potential available hosts. One consequence of over-dispersion is that a single heavily parasitized animal can serve as an important source of infection for an entire herd or population through translocation programs or normal migration or dispersal events.

Many parasite infections in domestic animals can be treated with anthelmintics or insecticides, while most infections in free-ranging wildlife cannot because the animals are not handled individually. Therefore, managers should recognize that their efforts are most effective when they are aimed at preventing the establishment of new parasites and controlling the levels of existing infections.

Internal Parasites

A large number of internal parasites are found in both domestic livestock and wild ungulates including the gastrointestinal nematodes and tapeworms, lungworms, and liver flukes. Although some species of parasites in each of these groups clearly are shared by both domestic and wildlife hosts, most internal parasites are relatively host-specific and are not likely to be of major concern to the range manager.

Gastrointestinal Nematodes. Several genera of trichostrongyle nematodes (e.g., *Ostertagia*, *Haemonchus*) are found in the gastrointestinal tract of domestic and wild ruminants. These parasites actually exist in 2 separate populations: those inside the host and those on the pasture. Infected animals pass eggs in their feces and in warm, moist conditions large numbers of infective larvae can accumulate on pasture. Infections are acquired by ingesting larvae with grass and typical clinical signs include diarrhea, weight loss, and anemia. Because these parasites have a direct life cycle, clinical disease is most likely to be seen when large numbers of animals share a restricted grazing environment. If necessary, the most effective way to manage these parasites is through a preventive approach. Fecal examinations will reveal the approximate numbers and types of internal parasites, and it is important to examine and treat new animals before they are added to or mixed with other captive or free-ranging animals. Treatment of internal parasites in penned exotic and domestic animals is relatively easy, but treatment is not effective if the pastures have already been contaminated with eggs and larvae, and treatment of free-ranging animals is impractical if they are widely dispersed and cannot be attracted to medicated feed.

Lungworms. Lungworms (*Protostrongylus stilesi*) are one of the most important parasites of Rocky Mountain bighorn sheep, but it apparently does not cause a problem in domestic sheep (Davis et al. 1981*b*). Lambs are infected in utero by transplacental migration of larvae and heavy infections can cause high morbidity and mortality. This highly pathogenic lungworm is a particular problem among high density bighorn sheep populations that live in mesic environments supporting large populations of the snail intermediate host. The anthelmintic fenbendazole is an effective treatment and treatment may help reduce morbidity and mortality in those situations where a substantial proportion of the population can be attracted to medicated feed. Other genera (e.g. *Dictyocaulus*, *Muellerius*) and species (*P. rushei*) of lungworm can infect bighorn sheep and domestic sheep or cattle (*Dictyocaulus* only), but infections with these lungworms do not appear to result in significant interspecies management problems.

Liver Flukes. Liver flukes (*Fasciola hepatica* and *Fascioloides magna*) are important internal parasites of domestic and wild ruminants in those areas of the Northwest, Midwest, and Southeast, where the intermediate host, an amphibious snail, is found (Davis et al. 1981*b*). Both parasites are transmitted by accidental ingestion of infective stages (metacercariae) present on vegetation. Both parasites are difficult to impossible to control in free-ranging animals. *Fasciola hepatica* differs from *F. magna* in that eggs are passed by infected hosts regardless of whether they are a domestic or wild ruminant. Domestic livestock are dead-end hosts for *F. magna* and do not pass eggs in their feces. Mule deer and white-tailed deer are the normal hosts for this parasite and infections in domestic livestock (sheep, goats, cattle) only occur when infected deer are present in the same environment. Both of these parasites can cause morbidity and mortality in domestic and wild animals and clinical signs include weight loss, diarrhea, and anemia. Although effective drugs are available for use in domestic livestock, it is impractical to treat these infections in free-ranging wildlife. The presence of only a few untreated animals is sufficient to cause the persistence of infective stages in the environment. Attempts to eliminate water sources where the snail intermediate hosts occur, or restrict access of deer to these sites, are typically impractical and unrewarding.

External Parasites

A large array of arthropod ectoparasites, including ticks, mites, flies, lice, and fleas, are found in or on the skin of mammalian hosts. Many of these parasites can cause disease in domestic and wild animals through their bloodsucking activities and by their ability to serve as vectors of viruses, bacteria, and protozoa. Examples of some of the important tick and fly vectors (and their respective pathogens) include the ticks *Dermacentor* spp. (*Anaplasma* [anaplasmosis]), *Ixodes* spp.(*Borrelia* [Lyme disease, borreliosis]), *Boophilus* spp. (*Babesia* [Texas cattle fever]), and gnats such as *Culicoides* spp. (BT, EHD) and tabanid flies (*Elaeophora schneideri* [arterial worm]). The relative importance of several emerging vector-pathogen relationships to rangeland management is currently unknown, but the fact that arthropod vectors and their associated pathogens can be found on domestic livestock and wildlife suggests that the situation should be closely monitored. These emerging potential problems include the tick-transmitted diseases of borreliosis, anaplasmosis, and babesiosis in bighorn sheep and deer.

Boophilus. The federal eradication program for cattle fever ticks (*Boophilus annulatus* and *B. microplus*) provides an instructive example of the problems that can arise due to parasite interactions between domestic cattle and free-ranging wildlife. *Boophilus annulatus* and *B. microplus* are vectors of the blood protozoan, *Babesia bigemina*, the causative agent of Texas cattle fever, in North America. In 1906 the federal government instituted the United States Cattle Fever Eradication Program because substantial economic losses (>40 million dollars annually) were being sustained by the cattle industry in the Southwest due to infections of *B. bigemina*. The program, which was highly successful, focused on cattle dipping to kill *Boophilus* ticks since these parasites are one host ticks. However, *B. microplus* and *B. annulatus* occur on cattle and white-tailed deer, and the only way to eradicate ticks on deer was through depopulation. Therefore, in a controversial program, thousands of deer were killed during the early 1900's to eliminate the wildlife host for *Boophilus* spp. ticks. Texas cattle fever no longer causes a significant problem in the U.S. However, deer populations are much higher now than in the early part of this century and outbreaks of cattle fever, or other shared infectious diseases, may necessitate a reexamination of the need for deer depopulation.

***Psoroptes* spp.** Psoroptic scabies is an ectoparasitic disease of domestic and wild ungulates caused by mites of the genus *Psoroptes*. Like Texas cattle fever, scabies was a serious problem in the livestock industry that warranted a federally subsidized eradication program that was initiated during the turn of the century. The program consisted of isolating and treating infested domestic livestock and culminated in the 1973 federal acknowledgment that sheep scabies had been eradicated from the United States. Many observers are convinced that the introduction of domestic sheep and cattle onto bighorn sheep ranges in the late 1800s and early 1900s precipitated bighorn die-offs. However, it is of interest that bighorn sheep scabies is still relatively common in the western states while domestic sheep and cattle scabies has decreased substantially (Boyce et al. 1991). These findings raise questions regarding the host specificity and species identification of *Psoroptes* spp. mites. During an outbreak of bighorn scabies in New Mexico, several unsuccessful attempts were made to transmit *Psoroptes* spp. mites from bighorn sheep to cattle and domestic sheep. However, other researchers have been successful in transferring mites between host species, and mule deer sympatric with mite-infested bighorn sheep have been found to be infested with mites that are indistinguishable from those found on the bighorn sheep.

Diagnosis of *Psoroptes* spp. infestations can be accomplished by microscopic examination of skin scrapings and ear swabs or by immunodiagnosis (Boyce et al. 1991). The advent of safe and effective antiparasitic compounds such as ivermectin has played a major role in reducing the prevalence of scabies in domestic livestock. These compounds have also been used to treat free-ranging bighorn sheep and remote delivery devices (i.e., biobullets) may be used in the future to treat selected free-ranging populations. However, the potential of these mites to infest different host species may preclude any control strategy that focuses on treatment as a management tool.

SUMMARY

Livestock and wild ungulates have shared the western rangelands for >100 years and will continue to do so. Under some circumstances livestock and wild ungulates share some of the same diseases and parasites. Sharing the same diseases and parasites can be a problem when diseases and parasites are harmful to the host or the host population, and when diseases and parasites are commonly exchanged between wildlife and livestock. The challenge is to manage wildlife and livestock to minimize negative interactions caused by their diseases and parasites.

In general, the healthier the livestock and wildlife populations are, the less risk there is of disease and parasite transmission. Good nutrition, reasonable stocking rates, and maintenance of animals at or below carrying capacity are the obvious basics. In general, healthy resident livestock are less likely to harbor pathogenic strains of disease agents than are animals brought from long distances or a variety of locations. Separating newly purchased stock or wildlife from a herd for a period of observation before introduction is a good practice. Preventive medical programs including routine locally appropriate vaccination, prophylactic antibiotic treatment, nutritional supplementation, and worming and parasite control programs will be an economically sound investment for the stockman or game rancher and may have benefits for wildlife. Serologic and pathogen isolation surveillance of free-ranging wildlife, particularly animals being relocated, should be routine. All these programs help insure healthy wildlife and livestock.

Although there may be some circumstances where separation of wildlife and livestock is advantageous, the existence of shared diseases and parasites should not be routinely used as an argument for exclusion of either wildlife or livestock from grazing lands. The key to minimizing the potential negative impacts of diseases and parasites that may be shared by wildlife and livestock is a knowledge of their existence and a willingness on the part of wildlife and livestock interests to take appropriate management actions.

LITERATURE CITED

Boyce, W. M., D. A. Jessup, and R. K. Clark. 1991. Serodiagnostic antibody response to *Psoroptes* infestations in bighorn sheep. J. Wildl. Dis. 27:10-15.

Callan, R. J., T. D. Bunch, G. W. Workman, and R. E. Mock. 1991. Development of pneumonia in desert bighorn sheep after exposure to a flock of exotic wild and domestic sheep. J. Am. Vet. Med. Assoc. 198:1052-1056.

Castro, A. E., and W. P. Heuschele. 1992. Veterinary diagnostic virology. Mosby-Year Book, Inc., St. Louis, Mo. 600pp.

Davidson, W. R., F. A. Hayes, V. F. Nettles, and F. E. Kellogg. 1981. Diseases and parasites of white-tailed deer. Heritage Printers, Inc., Charlotte, N.C. 458pp.

Davis, J. W., L. H. Karstad, and D. O. Trainer. 1981*a*. Infectious diseases of wild mammals. Second ed. Iowa State Univ. Press, Ames. 446pp.

———, ———, and ———. 1981*b*. Parasitic diseases of wild mammals. Second ed. Iowa State Univ. Press, Ames. 446pp.

Drew, M. L., D. A. Jessup, A. A. Burr and C. E. Franti. 1992. Serologic survey for brucellosis in feral swine, wild ruminants, and black bear of California, 1977 to 1989. J. Wildl. Dis. 28:355-363.

Foreyt, W. J. 1989. Fatal *Pasteurella haemolytica* pneumonia in bighorn sheep following direct contact with normal domestic sheep—an experimental study. Am. J. Vet. Res. 50:341-344.

———, and D. A. Jessup. 1982. Fatal pneumonia in bighorn sheep following direct association with domestic sheep. J. Wildl. Dis. 18:163-168.

Jessup, D. A. 1993. Translocation of wildlife. Pages 493-499 in M. E. Fowler, ed. Zoo and wild animal medicine. Third ed. W. B. Saunders Co. Philadelphia, Pa.

———, J. DeForge, and S. Sandberg. 1991. Biobullet vaccination of captive and free-ranging bighorn sheep. Pages 428-434 *in* L. A. Renecker and R. J. Hudson, eds. Wildlife production: conservation and sustainable development. Univ. Alaska, Fairbanks.

Miller, M. W., J. M. Williams, T. J. Schiefer, and J. W. Seidel. 1991. Bovine tuberculosis in a captive elk herd in Colorado: epizoology, diagnoses, and management. Pages 533-542 *in* Proc. of the 95th U.S. Animal Health Association, San Diego, Calif.

Onderka, D. K., S. A. Rawluk, and W. D. Wishart. 1988. Susceptibility of Rocky Mountain bighorn sheep to pneumonia induced by bighorn and domestic livestock strains of *Pasteurella haemolytica*. Can. J. Vet. Res. 52: 439.

———, and W. D. Wishart. 1988. Experimental contact transmission of *Pasteurella haemolytica* from clinically normal domestic sheep causing pneumonia in Rocky Mountain bighorn sheep. J. Wildl. Dis. 24:663-667.

Silflow, R. M., W. J. Foreyt, S. M. Taylor, W. W. Laegreid, H. D. Liggit, and R. W. Lied. 1989. Comparison of pulmonary defense mechanisms in Rocky Mountain bighorn (*Ovis canadensis canadensis*) and domestic sheep. J. Wildl. Dis. 25:514.

Thorne, E. T., N. Kingston, W. R. Jolley, and R. C. Bergstrom. 1982. Diseases of Wyoming wildlife. Second ed. Wyo. Game and Fish Dep., Cheyenne.

———, M. W. Miller, D. A. Jessup, and D. L. Hunter. 1992. Disease as a consideration in translocation and reintroducing wild animals: Western State Wildlife Management Agency Perspectives. Pages 18-25 *in* Proceedings of the joint conference of the American Association of Zoo Veterinarians and the American Association of Wildlife Veterinarians. Am. Assoc. of Zoo Vet.

Political and Economic Issues in Conservation of Rangeland Wildlife

Chapter 25
WILDLIFE LAW: JUDICIAL AND FISCAL RESPONSIBILITIES

Wendell G. Swank

INTRODUCTION

In law, wildlife is that type of property to which all citizens have equal rights and privileges held in common for their joint use. Such property and rights are referred to as "common," and is founded in English common law. In the United States, and most other countries, jurisdiction over the wildlife resource is assigned by law as the responsibility of governments, and everything pertaining to wildlife is specified in state or federal law, and frequently in both.

The origin of our laws is English law, and like English law our laws are formed from several sources. Our Constitution is the basic foundation for all laws in the Country and all laws must comply with the provisions of this historic document. Statutory laws are those made by an elected body, such as a state legislature or the United State Congress. Such bodies are usually not in continuous session nor can they anticipate problems arising from the administration of the laws that they make, hence government administrative entities and administrators are permitted by statutes to make rules and regulations to carry out the provisions of the statutes. Such rules are regulatory laws, and have the full force of the statutes from which they are derived.

Laws are not infrequently contradictory, often vague, subject to various interpretations and perceived intent, hence are challenged in the courts. Decisions of the courts serve as a guide as to how our laws are administered and enforced and have a tremendous influence on the entire legal system.

Wildlife laws, and the legal system has been greatly shaped by what we commonly refer to as politics, but in reality is democracy in action. Laws are dynamic, and those governing wildlife have changed dramatically since 1950 (Matthews 1986). Because of dramatic changes in wildlife habitat, competition for the use of natural resources, and increased interest by a larger segment of the public, even greater changes will probably occur in the future.

Most current government wildlife administrators came up through the ranks in a biological capacity, and although they have a good basic knowledge of biological implications in the management of the resource, it was not until they reached the administrative level that they become aware of the importance of wildlife law. Moreover, through public demand both federal and state laws have progressively given citizens more influence on the management of natural resources through administrative procedure regulations, and government administrators are constantly challenged by the general public and special interest groups regarding legal decisions. Some government administrators, particularly at the higher level, almost daily face possible litigation in response to their actions or inactions.

This chapter covers a brief history of the development of wildlife laws in this country, a discussion on the evolution of laws that govern the division of authority over wildlife between various governmental entities, laws that have had a profound influence on management of rangelands and rangeland wildlife, the administrative structure of most of the public agencies responsible for rangeland wildlife management, and a short review of the more important laws affecting management of wildlife in rangeland country.

WHO OWNS THE WILDLIFE ?

In the United States the people of individual states own the wildlife. To understand how this evolved it is necessary to review the history of this country. At the successful close of the Revolutionary War all of the rights formerly held by the government of England devolved to the original 13 colonies. Those Colonies were independent nations during that period from the conclusion of the Revolutionary War and signing of a treaty in 1783 until the formation of the United States through the Constitution and its ratification of the ninth State, New Hampshire, in 1788. Thus the Colonies acted independently without a union for about 5 years. The rights of the states to wildlife were first clarified in the case Martin versus Wadell in 1842 (Cases cited are listed in Table 1). Martin wanted to exclude others from taking oysters from certain mudflats in the Raritan River in New Jersey. He claimed ownership of the riparian and submerged lands, tracing his title to a grant in 1664 from King Charles II to the Duke of York, which was purported to convey "all lands, islands, soils, rivers, harbors, mines, minerals, quarries, marshes, waters, lakes, fishing, hawking, hunting and fowling" within certain metes and bounds. Chief Justice Taney ruled that the lands and waters in question were held by the King as a public trust, and that it "must be regarded as settled in England against the rights of the King since Magna Charta to make a private grant of such lands and waters," that is, the King was without power to abridge "the public common piscary," which translated means the right of everyone to fish public waters. Taney went on to state: "When the people of New Jersey took possession of the reins of government following the American Revolution they took into their own hands the power of the sovereign, and the prerogatives and regalities which before belonged either to the Crown or the parliament became immediately and rightfully vested in the state."

Several other court cases corroborated the jurisdiction of the states over wildlife, the most enlightening being that of Geer versus Connecticut in 1896. In that case Justice White, writing for the majority, reviewed the history of ownership of game from early Greek through Roman law, through the civil laws of Europe, to common law in England, and concluded that the states had the right to control and regulate the common property in game, which right was to be exercised "as a trust for the benefit of the people." The Geer decision shaped and dominated wildlife management, wildlife law, and the question of jurisdiction over wildlife for many years.

The writers of the Constitution were hesitant about creating a central government with an abundance of power, because they had just successfully thrown off the yoke of such a government. They demonstrated their hesitancy in Article X of the Amendments to the Constitution by stating that "powers not delegated to the Federal Government by the Constitution are reserved by the States, or to the people." Hence the States exerted complete jurisdiction over the wildlife resources of the country without Federal intervention until the passage by Congress of the first federal law pertaining to wildlife, the Lacey Act, in 1900 (Statutes cited are listed in Table 2).

In most instances the people of the individual states still own the wildlife, and the state governments are custodians of the wildlife and act for the people. The Federal Government, through acts of Congress, has gradually exerted more jurisdiction, using certain provisions of the Constitution to justify their action, but this has not been without considerable opposition from the states (Groves 1965, 1966: Olds 1965). The state wildlife agencies have relied on Article X of the Constitution to guard their jurisdiction, and the federal government has been judicious to assure that laws affecting wildlife have been in accordance with provisions of the Constitution. In some instances the initial positions taken by federal agencies have been modified in Congress through the political process to remove opposition from the state agencies.

Laws passed by Congress since 1900 giving jurisdiction over wildlife to federal agencies fall under 3 articles of the Constitution: the interstate commerce clause, the treaty clause, and the property clause. The courts have held that these provisions of the Constitution give the federal government jurisdiction to manage wildlife that supersede that of the states. The interstate commerce clause states that "Congress shall have power to regulate commerce with foreign nations, and among the several states, and with Indian tribes." The Lacey Act falls under this clause, but its major emphasis when passed was to prohibit the introduction, into the United States, species of wildlife thought

TABLE 1. Cases cited in the chapter entitled "Wildlife law: judicial and fiscal responsibilities."[a]

Geer v. Conn., 161 U.S. 519 (1896).
Hunt v. United States, 278 U.S. 96 (1928).
Iron Crow v. Oglala Sioux Tribe, 231 F.2d 89 (1956).
Klepp v. N.M., 426 U.S. 529 (1976).
Martin v. Wadell, 41 U.S. 367 (1842).
Mescalero Apache Tribe v. State of N.M. Dept. of Fish and Game, 451 U. S. 2378. (1983).
Mo. v. Holland, 252 U.S. 416. (1920).
United States vs. Mont. 604 F.2nd 1162 (9th Cir. 1979), cert. granted, 445 U.S. 960. (1980).
White Mountain Apache Tribe v. State of Ariz. Dept. of Game and Fish, 649. F.2nd 1274. (1981).
Worchester v. Ga., 31 U. S. 483. (1832).

[a]Citations follow the 1980, A uniform system of citation, twelfth ed., Harvard Law Review Association, Gannett House, Cambridge, Mass..

TABLE 2. Statutes cited in the chapter entitled "Wildlife law: judicial and fiscal responsibilities."[a]

Administrative Procedure Act of 1946, 5 U.S.C. 551-559; 701-706. (1977).

Classification and Multiple Use Act. Pub. L. 88-607, Repealed 1970.

Conservation Programs On Public Lands. 16 U.S.C. 670g-670o (1988).

Duck Stamp Act, 16 U.S.C. 718-718j (1988).

Endangered Species Act of 1973, 16 U. S. C. 1531-1544 (1988).

Federal Aid in Fish Restoration Act, 16 U.S.C. 777-777e (1988).

Federal Aid in Wildlife Restoration Act. 16 U.S.C. 669-669i (1988).

Federal Land Policy and Management Act of 1976, 43 U.S.C. 1701-1784 (1986).

Fish and Wildlife Conservation Act of 1980, 16 U.S.C. 2901-2912, (1988).

Forest Service Organic Act of June 4, 1897. 30 Stat. II, 215, codified at 16 U.S.C. 475. (1988).

Forest and Rangeland Renewable Resources Planning Act of 1974, 16 U.S.C. 1600-1614. (1988).

General Mining Law of 1872, R.S. 2319 ,c. 152, 17 Stat. 91 (1872), codified at 30 U.S.C. 22-24, 26-30, 33-35, 37, 39-42, 47 (1986).

Homestead Act of 1862, R.S. ss2289 derived from Acts of May 20, 1862; codified at 43 U.S.C. ss161. Repealed by Pub. L. 94-579, Oct. 21, 1976, 90 Stat 2787.

Lacey Act, 16 U.S.C. 701-708(in part), (1988).

Marine Mammal Protection Act, 16 U.S.C. 1361, 1362, 1371-1384, 1401-1407. (1988).

Migratory Bird Treaty Act, 16 U.S.C. 703-712, (1988).

Multiple-Use Sustained Yield Act, 16 U.S.C. 528-531. (1988).

National Wildlife Refuge System Administration Act, 16 U.S.C. 668dd-668ee. (1988).

National Environmental Policy Act. 42 U.S.C. 4321-4370b (1990).

National Forest Management Act of 1976, Pub. L. 94-588, 90 Stat. 2958.

Public Rangeland Improvement Act. 43 U.S.C. 1739, 1751-1753; 1901n-1908 (1986).

Reorganization Plan No. II of 1939. Fed. Reg. 2731, 53 Stat. 1433.

Stockraising Homestead Act, c 9, 39 Stat. 862, (1916) (repealed 1976).

Sikes Act. 16 U.S.C. 670a-670e (1988).

Taylor Grazing Act, June 28, 1934, ch. 865, 16. 48 Stat. 1275.

Whaling Convention Act, 16 U.S.C. 916-916l (1988).

Wilderness Act, 16 U.S.C. 1131-1136 (1988).

Wild Free-Roaming Horses and Burros Act. 16 U.S.C. 1331-1340 (1988).

[a] Citations follow the 1980, A uniform system of citations, twelfth ed., Harvard Law Review Association, Gannett House, Cambridge Mass.

to pose potential harm to native species and to domestic animals. In addition, it prohibits transportation between states of wildlife taken in contravention of the laws of states. This provision was made a part of the law to curb the commercialization of wildlife, which in 1900 was considered a major cause for the reduction of wildlife populations. The right of the federal government to exert control over wildlife in interstate commerce has seldom been tested in the courts.

The property clause of the Constitution states that "Congress shall have the power to dispose of and make all needful rules and regulations respecting the territory or other property belonging to the United States." This provision gives the federal government the authority to regulate the taking and management of wildlife in national parks, in wildlife refuges, and on other federal lands. The courts have also ruled that the federal government may reduce wildlife populations on federal lands to prevent damage to federal property (i.e., the habitat). (Hunt vs. United States, 1928). Primarily on the basis of the decision in Kleppe versus New Mexico (1976), Bean (1977) and Lund (1980) argue persuasively that under the property clause Congress can pass laws, which on federal lands have paramount authority, and can completely supplant the wildlife laws of the states.

The treaty clause of the Constitution states that only the federal government may make treaties with foreign nations, and that all treaties made under the authority of the United States shall be the supreme law of the land. The first treaty made by the United States affecting wildlife was the Migratory Bird Treaty Act of 1918 with Great Britain acting for Canada. That treaty gave the federal government the power to regulate the taking of all migratory game birds, and that authority was upheld in Missouri versus Holland (1920), and numerous other court cases. Many laws under this clause have followed, which now include the Endangered Species Act, the Marine Mammal Protection Act, the Whaling Convention Act. and the many treaties governing fishery management in the oceans and the Great Lakes.

Some proponents of more federal control have contended that Article 1, Section 8 (18) of the Constitution gives Congress the authority to further take from the states the responsibility for management of wildlife. That provision states that (Congress may)" make all laws which shall be necessary for carrying into execution the foregoing powers, and all other powers vested by this Constitution in the Government of the United States, or in any department or officer thereof." No federal wildlife law has been passed under this clause, hence the question has remained untested in the courts.

The state wildlife agencies and the Federal Government agencies concerned with wildlife were progressing toward a legal and legislative confrontation regarding the jurisdiction over resident wildlife in the mid-1960s (Groves 1965, Gottschalk 1975, Olds 1965). Officers in the Federal Government, particular those in the Department of Interior, were assuming a more aggressive role in the management of wildlife, hence wildlife officials of the states, particular the public land states in the West, felt that their traditional position was being threatened. Several court cases ensued and through the instigation of the wildlife agencies of the states and the International Association of Fish and Wildlife

Agencies a bill was drafted for introduction in Congress which stated, in part "it is hereby declared to be the intent of the Congress that no provision of any Act shall be construed or implemented in any manner as to displace, preempt or deprive the several states of their primary and historically recognized authority to control, regulate and manage fish and resident wildlife in and on any territorial boundaries, including all Federally owned lands (Groves 1966)." The proposed legislation went through several revisions, and numerous bills were introduced in the House and the Senate without the official endorsement of the International Association. In the meantime numerous meetings were held and a dialog continued between the International Association and the Secretary of Interior (Olds 1968). Senate Bill 1232 that contained the above clause passed the Senate on 8 December 1969. The International Association had ceased to lobby for the bill because of ongoing negotiations with Federal officials that culminated in an agreement between Secretary of Interior Walter Hickel and the International Association on a policy statement outlining the relationship between the wildlife agencies of the several states and the agencies within the Department of Interior in September 1970 (Woodward 1970).

Every succeeding Secretary of the Interior has followed this policy setting forth this relationship, and the Conservation Program on Public Lands Act, Pub. L. 93-452 of 2 October 1974, provided for cooperative agreements between the Federal agencies in the Departments of Defense, Agriculture, and Interior and the state wildlife agencies. The law specifically states that hunting, trapping and fishing of resident fish and wildlife on public lands shall be conducted in accordance with applicable regulations of the state in which such land is located. This language assured officials of the state wildlife agencies that their traditional function and position was safeguarded, and that the threat of the federal agencies imposing special licensing fees that would jeopardize their source of income was negated.

Most officials in the federal government and most members of Congress respect the traditional role of the states in the management of resident fish and wildlife, and know that federal agencies do not have sufficient personnel to provide an effective protection program or management program for the wildlife of the nation. The U. S. Forest Service issued its policy (W-2) in 1941 that outlines its cooperative program between the several states and the Forest Service (Cliff 1968). That policy states that

> The Chief of the Forest Service, through the Regional Foresters and Forest Supervisors, shall determine the extent to which National Forests or portions thereof may be devoted to wildlife production in combination with other uses and services of the National Forests, and in cooperation with the fish and game departments or other constituted authority of the States concerned, he will formulate plans for securing and maintaining desirable populations of wildlife species, and he may enter into such general or specific cooperative agreements with appropriate State officials as are necessary and desirable for such purposes. Officials of the Forest Service will cooperate with State game officials in the planned and orderly removal in accordance with requirements of State laws of the crop of game, fish, furbearers, and other wildlife on National Forest lands.

In the Department of the Interior a policy issued by the Secretary in 1983, that is still in effect, can be summarized as follows:

1. In Reclamation withdrawal lands the responsibility for the management of fish and wildlife shall rest with the states.
2. With respect to Bureau of Land Management (BLM) lands Rasmussen (1968) stated that the BLM recognizes and supports the position that the states have the primary responsibility for regulating the public's use of wildlife, and that BLM's primary responsibility is providing wildlife habitat. He went on to say that prior to 1964 there was no mandate for wildlife management on BLM lands, however the Classification and Multiple Use Act of 1964, now incorporated in the Federal Land Policy and Management Act of 1976, called for the retention, classification and management of public lands for fish and wildlife development and utilization. The BLM has developed cooperative programs for the management of the habitat with the states.
3. In the National Wildlife Refuge System Administration Act of 1966, Congress explicitly provided that regulations permitting hunting and fishing of resident fish and wildlife within the system shall be, to the extent practicable, consistent with state fish and wildlife laws and regulations. The act goes on to say that nothing in the Act shall be construed as affecting the authority, jurisdiction, or responsibility of the several States to manage, control, or regulate fish and resident wildlife under state law or regulation in any area within the system.
4. On units of the National Park System, in general, consumptive wildlife use is prohibited, however legislation has authorized limited hunting, trapping or fishing activity within certain areas. Such activities shall be in accordance with applicable federal and state laws. Fishing, which is generally permitted in the National Park System, shall be in accordance with applicable state and federal laws.

Most of the states have entered into cooperative agreements for the management of wildlife on lands within the Agriculture, Defense, and Interior departments. In general those agreements confirm that the states have jurisdiction over the wildlife resources of the state, except where federal laws specifically have assumed control over the resource, such as under the Migratory Bird Treaty Act and the Endangered Species Act. The states are responsible for promulgating regulations for hunting and fishing, and take the major responsibility for protecting the wildlife from illegal taking. Federal Agencies take the role of the landowner, and point out that they are responsible for management of the habitat that is essential to the well being of the wildlife. Both agencies agree to consult with and keep their cooperator informed regarding any plan or action that may have an effect on their responsibility or operations. As an example, wildlife departments will consult with the federal agencies in the formulation of hunting seasons and regulations prior to adoption. The state agencies will get permission from the federal agency concerned before planning and making installations such as water catchments or spring improvements.

The jurisdictional status of Indians and other native people over wildlife is quite complicated, and numerous court cases during the past decade involving state, federal and Indian agencies have kept the status indecisive. Court decisions have, however, indicated some trends (Dein 1978). Treaties were made with Indians as settlement moved westward, and over 400 treaties were consummated between 1778 and 1871. Power of Congress to make treaties with Indians was terminated in 1871, but existing treaties were to continue in force until consummated by a congressional act (U. S. Code, Title 25, Sect. 71. 1970), Each treaty is a separate legal contract between the tribe and the government of the United States. Most treaties were written in general terms in vogue at that specific time and leave much to be desired concerning specifics. Essentially the courts have ruled that Indian tribes reserved certain rights in their treaties with the United States, among which was the right to take wildlife on their reservations and in some cases "on unoccupied open land," and in their "usual and accustomed places." The tribes are "quasi-foreign entities possessing all inherent rights of sovereignty excepting where restrictions have been placed thereon by the United States itself" (Iron Crow vs. Oglala Sioux Tribe 1956). Because Indian tribes are foreign entities the treaties between those tribes and the United States fall under Article IV of the Constitution, hence take precedent over the laws of the individual states. Moreover, the courts have ruled that if the treaties are vague or unclear, the native people shall have the benefit of the doubt because they were not versed in the English language, and may have been deceived by those making the treaty for the United States Government (Worchester vs. Georgia 1832).

Traditionally, most state wildlife agencies have recognized the right of members of Indian tribes to take wildlife on their reservation free of state regulations. The taking of wildlife by persons not members of the tribe of the reservation on which the wildlife was taken is more complicated. Courts have held that states and Indian tribes share interest in wildlife that move off and on a reservation, and the states may impose restrictions on taking if the resource is placed in jeopardy (United States vs. Montana 1979). However, the burden of proof lies with the state. It has also been held by the courts that non-Indians may hunt on reservations without a state license as long as reservation regulations are not more liberal than state regulations. The Ninth Circuit Court held that the Arizona Game and Fish Department could not require non-Indians hunting on the White Mountain Apache Reservation to have a state license (White Mountain Apache Tribe v. State of Arizona, Dep. Game and Fish 1981). Two years later the Supreme Court held that New Mexico could not enforce state regulations that required a state license of non-Indians hunting and fishing on the Mescalero Apache Reservation (Mescalero Apache Tribe vs. State of New Mexico, Dep. Fish and Game, 1983). Over the years Congress has swayed back and forth between a policy of assimilation of native peoples versus a policy of self determination, and the present policy of self determination has influenced the courts to repeatedly lean in favor of the native people. As a consequence the states have been less aggressive in their attempts to apply state regulations on Indian reservations, but are still involved in complicated legal questions regarding off-reservation fishing by Indians. Current federal regulations prescribe seasons and bag limits for the hunting of migratory birds on certain federal Indian reservations and Indian territory, but do not cover resident game species (Code of Federal Regulations, Title 50, 20.110(a)-(j)).

So, who owns the wildlife? Unquestionably the people of this country own the wildlife, but the more germane point is one of jurisdiction. The state wildlife agencies have the primary responsibility for management and protection of wildlife. This has been recognized by Congress and is stated in federal laws and in policies by federal agency administrators. Where there is a conflict of interest on federal lands or where international treaties are involved, federal laws may be invoked that preempt state laws. On Indian reservations tribal wildlife laws apply, except for wildlife that move off and onto the reservation, then the states have the ultimate responsibility for conservation of the wildlife resource, but the courts have held that proof of necessity for restraint rests with the state agency.

STATE WILDLIFE AGENCIES

As previously mentioned, state wildlife agencies have primary responsibility for management and protection of the wildlife in the United States. Collectively the state agencies have considerably more resources and personnel than do the federal agencies, and they have closer ties to the public because they have local representatives in almost all communities.

State wildlife agencies come under the jurisdiction of the legislatures and the laws of the individual states, hence there is considerable variation from state to state in administrative structure (Wildlife Management Institute 1977). Most departments operate under policies established by a board or commission. Members of Commissions are not wildlife professionals. Their function is to represent the general public in wildlife affairs, establish regulations for the protection and management of the wildlife, serve as a buffer between the professional staff of the wildlife agency and the political realm within state government, and to implement legislative statutes through commission regulations. Members of these commissions receive no pay but usually are reimbursed for their actual expenses associated with their duties.

State wildlife agencies are administered by a director, who in most states is appointed by the commission. The term of office for the director may be for an established period set by law, or it may be for no set period, and the director serves at the will of the commission. In a few states the director is appointed by the governor, serves at the will of the governor, and usually goes out of office with the governor.

Personnel within the departments are usually appointed and dismissed by the director; however state civil service or personnel regulations are usually prescribed and adhered to in personnel matters. Turnover of personnel in state wildlife departments at the lower and middle levels is usually low, but turnover of directors is relatively high. The average term of directors from 1955 to 1981 for 46 state wildlife agencies was 7.1 years, with terms getting shorter in more recent times (Jahn 1982). Morale within departments greatly influences turnover, and fluctuates with the quality of the leadership and the working relationship between the department, particularly the director, and the commission. One frequent problem is that commissioners are reluctant to act as a body in giving direction to the department through the director, and have a tendency to become involved in the day to day operation of the organization. This results in a diversification of leadership, confusion as to who is in charge, and indecision as to whose direction to follow. This is usually followed by low personnel morale and an exodus from the organization.

Below the commission and director level, departments take on many different organizational structures. Department responsibilities vary considerably from state to state. In some states wildlife responsibilities come under an organization that is responsible also for forests, parks, and in some even water, minerals, air quality and other environmental resources. Of 46 states responding to a questionnaire in 1982, 24 were part of a larger agency and 22 were separate (Jahn 1982). About half of the state wildlife agencies are also responsible for enforcement of the boating laws. Many states have a separate enforcement division, and in a few, enforcement of the wildlife laws is the responsibility of the Department of Public Safety or state police rather than of the wildlife agency.

Funding for state wildlife organizations is provided primarily by license fees paid by those who hunt and fish, and by fines received from the courts from violators of game and fish laws (Jahn and Trefethen 1978). Licenses brought to the state wildlife agencies an income in excess of \$784 million in 1990 (Wildl. Manage. Inst. 1991). In fiscal 1978-79 licenses accounted for 58.3% of the total revenue to state wildlife agencies. The next most important source of revenue is provided from an excise tax imposed at the manufacturing level on firearms, fishing tackle, and archery equipment collected by the federal government and passed through to the states for management, research and land and water acquisition. These programs are in accordance with 2 federal laws, the Federal Aid in Wildlife Restoration Act and the Federal Aid in Fish Restoration Act. Funds from those programs contributed an additional 17.8% to the state wildlife revenue in 1978-79 (Glass 1980). In almost every state efforts have been made to get those who benefit from natural resources to financially support fish and wildlife programs, in addition to hunters and fishermen, but for the most part those have been unsuccessful. Revenue from special car license plates and check-offs from state income tax returns are 2 programs common to the states. These have helped to support management of so called non-game wildlife but revenue from these sources make up a small portion of state wildlife budgets. Proposed programs at both the state and federal level to obtain funds from bird watchers, hikers, campers, and other users of the wildlife resource to support so called non-game and endangered species have been tried for years, but have been ineffective for the most part (Heberlein 1991). At the federal level the Fish and Wildlife Conservation Act of 1980 was passed by Congress to financially support non-game wildlife species, but little or no funds have been appropriated by Congress to implement the Act. Some states get a portion of the money from excise taxes on certain items, others get a small portion of the general revenue, which usually is earmarked for a specific purpose and appropriated by the legislature, and some have been able to get bonds and additional funds specifically for capital outlay items such as purchase of lands for fish and wildlife. Such miscellaneous funding is helpful, but earmarked dependable funding is essential for the success of long-term wildlife management programs (Jahn and Trefethen 1978).

Traditionally, state wildlife agencies have concentrated their activities on protecting wildlife and regulating the legal taking of game species by establishing seasons and bag limits. This is still their major activity, but with more of their constituency becoming urbanized the states are increasing their efforts in such programs as watchable wildlife, environmental protection, rare species management, and wildlife education.

Monitoring huntable species of wildlife by various census techniques and measuring the influence of hunting is a major activity of most state wildlife agencies. Habitat evaluation, particularly that of large herbivores, also accounts for much of the efforts. Analyzing the combined information from hunts, census results, and habitat evaluation is an annual exercise and forms the basis each year for formulating proposed regulations for the upcoming fall hunts. Most state wildlife agencies must follow administrative procedure acts, and are required to hold meetings and give the public an opportunity to comment prior to establishing regulations for the taking of wildlife. This necessitates preparation of data to be presented and frequently involves making many contacts with the public even prior to these meetings, particularly if proposals are to deviate substantially from previous years. Briefing personnel from other natural resource management agencies is also necessary, as wildlife is frequently managed on a cooperative basis. State wildlife agencies own very little land, hence when establishing regulations for wildlife they must consider the views of private landowners and programs of state and federal agencies who control large blocks of wildlife habitat. Good relationships are essential because those who own the land control hunter access, and hunter access is vital to successful state wildlife programs.

State wildlife agencies also expend large sums of money developing and improving wildlife habitat on lands that they do not own; hence it is to the advantage of land managers and owners to cooperate with the state wildlife agency if they are interested in conservation and management of wildlife.

State wildlife agencies have many advantages for the management of the wildlife resources. They can act and respond more readily than most federal agencies; hence they can take advantage of opportunities and can alleviate problems more quickly. Because they are a local organization they are generally more responsive to the desires of local people. This may not in all instances be best for the resource, but in theory it is the way democratic governments should operate.

State wildlife agencies have, for the most part, a roster of very highly qualified personnel. State biologists and wildlife managers usually have ≥1 degree in wildlife, and they tend to stay with the agencies, hence have many years of local experience. Successful wildlife management requires an understanding of biological and social conditions, and such basic knowledge does not come without living and working in an area for several years.

There is an increasing criticism of state wildlife agencies for concentrating their efforts on huntable wildlife and neglecting other species. Perhaps this criticism is warranted, but considering the major sources of their funding, and the continued reluctance of the nonhunting segment of the public to contribute to the cost of management of the wildlife resources, their emphasis on huntable species is understandable. Another problem confronting all scientific wildlife management is the growing anti-hunting sentiment of the public. Most anti-hunting organizations want to concentrate the jurisdiction over all wildlife to a central point (i.e., the federal government), where they feel they have more of an opportunity to exert their influence to eliminate hunting as a recreational activity (Huey 1980). In addition to the legal constraints discussed above, the wildlife activities of the federal government, both in manpower and monetary resources, are much less than that of the collective states. Moreover, eliminating hunting would eliminate the revenue that comes to the states from licenses, and hunters have shown a willingness to pay for the costs associated with wildlife management. If that source of income were eliminated, the management of game and nongame species would become another burden on the back of the tax paying general public, and they have increasingly shown a reluctance to assume more taxes of any type.

RANGE WILDLIFE MANAGEMENT ON PUBLIC LANDS

Public Land Acquisition

At one time practically all of the land within the United States, except the original 13 colonies and the present state of Texas, belonged to the federal government. In 1781 the newly organized United States government took over all of the land between the Allegheny Mountains and the Mississippi River that formerly was claimed by the state of New York, and by 1802 all of the states that bordered the Atlantic Coast had relinquished their claims of their western lands to the federal government (World Book Encyclopedia 1956). The Federal Government in 1803 concluded the Louisiana Purchase with France that added all of the land from New Orleans west of the Mississippi River to the Oregon Territory, which about doubled the size of the country by adding 2.5 million km^2. The Treaty of Guadalupe Hildago concluded the Mexican War in 1847 and added 1.33 million km^2 which later became the states of California, Nevada, and Utah, and most of Arizona and New Mexico, and parts of Colorado and Wyoming. The Mexico-United States border was still in dispute, and was settled through the Gadsden Purchase in 1854, which added another 118,000 km^2 south of the Gila River in what is now Arizona and New Mexico. Oregon became a Territory of the United States in 1848, and this added another 670,000 km^2, which now make up the states of Oregon, Washington, Idaho, and parts of Montana and Wyoming. This concluded the acquisition of land in the Lower 48 states, but in 1867 Secretary of State William Seward committed "Seward's folly" by purchasing Alaska from Russia at a cost of \$7.2 million and adding 1.5 million km^2 of territory. The Federal Government was now land rich, but had little money, so Congress used gifts or grants of land to help settle and develop the country. The confusing land ownership pattern that exits in the western states today is the result of a public land disposal policy that had no central planning, no single administrative agency, and changed at the whim of Congress as each contingency came up.

Public Land Disposal

Public lands belonging to the United States have been disposed of for 4 primary purposes: homesteads to provide a living to an increasing population and to settle the "west," the exploration and development of minerals to increase the economic resources and income to the people and the nation, to railroad developers to facilitate and accelerate transportation into and across the vast expanse of the western mountains and prairies to the more productive coasts of California, Oregon and Washington, and to individual states to provide financing for public education.

The sale of public lands was administered by the General Land Office, established in 1812 as a unit of the Treasury Department, and transferred to the newly created Department of the Interior in 1849. Disposal of the public lands began in earnest with the passage by Congress of the Homestead Act in 1862 (Trefethen 1975). That law provided that any citizen could get title to 64.8 ha of any unoccupied and unwithdrawn public land after living on it and making minor improvements for 5 years. The amount of land/individual was later increased by an additional 64.8 ha in the lands of the arid west when Congress became convinced that additional land was needed there to support a family. The Stockraising Homestead Act of 1916 increased the land area to 259 ha (1 section) for small stockraisers. The General Land Office was established to administer the Homestead Act and to "dispose of the Federal lands in an orderly manner." Most of the land in what is now the "Plains States" was settled under the Homestead Law and in the western Public Land States much of the private land scattered throughout the National Forest System and Bureau of Land Management lands were settled under the various Homestead Acts.

The General Mining Law of 1872 authorized a person locating valuable mineral deposits on most federal lands to obtain a possessory interest of the land, including surface rights around the deposit of sufficient area to permit the development and processing of the minerals, and most, but not all of the minerals extending vertically downward into the earth. The locator, with a minimum of improvements, may apply for a patent and acquire a fee simple title to the land for a fee of \$6.20/ha. It is generally agreed by everyone, except the mining interests, that the mining law is outdated, and has been extensively abused. Much public land has passed into private hands for purposes other than mining. Summer home sites, land developers speculating on expansion of communities, potentially high recreational demand areas, and a variety of other possible land uses have been cited as common abuses. For many years there has been a tide of sentiment to tighten up the mining laws, but the mining lobby has thus far prevailed. From 1874 to June of 1992 the Department of Interior had issued 63,582 mining claim patents covering almost 1.3 million ha of once public lands, and as of June 1992 there were an additional 1.2 million claims on file putting an additional 9.7 million ha at risk (Horning 1992).

Railroad developers were granted odd numbered sections of public lands extending back from railroad right of ways 34 km on each side of the roadbed. This land was frequently sold by the railroad companies for the establishment of towns, to logging companies for the valuable timber resources, to individuals for farming or ranching, and for other purposes. Fortunes were made by some railroad developers, and fortunes were lost by others. Railroads changed the concept of transportation in America, making sport hunting and the wildlife resources of the West commercially available to the populations of the East. Those were exciting times, there was a race to see who would be the first to complete the transcontinental railroad link, and the momentum was moving westward. The Federal Government granted 524,000 km^2 of public lands to the railroads to stimulate this movement (Trefethen 1975).

In 1862 Congress granted each state 121 km^2 from the public lands for each representative and senator in Congress for the purpose of supporting education in the fields of agriculture and mechanical arts at the university level. The land given to the states would be sold or leased and the income would be used to support 1 university in each state, now known as land grant universities. In some states most, if not all of that land has been sold, but in others the states still retain a great amount of land. School lands are intermixed with private and public lands, and for the most part they are indistinguishable as to ownership from adjacent lands. State land departments vary considerably in their attitude regarding wildlife management; some devoting resources and modifying other land uses to benefit wildlife, while others totally ignore wildlife values. Under the University Land Grant program 48.6 million ha of public lands were given to the states (Trefethen 1975), of which 28.7 million were in the 11 western public land states (Plater et. al. 1992).

After 200 years into a program of disposing of the public lands the attitude of the people in the United State and that of Congress gradually changed as the value of those lands was recognized. A tremendous challenge that Congress faced was that of determining how the public lands were to be used, establishing priorities that would be acceptable to the many potential users of the those lands, and passing laws to carry out the will of Congress. This complex problem began with the formation by Congress of the Public Land Law Review Commission in 1964 and the publishing of its findings and recommendations (Public Land Law Review Commission 1970). Among important actions taken by Congress was a change in policy that public lands, where appropriate, were to be retained in government ownership, and that the Bureau of Land Management would be responsible for the management of those lands.

Bureau of Land Management and the Federal Land Policy and Management Act of 1976. The Bureau of Land Management (BLM) is responsible for the multiple use sustained yield management of about 110 million ha of so-called Public Lands, located primarily in the 11 western (public land) states and Alaska. Resources enumerated in the laws governing the activities of the BLM are livestock grazing, watershed protection, outdoor recreation, timber, industrial development, and fish and wildlife production (National Wildlife Federation 1992). The BLM is concerned also with the identification, classification, use and disposal of public lands, and the development, conservation, and use of the mineral resources on lands owned by the government.

The office of the Director of BLM and an advisory staff is located in Washington, D. C. There are offices of State Directors in all of the 11 western states except Washington, which is administered out of the Oregon state office. There is also a state office and Director for Alaska, and 1 collectively for all of the Eastern states that is located in Virginia. District offices administered by District Managers reporting to the State Director are located in strategic areas within each state.

The BLM was formed by consolidating the Taylor Grazing Service and the General Land Office during the Franklin Roosevelt administration under the Government Reorganization Plan II in 1939. The Grazing Service was established under the Taylor Grazing Act of 1934. The purposes of the Act, as stated by Congress, were to establish grazing districts on unappropriated, vacant, unreserved lands of the public domain, to regulate their use, to preserve the land and its resources from destruction, and to provide for improvement and development of the range. The law provided for the establishment of regulations by the Secretary of the Interior for cooperation with stockmen, state land officials and state wildlife agencies. After consolidation under BLM the responsibilities of the Grazing Service and the General Land Office continued almost unchanged. The BLM allocated most of the range resources to livestock, and district BLM employees were considerably under the influence of local range advisory boards, until 1975 composed of from 5 to 12 local stockmen and 1 wildlife representative. The law stipulated that it in no way restricted the right to hunt and fish on lands within grazing districts, and permittees had no right whatsoever to interfere with hunting and fishing. However, it was the contention by BLM that it had no mandate for wildlife use of the range resources until the passage of the Federal Land Policy and Management Act (FLPMA)(Berklund 1975).

In developing policy and making recommendations for management of the public lands many of the recommendations of the Public Land Law Review Commission were enacted into law, but some were rejected, primarily because they were not accepted by the general public and constituencies of special interest groups. Potential impacts of the proposed recommendations that would occur on fish and wildlife resources were formulated and made available to Congress by the International Association of Game, Fish, and Conservation Commissioners, which set up a special committee to follow the work and recommendations of the Land Law Review Commission at its inception (International Association 1971). The Association endorsed or slightly modified many of the recommendations made by the Commission, and most were enacted into the Federal Land Policy and Management Act. However, several notable recommendations by the Commission were rejected by the Association and are not included in the Act. Among those rejected by Congress that would have had an impact on wildlife were that public lands suitable for intensive grazing and farming be sold at market value; that lands within the National Forest System and BLM lands be classified and allocated to their highest use, with economic return to the Treasury the dominant deciding factor; that in the allocation of forage, wildlife would have a subordinate rather than an equal role with cattle; and that a federal land use fee be charged for hunting and fishing on public lands.

The enactment of the FLMPA was a very important event for wildlife on BLM lands. Congress declared that it would be national policy that public lands would be retained rather than disposed of, which had been the national policy up to that time. The law provided for the sale only of particular parcels of public lands if such disposal would better serve the national interest than retention of the lands by the government. Examples of such lands to be disposed of would be tracts that were difficult to manage because of their location, tracts that were acquired for a specific purpose that was no longer required, and tracts for expansion of communities and economic development where no private lands are available.

The law provided for the establishment of goals and objectives, and the planning for management and use of the land on the basis of multiple use and sustained yield of the renewable resources in perpetuity. It defined multiple use as the combination of uses that will best meet the present and future needs of the American people. The uses would include, but not be limited to recreation, range, timber, minerals, watershed, wildlife and fish, and natural scenic and historical values.

In the planning process major or principal uses were defined as domestic livestock grazing, fish and wildlife development and use, mineral exploration and production, rights-of-way, outdoor recreation, and timber production, and that any proposed land use that specifically excluded any of these uses would be subject to review by the Secretary of Interior.

The law called for the preparation of an allotment management plan for the grazing of livestock on BLM and National Forest lands. The plan would prescribe the manner in, and extent to which livestock operations would be conducted to meet the multiple-use, sustained yield objectives. The plan would be prepared in careful and considered consultation, cooperation. and coordination with lessees, permittees, and landowners involved. Leases in the western states would be for a term of 10 years, and where the permittee or lessee was in compliance with terms of the allotment plan, the holder would have first priority for the new permit or lease.

Congress would set the price for grazing leases, and directed that 50%, or $10 million/year, whichever is greater, of money received for grazing of domestic livestock on BLM and Forest Service lands in the 16 contiguous western states be made available for on the ground range rehabilitation in the BLM district or the national forest where the money was derived, and that the remaining one-half be used for range improvements where the Secretary directs. Range improvements with the above funds included enhancement for fish and wildlife habitat. The revenue for habitat improvement is the net amount after deducting 12.5% given state governments in lieu of taxes as provided under the Taylor Grazing Act.

In the FLPMA the statement is made that nothing in the Act shall be construed as authorizing the Secretary concerned to require federal permits to hunt or fish on public lands or Forest Service lands and adjacent waters, or enlarging or diminishing the responsibility and authority of the states for management of resident fish and wildlife. The law does, however, provide that the Secretary concerned may designate on BLM and Forest Service lands where, and establish periods when, no fishing and hunting will be permitted, for reasons of public safety, administration, or compliance with applicable law. Except in emergencies the Secretary concerned must consult with the state game and fish agencies before making any regulations relating to hunting and fishing on public lands.

To carry out the policies set forth by Congress the BLM is directed to develop, in cooperation and consultation with state wildlife agencies, comprehensive plans for the conservation and rehabilitation for fish and wildlife resources on public lands. The BLM and the several states have entered into Memoranda of Understandings to put into effect provisions called for in the FLPM Act, the Public Rangelands Improvement Act, the Endangered Species Act and the provisions of the National Environmental Policy Act. In general these Memoranda recognize the state wildlife agencies as having the responsibility for managing the wildlife on BLM lands and the BLM is responsible for management of the habitat. The BLM agrees to regulate the use of public lands and the state agency agrees to regulate wildlife populations in a manner that will ensure the productivity for future wildlife populations.

The BLM further agrees to advise the state agency of anticipated management measures that may affect wildlife production and carrying capacity, to assist in law enforcement by reporting violations and to make available to the state agency facilities that may be needed to carry out joint wildlife programs on BLM lands. The state agency agrees to notify the BLM of fires or other disturbances on public lands and to render assistance, if appropriate, and to advise the BLM of anticipated harvest or other wildlife management measures that may affect wildlife production or carrying capacity.

The state agency and BLM agree to resolve conflicts through administrative channels, to exchange information relating to range, wildlife and outdoor recreation, to meet at least twice annually to discuss developments, harvest recommendations, and to review accomplishments, to cooperate in resolving public access to public lands, and to cooperate in the preparation and release of publications or news releases about cooperative wildlife management activities.

Wildlife Law and Policy on United States Forest Service Lands. The United States Forest Service (USFS) is responsible for the administration of Forest Service lands under the principles of multiple use and sustained yield of the basic products and services for those lands; namely timber, range, watershed, recreation, and fish and wildlife. The USFS administers 154 national forests, 19 national grasslands, experimental forests and other lands aggregating about 75,506,000 ha (Public Land

Law Review Commission 1970). Administratively, the USFS is in the Department of Agriculture, with the Chief Forester and his staff located in Washington D. C. (National Wildlife Federation 1992). The United States is divided into 9 Forest Service Regions with Regional Foresters and their staffs strategically located to meet the needs of the organization. National forests are headed by Forest Supervisors, also with staff advisors, and forests are divided into Districts headed by District Forest Rangers. Responsibility emanates from the Chief Forester down through the system to the District Rangers who are responsible for decision making at the local level in conformance with policies established at various levels within the system.

The USFS, according to its Organic Act, was established to secure favorable conditions of water flow, and a continuous supply of timber for use and necessities of the citizens of the United States. No reference was made to the management of wildlife or other resources of the forests, and a mandate by Congress for management of national forests for other uses did not exist until the passage of the Multiple-Use, Sustained Yield Act in 1960. That Act stated that national forests are established and shall be administered for outdoor recreation, range, timber, watershed, and wildlife and fish purposes. It was the policy of the USFS almost since its beginning, however, to consider wildlife a part of the forest and managed on the principles of multiple use. The philosophy guiding this management policy was contained in the initial instructions in 1905 from the then Secretary of Agriculture Wilson to Chief of the Forest Service Gifford Pinchot that the national forests shall be managed from the standpoint of the greatest good to the greatest number of people in the long run (Trefethen 1975: 105). That management philosophy was adopted and still serves as the motto of the Forest Service. Accommodating wildlife and producing game for recreational use was one of the services provided for the people.

Congress expanded management and administration of forest service lands by passing the Forest and Rangeland Renewable Resources Planning Act of 1974 and the National Forest Management Act of 1976 that amended and was made a part of the Forest and Rangeland Renewable Resources Planning Act. These acts call for the establishment of objectives for the major USFS programs and the preparation of extensive plans for management of national forests by land management units in order to determine multiple-use and sustained yield relationships. The major thrust of the USFS programs, as viewed by Congress, still appears to be timber production as the Act states that "It is the policy of Congress that all forested lands in the National Forest System shall be maintained in appropriate forest cover with species of trees, degree of stocking, rate of growth, and condition of stand designed to secure the maximum benefits of multiple use sustained yield management." The Act calls for the Secretary of Agriculture to submit a proposed budget to the President with an estimate of the sums needed for "reforestation and other treatments, for securing seed, grow seedlings, prepare sites, plant trees, thin, remove deleterious growth and underbrush, build fences to exclude livestock and adverse wildlife from regeneration areas, and otherwise establish and improve growing forests to secure planned production of trees and other multiple use values." The Act, does however, require assurances from the Secretary of Agriculture that the plans shall "provide for multiple use and sustained yield of the products and services obtained therefrom in accordance with the Multiple-Use Sustained Yield Act, and in particular include coordination of outdoor recreation, range, timber, watershed, wildlife and fish, and wilderness." Plans are to be revised as required by changing conditions, and all plans are to comply with the requirements of the National Environmental Policy Act (NEPA) of 1969. including the preparation of environmental impact statements where appropriate.

Certain restrictions on timber harvest are called for where soil, slope, or other watershed conditions would be irreversibly damaged, and provisions are to be made for protection of streams, stream banks, lakes, and wetlands where harvests are likely to adversely affect fish habitat. Clearcuts and other cutting-regeneration procedures that produce even-aged stands are to be made in blocks, patches and strips blended to the natural terrain, size limits are to be established after appropriate public notice and local hearings and review by the Forest Service officer 1 level higher than the normal decision making officer, and such cuts are to be in a manner consistent with the protection of soil, water, fish, wildlife, recreation and aesthetic resources, and the regeneration of the timber resource.

The law calls for the development of plans for the management for each unit of the National Forest System to be made by an interdisciplinary team based on inventories of the forest resources, and incorporated into 1 document or set of documents. The plans, along with supporting data and maps are to made available to the public for examination, review, and participation at locations convenient to the affected unit at least 3 months before final adoption, and the Secretary, through National Forest officers, shall publicize and hold public meetings to foster public participation in the review of such plans.

Under the law the Secretary is to establish standards that will insure that trees in the National Forest System have generally reached their mean annual increment of growth, and that cuttings in exception to these standards will be made only after consideration has been given to multiple uses of the forest including, but not limited to recreation, wildlife habitat, and range, and after completion of the public participation process called for above.

The USFS, the public and the courts have wrestled with the provisions of the Multiple-Use Sustained Yield Act and the Forest and Rangeland Renewable Resources Planning Act, mostly as these acts apply to a "balanced" use. Obviously all uses cannot be equal on every area, and the courts have ruled that the provisions of the above 2 acts have been fulfilled as long as the planners have "considered" all of the various uses when decisions were made (Bean 1977). Litigants against Forest Service action have been more successful and more numerous using provisions of the Wilderness Act, the NEPA, the Endangered Species Act and the Administrative Procedure Act (United State Code, Title 5, 1988) rather than the Multiple-Use Sustained Yield Act and the Forest and Rangeland Renewable Resources Planning Act.

Plans for the management of wildlife on USFS lands are usually generated cooperatively by the USFS and the affected state wildlife agency, and such cooperative programs are covered in formal cooperative agreements. In the agreements the USFS recognizes the state wildlife agency as the organization primarily responsible for the control, propagation, protection and management of wildlife populations, The state agency recognizes the USFS as the agency responsible for management of the wildlife habitat on USFS lands and for the preparation of forest management plans as specified in the National Forest Management Act of 1976, and in compliance with the NEPA, the Endangered Species Act, the Wilderness Act and other relevant Federal legislation. The USFS further agrees to maintain favorable wildlife habitat on National Forests, to issue special use permits to the state wildlife department for structures or improvements for wildlife management and management activities, to invite the department to participate in USFS plans, including the NEPA process, and to allow access by all state department employees to areas and facilities necessary to accomplish mutually agreed upon objectives.

The state wildlife agencies agree to take primary responsibility for management of wildlife populations and enforcing applicable laws on USFS lands, to assist in the suppression of fires on Forest Service lands, and to submit environmental documents for approval prior to undertaking any project on Forest Service lands. The department further agrees to consult with Forest Service employees on candidates for endangered or threatened status of animals or plants, to participate fully in the NEPA process regarding possible impacts to wildlife, to keep the USFS informed and allow input on proposals for hunting, fishing and trapping regulations, to notify the Forest Service of changes in wildlife and watercraft laws, and neither make nor sanction the release of any wildlife that may affect forest management until a joint agreement has been approved.

Both agencies agree to share and exchange information related to wildlife and habitat studies, to meet at periodic intervals, to resolve conflicts through appropriate channels, and that publicity on joint wildlife projects give recognition to the nature of the work undertaken by each agency.

Wildlife Programs On United States Fish and Wildlife Lands. The U. S. Fish and Wildlife Service (USFWS) is the agency at the federal level with primary responsibility for the conservation and management of migratory birds and endangered species, marine mammals, and certain sport fisheries. The USFWS operates a system of wildlife refuges for migratory birds and other wildlife species, manages hunting of migratory game bird species through regulations, enforces federal regulations prescribed by federal laws, conducts research relating to fish and wildlife problems, administers federal aid fish and wildlife programs to the states, and serves as the leading federal agency in international conventions for wildlife conservation (National Wildlife Federation 1992).

The USFWS is in the Department of Interior, and the Director's office, along with administrative and advisory staff, is located in Washington, D.C. There are 7 regional offices located in different geographical areas of the country and 1 research and development regional office located in the Washington, D.C., area. Each region is headed by a regional director along with administrative and advisory staff. There are also several research and development centers, and the USFWS takes the major role in the Cooperative Fish and Wildlife Research Units located at many state universities.

The USFWS had its beginning in 1885 when Congress passed a law authorizing funds in the budget of the Department of Agriculture to establish a Division of Economic Ornithology and Mammalogy for studies on birds and mammals and their relation to agriculture (Trefethen 1975). Congress expanded the scope and renamed the organization the Division of Biological Survey in 1896. In 1939 the Bureau of Fisheries in the Commerce Department and the Bureau of Biological Survey in Agriculture were transferred to the Department of Interior, and under Federal Reorganization Plan III in 1940 the 2 bureaus were combined and renamed the USFWS. In 1970 commercial fisheries functions were transferred back to Commerce, and in Interior the Bureau of Sports Fisheries and Wildlife was created. However, under Pub. L. 93-271 passed by Congress in 1974, the present organizational and administrative structure was specified and the agency was again renamed the USFWS.

One of the major functions of the USFWS is to manage North American migratory bird species as provided in agreements between the United States, Canada, Mexico, Japan, and Russia. A major activity of the USFWS in the United States is providing resting areas, food, and protection from hunting and harassment for migratory waterfowl during their migration and on their nesting and wintering grounds. The creation of refuges for waterfowl by the United States had begun a long time before the consummation of the Migratory Bird Treaty Act in 1918. President Theodore Roosevelt created the first national wildlife refuge, Pelican Island, along the east coast of Florida in 1903, primarily to curtail the decimation of an egret (*Bubulcus spp.* and *Egretta spp.*) colony by plume hunters. This was followed rapidly by the withdrawal of public lands for numerous wildlife refuges for the protection and propagation of a variety of species, and now the National Wildlife Refuge System consists of over 272 different units encompassing an area of about 14 million ha operating in all 50 states (National Wildlife Federation 1992).

One of the major sources of funds for the acquisition of lands and waters for the National Wildlife Refuge System was created by the Duck Stamp Act. The origin of the Act began in Congress 10 years before it passed with the introduction of legislation with the avowed purpose of perpetuating the great North American waterfowl population However proponents of the objective could not agree on the methods, and the proposal died for lack of unified support within Congress (Trefethen 1975). The great drought of the 1930's decimated the waterfowl population as the fall flights from the nesting grounds of the prairie pothole region decreased to a trickle and birds on the wintering grounds encountered a shortage of food as they were forced to concentrate along the coasts for lack of inland waters. This plight of waterfowl was not lost on Congress, and the decade old concept of waterfowl hunters paying for the acquisition of waterfowl habitat was resurrected and became law as the Migratory Bird

Hunting (Duck) Stamp Act on 16 March 1934. Revenue derived from those stamps, over the amount paid the United States Postal Service for printing and selling the stamps, is deposited in the Migratory Bird Conservation Fund and used for the acquisition of areas for migratory bird refuges and waterfowl production areas. The original law provided for Duck Stamp funds to be used for management and enforcement of regulations on refuges and land acquisition. Over time the USFWS used an increasing amount of funds from duck stamps for management and enforcement, but those contributing the money deemed acquisition to be of paramount importance and through public pressure on Congress the law was amended in 1958 to confine funds from duck stamps to acquisition. The Duck Stamp program has been of tremendous benefit to the National Wildlife Refuge System as it has provided an assured source of funds, contributing millions of dollars since 1934 for the acquisition of wetlands and the management of waterfowl and other wildlife species that benefit from the program.

Refuges are administered under the provisions of the National Wildlife Refuge System Administration Act of 1966 and amendments thereto. National Wildlife Refuges are established for the restoration, preservation, development and maintenance of wildlife and wetlands habitat; for the protection and preservation of endangered or threatened species and their habitat; and for the management of wildlife and wildlands to obtain the maximum benefits from these resources (Code of Federal Regulations, Title 50 25.11(b).

The National Wildlife Refuge System Administration Act permits the Secretary of Interior to acquire, transfer, lease, and dispose of lands within the USFWS, prescribes prohibited and permitted activities on wildlife refuges, states that the Act does not permit the Secretary to regulate hunting and fishing of resident wildlife on lands not within the system, and that hunting and fishing of resident wildlife, to the extent practicable, shall be consistent with state laws and regulations. The Act provides for the use of refuge areas for public recreation to the extent practicable and when not inconsistent with the primary objectives and uses for which the areas were established, and provides for enforcement provisions and penalties. The Act provides that an area not to exceed 40% of a refuge at any one time may be open for the taking of migratory waterfowl. It also stipulates that prior approval must be secured from the Migratory Bird Conservation Commission for the disposal of any lands within the System and that funds derived from lands sold that were originally purchased with revenue from the Duck Stamp Act must be returned to the Migratory Bird Conservation Fund.

The public has an opportunity to participate in the rule making process for the management of refuges in accordance with the provisions of the Administrative Procedure Act. Rules pertaining to closing and opening refuges, and for governing entry to and uses and activities permitted on refuges are also given in Title 50, Code of Federal Regulations (50 C.F.R.). Various recreational activities are permitted on refuges as long as they are not in conflict with the primary objectives for which the area was established. Entry permits and fees may or may not be required, depending on facilities available, degree of public use, time of year, and planned activity of the visitor. State laws apply to motor vehicle use and state laws and Coast Guard regulations apply to boat use. Trespassing domestic livestock may be a problem and may be impounded and disposed of according to the provisions of state law. Dogs and cats running at large may be disposed of in the interest of public safety and protection of wildlife. Use of natural resources on refuges for economic return such as livestock grazing, timber, firewood, cultivating, and sand and gravel, may be permitted, depending on the purposes for which the refuge was established. The production and harvesting of hay and grain may be on a share-in-kind basis when such activities are of benefit to wildlife and the extraction of minerals, including oil and gas, are permitted under certain conditions.

Feral animals on refuges that have reverted to the wild may be taken by authorized federal and state personnel or by private persons operating under a permit. Species of wildlife determined to be detrimental or surplus may be donated to public agencies or institutions, destroyed by authorized federal or state personnel, or trapped or hunted by the general public. The opening of a refuge to hunting must be preceded by a determination by the Secretary of Interior that hunting will be compatible with the principles of sound wildlife management and in the public interest. Refuge-specific hunting regulations are initially published in the daily issue of the Federal Register, and subsequently published in 50 C.F.R. sec. 32. The number of Federal refuges that provide hunting varies from year to year, but in 1991 approximately 200, or 75% of the Federal Refuges were open for migratory game birds, upland game, and big game hunting.

Most Federal Refuges in 1991 provided migratory game bird hunting, and on most refuges the use of non-toxic shot was required. On many the number of shells permitted for each day's hunting was stipulated on the theory that this would reduce random shooting, hence crippling loss.

Because hunting of resident game species on Federal Refuges is in compliance with the regulations of the individual states there are great variations nation-wide. Federal refuges provide hunting opportunities for many popular small game species and some of the best hunting in the country for large game.

Birding has always been a major activity on wildlife refuges, but as the Nation's population has becomes more urban the participation in this activity has climbed significantly, and use of refuges as favorite birding spots has increased. The large concentrations of waterfowl found on refuges is in itself a great attraction, but the combination of facilities available specifically for bird watching and bird photography, the knowledge about birdlife inherent in personnel at those refuges, and their assistance to birders in locating specific species adds to the popularity of refuges among birders. The increased emphasis of endangered species management on national wildlife refuges is also a factor. Aransas Wildlife Refuge, Texas, is a great attraction because it is the wintering ground for the endangered whooping crane, and a refuge specifically for the endangered Attwater's prairie chicken (Tympanuchus) is in operation at Eagle Lake, Texas. Other very popular birding refuges are Great Swamp in New Jersey, Cape Romain in South Carolina, and Santa Ana and Laguna Atascosa in Texas (Nelson 1975).

Wildlife Conservation on Military Reservations. The Department of Defense has under its administration about 12 million ha in the United States, much of it prime wildlife habitat. Prior to the 1950s there were no laws or direction from Congress regarding Department of Defense wildlife programs or public access to wildlife on military lands. Wildlife programs, if they existed at all, were dictated by the base commander, and often in direct conflict with state laws (International Assoc. 1950, 1953). In a response to this criticism Congressman Robert Sikes of Florida in 1950 introduced a bill in Congress to create a cooperative program between the Florida Game and Freshwater Fish Commission and Elgin Air Force Base for the management of wildlife and wildlife habitat on the base. The wildlife program at Elgin proved to be very successful, benefiting wildlife, and receiving approval of the military establishment and the Florida wildlife personnel. In an attempt to resolve jurisdictional conflicts between state wildlife agencies and military base commanders several bills were introduced in Congress over the years to clarify jurisdiction and provide a national program for wildlife management on Department of Defense lands. In 1958 Congress passed P.L. 85-337 giving state wildlife agencies access to military reservations to carry out certain management functions, and in 1960 Congressman Sikes, drawing on his earlier success with the program at Elgin Air Force Base, succeeded in getting Congress to pass a more comprehensive law, Pub. L. 86-797, Fish and Wildlife Conservation on Military Reservations, commonly know as the Sikes Act. This law authorized the Secretary of Defense to carry out programs for management of the fish and wildlife resources on military reservations according to a plan developed cooperatively between the Secretary of Defense, the Secretary of the Interior (represented by the USFWS) and the appropriate state wildlife agency in which the military base is located. The law provided for fish and wildlife habitat improvements, as called for in the plan, to be paid for out of revenue from State issued special permits sold to those who hunted and fished on the reservation. The permits were to be sold by the Commanding Officer of the base, acting as agents for the state wildlife organization, and the funds were to be used only on the reservation where collected.

An extensive amendment was made to the Sikes Act in 1974 (Pub. L. 93-452), providing for range rehabilitation to support wildlife, and control of off-road vehicles to be included in the plans, for the inclusion of recreational facilities, and an authorization for appropriations out of general revenue not to exceed $1.5 million/fiscal year to carry out the provisions of the law.

Special management provisions for endangered and threatened animals and plants on military reservations were specified in an amendment to the Act in 1982 (Pub. L. 97-396), and amendments made in 1986 specified that Department of Defense lands would be managed to provide for sustained multipurpose uses, and that public access to the extent possible would be provided as long as the military mission of the base was not affected. In addition, fish and wildlife management functions would be performed by Department of Defense personnel professionally trained to perform those functions. Sale of products from the land or leasing of the land for other uses would be permitted only when such function would not interfere with natural resources or multiple use programs. This amendment was made to provide wildlife programs equal status with other natural resources activities on military bases.

The Sikes Act has made a difference in the management of wildlife resources on military bases. The recognition of the states as the primary agency responsible for resident fish and wildlife management and the requirement for those who hunt and fish on military reservations to comply with state regulations have defused the contentious issues that existed prior to the passage of the Act. Development and implementation of plans cooperatively generated have improved working relationships between state wildlife personnel and Department of Defense personnel, to the benefit of the wildlife resource. To some extent, however, the success of the program is still dependent on the decisions of the individual base commander. Almost any program requiring use of the land can be curtailed or eliminated with the assertion that there is interference with the assigned military mission. Moreover, funding has been a continuous problem. Congress, in the 1982 amendment provided for authorized funding to support wildlife programs on military bases through 1993, but actual appropriation of funds has been sporadic or non-existent, and wildlife programs have had to operate on money generated from the sale of permits, or use funds from other sources. This has contributed to instability and lack of progress for wildlife programs on some bases.

MISCELLANEOUS FEDERAL LAWS APPLICABLE TO RANGELAND WILDLIFE

The brief description of the some of the activities of the USFWS, USFS, BLM, and the Department of Defense covers only a portion of laws applicable to the responsibilities of the federal agencies for the management of rangeland wildlife. Those interested and concerned with the management of wildlife on federal lands should review appropriate current titles of the United States Code (USC), particularly Title 7, Agriculture, Title 10, Armed Forces, Title 16, Conservation, Title 33, Navigation and Navigable Waters (wetlands) and title 43, Public Lands. Annotated versions of the U.S.C. are published by several companies, and have the advantage of giving an exact duplication of the U.S.C. with up to date annotations of interpretations of the law by the courts.

In the use of the U.S.C. one must also consult the appropriate Code of Federal Regulations (C.F.R.), which spell out the details and current regulations prescribed by the federal agency responsible for carrying out the law. For example, Title 50 C.F.R. is devoted to regulations for fish and wildlife. Each volume of the C.F.R. is published at ≥ 1 time a year, hence is more current than the U.S.C. that is published every 4 years. These publications are available at many law firms, at most universities that have a school of law, and at universities that are official depositories of federal publications.

The Endangered Species Act

Passing The Endangered Species Act (ESA) by Congress in 1973 greatly modified management practices carried on by natural resource management agencies. The Act also limited wildlife habitat management options, particularly on federal lands. The ESA was passed by Congress in 1973 in response to the accelerated rate at which various species of fish, wildlife and plants were becoming extinct and in danger of or threatened with extinction. The responsibility for implementing the provisions of the Act rests primarily with the Secretary of the Interior acting through the U. S. Fish and Wildlife Service. The Secretary is charged with preparing a list of endangered species and threatened species, and making a review and possible revision of such lists ≥5 years. The Secretary may enter into cooperative agreements with state wildlife agencies to carry out provisions of the Act for species that are endangered within the respective state after determining that the state agency has a program that is adequate to assume the responsibilities called for in the Act. All federal agencies are to abide by the provisions of the Act, and are charged with implementing procedures that may be necessary to bring the threatened species to a level where special conservation measures are no longer necessary. Recovery plans are to be formulated for each listed species using the expertise of personnel from any source, and keeping state conservation agencies, county officials, and the general public informed of proposed actions. The Secretary, through regulatory procedures, is authorized to designate habitat considered to be critical for the survival of the species and "each Federal agency shall, in consultation with and with the assistance of the Secretary, insure that any action authorized, funded, or carried out by such agency, is not likely to jeopardize the continued existence of any endangered or threatened species or result in the destruction or adverse modification of habitat of such species which is determined by the Secretary, after consultation as appropriate with affected States, to be critical, unless such agency has been granted an exemption for such action by the Committee pursuant to subsection (h) of this section" (16 United States Code, Sect. 1536 (a)(2). The Committee referred to above is formally known as the Endangered Species Committee, and is composed of the Secretaries of Interior, Agriculture, Army, Chairman of the Council of Economic Advisors, the Administrator of the Environmental Protection Agency, and the Administrator of the National Oceanic and Atmospheric Administration. The Act prescribes procedures that the Committee must follow in making decisions for an exemption requested by a federal agency. The decision made by the Committee shall be considered final by the federal government. Procedures for an appeal through the courts from any decisions made by officials of the Federal Government are covered in the Act.

The National Environmental Policy Act

The NEPA was passed by Congress in 1970 (Pub. L. 91-190). Section 101 of the Act expresses the concern of Congress, and states (in part) "The Congress, recognizing the profound impact of man's activities on the interrelationships of all components of the natural environment . . . , declares that it is the continuing policy of the Federal Government . . . to create and maintain conditions under which man and nature can exist in productive harmony, and fulfill the social, economic, and other requirements of present and future generations of Americans." The major action mechanism of NEPA is contained in Sect. 102 that states that ". . . all agencies of the Federal Government shall . . ."

> include in every recommendation or report on proposals for legislation and other major Federal actions significantly affecting the quality of the human environment a detailed statement by the responsible official on—
> (i) the environmental impact of the proposed action,
> (ii) any adverse environmental effects which cannot be avoided should the proposal be implemented,
> (iii) alternatives to the proposed action,
> (iv) the relationship between local short-term uses of man's environment and the maintenance and enhancement of long term productivity, and
> (v) any irreversible and irretrievable commitments of resources which would be involved in the proposed action should it be implemented.

This "stop and think" provision of the NEPA, the Federal Administrative Procedure Act (5 U.S.C. 551-559) that requires information be made available to the public on almost every action proposed or taken by government officials, and the potential threat of litigation by citizens and citizen-backed organizations have changed the way in which business is conducted by most federal agencies that deal with natural resources. The preparation of an environmental impact statement is time consuming and costly, and many government officials attempt to short-cut that requirement by declaring that proposed action is not a major action, or that the human environment will not be significantly affected (Plater et al. 1992:615-621). In spite of the many decisions handed down by the courts focusing on the intent of Congress the determination of "major and significant action" is still elusive (Findley and Farber 1985:113-131). What is major and significant are both subjected to the judgment of individuals, and are the focus for most NEPA court cases. Undoubtedly, NEPA has been beneficial in protecting the environment in the review process to assure that action taken would minimize the adverse environmental effects, and in the many projects that have not been proposed because they would have been terminated before they were started. On the other hand, fulfilling the requirements of NEPA and attempting to appease some environmental activists has placed a burden on government agencies, delaying the implementation of worthy projects and costing the American taxpayer money for government personnel to prepare environmental impact statements and environmental assessments. The jury is still out on long-term public benefits versus liabilities of NEPA, and the real worth of the law may never be known.

The Wilderness Act

The Wilderness Act designates certain federally owned lands as wilderness areas, and was passed by Congress in 1964. The stated purpose of the Act is . . . "to secure for the present and future generations the benefits of an enduring resource of wilderness, where people may enjoy areas retaining their primeval character, unimpaired by the activities of man, and where man himself is a visitor who does not remain." Areas within the National Forest System managed as "wilderness, "wild" or "canoe" prior to the Act were automatically included as wilderness areas under the Act, and roadless areas in the National Parks System and the National Wildlife Refuge System came under review and were considered by Congress for Wilderness designation within 10 years after passage of the Act. Recommendations for wilderness status for areas are made by the President but must be created by an act of Congress, and the areas remain under the administration of the federal agency under which they were administered prior to designation as wilderness. Jurisdiction of resident fish and wildlife was not affected by wilderness status, livestock grazing was retained, and laws pertaining to mineral exploration and development were extended unaltered to 1 January 1984, at which time the appropriation to mineral rights in wilderness areas was withdrawn.

The prohibition of the use of any motorized vehicle, boat or aircraft in wilderness areas is their primary distinction, and this has advantages and disadvantages for wildlife managers. Any activity that requires access such as surveys, gathering research data, wildlife transplants, and controlling over-populations through hunting requires more time, effort and resources. On the other hand these are rather unique areas where wildlife can be studied unaltered by extensive activities of man. They can serve as control areas to evaluate management results on more accessible sites, such as controlled burns, timbering, various wildlife harvest intensities and techniques, and other conventional man-induced activities.

Wild Free-Roaming Horses And Burros Act

The Wild Free-Roaming Horses And Burros Act (WFRHBA) was passed by Congress in 1971. The history of providing protection and management of the wild horses and burros of the West is a unique example of 1 person goading Congress to take action. Velma B. Johnson, otherwise known as "Wild Horse Annie," walked the halls of Congress until she succeeded in getting Public Law 86-234 passed by Congress and signed into law by President Eisenhower on 8 September, 1959. That law prohibited the use of aircraft or motorized vehicles in rounding up or hunting wild horses and burros on federal land. Nevertheless, the practice of using vehicles and aircraft to round up these animals continued. They were processed for pet food or fertilizer, or they were killed just to reduce competition for forage with cattle. To give more protection to the mustangs Velma Johnson and other wild horse enthusiasts persisted and finally got Public Law 92-195 (WFRHBA) passed and signed on 15 December, 1971 giving specific legal protection to horses and burros ranging on BLM and USFS lands. Under protection, wild horses and burros increased on public lands, and that created management problems for the federal agencies. To assist in roundups and transportation of captured animals the Act was amended in 1976 as a portion of Public Law 94-579 (Federal Land Policy and Management Act) to permit federal agencies to use helicopters and motor vehicles. Two years later Congress, recognizing that the western ranges were deteriorating and forage had to be apportioned if an ecological balance was to be maintained, passed Public Law 95-514, the Public Rangelands Improvement Act, setting up a program for disposal of excess wild horses and burros, and directed the Secretary, ". . . that when an overpopulation (of horses and burros) exists on a given area of public lands . . . to immediately remove excess animals from the range so as to achieve appropriate management levels. Action shall be taken, in the following order and priority, until all excess animals have been removed so as to restore a thriving natural ecological balance to the range, and protect the range from deterioration associated with overpopulation."

The results of the program emanating from the FWRHBA have been controversial, not as successful as anticipated by Congress, and extremely costly to the American taxpayers (Eagle et al. 1992). The major task has fallen to the BLM, who had 38,000 of the 39,000 feral horses on federal lands in 1988, 27,000 of them in Nevada. The BLM found places for 9,500 horses through its adoption program in fiscal 1985 at a cost of $3.5 million, but due to public pressure the "humane destruction in a cost efficient manner" clause of the WFRHBA was not invoked and excess horses were maintained in corrals costing an additional $6 million (Eagle et al. 1992, from Boles 1986).

Experimental hormonal reproduction control studies have reduced foaling rates of mares from ≥42 to ≤11% for 2 years (Eagle et al. 1992), but the influence of the reduced foaling rates on the population, cost efficiency for any reduction that may occur, and prospects for improved range conditions are yet to be determined.

SUMMARY

Wildlife in the United States is public property, and is held in trust and managed by government agencies according to the laws of the country. This jurisdiction was inherited by the original 13 colonies from the British Government and descended to the individual states as they joined the Union. In 1900 Congress passed the first federal law, the Lacey Act, which made it a federal offense to transport between the states game taken contrary to the laws of the state in which the game was taken. This was followed by the Migratory Bird Treaty Act in 1918 that gave the federal government jurisdiction over birds that migrated from Canada into or through the United States and the court case, Hunt versus the United States in 1928 clarified the jurisdiction of the federal government over wildlife and other natural resources that occurred on government lands. The basis of the role of the federal government to preempt the traditional jurisdiction of the states over wildlife is based on the United States Constitution, particularly the treaty clause, the property clause, and the interstate commerce clause.

The jurisdiction of Indian tribes over wildlife on their tribal lands supersedes that of the states, and in some instances hunting and fishing rights off reservations were retained by treaty. Federal laws apply on Indian reservations unless specifically exempt by treaty.

The various responsibilities and functions of the federal and state agencies are clarified by federal and state laws, policy directives, and cooperative agreements. State wildlife agencies have the major role in the management of wildlife. Collectively, they have more personnel, more funds, and are organized to serve local constituencies and respond to local problems. On federal lands the federal agencies assume major responsibility for management of the habitat, and the involvement of management of wildlife populations varies with the agencies. The degree of cooperative federal-state wildlife programs vary between federal agencies concerned and within the various states. In general there are more federal-state cooperative programs in the western states because that is where most of the federal land occurs.

More emphasis on non-game wildlife is occurring in state and federal agencies. This is in response to a growing urban constituency and increased interest in watchable wildlife. However, major funding for wildlife management, particularly for state programs, is still provided by people who buy licenses to hunt and fish.

Rangeland wildlife are those animals that occupy extensive grasslands or grazing country, and to a great extent that means the western public land states. At different times in the history of the country the government of the United States (the people) owned all of the land west of the original 13 colonies. The land tenure, pattern of ownership, uses, revenue produced by that land, and the politics that shaped government laws and policies that greatly influences the management of rangeland wildlife is tied to events of the past. To understand conditions as they are today it is necessary to know those events.

The most productive lands were homesteaded, hence we find private lands surrounded by national forest. Large blocks of land are owned by timber companies that were purchased from those that received land to finance the building of the railroads from the Mississippi River to the Pacific. State land given to the state to finance public education is intermixed with federal land. Privately owned small tracts, and some not so small, are lands acquired under the mining laws that encouraged mineral exploration and development in the West. National parks were established in those areas that had great attractions, and national forests were mandated for those areas covered by trees and those forming important watersheds. Land that was left over and no one particularly wanted became the public domain, and today are BLM lands.

The BLM manages the federal lands under its jurisdiction, and the wildlife that occupies that land, primarily as specified in the Federal Land Policy and Management Act. Raising of domestic livestock and providing habitat for wildlife are 2 major uses of BLM lands, although many other forms of outdoor recreation are gaining in importance, particularly closer to large metropolitan centers. The lands are managed on a multiple use sustained yield basis, defined as the combination of uses that will best meet the present and future needs of the American people.

Congress established national forests to secure favorable conditions of water flow and a continuous supply of timber. Providing habitat for wildlife and other uses was not mandated by Congress until 1960; however, national forest administrators have considered wildlife an important component of the forest since inception. National forests are managed principally under the provisions of the Forest and Rangeland Renewable Resources Planning Act and the National Forest Management Act. Although Congress has called for the national forests to be managed on a multiple use sustained yield basis, in the view of the general public, the production of timber still seems to receive emphasis over other uses, particularly during the past decade. This has resulted in numerous legal actions taken and threats of legal action by conservation organizations, including state wildlife agencies, alleging that the Forest Service has not followed procedures called for in the Environmental Policy Act, the Endangered Species Act, and the Administrative Procedure Act.

The Forest Service and state wildlife organizations manage wildlife on USFS lands through cooperative agreements, the state organizations taking the lead on establishing hunting seasons and providing protection against illegal taking, and the USFS concentrating on habitat management.

The USFWS is responsible for the management of over 272 National Wildlife Refuges encompassing about 14 million ha. Most areas are managed for migratory bird species, particularly waterfowl, but many of these areas also have good populations of rangeland wildlife. Areas for some refuges have been acquired specifically for the management of rangeland wildlife species, particularly in the southwestern deserts. Federal refuges are managed on a multiple use basis, but activities must be compatible with the use for which the refuge was established. Wildlife viewing, hunting, camping, fishing, picnicking, and boating are major uses of refuges.

The Department of Defense administers about 12 million ha of public lands, much of it prime wildlife habitat. Wildlife programs on military reservations are carried on under plans developed cooperatively by the Secretary of Defense, the Secretary of Interior, represented by the USFWS, and the respective state wildlife agencies. Revenue from permits to hunt is used for wildlife management on the base where collected. Special management provisions are made for rare and endangered species, and the areas are managed for sustained multipurpose uses. Management functions are performed by professionally trained Department of Defense wildlife personnel.

The Endangered Species Act of 1973 greatly altered wildlife and land management practices by both federal and state natural resource agencies. Federal agencies must insure that any action taken is not likely to jeopardize an endangered species or result in the destruction of habitat of any listed species. The USFWS is required to prepare a list of rare and endangered species and prepare recovery plans for those species using personnel from any source.

The National Environmental Policy Act of 1970 requires all federal agencies to prepare environmental impact statements or environmental assessments prior to taking any action that may affect the human environment. Information on proposed action must be made available to the general public, along with an assessment of the effect on the environment should the proposed action be implemented, and possible alternatives to the proposed action. The NEPA has invoked extensive litigation between federal agencies and various conservation and other organizations. It has prompted federal agencies to conduct more in-depth analysis of proposed actions which has been beneficial to the environment, but it has been costly to the American taxpayer and delayed implementation of worthy projects.

The Wilderness Act of 1964 made provisions for establishing areas of federal land that would retain wilderness characteristics: no roads, no motorized vehicles, no development. These areas provide an opportunity to study and manage wildlife free of extensive human disturbance, but necessitate more resources to conduct wildlife inventories and make wildlife habitat improvements.

Congress, by passing the WFRHBA of 1971, gave these feral animals on Bureau of Land Management and Forest Service lands status equal to that of native wildlife. The BLM program of rounding up feral horses for adoption, then feeding those that are surplus to the adoption program cost the American people about $10 million in 1985. Experimental hormonal studies to decrease productivity are underway, but their costs and effects on the population are yet to be determined.

LITERATURE CITED

Bean, M. J. 1977. The developing law of wildlife conservation on national forests and natural resource lands. J. Contemporary Law 4:58-77.

Berklund, C. 1975. Land management for the conservation of wildlife. Proc. Int. Assoc. Fish and Wildl. Agencies 65:83- 87.

Boyles, J. S. 1986. Managing America's wild horses and burros. J. Equine Vet. Sci. 6:261-265.

Cliff, E. P. 1968. State-Federal cooperation in resource management. Proc. Int. Assoc. Game, Fish, and Conserv. Comm. 58:71-74.

Code of Federal Regulations, Title 50. 1991. Wildlife and Fisheries, Parts 1-199. U.S. Gov. Printing Off., Washington, D.C. 710pp.

Dein, J. G. 1978. State jurisdiction and on-reservation affairs: Puyallup Tribe v. Dep't of game. Environ. Affairs. 6:535-564.

Eagle, T. C., E. D. Plotka, R. A. Garrott, D. B. Siniff and J. B. Tester. 1992. Efficacy of chemical contraception in feral mares. Wildl. Soc. Bull. 20:211-216.

Findley, R. W. and D. A. Farber. 1985. Environmental law, cases and materials. 1985. West Publ. Co., St. Paul, Minn. 813pp.

Glass, J. H. 1980. Fish and wildlife agency funding. Proc. Int. Assoc. Fish and Wildl. Agencies 70:102-112.

Gottschalk, J. S. 1975. State-federal partnership in wildlife conservation. Pages 290-301 *in* H. P. Brokaw, ed. Wildlife and America. Counc. Environ. Quality, U.S. Gov. Printing Off., Washington D.C.

Groves, F. W. 1965. Report of the federal-state rights relations committee. Proc. Int. Assoc. Game, Fish and Conserv. Comm. 55:19-31.

———. 1966. Report of federal-state rights relations committee. Proc. Int. Assoc. Game, Fish and Conserv. Comm. 56:170-182.

Heberlein, T. A. 1991. Changing attitudes and funding for wildlife: preserving the sport hunter. Wildl. Soc. Bull. 19:528-534.

Horning, J. 1992. Selling our land cheaply. Natl. Wildl. Fed. EnviroAction. June 1992:10.

Huey, W. S. 1980. International treaties from the state agency viewpoint. Proc. Int. Assoc. Fish and Wildl. Agencies 70:20-27.

International Association of Game, Fish and Conservation Commissioners. 1950. Report of Resolution Committee: Hunting on military reservations. Proc. Int. Assoc. Game, Fish and Conserv. Comm. 40:157-158.

———. 1953. Report of Resolution Committee: Military establishments. Proc. Int. Assoc. Game, Fish and Conserv. Comm. 43:159-160.

———. 1971. Public land policy impact on fish and wildlife. Int. Assoc. Game, Fish, and Conserv. Comm. Washington, D.C. 80pp.

Jahn, L. R. 1982. Results from questionnaire on U. S. state and territory fish and wildlife agency directors or administrators. Proc. Int. Assoc. Fish and Wildl. Agencies 72:178-183.

———, and J. B. Trefethen. 1978. Funding wildlife conservation programs. Pages 456-470 *in*. H. P. Brokaw, ed. Wildlife and America. Counc. Environ. Quality, U.S. Gov. Printing Off., Washington D.C.

Lund, T. A. 1980. American wildlife law. Univ. California Press, Berkeley. 179pp.

Matthews, P. M. 1986. Who owns wildlife? Wildl. Soc. Bull. 14:459-465.

National Wildlife Federation. 1992. Conservation directory. Natl. Wildl. Fed., Washington D.C. 398pp.

Nelson, W. O., Jr. 1975. Nongame birds in current policies and decision making of the U. S. Fish and Wildlife Service. Pages 323-327 *in* Proc. symposium on management of forest and range habitats for non-game birds. U.S. For. Serv. Gen. Tech. Rep. WO-1, Washington D.C.

Olds, N. V. 1965. Report of the legal committee. Proc. Int. Assoc. Game, Fish and Conserv. Comm. 55:101-107.

———. 1968. Report of the legal committee. Proc. Int. Assoc. Game, Fish and Conserv. Comm. 58:103-106.

Plater, Z. J. L., R. H. Abrams, and W. Goldfarb. 1992. Environmental law and policy. West Publ. Co., St. Paul, Minn. 1039pp.

Public Land Law Review Commission. 1970. One-third the nation's land. U.S. Gov. Printing Off. Washington D.C. 342pp.

Rasmussen, B. L. 1968. Federal and state roles on the public domain. Proc. Int. Assoc. Game, Fish, and Conserv. Comm. 58:74-77.

Trefethen, J. B. 1975. An American crusade for wildlife. Winchester Press, New York, N.Y. 409pp.

United States Code. 1988. Title 5, government organization and employees. Office of Law Revision Counsel, House of Representatives, U. S. Gov. Printing Off. Washington, D. C.

———. 1988. Title 16, Conservation. Office of Law Revision Counsel, House of Representatives. U. S. Gov. Printing Off., Washington, D.C. 4307pp.

Wildlife Management Institute. 1977. Organization, authority and programs of state fish and wildlife agencies. Wildl. Manage. Inst., Washington, D.C. 39pp.

———. 1991. License sales hold steady. Outdoor News Bull. 45:3.

World Book Encyclopedia. 1956. Public lands, America land gains. 13:6640.

Woodward, H. R. 1970. President's address. Proc. Int. Assoc. Game, Fish and Conserv. Comm. 60:13-18.

Chapter 26
ECONOMIC VALUATION OF RANGELAND WILDLIFE

Phil R. Ogden

INTRODUCTION

Economic values for wildlife resources have been reported in various literature sources and have been referenced by numerous authors including Bryant (1989). He cited data showing that hunters spent >$775 million in the western United States in 1985 and >$6.0 billion for big game hunting across America. Economic values for wildlife resources and the amount of expenditures for recreation related to wildlife obviously are very large.

Economic values of wildlife for specific species and situations may vary greatly depending on many factors including methods of analysis and assumptions. Shalloof (1985) determined a value of $328/elk (*Cervus elaphus*) in northeastern and central Oregon using the travel cost method of analysis. Martin et al. (1985) calculated the marginal net benefit/elk at the 1979 herd size for an area in northern Arizona as $106.00. This value was determined using a marginal willingness to pay analysis with number of permits held constant. If the herd, however, were doubled with number of permits held constant, the marginal net value dropped to $6.35/elk. These authors pointed out the problems of valuing wildlife for use in analyses to arrive at efficient multiple use decisions.

The purpose of this chapter is to provide methods and describe information needs for analysis and interpretation of data that may be used in economic analyses of wildlife management alternatives. The discussion is intended to be used to help identify the questions to be asked and the kinds of answers that are necessary to estimate the economic benefits and costs for wildlife and/or habitat management alternatives; answers from which legitimate comparisons of alternatives may be made. Included in this chapter are sections on production functions, valuing wildlife, determining costs, evaluating the degree of competition among resource uses, evaluation of benefits and costs over time, and advantages and disadvantages of commercializing wildlife on private land.

A function of economics is to provide information and methods to assist with the efficient allocation of scarce resources among competing uses; and if this efficient allocation is to assist with the solution of resource allocation conflicts, allocations must be equitable (i.e., the benefits and costs of efficient allocation are fairly distributed). Thus, for economic analyses, not only must the dollar values for benefits and costs of resources, products, and activities be identified, the entities must be identified to whom costs and benefits accrue. Both monetary and nonmonetary costs and benefits must be considered.

The degree to which economic evaluation is used in the decision making process regarding rangeland wildlife may vary greatly. There is little doubt, however, that the long-term sustainability of any policy or action will depend greatly on how efficiently and equitably human needs are satisfied (Chandler 1986). Budgets for rangeland resource management often constrain management activities, and projects that favor rangeland wildlife must compete for available dollars.

PRODUCTION FUNCTIONS

All data necessary for detailed production economic analyses for natural resource decisions often are not available. Plant and animal population growth and production, however, provide a common basis for communication among biologists, range managers, and livestock producers.

As an initial step in the economic valuation of wildlife, it is essential to have an understanding of the products or activities to be produced by wildlife and the relationship of different levels of inputs to different levels of outputs of the products and activities. The products and activities (i.e., outputs) derived from wildlife may include the number of animals available for harvest, number of trophy animals available for harvest, hunter days, number of recreation days, number of observations, or number of licenses sold. The fixed input for production of the wildlife outputs often is the area and condition of habitat at a specific time within the management unit being analyzed. The wildlife population expressed as number of animals/unit area is a variable input.

A sigmoid curve represents the general shape of a production function for products derived for a specified unit of fixed input and changing levels of a variable resource. Workman (1986) refers to this curve as the total physical product curve (Fig. 1a). The total product curve (Fig. 1a) represents the total product as a function of animal population expressed as number of animals/unit area for a management unit. The biological relationship of product output to number of animals/unit area for this general curve is that, with very few animals/unit area, rate of recruitment may be low and the product response is slow (i.e., due to poor breeding success or high relative predator losses). Total production improves with an increase in animal numbers/unit area and then decreases with excessively high numbers because of biological competition that limits breeding efficiency or some other factor that affects quantity of animals available for harvest or observation. The average product curve (Fig. 1b) is obtained by dividing the numbers on the total product curve by each of their variable input numbers on the X-axis. The average product curve provides the product amount produced/unit of variable input; e.g., the product/individual animal (Fig. 1b).

The sigmoid total product curve (Fig. 1a) is the relationship of product to a variable stocking rate of a wildlife species on a fixed area of habitat of a specific quality. This function represents possible population numbers as variable inputs at a point in time and should not be confused with the similarly shaped sigmoid curve of a population buildup over time. The product value curve of wildlife/unit of habitat is determined by multiplying the dollar value/unit of product times the quantity of the total product curve (Fig. 1a). The economic value of wildlife/individual animal is obtained by multiplying the dollar value of the product in question times the appropriate number on the average product curve (Fig. 1b). Realistic dollar values/ unit of product, therefore, must be determined. Product values may be assumed to be constant over the range of alternatives compared, if the alternatives do not have a major effect on the supply and demand relationships of the product. Thus, value product curves may have the same general shape as the total and average product curves (Fig. 1) but with the units on the Y-axis changed to dollars.

The economically most efficient population level is attained at the point on the total product value curve where marginal returns are equal to total costs to attain the marginal return. If all costs are fixed costs (i.e., do not change with number of animals/unit area), the economically most efficient production level is at maximum total production. This is the point at which line Y' meets the total product curve (Fig. 1a). When all costs are variable costs associated with each increase in product, the economically most efficient population will be at the point on the total product value curve where the slope of the curve becomes less steep. This corresponds to the peak average product value as shown by line Y (Figs. 1a and 1b). The economically most efficient level of production, therefore, is within the range between maximum individual animal production and total population production.

Alternative management options may hypothesize changed habitat conditions; thus, product population responses must be represented by different production functions. The question to be asked in these cases is what is the most efficient management alternative? The answer is found by maximizing the economic return at the margin for the different production functions. Marginal analyses consider the economic benefits and costs of the last incremental increase in product hypothesized for each alternative. Sustainability of the chosen animal populations used for a specific comparison is a question that also should be asked, and the economic evaluation should be projected over the life of the planning period. This evaluation must consider costs and returns over time as discussed later in this chapter.

VALUING WILDLIFE PRODUCTS

Wildlife has economic value based on many different products accruing to different segments of society, depending on who derives the benefit. If economic analyses are comparable for efficient and equitable allocation of resources, values should be derived for different alternatives at equivalent points in the production-consumption process. The value of pronghorn antelope (*Antilocapra americana*) based on income to a game and fish department from sale of hunting tags does not provide a value that is comparable to the value to hunters based on their willingness to pay travel expenses for the experience of hunting for antelope.

Market Values

For a game and fish department, the product for a species of wildlife may be the hunting tags sold. The value/tag is the price paid for the tag. These prices generally are not market prices, but administered prices set by commissions. The values, however, are real and represent income to the department. These may be appropriate values for determining value product curves for comparing various management options where the question asked is: what is the economic return to the department for tags sold for various management options?

A high value product (i.e., activity) of a bighorn sheep (*Ovis canadensis*) or elk population might be the trophy hunt experience. If the trophy animals are found on private lands or Indian Trust Lands where access may be controlled and charged for, then the value of the trophy hunt experience may be valued on the market value for which the permission to hunt on the land is sold. The product unit valued may be hunter days or hunting

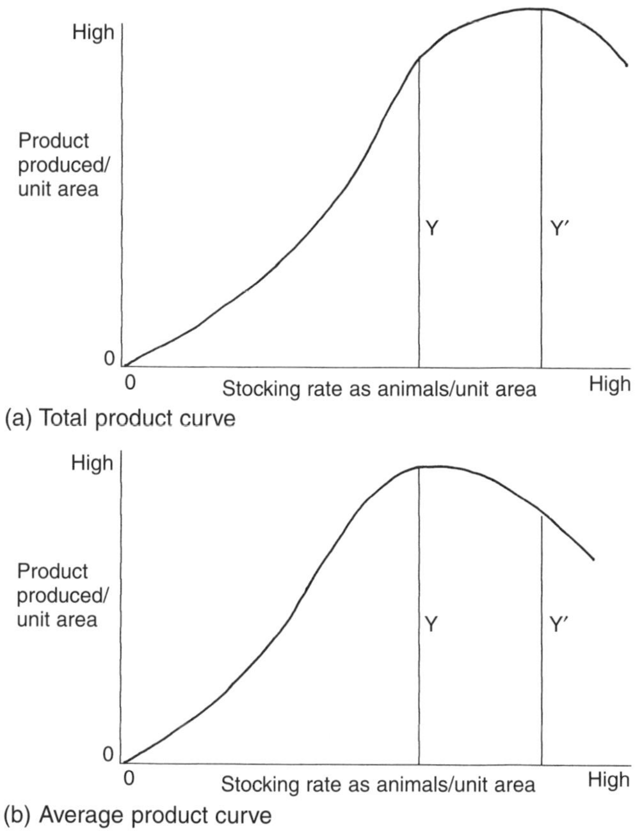

FIGURE 1. General production function curves.

permission tags sold. Here the value accrues to the land owner and may be an appropriate value on which the land owner may compare alternative management strategies for the land in question. Similarly, access to private lands may be charged for observation, hunting, or fishing. Market values may determine the income to private land owners for such activities. For a rancher on public lands, the product may be guide service days for which a market value may be determined, a value which can be compared with other alternatives for investment of ranch labor, time, and capital for specific situations.

For a local economy, the value of a specific wildlife population may be determined by the amount of money spent in local businesses with a specified wildlife population versus an alternative population. This may require surveys to determine anticipated income to businesses that may be associated with the wildlife populations to be compared. These values may be determined as a result of consumptive or nonconsumptive wildlife use. With proper design and conduct, such studies provide values of wildlife resources for the local economy or economy at any level for which the data are gathered. For direct value comparisons, a similar type of survey would need to be conducted for livestock grazing at stocking rates allowed with each wildlife population alternative.

Valuation When Market Values Are Not Available

For many wildlife valuation problems, market values do not exist. As an example, there is no market value for a recreation day to hunt, fish, or observe wildlife on public lands where no fee is charged. Randall (1987) discussed a number of methods used by economists when no market value exists. Economists may arrive at a contingent valuation by creating hypothetical or experimental markets, or collect data by surveys with questions such as willingness to pay or willingness to sell. Economists also may develop values using weak complementary or implicit price methods.

Lambert (1983) provided an example of how the value of a hunter day may be determined by utilizing sample survey data for a hunter population. Hunter populations have different levels of willingness to pay for a hunter day experience and also have differing costs/hunter day. An estimate of the total hunter population for the management unit must be determined and from sample survey data from this hunter population, a demand curve and average costs/hunter day can be determined (Fig. 2). Current management of the management unit provides for an existing number of hunter days (Fig. 2). The value of the hunting experience for the hunter population utilizing the management unit is the area bounded by the demand curve, the cost curve, and the existing hunter day line (Fig.2). This value is identified as consumer surplus. The value of a hunter day to evaluate management changes that may result in additional availability of hunter days is the consumer surplus/hunter day at the margin (value of an additional day beyond existing). Changes in management also may result in changes in quality of hunter days that may change the demand curve. In this case,

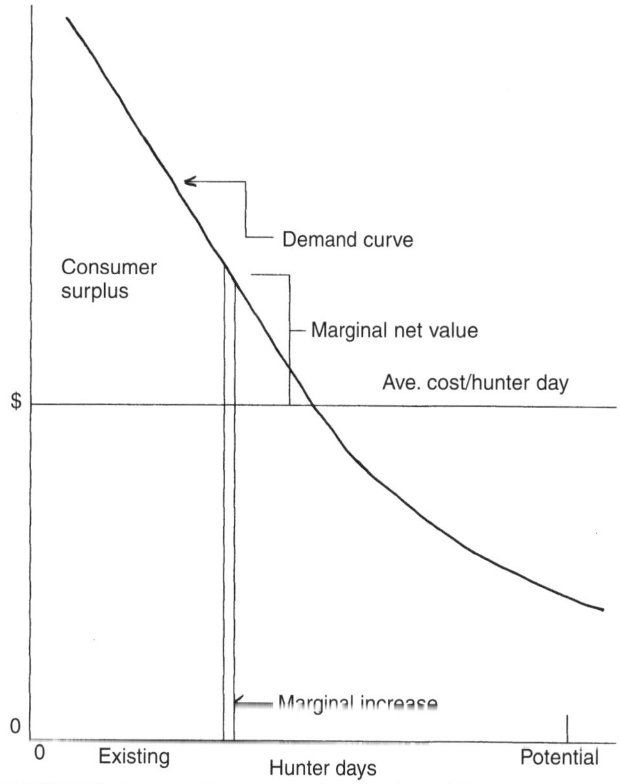

FIGURE 2. Demand curve and associated information derived from survey of hunter willingness to pay and travel costs for a specified management unit.

the economically efficient solution would be determined by comparing the marginal value for the different demand curves.

Consumer surplus represents value to the population sampled and can be applied to wildlife observation days and hunter days. The values developed are dependent on the population included in the survey and care to design a survey free from questions that might elicit biased responses.

Value Curves

All of the valuation methods discussed above are appropriate to answer specific questions for specific recipients of the values from wildlife products. The valuation that is used for any specific question is dependent on the comparisons to be made and the questions to be answered.

The value/unit of product determined by any of the valuation methods discussed above may be used as a factor by which numbers along the total product curve (Fig. 1) may be multiplied to obtain a total gross value curve (Fig. 3). Each point on the total value curve provides an estimate of the gross return/unit of habitat. The average product value curve, the total value curve divided by the input variable (no. animals), provides gross return/individual animal.

COSTS OF PRODUCTION

Total or gross values must be corrected to net values for economic analyses. That is, the fixed and variable costs of the production must be subtracted from the total income to determine net income (Fig. 3). Fixed costs are those costs that are independent of the numbers of wildlife. For instance they may represent a base level of costs associated with licenses and law enforcement for a game and fish department regardless of the wildlife population. For a land owner, they may represent the base level of water development and input to maintain the habitat to support a basic wildlife population. The variable costs are the costs associated with each unit increase in product output. For example, this may be supplemental water or salt, increased law enforcement, predation control, or even certain habitat improvement costs if these costs are associated with each unit of increasing number of animals.

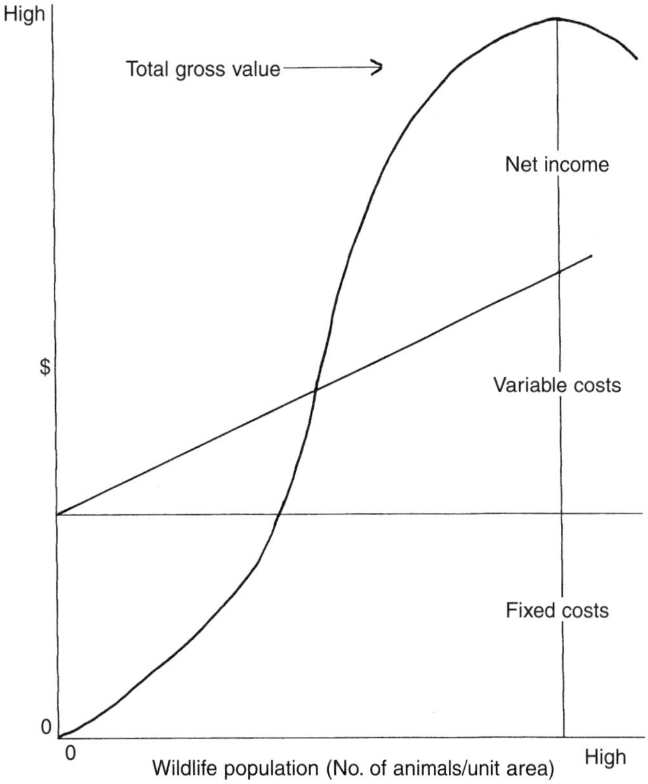

FIGURE 3. Total gross value curve and variable and fixed costs for various stocking rates of a wildlife species.

The maximum net income is the point at which the slope of the sum of the fixed and variable cost lines is tangent to the total gross value curve (i.e., where the slopes of the gross return and total cost lines are equal). This also is the point at which marginal costs equal marginal returns. The net value of the last increment of product produced whether at maximum net return or not, is the marginal net value of a product and is an appropriate economic value for comparing alternative courses of action (Verburg et al. 1987).

Costs for wildlife populations may not be borne by the same entity that derives the benefit. For instance, the economic value from increased elk may accrue to a game and fish department or to hunters or to the local economy, but some of the costs of an increased elk population may be induced costs of reduced livestock numbers by the livestock permittees. Thus, the cost of increased numbers of elk may be of little cost to a game and fish department, but the costs may be great to a livestock operation on the same management unit. The benefits and costs should be identified and clearly stated for each of the parties affected by resource management alternatives if a clear picture of the effects of alternative management is to be developed.

COMPETITION FOR RESOURCES

The valuation of the benefits and costs of wildlife products discussed above considers only a single product from the rangeland resource. The total gross value curve (Fig. 3) is based on a relationship of numbers of a specific wildlife species/unit of rangeland. Most management alternative decisions involving wildlife on rangelands affect many products from the rangeland resource. In addition to the single product economic evaluations as discussed above, major information needed to make efficient and equitable economic decisions is identification of the nature of the product competition for the resources of the rangeland.

The competition among products for rangeland resources may be supplementary, complementary, competitive, or antagonistic (Fig. 4). Workman (1986) referred to curves that identify the competitive relationship between two products derived from a single resource as transformation curves and discussed the biological and economic interpretation of these curves. Often in economic analyses, an assumption is made that products are directly competitive for resources (Fig. 4c). For competitive or antagonistic products, maximum economic benefit will be derived from one of the products, the specific one being dependent on the relative value of the products (i.e., the slope of the price line; Figs. 4c and 4d.). If products are completely supplementary, that is inhabit completely different niches on a unit of rangeland, the net value of these products to the range unit are additive through the range where they are supplementary (Fig. 4a). Regardless of the economic value of the individual products, the most efficient allocation of resources is to maximize both populations within the limits of the habitat for sustainable production (Fig. 4a). The maximum production of each product for maximum economic return of complementary products is the point at which the product price line is tangent to the transformation curve (Fig. 4b).

Products may be antagonistic if there is a relationship that any number of a species has a detrimental effect on another species (Fig. 4b). For a particular habitat, the population of coyotes (*canis latrans*) or bobcats (*Lynx rufus*) may be antagonistic to antelope, as the habitat may not support enough of the antelope population to accommodate depredation. The nature of competition for fixed habitat resources among variable resources and products should be determined and accounted for in economic analyses.

CONSIDERING COSTS AND RETURNS OVER TIME

For benefit/cost (B/C) analyses, costs and returns for products under consideration must be determined for the management alternatives and then projected over an appropriate planning period. Money received or spent in the future is not worth as much as the value today because of risk and the opportunity for use or investment of the money for a benefit if available today (opportunity cost). Interest rates reflect the relative value of money today versus the future. Money in the future, therefore, is discounted to present value by appropriate factors based on an appropriate interest rate. The present value of $1.00 received at a given time at various interest rates (Table 1) differs from the present value of $1.00 received annually over a given period of years (Table 2). The values of Table 2 are accumulative values of Table 1.

The values in Table 1 are used as multipliers to adjust one-time costs or returns in a given year to present value. Table 2 values are used as multipliers to convert a flow of costs or returns over more than 1 year to a present value. Tables 1 and

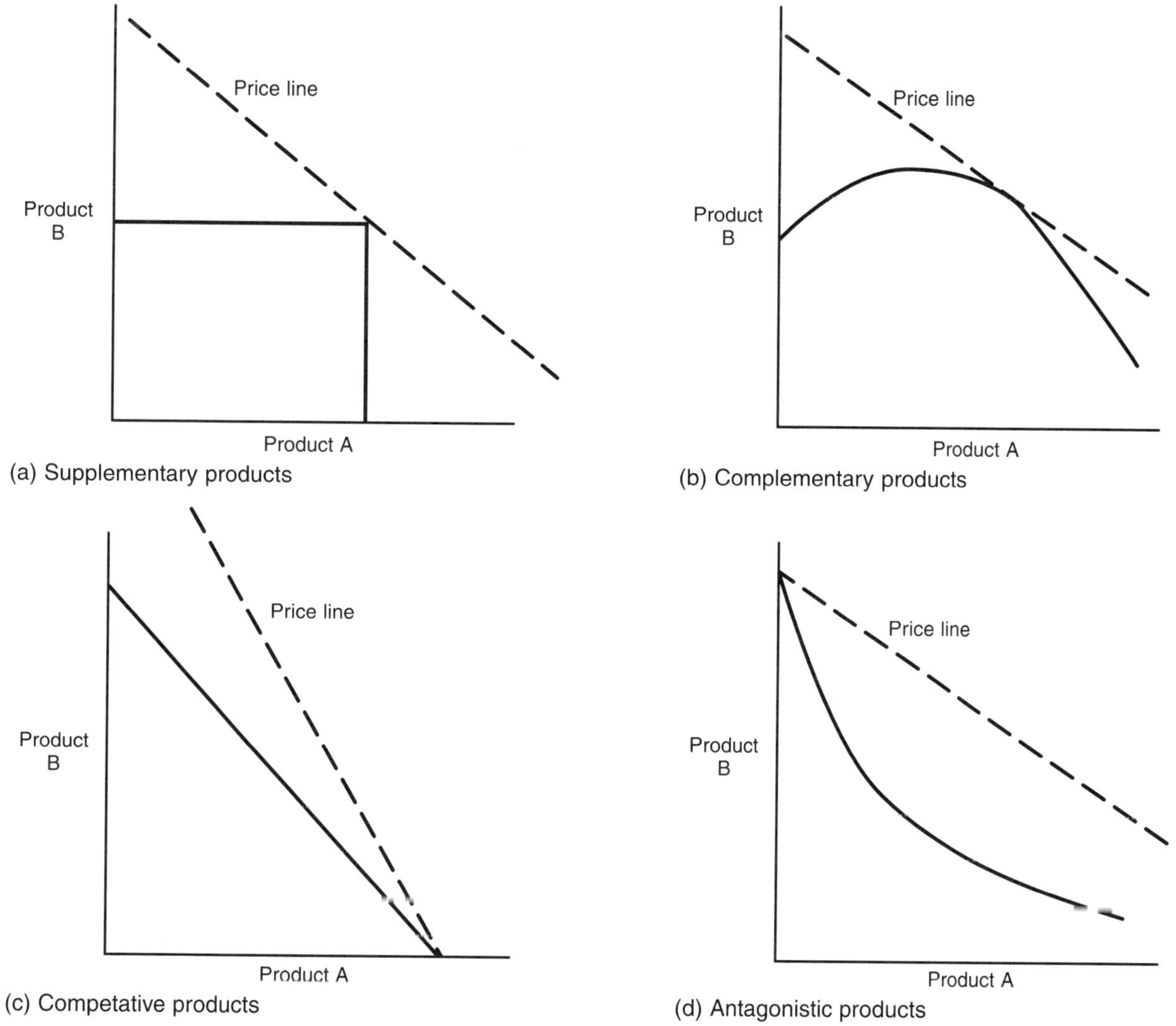

FIGURE 4. Transformation curves illustrating types of product competition for resources.

TABLE 1. The present value of $1.00 due in a specific future year at interest rates of 4-14%

	% Interest										
Year	4	5	6	7	8	9	10	11	12	13	14
1	0.96	0.95	0.94	0.93	0.93	0.92	0.91	0.90	0.89	0.88	0.88
2	0.92	0.91	0.89	0.87	0.86	0.84	0.83	0.81	0.80	0.78	0.77
3	0.89	0.86	0.84	0.82	0.79	0.77	0.75	0.73	0.71	0.69	0.67
4	0.85	0.82	0.79	0.76	0.73	0.71	0.68	0.66	0.64	0.61	0.59
5	0.82	0.78	0.75	0.71	0.68	0.65	0.62	0.59	0.57	0.54	0.52
6	0.79	0.75	0.70	0.67	0.63	0.60	0.56	0.53	0.51	0.48	0.46
7	0.76	0.71	0.67	0.62	0.58	0.55	0.51	0.48	0.45	0.42	0.40
8	0.73	0.68	0.63	0.58	0.54	0.50	0.47	0.43	0.40	0.38	0.35
9	0.70	0.64	0.59	0.54	0.50	0.46	0.42	0.39	0.36	0.33	0.31
10	0.68	0.61	0.56	0.51	0.46	0.42	0.39	0.35	0.32	0.29	0.27
11	0.65	0.58	0.53	0.48	0.43	0.39	0.35	0.32	0.29	0.26	0.24
12	0.62	0.56	0.50	0.44	0.40	0.36	0.32	0.29	0.26	0.23	0.21
13	0.60	0.53	0.47	0.41	0.37	0.33	0.29	0.26	0.23	0.20	0.18
14	0.58	0.51	0.44	0.39	0.34	0.30	0.26	0.23	0.20	0.18	0.16
15	0.56	0.48	0.42	0.36	0.32	0.27	0.24	0.21	0.18	0.15	0.14

2 are year end values and may be used to convert benefits received directly. A general convention in benefit/cost analyses is that benefits are considered as received at year end and costs are considered as beginning of year. For costs, therefore, discounting with Tables 1 and 2 is accomplished by using the year end value of the year prior to the year of the cost as the appropriate multiplier. For beginning of project costs (year 0), the multiplier is 1.00 for all interest rates.

The B/C ratio for any project is the sum of all benefits discounted to the present divided by the sum of all costs discounted to the present. Benefit/cost ratios of <1.0 show a negative return to the investment and ratios >1.0 indicate that there is a profit on the project in addition to covering interest charges. A B/C of 1.0 is the point at which there is no profit or loss; project returns and costs are balanced to pay only the rate of interest.

ADVANTAGES AND DISADVANTAGES OF COMMERCIALIZING WILDLIFE ON PRIVATE LAND

Associating a monetary value with wildlife is well established through license fees, trapping, guide service, tourism, and habitat improvements. The commercialization of wildlife for profit on private lands through ecotourism, game ranching, and other enterprises, however, is a matter of much debate (Hawley 1993). The debate centers around the question of whether profitization of wildlife is an aid to conservation or results in wildlife exploitation.

Gilbert (1993) stated that commercialization of wildlife currently is contributing to multiple use and ecologically based management of species. Game farming supplies sport and meat, zoological parks provide endangered species rehabilitation, and theme parks and destination resorts provide recreational viewing. He suggested that public agencies have exploited wildlife commercially; perhaps private enterprises should have like opportunities. Teer (1993) argued that the needs of people and the needs of nature must both be considered and that commercialization of wildlife provides an incentive for conservation.

Geist (1993), however, provided a list of 10 potential threats to wildlife that would result from the abandonment of current management of wildlife for a shift to commercialization. He described current wildlife management in North America as being based on 4 fundamental policies: public ownership, denial of economic value to dead wildlife, allocation of surplus wildlife by law, and nonfrivolous uses of wildlife. His 10 potential threats are: (1) the legal trafficking in wildlife complicates policing of wildlife laws, (2) a danger of genetic pollution from introduction of exotics, (3) the extinction of native wildlife by feralization of exotics, (4) threat of diseases, (5) extinction of predators, (6) encouraging trophy hunting for the few as opposed to hunting for the general public, (7) poor conservation of wildlife by stressing production of favored species and elimination of less favored, (8) paid hunting discriminating against modest income hunters resulting in a loss of support for hunting, (9) paid hunting fostering the concept of conspicuous consumption of wildlife by the elite, and (10) economic efficiency leading to depletion of wildlife populations and threatening the service and manufacturing industries that depend on the wildlife populations.

Neave (1993) addressed issues of commercialism in the context of societal goals. He concluded that, as evidenced by current environmental, ethical, and humane concerns and activities, society wants wildlife to be managed in natural habitats on a landscape basis. Commercialization, in his opinion, promotes scarcity of wildlife for economic gain of the few as opposed to

TABLE 2. The present value of $ 1.00 received annually for years and interest ratesof 4-14%

	% Interest										
Year	4	5	6	7	8	9	10	11	12	13	14
1	0.96	0.95	0.94	0.93	0.93	0.92	0.91	0.90	0.89	0.88	0.88
2	1.89	1.86	1.83	1.81	1.78	1.76	1.74	1.71	1.69	1.67	1.65
3	2.77	2.72	2.67	2.62	2.58	2.53	2.49	2.44	2.40	2.36	2.32
4	3.63	3.55	3.46	3.39	3.31	3.24	3.17	3.10	3.04	2.97	2.91
5	4.45	4.33	4.21	4.10	3.99	3.89	3.79	3.70	3.60	3.52	3.43
6	5.24	5.08	4.92	4.77	4.62	4.49	4.35	4.23	4.11	4.00	3.89
7	6.00	5.79	5.58	5.39	5.21	5.03	4.87	4.71	4.56	4.42	4.29
8	6.73	6.46	6.21	5.97	5.75	5.53	5.33	5.15	4.97	4.80	4.64
9	7.43	7.11	6.80	6.51	6.25	5.99	5.76	5.54	5.33	5.13	4.95
10	8.11	7.72	7.36	7.02	6.71	6.42	6.14	5.89	5.65	5.43	5.22
11	8.76	8.31	7.89	7.50	7.14	6.80	6.49	6.21	5.94	5.69	5.45
12	9.38	8.86	8.38	7.94	7.54	7.16	6.81	6.49	6.19	5.92	5.66
13	9.99	9.39	8.85	8.36	7.90	7.49	7.10	6.75	6.42	6.12	5.84
14	10.56	9.90	9.29	8.74	8.24	7.79	7.37	6.98	6.63	6.30	6.00
15	11.12	10.38	9.71	9.11	8.56	8.06	7.61	7.19	6.81	6.46	6.14

abundance for society. Thus, commercialization of wildlife does not meet societal goals.

Biological and social issues and concerns must be considered along with the economic questions when considering commercialization of wildlife on public or private lands. On private lands, however, the economic concerns become a greater factor than on public lands, as what happens to the wildlife habitat on private lands is more likely to be influenced by economics than it is on public land. Answers to the questions posed at the beginning of this chapter are the appropriate economic contribution to answering the question of whether to commercialize or not commercialize wildlife on private property. Is it efficient? Is it equitable? Methods to help answer these questions have been discussed in this chapter, and a planned sequence of economic analyses should be accomplished to develop data to answer these questions as part of a decision model just as for other management alternative decisions.

For wildlife products that are readily marketable and are supplementary or complementary to livestock grazing on private land, there is an opportunity to supplement ranch income with income from wildlife. To answer the question of efficiency for the ranch operation, marginal net income or benefit/cost analyses may provide the mix of wildlife production and livestock production that maximizes ranch profit. The equitable question would be addressed by evaluating the benefits and costs to the local economy, to game and fish departments, and other affected interests.

Private land owners may be hostile toward hunters and state agencies that profit from wildlife because of damage to private resources. Such a case was reported for Montana land owners who suffered losses from deer (*Odocoileus* spp.) populations (Blood and Baden 1984). Improvement of the equitable distribution of income from the deer population in this situation may improve the opportunities for management that favors the deer population. Staniford (1989) found that fee hunting added to ranch profits in the oak (*Quercus* spp.)-annual grass range in California. He suggested that as hunting quality increased there was less tree harvesting. He also suggested that fees may reduce pressure to subdivide private land due to increased income and spreading out risk of production. Bryant (1989) concluded that for many ranches in Texas, fees from hunting leases were being put back into the land for habitat improvement. There are many examples, especially in Texas, where wildlife species have benefited by commercialization. Wildlife species populations, however, may suffer under wildlife commercialization if they are not readily marketable and are antagonistic or competitive products of wildlife that are commercialized. Generalizations should be avoided, and each management situation should be decided on the best information available applied to the decision in question.

SUMMARY

Economic analyses that will provide useful data to assist with rangeland wildlife management decisions that are efficient and equitable must include the following considerations:

1. Identify appropriate production functions for the wildlife population in question and appropriate transformation curves to document the degree of competition for the fixed habitat resource among competing variable resources and products.
2. Identify product benefit values for alternative products and recipients of the benefits.
3. Identify costs of producing alternative products and identify to whom cost accrue.

4. Conduct benefit/cost analyses based on present values of flows of costs and returns over an appropriate planning period for each of the major entities which receive benefits or incur costs as a result of proposed management changes.

Seldom do range resource managers have the luxury of having specific data to complete detailed analyses for each of the steps as outlined above. Managers should, however, include the above considerations in decision models using the best available data. This includes making decision regarding commercializing wildlife on private land.

LITERATURE CITED

Blood, T., and J. Baden. 1984. Wildlife habitat and economic institutions: feast or famine for hunters and game. West. Wildlands 10:8-13.

Bryant, F. C. 1989. Economic implications of wildlife. Proc. West. Sect. Am. Soc. Anim. Sci. and Western Branch Can. Soc. of Anim. Sci. 40:500-502.

Chandler, W. U. 1986. The changing role of the market in national economies. Worldwatch Pap. 72. 57 pp.

Geist, V. 1993. Great achievements, great expectations: successes of North American wildlife management. Pages 47-72 *in* A. W. L. Hawley, ed. Commercialization and wildlife management: dancing with the devil. Krieger Publ. Co., Malabar, Fla.

Gilbert, F. F. 1993. The vision: wildlife management in North America. Pages 23-33 *in* A. W. L. Hawley, ed. Commercialization and wildlife management: dancing with the devil. Krieger Publ. Co., Malabar, Fla.

Hawley, A. W. L. 1993. Introduction. Pages 1-4 *in* A. W. L. Hawley, ed. Commercialization and wildlife management: dancing with the devil. Krieger Publ. Co., Malabar, Fla.

Lambert, D. K. 1983. Costs and benefits of BLM range improvements: private and public perspectives. Oregon State Univ. Ext. Serv. EM 8244, Corvallis. 34pp.

Martin, W. E., D. C. Cory, and M. A. Helfrich. 1985. Valuing wildlife for efficient multiple use: an Arizona case study. Univ. Arizona, Dep. Agric. Econ. Rep. 35. Tucson. 43pp.

Neave, D. J. 1993. Who is accountable; the ethics of conservation. Pages 85-101 *in* A. W. L. Hawley, ed. Commercialization and wildlife management: dancing with the devil. Krieger Publ. Co., Malabar, Fla.

Randall, A. 1987. Resource economics: an economic approach to natural resource and environmental policy. Second ed. John Wiley and Sons, New York, N. Y. 434pp.

Shalloof, F. M. 1985. Impact of various factors upon benefits from big game hunting estimated by the travel cost method. Ph.D. Thesis, Oregon State Univ., Corvallis. 152pp.

Staniford, R. B., IV. 1989. A bioeconomic model of California's hardwood rangelands. Ph.D. Thesis, Univ. California, Davis. 143pp.

Teer, J. G. 1993. Commercial utilization of wildlife; has its time come? Pages 73-83 *in* A. W. L. Hawley, ed. Commercialization and wildlife management: dancing with the devil. Krieger Publ. Co., Malabar, Fla.

Verburg, E. A., J. J. Charbonneau, W. R. Mangun, and L. G. Llewellyn. 1987. The importance of fish and wildlife values to the profession. Pages 49-59 *in* D. J. Decker and G. R. Goff, eds. Valuing wildlife: economic and social perspectives. Westview Press, Boulder, Colo.

Workman, J. P. 1986. Range economics. Macmillan Publ. Co., New York, N. Y. 217pp.